C0-ASZ-314

Handbook of Automotive Engineering

Other related resources from SAE International:

Automotive Engineering Fundamentals
By Richard Stone and Jeffrey K. Ball
Order No. R-199

BOSCH Automotive Handbook, Sixth Edition
Order No. BOSCH6

2005 SAE Handbook 3-Volume Set
Order No. 2005HBST

2005 SAE Handbook on CD-ROM
Order No. HBKCD2005

Advanced Vehicle Technology, 2nd Edition
By Heinz Heisler
Order No. R-337

Handbook of Automotive Engineering

Edited by Hans-Hermann Braess
and Ulrich Seiffert

Translated by Peter L. Albrecht

Warrendale, Pennsylvania USA

For permission and licensing requests, contact:

SAE Permissions
400 Commonwealth Drive
Warrendale, PA 15096-0001 USA
E-mail: permissions@sae.org
Tel: 724-772-4028
Fax: 724-772-4891

Library of Congress Cataloging-in-Publication Data

Vieweg Handbuch Kraftfahrzeugtechnik. English
Handbook of automotive engineering / edited by Hans-Hermann Braess and Ulrich Sieffert; translated by Peter L. Albrecht.
p. cm.
Translation of: Vieweg Handbuch Kraftfahrzeugtechnik. 2nd ed.
Includes bibliographical references and index.
ISBN 0-7680-0783-6
1. Automobiles—Design and construction. I. Braess, Hans-Hermann. II. Sieffert, Ulrich. III. Title.

TL240.V48 2005
629.2′3—dc22 2004061399

SAE
400 Commonwealth Drive
Warrendale, PA 15096-0001 USA
E-mail: CustomerService@sae.org
Tel: 877-606-7323 (inside USA and Canada)
724-776-4970 (outside USA)
Fax: 724-776-1615

ISBN 0-7680-0783-6

SAE Order No. R-312

Translated by Peter L. Albrecht.

Printed in the United States of America.

Originally published in the German language by Friedr. Vieweg & Sohn Verlagsgesellschaft mbH, D-65189, Wiesbaden, Germany, as “Hans-Hermann Braess und Ulrich Seiffert (Hrsg.): *Vieweg Handbuch Kraftfahrzeugtechnik 2. Auflage* (2nd edition)”
ISBN 3-528-13114-4

Preface

This latest edition of the *Handbook of Automotive Engineering* is the successor to a well-known German-language handbook published by Professors Heinrich Buschmann and Paul Koessler, and familiar to generations of automotive engineers. The first edition was published in 1940; the last, in 1973.

This new volume comprehensively describes the fascinating world of the automobile and its development. More than 40 well-known authorities from the automotive and supplier industries were engaged as contributors. This volume also serves as a "time capsule" in that it describes the current state of the art and rapid development of the automobile. Although based primarily on German authors, the global nature of the automobile industry ensures that its contents have worldwide applicability.

Beginning with the need for mobility, resulting requirements and conflicting goals are defined; combined with physical and engineering principles, these establish the boundary conditions for modern vehicles. The design process, as a major factor in winning customers, in the purchase decision, and in customer acceptance, is described in detail. The chapter covering vehicle concepts and package illustrates how a multitude of different overall concepts and variations may arise as a consequence of differing priorities. Related specific concepts, including electric drive, fuel cells, hybrid and gas turbine propulsion are described.

The chapter on "classical" powertrain systems occupies a prominent portion of this book. Modern internal combustion engine technologies (both gasoline and diesel) determine the course of development for the foreseeable future. It is evident that both engine types have very high potential for further development. Other important topics include exhaust gas aftertreatment, turbo- and supercharging, and optimization of auxiliary devices. Transmission systems exhibit increasing variety, as demonstrated by automatically shifted manual transmissions, CVTs, and automatic transmissions with a greater number of gears. Opportunities and risks for the two-stroke engine are analyzed and discussed. In the long term, great importance must be given to additives and alternative fuels, as well as alternative energy sources, whose merits and disadvantages are examined and compared.

Vehicle bodywork is becoming an increasingly demanding and complex field, as shown by the number of topics treated herein. These include the fundamentals of unit bodies, space frame technology, convertibles, ergonomics, comfort, vehicle safety, communications and navigation systems.

The section on suspensions includes tire and brake systems and demonstrates the incredible progress achieved in acoustics, road holding, comfort, and safety.

Electronics play an increasingly vital role in all areas of the vehicle. All future innovations will include electronic components and systems. This suggests, of course, that virtually all vehicle functions and systems will include electronic components, e.g., steer-by-wire, brake-by-wire, and drive-by-wire.

In the past decades, increasing demands have led to significant increases in vehicle weight. Future materials engineering, manufacturing technology, and structural design will in particular need to address requirements for light weight as well as recycling aspects. With increasing complexity of the vehicle itself, its development, the relationships between vehicle manufacturer and suppliers, and global production, it becomes evident that optimization of the product creation process plays an ever more important role. Reduction of development costs and time, with demands for increasing quality, lead to systematic application of computation, simulation, testing, and quality assurance methods, including virtual reality methods. All participants in the product creation process must work together from the very start of the project.

The *Handbook of Automotive Engineering* is a classical technical reference book, covering all areas of automotive research and development. Its content is equally interesting to the engineer and automotive specialist who seeks to expand his knowledge, as well as technically interested persons, students, and non-technical laymen who wish to learn more about the modern automobile and its systems.

The authors of this edition are all experts in their respective fields, and members of the German automotive and supplier industries. We would like to thank all of these authors for their contributions, as well as SAE, publishers of this handbook.

Grünwald/Braunschweig, March, 2004

Hans-Hermann Braess
Ulrich Seiffert

Contents

Note to the user:

Bibliographical references which are not cited in the text indicate additional literature. This may serve to provide the reader with more in-depth information on the material covered in the respective section.

1 Mobility

1.1 Introduction

As life requires movement, its opposite is equated with lack of motion, rigidity, lifelessness, indeed death. Matter itself is forever in motion. Galileo pronounced, "There is, in nature, perhaps nothing older than motion." Pascal, in his *Pensées*, wrote, "Our nature consists in motion; complete rest is death." Not surprisingly, since time immemorial, imprisonment has been considered one of the harshest punishments that could be meted out to man or beast. Imprisonment represents not only loss of freedom of movement, of the ability to leave one's lodgings, of travel or flight, but also considerable sensory deprivation and, therefore, a lack of new perceptions and communication of such perceptions.

Short of speculating on the Big Bang and the origin of the universe, it may be said that motion, along with time and space, is a cosmic absolute. The ability to move emancipated animals from the vagaries of nature liberated them from the bonds of a fixed location. Motion is a basic requirement for survival of species and therefore one of the principles of evolutionary success. According to Biblical scriptures, human mobility began with Adam and Eve being cast out of the Garden of Eden.

The fact is that humans have always sought to move faster, farther, and carry greater loads beyond the limitations of their own muscles—and, if possible, without any physical exertion on their part. For this reason, humans have always eagerly embraced the opportunities presented by technology. In ancient times, humans employed the ship, propelled by wind and human muscle power. The invention of the wheel led to the wagon, for centuries drawn by animal power, until the railway initiated a revolution in transportation about 160 years ago. Once again, it was shown that the progress of civilization and economic and cultural advancement are inseparable from mobility.

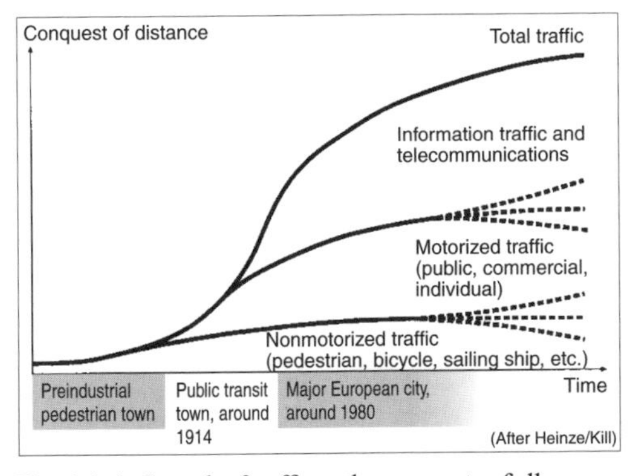

Fig. 1.1-1 Growth of traffic as the aggregate of all means of overcoming distance.

Beginning more than 110 years ago, the development of the automobile saw the creation of a means of transport which, thanks to its individual and flexible character, has remained the most commonly used method through the present day. Only the automobile is capable of traveling to virtually any desired destination at any time. This second transportation revolution made possible both personal individual mobility and virtually every form of freight transport in nearly every corner of the earth (Fig. 1.1-1, from [1], Fig. 1.1-2, from [2]).

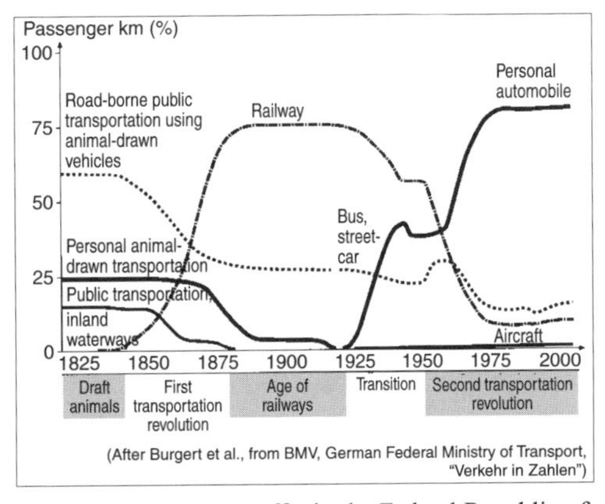

Fig. 1.1-2 Passenger traffic in the Federal Republic of Germany since 1820.

At present, more than 80% of German passenger miles are attributable to the personal automobile (Fig. 1.1-3).

Despite intensive efforts to improve public transportation, especially in urban areas (see Chapter 11), it may be assumed that the passenger car, thanks to its broad application (Fig. 1.1-4) as well as its unique advantages, will continue to be of major significance for the foreseeable future.

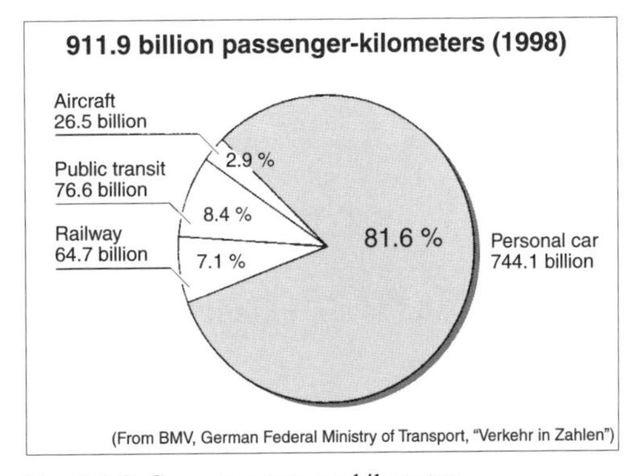

Fig. 1.1-3 German passenger-kilometers.

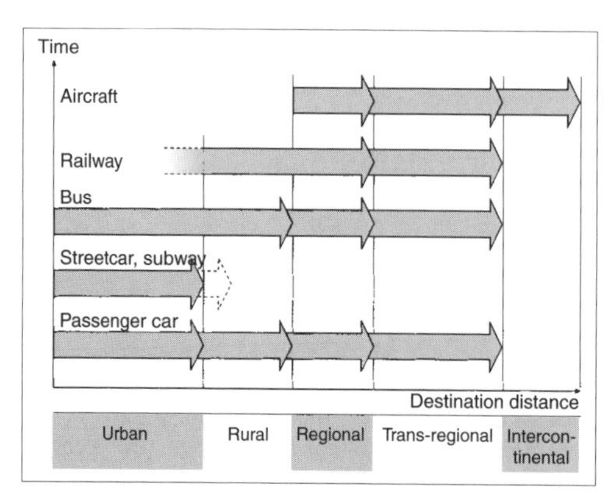

Fig. 1.1-4 Cooperation between major transportation systems.

One might almost say that at present, the great success of the automobile must also be regarded as its greatest problem. This leads to frequent discussion regarding regulatory efforts to restrict unlimited use of the automobile.

Currently, more than 500 million automobiles exist worldwide, even though mass motorization of several large and many small nations has not yet begun. Motor traffic has assumed a prominent place in the engineering world, as well as in general society, thanks to its advantages as well as disadvantages, particularly its appetite for natural resources, safety concerns, and environmental impact. It is therefore the primary mission of automotive and transportation engineering to minimize the disadvantages of global motorization while retaining and indeed reinforcing the automobile's advantages. This also suggests that a modern auto manufacturer can no longer merely offer independent technological products; vehicles as well as the transportation system—the road network, including parked vehicles—must be integrated into mobility concepts which encompass all transportation modes.

A necessary precondition for anyone involved in the automobile and vehicular traffic is familiarity with a wide range of disciplines. For example, mechanical engineering, manufacturing technology, chemistry, recycling, electrical engineering, electronics, data processing, ergonomics, biomechanics, transportation technology, and communications technology as well as sociology and ecology (effects on the inhabited and uninhabited environment) must be integrated, directly or indirectly, in the development of new vehicles.

1.2 Causes and forms of mobility

1.2.1 Definitions

In general, mobility represents the conquest of space, the reaching of destinations; to a certain extent, mobility is also understood as the ability to change location. For sociologists, mobility is also the movement of persons between social strata, while for the psychologist it is spiritual movement (cf. [3]).

Generally, three indicators serve to quantify physical mobility:

- The number of trips per person per unit of time
- The distance covered per person per unit of time
- The time per person per trip

The total of all movements of persons and goods is termed *traffic.*

The availability of vehicles is generally represented by the degree of motorization (number of vehicles per 1,000 inhabitants), a quantity largely dependent on economic factors (see, for example, [4] and Fig. 1.2-1).

Because often more than one person is traveling in a vehicle, a distinction must be made between "passenger-" or "person-kilometers" and "vehicle-kilometers." Transportation tasks for persons, goods, and information are extraordinarily complex (Fig. 1.2-2, from [5]). Different destinations, distances, and time constraints are but a few of the factors to be considered.

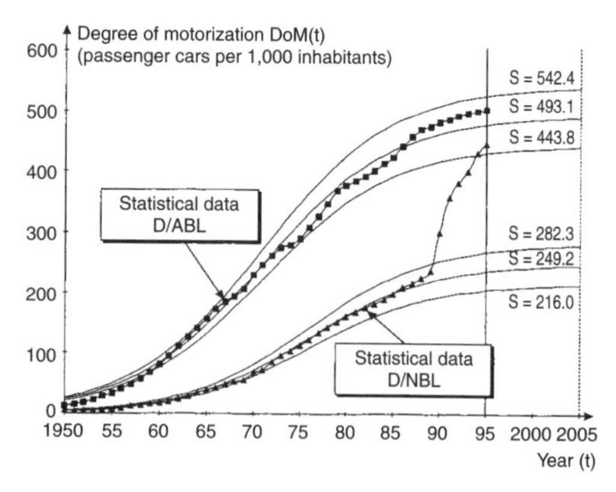

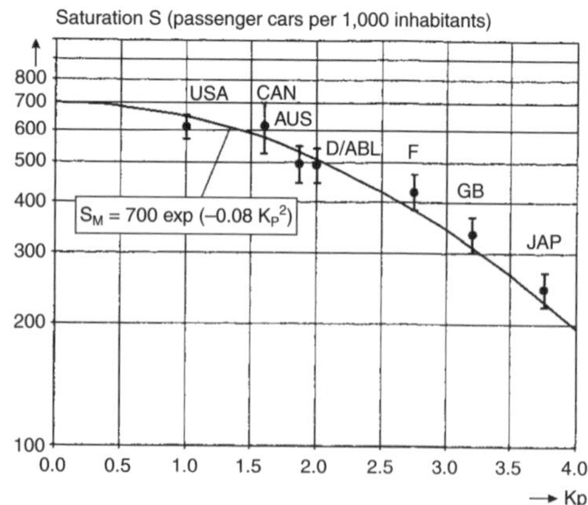

Fig. 1.2-1 Development of motorization in the former West and East Germany, saturation levels of motorization as a function of per capita income relative to fuel price. D/ABL, original West German states; D/NBL, original East German states.

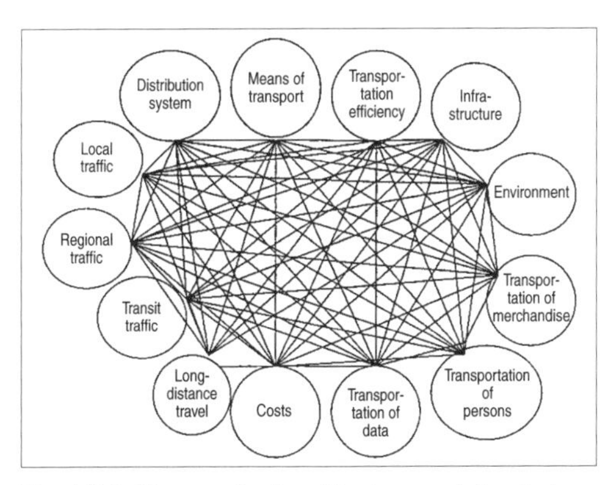

Fig. 1.2-2 The complexity of the transportation task.

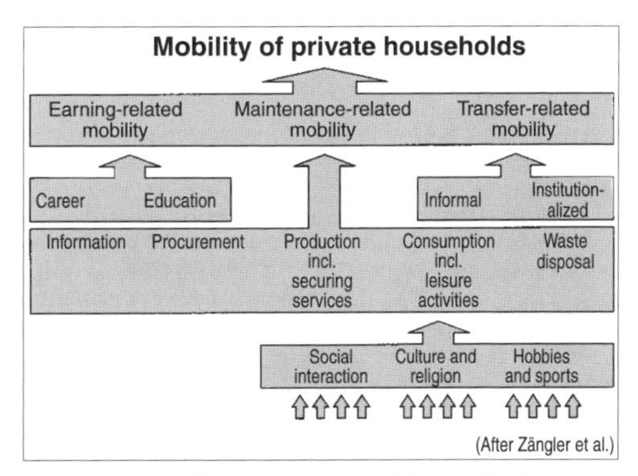

Fig. 1.2-3 Traffic-generating activities of private households.

Between departure point A and destination B, personal mobility is seldom conducted in a "unimodal" form (using only a single mode of transportation) or without interruption. Typical examples are:

- Walking from house to garage, driving from the garage to the train station, taking an express train from there to another city, where . . . In other words, a "chain" employing various transportation modes.
- Driving the car to the office, then driving to the supermarket after work, then driving to the movie theater, then driving home again. In other words, a chain with intermediate destinations.

In addition, often baggage and passengers must be carried along. Even these few examples underscore how mobility can assume a diversity of forms and how these forms impose different requirements on various transportation methods and their mutual interactions.

1.2.2 Activities determine mobility

Human age and the unique activities of various stages of human life influence the characteristics of mobility. Other major influences include the structures of housing, economy, and leisure time [6].

In Germany, about 88% of personal transportation mileage is traceable to the social and economic contexts of private households [7]; Fig. 1.2-3 schematically illustrates the activities which generate traffic. The remaining 12% of personal mileage is attributable to business and job-related activities. Of particular interest is leisure travel, which has been generally increasing for years (Fig. 1.2-4).

In addition, by virtue of unusual destinations, number of passengers, luggage, and leisure equipment, leisure travel imposes special demands on the transportation modes suited or preferred for this activity [8]. Tourism has become a major economic factor, particularly in the United States, where it has become the most significant source of employment in 32 states [9].

Although pedestrian traffic predominates for short

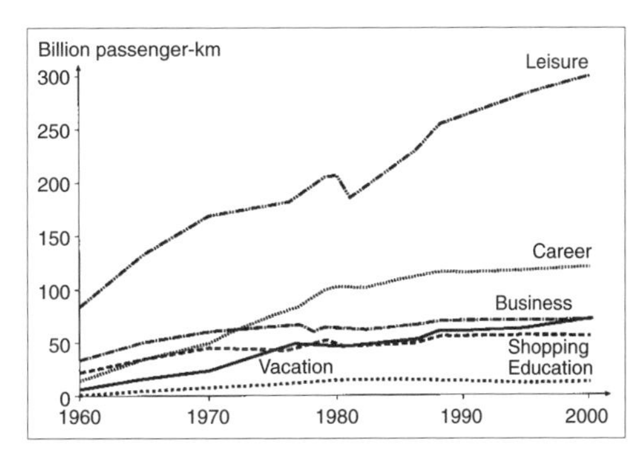

Fig. 1.2-4 Passenger car mileage by mission.

distances (Fig. 1.2-5, for example, from German data), motorized individual traffic is employed for short distances as well, such as for shopping trips or in inclement weather.

As already indicated, the particular advantages of the automobile (Figs. 1.2-6, 1.2-7 from [5]) give rise to the overall high proportion of motorized individual traffic. With increasing city size, however, public transportation bears a significantly higher proportion of traffic,

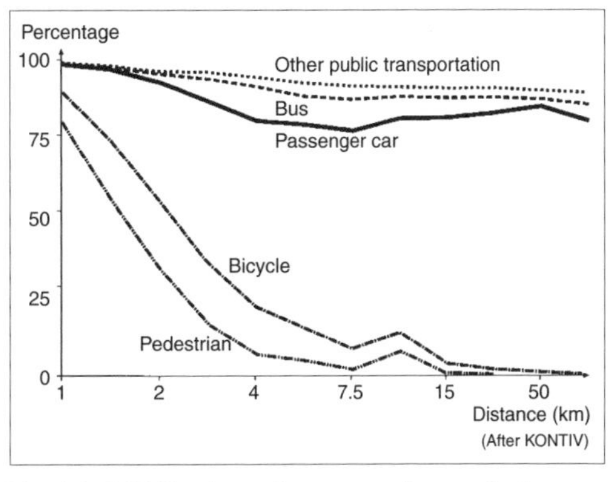

Fig. 1.2-5 Utilization of transportation methods as a function of destination distance.

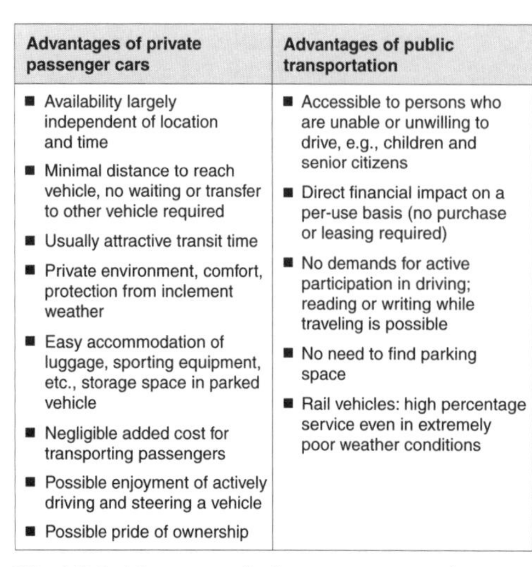

Advantages of private passenger cars	Advantages of public transportation
■ Availability largely independent of location and time ■ Minimal distance to reach vehicle, no waiting or transfer to other vehicle required ■ Usually attractive transit time ■ Private environment, comfort, protection from inclement weather ■ Easy accommodation of luggage, sporting equipment, etc., storage space in parked vehicle ■ Negligible added cost for transporting passengers ■ Possible enjoyment of actively driving and steering a vehicle ■ Possible pride of ownership	■ Accessible to persons who are unable or unwilling to drive, e.g., children and senior citizens ■ Direct financial impact on a per-use basis (no purchase or leasing required) ■ No demands for active participation in driving; reading or writing while traveling is possible ■ No need to find parking space ■ Rail vehicles: high percentage service even in extremely poor weather conditions

Fig. 1.2-6 Advantages of private cars compared to public transportation.

thanks to more capable subway and light rail systems providing easy access to city centers (Fig. 1.2-8, from [10]).

Besides the already addressed goal-oriented characteristics of personal mobility, one may also consider so-called "experience mobility" (Fig. 1.2-9, from [11]). Often coupled with leisure and vacation mobility, experience mobility usually leads to particularly unusual vehicle concepts and equipment features (see Section 4.2).

Commercial transportation also places special demands on passenger cars and their derivatives, for example with regard to cargo space and weight, economy, or configurability.

An important special condition of mobility and traffic is so-called "stationary traffic," which in the case of motorized individual traffic becomes a factor before and after every mobility event and becomes individually important if there is a shortage of private or public parking space. Insufficient parking space also has collective consequences (obstruction of traffic, the cost of traffic jams, etc.; see [12, 13]).

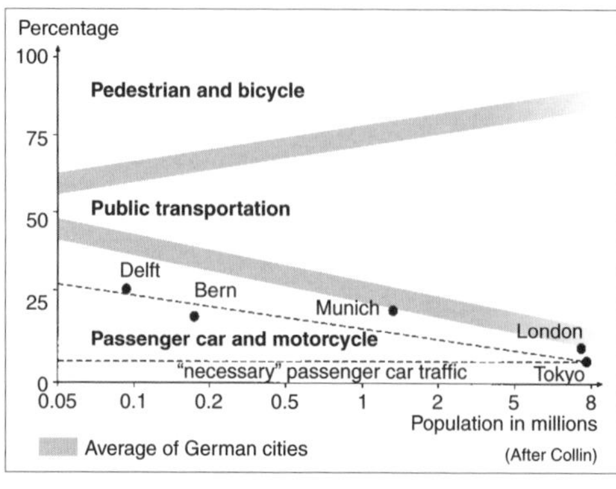

Fig. 1.2-8 Utilization of transportation modes for travel to city centers.

	Goal-driven mobility	Experience-driven mobility
Definition	Subsidiary needs derived from a primary goal; goal-oriented	Original/task-appropriate, independently formulated requirement: route oriented
Need	Rational utilitarian needs	Emotional need for recognition
Function	The automobile as means of transportation	The automobile as a means to an experience
Goal	"Value for money": Cost of ownership	"Value for money": Sheer driving pleasure
Features	Minimal experience value, primarily rigidly defined travel times and distances	High experience value, joy of driving, the pleasure of being on the road
		(After Weinhold)

Fig. 1.2-9 Comparison between goal- and experience-oriented mobility.

1.2.3 Transportation systems for freight traffic

For freight traffic, transport quality and economy are primary considerations in selecting the preferred means of transport. At this point, it suffices to say that here, too, road transportation offers unique advantages (e.g., [14] as well as Fig. 1.2-10, from [5]).

People	Private vehicles	Public transit	Aircraft	Rail
Safety	◆	●	●	●
Comfort	●	○	◆	◆
Speed	◆	◆	●	◆
Flexibility/transport frequency	●	○	○	○
Environmental impact	◆	◆	○	◆
Independence from weather conditions	◆	◆	◆	●
Availability	●	◆	○	◆
Consumer cost	●	○	●	◆

● high ◆ medium ○ low

Fig. 1.2-7 Qualitative evaluation of various human transportation systems.

Goods	Railway	Road traffic	Pipeline	Aircraft	Ship
Mass transport	●	○	●	○	●
Speed	◆	◆	○	●	○
Flexibility	○	●	○	○	○
Weather independent	●	◆	●	○	◆
Environmental impact	◆	○	●	○	◆
Transport frequency	◆	●	●	○	○
Variability	◆	●	○	○	○

● high ◆ medium ○ low

Fig. 1.2-10 Qualitative evaluation of various freight transportation systems.

1.2.4 Some marked aspects of mobility

While passenger cars are normally designed for as wide a variety of applications as possible ("encompassing all demands," see Chapter 2), there are special forms of mobility which may lead to constraints or very specific demands on their associated vehicle concepts. These include:

- Limited travel in urban areas (with distinct "city cars")
- Carrying heavy loads by means of trailers (boats, horses, etc.)
- Travel on unimproved roads and off-road
- Mobility of "younger" senior citizens (possibly with special senior citizen vehicles?)
- Travel in armored security vehicles

1.3 Stress points and disadvantages of mobility

Mobility is a prerequisite for economic well-being; without mobility of people and goods, it would be impossible to create value comprehensively in industry, trade, or the service sector. Similarly, mobility is a prerequisite for individual well-being. As in other sectors, mobility on a massive scale has its associated problems.

Collective traffic problems have always existed, above all when many people live and move in a limited space. For example, in an effort to combat noise, Julius Caesar issued an edict which banned military chariots from the streets of Rome at certain times of the day. And before the invention of the automobile, major cities were in danger of being strangled by horse-drawn vehicles (Fig. 1.3-1). Berlin in 1875—in other words, before the advent of the automobile—experienced more traffic deaths than it does at present.

Fig. 1.3-1 Even in the prehistory of the automobile, traffic conditions could be somewhat less than pleasant.

Today, with a global population of more than 6 billion, the consequences of personal and freight traffic have become a worldwide problem for urban areas, judicial districts, and indeed for the planet itself.

The direct negative effects of traffic are generally considered to be:

- Emissions
- Accidents
- Consumption of natural resources

Like other consequences of human activity, these may be attributed to various segments of individual means of transport (Fig. 1.3-2, from [15]).

Additionally, indirect effects on the general public are termed "external effects." Usually, these are only considered as external costs (e.g., [16]).

Business and economic drawbacks caused by traffic congestion and the resulting effects, such as loss of working or leisure time or failure to meet important deadlines and appointments, are discussed less often. Based on a German study [13], the costs of traffic congestion for the European Union are estimated at approximately 2% of EU gross domestic product.

Some effects, such as consumption of natural re-

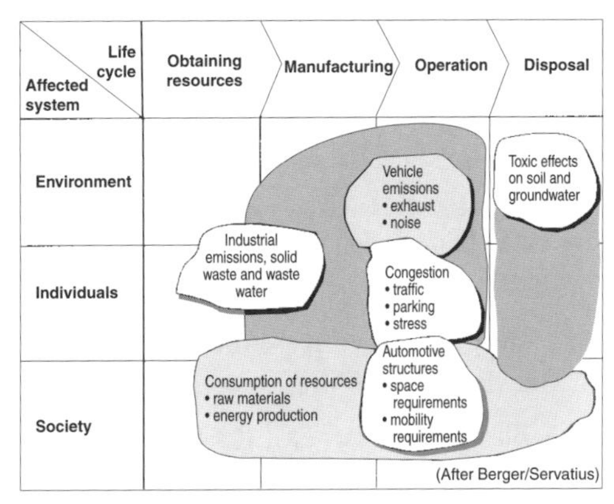

Fig. 1.3-2 Negative effects of the automobile.

• Objective limits	– Subjective limits
• Defined, rigid limits	– Expandable limits
• Technological limits	– Ecological limits
• Critical limits	– Noncritical limits
• Economic limits	– Social limits
• Availability limits	– Distribution limits
• ...	– ...

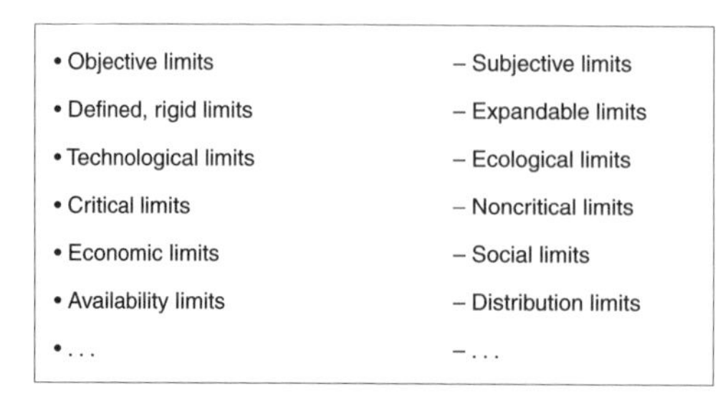

Fig. 1.3-3 Categories of traffic-related impact limits.

- Freedom from breakdowns and accidents
- Finding destinations in unfamiliar territory, assistance in finding parking space
- Avoiding unexpected delays (traffic congestion, etc.)
- Avoiding or mastering critical situations
- Mastering unavoidable accident situations
- Minimizing inconveniences and unusual conditions (weather, etc.)
- Parking assistance
- ...

Fig. 1.4-1 What do drivers want?

sources or CO_2 emissions, affect the entire planet; others, such as pollutants, noise, or use of space, manifest themselves primarily or entirely on a local or regional scale.

Not all problem areas have the same degree of criticality. For example, within the raw materials sector, ferrous materials are readily available worldwide, while special materials such as noble metals for catalytic converters or fuel cells are in limited supply.

Certain impacts, such as engine exhaust emissions or the environmental effects of manufacturing processes, have been dramatically reduced in recent years. Application of new technologies, such as telematics, may further reduce or even eliminate other burdens. In order to ensure optimum application of means to achieve these ends, it is necessary to have as complete a knowledge as possible of all resulting effects as well as the limits of the associated impacts (Fig. 1.3-3, from [17]). This, however, presents a great challenge not only for various scientific disciplines but also for legislators, industry, professional associations, and society as a whole. Enduring solutions are needed that will ensure the indispensable advantages of mobility while keeping negative consequences and disadvantages within tolerable limits. Without doubt, one aspect must be assigned a tolerance of zero: every traffic fatality is one too many.

It should be noted that traffic problems cannot be solved by better automobiles alone, nor can they be solved without the automobile. What is needed is a traffic management system which encompasses the entire traffic field, including four sectors:

- Traffic demand and supply management
- Traffic space management
- Traffic technology management
- Traffic organization management

All participants and responsible parties must be involved in this management system (see Chapter 11).

1.4 Mobility-relevant demands on the automobile

1.4.1 Basic demands

Since the introduction of motor vehicles in the late 1880s, the fields of automotive and traffic engineering have achieved a multitude of readily apparent, fundamental improvements in every aspect. Nevertheless, a ceaseless demand for continued improvements is matched by continued technical progress. Demands resulting from mass motorization were mentioned in section 1.3; moreover, every driver has his or her own expectations for future vehicles (for examples, see Fig. 1.4-1).

An effort to present an overall view of such demands is shown schematically in Fig. 1.4-2. This indicates that every category, from the worldwide distribution of customers and markets to the various economic sectors (such as vehicle manufacturers, suppliers, energy providers, and repair shops, for example), is more or less integrated in a network with common technological goals.

Another perspective originates in demands on traffic movement, concentrates on measures to improve traffic processes, in particular by means of internal and external driver-assist systems, and encompasses the responsible parties and participants (Fig. 1.4-3, from [18]).

Additional basic demands for mobility and traffic result from encroaching issues, such as "securing the energy supply" or "securing the supply of raw materials," that ultimately merge with the task of "sustainable mobility"—engineering development which can be maintained into the future (see Chapter 11).

Fig. 1.4-4 indicates one possible means of quantifying and representing demands imposed on vehicle con-

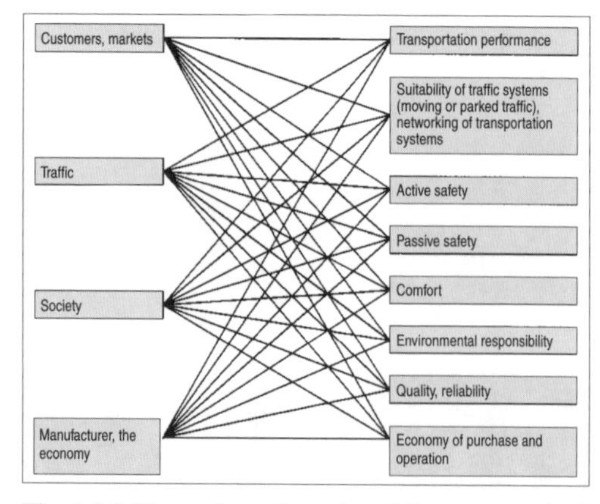

Fig. 1.4-2 Demands on the automobile—a networked system.

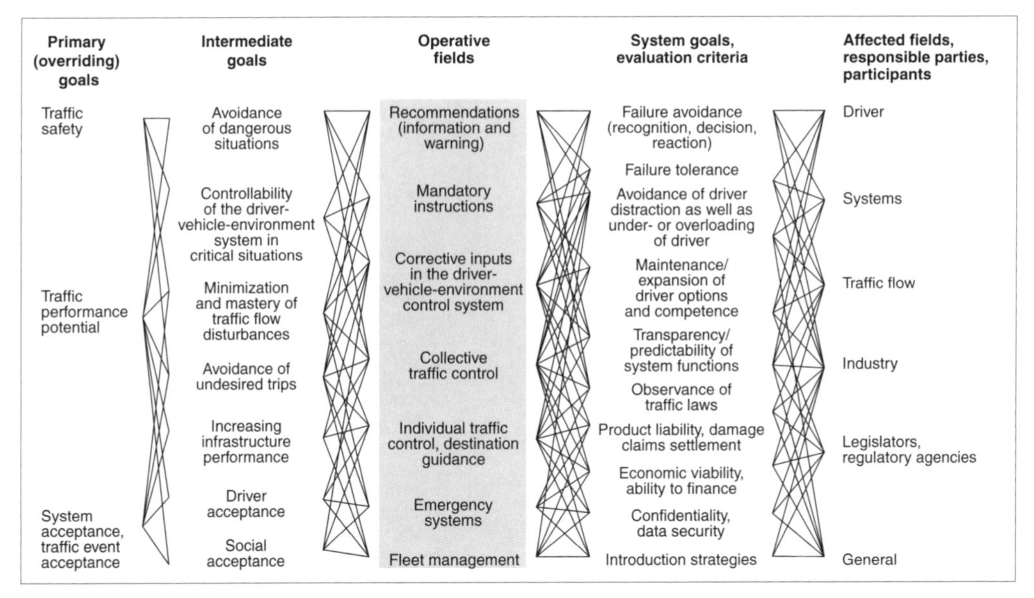

Fig. 1.4-3 PROMETHEUS goals and operative fields.

cepts, with range and maximum possible speed as the coordinate axes [19]. Other representational forms employ portfolios with various differentiating features, such as comfort, safety, and sportiness.

1.4.2 Special requirements

Despite the increasing importance of public transportation, motorized individual transportation will remain unavoidable in city centers. As a result, efforts to create dedicated city-friendly vehicle concepts (not a new idea in itself) continue to be made. Associated requirements include, above all, low exhaust and noise emissions as well as minimal parking space requirements. These goals can lead to special packaging concepts as well as unique powerplant configurations (e.g., electric cars); see Section 4.3.

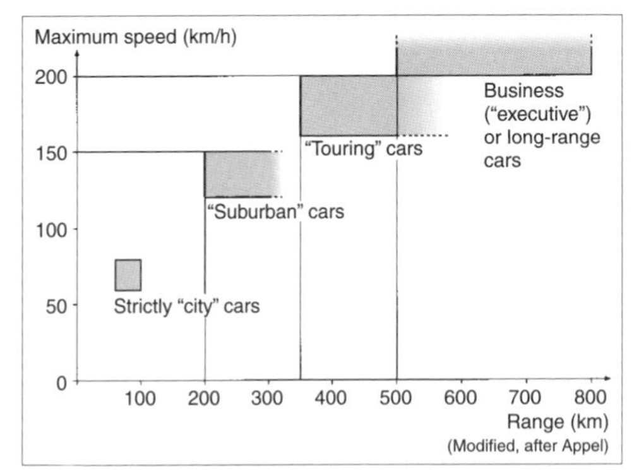

Fig. 1.4-4 Passenger car concepts: transportation properties and operating modes.

We fully concur with the following statement (from [20]): "Regardless of whether the topic is 'pure city cars' or 'city-capable cars': urban vehicle concepts will be successful only when the cars in question meet the actual demands of their users, and when they are successfully integrated into more comprehensive city traffic concepts."

While special conversions are offered to physically handicapped persons, the question of dedicated vehicle designs to meet the ergonomic needs of senior citizens is usually met with rejection rather than enthusiasm.

Although certain properties are preferred for taxi service (ease of entry, small turning circle, etc.), with the exception of London taxis, special vehicle concepts dedicated to this mission have been less than successful. On the other hand, off-road vehicles for extreme service, as well as derivatives which exhibit acceptable driving characteristics on paved roads, have achieved worldwide success for a variety of reasons.

It cannot yet be determined whether new forms of mobility, such as mobility leasing or car sharing (e.g., [21, 22]), will lead to special vehicle concepts.

References

[1] Heinze, G.W., and H.H. Kill. "Evolutionsgerechter Stadtverkehr." VDA Publication No. 66, 1991.
[2] Burgert, W., et al. "Tendenzen im Karosserieleichtbau." VDI-Bericht 1256, 1996, pp. 29–50.
[3] Gleich, M. *Mobilität—Warum sich alle Welt bewegt*, Hoffmann und Campe Verlag, 1998.
[4] Strobel, H., and H.-H. Braess. "Der Kraftstoffpreis—Eine verkehrspolitische Einflussgröße?" ATZ 1997, pp. 74–90.
[5] Seiffert, U. "Mobilität—Gesellschaftliche Anforderungen und technologische Optionen der Zukunft." RWE-Zukunftstagung "Gesellschaft und Technik im 21. Jahrhundert." Essen, Aug. 22, 1998.

[6] Hautzinger, H., and M. Pfeiffer. "Gesetzmäßigkeiten des Mobilitätsverhaltens." Berichte der Bundesanstalt für Straßenwesen, Vol. M 57, 1996.
[7] Zängler, Th., et al. "Ein mikroökonomischer Ansatz zur Analyse der Mobilität privater Haushalte." VDI-Bericht 1372, 1998, pp. 1–11.
[8] Opaschowski, H.W. "Freizeit und Mobilität." Vol. 12 of a series on leisure time research, BAT-Freizeit-Forschungsinstitut, 1995.
[9] Klein, H. "Wie mobil wollen Menschen sein?" In J. Speidel (editor), *Mobilität und TeleKommunikation,* Hüthig-Verlag, 1998, pp. 61–69.
[10] Collin, H.Z. "Integration des ruhenden Verkehrs in die Verkehrsentwicklungsplanung." VDI-Bericht 817, 1990, pp. 273–298.
[11] Weinhold, H. "Das öffentliche Meinungsbild zum Themenfeld Mobilität im europäischen Vergleich." Hochschule St. Gallen, 1994.
[12] Steierwald, M. "Benutzerfreundliche Parkgaragen." Dissertation TU Munich, 1993.
[13] Frank, D., and J. Sumpf. "Abschätzung der volkswirtschaftlichen Verluste durch Stau im Straßenverkehr." Unpublished study by BMW AG, 1994, revised 1998.
[14] Schulz, J. "Bewertung des Güterverkehrs auf Straße and Schiene." FAT publication No. 125, 1996.
[15] Berger, R., and H.-G. Servatius. *Die Zukunft des Autos hat erst begonnen—Ökologisches Umsteuern als Chance.* Piper Verlag, 1994.
[16] European Community Commission. "Green Paper on the impact of transport on the environment—A community strategy for 'sustainable mobility.'" COM(92)46, Brussels, April 6, 1992.
[17] Braess, H.-H., and D. Frank. "Sustainable Mobility—Gedanken zu Möglichkeiten und Grenzen der Präzisierung eines unscharfen Begriffes." Prague, FISITA-Congress, Paper E 16.07, 1996.
[18] Braess, H.-H., and G. Reichart. "PROMETHEUS—Vision des 'intelligenten' Automobils auf der 'intelligenten' Straße?" ATZ 1995, pp. 200–205 and pp. 330–343.
[19] Appel, H. "Automobil und Verkehr—Ganzheitliche Lösungsansätze." In VDI-Bericht 1007, 1992, pp. 1–57.
[20] Hautzinger, H. "Neue Ergebnisse zur PKW-Nutzung: Konsequenzen für innovative Fahrzeug-Konzepte." In H. Appel (editor), *Stadtauto—Mobilität, Ökologie, Ökonomie, Sicherheit.* Vieweg-Verlag, 1995.
[21] Voy, Ch., and K.-O. Proskawetz. "Mobilitätsleasing—Vision und Ansätze zur Realisierung." VDI-Bericht 915, 1991, pp. 115–132.
[22] Bundesminister für Verkehr (German Federal Transportation Minister). "Car Sharing kann Innenstädte entlasten." Verkehrs-Nachrichten, September 1995, pp. 1–4.
[23] Hüther, K. "Auswirkungen des Tourismus auf die Verkehrsnachfrage und die Umwelt." Internationales Verkehrswesen 4, 1998, pp. 133–136.
[24] Müntefering, F. "Mobilität als Grundlage für Innovation und Beschäftigung." Zeitschrift für Verkehrswissenschaften, 1999, Vol. 2, p. 87 ff.
[25] Baum, H. "Beschäftigungseffekte des Verkehrs—Eine quantitative Abschätzung." Zeitschrift für Verkehrswissenschaften, 1999, Vol. 2, p. 131 ff.
[26] Waschke, Th., et al. "Perspektiven der mobilen Gesellschaft." AVL Conference "Motor und Umwelt 99," conference proceedings, pp. 67–86.
[27] Haas, H.-D., and E. Störmer. "Angebotsqualität bei ÖV-Unternehmen." Internationales Verkehrswesen 4, 1999, pp. 119–124.
[28] European Commission. "Paving the Way for Sustainable Mobility." Transport Research Conference, Lille, November 8–9, 1999.
[29] Zängler, Th.W. *Mikroanalyse des Mobilitätsverhaltens in Alltag und Freizeit.* Springer-Verlag, 2000.
[30] ifmo-Institut für Mobilitätsforschung (editor). *Freizeitverkehr—Aktuelle und künftige Herausforderungen und Chancen.* Springer-Verlag, 2000.
[31] Brackmann, D., et al. "Untersuchung der Auswirkungen ausgewählter politischer Entscheidungen auf Verkehr und Umwelt." ifmo-Studien—Institut für Mobilitäts-Forschung (editor), Berlin, 2000.
[32] Various authors. "Gesamtverkehrsforum 2000." VDI-Berichte 1545, 2000.
[33] Various authors. Professional Congress Mobility (Proceedings, World Engineers Convention), EXPO Hannover, June 19–21, 2000. VDI-Verlag.

2 Requirements and conflicting goals

2.1 Product innovation and prior progress

For more than a century, the automobile has served as a means of transportation for people, animals, and goods. Although it has retained its basic characteristics throughout that time—four wheels, gasoline or diesel engine, and a transmission to modify output torque—it has nevertheless undergone tremendous changes. These have been characterized by technological progress, global availability, the need to satisfy a demand for mobility, international competition, legislation throughout the world, and the fact that the automobile as a product must be sold to customers. This last requirement means that customer desires have a strong influence on the product development process and must be given due consideration. Demands on the automobile itself are filled with contradictions, yet it has always been possible to resolve these collectively. Fig. 2.1-1 [1] clearly shows these conflicting goals:

- High vehicle safety
- Comfort and security
- Highest quality (no breakdowns, high retained value, great appeal)
- Low exhaust emissions
- Low exterior noise levels
- High recyclability
- Adequate performance and large payload volume
- Low costs, wear, and maintenance requirements
- Integration in an overall transportation system

Improving the traffic situation is of great importance for the future, as the customer attributes as yet unsolved traffic problems to the automobile industry and expects that same industry to provide a significant contribution toward their solution. Customer response to surveys of attitudes toward the middle-class automobile are shown in Fig. 2.1-2. Safety, reliability, and freedom from breakdowns are given high priority, along with good fuel economy.

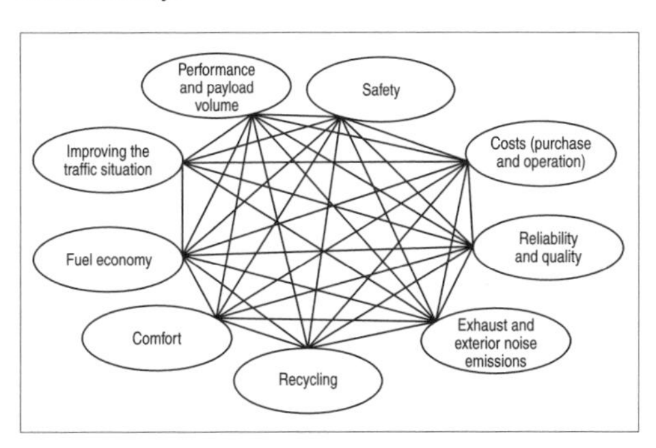

Fig. 2.1-1 Demands on the automobile.

Respondents answering "very important/important," in percent			
Safety	95	Fun to drive	51
Quality/reliability	93	Extensive appointments	50
Economy	85	Versatility, utility	48
Good fuel economy	84	Large trunk	44
Handling/roadholding	81	Interior design	42
Environmental compatibility	78	Styling, exterior design	40
Value	78	High engine output	25
Advanced technology	73	Automobile with personality and character	22
Comfort	62	Sportiness	18
Roomy interior	55		
Compact dimensions/ maneuverability	53		

Fig. 2.1-2 Relative importance of purchase criteria.

In response to customer demands, the variety of automotive offerings has expanded markedly in recent years. The range of models offered extends from extremely compact vehicles such as the "smart" [2], with an overall length of 2.5 meters, width of 1.515 meters, and height of 1.529 meters, to the long-wheelbase Mercedes S-Class [3], more than twice as long at 5.1 meters, 1.855 meters wide, and 1.444 meters high. Fig. 2.1-3 illustrates the size contrast between these extremes. Other segments also cover specific customer demands—for example, the large number of convertibles and roadsters such as the BMW Z3, Porsche Boxster, Audi TT Coupe and Roadster (Fig. 2.1-4 [4]), and multipurpose van and sport-utility segments. Fig. 2.1-5 is an example of a growth prognosis for vehicle classes with an overall length of less than about 3.5 meters. It shows that even now, customers are prepared to buy a special vehicle if it appeals to their inclinations and meets their expectations.

Prior to World War II, the development of the automobile as a means of transportation proceeded without significant attention to the environment, vehicle safety in terms of reducing the effects of accidents, or recycling. After the war, a multitude of smaller vehicles met immediate transportation demands. A distinct change occurred in the United States during the mid-1960s with

Fig. 2.1-3 Size comparison between Mercedes-Benz Smart and S-Class.

Fig. 2.1-4 Audi TT Roadster Quattro.

Proportion of small cars in overall market, 1992–2000

New vehicle registrations, western Europe (in millions)	13.8	11.7	12.2	12.3	13.0	13.6	14.0	14.4	14.5
Small car market share (percent)	3.3	4.2	4.3	4.3	4.1	5.7	5.9	6.6	6.9
Small car market share (millions)	0.45	0.49	0.52	0.52	0.53	0.77	0.82	0.95	1.0

The "small car" segment includes models such as Daihatsu Cuore and Move, Ford Ka, Seat Arosa, Fiat Selecto, MCC Smart, VW Lupo, Subaru Vivio, Rover Mini, Renault Twingo, etc.

Source: Marketing Systems

Fig. 2.1-5 Small car segment prognosis.

the introduction of safety and emissions regulations. Significant influence was exerted by consumer advocate Ralph Nader and his safety campaigns; this and purchasing requirements for government vehicles led to the Federal Motor Vehicle Safety Standards (FMVSS). This all-encompassing body of requirements includes all vehicle components and has undergone continuous expansion to the present day. Regulations for operation of motor vehicles have existed since the early 1900s in Germany, with the introduction of a requirement for motor vehicle liability insurance. International safety conferences, such as the ESV Conferences, helped to advance this important topic. At these conferences, accident researchers, mechanical and biomechanical engineers, and representatives of regulatory agencies meet to engage in intensive discussion of the latest vehicle safety research and information. Particularly in early years, the conferences included an innovative competition for the best engineering solutions for experimental vehicles or component systems. Vehicle safety may not only be influenced by vehicle manufacturers, but also to a very significant degree by customers and regulators. For example, road construction, signaling, driver education, restriction of alcohol and drug use, and legible signage have also made very positive contributions. Clear evidence is provided by the fact that in all (German) states, the number of fatal and injury accidents as a function of vehicle kilometers traveled has been considerably curtailed (Fig. 2.1-6; [5, 6]). This indicates how vital these activities have been and remain today. Differences by state should encourage "worse" states to imitate the "better" ones. This applies equally well to other transportation systems such as railways, public transit, and commercial airlines, which exhibit even greater safety per passenger-kilometer.

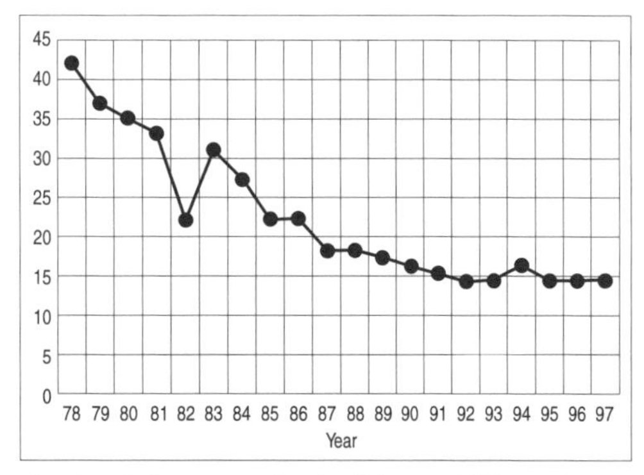

Fig. 2.1-6 German traffic fatalities per billion vehicle kilometers, from 1991 including the former East Germany.

The desire for change has not always originated with customers; often, society as a whole has brought about these changes by way of the political process. A typical example is California. Exhaust emissions regulations were introduced in response to the Los Angeles smog problem and poor air quality in other major cities.

In the mid-1960s, the United States placed limits on carbon monoxide (CO), oxides of nitrogen (NO_x), hydrocarbons (HC), and particulates. In the early 1970s, Europe restricted emissions of carbon monoxide, hydrocarbons, and particulates, and later oxides of nitrogen. In addition, vehicle manufacturers must answer the challenge that any additional components of exhaust emissions (those without specific restrictions) do not also represent a health hazard. The most significant technological breakthrough toward emissions reduction was achieved with the three-way catalytic converter and closed-loop mixture control (oxygen sensor or "lambda sonde") for gasoline engines in conjunction with unleaded gasoline. This applies to Europe as well, where high penetration of these concepts has been achieved in most countries. Measures were also implemented to reduce emissions from fuel tanks and lines. Evaporative emissions from parked as well as operating vehicles are controlled by means of appropriate sealing of tanks and lines and redirection of evaporative emissions to an active charcoal filter which is purged by the engine. Many gas stations draw vapor displaced during refueling operations through hoses at the fuel nozzle. For diesel engines, progress was at first achieved through internal engine measures, until the oxidizing catalytic converter, introduced in the late 1980s, permitted a greater degree of freedom in reducing emissions components. Since then, high-pressure fuel injection has resulted in continued success in diesel emissions control. Particulate emissions have been visibly reduced.

At present, intense effort is under way to improve both engine concepts. These may be divided into the following groups: minimizing cold-start emissions, further reduction of restricted exhaust components, and solving the oxides of nitrogen problem for direct-injection gasoline and diesel engines. California regulatory requirements, with that state's ULEV (Ultra Low Emission Vehicle), SULEV (Super Ultra Low Emission Vehicle), and ZEV (Zero Emission Vehicle) standards, and European targets embodied in Euro III and Euro IV will ensure that with the exception of carbon dioxide emissions, the automobile will completely disappear from any future discussions of environmental impact. Fig. 2.1-7 illustrates European measures since 1970 to reduce emissions and continued measures with Euro I through Euro IV in 2005. As a result of these measures, exhaust emissions per test kilometer will be reduced by an average of 97.5% to 2.5% of their pre-1970 levels.

External noise has also been reduced. The certification process includes simulated drive-by acceleration in second and third gear. Today, five times as many vehicles can perform this test (simultaneously), yet their combined sound emission will still be below the 82 dB(A) limit of 1975. With the current certification limit set at 74 dB(A), half of the noise is created by the interaction of tire and roadway; with a further reduction of the standard to 72 dB(A), this source will account for 75% of the generated noise. Obviously, the most important noise reduction activities will concentrate on reducing tire/roadway noise. External noise is gaining increased social importance; critics of the automobile often place it on the same level as urban parking and emissions problems.

It is obvious that the automobile has undergone positive changes in every important feature. This applies to its longevity—for example, galvanized bodywork or improved anticorrosion properties (corrosion protection for up to 12 years)—longer appeal and greater durability, increased torsional stiffness and the resulting freedom from rattles, improved ergonomics—seat design as well as location and function of control elements—and significant improvements in cabin noise and vibration. Customers pay special attention to noise and vibration behavior at idle as well as at speed. Many detail optimizations—decoupling, vibration dampers, even active control of air springs—have resulted in significant improvements for customers. This trend will doubtless continue.

A major contribution to these improvements is attributable to the body structure. Despite increased weight, the structure represents considerably better interior space utilization relative to the exterior footprint. Over the past decade, compact designs have seen a 10% improvement in this regard. This also applies to torsional stiffness; if one divides body weight by torsional stiffness and frontal area, modern bodies represent a better than 30% improvement.

Numerous support systems to assist in vehicle operation have been introduced to production cars. Fig. 2.1-8 gives examples of their areas of application.

For customers, a significant factor in the sales decision is fuel economy. Among vehicles offered in the Federal Republic of Germany by German manufacturers between 1978 and 1997, fuel use was reduced by approximately 26%, to an average of 7.3 L/100 km (32.2 miles per gallon). Although vehicle weights have (with a few exceptions) risen by an average of about 250 kg in the past few years, other innovations and applied measures have resulted in this improved fuel consumption.

Extensive detail work with regard to aerodynamic shaping has sharply reduced the drag coefficient (c_d), as well as the total aerodynamic drag. Compared to 1960,

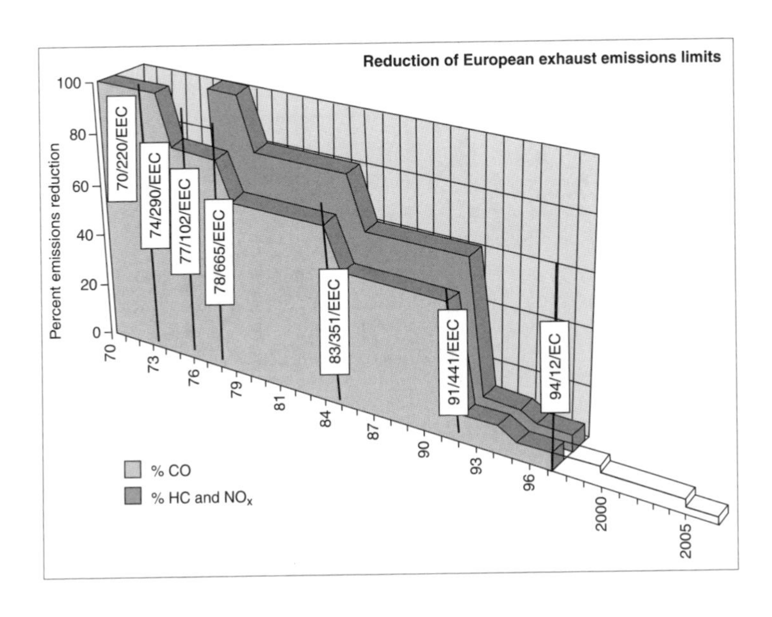

Fig. 2.1-7 Reduction of European exhaust emissions limits.

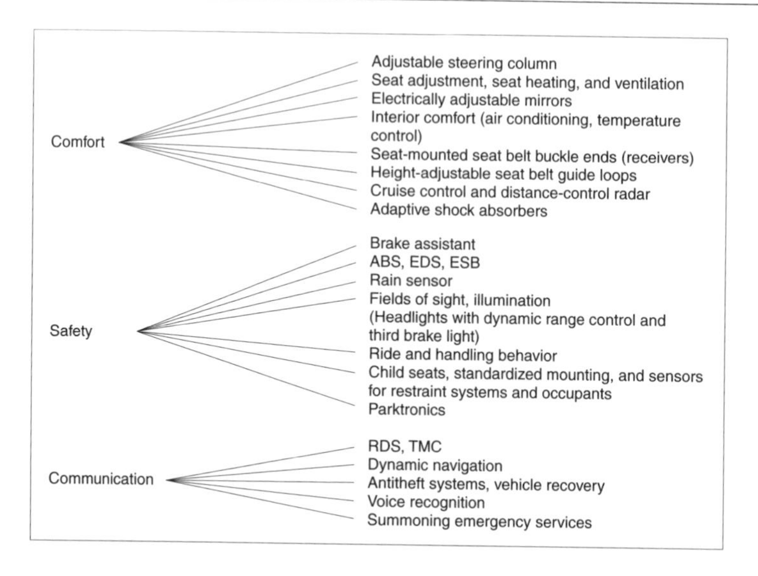

Fig. 2.1-8 Examples of vehicle support systems.

c_d has been reduced by about 40% to an average c_d of about 0.3 (see Fig. 2.1-9). Among exemplary modern vehicles, aerodynamic drag (the product of $c_d \times$ frontal area) is below 0.6 m^2. Despite the larger frontal areas resulting from the current shorter/higher design trend, there has been a steady reduction in drag coefficient and therefore aerodynamic drag. In current European Community test cycles, drag coefficient accounts for about 40% of fuel consumption.

The current tire generation, in part through the use of silica technology, has reduced the coefficient of rolling resistance from 0.02 in 1960 to the present value of about 0.008.

Automotive powerplants have made the greatest contribution to reduced fuel consumption and increased comfort. The gasoline engine has seen the introduction of multipoint fuel injection, solid-state ignition, multivalve engines, cylinder shutdown, lightweight design, optimized coolant flow, and mechanical supercharging for increased power or torque while still returning specific fuel consumption of 225 g/kWh. The diesel engine has been given direct fuel injection with variable supercharging to improve power and torque, common rail and unit injectors with injection pressures of nearly 1,600 bar and 2,000 bar respectively, and preinjection to reduce noise and emissions. The best engines with these features return specific fuel consumption numbers below 200 g/kWh. Transmissions have also been

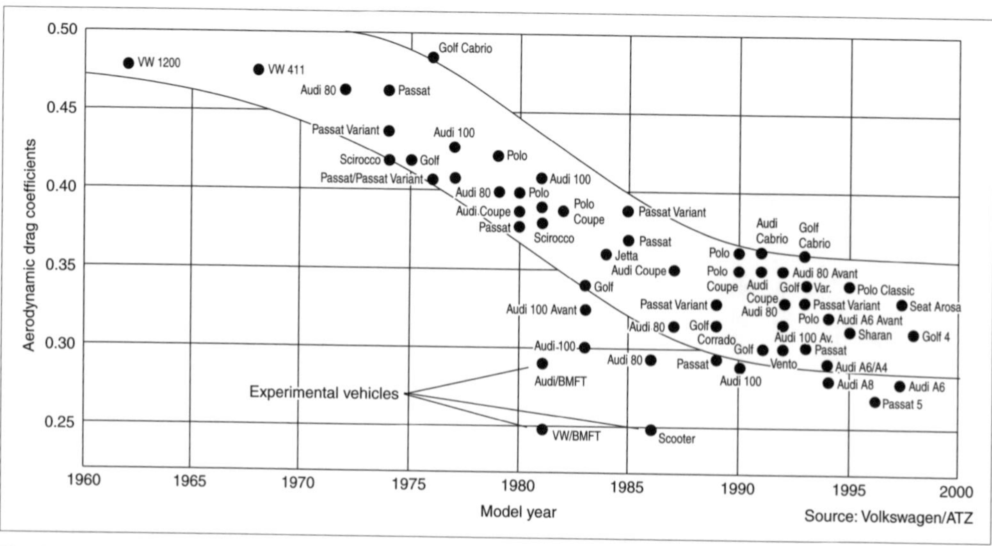

Fig. 2.1-9 Aerodynamic drag coefficients.

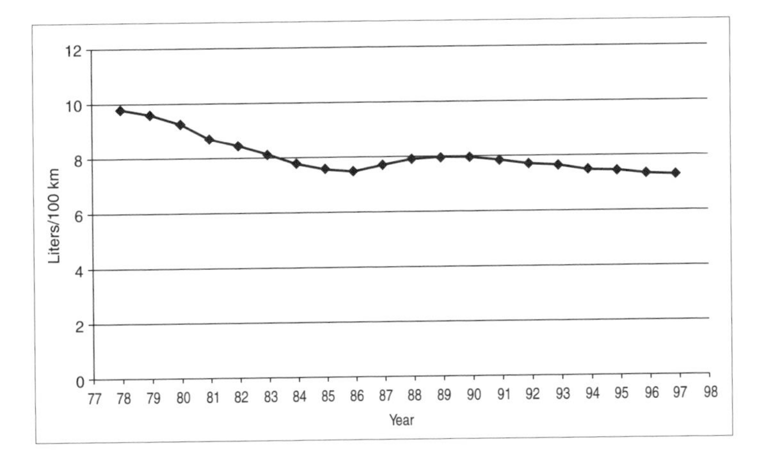

Fig. 2.1-10 Development of average German-made passenger car/wagon fuel economy.

improved with regard to torque capacity per unit weight and size. Among manual transmissions, five- and six-speed transmissions are the norm, while automatic transmissions have electronic controls and shifting with four-, five-, and six-speed automatics resulting in respectable fuel economy improvements. Recently introduced automated manual transmissions, in which gear selection may be made by the driver or by the automatic system, with the addition of engine shutdown at idle, represent a major contribution to reduced fuel consumption. Electronically controlled continuously variable transmissions will also carve out their own market niche.

Drivetrain improvements have more than offset the potential loss of fuel economy implied by increased vehicle weight. Fig. 2.1-10 illustrates the reduction in fuel consumption within the German automobile industry, plotted as average fuel economy of new models entering the market. Future goals with regard to further fuel economy improvements are just as ambitious. The German auto industry has voluntarily assented to meet the target set by the Association des Constructeurs Européens d'Automobiles (ACEA) to reduce fuel consumption below 5.7 L/100 km (assuming diesels account for 25% of the fleet) or 140 g CO_2/km as averages for new vehicles entering the market.

The VW Lupo/Audi AL2 holds the record in terms of fuel economy; when equipped with lightweight components, three-cylinder direct injection diesel engine with unit injectors, and automated manual transmission with idle shutoff, this fully capable four-seater returns about

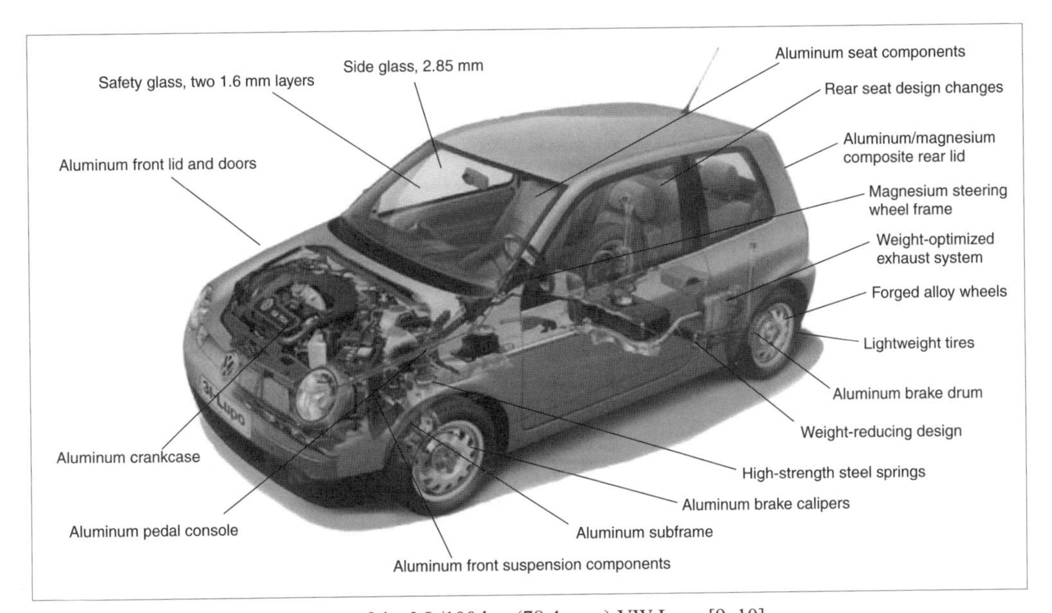

Fig. 2.1-11 Lightweight components of the 3 L/100 km (78.4 mpg) VW Lupo [9, 10].

80 g CO_2/km [7]. Fig. 2.1-11 illustrates the major components which contribute to its extreme fuel economy.

Another significant aspect is customers' indirect perception of improvements. Doubtless this includes acceptance of unleaded gasoline, the functionality of exhaust treatment concepts, and corrosion protection. This also applies to long-term qualities of installed electrical and electronic equipment, actuators, and sensors, which play a vital role in progress to date. Electronics have achieved enormous penetration in all vehicle components; Fig. 2.1-12 indicates that this trend will continue. Taking the example of feedback and control systems, Fig. 2.1-13 lists significant milestones on a timeline [8]; the history of continuous development is readily apparent. Electronics are increasingly being applied to monitoring the functionality of vehicle components, such as the airbag system, engine oil level and oil quality, and service intervals and the onboard diagnosis system (which monitors all components that might influence exhaust gas emissions).

Application of information and communications systems permits better integration of the automobile into other systems, through improved logistics and communication, dynamic navigation and destination guidance, and networking with other traffic systems. These also enable new communications possibilities for customers in their cars regarding, for example, general services, maintenance, and information on upcoming events. Customers expect:

1. Travel information
   Travel planning, choice of transportation mode, overall route planning, services, reservations and booking, route planning for pedestrians, tourist information
   Static route planning
   Road characteristics, parking availability, networking of transportation systems, road maps
   Personal communications
   Personal mailbox, ability to report road emergencies, summoning emergency services
2. Management systems in public transit systems
   Static and dynamic trip information, individual public transit travel planning, ticket sales or alternatives to conventional tickets
3. Parking space management
   Parking lot occupancy, dynamic parking space information, directions to parking, parking space reservation
4. Traffic information
   Navigation, individual and collective routing, dynamic accident information, general traffic conditions (congestion, road work, weather forecast, environmental conditions, special events, local restrictions, approach, transit)
5. Traffic demand management
   Carpooling, networking for freight service, city logistics

Along with changes in automotive technology have been changes in the relationship between customers and manufacturers. This is evidenced by a much larger selection of vehicle models, expanded services, innovative leasing offers, car sharing programs, and restructured sales organizations. Every company attempts to make its offerings as attractive as possible by means of a wide model range, either within the brand itself or by means of various nameplates within the same concern. This is also expressed through customer communications by

Present	Present and future
Starter	Electric steering assist
Fuel pump	Heated catalytic converter
Ignition	Electric engine coolant
Idle bypass valve actuator	Electric water pump
Throttle actuation	Emissions control air pump
ABS hydraulic pump	Active engine mounts
Front/rear windshield wipers	Active suspension
Headlight washing system	Tire pressure monitors
Front/rear electric windows	Brake assistant
Electric sunroof	Electric brakes
Electric door and trunk lid locks	Electric A/C compressor
Headrest and seat adjustment	Electric 5- or 6-speed mechanical transmission
Electrically adjustable mirrors	Heated windshield
Electric antenna	Improved night vision
Lights, headlights	Lane-change viewing device
Rear defroster	Parking assistant
Seat heat	Automatic distance control
Heated windshield washer nozzles	Intelligent airbag sensors
Heated oxygen sensor	Telephone/fax
Engine/transmission electronic control unit	Voice recognition
ABS and traction control	
Cruise control	
Steering	
Instruments	
Airbags and belt tensioners	
Air conditioning system	
Radio/CD/cassette/amplifier	
Infrared remote locking	
Navigation	
Antitheft system	

Fig. 2.1-12 Vehicle electrical and electronic systems (source: [11]).

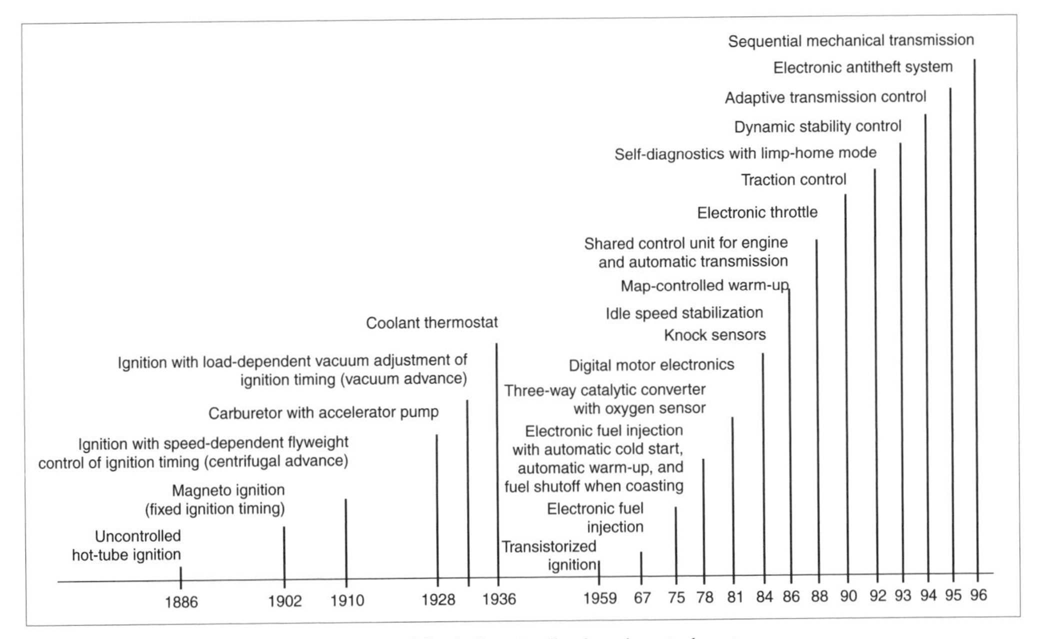

Fig. 2.1-13 Significant milestones in automobile design: feedback and control systems.

means of clinics, internal and external surveys, customer satisfaction indices, comparison tests of repair facilities, durability tests by consumer organizations (such as ADAC and TÜV in Germany or *Consumer Reports* in the United States), crash tests (such as Insurance Institute for Highway Safety, European and U.S. NCAP tests), and firsthand information from automotive magazines. In contrast to earlier times, today's customers have at their disposal outstanding product and brand information.

The aforementioned recyclability has been integrated into an already well-established recycling system. However, in the interest of preserving resources, it is necessary to channel recycled materials to new uses.

The automobile's importance in providing mobility for people and goods has made this industry an important economic factor throughout the world. With worldwide annual production of about 50×10^6 new cars, at about 25,000 DM (US$11,500) per car, the industry has gross revenues of 1,250 billion DM (US$575 billion). Added to this is indirect revenue from distribution and repair operations. The past 20 years have witnessed great challenges which the automobile industry has had to overcome: globalization; internal competition; increased productivity; continuous improvement processes; modularization; changing relationships with suppliers, scientific and technical facilities, and development partners; and a new orientation for the product development process. These changes include early inclusion of all departments involved in the product development process, a departure from sequential development to parallel processes which include external development by suppliers or outside development contractors. Shared components (within a company or between several companies) and platform, module, and common part strategies enabling a large number of model variations permit cost reductions and trimmed development capacities.

In the future, the automobile industry will have to have innovative solutions. Traditionally, the industry is prepared to embrace major changes only if customers are both available and willing to accept new engineering solutions. Fig. 2.1-14 illustrates two forms of innovation. The left side lists external factors which are difficult to influence, such as oil crises and regulatory activities, while the right side shows technology-driven changes such as microelectronics, fuzzy logic, and

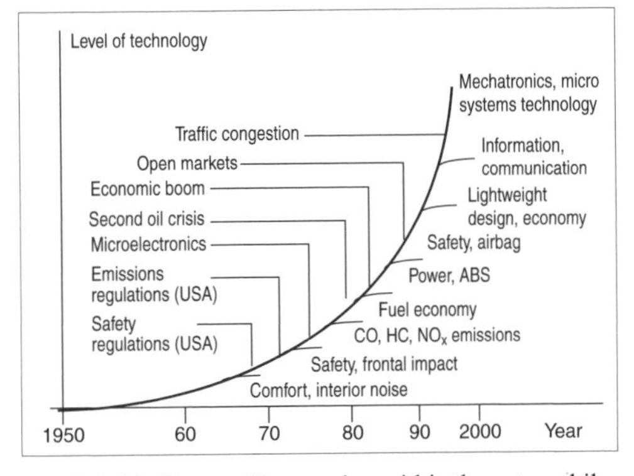

Fig. 2.1-14 Waves of innovation within the automobile industry.

neural controls. At the millennium, an increased pace of innovation is being demanded of the automobile industry. At the forefront of these innovations are new forms of gasoline and diesel engines, alternative powerplants, fuel cells and hybrids, accident avoidance by means of sensors and supported guidance systems including "electronic copilots," optimization of vehicle electrical and electronic devices, and lightweight design. The technological challenges are immense but achievable if all sectors involved—science, politics, and industry—are guided by achievable criteria and apply their energies to pull in the same direction.

References

[1] Smart press information. Micro Compact Car GmbH, Remmingen, 1998.
[2] Micro Compact Car GmbH. "Smart" press information, 1998.
[3] ATZ/MTZ. "Die neue S-Klasse." Wiesbaden, 1998.
[4] Audi AG press information.
[5] Seiffert, U. "Möglichkeiten und Grenzen der Erhöhung der Sicherheit im Kfz." ÖVK II/97, Technische Universität Wien.
[6] "Verkehr in Zahlen 1998." Deutscher Verkehrs-Verlag, German Federal Ministry of Transportation.
[7] Volkswagen AG press information. 1998 Paris Auto Salon.
[8] Reitzle, W. "Das Automobil: Zukunft durch Innovation and Faszination." ATZ/MTZ special publication, *Geschichte and Zukunft des Automobils.* Wiesbaden, 1998.
[9] Winterkorn, M. et al. "Das Dreiliter-Auto von Volkswagen—der Lupo 3 l TDi," ATZ Vol. 101 No. 6, Wiesbaden, 1999.
[10] Winterkorn, M. et al. "Das Dreiliter-Auto von Volkswagen—der Lupo 31 Tdi." ATZ Vol. 101 No. 7/8, Wiesbaden, 1999.
[11] Seiffert, U. "Die Automobiltechnik nach der Jahrtausendwende." Handelsblatt/ Euroforum, Munich, 1998.
[12] von Fersen, O. (ed.) *Ein Jahrhundert Automobiltechnik. Personenwagen.* VDI-Verlag, Düsseldorf, 1986.
[13] "Geschichte and Zukunft des Automobils." ATZ/MTZ special publication. Wiesbaden, 1998.
[14] VDI conference proceedings. VDI-Verlag, Düsseldorf.

2.2 Regulatory requirements

The following section will concentrate on European regulations. In the event that individual requirements vary globally, these will be treated in the appropriate chapters.

2.2.1 Certification for service on public roads

Section 1 paragraph 1 of the German motor vehicle code stipulates that motor vehicles and trailers operating on public roads must be approved for service by the appropriate licensing authority. Furthermore, §18 of the vehicle code—the Strassenverkehrs–Zulassungs–Ordnung (StVZO)—requires that any motor vehicle with a top speed of more than 6 km/h as determined by its configuration as well as its trailer(s) may only be operated on public roads if the type has been granted a use permit or European Community type approval and licensed for use by issuance of a motor vehicle or trailer license plate.

For vehicles produced or to be produced in series, the manufacturer may obtain a so-called General Use Permit (Allgemeine Betriebserlaubnis, §20 StVZO) or if possible an EC type approval. Prior to issuance of a General Use Permit, the manufacturer must guarantee that it will exercise the granted powers responsibly. In the case of an EC Type Certificate, the applicant must present proof to the licensing authority that a quality assurance system meeting that authority's requirements has been implemented. As of January 11, 1996, vehicles of class M_1 (see Table 2.2-2) will no longer be issued General Use Permits.

For class M_1 vehicles, Fig. 2.2-1 illustrates the regulatory processes while Table 2.2-1 provides an overview of the applicable legislative authority.

2.2.2 National and supranational legislative authorities

2.2.2.1 Traffic code and motor vehicle code (German StVZO)

Section 6 of the German traffic code empowers the Federal Ministry of Transportation, Building and Housing with approval of the German Senate (upper house of Parliament—the Bundesrat) to issue regulations and general management requirements covering the condition, equipment, examination, and identification of vehicles. Section 38 paragraph 2 and §39 of the Federal Emission Protection Law empowers the Federal Ministry of Transportation, Building and Housing and the Federal Environment, Environmental Protection and Reactor Safety Ministry with approval of the German Senate to issue appropriate regulations addressing environmental protection. The regulations, issued jointly by both ministries, are part of the StVZO. Construction and operational regulations for vehicles are addressed beginning with §30, "General Regulations regarding Vehicle Condition." According to §30 paragraph 1, vehicles must be built in such a manner that:

1. When used as intended in traffic, they do not harm any person(s) nor endanger, hinder or burden any person(s) or endanger any person(s) to any avoidable degree.
2. Occupants are protected from injury to the greatest possible degree, especially in the event of an accident, and the scope and consequences of injuries are minimized.

The paragraphs following §30 paragraph 1 regulate additional details.

2.2.2.2 European Community directives

The role of the European Community is in part an outgrowth of the Treaties of Rome (1957), specifically the treaty which formed the European Economic Community (EEC).

The main mission of the EEC Treaty is the establishment of a joint domestic market by dismantling trade

Table 2.2-1 Comparison of EC directives, ECE regulations, and German traffic codes

	EC Directive	ECE Regulation	StVZO
Requirements for active vehicle safety (accident prevention)			
Steering equipment	70/311/EEC	ECE 79	§38
Brake systems	71/320/EEC	ECE 13	§41
Replacement brake pads/shoes	71/320/EEC	ECE 90	§22
Equipment for acoustic signals	70/388/EEC	ECE 28	§55
Field of vision	77/649/EEC		§35b
Defrosting and defogging systems for glazing	78/317/EEC		§35b
Windshield wipers and washers	78/318/EEC		§40
Rearview mirrors	71/127/EEC	ECE 46	§56
Heaters (engine waste heat)	78/548/EEC		§35c
Gas heaters, auxiliary heaters			§22a
Installation of lighting and light-signaling devices	76/756/EEC	ECE 48	§49a, 53a
Reflex reflectors	76/757/EEC	ECE 3	§53
Clearance lamps, taillamps, stop lamps	76/758/EEC	ECE 7	§51, 51b, 53
Side marker lamps	76/758/EEC	ECE 91	§51a
Turn signal lamps	76/759/EEC	ECE 6	§54
Headlamps for high beam and/or low beam	76/761/EEC	ECE 1, 8, 20 ECE 37	§50
as well as their light sources			§22a
Gaseous discharge headlamps		ECE 98	
as well as their light sources		ECE 99	
Front fog lamps	76/762/EEC	ECE 19	§52
Rear fog light (fog taillamp)	77/538/EEC	ECE 38	§53d
Backup lamps (reversing lamps)	77/539/EEC	ECE 23	§52a
Parking lamps	77/540/EEC	ECE 77	§51c
Rear registration plate illumination devices	76/760/EEC	ECE 4	§60
Reverse gear and speedometer equipment	75/443/EEC	ECE 39	§39, 57
Interior equipment (symbols, warning lights)	78/316/EEC		§30
Wheel covers	78/549/EEC		§36a
Tire tread depth	89/459/EEC		§36
Tires and tire mounting	92/23/EEC	ECE 30	§36
Towing capacity, hitch vertical load	92/21/EEC		§42, 44
Towing equipment (trailer hitches)	94/20/EC	ECE 55	§43
Pedal arrangement		ECE 35	§30
Requirements for vehicle passive safety (injury mitigation)			
Interior fittings (protruding elements)	74/60/EEC	ECE 21	§30
Steering mechanism (behavior in event of impact)	74/297/EEC	ECE 12	§38
Frontal impact, occupant protection	96/79/EC	ECE 94	
Side impact, occupant protection	96/27/EC	ECE 95	
Seat belt anchorages	76/115/EEC	ECE 14	§35a
Seat belts and restraint systems	77/541/EEC	ECE 16	§35a
Seats, seat anchorages, head restraints	74/408/EEC	ECE 17, 25	§35a
Head restraints	78/932/EEC	ECE 17, 25	§35a
External projections	74/483/EEC	ECE 26	§30c
Fuel tanks and rear underrun protection	70/221/EEC	ECE 58	§47–47c
Liquefied petroleum fuel systems		ECE 67	§41a, 45, 47
Doors (locks and hinges)	70/387/EEC	ECE 11	§35e
Front and rear bumpers		ECE 42	
Rear-end collisions (not applicable to Germany)		ECE 32	
Child restraint systems		ECE 44	§22a
Safety glazing	92/22/EEC	ECE 43	§40
Electric propulsion (safety)		ECE 100	§62

Table 2.2-1 (Continued)

	EC Directive	ECE Regulation	StVZO
Requirements for emissions behavior			
Permissible sound level and exhaust system	70/157/EEC	ECE 51	§49
Replacement exhaust systems	70/157/EEC	ECE 59	§49
Pollutant emission from motor vehicle engines	70/220/EEC	ECE 83, 103	§47, 47c
Pollutant emission from diesel engines (smoke)	72/306/EEC	ECE 24	§47
Fuel consumption and CO_2 emissions	80/1268/EEC	ECE (84), 101	§47d
Radio interference suppression and electromagnetic compatibility (EMC)	72/245/EEC	ECE 10	§55a
Energy consumption of electric propulsion systems		ECE 101	
Miscellaneous			
Type approval of vehicles	70/156/EEC		§19, 20
Rear license plate (location and mounting)	70/222/EEC		§60
Devices to prevent unauthorized use of vehicle	74/61/EEC	ECE 18	§38a
Alarm systems and immobilizers	74/61/EEC	ECE 97	§38b
Manufacturer's plates and serial numbers	76/114/EEC		§59
Towing devices	77/389/EEC		§43
Engine full-load power measurement	80/1269/EEC	ECE 24, 85	§35
Masses and dimensions of class M_1 vehicles	92/21/EEC		§32, 34
Top speed (measurement)		ECE 68	§30a

barriers. This ensures, among other things, that vehicles which meet the unified technical standards can participate in unrestrained trade. High levels of safety and environmental protection were given priority in drafting these regulations (the so-called directives). Based on Article 94 of the 1957 Treaty and on a recommendation by the European Commission, the European Parliament has issued numerous directives for road vehicles. Today, Article 95a, in conjunction with Article 14 of the Treaty which created the European Community (in its October 2, 1997 version) form the legal basis for the directives. Organizationally, the directives consist of commission directives and individual directives. The three commission directives cover type approval for motor vehicles and motor vehicle trailers (70/156/EEC; at present only applicable to Category M_1 motor vehicles; see Table 2.2-1);

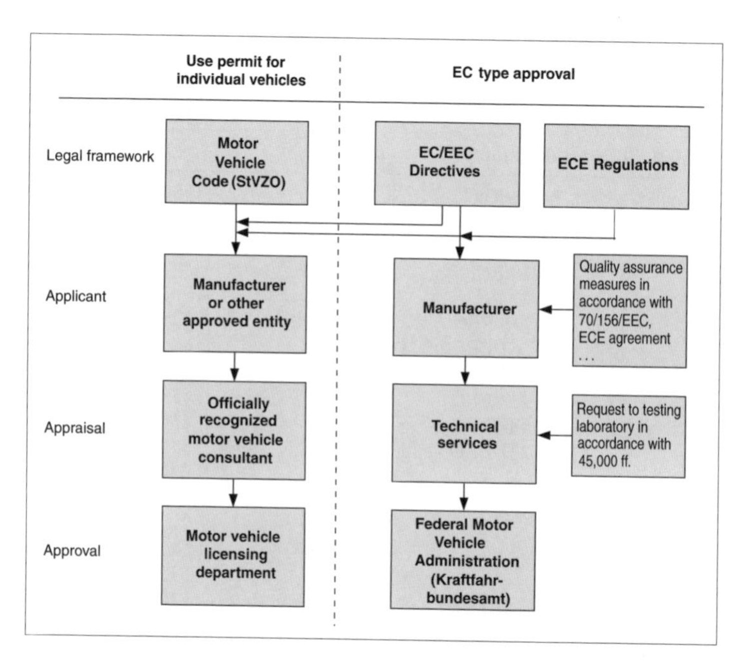

Fig. 2.2-1 Approval process for M_1 vehicles.

wheeled agricultural and forestry tractors (74/150/EEC); and two- and three-wheeled motor vehicles (92/61/EEC). Along with vehicles, type approval may also be granted to systems, components, and independent technical units.

In keeping with directive 70/156/EEC, last modified by directive 98/91/EC, motor vehicles and trailers subject to the aforementioned directives are divided into categories (Table 2.2-2). For off-road-capable motor vehicles, the letter G is appended to the M or N category designation. Motor homes are generally classified as M_1; these are special-purpose M_1 vehicles.

Depending on bodywork—ranging from sedan to wagon to multipurpose vehicle—M_1 vehicles are further differentiated. Multipurpose vehicles are capable of transporting passengers and baggage or goods in a single-volume interior. In order to qualify for an M_1 classification, these vehicles must fulfill additional criteria.

2.2.2.3 United Nations Economic Commission for Europe regulations

On the basis of an agreement signed March 20, 1958, "concerning the adoption and reciprocal recognition of approval for motor vehicle equipment and parts," the United Nations Economic Commission for Europe (ECE) is also involved with coordination of motor vehicle regulations. Significant parts of the agreement were changed in Revision 2, dated October 16, 1995. The title now reads "concerning the adoption of uniform technical prescriptions for wheeled vehicles, equipment and parts which can be fitted and/or be used on wheeled vehicles and the conditions for reciprocal recognition of approvals granted on the basis of these prescriptions." In the spirit of this agreement, the expression "wheeled vehicles, equipment and parts" refers to all wheeled vehicles, equipment and parts whose characteristics affect highway safety, the environment, and energy conservation. Accession to the agreement is voluntary.

Now UN members who are not also members of the ECE but are invited by the ECE as advisors, as well as regional economic organizations empowered by their members to exercise rights within the framework of the agreement, are also able to accede to the agreement.

The European Community acceded to the agreement on March 24, 1998; Japan entered on November 24, 1998.

Since 1958, more than 100 regulations have taken effect. The Federal Republic of Germany has adopted most of these. The agreement and its revisions have been given the power of law in Germany. This law empowers the Federal Ministry for Transportation, Building and Housing, after considering input from individual German states, to issue decrees implementing the regulations within the Federal Republic of Germany.

The European Community has recognized numerous ECE regulations as being equivalent to specific EC directives. Manufacturers may apply these within the framework of obtaining an EC-type approval.

2.2.2.4 Further measures to reduce trade barriers

Originating in transatlantic dialogue to strengthen economic ties and reduce trade barriers between Europe and the United States, a further agreement to codify global motor vehicle regulations was established within the ECE. This instrument is not intended to replace the 1958 ECE agreement, but rather to exist alongside it; it is designated a parallel agreement.

The basis of the 1958 agreement is mutual recognition of approvals based on ECE regulations. The objective of the parallel agreement is solely the formulation and establishment of global technical regulations regarding safety, environmental protection, energy efficiency,

Table 2.2-2 Vehicle classification

Category M	Passenger motor vehicles with at least four wheels
Category M_1	Passenger vehicles with no more than eight seats plus driver
Category M_2	Passenger vehicles with more than eight seats plus driver, not to exceed gross weight[a] of 5 metric tons
Category M_3	Passenger vehicles with more than eight seats plus driver and gross weight exceeding 5 metric tons
Category N	Motor vehicles having at least four wheels, or three wheels when maximum weight exceeds 1 metric ton, used for carrying goods
Category N_1	Vehicles used for the carriage of goods, maximum weight not exceeding 3.5 metric tons
Category N_2	Vehicles used for carrying goods, maximum weight exceeding 3.5 but not exceeding 12 metric tons
Category N_3	Vehicles used for carrying goods, maximum weight exceeding 12 metric tons
Category O	Trailers (including semi-trailers)
Category O_1	Trailers with maximum weight not exceeding 0.75 metric ton
Category O_2	Trailers with maximum weight exceeding 0.75 metric ton but not exceeding 3.5 metric tons
Category O_3	Trailers with maximum weight exceeding 3.5 but not exceeding 10 metric tons
Category O_4	Trailers with a maximum weight exceeding 10 metric tons

[a] Technically permissible total vehicle weight stated by the manufacturer. In place of the concept "total vehicle weight," some directives employ the term "maximum weight."

and antitheft protection. Specific methods used to implement the formulated regulations are left to the agreement's individual signatory parties. In this way, type approval processes as well as self-certification are possible. The global technical regulations are prepared by the same working groups involved in ECE regulations.

Along with the United States of America, the European Community, the Russian Federation, and Japan, member states of the European Community and others may also join the parallel agreement. The agreement took effect on August 25, 2000.

2.2.3 Accident prevention (active safety)

2.2.3.1 General

The present regulations regarding requirements for active vehicle safety do not represent a challenge for manufacturers of M_1 vehicles. Fostered by tests conducted by automotive magazines and automobile clubs, a healthy competition has existed in this field for many years, leading to continuous improvement of active vehicle safety as well as occupant comfort (which must be regarded as a related field). The standards established by this competition are considerably higher than those imposed by regulatory agencies.

Fig. 2.2-2 illustrates the fields encompassed by government regulation (with the exception of lighting technology).

2.2.3.2 Brake system

The brake system must enable application of service brakes, auxiliary brakes, and parking brakes. The service brakes must enable control of vehicle motion as well as safe, rapid, and effective deceleration of the vehicle from all speeds and load conditions and on hills or downgrades of any slope. The auxiliary brakes must enable stopping the vehicle within a reasonable distance: for example, in the event of failure of a main brake circuit. The parking brake must be able to hold the vehicle immobile even in the absence of the driver on hills and downgrades.

For the purpose of obtaining type approval, the vehicle's brake effectiveness is determined by actual road testing. The following tests are conducted:

- Normal test of brake effectiveness with cold brakes (Type 0)
- Testing of reduction in brake effectiveness (Type I) with repeated brake applications, followed by hot braking test

The values to be met are given by Table 2.2-3.

The effectiveness of the auxiliary braking system is determined by a Type 0 test, with the engine disengaged and with a starting speed of 80 km/h.

The braking distance may not exceed the value calculated from

$$s = 0.1v + \frac{2v^2}{150}$$

and the average deceleration may not be less than 2.9 m/s^2.

If the auxiliary braking system is foot-actuated, the specified braking action must be achievable with an actuating force of no more than 500 N.

The parking brake must be able to hold the fully laden vehicle on an 18% hill or downgrade. If the parking brake is hand-actuated, the actuating force may not exceed 400 N; force for foot-actuated parking brakes may not exceed 500 N.

For vehicles not equipped with an automatic antilock system, the front-axle coefficient of adhesion must exceed rear-axle adhesion under all load conditions.

Problems with the automatic antilock brake system must be indicated to the driver by means of an optical warning signal.

Antilock systems are divided into three categories. Category 1 is subject to the most stringent standards.

According to ECE Regulation 13 (ECE-R13), footnote 5, if antilock systems and/or functions such as traction control are installed, the manufacturer shall provide documentation which covers system layout, description of functions, and safety concept to the technical service performing the approval.

2.2.3.3 Driver's field of vision

The driver must have an adequate forward field of vision. With the vehicle installed in a three-dimensional

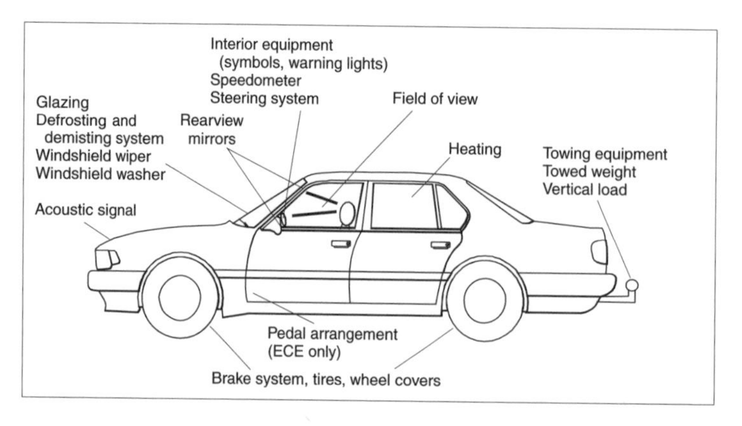

Fig. 2.2-2 Active safety.

Table 2.2-3 Brake efficiency requirements

	Brake test type	0-I
Type 0 test with disengaged engine	Test speed $\upsilon =$	80 km/h
	Braking distance s [m] $\leq$	$s = 0.1\upsilon + \frac{\upsilon^2}{150}$
	a_{ave}[a] $\geq$	5.8 m/s^2
Type 0 test with engaged engine	$\upsilon = 80\%$ of υ_{max} but $\leq$	160 km/h
	$s \leq$	$s = 0.1\upsilon + \frac{\upsilon^2}{130}$
	a_{ave}[a] $\geq$	5 m/s^2
	Pedal force $\leq$	500 N

[a] a_{ave} = average deceleration

coordinate system (Fig. 2.2-3), various reference points are used to determine whether an adequate field of vision is obtained.

Two of these reference points are the "H point" and the "SgRP" or "seating reference point." Both reference points also play a role in other directives.

The H point is the pivot center of the torso and thigh on the so-called H-point machine (Fig. 2.2-4), which is placed on the vehicle seats. The SgRP or manufacturer's design reference point is located for each seat by the manufacturer during the design process.

2.2.3.4 Tire chains

It must be possible to fit a type of tire chain to at least one wheel and tire combination of a permanently driven axle.

2.2.3.5 Lighting equipment

Only required or permitted lamps, luminescent materials, and reflecting devices may be installed (Fig. 2.2-5). Lighting equipment is subject to type approval, must be firmly attached to the vehicle, and must be capable of functioning at all times.

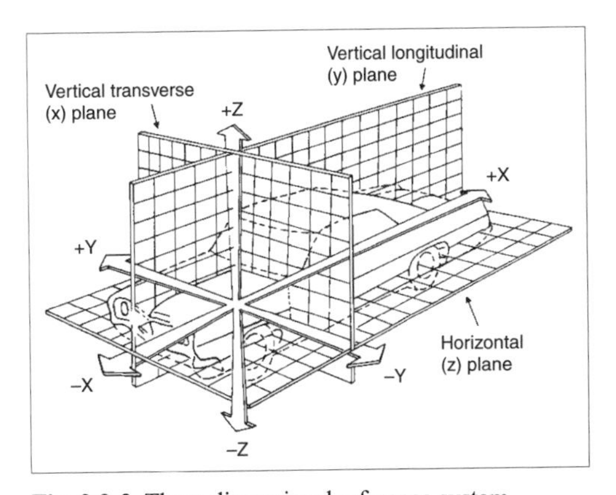

Fig. 2.2-3 Three-dimensional reference system.

2.2.4 Vehicle safety and injury mitigation

2.2.4.1 General

Comparative crash tests and dissemination of the evaluated results to the general public have fostered a rewarding competition among manufacturers for improved passive vehicle safety, benefiting occupant protection. In conjunction with this, the European New Car Assessment Program (Euro NCAP) should be mentioned. Its objective is to provide consumers with independent evaluation of vehicle safety levels. Test vehicles are awarded up to five stars. At present, the program includes the following tests:

— Frontal impact; 40% offset against a deformable barrier as described in 96/79/EC, but at 40 mph (64 km/h)
— Side impact, as described in 96/27/EC; results evaluated according to FMVSS 201
— Pedestrian impact with four impactors (leg, hip, child head, adult head) based on EEVC (Euro-

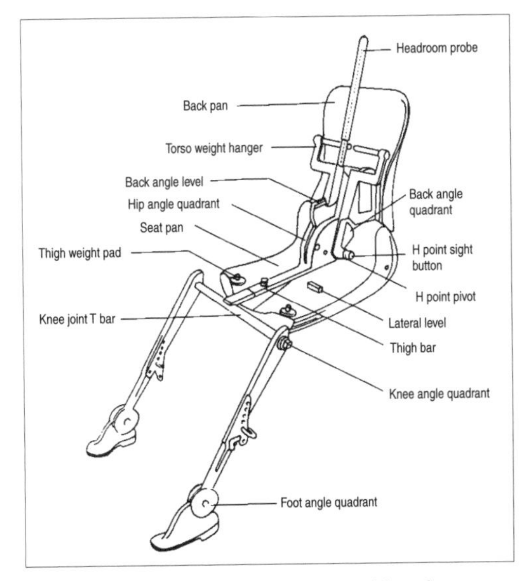

Fig. 2.2-4 Description of H-point machine elements.

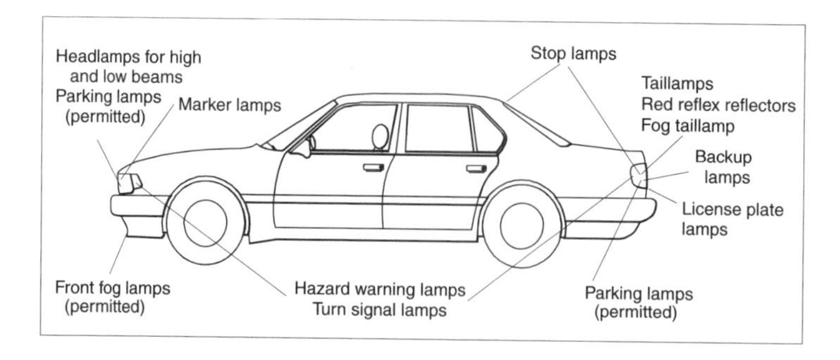

Fig. 2.2-5 Lighting equipment.

pean Enhanced Vehicle Safety Committee) guidelines

Fig. 2.2-6 provides an overview of elements addressed by regulations.

2.2.4.2 Occupant protection in frontal impacts

Directive 96/79/EC applies to category M_1 vehicles with a total permissible mass not exceeding 2.5 metric tons. Vehicles approved under this directive also meet requirements of the directive for steering system behavior during impacts with regard to steering wheel and column intrusion. The vehicle, fitted with test dummies on the front outboard seats, impacts against a transverse barrier at a speed of 56 km/h (35 mph). The front face of the barrier consists of a deformable structure. The vehicle must overlap the barrier face by 40% ± 20 mm. Injury criteria are determined on the basis of multiple transducers mounted in the test dummies; these criteria may not exceed certain maximum values. The criteria address head performance, neck injury, thorax compression, femur force, and tibia compression force.

2.2.4.3 Occupant protection in side impacts

Directive 96/79/EC applies to category M_1 and N_1 vehicles for which the R-point of the lowest seat is not more than 700 mm above ground level. A movable deformable barrier impacts the driver's side at a speed of 50 km/h (31 mph). A side impact dummy is installed in the driver's seat. No door may open during the test. After the impact, it must be possible without the use of tools to open a sufficient number of doors provided for normal entry and exit of occupants and, if necessary, tilt the seatbacks or seats to allow evacuation of all occupants and to remove the dummy from the vehicle.

Head, thorax, pelvis, and abdomen performance criteria may not exceed certain maximum values.

2.2.5 Requirements for emissions behavior

2.2.5.1 General

The regulatory requirements for emissions behavior, particularly exhaust gas behavior, constitute formidable challenges for vehicle manufacturers. Therefore, it is vital that emissions limits taking effect in the years 2000 and 2005 with stricter or additional requirements be established well in advance, enabling manufacturers to work toward concrete goals.

2.2.5.2 Sound emissions and exhaust system

Sound measurements are conducted on stationary as well as moving test vehicles. Stationary sound level is

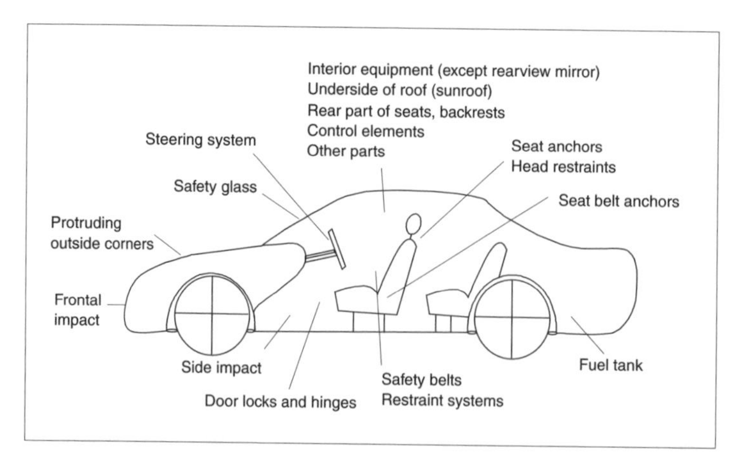

Fig. 2.2-6 Safety and mitigation of injuries.

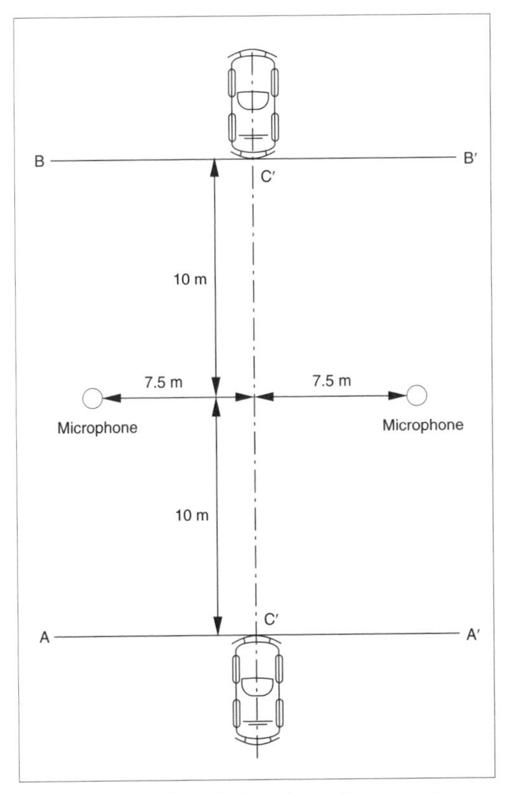

Fig. 2.2-7 Location of microphones for measuring maximum drive-by sound levels.

measured in order to determine a baseline for the same vehicle under traffic conditions.

To measure maximum drive-by sound levels, the vehicle is driven at a steady speed to line AA′ (Fig. 2.2-7). When the front of the vehicle reaches line AA′, the throttle control is depressed as quickly as possible. This throttle setting is maintained until the back of the car has reached line BB′. The throttle control is then released to its idle position as quickly as possible. During the vehicle drive-by between lines AA′ and BB′, the A-weighted maximum sound level in decibels is to be measured. For category M_1 vehicles, the maximum sound level of 74 dB(A)* may not be exceeded.

A special EC use permit may be obtained for exhaust systems or replacement muffler systems.

* **Note**: Higher limits for certain M_1 vehicles:
+1 dB(A) for diesel direct injection
Additionally +1 dB(A) for off-road vehicles of maximum weight >2 metric tons and <150 kW output or
Additionally +2 dB(A) for off-road vehicles of maximum weight >2 metric tons and >150 kW output
An idle period of 40 seconds is deleted with implementation of EURO 3 (as of January 2000).

2.2.5.3 Exhaust emissions

2.2.5.3.1 Emissions from motor vehicle engines

Vehicles powered by gasoline engines are subject to the following tests:

- Measuring average exhaust emissions after cold start (Type I) for vehicles with a maximum weight ≤3.5 metric tons
- Measuring carbon monoxide emissions at idle (Type II) for vehicles with a maximum weight >3.5 metric tons; as of EURO 3, also applies to vehicles ≤3.5 metric tons
- Measuring crankcase emissions (Type III)
- Measuring evaporative emissions (Type IV) for vehicles with a maximum weight ≤3.5 metric tons
- Durability of emissions-reducing components (Type V) for vehicles with a maximum weight ≤3.5 metric tons
- Measuring carbon monoxide and hydrocarbon emissions at low ambient temperatures after cold start (Type VI) for vehicles with ≤6 seats and ≤2.5 metric tons
- Onboard diagnostic system (OBD) test

As of 2003, new diesel-powered vehicle types with a maximum weight ≤3.5 metric tons must pass the Type I and Type V tests; diesels of ≤2.5 metric tons and no more than six seats must also pass the OBD tests.

For the Type I test, a driving cycle is conducted on a chassis dynamometer. The cycle consists of two parts: Part 1 (urban cycle) and Part 2 (suburban cycle). This is represented in Fig. 2.2-8.

The measured gaseous emissions masses (and in the case of diesel-powered vehicles, the mass of particulates) determined by each test may not exceed the limits, in g/km, given in Table 2.2-4.

The Type II test is to be conducted on gasoline-powered vehicles for which the previously mentioned Type I test does not apply. As of EURO 3 (January 2000), however, it is to be applied to all gasoline-powered vehicles.

A vehicle shall be deemed to pass the Type III test if gaseous emissions from the crankcase do not escape to the atmosphere in the course of any measurements. Measurements are conducted on the chassis dynamometer under three different operating conditions: for example, at idle or at a vehicle speed of 50 km/h.

Tests to measure pollutant emissions are complex and demand extraordinarily precise methodology. The seemingly most minor inattention during testing may necessitate abortion and repetition of the test.

An example from the Type IV test illustrates the effort required. The test is to determine the loss of hydrocarbons by evaporation from the fuel systems of gasoline-powered cars.

To measure evaporative emissions, the required equipment includes a gas-tight, rectangular measuring chamber with dimensions sufficient to enclose the test

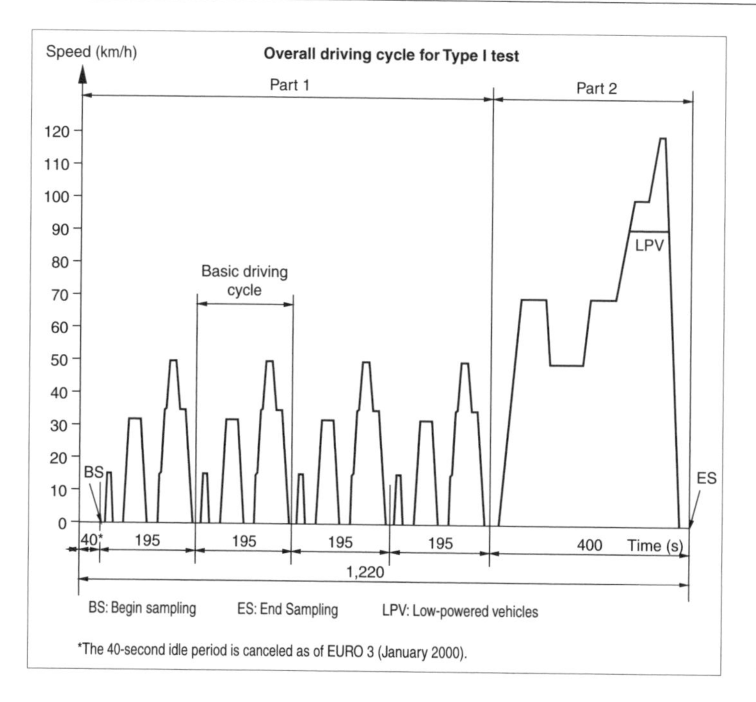

Fig. 2.2-8 Type I driving cycle.

Table 2.2-4 Type I emission test limits for M_1-class vehicles

Category	Class	Reference mass (RW) (kg)	Limit values								
			CO		HC		NO_x		HC + NO_x		PM[a]
			L_1 (g/km)		L_2 (g/km)		L_3 (g/km)		$L_2 + L_3$ (g/km)		L_4 (g/km)
			G	D	G	D	G	D	G	D	D
A (2000)											
M[b]	—	all	2.3	0.64	0.20	—	0.15	0.50	—	0.56	0.05
N_1[c]	I	RW ≤ 1305	2.3	0.64	0.20	—	0.15	0.50	—	0.56	0.05
	II	1305 < RW ≤ 1,760	4.17	0.80	0.25	—	0.18	0.65	—	0.72	0.07
	III	1760 < RW	5.22	0.95	0.29	—	0.21	0.78	—	0.86	0.10
B (2005)											
M[b]	—	all	1.0	0.50	0.10	—	0.08	0.25	—	0.30	0.025
N_1[c]	I	RW ≤ 1305	1.0	0.50	0.10	—	0.08	0.25	—	0.30	0.025
	II	1305 < RW ≤ 1,760	1.81	0.63	0.13	—	0.10	0.33	—	0.39	0.04
	III	1760 < RW	2.27	0.63	0.16	—	0.11	0.39	—	0.46	0.06

[a] For compression ignition engines.
[b] Except vehicles whose maximum mass exceeds 2,500 kg.
[c] And those Category M vehicles which are specified in Note b.
CO = Mass of carbon monoxide; HC = mass of hydrocarbons; NO_x = mass of oxides of nitrogen; HC + NO_x = combined mass of hydrocarbons and oxides of nitrogen; PM = mass of particulates; G = gasoline; D = diesel.

vehicle. The vehicle must be accessible from all sides. After closing, the chamber must be gas-tight and its inner surfaces impermeable to and nonreactive with hydrocarbons. In order to account for volume changes due to variations in chamber temperature, chambers with variable or fixed volumes may be employed. The interior wall temperature must be kept between 278 K (5°C) and 328 K (66°C) and, during the entire hot-soak test, between 293 K (20°C) and 325 K (52°C).

The air outside the chamber is to be monitored with a hydrocarbon analyzer of the flame ionization detector (FID) type. Fans or blowers must provide thorough mixing of the air in the chamber when sealed as well as adequate chamber ventilation when opened.

The test procedure is illustrated schematically in Fig. 2.2-9.

Evaporative emissions from each of these phases are calculated from the initial and final hydrocarbon concentration, air temperature and pressure, and net chamber volume. The total amount of emitted hydrocarbons is the sum of emissions during the tank heating and hot soak phases.

The Type V test represents an aging process over 80,000 km, conducted according to a predetermined test sequence on a test track, on the road, or on a chassis dynamometer. The manufacturer may employ predetermined aging factors in place of the aging test and the individual aging factors determined by this test, and employ these for the Type I test.

As of 2002, new gasoline-powered category M_1 vehicle types with no more than six seats and a maximum weight of no more than 2,500 kg are subjected to an additional test (Type VI) to measure exhaust CO and HC emissions at low ambient temperatures after a cold start. The test uses Part 1 of the driving cycle (Fig. 2.2-8), without the soon-to-be-deleted 40-second idle phase. At a test temperature of $-7°C$, CO may not exceed 15 g/km and HC is limited to 1.8 g/km.

As of 2000, new gasoline-powered M_1 vehicle types under 2.5 metric tons must in addition be fitted with an onboard diagnostic system (OBD) to monitor emissions. For new diesel-powered M_1 vehicle types under 2.5 metric tons and with no more than six seats, this went into effect in 2003.

2.2.5.3.2 Diesel engine emissions (diesel smoke opacity)

The light absorption coefficient of exhaust gases is measured continuously by an opacimeter. Measurements are conducted at both steady engine rpm and under "snap acceleration." The latter test is especially useful in establishing a baseline for later testing of vehicles in service. At idle, the accelerator control is rapidly and smoothly depressed to achieve the injection pump's maximum fuel injection quantity.

Light absorption by exhaust gas is measured at full load for six different engine speeds. The resulting absorption coefficient, as a function of air throughput, may not exceed certain specified limits.

2.2.5.4 Fuel consumption and CO_2 emissions

CO_2 and hydrocarbon emissions are measured during the test cycle illustrated in Fig. 2.2-8. Fuel consumption is calculated from HC, CO, and CO_2 emissions for city, suburban, and overall conditions. Limits are not specified.

2.2.5.5 Electromagnetic compatibility and radio noise suppression

In the spirit of directive 89/336/EEC regarding electromagnetic compatibility, "apparatus" is defined as all electrical and electronic appliances together with equipment and installations containing electrical and/or electronic components.

Such apparatus must be designed so that:

a. The electromagnetic disturbance it generates does not exceed a level which prevents radio, telecommunications equipment, and other apparatus from operating as intended.
b. The apparatus has an adequate level of intrinsic immunity to electromagnetic disturbances to enable it to operate as intended.

This is a so-called "horizontal" directive; with few exceptions, it applies to all apparatus, regardless of installation location.

If the protection requirements of directive 89/336/EEC are harmonized with individual (so-called "vertical") directives for specific devices, this directive does not apply to these devices and protection requirements. For example, directive 72/245/EEC is one such individual directive which sets more rigid requirements for electromagnetic compatibility and electromagnetic interference emissions from motor vehicles. Limits have been established for broadband and narrow-band electromagnetic interference emissions as well as invulnerability of vehicles to interference from electromagnetic fields. This directive applies to new vehicle types as of January 1, 1996, and to new vehicle registrations from October 1, 2002.

2.2.6 Miscellaneous requirements

2.2.6.1 Mounting rear license plate

For the rear license plate, a flat or nearly flat rectangular surface of the following dimensions is to be provided:

520 mm wide × 120 mm high or

340 mm wide × 240 mm high

The lower edge must be at least 0.30 m above the road surface.

2.2.6.2 Theft deterrence, immobilizer

An antitheft device to prevent unauthorized use shall be designed so that it must be deactivated in order to enable the engine to be started by means of the normal control and to enable the vehicle to be steered, driven, or moved forward under its own power. In addition, category M_1 vehicles must be fitted with an immobilizer. This system is intended to prevent the vehicle from being driven under its own power by unauthorized persons. This is achieved either by deactivation of at least two separate vehicle circuits that are necessary to permit operation under its own power (e.g., starter motor, ignition, fuel supply) or by interference by code of at least one control unit required for the operation of the vehicle.

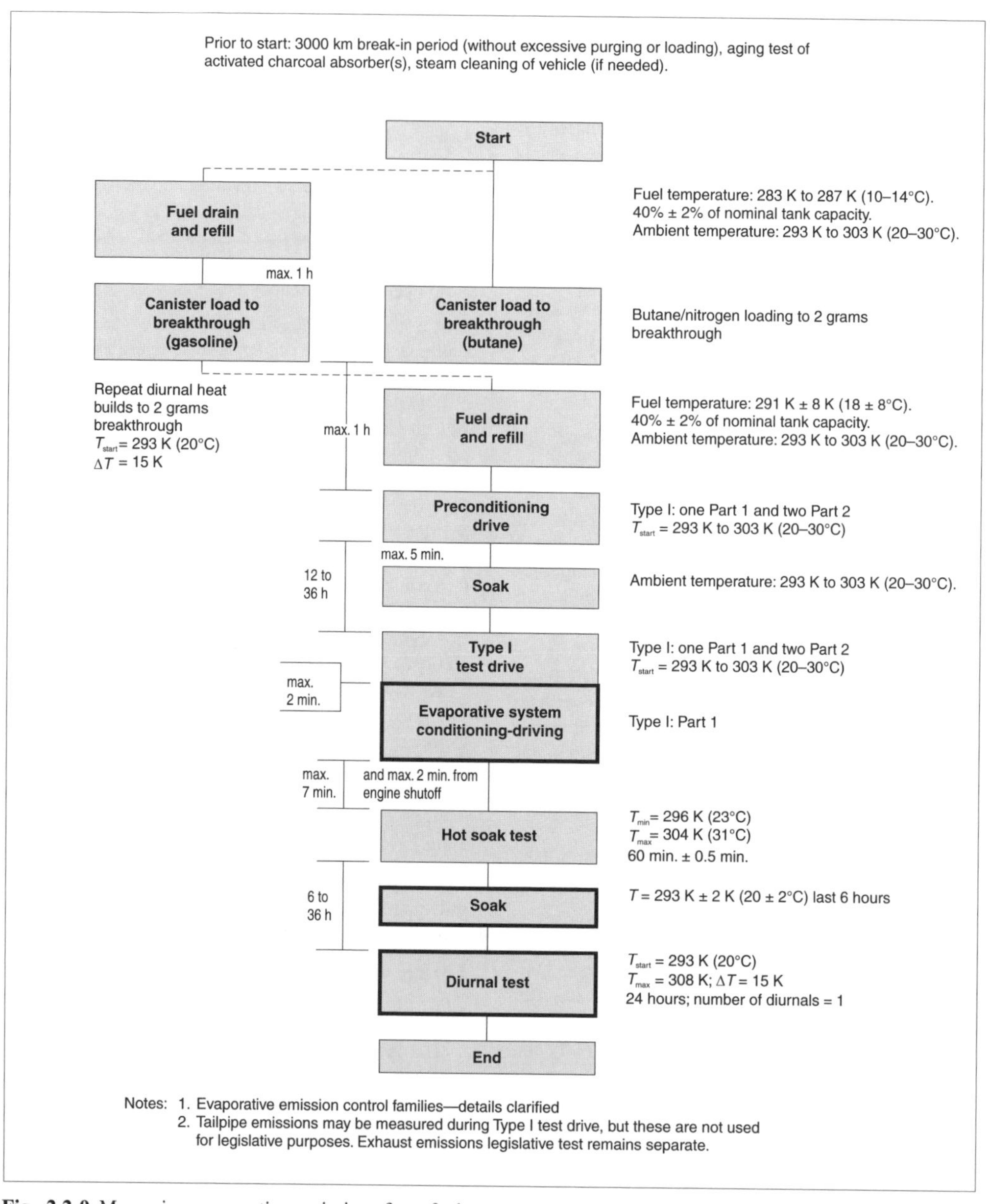

Fig. 2.2-9 Measuring evaporative emissions from fuel systems.

2.2.6.3 Manufacturer's plate, vehicle identification number

A manufacturer's plate must be firmly attached in a conspicuous and easily accessible position on the vehicle. The manufacturer's plate must include, among other things, the name of the manufacturer, the EEC type approval number, the vehicle identification number, and the maximum permitted laden weight of the vehicle. The vehicle identification number contains 17 characters and is also marked on the chassis, the frame, or the right side of the vehicle in such a way that it cannot be obliterated or deteriorate.

2.2.6.4 Engine power measurement

Measurements are to be conducted under prescribed conditions with a sufficient number of engine speeds to establish completely and accurately the full-load curve between the manufacturer's stated minimum and maximum engine speeds. One of the operating speeds shall be the rated speed. Relevant accessory equipment such as water pump, generator, and supercharging device are installed and operating; auxiliary equipment such as air conditioning compressor and air suspension compressor are removed.

2.2.6.5 Weights and dimensions of category M_1 vehicles

The sum of the technically permissible maximum axle loads may not be less than the technically permissible maximum laden mass of the vehicle and may not be less than the mass of the vehicle in running order (fuel tank 90% full) plus 75 kg per seating position. Permissible towing loads, including the static vertical load on the towing vehicle's coupling point, are subject to the following limits:

Trailer equipped with service braking system:
Permissible maximum mass is equal to the maximum mass of the M_1 vehicle (for off-road vehicles, 1.5 times the vehicle mass), but in no case more than 3.5 metric tons

Trailer not equipped with service braking system:
Half the vehicle mass in running order, but in no case more than 0.75 metric tons

2.2.7 Prognosis

Environmental protection will maintain its important position. As a result, the coming years will see increasing regulatory requirements for environmental responsibility of motor vehicles. The EC directive regarding "end of life" vehicles establishes measures intended primarily to prevent waste and, additionally, to promote treatment, recycling, and recovery utilizing scrapped vehicles and their components. The European Commission has drafted a proposal to change the framework directive 70/156/EEC, under which a minimum of 85% (by weight) of each vehicle issued a type approval and placed in service after January 1, 2005 must be reusable and/or recyclable, increasing to 95% as of January 1, 2015.

According to a study conducted by VBE and VDI-Gesellschaft Mikroelektronik (Verband Bildung und Erziehung/Verein Deutsche Ingenieure—Association for Education and Training, Association of German Engineers, Society for Microelectronics), the production value share of microelectronics in automobiles will rise from 22% in 1997 to 32% in 2002. From the mid-1980s, manufacturers began to complement or replace safety- and environment-related vehicle controls and devices with electronic components. Automatic antilocking systems, traction control, vehicle dynamics controls, and airbag and seat belt tensioning deployment systems all serve to increase vehicle safety. Engine electronics, oxygen sensors and mixture control, electronic diesel governors, and onboard diagnostics all benefit the environment.

Systems with electronic components networked with one another become ever more complex. Their failure may compromise safety or the environment. Consequently, regulatory requirements must be created for this field as well. Both the European Community (EC) Commission and the UN Economic Commission (ECE) are currently working on draft regulations for vehicle systems with electronic components. The first result is Supplement 5 to the 09 series of amendments to ECE Regulation No. 13 (braking). This will be followed by legislative work regarding "steer by wire."

2.3 New technologies

Chapter 2.1 showed that the automobile industry and its suppliers will have to apply innovative solutions where existing technologies prove inadequate.

New technology is only rarely an end in itself. More frequently, such technology is applied:

- When existing properties (customer relevant, regulation relevant) can no longer be improved or can only be improved to a limited extent using existing technologies and designs ("improvement innovations")
- When previously unrealizable properties must be achieved ("breakthrough innovations")

Often, new technologies serve to reach new strategic goals.

The variety of new technologies may be classified into four categories (Fig. 2.3-1).

The purpose of increased application of microelectronics is to adapt the vehicle and all its subsystems to all possible operating situations and to enable the vehicle to be networked with other traffic systems. The complete range of functions, from information acquisition to independent planning and execution of actions, often described as "technical intelligence," is available to achieve these ends (Fig. 2.3-2).

New technologies are not only applied to products, their manufacture, use, and deactivation, but also for the product creation process (from design to production and service; "process innovation," see Chapter 10). Not infrequently, product innovations imply accompanying process innovations. Examples include:

- New materials, joining/fastening, and surface treatment technologies
- Component integration
- Electrical and electronic systems
- Electronic data processing, software
- Rapid prototyping
- Virtual product creation process

The basis for new technology may also be found in new design concepts and more rational and more environmentally friendly manufacturing, as well as measures to reduce complexity in the overall chain of value creation (Fig. 2.3-3).

- Increased application of microelectronics, information, and communications technology for an "intelligent' automobile within an "intelligent" traffic system
- Improved and new materials, designs, and manufacturing methods
- Development of more resource- and environment-friendly recycling methods
- Development and introduction of additive power sources and powertrains

Fig. 2.3-1 Advances in automotive technology.

Definition:

The ability of a technical system to behave within its environment in such a way that a human, exhibiting the same behavior, would be credited with intelligent behavior.

Characteristics:

- The ability to receive, process, and implement information into behavior appropriate for the situation
- Information storage and retrieval
- Adaptability and ability to learn in the presence of changing conditions in systems and surroundings
- The ability to make appropriate independent decisions under changing conditions in systems and surroundings
- The ability to conduct independent planning of actions on the basis of experience and early indications of changes in systems and surroundings

Fig. 2.3-2 Intelligent technical systems—an avenue for long-term development.

It follows that decisions must be made early in the concept stage regarding which new technologies will be considered for which fields. It must be determined whether contingency solutions are available and applicable in the event important milestones are not achieved, particularly with regard to concept assurance, product assurance, and process assurance.

Examples of technologies that have improved on or expanded existing properties (see also 2.1) include:

- Closed-loop operation (oxygen sensor, lambda control) of three-way catalytic converter for gasoline engines for simultaneous oxidation and reduction of exhaust gas components
- Active engine mounts to achieve frequency-dependent optimum vibration damping and noise transmission
- Xenon headlights with greatly improved road illumination as well as other advantages
- "Tailored blanks"; i.e., steel sheet metal stock of varying thickness for optimum design of body shells
- Fiber optics, offering many advantages over copper wire
- Combined starter/alternator/vibration damper system to integrate components and functions
- Additional second onboard electrical system with 36/42 V for "power users" (motors, actuators)

Among the technologies that enable previously unrealizable properties are:

- Zeolite thermal storage, providing heat for engine coolant and vehicle interior immediately after a cold start
- Freely programmable valve trains, mechanically decoupled from the crankshaft
- New fuel injection systems with piezoelectric actuators, etc.
- New exhaust system concepts, including adsorber technologies, dielectric or plasma-based technological concepts
- New materials with "selectable" properties (e.g., electrochromic materials) or with special added properties (e.g., water-repellent glass)
- New design principles for human-machine interactions (e.g., voice recognition and speech capability)
- New telematics functions for a wide range of travel and personal services management including self-organizing communications networks ("Personal Travel Assistant")

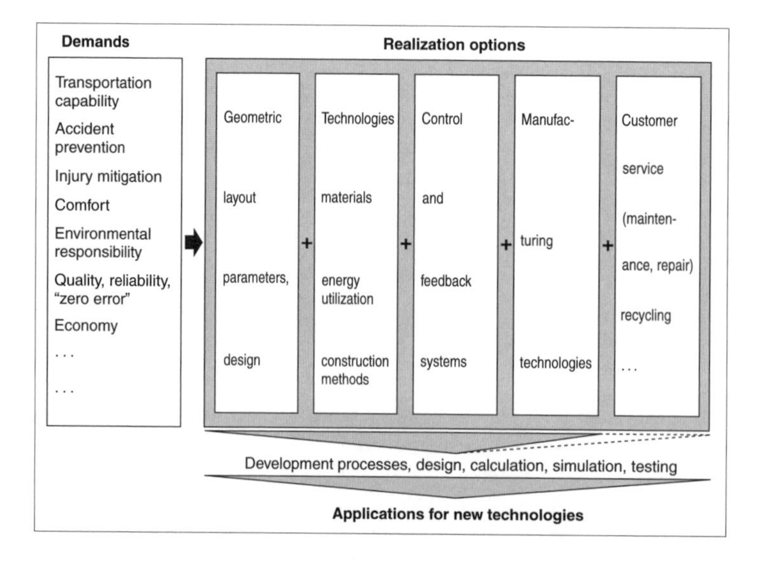

Fig. 2.3-3 Demands on the automobile and their realizability (simplified representation).

- Driver-assist systems for longitudinal and transverse vehicle guidance, responsive to traffic conditions and the path of the road (recognition of surroundings, evaluation of situations, selection and execution of actions)
- Fuel cells for onboard "direct generation" of electrical energy by means of tanked hydrogen, methanol, etc.

In addition to dependence on attainable technological progress, application of alternative or additive energy sources and drivetrains is also dependent on regulatory developments as well as traffic and infrastructure developments (Sections 4.3 and 5.8). For example, these factors will determine the market success of hybrid drivetrains, whose advantages are necessarily coupled with considerably greater complexity.

Automobiles are without exception driven under widely varying operating conditions. Electronic control systems are especially suited to mastery of the resulting conflicting goals. Examples include:

- Three-way catalytic converter with oxygen sensor and feedback mixture control
- Variable gas flow hardware (variable resonance manifolds, variable valve timing, etc.)
- Adaptive transmission controls
- Suspension systems with situation-dependent ride leveling and stabilization
- Controlled drivetrain mounts
- Seat occupancy detection and additional sensors for so-called "smart airbags"

Also of significance are combinations of existing systems to achieve improved or even new properties. An already classic example is the combination of engine management (fuel injection, ignition, etc.) with automatic transmission control to achieve comprehensive drivetrain management. A more recent development concerns headlights with variable light flux distribution that are controlled by the navigation system to aim into upcoming curves.

Beyond these, there are as-yet unrealized functions and properties for which new technologies must be invented. For example:

- The ability to perceive the tire/roadway traction coefficient of the road ahead
- The ability to "see around corners" from within the vehicle
- Self-healing paintwork (for minor scratches)
- Completely water-repellent glass (especially for windshields)

Not all concepts, much less ideas and efforts, lead to success. Fig. 2.3-4 shows the example of the Europe-wide PROMETHEUS research project [1], in which "filters" were normally applied to ideas, and how at the end, only a portion of research effort led to eventual success.

As illustrated by the cited examples and Fig. 2.3-5, new materials and electronic control systems play a major role in mastering the many conflicting goals found within the automobile. Software is becoming increasingly important. Advantages include flexibility in development and control of the number of variables (for example, programming at the end of the assembly line). As the classic methods for defining vehicle-specific technical demands on control systems and software are inadequate, formulation of the design targets must also include so-called "CASE tools"—Computer Aided Software Engineering—which enable exact descriptions of system behavior for programming purposes, free of logical errors and/or misinterpretation (see section 8.2).

New technologies should, at least in mass production, only be applied after all critical paths of the overall process chain have been mastered, from development through production all the way to recycling. In the process, demands for quality, reliability, and often economy frequently speak against new developments.

The possibility of introducing new technology despite these objections is often provided by "upper class" vehicles, to which special measures may be applied (for an added price)—or which initially appeal to special customers or markets.

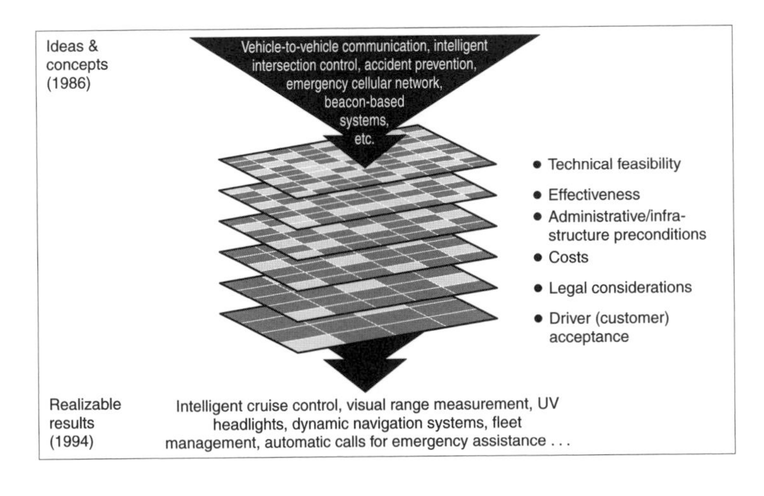

Fig. 2.3-4 The PROMETHEUS project, from idea to reality.

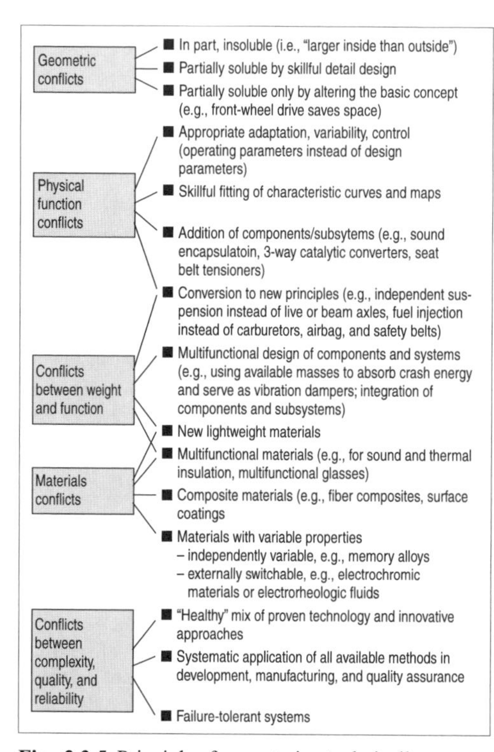

Fig. 2.3-5 Principles for mastering technically conflicting vehicle goals.

The production sector also derives significant benefits from new technologies, such as:

- New initial forming processes such as casting, near-net-shape component manufacturing
- New shaping processes, such as hydroforming and roll forming
- New fastening processes and combinations, such as clinch riveting
- New painting processes (environmentally friendly and energy saving)
- New nondestructive quality assurance processes

As already indicated, new technologies may also make the product creation process more effective and more efficient. Examples (for details see Chapter 10) include:

- Knowledge management and know-how recycling, with systematic perception of the available knowledge with regard to the overall task of automobile manufacturing.
- Systematic application of the most modern CA (computer assisted) methods, from 3D-CAD to CAE and CAM to CAQ (Fig. 2.3-6).
- Concepts such as the "digital car" (virtual prototypes) and "virtual reality" (virtual laboratory vehicles, for example, to optimize human-machine interaction).
- New experimental methods to shorten development time while simultaneously improving predictability

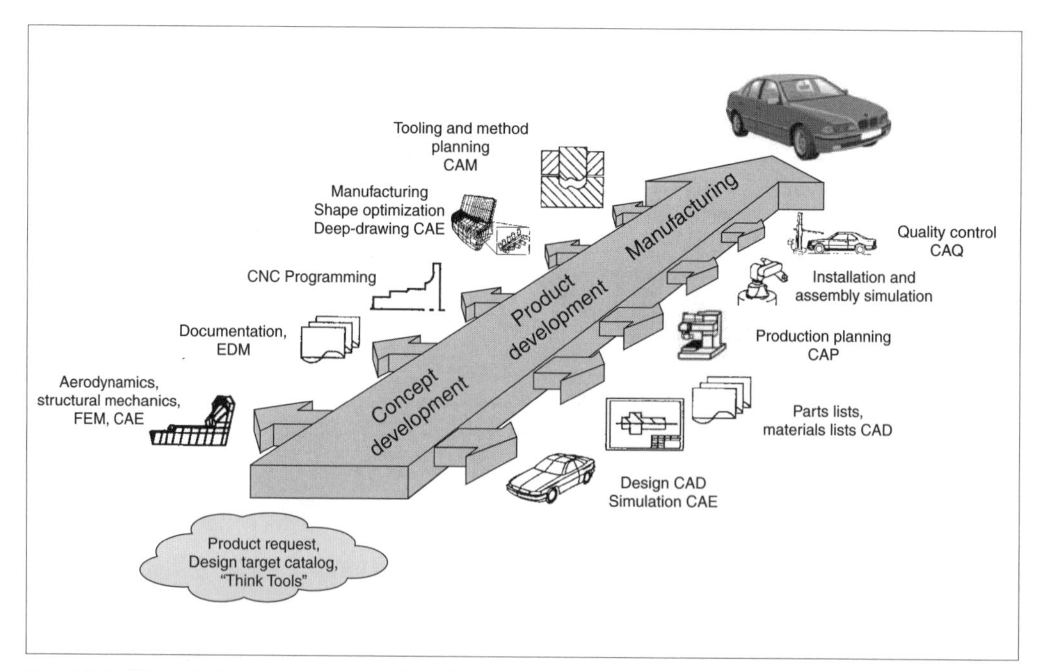

Fig. 2.3-6 CA methodology in the automobile industry.

for manufacturing of new developments. This includes methods such as "hardware in the loop" and rapid prototyping.

Yet alongside our appreciation of the potential offered by new technologies, we must not ignore their limitations, as illustrated by the following examples:

- Even the most intelligent materials and sophisticated control systems cannot negate basic physical and chemical laws, such as the laws of thermodynamics.
- Conflicting geometrical goals can be solved only partially by different design principles. "Negative material thickness" will probably remain a dream for the foreseeable future; there will never be an automobile whose interior is larger than its exterior.

Because increasingly complex technology inevitably leads to longer development times and usually leads to higher costs and other disadvantages, one might well formulate a desirable philosophy of technology as follows: "From the primitive to the complex to the ingeniously simple"—the last of which has an especially favorable relationship between utility and effort.

References

[1] Braess, H.-H., and G. Reichart. "PROMETHEUS—Vision des 'intelligenten Automobils auf intelligenter Straße?'—Versuch einer kritischen Würdigung." ATZ 1995, pp. 200–205 and 330–343.

[2] Seiffert, U. "The Automobile in the Next Century." Paper K 0011, FISITA Congress, Prague, 1996.

[3] Braess, H.-H. "Das Automobil im Spannungsfeld zwischen Wunsch, Wissenschaft und Wirklichkeit." 2. Stuttgarter Symposium Kraftfahrwesen und Verbrennungsmotoren, Feb. 18-20, 1997. Conference proceedings, Expert Verlag, 1997, pp. 786–801.

[4] Various authors. "Geschichte und Zukunft des Automobils." Special issue, "100 Jahre ATZ," 1998.

[5] Nord, D.K., and H.L. Fields. "A Vision of the Future of Automotive Electronics." SAE 2000-01-1358.

[6] Peters, D., et al. "Car Multimedia—multimediale Mobilität für das 21. Jahrhundert." VDI-Berichte 1547, 2000, pp. 771–786.

[7] Seiffert, U. "Die Automobiltechnik im neuen Jahrtausend." Vision Automobil conference, Euroforum, Munich, May 22–23, 2000.

[8] Konitzer, H. "Rahmenbedingungen für die technische Entwicklung des Automobils und des Verkehrs im nächsten Jahrtausend." Vision Automobil conference, Euroforum, Munich, May 22–23, 2000.

[9] Warnecke, H.J. "Innovationen und Entwicklungen für die Automobilindustrie und ihre Produkte." Vision Automobil conference, Euroforum, Munich, May 22–23, 2000.

[10] Naab, K. "Automatisierung bei der Fahrzeugführung im Straßenverkehr." Automatisierungstechnik, 2000, pp. 211–223.

[11] Fuchs, M., and J. Ehret. "Automotive Electronics—A Challenge for Systems Engineering." SAE 2000-01-CO48.

3 Vehicle physics

3.1 Basics

For the vehicle developer, vehicle physics represent a network of physical and technological requirements.

The following two illustrations show, for example, the various component-specific [1] demands on the body structure (Fig. 3.1-1) and for individual subsystems [2]. The basic physical laws and their effects on various disciplines must be taken into consideration during the design of the car. One concrete example is the vehicle's onboard consumption of electrical energy. This may amount to the equivalent of up to 3 liters/100 km: that is, without reducing electrical demand and/or improving efficiency of the electrical generation process and of the vehicle's electrical consumers, a vehicle which normally consumes 3 liters of fuel per 100 km would be unable to move. Vehicle physics therefore demands a network view of all requirements within the product creation process. It is not optimization of an individual property or quantity but rather the performance of the total system which determines product success.

3.1.1 Definitions

Although specific definitions will be developed in individual chapters, the following is intended to provide an overview.

The Directives of the European Community [3, 4] define various vehicle categories. In general, these are classified as:

— Road vehicles
— Nonself-propelled trailers
— Vehicle combinations

These groups are further subdivided; for example, the subgroup of road vehicles known as motor vehicles includes

— Motorcycles (single-track, two-wheeled vehicles), such as motorcycles, motor scooters, and motorized bicycles.
— Motorcars (multi-track vehicles). These are further divided into passenger cars and commercial vehicles.

The passenger car category includes vehicles capable of carrying a maximum of nine persons, sedans, coupes, convertibles, wagons, commercial wagons, special passenger cars such as motor homes, multipurpose vehicles (MPVs), and off-road vehicles. Commercial vehicles include vehicles for the transportation of persons and primarily goods; examples are motor buses (more than nine persons plus luggage), minibuses (maximum 17 persons), city motor buses, intercity motor buses, travel coaches, articulated buses, and special buses. Trucks, intended for transport of goods, include multipurpose trucks for general transportation tasks and special-purpose trucks.

Vehicles intended for towing trailers or other devices include machines equipped with towing hitches or fifth wheels, and tractors.

Trailers include those with steering and nonsteering axles. Vehicle combinations include all truck tractors

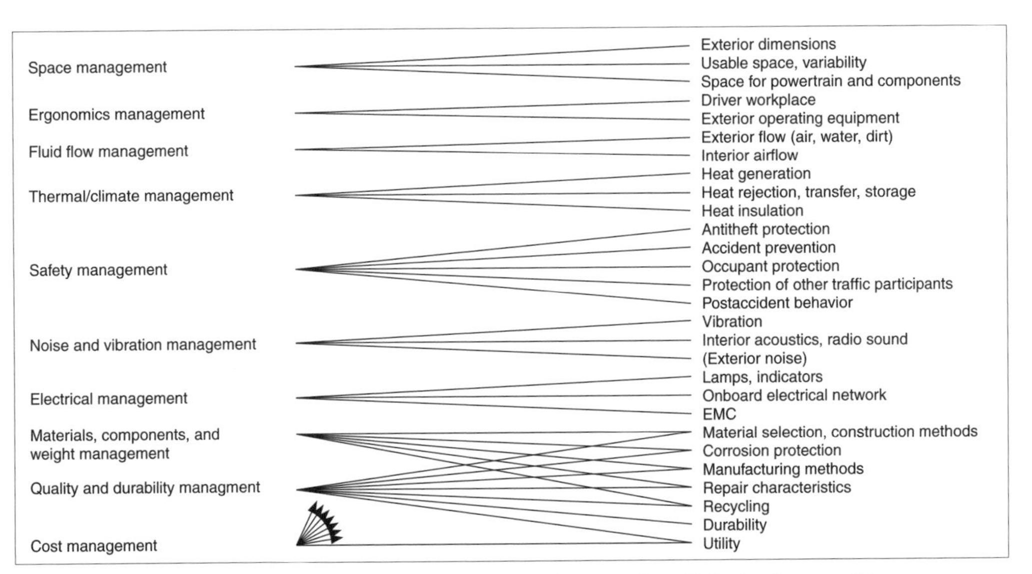

Fig. 3.1-1 Body-structure-based physical and engineering requirements for vehicle development [1].

Components, critical design parameters / Requirements; subdisciplines of vehicle physics	Body							Drivetrain				Chassis and suspension						Vehicle-wide subsystems							
	Body shell structure	Body shell exterior surfaces	Cooling air entry area	Brake cooling entry area	Heating, ventilation, air conditioning	Body equipment	Sound insulation	Engine	Auxiliary equipment	Power transmission	Powertrain	Front suspension	Steering system	Rear suspension	Spring system	Brake system	Wheels and tires	Engine compartment	Sound encapsulation	Exhaust system	Cooling system	Engine mounts	Fuel system*	Vehicle electrical system	Central hydraulic system
Aerodynamics	X	XX	XX	X	X	X**		XX	X	X		X		X		X	X	X	X	X	XX				
Thermodynamics and heat transfer		XX	XX	XX	X	X		XX	XX	X		X	X	X		XX	X	XX	XX	XX	XX	X	XX	XX	X
Fuel-dependent factors								XX										X		X			XX		
Exterior noise			X					XX	XX	XX	XX					X	XX		XX	XX	X	XX	X Pump		X Pump
Interior noise	XX	XX			XX	X	XX	XX	XX	XX	XX	XX	X	XX	XX		XX	XX	XX	XX	X	XX	X Pump	X	X
Vibration technology	XX	XX				X	X	XX	X	X	X	XX	XX	XX	XX	XX	XX			XX		XX		XX	

XX Strongly networked
X Weakly networked

**Exterior parts

*The fuel system includes:
- fuel tank, filler neck
- fuel pump(s)
- activated charcoal canister
- onboard system (refueling emissions)
- fuel rollover valve

Fig. 3.1-2 Vehicle physics as a network of functions incorporating systems, devices and components [2].

and trucks designed to tow trailers and passenger cars and commercial vehicles fitted with trailers.

These are further divided into categories L, M, N, and O.

L are motor vehicles with fewer than four wheels

M are motor vehicles used to transport persons, with at least three or four wheels and a total weight >1 metric ton.
- $M_1 \leq 9$ persons
- $M_2 > 9$ persons, <5 metric tons total weight
- $M_3 > 9$ persons, >5 metric tons total weight

N are motor vehicles for transporting goods, with at least three or four wheels and a total weight >1 metric ton
- $N_1 \leq 3.5$ metric tons total weight
- $N_2 > 3.5$ metric tons and ≤ 12 metric tons total weight
- $N_3 > 12$ metric tons total weight

O designates trailers or semitrailers
- O_1 is single-axle trailers ≤ 0.75 metric tons total weight
- $O_2 > 0.75$ metric tons and ≤ 3.5 metric tons total weight
- $O_3 > 3.5$ metric tons and ≤ 10 metric tons total weight
- $O_4 > 10$ metric tons total weight

3.1.2 Driving resistance and propulsion

Total driving resistance

The total driving resistance (Fig. 3.1-3) is calculated from:

$$F_{tot} = F_{Ro} + F_{Ae} + F_G$$

The power associated with this resistance is $P_{tot} = F_{tot} \cdot v$.

Rolling resistance

Rolling resistance is the result of deformation work on tires and roadway.

$$F_{Ro} = f \cdot G = f \cdot m \cdot g$$

Deformation resistance of the surface only becomes significant in off-road situations; on soft ground, this can exceed 15% of the vehicle weight. On surfaced roads, rolling resistance is almost exclusively the result of deformation work on tires (flexing). Determinants are flexing amplitude (compression distance, wheel load, tire internal pressure) and flexing frequency (vehicle

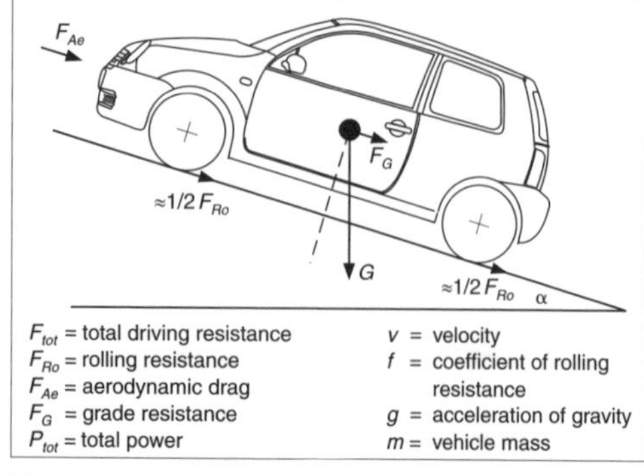

Fig. 3.1-3 Total driving resistance.

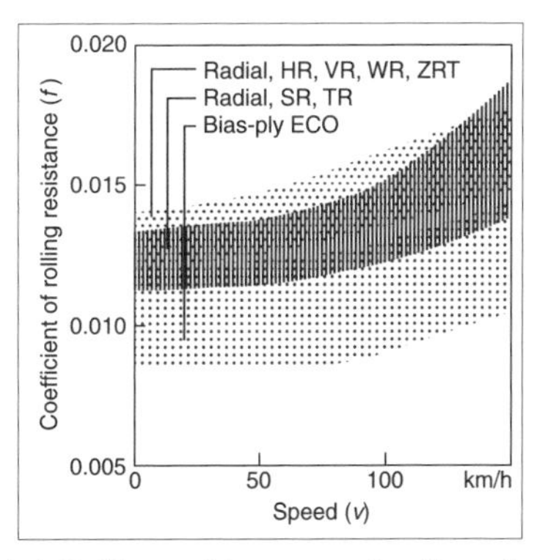

Fig. 3.1-4 Rolling resistance as a function of vehicle speed [3].

speed). Drivetrain friction increases rolling resistance. In the low-speed regime, new low–rolling-resistance tire designs achieve rolling resistance coefficients of 0.008. At 150 km/h, coefficients of 0.017 may be achieved. Fig. 3.1-4 illustrates dependence of rolling resistance as a function of vehicle speed. Because F_{Ro} is defined as acting on the wheel in a longitudinal direction, it must be differentiated from resistance arising from side forces (tire toe-in resistance). In cornering, rolling resistance increases with increasing slip angle (cornering resistance [5]).

Aerodynamic drag

Aerodynamic drag, F_{Ae}, is calculated from the following equation:

$$F_{Ae} = c_d A \cdot \rho \cdot \frac{v^2}{2}$$

where

ρ = air density
v = air velocity
A = cross-sectional area
c_d = aerodynamic drag coefficient

Drag coefficients for passenger cars range from c_d = 0.25 to 0.4. For trucks, c_d = 0.4 to 0.9. The cross-sectional area A of a passenger car ranges from 1.5 to 2.5 m^2 and from 4 to 9 m^2 for trucks. Drag is the result of airflow around and through the vehicle. This has been considerably reduced in recent years through extensive research and development work. At higher speeds, aerodynamic drag determines driving resistance and therefore becomes the dominant factor in fuel consumption. Oblique relative wind at an angle ε to the vehicle longitudinal axis results in different drag coefficients $c_T(\varepsilon)$. The same cross-sectional area and an oblique relative wind v_A results in

$$F_{Ae} = c_T \cdot A \cdot \frac{\varepsilon}{2} v_A^2$$

Aerodynamic drag power is

$$P_L = F_L \cdot v$$

Drivetrain drag

Drivetrain drag is $F_D = (1 - \eta)P/v$; it includes mechanical losses from the engine, through the transmission, and to the wheel hubs ($\eta = \eta_1, \eta_2, \eta_3, \ldots, \eta_n$), with power P and velocity v.

Grade resistance

Grade resistance $F_G = m \cdot g \sin \alpha$ where m is mass; grade power is

$$P_G = F_G \cdot v$$

Acceleration resistance

$F_{acc} = m_{eff} dv/dt$. If we neglect rotating masses with small moments of inertia on rotating shafts and in the transmission, and apply constant rotational energy ($I \cdot \omega^2$ = const),

$$m_{eff} = m + \frac{I_W + i^2 I_{eng}}{r_{stat} \cdot r_{dyn}}$$

where I_W and I_{eng} are the moments of inertia of the wheels and engine, i is the overall gear ratio, and r_{stat} and r_{dyn} are the static and dynamic radii of the tires.

Tractive force utilization

With a given tractive force F_x at the wheels for a given acceleration and grade,

$$F_{acc} + F_G = F_x - (F_R \text{ and } F_L)$$

Tractive force diagram

Given an engine map $M(n)$ expressing torque as a function of engine speed and taking into account internal resistance F_i, we can express tractive force available in various gears as a function of vehicle speed $F_x(v)$. The full-load curves should approach as closely as possible and without significant gaps the limiting hyperbola defined by $F_x = P_{max}/v$. The area on the other side of this limiting hyperbola represents the sum of all driving resistances, $\Sigma F_R(v)$. Fig. 3.1-5a is an example of such a diagram; Fig. 3.1-5b is a tractive power diagram based on the same data. The operating points and reserves available for climbing and acceleration may be read from the curves. One may also design to optimize fuel economy by plotting the lines of constant specific fuel consumption on the tractive force diagram. Unfortunately, in the real world, customers will often drive in gears offering less than optimum fuel economy.

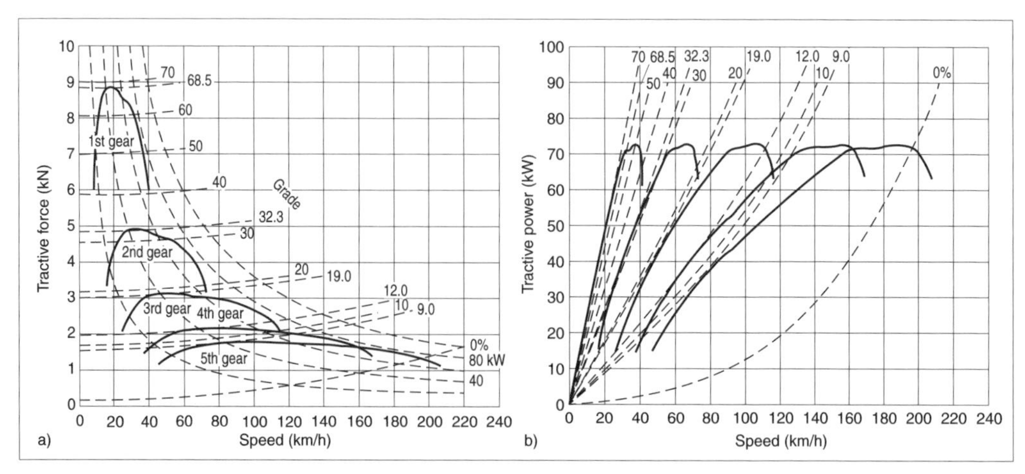

Fig. 3.1-5 Tractive force (a) and tractive power (b) diagrams.

3.1.3 Factors affecting fuel consumption

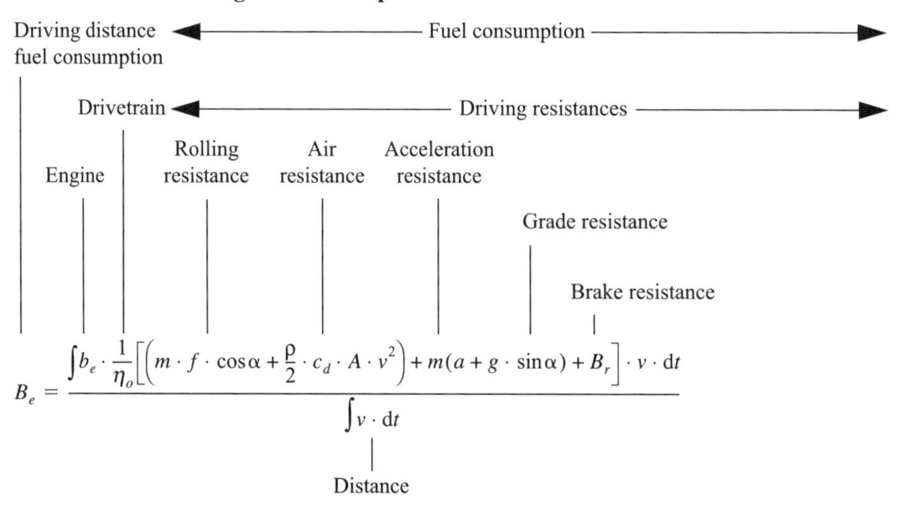

$$B_e = \frac{\int b_e \cdot \frac{1}{\eta_o}\left[\left(m \cdot f \cdot \cos\alpha + \frac{\rho}{2} \cdot c_d \cdot A \cdot v^2\right) + m(a + g \cdot \sin\alpha) + B_r\right] \cdot v \cdot \mathrm{d}t}{\int v \cdot \mathrm{d}t}$$

In addition, fuel consumption at idle, and the efficiency of electrical consumers, must also be considered.

Quantity		Units
B_e	point-to-point fuel consumption	g/m
η_O	overall drivetrain efficiency	–
m	vehicle mass	kg
f	coefficient of rolling resistance	–
g	acceleration of gravity	m/s²
α	grade	°
ρ	air density	kg/m²
c_d	drag coefficient	–
A	frontal area	m²
v	vehicle speed	m/s
a	acceleration	m/s²
B_r	brake resistance	N
t	time	s
b_e	specific fuel consumption	g/kWh

3.1.4 Dynamic forces

Under acceleration and braking, inertia forces result in dynamic load transfer ΔF on the vehicle's axles as shown in Fig. 3.1-6.

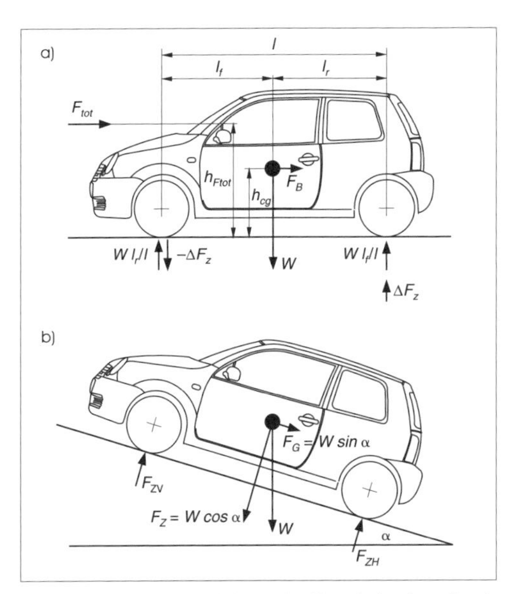

Fig. 3.1-6 Static (a) and dynamic (b) axle loads on level surface and on grade.

Vehicle dynamics and vehicle behavior

On a horizontal surface, Fig. 3.1.-6 a, the vertical forces change by

$$|\Delta F_z| = m \cdot \frac{\mathrm{d}v}{\mathrm{d}t}\frac{h_{cg}}{l}$$

where

ΔF_z = change in vertical forces
m = vehicle mass
$\mathrm{d}v/\mathrm{d}t$ = vehicle acceleration or deceleration
h_{cg} = height of center of mass
l = wheelbase

Accelerated motion results in a change in ΔF_z; under acceleration, this increases load on the rear axle and under deceleration or braking increases load on the front axle. These pitch changes must be taken into consideration in the suspension design. Subjectively, longitudinal oscillations while in motion are particularly annoying; these are usually accompanied by pitch movements. As a result, suspensions are usually mounted quite softly in a longitudinal direction, while higher stiffness is maintained in the remaining degrees of freedom.

In steady-state motion, the total driving resistance F_{tot} acts at a point located h_{tot} above the roadway; this results in

$$\Delta F_z = F_{tot} \cdot \frac{h_{Ftot}}{l} \text{ = change in vertical forces}$$

On grades, the weight components must be taken into consideration. From Fig. 3.1-6, the front axle load is

$$F_{zF} = \frac{W}{l} \cdot (l_R \cdot \cos\alpha - h_{cg} \cdot \sin\alpha) \pm \Delta F_z$$

The rear axle load is

$$F_{zR} = \frac{W}{l} \cdot (l_F \cdot \cos\alpha - h_{cg} \cdot \sin\alpha) \pm \Delta F_z$$

3.1.5 Motion of the vehicle body

Because of the importance of vehicle dynamics, some concepts related to ISO 8855 [6] should be defined.

Components of linear motion

Vehicle velocity $\vec{v}$: a vector quantity describing the origin of the vehicle-based coordinate system with respect to the fixed coordinate system
Longitudinal velocity v_x: component of vehicle velocity in the x direction
Transverse velocity v_y: component of vehicle velocity in the y direction
Vertical velocity v_z: component of vehicle velocity in the z direction
Horizontal velocity v_h: resultant of longitudinal velocity v_x and transverse velocity v_y
Vehicle acceleration $\vec{a}$: vector quantity, describing the acceleration of the origin of the vehicle-based coordinate system with respect to the fixed coordinate system
Longitudinal acceleration a_x: component of vehicle acceleration in the x direction
Transverse acceleration a_y: component of vehicle acceleration in the y direction
Vertical acceleration a_z: component of vehicle acceleration in the z direction
Centripetal acceleration a_c: component of vehicle acceleration in a horizontal normal direction to the horizontal velocity
Tangential acceleration a_t: component of vehicle acceleration in the direction of the horizontal velocity v_h
Horizontal acceleration a_h: resultant of the longitudinal acceleration a_x and the transverse acceleration a_y or centripetal acceleration a_c and tangential acceleration a_t

Note that it is also possible to resolve velocity and acceleration into components $v_{x,v}$, $v_{y,v}$, and $v_{z,v}$ and $a_{x,v}$, $a_{y,v}$, and $a_{z,v}$ in the vehicle-based coordinate system (X, Y, Z).

Components of rotational motion

Angles:

Yaw angle Ψ. Angle (X_E, X) resulting from a rotation around the Z_E axis.
Pitch angle θ. Angle (X, X_V) resulting from a rotation around the Y axis.
Roll angle ϑ. Angle (Y, Y_V) resulting from a rotation around the X_V axis.

Sideslip angle (attitude angle) β. Angle between the X axis and the direction of horizontal velocity, resulting from a rotation around the Z axis. It may be determined from the longitudinal velocity v and the transverse velocity v_y:

$$\beta = \arctan\frac{v_y}{v_x}$$

Angular velocities:

Note that angular velocities may be defined in the vehicle-based coordinate system as well as the horizontal coordinate system.

yaw velocity: $\frac{d\Psi}{dt}$

pitch velocity: $\frac{d\theta}{dt}$

roll velocity: $\frac{d\vartheta}{dt}$

Angular accelerations:

yaw acceleration: $\frac{d^2\Psi}{dt}$

pitch acceleration: $\frac{d^2\theta}{dt}$

roll acceleration: $\frac{d^2\vartheta}{dt}$

3.1.6 Forces and moments

External forces acting on the vehicle at any given time may be expressed as a force vector $\vec{F}$ and a moment vector $\vec{M}$.

Components of force $\vec{F}$

Longitudinal force F_x: component of the force in the X direction

Transverse force F_y: component of the force in the Y direction

Vertical force F_z: component of the force in the Z direction

Components of moment $\vec{M}$

Yaw moment M_z: component of the moment in the Z direction

Pitch moment M_y: component of the moment in the Y direction

Roll moment M_x: component of the moment in the X direction

3.1.7 Chassis

Wheelbase

Distance between tire contact patches on the same side of the vehicle, projected on the X axis (with the vehicle on a horizontal surface).

Note: on vehicles with more than two axles, wheelbases are given between successive pairs of wheels (from frontmost to rearmost wheel). The overall wheelbase (right or left) is the sum of these distances.

Track b

The distance between tire contact patches of both wheels of a single axle, projected on the YZ plane (with the vehicle on a horizontal surface). For dual wheels, track is the distance between points midway between the tire contact patches of both double-wheel units. Note: wheelbase and track generally do not remain constant, due to the effects of jounce and rebound.

3.1.8 Steering

Static total toe-in angle d

Sum of the static toe-in angles of the left and right wheels with the steering in its straight-ahead position. The static toe-in angle is the angle between the X axis and the X_w axis. It is positive when the intersection of the X_w and X axes lies ahead of the tire contact patch.

Static toe-in

Static toe-in is the difference in distance between the rim flanges behind and ahead of the wheel centers on one axle, measured at the elevation of the wheel centers under static reference conditions. It is positive when the distance between the rim flanges is greater behind the axle than ahead of it and negative when the distance at the rear is less than at the front.

Ackermann angle σ_A

Arctan of the wheelbase divided by the track radius of the rear axle midpoint at very low vehicle speed. The Ackermann angle is often used to define over- and understeer.

Steering wheel angle σ_H

Angular displacement of the steering wheel from its straight-ahead position. It is considered positive when the vehicle is in a left turn.

Steering wheel moment M_H

The moment applied by the driver to the steering wheel about its axis of rotation. It is considered positive when the vehicle is in a left turn.

Steering ratio i_s

Steering wheel position in relation to change in the average steering angle of a steered pair of wheels with an unloaded steering system and the vehicle under static reference conditions.

3.1.9 Wheels and tires

Slip angle a

Angle between the X_w axis and the tangent of the tire contact patch path. It is considered positive to the left.

Wheel camber angle ε_w

Angle between Z_w axis and the wheel plane. It is considered positive to the right.

Tire rolling characteristics

- Static rolling radius r_{stat}
  Distance between wheel center and tire contact patch under specific conditions of wheel load, tire inflation pressure, and dimensions (width and profile) of the rim on which the tire is mounted
- Dynamic circumference C_R
  The ground distance covered during one wheel rotation under specific conditions of wheel load, tire inflation pressure, temperature, speed, etc.
- Dynamic rolling radius r_{dyn}

$$r_{\text{dyn}} = \frac{C_R}{2\pi}$$

References

[1] Braess, H.-H. "Die Karosserie—Typisches Beispiel für Zielkonflikte und Zielkonfliktlösungen für Automobile." VDI-Bericht 968 (1992), *Entwicklungen im Karosseriebau*, Düsseldorf.

[2] Braess, H.-H. unpublished reference material.

[3] Bosch. *Kraftfahrtechnisches Taschenbuch*; 22nd German edition. VDI-Verlag, Düsseldorf, 1999. *Bosch Automotive Handbook*, 5th English edition. Robert Bosch GmbH, 2000. Distributed by SAE.

[4] Seiffert, U. *Kraftfahrzeugtechnik*, Dubbel, 19. Auflage, Springer Verlag, Heidelberg, 1997.

[5] Braess, H.-H., and R. Stricker. "Eigenlenkverhalten, Kurvenwiderstand, Kraftstoffverbrauch—Ein weiterer Aspekt des Fahrzeugkonzeptes und der Fahrwerksabstimmung." VDI-Bericht 418, 1981, *50 Jahre Frontantrieb im Serienautomobilbau*, pp. 275—280.

[6] ISO 8855 International Standard, "Road vehicles—Vehicle dynamics and road-holding ability—Vocabulary," April 1992.

3.2 Aerodynamics

3.2.1 Basics

Drag experienced by a body depends upon its shape, its size, and the medium through which it is moving. In the case of a passenger car, the medium is air, which may be regarded as incompressible in the normal range of speeds encountered. Its material properties are defined by its density ρ and kinematic viscosity ν, which in turn are functions of air pressure and temperature.

Under standard conditions, $P = 1{,}013$ mbar and $T = 0°C$,

$$\rho = 1.252 \text{ kg/m}^3 \qquad \nu = 1.373 \times 10^{-5} \text{ m}^2\text{/s}$$

Density ρ is air mass per unit volume, while viscosity ν is the ability to transfer stresses between layers of air: that is, the physical cause of frictional resistance. In heat transfer from radiators and brakes, thermal conductivity of air must be considered:

$$k = 0.0242 \text{ J/m} \cdot \text{s} \cdot \text{K}$$

Aerodynamic drag of a vehicle results from relative motion between the vehicle surface and surrounding air. Accordingly, it is physically irrelevant (ignoring effects such as rotating wheels, boundary layer at the road surface, etc.) whether a body is moving through air or a stationary body is subjected to a stream of air moving at the same velocity (as would be the case of a model in a wind tunnel). The front of the vehicle must push air aside, resulting in a pressure increase. At the rear, the air molecules cannot flow together smoothly, which causes a drop in pressure. The sum of these pressures results in the form drag (or pressure drag) of a vehicle. Friction between the body surface and air, a viscous medium, results in skin drag (or friction drag). The creation of vortices results in induced drag. For a passenger car, friction drag accounts for approximately 10 percent of total drag. The other two components are dependent on vehicle shape. For example, a "hatchback" vehicle with a steep rear cutoff generates a large region of separated flow, and therefore high form drag, while a "fastback" vehicle may have lower form drag but greater vortex generation and therefore greater induced drag. An additional drag component results from flow resistance through radiators, body gaps, and ventilation systems. Important areas affected by aerodynamics are shown in Fig. 3.2-1.

The aerodynamic quality of a vehicle is indicated by its drag coefficient, c_d, a dimensionless resistance determined from the equation

$$c_d = \frac{F}{A \cdot q}$$

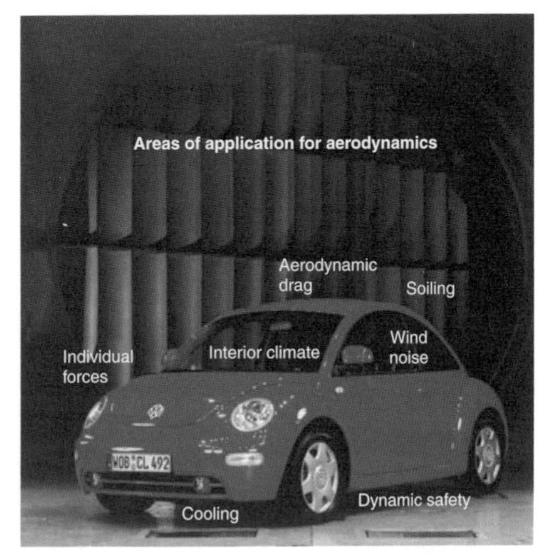

Fig. 3.2-1 Areas of application for aerodynamics.

where

F [N] = drag
A [m²] = reference area (for automobiles, the projected area normal to the direction of travel)
q [N/m²] = static pressure component = $\rho \cdot \upsilon^2/2$
υ [m/s] = vehicle or relative wind velocity

Coefficients of lift and lateral force are structured analogously. For wind tunnel measurements, a coordinate system, shown in Fig. 3.2-2, has been defined. Usually, moments about axes can be measured directly in the wind tunnel. Vehicle wheelbase is used as the reference length for moment coefficients.

Aerodynamic forces are usually determined in wind tunnels. To determine aerodynamic coefficients for production vehicles, actual vehicles are tested in correspondingly large wind tunnels. During the development process, scale models are also employed. It is important to recognize whether experimental results on small, geometrically identical models are valid for full-scale vehicles. Such a conclusion is justified if the flows are mechanically similar and the dimensionless parameter Re = $\upsilon_\infty l/\upsilon$ (Reynolds number) is kept the same for both cases. For wind tunnel measurements, this means that speed must be increased by the scale factor. On the other hand, because the Mach number, $Ma_\infty = \upsilon_\infty/a_\infty$, must not be allowed to become too large (in order to avoid compressibility effects), model studies are conducted at wind speeds between 60 and 80 m/s with models no smaller than 1/5 scale. This scale is not always sufficiently large for the design of curves and radii.

It has been known since the mid-1930s that the nozzles of automotive wind tunnels should be approximately rectangular, with test section floors connected directly to nozzles and collectors. For nearly as long, it has been recognized that simulation using a moving ground plane is the physically more correct method; however, this requires considerably greater effort. For the same reason, boundary layer control (blowing or suction) is also applied to improve simulation, usually as a combination of both methods. To this day, these techniques have been used primarily in testing race cars, due to their low ride heights. The benefit of using these methods for passenger cars is debatable, and their use is a reflection of company policy.

3.2.2 Areas of application

3.2.2.1 Aerodynamic drag/vehicle performance

Aerodynamics, or more precisely aerodynamic drag, is one of the factors influencing vehicle performance and fuel economy. Fig. 3.2-3 shows the results of fuel economy calculations using measured input parameters for a typical midrange passenger car and the resulting effects of aerodynamic drag on fuel consumption and top speed. The trends indicated may be applied to other vehicles.

From this behavior, it is apparently desirable to develop vehicles with low aerodynamic drag and low drag coefficient c_d. Fig. 3.2-4 illustrates how average c_d values have changed over the course of time. This diagram is not so much an indicator of engineering knowledge as it is of consumer acceptance of more aerodynamic vehicle shapes. The difference between technologically possible experimental vehicles and actual production cars underscores this relationship. Dependency on customer tastes makes it difficult to apply experimental results in extrapolating this diagram into the future.

Fig. 3.2-5 is a histogram showing the range of c_d values found on modern vehicles. The illustration mainly incorporates passenger cars sold in Germany. All vehicles were measured in the same wind tunnel under identical

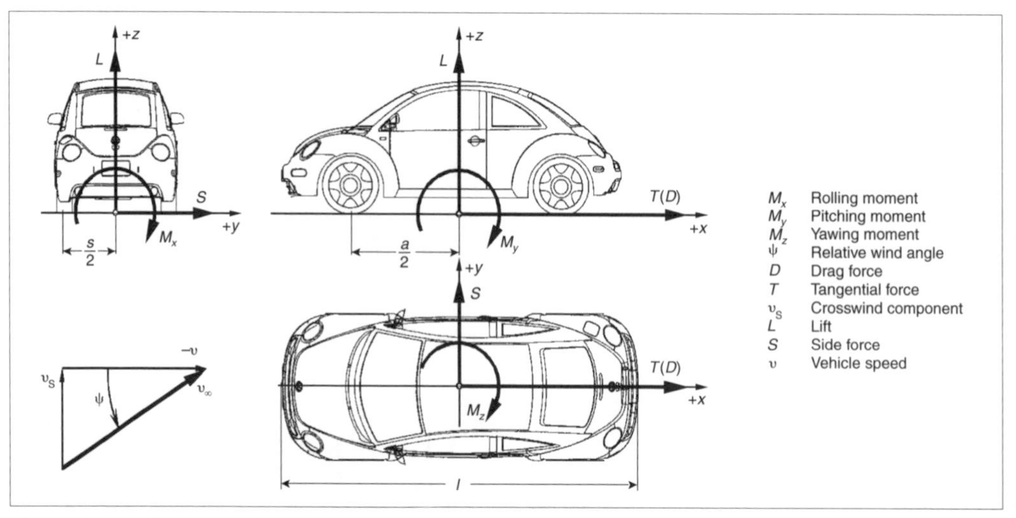

Fig. 3.2-2 Coordinate systems, forces, and moments.

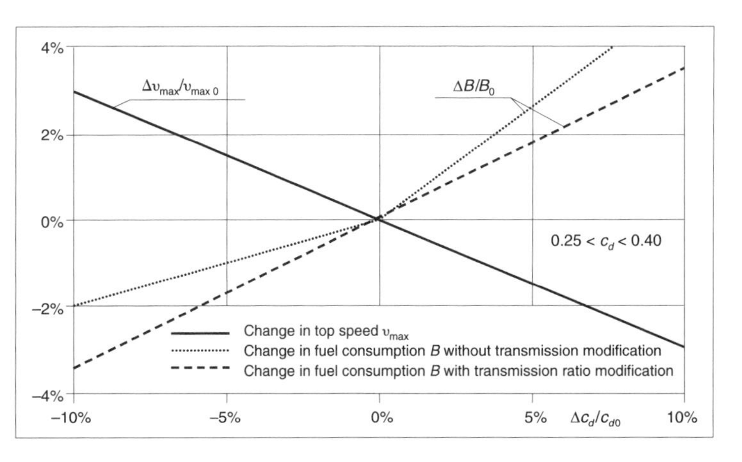

Fig. 3.2-3 Effect of c_d on vehicle performance and fuel consumption.

conditions. The lowest c_d values are almost without exception generated by "notchback" sedans. The higher values are returned by older vehicles and those which, due to the specific nature of their design, make limited or no compromise in favor of aerodynamics.

In the past, the literature [1] presented two techniques for developing vehicles with low c_d:

- Shape optimization, beginning with a basic low-drag body, progressing to a basic shape and basic model, thence to the production vehicle
- Detail optimization, progressing from an unmodified model to an acceptable production vehicle

This implies that shape optimization in principle produces lower drag coefficients. Today, apart from specialty vehicles, detail optimization is the method of choice. It should be noted that in the past few years, experience has led designers to begin with models of considerably lower c_d than was the case when these techniques were first developed. Detail optimization demands extensive wind tunnel testing. Computational methods have only a limited ability to present the results of minor modifications with sufficient accuracy and speed [2]. According to individual company philosophies, detail optimization is only applied to full-scale models or initiated even at smaller scales and later carried on with full-scale models. The result of this method depends on how far developers and designers are prepared to go in resolving conflicting goals in favor of aerodynamics and may indeed be comparable to the process of shape optimization.

To develop a vehicle with a low c_d, all drag-influencing body parameters must be optimized. The literature

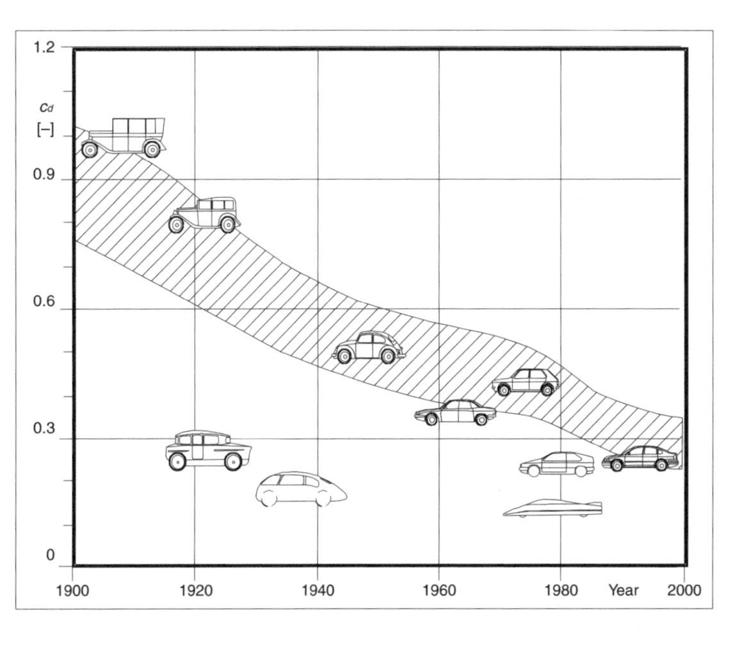

Fig. 3.2-4 Historical development of drag coefficient c_d.

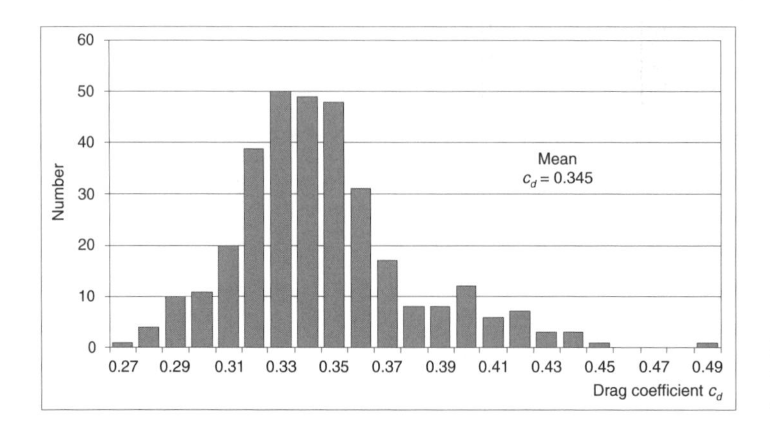

Fig. 3.2-5 Distribution of c_d values for 329 passenger cars, 1998.

presents a multitude of parameter variations. It should be noted, however, that the results are only applicable to the models examined and cannot be applied wholesale to other designs. The effects may generally not be superposed. Consequently, tests of models can only provide reliable results if all details, such as engine compartment and suspension, are accurately represented. Fig. 3.2-6, for example, shows the effects of rear body slope and tail length on c_d. This family of curves, however, is also strongly influenced by the radius of the transition between roof and backlight, the shape of the C-pillars, the shape of the trunk lid, the rear diffusor angle, and the rear diffusor length. Even the presence of a front spoiler has an effect. In principle, these interdependencies require that models be optimized through multiple iterations. For reasons of cost and time required, often only individual parameters are optimized in a single test sequence.

The conflicting goals of an aerodynamically desirable exterior shape and the wishes and demands of other development partners often shift attention to the floorpan, which the aerodynamicists may modify without interfering with other departments. The underside of the car accounts for roughly half of total drag. Of this, about 10% is created by open wheel wells and about 25% by the wheels. The actual floorpan itself accounts for about 15% of total drag. On modern cars, the floorpan is fairly smooth. There are unavoidable intrusions for the exhaust system, which must be cooled and given space for expansion; for suspension components, which must be permitted a certain freedom of motion; for the gap between fuel tank and floorpan, which in the event of a rear impact provides deformation space for the body without rupturing the tank; and so on. A conventional passenger car has a potential Δc_d = 0.01 to 0.02, which can be achieved with underbody cladding.

3.2.2.2 Dynamic safety

Airflow around a vehicle results in forces in all three directions and moments about all three axes of the vehicle-based coordinate system (see section 3.1). While drag, the force in the longitudinal direction, affects performance, the other forces and moments affect vehicle handling and safety. Forces become unsymmetrical in crosswinds and under the airflow-induced influence of other vehicles. On the part of the vehicle itself, handling and safety are influenced by:

— Suspension
— Center of mass location

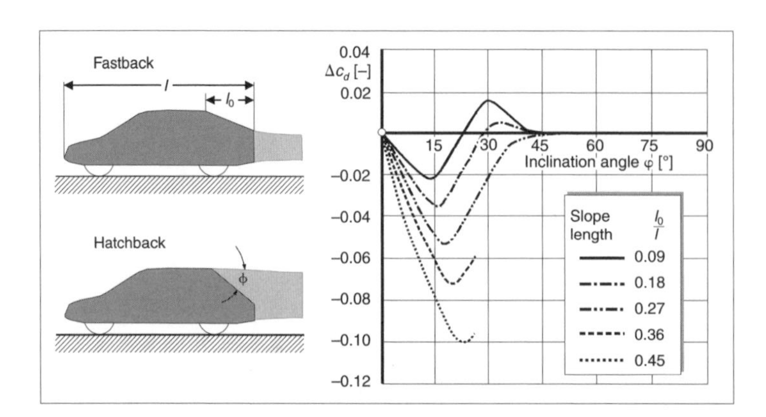

Fig. 3.2-6 Influence of rear body design on drag coefficient c_d.

— Aerodynamics
— Propulsive power

Prior to 1930, due to modest achievable speeds and lack of interest, the effects of aerodynamics on handling were of secondary importance. The 1930s saw increasing development of low-drag vehicles, but these exhibited excessive yaw moment gain with increasing yaw angle, inadequate suspensions, and often mass centers located far aft as a consequence of their package concepts. At the same time, construction of new "autobahns" permitted higher speeds, and for the first time, the phenomenon of crosswind sensitivity in conjunction with streamlining was encountered [3]. As a group, low-drag vehicles quickly garnered a reputation for crosswind sensitivity. Continued suspension development and a shift to front-mounted engines have reduced this phenomenon to the point where today, little effort is usually expended in finding shapes with low yaw moments.

On the other hand, modern vehicles achieve speeds at which front and rear lift become significant. As of about 160 km/h (100 mph), the following criteria are affected by lift to a greater or lesser degree:

— Reaction to steering inputs
— Cornering behavior
— Lane change behavior
— Power-off effect
— Steering feedback at high speeds
— High-speed instability (yawing)

In summary, it is desirable to have a low lift coefficient over the rear wheels and a front-end lift coefficient not significantly lower. Acceptable values of these coefficients are manufacturer-specific and are also influenced by suspension layout. In this way, unfavorable lift distribution may in part be compensated by suspension modifications (see Chapter 7).

Like aerodynamic drag, the other aerodynamic forces and moments, such as airflow around wheels and through the engine compartment and so on, are above all influenced by the exterior shape. Modifications to this must be introduced at an early stage of the design process. At this point, driveable models are certainly not available; wind tunnel tests must be used to evaluate designs. These provide quasi-static results: that is, the effects of gusts and such are not accurately reproduced and are not yet related to other vehicle dynamic parameters. In general, experience gained with previous production models is used to apply wind tunnel results to handling.

3.2.2.3 Wetting and soiling

A moving car is often subjected to nonhomogeneous—that is, particulate-carrying—flows. These particles range from gas and dust to water, insects, and rocks. In keeping with their history, the trajectory of these particles at first differs from the airstream lines. These differences lead to relative velocities, which in turn give rise to forces, which affect the trajectory of these particles as a function of their density. While exhaust gases, after a short transition period, move almost identically to the airstream, rocks continue on their paths almost without effect. Accordingly, in considering soiling, the density of particles must be considered:

- Gaseous materials generally follow the airstream. Exhaust gases are of particular interest, as these should not be allowed to enter the cabin. Because the concentration of exhaust gases increases near the ground, air inlet openings should be located as high as possible. For this reason, air inlets are generally located on the cowl just ahead of the windshield.

The vehicle is also subject, however, to its own exhaust train. Fig. 3.2-7 shows rear-end airflow typical of a hatchback. Characteristically, a vortex forms behind the rear bumper, which transports air and material carried along with it about a half meter to the rear and then forward again to the rear hatch. Above this, a vortex rotates in the opposite direction. These two vortices have a mutual mixing zone in which particles are transferred. The upper vortex then carries these to the rear edge of the roof. If exhaust gas is introduced into the lower vortex, the two vortices will distribute this across the entire rear of the car. Because certain driving conditions are accompanied by significantly lower pressure in the cabin, exhaust gases could be drawn into the car. The effect increases as the square of vehicle speed. Countermeasures include complete sealing, especially at high speed, and a suitable choice of exhaust tip so that exhaust gases are not admitted to this vortex structure in the first place. Downward-pointing exhaust tips have proven beneficial. Even better are mounting locations which blow exhaust gases behind a rear wheel.

- Dust follows the behavior described in the previous paragraph and may therefore also be drawn into the vehicle. Dust thrown under the center of the car by the wheels is deposited on the rear of the car by the trailing vortices. The only way to prevent its being drawn into the car is a suitable sealing system. Dust from the front wheel wells is also thrown at the door

Fig. 3.2-7 Airflow behind a hatchback car.

gaps, especially those just above the doorsills. In the event of significantly lower pressure inside the car, for example with a lifted tilting/sliding sunroof, dust may pass through the door gaps and enter the door shell through door drain holes and from there enter the cabin through the window channels. As the pressure distribution along the outer door gaps varies, flow is established in the door pockets. In this way, dust from below the rear doors—for example, from the rear wheel wells—can be drawn into the door pockets and from there be transported forward to the A-pillars behind the front wheel or upward to the gap at the top of the C-pillars. Only a continuous, unbroken seal system can reliably prevent entry of this dust.

- Water droplets are differentiated by their size and origin. On wet roads, droplets hurled under the car by the wheels behave in part like exhaust gas and dust. Like dust, lighter droplets are transported up to the top of the backlight; medium droplets only reach the lower rear of the car. The heaviest droplets cannot follow the airstream and remain in the wake of the car. Mudflaps behind the wheels increase the spray effect toward the vehicle centerline and are likely to cause even greater soiling of the rear of the car.
- In keeping with their weight and inertia, raindrops are nearly immune to influence from the vehicle airflow. From energy considerations alone, blowing the windshield free of raindrops is impossible. To prevent water swept by the wipers from the windshield to the A-pillars from reaching the side window and thereby obscuring the view of the outside mirrors, design solutions to catch this water are generally applied. Another option lies in the choice of door mirror mount; mirrors attached directly to the door enjoy a greater shielding effect than those on a separate pedestal. Unfortunately, measures to reduce side glass soiling usually lead to higher aerodynamic drag and increased wind noise, resulting in conflicting goals to be solved.

On nearly all cars, conical vortices are generated along the side glass behind the A-pillars, rotating along the glass surfaces toward the A-pillars. Their intensity is so great that frequently, even at speeds as low as 100 km/h (62 mph), water coming past the A-pillars is forced upward and then rearward along the upper window frame. As a result, the problem of side glass soiling only arises in the speed range from 50 to 100 km/h.

- A special problem is posed by water trailed behind the wiper blades. On a number of vehicles, the driver-side wiper reverses its motion when it is near and parallel to the A-pillar. At this point, a relatively large amount of water is found alongside the wiper blade, and the airflow over the blade produces a vortex which, near the glass surface, rotates toward the wiper. This vortex transports some of the water back toward the wiper blade, so that in reversing its motion, the blade draws a wake of water behind it. Solutions to this problem must be found on an individual basis.
- Heavy particles, such as rocks thrown up by the wheels, cannot be influenced by airflow measures.

3.2.2.4 Individual forces

Airflow affects not only the vehicle as a whole, but also individual components. The pressure distribution results in areas of the surface (i.e., side glass) with local pressures and loads not directly attributable to drag, lift, or side force. Fig. 3.2-8 shows typical pressure distributions along the longitudinal sections of three different vehicle types:

— Vans
— Notchback sedans
— Sports cars

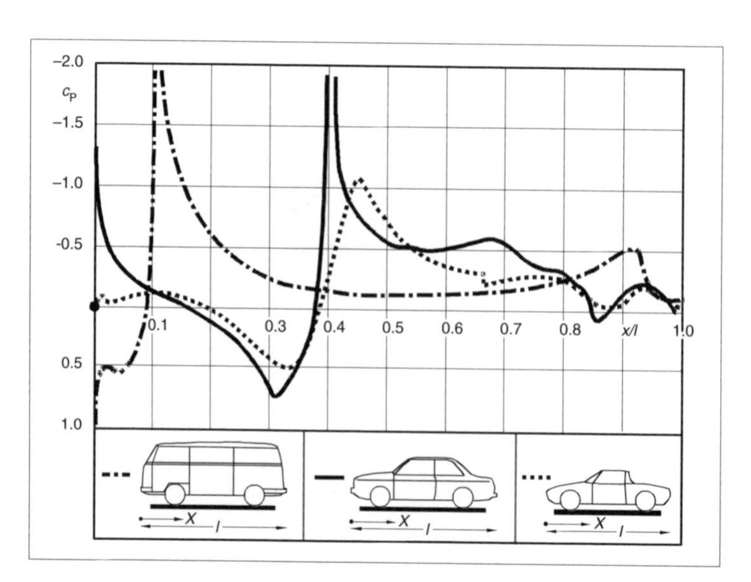

Fig. 3.2-8 Pressure distribution along longitudinal sections.

The maximum underpressure acting on the front of the passenger car can, for example, exert a force tending to lift the front hood. At a speed of 200 km/h (124 mph), the upward force at the front of the hood may, depending on the shape of the vehicle nose, amount to 300 to 500 Newtons. Force on the side glass immediately behind the A-pillars is about half as great, but may reach the same levels in crosswind conditions. These forces move components and may open sealing systems or, as a result of flow separation, lead to increased aerodynamic drag. Functions may also be compromised; for example, it may not be possible for lowered side glass to reenter the window channels during the closing process.

Pressure loads may cause dents in front spoilers, tear away rear wings, disturb outside mirror adjustments, etc. These may not be due to steady-state forces alone. Periodic flow separation can excite harmonic effects which reduce component life or affect function. At high speeds, vortex shedding from the mirror housings can often excite undesirable vibration of the mirror glass. Vibrations of the front hood have also been observed at high speeds.

Shape changes in response to aerodynamic loads often lead to "edgy" exterior contours which increase pressure loads. Often, oscillating components are themselves responsible for generating periodic pressure loads, thereby making the situation worse.

These pressures and forces should be identified at an early stage through wind tunnel testing or computational methods so that they may be minimized through form changes or incorporated in the design process.

3.2.2.5 Cooling and component temperatures

The purpose of the cooling system is to cool the engine and other components under any and all operating conditions in order to avoid functional problems or damage. Further, the heating system should function well, wear should be minimized, and fuel economy and power should be optimized (for details, see Section 3.3).

Heat rejection occurs either directly from the component to the surrounding air—e.g., from the brake discs—or by means of a coolant (including oil) and a radiator. From an aerodynamic standpoint, charge air coolers (intercoolers) are no different from water radiators. In all cases, an internal airflow is required. Energy is required to achieve internal airflows. This may be provided by blowers, which are especially necessary at low speeds or when the vehicle is stationary. At higher speeds, most vehicles employ the pressure differential between the vehicle front and underside to direct air through radiators or through the engine compartment. Because of greater resistance, internal flows fundamentally produce greater losses than the same volume flowing around the car. This manifests itself as an increase in aerodynamic drag and/or a higher c_d. Therefore, in the interest of improved fuel economy, the cooling-air component of drag and hence the cooling air volume flow should be kept to the minimum possible. Measures to this end include:

- Ducting the airflow from its entry to the radiator
- Positioning inlets and passages so that the radiator experiences uniform flow
- Avoiding obstructions and breaks in the air ducting
- Appropriate sizing of the radiator cross section
- Employing radiators with high specific heat-transfer performance

To cool individual components, an airstream is often directed at the part in question. Sealed channels are much better suited to this purpose because open streams often have only a limited range due to the available pressure drop and may be deflected by other flows.

3.2.2.6 Interior climate

The purpose of the heating, ventilation, and climate control system of a vehicle is to provide a comfortable and safe environment for the occupants to the highest possible level, insofar as this may be influenced by this system (see Section 6.4.3). In the relatively small volume represented by the cabin of a passenger car, the occupants are affected by the following factors:

- Temperature and temperature stratification
- Airflow speed and distribution
- Humidity
- Radiation from components
- Direct insolation (solar radiation)
- Contents of air

These parameters are in part influenced by airflow within the vehicle. Airflows are in turn dependent on the settings for blowers, vents, and so on; location of inlet and exhaust ducts; and the locations of openings which were unanticipated from a functional or design standpoint (in other words, leaks).

Flow around the body produces a pressure distribution on the surface, which may be employed in selecting locations for openings. The preferred locations for exit openings are in low-pressure regions. Care must be taken to ensure that hazardous substances are not drawn into the cabin in the event of a strong low-pressure condition, such as with opened side windows. Objectionable noise levels may result if the exit openings are located too close to the occupants' ears. To minimize the necessary blower power, inlet openings are often located in regions of high pressure. On the other hand, it is advantageous to have heater operation independent of vehicle speed. For this reason, a location with minimal overpressure would be preferred (Fig. 3.2-9).

3.2.2.7 Wind noise

Vehicle occupants are subjected to a variety of noises, some of which are generated outside the vehicle and may be reduced by insulation. However, the more important group of noises is generated by or within the vehicle itself. Apart from desired sounds such as radio, warning buzzers, and so forth, three sources may be identified: engine, rolling, and wind noises. As a result

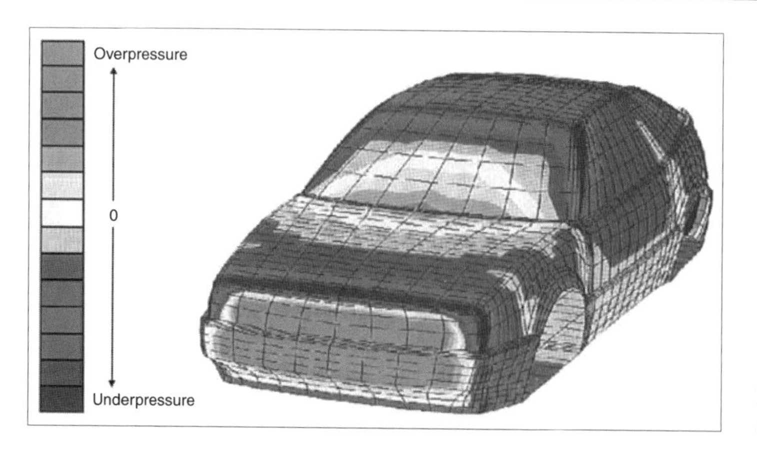

Fig. 3.2-9 Surface pressure distribution (computed).

of the successful reduction of engine and rolling noises over the past few years, wind noises can no longer be ignored (see section 3.4). Wind noises are caused by the following:

- Leaks, primarily in door and window seals, but also through openings in sheet metal. Due to the displacement of air, the outside of a vehicle body often exhibits appreciable areas of low pressure. The regions immediately behind the A-pillars are especially susceptible. Depending on the heating and ventilation operating mode, over- or underpressures within the vehicle are relatively small. If the seals are not seated properly, there will be a flow from inside the cabin to the outside. This flow results in noise, most apparent in the 4 to 10 kHz frequency range. As a result of the underpressure forces against the side windows of the front doors, at high speeds the door frames may move outward as much as 2 mm. If the door seal system is incapable of sealing in this situation, the result will be especially loud flow noises throughout the entire relevant frequency range.
- Cavity resonance excited by airflow. Air flowing across open door and lid gaps results in flow noise from about 500 Hz to 3 kHz. The cross section of channels behind the gaps also determines their frequency range. The noise volume is especially sensitive as to whether the back edge of the gap projects into the flow or deflects away ("fish scale effect"). An appropriate choice of seals can eliminate this noise, but a suitable "fish scale effect" is only slightly less desirable.

This family of noises also includes sunroof booming, typically about 20 Hz and often encountered at speeds between 40 and 60 km/h (25 to 37 mph). Air flowing over the open roof effectively turns the entire cabin into a Helmholtz resonator. As a remedy, wind deflectors, tailored to the specific vehicle design, may be installed. Perforated or serrated wind deflectors are generally more effective than smooth devices, but at higher speeds generate undesirable noise of their own. The length of the roof cutout is also a significant factor, to the point where some manufacturers sacrifice sunroof travel in order to have a smooth wind deflector.

Booming is also observed with open side windows, especially rear windows; to date, there is no technical solution. Occupants may, however, avoid this situation by also partially opening a front window in addition to the opened rear window.

- Flow separation along a contour. Separation results in vortices and turbulent pressure fluctuations, which impinge on the vehicle surface and excite vibrations. These fluctuations are imparted to the vehicle occupants as noise. Most noteworthy are A-pillar vortices, first because they are the most energetic vortices and second because they are located close to the driver. Still, their influence is considerably less than that of flow across body gaps and inadequate seals. These noise effects may be minimized by shaping the A-pillars to reduce turbulence or by increasing side glass insulation.

Due to the nature of the generating fluctuations, noise generated by flow separation at the back of the vehicle is in the 20 Hz range but is only noticeable in special cases, such as convertibles with soft plastic rear windows offering little insulation.

- Separation from protruding parts such as antennas, suspension components, mirrors, and wipers. Vortex streets are generated behind mast antennas, generating individual high-intensity tones. Countermeasures include angling antennas by at least 45° and applying a spiral around the antenna itself. This eliminates the separate tones, replacing these with a less noticeable general sound level.

  For functional reasons, suspension components and wipers cannot be shaped to give optimum airflow; therefore, they should be located out of the direct airstream. Wipers, for example, may be "parked" under the front hood.

Outside mirrors are often regarded as sources of noise as the driver perceives noise coming from their direction. They are not always the actual source, however;

A-pillar vortices and seals may also play a role. Mirror housings themselves or flow separation from them is less likely to be the actual source of noise than accelerated flow through a too-narrow gap between housing and side glass, flow around the housing base, and poor sealing between the base and door or between the base and vehicle interior.

- Turbulent oscillations in the airflow near the vehicle: in other words, the boundary layer. Although this component presents an insurmountable lower limit, it is of secondary importance on motor vehicles.

The wide variety of noise sources means that aerodynamic optimization will only rarely improve the vehicle's noise situation. By far the more important noise sources are inadequate sealing systems, which have little effect on aerodynamics and therefore are little influenced by aerodynamic optimization.

3.2.3 Aerodynamics in the overall development process

Aerodynamic requirements have a direct effect on the exterior vehicle shape. Accordingly, aerodynamic development must operate parallel to the design phase [4]. Whether design models are tested in the wind tunnel or additional wind tunnel models are employed depends on custom and practice within individual firms. In the second case, it is critical that parallel models are built in a timely manner and continuously updated to reflect the current design status. In general, there is little point in creating aerodynamic studies which deviate too far from the actual development goals and are therefore ignored by other development departments. The exception to this is special development projects which place a high priority on extremely low c_d values.

Aerodynamic parameters required for driving safety should be defined prior to the beginning development work and checked and assured during aerodynamic development. It should be expected that there will be resulting changes to the vehicle shape. After the fact, these coefficients can only be modified by spoilers and other add-on components, which often leads to problems with regard to acceptance by designers, engineers, financial planners, and, above all, customers.

Optimizing the cooling air component of airflow also begins during the design phase as cooling air inlets are created. The process continues with testbeds in which new details representing the front of the new design are tested on current, similar production models. This is followed by the prototype phase, in which experimental results necessitate further modifications. Experience with previous models accelerates the development process. Wetting and soiling are normally examined in the prototype phase, but some decisions, such as use of gutters along the A-pillars, must be reached at an even earlier stage. Functional tests in the wind tunnel are also conducted during the prototype phase.

With the short development times common today, there is a danger of aeroacoustic development coming too late in the process. Meaningful test results cannot be expected until substantially production-like prototypes, with actual seals and complete interior equipment, are available. By this time, however, development has progressed to the point where major modifications are no longer possible prior to the onset of production. Therefore, aeroacoustic development must also begin during the design phase, even though it would only be possible to draw on experience gained in developing predecessor models.

References

[1] Hucho, W.H. *Aerodynamics of Road Vehicles.* Fourth edition. Society of Automotive Engineers, 1998. ISBN 0-408-01422-9.

[2] Hupertz, B. "Einsatz der numerischen Simulation der Fahrzeugumströmung im industriellen Umfeld." Dissertation, TU Brunswick, 1997, and MR-Forschungsbericht 98-03, Institut für Strömungsmechanik, TU Brunswick.

[3] Huber, L. "Die Fahrtrichtungsstabilität des schnellfahrenden Kraftwagens." DKF Heft 44, 1940

[4] Hucho, W.H. "Design und Aerodynamik—Wechselspiel zwischen Kunst und Physik." In *The drive to design—Geschichte, Ausbildung und Perspektiven im Autodesign.* Edited by R.J.F. Kieselbach. Avedition, Stuttgart, 1998.

3.3 Heat transfer and thermal technology

3.3.1 Cooling of internal combustion engines

In cooling combustion engines, several different media, such as coolant, oil, engine oil, transmission oil, charge air, and exhaust must be considered (see Section 5.1).

In ensuring the serviceability of the engine, cooling the medium which carries heat away from the engine is of primary importance (Fig. 3.3-1). Furthermore, part of the engine heat loss is taken up by lubricating oil. For high-powered engines, cooling the oil pan is insufficient

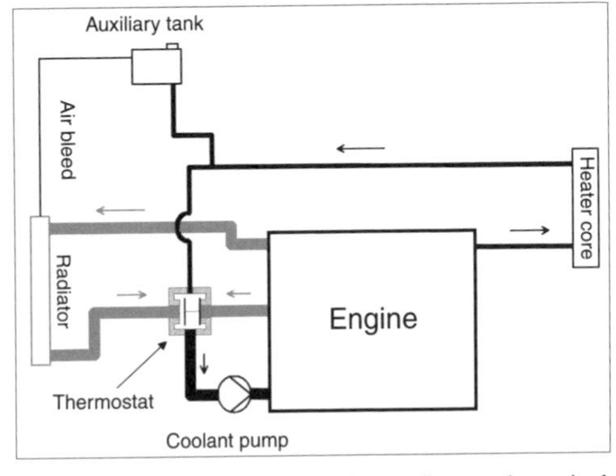

Fig. 3.3-1 Cooling system with overflow tank, typical of an upper midrange automobile.

to avoid exceeding maximum allowable oil temperature, and an engine oil cooler is required.

Modern supercharged diesel engines in both passenger cars and commercial vehicles also employ charge air intercoolers. Increased air density achieved through cooling results in improved cylinder filling and higher power output. In addition, lower temperature reduces thermal loading of the engine and leads to lower NO_x concentration in the exhaust gases.

Recently, exhaust coolers [1] have found application on diesel engines, which will be subject to ever-tighter emissions restrictions in the future. These limits can be met with little fuel economy impact by means of exhaust cooling in addition to the exhaust gas recirculation (EGR) systems familiar on passenger cars. By cooling and mixing incombustible exhaust components with charge air, combustion temperatures—and therefore exhaust NO_x concentrations—are reduced.

Vehicles fitted with automatic transmissions also require transmission oil cooling, either by means of coolant or air. Additionally, vehicles with air conditioning incorporate a condenser as part of the refrigeration system.

Ever-increasing demands with respect to fuel economy, exhaust emissions, longevity, ride comfort, and packaging have led to modern internal combustion engine cooling systems which, with very few exceptions, exhibit the following characteristics:

- Water cooling of the engine, with forced coolant circulation by means of a belt-driven centrifugal pump.
- Cooling system operation with up to 1.5 bar gauge pressure.
- Use of a mixture of water and antifreeze, usually ethylene glycol, in a 30–50% volume concentration.
- Corrosion-resistant aluminum alloys as the predominant material for radiators. Coolants contain additional corrosion inhibitors to protect aluminum radiators.
- Plastic as the predominant material for radiator tanks, fans, and fan shrouds.

Modern development trends are directed toward:

- Control inputs by means of fan drive and coolant thermostat
- Premounting all cooling components in the front end area in a single functional unit, the so-called "cooling module"

Along with numerous development activities toward even more compact, lighter, and more efficient components, the cooling system—especially when electronically controlled—continues to gain importance with respect to achieving the previously stated requirements.

3.3.1.1 Radiator layout

The basic goal in designing a cooling system is to provide the required cooling capacity using the most compact, lightest, and most economical radiators within the available space. To this end, an optimization process must be carried out with regard to location and dimensioning of the modular heat transfer unit, the choice of radiator tube and fin geometry, the fan's power draw, and fitting to vehicle-specific boundary conditions, often including c_d and crash behavior.

Analytical programs for calculating heat transfer by one-dimensional streamline theory are commonly used to assist in system layout. Given radiator geometry, heat transfer, conductivity, and pressure loss conditions as well as the fluid flows, input of entry pressures, and temperatures provides information regarding these same properties on the heat transfer unit's exit side. Supported by empirical data from years of test experience with a wide range of radiator variations, these algorithms, within the framework of similitude theory, permit very precise predetermination of any desired fin/tube configuration in any desired size for any desired operating conditions.

Today, applications almost universally require design of entire cooling modules with full or partial overlap of heat exchangers, blowers, and shrouds. Accordingly, for these modules so-called topological models are created with multiple stream paths, each of which may in turn be subjected to streamline theory calculations. The simulation algorithms take into account the mutual effects of various components [2].

Ultimately, this design aid is supplemented with elements such as relative wind, blowers and all of the vehicle's pressure drops such as the radiator grille and flow through the engine compartment. This permits iterative calculation of the vehicle's cooling airflow and, consequently, the cooling system's thermodynamic properties. Coupled with broad experience from wind tunnel tests of cooling capacity, one has a rapid, highly reliable simulation aid which greatly reduces the need for actual vehicle testing.

The near future will see a combination of analytical one-dimensional methods with numerical three-dimensional CFD (computational fluid dynamics) methods which will provide detailed descriptions of highly complex cooling airflow through the engine compartment.

Customarily, cooling system layout requires determination of several thermally critical vehicle operating conditions, such as "maximum level speed," "rapid ascent," or "slow ascent with trailer." Applications are differentiated between European service and operation in hotter climates. In every case, there are preset values of vehicle speed, ambient air temperature, heat rejection rates, and targets for maximum permissible coolant, charge air, and oil temperatures. Performance loss as a result of aging is indirectly addressed by higher targets. Typical rules of thumb and targets for the most common cooling configurations are summarized in Table 3.3-1.

Bandwidths for various operating conditions, from the lowest-powered passenger cars to the most powerful commercial vehicle powerplants, are:

Table 3.3-1 Typical targets and rules of thumb for cooling system layout

	Passenger car	Commercial vehicle (Euro3)
Maximum heat rejection rate from coolant for:		
Gasoline engine	$Q_{coolant} = 0.5$ to $0.6\ P_{mech}$	
Indirect injection diesel	$Q_{coolant} = 1.0\ P_{mech}$	
Direct injection diesel	$Q_{coolant} = 0.65$ *to* $0.75\ P_{mech}$	$Q_{coolant} + Q_{intercooler} = 0.65\ P_{mech}$
Maximum permissible temperature difference between coolant inlet temperature and ambient temperature	Approx. 80 K	Approx. 65 K
Maximum permissible temperature difference between charge air at intercooler exit and ambient temperature	Approx. 35 K	Approx. 15 K

Maximum coolant temperature	100–120°C
Maximum coolant flow rate	5,000–25,000 L/h
Maximum charge airflow	0.05–0.6 kg/s
Maximum charge air inlet temperature	110–220°C (at 25°C ambient temperature)

3.3.1.2 Radiator construction

Based on the above data, different performance requirements and different packaging demands for water, oil, and charge air heat transfer devices have led to a variety of different radiator designs [3, 4]. One example is shown in Fig. 3.3-2.

System thickness (dimension in the direction of cooling airflow) for coolant radiators ranges from 14 mm for the smallest passenger cars to 55 mm for the largest commercial vehicle radiators; frontal areas range from 15 dm^2 to 85 dm^2. Charge air coolers range from about 30 mm to more than 100 mm system thickness, with frontal areas from 3 dm^2 for passenger cars to 80 dm^2 for commercial vehicles. In passenger cars, charge air coolers can be mounted in a variety of configurations: as a large area in front of the coolant radiator, long and slim below or alongside the radiator, or located well away from the engine coolant module—for example in a wheel well; hence the large range of system thickness.

The primary determinant of radiator performance is design of the fin and tube geometry, the so-called radiator core. Distinction is made between mechanically joined and soldered systems. Mechanically joined fin and tube systems (Fig. 3.3-3) consist of round or oval tubes inserted in stamped fins. Expanding the tubes joins them to the fins. These systems typically cover the lower end of the performance spectrum, but with improved expansion methods and ever-smaller oval tubes, they begin to approach the performance of soldered systems.

In their modern form, soldered systems, consisting of tinned flat tubes and rolled corrugated fins (Fig. 3.3-4), are typically made with only one tube accounting for the entire system thickness. For improved rigidity, this may have stiffening ribs or a folded configuration. In either case, the tubes are located by so-called headers in the tanks, which distribute coolant or charge air to the tubes.

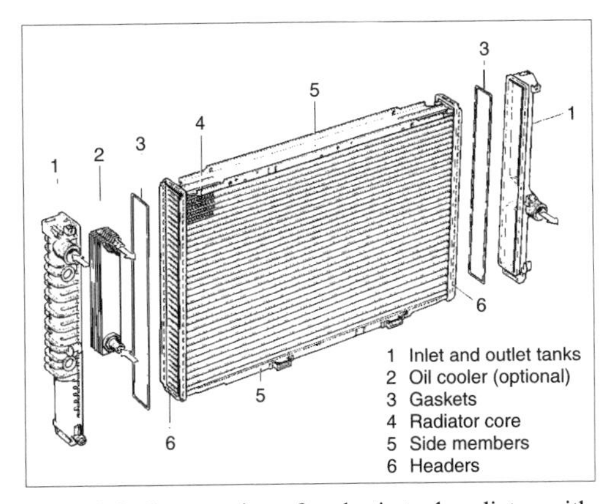

Fig. 3.3-2 Construction of a plastic tank radiator with an aluminum core.

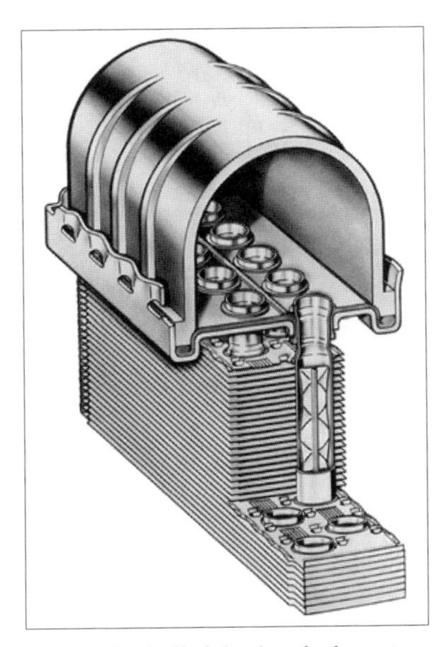

Fig. 3.3-3 Mechanically joined oval tube system.

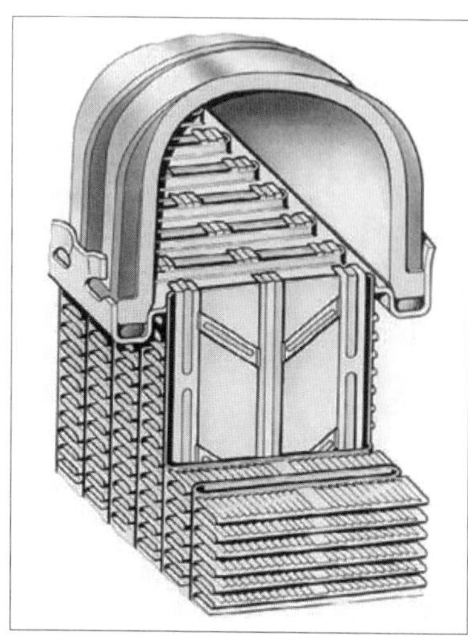

Fig. 3.3-4 Brazed flat-tube heat exchanger.

In passenger cars, engine oil coolers are preferentially located close to the engine. Oil is cooled by heat transfer to the engine coolant (Fig. 3.3-5) using designs such as aluminum heat exchangers in shell and tube, stacked plate, or flat-tube configurations. Passenger cars with automatic transmissions may employ air-cooled flat-tube heat exchangers, or they may have very slim, extended flat-tube exchangers built into the tanks of engine coolant radiators, where they transfer heat to engine coolant.

Charge-air coolers for supercharged diesel engines are, almost without exception, brazed aluminum flat-tube heat exchangers cooled directly by air. In some cases, charge air is cooled close to the engine by coolant, refrigerated to a very low temperature in a separate circuit.

On commercial vehicles, exhaust coolers are subjected to very high temperatures and extremely corrosive conditions, which dictate the use of stainless steel. Laser welding or nickel brazing are commonly used as joining methods. These coolers are of shell and tube construction. The tubes carrying exhaust gases may be simple round tubes or tubes employing special performance-enhancing yet soot-resistant measures.

Fig. 3.3-5 Stacked plate oil cooler.

3.3.1.3 Fans and fan drives

Modern fans for engine cooling systems are almost without exception axial flow designs made of composites or plastic. Depending on operating conditions in the vehicle, shroud rings and inlet nozzles at the blade tips ("jet ring fans") may be added to the axial blades. Other typical fan features might include sickle-shaped blades and uneven blade spacing. These could benefit fan efficiency and noise emissions.

In passenger cars, fans, either singly or paired, are usually installed to work in suction, with maximum diameters of about 500 mm. Except on the most powerful engines, electric motors are used to drive the fans. These consume up to 600 W of electrical power, with staged speed variation by means of ballast resistors or continuous speed variation with brushless electric motors. The most powerful passenger cars as well as all commercial vehicles employ viscous fan drives. Input speed, dictated by the crankshaft or otherwise mechanically transmitted from the engine (usually through the water pump), is transferred to the secondary side of the viscous coupling by fluid (oil) friction. The secondary side is connected to the fan. By means of varying the amount of oil filling, fan speed may be controlled over a range from idle speed to just under the input speed. Commercial vehicles employ fans up to 750 mm in diameter, consuming up to 30 kW of power.

3.3.1.4 Cooling modules

Cooling modules are constructive units which consist of various components, including a fan unit with drive and possibly including air conditioning components (Fig. 3.3-6). The modular approach, which has been gaining acceptance since the late 1980s, offers a number of technical and economic advantages [5]:

- Optimum layout and matching of components.
- This results in improved efficiency as installed in the vehicle or smaller, more economical components.
- Less logistic effort on the part of vehicle manufacturers in terms of development, testing, and assembly.

Normal road vehicles almost exclusively use body-mounted cooling modules attached to the vehicle's existing longitudinal and transverse members. Usually one of the heat transfer units serves as a load-bearing element for the module; other components are attached to its coolant or air tanks and side members by means of latching elements (snap assembly), clamps, or clips. The greater the number of components contained in a cooling module, the greater the advantage in using a separate carrier frame for all module components.

3.3.1.5 The overall engine cooling system

The field of engine cooling includes heat sources (among others, the engine and transmission), heat exchangers, and of course a heat sink—cooling air—to which all heat must be transported. Conventional cooling systems

Fig. 3.3-6 Top: cooling module for commercial vehicle with fan and viscous coupling. Bottom: cooling module for passenger car with electric fans.

customarily employ the following fluid stream control inputs:

A thermostat, whose wax element responds to the temperature of surrounding coolant and either directs coolant flow through the radiator or through a bypass. In this way, with very low coolant temperatures, unneeded additional cooling can be largely avoided, while at high temperatures, maximum cooling is provided.

Electrically driven fans may be driven at various discrete speeds in response to coolant temperature in radiator tanks or with continuously variable speeds.

Fans fitted with viscous clutches are controlled by varying the amount of oil, which in turn varies fan speed as a function of cooling air temperature upstream of the clutch. Air is heated after passing through a hot heat exchanger. This indicates a need for more cooling, and a bimetallic thermostatic element activates the fan.

All other systems are designed to meet worst-case operating conditions but are otherwise not actively controlled. The coolant pump is belt-driven from the crankshaft, charge air cooling is almost always uncontrolled, oil cooling is only partially thermostatically controlled.

Historically such cooling systems have been completely adequate and are characterized by highly reliable operation. In the future, however, this system, just like many other vehicle systems, will be electronically controlled. A network of sensors will monitor the thermal situation of engine and cooling system. An engine cooling control unit will employ control concepts to modulate fluid transport components (fans, pumps) and control components (valves, flaps, slats) in order to achieve demand-based cooling, thereby saving energy required to drive auxiliary devices, have a favorable impact on exhaust and noise emissions, and, in the interest of increased comfort and reduced engine wear, shorten the engine warm-up phase. For this reason, all transport and control components will have to be electrically controlled; at present, this is already the case for electric fans. Thermostats will be controllable by electrically heating their wax elements. This will permit thermostat settings independent of the actual coolant temperature, instead matching predetermined points on a map in computer memory. Viscous clutches will soon be electrically controllable, by means of an electromagnetically actuated valve replacing their present bimetallic controls. Other auxiliary drives will also be electronically controlled. In addition, more control components will enter service [6, 7].

3.3.2 Passenger compartment heating and cooling

Climate control of the passenger compartment serves several purposes:

- Guarantee clear outward vision
- Provide the driver with a less fatiguing environment
- Protect the occupants from unpleasant odors and irritating substances
- Create a comfortable climate for all vehicle occupants

Climate control serves both comfort and safety; the ability of a driver to concentrate is considerably greater in a comfortable environment than in hot or cold surroundings. In addition, glass areas are kept free of ice and condensation. In many countries, this task is mandated by appropriate regulation [8].

Issues of occupant comfort and climate control operation will be treated in detail in Section 6.4.3. However, for a climate control system to fulfill its function, it must be adequately supplied by the heating and refrigeration circuits. The components of these circuits and their interaction with the vehicle will be treated here.

Vehicle climate control is achieved by a flow of treated air, symbolized by arrows in Fig. 3.3-7, which enters the cabin through vents in the dashboard, in the footwells, and, in some cases, in the rear seating area. The climate control unit (3) is mounted below the instrument panel. With the help of a radial blower (1), outside air is drawn from just below the windshield, through a filter (2), and fed to the evaporator (4) of the climate control unit (3), where the air is cooled and thereby dried. From there, the air passes to the heater core (5), where it is warmed.

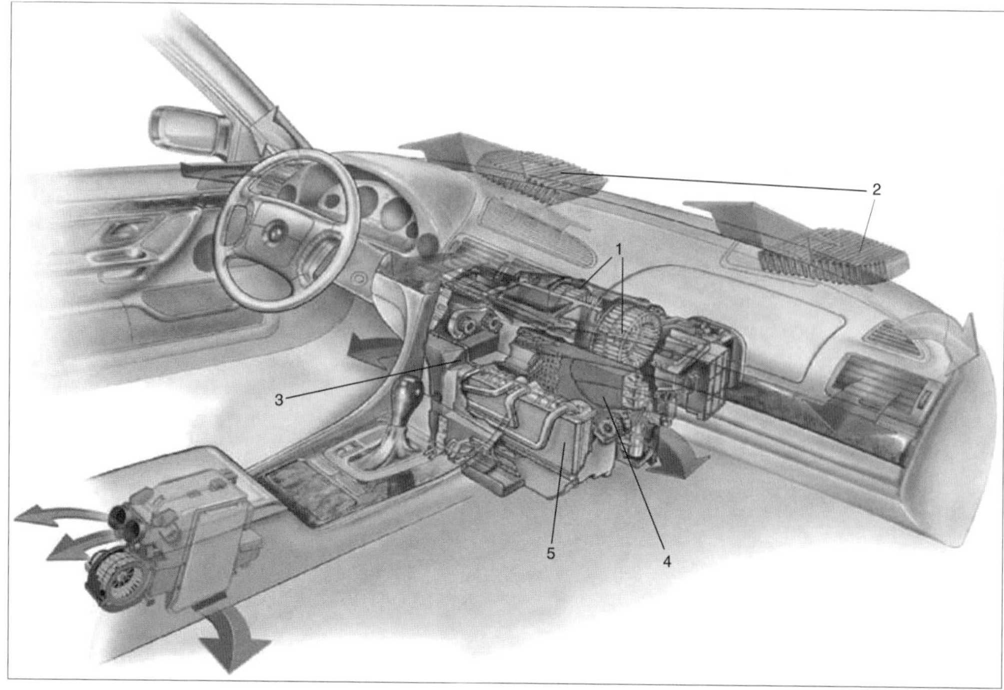

Fig. 3.3-7 The climate control system of a luxury vehicle.

3.3.2.1 The heating function and its components

Passenger cars are generally heated with waste heat from their combustion engines. After passing through the engine block, a portion of the coolant flow is diverted to the heating circuit, where it flows through the heater core located within the climate control unit and gives up heat to air bound for the passenger compartment.

The design principles of heater cores are similar to those of coolant radiators, already described in section 3.3.1.2. Fig. 3.3-8 shows a typical heater core, consisting of an assembly of flat tubes. Its heating capacity as a function of mass airflow is shown in Fig. 3.3-9 for three different coolant flow rates. The values given apply for an inlet air temperature of −20°C and coolant temperature of +80°C. For other inlet temperature differences, performance will vary accordingly.

To control the heat admitted to the passenger compartment, water valves may be installed (electrically modulated valves or continuously adjustable valves) in order to vary the coolant flow and so adjust heating performance (water-side temperature control). Alternatively, so-called air-side temperature control may be employed, in which the heater core is always subjected to its maximum flow. Heating performance in this case is controlled by mixing warm and cool air upstream of its entry to the passenger compartment (see also Section 6.4.3).

Fig. 3.3-8 Brazed flat-tube heater core.

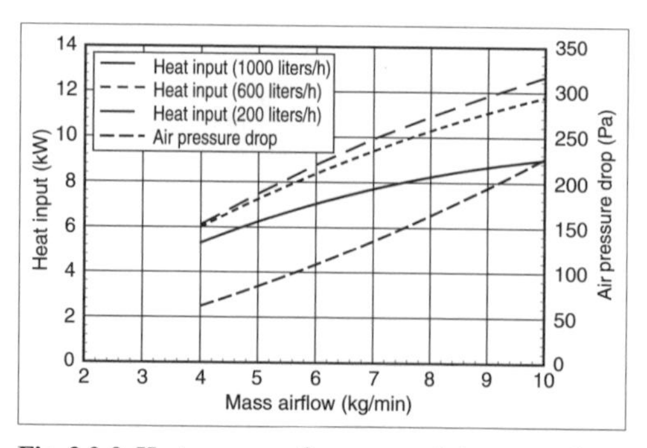

Fig. 3.3-9 Heater core performance and air pressure loss.

Because it is directly coupled to the engine cooling system, both the heater core feed temperature and the pressure differential, which drives the mass flow rate of coolant through the heating system, vary with engine speed and load. Regardless of vehicle size, the coolant circuit layout should provide a coolant flow rate of at least 600 L/h through the heater core inside the passenger compartment to ensure adequate heat transfer between the coolant and air. If for some design reason the water pump cannot supply an adequate coolant mass flow rate through the heater circuit, electrically driven auxiliary water pumps may be applied (with typical feed rates of about 1,000 L/h with 1,000 mbar pressure head). These also prevent a drop in heater performance at engine idle.

As a result of fuel economy measures applied to both engines and vehicles, problems have arisen in providing heat to the passenger compartment, whose heat demand at an ambient temperature of $-20°C$ amounts to about 7 kW at thermal equilibrium. On passenger cars and light trucks fitted with modern direct-injection diesels or gasoline powerplants, engine waste heat is insufficient to ensure rapid heating. As a solution, several auxiliary heater concepts [9] are available. Active systems cover the gap between heat available in engine coolant and the heat demand of the passenger cabin by adding an additional heat source: that is, by application of additional primary energy. This might take the form of a fuel-fired auxiliary heater (Fig. 3.3-10), air-side electric auxiliary heater (PTC heater, Fig. 3.3-11) or coolant-based electric auxiliary heating. Table 3.3-2 provides an overview of various auxiliary heating systems.

All electric auxiliary heaters are effectively indirect heaters, as they place an additional load on the engine's electrical generating system. As an approximation, the same amount of heat energy which is applied electrically is also indirectly rejected to the engine coolant.

In addition to these active systems, passive systems are under consideration. These recover heat—for example, from the exhaust system with the aid of an exhaust gas heat exchanger, or from cooling of ancillary devices such as the generator, or from recirculating already heated cabin air. All such systems are currently still under development.

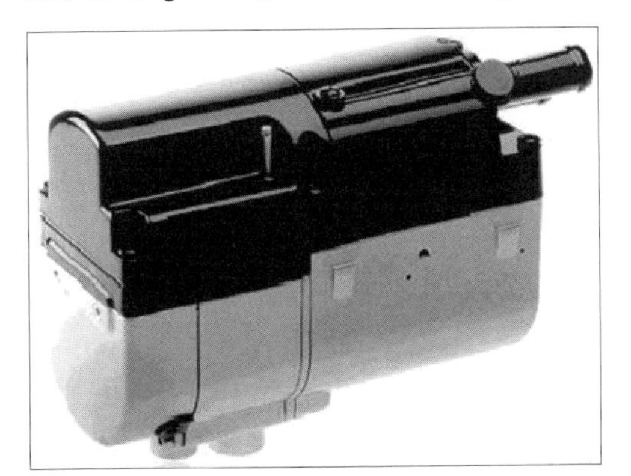

Fig. 3.3-10 Fuel-fired auxiliary heater.

Fig. 3.3-11 PTC heater.

3.3.2.2 Operation of the air conditioning system and its components

Vehicle air conditioning systems are based on the vapor-compression refrigeration cycle and are similar in principle to the operation of a household refrigerator. Air for the passenger compartment is cooled on the outer surface of the evaporator (Fig. 3.3-13). Moisture is condensed out of the air in the process. Inside the evaporator, a refrigerant, generally CFC (chlorinated fluorocarbon)-free R134a, at low pressure and at a temperature below ambient, vaporizes, thereby absorbing heat from the air. A compressor raises the vaporized refrigerant to a higher pressure. Heat carried by refrigerant vapor is released at higher temperature (above ambient) as it passes through the condenser, located in the engine compartment ahead of the engine coolant radiator. The refrigerant returns to a liquid state in the condenser and passes to the receiver-drier, which stores a supply of refrigerant to cover a range of operating conditions. From the receiver, refrigerant passes through an expansion valve, where it again expands to low pressure and low temperature, and then to the evaporator, where the the closed cycle repeats.

For the engine compartment package, the type of expansion device is of major importance. By regulating the refrigerant mass flow rate, a thermostatic expansion valve ensures that only vapor is present at the evaporator outlet. This not only provides for good evaporator utilization under all operating conditions, but also protects the compressor from so-called "hydraulic shock." In these systems, the receiver-drier is installed in the engine compartment between the condenser and expansion valve.

As an alternative to the thermostatic expansion valve, an orifice tube (actually a capillary tube) is often used. As the orifice tube does not have a variable cross section, there is a possibility of liquid refrigerant leaving the evaporator. To protect the compressor, a system using this design must also include an accumulator which separates

Table 3.3-2 Overview of auxiliary heating systems

Auxiliary heating system	Description	Advantages
Fuel-fired auxiliary heater	Combustion of fuel produces additional heat, which is transferred directly to the coolant. The fuel-fired heater is integrated within the engine coolant circulation system	Easily upgraded to preheater
Electric PTC heater in airflow (Positive Temperature Coefficient; electrical resistance rises with temperature)	Heat is transferred directly to air flowing into passenger compartment. The heating elements' PTC characteristic prevents unacceptably high temperatures	Rapid response; additional heating effect from greater engine load through electrical generating system; additional heating effect may be achieved with water-cooled generators
Electric heater in coolant	Heat is transferred to engine coolant	Additional heating effect from greater engine load through electrical generating system; additional heating effect may be achieved with water-cooled electrical generators
Heat pump	The air conditioning system is operated in heat pump mode; requires complex shrouding and additional valves	Small additional space requirement
Heat recovery from exhaust gas	Heat is transferred to engine coolant; added backpressure in exhaust system and additional heat input to engine coolant system must be taken into consideration	Virtually no additional primary energy demand
Closed air circulation with air dryer	Already warmed air from the passenger compartment is dried by absorbers and recirculated; requires complex air ducting	Very slight additional primary energy demand

and stores excess liquid refrigerant. An accumulator occupies about twice the volume of a receiver-drier.

The condenser and ventilation of the condenser are of major significance in fine-tuning the vehicle's overall heat management. For one thing, the heat rejection of the condenser must be considered in designing the engine cooling system. For another, good condenser ventilation is of extreme importance for the output and efficiency of the air conditioning system. In integrating the condenser in the vehicle, care must be taken to eliminate reverse flow of cooled air. The condenser itself may be a mechanically joined round-tube system or a brazed assembly of flat tubes and fins (Fig. 3.3-14).

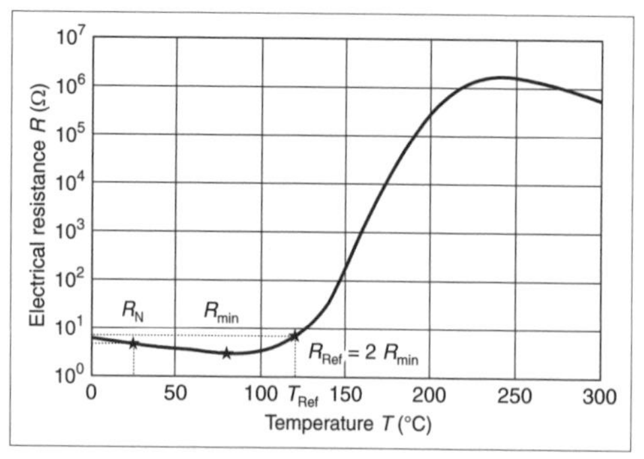

Fig. 3.3-12 Temperature-resistance curve of PTC heating element.

The evaporator is usually located inside the air conditioning unit. Along with high "performance density" to meet packaging requirements—for example, using a modern plate evaporator (see Fig. 3.3-15)—it is important to ensure that condensed water can drain easily. To this end, special coatings are employed, with the added benefit that they inhibit growth of bacteria and microorganisms which may lead to unpleasant odors.

The working fluid of a passenger car air conditioning system is generally R134a. It is a nonflammable, nontoxic fluorinated hydrocarbon compatible with many plastics and metals. Alternative refrigerants for motor vehicle air conditioning systems, such as CO_2, are under

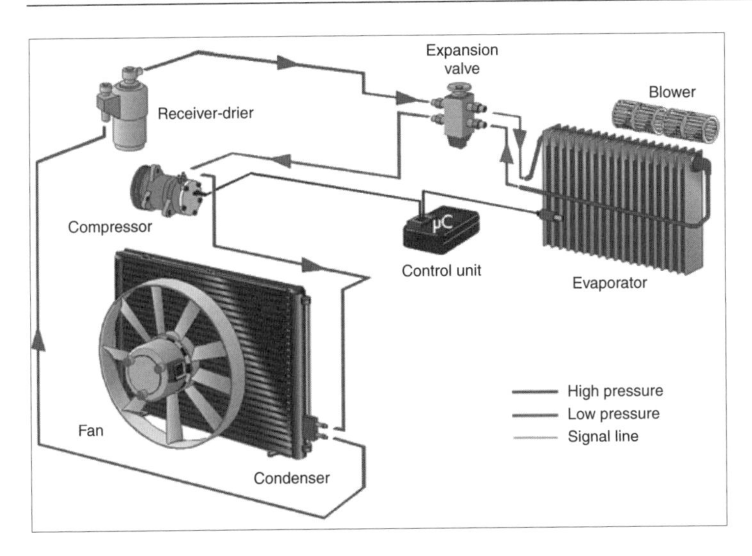

Fig. 3.3-13 Schematic of automotive air conditioning refrigerant cycle.

development [10]. About 10–20% oil is added to lubricate the compressor; the resulting performance losses are unavoidable [11].

The amount of refrigerant contained in the air conditioning system differs from car to car and is largely determined by the interior volume of those components which carry refrigerant in its liquid state. Typical values are 600 to 900 g of R134a.

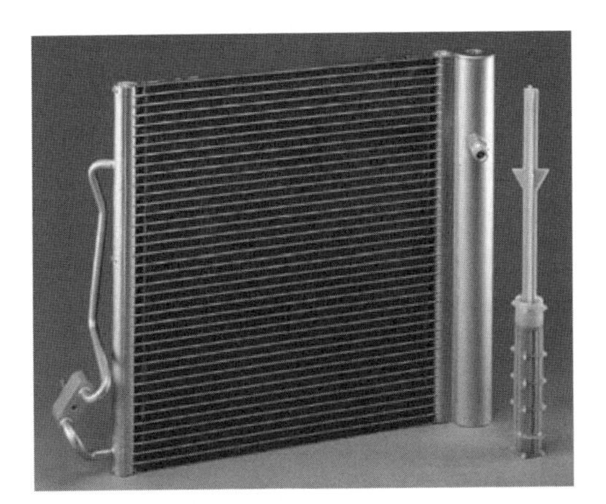

Fig. 3.3-14 Flat tube condenser with integral receiver-drier.

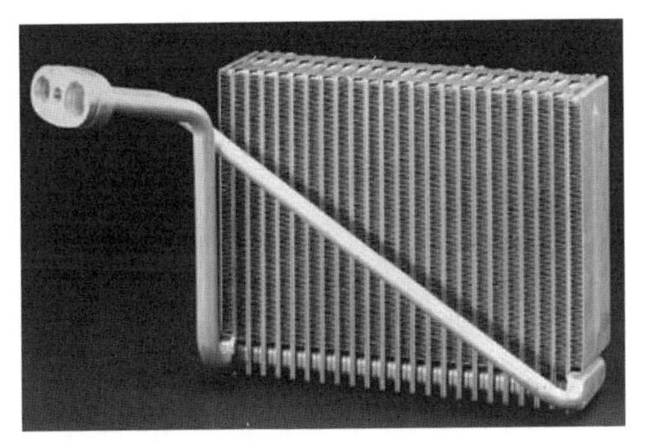

Fig. 3.3-15 Plate evaporator.

3.3.2.3 Compressors and control of cooling performance

Of all the components in the air conditioning system, the compressor plays a determining role for the package inside the engine compartment, engine management, and vibration characteristics. Along with its operating principle, the condenser's control method affects the overall behavior of the vehicle. The following is restricted to the most significant compressor designs found in modern automotive air conditioning systems [12].

A *rotary vane compressor* uses a rotor fitted with radially movable vanes turning in an asymmetrical cylinder. A compression space is formed between pairs of vanes. Rotary vane compressors have the following advantages: high delivery rates at medium speeds, unaffected by hydraulic shock, compact dimensions, few moving parts, and low weight. Furthermore, they have no oscillating masses. Disadvantages include noise at low speeds.

Scroll compressors also use rotary motion to compress the working medium. Scroll compressors consist of a fixed spiral and a second rotating eccentrically mounted spiral. The spirals touch at two or three points, forming pockets. Scroll compressors have the following advantages: high efficiency at medium speeds, ability to withstand high rotational speeds, uniform torque requirements, unaffected by hydraulic shock, compact dimensions, quiet operation, and no oscillating masses.

Swashplate compressors, which currently dominate the market, use an angled round disc driven by the compressor shaft. Pistons bearing on this disc are moved back and forth in their bores, compressing the refrigerant. Intake and exhaust of the compressed gas are accomplished by valves contained in bores in the cylinder head. In principle, the displacement of such compressors may be altered by changing the inclination of the swashplate relative to its shaft. Advantages of swashplate compressors include good efficiency at low speeds, easily achieved power regulation, and good performance through the controlled range.

Air conditioning systems not equipped with variable-displacement compressors meet changing cooling demand by means of duty cycling: that is, periodically engaging the unit by means of a magnetic clutch. The engine management system must compensate for this; otherwise, turning the compressor on and off will have a negative effect on vehicle behavior (a perceptible jerk as the compressor engages).

Many current European midrange and luxury cars employ performance-controlled variable-displacement compressors [13]. Their primary advantage lies in compensating for engine speed variations caused by traffic conditions. For a given pump head-capacity characteristic—the so-called control curve—compressor displacement is altered accordingly. This also eliminates annoying jerks caused by clutch engagement or disengagement.

A further development of automatic displacement control is external control to vary the displacement volume. A solenoid valve replaces the usual mechanical control valve. This permits altering the pump delivery indirectly by changing the control curve by means of an external electrical signal. External control is a prerequisite for clutchless operation. In addition, this system makes it possible to achieve evaporator temperature control.

3.3.2.4 Air conditioning system layout

A vehicle's requirement for interior heating or cooling depends on both vehicular and climatic boundary conditions:

Vehicular boundary conditions

- Mass and heat capacity of installed components and enclosing surfaces.
- Heat insulation of enclosing surfaces; e.g., roof, floorpan, firewall.
- Radiation properties of glazing.
- Dimensions and angles of glazing and cabin.
- Ventilation and airflow through cabin, air leakage through body.
- Airflow around vehicle; air intake conditions and external heat transfer vary with speed.

Climatic boundary conditions

- Outside air temperature
- Relative humidity
- Insolation (solar radiation)

The determining factor in vehicle climate control system layout is system startup, which requires several times as much performance as maintaining a steady temperature. A vehicle air conditioning system is normally sized so that a comfortable temperature may be attained in a short time even after an extended period of nonoperation. In the summer, under full sun, temperatures in a parked vehicle may reach 70°C (158 °F). A typical design condition is ambient temperature 40°C, 40% relative humidity, and 1,000 W/m^2 insolation.

In winter, the vehicle interior will cool to the ambient outside air temperature; −20°C may be assumed as a design condition. Heating, with an optimum airflow rate of 5 kg/min, which is below the maximum possible rate, proceeds as shown by the top curve of Fig. 3.3-16. The lower curve emphasizes the uncomfortable situation presented by vehicles with direct-injection diesel engines lacking auxiliary heaters.

In addition to climatic conditions, driving situations are specified. One typical situation is driving at constant speed on a level road in third gear, at either 32 km/h or 50 km/h (20 or 31 mph). Usually, climate control system behavior is also noted at engine idle as, due to low engine speed driving either the water pump or air conditioning compressor, this presents the most unfavorable situation for heating as well as cooling the vehicle.

A typical cooling curve for operation in 40°C ambient air with the evaporator blower at maximum speed achieves 30°C after 20 minutes. After 60 minutes, the temperature in the passenger compartment has been reduced to 23°C. With maximum air conditioning output, the curve of temperature versus time initially shows a steep gradient, which transitions into an equilibrium value; see Fig. 3.3-17. At the beginning of the cooling phase, the evaporator may be rejecting as much as 8 kW of heat, dropping to about 2.5 kW near equilibrium. The cabin airflow rate for these situations is between 7 and 10 kg/min.

To evaluate an air conditioning system, the following parameters are of significance:

- Cooling output
- Efficiency
- Noise/vibration
- Control response
- Durability

These parameters are influenced by the components themselves as well as the system environment: that is, the routing and properties of the refrigerant lines, the ventilation of the condenser, as well as the system pressures and temperatures determined by operating conditions. This means that ultimately, a refrigeration circuit can only be evaluated when installed in a vehicle.

3.3.2.5 Additional fuel consumption due to air conditioning system

A vehicle's additional fuel consumption due to air conditioning [14] is composed of 1) engine output required to drive the compressor, 2) added fuel consumption due to increased vehicle weight, and 3) the necessary additional electrical power to drive fans and blowers.

In air conditioning system layout, the most extreme boundary conditions determine the required performance and, therefore, the size of components in the heating as

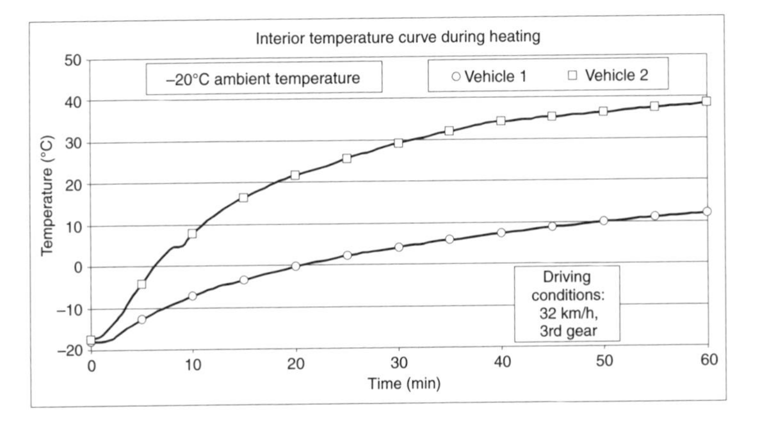

Fig. 3.3-16 Heating curve of a luxury car.

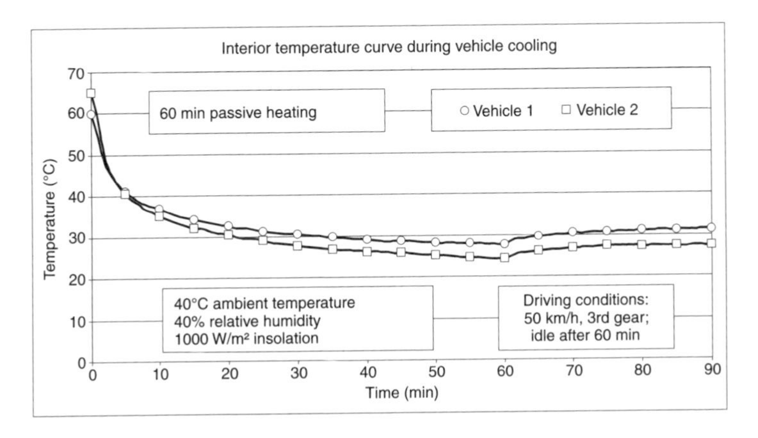

Fig. 3.3-17 Cooling curve of a luxury car.

well as cooling circuits. For air conditioning fuel consumption, the most common operating conditions—namely part load—are determinant. Consequently, added fuel consumption is determined as an annual average, with the aid of load profiles which give frequency-weighted fuel consumption for multiple operating conditions at various ambient temperatures and driving conditions [15].

Using one specific vehicle as an example, the compressor power draw is averaged over a load profile consistent with a central European climate and multiplied by the vehicle's annual operating time. Operational hours are set at 462 hours per year, determined on the basis of an assumed 15,000 annual kilometers and an average speed of 32.5 km/h (20.2 mph) for the established driving cycle. For the specific vehicle chosen for this example, the average annual fuel consumption dedicated to driving the compressor amounts to 68 L. Added to this are power draw for electrical items such as blower and clutch, amounting to about 13 L/year; the 12 kg of increased vehicle weight account for approximately 12 L/year. In total, an air conditioning system typically consumes 93 L/year of fuel. Based on annual fuel consumption and the annual mileage, the distance-specific fuel consumption works out to an annual average of about 0.62 L/100 km. A different vehicle utilization profile or climatic and vehicle-specific boundary conditions different from those of the chosen load profile will of course alter the fuel consumption.

References

[1] Lutz, R., and J. Kern. "Ein Wärmeübertrager für gekühlte Abgasrückführung." 6. Aachener Kolloquium Fahrzeug und Motorentechnik, 1997.

[2] Eichelseder, W., R. Marzy, I. Hager, and M. Raup. "Optimierung des Wärmemanagements von Kraftfahrzeugen mit Hilfe von Simulationswerkzeugen." Tagung Wärmemanagement, Haus der Technik, Essen, 9/98.

[3] Kern, J., and J. Eitel. "State of the art and future development of aluminium radiators for cars and trucks." SAE 931092.

[4] Kern, J. "Neue Konstruktion gelöteter Ganzaluminiumkühler für Kfz." ATZ 100, 9/98.

[5] Löhle, M., H. Kampf, and J. Kern. "Integrierte Systeme und Produkte zur Motorkühlung und Innenraumklimatisierung." ATZ special issue "Systempartner 98."

[6] Martin, M. "Elektronisch geregelte elektromagnetische Visco-Lüfterkupplungen für Nutzfahrzeuge." ATZ 95, 5/93.

[7] Ambros, P. "Beitrag der Motorkühlung zur Reduzierung des Kraftstoffverbrauchs." Tagung Wärmemanagement, Haus der Technik, Essen, 9/98.

[8] European Council Directive 78/317/EEC or U.S. safety standard FMVSS 103.

[9] Flik, M., M. Löhle, H. Wilken, M. Humburg, and L. Vilser. "Beheizung von Fahrzeugen mit verbrauchsoptimierten Motoren." Presented at the Vienna Engine Colloquium (Wiener Motorenkolloquium), 1996.
[10] Reichelt, J. *Fahrzeugklimatisierung mit natürlichen Kältemitteln auf Straße und Schiene.* Heidelberg, 1996.
[11] "Condensers of Automotive Air Conditioning Systems with a 3-E Cell Model." VTMS Conference, Indianapolis, 1997.
[12] Colinet, M. *La Climatisation.* Boulogne, 1987.
[13] Reichelt, J. *Leistungsgeregelte Verdichter zur PKW-Klimatisierung.* Karlsruhe, 1987.
[14] Taxis-Reischl, B. "Energieverbrauch von Klimaanlagen und Wege zur Verbrauchsreduzierung." ATZ 9, 1997.
[15] Flik, M., H. Kampf, A. Kies, and B. Taxis-Reischl. "Lastprofile zur Berechnung des Energieverbrauchs von Klimatisierungssystemen in Fahrzeugen." Presented at VDI conference "Simulation und Berechnung im Fahrzeugbau," Würzburg, 1996.

3.4 Vehicle acoustics

3.4.1 Introduction

Over the past two decades, the process of motor vehicle acoustic optimization has evolved from a more or less empirical approach to a highly precise process. The driving forces behind this evolution were, among others, increasing customer expectations for comfort as well as demands for ever-shorter development times prompted by competition between vehicle manufacturers. The following statistics underscore progress in the field of acoustics: in just the past ten years, interior noise levels at steady-state autobahn speeds have been reduced by about 50%, even as development times were cut in half. In particular, modern luxury sedans have achieved a hitherto unknown level of acoustic comfort, with noise levels around 60 dB(A) at steady-state speeds of 100 km/h (Fig. 3.4-1).

It is becoming ever more apparent that low noise levels alone are insufficient grounds for customer acceptance of a vehicle. Moreover, it can be shown that below a certain noise level, vehicles suffer a purely subjective loss of dynamics, an effect which may be irritating, at least for performance-oriented customers. In this regard, it is more than mere conjecture to say that an absolutely silent vehicle would also represent a soulless vehicle.

On the basis of these relationships, the acoustic engineer faces a new challenge. Along with criteria related to conventional topics in objective sound-level acoustics, he or she must now also add subjective "sound design" as a point of emphasis for the work. The acoustic engineer must ensure that every vehicle type is associated with an appropriate, well-defined sound package which meets customer expectations. Accordingly, a roadster must have sporty, performance-oriented sound characteristics, with plenty of driver feedback. A luxury sedan represents the other extreme: the desired sound ambiance, inside as well as outside the vehicle, is expected to impart a sense of superiority, of nobility.

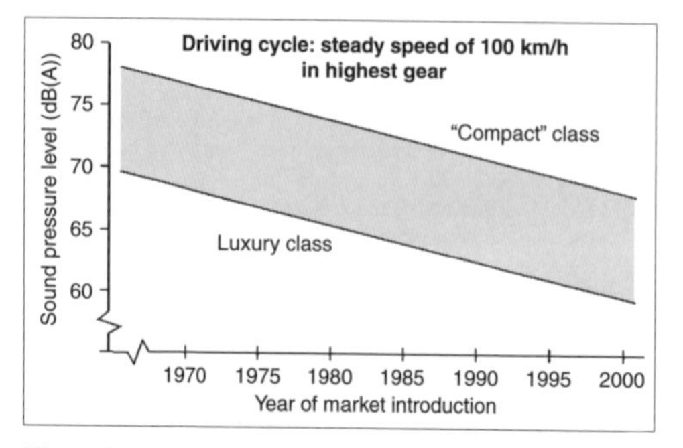

Fig. 3.4-1 Noise levels for production vehicles.

If such aspects of subjective acoustics are applied consistently across a manufacturer's entire model range, the result, just as in the case of geometric design, forges a (new) acoustic brand identity. Experience has shown that such acoustic fine-tuning measures are received favorably by customers as well as the press and are recognized as a differentiating feature, often influencing the purchase decision and making the brand more competitive in the marketplace.

Practical realization of an all-encompassing, brand-identifying acoustic design requires complete mastery of the acoustic development process, beginning with a new vehicle model's earliest phases and continuing to the very end of its series production life. Along with highly qualified personnel, this predicates access to powerful testing and computing facilities for development and qualification of new acoustic concepts, production development of vehicle projects, as well as quality assurance in the manufacturing plant. Furthermore, because of vehicle-encompassing considerations, it is vital that acoustic activities be given high priority and a high degree of acceptance within an organization; otherwise, implementation of comprehensive overall acoustic concepts can only occur to a limited degree.

The following sections examine several selected topics in the field of automotive acoustic development. Consideration is given to issues specific to this discipline as well as questions involving the entire vehicle, which have varying degrees of relevance at different times during the vehicle development process.

3.4.2 Target definition

At the very beginning of vehicle development, it is necessary to formulate an exact definition of its acoustic position. Only then it is possible to incorporate in the overall vehicle package those design considerations necessary to meet its acoustic requirements. It is vital to recognize that the packaging phase at the beginning of any new vehicle development program rigidly defines the vehicle's possible acoustic potential. Later improvements, usually associated with specific space restrictions, generally fail as a result of insufficient space available for implementation.

Qualitative acoustic positioning may be achieved by driving comparison of vehicles from a firm's own production as well as competitors' models. Experience has shown that groups of persons taking part in such driving comparisons are able to describe significant acoustic properties of the various vehicles relatively precisely. Subsequently, for objective purposes, characteristics which are deemed interesting may be examined in detail by means of sound level and/or subjective acoustic criteria. If deemed necessary, acoustic design measures which give rise to specific relevant phenomena may be analyzed.

Once the acoustic positioning of a vehicle is qualitatively defined, it is possible to specify quantitative global "boundary values" for the overall vehicle. Yet the catalog of acoustic targets encompasses considerably more than just these "zero tier" specifications. It is advantageous to establish a hierarchical structure for the acoustic targets catalog (Fig. 3.4-2). First, acoustic specifications are established for the three primary first-level subsystems: body, powerplant and driveline. Subsequent Tiers 2, 3, etc., describe in steadily increasing numbers and in ever finer detail the properties of specific subcomponents.

Initial establishment of such a target catalog requires considerable effort. One reason is that a definite relationship must be established between diverse customer-relevant noise and vibration phenomena and various vehicle components (Fig. 3.4-3). This requires detailed knowledge of various excitation mechanisms (Fig. 3.4-4) as well as the transmission paths associated with such phenomena: for example, by means of transfer path

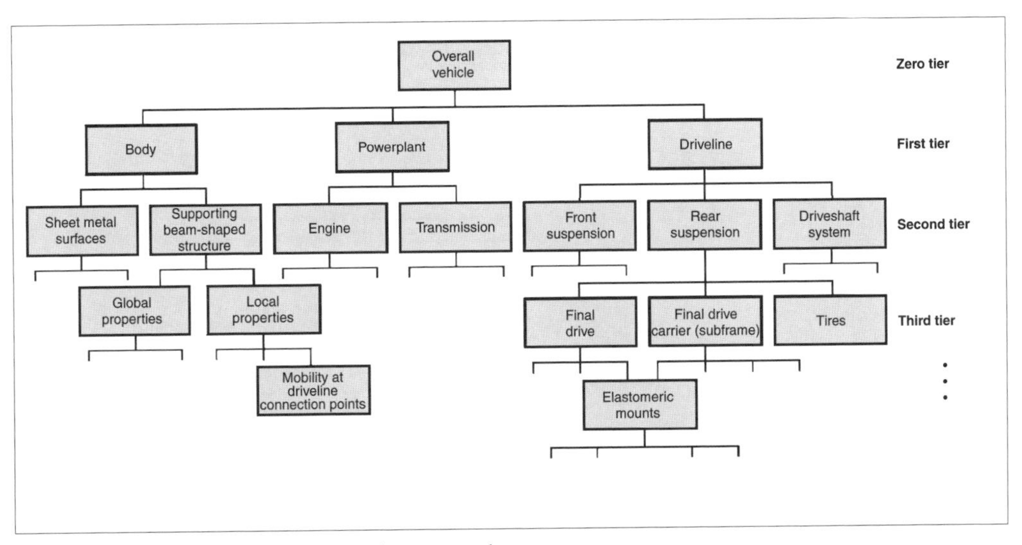

Fig. 3.4-2 Hierarchical structure of acoustic target catalog.

	Body	Engine	Transmission	Chassis	Rest of driveline	Auxiliary equipment
Engine noise	●	●		●	●	●
Transmission			●	●	●	
Wind noise	●					
Rolling noise	●			●		
Brake squeal	●			●		
Squeaks/rattles	●			●		
Idle noise	●	●	●			●
⋮						
Exterior noise	●	●	●			●
⋮						
Shunt vibrations		●		●	●	
⋮						
"Tingling"	●	●	●		●	●

Fig. 3.4-3 Component contributions to typical noise and vibration.

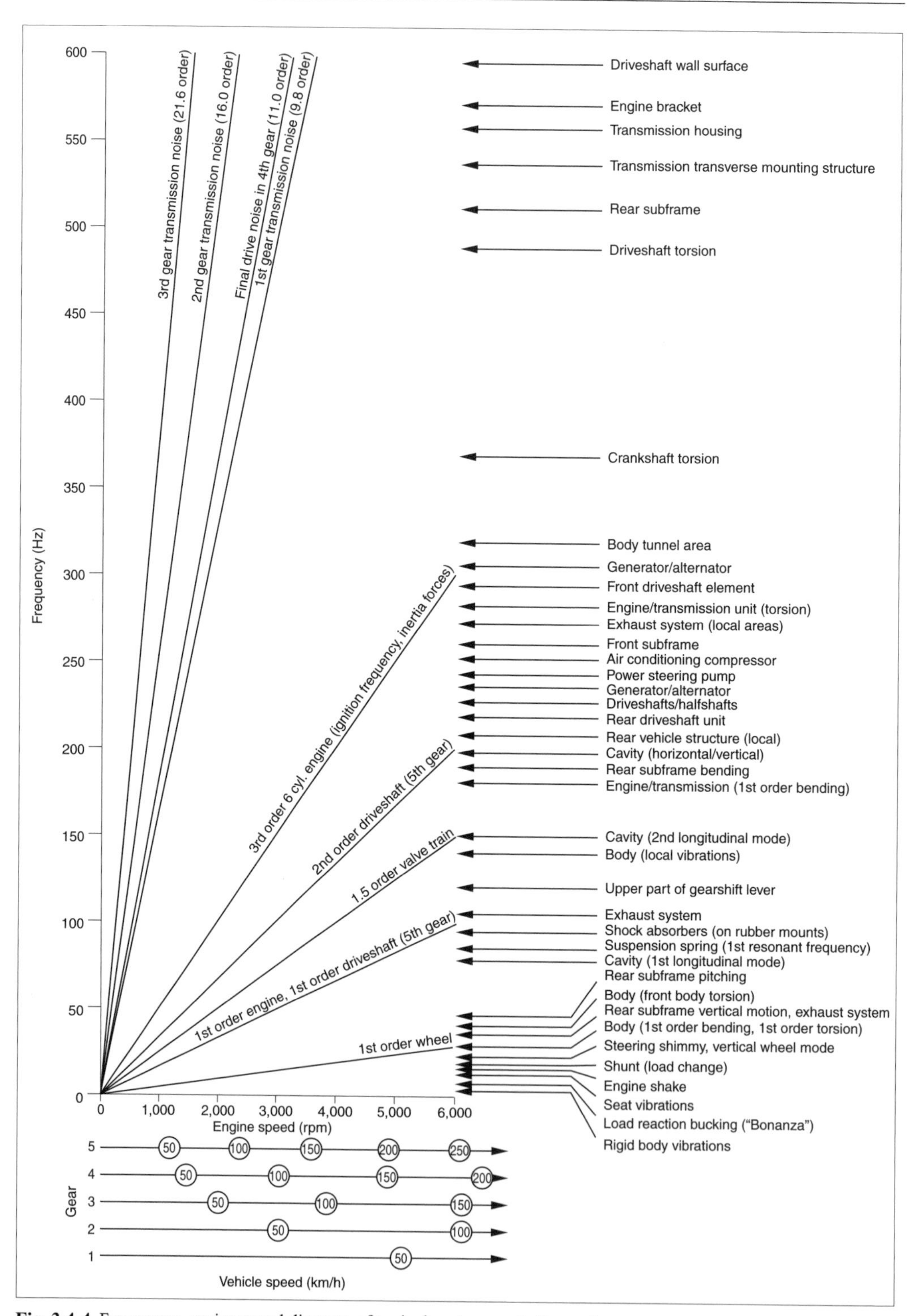

Fig. 3.4-4 Frequency–engine speed diagram of typical system excitation and resonant systems.

analysis methods, which will be discussed in section 4.2. On the other hand, definition of any objective target requires specification of an associated standard test. Based on such test standards, it becomes possible to carry the total acoustic development process to a successful conclusion, accompanied by continuous testing down to the component level.

It should also be mentioned that due to the myriad goal conflicts between the various disciplines involved in functional vehicle integration, specification of acoustic target values cannot be considered independently of the demands of these individual engineering specialties. What is important in this regard is maintaining a balanced view of overall vehicle qualities; in the final analysis, only a vehicle which demonstrates balance in all disciplines will garner a high degree of customer acceptance.

3.4.3 Component qualification

Acoustic qualification of individual components and subsystems separate from the total vehicle environment is an important element of an advanced acoustic development process. As a result of a steady reduction in product development time, early availability of actual prototypes is becoming increasingly difficult. It would be unrealistic to engage in development processes which concentrate on basic optimization work during the prototype test phase, as was once standard practice. This illustrates the significance of early application of calculations using virtual prototypes, so-called "digital cars," as well as qualification of individual components, which are available as geometrical data as well as hardware long before the overall vehicle is available. In general, based on this material, CAE and/or CAT studies [1, 2] can yield significant conclusions regarding functional integrity. To illustrate this process, four selected examples will be presented.

3.4.3.1 Propulsion unit

The propulsion unit, normally consisting of conjoined combustion engine and transmission components, represents the vehicle's main source of acoustic and vibratory excitation. Consequently, acoustic optimization of the propulsion unit is of primary importance. If this system's vital acoustic properties are ignored during development, no satisfactory result should be expected for the vehicle as a whole. This statement takes on added significance when one considers that a propulsion unit is generally installed in more than one vehicle model; as a result, weaknesses in the propulsion unit's acoustic concept result in an entire model range of vehicles with more or less strongly defined acoustic shortcomings. Acoustic layout of propulsion components is an increasingly demanding assignment, as the current trend toward ever lighter units with greater system complexity often runs counter to ever more stringent acoustic demands.

The prime directive for realizing a propulsion unit with desirable acoustic behavior is minimization of both structure-borne noise as well as airborne noise. Increasingly successful acoustic optimization permits increasingly precise and uncompromising acoustic layout of the complete vehicle. At this point it should be mentioned that the acoustic properties of a power unit may be completely defined merely by specifying target values for two quantities: structure-borne and airborne noise. In order to reduce the structure-borne noise level at the interfaces between the power unit and its immediate vehicular surroundings, special attention should be given to minimizing external forces resulting from moving masses and to the combustion process [3, 4].

Optimization of the crank train from a vibration standpoint is a significant element in the acoustic layout of a power unit. Parameters exerting major influence on dynamic properties include cylinder configuration (inline or V engine), firing order, crankshaft bearing concept, the number of counterweights, the fundamental frequencies for crankshaft bending and torsion as installed in the crankcase, as well as the ratio of connecting rod length to crank radius. To achieve an appropriate acoustic concept, it is important to consistently take cognizance of the necessary boundary conditions (geometry and space requirements) even during the earliest development stages. If, despite all efforts, it is not possible to achieve satisfactory results from a vibration standpoint, one may, for example, apply balance shafts to remedy the situation (Fig. 3.4-5).

On more recent engine families, the valve train has proven to be of increasing importance in terms of its influence on vibration. This is caused by the increasing complexity of this system. Fuel economy and emissions requirements lead not only to requirements for multivalve technology but also to ever more complex (variable) valve timing systems. In most cases, this results in greater moving masses in the valve train, the effects of which on dynamic engine properties in turn require careful analysis and, if necessary, compensation

Fig. 3.4-5 Four-cylinder engine with balance shaft module.

by means of counterweights. It should be noted that valve-train–induced vibrations generally take the form of frequencies of 0.5× the engine speed and multiples thereof. Combined with vibrations originating in the crank train and combustion processes, with frequencies that are whole multiples of engine speed, the resulting total body noise characteristic gives a subjectively harsh noise impression.

An important factor in sound transmission from a powertrain is the configuration of engine and transmission mounts. These must take into account both rigid body and elastic properties of the power unit. To avoid "booming" noise phenomena inside the vehicle, it is vital that the entire engine-transmission unit possess adequate dynamic stiffness. If the fundamental resonant frequencies in bending and torsion are too low, excitation of these resonant modes by internal engine forces will almost certainly transmit low-frequency vibration to nearby body and drivetrain components. To reduce significantly such phenomena, it is necessary on four-cylinder engines, for example, to achieve a resonant frequency level in excess of 200 Hz for first-order bending of the power unit. As a result, in the engine speed range up to 6,000 rpm, most of the engine excitation, combined with the engine's second-order harmonic, does not experience any additional reinforcement by elastic resonance of the power unit.

Great care must also be taken in determining acoustically desirable locations for power unit supporting brackets (engine and transmission mounts). The important point in this regard is that these elements are placed at structurally very stiff locations on the engine-transmission unit, so that base excitation resulting from the power unit's local compliance is minimized. Furthermore the pickup points must be located in areas exhibiting the smallest possible vibration amplitudes in the power unit's fundamental resonance modes. Finally, it is important to achieve the dynamically stiffest possible design of the support brackets themselves. This may be realized, for example, by geometrically short support bracket designs, which, aside from stiffness aspects, also offer advantages in terms of the levels of vibration imparted to surrounding areas. This offers added advantages in optimizing the power unit mountings with regard to engine shake. A limiting condition in achieving the smallest possible mounting footprint in the transverse direction is imposed by the ability to resist (static) engine torque, which requires a certain minimum spacing between engine mounting points. This leads to the condition that for powerful engines, an acoustically advantageous, narrow engine mount footprint cannot always be achieved.

The airborne noise emitted by a power unit is largely produced by sound radiation from vibrating surfaces. Obviously, for a given design, a higher level of internal structure-borne noise produced within the unit will result in a higher level of vibration for these surfaces. Consequently, achieving good structure-borne noise qualities for a power unit is a prerequisite for minimizing airborne noise emissions. Of added significance is the structure-borne noise level produced by combustion forces within the engine system. An important criterion for achieving good acoustic properties is the realization of a "soft" combustion process. In this regard, diesel engines, with their high compression ratios and sharp combustion pressure spikes, are at a distinct disadvantage. If soft combustion can be achieved by means of so-called pilot injection, as has been done on recent direct-injection diesel engines, the power unit's acoustic properties may be improved significantly.

A further means of potentially reducing combustion-generated sound radiation lies in reducing structural sound transmission through paths which are subject to excitation by combustion forces (Fig. 3.4-6). Furthermore, sound radiation can be decreased by reducing vibration of the power unit's outer wall surfaces. Specifically designed (local) reinforcing ribs and/or domed or crowned surfaces, among other measures, will generally improve sound radiation characteristics. Of continued importance is reduction of surface vibration amplitudes by means of damping measures, which may take the form of metal sandwich structures or application of bituminous materials (for example, in the oil pan area). It should also be mentioned that increasing importance is being given to sound insulation systems, for example in the form of (plastic) covers lined with absorbing materials and applied to specific locations responsible for significant sound generation. Typical examples include sound-damping elements in the form of decoupled covers immediately adjacent to the engine, such as those found in the vicinity of fuel injectors.

Finally, mention must be made of the major acoustical significance of power unit peripherals. These include power unit intake and exhaust systems as well as engine-driven accessories. Obviously, all of these systems and components require acoustical optimization in conjunction with the acoustic properties of the power unit; otherwise, these subsystems might well become primary noise sources under various driving conditions. Examples include airborne noise produced by a generator fan at high rpm, as well as that of an engine fan at idle.

The acoustic layout of intake and exhaust systems is often difficult, due to an entire series of conflicting goals which must be accommodated. For engine power output reasons, low exhaust back pressure is desirable, coupled with large-diameter exhaust pipes and the resulting negative effects on the exhaust system's acoustic effectiveness. This negative effect may be compensated for by greater muffler volume, but the required space and mounting considerations may not be realizable. In the case of such serious target conflicts, new solutions may be needed to relax constraints on the overall optimization process. In the example cited here, one possible solution is a switchable or variable-geometry exhaust system which, depending on operating parameters, switches between single- and dual-port operation (Fig. 3.4-7).

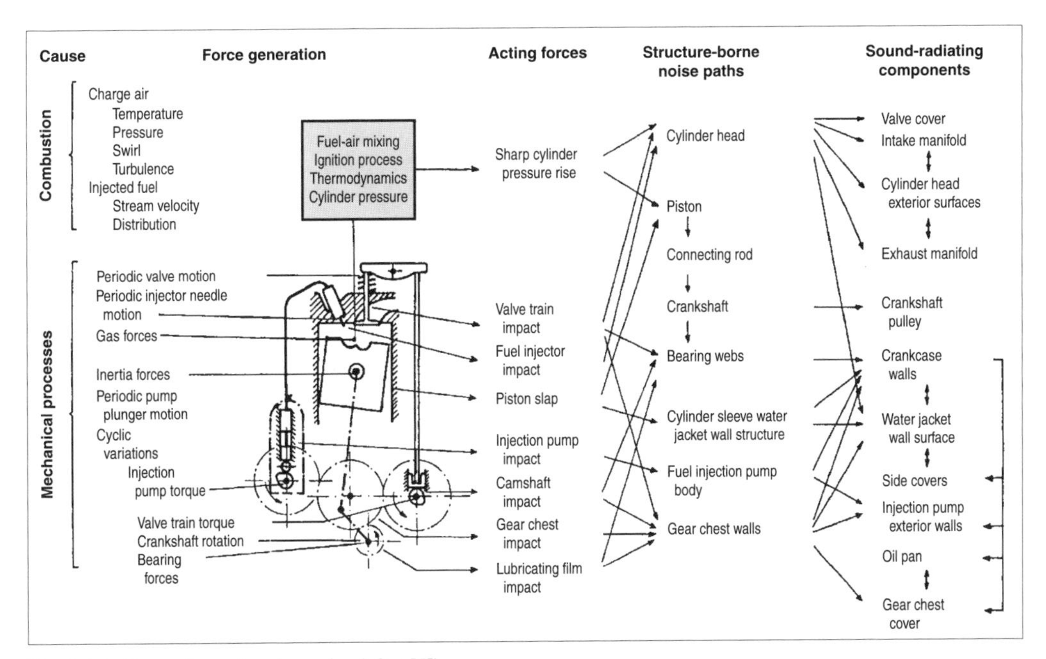

Fig. 3.4-6 Generation of engine noise (after [5]).

Engine-driven accessories can have a major influence on a power unit's overall acoustic properties. Such accessories include, for example, a water-cooled generator, air-conditioning compressor, power steering pump, and water pump. Once the power unit is installed, these auxiliary devices are connected to the body. For example, the air conditioning compressor is connected to the body-mounted evaporator and condenser elements by means of high- and low-pressure refrigerant hoses; by means of its hydraulic lines, the power steering pump is connected to the steering system, which is attached to the steering column. From these relationships it follows that a multitude of structure-borne noise paths, originating from engine-mounted accessories, lead into the body structure. With a view to realizing good acoustic properties in the vehicle as a whole, it becomes advantageous to minimize the noise generated by these accessories, as well as to minimize transmission of structure-borne noise from the power unit into these auxiliary devices. Absolutely rigid connections between accessories and the power unit (to eliminate transmission of low-frequency resonances) as well as elastic decoupling via isolation elements (to reduce high-frequency noise transmission) have proven to be important measures in achieving this goal. The subject of noise transmission from accessory systems will be discussed in detail in section 3.4.3.4, using the steering system as an example.

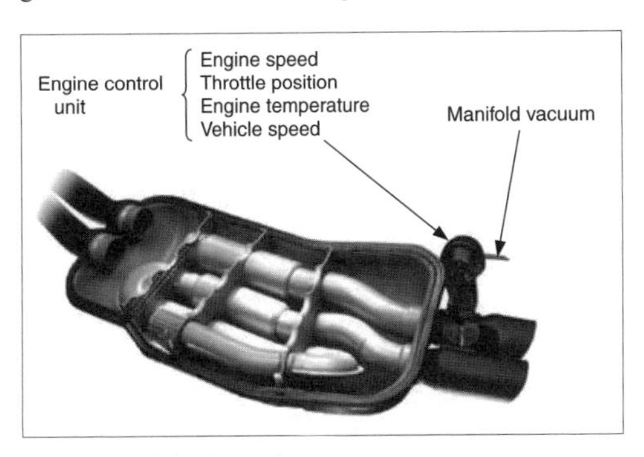

Fig. 3.4-7 Adaptive exhaust system.

3.4.3.2 Driveline

"Driveline" is understood here to mean the entire system mounted below the body and comprising the front and rear suspensions including wheels, the engine-transmission unit with exhaust system, and the entire driveshaft mechanism. In the vehicle, the driveline is connected to the body by means of elastomeric mounting elements (Fig. 3.4-8). From this relationship we get the qualitative result that the higher the structurally transmitted sound level in the driveline system, the higher the noise level in the vehicle as a whole.

A combined experimental–computational approach [6] has proven highly reliable and effective for driveline acoustic optimization. This approach is based on a fully assembled driveline, in its proper geometric orientation, on a chassis dynamometer, as well as a combination of multibody and finite element models (Fig. 3.4-9) to mathematically describe this structural system. This method employs the following system assumptions: at

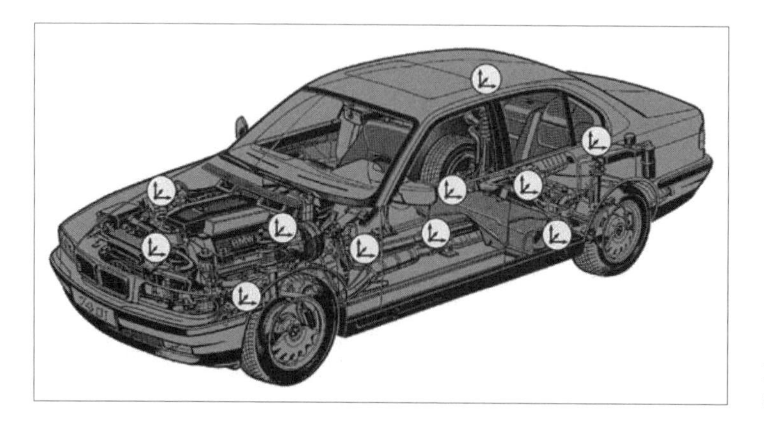

Fig. 3.4-8 Driveline-body coupling points.

the body side of its elastomeric connecting points, the driveline is rigidly mounted with respect to its environment; and individual drive wheels are powered by separate dynamometer rollers.

The advantage of a combined experimental–computational approach lies in the fact that characteristic system parameters which are difficult or impossible to calculate may be determined experimentally and then used as a data set in computations. One example of such a parameter is the system's resultant excitation vector under specific operating loads, as well as the (nonlinear) spring/damper characteristics of various joining elements such as driveshaft isolation dampers and slip splines. Experimental determination of such parameters is vital to reconciling the mathematical model with the actual experimental assemblage. The mathematical system description may be regarded as satisfactory if all major structural sound transmission paths exhibit identical behavior in both experimental and computational models.

An added advantage of a combined experimental–computational approach is that undesirable vibrational behavior detected experimentally may be precisely analyzed and optimized by the (appropriately configured) mathematical model. A typical optimization problem addresses the layout of mounting elements between drivetrain and body structure. The question is which (dynamic) parameters these elements must exhibit in order to minimize structural sound transmission to the body under specific operating conditions. Solution of this problem is extremely difficult, as it encompasses finding an optimum layout for about 15 mounting points. For each mounting point, the three translation coordinates alone imply three associated stiffness parameters; it is apparent that just the stiffness optimization of the overall mounting system must involve simultaneous solution of 45 variables. It goes without saying that such multidimensional problems cannot be solved intuitively, but rather that an optimum solution may only be arrived at by computational means.

Without going into detail, for the sake of more

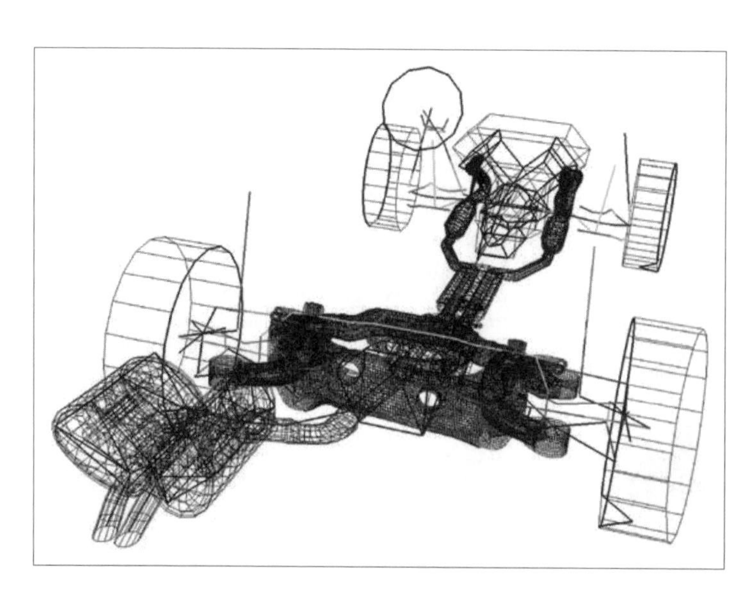

Fig. 3.4-9 Computational model of a driveline structure.

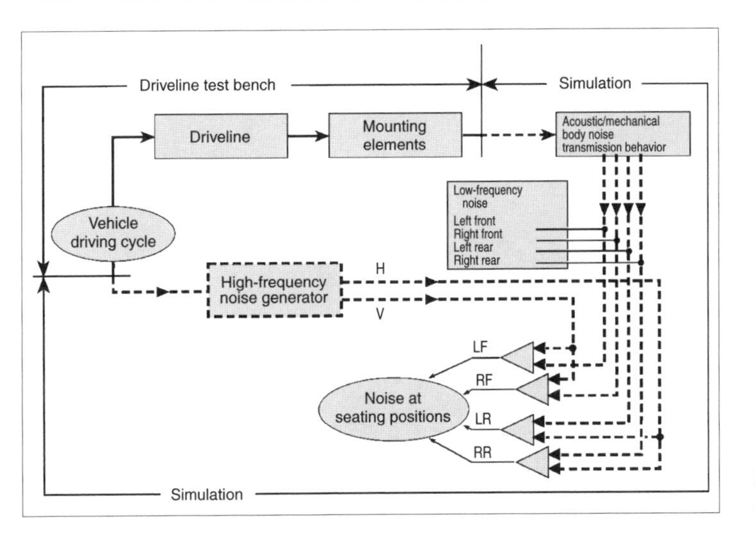

Fig. 3.4-10 Noise simulation on driveline test bench.

complete understanding it should be mentioned that the "driveline system" is well suited for analysis of an entire series of additional vibration-relevant questions: for example the topic of driveline shunt under load change, driveline torsional vibrations, engine and rear axle mounting, layout of driveshafts and halfshafts, as well as transmission of rolling noise, for which the dynamometer rollers are fitted with a rough surface.

Construction of experimental driveline rigs has proven to be an important tool for conducting early sound engineering studies. These are based on the methods discussed in section 3.4.4.2 for structural-acoustic transmission path analysis.

Assuming that the body's noise transmission behavior is known, as defined by a matrix of vibro-acoustic frequency response spectra between the sound pressure distribution at selected points in the passenger compartment and the input force at the various interface points between the driveline and the body (for example, from investigations of a predecessor vehicle model or from a data bank), the time-variant behavior of vehicle interior noise in response to specific excitation from the driveline may be determined. For sound engineering investigations, these signals may now be passed to an audio system, thereby making the vehicle interior noise audible (Fig. 3.4-10). What is always fascinating in this process is that an important part of vehicle noise under various operating conditions may be made available for subjective evaluation—without the existence of an actual vehicle.

3.4.3.3 Body

Just twenty years ago, the basic engineering of an automobile body was based primarily on static and vehicle dynamic criteria as well as crash requirements. The primary goal was to create a body structure with the required static stiffness properties for specific bending and torsional loads. Over time, with the introduction of computational finite element and experimental modal analysis techniques to automotive design, dynamic aspects gained increasing importance alongside static stiffness criteria. However, dynamic structural design was only rarely the driving factor in body development. Finite element methods were primarily, and with great success, used to reduce body weight and to improve crash behavior.

Today, the situation is quite different. Ever-increasing demands for greater comfort have elevated dynamic design criteria as the primary elements of modern body engineering [7]. It must be remembered that the body structure is only one link, in the overall chain of cause and effect, between the primary drivetrain excitation and vibration and acoustic impact on vehicle occupants. Consequently, body engineering itself cannot affect such excitation. On the other hand, it is possible to realize a body layout optimized for vibration transmission and therefore able to exert a positive effect on the noise transmission paths for diverse excitation mechanisms [8, 9].

In this regard, for example, tuning resonant frequencies of fundamental global body vibrations in keeping with critical drivetrain excitation is of great significance. These vibrations include, among others, engine excitation at idle as well as excitation from wheel imbalance forces at higher speeds. As exemplified in Fig. 3.4-11 for the case of a six-cylinder vehicle, minimizing the transmission of such vibration across a large speed range can be achieved if the resonant frequencies for the vehicle body's first-order bending and torsional vibrations are restricted to a defined frequency window. In the example shown, this window extends from 27 to 33 Hz. It is apparent that lowering this frequency window would result in increased excitation from first-order wheel forces, even at relatively low speeds; on the other hand, raising the resonant frequencies would result in a perceptible increase in idle vibration with third-order engine forces at the ignition frequency. It can be shown that the desirable resonant frequency range of a

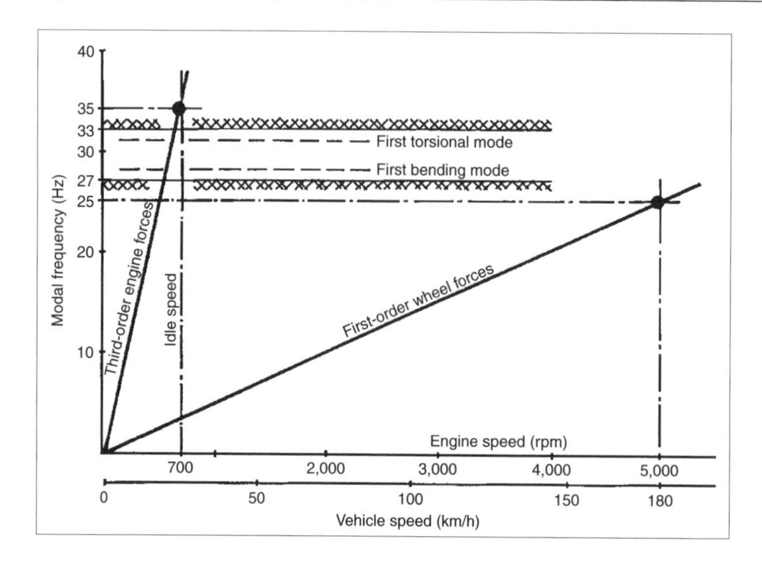

Fig. 3.4-11 Modal frequency (Hz) criteria for vehicle layout.

six-cylinder vehicle applies equally well to four- and eight-cylinder vehicles.

Experience has shown that this dynamic criterion for basic body layout results in outstanding vibration comfort properties of the vehicle as a whole. Furthermore, it has been determined that such a dynamic design philosophy automatically leads to static body parameters which far exceed the demands of vehicle dynamics. Another feature is that a vehicle designed for such outstanding stiffness properties imparts a subjective impression of great solidity, even after only a short drive. This may be attributed to specific material combinations at the interfaces between interior and chassis components which result in a marked reduction in body squeaks and rattles that would otherwise give customers a very negative impression.

It should be pointed out that the aforementioned body modal frequency targets refer to the configuration of a completely assembled vehicle, not merely the raw body shell structure. As the body-in-white represents only about 20% of the total vehicle weight, its resonant frequency level as an individual component is appreciably higher than when it is part of a complete vehicle. This relationship results in a significant challenge for engineers and designers tasked with the body development process. Their assignment is to realize a body-in-white design in the early stages of new vehicle development which meets all dynamic requirements of the later overall vehicle. In carrying out such "target-specific body design," specially tailored techniques of mass simulation, combined with dynamic structural optimization processes, have proven to be efficient tools for this interesting and demanding assignment. In addition, the body-in-white weight of each successive model could be reduced, despite markedly increased (dynamic) requirements.

Along with global body requirements, an entire range of local criteria must be taken into consideration in achieving good comfort properties. These include targeted dynamic layout of the structural assembly (consisting of sheetmetal, glass, and plastic surfaces) that encloses the vehicle interior. Care must be taken that the fundamental modal frequencies in bending of surrounding surfaces are not in the vicinity of the lowest cavity resonant frequencies; otherwise, undesirable coupling effects between structure and cavities might arise (Fig. 3.4-12). Obviously, such resonant frequency

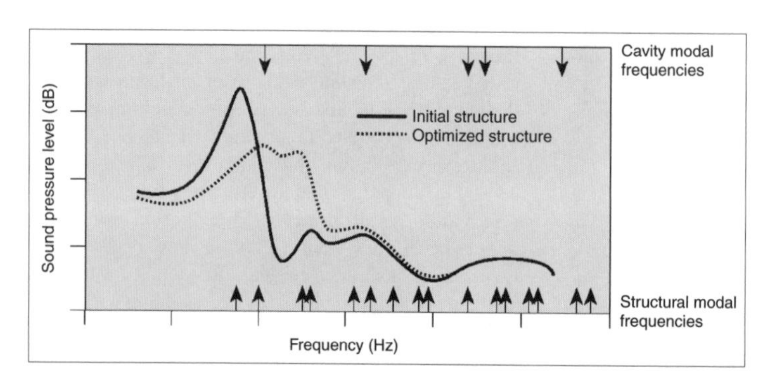

Fig. 3.4-12 Rolling noise frequency spectrum.

determination must be carried out with an eye to the later actual mass loading (including sound insulation) of individual surfaces in the completed vehicle. Shape optimization by means of reinforcing ribs and/or surface crowning provides a highly effective approach to targeted frequency tuning of boundary surfaces.

An additional local condition is limited mobility at all interface points between the body and drivetrain. This is an absolute requirement for achieving good isolation of structure-borne noise by means of elastic elements joining drivetrain and body. Experience has shown that to achieve satisfactory isolation properties, mobility of mounting elements must be a factor of three larger than the corresponding body mobility. This rule of thumb applies to all three translation coordinate axes of an interface point.

A low energy level for the total structural sound input is significant insofar as, for reasons of energy equilibrium, any structural sound energy fed into the body structure can only be eliminated by structural damping of induced body vibration. It follows that a high level of structural noise transmission must lead to a high level of global and/or local body vibration, combined with the resulting negative acoustic properties.

Furthermore, these considerations explain why local efforts at stiffening very strongly excited sheet metal structures often do not result in acoustic improvements. The reason is that such sheet metal expanses, whose vibration has been reduced by stiffening measures, eliminate less energy through their damping properties. As a result, by this diversion of structural sound paths, other structural components are excited to higher levels and proceed to radiate more noise of their own. In such problem situations, it is often more effective to apply additional damping (for example, in the form of bituminous coatings) on especially strongly vibrating surfaces, as this will permit drawing additional vibration energy out of the overall system.

One discipline directly related to the topic of body acoustics concerns itself with the layout of sound insulation systems (Fig. 3.4-13). Creating an appropriate sound insulation concept is of primary importance not only in achieving a competitive acoustic level (Fig. 3.4-14) but also in realizing a defined sound quality. Of importance in this regard is the specific application of sound insulation properties to various problem situations. Among other things, these problems include the typical sound insulation behavior of a body design, characterized by modal properties in the low frequency range, sound insulation properties in accord with the mass law in the middle frequency range, and coincidence effects at high frequencies [10]. A typical problem in the medium frequency range is noise generated by water and pebbles against the underbody and wheel wells. If adequate noise reduction is not achieved by means of sound insulation measures, the result is a distinctly perceptible noise characteristic, readily apparent to vehicle occupants, suggesting the impression of a low-quality, "cheap" vehicle.

Precise knowledge of structure-borne and airborne noise transmission paths in the complete vehicle under various operating conditions is of primary importance in the layout of sound insulation systems; this will be treated in greater detail in Section 3.4.4.2. This information is of importance in defining the properties of sound insulation packages. Depending on specific problems at hand, it must be decided whether systems for sound damping, isolation, or absorption are to be applied. Additionally, in solving specific problems, the acoustic efficacy of sound insulation measures may be tailored for especially critical frequency ranges. An example of this is provided by the very different frequency-dependent absorption properties of porous absorbers, foils, and so-called spring/mass systems [10].

The most important property of a sound insulation system is its "acoustic impermeability"; sound present

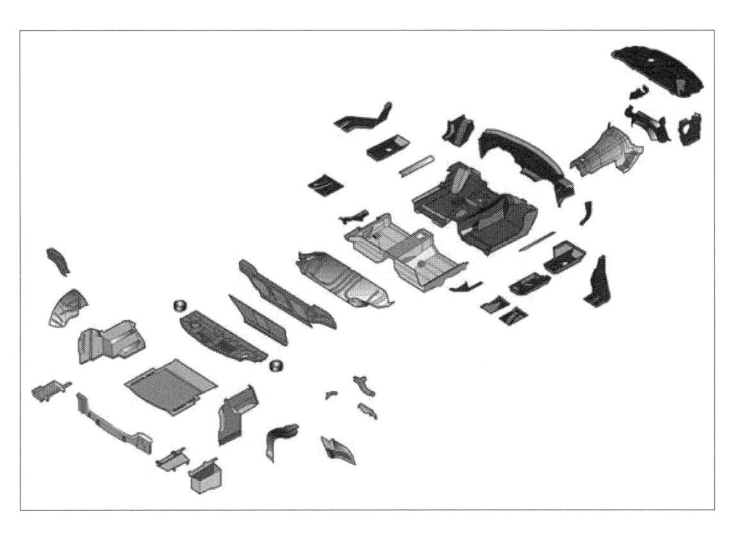

Fig. 3.4-13 Auto body sound insulation elements.

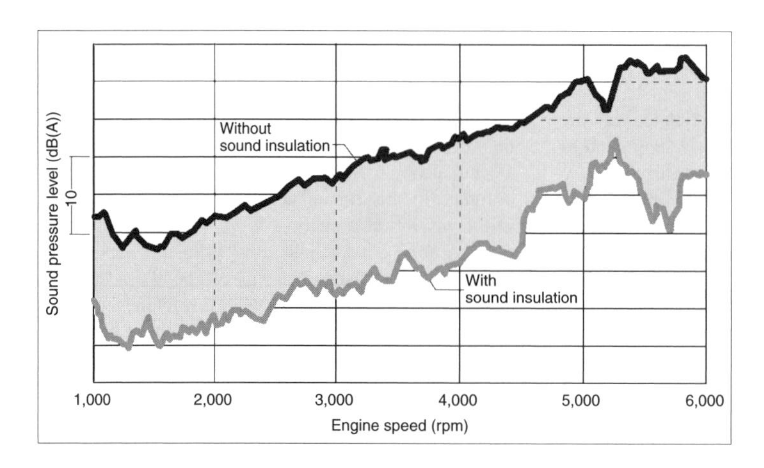

Fig. 3.4-14 Reduction of total sound pressure level by means of sound insulation measures.

outside the cabin must not enter this space through direct airborne transmission. This requires sealing of all actual existing "holes" between the cabin and its surroundings. Examples of such "holes" include openings in the firewall to admit heating and air conditioning system components, for the steering column, for the wiring harness and for the pedal linkages. All such openings must be sealed airtight by means of sound insulation measures. As movable elements such as the steering column system are sealed, demanding sound insulation requirements may result in quite complicated systems: such as, for example, a system in the form of sleeves with multiple levels of sealing (Fig. 3.4-15).

Not to be ignored in defining the sound insulation package—the so-called scope of secondary measures—is a series of boundary conditions such as cost, weight, and package restrictions. Experience has shown that the smaller the space available for sound insulation measures, the heavier and more expensive the resulting systems. This conclusion shows how important it is to establish the sound insulation concept and thereby its associated space requirements early in the packaging phase of a new vehicle concept.

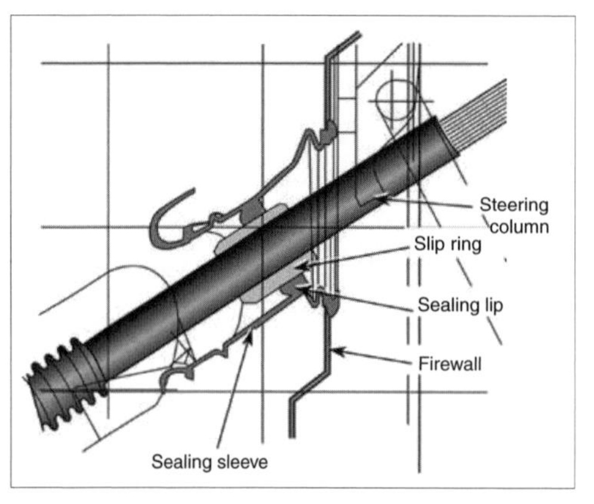

Fig. 3.4-15 Acoustic sealing of steering column opening in firewall.

3.4.3.4 Steering system

As a result of steady improvement in vehicle acoustic quality, noise from subsystems and auxiliary devices becomes more apparent. Examples of this phenomenon include windshield wiper systems, the fuel supply system, the heating and air conditioning system, the ABS system, power steering, and ride level control as well as a multitude of electrically actuated functions.

It may be asserted that an acoustically competitive automobile, at least in the midrange and luxury class, may only be achieved by consistent optimization of these auxiliary devices. Based on the many components and systems which must be considered, this requires a process which enables acoustic qualification of auxiliary devices in terms of components or systems. We will use the steering system as an example of this process.

For acoustic studies, the entire steering system, as shown in Fig. 3.4-16, is mounted on a vibration isolating foundation in its designed orientation. The system is operated by electrically powering the power steering pump. All customer-relevant operating conditions for the system may be represented as functions of input and output signals at the steering box. Thanks to ready accessibility of the components when mounted in this test configuration, it is possible to analyze in detail and optimize both the structural sound transmission and airborne sound behavior of the overall system. Among other things, this process takes into consideration the airborne noise radiation and structural noise excitation of the power steering pump; the transmission behavior of the hydraulic lines with regard to their body, fluid, and tubular noise transmission characteristics (Fig. 3.4-17); and the establishment of effective system isolation to be applied later in the complete vehicle.

It is readily apparent that such studies at the component and system levels represent an enormous expenditure of effort. On the other hand, they are indispensable

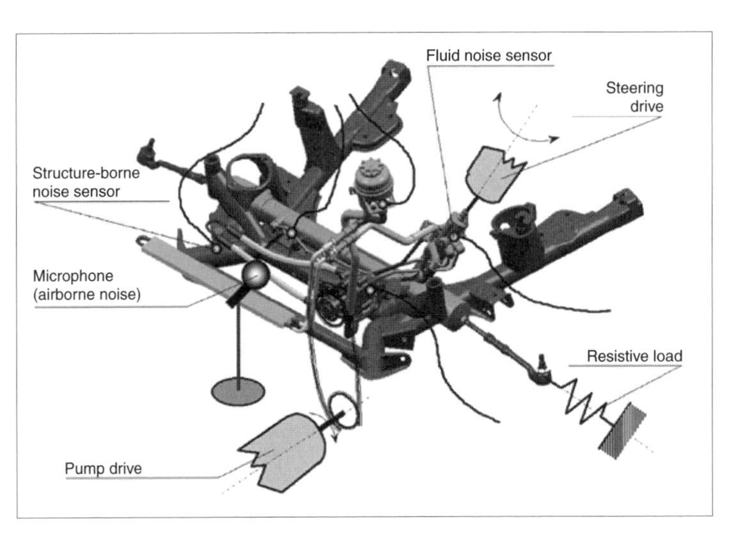

Fig. 3.4-16 Component testing setup for acoustic qualification of steering system.

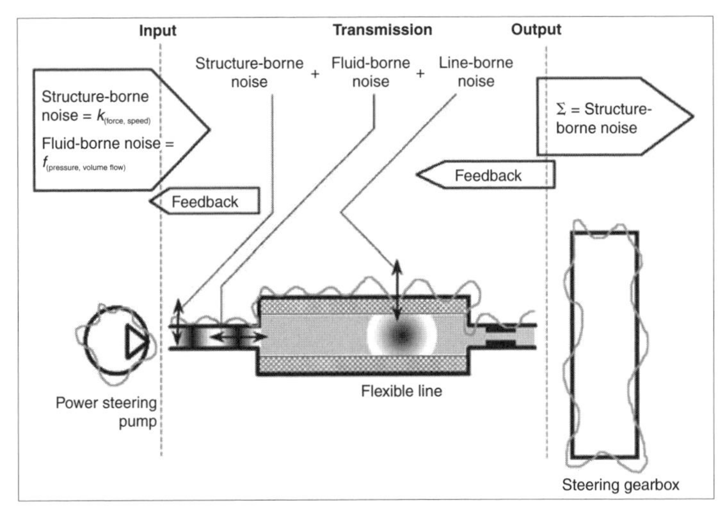

Fig. 3.4-17 Noise transmission mechanisms in a system of fluid-filled lines.

in achieving the necessary concept assurance during development of a new vehicle and in avoiding costly last-minute changes before the vehicle enters production.

3.4.4 Overall vehicle optimization

In an advanced development program, vehicle prototypes are not available until a point relatively late in the process. It is obvious that at such a late juncture, only finishing touches may be applied to the vehicle. The number and degree of acoustic problems detected on prototype vehicles is a very objective measure for evaluating the quality of the acoustic development process. The fact that a single serious fault, which can no longer be eliminated at such a late stage, may have a deciding effect on overall acoustic quality (and therefore market success) of the vehicle once again underscores the real significance of an acoustic target catalog that accompanies the entire vehicle development process.

The final development phase, which ends as the vehicle enters production, is marked by intense effort on the part of those engaged in the acoustic process; only consistent elimination of all acoustic defects detected on prototypes and preproduction vehicles will result in a comfortable vehicle meeting customer expectations. Step by step, various acoustic phenomena are analyzed in minute detail and repeatedly optimized through detailed tuning work. The following sections examine several topics in overall vehicle acoustic optimization.

3.4.4.1 Aeroacoustics

In general, wind noise (see also Section 3.2.2.7) represents a very irritating component in sound engineering studies. Aside from its level, this noise is particularly

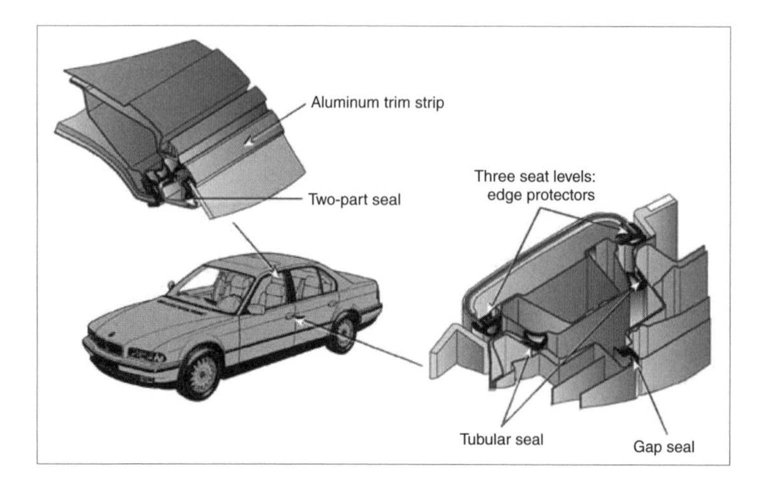

Fig. 3.4-18 Door seal system.

annoying because it cannot be influenced by the driver. Moreover, sources of high-frequency aeroacoustic noise components are often easily located and may be readily traced to systems which are deficient in this regard. Such imperfections give customers an impression of poor quality.

Consequently, the goal of aeroacoustic development must be to achieve a low wind noise level as well as a sound aspect free of tonal and periodic components. Available tools include complex experimental facilities in the form of aeroacoustic wind tunnels as well as modern analysis methods based on the use of artificial acoustic head technology for subjective noise evaluation.

A basic condition for achieving good aeroacoustic properties is the acoustic impermeability of the entire passenger cell enclosure. Every opening requires careful sealing to prevent any direct passage of airborne noise from exterior to interior. This in turn results in high acoustic demands on the door sealing system. Along with absolute impermeability, this system, because of its considerable sealing length, must also provide a high degree of acoustic effectiveness within the gap. To ensure this, modern designs generally employ multiple sealing systems (Fig. 3.4-18), in which sealing between the body and doors is accomplished by two or even three successive layers. Because all possible tolerances for every sealing layer—and for every vehicle built—must be considered to ensure absolute sealing, the required design and manufacturing effort is enormous.

Experience has shown that on occasion, quite unusual aeroacoustic transmission paths can play a significant role. For example, in one development case, an airborne noise path was traced to a narrow gap between an exterior door handle and the body, through the cavity between the door exterior shell and interior panel, and exiting through the interior door handle opening. Obviously, identification and rapid elimination of such phenomena demand extensive experience.

Depending on the problem at hand, it may be expedient to attack the actual noise source, in addition to optimizing airborne noise transmission paths. Especially strong aeroacoustic excitation results from, among other things, local vortex generation at locations on the body exterior exhibiting unfavorable aerodynamic flow. In this regard, flow around the A-pillars and outside mirrors has proven extremely critical (Fig. 3.4-19). In cases of unfavorable airflow, a situation may arise in which the (front) side windows are subjected to high pressure fluctuations arising from the vortex wake of the resulting separated flow. Due to the limited insulating properties of glass, the result is high sound pressure levels in the vehicle interior. It is possible to reduce the aeroacoustic excitation in this area by addressing the geometric shaping of the A-pillar rain gutter and the outside mirror.

It is a commonly held belief that wind noise primarily affects vehicle acoustics in the higher frequency ranges. Recent extensive research [11] has led to the realization that aeroacoustic phenomena may also have an appreciable effect on a vehicle's low-frequency acoustic properties. It was determined that aeroacoustic excitation of the heating and air conditioning inlet and

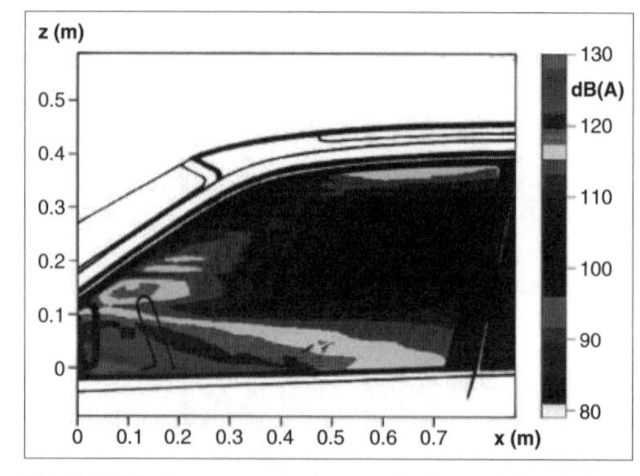

Fig. 3.4-19 Aeroacoustic impact in front side window area.

Fig. 3.4-20 Aeroacoustically excited vibration of vehicle exterior skin.

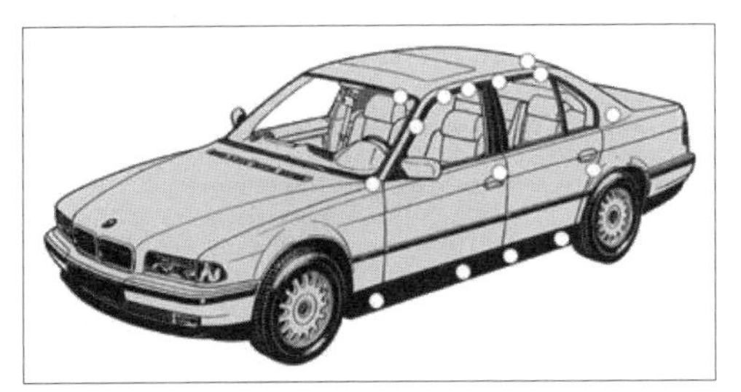

Fig. 3.4-21 Foam barriers within the body structure to suppress airborne noise transmission.

exhaust vents could excite strong infrasound within the vehicle interior. Although this sound pressure loading cannot be perceived acoustically by the vehicle occupants, it may lead to discomfort or even nausea.

Furthermore, it was shown that turbulence within the airflow boundary layer could excite low-frequency resonant vibration of the vehicle skin (Fig. 3.4-20). As a result of this vibration of sheetmetal surfaces, some of whose areas are quite extensive, low-frequency sound pressure may be induced in the vehicle interior via vibro-acoustic coupling phenomena. For the vehicle occupants, the character of the accompanying noise is comparable to rolling noise. This explains why low-frequency aeroacoustically induced noise is often erroneously interpreted as rolling noise. Measures to reduce this noise component include damping and/or stiffening critical sheetmetal areas, as well as location of barriers to airborne sound within the body structure (Fig. 3.4-21).

3.4.4.2 Transmission path analysis

Transmission path analysis is an important tool for low-frequency acoustic optimization of the overall vehicle. This methodology permits detailed tracing and analysis of noise paths—for example, from an originating noise source outside the passenger cell to a defined reference point within the cabin. In the process, various methods of analyzing structural and airborne noise transmission paths [11] are employed.

Fig. 3.4-22 illustrates an example of a block diagram for rolling noise transmission within a vehicle. Excitation occurs at the tire/roadway contact area. Structural noise originating there is then introduced into and distributed throughout the body structure through the drivetrain and elastomeric mounting elements at their coupling points to the body. This excites vibration in the vehicle interior boundary areas, which in turn give rise to interior noise though vibro-acoustic coupling. Using CAT (computer-aided testing) techniques, it is now possible to determine the transmission behavior of individual block elements indicated in Fig. 3.4-22. Examples include the contributions to vehicle interior rolling noise by individual power inputs at the coupling points between rear suspension and body, shown in Fig. 3.4-23. For the exact same operating condition, Fig. 3.4-24 shows the contribution of various interior boundary areas to

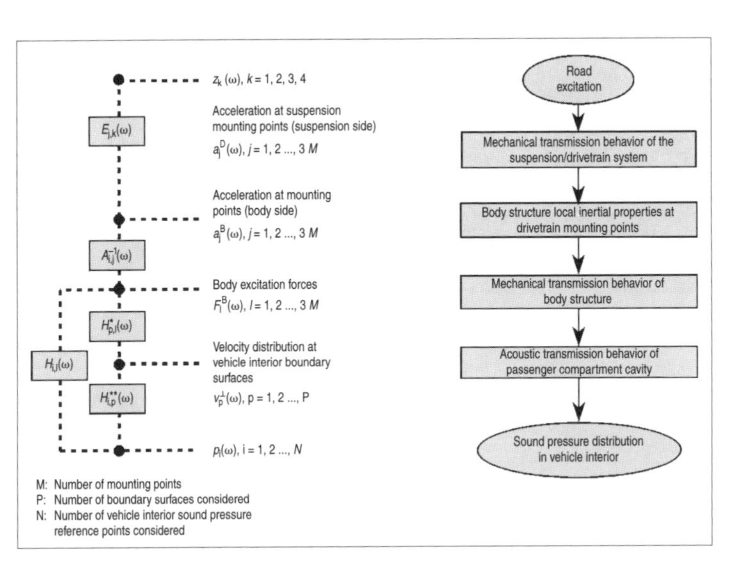

Fig. 3.4-22 Block diagram for structural sound transmission of rolling noise.

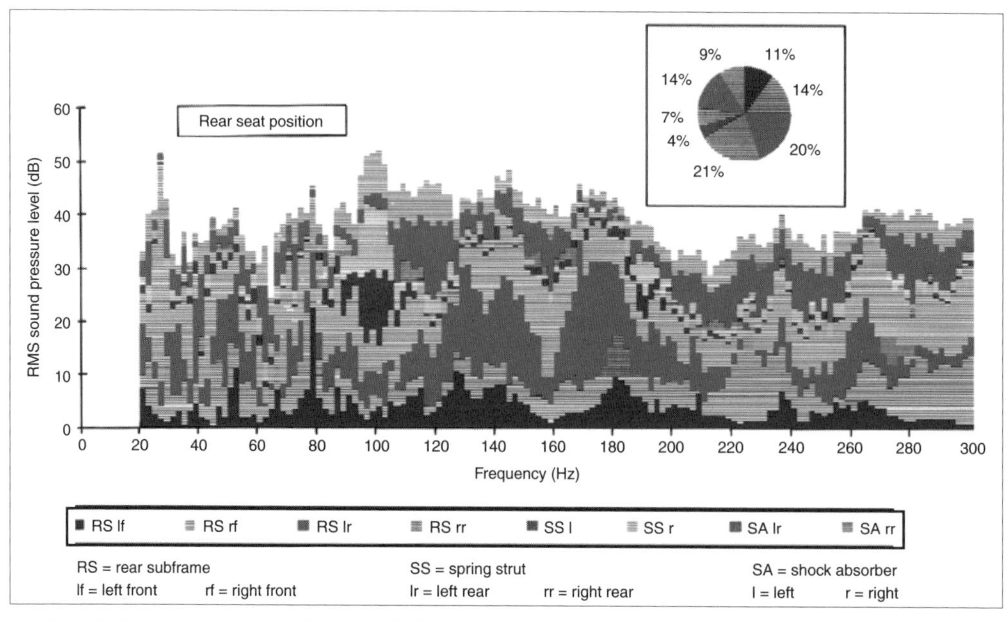

Fig. 3.4-23 Structural sound contribution of individual rear axle system connecting points.

noise generation. By means of this information, it is possible to identify positively acoustic weaknesses within the overall vehicle system and to appraise the improvement potential of various modifications.

Similar CAT methodology exists to determine airborne noise behavior of the overall vehicle. This considers direct noise transmission paths through acoustically open holes and gaps as well as secondary air transmission via boundary surfaces and their limited acoustic effectiveness. For vehicle acoustic engineers, the important question is how airborne noise enters the passenger compartment under the relevant vehicle operating

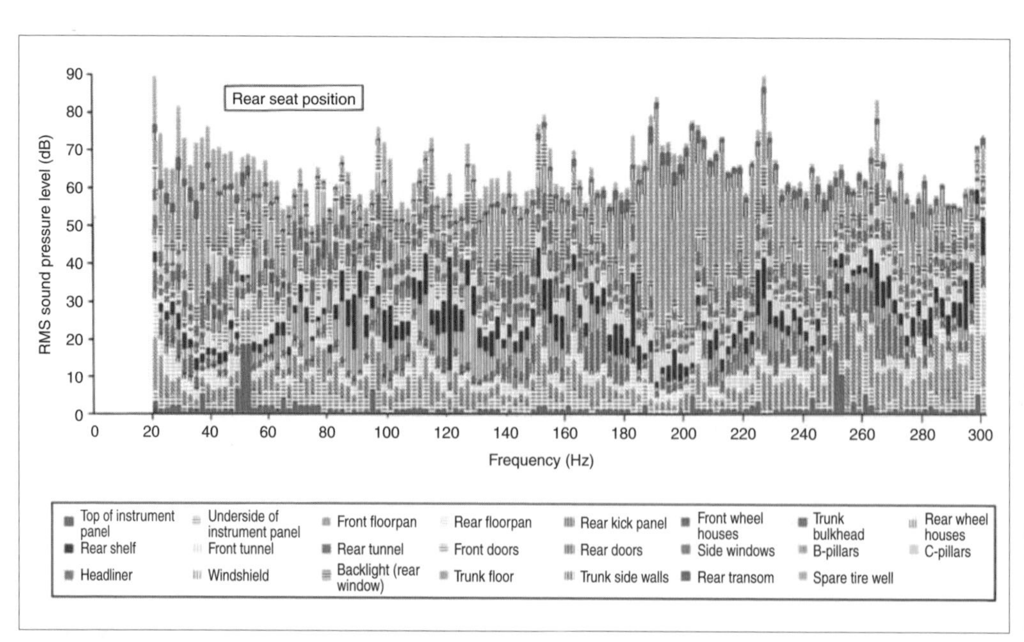

Fig. 3.4-24 Contribution of individual vehicle interior boundary surfaces to interior noise.

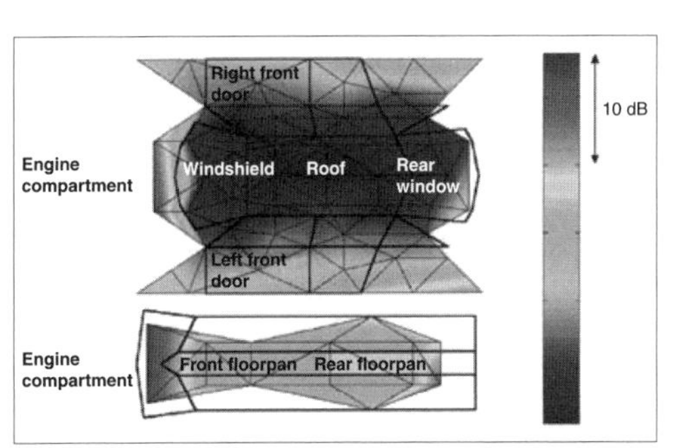

Fig. 3.4-25 Significance of passenger compartment boundary areas with respect to airborne noise transmission.

conditions. As shown in Fig. 3.4-25, representation of acoustic contributions through various cavity boundary surfaces provides insight into this process. From this illustration, it may be seen that an overall improvement in airborne noise transmission may be realized by additional insulation in the lower firewall and left rear floorpan areas (near the exhaust muffler). On the other hand, sound insulation efforts in the roof area may be curtailed with no negative acoustic impact. These relationships show the significance of such detailed analysis; only by such methods can various demands for acoustic quality, cost, and weight be balanced objectively.

3.4.4.3 Vibration comfort

Acoustic and vibration comfort of a vehicle are two closely related disciplines. Conflicting goals between both fields generally result from different demands with regard to vibration behavior and structure-borne noise insulation properties of various components. This relationship was already mentioned in Section 3.4.3.2, illustrated in the layout of drivetrain mounts. To examine this situation in greater detail, the following will use the topic of engine and transmission mounting as an example.

For reasons of vibration comfort, it is desirable to have a very rigid connection between the power unit and the body, which requires engine mounts exhibiting stiff, highly damped characteristics. A correspondingly optimized mounting system ensures that when a car is driven over irregular surfaces, the resulting so-called engine shake vibrations remain negligible and therefore provide little excitation of the body structure. On the other hand, stiff engine mounts transmit essentially unfiltered high-frequency structure-borne noise from the power unit into the body system, leading to high noise levels inside the passenger compartment.

Reconciling these conflicting goals requires very careful design of engine mounts to meet acoustic as well as vibration criteria in the best possible way. Application of hydraulic engine mounts has proven to be a highly effective measure; in contrast to conventional elastomeric elements, hydraulic mounts make possible the re-

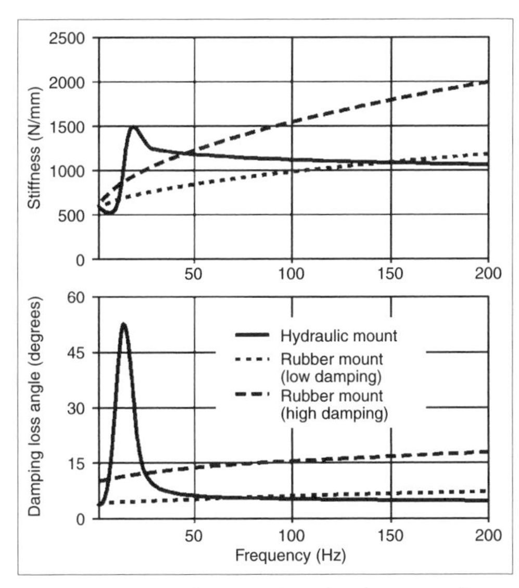

Fig. 3.4-26 Dynamic properties of various mounting elements.

alization of very different dynamic stiffness in a low as well as a high frequency range (Fig. 3.4-26). In connection with these, potential future application of active mounting elements, currently still under development, should also be mentioned. As elements of a control circuit, they permit either additional damping at low frequencies or reduced stiffness at high frequencies.

Experience has also shown that semiactive, controlled-stiffness mounting elements can result in a perceptible increase in acoustic and vibration comfort. These components, switched by means of an external control signal, can act as stiff or soft mounts. These vibration insulation elements are primarily installed in diesel-powered vehicles with their typical high degree of engine rotational irregularity at idle. To reduce passenger compartment vibration, the mounting elements are switched to "soft" at idle. While in motion, they are switched to "stiff" mode, with a positive effect on engine shake.

A further possibility for goal conflict resolution may be provided by a subframe design, which provides double insulation of structure-borne sound from the power unit (Fig. 3.4-27). In realizing such an acoustically advantageous structure, care must be taken that the additional six degrees of rigid-body freedom provided by a subframe design do not result in decreased vibration comfort.

It goes without saying that all mounting elements between powertrain and body should undergo an acoustic–vibrational optimization process. This process may result in installation of hydraulic mounts in front suspension link attachment points to reduce steering wheel rotational vibration, as well as vibrationally advantageous implementation of hydraulic mounts for rear subframe and final drive. Furthermore, mounting elements with differing dynamic stiffness characteristics

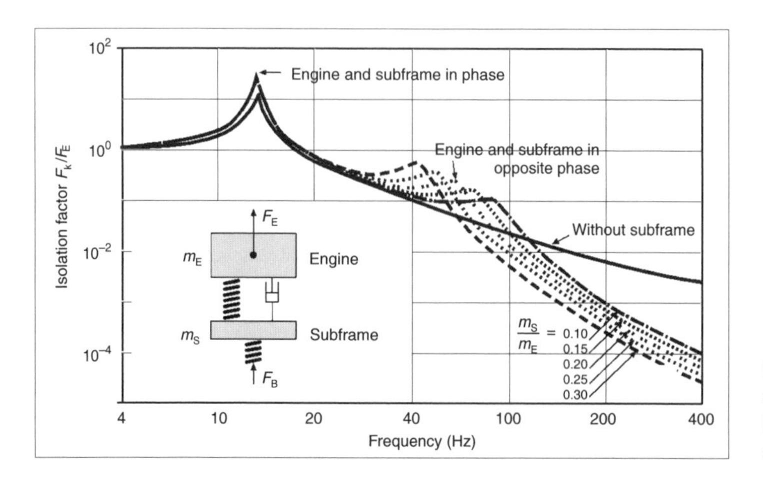

Fig. 3.4-27 Effect of subframe design on structure-borne noise behavior. F_B force to the subframe.

in various directions may be employed in the rear suspension support system—for example, to decouple conflicting ride comfort and handling targets. Such mounts may also be used around the muffler to fulfill various acoustic and durability requirements.

A further strongly negative effect on a vehicle's vibration comfort may be caused by so-called load-change vibrations, which may appear in various forms [12]. Load-change bucking is a particularly critical manifestation, as is the (large) jerk experienced during load change. Both phenomena appear as a result of steplike transient torque changes in the driveline in response to throttle pedal movement.

In the case of load-change bucking, also known as the "Bonanza effect" (from the opening title of the *Bonanza* television series, with the Cartwright clan bouncing up and down in their saddles), the participating single-mass vibration system may be described as follows. The automobile body, as a rigid body, acts as a translatory mass; the "spring" is formed by elasticities in the drivetrain system, from the tire/road contact area all the way to the mounting elements between body and drivetrain. Accordingly, it is to be expected that load-change bucking is dependent on the selected transmission gear ratio and will appear at various resonant frequencies. Particularly annoying are bucking oscillations at low engine speed in first and second gear, with frequencies between 1.5 and 4 Hz. The reason these oscillations are so annoying is not the amplitudes of vehicle acceleration perceived over an extended period of time, but rather their unfavorable "ringing" (Fig. 3.4-28).

Load-change jerk with the vehicle in motion appears in the higher gears. This vibration phenomenon is characterized by a subjectively perceived jolt in the vehicle's longitudinal acceleration accompanied by a single, dull thud; it occurs in response to steplike increases or decreases in drivetrain torque. As load-change vibration can be directly attributed to rigid-body dynamics of the driving axle, these jerks (in contrast to bucking) occur at higher frequencies. The primary annoyance is the

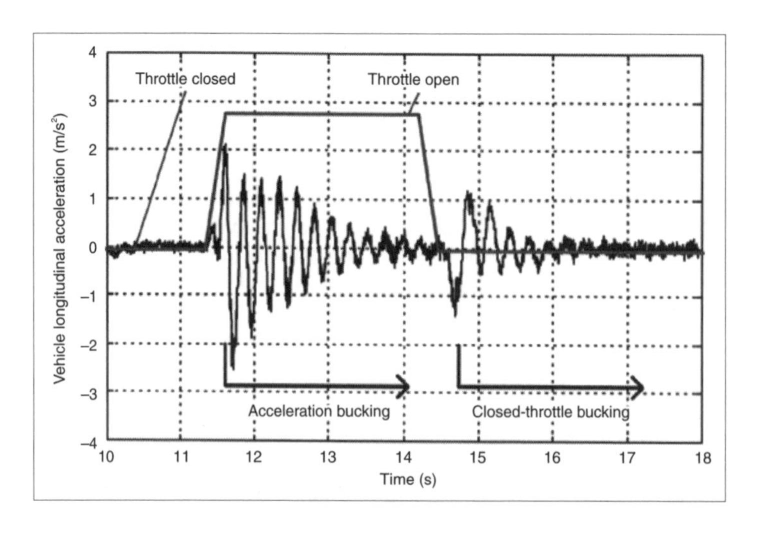

Fig. 3.4-28 Load-change oscillation over time.

amplitude of the initial event, less so the subsequent ringing behavior, as this vibration is generally well damped.

Extensive testing has shown that load-change vibration may be addressed primarily by modifications to the electronic engine management system, as this has a direct influence on excitation of the vibratory system. By contrast, modification of the transmission chain, represented in the form of structural countermeasures undertaken on the drivetrain system, have proven less effective. In connection with this, it should be mentioned that despite widely held belief to the contrary, an engine torque response characteristic which favors load-change behavior does not necessarily act counter to good throttle response or emissions behavior. Moreover, intelligent intervention in the engine management system permits fulfilling both criteria.

3.4.5 Exterior noise

The only existing regulations for vehicle acoustic properties which need to be considered in meeting certification requirements and which at present require continuous monitoring in production involve exterior noise. Along with requirements for idle noise, which are almost always easily met, the mandated drive-by test represents a real challenge. The test procedure is described by the international norm ISO 362 (see also section 2.2).

As shown in Fig. 3.4-29, regulatory agencies have steadily lowered the levels of vehicle exterior noise permissible for type certification. This illustrates that today, ten vehicles are allowed to emit as much noise as a single vehicle of 1980. Achieving the currently mandated European exterior noise limit of 74 dB(A) has required extensive development effort on the part of all automobile manufacturers. Improved exhaust silencer systems, acoustically optimized intake tracts, and engine compartments with partial acoustic encapsulation were all required to meet regulatory standards. As a result of these extensive and expensive efforts undertaken on the vehicle itself, we now have the situation that, at least in the case of more powerful vehicles, the maximum sound pressure levels in the driveby test are dominated by tire/road noise (Fig. 3.4-30). This is all the more remarkable because the tire noise sector has itself undergone considerable improvement over the past ten years.

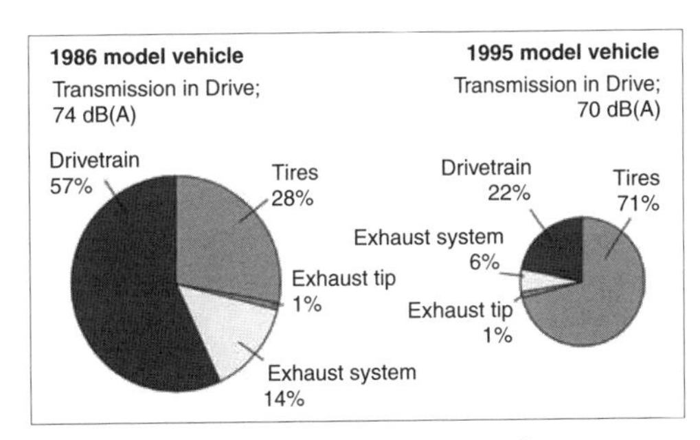

Fig. 3.4-30 Exterior noise components for high-performance vehicle.

Even though regulatory requirements represent an unyielding limit for exterior noise generation, many manufacturers undertake additional efforts to improve exterior noise. Examples include extensive efforts toward encapsulation of diesel-powered vehicles. The goal is to reduce noise emissions for these vehicles, which are characterized by their "knocking" during cold starts. As shown in Fig. 3.4-31, such encapsulation measures often extend deep into the vehicle concept and require very careful attention to other functional requirements—for example, cooling performance. This example shows that vehicle manufacturers are intensively involved with problems related to social acceptance of their products.

Finally, there is the problem of noise emissions from parked vehicles. Noises generated in this "operating condition" (by slamming doors, for example) may be regarded as environmentally unfriendly. To counteract this, some manufacturers have introduced so-called "soft close" systems in the form of electric closing assists to enable soundless closing of doors. Due to the high costs of providing such systems while meeting all important safety aspects (such as pinching hands or fingers), soft close mechanisms have so far only found application in the luxury market segment.

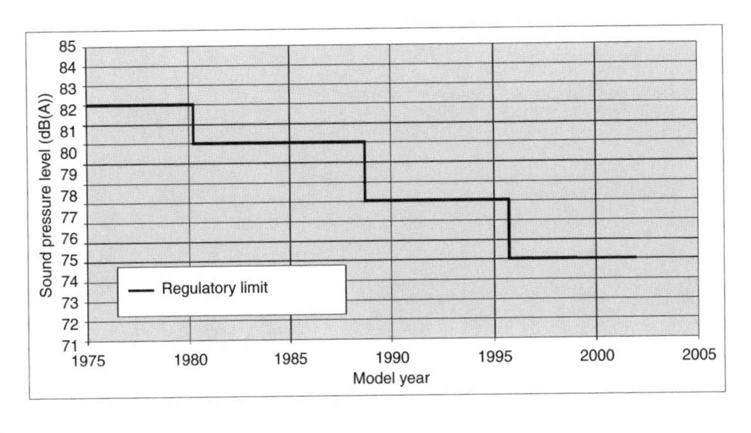

Fig. 3.4-29 Historical development of European vehicle exterior noise limits.

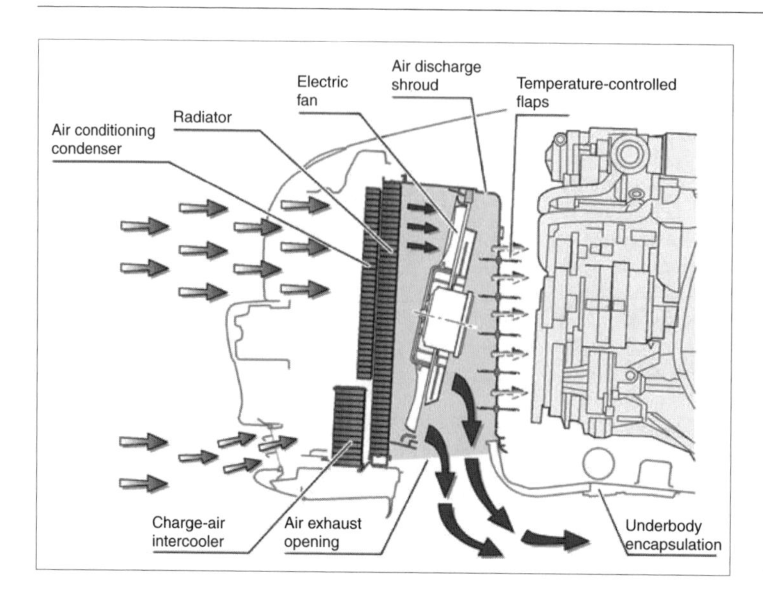

Fig. 3.4-31 Cooling module as a component for blocking exterior noise generation.

Despite these wide-ranging developments in large and small details, the topic of exterior noise has not yet been addressed to complete satisfaction. Reasons include that acoustic improvements applied to vehicles are in part offset by steadily increasing traffic. For practical purposes, the public has not experienced any drastic reduction in exterior noise burdens. It remains questionable whether even tighter drive-by noise limits can ever achieve such a goal. It would most certainly be more effective to address this problem by concentrating on other, more important topics.

These include reduction of tire/road noise [13], the dominant source of noise in suburban traffic. The noise level of vehicles entering urban areas at 60 to 120 km/h (37 to 75 mph) or driving past at steady speeds is all too clearly characterized by the high-frequency "singing" of tire/road noise. The objective is to introduce additional improvements in close cooperation between tire manufacturers and highway engineers (see Section 7.3). It should be noted that there is no simple solution to this extremely important problem; along with acoustic requirements, vehicle tires as well as road surfaces must satisfy a multitude of additional criteria such as long life and good hydroplaning properties.

Reduction of vehicle-generated urban noise requires specific traffic flow control. Every acceleration of a string of vehicles at a traffic signal is associated with noise emissions. The objective must be to provide possibilities enabling vehicles to transit cities at constant (possibly reduced) speeds. Creation of so-called "quiet traffic" zones, in extreme cases with cobblestone paving combined with speed bumps, as well as meandering traffic paths, does not lead to solution of this problem. Due to the resulting general nonsteady-state driving conditions in low gears, such measures actually produce higher noise burdens (Fig. 3.4-32).

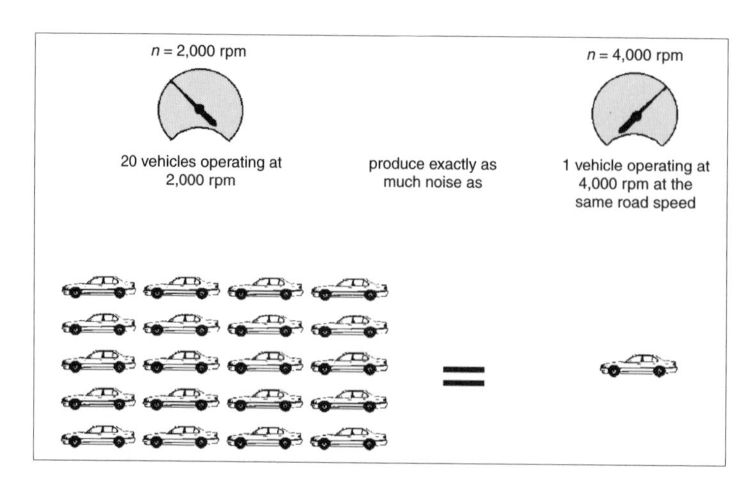

Fig. 3.4-32 Relationship between vehicle operating condition and exterior noise.

3.4.6 Quality assurance

Customers have little interest in how much effort went into development of a vehicle; they judge their own vehicle on the basis of its actual characteristics. For this reason, it is of utmost importance that every delivered vehicle completely meet or even exceed ever more demanding customer expectations.

Continuous production control within manufacturing plants is required in order to meet these demands. Special efforts are necessary at production startup of a new vehicle model. The objective is to assure that the acoustic quality realized by individual prototype vehicles is matched by production models. To this end, at the beginning of production, close cooperation is required between acoustic specialists in the development departments and quality assurance personnel in the plants. Only in this way is it possible to quickly discover, analyze and eliminate acoustic deficiencies of individual vehicles.

In most cases, this means dealing with seemingly minor issues that nevertheless have an enormous effect on the acoustic behavior of individual vehicles. Examples include effects of assembly methods, which only become apparent under production conditions after development work has been completed. For instance, stressed installation of the exhaust system, or even stressed fastening elements for hydraulic lines, can have a disastrous effect on vehicle acoustics. At that point, it is imperative to continuously identify and eliminate new acoustic problems. It is important that all participants in the production startup process, from company management to line worker, clearly recognize the importance of this task.

Later, even as the production crest is reached, it remains a lofty goal to maintain production quality under all circumstances. By means of specific production checks, it is possible to recognize early indications of diminishing acoustic performance caused, for example, by changes in the manufacturing process and provide timely countermeasures. For these checks, detailed acoustic analysis in sound chambers and/or in the test driving department are performed on selected production examples. The wealth of data acquired by these tests permits determining whether an acoustic problem only occurs on a certain model variation or on vehicles assembled on a specific production line. These data are a prerequisite in enabling generally rapid elimination of acoustic problems soon after they are recognized.

For acoustic specialists, this information from the plant floor is of great significance. By means of this feedback process, likely improvements are clearly identified and may be considered in establishing acoustic goals for subsequent vehicle models. One emphasis of these activities is realization of robust acoustic concepts which are insensitive to manufacturing and assembly methods, with special focus on achieving good long-term quality.

References

[1] Peter, W. "Berechnung im Automobilbau." ATZ Automobiltechnische Zeitschrift 90 (1988), Vol. 6, pp. 319–325.

[2] Wisselmann, D., and R. Freymann. "Der Einsatz von CAE/CAT-Verfahren in der Fahrzeug-Akustikentwicklung." VDI-Berichte No. 1491, 1999, pp. 299–320.

[3] Maass, H., and H. Klier. *Kräfte, Momente und deren Ausgleich in der Verbrennungskraftmaschine*. Springer-Verlag, Vienna and New York, 1981.

[4] Klingenberg, H. "*Automobilmesstechnik/Band A: Akustik*." Springer-Verlag, Berlin, Heidelberg, New York, London, Paris, Tokyo, 1988.

[5] Thien, G.E. "Geräuschquellen am Dieselmotor." VDI-Berichte No. 499, 1983, pp. 107–119.

[6] Strohe, M. "Powertrain Test Bench as an Innovative CAT Development Tool." Euro-Noise 98 conference proceedings, Vol. 1, 1998, pp. 471–476.

[7] Freymann, R. "Strukturdynamische Auslegung von Fahrzeugkarosserien." VDI-Berichte Nr. 968, 1992, pp. 143–158.

[8] Tönshoff, H.-K., and W. Kiehl. "Aspekte des lärmarmen Konstruierens mit Blech." VDI-Berichte No. 523, 1984, pp. 19–40.

[9] Cremer, L., and M. Heckl. *Körperschall*. Springer-Verlag, Berlin, Heidelberg, New York, 1982.

[10] Fasold, W., W. Kraak, and W. Schirmer. *Taschenbuch der Akustik*. VEB Verlag Technik Berlin, Teil 1, 1984.

[11] Freymann, R. "Acoustic Applications in Vehicle Engineering." In: *Fluid-Structure Interactions in Acoustics*. Edited by D. Habault. CISM Lectures No. 396, Springer-Verlag, 1999.

[12] Bencker, R. "Simulationstechnische und experimentelle Untersuchung von Lastwechselphänomenen an Fahrzeugen mit Standardantrieb." Dissertation, TU Dresden, Hieronimus Buchreproduktions GmbH, Munich, 1998.

[13] Tonhauser, J. "Einfluss des Reifen-/Fahrbahngeräuschs auf das Außengeräusch von PKW—Stand der Technik." Zeitschrift für Lärmbekämpfung, Vol. 43, Issue 6, 1996, pp. 158–163.

[14] Henn, H., G.R. Sinambari, and M. Fallen. *Ingenieurakustik*, 2nd edition. Friedr. Vieweg & Sohn Verlagsgesellschaft mbH,. Brunswick/Wiesbaden, 1999.

[15] Meyer, E., and E.-G. Neumann. *Physikalische und Technische Akustik*, 3rd edition. Friedr. Vieweg & Sohn Verlagsgesellschaft mbH, Brunswick, 1979.

[16] Zwicker, E., and H. Fastl. *Psychoacoustics*, 2nd edition. Springer-Verlag, Berlin, Heidelberg, New York, 1999.

[17] Lyon, R.H. *Machinery Noise and Diagnostics*. Butterworth Publishers, Boston, London, Durban, Singapore, Sydney, Toronto, Wellington, 1986.

[18] Heckl, M., and H.A. Müller. *Taschenbuch der Technischen Akustik*, 2nd edition. Springer-Verlag, Berlin, Heidelberg, New York, London, Paris, Tokyo, Hong Kong, Barcelona, Budapest, 1995.

[19] Olson, H.F. *Music, Physics and Engineering*, 2nd edition. Dover Publications, Inc., New York, 1967.

[20] Crocker, M.I., and N.L. Ivanov. *Noise and Vibration Control in Vehicles*, Interpublish Ltd., St. Petersburg, Russia, 1993.

[21] Zwicker, E., and M. Zollner. *Elektroakustik*, 2nd edition. Springer-Verlag, Berlin, Heidelberg, New York, London, Paris, Tokyo, 1987.

4 Shapes and new concepts

4.1 Design

4.1.1 The definition of design

The linguistic origin of the word "design" lies in the Latin *designare*, to denote, designate, appoint . . . to mark out, to trace out. In modern Anglo-Saxon or Romance language usage, the term is applied to nearly every creative process (for example, hair design for barbers and hairdressers, light design for illumination, fashion design for clothing, software design for development of computer programs, etc.) The *Oxford English Dictionary* gives one definition of design as a "sketch or plan for a future product" and defines industrial design as "the art of making designs for objects which are to be produced by machine." In any case, it should be noted that in English usage, design has a much broader connotation and may be used to signify the entire range of development (including pure engineering development), while in German, the concept always implies a strictly constructive or creative component.

Difficult as it is to define precisely the concept of design, it is equally difficult to circumscribe the designer's job description. In the automobile industry alone, the task of a so-called "designer" may range from purely aesthetic or artistic activity to scientific work.

Today, an automobile designer's assignment encompasses not only functional shaping and pure styling; the task is much broader, and subject to numerous constraints such as regulatory requirements, product specifications, and so on. The most important function of design, however, is to imbue the product with a soul. This soul does not arise from engineering development nor from sober packaging and certainly not from ornamentation; the soul of a product can only be created by a designer who knows how to combine aesthetics, style, and emotion in a suitable form. "Form follows emotion" [1] may therefore be taken as a basic law of automobile design, to be achieved while solving various conflicts as indicated by Fig. 4.1-1.

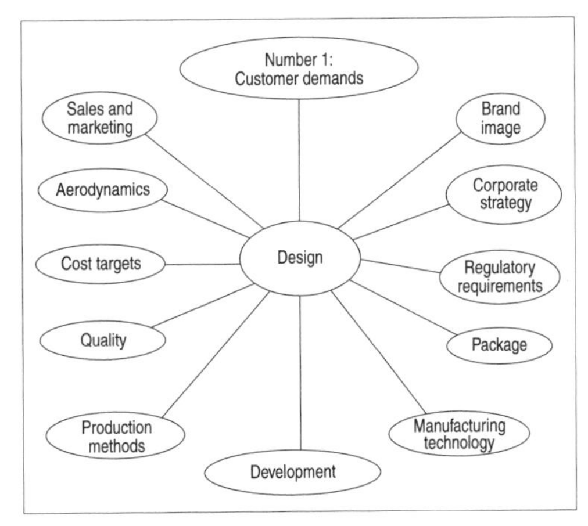

Fig. 4.1-1 Conflicting design goals.

4.1.2 The growing importance of design

As early as the 1920s, Alfred P. Sloan, the chairman of General Motors at the time, recognized the commercial significance of attractive design and founded the automobile industry's first design studio, under the leadership of Harley Earl. The automobile was developing away from its earlier position as a means of transportation, becoming instead an article of fashion. A car's dynamic and emotional qualities have always given this product a certain fascination. These emotional components, interpreted in three dimensions by talented designers, have created a steady stream of desirable products and design milestones.

In Harley Earl's time, factors such as reliability, performance, quality, and, of course, price counted as the most important purchase criteria for most drivers. Today, the situation is very different: there really are no bad cars anymore. Minimum requirements for reliability and safety are met, there is an abundant range of choices, and prices offer something for any budget. Manufacturers are left with design and brand as competitive advantages and means of differentiating their products from those of the competition.

4.1.2.1 Design and platform/shared parts concept

Currently, the best example of this is Volkswagen, the souls of whose products are formed exclusively by specific body and interior designs. From an engineering standpoint, these vehicles, despite very different vocabularies of form and strongly differentiated images, are virtually identical except for fine-tuning; they are based upon a platform which has proven itself in millions of examples. The shared parts concept has been applied to drivetrains, through all subassemblies, down to the last bolt, but by means of design, each achieves a unique product language. By this strategy, Volkswagen hopes to achieve the greatest possible market penetration for various corporate brands with a minimum of engineering development and investment. Fig. 4.1-2 shows several versions built on a shared platform, with appropriately modified underpinnings.

In the struggle for market share, manufacturers must satisfy customer demands in an ever-increasing number of market segments. Parallel to customers' lifestyle demands, which seem to become more complex with each passing day, and the growing range of manufacturers' offerings, the soul of the product and therefore its design

Fig. 4.1-2 Different vehicles on a single platform.

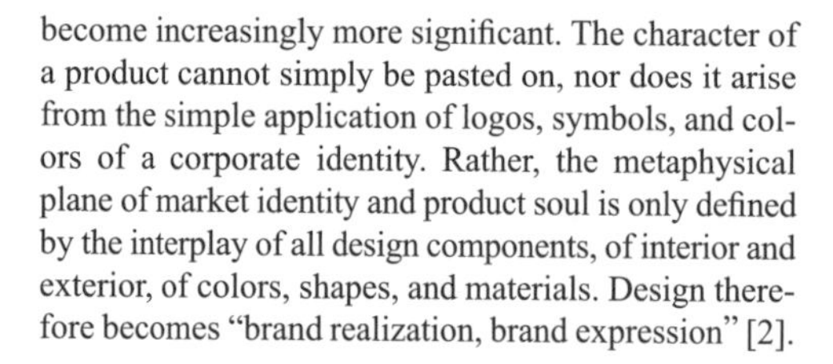

become increasingly more significant. The character of a product cannot simply be pasted on, nor does it arise from the simple application of logos, symbols, and colors of a corporate identity. Rather, the metaphysical plane of market identity and product soul is only defined by the interplay of all design components, of interior and exterior, of colors, shapes, and materials. Design therefore becomes "brand realization, brand expression" [2].

4.1.2.2 Design and brand image

"Products are made in factories, brands are bought by customers." To this quote from Bill Webb [3] we may add "people buy image." Brand image is something with which we like to identify. It fits our lifestyle and personality. By buying a branded article, we also acquire its image. Naturally, the appearance—that is, the form of the product—plays a very decisive role. Clever advertising and critical media reports may influence the purchase decision, but product design is determinant: it must validate the promises of advertising, media, and image, and it must not disappoint or deceive by a contradiction in brand identity.

The growing importance of brand and image can be observed everywhere in the automobile industry. Ford's acquisition of Jaguar was a clear example of purchasing a brand as well as its image, as were Volkswagen's acquisitions of Bugatti, Lamborghini, and Bentley. For all of these historically significant, image-laden brands, it is the task of design not only to prevent a break in each individual brand image but also to continuously cultivate the brand appearance and therefore its image, thereby further developing the brand identity.

4.1.2.3 Design as a marketing tool

In the past, decisions regarding which new products would be created and how they would look were made primarily on the basis of sales figures of previous model variations, dealer opinion, and available competing products. Today, neither design nor marketing departments are satisfied with such information. Even though it was once convenient for engineering to fall back on the old, familiar, and proven solutions, this path is neither goal-oriented nor sufficiently forward-looking for the application of design as an instrument of marketing [4]. Market demands—that is, customer needs, both immediate and in the near future—are the deciding factor in determining economic success. Experience has shown that designers, with their emotionally driven sense of form and color, more nearly reflect customer tastes and are therefore more closely attuned to future trends. After all, it is their profession to discern future style trends from social and cultural streams and, at the proper time, reflect these in automobile design.

4.1.2.4 Trend research

Trend research is the most important tool for detecting streams and developments which are suitable for precipitation in new designs. Today, trend research is part of the established repertoire of both design and marketing departments. Its goal is to forecast trends, make predictions, and carry out a form of creative prognosis. It is clearly different from "clinics," which began in the 1970s as a means of probing customer tastes: future vehicle models were shown to a selected audience, whose opinions were solicited. The results of such processes, still practiced today, are anything but reliable; examples abound of products which tested positively in clinics but then, ultimately, became marketplace flops.

Trend research, established as a nearly scientific discipline, is therefore an indispensable source of information for all manufacturers. In a market situation in which brand and image are so decisive for market success, it is especially important to determine the taste of target groups. These have long since expanded beyond one-dimensional structures; today, they are multilayered. It has proven worthwhile to analyze the target customer group not purely demographically but rather according to social background, lifestyle, values, and ambiance, and to illustrate these by means of image boards.

4.1.2.5 Conflicting goals

Automobile companies are molded by engineers—and often led by them. If design is given too little credence, conflicts are preprogrammed, especially because engineers are constantly under pressure to meet tougher technical demands with increasingly shorter development time at ever-lower cost. In such situations, it is of course easier to fall back on existing concepts than to act on designers' progressive proposals.

Not all engineers have a feel for the enormous potential for ideas that exists in a design department. However, when the engineers' innovation is successfully combined with designers' creativity, imaginative solutions are possible. A large portion of the credit must go to the studio engineer, whose role and significance are often misunderstood and/or underestimated. The studio engineer is one who works for the design department (the studio) and supports designers in their daily tasks. He feeds them information on new developments and technologies and ensures that design proposals meet production requirements and regulatory demands. This demands a great breadth of experience because, for example, the studio engineer must be familiar with all regulatory requirements for all disciplines and in various destination countries as well as their technical solutions.

Within the corporation, the studio engineer assumes the function of a technical advocate. He defends new, unconventional ideas from the design department against objections from the engineering departments. With the aid of sketches, models, and, if necessary, actual solutions presented by market competitors, he attempts to convince engineering departments that a solution proposed by the design department can be done (and is affordable). It is not uncommon for the studio engineer to propose new engineering ideas to designers, which in turn permits new design studies in which the aforementioned conflicting goals of Fig. 4.1-1 must be taken into consideration.

4.1.3 Exterior and interior design, color and trim, advanced design

At present, most manufacturers' design departments are divided according to areas of responsibility but work closely with one another on their respective projects in the sense of simultaneous design. Among designers, it is customary to move between various departments. One exception to this is the field of color and trim, where the required special knowledge of plastics, textiles, paint, and manufacturing techniques limits free movement between departments.

Exterior design

When people talk about automobile design, they usually mean exterior design. Accordingly, this segment represents the crowning discipline in the eyes of designers. The fact is that the appearance of a car communicates its character and therefore the soul of the product. By means of interplay between proportions and lines, highlights and graphic elements, a design may express sportiness or elegance, comfort or dynamism, avant garde or tradition. This serves to reinforce the brand image, and a brand-specific shaping of front and rear permits linking it to the corporate identity. In terms of the design process, exterior design determines development direction and timing.

Excursion: wheel design

Wheels are the legs of a car and provide the only contact between vehicle and road. They give the car its "stance." Relative proportions such as track, wheelbase, wheel diameter, and width are fundamental to a vehicle's appearance. If these proportions are right, then the "stance" of a vehicle imparts an image of power and security; if not, it can easily give rise to impressions of imbalance, instability, or weakness.

Wheel design plays a significant role; it permits expressions of sportiness or elegance, ruggedness (4×4 character) or luxury (sedan). Moreover, wheels—like headlights and taillights—are highly decorative elements of a vehicle design. However, in contrast to headlights or taillights, wheels are a design element on which customers can apply the wide range of offered accessories to customize their car according to their own individual taste and wallet, thereby establishing their own personal accents.

This aspect plays an ever more important role in individualizing the product and for many drivers still represents the actual status symbol. Therefore, no car manufacturer's options list and no enterprising dealer fail to offer a wide selection of wheels. For dealers, wheel design will become even more important in the future not only as a means of differentiation but also as a decorative element. The trend toward even larger wheel diameters continues unabated.

Nevertheless, wheel design is subject to strong fashion currents. Even Porsche's beloved Fuchs wheels of the 1970s no longer fit under the fenders of a 911/Type 996. The Alpina wheel, unchanged since the 1970s, is a notable exception. The first light alloy wheels were offered in the 1920s by Bugatti. For motorsports, these represented a logical development of the wooden-spoked wheels, which traced their origin to horse-drawn coachbuilding practice.

On production cars, the aluminum wheel finally replaced the sporty steel-spoked wire wheel in the late 1960s (Porsche 911). Since then, we have seen wheels made of magnesium and various aluminum alloys, stamped steel, riveted carbon fiber, and the first carbon composite solutions.

Interior design

The importance of interior design is often underestimated. To a much greater degree than the exterior, the interior may significantly influence the buying decision, and interior design determines the satisfaction with the acquired product. It has been shown that the purchasing decision may also be steered just by color combinations, fabric and material quality, or the smell of leather.

Interior design must harmonize with the exterior language, in form, color, and materials. The interior designers' task is all the more complicated because they must deal with physiological problems (such as ergonomics; see Section 6.4.1), the visual significance of various colors and shapes, production technology for various materials, and the regulatory and aesthetic demands of different markets. For this reason, the interior designer must intensify his relationship with the product. While the exterior message is still purely visual ("love at first sight"—or not), all senses are called into play in perception of the interior: sight, sound, smell, touch, and, in a different sense, taste.

Interior design affects the sense of well-being inside the car. This discipline also includes designing the seats and seating position, upon which comfort and ergonomics are largely dependent. Interior design also conveys perceptions of the vehicle interior: does one feel constrained or lost or that one is in restricted or expansive, spartan or luxurious, repelling or inviting surroundings? All of these are directly influenced by interior design. Because all senses are addressed by interior design, all elements—such as steering wheel, shift lever, door handles, switches, etc.—deliver a message which is just as important as the visual impression. All of these impulses produce the overall impression of a vehicle; they communicate values such as quality, functionality, modernity, solidity, and tradition, and they transform a journey in the car into a pleasurable experience—or a nightmare.

Color and trim

Colors, materials, and patterns developed by color and trim designers must harmonize with the vehicle interior as well as its exterior and should also address the taste of the target market. Customers experience exterior and interior colors, seating materials, and leather firsthand, and therefore these strongly affect their relationship to the car. Within cost limits, an attempt is made to satisfy customers with a limited number of appointment levels. Variation possibilities for selecting individual features and options of production vehicles are also determined by color and trim departments.

In a lengthy process with at least two years' lead time, paints and fabrics are developed to match their application to and within the car. From a materials standpoint, the primary considerations are quality and durability. This demands extensive specialized knowledge on the part of the designer, including familiarity with the latest production processes. Cooperation with paint and fabric suppliers plays a vital role. The creativity of a color and trim designer lies not only in creating new patterns and material and color combinations, but also in proper timing. The designer observes future color and fashion trends and attempts to create designs that will be in fashion three years down the road.

In the future, customers will become increasingly more demanding even for economically priced vehicles. New developments indicate what color and trim designers must achieve in interior design, both in terms of quality and with regard to ambiance: a high-tech look with aluminum and Kevlar or a club atmosphere with leather and walnut.

Advanced design

Sometimes also termed "concept design" and almost always given its own department by manufacturers, the responsibility of this team is exploring ideas for future development. Within the normal design process, this is usually impossible or can only be integrated with great difficulty. Because the everyday design process is usually based on precise technical requirements and a tight timetable, both of which must be meticulously maintained, it offers too little leeway for experiments or risky ideas.

Yet to secure success in the future, a company needs bold ideas for new products, as exemplified by the

Fig. 4.1-3 Bugatti EB 18/3 "Chiron" design study.

Mercedes-Benz A-Class. In design terms, the advanced or concept studio represents the counterpart of the engineering research department. Its designers are fed with information from trend research and encouraged to be as creative as possible. Their fields of endeavor include new exterior and interior design themes and new materials, colors, and textiles, but they also include completely new products. The results of these studios often appear as concept studies or show cars at the major auto shows and are often used to gauge market reaction (see Fig. 4.1-3). If the reaction is positive, "tamer" production versions will roll off the assembly lines years later. Beyond developing new product ideas, advanced design also has a second, equally important function: it serves as exercise and as motivational training for designers. Without an opportunity to seize upon new impulses and work somewhat more freely, a design department would gradually lose the necessary creativity and stimulation.

4.1.4 The design process

Beginning in the 1970s, the computer permanently changed the automobile product creation process. Of course, development of digital technology did not come to a halt at the design deparment's door. Although the computer's influence has yet to alter the actual design process—that is, the systematic progression of the design—individual design steps may now be carried out digitally instead of using classic methods, such as paper, marker, tape, and so on.

4.1.4.1 The classic design process

Assuming that corporate strategy, brand image, and design identity are already known and serve as general guidelines for the design team, the classic design process is as follows.

Design brief

Every company uses a different designation for this information; often it takes the form of a target catalog, for example, sometimes called a PPS (Product Program Submission) in English. This contains all relevant information required to define the product (e.g., vehicle type) as well as a detailed development plan extending to the beginning of sales. The following are of particular interest to design departments:

- Product definition and positioning (example: a small two-seater)
- Target customer group and segment (singles, midrange price)
- Design character (sporty/dynamic)
- Main competitors (benchmarking)
- Schedule
- Engineering and safety goals (wheel/tire innovation, number of airbags, etc.)
- Technical and equipment specifications (two-engine versions, two interior trim levels)
- Information on production process and methods (instrument panel installation through windshield opening, size of spray booth, conveyor pickup points on body)
- Main markets and sales volume

Package data

This information is decisive for exterior as well as interior design. It is usually based on an established (existing) platform, on which drivetrain and main assemblies—such as cooling system, fuel tank, etc.—are placed (see Section 4.2). Occupant seating positions and, therefore, H-point and sightlines are established by the package. Design must work with these initial so-called "hard points" to develop the best and most skillful utilization within the given vehicle dimensions. Simultaneously, however, the package also affects vehicle proportions: for example, height, width, length, windshield rake, overhangs—in other words, specific design-relevant factors.

Concept phase

With design brief and packaging information in hand, the design team can begin work, first in the form of brainstorming, then with sketch pad and marker. At this stage, key concepts or so-called image boards from trend research are helpful. They interpret the target group's lifestyle, reflect the image and character of the product, or employ other products to exhibit a certain desired vocabulary of form.

Ideation

The first scribbles and sketches are committed to paper during this phase. In addition to the role of creative talent, drawing skill also plays a part. A designer who can more easily visualize his ideas on paper and experiment with them achieves better results more quickly. Fig. 4.1-4 shows one example.

Theme refinement

In theme refinement, work continues on ideas selected during the ideation phase. Designers go into more detail and attempt to define shapes and lines more precisely by means of more realistic renderings. This is usually

Fig. 4.1-4 Design sketch.

followed by the first presentation to management—for example, in the form of so-called orthographic projections (side, front, rear, and top view) and perspective representations. In cooperation with management, a choice is made, channeling the design in one specific direction, or at least reducing the choices to a few favorites.

The same design process also applies for interior design, but usually after a short delay. Naturally, the exterior shape is a determining factor in interior design, dimensionally as well as in terms of theme (vocabulary of form).

Tape drawing

The tape drawing (Fig. 4.1-5) is a technique employed almost exclusively by the automobile industry. It has immense significance in translating two-dimensional ideas into three dimensions. Usually working in full scale, designers and studio engineers create a type of technical illustration using various widths of black crepe tape applied to a wall-sized sheet of grid or graph paper. The tape drawing is based on a dimensional drawing which specifies hard points. Because the crepe tape is easily streched, removed, and reapplied, this somewhat unusual method permits efficient experimentation with lines and curves. One rapidly gets an impression of how the full-scale model would look, and a very realistic representation (a rendering) can be done with the aid of an airbrush.

The tape drawing is an outstanding tool for controlling the design's proportions and main lines. Later, the quality of design achieved in tape is immediately reflected in the model. Now, for the first time, the studio engineer can check engineering functions on the basis of the tape drawing: for example, whether side windows can roll down, whether doors can open freely, and whether regulatory requirements (heights of bumpers, light units, etc.) are met.

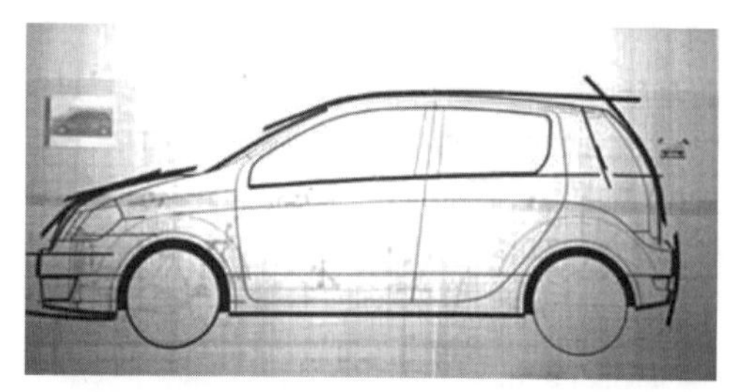

Fig. 4.1-5 Tape drawing.

Model phase

As soon as the designers (responsible for shape and proportion) and studio engineers (realisability) are satisfied with the tape drawing, work can commence on the model. Using a coordinate measuring machine, modelmakers transfer dimensions from the tape drawing to the model. This is usually made of clay (Plasticine™), but sometimes Epowood (polyurethane resin) or plaster is employed (Fig. 4.1-6). The model is built on a rigid steel armature whose substructure is attached at precisely defined points to a steel surface plate, ground flat to hundredths of a millimeter, and painstakingly leveled. This process is absolutely necessary in order to guarantee the required dimensional accuracy, even if the model is repeatedly removed from the surface plate for presentations.

Work on the ergo buck (also called an ergonomics buck or a seating buck) begins simultaneously with the exterior model work (Fig. 6.4-1). The purpose of this exercise is to obtain the first impressions of the interior: for example, entry and egress, seating position, and view through the windshield. Results of ergonomic tests in the seating buck may lead to changes to the tape drawing and therefore to changes in the model. Even during the modeling phase, necessary or desired changes to the model must flow back to the tape drawing. This ensures that in the tape drawing, a current representation of the actual engineering state of the model is always available to designers, engineers, and modelmakers.

The modeling phase is the most exciting part of the design process. It is here that ideas are converted into three-dimensional reality. First, several modelers transfer the dimensions of main design lines from the tape drawing to a huge lump of clay and sculpt the surfaces

Fig. 4.1-6 Work on a clay model.

between them. Bit by bit, pieces appear—hood, roof, windshield, and doors—at last resulting in a complete car. The first three-dimensional results are visible just a few weeks after the design brief. At the earliest opportunity, these are evaluated outside the design studio in open air. This perspective always provides new realizations: the roof is too round, the windows are too flat, the entire thing looks too heavy . . . The tape rolls are brought out again, new lines are sought, new proportions considered.

As the model develops, evaluations come ever closer to reality. Dynoc film is applied to the surface to allow better evaluations based on light reflections. For the final management hurdle—the so-called signoff or design freeze—the model is painted, lighting modules are installed in the form of hard components, and the structure above the beltline is fitted with clear glazing.

Finally, feasibility studies are initiated after the design freeze. At their conclusion, a new, digitally measured verification model is created. Tooling is constructed and outsourced parts are manufactured on the basis of this reference model.

4.1.4.2 The digital design process

The digital design process uses a new tool—the computer—to enable a new means of communication. Instead of paper and pencil, we now work in binary code. The capabilities of this new form of data communication are redefined on an almost weekly basis. In the future, the computer will certainly revolutionize the entire product creation process (see [5]). At present, however, it would be more accurate to refer to this as a semidigital design process. Although completely digital design has been attempted, virtual reality has not yet given entirely satisfactory results. Also, there has been reluctance to make million-dollar decisions without first evaluating an actual full-scale model. This is also true throughout the entire spectrum of engineering design; although CAD, CAE, and CAM have long been regarded as standard tools, these have reduced but not yet eliminated the need to construct actual prototypes.

The digital design process differs significantly among various manufacturers. Fig. 4.1-7 shows process differences. Although individual steps follow those of the classic design process, differing only in the aids employed, every manufacturer has different demands for computer-aided design and expects different advantages: reaching goals more quickly, having more alternatives, achieving simpler data transfer, and so on. In order to permit the semidigital design process to function at all, every possible output format must be available, including printers, plotters, milling machines, stereolithography (SLA), computer projection and virtual reality (VR) systems. Binary codes can be fully used and optimum design results achieved only if these preconditions are met.

Design brief, package data, concept phase

The most important steps follow those of the classic process: the design brief must clearly define the program goals, the package must precisely establish data, and the concept phase must correctly interpret the specified engineering and form requirements.

Ideation

Even in the digital process, actual design work begins with hand sketches. Although a computer may be employed for this highly creative and intuitive phase, most designers still prefer classical methods for their first scribbles. Part of the appeal lies in the spontaneity and speed which can be achieved in sketching with pen or pencil. In this early design stage, individual preferences (such as how the paper is turned, how the pencil is guided, a favorite working location to put one's ideas on paper) make working on a computer cumbersome, indeed counterproductive. Usually, though, after a few days, the ideas can be further developed on the computer using a two-dimensional drawing program (such as Studio Paint, for example).

3-D design and modeling

Based on uniformly applied digital data, this phase blends theme refinement, tape drawing, and modeling. As a designer refines his concept on the computer monitor, changes are automatically passed to the data file, which forms the basis for the model (and later, the production tooling). With the help of modeling software, the designer can now build his own proportional models in the computer using digital processes. Based on main design lines, he can shape surfaces and quickly get an impression of which forms and proportions may be achieved on the predetermined package. Naturally, three-dimensional package data may now be called up in the computer (Fig. 4.1-8), while in the classic design process these were only captured two-dimensionally in the tape drawing.

This, above all, is the main advantage of 3-D modeling over the classic design process: in the tape drawing, the interplay between four main viewing directions is often difficult to imagine, while the digital process offers not only the computer's very flexible representation of a wire frame model, but also, at the touch of a button, enables viewing from all possible angles. If the computer is allowed to stretch surfaces between design lines, the shape may be examined in more detail and in various lighting situations. Simultaneously, the designer is not only working with mathematical precision—all surfaces fit together precisely—but also more efficiently, as he can evaluate the shape and proportions of his concept and quickly and easily implement changes at any time in a three-dimensional representation.

Three-dimensional data stored in a computer also ease the task of studio engineers. Full-scale data of the latest model iteration are retrievable at any time and are

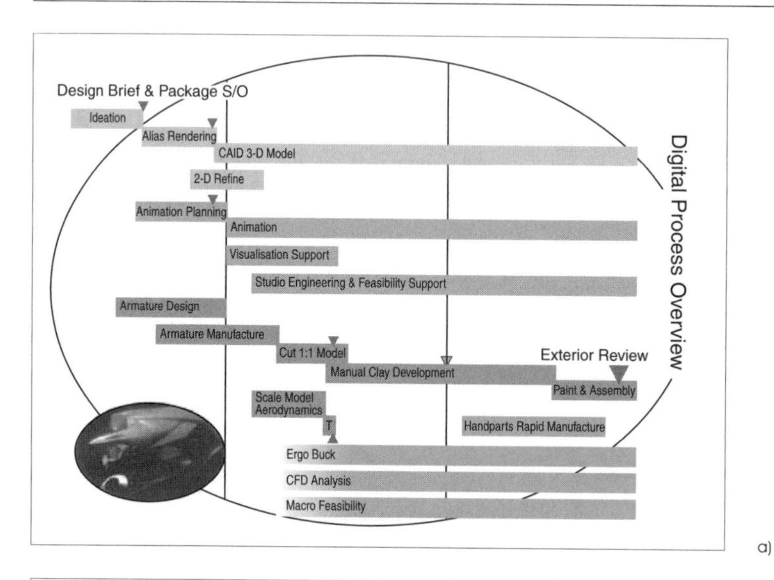

a)

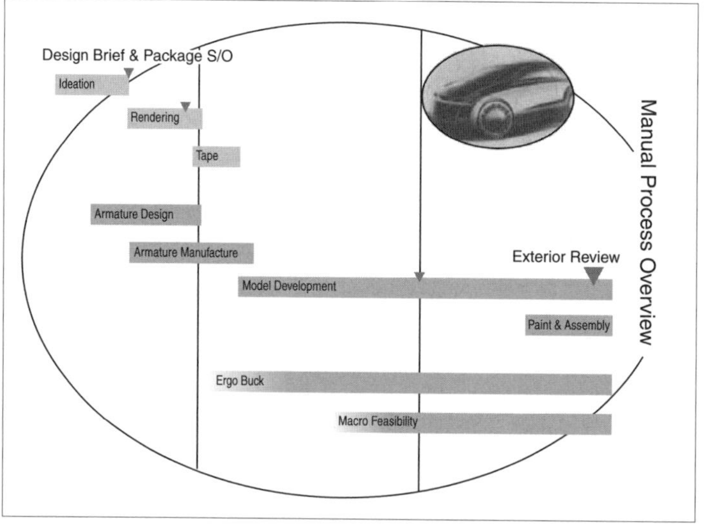

b)

Fig. 4.1-7 Comparison of digital (a) and manual (b) design processes.

compatible with CAD software such as Catia™. Compared to information in a tape drawing, digital files save time and help avoid mistakes. The same is true of computer-controlled milling of advanced models, which can be accomplished rapidly (in about two days) with the aid of digital data, and, later, development of tooling, which is also based on stored data files.

Ideally, computer models would be constructed by

Fig. 4.1-8 Computer-based three-dimensional model.

the designer himself, but as in the classical process, after a certain time, a digital modeler and his team assume responsibility for generating surfaces. Evaluation of the digital model should be done as often as possible and in as large a scale as possible (up to full-scale projections). Although some aids, such as environmental simulation, simplify evaluation of a digital model, design critiquing methods in the digital process must also be learned and developed. While the trained computer user can certainly do a passable evaluation of his own model, each designer is working alone, in front of his monitor, and is no longer in visual contact with other members of the team. The individual is separated from continuous team feedback, which otherwise makes up a large part of the studio atmosphere.

In addition, visual information gleaned from a clay model has an entirely different quality. Therefore, at least until design teams are fully trained in the digital process and their feedback tuned accordingly, every opportunity should be used to see the digital model in three-dimensional form. Even the tape drawing may be an important part of the digital process: a few crepe lines on full-scale plotter output of a digital model may provide entirely new insights. In any case, it makes sense to enter reality as soon as possible and have at least a small section of the model—for example, a door surface—milled. Precisely because conversion of aliased data into milled models may be accomplished so quickly, every opportunity should be taken to check the current state of the design in actual three-dimensional reality.

Based on digital data, the design may be developed in the interplay between virtual and real modeling. Improvements to the milled model may be added manually but then must be digitized and fed back to the computer model, analogous to the tape drawing. In the digital design process, this task is greatly simplified by the fact that milling and coordinate measuring machines are directly linked to the computer; manually altered surfaces can be automatically scanned and the corresponding files digitally modified. This interface between milling machine and computer model is vital, as it ensures that the most current database is available for market research (feasibility).

Moreover, it is of course possible to employ the digital model in a variety of ways, for example in aerodynamic investigations [6], for crash simulation (with the addition of FEM), and in animation. In this way, design evaluation of a digital model on a computer monitor is greatly improved if it simulates its environment. Normally, a digital image of the design studio itself is built in order to evaluate the model in a situation resembling that of the classical process. Also typical are street scenes; it is of course possible to "drive" the digital model inside a parking structure or any other desired environment. For management, this provides a new decision-making aid and interior design benefits if it can be seen how the cupholder deploys or how the rear seatback behaves when folded forward.

4.1.4.3 The digital process—strengths and weaknesses

Advantages and strengths:

- Data are mathematically exact and may be employed in many different ways: for example, for feasibility studies, simulations, animations, rapid prototyping, and virtual reality.
- Simple, rapid worldwide data transfer to other departments and suppliers.
- The engineering department can start feasibility studies much sooner, thereby achieving shorter development time.
- The digital process enables creation of additional digital models and therefore greater variety and possible variations, which can be quickly realized in the computer.
- There is greater efficiency; i.e., any designer working in a computer produces more in any given period of time than a designer using classical methods.
- Higher quality is possible because the data is mathematically precise and therefore hard parts fit perfectly.

Disadvantages and weaknesses:

- Evaluation of digital models is more difficult than conventional clay models because the computer monitor displays on only a reduced scale. Furthermore, context and environmental cues are limited: e.g., the effects of light and shadow.
- The infrastructure required for the digital design process is complex and costly.
- Development of a design on the monitor screen largely takes place in isolation; i.e., the designer is removed from his team's continuous feedback.
- At present, the digital process is still in the experimental stage; that is, reliable long-term experience is lacking, development is proceeding rapidly and continuously, and both software and hardware quickly become obsolete.
- As in all computer-supported processes, the computer may crash, and individual functions may exhibit bugs.

4.1.5 Prognosis

In the future, with growing significance of brand and image, the significance of design will also increase. However, design development time will continue to decrease: the vitality of the market and growing competition for customers will force companies to react more rapidly to the market and bring their products to market even sooner. This in turn will demand continuous refinement of the design process in order to achieve higher quality in ever-shorter time periods. In its results, design is doubly challenged: to provide products with an unmistakable identity and to achieve what was once described by Bruce Springsteen: "The whole idea is to deliver what money can't buy."

References

[1] Tucker Viemeister, founder of Smart Design, in *Design Week*, May 29, 1998, p. 14.
[2] Richard Murray, partner of Williams Murray Banks, in *Design Week*, March 27, 1998, p. 48.
[3] Bill Webb, director of the Institute of Retail and Distribution Management, speech at a seminar on "Brands as Lifestyles—The Brand Experience" at Design Show '97, Business Design Center, London, October 30, 1997.
[4] "More style gives us more ways of expressing the personality of a product, which, in turn, has a better chance of finding a meaningful place in our lives." Janice Kirkpatrick, in *Design Week*, February 20, 1998, p. 10.
[5] Multiple authors. "Neue Generationen von CAD/CAM-Systemen," VDI-Bericht 1357 (1997).
[6] Hucho, W.-H. "Design und Aerodynamik—ein Wechselspiel zwischen Kunst und Physik." 3. Stuttgarter Symposium Kraftfahrwesen und Verbrennungsmotoren, Expert-Verlag 1999.

4.2 Vehicle concepts and packages in passenger car development

4.2.1 Development process

4.2.1.1 Product requirements and the target catalog

In any newly defined vehicle, basic requirements are defined in a target catalog. These documents describe various basic requirements of the new product, as seen by:

- Marketing (production numbers, retail prices, markets)
- Development (process, special considerations)
- Product description
- Production
- Finance and company management

The product description segment, established in coordination with all internal corporate groups which exert influence on the product, defines the salient characteristics of a new vehicle. The following methodological sequence in establishing the product description has proven effective:

- Vehicle class (size class, e.g., "compact")
- Vehicle variations (e.g., four-door sedan, five-door station wagon)
- Powertrain assignment (engine program, transmission choices)
- Major vehicle dimensions
  - Exterior dimensions (wheelbase, length, overhangs, width, height, tracks)
  - Interior dimensions (length, width, height for seating positions, usable spaces)
- Technical description
  - Body style, variant concept
  - Engine versions and equipment (i.e., power and export market variations)
  - Transmission types (torque classes, automatic transmissions)
  - Chassis (suspension, wheels and tires, steering, control systems)
  - Technical equipment (e.g., air conditioning, electronics, fuel system)
- Technical specifications
  - Weights, payloads, towing loads
  - Performance
  - Fuel consumption and exhaust emissions targets

The product description forms the basis of the technical conception of a vehicle. During the conception phase, details are added and modified to establish a target catalog, which in turn forms the basis of vehicle development culminating in start of production.

In the establishment phase of the basic product parameters, one must differentiate between development of an essentially new vehicle or of a successor to an existing product. A generally new vehicle design emphasizes placement of the vehicle within the expected competitive market segment. For a successor model, greater technical emphasis is placed on rational development (for example, interior dimensions) and elimination of possible weaknesses, as well as refinement of carryover components from the predecessor model.

4.2.1.2 Establishing the vehicle concept

In the course of development, the vehicle concept is established based on the product description. The first layouts are created to assure the vehicle's buildability, establish major interior and exterior dimensions, indicate major component locations, and show early conflicts. The objective of this first design phase is confirmation of anticipated dimensions (interior and exterior) and component selection (engine, transmission), as well as vehicle variants, to enable initiation of concrete styling work.

4.2.1.3 Styling/hard point package

Requirements for components and main body structures, for occupants, and for useful volumes constitute a "surface" (even before an exterior skin is considered), a so-called "hard point package," which represents a basis for the first concrete interior and exterior styling renderings. Stylists must consider these engineering space requirements. Naturally, this phase is especially marked by conflicts between engineering and styling development goals, which require corresponding expenditure of effort in reconciling these conflicting goals by applying corrections and compromises.

4.2.1.4 Package

In the course of the concept phase and subsequent development, the vehicle conceptual design is corrected, optimized, and increasingly detailed. This more precise layout of overall vehicle dimensional definition is termed the "package."

This package undergoes constant refinement during the entire development period; after the beginning of series production, it supports product development activities or facelifts. Packages for vehicle variants, such as wagon or convertible models, are created on the basis of such a basic package.

The vehicle package forms a reference for all branches of development activity and, for this reason, always represents the current state of the project.

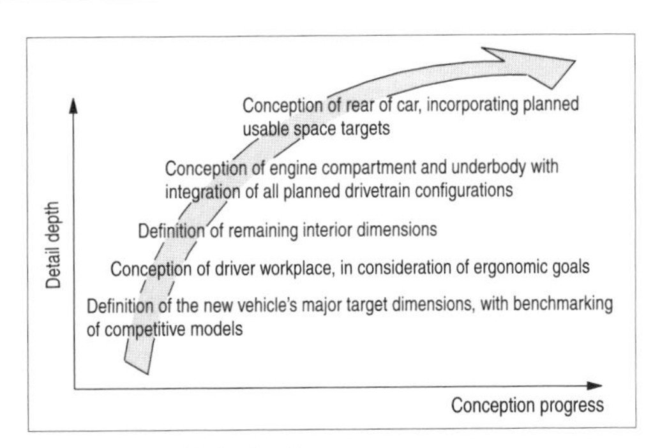

Fig. 4.2-2 Vehicle development concept sequence (example: a front-engined vehicle).

4.2.2 Vehicle conception as a process

4.2.2.1 Sequence

Initially, the development of a successor vehicle requires definition of major properties and specifications which differentiate it from its predecessor.

The new vehicle is conceived on the basis of these characteristics, which necessitate altered dimensional definitions. As a rule, a certain degree of carryover (COP; carryover parts) is given as a design goal in order to reduce development and investment expenditures.

Where a successor vehicle concept draws its design inspiration from various features (see Fig. 4.2-1), new vehicle development proceeds in a definite sequence. Consider the example of a front-engined vehicle (see Fig. 4.2-2).

A new vehicle is generally created from the inside out. Interior studies, with space and ergonomic investigations, come first in the sequence of concept development. Somewhat later, these are followed by layouts for the powertrain area (engine arrangement, transmission, auxiliary devices, front suspension, steering system, consideration of safety features such as body structures, crash deformation zones). Conceptual work in the underbody area concentrates on transmission and powertrain, exhaust systems, and layout of lines and body structure. At the rear of the car, main emphasis is placed on layouts for body structure, rear suspension, fuel tank, exhaust systems, and luggage space optimization. Early investigations of variants are conducted; for example, different rear body configurations or number of doors.

Often, in creation of the vehicle styling—which usually begins shortly after the start of concept work—conflicts arise within engineering specialties as well as between engineering and design departments. These are handled iteratively [1]. By means of new data and process relationships (see below) as well as project management capable of taking these procedures into account and optimizing them, the resulting time lost in the development process can be kept to a minimum or even eliminated.

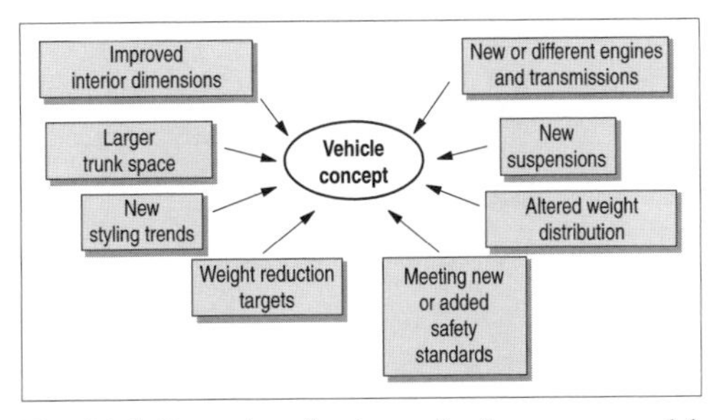

Fig. 4.2-1 Examples of major goals of successor model development.

The vehicle concept layout represents the germ of its engineering development. Throughout the entire development period, the package arising out of the concept design represents the central core for engineering information. Aside from tradeoffs between styling and engineering, the package serves above all as a source of basic reference data for the traditional development specialties—body, chassis, powertrain, and electrical/electronic systems, as well as tuning of their mutual interfaces. During the development process, optimization within engineering disciplines free of styling influence takes place, necessitating modification of the vehicle concept and package. The package always represents the current state of development, throughout the entire development process.

4.2.2.2 Package forms and product data management

In this age of electronic data processing, the exchange of product data and data communication during vehicle conception is undergoing a change. Today, with the exception of the first pen strokes of the creative process, vehicles are exclusively developed using CAD processes.

In earlier times, the actual drawing of the overall vehicle still carried decisive significance. Today, this has evolved into a means of managing design data. Clearly delineated responsibility for data, hierarchical levels, and release processes, as well as reading and writing approval for CAD files with controlled retrieval and storage responsibility, are defined. For the most part, this system affects design data for all components which possess an interface to others and therefore have an effect on the vehicle as a whole. The rationale of this management, which assigns a central role to the package sector, is assurance of an optimum development process with a minimum of development loops and the best possible quality of prototypes as well as production vehicles. The required management of this product data, which affects data processing throughout the company, is called Product Data Management (PDM) [2].

It is a given that among various carmakers, there are differences in the depth of involvement and degree of authority given to the package sector. The package database enables derivation of special layouts, such as an overview of electrical, cooling, or heating lines, visualization of assembly sequences, etc.

4.2.3 Layout sectors and factors influencing the vehicle package

The layout sectors of a vehicle may be roughly divided into:

- Interior
- Front end (mostly engine compartment)
- Rear end (mostly luggage space)
- Underbody

Within the project organization, these departments are often represented by corresponding teams. The main influences on a new vehicle package are manifold and range from the choice of vehicle concept and the actual layout sectors through strategic requirements to targets for the manufacturing and customer service branches. Regulatory standards and requirements imposed by consumer organizations as well as by the manufacturer itself round out the spectrum of requirements.

4.2.3.1 Definition of dimensions

In order to permit comparison of major vehicle dimensions, both internally and externally, unified definitions have been created. In the United States, these are encompassed in SAE standards; in Europe, they are codified by the "European Car Manufacturers Information Exchange Group" (ECIE) [3].

The following illustrations (source: ECIE) show a selection of major interior and exterior dimensions.

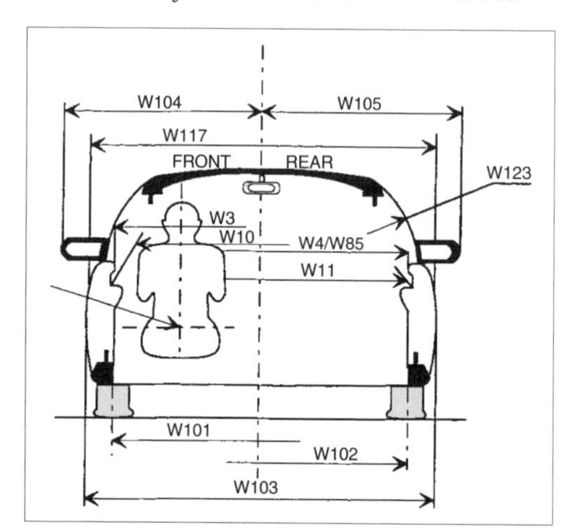

Fig. 4.2-4 ECIE width dimensions.

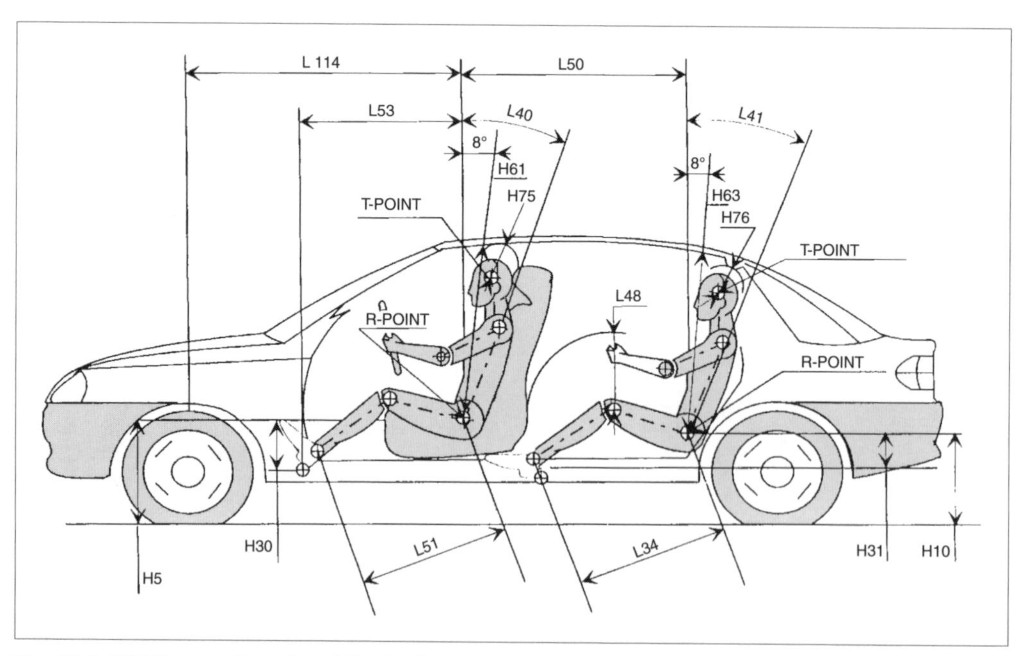

Fig. 4.2-3 ECIE interior dimensions (side view).

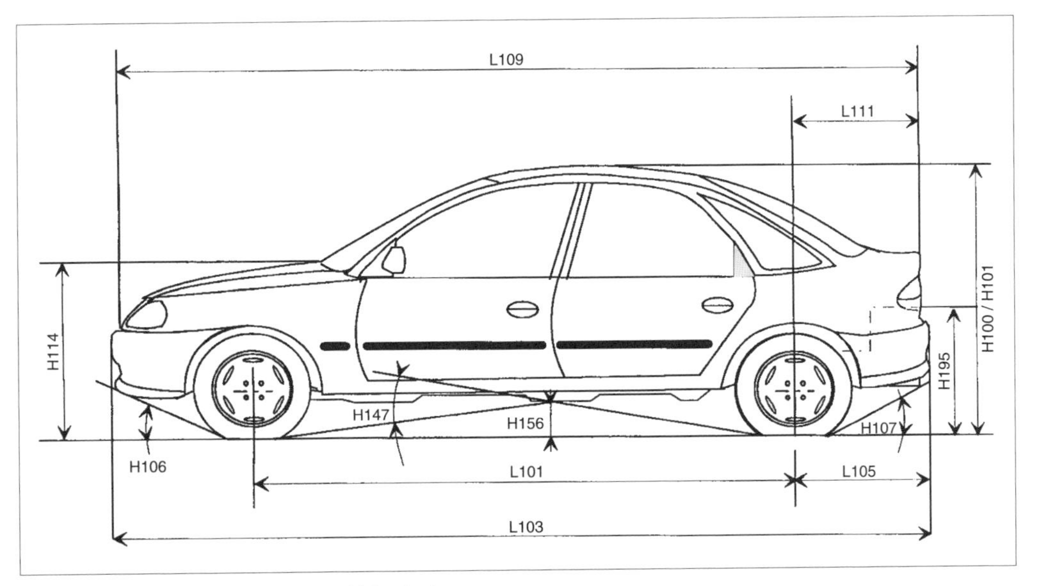

Fig. 4.2-5 ECIE exterior dimensions (side view).

4.2.3.2 Vehicle concept

The greatest influence on the form of the vehicle package is exerted by the choice of the vehicle concept, with its major characteristics:

- Powerplant location (front, rear, mid-engine)
- Drivetrain concept (front, rear, all-wheel drive)
- Powerplant orientation (transverse, longitudinal)
- Number of seats
- Comfort characteristics (e.g. leg room)
- Storage capacities

Engine locations found in passenger car design are:

- Front engine
- Mid-engine
- Rear engine
- Mid-engine configurations

Underfloor engines are rare in passenger car design (example: see Fig. 4.2-11), but are often employed in commercial vehicles.

4.2.3.2.1 Front-mounted engine

Characteristics

This engine location represents nearly the entire passenger car market. Engine and transmission are conjoined and mounted ahead of the passenger compartment (either longitudinally or transversely; in automotive parlance sometimes referred to as north-south or east-west mounting). The radiator and air conditioning condenser are mounted ahead of the engine at the front of the vehicle. Drivetrains may be front-, rear-, or all-wheel drive.

Transverse mounting is used only on front-wheel drive vehicles. For rear-wheel drive, longitudinal mounting is employed exclusively due to its simpler drivetrain and vibration advantages.

Advantages

This compact package permits short lines to all auxiliary devices, radiators, and heat exchangers. Powerplant noise can be well isolated from the cabin by means of the front firewall. In the case of a frontal accident, early contact of the powerplant block with the crash partner relieves loading of the body shell structure by powerplant inertia forces. The configuration provides adequate space for the exhaust system (particularly mufflers and catalytic converters) and fuel tank in the underbody area and at the rear of the vehicle.

Combined with front-wheel drive, the entire powertrain may be realized as a compact preassembled unit including the front axle. This permits, in addition to a flat tunnel, sufficient front-axle loading for good traction. The advantage of rear-wheel drive compared to front-wheel drive lies in its increasing traction potential with increasing rear-axle loading under acceleration or while ascending grades.

Disadvantages

Transverse engine installations, with few exceptions, are limited to a maximum of six cylinders. Transmission size is also severely restricted by installed length limitations (direct effect on vehicle width). Efforts to achieve more powerful yet very compact powerplants in transverse as well as longitudinal installations (crash length) underscore this problem. An additional disadvantage of front-wheel drive is reduced traction potential with increased rear-wheel loading due to dynamic load transfer under acceleration and in ascending grades.

With relatively modest effort, it is possible to produce an all-wheel drive variant based on a front-wheel drive design with a longitudinal engine installation by adding a transfer case to the transmission and, without any additional detours, extending the drivetrain to the rear axle [4].

The rear-axle loading required for rear-drive vehicles leads to a front axle mounted well forward in comparison to the engine/transmission unit. Still, for rear-wheel drive, it is seldom possible to achieve

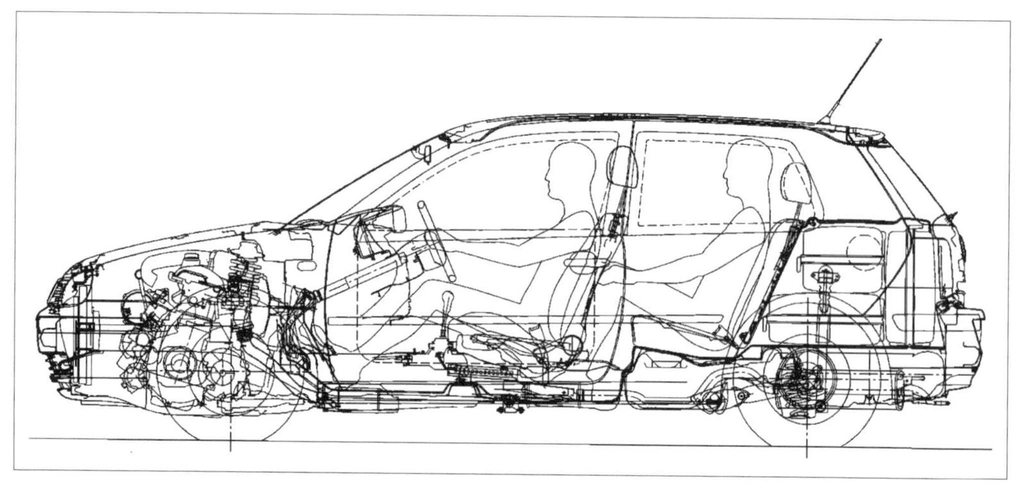

Fig. 4.2-6 Volkswagen Golf (transverse front engine, front-wheel drive).

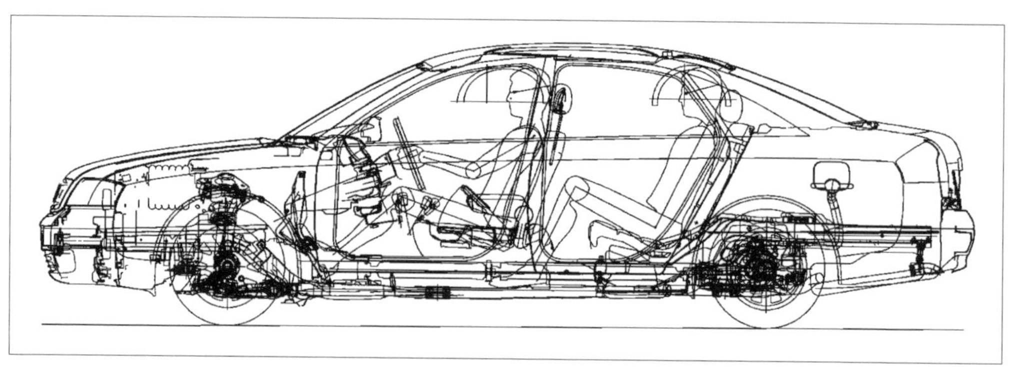

Fig. 4.2-7 Audi A6 (longitudinal front engine, front-wheel drive).

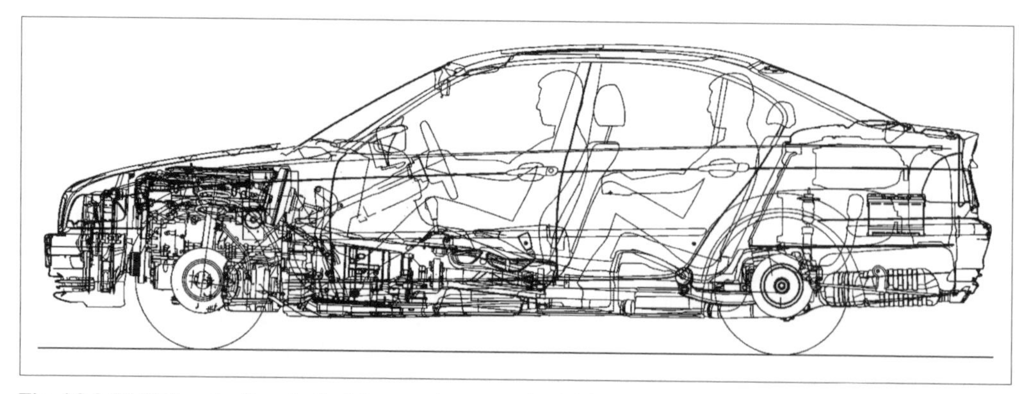

Fig. 4.2-8 BMW 3-series (longitudinal front engine, rear-wheel drive).

more than 50% rear weight distribution (in unladen condition).

A comprehensive comparison of advantages and disadvantages of front-wheel drive may be found in [5]. Development of representative examples of the "classical layout"—front engine and rear-wheel drive—is examined in [6] and [7].

Front engine examples

Transversely mounted front engine, front-wheel drive: Volkswagen Golf; see Fig. 4.2-6.

Longitudinally mounted front engine, front-wheel drive: Audi A6; see Fig. 4.2-7.

Longitudinally mounted front engine, rear-wheel drive: BMW 3-series; see Fig. 4.2-8.

4.2.3.2.2 Rear-mounted engine

Characteristics

Rear-mounted engines were more commonly employed in the past (e.g., VW Beetle, Renault, Fiat) than they are today. Rear-mounted installation is analogous to front-wheel drive designs, in that engine, transmission, and, in this case, rear axle are installed as a preassembled unit at the rear of the car. The transmission is mounted ahead of the longitudinal engine. A modern vehicle with this configuration is the Porsche 911 Carrera [8].

Advantages

Engine location behind the rear axle results in high rearward weight distribution (>60%) and therefore outstanding traction, which increases under acceleration and in ascending grades, and traction is still very high with a loaded vehicle interior or front trunk. No heat loading of the interior occurs through thermal radiation from the powerplant. It has a flat tunnel (no driveshafts or exhaust pipes). Longitudinal installation of the powerplant permits simple realization of all-wheel drive to the front axle.

Disadvantages

High rear-axle loading necessitates sophisticated suspension concepts (above all for the rear axle) to achieve good vehicle dynamics. Long lines result in the case of water cooling with front-mounted radiators as well as for heating and air conditioning systems. Body variability at the rear of the car is very strongly restricted by powerplant space requirements. Wagon variants offering competitive space are not possible.

Rear engine example

Rear-mounted engine, rear-wheel drive: Porsche 911 Carrera [8], Fig. 4.2-9.

4.2.3.2.3 Mid-mounted engine

Characteristics

A classic sports car configuration with engine located ahead of the rear axle. The engine may be mounted longitudinally (transmission behind engine) or transversely (analogous to transverse front engine). Due to the space requirements of the engine-transmission unit, only two-seat designs are practical. Modern monoposto (single-seat) race cars (e.g., Formula 1) are exclusively mid-engined.

Advantages

Relatively high rear weight distribution (>52%) due to the engine location ahead of the rear axle provides very good traction characteristics, traction that increases under acceleration and in ascending grades and neutral with increasing vehicle load. This is a vehicle concept with optimum vehicle dynamic potential thanks to balanced weight distribution. Low heat loading of the interior is due to thermal radiation from the powerplant. Generally a front luggage compartment is realizable. An additional trunk at the rear is possible; and the tunnel is flat (no driveshafts or exhaust pipes).

Disadvantages

For water cooling, front-mounted radiators and heat exchangers require long lines for coolant, heating, and air conditioning. Body variability at the rear and interior is strongly restricted by powerplant space requirements.

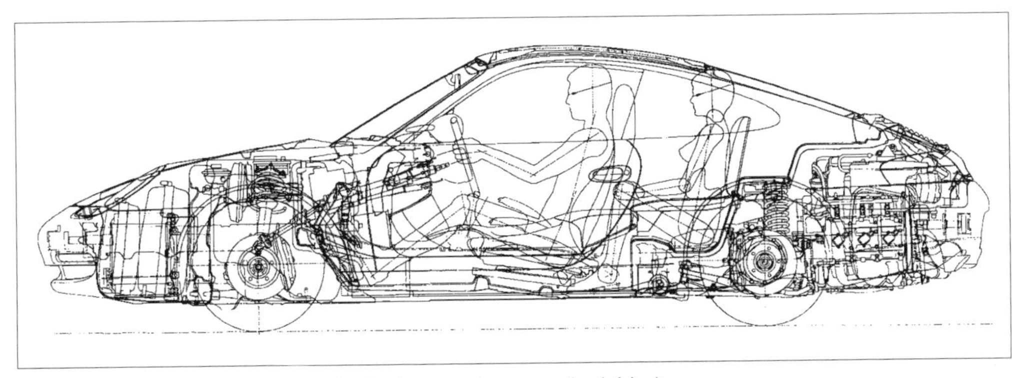

Fig. 4.2-9 Porsche 911 Carrera (longitudinal rear engine, rear-wheel drive).

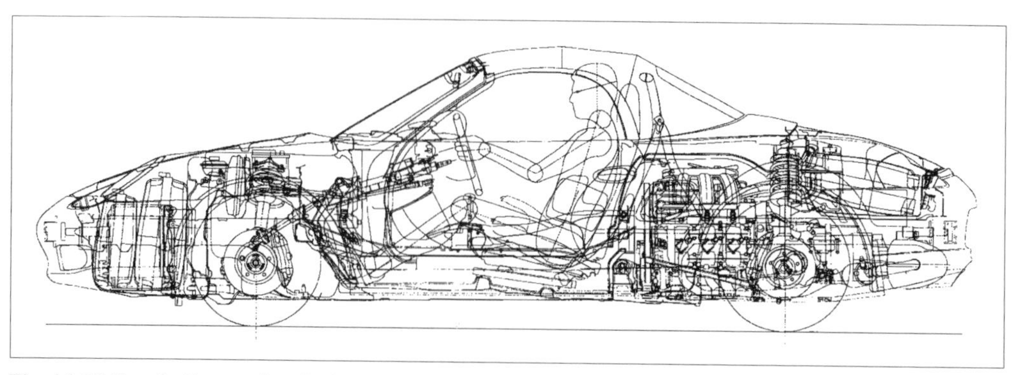

Fig. 4.2-10 Porsche Boxster (longitudinal mid-engine, rear-wheel drive).

Therefore, this configuration is almost exclusively limited to two-seat vehicles (four-seat capability leads to a very large wheelbase). With longitudinal mounting, placing the transmission behind the engine requires long shifter cables and makes all-wheel drive impossible. Even with a transversely mounted powerplant, this would be very difficult to realize.

Mid-engine example

Longitudinal mid-mounted engine, rear-wheel drive: Porsche Boxster [9], Fig. 4.2-10.

4.2.3.2.4 Special variations on engine location

Even today, efforts to differentiate new products within an existing market segment give rise to new vehicle concepts. These attempt to achieve market advantages by means of their unique, specific product features.

Front engine/underfloor concept

By spatially superimposing the major length-determinant elements such as powerplant and passenger compartment, it is possible to realize an extremely short overall vehicle length in comparison to market competitors. However, total required technical space as well as useful volume cannot be reduced; therefore, such a concept results in greater vehicle height as a typical characteristic of this class. It is possible to achieve a very compact and space-saving engine location by consistent development of a very flat engine and transmission that fit within the available space between firewall and front axle. Deflection of the powertrain below the floorpan provides the necessary deformation length required for frontal crash protection.

Front engine/underfloor concept example

Front transverse underfloor engine, front-wheel drive: Mercedes-Benz A-Class [10], see Fig. 4.2-11.

Rear engine/underfloor concept

This concept uses an unconventional solution to address the frequent urban problem of inadequate parking

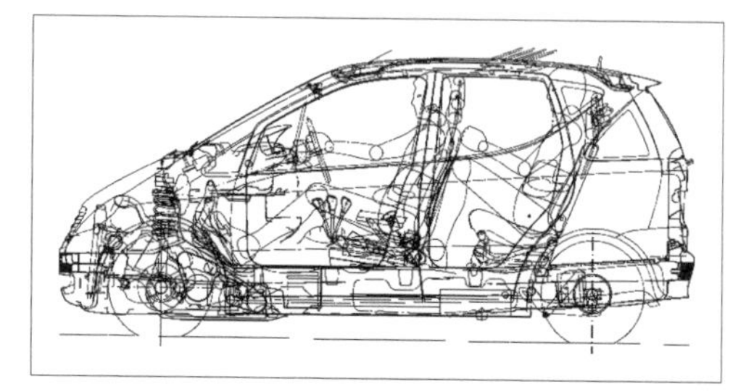

Fig. 4.2-11 Mercedes-Benz A-Class (transverse front underfloor engine, front-wheel drive).

space. The transverse engine, transmission, and rear axle are combined into a very compact assembly. Elimination of any rear seating provision and the resulting strictly two-seat configuration permits acceptable interior space for two occupants, despite an extremely short overall vehicle length. But here, too, the short vehicle length could only be achieved by "building upward," giving a greater vehicle height than typical of the class.

Rear engine/underfloor concept example

Rear transverse underfloor engine, rear-wheel drive: Smart, see Fig. 4.2-12.

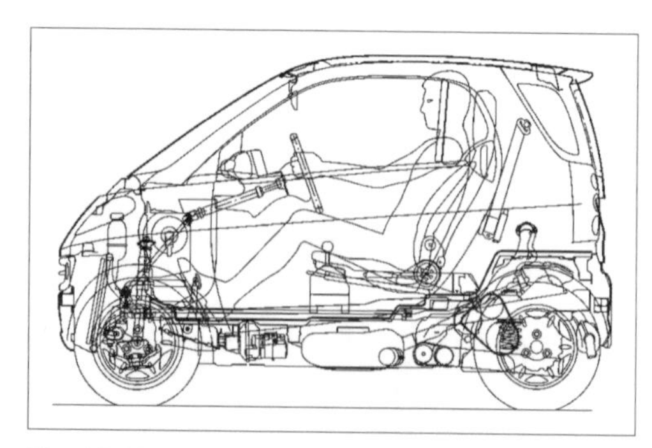

Fig. 4.2-12 Smart (transverse underfloor rear engine, rear-wheel drive).

4.2.3.2.5 Comparison of vehicle concepts

The major advantages and disadvantages which characterize a vehicle concept are outlined in the following table (Table 4.2-1), ranging from ++ for very good or very suitable, to −− for very poor or unsuitable. The overall evaluation of a concept is dependent on its intended application.

An attempt to influence the vehicle concept predominantly by means of weight and fuel economy is described in [11]. The various major vehicle specialties such as body, equipment, suspension, and drivetrain have a direct or indirect effect on vehicle weight [11, 12]. Special attention should be given to the problem of mutual influences of various vehicle component groups and designs and their effects on weight [12, 13].

4.2.3.3 Vehicle package layout domains

4.2.3.3.1 Interior

The majority of present-day vehicles have four or five seating positions, of which the fifth position, generally a central position on the rear seat, makes concessions in terms of comfort (seat width and cushion design) and safety (often with only two-point safety belts).

Increasingly, special configurations are coming into use. So-called vans, with emphasis on second or possibly third seating rows in the form of individual seats in place of bench seats, offer seats which are individually removable or individually mountable by means of suitable attachment concepts. Generally, these also require design of a flat vehicle floor, whose utility is not interrupted by seat consoles or a central tunnel. A disadvantage is the resulting increase in vehicle height.

Seats are designed in accordance with ergonomic criteria. These are based on an "anthropometric atlas" which provides insight into body proportions of the human population. Along with the worldwide range of body dimensions, consideration must be given to the increase in body size with successive generations. This leads to the result that new vehicles must above all offer considerable increases in leg- and headroom.

The limits of design are generally defined to encompass range from the so-called 5% female (only 5% of the female population is shorter), to the 95% male (only 5% of the male population is taller). This range must be considered in meeting various regulatory requirements, such as, for example, the achievable forward sight angles (ground visibility, traffic signals). The requirements for children and adolescents are met by fitted, sometimes integrated child safety seats and restraint systems.

Until recently, design layouts were done manually on drawings of the overall vehicle with the aid of scaled templates. Today, in the age of CAD development, this method has been replaced by appropriate 2-D modeling. Today's state of the art is the so-called "RAMSIS" system [14], a three-dimensional CAD model of the human form, developed under contract to the German automobile industry. RAMSIS permits ergonomically correct workplace design with simultaneous evaluation of achieved comfort criteria. The goal of such investigations is a low-fatigue seating position with ergonomic pedal and steering wheel configurations. In addition, the driver's reach zones and rear-seat space conditions are defined (see Section 6.4.1).

Despite all such nearly perfect theoretical layouts, automotive developers check design data by means of a seating buck. This is a reconfigurable mockup in wood, possibly with the assistance of shaped pieces, permitting verification or, if needed, correction of important

Table 4.2-1 Comparison of vehicle concepts

	Front engine Front drive	Front engine Rear drive	Mid engine Rear drive	Rear engine Rear drive
Traction, unladen[a]	+	−	+	++
Traction, laden[a]	−	+	+	+
Suspension conceptual demands	+	0	−	−−
Interior space	++	++	−−	0
Luggage space	++	+	0	−−
Rear bodywork variability	++	++	−	−−
Vehicle length requirement[b]	++	0	− (0)	0
Body structure loading in frontal impact	++	++	−−	−−
Thermal effects on interior	−	−−	0	+
Noise effects on interior	+	+	0	+
Suitability for all-wheel drive[c]	+ (++)	−	−−	++
Overall weight	++	0	+	+
Hose and line lengths	++	++	−	−−
Manufacturing costs	++	+	+	+

[a] Exact evaluation requires consideration of grade, traction coefficient, vehicle loading.
[b] (0) For transverse instead of longitudinal mid-engine installations.
[c] (++) For longitudinal instead of transverse front engine installations.

interior dimensions. This static—nonmoving, that is—seating buck is gradually replaced by a dynamic seating buck. This permits construction on an appropriate floorpan with drivetrain and a partial representation of exterior styling to enable realistic evaluation of critical visibility and seating relationships while the vehicle is in motion, even if only at low speeds.

4.2.3.3.2 Luggage compartment

An important sales criterion for customers, along with interior space, is the amount of available luggage space. Under VDA norms, the available volume is specified in liters (as determined using "luggage" consisting of rectilinear boxes with specifically defined dimensions) and maximum cargo height. The available variability usually found on modern vehicle concepts permits reporting of various achievable volumes. These often include additional information such as the linear dimensions of the largest stowable object for wagons.

4.2.3.3.3 Body structure

The possible arrangements of vehicle components and interior space as well as overall vehicle dimensions are strongly influenced by space requirements of the load-carrying body structure.

The main tasks of this load-carrying structure are:

- Transfer of operating forces to the body
- Realization of high body stiffnesses (bending, torsion, etc.)
- Assurance of the required crash deformation zones and required crash behavior

Above all, crash requirements imposed on the load-carrying structure have expanded greatly in recent decades. A variety of different tests imposed by regulatory agencies and consumer organizations must be met; export to the American market nearly doubles the number of tests required and in part results in conflicting structural design goals. Offset frontal crash tests impose high demands on the body structure and its development (see Section 6.5). Due to their great significance, the cross sections of load-carrying members and their location and nodes represent major boundary conditions for the package concept, affecting overall vehicle design.

4.2.3.3.4 Engine compartment

Increasing demands for reduced vehicle emissions, on the one hand, and efficiency and reduced maintenance for combustion engines, on the other, have led to ever greater complexity within engine compartments [15]. Increased product content such as ABS systems, headlamp washer systems, and so on make matters even more complicated.

Above all, these demanding packaging conditions, as well as engine encapsulation—most often employed to reduce noise emissions of diesel engines—represent thermal challenges in which limiting temperatures of systems and components must be considered (see Section 3.3). Specific ventilation considerations and airflow through the engine compartment must be taken into account, including exhaust air from the air conditioning condenser, which then flows through the engine coolant radiator located immediately behind and continues on to ventilate the engine compartment [4].

One major focus of engine compartment design is optimization of crash behavior. This requires minimization of so-called "compressed lengths" in the engine compartment. Along with the need for the shortest possible length of the engine itself, designers must take care, by means of suitable location or deliberate failure points, that auxiliary devices do not increase this compressed length.

The space demands of the air filter housing in the engine compartment play an important role, especially in the case of large-displacement engines. As a rule of thumb, the required volume may be estimated at about 1 L per 10 kW of engine output; the actual required volume will be affected by the engine concept and the specific oscillatory behavior of its induction airflow.

4.2.3.3.5 Exhaust system, drivetrain, and underbody

The drivetrain, exhaust system, lines and hoses, fuel tank, and rear axle determine the design of the vehicle underbody.

The exhaust treatment system represents the most significant part of the exhaust system; to reach its operating temperature as quickly as possible after a cold start, at least certain parts of the system must be located as close to the engine as possible. Exhaust piping with minimized throttling losses (for V and horizontally opposed engines, these generally employ dual exhausts [8, 9]) and sufficient muffler (silencer) volume are required to achieve low exhaust back pressure and thereby higher engine efficiency and output. Mufflers represent the largest volume requirement in the tunnel and rear body area. To meet present-day regulations for traffic noise limits of 74 dB(A), acceptable back pressures imply roughly the following overall silencer volumes (including catalytic converter):

- Normally aspirated engines: approx. 0.20–0.23 L/kW
- Turbocharged engines: approx. 0.16 L/kW

The lower volume requirement for turbocharged engines is the result of noise reduction by the turbocharger itself. The actual requirement is additionally dependent on the engine concept (e.g., inline four-cylinder or six-cylinder horizontally opposed engine), whose specific gas flow profile might demand a greater or lesser volume.

The drivetrain of the "classic" front engine/rear drive layout is characterized by the transmission located at the front of the tunnel and by the final drive (rear axle/differential). These components are joined by a driveshaft, whose diameter must be optimized for torsional vibration and, due to elasticity considerations, must not exceed a certain length. To stay below this maximum length, the driveshaft may be divided by a central bearing or supplemented by a "long-necked final drive," which moves the rear driveshaft flange forward in the tunnel.

To reduce aerodynamic drag on more expensive vehicles or vehicles capable of high speeds, an underbody cover (cladding) may be applied in order to achieve a relatively flat underbody. Side effects of such elements include better protection for lines and hoses but generally somewhat worsened thermal conditions.

4.2.3.3.6 Fuel tank, lines, and spare tire

The tank location is strongly influenced by necessary crash protection measures. Characteristically, for front-engined vehicles (for example, see Fig. 3.2-1), this is placed in a protected area ahead of and near the rear axle. A goal conflict arises in the case of rear-wheel-drive vehicles, in which a tank location behind the axle would be preferable for achieving higher rear axle loading. Modern tank location in the crash-protected area and customer desires for extended cargo carrying capability (i.e., fold-down seats or ski pass-through) result in some tank variations with reduced capacity [6].

For mid-engined and rear-engined vehicles manufactured by Porsche (see Fig. 4.2-9 and 4.2-10), the tank is located in a crash-protected area behind the front axle.

Production- and safety-relevant criteria determine the line package for fuel and hydraulic lines and electrical wiring. Crash-safe, noncrossover routing must be ensured by secure, unambiguous rapid connections which guarantee high production quality. Minimization of the number of fuel line connections, combined with materials permitting the lowest possible fuel diffusion, are necessary for reduction of hydrocarbon emissions. For the same reason, emissions (charcoal) canisters, with volumes ranging from 1.5 to about 5 L depending on fuel tank volume and refueling emissions system, are used for intermediate storage of fuel vapor released during vehicle operation (e.g., fuel tank warming) and, in the United States, also during refueling (see Section 7.6).

The spare tire constitutes a major space-consuming element. For front-engined vehicles, this is located at the rear of the car, consuming anywhere from 50 L (small, high-pressure "space saver" temporary spare) to about 80 L (fully functional replacement wheel and tire). For this reason it is often replaced by space-saving tire repair and inflation systems.

An example of complex underbody packaging is represented by the all-wheel-drive Porsche 911 Carrera 4 (Fig. 4.2-13). Its front axle is driven by a driveshaft.

4.2.3.3.7 Dimensional comparison of various vehicle classes

Table 4.2-2 compares several significant exterior and interior dimensions of various representative vehicle classes (all values approximate).

Analysis of the data in that table clearly shows that vehicle classes are predominantly defined by:

- Front shoulder room (W3)
- Rear legroom (L51)
- Trunk volume

The other dimensions are primarily the result of basic ergonomic requirements which must be met within any vehicle class.

Compared to other classes, sport utility vehicles (SUVs, off-road vehicles) are primarily defined by their considerably greater ground clearance (at least 200 mm at normal ride height, compared to 130 to 150 mm for passenger cars) and greatly increased approach and departure angles (ramp angles; up to 40° front/30° rear).

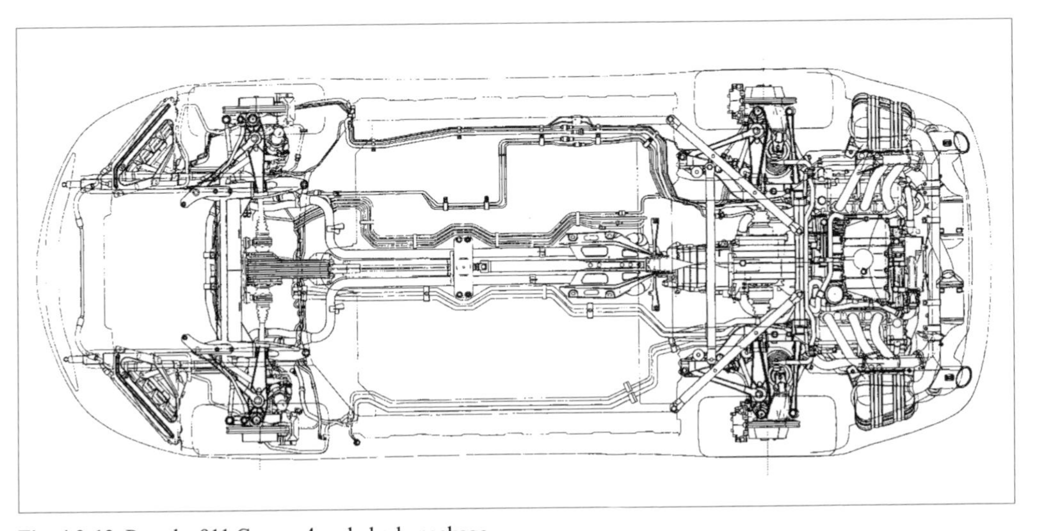

Fig. 4.2-13 Porsche 911 Carrera 4 underbody package.

Table 4.2-2 Dimensional comparison of various vehicle classes

Criterion	Vehicle class					
	Compact	Lower mid-range	Mid-range	Upper mid-range	Upper range	Vans
Exterior dimensions (mm)						
Length (L103)[a]	3,600–3,800	3,800–4,400	4,300–4,700	4,300–4,700	4,700–5,100	4,500–4,800
Wheelbase (L101)	2,350–2500	2,400–2,700	2,500–2,700	2,500–2,700	2,700–3,000	2,700–3,000
Width (W103)	1,550–1,650	1,670–1,740	1,670–1,770	1,670–1,770	1,800–1,900	1,750–1,900
Height (H100)	1,350–1,480	1,330–1,440	1,360–1,430	1,360–1,430	1,400–1,500	1,650–1,800
Interior dimensions (mm)						
Legroom, front (L34)	960–1,080	970–1,080	1,000–1,100	1,000–1,100	1,000–1,100	970–1,080
Headroom, front (H61)	920–1,000	940–1,010	950–1,010	950–1,010	980–1,020	1,000–1,050
Shoulder room, front (W3)	1,280–1,360	1,340–1,440	1,340–1,460	1,340–1,460	1,450–1,500	1,500–1,650
Seat spacing front-rear (L50)	680–760	670–790	730–830	730–830	840–950	850–900
Legroom, rear (L51)	730–920	760–880	750–920	750–920	900–1,000	800–900
Headroom, rear (H63)	900–970	900–980	910–980	910–980	950–990	950–1,000
Luggage capacity (liters)						
Trunk volume	200–460	240–550	330–550	330–550	500–600	250–2,500
Representative vehicles (2000)	VW Polo 3	VW Golf 4	Audi A4	BMW 5 Series	Mercedes S Class	VW Sharan

[a] Dimension designations in parentheses according to ECIE practice. Source [16]. See also SAE J1100.

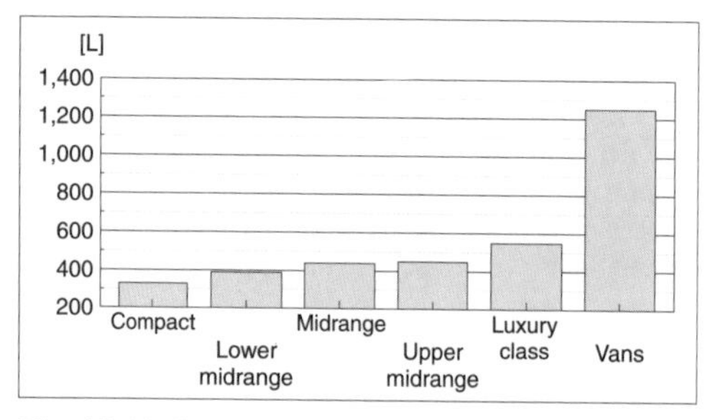

Fig. 4.2-14 Comparison of vehicle classes by trunk volume (average values, not including station wagons; vans not including third seat row).

Various forms within the SUV class are differentiated by the same main criteria as passenger cars.

4.2.3.4 Regulatory requirements

In the conception of a vehicle, various country-specific regulatory requirements must be met. (See Section 2.2).

4.2.3.4.1 Direct influences on the vehicle concept

Detailed regulatory requirements have been established for such features as:

- Bumper height
- Installation and location of lamps
- Wiper fields
- Driver's field of view
- Interior dimensions (e.g., pedal cluster)
- License plate location

These exert a direct influence on the vehicle design because they must be considered as firmly established dimensional requirements.

4.2.3.4.2 Indirect influences on the vehicle concept

Not only do regulatory requirements affect vehicle concept. Additionally, demands from consumer organizations, the general state of the engineer's art, and in-house requirements influence vehicle concepts, above all in terms of safety standards.

These myriad requirements result in design characteristics of which the following examples are but a tiny fraction:

- Assurance of adequate crash deformation zones
- A vehicle package that permits body shell structures suitable for achieving optimum load paths
- A vehicle package which locates expensive components in relatively crash-protected areas (for purposes of insurance costs)
- Function-optimized location of catalytic converters
- Consideration and optimum location of safety-related equipment such as
  - Airbags (front, side, roof)
  - Safety belts (belt fields)
  - Kneebar
- Space consideration for activated charcoal canister

4.2.3.5 Customer service and manufacturing requirements

4.2.3.5.1 Production and modularization

An assembly process based on a rational manufacturing concept whose goals are the highest production quality with the shortest manufacturing time affects the vehicle design: for example, the direction of installation (e.g., is the engine installed from above, below, or the front?) must be taken into consideration. Just as significant as physically large components, such as the powertrain, are installed modules whose possible installation in the vehicle determines their feasibility.

The trend toward greater modularization of the vehicle—that is, creation of larger preassembled component groups—results from the following:

- Freeing the vehicle assembly line of labor-intensive processes (resulting in, e.g., reduction in manufacturing time, improved quality)
- Simplified procedures on the assembly line
- Creation of variants and their pretesting away from the vehicle assembly line
- Reduction in losses from process cycle times
- Possibility for outsourcing larger and more complex items for various reasons (manufacturing costs, production know-how, quality, logistics, etc.)

Examples of larger modules include:

- Front end (e.g., front bumpers with radiator, headlamps, front valence crossmember, various other installed components, etc.)
- Cockpit (assembly crossmember, for example, including pedal cluster, steering column, heating system, airbag, instruments, electronics, trim, etc.)
- Front and rear axles
- Drivetrain, possibly including axle(s)

Fig. 4.2-15 shows the assembly sequence of a cockpit (source: Porsche AG); its final assembly state is shown in conjunction with a "handling traverse" with whose assistance the preassembled cockpit is inserted through the door opening and installed.

4.2.3.5.2 Customer service

Just as the design targets originating from the production sector are in place from the very beginning of vehicle development, similar requirements from the customer service sector also exert their influence on the development process.

Demands influencing the vehicle concept are primarily in the form of maximum permissible times for replacement of vehicle components (examples: parts of the exhaust system, heater blower motor, oil filter, starter, etc.) as well as minimization of repair costs. During concept development, these requirements give rise to removal and installation tests of components or assemblies and theoretical determination of time requirements.

The sole purpose of this optimization is reduction of service, repair, and insurance costs for the customer

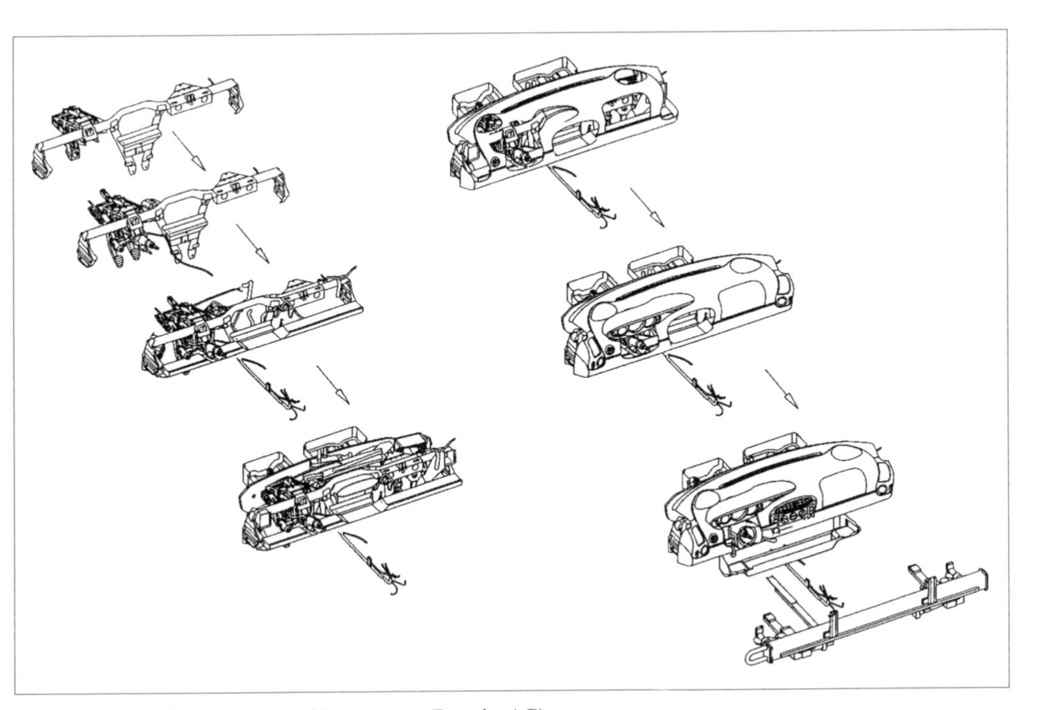

Fig. 4.2-15 Vehicle cockpit assembly sequence (Porsche AG).

and thereby increased market competitiveness of the product.

4.2.3.6 Strategic influences

4.2.3.6.1 Vehicle platform

One of the important influences on the vehicle concept is exerted by defined development of a so-called "vehicle platform." Although the definition of a platform varies between automobile manufacturers, important features include:

- Significant parts of the floorpan, generally load-carrying structures including the front firewall and, at the rear, longitudinal and often package variables (e.g., tank location or spare tire for sedan and wagon)
- Drivetrain including cooling modules
- Suspension and steering
- Cockpit design, not including trim
- Due to the above features, generally also the seating position, or at least the front seating position

As different vehicle types are represented on a common platform, the overall body structure and therefore the visible shapes presented to customers (interior as well as exterior) are not included in the contents of a platform.

The goal of platform development is to achieve a maximum of different vehicle types and variations, sharing a common engineering basis, with minimum overall development effort and investment costs. Additionally, purchasing conditions and production costs are improved for high-volume platform components. This strategy is employed above all by automobile concerns which include several different marques, but also by smaller manufacturers seeking more cost-effective manufacturing of smaller vehicle production volumes [4, 15].

4.2.3.6.2 Vehicle variants

In order to serve as many market segments and reach as high a market penetration as possible with a single vehicle concept, modern vehicles are generally produced in a variety of body variations. For example, a concept may include two- and four-door sedans, wagon, and convertible.

In the past, variants were often developed only well after introduction of the first model of a given platform; today, variant development runs parallel or with only a slight delay. Reasons include:

- Possible consideration of variants in design and engineering of the base vehicle
- Component design taking into account variants
- Synergistic effects in development of different variants
- Greater likelihood of using common components in variants
- Lower overall development and investment costs as well as shorter development times
- Improved integration of variants in the production process

In order to realize benefits, vehicle variants must be considered on a conceptual basis. Different variants imply design differences at the rear of the vehicle, and these in turn demand corresponding body structures in keeping with the variants' specific requirements. In concept and packaging, space requirements such as convertible top stowage or suspensions (e.g., strut towers in view of wagon variants) must be taken into consideration and optimized.

4.2.3.6.3 Modular construction

In vehicle design, "modular construction" is taken to mean application of components or assemblies in a variety of vehicle lines and types. Using common parts in several different vehicles results in higher production volume of these components and therefore lower component costs. Use of these components for a given project is predefined in the target catalog, the engineering product description, or the compiled design goals and are to be considered accordingly in the vehicle concept and package.

"Major" modular components include:

- Powertrain (engines, transmissions)
- Suspension components or entire suspension assemblies (axles)
- Steering column
- Drivetrain components (driveshaft, differential)
- Radiator
- Auxiliary devices (alternator, air conditioning compressor)
- Heating and climate system
- Emergency or spare tire

In addition, many smaller components and functional groups, including locks, electrical components, and so forth, are installed as modular elements but do not affect the vehicle concept.

The following illustration, Fig. 4.2-16, schematically shows an example of a platform structure with three derivative vehicle types and their variations.

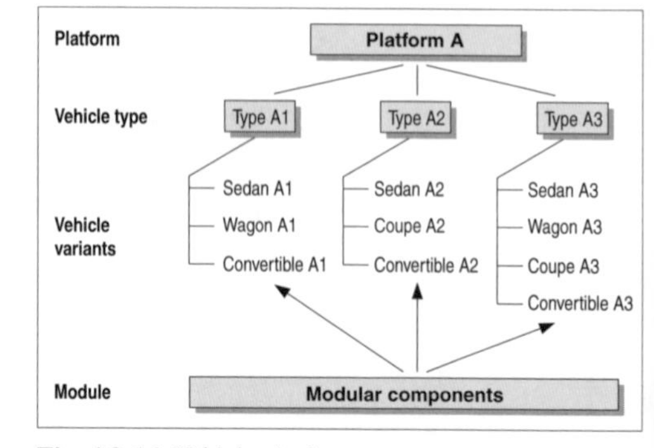

Fig. 4.2-16 Vehicle platform structure.

4.2.4 Vehicle development in the future

The recent past has been distinguished by the shift from manual design efforts to CAD technology.

In terms of the vehicle conception process, the future will, above all, be marked by:

- Further integration of software tools in design, styling, planning, and simulation
- Increased shift of testing and evaluation activities to the earlier, theoretical phase, with the help of simulation technology ("digital mockup") [17]
- Replacement of laboratory experimental constructs by corresponding studies in further-developed CAD processes ("virtual laboratory")
- The possibility of conducting "virtual product clinics" (in conjunction with computer aided styling [CAS]) to evaluate vehicle and styling features at a very early (and therefore cost-effective) phase of development

These improvements to the development process, steps along the path to "virtual product development," will result in a reduction in the number of development stages and prototypes and a further reduction in vehicle development time and cost, all while improving quality [2, 17, 18]. These trends will lead to more complex tasks for and increased importance of the "vehicle concepts and packaging" sector.

While these efforts show a dramatic change in the theoretical possibilities for vehicle development, we must not forget that simple laws of physics and especially mechanics, with their attendant influence on the future product, must be observed, especially in the vehicle conception phase. Some examples include:

- Estimation of vehicle weight and weight distribution, which leads to wheel and tire selection (load capacity)
- Performance calculation
- Rough estimation of fuel economy: the desired vehicle range results in required fuel tank capacity
- Determination of the required crash deformation dimensions from the permissible occupant acceleration limits

References

[1] Braess, H.-H. "Das Automobil im Spannungsfeld zwischen Wunsch, Wissenschaft und Wirklichkeit." 2. Stuttgarter Symposium Kraftfahrwesen und Verbrennungsmotoren 1997, pp. 786–799.

[2] Spur, G., and F.L. Krause. *Das virtuelle Produkt—Management der CAD-Technik.* Carl Hanser Verlag, Munich and Vienna, 1997.

[3] ECIE Übersicht Maßdefinitionen.

[4] "Der neue Audi A6." ATZ/MTZ special edition to ATZ and MTZ, 3/97.

[5] "50 Jahre Frontantrieb im Serienautomobilbau." VDI-Bericht 418, 1981.

[6] "Die neue S-Klasse." ATZ/MTZ special edition to ATZ and MTZ, 10/98.

[7] "Der neue 3er-BMW." ATZ/MTZ special edition to ATZ and MTZ, 5/98.

[8] "Der neue Porsche 911 Carrera." ATZ/MTZ special edition to ATZ and MTZ, 12/97.

[9] "Der Porsche Boxster." ATZ/MTZ special issue, 1996.

[10] "Die neue A-Klasse von Daimler-Benz." ATZ/MTZ special edition to ATZ and MTZ, 10/97.

[11] Braess, H.-H. "Zur gegenseitigen Abhängigkeit der Personenwagen-Auslegungsparameter Höhe, Länge und Gewicht." ATZ 81,1979, p. 9.

[12] Piëch, F. "3 Liter/100 km im Jahr 2000?" ATZ 94, 1992, p. 1.

[13] Braess, H.-H. "Negative Gewichtsspirale." ATZ 101, 1999, p. 1.

[14] RAMSIS, Forschungsvereinigung Automobiltechnik e.V., FAT-Schriftenreihe No. 123.

[15] Indra, F. "Package: Die Herausforderung für den Motorenentwickler." 19. Internationales Wiener Motorensymposium, Fortschrittsbericht VDI Reihe 12 No. 348, VDI Verlag, 1998.

[16] Porsche AG, Bereich Fahrzeugkonzepte und Package.

[17] Schweer, G., O. Tegel, M. Terlinden., and P. Zimmermann. "Digital Mock-Up und Virtual Reality—Wege zur innovativen Produktentwicklung bei Volkswagen." VDI-Jahrbuch 1998 (FVT), Systemengineering der Kfz-Technik.

[18] Heer, M., and D. Crede. "Berechnungskonzept zur virtuellen Konzeptabsicherung am Beispiel des Audi TT." VDI-Bericht No. 1398, 1998.

4.3 Innovative propulsion systems

4.3.1 Electric drive systems

Electric drives are employed in various transportation systems. The following section will examine electric drives for electric vehicles in the sense of electrically powered nonrail vehicles for transportation of persons and goods. These vehicles are propelled by electric motors and draw their power solely from an onboard traction battery.

Electric vehicles appear to have advantages in that they do not release pollutants at the point of use—no exhaust emissions, no release of fuels in refueling or fuel storage. This is independent of the age and condition of the vehicle. However, because electrical energy used to propel electric vehicles is generated in stationary powerplants, these generators may themselves produce emissions. The ability to reduce regional or global emissions by means of electric vehicles is therefore ultimately dependent on pollutant emissions by the generating facility.

One advantage is that electrical power generation provides an extensive, existing infrastructure—or one which may be quickly emplaced—for recharging of electric vehicles. Electric motors also exhibit high efficiency and offer the possibility of energy recovery by means of regenerative braking. And, electric vehicles have an added advantage in that they are very quiet at low speeds.

On the other hand, limited range presents a disadvantage for users of electric vehicles. Longer battery recharging times place a restriction on electric vehicle availability, except in specific cases which employ rapid charging or battery exchange.

Electric vehicles have been in service since the end of the nineteenth century, even before internal-combustion-powered vehicles. In the early twentieth century, electric propulsion was rapidly replaced by internal combustion engines. New impetus for development of modern electric vehicles arose from the oil crisis of the 1970s and California's ZEV (Zero Emissions Vehicle) regulations. In their original (1990) form, these regulations required emission-free vehicles for California beginning with the 1998 model year. Technically, this could only be achieved with battery-electric vehicles or hydrogen fuel-cell vehicles. In the years that followed, the California law was changed several times [1] and remains under discussion at this writing. In their most recent form (January 2001), the regulations permit other vehicles with very low emissions alongside pure "ZEV vehicles." ZEV regulations encompass the following vehicle categories:

ZEVs

Absolutely emission-free vehicles such as electric vehicles and hydrogen fuel-cell vehicles. Among electric vehicles, further distinction is made according to range and maximum speed.

PZEVs

"Partial Zero Emissions Vehicles" with combustion engines, extremely low pollutant emissions, and long-term emissions stability.

AT-PZEVs

These "Advanced Technology PZEVs" exhibit the same low emissions as PZEVs and additionally incorporate electric drive. In practice, these are extremely low emissions hybrid vehicles and possibly also fuel-cell vehicles with reformers.

These complex regulations envision various percentages of ZEVs, PZEVs, and AT-PZEVs in a manufacturer's fleet as a function of that manufacturer's market share. Beginning in 2003, large manufacturers are required to produce 10% of their offerings as "ZEV equivalent" vehicles for the California market, rising to 16% in 2018. Automakers are given a certain degree of flexibility with regard to the mix of ZEVs, PZEVs, and AT-PZEVs. Nevertheless, larger manufacturers must provide at least 2% ZEVs beginning in 2003.

4.3.1.1 Drive systems for electric vehicles

The overall electric vehicle drive system [2] encompasses:

- Traction battery with battery management and, typically, a charging device
- Electric motor with electronic controls (inverter) and cooling
- Transmission, if needed, including differential
- Power transmission to drive wheels

Furthermore, auxiliary devices such as power steering, power brakes, and the heating and climate control system must be adapted to electric drive. Rechargeable traction batteries require charging devices in the form of stand-alone (stationary) or onboard rechargers.

For power transmission to the drive wheels, the drivetrain may assume one of several configurations (Fig. 4.3-1) [3]. Usually, front or rear wheels are driven by a central electrical machine coupled to a generally required transmission and differential. Alternatively, tandem drive systems with two electric motors or electric wheel hub motors may be employed.

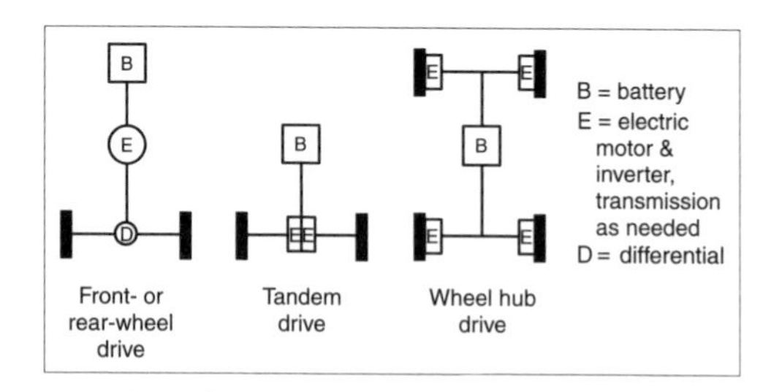

Fig. 4.3-1 Electric vehicle drivetrain configurations.

4.3.1.2 Electric motors for electric vehicles

Motors for use in vehicles must operate across a wide speed and torque range. Electric motors are nearly ideal prime movers for vehicles: they are quiet, operate at high efficiency, and exhibit attractive torque/speed characteristics. Their maximum torque is available at very low speeds, and even at high speeds they are still capable of producing adequate torque at nearly constant power across the speed range. Fig. 4.3-2 shows power and torque curves of an inverter-fed asynchronous motor for an electric vehicle. As electric motors may be operated in an overload condition for brief periods, additional torque is available for acceleration and short climbing grades. In electric vehicles, in contrast to combustion engines, a multistage transmission may be dispensed with. Usually, a simple single-ratio transmission is sufficient.

Various configurations of electric motors may be employed in electric vehicles:

- Direct-current (DC) motors
  - Series-wound DC motors
  - Shunt-wound DC motors

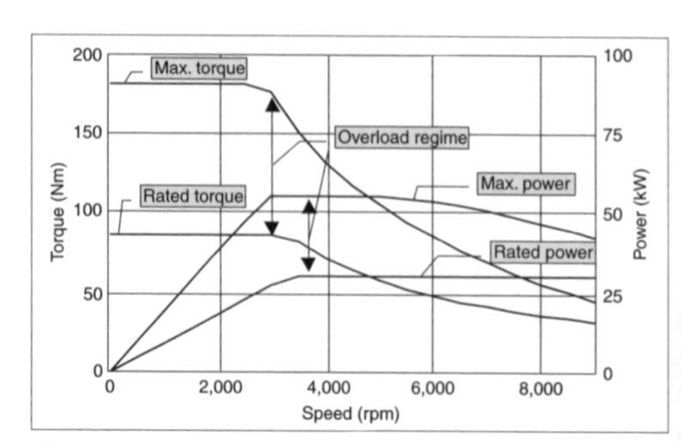

Fig. 4.3-2 Power and torque curves of an inverter-fed asynchronous machine.

- Alternating current (AC) motors
  - Asynchronous motors
  - Synchronous motors
    - Permanently excited synchronous motors
    - Electrically excited synchronous motors
- Special motors
  - Brushless DC motors
  - Transverse flux motors
  - Switched reluctance motors

Criteria for motor selection [4] include compact dimensions, low weight (high power density), high efficiency, simple controllability across a large speed and torque range, overload capability, low noise generation, low cost, and low maintenance requirements.

Direct-current motors

These motors were once the most common choice for electric vehicles. They are powered by direct current from the traction battery passing through a chopper. A principal distinction is made between series-wound and shunt-wound machines.

For practical purposes, among DC motors only the shunt-wound variety finds application in electric vehicles. In this configuration, magnetic flux is generated by exciter windings in the stator. Direct current is applied to the rotor by means of carbon brushes (commutator). Rotation of the rotor causes current in the armature to be polarized by the commutator. Armature and field excitation (chopper) are fed directly from the battery and serve to control armature and field winding current. As a class, direct-current motors have attained a high state of engineering refinement. Thanks to relatively simple motor control, these motors are very attractively priced. Technical drawbacks include the commutator and its brushes, which require maintenance. The maximum motor speed is limited to about 7,000 rpm by the commutator circumferential velocity. Also, efficiency and power density of direct-current motors are relatively limited.

Alternating-current motors

Alternating-current motors are finding increasing application in electric vehicles. In this case, the DC voltage provided by the traction battery is converted to variable-amplitude and variable-frequency alternating current by means of an inverter. The stator consists of a three-circuit winding, laid in slots. The stator winding generates a rotating electromagnetic field. Among AC motors, distinction is made between synchronous and asynchronous motors. The distinction is due to differences in rotor design and the corresponding rotation of the rotor either synchronously or asynchronously with the rotating stator field. For the asynchronous or induction machine, the rotor consists of a slip ring or squirrel-cage design. The slip-ring rotor has alternating-current windings whose connections to the outside are made by slip rings and carbon brushes.

For the squirrel-cage motor, slots in the rotor are filled with aluminum, copper, or bronze rotor bars and connected at their ends by shorting rings. Function of these machines is based on the fact that with different rotational speeds between rotor and field, voltages are induced in the rotor. Current flows through the short-circuited rotor bars, which by virtue of the surrounding magnetic field causes forces (torques) to be exerted on the rotor. Among synchronous motors, distinction is made between permanently excited and externally excited synchronous motors. In externally excited synchronous motors, the rotor is magnetized by DC current applied to its windings. By varying the rotor field (exciter current), it is possible to achieve an extended range of constant maximum power. In the permanently excited synchronous machine, the rotor field is built up by permanent magnets. No additional energy is required for the rotor's magnetic field; therefore, these machines exhibit high efficiency. Distinction is made between inside and outside rotor designs. So-called permanent magnet motors are included in the latter, which produce higher torque thanks to their higher number of poles.

Squirrel-cage alternating-current machines have the advantage of compact, rugged design (maintenance free). Maximum speeds of up to 15,000 rpm are possible. These also benefit from the higher efficiency of AC motors compared to DC motors. Permanently excited synchronous machines exhibit the highest efficiencies. One disadvantage of AC motors is their added control complexity, although this is being steadily decreased by continued development of power semiconductors.

Special motors

Special motors include brushless DC motors. In principle, these are permanently excited DC motors without commutators. Commutation is achieved electronically by means of an inverter, which feeds pulse-width–modulated DC current to the stator winding. Design of this type of machine is similar to that of a permanently excited synchronous machine.

Transverse flux motors [5] differ from conventional motors in the manner in which flux is channeled. Flux is directed transversely to the direction of motion, while current is directed in the direction of motion. This is achieved by a coaxial ring winding which conducts current in a circumferential direction. The ring winding is surrounded by a large number of pole pairs with axially oriented flux. Each winding requires its own stator/rotor system with its own controller. At least two systems are required for a single propulsion application. These machines are characterized by very good efficiency, high torque densities, and compact dimensions. For these reasons, transverse flux motors are suitable for wheel hub motors (direct drive, without transmissions). One disadvantage is their complex geometry, which, along with the necessary high-energy magnets, results in high manufacturing costs.

The class of special motors also includes switched reluctance motors [5, 6]. These are brushless, magnetless motors with differing numbers of poles in the rotor and stator. Each stator has its own exciter winding. Several stator windings are switched together electrically to form the north/south pole pair of a single phase. Each phase undergoes unipolar excitation—that is, only one current direction per winding—until rotor and stator poles are aligned. Sequential excitation of phases results in continuous motor rotation.

Thanks to their simple mechanical design, their lack of magnets and rotor windings, and their comparatively simple control structure, reluctance motors are robust and economical. Furthermore, they offer high peak torque and high efficiency. The cyclical nature of their torque development may cause problems in terms of acoustic noise generation, but this can be addressed by good mechanical design as well as specifically tailored electrical controls.

4.3.1.3 Inverters

In electric vehicles, inverters [7] serve to supply the drive motor with electrical power from the battery, to modulate this power in response to driver commands, and to enable storage of regenerated braking energy in the battery. Moreover, the inverter must ensure adherence to certain operational limits arising from limited battery voltage and maximum battery current, the tractive condition of the brake control system, and motor and inverter temperatures.

Driver commands are defined by accelerator and brake pedal travel, direction of travel (forward or reverse), and desired speed (cruise control). A vehicle control unit translates these commands into a target output torque. In the process, this unit also controls operational limits.

Depending on the details of individual electric motors, inverters of different configurations may be employed. DC machines are generally supplied directly from the traction battery through a DC controller. Synchronous or asynchronous motors require a symmetrical rotating field, which in turn requires alternating current. Among AC inverters, a principal distinction is made between inverters with a constant-voltage intermediate circuit (constant voltage type) and those with an alternating-current intermediate circuit (constant current type). Because of its simpler construction and better dynamics, the constant voltage inverter has become the predominant configuration for modern electric vehicle designs.

DC chopper converter

The basic principle of the DC chopper converter is that battery voltage is supplied to the drive in the form of pulses. Active power components may consist of switchable transistors (bipolar, IGBT, MOSFET) with an appropriate free-wheeling diode.

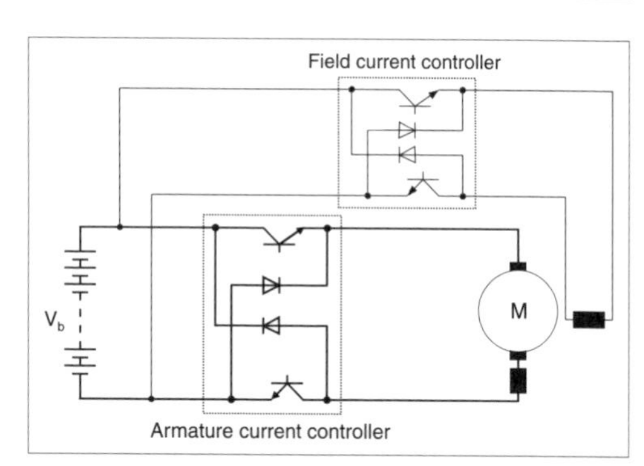

Fig. 4.3-3 DC chopper converter for externally excited DC machine.

Fig. 4.3-3 shows a converter for an externally excited DC machine. Armature and field current controllers are supplied directly from the traction battery. Armature control covers the basic constant-torque speed range. Higher speeds are possible by dropping the field current. In the weakened-field operating regime, the armature controller is fully bypassed.

Constant-voltage inverter (constant-voltage intermediate circuit)

The constant-voltage inverter (Fig. 4.3-4, represented schematically in conjunction with an externally excited synchronous machine) consists mainly of a pulse modulator which converts battery voltage into alternating voltage of variable amplitude and frequency. The capacitor is necessary to decouple the battery from the pulse modulator's higher harmonics. Recovery of braking energy (regenerative braking) is possible without any additional circuitry. The pulse modulator may be used with synchronous as well as asynchronous machines. Fig. 4.3-5 shows a block diagram of a typical alternating current drive with inverter. The modulator uses battery voltage to produce symmetrical alternating voltage for the connected synchronous or asynchronous machine. The objective of the control system is to use the machine to produce a desired torque as demanded by the control unit. To this end, the desired torque is first transformed into suitable parameters (e.g. winding currents). The current control then compares these to the output parameters of a machine model, which cal-

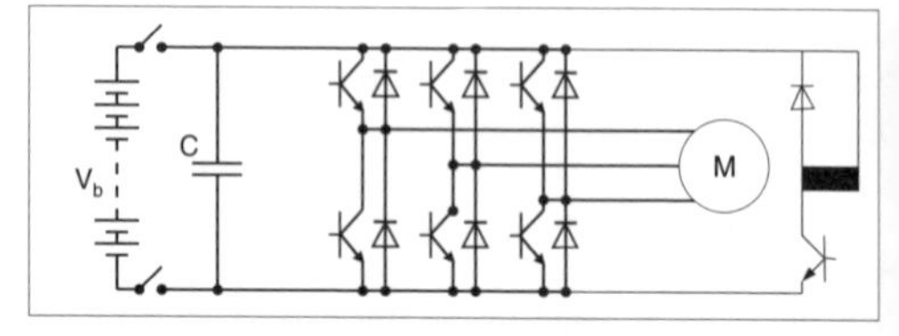

Fig. 4.3-4 Constant-voltage inverter.

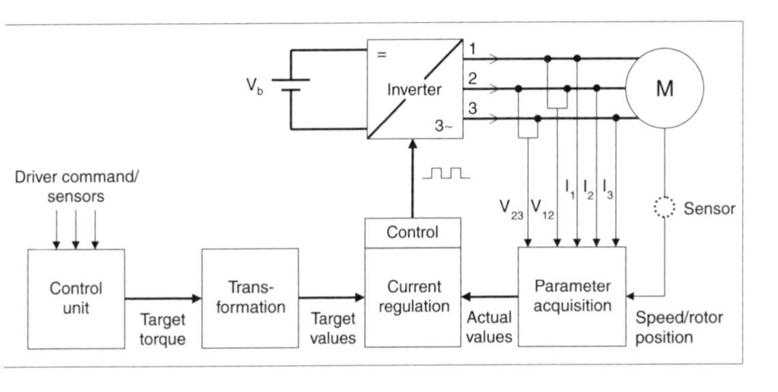

Fig. 4.3-5 Alternating-current drive with constant-voltage inverter.

culates the actual values by means of suitable measured parameters. These measured parameters might consist of machine voltages and currents, as well as speed or rotor position. Finally, the current control transmits the required control parameters to the master pulse generator, which switches the inverter's transistors on and off.

Two different techniques are employed to achieve current or torque control. One is control of winding currents by means of hysteresis tolerance bands, in which currents are kept within a tolerance band of the sinusoidal target currents. However, indirect current control with underlying voltage control (pulse-width modulation, or PWM technique) is becoming increasingly popular.

Table 4.3-1 gives an overview of motors for electric vehicles. It illustrates that no single type of machine meets all criteria to a high degree. Accordingly, every unique application requires an appropriate choice of suitable drive. Despite their high level of development, DC machines appear to be less suitable for modern electric vehicles, while asynchronous machines as well as switched reluctance machines offer an overall balance of desirable qualities. Asynchronous machines furthermore benefit from a high state of development, while switched reluctance machines and especially transverse flux machines still have much development ahead of them. The permanent-magnet-excited synchronous machine appears attractive for applications requiring high efficiency and compact design.

4.3.1.4 Traction batteries

The traction battery is the most important component of any electric vehicle. The energy content of the battery determines vehicle range, while the battery weight and its electric power output determine the electric vehicle's performance. Further, traction batteries should offer long service life and a high level of safety. Ultimately, battery costs are decisive in the marketing of electric vehicles.

Traction batteries are differentiated into primary and secondary systems. Primary batteries, such as the zinc-air battery, can only be discharged once. After being fully discharged, these batteries must be exchanged and refurbished. Their electrochemical reactions are effectively irreversible, while in the case of secondary batteries, repeated charging and discharging is possible.

A further differentiating characteristic is operating temperature. Distinction is made between batteries which operate at ambient temperatures and those which oper-

Table 4.3-1 Comparison of electric motors for electric vehicles

	DCM[a]	ASM[b]	EESM[c]	PMSM[d]	SRM[e]	TFM[f]
Efficiency	−−	+	+	++	+	++
Maximum speed	−−	++	+	+	++	−−
Volume	−−	+	+	++	+	−
Weight	−−	+	+	++	+	+
Cooling	−−	+	+	++	++	+
Manufacturing process	−	++	−	−	++	−−
Costs	−	++	−	−−	++	−−

[a] DC machine.
[b] Asynchronous machine.
[c] Externally excited synchronous machine.
[d] Permanent magnet excited synchronous machine.
[e] Switched reluctance machine.
[f] Transverse flux machine.

ate at elevated temperatures (high-temperature batteries). Among the ambient temperature batteries are lead-acid, nickel-cadmium, nickel-metal hydride, zinc-bromine and lithium-ion batteries. High temperature batteries operate at temperatures up to 350°C. These include lithium-polymer systems (approx. 80°C), sodium-sulfur (approx. 300°C) and sodium-nickel chloride (approx. 300°C).

Table 4.3-2 compares the most important electric vehicle battery types [8]. Listed are data for complete battery systems (including battery containment, battery management, and cooling). The indicated target values reflect auto manufacturers' acceptable specifications and performance requirements.

Energy density characterizes the energy capacity of the battery relative to the total battery weight or volume and is measured in Wh/kg and Wh/L respectively. It is customarily measured during a discharge period of two hours. The range of an electric vehicle is largely determined by energy density.

Power density characterizes the electrical power which may be drawn from the battery, based on total battery weight or volume, and is measured in W/kg and W/L respectively. Power density is dependent on the battery's state of charge and is usually referenced to 80% discharge. Power density determines the performance (maximum speed, acceleration) of an electric vehicle. For battery life, distinction is made between cycle life and calendar life. Cycle life indicates the number of possible charge/discharge cycles of a secondary battery until its capacity drops to 80% of rated capacity. Calendar life is based on practical experience and indicates battery life if it does not reach its cycling limit.

Battery costs are referenced to energy content and are indicated in Euros/kWh. These costs are dependent on materials and manufacturing costs as well as production numbers. The following table is based on production volume of 10,000 to 20,000 complete battery systems per year.

Lead-acid batteries

Lead-acid batteries have long been used in electric vehicles. Due to their low energy density, these vehicles are heavy and have limited range. In particular, the life expectancy of lead-acid batteries is unsatisfactory. Only their power density exceeds target values. Although their relatively low acquisition cost appears attractive, this is offset by their limited service life, and ultimately they have no cost advantages.

Nickel-cadmium batteries

Compared to lead-acid batteries, nickel-cadmium batteries exhibit higher energy density and cycle life. "Memory effect," which affects some types, may limit their usefulness. There are concerns regarding this battery type with respect to its use of cadmium, a poisonous heavy metal.

Nickel-metal hydride batteries

These batteries are environmentally responsible and have largely displaced nickel-cadmium batteries. In contrast to nickel-cadmium batteries, they display higher energy and power densities. Cycle durability, at over 1,000 cycles, is good, and calendar life of 10 years is very good. In terms of medium- and long-range cost targets, these are relatively expensive batteries. They are now employed in most modern electric vehicles.

Sodium-nickel chloride batteries

The class of high-temperature batteries includes sodium-nickel chloride batteries. Operating temperatures inside the battery range from 270 to 350°C. A double-walled, possibly evacuated container limits thermal losses. A temperature management system is necessary to maintain operating temperature. This battery type is characterized by high energy density and good performance and offers adequate service life. Battery costs are relatively high. Energy losses (thermal losses) are an added disadvantage.

Lithium-ion battery

Lithium ion batteries also offer high energy density and very high performance levels, but with limited calendar life. Their disadvantage is high cost. To date, only a few prototypes have been produced. Development of this battery type concentrates on reducing costs and increasing life expectancy.

Table 4.3-2 Comparison of battery systems for electric vehicles

Battery type	Energy density Wh/kg	Power density W/kg	Battery life		Costs EUR/kWh
			Cycles	Years	
Lead-acid	30–35	200–300	300–400	2–3	100–150
Nickel-cadmium	45–50	80–175	>2000	3–10	<600
Nickel-metal hydride	60–70	200–300	>1000	10	300–350
Sodium-nickel chloride	90–100	160	1000	5–10	<300
Lithium-ion	90–140	300–600	500–750	<5	300–600
Lithium-polymer	110–130	~300	<600	—	300
Zinc-air	100–220	~100	—	—	60
Target values	100–200	75–200	1000	10	100–150

Lithium-polymer

Lithium-polymer batteries are also members of the high-temperature class. Operating temperatures inside the battery range from 60 to 100°C. Their cells are manufactured as membranes; the electrolyte is also in membrane form. Development of this battery type is ongoing. Commercially usable batteries in large volume are not expected within the next few years.

Zinc-air battery

Zinc-air batteries are considered primary batteries—that is, not repeatedly rechargeable. This battery type exhibits the highest energy densities, with relatively low power densities. Because a dedicated infrastructure is needed for removal (exchange system) and regeneration or refurbishment, this system is suitable mainly for large fleet operators.

4.3.1.5 Battery chargers

Electric vehicles are generally recharged at night. Usually, this is done via onboard chargers (AC/DC converter). Distinction is made between conductive and inductive chargers. Conductive charging is likely to become predominant, as the California Air Resources Board (CARB) has declared this technology as standard [9]. Charging power is dependent on the available power supply (mains) voltage and the maximum permissible current. In order to recharge electric vehicles in five to six hours, 7 kW onboard chargers (230 V 30 A, single phase) are used.

4.3.1.6 Looking to the future

The development of electric vehicles has moved to the foreground, with a backdrop of tighter emissions regulations and requirements for Zero Emissions Vehicles. High-performance traction batteries and electric drive systems are applied to achieve respectable vehicle performance [10]. In view of acceleration ability and top speed, electric vehicles have proven themselves to be fully capable road vehicles. Operating ranges of about 200 km allow a wide variety of applications.

Nevertheless, in the eyes of the consumer, the limited range and increased weight of electric vehicles presents an imminent handicap, particularly when combined with high acquisition costs.

Offsetting these drawbacks are the incontrovertible advantages of electric vehicles for environmentally responsible transportation, opening significant long-term possibilities for this means of propulsion.

References

[1] Shulock, C. "The California ZEV Program—Compliance Scenarios." EVS 18, October 20–24, 2001, Berlin.

[2] International Energy Agency. "Electric Vehicles: Technology, Performance and Potential." OECD Publications, 2 rue Andre-Pascal, 75775 Paris Cedex 16. ISBN 92-64-14015-8-No. 46876. 1993.

[3] Wallentowitz, H. "Einsatz von Elektrofahrzeugen." BMV-Forschungsvorhaben FE-Nr. 70466/95, ika Bericht 6062.

[4] Reckhorn, T., A. Dietz, and K. Müller. "Drive Systems for Electric, Hybrid and Fuel Cell Vehicles." EVS 18, October 20–24, 2001, Berlin.

[5] Altendorf, J., T. Balimann, R. Inderka, and A. Lange. "OKOFEH—A Research Project for Initiating Further Development of Drive Systems for Electric, Hybrid, and Fuel Cell Vehicles." EVS 18, October 20–24, 2001, Berlin.

[6] Iderka, R., J. Altendorf, L. Sjöberg, and R. De Doncker. "Design of a 75 kW Switched Reluctance Drive for Electric Vehicles." EVS 18. October 20–24, 2001, Berlin.

[7] FAT-Schriftenreihe No. 104. "Antriebe für Elektrofahrzeuge."

[8] Anderman, M., F. Kalhammer, and D. MacArthur. "Advanced Batteries for Electric Vehicles: An Assessment of Performance, Cost and Availability." State of California Air Resources Board, Sacramento, June 2000. Available at www.arb.ca.gov/msprog/zevprog/2000review/btapreport.doc.

[9] Sweigert, G. "Standardization of Charging Systems for Battery Electric Vehicles." EVS 18, October 20–24, 2001, Berlin.

[10] Hoehl, K., and E. Wüchner. "Das Elektroauto—Zukunft oder Sackgasse?" DI Berichte 1292, 1997.

4.3.2 Fuel cell propulsion for mobile systems

The conventional combustion engine, in its current high state of development, is at present the most compact automotive propulsion system available. The maximum efficiency of passenger car diesel engines exceeds 40%. However, the average overall efficiency for typical driving cycles, from the release of primary energy to conversion into mechanical work to transfer to driven wheels, is little more than 20%. Electric propulsion systems with suitable batteries for energy storage may offer compact solutions for urban vehicles, but to date, greater range is not possible due to correspondingly heavier batteries. Moreover, with such vehicles, losses and emissions resulting from the conversion from primary to useful energy are merely relocated from the road to the electrical generating facility. In contrast, fuel cells combined with electric propulsion offer a new propulsion concept with considerable advantages:

- Absolutely no harmful vehicle emissions from hydrogen fuel-cell systems.
- Methanol fuel-cell systems have extremely low hydrocarbon emissions and, compared to combustion engines, considerably lower carbon dioxide emissions and elimination of NO_x and CO emissions.
- Minimal emissions throughout the overall energy conversion chain.
- Appreciably higher overall efficiency relative to the primary energy source.
- Long service life.
- Fully compatible with energy sources such as natural gas and renewable resources, which are currently being used only on an experimental basis for transportation applications.
- Vehicle range and refueling methods comparable to current vehicles fitted with combustion engines.

These advantages would make a major contribution toward reducing traffic-related pollutant emissions and consumption of energy resources. Accordingly, many firms, including nearly all vehicle manufacturers (e.g., DaimlerChrysler, General Motors, Toyota, Ford, Honda, Nissan, and others) and a multitude of research institutions throughout the world are engaged in independent fuel-cell research projects.

The most promising fuel-cell type for mobile systems appears to be the polymer electrolyte membrane fuel cell.

Polymer electrolyte membrane (PEM) fuel cell

The strengths of the PEM concept in comparison to other fuel-cell types arise from the more easily managed proton-conducting polymer film (compared to alternative fuel-cell electrolytes), a working temperature below 100°C, and very high power densities. Alternative types are primarily used in stationary energy production applications.

A PEM fuel cell consists of an electrolyte membrane, catalysts and electrodes applied to both sides of the membrane, and so-called bipolar plates which admit gases and conduct away the generated electrical current. The membrane is only a tenth of a millimeter thick, made of a sulfonated fluorocopolymer—essentially a solid electrolyte which separates the reacting gases, oxygen and hydrogen—and only permits protons (H+) to pass through. These are produced by oxidation of hydrogen at the anode, releasing electrons:

$$H_2 \Rightarrow 2H^+ + 2e^-$$

Simultaneously, at the cathode on the opposite side of the electrolyte, reduced atmospheric oxygen reacts with protons which have passed through the electrolyte, producing water vapor:

$$\tfrac{1}{2} O_2 + 2H^+ + 2e^- \Rightarrow H_2O$$

The resulting potential difference between the two electrodes is the driving force behind the fuel-cell reaction and may be converted into work in an external electrical circuit.

The membrane's polymer chains contain a large number of active centers which give the membrane its high ionic conductivity; values of 0.1 per ohm-cm are typical. The membrane thickness is between 50 and 150 μm. In order to accelerate electron reactions, both sides of the membrane are coated with a thin layer of noble metal catalyst—for example, platinum—which in turn is coupled to a porous support structure. This structure, the so-called MEA (membrane electrode assembly), forms the core of the cell. Only a few years ago, 4 mg of catalyst per square centimeter were required. Recent research indicates that as little as 0.1 mg may be sufficient, with no loss of performance. This would drastically reduce costs and improve the economic viability of fuel cells.

The bipolar plates uniformly distribute the reactants—oxygen, or air, and fuel—to the electrodes and carry off electrical current and generated heat. They enclose the membrane electrode assembly from both sides and together with the MEA comprise an individual cell, the smallest possible configuration of a fuel cell. They also distribute coolant to keep operating temperatures at the desired level and remove the resulting water. The plates also divide reaction gas and coolant channels. Finally, these plates permit several individual cells to be coupled electrically. The result is a stack mounted between two end plates. The number of components or the effective total electrode area depends on voltage and current requirements. At present, specific output of up to 1 kW per kilogram, with power density as high as one watt per square centimeter at current densities of up to two amperes per square centimeter, is achievable. The voltage of such individual cells operating at temperatures of about 80°C and pressures between 2 and 5 bar ranges between 1.0 volts at idle and 0.5 volts at maximum current draw. Even today, with fuel cell technology at an early development stage, the measured efficiency of an elementary fuel cell (see Fig. 4.3-8) falls between 55 and 80% in the typical operating range between 0 and 0.7 A/cm^2; in the driving cycles commonly encountered in everyday operation, it is important that the higher efficiency be reached under part-load conditions.

Efficiency losses appear in the form of ohmic losses due to the finite conductivity of the membrane, catalyst

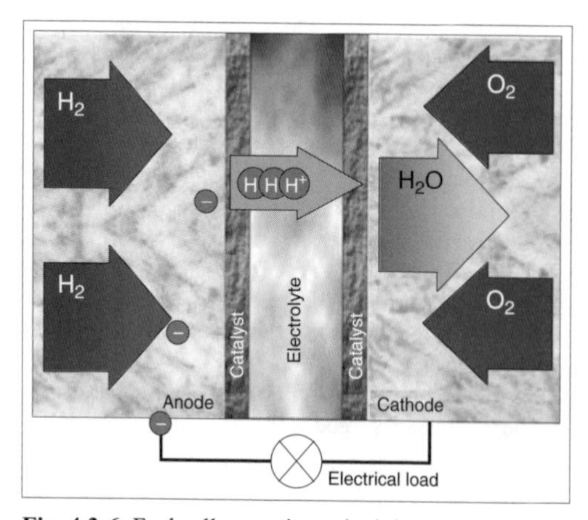

Fig. 4.3-6 Fuel cell operating principle.

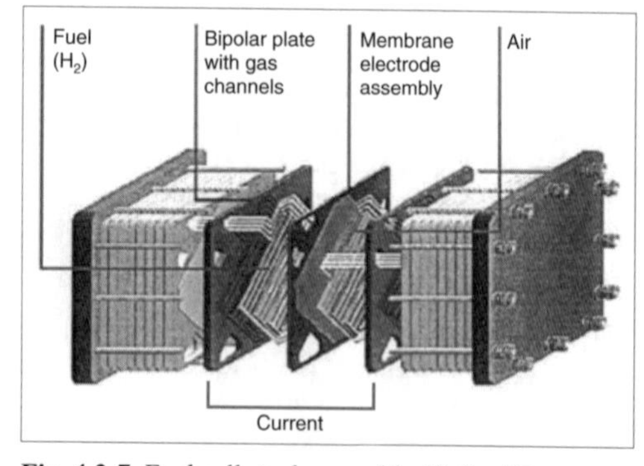

Fig. 4.3-7 Fuel cell stack assembly (Ballard Power Systems).

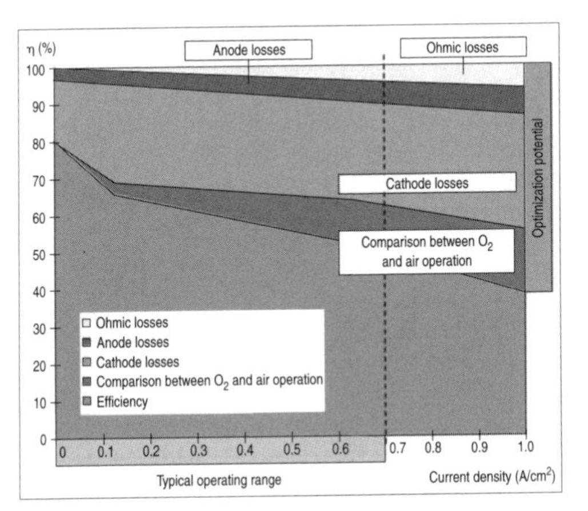

Fig. 4.3-8 Fuel-cell efficiency.

surfaces, and bipolar plates. Conversion of reactants at the electrodes leads to anode and, to an even greater extent, cathode losses which, with the present state of the art, suggest that there is considerable potential for improvement. In mobile applications, substitution of ambient air for pure oxygen is another significant source of losses.

Increasing operating pressure above normal results in lower losses but requires additional power to drive the corresponding compressor; the optimum tradeoff must be found for any given overall system.

Because the effective efficiency of a fuel cell—that is, the efficiency during current draw—is defined by the ratio of measured terminal voltage relative to theoretical cell voltage (1.23 volts in the case of hydrogen-oxygen cells), efficiency is directly related to voltage response to current draw.

While the efficiency of combustion engines is basically limited by the thermodynamics of the Carnot cycle, such a basic limitation is currently not known to exist for fuel cells. We may expect further reduction in losses with continued development. As for future development, it appears that the potential for improvement of fuel cells is appreciably greater than the remaining optimization opportunities for combustion engines.

The following table lists typical performance data for a modern fuel cell stack manufactured by the Canadian company Ballard Power Systems, as installed in the Necar II (New Electric Car), a small fuel-cell–powered van based on the Mercedes-Benz V class, and in the Nebus (New Electric Bus), a fuel-cell–powered transit bus based on the Mercedes-Benz 0405.

Table 4.3-3 PEM fuel-cell specifications

Operating temperature	80°C
Operating pressure	2–3 bar
Current density	0–15 A/cm²
Cell voltage	1.0–0.5 V
Power density (stack)	
Relative to weight	9.7 kW/kg
Relative to volume	1.0 kW/L

Hydrogen supply peripherals

The stack is completed through the addition of other components which, among other things, supply the cells with gases and provide a cooling function. The engineering configuration of these peripheral components as well as all specifications of the overall system are strongly dependent on the choice of process gases. The simplest variant uses hydrogen and oxygen stored in pressurized containers. Major additional components include pressure regulators and recirculation systems for process gases, condensers to remove byproduct water, and a cooling circuit. Such systems are best known for their application in the Gemini and Apollo space programs and in the Space Shuttle. In manned space flight applications, hydrogen and oxygen are readily available for propulsion and life support, and neither cost nor overall primary energy consumption carry as much importance as on Earth. Moreover, such systems, given that they do not require complex storage systems, permit power densities of more than 1 kW per kilogram, with system efficiencies of more than 60%. For road vehicle applications, important questions include those of cost-effective hydrogen generation and its storage on board the vehicle. The most familiar hydrogen generation process is certainly electrolysis. Although modern installations can make use of 80% of the current used to dissociate water, the overall efficiency, including use of primary energy to generate electricity, is only about 25% (in the case of the German energy economy). At present, the cheapest and therefore most common method extracts hydrogen from natural gas, predominantly composed of methane (CH_4). The resulting hydrogen fuel is cleaned after reduction. At over 70%, the overall efficiency of this method is considerably higher than that of hydrogen by electrolysis.

Several technologies are suitable for carrying sufficient hydrogen aboard a vehicle:

- Pressurized tank storage requires a large storage volume, at least 50 L of tank volume per kilogram of hydrogen, at a pressure of 220 bar, as well as energy needed to compress the gas (the driving range of a mid-range car is about 100 km per kg of stored hydrogen).
- Cryogenic storage—that is, of liquid, refrigerated hydrogen—permits higher storage density, but liquefaction consumes about 30% of the energy content; moreover, the hydrogen evaporation rate is about 2% per day.
- Chemical storage in the form of metal hydrides permits high storage density relative to the occupied volume, but in terms of weight its performance is rather meager; the refueling process is lengthy and tanks cannot be refilled and emptied an unlimited number of times.

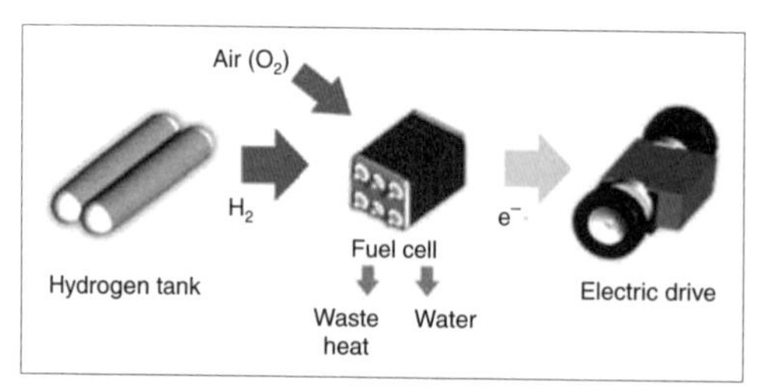

Fig. 4.3-9 Hydrogen fuel-cell propulsion.

The necessarily heavy and, above all, voluminous tanks for hydrogen as well as oxygen plus the lack of a hydrogen supply infrastructure limit widespread application of fuel-cell–powered vehicles. PEM cells may be operated with air, eliminating the need to generate and store oxygen, but operation at lower atmospheric partial pressures causes a significant drop in power density. Compensation by means of compressed air, however, consumes part of the generated electrical energy in proportion to the compressor's power requirement. For this reason, a compromise is generally made in using relatively low pressures. While this does not change anything on the hydrogen side or in cooling, on the air side the pressure regulator is replaced by a compressor plus some peripheral components including air filter and muffler. Hydrogen/air units presently exhibit system efficiencies of about 50% and system power densities of up to 0.3 kW per kilogram.

Methanol fuels for individual transportation

If fuel cells are to be operated on more readily accessible, more easily stored fuel such as methanol, this hydrocarbon fuel must first be reformed to produce a hydrogen-rich gas. Fuel and water, stored in unpressurized tanks, are metered to an evaporator. The superheated fuel/steam mixture is passed over a catalyst, the so-called reformer, where heat is added, and, in the case of methanol, the main products are hydrogen and carbon dioxide, from the reaction

$$CH_3OH + H_2O \Rightarrow 3H_2 + CO_2$$

The remaining trace of carbon monoxide (CO), amounting to 1–2%, must be removed before the gas passes to the PEM fuel cell, as the catalyst has a limited tolerance for this constituent. Possible strategies include CO methanization, selective CO oxidation, and membrane separation processes. Finally, the combustion gas is fed to the anode, where most of the hydrogen is converted. The waste gas, still containing hydrogen, is used to generate heat for evaporation and the energy-intensive reformation process.

Methanol is also obtained from natural gas or, alternatively, from coal gasification; the efficiency of this preliminary step is about 70%, based on the amount of primary energy used.

In the long term, methanol production from biomass may become interesting, as this will represent a large-scale renewable energy source. The special advantages of methanol compared to hydrogen are, first, high storage density of the liquid hydrocarbon and its ease of handling; and second, with only minimal modifications, the existing network which supplies gasoline stations could be used for transportation, storage, and distribution of methanol fuel.

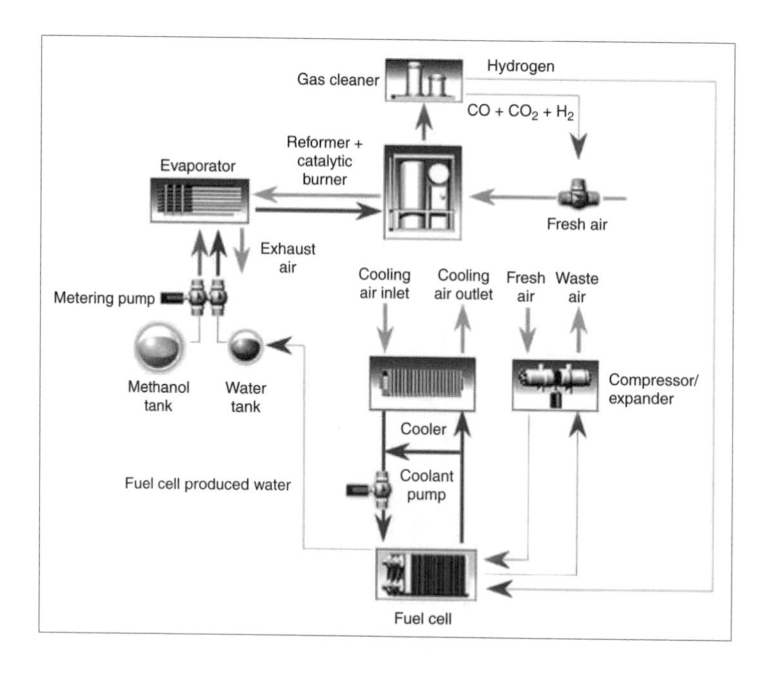

Fig. 4.3-10 Methanol fuel-cell system and hydrogen fuel cell.

The state of fuel-cell–vehicle technology

At present, nearly all fuel-cell technologies and components require further development in order to increase individual and overall efficiency, to simplify the units, to make them suitable for volume production, and to reduce costs. Engineering challenges include improvement of cold-start behavior for methanol fuel-cell vehicles and assurance of operating capability at high and low temperatures.

Even now, prototypes are demonstrating the feasibility and desirable properties of fuel-cell systems in road vehicle propulsion applications. These experimental vehicles range from the hydrogen-powered Necar minivan, based on the Mercedes-Benz MB 100 of 1994, and the Necar II, based on the Mercedes-Benz V class of 1996, to the Necar 3, employing methanol reformation and based on the 1997 Mercedes-Benz A class. Also completed in 1997 was an urban transit bus, named Nebus, employing a hydrogen fuel cell system.

4.3.2.1 Necar vehicles

Table 4.3-4 outlines the rapid sequence of events in the development of fuel-cell–propulsion technology. Necar I may be regarded as a rolling laboratory, packed with components for generating electricity, with space left over for only two drivers. Just two years later, Necar II represented a vehicle with the same number of seats as a standard production Series V van. Only the space dedicated to the fuel cell stacks, behind and below the rear bench seat, resulted in a moderate restriction of available cargo space. Along with significant reduction in the size of the fuel cell stacks, mounting the hydrogen tanks on the vehicle roof provided additional space reserves. The fuel-cell size reduction was accompanied by weight reduction from 21 kg/kW to only 6 kg/kW. Despite larger tanks and correspondingly greater range, the total weight of Necar II was nearly one metric ton less than that of Necar I.

The Necar II's energy demand, as determined in the NEDC (New European Driving Cycle), already demonstrates a higher system efficiency than even the best combustion engines currently available in production vehicles. The Necar 3, presented by Daimler-Benz in 1997, represents the next step toward a production passenger car powered by fuel-cell technology; it is the world's first vehicle to offer onboard methanol-to-hydrogen reformation. Its 38 L unpressurized fuel tank provides enough energy for a range of about 400 km.

The sandwich design of the A-Class floorpan makes this vehicle especially attractive for operation with alternative propulsion systems. The packaging concept of the Necar 3, for example, has the fuel-cell stacks built directly into spaces in the floor. The gas generator system is mounted behind the two front seats; in the Necar 3, the generator's size limits the number of occupants to only the driver and one passenger. This system, however, promises future possibilities for component downsizing.

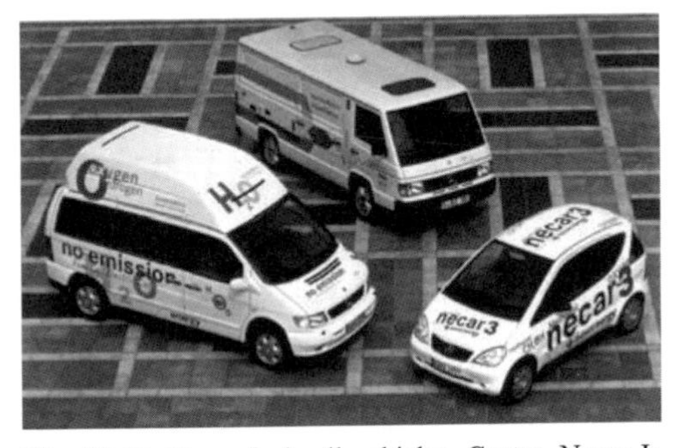

Fig. 4.3-11 Necar fuel-cell vehicles. Center, Necar I; left, Necar II; right, Necar 3.

Table 4.3-4 Specifications of Necar fuel-cell vehicles

	Necar I	Necar II	Necar 3
Fuel-cell system			
Power output	50 kW from 12 stacks	50 kW from 2 stacks	50 kW from 2 stacks
Power density	21 kg/kW 48 W/kg	6 kg/kW 167 W/kg	15 kg/kW 66 W/kg
Voltage level	130–230 V	180–240 V	185–280 V
Fuel tank system			
Type	Pressurized hydrogen tank, fiberglass-sheathed aluminum	Pressurized hydrogen tank, carbon-fiber–reinforced plastic	Unpressurized stainless steel tank
Volume	150 L	2 × 140 L	38 liters
Pressure	300 bar	250 bar	
Drive system			
Electric drive	30 kW	33 kW continuous 45 kW maximum	33 kW continuous 45 kW maximum
Top speed	90 km/h	110 km/h	120 km/h
Range	130 km	>250 km	400 km
Max. gross weight	3,500 kg	2,600 kg	1,750 kg

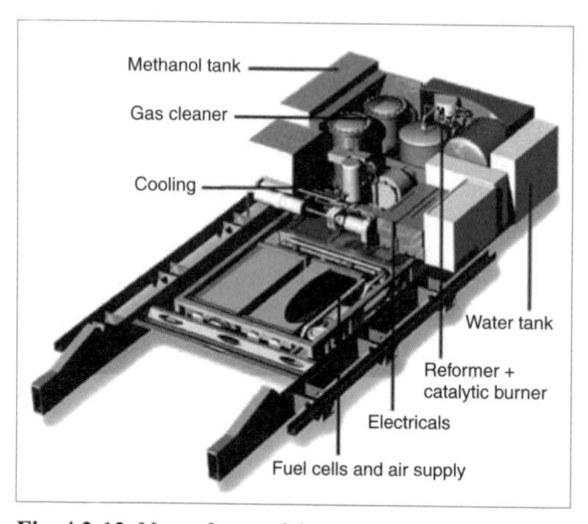

Fig. 4.3-12 Necar 3 propulsion system packaging.

Mechanical integration of optimized components will enable appreciable reduction in space requirements to the point where the passenger compartment will no longer show any significant restrictions.

4.3.2.2 Nebus fuel-cell bus

Development of fuel-cell–powered buses extends propulsion system development to higher power levels. The Nebus is powered by a hydrogen fuel-cell system combined with two electric wheel hub motors made by Zahnradfabrik Friedrichshafen (ZF). These AC asynchronous motors permit direct transmission of electrical energy to the driven wheels, eliminating the need for transmission and driveshafts. Correspondingly, the Nebus was designed as a low-floor bus, permitting stepless entry for riders. Conventional diesel–electric buses in regular service have demonstrated simple maintenance requirements and yielded overall positive experiences, underscoring the advantages of this electric motor technology. The two hub motors are designed for continuous output of 50 kW each, but permit maximum power of 75 kW and give the bus a top speed of 80 km/h.

The fuel-cell system consists of ten stacks identical to those of the Necar II, each providing 25 kW. With a total output of 250 kW, this system permits a fivefold improvement in fuel-cell power density compared to the first Necar of 1994 and correspondingly a quite compact installation. The desired operating voltage and current are obtained with groups of five stacks operating in series and the two stack groups in parallel. The achieved voltage range of 440–720 V is ideally suited to powering the wheel hub motors. The entire fuel-cell system can be installed and mounted in a frame which fits in place of a conventional diesel propulsion system. The required hydrogen is carried in seven fiberglass-wrapped aluminum pressure tanks (300 bar) mounted on the roof of the bus, providing a 250-km range. Although normally no refueling is required during a routine day of transit bus operation, the refueling time of 15 minutes is satisfyingly brief.

Fig. 4.3-13 Hydrogen fuel-cell transit bus.

Table 4.3-5 Specifications of the Nebus fuel cell bus

	Nebus
Fuel cell system	
Power output	250 kW from 10 stacks
Power density	5.6 kg/kW 178 W/kg
Voltage level	440–720 V
Fuel tank system	
Type	Pressurized hydrogen tank, fiberglass-sheathed aluminum
Volume	1,050 L
Pressure	300 bar
Drive system	
Electric drive	2 × 50 kW cont. 2 × 75 max.
Top speed	80 km/h
Range	250 km
Max. gross weight	18,000 kg

Besides quiet and virtually pollution-free operation, the advantages of fuel-cell propulsion include the previously mentioned high efficiency, which manifests itself as good fuel economy. The results of a typical urban driving cycle are shown in Fig. 4.3-14.

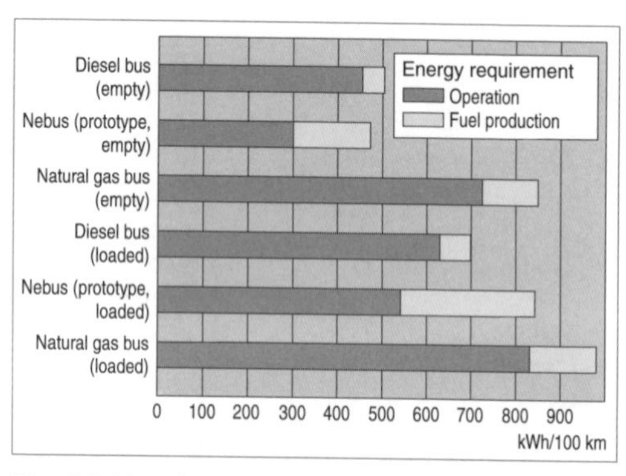

Fig. 4.3-14 Nebus energy requirements as measured in a typical urban driving cycle. Note that data for energy required to produce fuel are estimates and may differ according to method.

The Nebus's actual operational energy demands are especially attractive when operating in an empty condition. In comparison to the most efficient conventional combustion engine—the diesel—Nebus still enjoys a 50% fuel economy advantage. With increasing vehicle load, this advantage diminishes, as the efficiency of a combustion engine rises as it approaches its rated output, while the efficiency of Nebus drops. The result is a decreased fuel economy advantage for the fully laden Nebus. The natural gas bus, included for comparison purposes, is in both situations less economical than the Nebus. If, however, we take into consideration the entire energy chain including fuel production, the picture worsens because petroleum refining is a much more highly optimized process, with correspondingly higher efficiency, than either natural gas production or hydrogen regeneration. As a result, when operating empty, the Nebus has only a slight fuel economy advantage compared to diesel power and at full load is less economical than the diesel. However, it must be considered that hydrogen production offers the potential of significant improvements so that in future, its production costs may be greatly reduced. Furthermore, hydrogen offers the possibility of regenerative production from renewable resources.

4.3.2.3 Pollutant emissions from fuel-cell propulsion systems

Fig. 4.3-15 shows the relative pollutant emissions from vehicles with diesel engines and from various fuel-cell systems. Diesel emissions are normalized at 100%. For the case of hydrogen fuel-cell propulsion, the illustration represents fuel production using regenerative methods. Large-scale production of methanol is normally based on a process using natural gas reformation. Complete freedom from emissions is possible with purely regenerative hydrogen production, which may present an economically viable alternative within a few years. Emissions for large-scale production of hydrogen from natural gas are comparable to those with methanol production from natural gas. In both cases, emissions are produced almost entirely during fuel production and only in trace amounts during vehicle operation. With more efficient production methods, as are already being applied in the most modern plants, these emissions can be reduced even further. As a possible future alternative, methanol may conceivably be produced by regenerative processes—for example, from biomass.

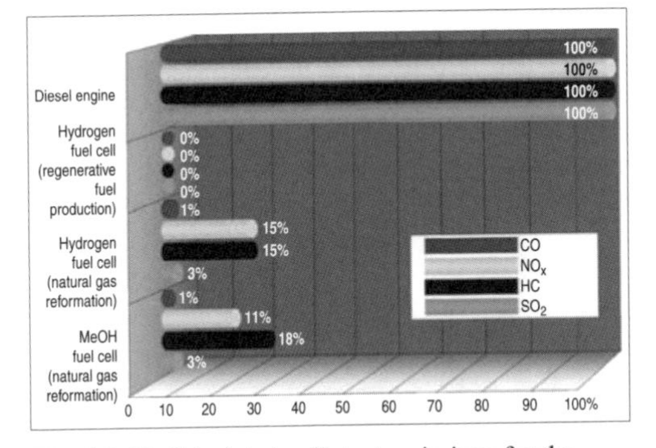

Fig. 4.3-15 Calculated pollutant emissions for the entire process chain from well to wheel. Fuel-cell vehicles compared to corresponding Euro2 diesel combustion engine vehicles (in both cases including emissions from fuel production). Note that for fuel-cell vehicles, emissions are almost entirely the result of fuel production. For methanol, manufacturing-related methane emissions are not taken into account due to strong dependence on process-specific differences.

4.3.2.4 Future outlook

In view of achieved development successes, planned 2004 market introduction of fuel-cell vehicles in limited numbers for fleet applications will be realized. Along with engineering improvements to the propulsion system, particularly in the areas of performance, reliability, and dynamics, special attention is being directed toward reducing propulsion system costs and developing the fuel and service infrastructure.

Beyond these, development projects are under way whose successful outcomes are as yet unforeseeable and whose introduction is surely several years beyond introduction of the technology described here. Among these are gasoline reformation and direct methanol fuel cells.

Onboard hydrogen generation from gasoline-based fuels instead of methanol makes use of the existing fuel infrastructure, thereby offering advantages compared to a newly created methanol infrastructure. However, it should be noted that fuels currently available for combustion engines are not suitable for reformation in their current form.

Furthermore, it is not possible by this means to reduce the dependency on petroleum. In contrast to methanol-based approaches, this form of hydrogen generation produces, among other things, oxides of nitrogen. It is also accompanied by basically less favorable reformability of the complex hydrocarbons, which results in lower efficiencies and higher system weights.

At present, the direct methanol fuel cell is still in the experimental stage, but it offers an interesting opportunity to generate electricity directly from methanol without first converting to hydrogen. Currently, demonstrated efficiencies are far from being competitive with the technology already presented by onboard methanol reformation.

References

[1] Mok, P.P., and A. Martin. "Automotive Fuel Cells—Clean Power for Tomorrow's Vehicles." SAE Conference, Detroit, 1999.

[2] Kordesch, K., and G. Simader. *Fuel Cells and their Applications* (ISBN 3527285792). John Wiley and Sons, 1996.

[3] Prater, K.B. "Solid Polymer Fuel Cells for Transport and Stationary Applications." Ballard Power Systems, *Elsevier Journal of Power Sources* 61, 1996.
[4] "The Promise of Methanol Fuel Cell Vehicles." American Methanol Institute, 1998.
[5] Atkin, G., and J. Storey. "Electric Vehicles—Prospects for battery-, fuel cell- and hybrid-powered vehicles." *Financial Times Automotive*, 1998.
[6] Willand, J., R. Krauss, A. Rennejeld, and K.-E. Noreikat. "Versuch und Berechnung bei der Entwicklung eines Brennstoffzellen-Elektrofahrzeugs." Stuttgarter Symposium Kraftfahrwesen und-Verbrennungsmotoren, Stuttgart, 1997.
[7] Wallentowitz, H., and K. Wittek. "Konzeptstudie für ein Methanol-Brennstoffzellenfahrzeug auf Basis der Mercedes-Benz-A-Klasse." Institut für Kraftfahrwesen, RWTH Aachen, Aachen, 2001.
[8] Gerl, B. "Innovative Automobilantriebe—Konzepte auf Basis von Brennstoffzellen, Traktionsbatterien und alternative Kraftstoffe." Verlag Moderne Industrie, Landsberg/Lech, 2002.
[9] Tamura, H., and M. Matsumoto. "Fuel Cell Vehicles: Technology Development Status and Popularization Issues." Nissan Motor Co., Convergence 2002, Detroit, 2002.
[10] McCormick, B., and E. Schubert. "Fuel Cells at General Motors." Convergence 2002, Detroit, 2002.
[11] Panik, F., and O. Vollrath. "Visions for the Fuel Cell Future." f-cell 2002, Stuttgart, 2002.

4.3.3 Hybrid drive

4.3.3.1 Scenario

"Hybrid drives" are defined as that class of propulsion systems which employ at least two different energy conversion processes and two different energy storage media.

With few exceptions, for practical purposes the concept currently implies energy conversion by means of combustion engines and electric motors and energy storage in the form of liquid fuels and batteries.

By virtue of optimized design and sophisticated interaction between drive components, hybrid propulsion systems may exhibit the following advantages:

- Reduced fuel consumption
- Lowest possible emissions
- Noise reduction
- Increased operational comfort
- Locally emission- and noise-free operation in sensitive areas

The extent to which these desirable properties are realized is strongly dependent on the hybrid concept and its engineering layout.

4.3.3.2 Concepts

The biggest problems regarding application of the hybrid concept are associated with locating the additional drive components within the vehicle. This may be especially difficult if the base vehicle was originally designed for conventional drive. Increased costs compared to conventional drives will put the brakes on rapid market penetration, even if hybrids are quite feasible from their engineering development standpoint.

Hybrid vehicles differ from pure electric vehicles in three important points:

1. Hybrid vehicles can deal with the existing infrastructure.
2. Their rapid refueling capability means unrestricted range and ready availability.
3. Performance is not dependent on the battery alone, permitting vehicle layouts with higher performance levels.

While electric vehicles are primarily operated as "second vehicles" for the reasons stated, hybrids are fully capable primary vehicles. They completely meet the human demand for individual mobility.

Hybrid vehicles may be classified in three groups: parallel hybrids, serial hybrids, and mixed hybrids.

Parallel hybrids

In parallel hybrids, the combustion engine and electric motor are able to drive the wheels simultaneously. This offers the possibility of operating entirely under combustion engine power, entirely under battery power, or in combination. This is a sensible approach in cases where pure electrical propulsion may be required of a vehicle, for example in entering or making deliveries within emission- and noise-sensitive areas. With regard to weight, space, and costs associated with the electric machine and battery, this approach seems all the more attractive if pure electric operation must take place with

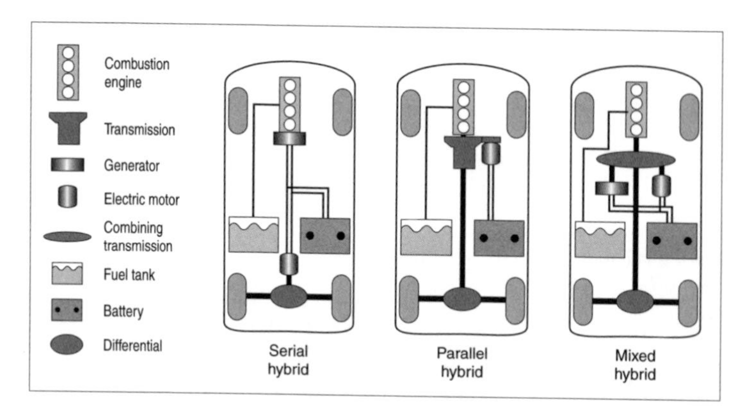

Fig. 4.3-16 Hybrid propulsion concepts.

limited performance and range. With battery recharging from the electrical grid, the parallel hybrid system's ecological impact is strongly dependent on the nature of the grid's electrical generating method.

With both drive systems operating during acceleration, the combustion engine may be a smaller design with associated economy benefits. Additional major economy improvement potential may be realized with regenerative braking and shutdown of the combustion engine under those operating conditions where it is at its worst—in stop-and-go operation, traffic jams, and motoring (coasting).

At the lower end of the spectrum of installed electrical performance, parallel hybrid concepts blend with so-called starter/generator concepts. In these, an electric machine operates in parallel with the combustion engine, but primarily to supply the onboard electrical network. In keeping with its limited electrical performance, its designed traction functions are at a much lower level.

Serial hybrids

In serial hybrids, the drive wheels are driven entirely by electric motors, with electrical energy supplied by a combustion engine and an onboard generator. The battery, as an energy buffer, permits complete temporal decoupling of the combustion engine from the drive. In this way, combustion engine operation may be set independently of the actual momentary driving demands and optimized for efficiency or emissions. Serial hybrids are therefore preferred for applications operating under restricted emissions regulations. Along with operating the combustion engine at its optimum point, fuel economy is also improved by regenerative braking. Any fuel economy advantages are, however, diminished because all of the combustion engine output must pass through the entire electrical efficiency chain.

When designed as a self-sufficient hybrid—which is by definition independent of the power grid—the combustion engine and generator output and subsequent efficiencies must be designed in keeping with the vehicle's maximum sustained speed. The electric motors must also satisfy requirements for maximum acceleration. Overall, in this type of hybrid, the total power of all installed components (engine, generator, electric motors) is higher than that of any other hybrid concept—with corresponding cost and weight disadvantages.

If emphasis is placed on purely electric propulsion, the term "range extender concept" may be used. In such

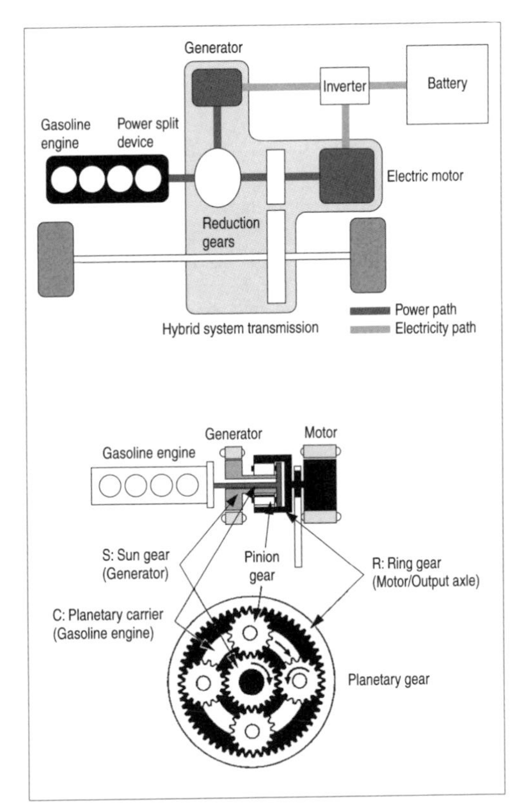

Fig. 4.3-18 The Toyota hybrid system.

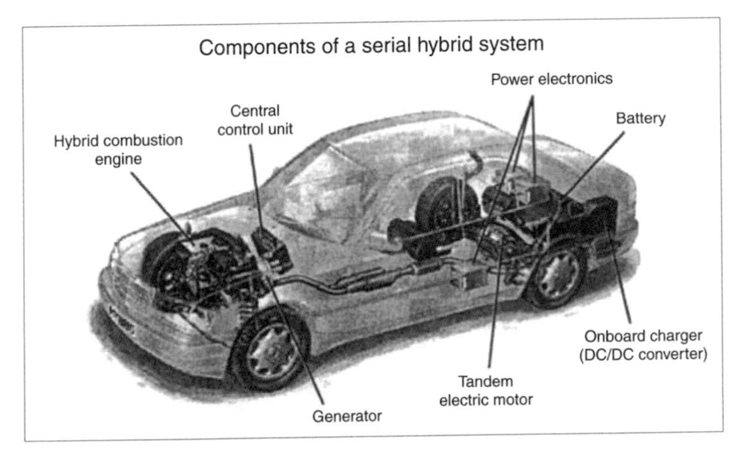

Fig. 4.3-17 Serial hybrid drive in a Mercedes-Benz C-class vehicle.

a system, a small combustion engine and generator only serve to extend battery range.

Mixed hybrids

Disadvantages of the basic parallel and serial hybrid concepts led to development of mixed hybrid systems. Equipped with combustion engines, electric machines, mechanical transmission components, clutches, free-wheeling units, and brakes, they can assume a wide variety of forms. Typical of the class is a serial hybrid concept with direct coupling of the combustion engine to the driven wheels. The term "power split hybrid" is used to describe power flow from the combustion engine reaching the driven wheels through various parallel paths. Power splitting may be implemented electrically, or it may be implemented mechanically by means of differential gearing. An example of such a concept is the Toyota Prius (Fig. 4.3-18), the world's first production hybrid vehicle, in production since late 1997.

Mixed hybrids are primarily characterized by the following advantages:

- The combustion engine may drive the wheels directly while operating under its optimum conditions.
- Discrete gears or transmission stages may not be required.

Disadvantages include increased system complexity, with corresponding control complexity and cost.

4.3.3.3 Components from the hybrid point of view

Heat engines

Among hybrids, methods of connecting the heat engine to the driven wheels vary from parallel hybrids, with their conventional layout using a transmission, to complete decoupling in serial hybrids. The engine concept may range from the conventional to steady-state stationary engines, operating at a single load and rpm point. In the latter case, new perspectives are opened for application of gas turbines and Stirling engines. In future, fuel cells may take over the function of supplying onboard electrical power.

Battery

Distinction must be made between two different battery concepts. If greater emission-free range is desired, high-energy batteries are used, bringing with them the same problems as found in electric cars: namely, weight, packaging, and cost. In the other concept, the battery only serves as a power buffer for regenerative braking and acceleration assistance or for smoothing the dynamic behavior of the combustion engine.

If we compare electrical or electromechanical energy storage media in terms of energy density divided by power density, our attention soon turns to supercapacitors and flywheels. These technologies, however, are still in their early development stages. Moreover, the required

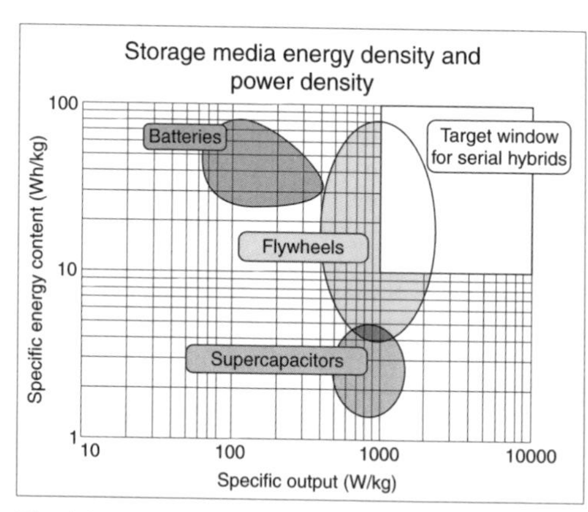

Fig. 4.3-19 Storage media energy and performance.

energy content is rather high for such systems, particularly for supercapacitors.

Lead/gel batteries ("gel cells") are inadequate for the described hybrid applications. Their charging and discharging performance is limited and drops even further at low temperatures. If the state of charge exceeds 50 or 60%, charging current must be drastically reduced to avoid gassing. Lead-acid batteries are highly developed and therefore very attractively priced.

The Na/NiCl battery is only suitable for "range extender" hybrid variants; otherwise, its specific output is too low. This battery must be constantly held at a high operating temperature, which appears practicable for electric vehicles which generally spend several hours per day connected to the power grid via a charging device. However, for hybrid vehicles whose stated concept predicates that they be independent of any charging infrastructure, such required continuous battery heating is unacceptable. At present, high-performance lithium batteries, capable of respectable energy content, have been announced. This type of battery uses rare, expensive cobalt; currently, intensive work is under way to replace this with alternative materials such as manganese oxide.

Alkaline batteries (nickel-cadmium and nickel-metal hydride) inherently offer high performance potential, even at relatively low charge. This is equally true for quick charging at high states of charge (load leveling, regeneration). For most hybrid applications, these are decisive parameters. Alkaline systems are relatively expensive due to their higher material costs.

Other important battery considerations are longevity, recyclability, and battery behavior in the event of operational problems and accidents.

Transmission

Transmission applications in hybrid vehicles range from conventional transmissions in parallel hybrids to total lack of a staged transmission in the case of serial or branched hybrids. In parallel hybrids, automation of shift processes as well as clutch actuation is recommended in order to reduce driver workload. Application of CVTs (continuously variable transmissions; see Section 5.4.5) to hybrids offers new possibilities.

Electric machines

Requirements for electric machines in hybrid service are comparable to those in a pure electric car, which have been examined in Section 4.3.1.

Bibliography

[1] "Hybridfahrzeuge, Zukunftstechnologien für die Praxis." TÜV Südwestdeutschland, Holding AG/Verlag.
[2] Friedrich, A. "Gegenüberstellung von PKW mit Verbrennungskraftmaschinen, Hybridantrieben und Brennstoffzellen aus Umweltsicht." VDI-Bericht 1418, 1998.
[3] Killmann, G. "Toyota Prius—Development and market experiences." VDI-Bericht 1459, 1999.
[4] Antony, P., M. Krämer, and J. Abthoff. "Der serielle Hybridantrieb von Mercedes-Benz." VDI-Bericht 1418, 1998.
[5] Boll, W., and P. Antony. "Der Parallel-Hybridantrieb von MercedesBenz." VDI-Bericht 1225, 1995.
[6] *Technology for Electric and Hybrid Vehicles.* SAE SP-1331.
[7] Seiler, J., and D. Schröder. "Hybrid Vehicle Operating Strategies." Electric Vehicle Symposium 15, 9/1998, Brussels.
[8] "Hybrid Electric Propulsion Systems." Gorham Intertech Consulting, 7/1995.
[9] Friedmann, S. "Entwicklung von Bewertungswerkzeugen für Hybridantriebe." VDI-Bericht 1378, 1998.

4.3.4 Stirling engines, gas turbines, and flywheels

4.3.4.1 Stirling engines

A Scottish inventor, Rev. Robert Stirling, described in 1816 the engine concept bearing his name.

The Stirling engine operates with continuous external combustion or heat input. A heat exchanger transmits thermal energy to the working fluid (usually helium) within the working cylinder. With the aid of a displacer piston, the working fluid is forced back and forth between a chamber at constant high temperature and another chamber at constant low temperature, which results in a cyclical variation of pressure inside the chambers. These pressure oscillations are converted to mechanical energy by a working piston and crank drive. A radiator removes the Stirling engine's waste heat. For higher thermal efficiency, a regenerator is located between the hot and cold chambers.

The theoretical Stirling cycle (a closed cycle with continuous heat input) is described by two isotherms and two isochors. Fig 4.3-20 shows the theoretical Stirling cycle on *p-V* and *T-s* diagrams. In a Stirling engine process, the cycle operates "clockwise," while for a refrigerator or heat pump, it operates in a "counterclockwise" direction.

The individual stages of the theoretical cycle are:

1–2: isothermal compression; after adiabatic compression, the working fluid is cooled to its original temperature by a heat exchanger, and heat is rejected to the environment or to a medium to be heated.

2–3: isochoric (constant volume) heat transfer; heat is supplied to the fluid by the regenerator.

3–4: isothermal expansion; after adiabatic expansion, the working fluid is heated to its original temperature, which requires heat supplied by an external, continuous combustion process. Useful work is done during this stage.

4–1: isochoric heat rejection; heat is transferred to the regenerator.

The ideal process shown here can only be achieved if working and displacer pistons move discontinuously.

The thermal efficiency of the ideal Stirling cycle is identical to that of the Carnot cycle; that is,

$$\eta_{th} = 1 - \frac{T_1}{T_3} = 1 - \frac{T_{\min}}{T_{\max}}$$

Working fluids for this closed working cycle are almost exclusively gases such as hydrogen, helium, nitrogen, air, or various compound media. The working fluid must have certain properties such as high specific heat, low density, low viscosity, and high thermal conductivity. Helium and hydrogen are well suited for this application. In practice, the average process pressure, which should be as high as possible for optimum power density, is between 2 and 20 Mpa (20–200 bar).

In practice, there are, among other things, deviations from the ideal Stirling process:

— Discontinuous piston motion is not achievable with kinematic drive systems.

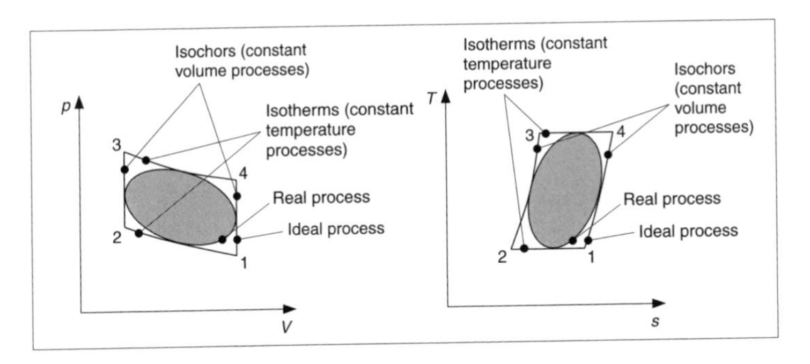

Fig. 4.3-20 The Stirling engine cycle on *p-V* and *T-s* diagrams.

- The presence of "dead space" due to heat exchanger and fluid passages.
- Heat input and rejection cannot be limited to cylinder walls.
- The regenerator volume can never be zero.
- The regenerator temperature is not constant over space and time.
- A heat transfer path exists between the hot and cold space.

The most common configurations of Stirling engines as powerplants are: *alpha machines*, designed as single-acting and double-acting machines. Instead of a displacer piston, alpha machines use a second working piston. *Beta machines* are those in which working piston and displacer piston occupy the same cylinder, with part of the cylinder traversed by both displacer and working pistons. *Gamma machines* are those in which working piston and displacer piston are realized in different cylinders.

The main components of Stirling machines are: working spaces and gas passages, working piston and displacer piston, drive system, regenerator, heat exchanger for rejecting heat to the cold side (cooler), heat exchanger for heat input on the hot side (heater), and combustion chamber with air preheater.

Drive systems are categorized as free piston, hybrid free piston, or kinematic power trains. Kinematic power train drives include, among others, crankshaft, crankshaft and crosshead, rhomboid drive, swashplate, wobble drive, rotary piston, slider crank, Carlquist, Ross, and Parsons drives. In every case, a drive system connects the motion of working and displacer pistons in such a way that their movements are coupled to one another.

Modern Stirling engines are double-acting, with, for example, four cylinders operating at different phase offsets. Because they exhibit good part-load efficiency compared to other combustion engines and their best-point efficiency is comparable to modern diesel engines, development of Stirling engines is ongoing. Particularly within the framework of hybrid combinations, Stirling engines open up a variety of new possibilities. By virtue of their continuous external combustion, Stirling engines have outstandingly low raw exhaust emissions, lower than those of gas turbines.

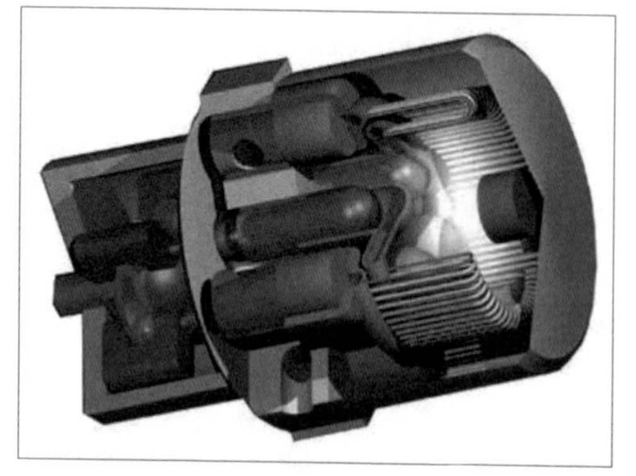

Fig. 4.3-21 Stirling engine (25 kW).

The most important advantages of Stirling engines are:

- Very low emissions of all regulated pollutants (HC, CO, NO_x), especially when combined with catalytic combustion chambers.
- Good efficiency at best operating point; variable displacement types even show good part-load efficiency.
- Better torque characteristics than internal combustion engines.
- Lower noise and vibration levels.
- Wide variety of possible heat sources.
- Multifuel capability.

These are countered by the following disadvantages:

- Less satisfactory throttle response than piston engines (not applicable to variable-displacement types)
- More complex power regulation
- Greater space requirements due to large heat exchangers
- High manufacturing costs due to cost-intensive construction

Vital statistics

Specific output	10–500 W/kg
Power density	50–500 W/L
Efficiency, part load	approx. 30%
best point	approx. 40%
Cost	50–1,500 €/kW
Service life	>11,000 h (operational)

Fig. 4.3.21 shows a cutaway of a Stirling engine.

4.3.4.2 Gas turbines

Gas turbines are combustion engines employing continuous internal combustion. Air required to oxidize the fuel undergoes the engine cycle's individual state changes in spatially separate sections such as compressor, combustion chamber, and turbine(s) which may be connected via diffusers and/or scrolls.

The operating principle of the gas turbine is as follows: atmospheric air, continually drawn in through a filter and silencer, is compressed to a working pressure of about 5 bar by a compressor, generally a centrifugal-flow compressor. Next it passes through a heat exchanger, configured as a recuperator or, preferably, as a rotating regenerator, where the air is preheated and directed to a combustion chamber. In the combustion chamber, gaseous, liquid, or emulsified fuel is injected, ignited, and burned with a portion of the airstream. By mixing with the remaining air, combustion products are cooled to about 1,300 K at the turbine inlet. The resulting combustion gases transfer their energy to one, two, or three turbine stages, mounted on one or more shafts (up to a maximum of three). The gases are expanded in the turbine; for example, in a two-shaft gas turbine, about two-thirds of the energy drop occurs in the turbine used to drive the compressor. The remaining energy may be used to drive the power turbine. The remaining hot combustion gases flow through a heat exchanger and so provide energy for preheating induction air. The power turbine rotational energy is drastically geared down (reduction gearbox), passed through a change-gear transmission, and directed to the driveshaft. The compressor turbine not only draws in air, but may also be used to drive auxiliary devices such as alternator or hydraulic pumps.

The usual gas turbine configurations for automotive applications are differentiated by the number of shafts and by individual components such as heat exchanger, intercooler, or secondary combustor (reheater) for improved thermal efficiency. In the case of the single shaft gas turbine, the gas generator section and power turbine are mounted on a common shaft. This configuration exhibits an unfavorable torque response when accelerating from idle, as the gas generator speed is always identical to driveshaft speed. This torque response is of course less problematical if the gas turbine is used to power the generator of a serial hybrid. In the case of a two-shaft gas turbine (Fig. 4.3-22), the gas generator shaft (with compressor and associated compressor turbine) is decoupled from the driveshaft and power turbine. The torque response of this configuration is noticeably better than that of the single-shaft gas turbine or gasoline engine. To reduce automotive gas turbine fuel consumption at part load and idle, as well as to improve acceleration, load regulation is achieved by controlling the working gas temperature and/or by means of variable-geometry turbine and compressor guide vanes. In the three-shaft gas turbine, compression includes intermediate cooling; and a second combustion stage is possible in the expansion section, which further reduces fuel consumption, but at the cost of increased complexity and greater manufacturing effort.

Possible fuels for gas turbines include diesel or gasoline fuels, but also alternative hydrocarbons, natural gas, gasified coal, and even coal dust. The combustion process takes place continuously with a great excess of air. Admixing cold air at the combustion chamber inlet keeps the combustion temperature at about 1,300 K, lower than the peak temperatures of combustion engines with discontinuous combustion. As a result, among the turbines currently available for automotive applications, CO_2 emissions from this continuous combustion process are higher than those of conventional combustion engines, but CO, HC, and, to some extent, NO_x emissions are considerably lower.

A vehicle propulsion system powered by a gas turbine offers emissions advantages plus additional benefits such as multifuel capability, attractive steady-state torque curve, relatively good torque response (dual-shaft gas turbines), low vibration levels, and long service intervals. Drawbacks include higher fuel consumption, large heat exchangers needed for higher thermal efficiency, as yet economically unavailable materials for combustion chamber components (high temperature ceramics such as Si_3N_4, SiC, and glass ceramics which cannot yet be made or worked in the required purity), excessive noise development, and limited suitability for lower-power applications, as well as less desirable throttle response compared to reciprocating piston engines.

Vital statistics

Specific output	300–500 W/kg
Power density	200–400 W/L
Efficiency, part load	10–15%
best point	25–40%
Cost	15–25 €/kW
Service life	2,000–4,000 h (operational)

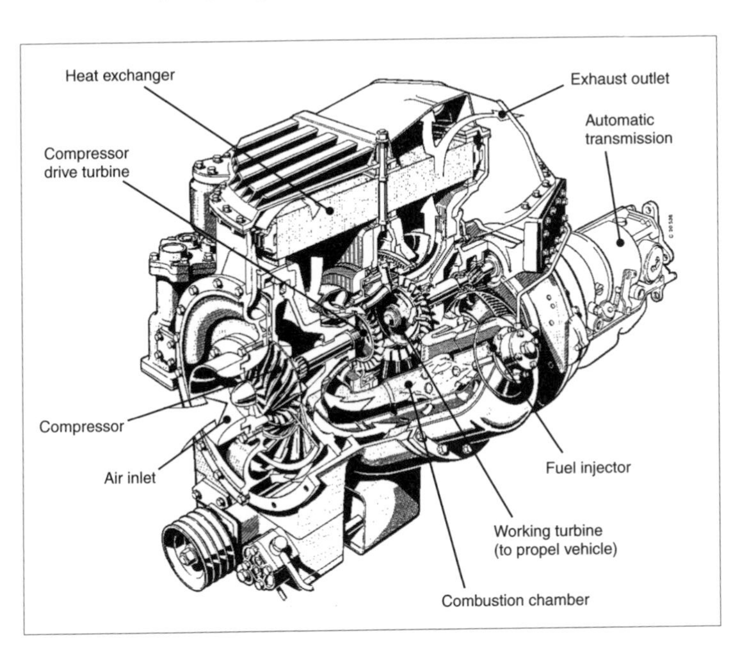

Fig. 4.3-22 Gas turbine (Daimler-Benz experimental passenger car).

4.3.4.3 Flywheels

Flywheels are among those mechanical energy storage media which employ kinetic energy (in this case, energy of rotation) to store energy. Flywheels are primarily used to smooth short-term load and power fluctuations, to achieve higher peak power, to achieve a transition during power interruptions, and to store otherwise unusable energy (e.g., regenerative braking—recovering kinetic energy through braking, and using this energy for the next acceleration event). In the past, flywheels consisted of steel discs; today, for storing large amounts of energy, flywheels are made of composite materials. Distinction must be made between purely mechanical flywheels as short-term energy reservoirs and electromechanical flywheels, in which a motor-generator unit stores or extracts energy from the flywheel (Fig. 4.3-23).

The energy stored in the flywheel, T, may be calculated from the moment of inertia I and the angular velocity ω: $T = \frac{1}{2} I \omega^2$. The rotational moment of inertia I is proportional to the mass of the object and the square of its radius of gyration, r, from the axis of rotation. For a cylindrical flywheel, $I = \frac{1}{2} m r^2$. Because flywheel storage media are usually not completely "discharged," given a minimum angular velocity ω_{min} and a maximum angular velocity ω_{max}, the energy content of a flywheel is

$$T = \frac{1}{2} I (\omega_{max}^2 - \omega_{min}^2).$$

The choice of flywheel material and geometric dimensions determines the storable energy per mass unit, also known as the specific energy density of the flywheel: $T/m = K_s(\sigma/\rho)$. The specific flywheel energy is dependent on the flywheel geometry, characterized by the shape factor K_s, and its material properties (yield strength σ and material density ρ). The value of the shape factor K_s ranges between 0 and 1 (a disc of uniform strength has a shape factor of 1, in practice not achievable; a double conical disc 0.81; an unperforated circular disc 0.61; a circular ring 0.5; and a perforated disc perhaps 0.31). High specific energy densities are achieved with high tensile strength combined with low density of the materials used.

In addition to the rotor, system components of a flywheel storage system include the housing, bearings, and hydrostatic or electric converter. While earlier rotors were almost exclusively made of high-strength rolled or forged steel or titanium alloys, modern flywheels with high energy densities may be realized using fiber-reinforced composite materials. Good specific strength, defined as the ratio of yield strength σ and material density ρ and expressed as a breaking length in km, is exhibited, for example, by materials such as alloy steel 18Ni-400 (specific strength = 35 km) and fiber composite materials S1014 glass-epoxy (92 km) and carbon fiber-epoxy (110 km). The housing is intended to serve two functions: fulfill a protective function in the event of a burst flywheel and permit operating the rotor in a low-pressure atmosphere (considerable reduction in aerodynamic friction losses). Flywheel bearings should introduce low friction losses, while meeting very high safety standards. Magnetic bearings (permanent magnets or electromagnets) could be used for high-speed flywheels; in such bearings, there is no mechanical contact within the bearing (no wear, no friction losses). The energy converter is responsible for adding or withdrawing kinetic energy from the flywheel, a task achievable only by means of changing the flywheel's rotational speed. This function may be achieved by stepless hydrostatic drives or electromechanical converters (motors) combined with power electronics.

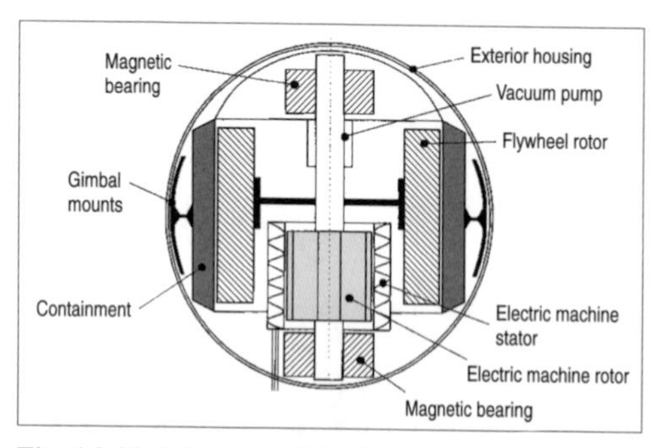

Fig. 4.3-23 Schematic of flywheel.

Flywheel storage makes sense in such vehicles whose operation includes frequent braking and accelerating phases (e.g., city buses and light rail vehicles in public transit service). Modern concepts are predicated on hybrid drives, in which flywheel storage is used alongside a conventional propulsion system based on an electric motor or combustion engine. The flywheel stores and utilizes otherwise lost braking energy. In actual hybrid buses, fuel savings of about 25% have been realized. Other advantages of a hybrid design are possible reduction of the main propulsion engine's rated power output, or a smaller battery power rating (battery load leveling) and more continuous engine operation. Flywheel storage combined with high-tech solutions (high-strength materials, vacuum housing, magnetic bearings, electronic control) may enjoy a bright future, even if the concept is restricted to specific market niches.

Vital statistics of electromagnetic flywheels

Specific output	500–4,500 W/kg
Power density	700–1,450 W/L
Specific energy	5–65 Wh/kg
Energy density	10–50 Wh/L
Efficiency (charging/discharging)	85– 95%
Cost	5,000–25,000 €/kWh
Service life	approx. 150,000 cycles

Bibliography

Stirling engines

[1] Förster, H.J., and K. Pattas. "Fahrzeugantriebe der Zukunft," Teil 1 und 2. Special edition of *Automobilindustrie*, Vogel-Verlag, Issues 3/1972 and 1/1973.

[2] Gelse, W. "Erfahrungen mit Stirlingmotoren bei DB im mobilen Einsatz." Presentation at the beim DB Technology Workshop "Stirlingmotor," 1994.

[3] Künzel, M. "Stirlingmotor der Zukunft." VDI-Verlag, Düsseldorf, 1986.

[4] Meijer, R.J. "Prospects of the Stirling Engine for Vehicular Propulsion." *Philips Tech. Rev.* 20, 1970, pp. 245–276.

[5] Meijer, R.J. "Der Philips-Stirlingmotor." MTZ 29, 1968, pp. 284–289; *Forschen, Planen, Bauen* (MAN house publication), May 1970.

[6] Peters, H. "Stirlingmotor—Stand der Technik und Anwendungsmöglichkeiten." Dissertation toward fulfilling requirements of state examination, Universität Bonn (Bon–1B-9632), 1996.

[7] Steinle, F. "Stirling-Maschinen-Technik: Grundlagen, Konzepte und Chancen." Verlag C. F. Müller, Heidelberg, 1996.

[8] Walker, G., et al. *The Stirling Alternative, Power Systems, Refrigerants and Heat Pumps*. Gordon and Breach Science Publishers, New York, 1994.

[9] Walker, G. "Stirling Powered Regenerative Retarding Propulsion System for Automotive Application." 5th International Automotive Propulsion Systems Symposium *Proceedings*, Detroit, April 1980.

[10] Werdich, M. *Stirling-Maschinen—Grundlagen, Technik, Anwendung*. Ökobuch Verlag, Freiburg, 1994.

Gas turbines

[11] Buschmann, H., and P. Koessler. *Handbuch für den Kraftfahrzeugingenieur.* "Gasturbinen," pp. 485–510. Deutsche Verlags-Anstalt, Stuttgart, 1991.

[12] Förster, H.J., and K. Pattas. "Fahrzeugantriebe der Zukunft," Teil 1 und 2. Special edition of *Automobilindustrie*, Vogel-Verlag, Issues 3/1972 and 1/1973.

[13] Förster, H.J. *Stufenlose Fahrzeuggetriebe in mechanischer, hydrostatischer, hydrodynamischer, elektrischer Bauart und in Leistungsverzweigung*. Verlag TÜV Rheinland, Cologne, 1996.

[14] Seiffert, U., and P. Walzer. *Automobiltechnik der Zukunft.* VDI-Verlag, Düsseldorf, 1989; published in English as *Automobile Technology of the Future*. SAE, Warrendale, PA, 1991.

[15] Walzer, P. *Die Fahrzeug-Gasturbine*. VDI-Verlag, Düsseldorf, 1991.

Flywheels

[16] Biermann, J.W. "Untersuchungen zum Einsatz von Schwungradspeichern als Antriebselemente für Kraftfahrzeuge." Dissertation, Rheinisch-Westfälische Technische Hochschule, Aachen, 1981.

[17] Kolk, M. "Ein Schwungrad-Energiespeicher mit permanentmagnetischer Lagerung." Report by Forschungszentrum Jülich, 1997.

[18] Reiner, G., and K. Reiner. "Energetisches Betriebsverhalten eines permanenterregten Drehmassenspeichers in Theorie und Praxis." VDI conference "Energiespeicher für Strom, Wärme und Kälte," VDI Berichte 1168, Leipzig, 1994.

[19] Sprengel, U. et al. "Positionspapier zur Energieversorgung in der Raumfahrt." Deutsche Forschungs- und Versuchsanstalt für Luft- und Raumfahrt (DFVLR), Stuttgart, 1986.

[20] von Druten, R.M. et al. "Design Optimization of a Compact Flywheel System for Passenger Cars." VDI-Bericht 1459 (1999), "Hybridantriebe," pp. 331–343.

5 Powerplants

5.1 Otto-cycle engines

This section will describe that class of powerplants which came to dominate the first century of the motor vehicle: reciprocating-piston combustion engines coupled to a speed/torque converter and a clutch for driving off or shifting gears.

An automotive powerplant must fulfill an entire series of operational demands:

- The vehicle must be able to start from rest and to maintain any desired speed up to a certain top speed.
- Output torque and speed must be capable of rapid modulation in order to permit dynamic changes in vehicle behavior.
- The energy carrier must provide high energy content within a small space and with light weight. It should be possible to achieve the desired vehicle range without interruption or refueling and without sacrificing significant payload or space.
- For mobile applications, masses and volumes should be kept as small as possible.
- The overall system should be capable of withstanding motion, shocks, and vibration.
- The powerplant should be available for operation on short notice.

Along with these basic technical requirements, vehicle manufacture and operation must also meet economic goals and, increasingly, ecological requirements and growing demands for operational comfort.

Of "alternative" powerplants, of which there has been no shortage over the years, none has been able to assert a position as a prime mover of motor vehicles. Their special advantages have always been countered by their own unique disadvantages—some technical, some economical—which resulted in unfavorable compromises within the total spectrum of properties that must be considered for a motor vehicle engine: technology, cost, immediate availability and range. In the face of changing boundary conditions–for example, regulatory requirements for emissions limits or, indeed, absolute lack of emissions at the point of vehicle operation—new yardsticks must be established for evaluating new or alternative propulsion systems.

The overwhelming, undisputed choices for motor vehicle powerplants are the *Otto-cycle engine* and the *diesel engine*, both reciprocating-piston internal combustion engines. The primary difference between these two concepts is *external ignition* in the case of Otto-cycle engines and *autoignition* in the case of diesel engines. An additional classic differentiating feature is the means of load regulation. In Otto-cycle (in this context generally taken to mean gasoline-fueled) engines, load is regulated by means of the inducted air–fuel mixture. A throttle plate (throttle butterfly) restricts the cross section of the intake tract, throttling the engine. The diesel engine draws its combustion air freely; only the mass of injected fuel is varied (although this tactic now also applies to the direct-injection gasoline engine; see Section 5.1.2). Because of the different ignition properties of gasoline and diesel fuels, the compression ratio of Otto-cycle engines is lower than that of diesels. The compression ratio limit is determined by the fuel's knock resistance; at present, gasoline-engine compression ratios are limited to approximately $r_c = 10–11$. For diesel engines, the compression ratio must be approximately $r_c = 17–24$ to make certain that temperature and pressure are sufficient to ensure autoignition of diesel fuel under cold starting conditions. Other engineering specifications and properties for Otto-cycle and diesel engines occupy their own typical ranges, but their domains overlap.

5.1.1 Otto-cycle engines, Part 1: Fundamentals

The designation "Otto engine" goes back to Nicolaus August Otto, who operated the first four-cycle engine in 1876 at the Gasmotorenfabrik Deutz AG (Deutz Gas Engine Works). Otto was granted a patent for this invention on August 4, 1877 [1]. Diesel engines also employ the four-stroke cycle, and both Otto-cycle and diesel engines may also operate on a two-stroke cycle (see Section 5.7). In this regard, the designation "Otto-cycle engine" as it is used today is not strictly a reference to the work of N.A. Otto. A more appropriate name would be the designation "SI engine" or spark ignition engine, signifying an engine whose combustion process is initiated by an electric spark.

5.1.1.1 Operating principles

In its working cycle, the engine employs combustion to release chemical energy bound in fuel, converting it to heat and mechanical work performed on a crankshaft. The theory of the working cycle is described by the field of study known as thermodynamics. One characteristic feature is "internal combustion;" that is, the combustion process takes place within a chamber bounded by the piston, cylinder, and cylinder head. If continuous work is to be done, this process must repeat cyclically. To this end, the working medium must be returned to its initial condition. Because of the previously completed combustion process, this is only possible within an enclosed space by replacing the combustion gases with freshly admitted fuel and air; this is termed the gas exchange process. It is customary to represent this process on a p-V diagram: that is, plotting cylinder pressure as a function

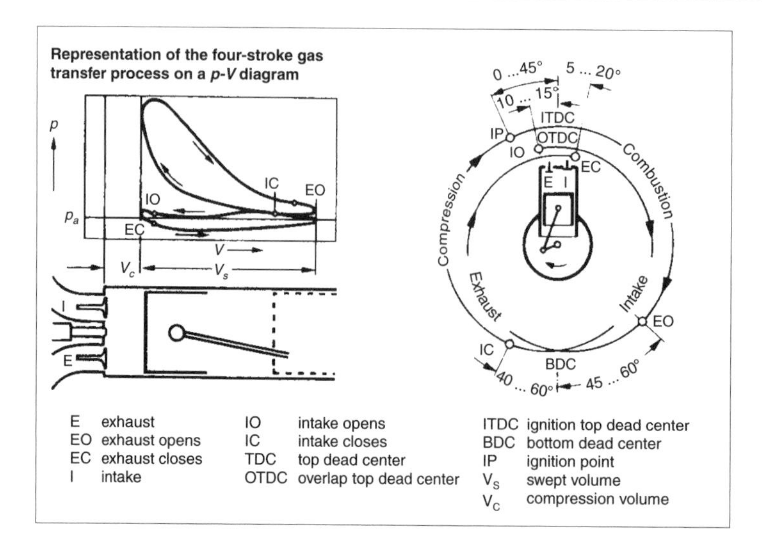

Fig. 5.1-1 Four-stroke Otto cycle.

of cylinder volume (Fig. 5.1-1) or as a function of crankshaft angle.

5.1.1.1.1 Four-stroke cycle

In the four-stroke cycle (Fig. 5.1-1), the piston, traveling from top dead center (TDC) to bottom dead center (BDC), draws fuel–air mixture into the cylinder through the open intake valve. Thereupon the intake valve closes, and as the piston returns from BDC to TDC, the contents of the cylinder are compressed. In the process, pressure and temperature rise in accordance with the physical properties of the mixture in the cylinder (air, fuel/fuel vapor, exhaust end gas). Shortly before TDC, ignition takes place, followed by combustion and expansion of the combustion gases as the piston again travels toward BDC. Thereupon the exhaust valve opens, the spent combustion gases are forced out as the piston again travels to TDC, and the cycle repeats.

From the *p-V* diagram, the work done by the cyclic process, W_c, may be determined from $W_c = \int V \cdot dp$ (if pressure is plotted as a function of crank angle, $W_c = -\int p(\alpha)dV$). V_s is the swept volume of the piston. V_c is the compression volume (the volume contained between the piston and cylinder head at top dead center), p_a is ambient (atmospheric) pressure, against which the spent combustion gases are exhausted and from which the engine draws its fresh combustion air. The work produced during expansion must be greater than the work required for gas transfer and compression.

5.1.1.1.2 Efficiencies

The quality or efficiency of a working cycle is customarily compared to what would be possible under theoretical or "ideal" conditions. Ideal, in this sense, signifies simplified conditions which permit mathematical analysis, assuming "reversible" processes (free of losses). The efficiency chain of an Otto-cycle engine may be assembled from the following steps (see also DIN 1940) (Fig. 5.1-2).

Thermal efficiency η_{th}

As a thermodynamic reference process, the Otto-cycle engine is compared to the so-called "constant volume process"—that is, isotropic compression ($1 \rightarrow 2$), isochoric (= constant volume) heat input (combustion, $2 \rightarrow 3$), isentropic expansion ($3 \rightarrow 4$), and isochoric return of the ideal working fluid to the process's original condition ($4 \rightarrow 1$). Boundary conditions are:

- No thermal or gas losses, no end gas
- Ideal gas with constant specific heats c_p and c_v, $k = c_p / c_v = 1.4$
- Infinitely fast (instantaneous) heat transfer
- No flow losses

Thermodynamic efficiency η_{th}, or the efficiency of an ideal "complete engine," is defined as

$$\eta_{th} = (Q_{in} - Q_{out})/W_{in} = 1 - (Q_{out}/Q_{in})$$

where Q_{in} = heat transferred in ($2 \rightarrow 3$)
Q_{out} = heat transferred out ($4 \rightarrow 1$)
and $Q_{in} = m \cdot c_v(T_3 - T_2)$
$Q_{out} = m \cdot c_v(T_4 - T_1)$.
It follows that

$$\eta_{th} = 1 - [(T_4 - T_1)/(T_3 - T_2)].$$

For an ideal gas undergoing a reversible adiabatic process, $T \cdot V^{k-1}$ = const. from State 1 to State 2 and from State 3 to State 4. It follows that the thermal efficiency of the constant-volume process

$$\eta_{th} = 1 - (T_4 - T_1), \quad \text{and since} \quad T_1/T_2 = (1/r_c)^{k-1}$$

it follows that the thermal efficiency of the constant-volume process is

$$\eta_{th} = 1 - r_c^{1-k}$$

where compression ratio $r_c = (V_c + V_s)/V_c$.

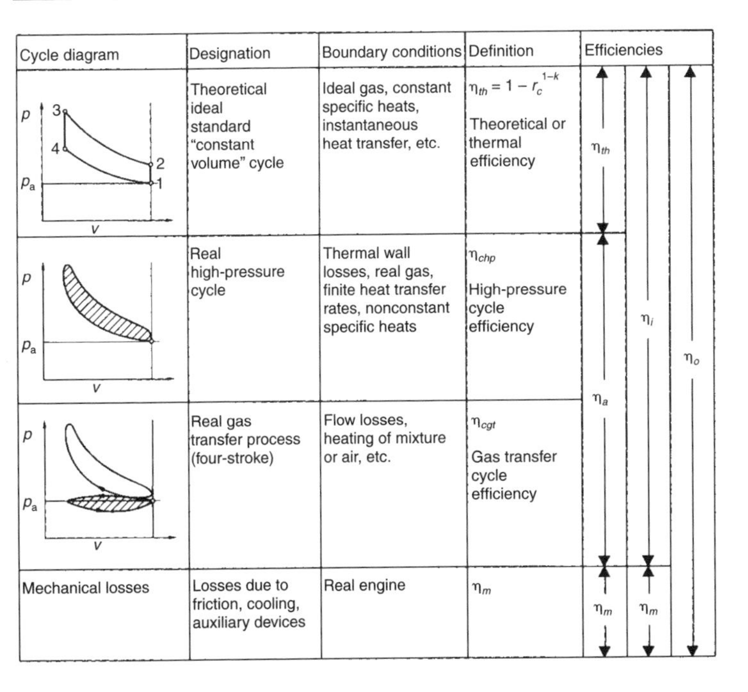

Fig. 5.1-2 Piston engine individual and overall efficiencies.

Availability conversion efficiency, η_a

This encompasses all differences between "ideal" and "real" cycles, in both high- and low-pressure processes: real working fluid, end gas trapped in the cylinder, thermal losses through cylinder walls, gas-flow losses, charging losses (volumetric efficiency), and real combustion processes.

$\eta_a = W_i/W_{th}$

where W_i = indicated work, calculated from the real pressure history, and W_{th} = work of the comparable ideal cycle = $Q_{in} - Q_{out}$.

Indicated efficiency, η_i

η_i is the product of (ideal cycle) thermal efficiency η_{th} and (real) cycle efficiency η_a, and tells us how much of the energy contained in the fuel is available in the form of indicated work.

$\eta_i = W_i/W_F$

where W_F = energy content of the added fuel
= $m_F \cdot Q_c$ = fuel mass × heat value ("heat of combustion").

Mechanical efficiency, η_m

Mechanical efficiency, η_m, encompasses all friction losses in the crank train and cylinder head, lubrication, coolant, and fuel pumping losses, etc.—In other words, all auxiliary devices necessary for engine operation.

$\eta_m = W_{eff}/W_i$

where W_{eff} = the effective work available at the flywheel.

Overall efficiency, η_o

The overall efficiency of an engine, the useful efficiency or effective efficiency η_o is the product of all individual efficiencies:

$$\eta_o = \eta_{th} \cdot \eta_a \cdot \eta_m = W_{eff}/W_F.$$

This represents the ratio of effective work available at the flywheel to the energy supplied by the fuel.

Normally-aspirated real automotive engines achieve overall efficiencies of up to $\eta_o = 0.3$ in their optimum operating range. In conventional Otto-cycle engine designs, this optimum operating range is near the middle of the engine speed range and at slightly less than full-load operation (Fig. 5.1-3).

5.1.1.1.3 Definitions and units

Engine power is determined according to standard procedures outlined in international standards such as DIN ISO 1585 or SAE J1349. These procedures include all auxiliary devices necessary for engine operation and are referenced to "standard conditions" of T_a = 298 K, p_a = 1 bar, and 30% relative humidity (DIN/ISO standard), or T_a = 25°C (= 298 K), p_a = 100 kPa (= 1 bar), and absolute dry air pressure of 99 kPa (SAE standard; see also SAE J3046/1). Reference to standard environmental conditions is important, because engine output varies according to atmospheric conditions. Power increases with denser air (higher barometric pressure, lower temperature). An altitude increase of 100 m reduces engine output by about 1%. Increased moisture content (higher relative humidity) also reduces engine output.

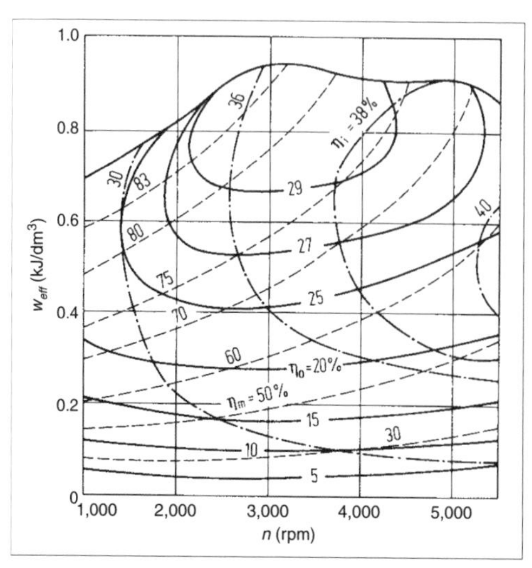

Fig. 5.1-3 Engine efficiency map for a typical Otto-cycle engine.

Reported parameters include maximum output at the flywheel, P_r, labeled "net brake power" in the standard, along with the associated engine speed n; and peak torque T_r and its associated engine speed (or speed range).

Engine output
$P_r = T \cdot \omega = 2\pi \cdot T \cdot n$,
or for P_r in kW, T in Nm, n in rpm,
$P_r = T \cdot n/9549$
$P_r = V_D \cdot bmep \cdot n/2$,
or for P_r in kW, $bmep$ in bar, n in rpm and displacement V_D of entire engine in liters,
$P_r = V_D \cdot bmep \cdot n/1{,}200$

Brake mean effective pressure
$bmep = 2 \cdot P_r/(V_D \cdot n) = 4\pi \cdot T/V_D$
or for P_r in kW, $bmep$ in bar, n in rpm and displacement V_D of entire engine in liters,
$bmep = P_r \cdot 1{,}200/(V_D \cdot n)$
or for $bmep$ in bar, T in Nm, V_D of entire engine in liters,
$bmep = 0.1257 \cdot T/V_D$

Torque
$T = V_D \cdot bmep/(4\pi)$
or for T in Nm, V_D of entire engine in liters, and $bmep$ in bar,
$T = V_D \cdot bmep/0.1257$

Specific output
$P_s = P_r/V_D$
in kW/L

Specific weight
(specific mass) $m_P = m/P_r$
in kg/kW, where m = engine weight according to DIN 70020-A

This is sometimes reported as the reciprocal, in kW/kg;
$m_P = P_r/m$

Brake specific fuel consumption ("bsfc") in g/kWh
$b_r = \dot{m}_f/P_r = 1/(Q_C \cdot \eta_O)$
$= (V_f \cdot \rho_f \cdot 3{,}600)/t_f \cdot P_r)$
where $\dot{m}_f$ = measured fuel consumption in g/h (or kg/h)
Q_C = heat value (heat of combustion) of fuel in kJ/kg
V_f = measured fuel volume in cm³
ρ_f = fuel density in g/cm³
t_f = measuring time in seconds

Efficiency
$\eta_O = P_r/(\dot{m}_f \cdot Q_C)$
$= 86/b_r$
if Q_C = 42,000 kJ/kg

5.1.1.2 Engine design and mechanics

5.1.1.2.1 Engine configurations

Motor vehicles may be powered by any of several engine configurations: inline engines, V engines, or horizontally opposed ("boxer") engines. Most recently, "W" engines have reappeared on the automotive scene (Fig. 5.1-4). Single- and two-cylinder engines are common on motorcycles, along with a few four-cylinder engines and in individual cases six-cylinder engines. Passenger cars are equipped with 3-, 4-, 5-, 6-, 8-, 10-, and 12-cylinder engines, and an 18-cylinder engine in W configuration has been presented in prototype form.

Motor vehicle *cylinder numbering convention* is standardized by DIN 73021 (see also SAE J824; these standards are not identical: for general applications and marine engines, ISO 1204 and 1205 specify reverse numbering order, beginning from the engine's output end). Seen from opposite the engine power output end (opposite the flywheel or clutch end), cylinders are numbered sequentially (1, 2, 3, etc.) as they are intersected by an imaginary horizontal plane, initially extending to the left from the crankshaft axis, which is then rotated clockwise around the engine axis. In the case of coplanar cylinders (cylinder banks), the cylinder nearest the observer is numbered "1" and the following cylinders in that plane are numbered sequentially.

Firing order is the sequence in which mixture in the cylinders is ignited. It is determined by the engine configuration, even firing sequences, ease of manufacturing the chosen crankshaft configuration, favorable crankshaft loads, etc. Firing order is specified beginning with cylinder number 1.

Inline engines (example 1 in Fig. 5.1-4, up to six cylinders) have one crankpin per cylinder and generally a single-piece cylinder head.

Horizontally opposed (also known as "flat" or "boxer") engines (example 4) with cylinder banks opposite each other also have one crankpin per cylinder and a single cylinder head per cylinder bank.

V engines (example 2) are characterized by two connecting rods sharing each crankpin, one rod from each cylinder bank. Again, each cylinder bank is surmounted by a single cylinder head. The angle between cylinder

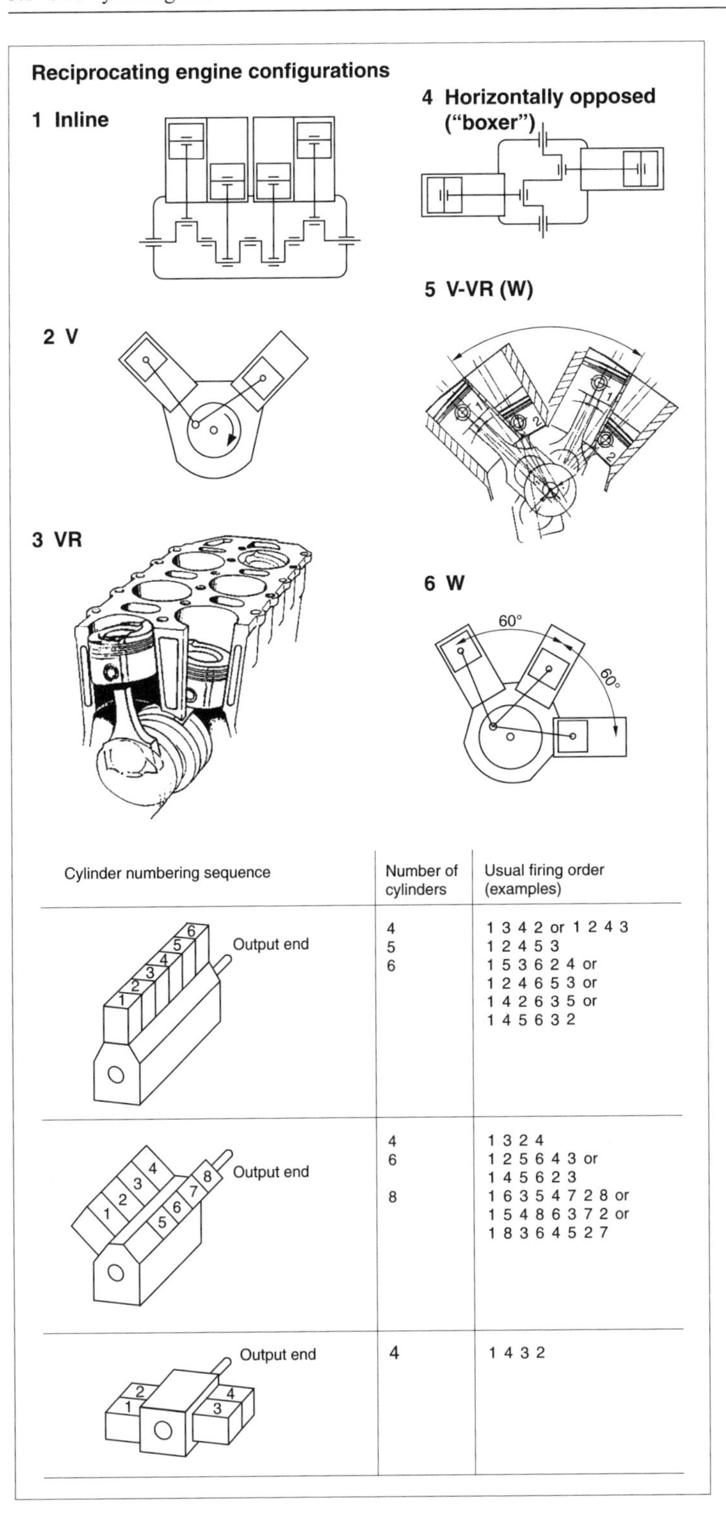

Cylinder numbering sequence	Number of cylinders	Usual firing order (examples)
Output end	4 5 6	1 3 4 2 or 1 2 4 3 1 2 4 5 3 1 5 3 6 2 4 or 1 2 4 6 5 3 or 1 4 2 6 3 5 or 1 4 5 6 3 2
Output end	4 6 8	1 3 2 4 1 2 5 6 4 3 or 1 4 5 6 2 3 1 6 3 5 4 7 2 8 or 1 5 4 8 6 3 7 2 or 1 8 3 6 4 5 2 7
Output end	4	1 4 3 2

Fig. 5.1-4 Automotive engine configurations, cylinder numbering, and firing order.

banks is 60°, 72°, or 90°, as determined by the number of cylinders (to achieve even firing) or by a common bank angle established for an engine family encompassing designs with varying numbers of cylinders. For an engine as shown in example 3, the V angle is so narrow (15°) that a single-piece cylinder head can be utilized; the staggered cylinders result in short overall length. However, the crankshaft has one crankpin per cylinder. On this basis, it would be considered an inline engine, and, indeed, the cylinders are numbered as for an inline engine. Consequently, this engine configuration is also known as a VR engine, a hybrid of an inline and V engine. (The designation VR comes from its V configuration and the German word *Reihenmotor*—literally "row engine"—for "inline engine"). One noteworthy characteristic is the "shortened" crank train of this design: planes through the axes of both cylinder banks intersect below the crankshaft axis. As a result, top dead center and bottom dead center are not 180° apart. For cylinders 1, 3, etc., the angular separation between TDC and BDC is <180° crank angle, for cylinders 2, 4, etc., it is >180° crank angle.

This VR engine configuration has recently also been presented as a V engine (V-VR or W engine, example 5), in which each crankpin carries two connecting rods, one from each cylinder bank. Cylinders are numbered sequentially in banks as in V-engine practice.

Most recently, another engine concept with cylinders in a W configuration has been presented, with three cylinder banks inclined 60° to each other (3 × 6 = 18 cylinders). Each crankpin carries three connecting rods, one per cylinder bank. The main advantage of this configuration is that a large number of cylinders can be arranged in a very compact form and provide very uniform, low-vibration engine operation.

At present, two opposing trends may be observed. Until now, by far the largest number of cars in Europe were powered by four-cylinder engines of approximately 1.4 to 2.0 liters displacement. Beyond these, there were several six-cylinder engines on the market, a few eight-cylinders, and very few 12-cylinder engines. Small vehicles with engines below 1.2 liters, as well as three-cylinder engines, were also exceptions. Since the mid-1990s, vehicles at the lower end of the market as well as vehicles in the luxury segment have become more numerous. On the one hand, "economy cars" with the lowest possible fuel consumption (3 L per 100 km, or 78 mpg) are being promoted, while at the other extreme, the market for powerful, comfortable luxury cars is experiencing worldwide expansion.

5.1.1.2.2 Crank train

The crank train serves to convert the linear motion of pistons and gas forces resulting from combustion in the cylinders into rotary motion and useful torque. As a result of cyclic operation and nonuniform motion, "internal" and "external" forces are generated by gas and inertia forces. Internal forces dictate design of pistons, connecting rods, crankshaft, and bearings. The free "external" forces produce forces and moments on the engine itself, which manifest themselves as vibrations and must be absorbed by the engine mounts.

From the crank train geometry (Fig. 5.1-5), for piston travel s_p measured from top dead center as the origin, and λ, the ratio of crank radius r to connecting rod length l:

$$S_p + l\cos\beta + r\cos\alpha = l + r;$$

solving for s_p

$$S_p + r(1 - \cos\alpha) + l(1 - \cos\beta)$$

Furthermore, $\overline{BD} = l\sin\beta = r\sin\alpha$, and therefore

$$\sin\beta = \frac{r}{l}\sin\alpha = \lambda\sin\alpha = \sqrt{1-\cos^2\beta}$$

$$\rightarrow(\cos\beta = \sqrt{1-\lambda^2\sin^2\alpha}).$$

Therefore, for piston travel s_p as a function of crank angle α and with crank radius to connecting rod length ratio $\lambda = r/l$,

$$s_p = r(1-\cos\alpha) + l(1-\sqrt{1-\lambda^2\sin^2\alpha}).$$

The square root may be replaced by the binomial expansion

$$1 - \tfrac{1}{2}(\lambda\sin\alpha)^2 - \tfrac{1}{8}(\lambda\sin\alpha)^4 - \tfrac{1}{16}(\lambda\sin\alpha)^6 - \dots$$

Terms of fourth and higher order are very small and may be ignored, so that piston travel may be expressed as

$$S_p = r(1 - \cos\alpha + \tfrac{\lambda}{2}\sin^2\alpha).$$

It follows that piston velocity $\dot{s}_p$ is

$$\dot{s}_p = \omega \cdot r\sin\alpha(1 + \lambda\cos\alpha), \quad \text{with} \quad \omega = \pi \cdot n$$

and piston acceleration $\ddot{s}_p$

$$\ddot{s}_p = \omega^2 r(\cos\alpha + \lambda\cos2\alpha)$$

with maximum value at top dead center

$$\ddot{s}_{ptdc} = \omega^2 r(1 + \lambda)$$

and at bottom dead center $\ddot{s}_{pbdc} = -\omega^2 r(1 - \lambda)$.

Acceleration as a function of crank angle α and piston travel s_p for various values of λ are also presented in Fig. 5.1-5. For $\lambda = 0$—that is, an infinitely long connecting rod—piston travel, velocity, and acceleration are purely sinusoidal.

From piston acceleration, it follows that the oscillating inertia force F_{osc} along the cylinder axis is

$$F_{osc} = m_p \cdot r\omega^2(\cos\alpha + \lambda\cos2\alpha)$$

where m_p is the reciprocating mass of the piston and connecting rod assembly, defined as

$$m_p = (1/3\, m_{conrod}) + m_{piston}$$

(1/3 connecting rod mass + entire piston mass).

The oscillating forces may be shown graphically if terms $\cos\alpha$ and $\cos2\alpha$ are plotted as a function of crank

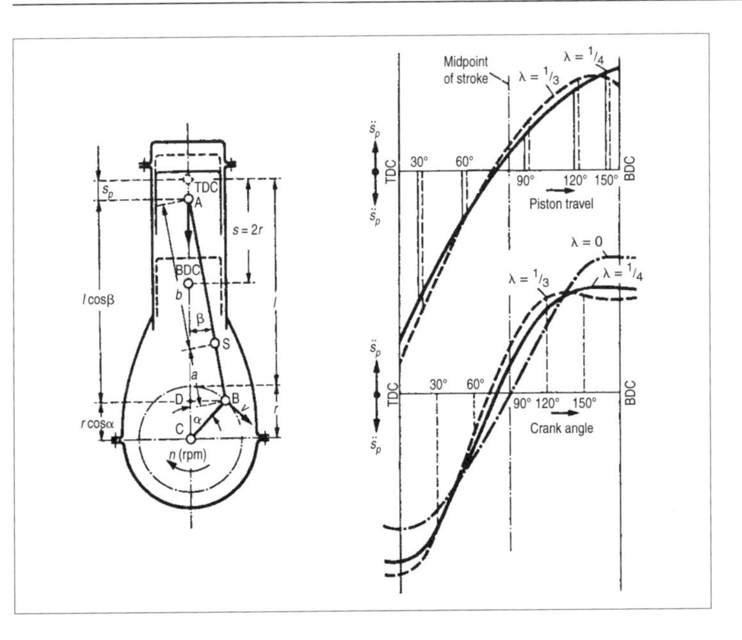

Fig. 5.1-5 Geometry and acceleration of crank train oscillating masses.

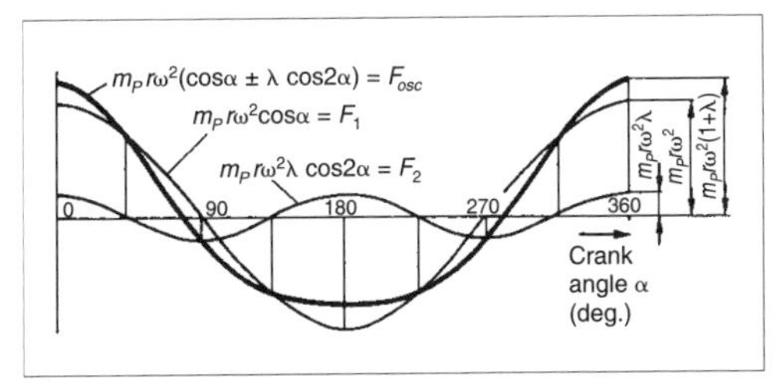

Fig. 5.1-6 Oscillating inertia forces resolved into base and harmonic oscillations.

angle (Fig. 5.1-6). The sinusoidal trace of force $F_1 = m_p \cdot r \cdot \omega^2 \cdot \cos\alpha$—that is, one complete cycle per crankshaft rotation—is overlaid with the sinusoidal force $F_2 = m_p \cdot r \cdot \omega^2 \cdot \lambda \cdot \cos 2\alpha$, with two complete cycles per crankshaft rotation. Therefore one speaks of first-order forces ($\sim\alpha$) and second-order forces ($\sim 2\alpha$). When these forces are superimposed, they reinforce at top dead center and partially cancel at bottom dead center.

The rotating mass is

$$m_R = 2/3 m_{conrod} + m_{\text{crank}}$$

where m_{crank} is the unbalanced rotating mass of the crankshaft with respect to the crankpin and therefore the rotating force on the crankpin

$$F_{rot} = m_R \cdot r \cdot \omega^2.$$

The magnitudes of forces F_{osc} and F_{rot} rise linearly with mass and quadratically with angular velocity (engine speed). Therefore, the objective is to design the lightest possible pistons, piston wrist pins, and connecting rods, and to achieve the smallest possible λ—that is, a relatively long connecting rod.

In the case of a multicylinder engine, inertia forces of individual cylinder units are superimposed. However, in the engine longitudinal direction, they operate in various planes, separated by the cylinder spacing. This may lead to amplification of exterior forces and moments or to reduction of exterior forces. One speaks of the engine's so-called "free" forces and moments. Some of the more common engine configurations and their associated first- and second-order free inertia forces and moments (not including counterbalancing measures) are shown in Table 5.1-1.

The gas force F_G acting on the piston is resolved into the force acting along the connecting rod, F_{Grod}, and the piston side thrust, F_{Gside} (Fig. 5.1-7). At the crankpin, the connecting rod force F_{Grod} in turn may be resolved into a radial force $F_{Gradial}$ and a tangential force F_{Gtan}. The tangential force acts at the crankpin offset r and transmits torque to the crankshaft. At crankshaft angle α, connecting rod angle β, and crank radius to connecting rod length ratio λ, the following may be derived analogously to the preceding for inertia forces:

Connecting rod force F_{Grod}

$$= F_G / \cos\beta$$
$$= F_G / \sqrt{1 - \lambda^2 \sin^2\alpha}$$
$$= F_G / [1 - ½(\lambda\sin\alpha)^2 - \ldots]$$

Piston side thrust F_{Gside}

$$= F_G \tan\beta = F_G \lambda \sin\alpha / \sqrt{\ldots}$$

Crankpin radial force $F_{Gradial}$

$$= F_G \cos(\alpha + \beta)/\cos\beta$$
$$= F_G (\cos\alpha - \lambda \sin^2\alpha / \sqrt{\ldots})$$

Crankpin tangential force F_{Gtan}

$$= F_G \sin(\alpha + \beta)/\cos\beta$$
$$= F_G [\sin\alpha + \lambda\sin\alpha\,(\cos\alpha / \sqrt{\ldots})]$$

Table 5.1-1 First- and second-order free forces and moments (without counterbalancing) for several common engine configurations

$F_{rot} = m_R \cdot r\omega^2$ $F_1 = m_p \cdot r \cdot \omega^2 \cdot \cos\alpha$ $F_2 = m_p \cdot r \cdot \omega^2 \cdot \lambda \cdot \cos 2\alpha$

Cylinder configuration	First order free inertia forces[a]	Second order free inertia forces[a]	First order free inertia moments[a]	Second order free inertia moments[a]	Firing interval
3-cylinder					
Inline, 3 crankpins	0	0	$\sqrt{3} \cdot F_1 \cdot a$	$\sqrt{3} \cdot F_2 \cdot a$	240°/240°
4-cylinder					
Inline, 4 crankpins	0	$4 \cdot F_2$	0	0	180°/180°
Horizontally opposed, 4 crankpins	0	0	0	$2 \cdot F_2 \cdot b$	180°/180°
5-cylinder					
Inline, 5 crankpins	0	0	$0.449 \cdot F_1 \cdot a$	$4.98 \cdot F_2 \cdot a$	144°/144°
6-cylinder					
Inline, 6 crankpins	0	0	0	0	120°/120°

[a] Without counterweights.

If the periodic gas forces and periodic inertia forces acting on the crank train components are now added together, we get the resulting piston force F_P, as shown in Fig. 5.1-7. In general, gas forces counteract inertia forces during the compression and power strokes. The representation of Fig. 5.1-7 is merely qualitative; the actual magnitudes of gas and inertia forces change considerably toward their maximum values, the former nearly linearly with engine load, the latter quadratically with engine speed. At low engine speed and full load, gas forces are dominant, while at high speeds, inertia forces predominate.

If an engine contains multiple cylinders and a crankshaft with crank throws offset relative to one another, individual cylinder torques overlap. Fig. 5.1-8 shows the resulting rotational force F_T for various engine configurations over a complete working cycle of 720° crankshaft rotation. For a four-stroke single-cylinder engine, rotational force is once again shown as the superposition of gas and inertia forces. The more cylinders are acting on a crankshaft, the less the degree of nonuniformity of rotational force, which approaches the mean rotational force F_{Tm}. The variation in rotational force during a complete cycle results in nonuniform rotational speed at the crankshaft output end. The degree of nonuniformity is defined as $\delta_S = (\omega_{max} - \omega_{min}) / \omega_{min}$. An energy storage device (a flywheel) must be used to reduce this nonuniformity to a level appropriate for the application (compromise between uniformity and throttle or acceleration response).

Free inertia forces and moments may be completely or at least partially compensated by rotating counterweights turning at crankshaft speed (for first-order

Table 5.1-1 (continued)

Cylinder configuration	First order free inertia forces[a]	Second order free inertia forces[a]	First order free inertia moments[a]	Second order free inertia moments[a]	Firing interval
6-cylinder (continued)					
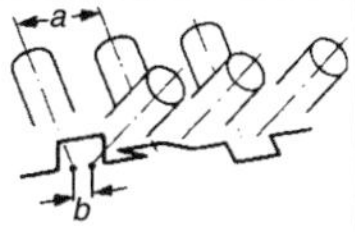 90° V, three crankpins	0	0	$\sqrt{3} \cdot F_1 \cdot a$[b]	$\sqrt{6} \cdot F_2 \cdot a$	150°/90° 150°/90°
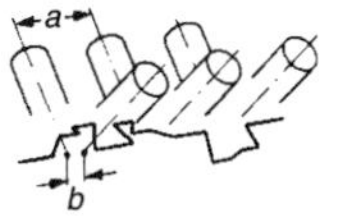 Even-firing 90° V6: three crankpins, crankpins offset 30°	0	0	$0.4483 \cdot F_1 \cdot a$ 	(0.966 ± 0.256) $\sqrt{3} \cdot F_2 \cdot a$ 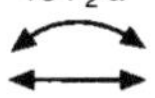	120°/120°
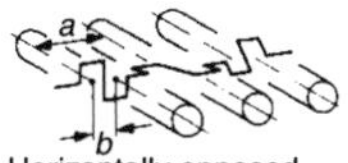 Horizontally opposed, six crankpins	0	0	0	0	120°/120°
 60° V, six crankpins	0	0	$3 \cdot F_1 \cdot a/2$	$3 \cdot F_2 \cdot a/2$	120°/120°
8-cylinder					
 90° V, four crankpins in two planes	0	0	$\sqrt{10} \cdot F_1 \cdot a$[b]	0	90°/90°
12-cylinder					
 60° V, six crankpins	0	0	0	0	60°/60°

[a] Without counterweights.
[b] May be fully counterbalanced with counterweights.

forces and moments) or at twice crankshaft speed (second-order), so that engine mounts need only absorb little or no inertia forces from moving masses within the engine. However, there remain other forces which must be absorbed by engine and transmission mounts: reaction forces from engine torque, gas forces acting through pistons and connecting rods, and inertia forces of the engine as a whole during vehicle acceleration in the longitudinal, vertical, and transverse planes.

5.1.1.2.3 Valve train

The valve train consists of inlet and exhaust valves, valve springs which close these valves, cam drive, and cam motion transfer elements. Conventional valve train configurations are shown in Fig. 5.1-9, along with a conventional valve timing diagram with associated valve velocity and acceleration curves. Modern motor vehicle design generally employs overhead valves (the valve heads hang "down" into the combustion chamber), abbreviated as OHV, with an overhead camshaft—OHC—or double overhead camshafts—DOHC. Valve configuration essentially determines the combustion chamber shape. Current designs include two, three, four, or five valves per cylinder, driven by one or two camshafts (whose duties are divided between intake and exhaust valves). In the case of an odd number of valves per cyl-

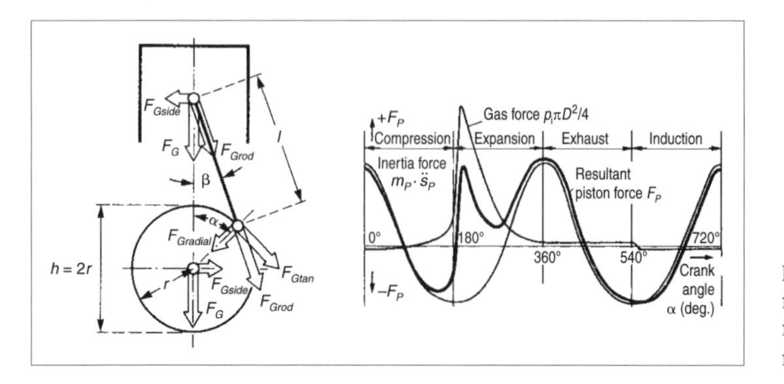

Fig. 5.1-7 Resolution of gas forces and resultant piston force F_P (sum of gas and inertia forces).

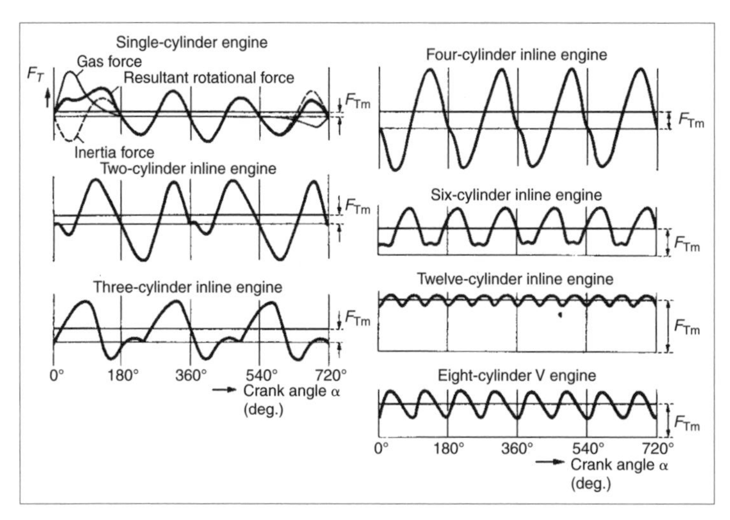

Fig. 5.1-8 Rotational force F_T over crankshaft rotation for various engine configurations.

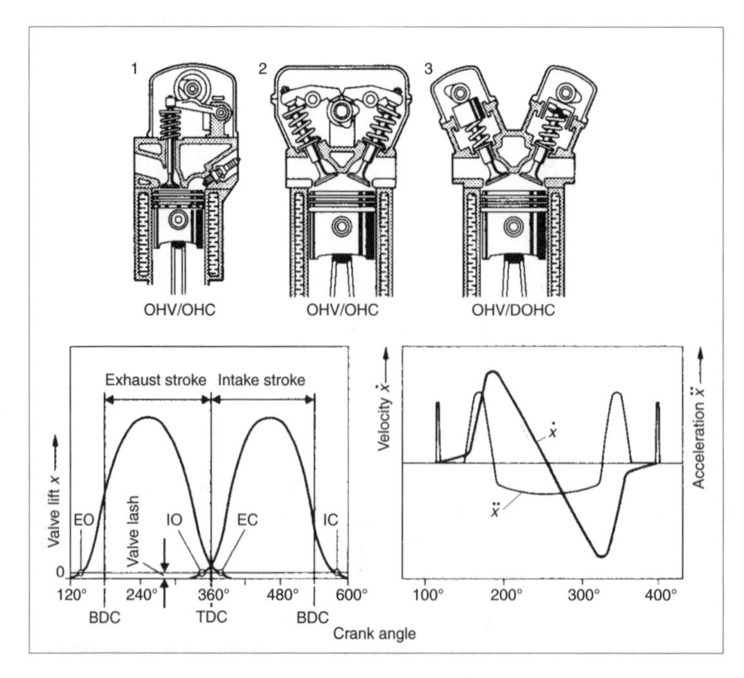

Fig. 5.1-9 Valve train configurations with valve timing diagram, valve velocity and acceleration curves. EO, exhaust opens; IO, inlet opens; EC, exhaust closes; IC, inlet closes.

inder, there is always one more intake valve than exhaust valve(s). Intake valves provide a larger flow cross section because cold intake air presents more flow resistance than hot exhaust gases.

Valve motion is transmitted from the cam lobe by means of finger followers (example 1), rocker arms (example 2), or bucket tappets (example 3). These intermediate elements transfer valve spring forces and forces from the sliding action of the cam lobe on their bearing surfaces. The valve stem should carry only minimal side forces; bucket tappets exhibit the ideal situation in that such tappets are constrained to move only in the direction of the valve axis.

Design of the cam lobe profile follows from the permissible Hertzian stress at the point of contact, the acceleration curve on the "ramp" sections of the profile leading to and from the lift section from the point where the valve lifts off its seat or reseats, as well as the amount and duration of valve lift—that is, the greatest possible opening time with the largest possible opening. Cam profiles are designed to have "smooth" acceleration curves in which the valve experiences steady acceleration without any discontinuities. Valve trains with roller followers require concave cam faces in order to achieve smooth valve motion. Grinding concave ("hollow") cam profiles is no longer a significant problem, as suitable band or disc grinding machines capable of generating the required radii are now available.

Dynamic forces impose limits on cam and valve lift. Positive contact must be maintained throughout the mechanical train which stretches from cam lobe to valve; up to the engine's design speed, no "jump" or "bounce" is permissible anywhere between the cam and valve. Due to the high contact stress at the cam lobe, special material treatment or surface coatings are required on both the cam lobe and its mating surface. These two surfaces must exhibit favorable wear and tribological properties in combination with each other. Despite all best efforts, the sliding action of finger, rocker, or bucket followers results in appreciable friction forces. Accordingly, needle roller bearing followers are finding increasing application at the cam lobe contact. This can result in considerable reduction in friction power. As shown in Fig. 5.1-10, at low engine speeds, roller followers may drop the friction mean effective pressure (*fmep*) to a third of its usual value and exhibit a very smooth characteristic across the entire engine speed range. At higher speeds, hydrodynamic lubrication at the sliding contact surface is so good that fmep values comparable to those of roller followers may be achieved. Because only the lower engine speed range has a significant influence on practical vehicle operation, friction reduction by means of roller followers has a marked effect on fuel economy.

Most valve trains are maintenance-free. Bucket tappets, finger followers, or rocker arms may incorporate hydraulic compensating elements, mounted at a support point or at the point of contact with the valve stem. Fed by engine oil, these continuously eliminate valve lash—the play between the cam base circle and the valve stem. This compensates for valve recession as a result of repeated seat impacts as well as for thermal expansion in both long- and short-term operation. Hydraulic compensating elements make the valve train quieter and maintenance-free. However, such compensating elements produce a light, continuous pressure of the lifter or rocker arm against the cam base circle, resulting in increased valve train friction. Some manufacturers are once again abandoning hydraulic compensating elements, instead employing valve seat materials with better long-term thermal stability; as a result, the preset valve lash hardly changes over extended periods of operation.

Camshafts are driven by a chain (single or duplex roller chain) or a timing belt, in either case driven at half crankshaft speed (for four-stroke engines). For fixed cam timing (no variable valve timing), the timing chain or timing belt passes over one or both cam timing sprockets. Loading is periodic, as camshaft drive torque is the sum of cyclic loading imposed on individual cam lobes.

In conjunction with the intake tract and exhaust system, the valves and valve train control the engine's gas exchange function. This is a highly dynamic process, affected as it is by flow cross sections and lengths as well as valve opening characteristics. As a result, cylinder filling can only be optimized over a limited engine speed range, unless variable valve geometry and timing are employed.

To better match the valve train to varying engine speeds (speed range <1,000–>6,000 rpm), the intake cam timing may be shifted to provide greater valve overlap at higher speeds—that is, the intake valve opens even earlier (while the exhaust valve is still open). Some manufacturers also vary valve lift in order to amplify the effect. The objective is to improve volumetric efficiency at low engine speed as well as higher rpms, thereby increasing engine torque. In interplay with volumetric efficiency, compression ratio, and ignition, the limiting factor is ultimately the engine's knock limit, which limits torque.

Additional comments regarding variable valve trains may be found in Section 5.1.2.3.

5.1.1.2.4 Gas exchange elements

"Gas exchange" is the replacement of combustion gases by fresh induction air or air/fuel mixture. This is controlled by the intake and exhaust systems, which, in conjunction with the valves and their opening characteristics, determine filling of the working cylinder. The quality of the gas transfer process is defined by volumetric efficiency—the relationship between m_i, the amount of fresh air actually inducted, and the theoretically possible induction charge, m_{th}, for a given displacement: $\eta_v = m_i / m_{th}$. The quantity of fresh induction charge, $m_i = m_a$ for internal mixture formation (fuel added directly to cylinder) and $m_i = m_a + m_f$ for external mixture formation—that is, the induction charge—consists of either air alone or air plus fuel. Only if as much air (and therefore oxygen) as possible enters the cylinder and remains there can a correspondingly large quantity of fuel be added or injected for "complete" combustion to produce high engine output.

Distinction is made between "crossflow" and "counterflow" gas exchange concepts. In the case of a counterflow cylinder head, both intake and exhaust ports are located on the same side of the head. Intake and exhaust gas streams flow in opposite directions. This principle is often encountered on classic two-valve engines with valves arranged in a straight row. In the case of a crossflow cylinder head, one side serves as the intake side, the other as the exhaust side. This layout may sometimes be found on two-valve engines, but it is a requirement for all multivalve (i.e., ≥3 valves per cylinder) engines. In crossflow heads, the "cold" and "hot" sides are separate, giving more freedom in arranging ports and passages and in fuel system layout.

Gas exchange events must take place in very short time spans, over approximately 180 to 240° of crankshaft rotation for intake and exhaust. This implies times of approximately 3 to 0.5 ms at engine speeds of 1,000 to 6,000 rpm respectively. Given fixed geometry for intake and exhaust tract lengths and diameters, this process can be optimized over only a limited engine speed range. Fig. 5.1-11 shows the brake mean effective pres-

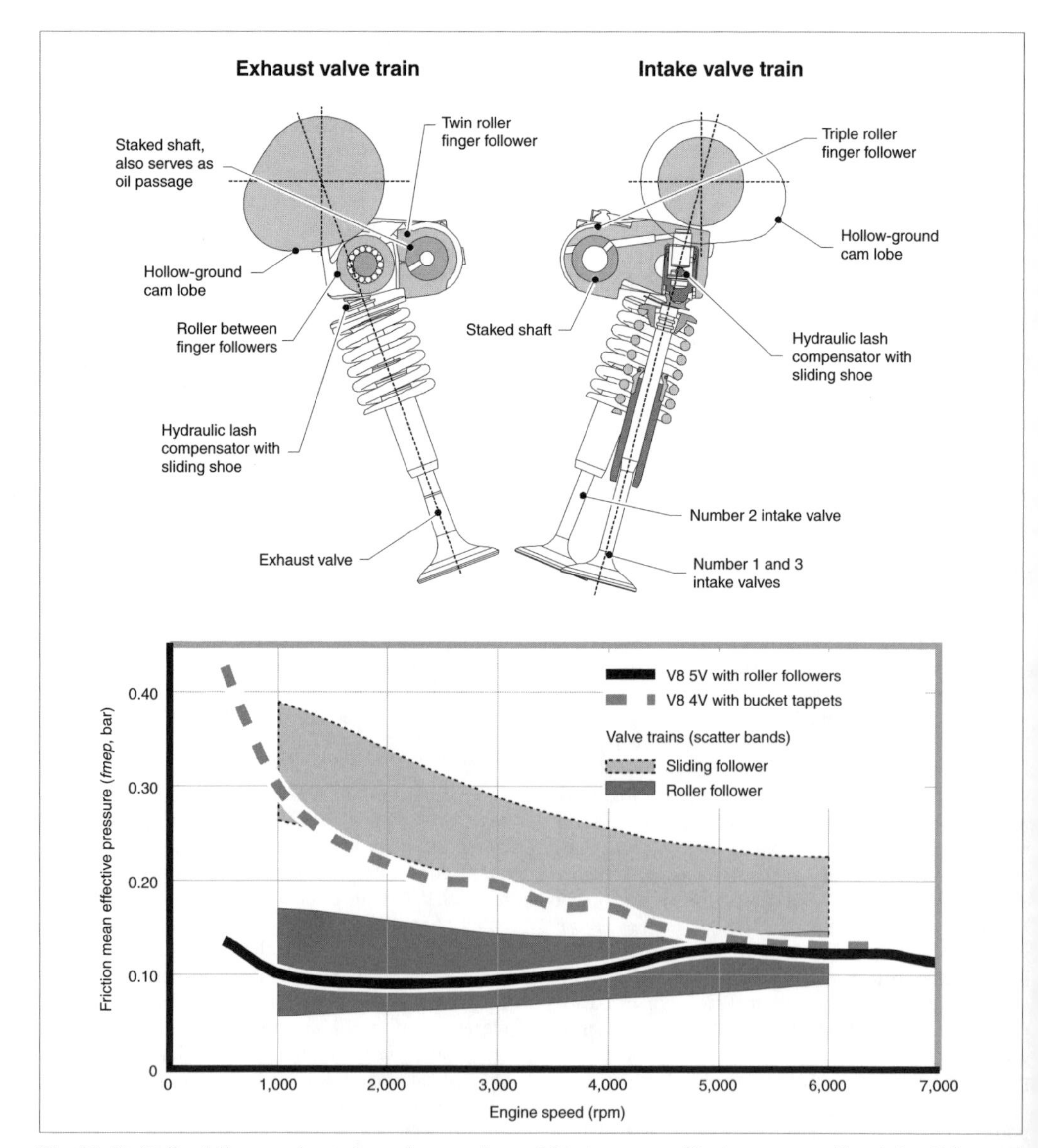

Fig. 5.1-10 Roller-follower valve train, and comparison of friction mean effective pressure (*fmep*) for sliding and roller followers (example: Audi V8 with five-valve technology).

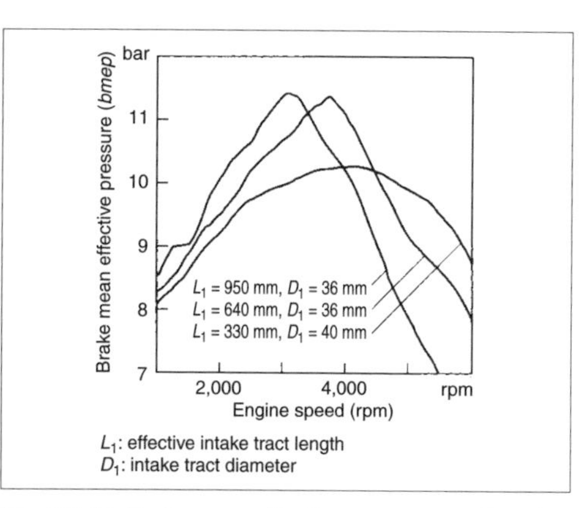

Fig. 5.1-11 Brake mean effective pressure as a function of engine speed for various intake tract geometries.

sure as a function of engine speed for a certain engine. Maximum torque is achieved at lower speeds with long, narrow intake tracts, while high maximum power is achieved with large, short passages. For ordinary cars, one would choose a layout giving high torque at low and medium speeds, while sports cars would be fitted with intake tracts giving a torque peak at higher speed and with greater maximum power.

In addition, the inner surfaces of the intake passages should be as smooth as possible and without any lips or edges to "trip" the gas flow. Sharp bends should be avoided. For multicylinder engines, it should be ensured that successive cylinders (in firing order) do not have negative mutual interactions. For these reasons, modern designs no longer use an intake manifold, instead employing a plenum and individual equal-length intake runners for each cylinder.

An increasingly common feature of current engine designs is variable-geometry intake systems. Flaps or rotary valves switch between two or even three different intake runner lengths. This permits a torque curve optimized across the entire engine speed range. The intake tract upstream of the throttle, including the intake duct and air filter, has a significant effect on gas flow. This part of the induction system must be designed with volumes and cross sections which damp out induction noise while minimizing pressure losses. Induction systems often employ the principle of the Helmholtz resonator, offering good sound damping across a wide frequency range.

In recent years, modular technology has been advanced to counter cost pressures and guarantee better production quality. An air intake module combines all components associated with combustion air supply and fuel injection (function and spatial arrangement) in a preassembled, prewired unit ready for testing, which is supplied to the engine assembly line. Fig. 5.1-12 shows an example of an intake module with a multipiece intake runner assembly in composite and aluminum, electronically controlled throttle butterfly, variable-geometry intake runner, fuel rail, and fuel injectors. In midsize and smaller engines, the intake module often also includes the air filter and ducting for prewarmed (cold-start) air.

On the exhaust side, the familiar thick-walled, heat-resistant cast iron exhaust manifold has become obsolete. Not only were such manifolds heavy, but they also absorbed a great deal of heat, which delayed catalytic converter lightoff. Modern tubular steel manifolds permit exhaust passage shapes and routings for better flow, they are lighter, and they extract less heat from the exhaust, allowing faster catalytic converter warmup. Often, exhaust manifolds are also fitted with sheet metal shrouds separated by an air gap, which acts as an insulator. New manufacturing technologies such as hydroforming have permitted development of products incorporating high

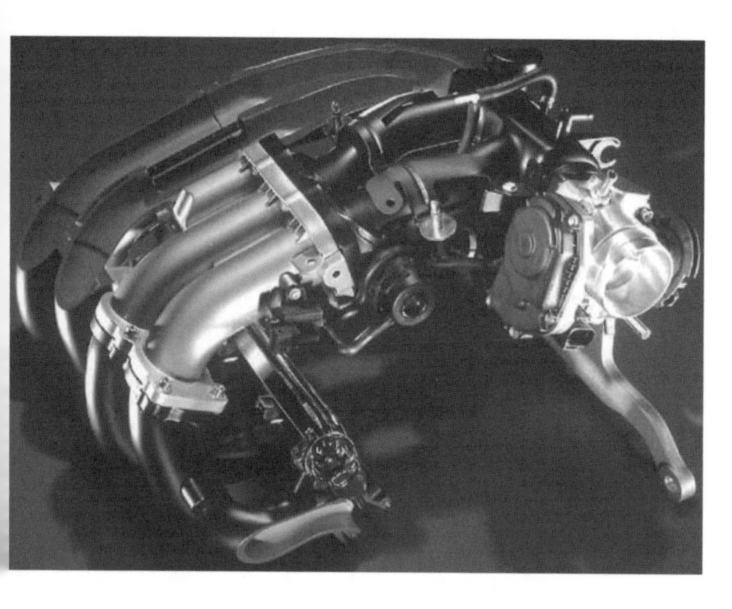

Fig. 5.1-12 Intake module (variable-geometry intake runners, electronic throttle, fuel injection system including wiring and hoses; Siemens).

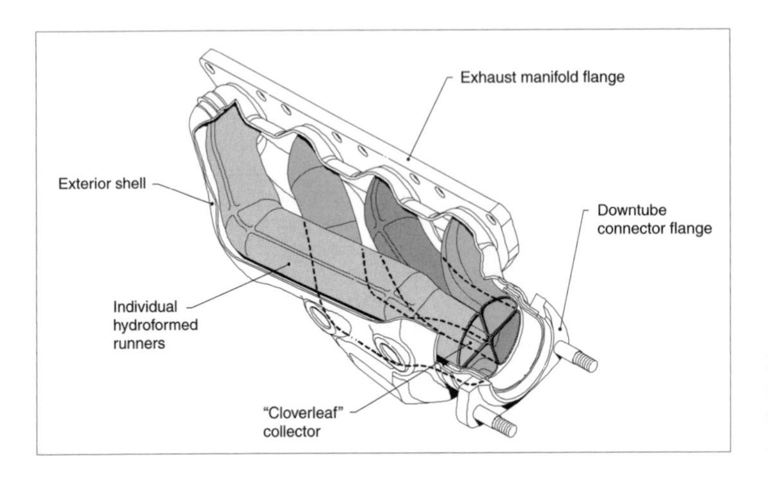

Fig. 5.1-13 Tubular steel exhaust manifold including air-gap insulation (example: Audi V8).

technical quality with low cost that are capable of withstanding severe thermal and vibration loading (Fig. 5.1-13).

Downstream of the exhaust manifold is the catalytic converter. On some vehicles, this is bolted directly to a flange at the manifold outlet, but most applications mount the converter some distance away, underneath the floorpan, in order to limit thermal loading of the converter at full load. Exhaust system dimensions also affect the gas exchange process. Exhaust back-pressure should be low, sound damping should be effective, and noise radiation from the system itself should be as low as possible. In order to prevent corrosion, stainless steel or aluminized steel is used.

Recently, intake and exhaust systems have not only been expected to provide noise insulation but have also been designed to integrate their associated noise with the vehicle's overall acoustic design. Psychoacoustic evaluation of vehicles demands not only compliance with low driveby noise levels, measured in dB(A) as dictated by test techniques and procedures, but also a sonorous, "agile," powerful, or even nostalgic sound image, according to the vehicle's brand philosophy. These developments, still in their early stages, are being advanced by cooperation between suppliers and vehicle manufacturers.

5.1.1.2.5 Materials

Engine design has recently witnessed introduction of an entire series of new materials: plastics with and without fiber reinforcement, new aluminum- and magnesium-based light alloys and composites, sintered metals, and improved ferrous materials. The driving force behind development and application of new materials is the desire to achieve improved functionality and weight savings. Product costs, including the material itself and the manufacturing process associated with any new material, must be financially attractive. A further significant consideration in selecting new materials is recyclability. Materials must not only be reintroduced to the raw materials cycle, but should themselves have high recycled material content.

Cast gray iron (GG25), sometimes with minor alloying elements, is the classic material for crankcases. It is cheap, easily cast and machined, exhibits good surface properties in association with pistons in the cylinder bores, is heat resistant, and has good sound and vibration damping qualities, but it is also heavy, with wall thicknesses of at least 3 mm. Some manufacturers now employ cast gray iron with vermicular graphite (GGV in German, or CG—compacted graphite iron—in English). It exhibits higher strength and can be cast in thinner sections to reduce component weight. This material is used not only for gasoline (Otto-cycle) engines but also is found on almost all new larger passenger-car diesel engines (V6 and V8).

Aluminum offers considerable weight reduction. Crankcases made of aluminum alloy AlSi9xx are fitted with cast-iron cylinder liners or Nikasil-coated (nickel-silicon compound) aluminum liners because the basic material is unsuitable as a sliding surface for aluminum pistons. On the other hand, crankcases made of the hypereutectoid alloy AlSi17xx, cast in low-pressure permanent molds, have suitable wear surfaces because silicon crystals which precipitate upon solidification can be exposed by chemical etching. In general, however, machining of this hard material is more difficult. At present, engine blocks are available in a more easily worked alloy, AlSi9xx, in which the cylinder bores are coated with ceramic particles or ferrous material using a laser or plasma spray coating. Other manufacturers produce thin-wall cylinder liners in either hypereutectoid alloys or a special spray process which ensures especially fine silicon grains distributed in the aluminum alloy. These liners are inserted in a pressure die-casting machine and embedded in the "standard alloy." In a different process, cylinder bores are locally enriched with silicon. In this casting process, a "preform" of ceramic fibers coated with silicon particles is inserted in the mold and embedded in "standard material," which penetrates

Preheating oven containing preforms

Lokasil cylinder surface produced during casting process by "local material engineering."

Scanning electron micrograph

R Profile
LC (M50) 0.80 mm
VER 2.50 mm
HOR 250.0 mm

Silicon crystals Aluminum

Example: Porsche Boxster

Ra = 0.15–0.25 μm, Rz = 1.0–3.0 μm, Rpk = 0.4–0.8 μm

Lokasil structure and surface finish

Fig. 5.1-14 Cylinder crankcase of aluminum alloy with Lokasil® process (example: Porsche).

the preform ("squeeze casting" or Lokasil® process, Fig. 5.1-14). In the cylinder wall area, this forms a morphology similar to that of a hypereutectoid alloy, which can be honed and etched or brushed to form a suitable cylinder bore surface.

Development has progressed to the point where aluminum crankcases may be employed not only for gasoline engines but also for diesels, which are subjected to very high mechanical stresses. Volkswagen, in its "Three Liter Car" (so called for its fuel consumption per 100 km, not its engine displacement, and designed for exceptionally light weight), employs a 1.2 L TDI (turbodiesel direct injection) engine whose crankcase is made entirely of aluminum; DaimlerChrysler designed its V8 CDI in aluminum with gray iron liners.

In all variations of aluminum-alloy crankcases, pistons are given a thin ferrous coating to facilitate cylinder break-in. Because aluminum is not as effective as gray iron in damping vibration, aluminum crankcase designs must pay particular attention to stiffness, ribbing, and good load transfer so that the aluminum engine does not become acoustically objectionable.

Crankshafts for engines with high specific output are forged of heat-treatable steel. More conventional engines usually employ spheroidal graphite cast iron crankshafts. Improved materials and manufacturing techniques now permit application of cast crankshafts for higher-performance engines, such as, for example, cast cranks based on nodular cast iron GGG70 with slight modifications. This alloy, combined with rolling the fillets at the transitions between crank cheeks and main bearings and crankpins, results in crankshafts with great long-term durability.

Connecting rods for standard duty are cast of GTS-70. Heavy duty connecting rods are forged from heat-treatable Ckxx steel. Connecting rods have also been made of sintered metal. The attraction of sintered rods is the ability to create the big-end bearing cap by simply "cracking" the sintered rod. The fracture provides a very precise fitting surface structure, eliminating the need for any special shoulder bolts or dowel sleeves, thereby significantly reducing the necessary machining effort. Because this process also results in a very precise exterior contour of the component, weight classification of connecting rods can be eliminated. In view of the significance of inertia forces, conventional connecting rod designs generally involve sorting and classifying all connecting rods by big-end and wristpin-end weights; all connecting rods installed in any given engine are of the same weight class. Cracking of sintered connecting rods is a good example of system optimization. The added cost of the connecting rod blank is more than made up by savings in machining and handling. Meanwhile, more brittle (less ductile) cast and forged steel materials (for example, C70S6) have been developed which allow good cracking, once again making cast and forged connecting rods competitive from a systems cost standpoint.

Pistons for passenger car engines are generally cast in permanent molds, while sports car and competition applications often employ forged pistons, in either case using aluminum alloys of the AlSixx or AlCuxx families. These alloys are especially heat- and wear-resistant, exhibit high thermal conductivity and hardness, and are easily cast. Cast pistons are heat-treated (solution heat-treating followed by artificial aging and hardening) for increased strength and hardness as well as dimensional stability under operating conditions. For improved break-in performance and added security against piston seizing, piston wear surfaces (piston skirts) are coated: graphite coatings for gray iron cylinders, ferrous coatings for aluminum cylinders.

Modern cylinder heads are without exception cast of AlSixx alloys. Because of the complex geometry of intake and exhaust ports, coolant jacket, and valve mechanism, these are made using low-pressure permanent mold or lost-foam casting methods. The material must exhibit good casting properties. Cooling requirements, especially in the area of spark plugs and exhaust valves, often demand narrow, constricted water passages which must not be allowed to sinter themselves to the sand core. For added strength at operating temperatures, highly stressed cylinder heads are later hardened by heat-treating.

Valves, or at the very least valve heads, are made of steel with high strength at elevated temperatures (e.g., NiCr20TiAl, or Nimonic). The valve stem is made of less highly alloyed steel and is friction welded to the head. Valve heads have either plasma-coated or hardened valve seat areas. Some high-performance engines employ sodium-filled exhaust valves. These use hollow valve stems partially filled with sodium. Molten sodium (melting point for sodium is a relatively low temperature of 97°C) is shaken up and down within the hollow stem by rapid valve motion, which transports some of the heat away from the hot valve head to the cooler stem, where it is conducted away through the valve guide. To date, ceramic valves have not yet gained acceptance, although this light material would appear to be very attractive. Questions remain with regard to assurance of repeatable good quality, and uncompetitive manufacturing costs have so far prevented application in mass-produced vehicles.

Camshafts demand a hard, wear-resistant cam lobe surface. Case hardening steels or nitriding steels are employed, or alternatively spheroidal graphite cast iron (e.g., GGG60), in which the cam lobe surface is hardened by a TIG (tungsten-inert gas) process. New variations include composite assemblies in which cam lobes, bearings, spacer sleeves, and other necessary parts are (economically) produced as individual components and then slipped into the required position on a carrier tube. The individual elements are then joined to the tube either by a brazing process or, more often, by high pressure expansion of the carrier tube. The objective must be to eliminate all additional finishing steps, although sometimes the cam lobes are given a final grind. Advantages include freedom of materials choice for individual components and considerable weight savings achieved by use of a tubular shaft instead of solid material.

Sintered metals (powdered metals) are preferred for components whose shape would require expensive manufacturing processes or which require special alloys. Sintered oil pump gears and timing belt sprockets are "finished parts" which do not require any additional machining or treatment. Connecting rods and cams are examples of sintered parts which need only minimal work or rework. Valve finger followers or rocker arms are sintered (or made of stamped sheet steel). Valve seats require special heat-resistant alloys which will prevent valve seat recession during service. Sintering permits these parts to be made dimensionally accurate, ready for assembly.

Magnesium is enjoying renewed interest as a lightweight material. The old problems of corrosion and lack of strength at elevated temperatures are nowadays addressed by high purity material as well as new alloys. In engine design, die-cast magnesium is preferred for induction runners, cylinder head covers ("valve covers"), general covers, and brackets.

Induction systems are made of sand or permanent mold aluminum castings or built up from die-cast aluminum or magnesium components. For further weight savings, glass-fiber–reinforced plastic (e.g., glass-fiber–reinforced polyamide such as PA6-GF30) may be selected. Given their complex spatial geometry, intake systems are made using either meltable cores or assembled from multiple components using friction welding. There are also mixed constructions using aluminum or magnesium along with plastic materials. The metal components usually contain the variable intake geometry elements. Plastics used for these applications have good thermal stability (up to 150°C) and good strength (with fiber reinforcement). Mounting points for threaded fasteners or especially thermally stressed areas (e.g., EGR port) may include metal inserts. All plastic components are marked with an identifying symbol according to VDA 260, in conjunction with DIN standards for designation and abbreviation of plastics, in order to ensure correct material classification in the recycling process (see also SAE J1344).

5.1.1.3 Engine management

Electronically controlled engine management systems found in modern engines are a further development of engine control systems which, in earlier times, were actuated directly by the driver or indirectly by means of mechanical systems. Advances in microelectronics and electromechanical sensors, along with regulatory requirements to reduce pollutant emissions, were on the one hand a necessary advance, and on the other hand a strong driving force in promoting adoption and application of electronic control systems, sensors, and actuators in engine development. Electronic engine controls permit comprehensive engine management with targeted, deliberate actions affecting engine operation, in consideration of a multitude of physical and mechanical quantities as well as desirable actions in the event of unanticipated events.

5.1.1.3.1 Fuel injection

During the transition phase from carburetors to fuel injection, a form of fuel injection was introduced at the exact same point where carburetors delivered fuel to the intake manifold—so-called "throttle body injection" (TBI) or "single point injection" (SPI). This, however, did little to alter the fact that intake manifold runners were wetted by fuel, which had a detrimental effect on throttle response due to considerable delays introduced by fuel wetting and evaporation processes in the manifold. As a result, multipoint injection (MPI) soon established itself. In MPI injection, each cylinder is fed by its own individual fuel injector, preferably located close to the cylinder head and aimed at that cylinder's intake valve(s) (Fig. 5.1-15). Injection is initiated by a solenoid in the injector, which lifts the injector needle to expose the spray orifice. The duration of solenoid activation and fuel pressure–manifold pressure differential determine the injected fuel quantity. For four-valve cylinder heads with two inlet ports, dual-stream injectors are usually employed, with one stream directed at each intake valve. In gang injection, all injectors are actuated simultaneously, once every 360° of crankshaft rotation; in the course of a single working cycle, 720 crankshaft degrees, two injection events take place, but in effect at different times in the cycle of individual cylinders. In individual fuel injection, also known as sequential injection, the injectors are actuated individually, so that each cylinder is provided with fuel at a specific point

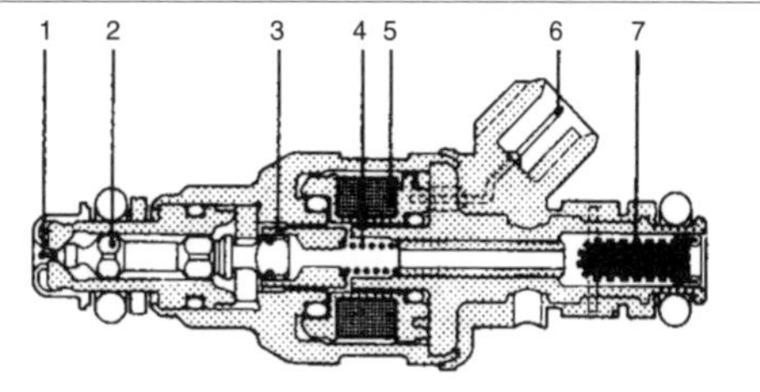

1 Pintle, 2 Valve needle, 3 Solenoid armature, 4 Closing spring, 5 Solenoid winding, 6 Electrical connection, 7 Fuel strainer

Fig. 5.1-15 Fuel injector (Bosch).

during its working cycle. This may take place shortly before intake valve opening, during valve opening, or in some combination, depending on engine load, speed, and temperature. Modern injection systems, combined with extensive testing of individual engine designs, permit very precise control of dynamic processes during load or speed changes, without injecting too much or too little fuel into the cylinder. On overrun (coasting), fuel injection is cut off entirely (overrun fuel cutoff).

Fuel is supplied to the injectors by means of fuel rails. For "top feed" injectors such as Fig. 5.1-15, the fuel rail is inserted into the tops of a row of injectors. "Bottom feed" injectors are inserted into a fuel rail integrated in the intake manifold near its cylinder head mating face. The injector body is sealed by O-rings above and below the fuel supply passages. Fuel pressure is typically 3–4 bar above actual manifold pressure and is provided by an electric fuel feed pump located in the fuel tank (see Section 7.6). A fuel pressure regulator at the end of the fuel rail responds to manifold pressure and fuel consumption, maintaining a constant pressure differential. The excess fuel quantity flows back to the tank via a return line, which also serves to bleed air bubbles and fuel vapor from the fuel lines. Some new systems operate without return flow; fuel pressure and quantity control are integrated in the pump, which provides only enough fuel to meet the fuel rail's instantaneous demand.

In addition to the previously mentioned "single stream" or "dual stream" variations, which may be skewed from the injector axis by several degrees for certain engine applications, there are deliberate differences in the degree of fuel atomization. The objective is to optimize the fuel spray for the engine type at hand and in particular to assure a consistent quality of the injected fuel stream. Fuel atomization may be improved by drawing a portion of the engine combustion air through the nozzle end of the fuel injector (air-shrouded fuel injectors). Air is diverted upstream of the throttle, using the pressure drop to the intake manifold; air and fuel are mixed at the point of injection to produce smaller fuel droplets.

5.1.1.3.2 Mixture formation

Mixture formation in Otto-cycle engines is controlled by fuel metering (quantity and composition), mixture generation, transport, and distribution. Mixture formation has always been a pivotal aspect of gasoline engine technical development. Ideally, the engine should be supplied with a homogeneous mixture of air (oxygen) and fuel, at the stoichiometric ratio, easily ignited by the spark plug and capable of rapid combustion. The quantity of air is controlled by the throttle, while the corresponding fuel quantity is controlled by the fuel injection system. The mixture should be distributed evenly among all cylinders.

The induction system is expected to provide even air distribution to the cylinders (see 5.1.1.2.4), while even fuel distribution is the province of each individual cylinder's fuel injector (in MPI systems). Mixture generation—that is, fuel vaporization and mixing with induction air—is affected by the fuel's vaporization (distillation) curve, temperature, pressure, airflow velocity and turbulence, degree of atomization, fuel concentration—and the available time.

The mixture is described by the relative air/fuel ratio, λ, the ratio of inducted air to the theoretical air requirement for complete combustion. Complete combustion of 1 kg of fuel requires about 14.6 kg of air, which would be $\lambda = 1$. In the case of excess fuel ("rich mixture"), $\lambda < 1$; in the case of excess air ("lean mixture"), $\lambda > 1$. The value of λ determines the operational behavior of the engine. Good torque and smooth engine operation are achieved with $\lambda = 0.9$ (Fig. 5.1-16), while best fuel economy is achieved at $\lambda = 1.1$ to 1.2. For $\lambda < 1$, however, HC and CO emissions rise; while oxides of nitrogen (NO_x) emissions rise for $\lambda > 1$.

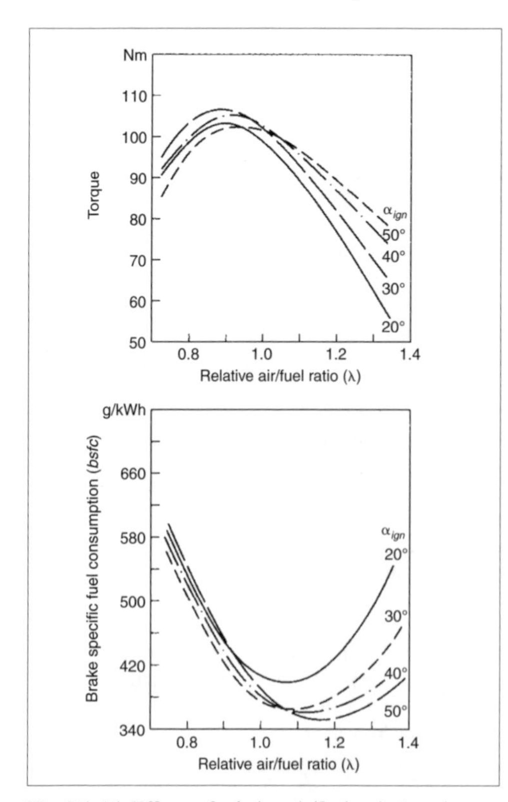

Fig. 5.1-16 Effects of relative air/fuel ratio λ and ignition timing α_{ign} on fuel consumption and torque.

The above considerations apply to overall air/fuel ratios, given homogeneous mixtures. There have been repeated attempts to reduce fuel consumption by reducing the amount of throttling in Otto-cycle engines. However, because mixture ignition by the spark plug requires λ between about 0.8 and 1.2, attempts were made to create a stratified charge in the combustion chamber—that is, deliberately producing a nonhomogeneous fuel distribution within the combustion chamber—with pure air in the zones along the cylinder walls. There have been various techniques which attempted by means of induction swirl or specifically directed turbulence to achieve operation with extremely lean mixtures (λ = 1.4 to 1.6). Some of these concepts for fuel-efficient engines were more or less successful but often exhibited dynamic behavior problems and therefore were never able to establish themselves.

These efforts were finally laid to rest by the introduction of three-way catalytic converters, which oxidize HC and CO and reduce NO_x. This necessarily mandates precise control of $\lambda = 1$. As a result of the introduction of catalytic converter technology, vehicle average fuel consumption rose by about 10% (see Section 5.1.1.3.5).

Under normal operating conditions, with a warm engine, mixture formation no longer presents any serious challenges. What remains critical is cold starting. In order to provide an ignitable mixture in the combustion chamber near the spark plug, even under unfavorable conditions (low temperatures, low airflow velocity and turbulence, cold components), it is necessary to inject extra fuel. This cannot be burned completely and leads to very high HC emissions. This is the most critical operating condition in meeting emissions standards.

5.1.1.3.3 Ignition

The ignition system ensures ignition of the fuel–air mixture in the combustion chamber by means of a high-tension discharge at the spark plug. Ignition initiation and progression are influenced by ignition timing, ignition energy, and spark duration.

Conventional coil ignition (CI), provided by means of contact breaker points, rotating ignition distributor, centrifugal spark advance, and vacuum advance or retard, is no longer used today. These systems exhibited too much contact wear and required too much maintenance.

Transistorized ignition (TI) replaces breaker points with an inductive pulse generator or a Hall effect generator, neither of which is subject to wear. Ignition distribution and timing are identical to that of the CI system. At present, TI is still in use.

High-tension capacitor discharge ignition (CDI), also known as thyristor ignition, employs ignition energy

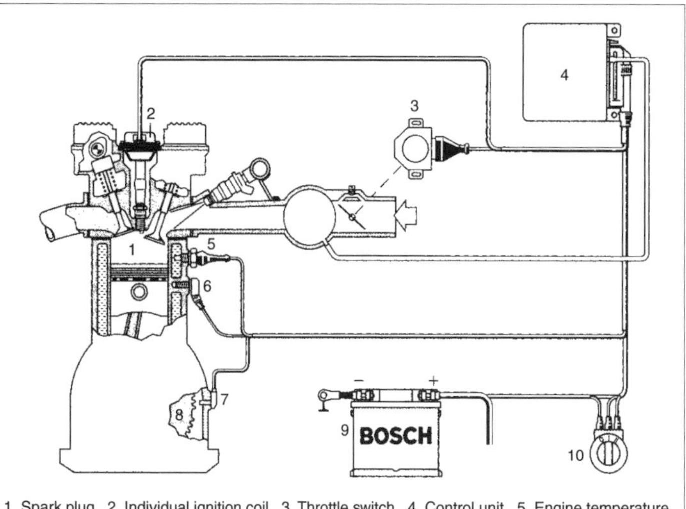

Fig. 5.1-17 Schematic of an electronic ignition system (BSI, breakerless semiconductor ignition; Bosch).

stored in a capacitor in place of inductive storage in a coil. When the capacitor is discharged, the resulting voltage is conducted to the ignition coil primary winding. Spark duration (0.1–0.3 ms) is appreciably shorter than for TI systems and does not always guarantee mixture ignition. For this reason, CDI is seldom used today.

Current practice is to use electronic ignition (SI, for semiconductor ignition) and distributorless electronic ignition (BSI, for breakerless semiconductor ignition), the former with a rotating mechanical high-tension distributor, the latter with solid-state high-tension distribution or individual spark coils (Fig. 5.1-17) to achieve completely maintenance-free ignition systems (with the exception of spark plugs).

For any given ignition system, ignition timing is that parameter which has been adjustable and variable (over a certain range) even on older ignition system designs. "Correct" ignition timing determines the combustion process. Early timing leads to detonation (combustion knock), while late timing causes long, drawn-out, ineffective combustion (Fig. 5.1-18). The ignition distributor matched ignition timing to engine speed and load by means of its centrifugal advance mechanism and a vacuum control unit containing a diaphragm sensitive to

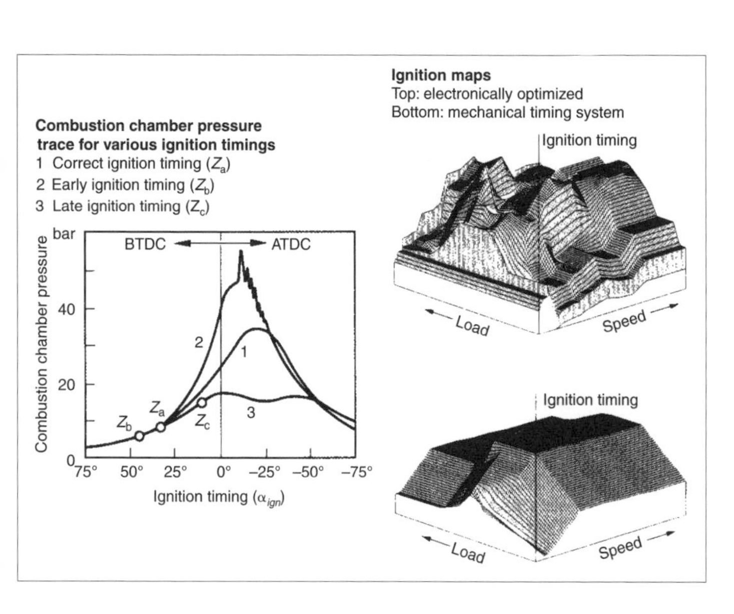

Fig. 5.1-18 Combustion chamber pressure as a function of ignition timing, ignition timing maps.

manifold vacuum. A typical ignition map for such a mechanical timing system is shown in Fig. 5.1-18. The map consists of large surfaces with limited changes in ignition timing; its response is slow compared to dynamic changes in engine operational parameters.

With electronic ignition, timing is controlled electronically in response to an entire series of parameters including engine load, engine speed, temperature, combustion knock, torque modification for transmission shifts and anti-slip control, etc. Fig. 5.1-18 also shows a representative ignition map for electronic ignition.

For stoichiometric mixtures, spark energy of about 0.2 mJ is sufficient to initiate combustion. However, for richer or leaner mixtures, the energy requirement climbs above 3 mJ. Sufficient spark plug extension into the chamber, greater electrode gap, and thin electrodes improve mixture ignition and provide smoother engine operation and lower HC emissions. Some vehicle manufacturers install two spark plugs per cylinder. This provides even greater assurance of mixture ignition and shorter flame paths.

The high-tension spark at the spark plug causes electrode erosion; both electrode gap and voltage demand increase. This may lead to irregular ignition or even misfire. The electrode erosion problem may be greatly reduced by highly wear-resistant surfaces—that is, high-temperature– and chemical-resistant platinum coatings on the surfaces where the arc is struck. With the introduction of such platinum plugs, spark plug replacement intervals could be extended to 60,000 or even 100,000 km. An additional optimizing measure is the design of the surface gap or surface air gap spark plug, in which the spark travels across the surface of the insulator tip and thence to the ground electrode (Fig. 5.1-19). This prevents spark plug fouling, which might otherwise occur under cold start conditions or with frequent short trips. Spark plugs are now available (e.g., Saab 95, Mercedes-Benz V12) with an included ion current detector, which permits direct detection of individual cylinder misfires, instead of the more complicated method using irregular crankshaft rotation (see engine controls, OBD; Section 5.1.1.3.6).

The ignition coil is not only a transformer but also an energy storage device in that it is an inductive voltage converter. The primary winding is fed by the vehicle's electrical network. The secondary winding delivers high voltage current with the required energy to the spark plug. The delivered voltage lies between 20 and 35 kV; spark energy is about 60 to 100 mJ; spark duration is about 2 ms. Three different coil configurations are currently in use (Fig. 5.1-20).

a) A single coil serving all cylinders. A high-tension lead connects coil and distributor. The distributor finger, rotating at camshaft speed, connects the cylinder to be ignited with the high-tension lead.
b) Double-spark coil with "solid state distribution." A coil is fitted with two high-tension outputs for two cylinders, ignited 360 crankshaft degrees apart (e.g., cylinders 1 and 4 of a four-cylinder engine). The rotating distributor is eliminated, and with it one more wear component. On the other hand, this system generates an unneeded spark during the valve overlap period. For four- and six-cylinder engines, two

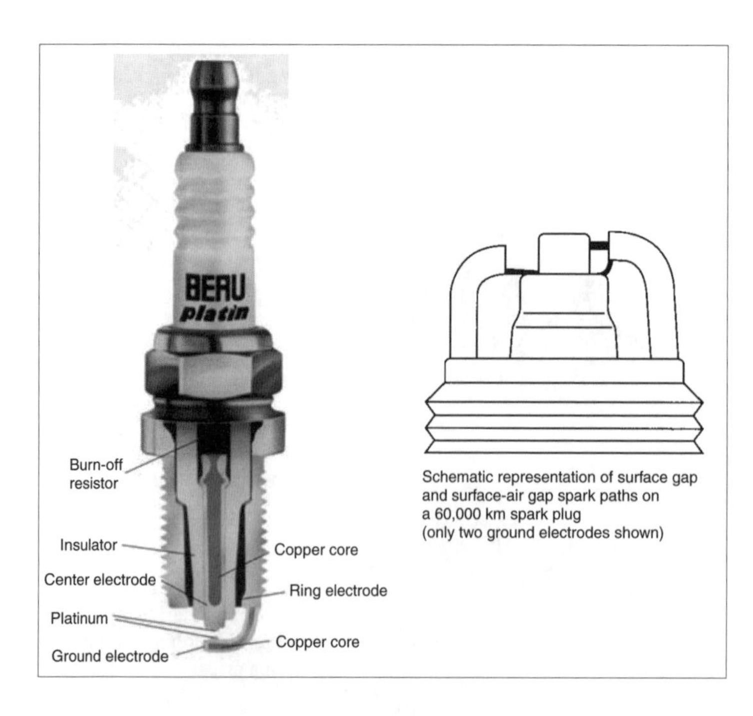

Fig. 5.1-19 Long-life spark plug and surface gap principle.

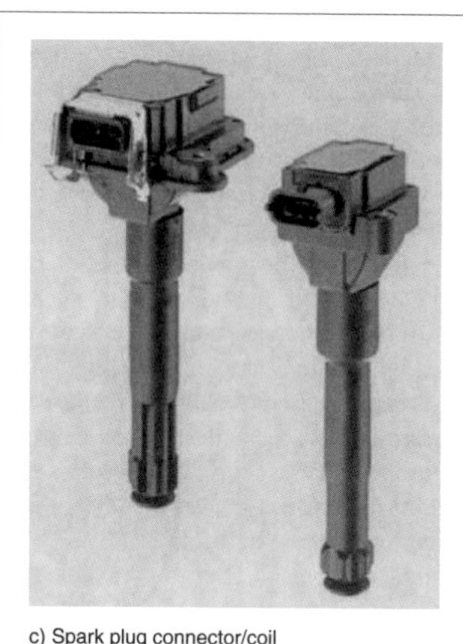

c) Spark plug connector/coil

a) Distributor coil

b) Six-cylinder ignition module (double spark concept)

Fig. 5.1-20 Ignition coil configurations.

or three double-spark coils are combined into a single package.

c) The single-spark coil serves a single cylinder and is mounted directly on the spark plug. This eliminates the high-tension leads. For electronic engine controls with knock sensors, each cylinder can be fired with its own individual ignition timing.

5.1.1.3.4 Combustion

The combustion process within the cylinder is strongly affected by the previously mentioned injection, mixture formation, and ignition processes and in turn affects exhaust gas composition. The objective of optimum combustion is to approach the ideal constant-volume process as closely as possible. Aside from the above factors, gas and combustion chamber wall temperatures play a significant role, as do chamber wall heat transfer and thermal conductivity and chamber geometry. For short flame travel, the combustion chamber should be as compact as possible, should not contain any narrow spaces which might quench the flame, and should not have any hot spots or deposits which might cause preignition and therefore lead to detonation ("knock").

Theoretically, given a stoichiometric relationship between hydrocarbons contained in the fuel and atmospheric oxygen, the only combustion products should be CO_2 and H_2O. However, because chemical reactions are also dependent on pressure, temperature, species concentration, and time, intermediate reaction products may remain; also, additional compounds may form through reaction with atmospheric nitrogen.

CO is formed during combustion with insufficient oxygen. CO is also formed if combustion is initiated late and if pressure and temperature drop during the expansion stroke, quenching the reaction and leaving a high concentration of CO in the exhaust.

In hot combustion gases, N_2 molecules break up, leading to the formation of nitric oxide, NO. The higher the maximum combustion temperature and the greater the oxygen concentration, the more oxides of nitrogen (NO, NO_2), collectively known as NO_x, are formed. With a higher concentration of inert gas in the form of exhaust in the combustion chamber, maximum combustion temperature is reduced, which counteracts NO_x formation. To achieve this, gas transfer may be controlled by means of valve timing, or exhaust gas, possibly cooled exhaust gas, is mixed with induction air.

Unburned hydrocarbons remain if fuel impinges on combustion chamber walls or deposits (either through injection or motion of the air–fuel mixture) or if the flame front cannot sweep through the entire combustion chamber because the mixture is no longer ignitable or is quenched in narrow gaps (piston top lands). Unburned fuel components and vaporized fuel are ejected with the exhaust gases as HC emissions.

In reality, the combustion process consists of an entire series of chemical reactions taking place simultaneously, sequentially, and with mutual interactions. At present, only partial processes have been described mathematically. In place of actual, not-fully-understood reactions, combustion models assume processes and models which are intended to describe combustion and the working cycle as realistically as possible.

Pressure history within the cylinder as a function of crankshaft angle can be measured relatively accurately by means of piezoelectric pressure transducers. Combustion models are simulated by computer programs whose input parameters are intake gas flow, heat transfer, etc. and which then represent burn rate and the mass fraction burning function. The mass fraction burning function represents the portion of burned fuel relative to the total inducted fuel quantity; burn rate is the derivative of this function with respect to crankshaft rotation. Ideally, combustion would not begin too far before top dead center, would exhibit smooth progression with maximum pressure at 8 to 12 crankshaft degrees ATDC, and would drop quickly thereafter, reaching zero at about 60 to 65 degrees ATDC. Deviations from this ideal situation indicate improper ignition timing, combustion knock, or delayed combustion.

Fig. 5.1-21 shows actual pressure traces from a 1.8-L normally aspirated engine operating in a low part-load regime. Note that although the engine as a whole is operating normally and smoothly, there is considerable cycle-to-cycle cylinder pressure variation. Therefore, a single cylinder pressure trace cannot be representative of a single operating condition; one must use a statistical mean trace as the "actual" cylinder pressure trace. Such a mean cylinder pressure trace was used to calculate the burn rate and mass fraction burning function shown in the lower diagram. The burn rate (from the statistically averaged pressure trace) is very uniform and falls into the optimum crank angle range. The mass fraction burning function documents that less than 100% of the inducted fuel undergoes complete combustion. On the one hand, this is a realistic observation, but on the other, the magnitude of the unburned fraction is not easily determined. The lower one goes into part-load operation—that is, the lower the amount of fuel made available for combustion—the greater the uncertainty in determining unburned fuel fraction.

This information is very useful to development engineers in optimizing the engine. Modern indicating and

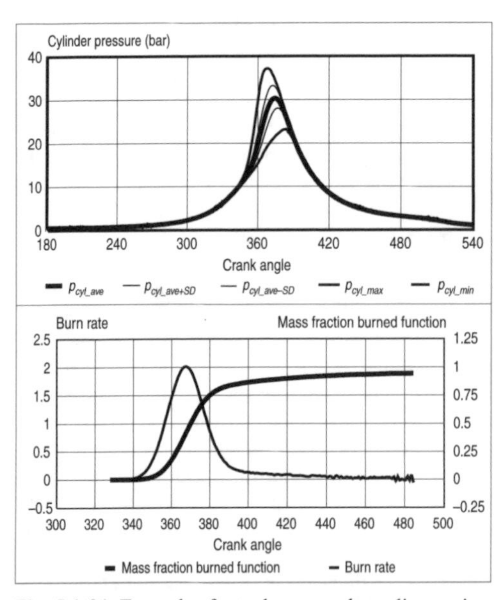

Fig. 5.1-21 Example of actual measured gasoline engine cylinder pressure traces and burn rate and mass fraction burned functions calculated from the mean pressure trace. SD, standard deviation.

evaluation systems essentially deliver this data online at the test cell. It is less important to obtain physically or thermodynamically "exact" data. Instead, these systems make possible a comparison within the framework of established boundary conditions, thereby fulfilling the requirements needed by developers in fine-tuning various parameters.

Two types of extreme combustion disturbances must be avoided at all costs: misfiring and knock.

Misfiring causes unburned gases to pass into the exhaust system, leading to unacceptable HC emissions and, as a result of reactions in the catalytic converter, to very high temperatures which can destroy the catalyst. In addition to poor mixture formation under particularly unfavorable conditions, misfiring may also be caused by injection or ignition system defects.

Combustion knock may result in piston or crank train damage. Even before the flame front has traversed the entire combustion chamber, at a speed of 25–30 m/s, the temperature and pressure rise in the chamber can cause as yet unburned areas of the charge to undergo reactions which propagate at speeds of up to 500 m/s. Temperature and pressure continue to rise rapidly and uncontrollably, creating high-frequency pressure waves (knocking or "pinging" noises). The associated mechanical and thermal overloads will cause damage if knocking operation continues over an extended period. The knocking tendency can be reduced by various measures: short flame paths originating from a centrally located spark plug, compact combustion chamber, high turbulence within the chamber, fuel with a higher octane number, avoidance of "hot spots" in the combustion chamber, lower compression and low induced charge temperature. Because engines should operate as close to the knock limit as possible for fuel economy reasons, modern electronic engine management systems include knock control. A knock sensor mounted against the cylinder wall detects the high-frequency vibrations typical of combustion knock; the management system retards ignition timing, and then gradually advances it until knock reappears briefly.

5.1.1.3.5 Exhaust emissions and catalytic converters

In complete combustion, ideally the only products should be carbon dioxide (CO_2), water (H_2O), and heat. In reality, however, combustion does not result in complete reaction of the entire fuel mass; intermediate products appear, and the presence of atmospheric nitrogen as well as trace contaminants in fuel (e.g., sulfur) also result in reactions.

Carbon dioxide emissions are largely proportional to fuel consumption. These emissions contribute to the "greenhouse effect" and therefore should be kept as low as possible. Several exhaust gas components are classified as "pollutants"; emissions of these substances are regulated.

Carbon monoxide (CO) is immediately highly toxic. It is mainly generated during engine idling. CO has long been used as a test parameter for correct fuel mixture adjustment. CO is not stable and eventually combines with atmospheric oxygen to produce CO_2.

In the presence of air, nitric oxide (NO) converts to nitrogen dioxide (NO_2), a gas which irritates human mucous membranes. Usually these compounds are grouped together as oxides of nitrogen (NO_x).

Exhaust gas contains many different forms of hydrocarbons (HC). Some of these are direct high-molecular-weight components of fuel, but most are partial and intermediate reaction products. These may also act as irritants, and some are considered carcinogenic.

Particulates encompass all materials (with the exception of water) which are entrained in exhaust gas under normal conditions as solid bodies (ash, soot) or as liquids.

These exhaust components are regulated on a nearly worldwide basis, either individually or as groups. However, there are significant local regulatory differences. The state of California sets the strictest standards, followed by the remaining United States; Europe and Japan have similar standards, followed in a weaker form by many other nations. There are very few nations which have not already issued emissions regulations or which are not about to do so.

In general, the situation for gasoline engines may be described as follows (Fig. 5.1-22): engines develop best power at $\lambda \cong 0.9$, but at that relative air/fuel ratio, HC and CO emissions are high. Best fuel economy is achieved at $\lambda \cong 1.1$, but under such conditions NO_x emissions reach their peak because this λ regime also results in the highest combustion chamber temperatures.

It would not be possible to meet existing emissions

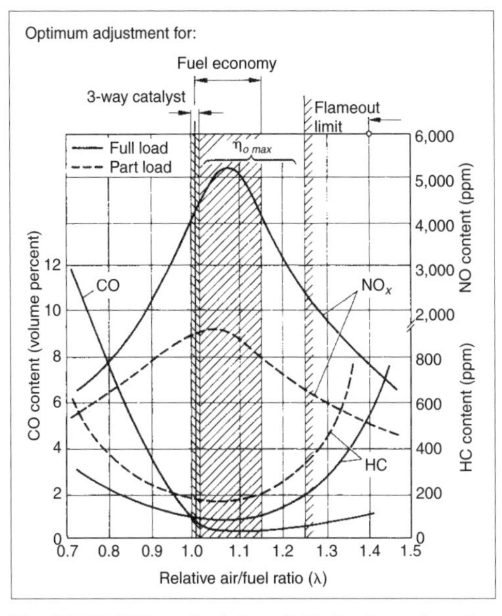

Fig. 5.1-22 Effect of relative air/fuel ratio and engine load on gasoline engine pollutant emissions.

standards for all controlled components (CO, HC, and NO_x) with engine-based measures alone. In the 1960s, it was necessary to introduce catalytic converters on cars built for the California market. Catalytic converters use three-way catalysts, which oxidize CO and HC and reduce NO_x (Fig. 5.1-23). Oxidation and reduction within a single device are only possible if the overall mixture is $\lambda = 1 \pm 0.005$. The term "λ concept" is sometimes used. Maintaining the correct air/fuel ratio is achieved by means of a "λ sensor" (oxygen sensor) of zirconium dioxide. This detects residual oxygen in the exhaust stream, with a step response in its output voltage signal at $\lambda = 1$. Because the goal is to keep the exhaust oxygen content at nearly zero, the control system must keep the engine operating in the vicinity of the sensor's step response in order to keep all three exhaust pollutants at their minimum values.

The catalytic converter (Fig. 5.1-23) consists of a substrate—a ceramic matrix or corrugated metal bundle—whose many small passages create a large surface area. This surface is given a "washcoat," a highly porous ceramic material which again increases the surface area for chemical reactions. Embedded in the washcoat are the actual catalyst materials which initiate the oxidation and reduction reactions. Catalytic materials consist of the noble metals platinum (Pt), rhodium (Rh), and palladium (Pd), which are used in various amounts and concentrations. Catalyst effectiveness is diminished or even eliminated by various substances including lead and lead compounds. For this reason, introduction of catalytic converter technology coincided with the introduction of unleaded gasoline.

Conversion rate, the measure of catalyst effectiveness, is strongly dependent on temperature. Below 250°C, virtually no reactions take place. Ideal conditions for good conversion and long service life are achieved at temperatures between 400 and 800°C. Temperatures over 1,000°C will destroy a catalytic converter. A converter mounted close to the exhaust manifold will start ("light off") faster, but at very high engine output it will be subjected to very high exhaust gas temperatures. Most vehicles therefore have their catalytic converter mounted below the floorpan at a certain specific distance from the engine.

The lower the raw emissions from the engine, the lower the postconverter emissions. Oxides of nitrogen formation are dependent on local maximum temperatures in the flame front. At part load, when the engine is operating in a throttled condition, these temperatures can be reduced by the addition of inert gas—in other words, exhaust gas; this will appreciably reduce NO_x concentration. As the inert gas component also reduces the average exhaust gas temperature and therefore may have a negative effect on the operating temperature or lightoff behavior of the catalytic converter, exhaust gas recirculation (EGR) is applied only during noncritical operating conditions, when it may have a positive effect on tailpipe emissions. During cold start, it may be

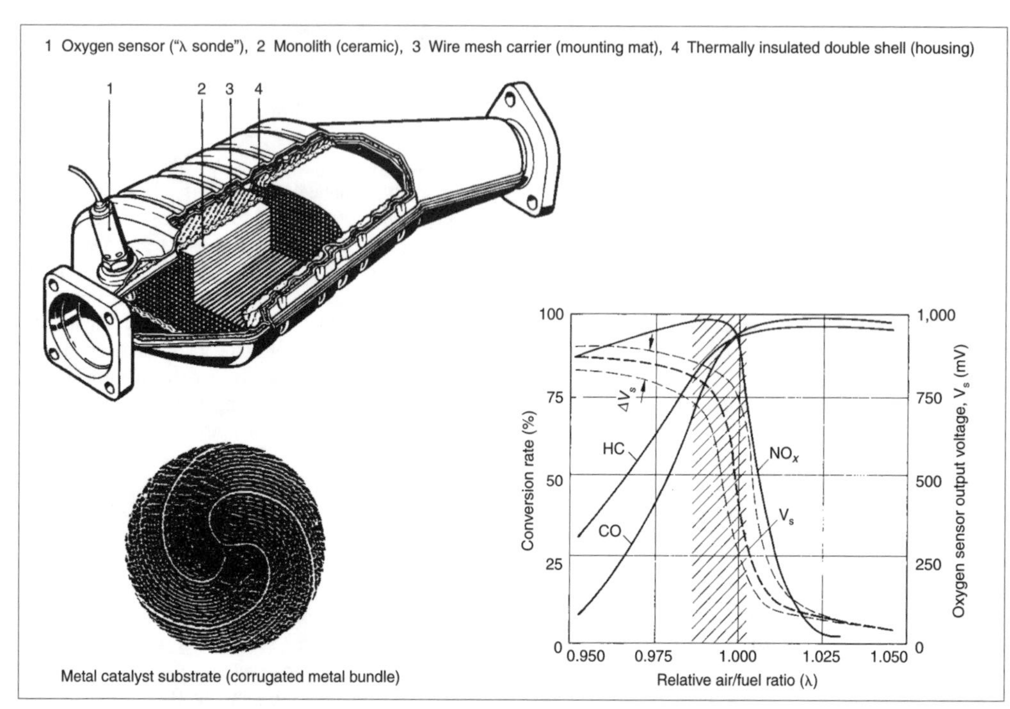

Fig. 5.1-23 Construction and operating principle of a three-way catalytic converter with oxygen sensor.

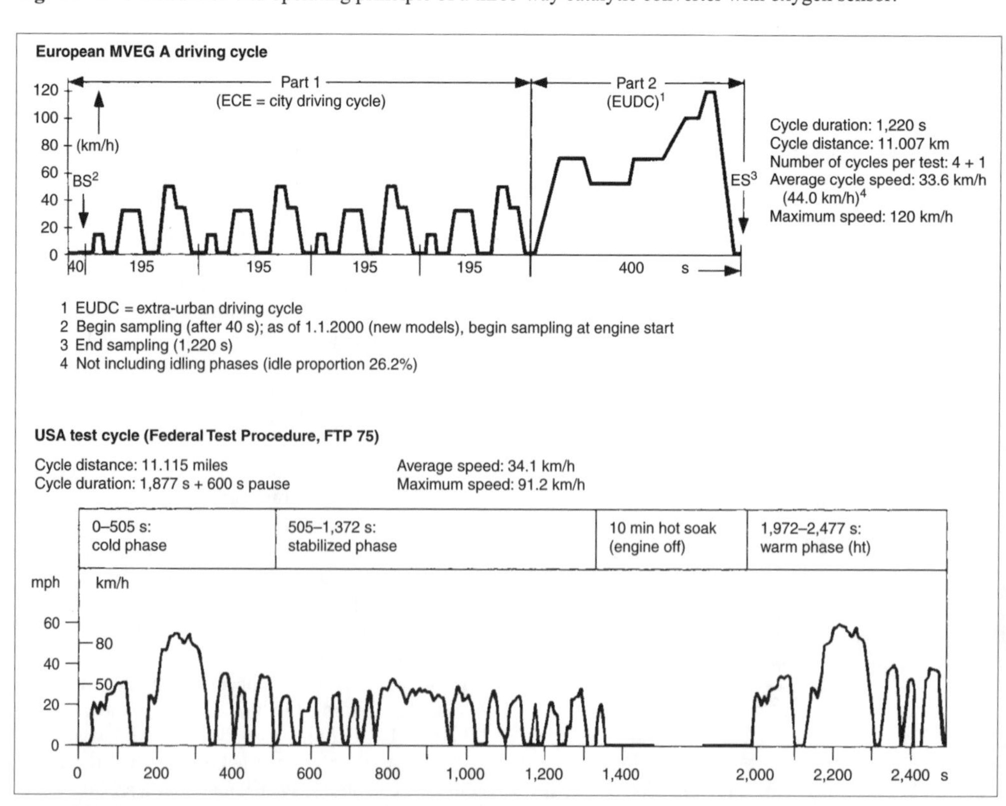

Fig. 5.1-24 European MVEG A and U.S. FTP 75 exhaust emissions test cycles.

necessary to pump air into the exhaust upstream of the converter (secondary air injection) in order to achieve afterburning of high hydrocarbon concentrations and simultaneously speed catalytic converter lightoff. This will greatly reduce the HC peak during cold start.

Emissions standards must be met under specific driving conditions in the course of a predefined test procedure. The test is conducted on a chassis dynamometer. The vehicle must be "conditioned"—that is, the vehicle history prior to the test guarantees the test's initial conditions. The inertial resistance of the dynamometer rollers is matched to vehicle weight. The vehicle must cover a defined (simulated) test course, at defined speeds and in specific gears (for manual-transmission vehicles), as specified in the European MVEG A test cycle or the American FTP75 test cycle (see Fig. 5.1-24).

Using the CVS (continuous volume sample) method, exhaust is collected in a plastic bag during the entire test period (U.S. test requires three bags) and then analyzed for HC, CO, NO_x, CO_2, and O_2. Emissions are evaluated by means of specified calculation operations and expressed in g/km (or g/mi), and fuel consumption in liters/100 km (or mpg, miles per gallon).

Table 5.1-2 shows European and U.S. 49-state exhaust emissions limits for passenger cars with gasoline engines. Shown are current and previous limits, and new limits which take effect in 2005. Reductions from stage to stage are so severe that an entirely new package of measures must be employed to meet each new set of emissions demands. These measures encompass sensors, electronic engine controls with intelligent control strategies, catalytic converter technology with expanded operating ranges, improved fuels, and new technologies permitting additional control of gas transfer mechanical components.

5.1.1.3.6 Engine control systems

An engine's ECU (electronic control unit) processes all incoming data to generate control commands as well as output data. Fig. 5.1-25 shows a schematic of an engine with attached engine control components. Fig 5.1-26 is a block diagram of an engine control system showing functional groups, with input parameters and output functions. Modern control units are equipped with 16-bit or sometimes 32-bit processors in order to handle the necessary information streams in a timely manner, even at high engine speeds and during dynamic events. At maximum speed, a six-cylinder engine allows only about 3 ms between ignition events.

Depending on vehicle equipment, there may also be interaction with other electronic control systems, such as automatic transmission, automatically shifted "manual" transmission, brake system, or climate control system. Various configurations are possible, with central or decentralized control systems. In the case of a central control system, a single electronic control unit serves

Table 5.1-2 European and American emissions limits

EUROPE MVEG A Test	Emissions limits for passenger cars (≤2.5 t, ≤6 persons), with gasoline engines			
		Stage II	Stage III	Stage IV
	Type certification issued after	1/1/1996	1/1/2000	1/1/2005
	First licensed after	1/1/1997	1/1/2001	1/1/2006
	CO	2.2	2.3	1.0
	HC	—	0.2	0.1
g/km	NO_x	—	0.15	0.08
	HC + NO_x	0.5	—	—
g/test	Evap.	2.0	2.0	
As of Stage III:	– MVEG A measurement including first 40 s – OBD			
USA	49 states + Canada	Emissions limits for passenger cars (≤12 persons) with gasoline engines		
FTP 75 test	1996–2002 model years			
		Standard	Clean Fuel	
g/mi	THC	0.41	—	Total hydrocarbon
	NMHC	0.25	—	Nonmethane hydrocarbon
	NMOG	—	0.125	Nonmethane organic gases
	HCHO	—	0.015	Formaldehyde
	CO	3.4	3.4	
	NO_x	0.4	0.4	
	Particulates	0.08	—	
g/test	Evap.	2.0	2.0	

New limits after 2003 model year are still under discussion; possible halving of CO and NO_x for clean fuel levels.

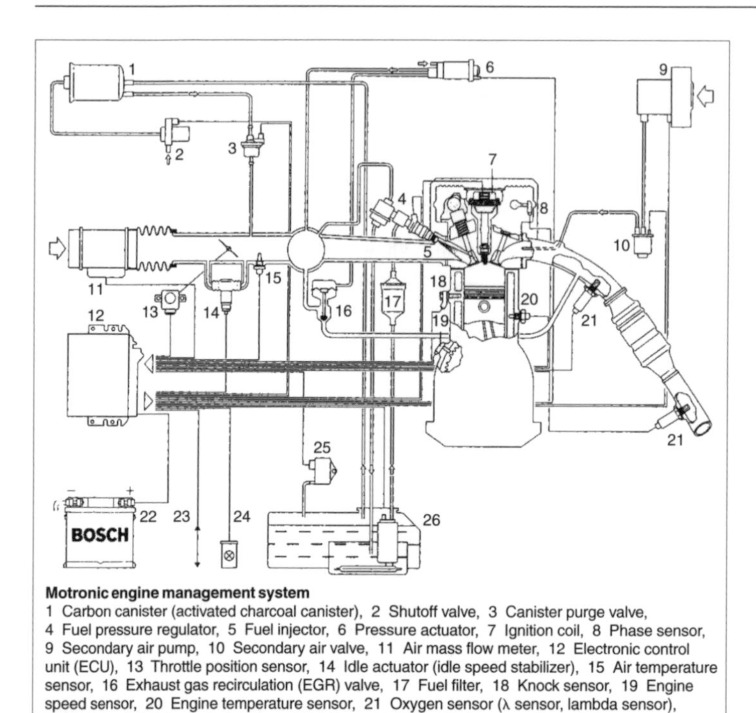

Fig. 5.1-25 Engine management system components (Motronic system).

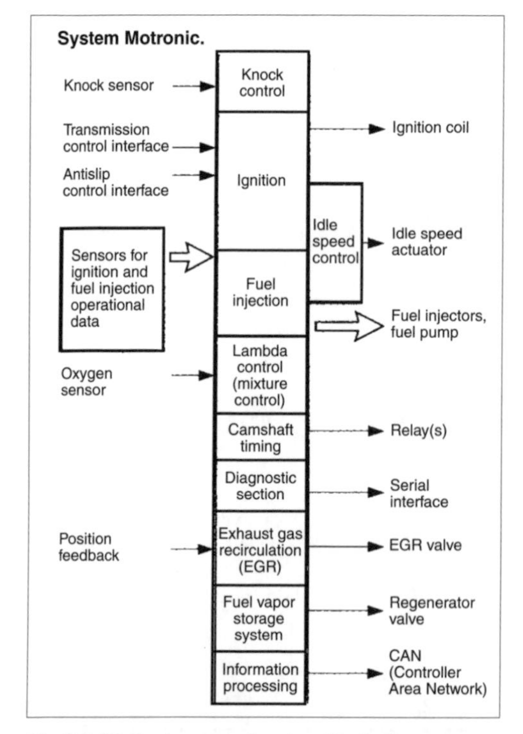

Fig. 5.1-26 Engine control system block diagram.

several components or systems (e.g., engine and automatic transmission). In decentralized control systems, the various components include their own electronic controls which serve that component's basic functions and provide exchange of control information for higher-order vehicle functions, such as between the antilock braking system and the engine and transmission control units (see Section 5.5.2). Fuel and exhaust systems are always included in the core of the engine control system.

Sensor signals are processed by signal conditioners and converted to a uniform voltage range. An analog-to-digital converter transforms these conditioned input signals to numerical data capable of being handled by the microprocessor. Digital microprocessor output must again be transformed and amplified to the power levels required by actuating devices. All necessary programs (connections) and maps (application data) are maintained in semiconductor storage.

In addition to the systems covered in previous sections—fuel injection (timing, duration), ignition (timing), emissions (λ control, exhaust gas recirculation —other parameters must also be considered:

— Knock control, such as retarding ignition timing if knock is detected
— Idle speed control, by means of ignition and fuel injection, under some circumstances treated differently under increased load, such as. air-conditioning compressor or power-steering pump

— Fuel tank venting—i.e., "discharging" the activated carbon canister
— Camshaft timing control as a function of engine speed and possibly engine load
— Boost pressure control of supercharged engines

An important element of modern engine control systems is their self-learning capability. Within preset limits, they are able to compensate for tolerances inherent in certain components. Examples include the λ map, basic idle air, and throttle angle. This self-learning capability permits reduction of complex adjustment and calibration operations in production as well as in customer service. A further example is using the knock sensor to modify the ignition timing map for various fuel grades.

Another vital question regarding engine management is "what happens if . . . ?"

For example, what happens if a sensor fails? This could be due to operational failure (sensor no longer senses), but could also result from a fault as simple as a broken electrical lead or disconnected plug somewhere in the wiring harness or at the sensor itself. Such failures must be anticipated and dealt with. All microprocessor-controlled systems include integrated self-diagnostics, which fulfill several tasks:

— Monitoring all components and systems. Limits for logical, rational values may be established and adherence to these limits monitored. Logical relationships are defined and monitored. Given the complexity of these systems, the causes of failures can be narrowed down, simplifying diagnosis.
— Protection of sensitive components in the event of malfunctions. For example, misfires may cause overheating, thereby endangering the catalytic converter. If multiple misfires are detected for a given cylinder, fuel injection for that cylinder may be shut off.
— Assuring a limp-home function. In the event of sensor or actuator failure, the system calls up substitute values to permit limited mobility. For example, if the engine load signal (air mass flow, manifold pressure) becomes unavailable, a replacement value is obtained from engine speed and throttle opening. If, for example, a temperature sensor fails, a representative replacement value from other operational data is obtained and substituted in calculations.
— Storage of detailed information. If the diagnostic system detects a malfunction, the electronic control unit stores the fault information, along with information on engine environmental and operational conditions, as a problem-specific code in a fault buffer. Because the driver may not even be aware of a system defect (not least because limp-home functions have been activated), U.S. regulations require that a warning light inform the driver of a recognized system fault if this may have an affect on vehicle emissions. The driver should then visit a repair facility at the earliest opportunity. Beginning in 2005, such regulations covering OBD (onboard diagnostics) will take effect in Europe as well.
— Querying the stored fault information by a repair facility via a standardized interface (interface protocol in accordance with ISO 9141 and 14230 standards). Stored fault codes help narrow the search for the causes of the malfunction. For highly complex systems, such codes are the only effective means of troubleshooting.

During the introduction of a new vehicle model, it is sometimes necessary to alter "maps" or even processor routines stored in the electronic control unit based on field experience and customer complaints. In earlier development states of electronic engine management systems, this was only possible through complete replacement of the control unit (sometimes, unfortunately, this included resoldering the EPROM by the unit's manufacturer). Modern practice is to install flash EPROMs (rewriteable electronic data storage components). If necessary, these may be reprogrammed and so permit adaptation to the engine application's most recent development stage. An added advantage is that in production, a single hardware component supplied by the control unit manufacturer may be loaded with specific data at the end of the assembly line, covering multiple engine and application variations. This streamlines logistics and ensures problem-free integration of the latest variations.

It should be cautioned that uncontrolled "chip tuning" may result in serious damage. The engine may be overloaded, and exhaust emissions limits may be exceeded. Furthermore, any modification of the operating mode approved and certified by the manufacturer would void the vehicle's type approval.

5.1.1.4 Cooling and lubrication

Heat losses in real thermodynamic processes result in heating of combustion chamber surfaces (cylinder head, piston, cylinder) and necessitate cooling to prevent component overheating, oil coking, and power loss due to reduced volumetric efficiency. Well-designed cooling and lubrication systems are prerequisites for optimum engine operation, keeping operating temperatures and friction of all moving parts within predetermined acceptable limits.

Air cooling is only rarely employed in modern automotive engine applications; increasingly, even motorcycles are liquid cooled. Although the cooling function of air-cooled systems was steadily improved and noise developed by the cooling blower greatly reduced by design measures, air cooling eventually reaches practical limits. Compared to liquid cooling, reduced heat transfer rates and the lower heat capacity of air require very large cooling areas, achieved by fins. Liquid cooling offers better preconditions for uniform temperature distribution within the engine and offers advantages in heating system design and control.

Liquid cooling for an engine/vehicle system generally

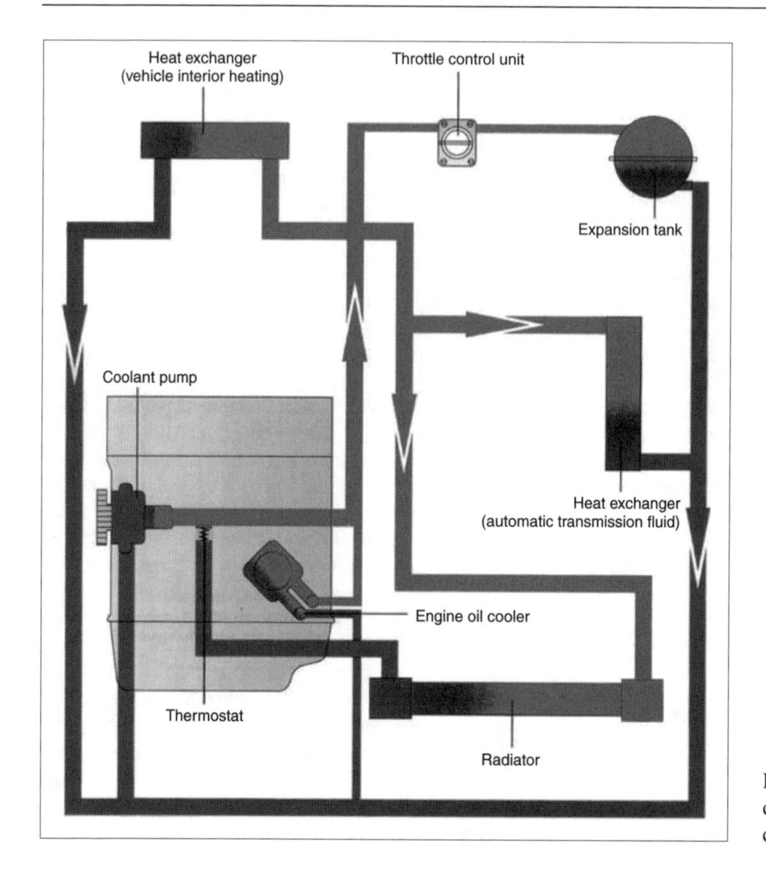

Fig. 5.1-27 Coolant circulation system for liquid cooling.

consists of two or three coolant circuits, selected automatically or manually (vehicle interior heating; Fig. 5.1-27; see also Section 3.3).

Coolant consists of a mixture of water and antifreeze (usually ethylene glycol) and, depending on application, specific inhibitors (for corrosion protection). The coolant absorbs heat within the engine and transports it to the radiator, which in turn rejects heat to the ambient air flowing through it.

The engine cooling system's task is to assure that sufficient heat is carried away from those areas subject to the greatest thermal loads—the cylinder head and its exhaust ports. Conversely, cooler areas should be cooled less, or even heated by the coolant, in order to achieve uniform temperature distribution within the component. This applies to the exhaust and intake side of the cylinder head, and also to individual cylinders regardless of their distance from the water pump. In general, water is pumped into the cylinder block, from which it flows to the cylinder head and exits at the other end of the engine. Flow through the engine is controlled by the area of openings in the cylinder head gasket; within the cylinder head, flow is deliberately directed at "hot spots." It is important that the flow path within the cylinder head eliminate any possibility of steam bubble accumulation, as this would result in a breakdown of heat transfer. The objective of achieving uniform temperature distribution within individual components has also led to reducing the extent of water jackets around cylinder walls; only the upper half or upper third of the area swept by the piston, where heat load is greatest, is surrounded by the water jacket.

In addition to the engine, heating system, and radiator, the cooling system of some vehicles may include an engine oil cooler, transmission oil cooler, and the (electrical) generator.

Ideally, in order to achieve faster warmup during cold start, coolant should not flow at all. But because the coolant pump is directly connected to the engine's rotating parts (via V-belt or poly-V/"V-ribbed" belt), a thermostat is used to short-circuit the flow through the radiator and thereby reduce the mass of water which must be heated by the engine. The thermostat opens smoothly, only after normal operating temperatures of 80–90°C are reached, directing coolant flow to the radiator. Another ideal situation would be to have coolant flow not only modified during startup, but also adjusted to match load after the engine has reached operating temperature. Output of the mechanical coolant pump is designed for full-load engine operation at low engine speeds. As a result, its output at any other engine operating points is too great and therefore associated with undesirable power losses. To date, actively controlled, electrically driven water pumps have not yet achieved any market penetration because peak-load operation requires large electric motors which, in the light of cost and efficiency analyses, no longer offer any economic or practical advantages. Efforts have been made to implement vapor cooling systems, which would use the heat of vaporization of water to reduce coolant flow to a fraction of that currently used in purely liquid-state cooling systems. Because this would also require a complete redesign of the vehicle heating and cooling systems, and because the problem of coolant component separation is still an open question, to date no one has stepped forward with a production application.

Oil used in the internal combustion engine's lubrication system must meet a series of demands: lubrication and cooling of all powerplant components and in some cases auxiliary equipment (e.g., exhaust turbocharger); transmission of forces to bearings, tensioners, adjusters, and actuators; vibration damping; removal of contaminants and neutralization of chemically active combustion products. Oil can perform all of these various functions only if it is suited to the specific tasks at hand and is available in sufficient quantities at those points within the engine where it is needed. Engine oil formulations are based on mineral oils and contain specific sets of additives, differing between manufacturers and intended applications. Fully synthetic oils are available on the market, offering greater resistance to aging, albeit at very high cost.

Most vehicle engines are equipped with pressure lubrication systems. The oil pump draws oil from the sump (oil pan) below the crank train and feeds it under pressure through an oil filter and, in some cases, an oil cooler and thence to the oil passages. The oil pump is generally a rotor (internal gear) pump mounted on the crankshaft or a gear or rotor pump mounted closer to the sump. It is designed to provide at least a minimum oil pressure and ensure lubrication even with a hot engine at idle speed (and after many hours of engine operation). To prevent oversupply with cold oil and at high engine speeds, oil pressure is limited by a pressure relief valve and a bypass circuit. Controllable oil pumps, whose volume flow is adjusted as a function of pressure,

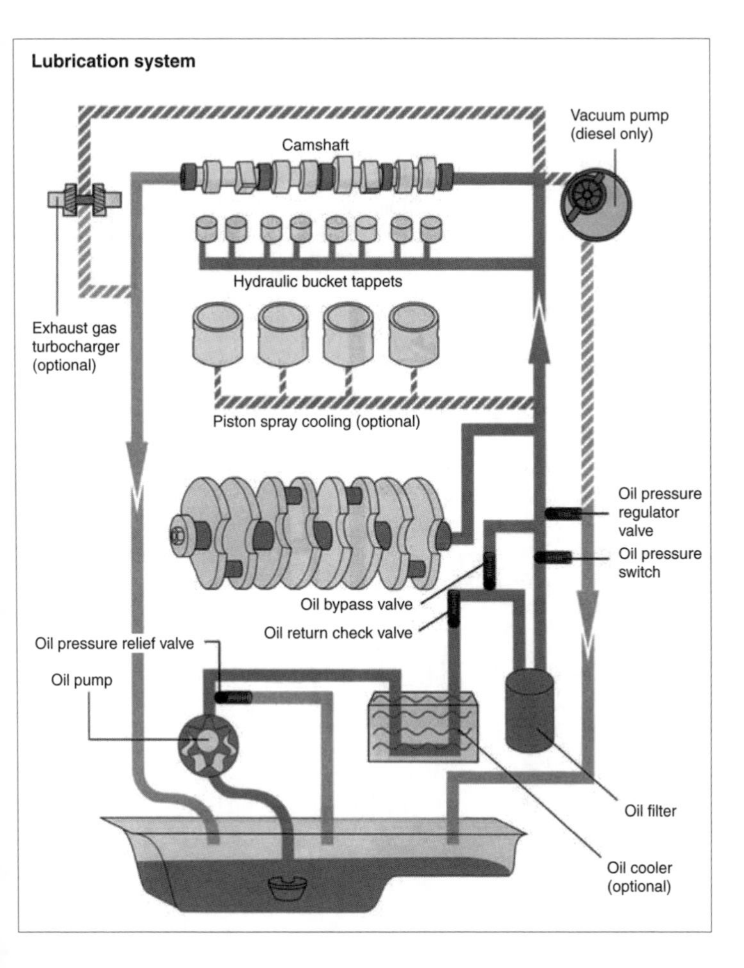

Fig. 5.1-28 Pressure lubrication system.

have been rarities due to their cost. It may be expected that they will see increased application if it becomes necessary to exploit even minor possibilities of improving engine mechanical efficiency.

All crankshaft bearings are supplied from a main oil gallery which runs along the entire length of the engine block (Fig. 5.1-28). Drilled passages in the crankshaft draw from this gallery and supply oil to the lower connecting rod bearings and possibly, by means of holes drilled through the connecting rod, to the wrist pin. Highly stressed engines also employ piston cooling by means of spray nozzles, mounted in the main oil gallery and aimed at the piston undersides. An oil passage leads up to the cylinder head to supply the camshaft bearings. Depending on design, the lubrication system may also supply hydraulic compensating elements (hydraulic lifters, lash adjusters) within bucket tappets or support elements in the cylinder head, and it may also provide pressurized oil for valve or camshaft timing mechanisms and oil spray for cam lobe lubrication.

Oil exiting from the sides of bearings and other oil spray and drips collect and flow back to the oil sump ("oil pan") through vent passages in the engine block. The oil sump serves both as a reservoir and a "settling tank" in which the returned oil is de-aerated and cooled (by the free airstream passing over the sump, possibly equipped with cooling fins). Oil may be cooled additionally by an oil cooler in the pressurized circuit. Such an oil cooler is either integrated within the engine's liquid coolant system or configured as an oil/air heat exchanger.

Dry-sump lubrication systems are found in only a few sports and off-road vehicles. In such cases, oil is drawn from a collection point below the engine by an (electric) scavenge pump and transported to a separate oil tank, from which it is drawn by the pressure pump. This ensures intensive oil cooling and good oil supply regardless of vehicle attitude, road grade, offroad operation, or dynamic (cornering, braking, acceleration) forces.

The oil filter is vital in ensuring engine operation and durability. It removes solid foreign particles from the engine oil (metal wear particles, dust, combustion products), thereby maintaining lubricant functionality between service intervals. Modern engine designs usually employ full-flow oil filters, in which contaminants are immediately caught by the filter as the entire oil stream passes through it continuously. A bypass valve ensures continued oil flow in the event of a clogged filter. If there is insufficient oil flow through the filter surface, resulting in higher pressure drop across the filter, the bypass opens, short-circuiting the filter but providing the necessary lubricant flow.

Recent filter designs use a split housing containing a replaceable paper filter cartridge, which can be removed and disposed of without dripping waste oil. Fig. 5.1-29 illustrates an oil filter module which integrates oil filter, oil cooler, and alternator bracket in a single unit. It is provided by the supplier, ready to install on the engine assembly line.

To monitor oil levels, some vehicles use electric sensors which activate a warning indication when the minimum oil level is reached. To reduce operating costs, oil consumption, and the amount of waste oil which must be processed, there is an increasing tendency to fully

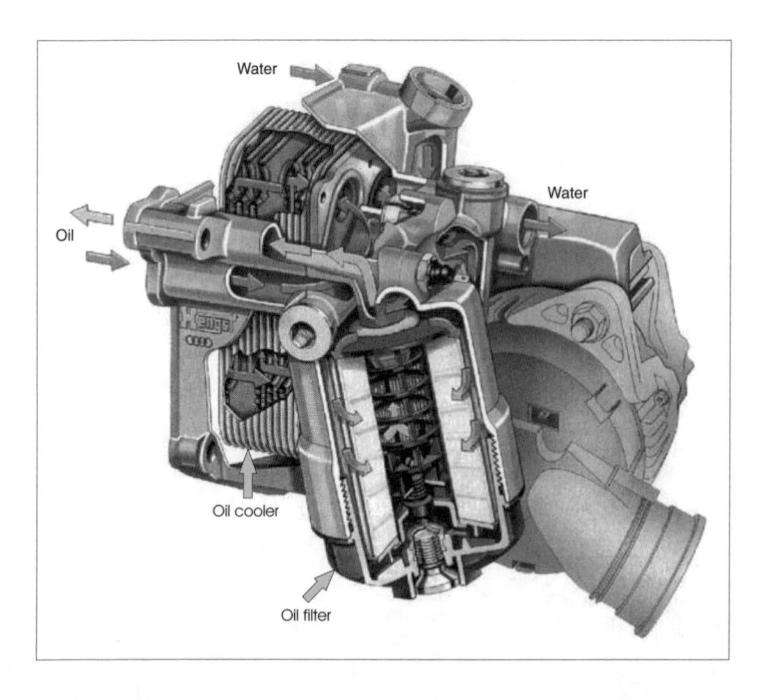

Fig. 5.1-29 Oil filter module (example: Audi V8).

utilize engine oil as long as possible without endangering the engine. Oil change intervals are then set not at predetermined distances (with included safety margins) or at specific time intervals (one or two years) but rather are determined by engine operating conditions and duration. Electronic engine controls are able to integrate engine loading as determined by fuel flow, operating temperatures, running times, oil replenishment, and similar factors, and indicate to the driver when an oil change is required. For average drivers with regular, moderate driving styles, this translates into considerably greater driving distances between oil changes.

5.1.1.5 Auxiliary devices and package

Several decades ago, opening a car's hood would reveal a large hole containing a black lump and a few tubes, pipes, and gadgets; besides these, there was a great deal of empty space, and a view of the road surface below. Today, even after removal of "designer covers" or sound insulation, a great deal of effort must be expended even to stick one's hand between components. The actual powerplant, the basic engine, has not gotten any larger, but expansive induction systems with their air filters and ducting, possibly also an intercooler, consume a great deal of space. Exhaust systems are often configured as a great convolution of tubes, sometimes with an exhaust-driven turbocharger added for good measure. Power steering and air conditioning compressor are considered almost standard on modern cars. The vehicle itself adds many safety and comfort elements which must be located somewhere in the engine compartment. As a result, space is in short supply and must be allocated carefully.

To drive auxiliary devices, modern designs almost exclusively rely on poly-V belts. These are plastic belts with a multiwedge profile molded on one or both sides. In contrast to the old V-belt, poly-V belts may be bent in either or both directions within their mounting plane. Idler or tensioning rollers can be interspersed with auxiliary device drive pulleys, thereby providing very compact installations (Fig. 5.1-30). The previously customary design practice of using multiple V-belts in various planes to drive individual devices or groups has now been reduced to a single poly-V belt operating in a single plane. Belt-driven devices include the water pump and alternator (generator). Usually, vehicle installations include a power steering pump and an air conditioning compressor. Some vehicles also have a belt-driven viscous-clutch cooling fan. Poly-V belts are sensitive to misalignment of the drive pulleys. Therefore, it makes sense, and contributes to achieving compact installations, if all auxiliary devices are mounted on a common carrier. This can be accurately manufactured to form a module which can be bolted to the engine as a single unit with a large mounting footprint.

The engine package also includes the starter, which engages the flywheel, as well as the induction and exhaust modules with associated components upstream and downstream of these, respectively. Their spatial arrangement is not only important in achieving a functionally optimized layout, but also represents components which must be considered in examinations of vehicle crash behavior. These represent rigid, nondeformable blocks which may limit body deformation, thereby endangering vehicle occupants. In such cases, other geometries must be chosen or the critical components must

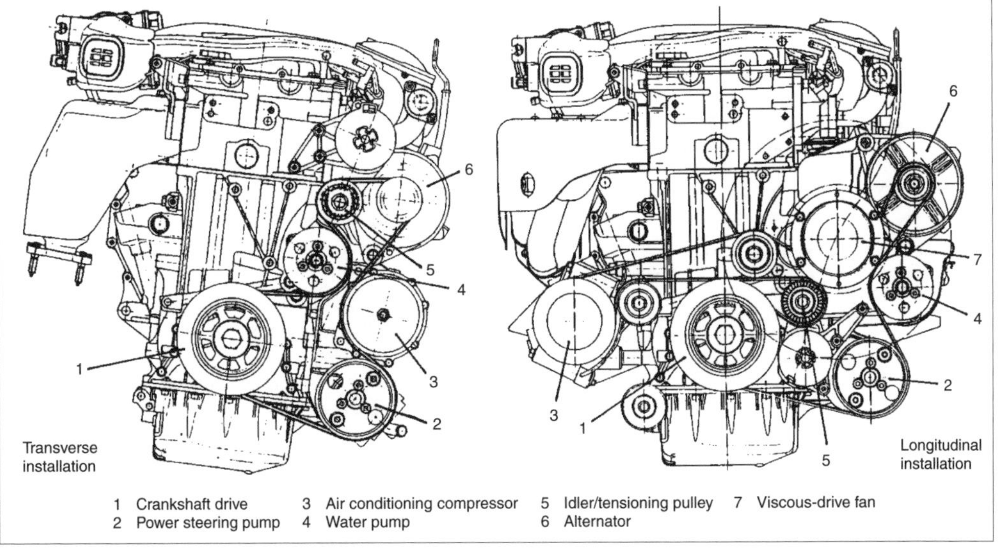

Fig. 5.1-30 Poly-V belt drive for auxiliary devices, longitudinal and transverse engine installations (example: Volkswagen five-cylinder VR5 engine).

be "defused" by designing them to break away or deform. The air mass flow sensor, located between the air filter and throttle butterfly, requires a certain minimum air duct length in order to measure undisturbed airflow values. The vacuum hose for the power brake booster, the hose to the activated charcoal filter (fuel tank venting), and the cable for throttle actuation (or electrical wiring for electric throttle) are routed on the engine's air induction side. On the exhaust side, the engine compartment may contain the catalytic converter and oxygen sensor, along with their heat shields. Electrical wiring, starter cable, hoses for coolant, air conditioning refrigerant, heating, and many more functions fill the engine compartment. These must all be arranged to avoid chafing, mutual electrical interference, excessive heat, and, if possible, remain readily accessible in the event of a repair.

Vehicle manufacturers attempt to address engine compartment complexity and packaging difficulty by combining as many items as possible into preassembled modules. This reduces assembly effort at the engine and vehicle production facilities to a process of coupling modules together.

References

[1] Goldbeck, G. *Kraft für die Welt, 1864–1964 Klöckner-Humboldt-Deutz AG*. Econ Verlag, Düsseldorf and Vienna, 1964.
[2] Dubbel, *Taschenbuch für den Maschinenbau*. W. Beitz and K.-H. Grote (editors), 19th edition. Springer Verlag, 1997.
[3] Buschmann, H., and P. Koessler (editors), *Handbuch für den Kraftfahrzeugingenieur*. 8th edition. Deutsche Verlagsanstalt, Stuttgart, 1973.
[4] Schäfer, F., and R. van Basshuysen. *Schadstoffreduzierung und Kraftstoffverbrauch von Pkw-Verbrennungsmotoren: Die Verbrennungskraftmaschine*. New Series, Vol. 7. H. List und A. Pischinger (editors). Springer Verlag, 1993.
[5] Hauschulz, G., H.-J. Heich, P. Leisen, J. Raschke, H. Waldeyer, and J. Winckler. *Emissions- und Immissionsmesstechnik im Verkehrswesen*. Verlag TÜV Rheinland, 1983.
[6] Scheiterlein, A. *Der Aufbau der raschlaufenden Verbrennungskraftmaschine*. H. List (editor), *Die Verbrennungskraftmaschine,* Vol. 11. Springer Verlag, Vienna, 1964.
[7] Maass, H. *Gestaltung und Hauptabmessungen der Verbrennungskraftmaschine*. H. List and A. Pischinger (editors). *Die Verbrennungskraftmaschine.* New Series, Vol. 1. Springer Verlag, Vienna, 1979.
[8] Maass, H., and H. Klier. *Kräfte, Momente und deren Ausgleich in der Verbrennungskraftmaschine*. H. List und A. Pischinger (editors). *Die Verbrennungskraftmaschine.* New Series, Vol. 2. Springer Verlag, Vienna, 1981.
[9] Gerigk, P., D. Bruhn, D. Danner, L. Endruschat, J. Göbert, H. Gross, and D. Komoll. *Kraftfahrzeugtechnik*. Westermann Schulbuchverlag, Brunswick, 1987 (1991).
[10] *Hütte. Die Grundlagen der Ingenieurwissenschaften*. H. Czichos (editor), 30th edition. Springer Verlag, 1996.
[11] Pahl, G., and W. Beitz. *Konstruktionslehre*. 4th edition. Springer Verlag, 1997.
[12] Baehr, H.D. *Thermodynamik*. 9th edition. Springer Verlag, 1996.
[13] Baehr, H.D., and K. Stephan. *Wärme- und Stoffübertragung*. 3rd edition. Springer Verlag, 1998.
[14] Köhler, E. *Verbrennungsmotoren (Motormechanik, Berechnung und Auslegung des Hubkolbenmotors)*. ATZ-MTZ-Fachbuch. Vieweg Verlag, 1998.
[15] Zima, S. *Kurbeltriebe (Konstruktion, Berechnung und Erprobung)*. ATZ-MTZ-Fachbuch. Vieweg Verlag, 1998.
[16] Bosch. *Kraftfahrtechnisches Taschenbuch*. 22nd edition. VDI Verlag, Düsseldorf, 1995. English edition: *Bosch Automotive Handbook*. 5th edition. Robert Bentley, New York, 2000.
[17] Company publications by Beru, Siemens, and Mahle.
[18] Forschungsvereinigung Verbrennungskraftmaschinen e.V. FVV. Schriftenreihe zu Forschungsvorhaben. FVV Frankfurt/ Main.

5.1.2 Otto-cycle engines, Part 2

5.1.2.1 Future solutions

Building on conventional gasoline-engine technology, future solutions must depart from traditional restrictions on external mixture formation, throttle control, and homogeneous cylinder charges. What then remains as the main differentiating feature between gasoline and diesel engines is the type of ignition, as expressed by the terms "spark ignition" (SI) and "compression ignition" (CI).

In evaluating future powerplant concepts, a wide variety of criteria must be considered. For consumers, reliability under all conceivable operating situations and throughout the vehicle's entire anticipated service life are taken as a matter of course. Ultimately, acceptance of a new propulsion concept is dependent upon its economic viability. This will be determined by manufacturing costs as well as operating costs. Fuel economy exerts a direct influence on overall operating economy but is also at the forefront of public discussion for reasons of natural resource conservation as well as formation of CO_2 emissions associated with fossil fuel combustion.

Increased regulatory activity, which imposes ever more stringent exhaust emissions standards on passenger car propulsion systems, is an outgrowth of the dramatic traffic density increase in urban areas. These regulations present additional important selection criteria for future passenger car propulsion systems. Noise emissions emanating from the propulsion system are also regarded as irritating, burdensome consequences of road traffic and must therefore be minimized. Noise emissions also have a direct influence on comfort as perceived by vehicle occupants. Efforts toward reducing vehicle weight and achieving the most compact dimensions result in demands for higher propulsion system power density. Dynamic operational behavior presents an emotion-based criterion as well as a requirement associated with the rational concept of "active safety."

In particular, recent significant progress in high-pressure direct-injection passenger car diesel engine development has emphasized the conceptual disadvantages of the conventional gasoline engine. In the conventional gasoline engine, control by means of throttling the intake airflow results in appreciable gas exchange losses. Concepts to emancipate the gasoline engine from throttling losses include gasoline direct injection with stratified charge (see Section 5.1.2.2) and variable valve actuation (see Section 5.1.2.3). Downsizing concepts also achieve a significant portion of their fuel economy advantage by shifting the engine operating point, thereby dethrottling the engine.

The present operating principle of homogeneous fuel–air mixtures as applied to conventional gasoline engines limits opportunities for achieving the thermodynamic advantages of lean operation within fuel flammability limits. One means of overcoming this disadvantage may be found in gasoline direct injection with stratified charge (see Section 5.1.2.2). However, such lean-burn engine concepts require new solutions for aftertreatment of oxygen-rich exhaust gas (see Section 5.1.2.2.5).

Limits on compression ratios imposed by knock tendencies at high load have a direct influence on ideal-process thermal efficiency. Customarily, a compromise must be struck between power density and part-load efficiency; however, this compromise may be avoided by means of variable valve actuation. In particular, when applied to supercharged gasoline engines with resulting downsizing effects, this concept permits realizing significant reductions in fuel consumption (see Section 5.1.2.4).

5.1.2.2 Direct injection

In direct injection of fuel into the combustion chamber, timing of the injection process establishes the time frame for mixture formation. Charge stratification realized by late injection permits good part-load operation, with generally very lean air/fuel ratios. The associated unthrottling of the engine as well as thermodynamic advantages arising from altered properties of the cylinder charge have a beneficial effect on fuel consumption. Early injection during the induction stroke permits a largely homogeneous cylinder charge. This operating mode is primarily used for high load and full load operation. The internal cooling effect of fuel vaporization during the induction process results in volumetric efficiency advantages and charge cooling, which reduces knock tendency, thereby permitting higher mean effective pressures.

5.1.2.2.1 Combustion system

The main purpose of a direct-injection combustion system is to ensure complete mastery of the engine process, utilizing concept-related advantages across as wide a range of engine operating map conditions as possible. In contrast to the conventional gasoline engine, late timing of direct cylinder injection in particular allows comparatively little time for mixture formation. Special effort must be made to ensure adequate mixture formation in the time available prior to ignition. At the moment of ignition, flammable mixture must be concentrated around the spark plug to ensure reliable ignition and as complete a combustion process as possible. From a thermodynamic standpoint, an insulating shell of air or end gas around this fuel-rich mixture zone would be advantageous, as this would permit effective minimization of thermal wall losses.

In addition, the combustion concept should also avoid any possible negative side effects of direct injection. Cylinder wall wetting by liquid fuel and the associated risk of washing off the lubricating oil film must be avoided in any case. Similarly, formation of lean mixture zones which undergo only partial combustion as a result of quenching effects, thereby raising hydrocarbon emissions, should be suppressed by appropriate measures. To avoid creating soot, mixture formation by the injected fuel must be supported to the point where no fuel droplets or areas of over-rich mixture remain by the time the flame front passes. Deposits of incomplete combustion products on the fuel injector should be minimized to avoid impairing engine operation. In addition, fuel spray should not be allowed to impinge directly on the spark plug, as this would result in unacceptable thermal shock.

To meet these combustion concept requirements, individual fuel spray characteristics, cylinder internal flow, and combustion chamber shape must undergo a careful mutual optimization process. The primary fuel injector design criteria are the volume enveloped by the fuel spray and its expansion over time and space (see also Section 5.1.2.2.3). Mixing of combustion air within the fuel spray is in large part attributable to secondary flow effects initiated by the fuel spray. In addition, the cylinder charge motion generated by the (air) induction process may be used to support the mixture formation process. Furthermore, charge motion can also be used to transport vaporized fuel, thereby controlling mixture concentration at the spark plug. The combustion chamber shape, in particular piston crown contour, must follow from these considerations. The combustion chamber shape may also be used to deliberately direct the fuel stream (see "Wall-guided systems" below). In addition, combustion chamber shape also affects charge motion and its subsequent decay during the compression process.

Direct-injection combustion systems may be classified according to the predominance of any of the aforementioned effects. Classification into spray-guided, wall-guided, and air-guided combustion systems has become generally recognized. Even if actual combustion systems are not always clearly assignable to one of these three classes, such classification nevertheless promotes better understanding of key processes.

Spray-guided systems

In spray-guided combustion systems, mixture formation and charge stratification are largely dependent on fuel spray characteristics. Since this combustion system does not incorporate any deliberate support for mixture formation by means of charge motion, and since the combustion chamber shape is designed for unimpeded fuel spray propagation, charge stratification is primarily driven by the mixture cloud formed by the fuel spray. The outer edge of the spray envelope forms a layer of flammable mixture, in which the ignition point should

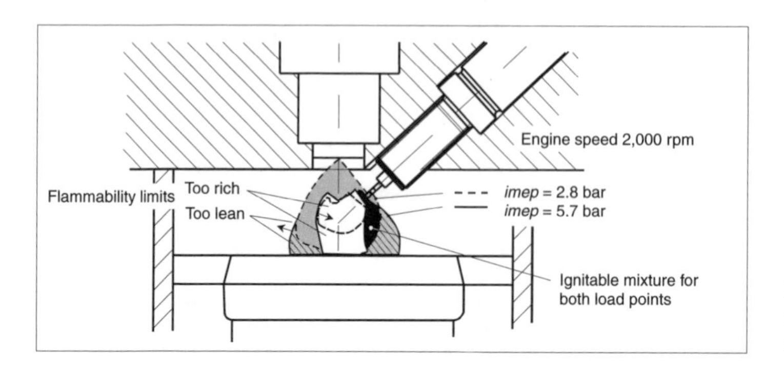

Fig. 5.1-31 Flammability limits for spray-guided mixture formation.

be located. Consequently, for spray-guided systems, fuel injector and spark plug are located in close proximity to one another to ensure that some flammable mixture is concentrated at the spark plug even for small injection quantities.

Basic investigations using a single-cylinder experimental engine fitted with a spray-guided combustion system [1] have shown that the ignition point must be located within a narrow zone at the outer edge of the spray envelope (Fig. 5.1-31). This places high demands on fuel injectors with respect to scatter. For these experiments, the engine was fitted with optical access to the combustion chamber, permitting use of laser-induced fluorescence (LIF) methods to determine local mixture composition during actual (fired) engine operation. This methodology has proven to be an extremely valuable tool in the highly complex development process of gasoline direct-injection systems.

A favorable air/fuel concentration at the "core" of the combustion chamber, surrounded by a thermally insulating layer of air or air/residual gas mixture, may be achieved with the fuel injector located in the center of the cylinder head and a steep inclination of the fuel spray near the cylinder axis. It should be pointed out that such spray-guided combustion systems exhibit very favorable part-load fuel consumption [2].

Central fuel injector location is also the preferred configuration for full-load operation with homogeneous, stoichiometric mixtures. This arrangement ensures good mixture formation with minimal cylinder wall wetting. However, in high part-load operation, the expanded fuel cloud cannot be adequately vaporized by induced turbulence alone. Therefore, purely spray-guided systems often exhibit serious mixture formation problems, associated with increased hydrocarbon and soot emissions, which greatly restrict the usable stratified charge operating range.

Wall-guided systems

In wall-guided combustion systems, mixture formation relies on a significant portion of the injected fuel being deposited on the combustion chamber walls, from which it vaporizes and mixes with air. The combustion chamber surface used to control mixture formation is usually the piston crown, which is given a special bowl shape. After the injection process, a fuel-rich multiphase mixture forms at the surface of the piston bowl, with a film of liquid fuel adhering directly to the surface. A short distance above the surface is a layer occupied primarily by vaporized fuel; this layer may also contain spray droplets or droplets splashed from or released by the wall.

The fuel-rich zone near the wall moves along the wall in response to the impulse of the impinging fuel spray. Appropriate geometric design of the piston bowl with respect to the spark plug promotes concentration of flammable mixture at the ignition point. However, motion of the fuel spray alone will not result in satisfactory operation. Therefore, a specific air charge motion is used to achieve timely mixture formation of the entire injected fuel quantity. This continuously provides fresh air to the fuel-rich wall zone and transports vapor-enriched mixture from the wall contour to the spark plug.

It is apparent that in addition to the combustion chamber shape, charge motion has a significant effect on the engine's operational behavior. Charge motion must be carefully tuned with regard to both intensity and configuration. In principle, for this type of charge motion, the familiar basic patterns of swirl and tumble enter consideration. In the reality of actual operating engines, one often encounters hybrid forms of these two different charge motion concepts, which are differentiated primarily by the orientation of their vortex axes.

Gasoline direct-injection engines introduced to the market by Japanese manufacturers may be regarded as typical representatives of wall-guided combustion systems. Mitsubishi's gasoline direct-injection concept, introduced in Japan in 1996 and in Europe the following year under the "GDI" appellation, is based on a four-valve engine with centrally located spark plug (Fig. 5.1-32). This four-cylinder DOHC engine (81 mm bore, 89 mm stroke) has a relatively high compression ratio (12:1, or 12.5:1 for European models), which is possible due to the internal cooling effect provided by direct fuel injection. The piston contains a pronounced bowl, which presents a "flat" aspect to the fuel injector, mounted on the intake side, and a "steep" aspect to the spark plug. To support mixture formation and transport, a tumbling motion is imparted to the charge in a

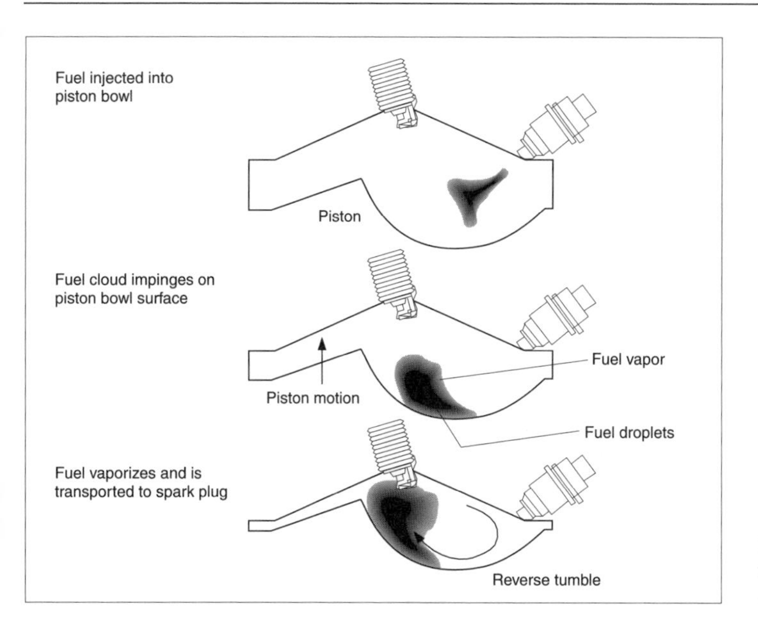

Fig. 5.1-32 Wall-guided reverse tumble combustion concept.

direction opposite that usually employed by tumble systems. This has come to be known as "reverse tumble." To create this charge motion, the intake ports are oriented nearly vertically, which, with the associated intake manifold, results in a relatively tall engine.

Detailed combustion analyses were conducted in the course of developing this engine concept [3]. Mixture formation processes as well as particulars of the combustion process in the stratified fuel–air mixture were closely examined with the aid of laser-optical methods. These examinations underscored the importance of charge motion for vaporizing fuel adjacent to the wall and for transporting the mixture toward the spark plug. The Mitsubishi GDI engine has also drawn the interest of many engine developers and scientists. Their work largely substantiates the manufacturer's own publications, even if the claimed fuel economy advantages have not been completely verified [4].

In developing this engine concept for the European market, several details of the combustion system, such as fuel injector inclination as well as combustion chamber geometry, have been matched to driving habits which are perhaps more performance-oriented than those of the Japanese market. The objective of these modifications has been, in particular, optimization for operation with homogeneous charges at higher loads, achieved by early injection during the intake stroke. Meanwhile, Mitsubishi has introduced other direct-injection engines with different numbers of cylinders and different dimensions but based on the same combustion system.

Meanwhile, other manufacturers have also introduced direct-injection gasoline engines to the Japanese market. The four-valve concept applied to Toyota Motor Co.'s 1.8-L gasoline engine [5] also relies on a wall-guided combustion system (Fig. 5.1-33). This four-cylinder engine displaces 1,998 cc (86.0 mm bore × 86.0 mm stroke) and has a moderate compression ratio of only 10:1. The fuel stream is directed at the edge of a piston bowl. Mixture formation is aided by swirl induced by deactivating one of the two intake ports, which is configured as a straight "fill" port, forcing air to enter through the remaining helical intake port. The swirl motion of intake air facilitates vaporization of fuel from the piston bowl wall as well as mixture transport to the spark plug. At top dead center, the centrally located spark plug extends into the piston bowl, whose edge is slightly constricted to form a lip.

Nissan Motor Co. has presented a direct-injected

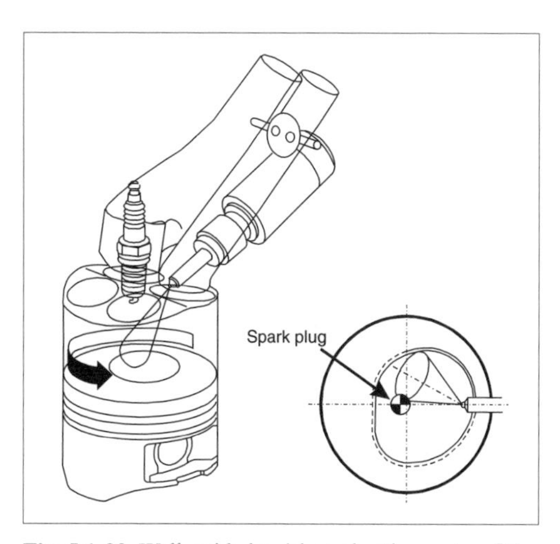

Fig. 5.1-33 Wall-guided swirl combustion system [5].

gasoline engine under the name NEO Di. This engine, bearing the designation VQ30DD, consists of a V6 DOHC engine displacing 2,987 cc (93.0 mm bore × 73.3 mm stroke) with four valves per cylinder and an 11:1 compression ratio. The engine operates with a swirl-based combustion system which Nissan calls NExT ("Nissan Exquisitely Tuned Combustion") [6], which also falls into the class of wall-guided systems. Charge motion is achieved by means of a port deactivation system using symmetrical intake ports. Closing the swirl control valve creates swirl to support mixture formation; an open swirl control valve takes advantage of moderate "tumble" motion of the charge to support homogeneous mixture formation. The comparatively simple form of the piston bowl is remarkable; it may be regarded as a pure tub-shaped bowl. Meanwhile, Nissan has also presented a four-cylinder version of this engine, displacing 1,769 cc (80.0 mm bore × 88.0 mm stroke).

The presented examples of actual engines with wall-guided mixture formation show remarkable differences, particularly in terms of piston bowl design. This variety is supplemented by even more examples to be found in the technical and patent literature. In view of production-relevant boundary conditions, the examples shown may be regarded as representative.

Air-guided systems

Air-guided combustion systems achieve mixture formation by means of interaction between the fuel spray and intensive, directional charge motion. Mixture formation takes place without any direct influence on the part of the combustion chamber walls. Rather, the objective of the combustion chamber design is to stabilize the flow induced in the intake charge during the induction stroke, thereby ensuring that air mixes with fuel after the subsequent injection process and that the mixture cloud thus formed is transported to the spark plug. Consequently, in contrast to spray-guided systems, the fuel spray cannot be aimed directly at the spark plug electrodes. This has advantages in terms of spark plug operating conditions. To differentiate air-guided systems from wall-guided systems, ideally liquid fuel does not reach the combustion chamber walls; instead, it vaporizes and mixes with air within the chamber.

Typical design features of air-guided combustion systems include relatively large spacing between fuel injector and spark plug, orientation of the fuel spray toward the point of ignition but without directly impinging on the spark plug, and a combustion chamber shaped to provide free spray propagation even with late injection. Fig. 5.1-34 shows this configuration as exemplified by a combustion system with tumble charge motion, developed by FEV Motorentechnik. Published research reports show that the basic mechanisms of mixture formation and transport as employed in air-guided combustion systems can be assured over an extended range of engine operating conditions [7]. Such systems permit good part-load fuel economy with high operational stability, as well as good full-load operation. Compared to other direct-injection concepts, emissions behavior shows very good hydrocarbon emissions with essentially soot-free part-load operation.

In order to analyze mixture formation processes, laser-optical and numerical methods were also applied during development of this combustion system [8]. Fig. 5.1-35 shows mixture distribution, as determined by computational fluid dynamics (CFD), at the end of injection and at ignition. It is apparent that charge motion directs the fuel spray toward the spark plug, thereby concentrating a flammable mixture at the plug. As this example demonstrates, numerical process simulation methods are a powerful tool, able to provide engine developers with good predictions of the effects of modifications or altered operating conditions. It is to be expected that this

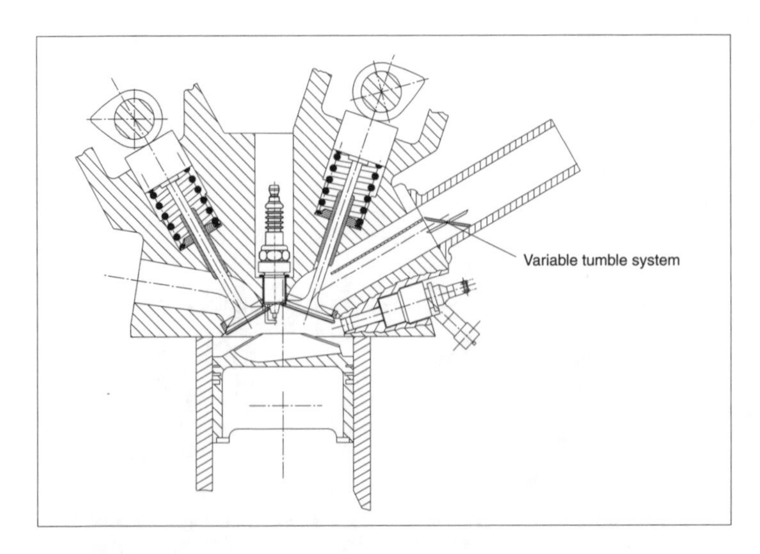

Fig. 5.1-34 Air-guided combustion system.

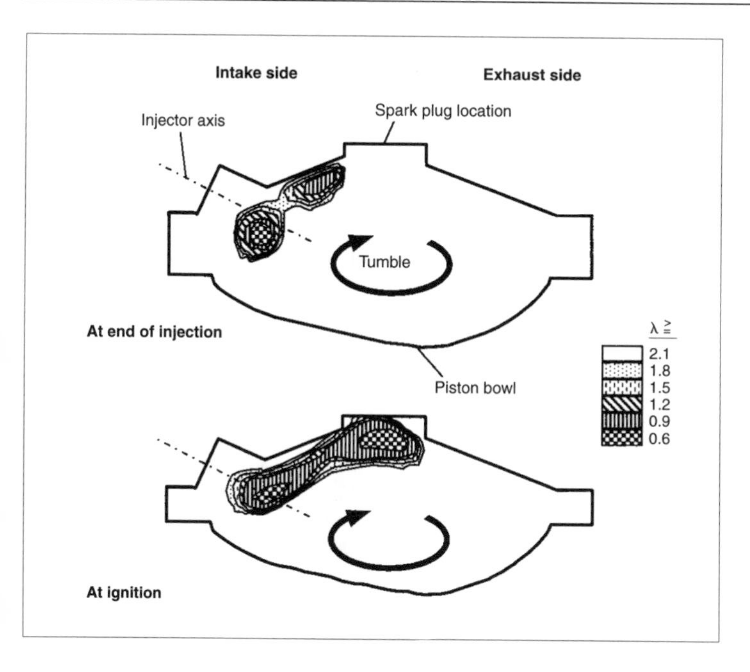

Fig. 5.1-35 Application of numerical methods to represent mixture distribution [8].

methodology will become an indispensable component of gasoline direct-injection development strategy.

Systematic comparison by Grigo [9] of the combustion systems described above, using a single-cylinder engine with consistently maintained boundary conditions, leads to conclusions regarding the potential of spray, wall, and air-guided combustion systems. With the aid of burning function calculations, Grigo was able to distinguish significant differences between these combustion processes (Fig. 5.1-36). Differences in elapsed time between end of injection and ignition are readily apparent. This time span is shortest for spray-guided systems, which may be interpreted as direct ignition of the injection spray. By contrast, for wall-guided systems, the extended mixture formation path along the piston bowl surface offers the longest available time frame for mixture formation. These systematic differences result in different ignition delays. The spray-guided system exhibits the longest ignition delay and therefore demands the greatest spark advance. Despite the brief mixture formation time available compared to the wall-guided system, the air-guided system, with its intense charge motion, achieves a high-quality mixture. Consequently, the air-guided system provides a shorter ignition delay, in contrast to the spray-guided system.

As a result of these comparative investigations, the following typical properties of the three combustion systems may be established:

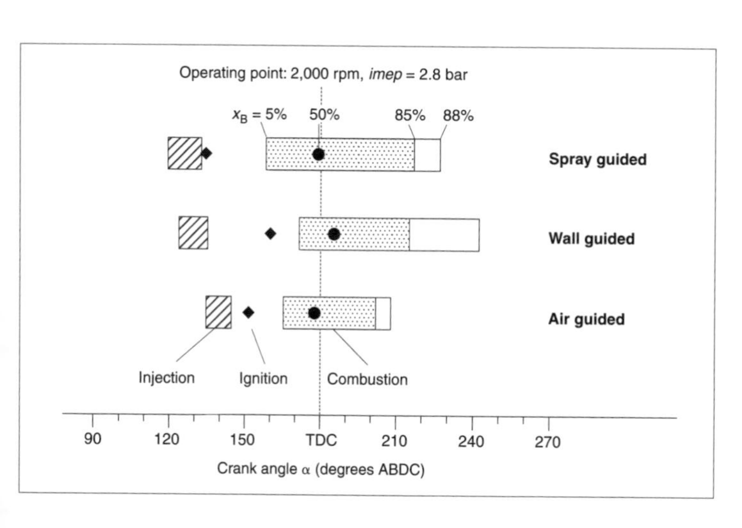

Fig. 5.1-36 Combustion progress in direct injection combustion systems [9].

- *Spray-guided combustion systems* exhibit a high degree of operational stability under part-load conditions with stratified charge. However, their operating principle results in high sensitivity to spray quality as delivered by the fuel injector. Under high part-load conditions, the spray-guided principle alone is not capable of assuring adequate mixture formation nor of avoiding extended zones of extremely rich mixture and their resultant significant soot emissions. However, with spray directed toward mid-cylinder and with compact combustion chamber design, the spray-guided principle offers advantages for full-load operation.
- *Wall-guided combustion systems* also offer a high degree of operational stability at part load. Often, however, the mixture formation process, based on interactions between fuel spray, piston bowl, and charge motion, does not proceed rapidly enough, resulting in drawn-out end of combustion and correspondingly high emissions of unburned hydrocarbons. Moreover, it should be noted that wall-guided mixture formation is strongly influenced by piston bowl surface temperature. For this reason, ability to achieve stratified charge operation immediately after cold start is limited. The complex shape of the piston bowl often leads to a convoluted combustion chamber, which runs counter to optimum full-load behavior. Similarly, the intensity of charge motion required for stratified charge operation has an effect on full-load operation. In this case, variable concepts (see also Section 5.1.2.2.2), with correspondingly lower volumetric efficiency losses, offer good possibilities for translating the conceptual advantages of gasoline direct injection into higher full-load torque output.
- *Air-guided combustion systems* for direct-injection gasoline engines offer the possibility of stable stratified charge operation across wide areas of the engine map. They exhibit the best potential for achieving good hydrocarbon emissions. At low engine speeds, however, there is a tendency that charge motion intensity may be insufficient to transport mixture to the spark plug. This may be countered by partial throttling, which in any case is necessary to achieve sufficiently high temperatures for catalytic exhaust gas aftertreatment (see Section 5.1.2.2.5). Like wall-guided systems, air-guided combustion systems require that the necessary charge motion intensity be achieved with as little loss in full-load volumetric efficiency as possible. Therefore, variable means of affecting charge motion are also preferred for these systems.

As already mentioned, not all combustion system concepts clearly fall into a single category. More often, an attempt is made to combine the most favorable characteristics of various systems in some advantageous manner. In the majority of cases involving development of direct-injected gasoline engines, consideration must be given to boundary conditions which may include incorporating existing components, models, and manufacturing facilities. Design details associated with combustion system choice are usually subordinate to these restrictions, which often represent considerable cost. The mission of combustion system development is then primarily one of finding and optimizing a suitable concept within the given boundary conditions.

Exhaust gas recirculation

Meeting exhaust emissions standards with gasoline direct-injection engines is especially problematical in terms of oxides of nitrogen because in lean operation, conventional catalytic exhaust treatment cannot achieve sufficiently high NO_x conversion rates. Even future solutions for treatment of oxygen-rich exhaust gas, discussed in Section 5.1.2.2.5, are by themselves insufficient to meet NO_x emissions limits. For this reason, internal engine measures to reduce NO_x raw emissions must be utilized. Exhaust gas recirculation (EGR) applied for this purpose increases heat capacity and reduces oxygen content, therefore reducing flame speed. Both result in lower peak temperatures during the combustion process.

Highly developed direct-injection combustion systems generally exhibit high tolerance for EGR. At part load, $n = 2000$ rpm, $imep = 2$ bar, EGR rates on the order of 30% may be regarded as typical. In evaluating such high (compared to conventional gasoline engine) recirculation rates, it must be remembered that the recirculated gas has a high oxygen content. However, exhaust gas recirculation reduces the engine's overall excess air ratio, which raises exhaust gas temperature. This is a welcome side effect, as otherwise temperatures at part-load operation would drop below the level needed for catalytic exhaust treatment. Moreover, it is often noted that higher mixture temperatures associated with exhaust gas recirculation have a positive influence on mixture formation.

For direct-injected gasoline engines, the preferred method of exhaust gas recirculation is external introduction of exhaust into the induction system. The engine must be slightly throttled to achieve the necessary low manifold pressures. Throttling is in any case necessary at low part load, in order to achieve a sufficiently high exhaust gas temperature for exhaust gas aftertreatment. The point of introduction should be located as close to the valves as possible, enabling rapid EGR rate adjustment when switching operating mode from lean stratified charge to homogeneous stoichiometric or rich mixtures, as required for regeneration of a catalytic converter with NO_x adsorber (see Section 5.1.2.2.5).

Alternatively, internal exhaust gas recirculation by means of variable valve timing may be considered. However, to achieve adequate exhaust gas recirculation rates, such systems require large shifts in camshaft timing and therefore pistons with deep valve pockets to ensure that valves do not contact pistons. This in turn often conflicts with piston crown design, which is largely driven by combustion system requirements.

5.1.2.2.2 Charge motion

The previously described combustion systems employed by direct-injected gasoline engines underscore that gas flow within a cylinder represents a key optimization point for many combustion system concepts. There is, however, no clear preference with regard to choice of charge motion configuration. Highly promising combustion system concepts have been demonstrated using swirl as well as tumble concepts. In many cases, charge motion cannot clearly be assigned to one or the other of these two basic forms, but rather represents a hybrid form of both. In particular, generating pure swirl flow is hardly realizable given the design constraints of modern gasoline engines; usually, "swirl" actually consists of rotational flow whose axis is angled within the combustion chamber.

Charge motion, required for part-load stratified charge operation, must exhibit sufficient intensity, depending on the choice of combustion system. Realization of such high flow intensity is generally only possible if intake runners are modified at the expense of flow capacity. This, however, implies losses in volumetric efficiency at full load, which cannot be tolerated. As a result, preference is given to concepts which permit variability with regard to charge motion and flow rates. Such concepts are primarily designed for good flow rates. Under part-load operation, the required charge motion is generated by switchable elements within the induction system.

The solution usually applied to multivalve engines with swirl systems is induction runner shutoff, in which one of the two induction passages is deactivated by means of a swirl control valve (see also Fig. 5.1-33). In such a system, dual intake runners must be used downstream of the swirl control valve. In view of sealing considerations at the swirl control valve, it is sufficient if the airstream through the deactivated passage is throttled to the point where it does not disturb the fresh air charge flowing into the combustion chamber tangentially through the remaining passage. This technology has already been extensively investigated on lean-burn engines with external mixture formation and successfully applied to series production vehicles. Alternatively, valve deactivation may be considered. However, this does not offer any fundamental advantages, and a concept decision would be based on other, perhaps ultimately cost-based, considerations. Other conceptual considerations—for example, combination with a cylinder shutoff system—may make valve deactivation a more attractive proposition.

To achieve variable charge motion in two-valve engines, partial activation of the intake runner may be considered. In such a system, the intake runner is divided until just upstream of the intake valve, with one of the passages deactivated by a flap valve. Alternatively, a deflector may be swung into the intake passage to produce a deliberate change in flow through the runner and asymmetrical flow distribution at the intake valve.

For tumble concepts, variable effects are also achieved by switchable partial deactivation of the intake tract. With symmetrical intake runner arrangements on multivalve engines, the siamesed runner is divided horizontally and one of the two resulting passages deactivated by means of a control valve or control spool. Tumble intensity is then determined by runner geometry and the orientation of the horizontal parting plane. Fig. 5.1-37 shows flow characteristics of such a variable tumble system within the scatter band of numerous intake runners examined by FEV Motorentechnik. It is apparent that this concept permits combination of high tumble intensity for stratified charge operation, as well as favorable flow rates for full-load operation.

5.1.2.2.3 Injection technology

In order to realize gasoline direct-injection concepts, present-day production engines employ high-pressure

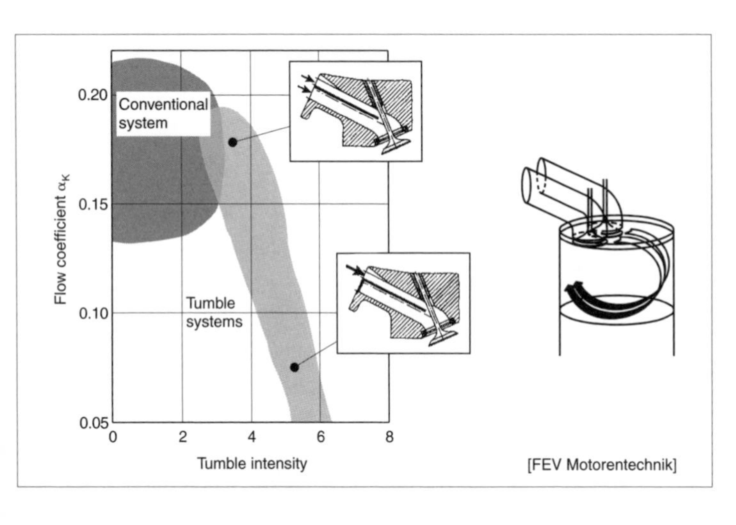

Fig. 5.1-37 Flow behavior of tumble systems.

liquid fuel injection. In addition, there exists the alternative of air-assisted injection, which uses air or exhaust gas to blow a metered, stored fuel quantity out of an injector.

High pressure liquid injection ("solid injection")

In high-pressure liquid injection, a high-pressure pump feeds fuel to a fuel distributor rail ("common rail") at a maximum system pressure on the order of 120 bar. Radial or axial, single- or multiple-plunger high-pressure pumps may be employed. These are driven either by the camshaft or from the engine timing drive. System pressure is maintained at a constant or map-controlled level by a pressure control valve. The applied rail pressure is the object of combustion process detail optimization and normally ranges between 40 and 120 bar. The fuel rail feeds one injector per cylinder, which injects fuel directly into the combustion chamber. The overall system architecture is shown schematically in Fig. 5.1-38.

The fuel injector is actuated electromagnetically, which permits map-controlled and individual-cycle modulation of injected quantity and injection timing. The characteristics of the fuel spray generated by the injector are primarily determined by flow rate, cone angle, and droplet size. Flow rate is customarily reported as volume at continuous injection against ambient pressure, with a fuel pressure of 100 bar. In designing an injector, consideration must be given to achieving the shortest possible injection duration for the required idling fuel quantity, as well as the necessary injection quantity for full-load operation at rated power.

Compared to manifold injection, the corresponding injection quantity for a direct-injection engine is concept limited; full-load injection can only take place over the duration of the induction stroke. Smaller spreads in injection quantity can in some cases be achieved by load-dependent control of fuel system pressure. The fuel spray cone angle is a significant component of combustion system design. Defining the actual spray cone angle is problematical from several standpoints; on the one hand, the spray shape is not a complete cone, and on the other, as a result of aerodynamic effects, spray propagation in air is sensitive to density and, therefore, pressure within the cylinder. For these reasons, agreement must be reached between system supplier and engine developers regarding spray cone angle and how it is determined.

The degree of atomization in the fuel spray is quantified by fuel droplet size distribution. This is expressed as a one-dimensional number, the Sauter mean diameter (SMD). This may be interpreted as the diameter of a droplet whose ratio of volume to surface area is representative of the entire spray. Typical values of Sauter mean diameter for series-production fuel injectors are less than 20 μm. Optimizing a fuel injector's hydraulic layout is the object of intensive research and development work, employing both optical methods to determine spray characteristics as well as numerical methods for spray modeling (Fig. 5.1-39). For the predominant inward-opening injector design, achieving the required cone angle is accomplished with a tangential fuel flow component, generated by appropriate geometry in the injector needle seat area. Depending on the execution of this swirl inducer, the flow pattern may not be fully developed for very short injection duration or at the beginning of an injection event. The result is a compact pilot jet with a small cone angle. However, this can be used to advantage within the framework of combustion system development to achieve charge stratification.

Operational stability of direct-injected gasoline engines is largely dependent on maintaining spray characteristics within narrow tolerances throughout the entire service life. For this reason, deposits on the combustion chamber side of the injector should be avoided. It has been shown that it is advantageous to have the injector face mounted flush with the combustion chamber wall, as this avoids turbulence or stagnant areas near the injector nozzle.

Air-assisted injection

Atomization of injected fuel can be achieved even with low injection pressures by means of air assistance. Fuel

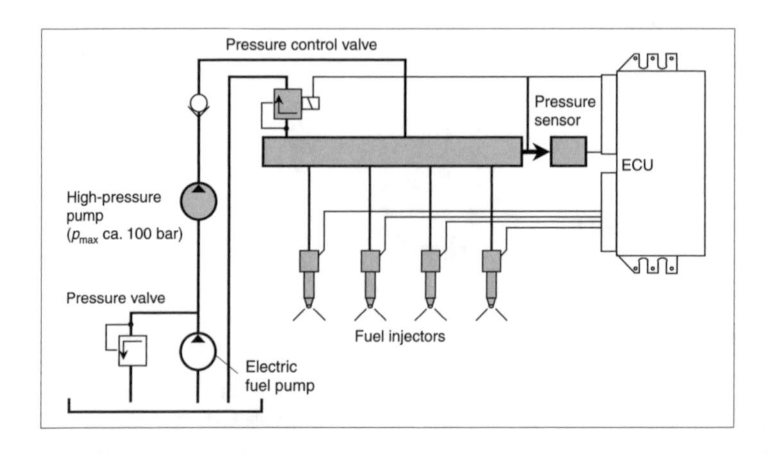

Fig. 5.1-38 High-pressure liquid injection. ECU, Electronic control unit.

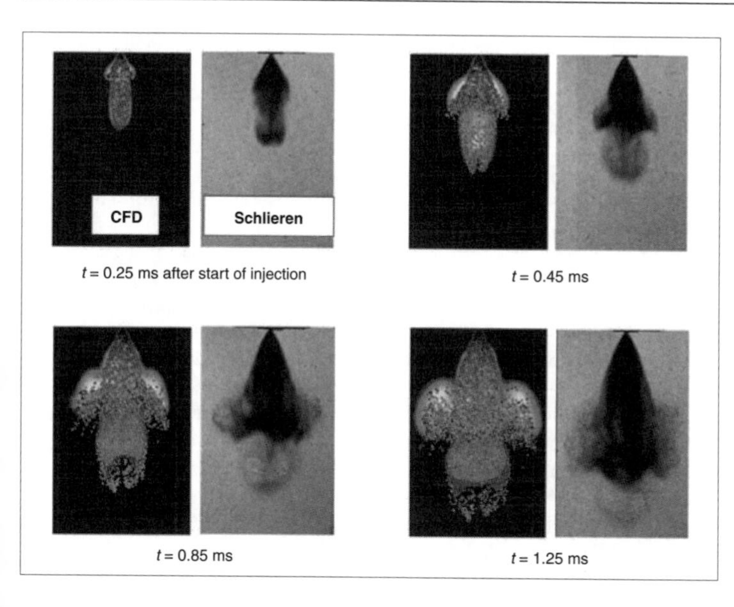

Fig. 5.1-39 Measured and numerically simulated spray cone development.

is metered into a prechamber, from which it is blown into the combustion chamber by a blast of compressed air or exhaust gas. For each working stroke, the amount of injected fuel and blown air represents an extremely rich mixture, so that even with this injection method, the combustion system must assure adequate mixing with combustion air already present in the cylinder. Because of system pressure limits, air-assisted injection systems do not offer the same degree of freedom with respect to injection timing as is available with high-pressure liquid injection.

System components for air-assisted injection were made available within the framework of two-stroke passenger car engine development and may also be found in several marine (boat) engine applications. Matching such systems to the requirements of four-stroke direct-injected gasoline engines is the object of Synerject LLC, a joint venture by Siemens Automotive and Orbital Engine Corporation. Central elements are a combined electromagnetically actuated fuel/air injector as well as a piston compressor to supply the required air.

As an alternative to supplying air by means of a compressor, it has been suggested that part of the cylinder charge could be diverted during the compression stroke, stored in an adjacent chamber, and used during the following stroke to blow the premetered fuel charge [10]. This eliminates the need for an additional compressor. However, such a system would involve a high degree of complexity.

5.1.2.2.4 Operating strategy and engine management

At part load, direct-injection gasoline engines operate with stratified charge. Under these conditions, fuel is not injected until the compression stroke. At higher loads, stratified charge would result in extended over-rich mixture zones; therefore, operation must shift to homogeneous mixtures, implying early injection during the intake stroke. Simultaneously, the amount of air drawn into the engine and the corresponding fuel quantity must be matched to one another. This results in functional demands on the engine management system which extend far beyond the scope encountered in conventional gasoline engines.

In addition to these two basic operating modes, additional modes may be of interest. Fig. 5.1-40 shows a sample operating strategy for a direct-injection gasoline engine on an engine map. For very low loads and at idle, partial throttling is generally necessary to keep exhaust temperatures at a level which enables catalytic exhaust gas aftertreatment. In the transition between lean stratified operation and homogeneous operation with $\lambda = 1$, it may be advantageous to operate with lean

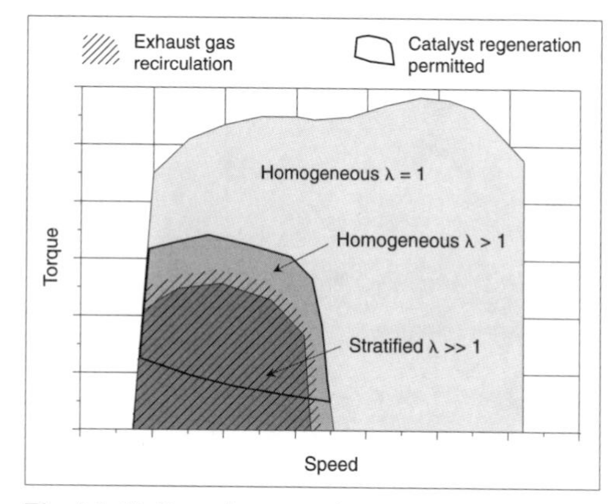

Fig. 5.1-40 Operating strategies.

mixtures in order to extend the benefits of lean operation to a greater range of engine operating conditions. Furthermore, full-load operation often calls for rich mixtures to limit exhaust temperatures (component protection). The engine management system also needs to be capable of applying its operating strategy under actual driving conditions. This calls for realization of hysteresis functions for transition between different operating modes as well as matching operation to operational boundary conditions—for example, cold start and warmup phases.

A further increase in complexity results from integration of the engine in the entire vehicle powertrain system, in particular as exhaust treatment equipment affects the engine control system. For example, use of an adsorber-type catalytic converter (see Section 5.1.2.2.5) during lean operation requires a brief shift to slightly below-stoichiometric operation. The frequency of this regeneration process depends on the adsorber's NO_x storage capacity and the engine's own NO_x emissions. It is the task of the engine management system to balance NO_x storage according to a certain "model" and, based on this model, to initiate regeneration as required [11]. Regeneration of the NO_x adsorber can be controlled with the aid of a NO_x sensor in the exhaust stream. Aging and sulfur poisoning of the adsorber can be monitored continuously in conjunction with the NO_x storage model. This forms a basis for appropriate initiation of the desulfurization procedure, which represents an operating phase with rich mixtures and high exhaust gas temperatures. Ideally, all of these dynamic processes should be applied in such a way that they cannot be perceived by the vehicle occupants. This requires rapid intervention in engine load regulation on the part of the engine control system; it is accomplished by means of an electrically actuated throttle (E-gas) as well as ignition timing changes.

Conventional engine management systems work on the basis of air quantity as a control input parameter. For the gasoline direct-injection systems described here, there is no definite relationship between air quantity and engine load, making such engine control systems unsuitable for this application. On the other hand, a torque-based functional structure permits representing the driver's desired torque in various operating modes [12]. Fig. 5.1-41 is a schematic of such a functional structure. Torque to be provided by the engine, as indicated by driver command and taking into consideration secondary effects from drivetrain and auxiliary devices, is calculated as the controlling parameter. Dynamic control inputs, such as antislip, vehicle dynamic controls, or other vehicle-based systems, enter into the torque demand calculation. Torque demand is calculated by the control unit, taking into consideration the engine's currently desired operating mode. The air quantity calculated from torque demand and selected operating strategy is then set by a throttle (E-gas) controlled by the engine management system.

5.1.2.2.5 Exhaust gas aftertreatment

Noble-metal catalytic converters, a proven technology for conventional gasoline engines, also achieve high conversion rates for oxidation aftertreatment of hydrocarbons and carbon monoxide emissions from lean-burn direct-injection gasoline engines. However, given an oxygen-rich exhaust stream, the reaction train of the reduction reactions used to address oxides of nitrogen is strongly restricted. Operating direct-injection gasoline engines with lean mixtures therefore requires new solutions for reduction of engine-emitted oxides of nitrogen. Many of these concepts are also applicable to diesel engines and will be pursued in parallel.

NO_x selectivity of catalytic converters may be significantly increased by choice of alternative catalysts. "Selectivity" is taken to mean the degree to which the catalytic converter, even with excess oxygen in the

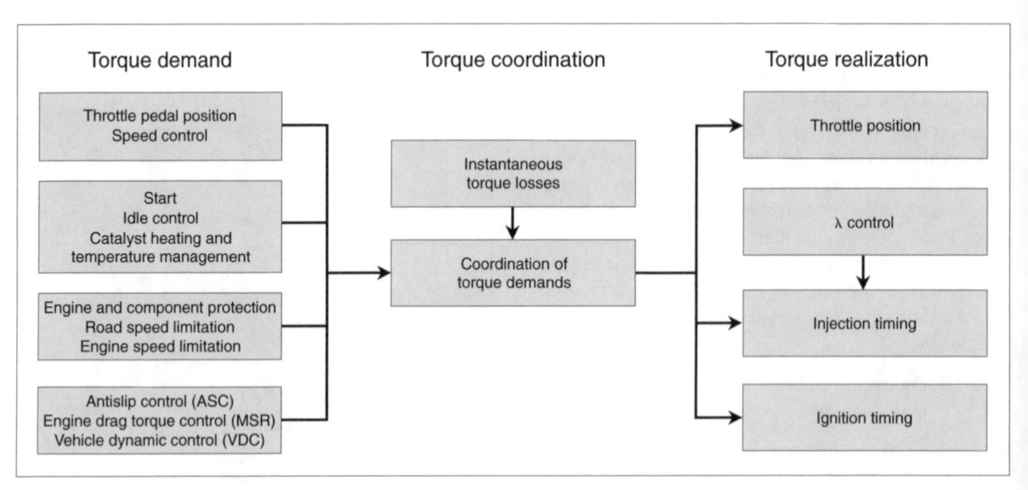

Fig. 5.1-41 Torque-based functional structure [1].

exhaust stream, is able to integrate oxides of nitrogen within the oxidation mechanism used to address hydrocarbons and carbon monoxide. To this end, indium has proven to be an interesting alternative. Strehlau et al report 90% NO_x reduction; however, this only occurs with an extreme exhaust C_1/NO_x ratio of 14:1 [13]. The required extremely high exhaust hydrocarbon concentrations represent a serious emissions problem during the warm-up phase. Furthermore, NO_x activity only occurs in a narrow temperature window, which means that the concept will hardly be capable of meeting future emissions standards.

A further possible means of NO_x reduction in oxygen-rich exhaust is use of an additional, selectively operating reducing agent. "Selective catalytic reduction" (SCR), extensively applied to large stationary power generating facilities, works with ammonia as a preferred reducing agent. For passenger car applications of this technology, urea, in the form of aqueous solutions or solid reactant, is under discussion [14]. This process is known for its great potential for NO_x reduction and has the added advantage of low sensitivity to catalyst poisoning, such as might result from sulfur contained in fuels. Potential problems include how the urea is handled as well as the need to carry urea on board as an additional operating material.

Probably the most promising means of NO_x reduction in oxygen-rich exhaust gas is NO_x storage technology. In this process, oxides of nitrogen emitted during lean operation are chemically stored, to be desorbed during below-stoichiometric operating phases and then reduced by a three-way catalyst as used in conventional gasoline engines [15]. This requires intermittent engine operation to ensure that the adsorber is always regenerated when NO_x storage capacity is exploited. The operating principle is shown in Fig. 5.1-42. In the first step, oxides of nitrogen in the exhaust, consisting mainly of NO, are oxidized by a noble metal catalyst to form NO_2. The NO_2 may then be stored (adsorbed) in the form of nitrate compounds in a storage element. The alkali metals and alkaline earth metals have proven to be especially effective as storage agents.

This functional principle is temperature dependent; newer developments show higher NO_x conversion rates across a sufficiently broad temperature range. If temperatures drop too low, effectiveness is significantly compromised by loss of catalytically supported NO_2 formation. If temperatures are too high, NO_2 dissociates to NO and can no longer be stored as nitrate. Furthermore, with increasing temperature, nitrate stability decreases and stored nitrogen is again desorbed. High converter temperatures should also be avoided to reduce thermal aging; this results in permanent loss of activity due to reduced dispersion of noble metal components and chemical reactions between adsorber material and the oxide substrate [13]. For this reason, adsorber catalysts are preferably located well downstream from the engine. It is especially advantageous to combine a downstream NO_x storage converter with an upstream oxidizing catalytic converter located close to the engine; the oxidizing catalyst lights off at low exhaust temperatures and shortly after cold start, to support oxidation of NO to NO_2.

Once the adsorber reaches its storage capacity, it must be regenerated. For this, the engine must be operated briefly with below-stoichiometric (rich) mixture, typically for only a few seconds. This results in spontaneous desorption of stored NO_2, and in the presence of rich-mixture exhaust components, the familiar three-way catalyst process takes place. The duration and frequency of this regeneration process have a direct effect on fuel consumption. A typical fuel consumption increase of 1–2% in the New European Driving Cycle (NEDC) has been reported.

The effectiveness of the regeneration process is in part dependent on whether rich-mixture exhaust components are immediately available for desorption and subsequent three-way reaction. As a result of the catalytic converter's oxygen storage capacity, at first part of the hydrocarbons and carbon monoxide are oxidized without enabling the regeneration process. Coatings within the upstream oxidation catalyst should therefore be designed for low oxygen storage capacity.

The alkali and alkaline earth metals used to store NO_x also absorb sulfur oxides in the form of sulfates which may result from combustion of sulfur-containing fuel. Thermal stability of sulfates is appreciably greater than that of nitrate compounds. Therefore, sulfur regeneration only takes place at higher temperatures than could be achieved in storage regeneration as described above. Over time, this leads to blockage of storage elements and loss of NO_x storage capacity. Demands for minimum sulfur content in fuels originate from this circumstance. Regulatory agencies have established a limit of 50 ppm, which, however, does not solve the problem of sulfur poisoning, but rather only diminishes it over a

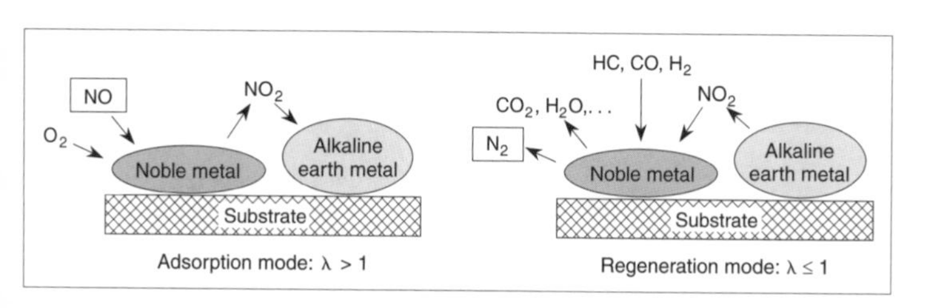

Fig. 5.1-42 NO_x adsorption.

longer time span. Investigations by Quissek et al. have shown that even the smallest sulfur concentrations eventually lead to deactivation of NO_x adsorbers [16].

It is therefore necessary to develop special regeneration strategies for desulfurizing the adsorber. The heretofore known regeneration procedures require operation with below-stoichiometric mixtures and high exhaust gas temperatures over an extended time period measured in minutes. This raises the question of how such conditions are to be met in every possible driving situation. Furthermore, this process results in added fuel consumption which partially offsets the advantages of lean burn direct-injection engines. The total consumption increase attributable to exhaust treatment may be pegged at 2–5%. This reduces the total realizable fuel economy advantage of gasoline direct injection to a bandwith of about 10–15%.

Due to high-temperature operation during desulfurization, sulfur tolerance of the entire system is directly related to the previously discussed thermal stability of the NO_x adsorber. This represents the main emphasis of further development work. Successful commercialization of NO_x adsorber technology will be largely dependent on whether additional improvements can be achieved. The main approach will be improved NO_2 formation in the lower temperature range as well as creation of a thermally stable junction between the active elements and washcoat material [15].

5.1.2.3 Variable valve timing

As does operation with lean mixtures, unconventional actuation of gas exchange elements also permits unthrottling the gasoline engine. Gas exchange work resulting from conventional throttle-based engine control represents a significant portion of gasoline engine cycle losses. Fig. 5.1-43 illustrates gas exchange work, measured by indicated cylinder pressure, as a percentage of total indicated work plotted on an engine map [17]. With decreasing engine load, these losses take on greater importance, which suggests a high potential for fuel economy improvement.

Beyond this, variable valve actuation offers added potential for improved operational behavior. This may be realized in terms of full-load torque response as well as idle quality, part-load fuel economy, and part-load emissions behavior.

5.1.2.3.1 Throttle-free load control

Gas exchange losses in conventional gasoline engines may be largely avoided if the inducted air charge can be controlled without throttling the intake airflow. With variable valve timing, this may be accomplished by early intake closing after the desired air mass has been inducted or by late intake closing after excess charge air has been displaced back into the intake tract (Fig. 5.1-44). The early-closing option requires mastery of very brief valve opening durations to achieve the minimum fresh air masses needed in zero-load operation. The late-closing strategy, on the other hand, is characterized by a tendency toward higher gas exchange losses as a result of repeated flow of the excess air mass for a given load condition.

Variable valve timing may also be employed to control residual gas mass (internal exhaust gas recirculation). In principle, three alternatives may be considered (Fig. 5.1-45). Early or extended valve overlap into the exhaust stroke forces exhaust gas into the intake tract ("intake valve recirculation") and dilutes the following fresh charge. This effect may also be used to improve mixture formation in the intake port.

In the case of "exhaust port recirculation," the valve overlap phase is shifted to the intake stroke. Consequently, during the first part of the intake stroke, fresh charge is drawn in through the intake valve while simultaneously exhaust gas is drawn back through the exhaust valve. An alternative to these exhaust gas recirculation methods is "combustion chamber recirculation," in which the residual gas fraction is determined by early exhaust valve closing. The exhaust gas remaining in the cylinder is compressed until the piston reaches TDC and then is expanded. After the cylinder again reaches atmospheric pressure, the intake valve is opened and the intake process is carried out according to the "early inlet closing" strategy (compare with Fig. 5.1-44).

For gasoline engines, throttle-free load control brings with it altered boundary conditions for mixture formation within the intake tract. Loss of low manifold

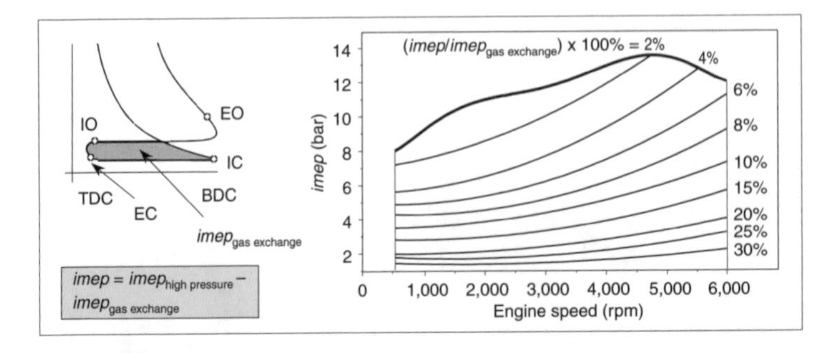

Fig. 5.1-43 Gas exchange losses for a conventional gasoline engine. IO, intake opens; EO, exhaust opens; IC, intake closes; EC, exhaust closes.

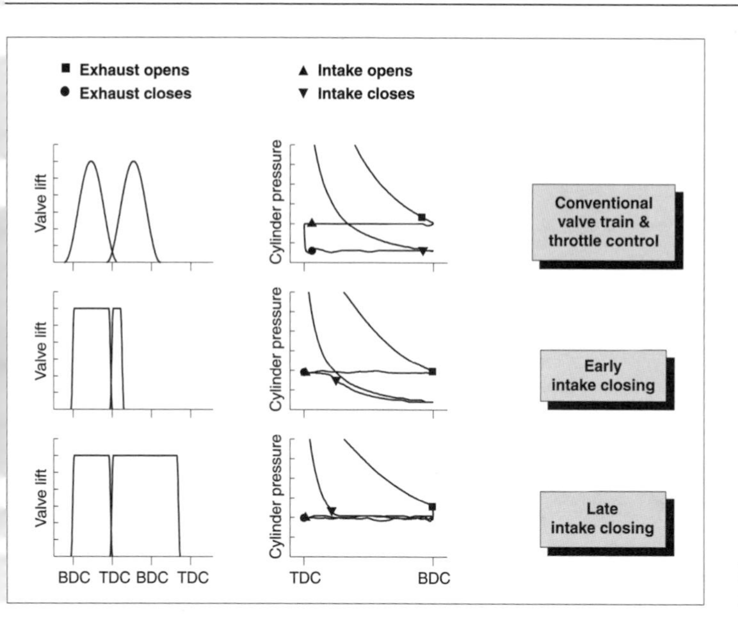

Fig. 5.1-44 Load control methods (part load).

pressure, which promotes gasoline vaporization, may result in mixture formation problems, especially during cold start and warm-up operation. Appropriate valve train variability, however, offers opportunities for improving mixture formation through other effects. For example, Fig. 5.1-46 shows a late intake opening process, examined by Salber [18]. This ensures that low manifold pressure is available when the intake valve opens (IO). Fresh charge enters the cylinder at sonic velocity, which yields a significant improvement in mixture formation. Reports indicate significant emissions improvements during cold start and warm-up. These emissions benefits, along with the fuel economy advantages of throttle-free load control, make the examined engine concept with electromechanical valve actuation (see Section 5.1.2.3.2) very attractive in view of future exhaust standards. In the larger view, throttling losses associated with this process are insignificant, as the system reverts to throttle-free operation after cold start.

Variable valve lift, a feature associated with several recently introduced systems (see Section 5.1.2.3.2), represents an additional means of improving mixture

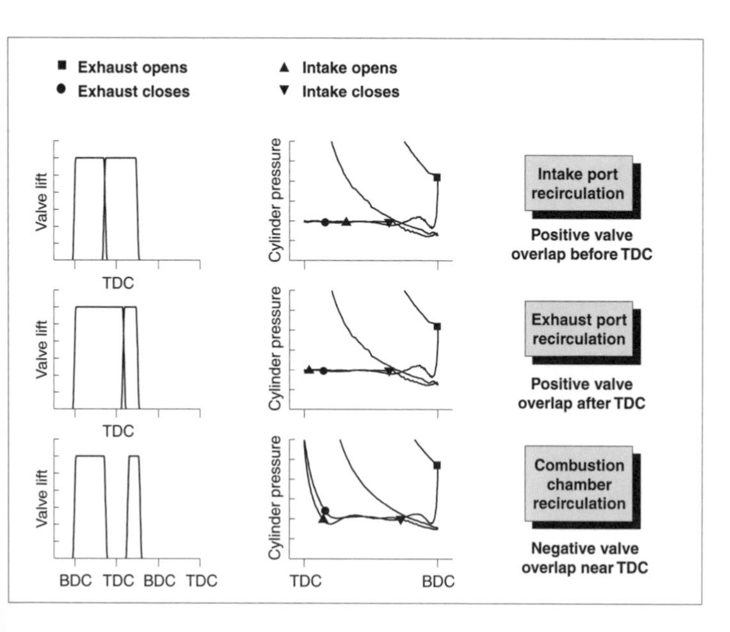

Fig. 5.1-45 EGR control methods.

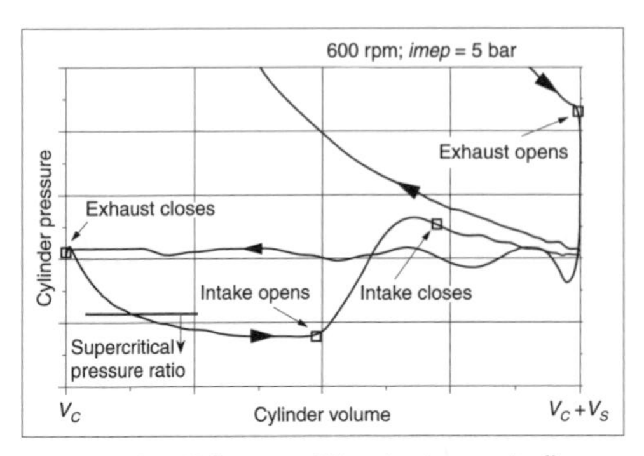

Fig. 5.1-46 *p-V* diagram of "late intake opening" strategy [18].

formation at low load by using the effects of small valve lift. A disadvantage is that this represents incomplete dethrottling of the engine.

Depending on the degree of variability achieved in the valve train, other potential improvements to engine operation may be achieved. Complete valve timing variability permits expansion of the usable torque range. The associated possibility of lowering idle speed results in additional fuel economy advantages. Moreover, full-load behavior can be optimized to minimize trapped residual gas at low speed, thereby achieving higher volumetric efficiency. Thanks to simultaneously reduced knock limitations, in particular with electromechanical valve actuation (see Section 5.1.2.3.2), this volumetric efficiency advantage can be translated into higher full-load torque. This permits installation of a taller (numerically lower) final drive ratio, which shifts the engine's operating point and leads to additional economy advantages.

Design solutions which permit complete deactivation of individual valves may also be employed to achieve map-controlled individual cylinder cutout. For the remaining active cylinders, this results in an operating point shift with correspondingly improved fuel economy. Valve control possibilities extend to cyclically intermittent operation, in which a selectable number of "idle" strokes is interspersed between any individual cylinder's four working strokes.

In combination with supercharging, variable valve control opens a series of interesting possibilities. For example, in addition to realizing the Miller process, it is possible to apply deliberate early exhaust valve opening to achieve pressure waves in the exhaust system, which, within the framework of a pressure-wave supercharging system, results in improved dynamic behavior.

This survey of possibilities for engine process and operational optimization associated with variable valve control makes no claims for completeness. Variable valve control in combination with gasoline direct injection (see Section 5.1.2.2) represents a very promising alternative toward achieving a highly efficient, emissions-optimized passenger car propulsion system.

5.1.2.3.2 Design solutions

The degree to which the previously described advantages of variable valve control may be achieved in practice is largely dependent on the extent to which the chosen design solution permits variability of valve timing and lift. With regard to design implementation of variable valve control, many concepts have been proposed and in some cases realized. The following overview of the most important concepts is divided into cam-actuated and direct-actuated systems.

Cam-actuated systems

By means of intervention in valve kinematics, cam-actuated systems permit variable changes to valve phasing, duration, and lift within certain limits.

Timing variation is usually obtained by means of rotating the camshaft relative to the crank train. Operation of such cam actuators relies on hydraulically shifted helical gearing (Fig. 5.1-47). Systems currently in production achieve rotation of about 20° (representing 40° of crankshaft rotation). Camshaft phasers were first offered in production by Alfa Romeo in 1983. In addition to two-stage phasers, which only permit a choice of either of two defined camshaft timings, continuously variable systems are becoming increasingly common [19, 20]. These permit shifting in unison all valves actuated by any given camshaft. In engines with double overhead camshafts (DOHC), this method permits altering the amount as well as phasing of valve overlap. Even though such systems permit improvements in part-load and full-load operational behavior, their limited adjustment range and lack of any means of modifying valve duration mean that throttleless load control is not possible without additional system modifications.

A further possibility is provided by switchable transfer elements between cam profile and tappet or follower. Switchable tappets or support elements permit shutoff of individual valves. Alternating operation with two different cam profiles has also been realized; in such systems, the camshaft carries two different cam profiles per valve, activating concentric portions of each switchable tappet. This enables both part-load and full-load optimized valve actuation, eliminating the otherwise necessary compromises in valve lift and timing. An example of

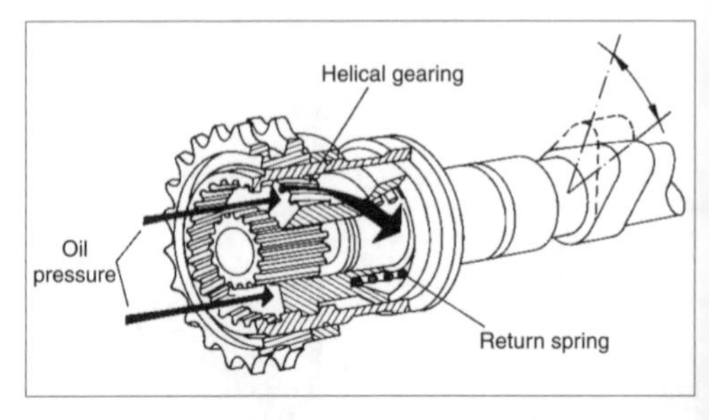

Fig. 5.1-47 Variable cam timing mechanism.

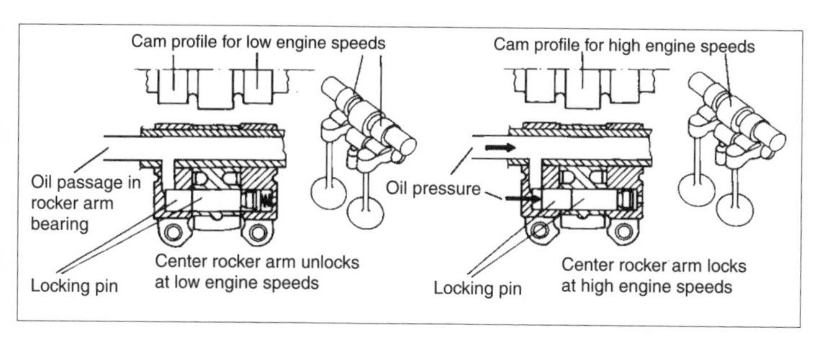

Fig. 5.1-48 Variable valve lift and timing (Honda VTEC).

this technology is Mitsubishi's production MIVEC engine, in which the shift mechanism is incorporated in the rocker arms. This enables switching between a valve lift/duration characteristic for low engine speeds and one for high speeds, as well as complete deactivation of individual valves to achieve cylinder cutout. Fig. 5.1-48 shows a similar design solution by Honda, which has been applied to various production engines under the VTEC designation.

If cam action is transferred to the valve by means of hydraulic elements, the valve may be closed early by relieving hydraulic pressure (Fig. 5.1-49). This decouples valve lift and closing timing from the cam profile. Hydraulic systems are applied to damp valve motion as the valve nears its seat.

A wide variety of design solutions has been proposed to continuously affect the relationship between cam lift and valve motion. Many of these concepts employ finger followers or rocker arms whose position may be varied by means of an eccentric. Depending on kinematic layout, this permits altering valve lift and duration (Fig. 5.1-50). The additional valve train components result in an increase in friction work. However, at part load, this is compensated by decreased valve lift, so that in practice it is even possible to realize advantages in this regard [21]. Combined with cam timing actuators, such systems provide a high degree of variability.

A further increase in variability follows from achieving valve motion by means of two camshafts, one for

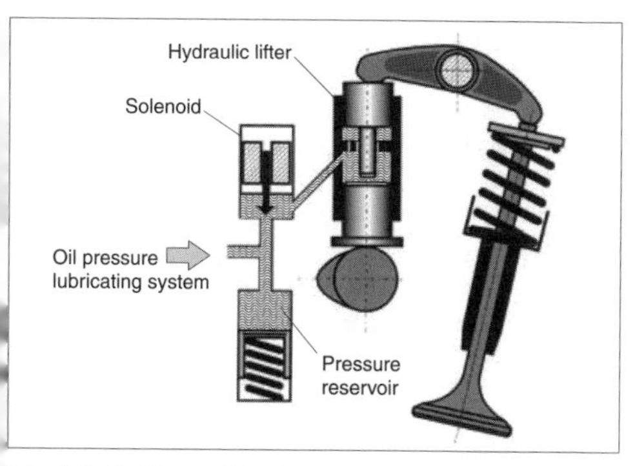

Fig. 5.1-49 Hydraulic valve control (source: Davenport et al. [25]).

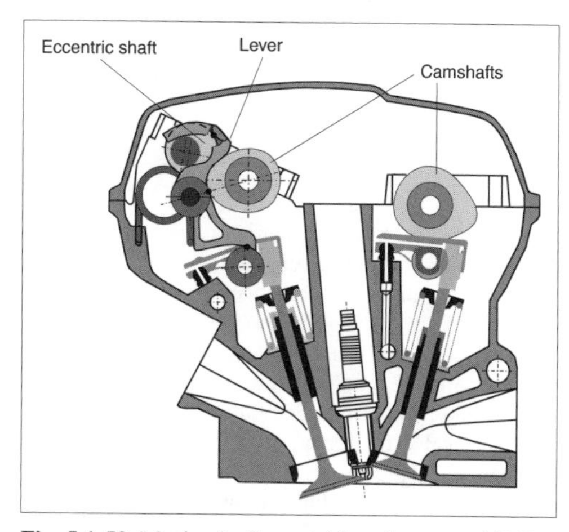

Fig. 5.1-50 Mechanically variable valve control [21].

valve opening and the other for valve closing. Valve lift of both camshafts is added mechanically by a lever mechanism. With independent phasing of both camshafts, it is possible to affect both opening and closing timing while also changing valve lift [22].

A further possibility for increasing variability of cam-controlled valve trains lies in design concepts which seek to directly influence valve lift. Systems employing axially movable "space cams" are familiar in large-engine practice. Systems have been designed for passenger cars that employ inner and outer cam "shells"; mechanical actuators permit rotating individual cams on the camshaft [23]. A similar system has been implemented in the series production Rover 1.8-L 16-valve engine (MGF 1.8i VVC), in which variable acceleration or deceleration of the intake valve motion is realized during valve opening and closing.

Directly actuated systems

The principal means of direct valve actuation are hydraulic and electromagnetic systems. Hydraulic concepts operate using energy stored in a hydraulic system. Valve actuation is by means of fast solenoids and hydraulic cylinders. Valve motion is variable in both timing and lift. Similarly, individual valve actuation is possible. However, the systems which have become available

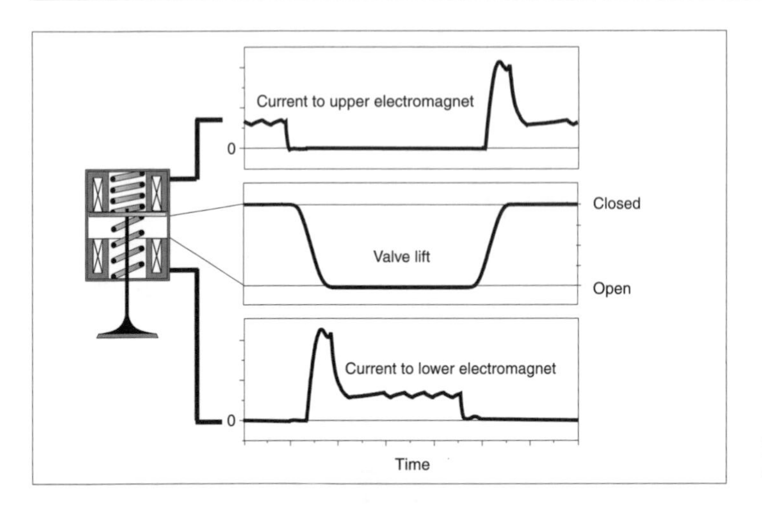

Fig. 5.1-51 Electromechanical valve actuation.

thus far exhibit limited dynamic range, which speaks against application at high engine speeds.

When activating gas exchange valves by means of magnetic force, it is especially advantageous if motion is actually accomplished by spring pressure and if magnetic force is used only to initiate and support valve opening and closing. This combination is termed "electromechanical valve actuation" (EMV). Fig. 5.1-51 illustrates schematically a valve actuation system based on this principle. The valve is activated by a tappet connected to a solenoid core. This core, combined with two springs, constitutes an oscillating spring and mass system whose maximum amplitudes represent the fully opened and fully closed valve positions. The oscillator's harmonic frequency largely determines the valve opening and closing time. The two electromagnets are used to interrupt the valve's harmonic motion at either extreme of travel. Without excitation by the electromagnets, system oscillations would damp out by friction, coming to rest in a position equivalent to half of maximum lift. Therefore, before starting the engine, the valves must be excited to their harmonic frequency—for example, by alternate application of the two electromagnets of each valve, until the closing electromagnet is able to "catch" the valve in its closed position.

The valve actuation history may be recognized in the typical current traces shown in the illustration. Beginning with the valve in its closed position, hold-shut current to the closing magnet is interrupted. Energy stored in the upper spring moves the valve until the mass system reaches the bottom of its travel. At this point, oscillation is interrupted by applying current to the lower electromagnet. Friction losses encountered during valve motion are compensated by briefly increasing current to the lower electromagnet. This "catching current" is applied just before the oscillating mass reaches the bottom of its travel. Closing the valve is a separate process, undertaken in reverse order, in which the lower spring initiates movement and the upper magnet is activated.

Because electromechanical valve action is no longer dependent on cam motion, but rather valve opening and closing is determined by a law which is constant over time, it follows that the valve flow cross section opens comparatively faster at low engine speeds. This results in a drop in flow velocity of the fresh charge and therefore reduced thermal losses. It follows that process temperatures at the end of compression are lower, which explains the significant decrease in knock tendency at low engine speeds. This results in appreciable increases in low-end torque.

The electrical energy demand of electromechanical valve actuation is in large part determined by the fact that magnetic force is only needed near the oscillating mass travel limits. It is precisely at those points where the electromagnets work at their highest efficiency, as the air gap between magnet and core is at its smallest. Furthermore, electrical generator efficiency has a decisive effect on the overall system energy balance. The resulting overall energy requirement must be compared to that of conventional valve actuating systems. Fig. 5.1-52 shows this comparison on the basis of equivalent friction *mep* of a 16-valve engine [18]. For a generator operating at 80% efficiency, electromechanical valve actuation exhibits part-load energy demand comparable to that of low-friction valve trains fitted with roller followers.

Application of electromechanical valve actuation to passenger car gasoline engines has far-reaching effects on overall engine design. Elimination of the mechanical valve timing mechanism permits a basic redesign of the overall engine concept, including auxiliary devices. High-performance generator systems—for example, integrated starter/generator units—have been developed to provide electrical energy for valve actuation. Rational extension of this concept implies that electrification of other auxiliary devices, such as coolant pump, power steering pump, and air conditioning compressor, be considered. Within the framework of appropriate map control

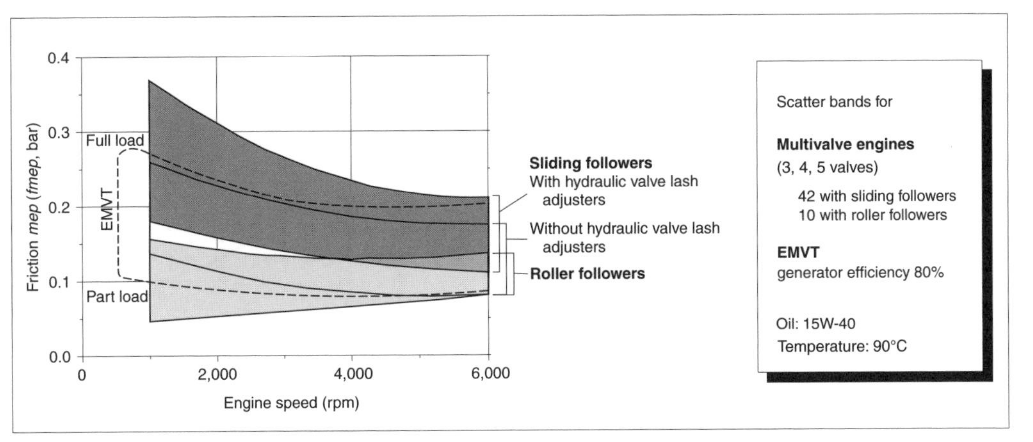

Fig. 5.1-52 Comparative energy demands for electromechanical valve actuation.

of these devices, other advantages may be integrated in the overall propulsion system concept.

Fuel economy improvements achieved with variable valve control in the European test cycle are strongly dependent on the realized variability. A fuel economy benefit of 20% may be attainable using electromechanical valve control combined with part-load cylinder cutoff or supercharging. In contrast to gasoline direct injection, these concepts do not carry any disadvantages with regard to exhaust emissions. On the contrary, variable valve control offers possibilities for reduced cold-start emissions and improved utilization of three-way catalytic converter technology due to faster catalyst light-off.

5.1.2.4 Downsizing and supercharging

While supercharging has found wide application on passenger-car diesel engines, among gasoline engines it is still limited to niche products. Most of these gasoline applications employ supercharging primarily to achieve highest possible performance numbers. Supercharging also offers the opportunity to achieve the required performance with the smallest possible powerplant. This concept, termed downsizing, offers definite fuel economy advantages, largely attributable to shifting the operating point on the engine map.

5.1.2.4.1 Operating point shift

Under steady-state conditions, the engine operating point corresponding to a specific vehicle performance point is determined by the engine speed required for that vehicle speed and the torque necessary to overcome driving resistance (aerodynamic drag and rolling resistance). This relationship is represented by the driving resistance curve plotted on the engine map. Fig. 5.1-53, using the example of a normally aspirated engine, shows that the driving resistance curve traverses areas of the engine map where specific fuel consumption is far above optimum levels. This is because the engine is throttled in part-load operation, and at low loads, friction losses consume a relatively high proportion of the engine's useful work.

As the illustrated example shows, reducing engine displacement shifts the engine operating point to an area of the engine map characterized by more desirable specific fuel consumption. In developing a combustion system, particular attention should be given to ensure that the region of good specific fuel consumption is not shifted to higher engine loads, as this would negate the advantages just described.

All efforts to improve torque, particularly at low engine speeds, benefit the engine's fuel consumption characteristics in actual operation, thanks to the possibility of using a taller final drive ratio. This permits an additional operating point shift. On the other hand, supercharged engines with correspondingly smaller displacement often exhibit a lack of drive-off torque, which necessitates shorter overall gearing. Development of supercharged engines optimized for fuel economy must pay special attention to this aspect.

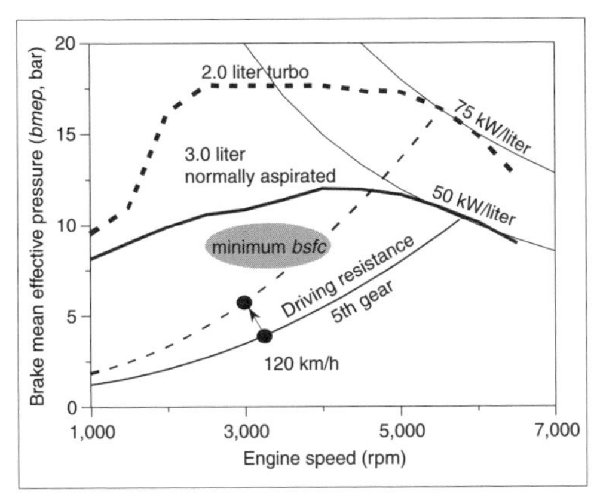

Fig. 5.1-53 Operating point shift.

An increasing number of automobile manufacturers offer an alternative supercharged engine developing identical power as a larger-displacement normally aspirated engine. Certified fuel consumption numbers clearly underscore the potential of this concept. However, it should be noted that to avoid combustion knock, supercharged engines must operate with lower compression ratios, which has a negative effect on thermal efficiency. Still, the trend for supercharged engines shows a steady increase in compression ratios. The associated knock problem at full load may be addressed by appropriate countermeasures. These include intensive charge air cooling as well as reduced valve overlap to reduce residual gas fraction in the cylinder.

Often, late ignition and the resulting losses in full-load fuel economy are accepted as a trade-off for improved part-load efficiencies. Recent investigations have shown that direct injection, with its associated charge cooling, exhibits great potential for minimizing this problem.

The topic of supercharging will be treated in depth in Section 5.3. In principle, the following supercharging methods may be applied:

- Exhaust gas turbocharging with and without variable turbine geometry (VTG), alternatively variable sliding ring turbine (VST), or with electrically driven auxiliary compressor.
- Mechanical supercharging; various apparatuses such as Roots blowers, screw compressors, and spiral compressors ("G-Lader") are available.

5.1.2.4.2 Variable compression ratio

From a thermodynamic standpoint, an engine's compression ratio is directly related to its process efficiency (Carnot cycle efficiency). However, for a gasoline engine, compression ratio is limited by the tendency toward combustion knock. This limitation is particularly applicable to the engine's full-load curve, while at part load, a higher compression ratio could achieve economy benefits without risk of knock. In supercharging an engine, increased knock tendency associated with higher process pressures and temperatures must be countered by reducing the compression ratio. This has a negative effect on the engine's thermodynamic efficiency, resulting in economy losses even at part load. This again negates some of the advantages gained by downsizing.

With variable compression ratio, it is possible to restrict compression ratio reduction to only the knock-sensitive full-load region. With an appropriately large compression ratio range, at part load it may even be possible to take advantage of the opportunity to increase compression ratio beyond levels usually found in conventional gasoline engine designs, generating additional part-load fuel economy benefits. Fig. 5.1-54 shows an example of fuel economy gains achieved with a variable compression ratio engine. The starting point was a supercharged engine with a 10:1 compression ratio.

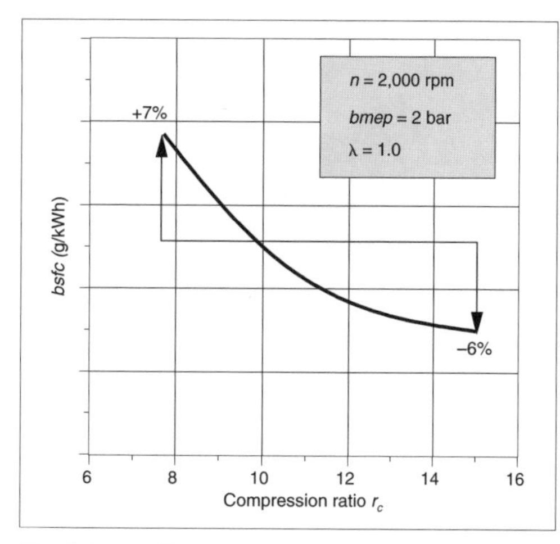

Fig. 5.1-54 Effect of variable compression ratio on part-load fuel economy.

Under the assumption of even higher supercharging to achieve pronounced downsizing effects, compression ratio had to be reduced by about two points. At the engine operating point in question, this would increase part-load fuel consumption by about 7%. This is avoided with variable compression ratio; moreover, this permits a compression ratio increase at part load, which in the case at hand creates a potential fuel consumption decrease of 6%. This demonstrates that variable compression ratio promises great potential, particularly for supercharged gasoline engines.

A multitude of design solutions has been proposed to realize controllable variable compression in reciprocating-piston engines. A number of criteria must be employed in evaluating these designs. Most important for engine processes is the combustion chamber shape, which should be as compact as possible throughout the entire range of variability and which should permit short flame paths. Compression variation should be rapidly implementable, so that compression ratio matching under actual driving conditions can take place dynamically. In concepts with variable-geometry crank trains, friction, including mechanical losses associated with compression ratio variation, must not present any significant disadvantages.

Fig. 5.1-55 shows an example of a crank train with a divided connecting rod whose connecting joint is guided by an eccentric arm. The fixed point of this eccentric arm is variable, which permits altering the crank train kinematics and thereby the compression ratio. With an infinitely long arm, this crank train would be kinematically identical to a crosshead design with variable connecting rod inclination. Comprehensive mechanical and thermodynamic development work has been carried out with this crank train, which has demonstrated a high potential for fuel consumption reduction in the European test cycle [24]. Compared to conventional gas-

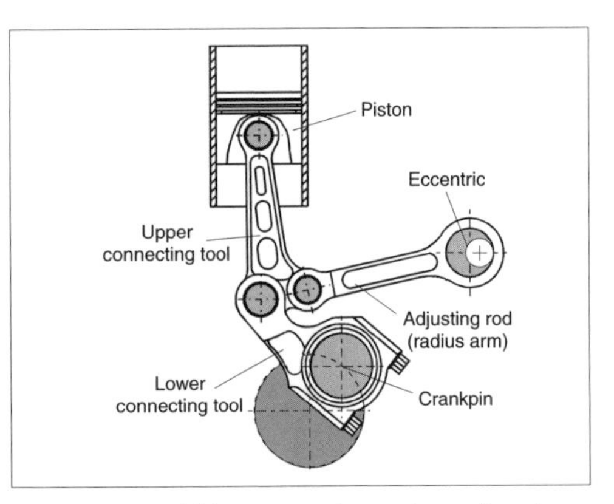

Fig. 5.1-55 Variable-compression reciprocating piston crank train.

oline engines, such an engine, highly supercharged, represents a potential fuel economy improvement of 29%, of which 17% is attributable to downsizing effects and 12% to variable compression ratio.

Pressure to achieve better fuel economy has greatly spurred gasoline engine development. At this juncture, it is still too early for accurate predictions about the extent to which direct injection, decreased displacement, fewer cylinders, cylinder deactivation, variable valve timing, and variable compression ratio, individually or in combination, may find application. Highly promising combinations include supercharging, variable compression ratio, and direct injection; and supercharging and electromechanical valve control. Both offer respectable fuel economy improvements of up to 30% compared to conventional gasoline engines.

References

[1] Maser, Mentgen, Rembold, Preussner, and Kampmann. "Benzin-Direkteinspritzung - eine Herausforderung für künftige Motorsteuerungssysteme." MTZ 58, 9/10, 1997.

[2] Karl, Abthoff, Bargende, Kemmler, Kühn, and Bubeck. "Thermodynamische Analyse eines DI-Ottomotors." 17. Internationales Wiener Motorensymposium, 1996.

[3] Kume, Iwamoto, Iida, Murakami, Akishino, and Ando. "Combustion Control Technologies for Direct Injection SI Engine." SAE 960600.

[4] Hohenberg. "Vergleich zwischen Direkteinspritzung und Saugrohreinspritzung am Mitsubishi GDI." 19. Internationales Wiener Motorensymposium, 1998.

[5] Sawada, Tomoda, Sasaki, and Saito. "A Study of Stratified Mixture Formation of Direct Injection SI Engine." 18. Internationales Wiener Motorensymposium, 1997.

[6] Liyama, Teruyuki, Muranaka, and Takagi. "Attainment of High Power with Low Fuel Consumption and Exhaust Emissions in a Direct Injection Gasoline Engine." FISITA World Automotive Congress, Paris, 1998.

[7] Wolters, Grigo, and Walzer. "Betriebsverhalten eines direkteinspritzenden Ottomotors mit luftgeführter Gemischbildung." 6. Aachener Kolloquium Fahrzeug- und Motorentechnik, 1997.

[8] Lang, Willems, and Grigo. "Rechnerische und experimentelle Optimierung am direkteinspritzenden Ottomotor." 7. Aachener Kolloquium Fahrzeug- und Motorentechnik, 1998.

[9] Grigo. "Gemischbildungsstrategien und Potential direkteinspritzender Ottomotoren im Schichtladebetrieb." Dissertation RWTH Aachen, 1999.

[10] Fraidl, Hazeu, and Carstensen. "DMI—ein Schichtladekonzept für zukünftige PKW-Motoren." 14. Internationales Wiener Motorensymposium, 1993.

[11] Schürz and Ellmer. "Anforderungen an das Motormanagement bei Anwendung von NO_x-Speicherkatalysatoren." 7. Aachener Kolloquium Fahrzeug- und Motorentechnik, 1998.

[12] Moser, Küsell, and Mentgen. "Bosch Motronic MED7—Motorsteuerung für Benzin-Direkteinspritzung." 19. Internationales Wiener Motorensymposium, 1998.

[13] Strehlau, Höhne, Göbel, v. d. Tillaart, Müller, and Lox. "Neue Entwicklungen in der katalytischen Abgasnachbehandlung von Magermotoren." AVL-Tagung Motor und Umwelt, 1997.

[14] Lüders, Backes, Hüthwohl, Ketcher, Horrocks, Hurley, and Hammerle. "An Urea Lean NO_x Catalyst System for Light Duty Diesel Vehicles." SAE 952493.

[15] Dahle, Brandt, Velji, Hochmuth, and Deeba. "Abgasnachbehandlung bei magerbetriebenen Ottomotoren—Stand der Entwicklung." 4. Symposium Entwicklungstendenzen bei Ottomotoren, Technische Akademie Esslingen, 1998.

[16] Quissek, König, Abthoff, Dorsch, Krömer, Sebbeße, and Stanski. "Einfluß des Schwefelgehaltes im Kraftstoff auf das Abgasemissionsverhalten von PKW." 19. Internationales Wiener Motorensymposium, 1998.

[17] Pischinger, Hagen, Salber, and Esch. "Möglichkeiten der ottomotorischen Prozessführung bei Verwendung des elektromechanischen Ventiltriebs." 7. Aachener Kolloquium Fahrzeug- und Motorentechnik, 1998.

[18] Salber. "Untersuchungen zur Verbesserung des Kaltstart- und Warmlaufverhaltens von Ottomotoren mit variabler Ventilsteuerung." Dissertation RWTH Aachen, 1998.

[19] Schmidt, Flierl, Hofmann, Liebl, and Otto. "Die neuen BMW-6-Zylindermotoren." 19. Internationales Wiener Motorensymposium, 1998.

[20] Steinberg, Lenz, Köhnlein, Scheidt, Saupe, and Buchinger. "A Fully Continuous Variable Cam Timing Concept for Intake and Exhaust Phasing." SAE 980767.

[21] Flierl, Klüting, Unger, and Poggel. "Drosselfreie Laststeuerung mit vollvariablen Ventiltriebskonzepten." 4. Symposium Entwicklungstendenzen bei Ottomotoren, Technische Akademie Esslingen, 1998.

[22] Kreuter, Heuser, and Reinicke-Murmann. "The Meta VVH System—A Continuously Variable Valve Timing System." SAE 980765.

[23] Hannibal and Bertsch. "VAST: A New Variable Valve Timing System for Vehicle Engines." SAE 980769.

[24] Habermann. "Untersuchungen zum Brennverfahren hochaufgeladener PKW-Ottomotoren." Dissertation RWTH Aachen, 2000.

[25] Davenport et al. "Ein hochflexibles Ventilsteuerungssystem und seine Einflüsse auf Ladungswechsel und Hochdruckprozess moderner Ottomotoren." Conference proceeding, Graz 1991.

5.2 Diesel engines

5.2.1 Definitions

Combustion engines

Combustion engines are defined as heat engines which produce usable work by means of discontinuous combustion of fuels within a working chamber whose volume is changed by the motion of a piston or slider. During the combustion process, such engines ignite and burn a flammable air/fuel mixture within a working cylinder. The released heat of combustion increases the pressure of precompressed gases. These produce mechanical work through action of the piston and crankshaft. Burned gases are exchanged for fresh air/fuel mixture after each working stroke.

Diesel engines

Diesel engines are defined as combustion engines in which liquid fuel, injected into the combustion chamber, ignites within the inducted air charge after compression raises air charge temperature to a level sufficient to initiate combustion.

Rudolf Diesel's idea was to create an especially economical combustion engine in which most heat losses were reduced to a minimum, first by cooling the burned gases to ambient temperature and second by limiting maximum combustion temperature by gradual introduction of fuel. A high compression ratio, which necessarily leads to autoignition, aids in achieving the most effective use of fuel. "The purpose of high compression is not autoignition, as is often wrongly assumed," writes R. Diesel [1]. "I sought to discover a process with the highest possible heat utilization, and this goal demanded highly compressed air." The inventor sought to achieve the highest possible compression pressure in his engine and maintain this maximum pressure by adding fuel during the working stroke. The objective was to achieve especially smooth power development.

5.2.2 Configurations

Diesel engine with divided combustion chamber

In the case of the *prechamber diesel engine* (Fig. 5.2-1a), fuel is injected into a chamber (prechamber) connected to the working cylinder by one or more relatively small passages. Directed air motion in the prechamber is not required. The *swirl chamber diesel engine* (Fig. 5.2-1b) uses a divided combustion chamber in which fuel is injected into a chamber which is connected to the working cylinder by a relatively large opening. A directed movement of air is generated in the prechamber during the compression stroke.

Direct-injection diesel engine

A diesel engine in which liquid fuel is directly injected into the main combustion chamber (there is no prechamber) is termed a *direct-injection diesel* (Fig. 5.2-1c).

One special combustion chamber configuration is represented by the MAN M system (Fig. 5.2-1d). Only about 5% of the fuel injected into the spherical combustion chamber is finely atomized; the remaining 95% impinges on the combustion chamber walls as a stream of liquid.

5.2.3 History of an invention

"The diesel smells like the future." This quote could well have been applied to Rudolf Diesel's vision, as it was his intention that his engine produce exhaust which was free of smoke and odor.

Born on March 18, 1857 in Paris to German parents, engineer Rudolf Diesel applied for a patent at the Imperial Patent Office in Berlin for a "new rational heat engine." He was issued patent number DRP 67,207 on February 23, 1893 for a "Working method and design for combustion engines" (Fig. 5.2-2). (The correspond-

PATENT-URKUNDE

№ 67207

UF GRUND DER ANGEHEFTETEN BESCHREIBUNG UND ZEICHNUNG IST DURCH BESCHLUSS DES KAISERLICHEN PATENTAMTES

an Rudolf Diesel, Ingenieur, in Berlin

EIN PATENT ERTHEILT WORDEN.

GESETZ v. 7. APRIL 1891

GEGENSTAND DES PATENTES IST:

Arbeitsverfahren und Ausführungsart für Verbrennungskraftmaschinen.

ANFANG DES PATENTES: 28. Februar 1892.

DIE RECHTE UND PFLICHTEN DES PATENTINHABERS SIND DURCH DAS PATENTGESETZ VOM 7. APRIL 1891 (REICHS-GESETZBLATT FÜR 1891 SEITE 79) BESTIMMT.

ZU URKUND DER ERTHEILUNG DES PATENTES IST DIESE AUSFERTIGUNG ERFOLGT.

Berlin, den 23. Februar 1893.

KAISERLICHES PATENTAMT.

Beglaubigt durch Frank

Bureau-Vorsteher des Kaiserlichen Patentamtes.

Fig. 5.2-2 Cover page of the February 23, 1893 patent certificate.

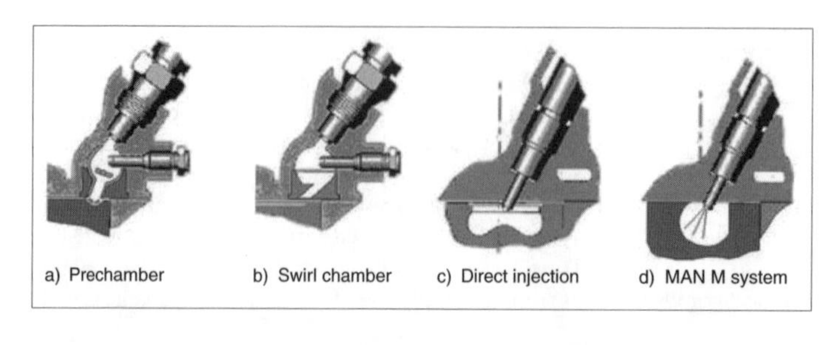

Fig. 5.2-1 Diesel combustion concepts.

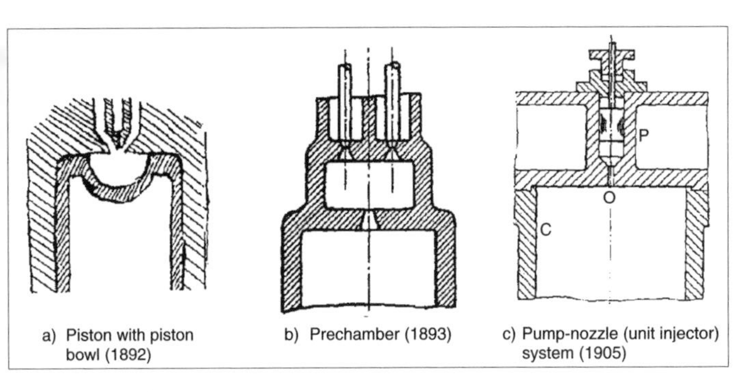

Fig. 5.2-3 Rudolf Diesel's proposals for combustion systems.

ing U.S. patent is 542,846, "Method of and apparatus for converting heat into work," July 16, 1895.)

Diesel, in his book *Die Entstehung des Dieselmotors*, wrote, "An invention consists of two parts: the idea and its execution" [1]. His above-average intelligence, his uncommon technical talent, and his ability to capture an idea and systematically translate it into physical reality were necessary components for the birth of the engine bearing his name. Another detail provides insight into Diesel's engineering empathy: although the current state of the art did not give any indication as to what fuel would prove most suitable, Diesel offered several suggestions regarding combustion systems (Fig. 5.2-3).

Diesel himself experienced only a few early mile-

Table 5.2-1 Diesel engine development mileposts

"High powered large diesel engines" development line	
1897	First operating diesel engine, with a maximum overall efficiency η_O = 26.2%, at Maschinenfabrik Augsburg (Germany)
1898	Delivery of first two-cylinder diesel engine, developing 2 × 30 hp at 180 rpm, to Vereinigte Zündholzfabriken AG of Kempten (Germany)
1899	Hugo Güldner's first two-stroke MAN diesel engine (not marketable)
1899	First noncrosshead diesel engine, Type W, by Gasmotorenfabrik Deutz
1901	First conventional (single-piece connecting rod) MAN diesel engine by Imanuel Lauster (Type DM 70).
1903	First installation of a two-cylinder four-stroke opposed-piston diesel engine, developing 25 hp, in a marine vessel (canal boat *Petit Pierre*) by the firm of Dyckhoff, Bar-le-Duc (France)
1904	First MAN diesel generating plant, with four 400 hp engines, goes online in Kiev (Russia)
1905	Alfred Büchi proposes utilizing exhaust energy for supercharging
1906	First reversible two-stroke marine diesel presented by the Sulzer brothers, Winterthur (Switzerland), developing 100 hp/cylinder (stroke & bore 250 × 155 mm)
1912	MS *Selandia* enters service, first oceangoing ship to be equipped with two reversible two-stroke diesels by Burmeister & Wain, each 1,088 hp.
1914	First test run of a double-acting six-cylinder two-stroke engine developing 2,000 hp/cylinder, by MAN of Nuremberg (stroke & bore 1,050 × 850 mm)
1951	First MAN four-stroke diesel engine (Type 6KV30/45) with high-pressure supercharging, η_O = 44.5% at $w_{O\max}$ = 2.05 kJ/L, $p_{cyl\max}$ = 142 bar, and P_A = 3.1 W/mm^2
1972	Largest two-stroke diesel engine to date (stroke & bore 1,800 × 1,050 mm, 40,000 hp) enters service
1982	Market introduction of super long stroke two-stroke engines with stroke/bore ratio = 3 (Sulzer, B&W)
1987	Largest diesel-electric propulsion system with new MAN–B&W four-stroke diesel engines and total power output of 95,600 kW enters service, propelling the *Queen Elizabeth 2*
1991/92	Experimental two-stroke and four-stroke engines by Sulzer (RTX54, with $p_{cyl\max}$ = 180 bar, P_A = 8.5 W/mm^2) and MAN–B&W (4T50MX, with $p_{cyl\max}$ = 180 bar, P_A = 9.45 W/mm^2
"High speed automotive diesel" development line	
1898	First run of a two-cylinder four-stroke opposed-piston engine ("5 hp buggy motor") by Lucian Vogel at MAN Nuremberg (experimental engine, not marketable)
1905	Experimental engine by Rudolf Diesel, based on a four-cylinder Saurer Otto-cycle engine with air compressor and direct injection (not marketable)
1906	German Reich Patent DRP 196,514 issued to the Deutz company for prechamber injection
1909	Basic patent DRP 230,517 issued to L'Orange (Germany) for prechamber

Table 5.2-1 (continued)

1910	British patent 1059 by McKechnie for high-pressure direct injection
1911	First compressorless Deutz diesel engine, Type MKV, enters production
1912	First diesel locomotive, with four-cylinder two-stroke V engine (1,000 hp) presented by Sulzer Brothers
1913	First diesel-electric railway locomotives, with Sulzer motors, enter service with Prussian and Saxon State Railways
1924	First diesel-engined commercial vehicles presented by MAN (direct injection) and Daimler-Benz AG (indirect injection into prechamber)
1927	Bosch begins production of diesel fuel injection systems
1931	Type approval testing of six-cylinder two-stroke opposed-piston Jumo 204 aircraft diesel engine, by Junkers Motorenbau GmbH. Output 530 kW (750 hp), specific output 1.0 kg/hp including speed reduction gearbox
1934	Four-stroke V8 prechamber diesel engine by Daimler-Benz AG for rigid airship LZ 129 *Hindenburg*, developing 1,200 hp at 1,650 rpm (specific output 1.6 kg/hp including transmission)
1936	First passenger car prechamber diesel engines by Daimler-Benz AG (installed in Mercedes Type 260 D) and Hanomag enter production
1953	First passenger car swirl chamber diesel engines introduced by Borgward and Fiat
1978	First passenger car diesel engine with exhaust-driven turbocharging enters production (Daimler-Benz AG)
1983	First high-speed, high-output diesel engine with twin supercharging by MTU enters production; $w_{O\max} = 2.94$ kJ/L, $p_{cyl\max} = 180$ bar, and $P_A = 8.3$ W/mm²
1987	First electronically controlled diesel fuel injection (BMW)
1988	First direct-injection passenger car diesel enters production (Fiat)
1989	First direct-injected, exhaust turbocharged passenger car diesel engine enters production at Audi (installed in Audi 100 DI)
1990	First catalytic converter-equipped passenger car diesel enters production (VW)
1992	First direct-injection, variable turbine geometry passenger car diesel engine enters production (VW)
1996	First direct-injection four-valve passenger car diesel (Opel Ecotec diesel engine)
1997	First supercharged passenger car diesel engine with high-pressure common rail direct injection and variable turbine geometry (Fiat)
1998	First passenger car diesel engine with unit injectors (VW)
2000	Particulate filter (Peugeot)

posts in the history of diesel engine development (Table 5.2-1), which grew to encompass large, powerful diesels as well as high-speed automotive engines. Diesel's suicide in September 1913 is attributed to mistakes, failed speculation, and an inventor's pride.

5.2.4 Diesel engine fundamentals

5.2.4.1 Introduction

Combustion engines are essentially energy conversion devices in which chemical energy bound in fuel is converted to mechanical energy—that is, useful work. In operation, energy released within the engine during combustion takes part in a cyclic thermodynamic process and is used to perform pressure/volume work. The energy balance of this conversion device (Fig. 5.2-4) is

$$E_{\text{fuel}} + E_{\text{combustion air}} + W_{\text{useful}} + \Sigma E_{\text{losses}} = 0.$$

From an economic standpoint, energy losses should be minimized. Today, however, this is no longer consistent with ecological demands that every conversion of matter and energy must take place with maximum efficiency and minimum environmental impact. In the past, this requirement spawned intensive research and development projects, and it continues to do so, transforming Rudolf Diesel's simple engine into the complex system it is today (Fig. 5.2-5). Not shown in the illustration are exhaust gas recirculation and charge air intercooling systems. An important feature is increased use of electrical and electronic components, as well as a shift from open control systems to closed-loop control. Furthermore, material applications and manufacturing effort should be minimized for competitive reasons.

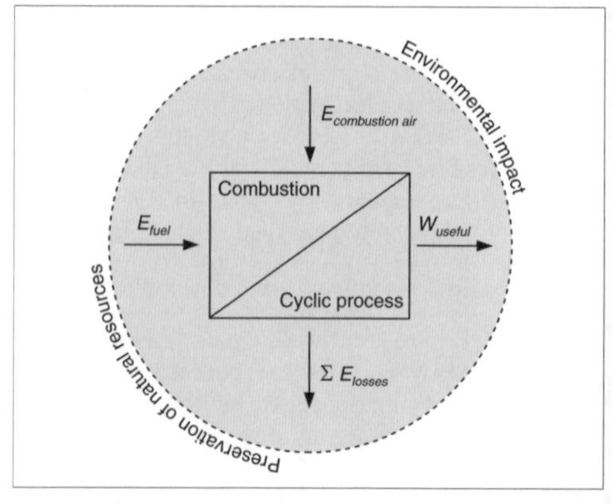

Fig. 5.2-4 Energy conversion in combustion engines.

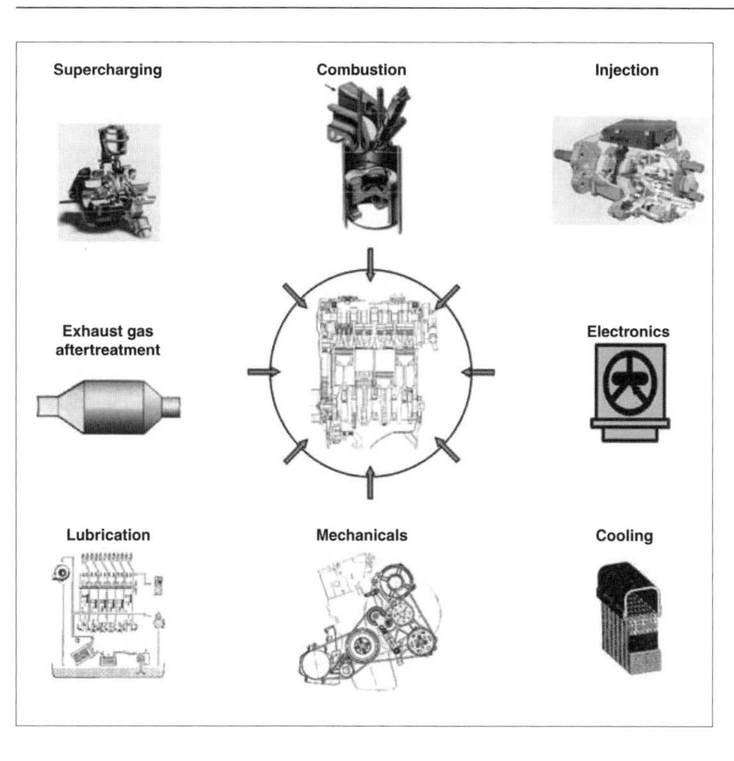

Fig. 5.2-5 Complexity of the modern diesel engine.

5.2.4.2 Comparison of engine combustion processes

Fuel, usually in liquid form, undergoes mixture formation prior to ignition and combustion. An ignitable mixture consisting of fuel vapor and air must be established. There are differences between these processes as they apply to diesel and Otto-cycle (gasoline) engines (Table 5.2-2).

In the diesel engine, fuel is injected into highly compressed, heated air shortly before top dead center (internal mixture formation). The classic Otto-cycle engine, on the other hand, employs external mixture formation: air and fuel are mixed outside the working cylinder over an extended time frame by means of a carburetor or manifold fuel injection.

After injection, the diesel engine contains a heterogeneous mixture of air, fuel vapor, and fuel droplets. By contrast, the cylinder of an Otto-cycle engine is filled by a heterogeneous air–fuel mixture. Otto-cycle ignition is initiated by an electrical discharge at the spark plug, given that the homogeneous mixture is within flammability limits (premixed flame). In the diesel engine, autoignition takes place, for which ignitable mixture is only required in a limited area (diffusion flame).

Because the Otto-cycle engine requires homogeneous ignitable mixtures, its power can only be modulated by varying the amount of admitted air charge (quantity change). The diesel engine operates with excess air. Engine load is established by means of the injected fuel quantity—in other words, the air/fuel ratio (quality change). As found on Otto-cycle engines, throttling, with its attendant significant losses, is not required.

Various fuel requirements arise directly from these different processes. Diesel fuel must be readily ignitable (high cetane number). Gasoline, on the other hand,

Table 5.2-2 Comparison of diesel and gasoline engine operating characteristics

Characteristics	Diesel engine	Classic gasoline engine
Mixture formation	Within cylinder	Outside cylinder
Mixture	Heterogeneous	Homogeneous
Ignition	Autoignition with excess air	External ignition within ignition limits
Air/fuel ratio	$\lambda_V \geq \lambda_{min} > 1$	$0.6 < \lambda_V < 1.3$
Combustion	Diffusion flame	Premixed flame
Torque modification method	Alteration of λ_V(quality change)	Mixture throttling (quantity change)
Fuel	Readily flammable	Not readily flammable

should resist uncontrolled autoignition; that is, it should be rather difficult to ignite (high octane number).

5.2.4.3 Diesel engine thermodynamics

State changes of an ideal gas

We will consider closed systems; that is, no interaction with any other bodies. Thermodynamic quantities describe macroscopic conditions of the bodies. The equilibrium condition of a homogeneous mass of gas, m, is defined by any two thermodynamic quantities (e.g., pressure, volume, temperature, internal energy).

The general state equation describes ideal gases:

$$p \cdot V = m \cdot R \cdot T$$

where p = absolute pressure in Pa, T = temperature in K, V = volume in m^3 and R = the gas constant in J/kgK.

The state of a gas may be represented on a p-V diagram. Special transformations may be calculated by holding certain thermodynamic quantities constant. Equations exist for isobars (constant p), isotherms (constant T), and isochors (constant V).

If a body is thermally isolated (i.e. there is no heat transfer between a gas and its surroundings), and its external conditions are changed sufficiently slowly, the process is termed adiabatic. Entropy remains unchanged; that is, the adiabatic process is reversible. Poisson's equation applies:

$$p \cdot V\kappa = \text{const.}$$

The isotropic exponent κ is the ratio of the specific heats at constant pressure (c_p) and at constant volume (c_v).

Ideal cycles and comparative cycles

In an ideal cyclic process, the gas undergoes a "closed" state change, so that after undergoing the cycle's quasistatic process, it is once again in its original state. According to the first law of thermodynamics (conservation of energy: the change in internal energy ΔU is equal to the sum of external heat added, ΔQ, and work, ΔW), it follows that heat converted in the cycle appears as mechanical work:

$$0 = \Delta U = \Delta Q + \Delta W.$$

In other words, the pressure and volume changes represent the theoretical useful work of the ideal process:

$$-\mathrm{d}A = p_e \cdot \mathrm{d}V.$$

For an ideal gas, the distinction between the system's external pressure p_e and internal pressure p_i may be ignored. The process takes place quasistatically, but the temperature of the heat reservoir, T_e, must be virtually identical to that of the system temperature, T_i.

For his invention, Rudolf Diesel considered an engine operating with a constant-pressure cycle (Fig. 5.2-6a). The cylinder contents are first compressed adiabatically along path 1 → 2. The subsequent combustion takes place at constant pressure and increasing volume (2 → 3), after which the heated air expands adiabatically (3 → 4). Finally, unused heat is rejected (4 → 1).

In the Diesel cycle p-V diagram, the combustion phase (2 → 3) is an isobar, as heat Q is added at constant pressure.

If these events are adapted to the realities of a heat engine, the ideal cyclic process becomes a combined constant volume–constant pressure process (also termed a limited-pressure, dual cycle, or Seiliger process). In this process (Fig. 5.2-6b), compression takes place along path 1 → 2, heat is added in part at constant volume (2 → 3) and in part at constant pressure (3 → 4). Expansion along path 4 → 5 does not proceed to ambient pressure; this would require an unrealistically long piston stroke. The Seiliger process represents the most general case of a comparative cycle, as it may be applied to actual engine processes. It encompasses both the limiting cases of the Diesel cycle constant pressure process and the constant volume process of the ideal Otto-cycle engine.

Assumptions for this model include:

- The charge represents an ideal gas.
- Combustion takes place according to an established law.
- Adiabatic processes (no heat transfer through walls).
- No friction in cylinder.
- No flow losses.

In the real four-stroke cycle (Fig. 5.2-7a), two closed "loops" are established, of which the high-pressure loop is normally positive and the gas exchange loop is usually

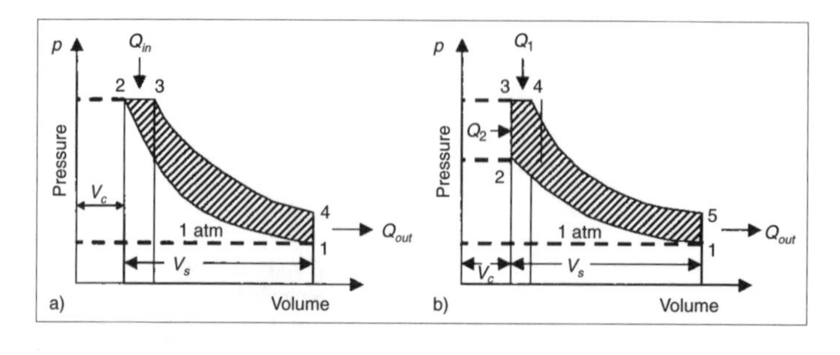

Fig. 5.2-6 a) Constant-pressure process; b) combined constant volume – constant pressure process. V_c, chamber volume; V_s, swept volume.

negative. For supercharged engines, the gas exchange loop may be positive.

The efficiency η_c of the complete engine and the overall efficiency η_O of the actual engine differ by the sum of individual losses in the latter. Determining magnitudes of these individual losses is the purpose of loss analysis. These losses are traceable to:

- Incomplete conversion of fuel
- Nonideal combustion process
- Heat losses to combustion chamber walls
- Leakage (blowby) past piston rings
- Spillover between main and auxiliary combustion chambers in divided-chamber engines
- Gas exchange
- Mechanical friction

Fig. 5.2-7b shows losses for a 1.5 L swirl-chamber diesel engine at 3000 rpm as a function of load (expressed as mean effective pressure, mep). Overall efficiency is represented by η_O; indicated efficiency of the entire process is η_i; indicated efficiency of the high-pressure loop alone is η_{iHP}; the efficiency of the comparative ideal-gas process is η_{ideal}; $\Delta\eta_m$ represents the loss in efficiency due to mechanical friction; $\Delta\eta_{gas}$ the gas exchange losses; $\Delta\eta_{spill}$ the "spilling" losses in divided-chamber engines; $\Delta\eta_{leak}$ represents leakage losses; $\Delta\eta_H$ represents loss of efficiency due to heat

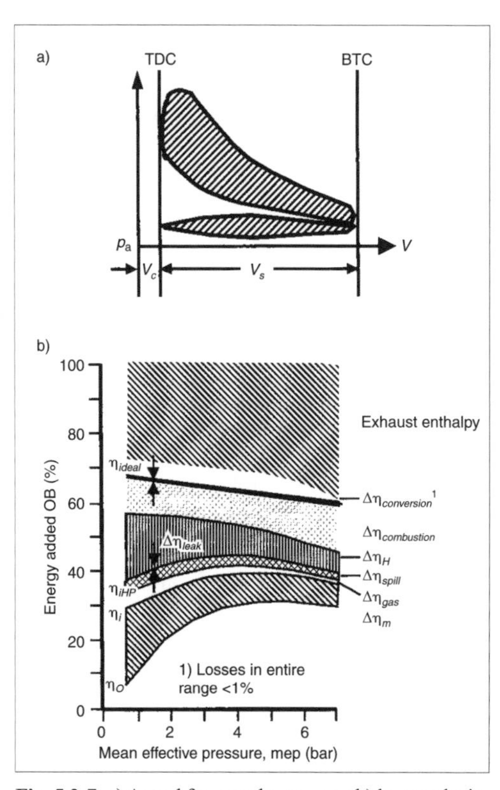

Fig. 5.2-7 a) Actual four-stroke process; b) loss analysis.

transfer; $\Delta\eta_{combustion}$ represents combustion losses, and $\Delta\eta_{conversion}$ the energy conversion losses. All of the listed efficiencies or efficiency losses are referred to the energy added to the system, Q_B. The following relation holds:

$$\eta_O = \eta_{iHP} - \Delta\eta_{conversion} - \Delta\eta_{combustion} - \Delta\eta_H - \Delta\eta_{leak} - \Delta\eta_{spill} - \Delta\eta_{gas} - \Delta\eta_m$$

Individual losses exhibit various load dependencies, in which efficiency losses due to combustion, heat transfer, and mechanical friction are far greater than those attributable to gas exchange, spilling processes, and incomplete conversion of fuel.

Limits of model calculations

In considering ideal processes, assumptions are made which, from the point of view of physics, cannot be applied to real processes. Table 5.2-3 indicates which individual components of thermodynamic models must be formulated to at least permit relative comparisons (e.g., in parameter studies). Demands for accuracy are less rigid here than, for example, design calculations for a supercharging system or cooling system.

5.2.5 Diesel engine combustion

5.2.5.1 General

In combustion engines, energy bound in supplied fuel is released by oxidation with combustion air. Real combustion processes, however, are incomplete and accompanied by losses. As a result, real combustion produces not only carbon dioxide (CO_2) and water vapor (H_2O), but also carbon monoxide (CO), hydrocarbons (HC), oxides of nitrogen (NO_x), and soot particles or particulate matter (PM).

The diesel engine combustion process is customarily divided into distinct phases: fuel injection, mixture formation, autoignition, and combustion with production of exhaust gases.

5.2.5.2 Injection and mixture formation

Air motion

Flow processes play a decisive role in mixture formation. These processes affect ignition delay, the amount of fuel impinging on the combustion chamber walls, and the combustion process insofar as these flow processes contribute to air utilization and the degree of charge burning. Distribution and movement of air and fuel must be matched to one another across the entire engine operating range. This is especially important in direct injection diesel engines, which lack the added impulse and intensive mixing of combustion gases and unburned charge ejected from prechambers or swirl chambers.

Swirl in direct injection diesel engines is generated by intake port geometry and intake valve seats. Spiral or tangential ports are customary (Fig. 5.2-8). The air-

Table 5.2-3 Comparison of model components in ideal and real processes

Model component	Ideal process	Real process
Material properties	Ideal gas c_P, c_V, K = constant	Real gas; composition changes during the process Material properties dependent on pressure, temperature, and composition
Gas exchange	Gas exchange represented as heat rejection	Mass transfer through valves; residual gas remains in cylinder
Combustion	Complete combustion in accordance with idealized laws	Different combustion processes are possible depending on mixture formation and combustion system; fuel sometimes only partially burned
Heat losses (wall losses)	No heat losses to walls	Wall losses present, are considered
Leakage	No leakage	Leakage present, partially considered

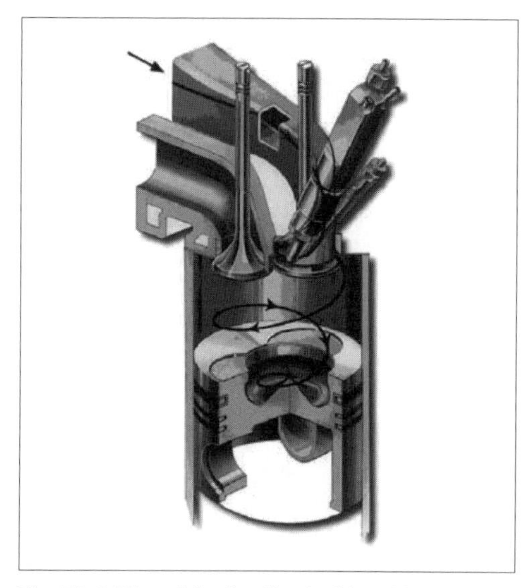

Fig. 5.2-8 Direct injection diesel with swirl port.

stream directed into the engine is given a rotary motion about the cylinder axis. Air swirl increases with increasing engine speed; for high-speed diesels, this increase is actually too rapid for effective mixture formation. However, acceptable compromises are possible.

As the piston approaches the cylinder head, rotary air motion is supplanted by quench motion, in which air flows out of the narrow space between cylinder head and piston into the piston bowl. At the beginning of the expansion stroke, flow direction reverses. This turbulent air motion aids mixture formation just before and after ignition. With the help of modern laser-doppler techniques, local flow conditions may be determined [2].

Injection and spray propagation

Fuel injection takes place near the end of the compression stroke. Air is highly compressed and correspondingly heated (30 to 60 bar, 300–400°C). The time history of the injection process is very strongly dependent on injection system design details. Important factors include pump plunger dimensions, fuel injection nozzle design details, and geometric relationships between injection line and delivery valve. Deliberate changes to the injection history and therefore the combustion process can be achieved by modifying one or more of these parameters.

At the beginning of injection, air already provides the environment necessary for ignition. Diesel fuel exits the injector nozzle primarily as a compact stream of liquid. In this phase, the proportion of fuel vapor is relatively low.

Droplets form as a result of cavitation at the nozzle tip, fluid body forces in the stream core, and external waves established in the spray envelope as a result of air drag. Individual fuel droplets deform and divide several times as they pass through the combustion chamber (interaction with air). There may also be mutual effects between droplets: these collisions may lead to breakup or coalescing of droplets. Furthermore, fuel may land on the combustion chamber walls. In order to establish locally ignitable conditions within a heterogeneous mixture of air and fuel droplets of varying sizes and distribution, fuel must first be heated by the compressed air charge.

As a result of heat transfer from hot air to liquid fuel, a fuel vapor layer forms around the individual droplet. This fuel vapor layer mixes with the surrounding air. The mixture is ignitable once the air/fuel ratio exceeds about 0.7. This explains the phenomenon of physical ignition delay.

5.2.5.3 Autoignition and ignition delay

One of the most important characteristics of the diesel engine combustion process is ignition delay—the time span between the beginning of injection and ignition (Fig. 5.2-9). Distinction must be made between a physical and a chemical component of ignition delay. The latter is that time period in which preflame reactions take place. Typically, ignition delay is about 1–2 ms. This time period sets the stage with respect to fuel consumption and emissions.

Parameters affecting ignition delay may be classified

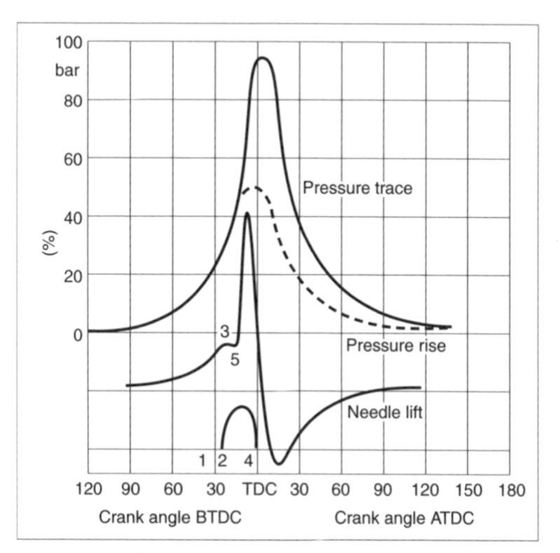

Fig. 5.2-9 Direct injection diesel engine ignition delay. 1, begin of delivery; 2, begin of injection; 3, ignition; 4, end of injection; 5, ignition delay.

as fuel-based or air-based. The most important fuel-based parameters are fuel quality, injection pressure, fuel temperature, injector nozzle geometry, and injection timing. Air-based parameters are air pressure and temperature in the combustion chamber, air charge motion (flow field), as well as reduction of fuel, air, and residual gas. From a design standpoint, these factors may be affected by:

— Intake port design
— Valve timing
— Combustion chamber shape
— Compression ratio
— Coolant temperature
— Supercharging
— Cold-start measures

Shorter ignition delay is achieved with increased:

— Fuel cetane number
— Fuel temperature
— Injection pressure
— Combustion chamber pressure
— Combustion chamber temperature

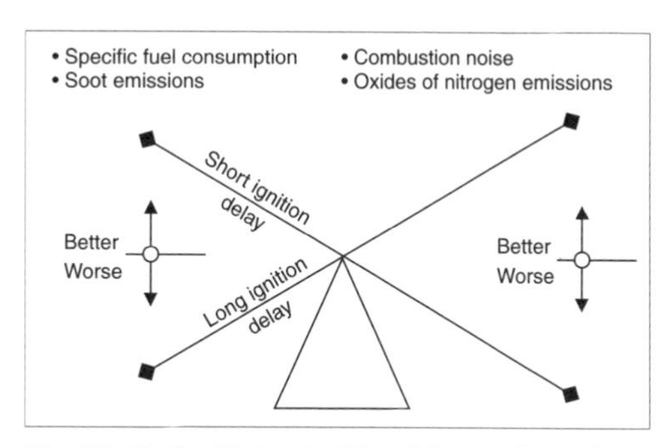

Fig. 5.2-10 Conflicting ignition delay goals in diesel engine combustion.

Ignition delay is also shortened by:

— Later injection timing (but still BTDC)
— More evenly, more finely divided fuel droplets
— Greater relative motion between fuel and air

Shorter ignition delay and the correspondingly smaller quantity of fuel injected during this period (Fig. 5.2-10) result in:

- Gentler pressure rise → lower combustion noise
- Lower peak pressure → lower combustion noise and reduced mechanical stress
- Lower peak temperature → lower oxides of nitrogen

Longer ignition delays result in higher combustion noise and more oxides of nitrogen.

Several of the influential parameters are also dependent on the engine's operating point. The search for an acceptable compromise across the entire operating range represents a central task in optimization of a combustion process.

The diesel combustion process

Diesel engine combustion is governed by diffusion. The injection phase often extends past the ignition point (see Fig. 5.2-11), allowing already established inhomogeneities to continue. In addition to charge inhomogeneities, there are also temperature inhomogeneities. In order to guarantee an ample supply of oxygen to support combustion, the diesel process must operate with excess air. The time history of energy conversion (com-

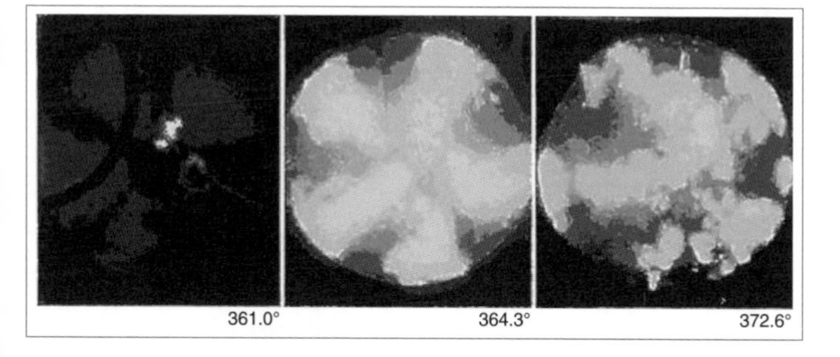

Fig. 5.2-11 Autoignition and combustion in Volkswagen 1.9-L TDI engine.

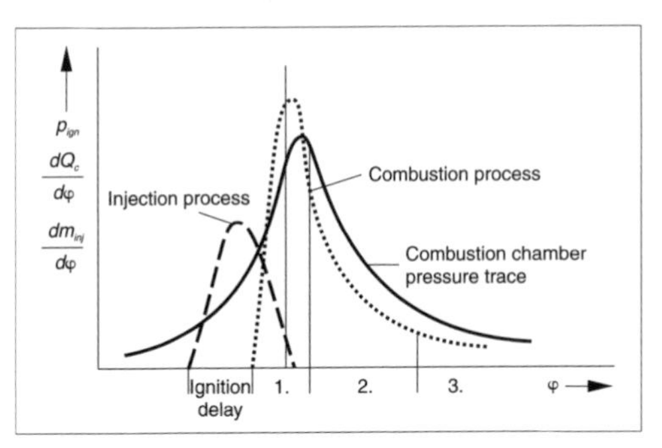

Fig. 5.2-12 Direct-injection diesel ignition delay. Q_c, heat released in combustion; m_{inj}, mass of fuel injected.

bustion progression) may be divided into three phases (Fig. 5.2-12):

1. The mixture, which has achieved a flammable condition during ignition delay, is thermally ignited. By the end of this phase, most of the fuel injected during ignition delay has been burned. The governing factor is the fuel's chemical energy.
2. Fuel injected well after the start of injection is mixed and burned. The combustion process is governed by the rate of mixture formation. Along with the velocity field, the temperature field is important.
3. The final phase of combustion is marked by relatively slow conversion of the last of the injected fuel. Air motion, temperature, and excess air decrease.

The first phase of the diesel combustion process is of decisive importance for combustion harshness, noise, and NO_x emissions. Multistage injection systems, offering features such as separate preinjection, provide a larger envelope for mixture formation and combustion.

The third phase affects fuel consumption and emissions formation, especially particulates. The goal is to achieve as early and quick an end of combustion as possible.

The described diesel engine combustion processes are shown schematically in Fig. 5.2-13.

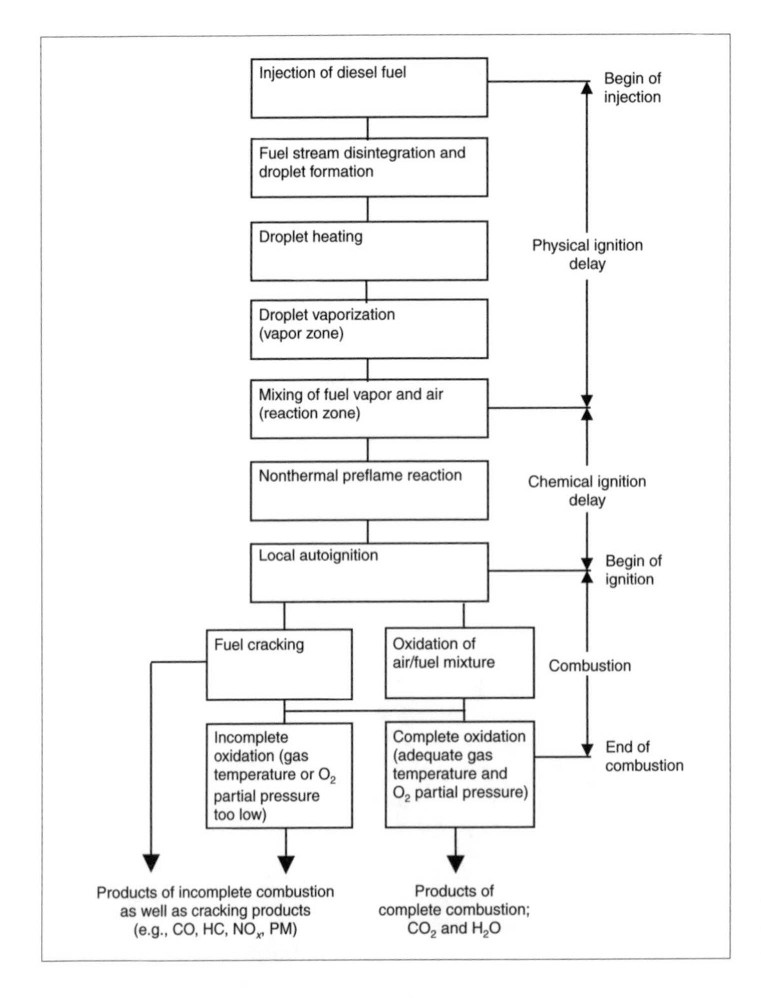

Fig. 5.2-13 Sequential representation of diesel engine mixture formation and combustion [4].

5.2.5.4 Exhaust emissions

Complete combustion of diesel fuel, containing sulfur, results in end products carbon dioxide (CO_2), water (H_2O), and sulfur dioxide (SO_2). Due to the short time available for combustion and varying degrees of mixture formation, temperature distribution, and air concentration, locally incomplete burning may take place.

Fig. 5.2-14a shows the region of soot formation on a *T*-λ diagram in relation to mixture and burned fraction, near top dead center. Burned material with air concentrations below λ = 0.5 must contain soot. In addition, NO components formed within 0.5 ms are indicated. The typical "scissors effect" forcing a compromise between soot and NO_x emissions is discernible. If soot and NO_x formation are to be avoided, the mixture should be λ = 0.6–0.9.

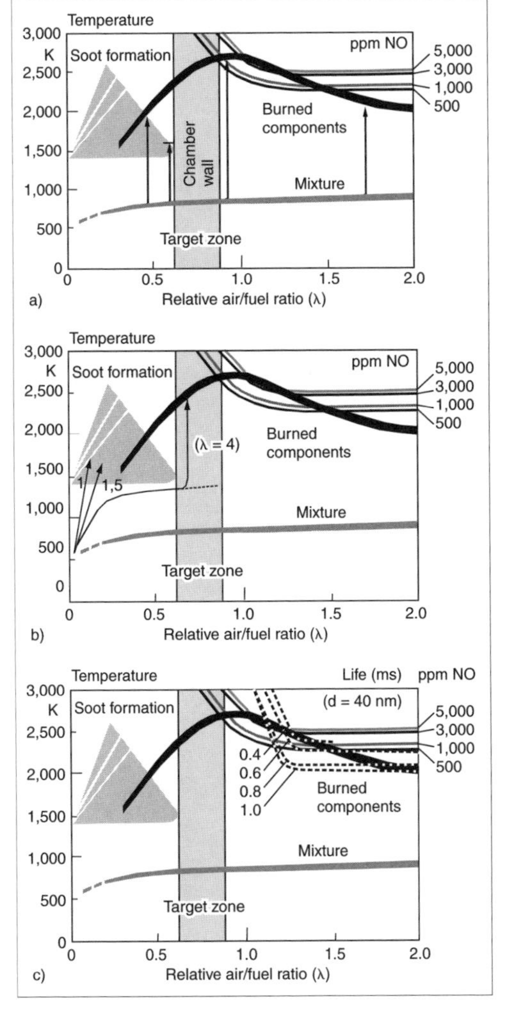

Fig. 5.2-14 a) First phase of diesel combustion (premix combustion); b) second phase of diesel combustion; c) third phase of diesel combustion.

Due to the inhomogeneity of mixture formation in the first combustion phase (premixed combustion), production of primary soot and oxides of nitrogen may be assumed [3]. In the interest of low-emissions combustion, the first combustion phase should take place with a minimum amount of mixture, if possible with λ = 0.6–0.9. The soot formation regime may shift to higher λ, for example if temperature drops sufficiently through cylinder wall cooling to enter the soot formation regime.

During the second combustion phase, injected fuel is mixed with air and combustion gas (Fig. 5.2-14b). This results in various compositions (λ). Mixtures of combustion gases and low λ may lead to secondary soot formation. Mixing fuel with hot, oxygen-poor exhaust should be avoided. Adequate fresh air should be supplied.

In the third combustion phase (Fig. 5.2-14c), combustion gases lean out after the end of injection. The regime in which soot particles burn with oxygen partially overlaps the regime of intensive NO_x formation. It is therefore advisable to limit formation of primary and secondary soot as much as possible and not count on oxidation toward the end of combustion.

Formation of oxides of nitrogen

Nitrogen is the primary component of air. Nitric oxide is formed at high temperatures during the combustion process (endothermic reaction) and may be represented in simplified form by the Zeldovich mechanism:

$$N_2 + O \leftrightarrow NO + N$$
$$O_2 + N \leftrightarrow NO + O$$
$$OH + N \leftrightarrow NO + H$$

These reactions proceed relatively slowly. However, temperature, pressure, and concentration fields change rapidly and strongly during combustion. This leads to nitric oxide concentrations which lie below those that would be established in thermal equilibrium. Nitric oxide emissions may be minimized in diesel combustion by limiting combustion temperature (late injection, charge air cooling) and lowering oxygen concentration (exhaust gas recirculation).

Soot formation and particulate emission

Soot forms as a result of molecular processes during combustion.

Fuel molecules are first consumed by oxidation. This produces ethyne (acetylene), the starting point for formation of higher hydrocarbons and aromatics. The latter undergo planar growth by an H abstraction/ethyne addition mechanism (Fig. 5.2-15). Spatial growth occurs through agglomeration of larger polycyclic aromatic hydrocarbons. Volume of the resulting aggregates increases through further coagulation and through surface growth. For the latter, a mechanism analogous to planar growth is often cited. As the process continues, development of particle size is largely determined by

Fig. 5.2-15 H abstraction/ethyne addition mechanism for planar growth of polycyclic aromatic hydrocarbons.

soot particle coagulation. Oxidation is the dominant process for the final phase of diesel engine combustion, in which the previously formed soot particles are introduced into oxygen-rich areas by mixing combustion products with combustion air. Overall, the following processes must be considered in soot formation:

$$(dN/dt)_{\text{overall}} = (dN/dt)_{\text{growth}} + (dN/dt)_{\text{coagulation}} + (dN/dt)_{\text{condensation}} + (dN/dt)_{\text{surface growth}} + (dN/dt)_{\text{oxidation}}$$

where N = the number of particles.

A full analytical description is not possible at present, as some of these processes are thoroughly understood while others are only known phenomenologically. The main difficulties include the large number of chemical reactions involved and their dependence on high pressures and temperatures, as well as flow and mixing fields. The corresponding reaction times are listed in Table 5.2-4. Insufficient time is available during the diesel engine cycle to permit formation of all of the larger particles. Larger particles continue to form even after their components have left the combustion chamber.

As described, soot particles are created through

Table 5.2-4 Typical times for soot formation [6]

Diesel combustion			
– Nucleation		0.001	ms
– Coagulation		0.05	ms
– Chain formation		few	ms
– Condensation		few	ms
– Oxidation		4	ms
• Turbulence duration		1	ms
• Laminar flame			
– Premixed	Nucleation	2–3	ms
	Overall formation process	10–30	ms
– Diffusion	Oxidation > 10 ms > formation		

molecular processes (up to ~10 nm), which then join to form primary particles (10–50 nm; see Fig. 5.2-16). Diesel particles in the context of emissions regulations are exhaust components which are measured by collection on a filter. These soot kernels adsorb organically soluble unburned hydrocarbons, sulfates, metal oxides, and other residual compounds. The soot kernel, complete with its adsorbed components, is termed a particle. Particles from diesel engine combustion represent only a portion of total particulate emissions (Fig. 5.2-17). Diesel engine particulate emissions may be reduced by:

- Improved combustion within the engine
- Improved fuel quality
- Exhaust gas aftertreatment (e.g., filtering)

Nonrestricted emissions

Exhaust emissions in the form of hydrocarbons (HC), carbon monoxide (CO), oxides of nitrogen (NO_x), and particulates are limited by regulations. With the exception of CO, these are expressed as total quantities of all related species (i.e., "HC" includes all hydrocarbon species, "NO_x" is NO + NO_2, etc.).

A differentiated composition for hydrocarbons was first applied in the United States for the 1994 model year, with a limitation on nonmethane hydrocarbons (NMHC) for passenger cars. This reflected the special status of methane, which is fairly unreactive in terms of atmospheric chemistry. Along with methane-free hydrocarbon measurements, a formaldehyde limit was established for so-called clean fuel vehicles.

California passenger car emissions standards, which place special emphasis on reducing ozone precursors, go one step farther and evaluate total NMHCs for their reactivity. The evaluation criterion is the so-called maximum ozone formation potential of these differentiated hydrocarbons. The resulting summary NMHC value is used for certification as NMOG—nonmethane organic gases.

None of the other substances present in combustion

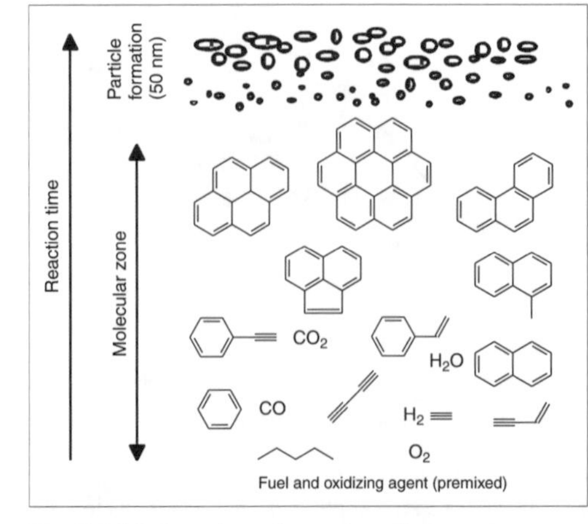

Fig. 5.2-16 Overview of soot formation reactions.

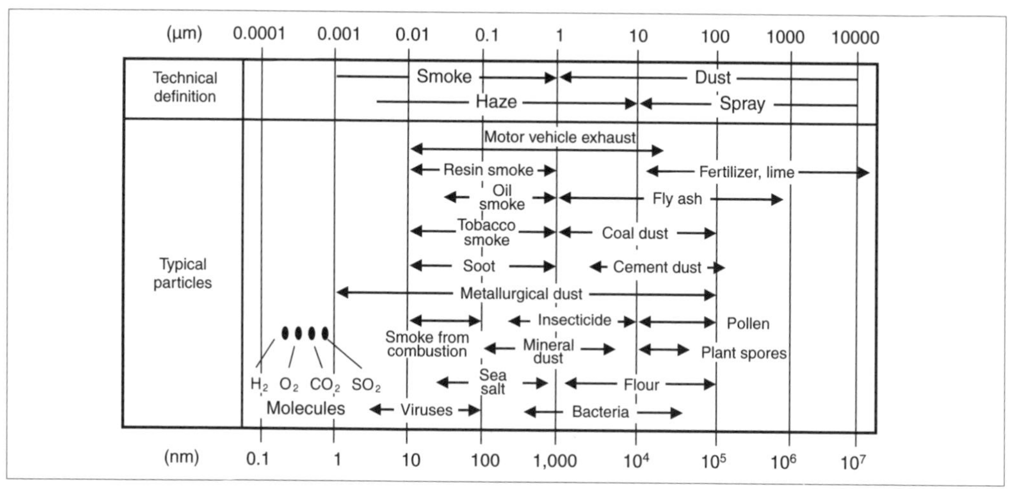

Fig. 5.2-17 Sources and magnitudes of atmospheric particulates [12].

engine exhaust are restricted by concrete emissions limits, nor have appropriate procedures for determining such emissions data been established. All of these substances may be encompassed by the concept of *unregulated exhaust components*. This definition includes such components as those whose emissions levels are indirectly limited by fuel composition regulations.

5.2.6 Injection system design characteristics

Requirements and principles

A precondition for efficient combustion is good mixture formation. The injection system plays a central role in this process. Fuel must be injected in the correct quantity, at the proper time, and at high pressure. Even slight deviations from the ideal will result in higher exhaust emissions, objectionable combustion noise, or high fuel consumption. Important for the diesel engine combustion process is a short ignition delay, the time between start of injection and onset of pressure rise in the combustion chamber. If a large fuel quantity is injected during this period, the result will be a sudden pressure rise and therefore loud combustion noise.

To achieve as "soft" a combustion process as possible, a small amount of fuel is injected at low pressure prior to the main injection event. This is termed *pilot injection*. Burning this small fuel quantity raises pressure and temperature in the combustion chamber.

This achieves the precondition for rapid ignition of the main injection quantity, thereby reducing ignition delay of the main fuel charge. Pilot injection and an injection pause between pilot and main injection result in gentle rather than abrupt pressure increase with attendant lower combustion noise and reduced oxides of nitrogen emissions.

The purpose of the *main injection event* is to support good mixture formation, so that fuel may be burned as completely as possible. High injection pressure results in very fine fuel atomization, permitting good mixing of fuel and air. Complete combustion leads to reduced pollutant emissions and high power output.

At the *end of injection* it is important that injection pressure drop rapidly in order to close the injector needle as quickly as possible. This prevents fuel from entering the combustion chamber at low pressure and therefore in the form of large droplets. These droplets would not burn completely, resulting in increased emissions (see Fig. 5.2-18).

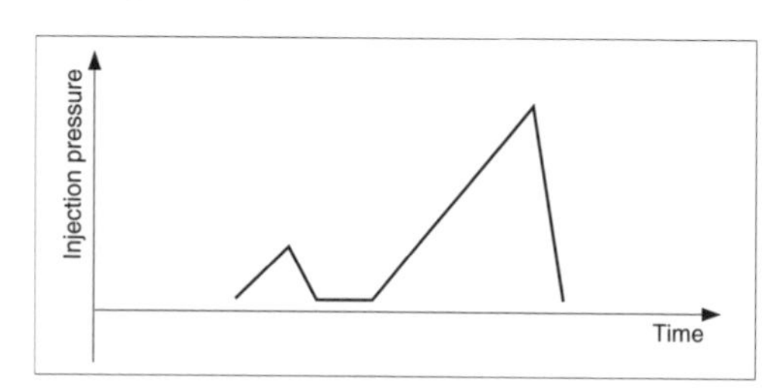

Fig. 5.2-18 Engine requirements for injection history: low pressure for pilot injection, followed by a pause, and then rising pressure during main injection, followed by a rapid end of injection.

It is the task of the injection system to deliver fuel from the fuel tank to the engine and inject it into the combustion chamber at high pressure and at the correct time. The system consists mainly of a fuel transfer or feed pump with fuel filter, an injection pump to achieve high pressure, injector nozzles (or injector valves), and the injection lines joining pump and injectors (Fig. 5.2-19).

5.2.6.1 Distributor injection pump

The distributor injection (VE) pump was developed for small, high-speed diesel engines (Fig. 5.2-20). This compact design permits a versatile, high-performance

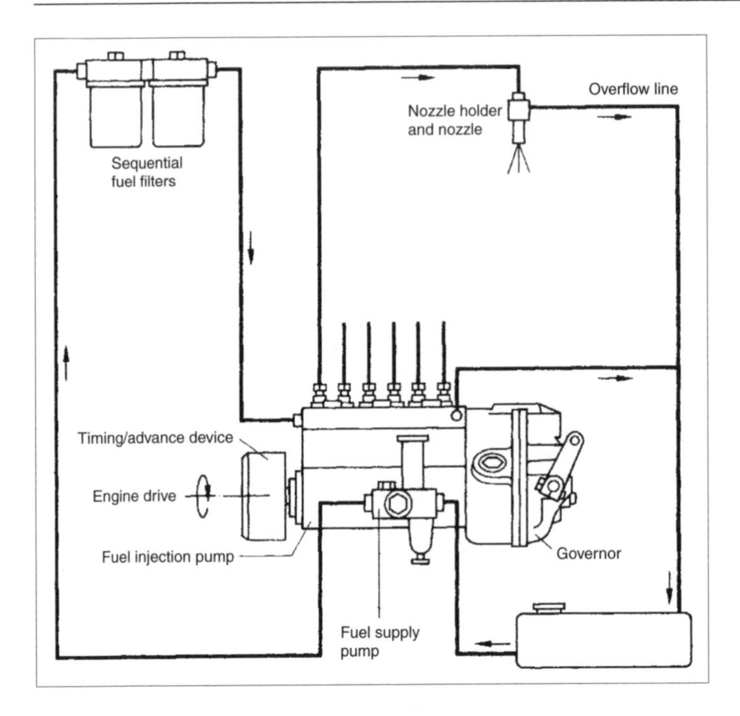

Fig. 5.2-19 Diesel fuel injection system principle.

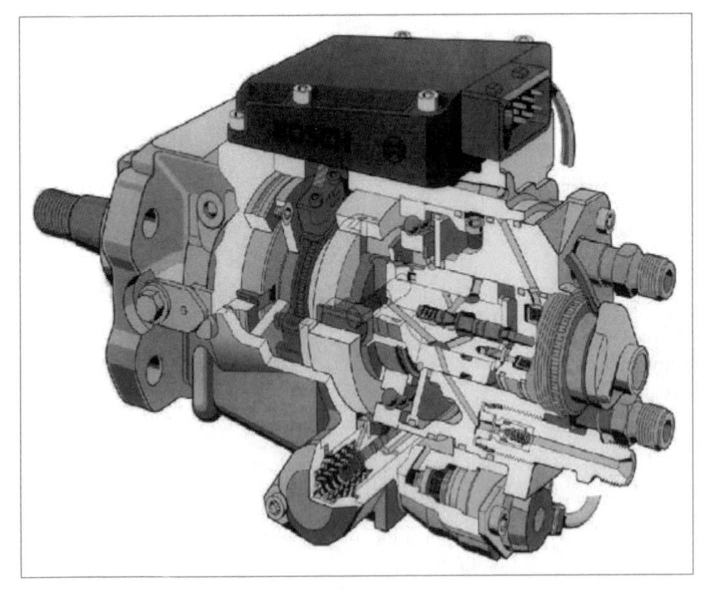

Fig. 5.2-20 Bosch VP44 radial plunger distributor injection pump with integral control unit.

injection system with low weight and small installed volume. The number of cylinders is generally limited to no more than six.

Typically, the VE pump consists of the following assemblies:

- High-pressure pump and distributor
- Speed and injection quantity governor
- Timing device
- Low-pressure supply pump
- Electromagnetic fuel shutoff valve
- Engine-specific add-on modules

Distributor pumps may be divided into two groups, differentiated by their basic operations: axial plunger pumps and radial plunger pumps. In axial pumps, rotary motion of the input shaft is converted into rotary and reciprocating motion of the distributor plunger. The reciprocating motion generates the required fuel pressure and volume, while the rotary motion distributes fuel to the individual engine cylinders. Only one plunger is needed. By contrast, in the case of radial plunger pumps, each engine cylinder or pair of cylinders is supplied by a separate plunger. Radial plunger designs are characterized by separation of functions between high pressure generation by the plungers and distribution by the axial distributor shaft. Pumps are lubricated by the fuel itself.

The second generation of VE pumps introduced electronic controls, eliminating the need for various individually tailored control groups. Injection quantity and timing were governed by means of sensors and by data and maps stored in the control unit.

The latest generation of distributor pumps employs electromagnetic control of injection quantity and timing. The axial piston version achieves nozzle pressures of up to 1,400 bar.

New solenoid-controlled radial plunger pumps achieve nozzle pressures of up to roughly 1,600 bar, with a potential of over 1,800 bar. In these pumps, fuel is pressurized by two to four plungers actuated by a very rigid ring cam. This pump configuration can serve a maximum of six cylinders. Separate pilot injection at low engine speeds, previously achievable only through the use of a dual-spring nozzle holder, is now made possible by the solenoid valve. However, this system does not permit postinjection. Its special advantages are:

- High fuel quantity dynamics (cylinder-specific fuel quantity control)
- High precision
- Fuel rate control by means of variable port closing (beginning of injection)
- Pilot injection (time-separated only at low engine speeds)

5.2.6.2 Inline injection pump

With a layout similar to that of inline engines, separate plunger elements, one for each engine cylinder, are arrayed in line in a common housing. The principal operating characteristics are:

- Fuel drawn from the tank by a separate supply pump (also called feed pump, lift pump, or transfer pump) enters the pumping element at bottom dead center through lateral ports and is subsequently compressed.
- Control of injected quantity is achieved by varying the plunger's effective stroke.
- The plunger, actuated by a camshaft, forces fuel through a delivery valve at the upper end of the pumping element and into the injection line to the engine.
- At the engine cylinder, fuel in the form of a finely atomized spray enters the combustion chamber through an injector nozzle.
- Subsequently, the pump plunger is returned to its original position by a spring, and the delivery valve, also spring-loaded, closes the injection line.

An inline injection pump (Fig. 5.2-21) includes a governor to control injected quantity, a supply pump to draw fuel from the tank, and a timing device for engine-speed–dependent adjustment of injection timing.

Both fuel injection pump and governor are lubricated by means of connections to the engine lubricating system.

Most manufacturers of passenger car diesel engines prefer VE (distributor) injection pumps.

5.2.6.3 Injector nozzle and nozzle holder

Without question, the injector nozzle plays a key role in diesel engines. It atomizes and distributes fuel in the diesel engine combustion chamber and influences the nature of the injection process. In this way, the nozzle helps to define how thoroughly fuel supplied by the injection pump is burned and, consequently, how effectively the energy content of the fuel is extracted. This task is by no means a simple one, especially when one considers that the nozzle must open as many as 2,500 times per minute and, in the case of direct-injection engines, distribute fuel in the combustion chamber at pressures of up to 2,000 bar.

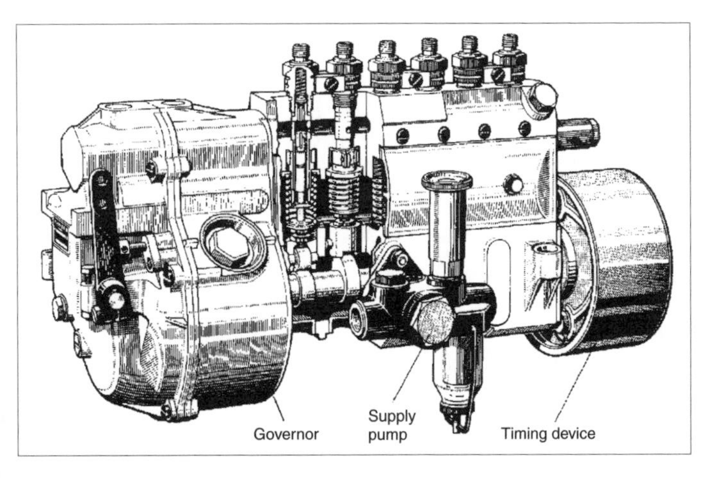

Fig. 5.2-21 Cross section of an inline injection pump.

All diesel injector nozzles in use today employ hydraulically actuated needles (valves). Injectors consist of a nozzle holder and a nozzle body in which a spring-loaded needle is fitted to tolerances of less than a thousandth of a millimeter. Pressure is applied to the needle seat, in the central section of the needle, by fuel which flows into the nozzle through the nozzle body. Once force on the needle seat exceeds spring preload, the needle lifts off its seat, opening at its lower extremity nozzle orifice(s) aimed into the combustion chamber. Fuel leak-off past the top of the needle is directed back through the nozzle holder and returns to the fuel tank via a separate overflow line. Quality of a fuel injection nozzle is largely determined by the nozzle orifice(s).

Distinction is made between pintle nozzles (Fig. 5.2-22a) and hole-type nozzles (Fig. 5.2-22b). Pintle nozzles are best suited for the prechamber and swirl chamber diesels found in passenger cars. The tip of a pintle needle has a profiled protrusion which extends through the orifice and provides a comparatively tight spray cone. A throttle (or delay) pintle nozzle employs special pintle dimensions. In this type, the shape of the pintle is specifically designed to modify the spray cone, matching it to engine requirements. At small needle lifts (in other words, at the beginning of injection), the spray cross section is small. The resulting spray (pilot injection effect) slows combustion pressure rise and reduces combustion noise.

Hole-type nozzles, on the other hand, are primarily suited to direct-injection diesels. The injected fuel spray forms a wider fan than with the pintle nozzle, resulting in fine atomization and thorough mixing with combustion air—a task accomplished in pre- and swirl-chamber engines by more intense air turbulence. Hole nozzle spray orifices are arranged in a cone at the nozzle tip. Distinction is made between single- and multihole nozzles, whereby the latter may have as many as twelve orifices. For low hydrocarbon emissions, it is important to minimize the fuel-filled volume between valve seat and orifice openings to the combustion chamber. "Valve-covered orifice" (VCO) or sac-hole nozzles, in which the seated needle covers the orifices, are employed to achieve this.

Today, the highest precision is called for in optimizing and locating nozzle holes. Flow conditions and pressure distribution are examined with three-dimensional computational models (Figs. 5.2-23, 5.2-4, 5.2-25). Important parameters for nozzle geometry are: number of holes, entry conditions (sharp- or round-edged), and length–diameter ratio. New manufacturing methods (e.g., hydroerosive forming) increase nozzle flow, tighten flow tolerances, and eliminate abrasive wear at the nozzle hole entry. If the hole entry is rounded, spray penetration and droplet size increase, which permits higher injection pressure.

The nozzle is mounted in a nozzle holder, which serves to attach the nozzle to the cylinder head and, with the nozzle, forms a nozzle and holder assembly (also known as an injector). Generally, nozzle holders contain a nozzle spring and some form of needle lift limitation. The dual-spring nozzle holder employs two springs (Fig. 5.2-26). The first compression spring (3) initially limits needle lift to the prelift, H_1 (pilot injection). Once pressure exceeds a level preset by the second compression spring (6), the needle is lifted further, up to its maximum lift H_2 (main injection). The injection

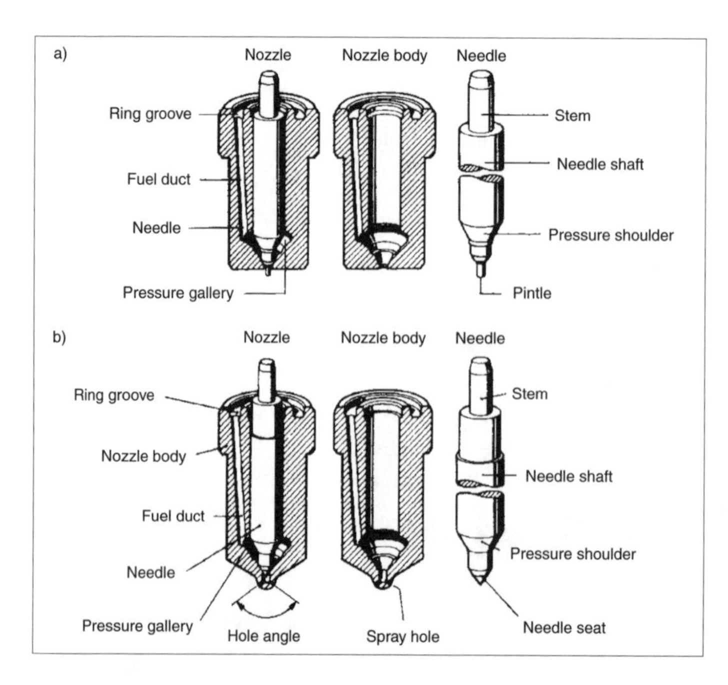

Fig. 5.2-22 a) Pintle nozzle for pre- and swirl-chamber engines; b) hole-type nozzle for direct-injection diesel engines.

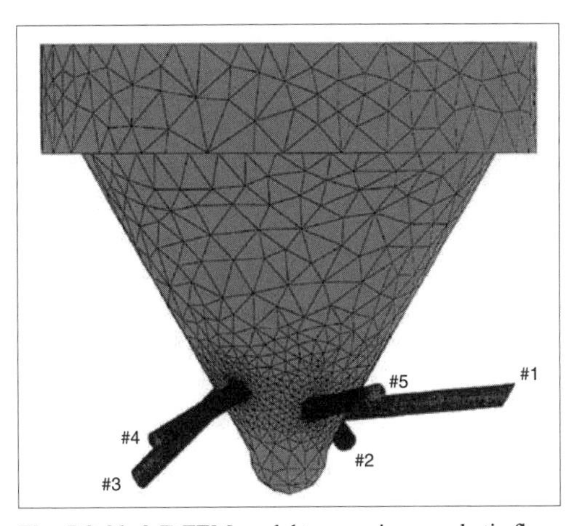

Fig. 5.2-23 3-D FEM model to examine nozzle tip flow conditions.

characteristic provided by such a dual-spring nozzle holder results in decreased combustion noise.

5.2.6.4 Unit injector

The unit injector is a system capable of meeting the most difficult diesel fuel injection demands.

Rudolf Diesel himself considered combining fuel injection pump and nozzle into a single unit (Fig. 5.2-27) in order to eliminate high-pressure lines and thereby achieve high injection pressure. However, he lacked the technical means to realize this idea. Since the 1950s, diesel engines for commercial vehicles and marine applications have been equipped with mechanically controlled unit injectors.

Volkswagen, in cooperation with Robert Bosch GmbH, was the first to develop successfully a passenger car diesel engine with a solenoid-actuated unit injector system. This engine meets all demands for higher

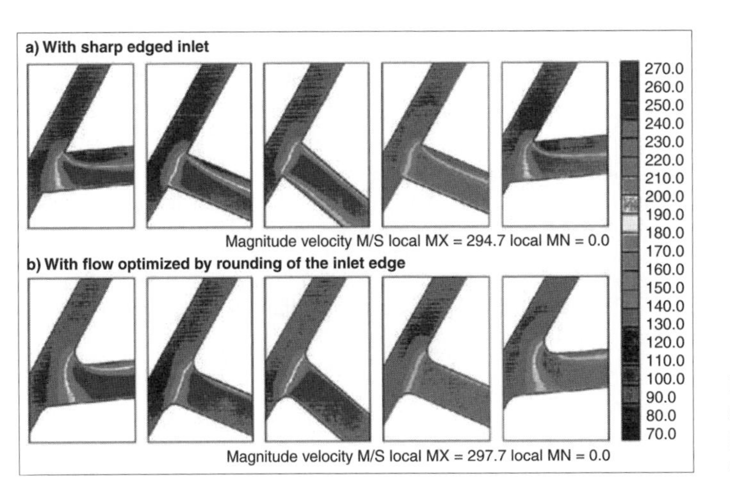

Fig. 5.2-24 Flow conditions in nozzle holes a) with sharp-edged inlet and b) with rounded inlet edge.

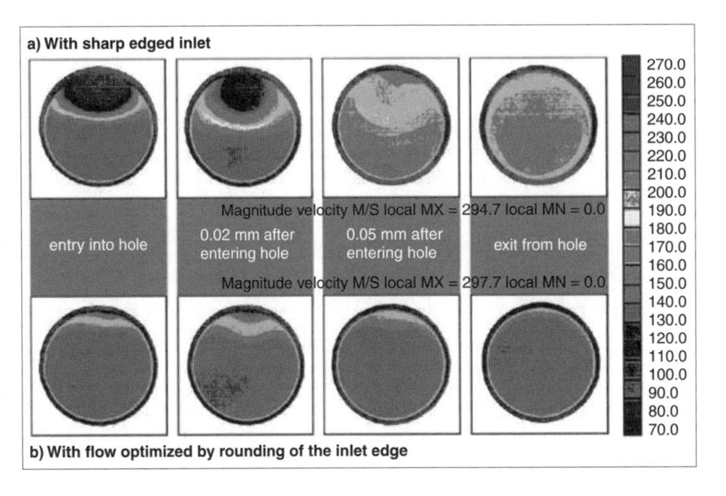

Fig. 5.2-25 Flow conditions at various nozzle holes a) with sharp-edged inlet and b) with rounded inlet edge.

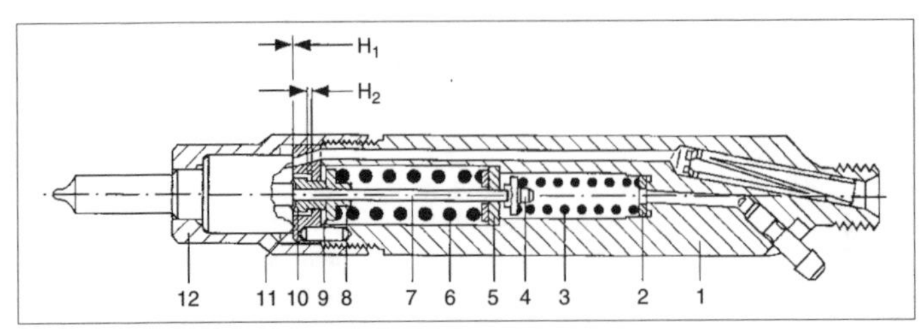

Fig. 5.2-26 Two-spring nozzle holder (Type KBEL, Robert Bosch Corp.). 1, Nozzle holder body; 2, pressure-adjusting shim; 3, pressure spring 1; 4, pressure pin; 5, guide sleeve; 6, pressure spring 2; 7, pressure pin; 8, spring seat; 9, pressure-adjusting shim; 10, stop sleeve; 11, intermediate element; 12, nozzle retaining nut.

Fig. 5.2-27 A scene such as this may have inspired Rudolf Diesel to consider the unit injector.

power output while simultaneously reducing environmental impact. It is a step toward a future envisioned by Rudolf Diesel—that someday the engine bearing his name should be "free of smoke and odor" [1].

A unit injector consists of an injection pump, control unit, and fuel injector nozzle combined in a single assembly (Fig. 5.2-28). Each engine cylinder is fitted with a single unit injector. This eliminates high-pressure injection lines such as those found on distributor or inline injection pumps; the high-pressure volume that must be compressed is much smaller.

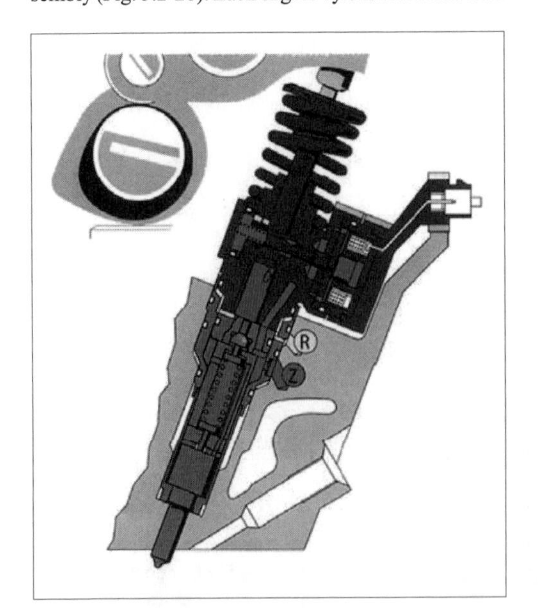

Fig. 5.2-28 Schematic representation of a unit injector with actuating cam.

Like a distributor or inline injection pump, a unit injector must fulfill the following requirements:

- Generate high pressure for injection
- Inject the right amount of fuel at the right time

To drive the unit injectors, the engine camshaft has one additional cam lobe per cylinder. Roller rocker arms activate the unit injector plungers. The injection cam profile has a steep opening flank and a flat closing flank. This depresses the pump plunger at a high rate, rapidly building up high injection pressure. After injection, the plunger returns slowly and steadily, permitting fresh fuel to refill the unit injector's high pressure space without forming bubbles.

In comparison to distributor injection pumps, diesel engines equipped with unit injector systems exhibit the following advantages:

- Low combustion noise thanks to separate pilot injection
- Low fuel consumption
- Low exhaust emissions
- High power output

These advantages are achieved through

- High injection pressure, currently as high as 2,050 bar, with a goal of over 2,200 bar at the nozzle
- Precise control of the injection process
- Pilot injection

Thanks to injection pressure as high as 2,050 bar and the resulting good combustion, VW's engine develops 285 Nm of torque at just 1,900 rpm. Its maximum output of 85 kW is produced at 4,000 rpm. Compared to the 1.9-L 81 kW engine, equipped with a distributor pump, the unit injector engine produces about 21% more torque.

5.2.6.5 Accumulator or common rail injection system

Another high-pressure injection system with great future potential is the accumulator or common rail system (Fig. 5.2-29). In this system, pressure generation and injection functions are decoupled from one another. Fuel for individual cylinders is drawn from a common accumulator, which is constantly held at high pressure. Accumulator pressure is generated by a high-pressure radial plunger pump and may be altered independently of operating conditions.

Each cylinder is fitted with a solenoid-actuated injector. Injection quantity is determined by the injector discharge cross section, solenoid opening duration, and accumulator pressure. At present, system pressures of up to 1,400 bar are being achieved; future systems with 1,500–1,600 bar and even 1,800 bar are under development. Injection quantity and injection timing as well as pilot and afterinjection are controlled by extremely fast solenoids. In the future, control will be via piezoelectric elements which permit extremely short switching times of less than 100 picoseconds. The goal is to achieve injection quantities well below 1 mm^3.

Functional separation of pressure generation and injection permits better control of the injection process and therefore of combustion. Injection pressure may be chosen freely from any value on the control unit map. Pilot injection is possible. At present, it is more difficult to match the common rail injection process to market demands. One advantage is that common rail systems may be installed on existing engine designs without cylinder head modification.

5.2.6.6 Injection system simulation

In addition to traditional simulation of injection hydraulics, mastery of highly stressed injection systems requires a thorough analysis of fuel behavior and of the overall energy conversion process. A cyclic process takes place in the high pressure circuit, shown schematically in Fig. 5.2-30.

Theoretical input is the work done to the plunger. Output may be regarded as the energy of the injector spray. Major losses occur during delivery cutoff. For high-pressure injection systems, changes in density, speed of sound, compressibility, and viscosity must be taken into account. Furthermore, various degrees of component expansion caused by high temperatures must be considered [4].

For development work, numerical solutions are usually determined by one-dimensional programs based on streamline theory. Three-dimensional effects are given weight by application of empirical corrections and coefficients—for example, for frictional losses in injection lines or flow losses at changes of cross section. In this way, dynamic processes in complete systems can be effectively simulated, and even if multiple variations

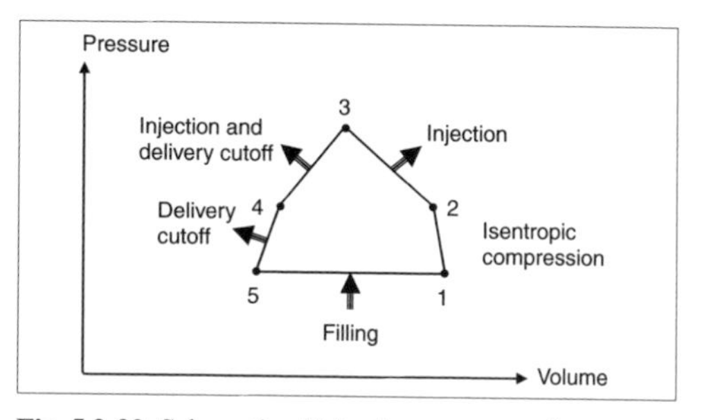

Fig. 5.2-30 Schematic of injection system cyclic process.

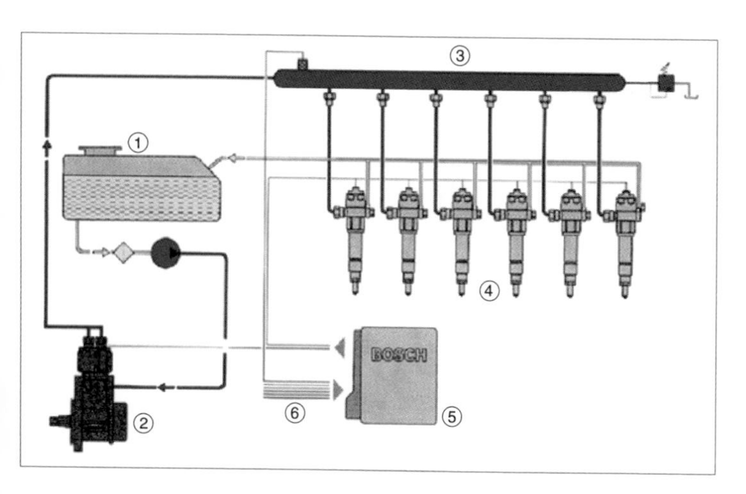

Fig. 5.2-29 Schematic representation of accumulator (common rail) injection system, with fuel tank (1), high pressure pump (2), accumulator (common rail) (3), injectors (4), electronic control unit (5), and sensors (6).

are examined, the required effort remains economically manageable.

5.2.7 Combustion system design characteristics

Modern diesel engines face increasingly stringent demands with regard to performance, fuel economy, and exhaust and noise emissions. Because of high engine speeds in passenger car diesels, only a very short time is available for the combustion process. A prerequisite for meeting these requirements is good mixture formation. For this, engines must be equipped with high-performance injection systems which provide high injection pressures for very fine fuel atomization and which allow precise control of injection timing and injected fuel quantity as well as appropriate combustion chamber design. Although classic diesel engine combustion systems, divided into indirect (Fig. 5.2-1 a and b) and direct (Fig. 5.2-1 c and d) injection systems, are employed, passenger car diesels also exhibit their own unique characteristics.

5.2.7.1 Prechamber systems

At the beginning of its development history, a detour by way of air injection helped diesel engines achieve growing popularity. Simultaneously, however, disadvantages of this complex, rather uneconomical process became ever more obvious. Instead of mixing fuel with blown-in air, attempts were then made to agitate combustion air and to create an ignitable mixture through other means.

In prechamber systems (Fig. 5.2-1a), fuel is injected under relatively low pressure by a flatted pintle nozzle into a rotationally symmetric prechamber arranged concentrically to the cylinder axis, where precombustion is initiated. The prechamber accounts for about 40% of the compression volume. Precombustion raises pressure in the prechamber to the point where partially burned charge is ejected at high velocity through the throat into the main combustion chamber.

The relatively long throat ends in several combustor holes whose angular orientation and diameter are carefully chosen to engulf as much of the main combustion chamber air as possible. The total combustor hole cross section, relative to the theoretical piston surface, amounts to about 0.5%.

An impact body located in the center of the prechamber breaks up the impinging fuel stream and promotes intensive mixing with air. It also serves to impart some swirl to air forced from the cylinder into the prechamber during the compression stroke.

Thanks to its relatively gentle pressure rise, prechamber diesels are of great interest for passenger car applications. "Soft" combustion results in lower noise, NO_x and particulate emissions.

In keeping with the required valve cross sections, two-valve engines generally require compromises to avoid the combustion disadvantages of an eccentrically located prechamber. Four-valve technology enables improved combustion chamber shapes. The resulting central prechamber location promotes very even mixture distribution in the main combustion chamber, with resulting good air utilization.

A decrease in soot formation may be observed in conjunction with an increased number of combustor holes, with a total of nine combustor holes (eight radial, one axial). The holes are dimensioned to ensure that kinetic energy of the gas jets entering the main combustion chamber provides good, uniform mixture formation and distribution of the prechamber contents.

5.2.7.2 Swirl chamber systems

Like prechamber systems, swirl chamber systems (Fig. 5.2-1b) are divided-chamber diesel combustion systems. Swirl chambers represent the most common type found in modern passenger car diesel engines. Advantages include high-speed capability (up to 5,000 rpm), relatively good emissions behavior, and significantly lower pressure rise, an important consideration for noise emissions.

Fuel injection occurs by means of a throttling pintle nozzle, at relatively low pressure, injecting into a spherical chamber. The nozzle is situated to aim the fuel stream perpendicular to the rotational axis of the swirl formed during the compression stroke. The stream penetrates the swirl and impinges on a hot zone on the opposite wall of the chamber. To achieve complete mixture formation within the swirl chamber at all engine speeds and loads, the shape and orientation of swirl chamber, nozzle, and glow plug must be carefully optimized to one another. Current knowledge and experience indicate that the optimum chamber size is 50% of the total compression volume. Changes in swirl chamber volume produce opposing effects with respect to NO_x emissions and noise on the one hand and HC, CO, and particulate emissions on the other.

It is preferable that the flame port—that is, the tangential passage connecting the swirl chamber to the cylinder—be configured so that its flow cross section represents about 1 to 2% of the piston area. This serves to produce swirl in air streaming into the chamber.

The main combustion chamber, in the piston crown, is flat or hourglass-shaped. Combustion gases, tightly confined to the piston crown, result in high thermal loading of the piston.

Swirl chamber designs produce an overall efficiency of about 36%. Under actual operating conditions, frequent part-load operation has a pronounced effect on fuel consumption. Swirl chamber engines exhibit uncommonly low fuel consumption across a relatively large load range.

5.2.7.3 Direct injection systems

The direct injection (DI) concept (Fig. 5.2-1c) was originally applied to stationary and commercial vehicle engines of all sizes. Since 1988, it has made inroads in passenger car diesels (Table 5.2-1). In the near future, direct-injection diesels will displace divided-chamber concepts as the preeminent configuration on passenger cars.

DI engines address the demand for passenger cars with low fuel consumption. Thanks to their lower thermal losses, direct injection engines represent the most economical form of diesel engine, and they surpass all other combustion engines in this regard. Their primary drawbacks are noise and emissions problems.

To achieve the best possible thermodynamic efficiency, high demands are placed on this system's fuel and air supply. These must be met in order to carry out in succession the required processes of fuel atomization, heating, vaporization, and mixing in the brief time available. To master this assignment, the following are required:

- High injection pressures in excess of 1,000 bar to achieve short injection duration.
- Higher degree of atomization: small nozzle holes (<0.2 mm) ensure better atomization and thereby better mixing of fuel and air.
- Multihole nozzle for more uniform spatial distribution of fuel within the combustion chamber.
- Deliberate air motion (swirl) during intake and compression strokes supports mixture formation.
- A deep piston bowl serves as combustion chamber.
- A glow plug extending into the combustion chamber aids in cold starting.

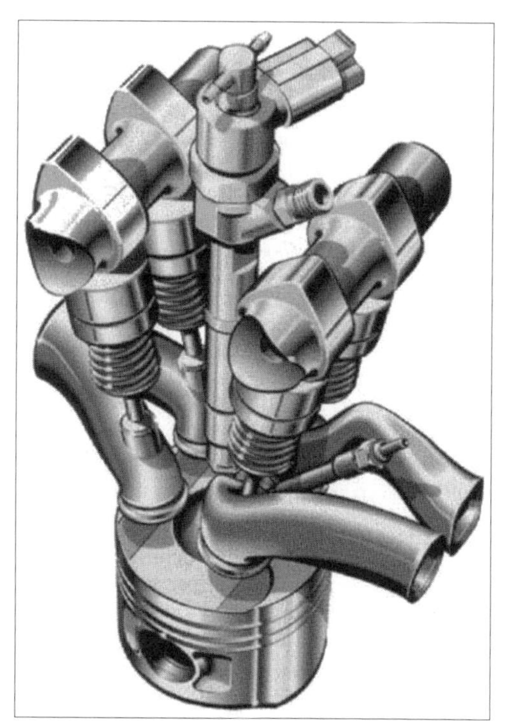

Fig. 5.2-31 Four-valve technology. Combustion chamber with intake and exhaust runners and centrally located injector (Daimler-Benz OM611).

Modern direct-injected engines have rated speeds up to 4,500 rpm. They exhibit a maximum overall efficiency of 43%. By adding additional features such as

- Dual-spring nozzle holders
- High pressure injection system
- Controlled, cooled exhaust gas recirculation
- Oxidizing catalytic converter
- Additional exhaust gas aftertreatment for NO_x and particulates

exhaust emissions and combustion noise can be reduced to low levels. The potential for further development with regard to engine improvements is by no means exhausted. For example, central, vertical nozzle installations may avoid tolerance-induced deterioration of operational parameters. Four-valve technology (Fig. 5.2-31) is a prerequisite for vertical nozzle installations.

In four-valve designs, the two intake runners usually exhibit differing configurations. One port is spiral-shaped, to produce strong swirl at part load. The other port is shaped to produce strong tangential flow. It is possible to shut down one runner, depending on engine operating point, which may help realize benefits in terms of smoke emissions.

Combustion noise, clearly audible during acceleration due to the harsh nature of full-load combustion, can be greatly reduced by injecting a small portion of finely atomized fuel into the combustion chamber. Such pilot injection requires a nozzle with variable orifice cross sections, only achievable with considerable cost and effort.

Another source of potential improvement is shaping intake runners to induce swirl. As the need for air charge rotational intensity varies with engine speed, variable-swirl intake ports may offer significant benefits.

The M-system, developed by MAN of Augsburg (Fig. 5.2-1d) operates with a completely different concept. In this system, the objective is to allow only the smallest possible portion of injected fuel to undergo autoignition. The larger portion (95%) is injected against the combustion chamber wall.

Combustion is initiated by autoignition of air-atomized fuel; this and hot rotating air within the combustion chamber remove and ignite fuel in layers from the combustion chamber wall.

The resulting "soft" extended combustion process leads to low soot and noise emissions. Volatile fuels with poor autoignition properties may be utilized (multifuel engines).

Compared to other direct-injection engines, strong turbulence in the combustion chamber leads to increased flow losses, expressed as higher fuel consumption. Moreover, pistons are subject to greater thermal loads.

The M-system is not applied to passenger car engines.

5.2.7.4 Crank train and cylinder head

The crank train (Fig. 5.2-32) converts the energy released in combustion into useful torque. Its major components are pistons (Fig. 5.2-33), connecting rods, and crankshaft. In contrast to Otto-cycle (gasoline) engines, which may be operated at speeds up to 7,000 rpm or more, diesel engine crank trains need only withstand 5,000 rpm at most. However, they are subjected to considerably higher combustion pressures and temperatures. Consequently, stronger designs are required for components which play a major role in transferring forces (piston pins, connecting rods, crankpins, and bearings), and their increased thermal loading must be transferred away.

Pistons must be designed to withstand higher temperatures and pressures. Greater demands are placed on piston rings with regard to sealing between combustion chamber and crankcase. Furthermore, oil control rings must limit consumption of lubricating oil. Special care must be given to design of the fire ring (top ring land), the section of piston between the piston crown and the top piston ring, in order to prevent coke buildup in ring grooves. Coking of lubricating oil can lead to "baking" of the piston rings. To avoid this, piston rings must also transfer heat as rapidly as possible to the cylinder wall and thereby to the coolant.

Compared to Otto-cycle (gasoline) engines, diesel engines typically exhibit longer piston strokes. With much smaller compression volume (about one third that of gasoline engines), dimensional deviations of crank train components have a much greater effect on compression ratio. This places higher demands on manufacturing tolerances in diesel engines.

Another advantage of long-stroke engines with small

Fig. 5.2-33 VW 1.9 L TDI piston.

cylinder bores is that the necessarily smaller combustion chamber surfaces lead to reduced thermal losses and permit the optimum compression ratio to be approached. Long-stroke engines provide easy cold starting and good idle, with lower compression ratios than required for short-stroke engines, whose compression ratio must be increased (at some cost in economy) for this purpose.

There are many differences between diesel and gasoline cylinder heads. In prechamber and swirl chamber engines, combustion chambers are located within the cylinder head. This results in extraordinarily high thermal loading (temperatures up to 900°C), which place especially high demands on cylinder head design. Combustion chamber location may result in uneven thermal expansion and result in cylinder head distortion.

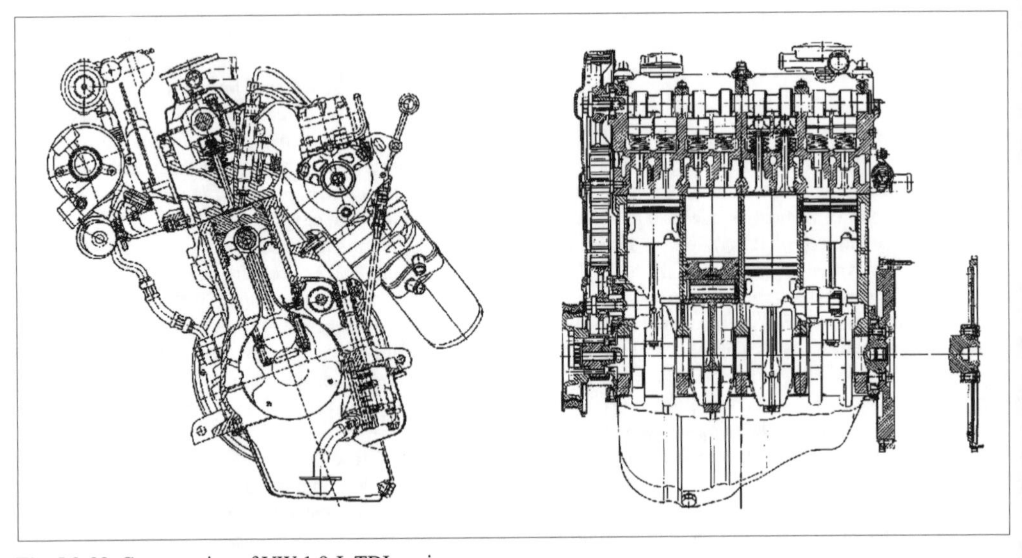

Fig. 5.2-32 Cross section of VW 1.9-L TDI engine.

Moreover, directing coolant to the combustion chambers represents a design challenge.

Furthermore, a diesel cylinder head must provide space for injector nozzles and glow plugs.

5.2.7.5 Supercharging

The idea of filling a diesel with precompressed air was tried experimentally by Rudolf Diesel himself. The breakthrough in supercharging was achieved in 1905 by Alfred Büchi, who seized upon the idea of using energy present in the exhaust stream. Even in his early work, Büchi proposed using a charge-air intercooler to improve thermal efficiency.

At barely 5,000 rpm, even modern diesel engines have limited speed capability. Increased power (greater air quantity permits greater injected fuel quantity) is only possible by means of increased displacement or turbocharging. In addition, supercharging leads to increased economy, since the turbocharger utilizes the exhaust-side pressure and temperature gradient relative to atmospheric conditions to precompress combustion air. So-called gas exchange work (that portion of the cycle in which the engine draws its combustion air against flow resistance upstream of the cylinder) is eliminated.

A compound machine consisting of an exhaust turbocharger and piston engine exhibits improved efficiency for several reasons. First, higher working pressures improve exhaust gas energy utilization, and elimination of negative pumping work improves internal efficiency, which reflects thermodynamic utilization of the supplied energy. Second, the relationship between mechanical losses (mostly frictional losses) and overall output is improved, which leads to higher mechanical efficiency.

Like all flow machines, the exhaust turbocharger is optimized for a specific engine operating point. Due to the large speed range of passenger car diesel engines, full-load boost pressure, left to itself, would rise considerably, leading to undesirable loading of the crank train. Therefore, boost pressure must be limited, which is accomplished by means of a wastegate (bypass valve). Matching turbocharger air supply to an engine's actual air requirement, a function of its operating point, may also be accomplished by means of variable turbine geometry (VTG, Fig. 5.2-34).

Other possibilities are offered by sequential turbocharging (several turbochargers operating in parallel), pressure-wave supercharging (COMPREX), and mechanical supercharging.

5.2.7.6 Life expectancy

Diesel engines are heavier and more generously dimensioned than Otto-cycle engines. They must be designed to cope with higher pressures and temperatures and steeper pressure gradients. Despite these conditions, diesels exhibit outstanding life expectancy, as they operate at lower speeds than gasoline engines, resulting in reduced wear of moving parts. In addition, although diesel engines are designed for full-load performance,

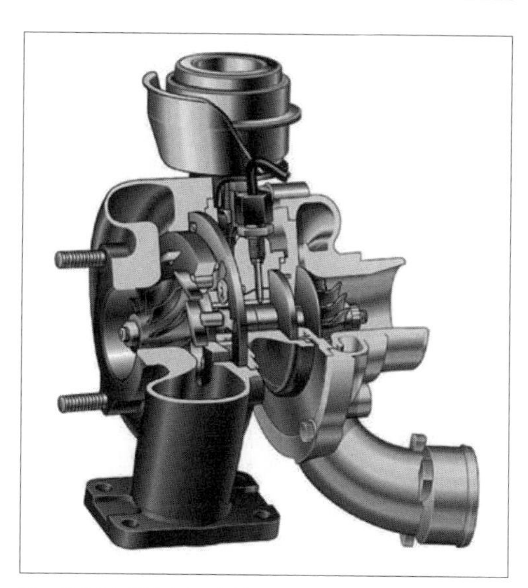

Fig. 5.2-34 VTG (variable turbine geometry) turbocharger.

the driving style of diesel drivers also has a positive effect on powerplant longevity, as they are apt to use full load less frequently than their counterparts driving gasoline engines.

5.2.8 Qualitative evaluation of combustion systems

Various combustion processes may be described on the basis of differences in combustion chamber geometry, fuel injection, and mixture formation, as well as load considerations.

In comparing diesel engines with direct or indirect injection, the direct-injection engine wins convincingly, with up to 15% lower fuel consumption. This results from the higher portion of its combustion process occurring at constant volume (see Section 5.2.4.3), lower heat losses, and elimination of flow losses between prechamber/swirl chamber and main combustion chamber.

Divided-chamber diesel engines exhibit their own unique advantages. Their shorter ignition delay and "staged" combustion reduce combustion noise, emissions of regulated exhaust pollutants, and exhaust odors.

Compared to gasoline engines, diesels offer clear benefits in terms of specific fuel consumption, especially at part load, and in raw emissions of CO, HC, and NO_x. The higher thermal efficiency of diesels is attributable to higher compression ratio, excess air (especially at part load), and lack of induction air throttling. The relatively low raw emissions of diesels result from the high relative air/fuel ratio (for low CO, NO_x) and the resulting lower peak combustion temperatures (low NO_x), as well as elimination of fuel condensation during induction, freedom from ignition misfire, absence of extreme

Table 5.2-5 Qualitative evaluation of various combustion systems as exemplified by motor vehicle engines

Property	Type of combustion system			Gasoline engine (homogeneous mixture)
	Diesel engine with:			
	Swirl chamber	Prechamber	Direct injection	
Specific torque	Reference values	–	++	0
Specific fuel consumption		–	+++	– –
Pollutants				
Raw emissions				
Carbon monoxide		0	–	+
Hydrocarbons		0	–	+
Oxides of nitrogen		0	– –	++
Soot particles		+	–	+++
Combustion noise		–	– – –	++
Exhaust odor		–	– –	++

0 about the same
+ slightly better
++ better
+++ significantly better
– slightly worse
– – worse
– – – significantly worse

cycle-to-cycle combustion variation, and quenching effects in narrow combustion chamber spaces (HC).

On the other hand, the gasoline engine has advantages over the diesel in terms of specific output, combustion noise, and particulate emissions, as well as better preconditions for exhaust gas aftertreatment.

Overall, it is apparent that thanks to their low fuel consumption, unmatched by any other combustion engine, diesels represent a very significant contribution toward extending oil reserves as well as reducing the "greenhouse effect." With regard to exhaust and noise emissions, the internal mixture formation of diesels creates a thermodynamic conflict which, despite all best efforts using internal engine measures, does not appear to be resolvable without exhaust gas aftertreatment.

5.2.9 Exhaust gas aftertreatment

Ever-tightening exhaust emissions limits and demands for low-emission vehicular operation have led and continue to lead to development of exhaust treatment systems appended to combustion engines.

5.2.9.1 Oxidizing catalytic converter

Reducing CO, HC, and PM emissions by means of oxidizing catalytic converters has been applied to diesel passenger car engines since 1990 (Table 5.2-1). As soon as light-off temperature is reached, conversion rates in excess of 80% can be achieved. Along with reduction of gaseous CO and HC emissions, hydrocarbons adsorbed by soot particles are also converted. There is, however, a danger that at high temperatures, the combustion product sulfur dioxide may form sulfates, and under the conditions specified for particulate measurement (52°C), these in turn may lead to condensation of acids. These are evaluated as particulate mass.

Depending on catalytic coatings, converter location, and therefore temperature levels at the converter, the following reductions are achievable:

- HC reduction of about 50–80%
- CO of about 40–70%
- NO_x of about 10–20%
- Particulates of about 30–40%

5.2.9.2 Diesel particulates

Diesel particulates are extremely small. At present, measurements of size distributions (Fig. 5.2-35) have not yet been standardized, and therefore results are not always comparable.

High nanoparticle counts have been observed, but these are almost exclusively attributable to droplets which do not possess carbon cores. Measurements which determine particle size on the basis of particle mass find most of these particles to range in size from 0.1 to 1 μm.

After leaving the exhaust tract, diesel particulates react with dust particles and other components of ambient air, thereby changing size and composition. Processes occurring in the atmosphere are extremely complex and not yet fully understood.

New diesel engine designs emit significantly less soot (on a mass basis) than older concepts. This is mostly attributable to better mixture formation. Examinations of size distributions for various vehicles employing different engine technologies show differences only in the maximum values of particle distribution (Fig. 5.2-36). Note that the particle size peak does not shift.

If particulate emissions are to be reduced even further, high separation rate particulate filters must be employed. Filters have been developed based on ceramic monoliths (Fig. 5.2-37), ceramic fibers, and metal substrates. Efficiencies of over 90% are achievable.

One technical problem to be solved is filter regener-

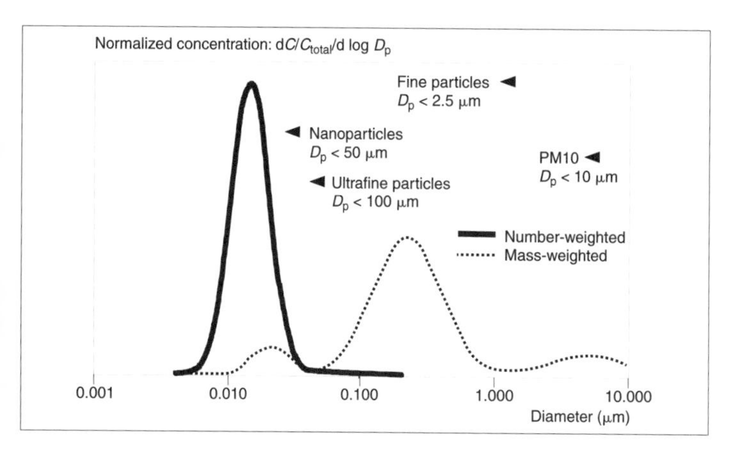

Fig. 5.2-35 Typical diesel particulate counts and mass distributions (source: [5]).

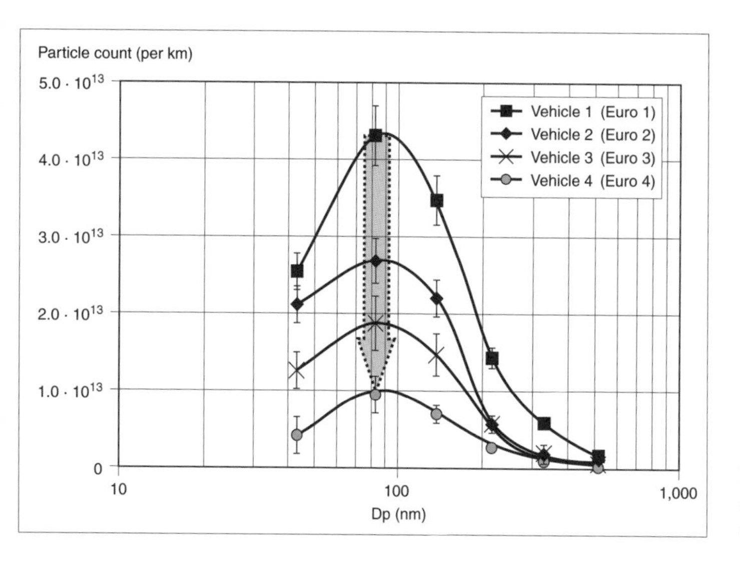

Fig. 5.2-36 Particle size distribution for various exhaust concepts.

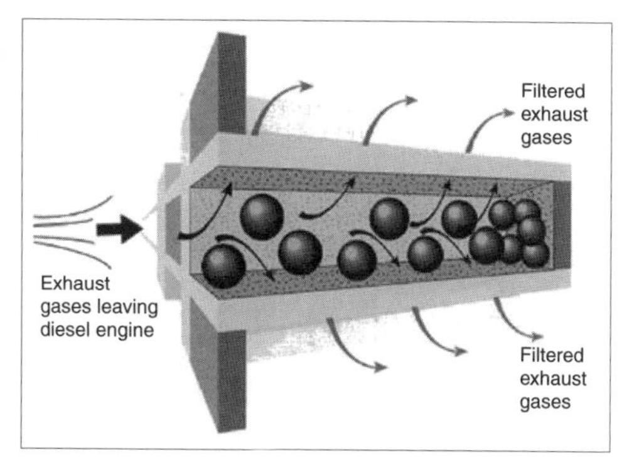

Fig. 5.2-37 Operation of a ceramic monolith whose passages are alternately closed on one end or the other. Exhaust gases pass through the walls, while particulates are filtered out.

ation. Particulate deposits rapidly plug the filter, which reduces engine output and increases fuel consumption. Temperatures in excess of 600°C are required to burn off soot; such temperatures can only be achieved under full load operation. Work is proceeding on various strategies to lower soot ignition temperature, including:

- Filter regeneration without application of additional energy
  - Catalytic coatings
  - Additive-supported regeneration
- Filter generation by means of added thermal energy
  - Electric heating
  - Diesel combustor

The systems outlined above operate discontinuously. Application to production passenger cars has yet to be achieved. By contrast, the CRT ("continuously regener-

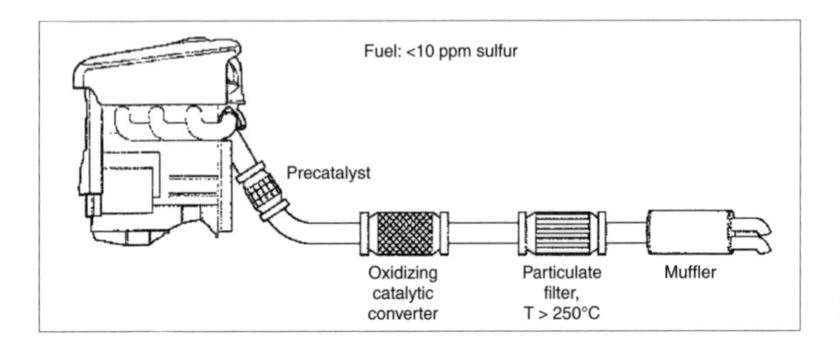

Fig. 5.2-38 CRT (continuously regenerating trap) particulate filter system.

ating trap") system ensures continuous burnoff. CRT systems consist of a catalyst/soot filter unit (Fig. 5.2-38). Vehicle exhaust gases first flow through an oxidizing catalytic converter. CO and HC are almost completely oxidized in a temperature range between 200 and 600°C. NO is converted to NO_2. In the soot filter, this NO_2 combines with carbon compounds to form CO_2. For proper operation, sufficient NO_2 must be available (NO_2/particulate ratio). NO_2 is in turn reduced to NO. These reactions take place between 200 and 500°C. The system reaches equilibrium in this temperature range, so there are no temperature spikes.

One important prerequisite for application of CRT systems is diesel fuel sulfur content below 10 ppm. The equilibrium temperature for soot deposition and burnoff is affected by fuel sulfur content (Fig. 5.2-39). SO_2 in the exhaust inhibits NO_2 formation so that insufficient oxidizing material is available. In addition, sulfur may result in sulfate formation, which would irreversibly block the filter surface.

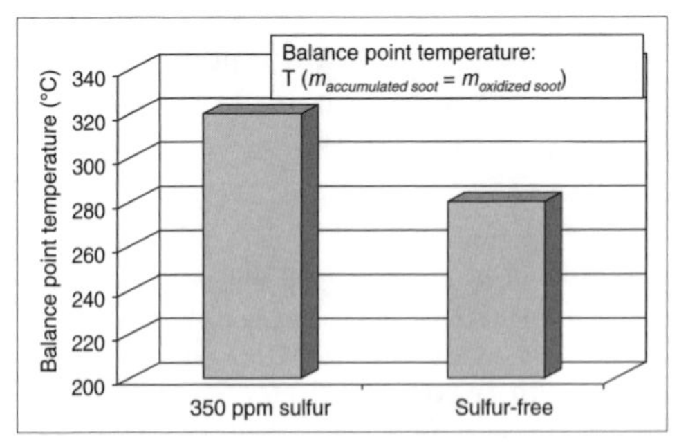

Fig. 5.2-39 Effect of sulfur content on CRT system equilibrium temperature.

5.2.9.3 Denitrogenation (DeNOx)

Since the mid-1980s, a search has been under way for NO_x catalytic converters that can selectively convert oxides of nitrogen using hydrocarbons or CO, even in the presence of excess oxygen. Various methods are known for exhaust gas aftertreatment of oxides of nitrogen. Vehicle-capable processes must produce gaseous nitrogen or gaseous water vapor emissions. The following practical means of reducing NO_x are possible:

- NSCR technology (nonselective catalytic reaction)
- SCR technology (selective catalytic reaction)
- NO_x storage catalyst

Vehicular applications of nonselective catalytic processes attempt to use unburned hydrocarbons (from fuel) and a suitable catalyst to reduce oxides of nitrogen. For this, metal substitution zeolites (e.g., copper zeolites) or noble metal catalytic coatings are employed. The following reactions take place:

$$4NO + 2CH_2 + O_2 \rightarrow 2N_2 + 2H_2O + 2CO_2$$

$$2CH_2 + 3O_2 \rightarrow 2H_2O + 2CO_2$$

NSCR catalysts have been examined extensively. In the test-relevant temperature range, the noble metal catalytic conversion process exhibits its maximum within an extremely narrow temperature window (Fig. 5.2-40). The difference in NO_x conversion rate between the two processes leads to the conclusion that different reaction processes are taking place. These processes have yet to be fully explained. In actual vehicle operation, lower conversion rates than those observed in the laboratory are realized due to higher spatial velocities and water and sulfur content. Conversion rates of up to 30% in the U.S. test procedure and up to 25% in the new European test cycle are observed with fresh catalysts. Moreover, long-term durability of several substituted Cu types is

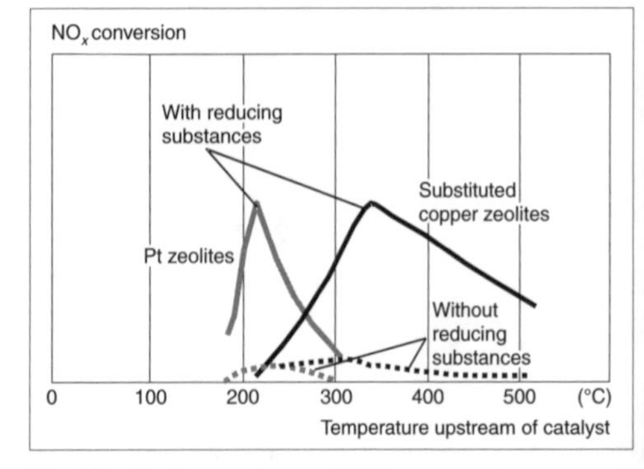

Fig. 5.2-40 Comparison of NO_x conversion rates of various catalysts with and without additional reducing substances (source: Daimler-Benz).

still unsatisfactory. Over time, fuel sulfur content reduces effectiveness by about 15%.

Reduction of nitrogen oxides by means of SCR catalysts is known from stationary powerplant practice. Ammonia (NH_3) is injected upstream of the catalytic converter to act as a selective reducing agent. Hydrogen bound in ammonia reacts with free oxygen and oxygen bound in nitrogen oxides to produce water and molecular nitrogen. In simple terms,

$$4NH_3 + 4NO + O_2 \rightarrow 4N_2 + 6H_2O$$

$$8NH_3 + 6NO_2 \rightarrow 7N_2 + 12H_2O$$

Transferring this technology to motor vehicles is nontrivial; direct application of hazardous NH_3 in gaseous form is unacceptable for safety reasons. Instead, other reducing agents such as urea, $(NH_2)_2CO$, must be used. NH_3 may be extracted from urea by hydrolysis:

$$(NH_2)_2CO + H_2O \rightarrow 2NH_3 + CO_2$$

Vehicle applications require a metering unit, a reducing agent tank, a control unit and hydrolysis/SCR catalytic converter. An air-atomizing nozzle introduces urea solution to the exhaust system as a function of vehicle load. Because of possible time lag in NH_3 generation during actual vehicle operation, active control of the NH_3 supply rate is necessary. Usually, an oxidizing catalyst is added downstream. Efficiency may be increased even further if the NO/NO_2 ratio is adjusted by means of an upstream oxidizing catalyst. Highly active oxidizing catalysts increase sulfate emissions, and therefore particulate emissions, making low-sulfur fuels mandatory for these systems as well. For cold-weather operations below −10°C, the reducing agent must be heated.

Unfortunately, this technology has serious disadvantages in passenger car operation, particularly for the driver. Adapting the system for nonsteady-state operation presents significant challenges. Generally very low exhaust temperatures imply catalysts with corresponding low-temperature activity; these are not yet ready for production. One disadvantage for customers is that it is not possible to carry sufficient urea on board to permit refilling only at scheduled vehicle service intervals. In the new European test cycle, NO_x reduction of 60–70% may be achieved.

If it becomes possible to carry the urea reducing agent aboard the vehicle not as a liquid solution but rather as a solid and to meter the generated ammonia into the exhaust stream, it may be possible to greatly diminish the disadvantages imposed by this additional operating material.

It is likely that NO_x storage catalytic converters will enjoy much greater distribution, as they promise overall higher NO_x conversion rates across a wider vehicle operating range. However, with the current state of the art, their application again predicates sulfur-free fuels.

The operating principle of NO_x storage catalytic converters and the reason for their sensitivity to sulfur are

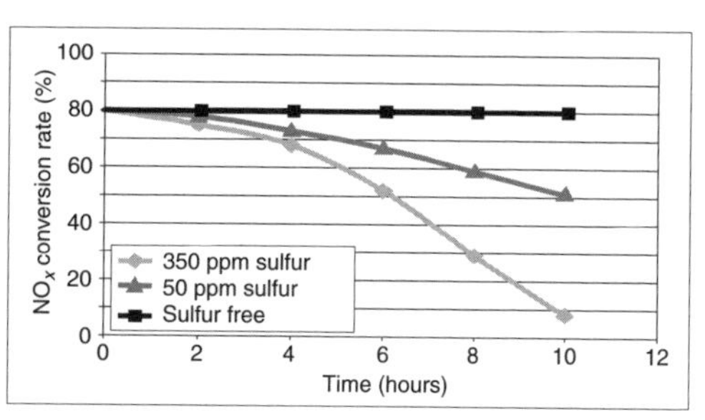

Fig. 5.2-41 Effect of sulfur content on change in NO_x conversion rate over time.

shown in Fig. 5.2-41. The converter consists of two main elements: a combination of noble metals (e.g., Pt/Rh) and the actual NO_x storage material, a member of the alkaloid or earth alkali metals or the rare earths, in the form of oxide or carbonate—here, for example, barium carbonate.

During the NO_x storage phase (Fig. 5.2-42), NO_x from the engine, present mainly as NO in lean exhaust gas, is converted to NO_2 by the catalyst's platinum component. The storage component reacts with this to form barium nitrate. Simultaneously, SO_2 in the exhaust gas is oxidized to SO_3 and also reacts with the barium carbonate to form sulfate. During the NO_x storage process, SO_2 conversion has not one but two negative effects: first, it competes with NO for adsorption space on the platinum, thereby hampering NO_2 formation, and second and more decisively, it consumes barium carbonate, the storage medium.

Once all available barium carbonate is converted to nitrate, the storage catalyst must be regenerated. The engine is briefly run with a rich mixture. The resulting excess of reducing exhaust gas constituents (H_2, CO, and HC) reconvert barium nitrate to carbonate or oxide, while the released NO_x reduces to molecular nitrogen N_2 on the noble metal catalyst. The sulfate exhibits considerably better temperature stability than the nitrate, and in this regeneration cycle ($\lambda < 1$, $T > 650°C$, $t > 2$ min) undergoes little or no conversion. It therefore remains in the catalyst, building up during every lean/rich cycle, and increasingly restricts the amount of barium carbonate available for NO_x storage as well as NO_2 accessibility to storage material.

Storage catalyst applications on diesel engines must overcome major hurdles. The main cause of these obstacles is the fact that diesel engines operate with excess air, and therefore offer no operating conditions for regeneration. The "rich" air/fuel ratio must be achieved by appropriate measures (increased exhaust gas recirculation rate, throttling, additional fuel injection upstream of the catalyst, altered engine fuel injection). The second problem is low exhaust gas temperature, above all in direct-injection engines. This limits storage effectiveness and especially the choice of possible regeneration strategies.

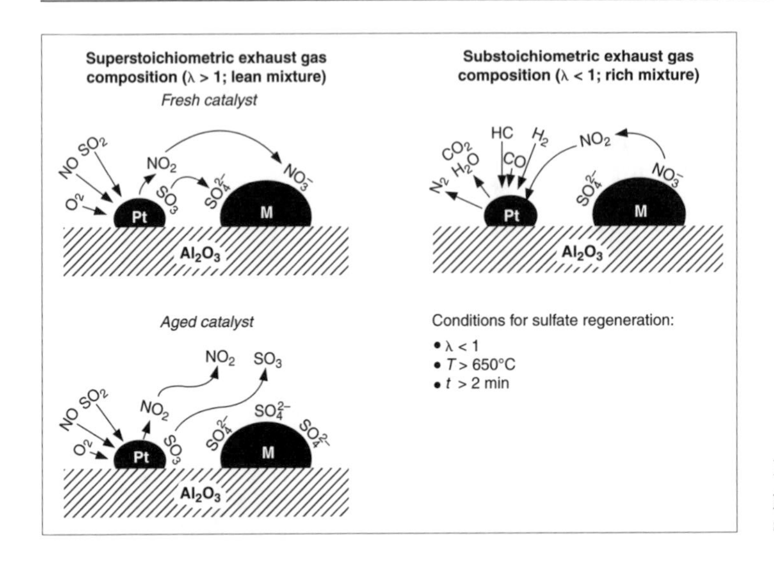

Fig. 5.2-42 NO_x storage process and effect of fuel-borne sulfur.

Initial experiments showed conversion rates (with catalyst in new condition) ranging between 35 and 65%. Mutual interactions with other exhaust emissions and driveability problems may arise. NO_x storage catalytic converters always require an additional oxidation catalyst downstream and sometimes a precatalyst mounted upstream (Fig. 5.2-43).

5.2.10 Simulation of diesel engine combustion

Zero-dimensional single-zone models represent the simplest computational models of nonsteady-state diffusion-controlled diesel combustion processes. In these, the combustion chamber is described by the thermodynamic model of an ideal mixture. There is no spatial resolution of pressure, temperature, or composition. Conclusions regarding pollutant formation are not possible.

In multizone models, a zero-dimensional model is calculated separately for each zone. This permits determination of temperature and composition over time for each zone.

In the case of zero-dimensional models, heat release is derived from experimentally determined equivalent combustion processes. By contrast, modeling spray propagation, vaporization, mixture formation, ignition, and combustion leads to quasidimensional or phenomenological multizone models. These permit actual calculation of combustion processes on the basis of injection history.

The bases for calculating internal engine processes using three-dimensional models are the conservation equations for momentum, mass, materials, enthalpy of turbulent energy, and turbulent dissipation. Modeling the fuel spray begins with events in the nozzle bore, since cavitation within the nozzle plays an important role in stream constriction and therefore stream exit velocity. For fuel spray simulation, the "discrete droplet" model has established itself (Fig. 5.2-44). The injection stream is approximated by statistical drop packets, each of which represents a group of droplets with identical thermodynamic properties and conditions. The injection process becomes calculable only by means of this simplification. Droplet packets undergo all physical processes experienced by droplets within the actual engine.

Atomization models describe shearing of primary

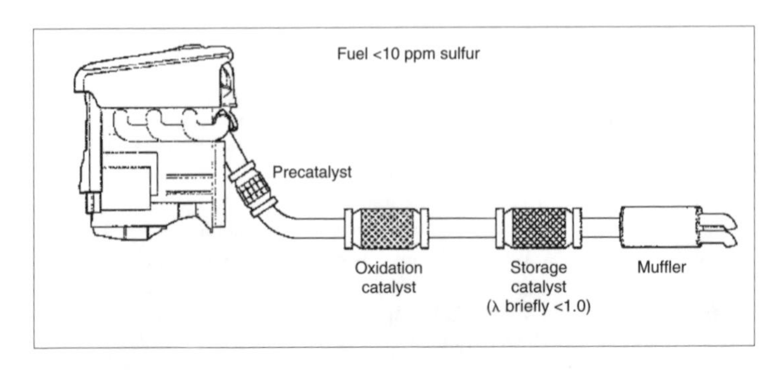

Fig. 5.2-43 NO_x storage catalytic converter system.

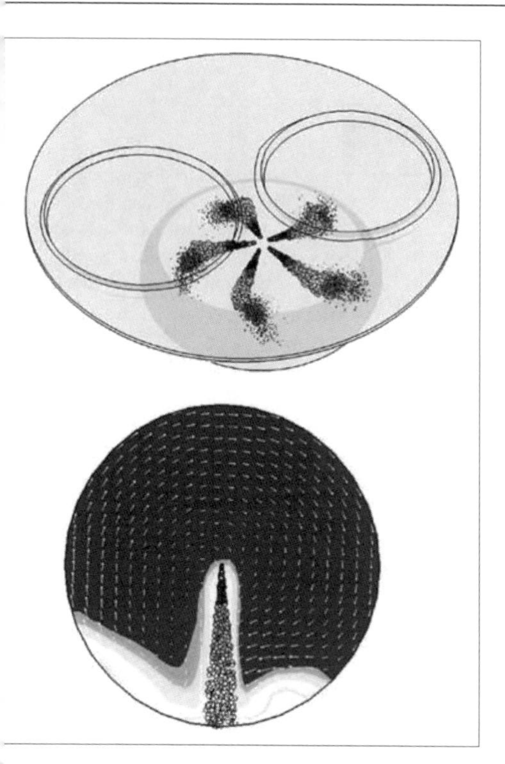

Fig. 5.2-44 Fuel spray of a five-hole nozzle in a direct-injection piston bowl.

droplets from the optically dense, liquid stream core after the stream exits the nozzle bore. "Breakup" models deal with disruption of primary droplets into secondary droplets. Further models provide information on various interactions (droplet to droplet, droplet to vapor phase, droplet to combustion chamber wall). Autoignition reactions begin as soon as fuel is vaporized and the fuel, in its vapor phase, mixes with the oxidizer. Ignition is calculated locally (in each grid cell) using a reduced chemical kinetic model, which defines reaction progress as a function of initial concentrations of fuel, oxidizer, radicals, and local gas temperature. As soon as ignition takes place, the remainder of the reaction process is simulated by a simpler combustion model (Fig. 5.2-45).

In combustion engines, nitric oxide (NO) is formed in three different ways: along with thermal NO formation, rapid formation of prompt (Fenimore) NO may take place in the flame front, where local temperatures may reach 2,800 K. In addition, nitrogen contained in fuel may be oxidized to form NO. In diesel engines, 90–95% of oxides of nitrogen are attributable to thermal NO, 5–10% to prompt NO, and less than 1% to fuel NO. Modeling thermal NO, formed at temperatures above 2,000 K, can be described by the Zeldovich mechanism:

$$N_2 + O \leftrightarrow NO + N_2$$

$$O_2 + N \leftrightarrow NO + O_2$$

$$OH + N \leftrightarrow NO + H$$

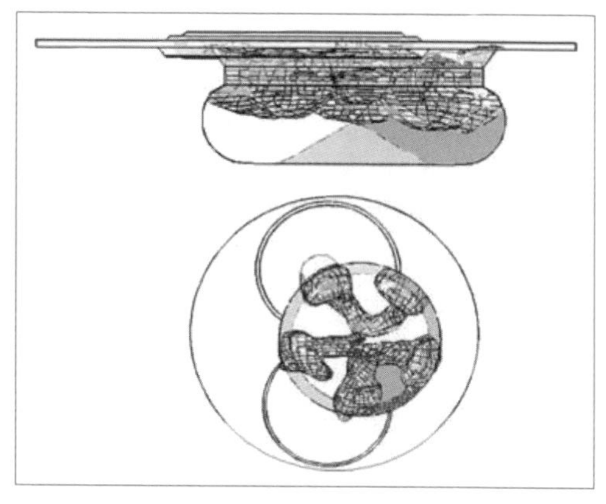

Fig. 5.2-45 Combustion simulation in a direct-injection diesel engine with a five-hole nozzle.

The major factors are local temperature field, local mixing field, and local flow field (residence time of the mass element being considered within the observed volume element).

The exact mechanism of soot formation is not yet fully understood. Therefore, semiempirical models are employed to simulate soot formation. These are dependent on vaporized fuel, oxidizer, pressure field, and temperature field. The current state of modeling is still unsatisfactory.

Sufficiently exact modeling of diesel engine mixture formation, combustion, and emissions formation is not yet possible because vital partial processes are not understood or not yet modeled in sufficient detail. Consequently, phenomenological models remain important. However, in the long term, CRFD (computational reactive fluid dynamics) programs have greater potential.

5.2.11 Diesel fuels

The quality of diesel fuels has a direct effect on emissions of all diesel vehicles independent of their intrinsic emissions levels. An improvement in fuel quality will result in an immediate emissions reduction for all existing vehicles.

A test program conducted by various manufacturers using passenger cars meeting emissions level II has demonstrated the potential benefits of Swedish Class 1 diesel fuel compared to conventional German commercially available diesel fuel (Fig. 5.2-46) [7].

The reduction potential (nearly 50%) is readily apparent.

The EPEFE program, part of the European automotive/oil program [8], systematically examined the effects of various fuel components on emissions behavior of modern engines. Specifications are shown in Table 5.2-6.

Here, too (see Fig. 5.2-47), we can see the great emissions reduction potential for vehicles employing modern technology (EU II and better). These realizations have lent support to the ACEA (Association des

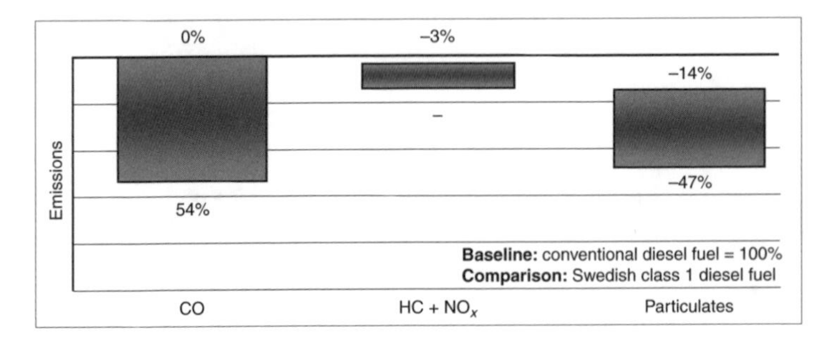

Fig. 5.2-46 Emissions reduction through use of higher-quality fuel (Swedish Class 1 diesel compared to German commercial diesel; 1999).

Table 5.2-6 Specification of fuels examined by EPEFE (D I – D V)

	Fuel scenarios					
	Base	D I	D II	D III	D IV	D V (local)
Cetane	51	53	54	55	58	58
Density (kg/m³)	843	835	831	828	828	820
Aromatics (%)	9	6	4.5	2.2	2.2	0.7
T95 (°C)	355	~350	~345	~340	~340	~310
Sulfur (ppm)	450	300	200	50	50	50

Constructeurs Europeens d'Automobiles) Fuel Charter [8]. The most important realizations are:

1. Cetane number
   An increase in cetane number to at least 58 and cetane index to at least 54 is required for improved ignitability and therefore cold starting properties.
2. Density
   Limitation (minimum and maximum values) is required for engine optimization. Density has a direct effect on exhaust emissions. The range from 820 to 840 kg/m³ is regarded as necessary.
3. Aromatics
   Multiringed aromatic hydrocarbons have a significant effect on diesel particulate formation. Therefore, polycyclic aromatics (PAHs) should be limited to no more than 1%, and total aromatics to a maximum 10% by weight.

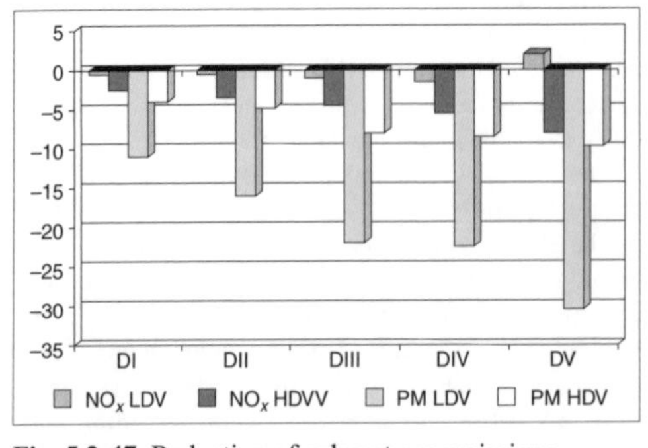

Fig. 5.2-47 Reduction of exhaust gas emissions through use of EPEFE fuels.

4. Sulfur
   Fuel-borne sulfur contributes to particulate formation. Emissions of all vehicles already in the fleet can be reduced through use of low-sulfur fuel grades (see Fig. 5.2-41). Sulfur also affects cold-start behavior of oxidizing catalysts; to achieve full performance, higher exhaust temperatures are required (see Fig. 5.2-48); that is, the cold start phase is extended, thereby increasing cold-start emissions [9].

Along with these effects which influence present-day technology, sulfur reduction is an imperative prerequisite for introduction of future exhaust gas after-treatment technologies such as $DeNO_x$ catalysts. Sulfur poisons these storage systems (Fig. 5.2-42). Incomplete regeneration is only possible with $\lambda = 1.0$ and high temperatures (>650°C). This places greater demands on catalyst material, requires added control complexity, and leads to higher fuel consumption and exhaust emissions. Therefore, a limit of 10 ppm is necessary.

5. Distillation end point
   Distillation end point is a marker for content of high boiling-point hydrocarbons (mainly PAHs), which are difficult to burn and contribute significantly to particulate formation. Therefore, the distillation end point should be lowered to 350°C.

5.2.12 The future of the diesel engine

Conventional internal combustion engines (gasoline and diesel) will remain the dominant concepts for the next 10 to 20 years. The high-speed, high-output diesel

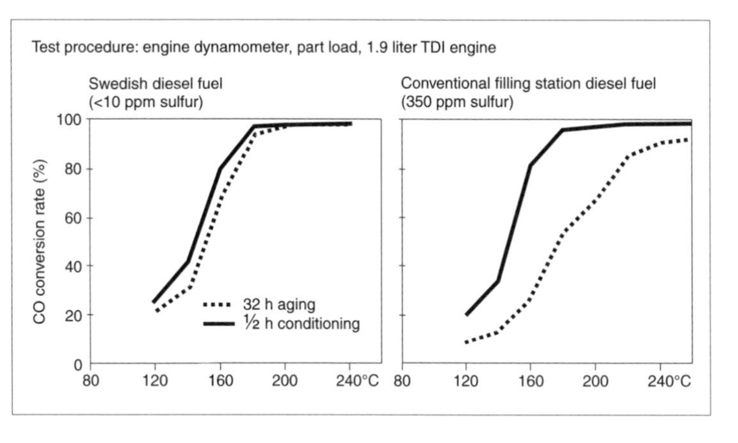

Fig. 5.2-48 Catalytic converter lightoff behavior.

engine has achieved a high state of development and withstands comparison to the gasoline engine. Development in recent years has been marked by

- Emissions requirements
- Improved fuel economy
- Noise standards
- Political discussion
- Cost reduction

Diesel or gasoline?

Compared to gasoline engines, the diesel engine has undisputed advantages in terms of fuel economy (and therefore in CO_2 emissions) and produces higher torque, especially at low engine speeds. All experts are of the opinion that fully usable vehicles with CO_2 emissions below 90 g/km are only possible with diesel engines. In other respects, the diesel engine is inferior to gasoline engines, which operate at higher engine speeds: specific output, combustion noise, and exhaust gas aftertreatment (three-way catalytic converter).

These comparisons have entered the discussion as a result of the potential presented by direct-injection (DI) gasoline engines.

The fuel economy advantage of the diesel engine compared to the gasoline engine will diminish. The DI gasoline concept means that the old gap between prechamber diesel and gasoline engine is restored, as both concepts have experienced about the same fuel economy improvement through introduction of direct-injection technology. However, even in the future, extremely economical vehicles will continue to rely on diesel power.

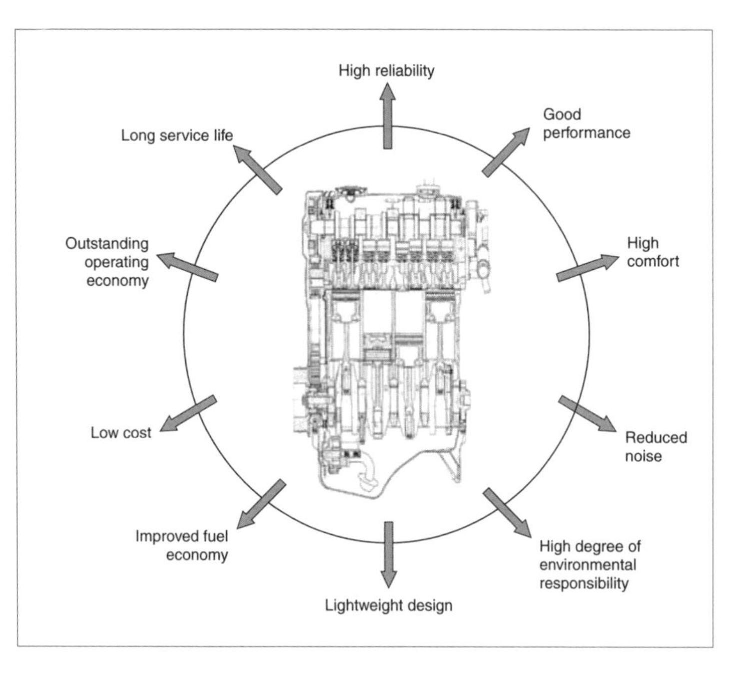

Fig. 5.2-49 Demands on future diesel engines.

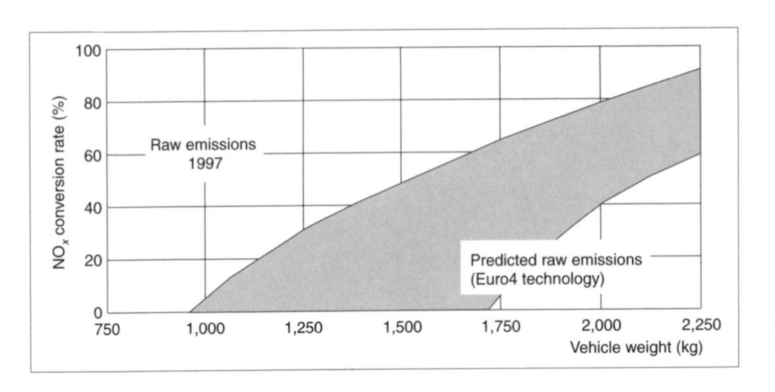

Fig. 5.2-50 NO_x conversion rates required to meet Euro4.

In exhaust gas reduction, DI gasoline engines encounter problems similar to those of diesels, as both DI concepts must be equipped with complex catalyst systems which are not yet fully developed.

At low engine speeds, diesel combustion noise is still greater than that of gasoline engines. From an overall vehicle standpoint, the relative noise levels of these two engine types are converging. In summary, the answer to the question of which combustion concept will earn the right to a future existence may well be "both." The DI gasoline engine will not replace the DI diesel. Moreover, it may be assumed that both concepts will see continued advances in the fields of fuel economy and exhaust gas improvement.

Further potential of the diesel engine

Development of direct-injection diesel engines will continue. Exhaust gas emissions reductions and their assurance in the face of vehicle aging will be the dominant focus of development, along with tighter fuel consumption scenarios. From these we may derive future demands on diesel engine technology (Fig. 5.2-49).

Modern diesel engines are characterized by multivalve technology, high-pressure fuel injection, exhaust turbochargers with variable turbine geometry, dual-spring nozzle holders, controlled exhaust gas recirculation, oxidizing catalytic converters, and fully electronic engine management.

From the present standpoint, exhaust emissions of NO_x and particulates represent the greatest problem of the diesel engine. In order to meet future exhaust emissions standards and other demands such as reduced fuel consumption, considerable effort will be required in engine development, exhaust gas aftertreatment, vehicle mass reduction, and fuel quality improvement.

In meeting these requirements, engine-side measures are in principle preferred over exhaust gas aftertreatment. The greatest potential improvement may be achieved by multivalve technology and continued development of high-pressure fuel injection. In addition to these optimizations, it will be necessary to achieve improved combustion chamber shape and swirl, and reduced friction on all engine components including pistons and piston rings, crankshaft and connecting rod bearings, oil pump, valve train, and auxiliary devices.

Although the VW Lupo 3L represents the first diesel-powered vehicle capable of meeting European Union regulations for 2005 (Euro4) without NO_x or particulate aftertreatment, most vehicle classes will not meet future standards without aftertreatment, especially in view of further fuel consumption reductions (Fig. 5.2-50).

In addition to oxidizing catalysts, NO_x storage catalysts exhibit the greatest promise for future passenger car applications. Depending on vehicle weight and achievable raw emissions, NO_x conversion rates in excess of 80% are deemed necessary. Virtually sulfur-free fuel is mandatory for such installations, as sulfur eliminates NO_x storage capacity. The SCR process is currently regarded as too complex for passenger car applications.

A further possible means of exhaust gas aftertreatment is the particulate filter. The particulate filter system,

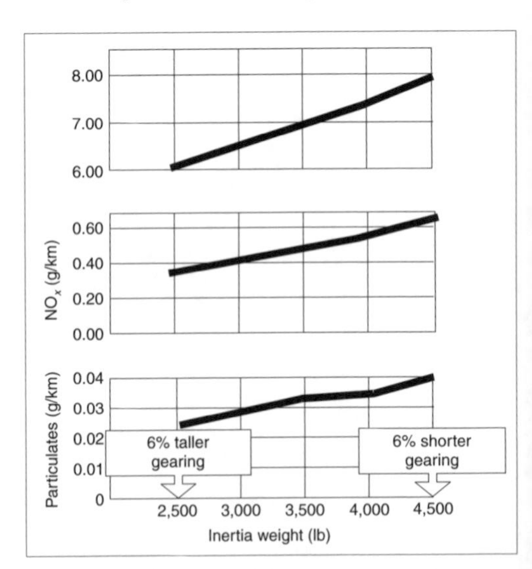

Fig. 5.2-51 Effect of vehicle weight on emissions and fuel consumption (extrapolated for new European driving cycle).

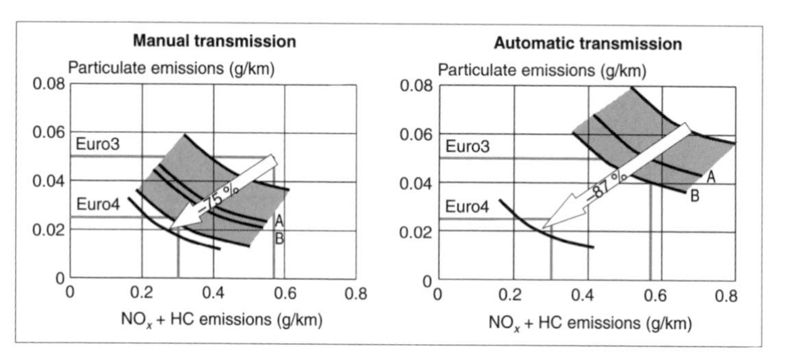

Fig. 5.2-52 Demands on diesel engines for meeting Euro4 standards. Present state of development (A and B represent comparable concepts) with required reductions for vehicles with different transmissions (manual and automatic).

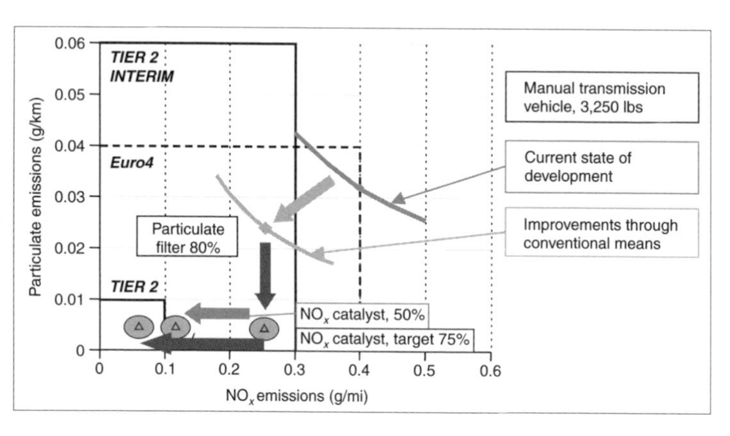

Fig. 5.2-53 Potential diesel engine emissions reduction; example using a Mercedes-Benz A-class vehicle.

with additives acting as a means of regeneration, is also regarded as too complex. Particulate aftertreatment by means of the CRT system appears to offer a more convenient option. However, this, too, requires sulfur-free fuel.

A reduction of only oxides of nitrogen or particulate emissions by means of aftertreatment permits improvement of both components. Which of the possible systems (particulate filters or NO_x catalyst) might reach mass production first and meet demanding customer expectations with regard to reliability and cost remains an open question.

In addition to combustion and friction optimization as well as exhaust gas aftertreatment, vehicle weight should be reduced to achieve further emissions and fuel consumption improvements. The potential for improvement is shown in Fig. 5.2-51.

Introduction of higher quality fuel similar to Swedish Class 1 fuel would not only result in considerable reduction of emissions from the existing diesel vehicle fleet; sulfur-free fuel would also ensure long-lasting effectiveness of new diesel exhaust gas aftertreatment technologies.

Considering these various aspects, the potential for emissions reduction may be represented on a graph of PM vs. NO_x. Fig. 5.2-52 shows the current state of development (A and B are mutually comparable concepts) and reductions, for both manual and automatic transmission vehicles, required to meet European regulations for 2005 (Euro4). According to these estimates, reductions of up to 75% are needed for passenger cars with manual transmissions. By contrast, demands on vehicles with automatic transmissions are even more stringent (87% reduction). In order to remain in the American market, diesel engines must meet Tier 2 standards proposed by the U.S. Environmental Protection Agency. As Fig. 5.2-53 shows, this will require considerable effort and with the current state of the art is only possible, if at all, with small cars using manual transmissions. Particulate and NO_x exhaust gas aftertreatment with about 80% effectiveness would be required.

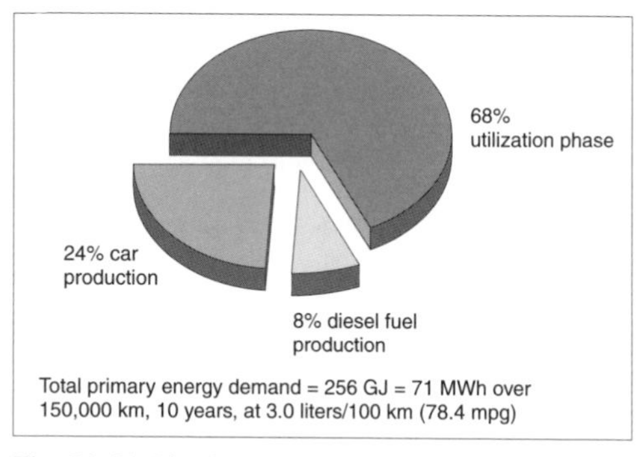

Fig. 5.2-54 Pie chart for total energy consumed by a "three-liter car."

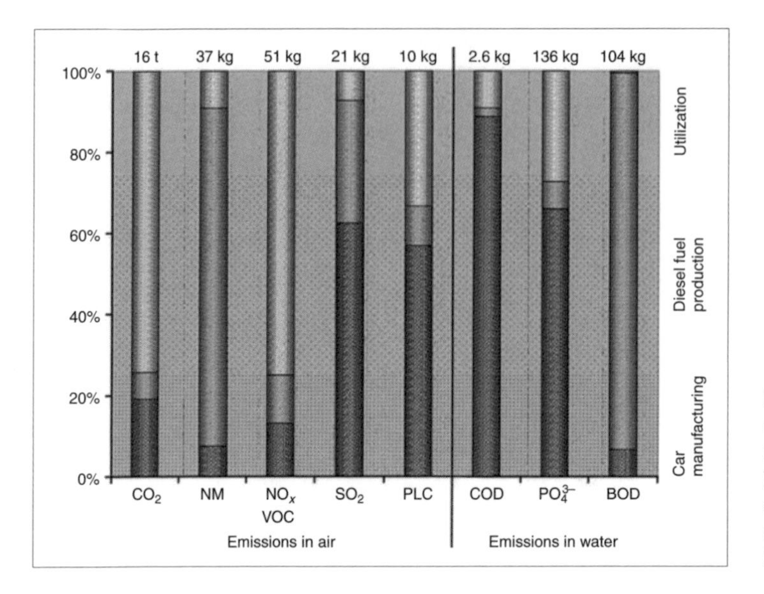

Fig. 5.2-55 Selected emissions for a "three-liter car." PLC, particulates; NM, Nonmethane volatile organic compound; COD, chemical oxygen demand; BOD, biological oxygen demand.

At present, the following disciplines offer potential diesel fuel consumption reductions:

- Combustion, thermodynamics
- Friction power (engine and vehicle)
- Auxiliary devices
- Drivetrain management
- Vehicle weight

Increasingly, environmental friendliness of diesel engines is evaluated as part of a larger picture. If one considers material and energy used in production and operation of a diesel engine, it becomes apparent that seen over the course of its service life, the diesel engine is a relatively environmentally friendly unit. This is clearly demonstrated by the material balance (according to ISO 14040/41) of the first "three-liter car": the VW Lupo 3L is equipped with a 45 kW unit injector diesel engine, the first to meet 2005 EU exhaust standards, while consuming only 2.99 L per 100 km (78.7 mpg), corresponding to 81 g/km of CO_2 emissions [10].

Energy consumption is the sum of fuel consumption, energy required to produce fuel, and energy required to produce materials and manufacture the car itself (Fig. 5.2-54). CO_2 emissions are largely determined by energy production; during the utilization phase (150,000 km, 10 years), this consists of combustion of fuel (Fig. 5.2-55). Hydrocarbons are mostly produced by the oil refining process. Engine exhaust accounts for approximately half of SO_2 emissions during the utilization phase, assuming a fuel sulfur content of 100 ppm. The particulates column encompasses various types of dust created in manufacturing and operation, but does not include tire wear. Emissions in water result from car washing and manufacture of spare parts. Compared to the total amount of wastewater, these emissions are very low.

A favorable overall materials balance is the result of:

- Use of nonhazardous materials (iron, steel, aluminum)
- Manufacturing process
- Manageable environmental requirements
- High materials recyclability
- Long service life
- High thermal efficiency

It may be concluded that the modern diesel engine is powerful, economical, environmentally friendly and conserving of resources. It opens a series of possibilities, leading the way into the next millennium economically as well as ecologically. The diesel engine has earned its place alongside the gasoline engine in the future of the automobile.

References

[1] Diesel, R. *Die Entstehung des Dieselmotors.* Springer Verlag, 1913.
[2] Arcoumanis, C., and K.-P. Schindler. "Mixture Formation and Combustion in the DI Diesel Engine." SAE 972681, 1997.
[3] Pischinger, F., et al. "Grundlagen und Entwicklungslinien der dieselmotorischen Brennverfahren." VDI-Berichte 714, 61, 1988.
[4] Engeler, W., et al. MTZ 58, 11, 1997, pp. 670–675.
[5] Kittelson, D.B. "Engines and Nanoparticles: A Review." J Aerosl Sci 29, 5/6, 575–588, 1998.
[6] Schindler, K.-P. "Why do we need the diesel?" SAE 972684, 1997.
[7] EPEFE (European Programme on Emissions, Fuels and Engine technologies), Final Report, 1998.
[8] World-Wide Fuel Charter, ACEA, Brussels, 1996.
[9] Quissek, F., et al. "The Influence of Fuel Sulphur Content on the Emissions Characteristics of Passenger Cars." 19. Internationales Wiener Motorensymposium, May 7–8, 1998.
[10] Dick, M. "Der 3-1-Lupo—Technologien für den minimalen Verbrauch." VDI-Bericht 1505, Düsseldorf, 1999.
[11] Baum, F. *Luftreinhaltung in der Praxis.* Springer-Verlag, 1988.

General References

[12] Mollenhauer, K. *Handbuch Dieselmotoren.* Springer-Verlag, Berlin, Heidelberg, New York, 1997.
[13] von Fersen, O. *Ein Jahrhundert Automobiltechnik* VDI-Verlag GmbH, Düsseldorf, 1986.
[14] Pölzl, H.-W., et al. "Die evolutionäre Weiterentwicklung des Automobils. Der neue V8 TDI- Motor mit Common Rail." Eurotax International AG, 11/1989.
[15] Hack, G. *Der schnelle Diesel.* Motorbuchverlag, Stuttgart, 1985.
[16] Bauder, R. "Die Zukunft der Dieselmotoren-Technologie." MTZ 59, 7/8, 1998, X-XVII.
[17] Basshuysen, G., et al. "Zukunftsperspektiven des Verbrennungsmotors." 60 Jahre MTZ, Sonderheft, 1999.

5.3 Supercharging

5.3.1 Basic principles and objectives

Supercharging is understood to mean increasing cylinder filling for the purpose of improved power output. This discussion will not consider supercharging by means of resonance effects in the intake tract, a technique which nevertheless may improve power output by as much as 8%. We will instead focus on increased cylinder filling by means of discrete components attached to the engine which employ exhaust gas energy or are mechanically driven by the engine itself. To keep the scope of this section within reasonable limits, we will consider only such systems and components that may be found in significant numbers in roadgoing passenger cars.

Supercharging offers a tool for improving passenger car gasoline or diesel engine power output by as much as 50%. This increased power may be employed for:

1. Improved vehicle performance
2. Reduced fuel consumption in short trips or for "intermediate" engine layouts

Supercharging is usually chosen as a means toward improved vehicle performance if there is insufficient space within the vehicle for installation of a larger displacement (and therefore physically larger) engine. Moreover, it may be economically advantageous to achieve the desired power increase by means of supercharging an existing engine that is already being installed in the vehicle, instead of developing and producing a completely new powerplant. Finally, marketing considerations may call for supercharging. At present, consistent application of supercharging to reduce fuel consumption in everyday operation—which is dominated by short-range operation—has yet to be realized in series production vehicles, although its potential has been demonstrated repeatedly in experimental vehicles. This application also involves increasing engine output. Better performance is achieved by simultaneously reducing engine displacement. Fig. 5.3-1 shows characteristic horsepower curves for a powerful, large-displacement normally aspirated engine and a small-

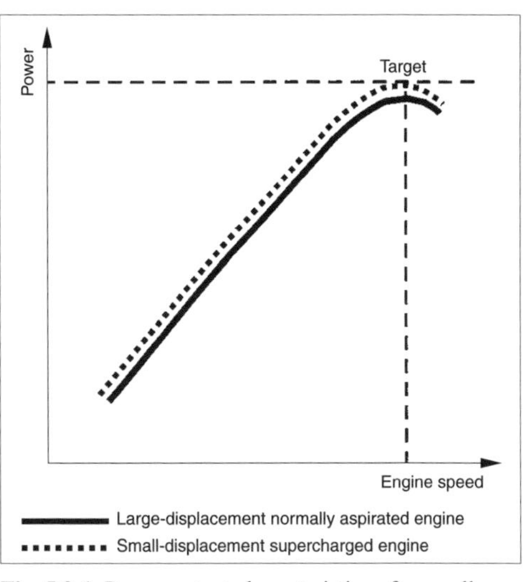

Fig. 5.3-1 Power output characteristics of normally aspirated and supercharged engines.

displacement supercharged engine. The characteristic power curves should be identical in both cases. The exact nature of the supercharging system used to achieve this output curve is immaterial at this point.

Fig. 5.3-2 illustrates the dependence of mean effective pressure on engine speed for both engine concepts. Supercharged mean effective pressure is related to normally aspirated mean effective pressure as the ratio of normally aspirated displacement to supercharged displacement. For illustrative purposes, it is assumed that the lines of constant specific fuel consumption are identical for supercharged and normally aspirated engines. If, in addition, given identical vehicle performance, both vehicles are identical (drivetrain, vehicle configuration), then identical road-load operating points will lie on the same abscissa. As the diagram shows, the operating point

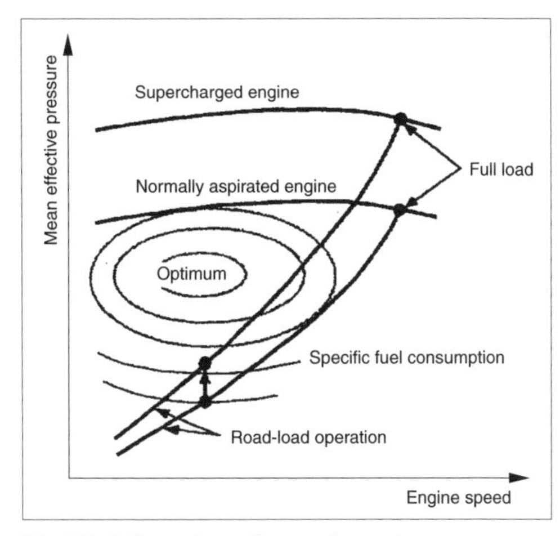

Fig. 5.3-2 Operating points on the engine map.

in the low part-load range moves toward better specific fuel consumption. In this way, fuel economy improvement of as much as 10% may be realized in short-range operation. However, Fig. 5.3-2 also shows that less favorable conditions may be encountered at higher engine speeds.

5.3.2 Systems

To develop power, an engine must liberate energy contained in fuel. In practice, it is a relatively simple matter to increase fuel flow to the engine by means of higher-capacity pumps or fuel injection equipment. However, because fuel can only burn with the supplied air in a certain specific ratio, it is not possible to achieve higher power output by merely increasing fuel flow. It is necessary to provide a supercharging system. And because all supercharging systems draw and compress atmospheric air before supplying this to the engine, the combustion air undergoes undesirable warming. This may result in charge air temperatures of up to 150°C. Because the temperature of air entering the engine at full load should not exceed 50°C, it is necessary to add more or less voluminous charge-air coolers to the vehicle. These charge-air coolers, generally fabricated of light-alloy sheet metal, will not be examined here in greater detail. We should, however, consider the basic engine, an important factor in all supercharging systems. Adding a supercharger is by no means as simple as it first appears. For example, the basic engine must be modified to accommodate higher thermal and mechanical loads. Typically, affected components include crankshaft bearings, pistons, cylinder head gaskets, and exhaust valves. Of course, ignition and mixture formation must be tuned to meet extremely high demands for driveability, reliability, and, above all, exhaust and noise emissions. In practice, this implies lengthy and therefore expensive development. The following sections examine three different supercharging systems; more than 100,000 examples of each of these have reached the market.

5.3.2.1 Exhaust gas turbocharging

The schematic layout of an exhaust gas turbocharger system [1] is shown in Fig. 5.3-3. Exhaust gas drives a radial turbine, which in turn drives a coaxially mounted radial compressor. In diesel engine applications, the throttle shown in Fig. 5.3-3 is deleted. The necessary charge air intercooler in the airflow to the engine is not shown. As the engine should develop high torque at low rpm for good driveability, it is necessary to use small turbines and compressors. Maximum compressor efficiency is about 75%, while turbine efficiency is about 50%. The combined efficiency of 0.75×0.50 implies that at most, 37.5 percent of exhaust energy may be recovered to compress intake air. The requirement for very small turbines in correspondingly small turbine housings leads to exhaust gas choking at high engine speeds. Because higher exhaust gas backpressure has a negative effect on fuel consumption and component life, excess exhaust must be diverted around the turbine by means of a bypass valve (wastegate). Boost pressure is maintained at the desired level by active exhaust bypass control. A new type of exhaust control is covered in Section 5.3.3.1. Mass airflow is monitored and evaluated to maintain the correct air/fuel ratio. The exhaust gas turbocharging system and components of the Audi V6 TDI engine [2] are shown in Fig. 5.3-4. In principle, there is no significant difference between supercharging system components of turbocharged diesel and

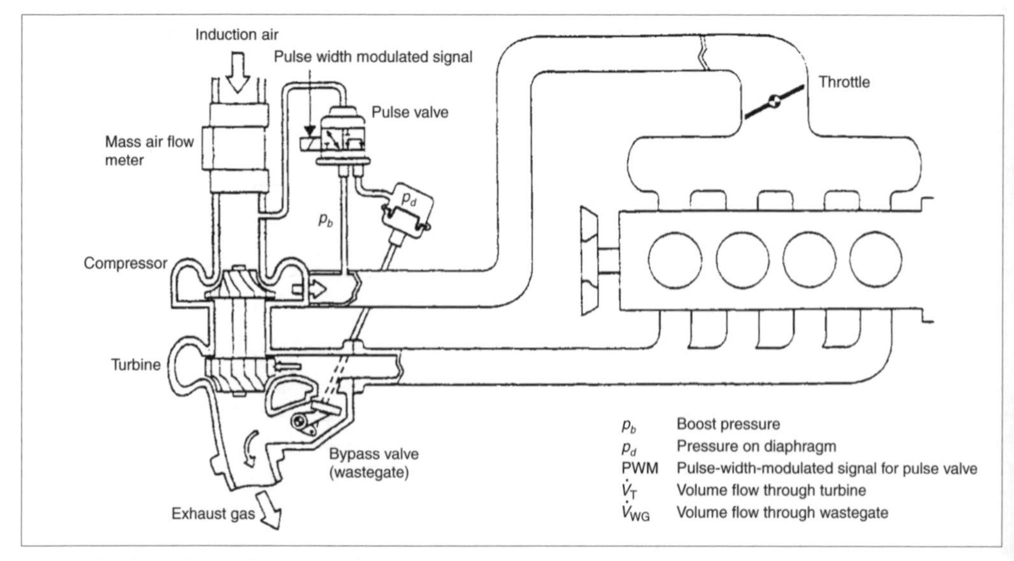

Fig. 5.3-3 Schematic of exhaust gas turbocharging system.

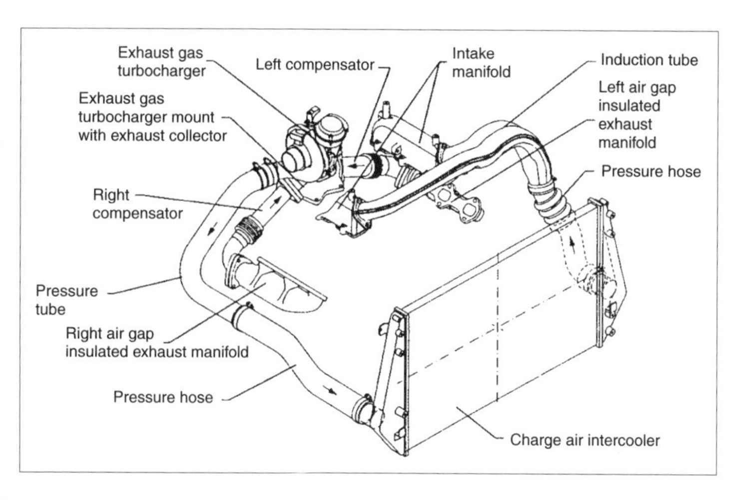

Fig. 5.3-4 Audi A6 TDI exhaust gas turbocharger.

turbocharged gasoline engines. For turbocharged gasoline engines, however, exhaust temperature at rated output is about 300°C higher. Materials considerations limit the operating temperature in that regime to approximately 1,050°C.

5.3.2.2 Pressure wave supercharging

The arrangement of a pressure wave supercharging system [3] is shown schematically in Fig. 5.3-5. A drum-shaped cellular wheel driven by the engine consumes relatively little power. It spins at three to four times engine speed. The ends of the cells are open; as the rotor spins, these are periodically closed by pockets in the exhaust gas and intake air distributors which are contained in the stationary housing elements. Impact of exhaust gas flowing into any given cell results in a pressure wave which compresses and accelerates the intake air charge. As the cellular rotor turns, the end of the cell is opened and charge air is forced into the engine. Thanks to its large number of cells, the cellular rotor supplies charge air with little pressure fluctuation. The arrangement of individual components for Mazda's diesel engine fitted with Comprex pressure wave supercharging [4] may be seen in Fig. 5.3-5a.

5.3.2.3 Mechanical supercharging

A typical mechanical supercharging system is shown schematically in Fig. 5.3-6, which illustrates the component layout of a Volkswagen G-Lader engine [5]. The positive displacement supercharger (compressor) is belt-driven from the crankshaft. The drive ratio between crankshaft and supercharger and the supercharger's air capacity is selected so that the required full-load supercharging airflow rate is always available throughout the entire engine speed range. To minimize drive losses at part load, surplus air is returned to the compressor intake tract by means of a bypass valve. At very low part throttle—for example, at idle—the compressor is not

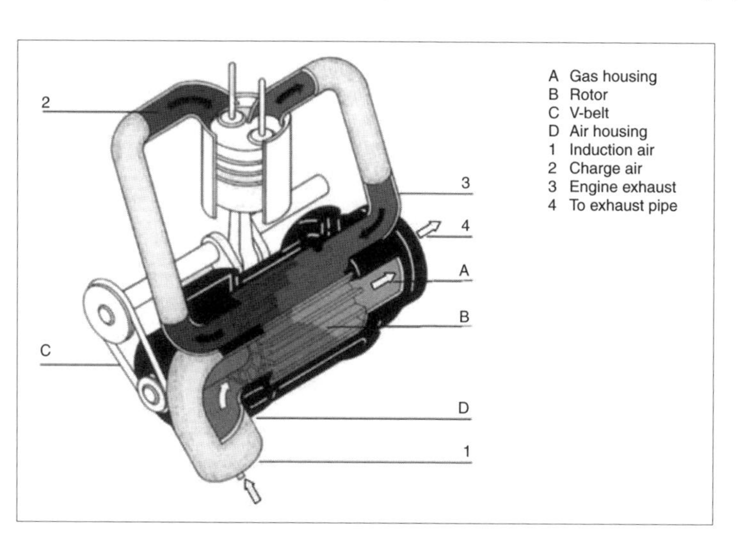

Fig. 5.3-5 Schematic of pressure wave supercharger on engine.

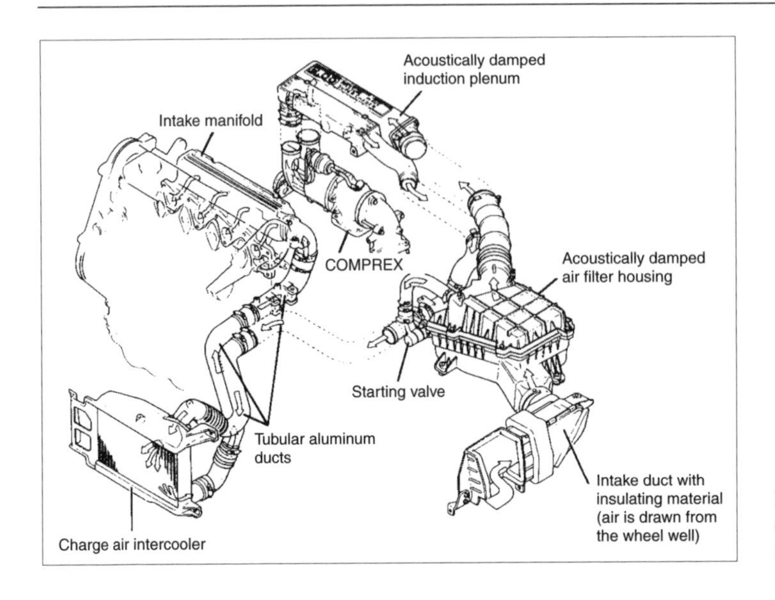

Fig. 5.3-5a Airflow components of a Comprex supercharging system.

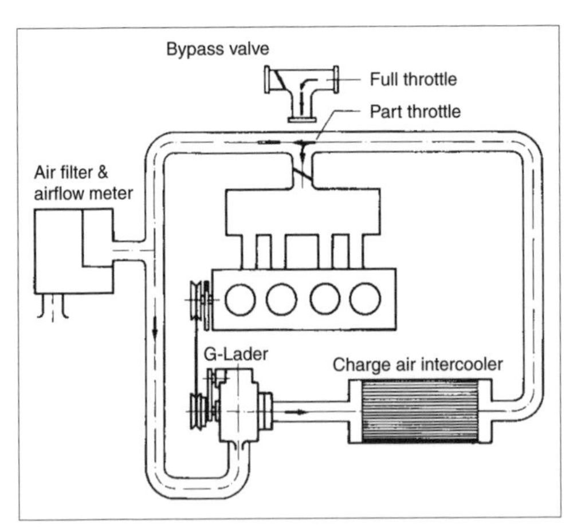

Fig. 5.3-6 Schematic diagram of mechanical supercharging.

needed at all. Under such conditions, power to drive the compressor can be eliminated entirely by means of a supplementary clutch: for example, on the compressor input shaft. The decision whether to include a clutch for the compressor input side is determined by considerations of fuel consumption versus system cost. Activation of the aforementioned bypass valve may be achieved by means of the throttle linkage (wide-open throttle implies completely closed bypass valve) or, much more preferably, by an engine-map–controlled electric or electronic actuator. Except for the presence of a throttle in the case of a gasoline engine, there is no difference between mechanically supercharged diesel and gasoline engines. Specific compressor types may require a muffler at the compressor inlet and/or outlet (not illustrated in Fig. 5.3-6). Compressor location on the Mercedes-Benz M111 engine [6] is shown in Fig. 5.3-7. Fig. 5.3-8 illustrates combustion airflow in this supercharged engine [7].

5.3.3 Components

Performance and manufacturing costs are largely determined by supercharger/compressor properties. The following sections will examine three different production configuration variations.

5.3.3.1 Exhaust-driven turbocharger

Fig. 5.3-9 illustrates a modern exhaust gas turbocharger [8]. In contrast to the exhaust gas turbocharger of Fig. 5.3-3, this does not employ an external bypass valve to control turbocharger boost. Instead, in this case, a system of at least ten movable guide vanes at the entrance of the radial turbine allows processing of the exhaust gas stream in keeping with the engine operating point. This means of controlling exhaust gas flow permits rapid acceleration of rotating parts, thereby realizing rapid boost gain, improved matching of the turbocharger with respect to exhaust gas behavior at mid-range engine speeds, and higher rated power with reduced fuel consumption. The drawbacks of variable turbine inlet geometry are appreciably higher manufacturing costs. Furthermore, application of this concept to very small turbochargers (i.e., on engines in the 1-L class) appears impractical due to relatively high flow losses (the problem of relative tip clearance). Application of variable turbine inlet geometry to gasoline engines with very high exhaust gas temperatures will probably necessitate adoption of ceramic materials for guide vanes. Without exception, turbocharger shafts are carried by journal bearings fed by the engine oil system. Lubricating oil de-

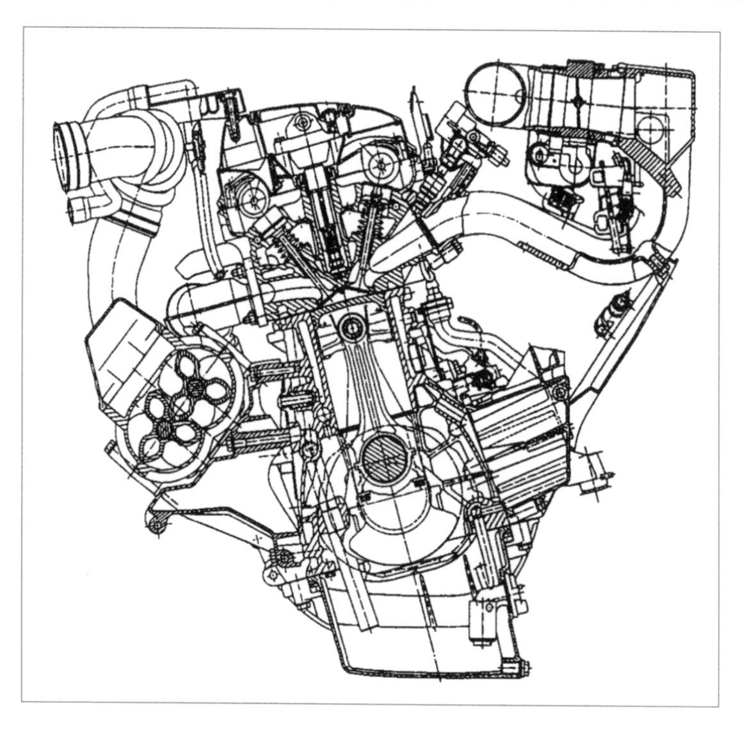

Fig. 5.3-7 Supercharger location on the Mercedes-Benz M111 engine.

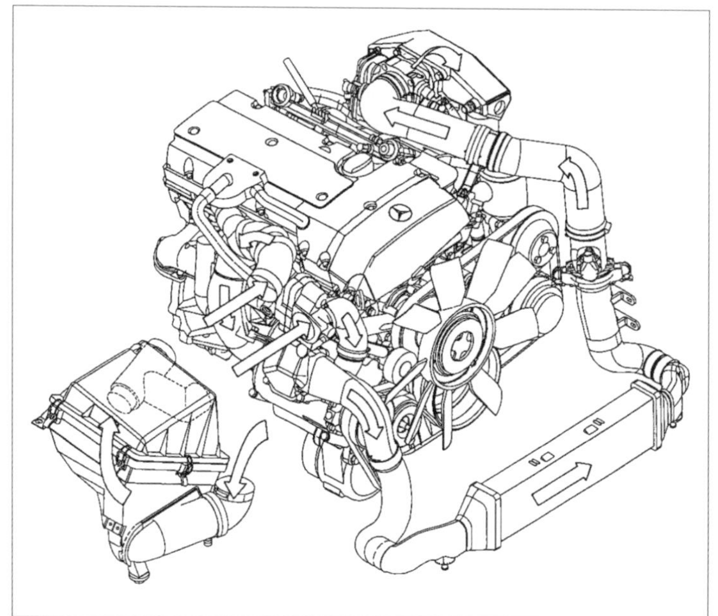

Fig. 5.3-8 Supercharged engine combustion airflow.

mand of a passenger car exhaust turbocharger amounts to approximately 1.5 L/min. Because of the intense exhaust heat found in gasoline engine turbocharger applications, which could lead to coking of the lubricating oil, the bearing housing is sometimes cooled by water from the engine cooling system. The cast radial compressor wheels require high-strength lightweight alloys to withstand operational speeds of up to 130,000 rpm.

The radial turbine wheel is produced by investment casting and consists of a superalloy exhibiting high strength at elevated temperatures. Application of ceramic

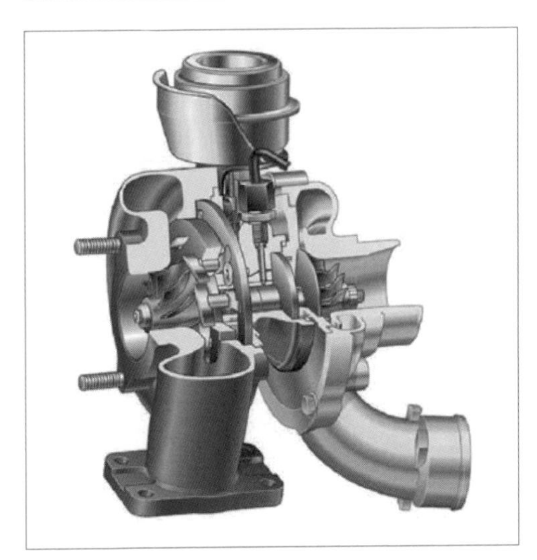

Fig. 5.3-9 Exhaust gas turbocharger with variable inlet geometry turbine.

materials may yield advantages in terms of rotational inertia (turbocharger response) and costs.

A further design solution may be provided by electrically assisted turbochargers; in the event of insufficient exhaust gas energy, an electric motor makes up the shortfall.

5.3.3.2 Pressure wave supercharger

The main component of the Comprex pressure wave supercharger, manufactured by BBC Brown Boveri, now Asea Brown Boveri (ABB) [4], is shown in Fig. 5.3-10. The cellular rotor has two concentric sets of 34 cells, with wall thicknesses of 0.6 mm. The cells are of varying sizes to eliminate howl during operation. The rotor is made of a special alloy which ensures minimal changes in length even with extreme temperature changes. The cellular rotor is a lost wax casting.

Fig. 5.3-10 Cellular rotor of a pressure wave supercharger.

5.3.3.3 Compressor

Compressors for engine supercharging consist of devices operating on the positive displacement principle; exhaust gas turbochargers rely on aerodynamic processes, while pressure wave superchargers employ an exchange of pressure energy between exhaust gas and charge air. A number of different compressor designs are described in [9]. In keeping with the selection criteria for supercharging systems treated herein, two design configurations will be described in greater detail.

5.3.3.3.1 Roots supercharger

Fig. 5.3-11 shows a Roots supercharger made by Eaton Corporation (USA), as installed on the aforementioned Mercedes-Benz M111 engine [7]. This device is of the rotary piston configuration. To reduce pressure fluctuations while pumping charge air, its two rotors, synchronized to one another by gears, resemble a three-leaf clover in cross section. The rated speed of this Roots supercharger is on the order of 10,000 rpm, permitting use of ordinary roller bearings on the rotors. Charge air is pumped between the light-alloy housing and the spaces between the "clover leaves," also of light alloy. The rotors are helical to reduce charge air pressure fluctuations. To achieve the best possible efficiency, these rotors must operate with the smallest possible clearances between one another and their housing. Because of the minimal lubricating oil requirements of roller bearings, gears, and radial shaft seals for sealing the pressure chamber against atmosphere, the Roots supercharger does not necessarily have to be connected to the engine lubrication system. Fig. 5.3-11 also shows an electromagnetic clutch that permits decoupling the supercharger under circumstances in which driving it might be undesirable.

5.3.3.3.2 Scroll compressor

The operation and construction of a scroll compressor will be described using the example of the Volkswagen G-Lader [10]. Fig. 5.3-12 illustrates its functional principle. Inlet *I* is connected to the intake air filter, and exhaust *E* is connected to the charge air intercooler. Inside the housing, a displacement rotor describes a biaxial translational motion; all points on the displacement rotor undergo rotary motion in a single plane. Housing and rotor have similar spiral shapes. The displacement rotor divides the housing chamber into an inner and outer working space. The size of these spaces changes as a function of displacement rotor position. Using the inner working space as an example, the pumping process is illustrated in four phases. In phase 1, the inlet to this working space is open. As the driveshaft and its eccentric turn 90° clockwise, the displacer moves toward the right, enlarging the working space and thereby drawing in air. After another 90° of rotation, the displacer is so close to the housing at the inlet that the inner working space is sealed. During the next phase, the sealing point moves toward the exhaust, forcing the trapped air

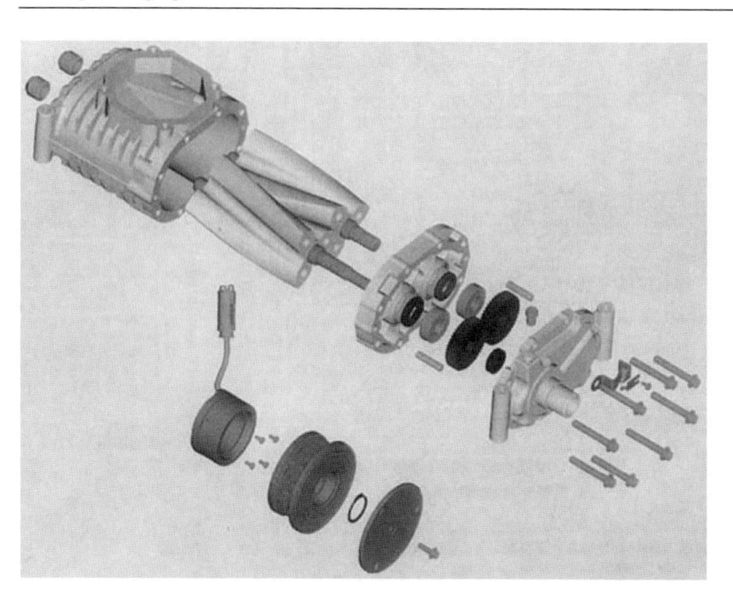

Fig. 5.3-11 Roots supercharger (Eaton Corp.)

volume toward the charge air intercooler. During the last phase of the exhaust process, the intake side of the working space again opens to repeat the process. This same process takes place, with a 180-degree phase difference, in the outer working space. Fig. 5.3-13 shows a partially sectioned G-Lader. The magnesium displacer, with a total of eight working spaces, operates between the two light-alloy housing halves. A secondary shaft running in permanently lubricated roller bearings is driven by a timing belt from the input shaft and ensures the displacer's twin-crank parallelogram motion. The input shaft bearings are fed with engine oil. Because the displacer operates eccentrically on the input shaft, it is necessary to add counterweights to balance inertia forces. Despite its rated speed of 10,000 rpm, because all points on the displacer describe very small circles, relative speeds between displacer and housing are very small, so that face seals may be used to improve pumping efficiency. A further increase in efficiency (particularly in the compressor's lower operating range) is possible by providing the displacer vanes with an elastic coating. Thanks to its operating principle, the scroll compressor has a disc-shaped form factor with relatively short axial length. As Fig. 5.3-14 shows, it achieves good efficiencies [11]. The full-load air demand curve for a 2.3-L gasoline engine is indicated on the graph. The compressor exhibits good efficiencies throughout the engine speed range.

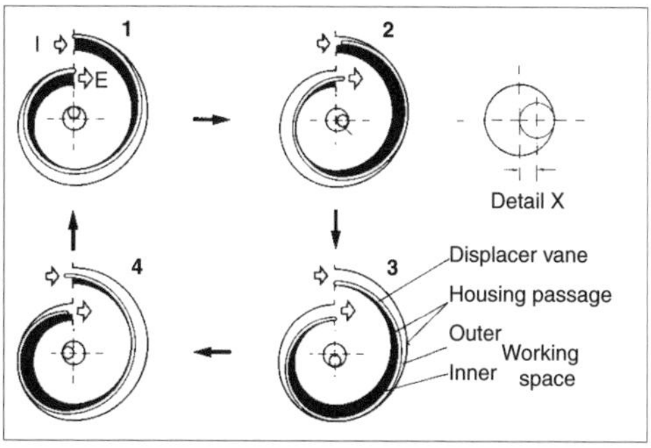

Fig. 5.3-12 Operating principle of scroll compressor.

5.3.4 Application problems

When a particular supercharging system is being selected, manufacturing and development costs are of primary concern. Costs of the supercharger itself play an important role but need not be decisive. It is quite possible that more favorable costs for a given supercharger may be more than offset by other, more expensive components made necessary by a particular choice of supercharging system.

With regard to operation and type of supercharging system, the following will compare exhaust gas turbocharging and mechanical supercharging systems. Several factors must be considered. First, the installation situation must be noted. Can a compressor, with its belt drive plane requirement, be installed in the available space? Or is an exhaust gas turbocharger unavoidable merely from packaging considerations? For airflow corresponding to the given engine output, the volume of a turbocharger is much smaller than that of a compressor, whose rated speed is about 1/10 that of the exhaust turbocharger. Given that costs and installation problems do not automatically exclude compressors from consideration, potential problems with operational behavior of the combined engine-supercharging system and its exhaust gas quality assurance must be addressed. Assuming that the driver is not expected to make allowances for any particular property of the supercharging

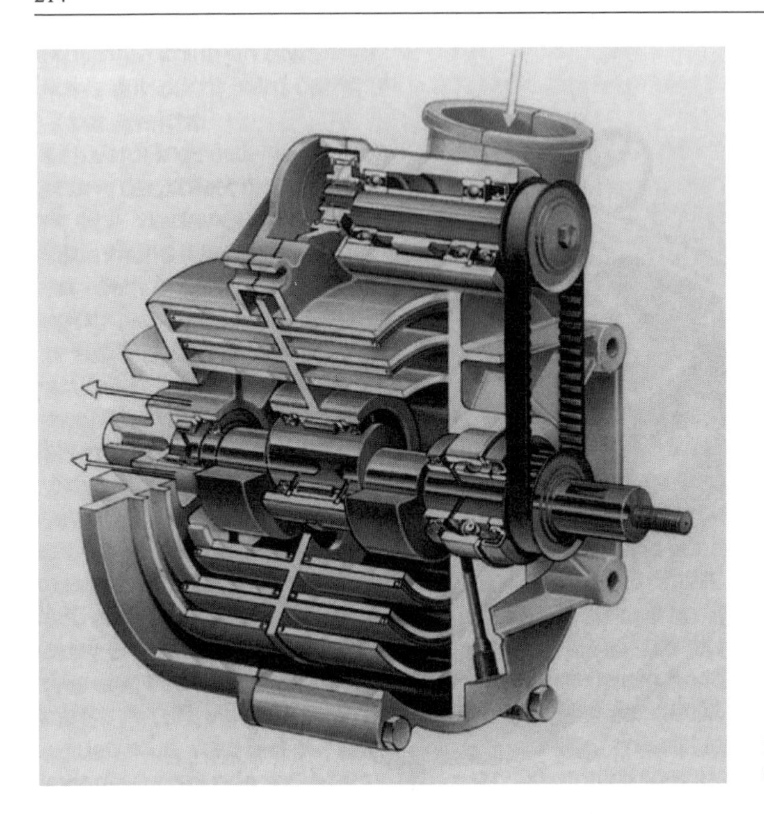

Fig. 5.3-13 Volkswagen G-Lader.

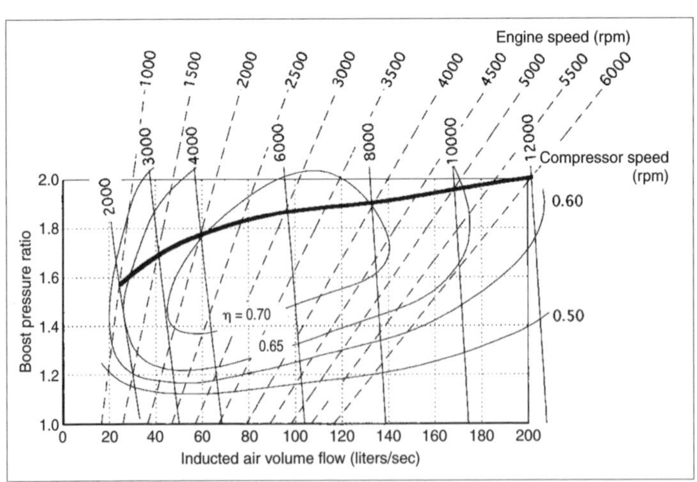

Fig. 5.3-14 Scroll compressor efficiency map.

system, compressor systems have many advantages, including very good performance characteristics at lowest engine speeds and transient response comparable to large-displacement normally aspirated engines. In a compressor system, charge air is always instantly available. There may, however, be noise problems, as charge air pressure fluctuations are appreciably greater than those found in turbocharged systems. Despite all of the advantages offered by turbocharging, the combination of the engine, itself a positive displacement machine, with a free-running flow machine results in less than satisfactory behavior at low engine speeds—above all in small-displacement engine applications. Throttle-controlled gasoline engines are particularly affected when turbocharger speed drops excessively in closed-throttle operation (e.g., coasting, overrun), so that in transitioning back to higher engine loads, delays in pressure buildup may become apparent ("turbo lag"). The diesel engine, which employs mixture regulation instead of combustion air throttling, exhibits markedly better response, as does the unthrottled gasoline engine. Because turbocharger miniaturization leads to much lower overall efficiencies, one basic conclusion becomes apparent: the smaller the base engine displacement and the greater the proportion of low engine speed operation, the more likely it is that mechanical supercharging will come into consideration.

With regard to exhaust gas quality assurance, given a choice between exhaust-driven turbochargers and mechanical compressors, there are no definitive reasons to exclude one system in favor of the other. Turbocharging offers the advantage that at rated engine power, energy extracted from the exhaust stream by the turbine tends to reduce thermal loads on exhaust gas aftertreatment components. Mechanical supercharging systems permit application of low-mass exhaust systems, which will warm up more quickly after cold start. The compressor may also be employed as a secondary air pump for exhaust gas treatment.

Summary

Cost and driveability arguments determine the choice between different supercharging systems. One other factor must be considered in a decision to employ a supercharged engine: the number of cylinders is a dominant marketing argument. A successful supercharged engine should have as many cylinders as possible.

References

[1] van Basshuysen, R., and F. Schäfer. *Shell Lexikon Verbrennungsmotoren*, a supplement to ATZ and MTZ. Chapter 13.

[2] Bach, M., R. Bauder, L. Mikulic, H.-W. Pölz1, and H. Stähle. "Der neue V6-TDI-Motor von Audi. Teil 1: Konstruktion." MTZ Motortechnische Zeitschrift 58, 1997, 7/8.

[3] BBC Brown Boveri. "Der Druckwellenlader Comprex." Brown Boveri Technik 8-87.

[4] Tatsutomi, Y., K. Yoshizu, and M. Komagamine. "Der Dieselmotor mit Comprex-Aufladung für den Mazda 626." MTZ Motortechnische Zeitschrift 51, 1990, 3.

[5] Seiffert, U., and P. Walzer. *Automobiltechnik der Zukunft*. VDI-Verlag, Düsseldorf, 1989, ISBN 3-18-400836-3; published in English as *Automobile Technology of the Future*. SAE, Warrendale, PA, 1991, ISBN 1-56091-080-X.

[6] Hass, A. 5. Aachener Kolloquium "Fahrzeug- und Motorentechnik 1995." MTZ Motortechnische Zeitschrift 57, 1996, 7/8.

[7] Hüttebräucker, D., C. Puchas, W. Fick, and K. Joos. "Entwicklungskonzept des Mercedes-Benz-Vierzylinder-Ottomotors mit mechanischer Aufladung für die C-Klasse." MTZ Motortechnische Zeitschrift 56, 1995, 12.

[8] Willmann, M. "Aufgeladene TDI-Motoren von Volkswagen." 6. Aufladetechnische Konferenz 1./2, Dresden, October 1997.

[9] von Fersen, O. (Editor). *Ein Jahrhundert der Automobiltechnik, Personenwagen*. VDI Verlag, ISBN 3-18-400620-4, 1986.

[10] Wiedemann, B., H. Leptien, G. Stolle, and K.-D. Emmenthal. "Development of Volkswagen's Supercharger G-Lader." SAE 860101.

[11] Spinnler, F.W., and R.W. Kolb. "ECODYNO, ein neuer Kompressor nach dem Spiralprinzip." 6. Aufladetechnische Konferenz, 1/2, Dresden, October 1997.

5.4 Drivetrain

5.4.1 Overview

5.4.1.1 Introduction

In a vehicle drivetrain, the transmission is no less important than the combustion engine. Only with proper interplay of both units and by selection of suitable components is it possible to drive off from a standstill, move forward or in reverse, and overcome varying driving resistances. The vehicle drivetrain concept encompasses all driveline components which transmit engine output to the driven wheels.

If we consider the passenger car, we find manual and automatic transmissions represented in varying degrees depending on geographical region. Continuously variable transmissions are in their infancy, but they show increasing promise; see Table 5.4-1. Automated manual transmissions are not relevant in the period under consideration, 1990–1998.

Essential demands on drivetrain components include simple operation, minimal losses, small installed volume and low weight, and high reliability and life expectancy, as well as low cost.

Table 5.4-1 Distribution of passenger car transmission types

	Total passenger car production (millions)		Manual transmissions (%)		Automatic transmissions (%)		Continuously variable transmissions (%)	
Region	1990	1998	1990	1998	1990	1998	1990	1998
Worldwide	35.5	37.7	58.1	54.7	41.6	44.7	0.3	0.6
North America	7.7	8.0	14.5	12.3	85.5	87.6	0	<0.1
Japan	9.9	8.1	34.6	15.5	64.6	82.4	0.8	2.1
Western Europe	13.1	14.8	90.7	86.7	9.0	13.1	0.3	0.3
Germany	4.7	5.3	83.3	76.4	16.7	23.5	<0.1	0.1

5.4.1.2 The tasks of a transmission

Worldwide, gasoline and diesel engines have come to dominate vehicle propulsion systems. This situation will continue for the foreseeable future. Both propulsion systems are characterized by the following:

- They only operate across a specific engine speed range, between idle and maximum rpm. Vehicle propulsion is not possible from a nonoperational (nonturning) engine.
- Engine torque alone is inadequate for rapid acceleration or overcoming significant grades. This might be possible given extremely overdimensioned engines, but the resulting powerplant would be far too large, too heavy, and uneconomical.
- Such combustion engines only turn in one direction. Forward and reverse motion would not be possible.

The transmission compensates for these engine weaknesses, and with its various gear ratios (or with continuous variability) combined with some sort of drive-away (startup) element, the transmission ensures that tractive effort is matched to available torque. Transmissions offer suitable gear ratios for acceleration, climbing grades, and economical cruising, and they also make reversing possible.

The driving condition of a vehicle is described by the so-called driving equation. This establishes an equilibrium between the driving resistance to be overcome and the driving torque available from its drivetrain. It should be noted that in the case of speed changes, the quasistatic torque developed by the engine is reduced or increased by its rotating masses. The overall drive ratio (i/r) and overall driveline efficiency η_{ovl} is interposed between transmission input and driven wheels. Therefore, the driving equation may be written as

$$\underbrace{(T_{eng} - J_{eng} \cdot \dot{\omega}_{eng})}_{\text{Drive torque}} \, {}^{i}\!/\!_{r} \cdot \eta_{ovl} = F_d = \underbrace{m \cdot g \cdot f \cos\alpha}_{\text{Rolling resistance}}$$

$$+ \underbrace{c_d \cdot (\rho/2) \cdot A \cdot v^2}_{\text{aerodynamic drag}} + \underbrace{F_B}_{\text{brake force}} + \underbrace{m \cdot g \cdot \sin\alpha}_{\text{grade drag}} + \underbrace{m \cdot \kappa \cdot a}_{\text{acceleration drag}}$$

The drive ratio i/r is the ratio of tractive effort at the wheels to torque at the transmission input shaft. If there is no slip in the drivetrain, i/r represents the ratio of angular velocity at the transmission input shaft to vehicle speed.

The drive ratio i/r is composed of the transmission's variable gear ratios and constant ratios within the drivetrain, such as the final drive ratio.

Acceleration drag incorporates not only vehicle mass, but also, by inclusion of the factor κ, rotationally accelerated masses between transmission input and road wheels. κ is defined as

$$\kappa = 1 + \frac{1}{m}\sum^{n} J_n \cdot i_n/r$$

In order to lay out a transmission, it is necessary to establish limits for the drive ratio (i/r). The (numerically) highest ratio $(i/r)_{max}$ is used for climbing steep grades, rapid acceleration, and, generally, for starting from rest. $(i/r)_{max}$ may be calculated from the road load equation, excluding aerodynamic drag and assuming good traction between tires and road surface. Maximum grade climbing ability is calculated by setting acceleration equal to zero in the equation; maximum acceleration, on the other hand, is calculated by setting grade equal to zero.

The calculated values $(i/r)_{max}$ for these two criteria differ because in maximum acceleration, the engine's rotating masses must be taken into account. Whether a vehicle is front-wheel, rear-wheel, or all-wheel drive also plays a role.

Questions to be answered in establishing the (numerically) lowest gear ratio $(i/r)_{min}$ are:

1. Is the vehicle expected to reach its terminal velocity on level road in that gear (application of maximum engine output, grade = 0, acceleration = 0)?
2. Are good hillclimbing ability and acceleration reserves in top gear important, or should fuel consumption be as low as possible?

The conditions cited under the first question permit calculating the overall ratio $(i/r)_0$ for maximum speed from the road load equation.

Deviations from $(i/r)_0$ are defined by the so-called overdrive factor, φ:

$$(i/r)_{min} = \varphi \cdot (i/r)_0$$

φ can assume the following values (see Fig. 5.4-1):

- $\varphi = 1$: Terminal velocity = maximum speed in gear, moderate climbing and acceleration reserves, relatively high fuel consumption.
- $\varphi > 1$: Climbing and acceleration reserves increase, fuel consumption increases, transmission spread is reduced.
- $\varphi < 1$: Climbing and acceleration reserves decrease, fuel consumption drops, transmission ratio range is increased.

The transmission ratio range is defined as

$$I = (i/r)_{max}/(i/r)_{min}$$

Ideally, this range would be covered by some form of continuously variable transmission. However, it has become common practice to employ transmissions of various configurations containing sets of discrete gears, which offer the advantages of compact dimensions, light weight, high efficiency, and manageable technology. At present, manual transmissions for passenger cars overwhelmingly use five discrete gears; sporty vehicles are often equipped with six-speed transmissions. Passenger car automatic transmissions typically use four or five gears; six-speed automatic transmissions are available, and seven-speed automatics have been announced. Gear staging is generally progressive: that is, the lower gears (numerically high gear ratios) exhibit larger "jumps" between gears. In higher gears, which see more service,

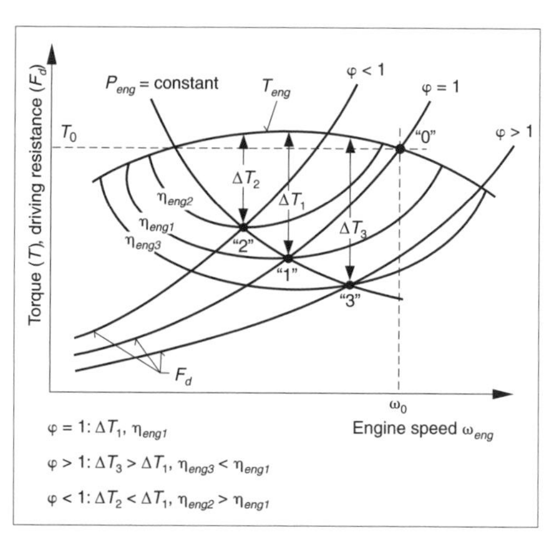

Fig. 5.4-1 Engine map and road load curve as a function of overdrive factor.

jumps are considerably smaller. Fig. 5.4-2 illustrates the interplay between engine and vehicle in the case of a progressively staged five-speed automatic transmission.

5.4.1.3 Drivetrain construction and components

The following components are necessary in order to transmit engine torque to the driving wheels:

— Mechanical clutch for starting from rest and separating engine and drivetrain, or a hydrodynamic torque converter
— Transmission, either with discrete gear ratios or continuously variable

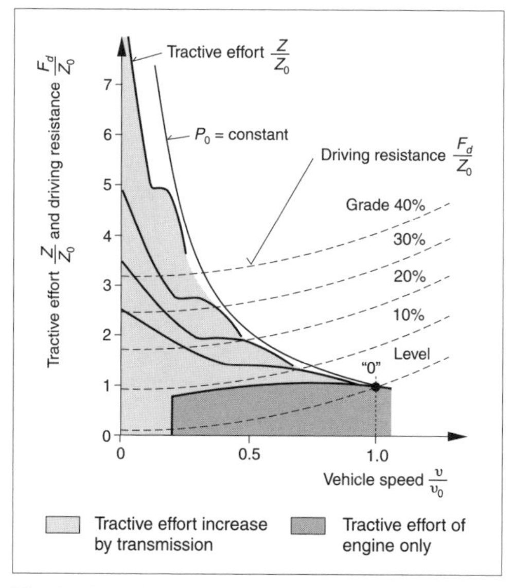

Fig. 5.4-2 Performance map of five-speed automatic transmission.

Table 5.4-2 Drivetrain configurations

Configuration	Engine location and orientation	Driven axle(s)
Standard	Front, longitudinal	Rear
Front	Front, longitudinal or transverse	Front
Mid	Mid (between axles)	Rear
Rear	Rear (behind rear axle)	Rear
All-wheel	Front, longitudinal or transverse; also mid or rear	Front and rear

— Final drive with fixed gear ratio
— Differential(s) for driven axle(s) and possibly for all-wheel drive
— Shafts and joints to transmit power through the drivetrain to the driven wheels
— Actuating elements for clutch, transmission, and possibly all-wheel drive activation

Table 5.4-2 lists the possible passenger car drivetrain configurations.

5.4.1.4 Final drive

Final drives are tasked with matching the differing rotational speeds of engine and wheels. Actual final drive design details depend upon engine location and orientation. Transverse engines require a cylindrical gearset; longitudinal engines require hypoid bevel gearing.

The fixed gear ratio i_f of a final drive is calculated from

$$i_f = (i/r)_{\min}/(i_g)_{\min}$$

where $(i_g)_{\min}$ is the lowest speed ratio of the associated discrete gear or continuously variable transmission.

Typical values for final drive ratio fall in the range of approximately 2.6 to 4.5.

5.4.1.5 Differential

Differentials are employed to compensate for unequal wheel rotational speeds of driven axles encountered in cornering or caused by differing tire dynamic rolling radii (manufacturing tolerances, tire pressure differences). A differential simultaneously prevents forced slippage while transmitting torque.

Differential and final drive are combined to form the final drive unit. Fig. 5.4-3 illustrates a rear-axle final drive unit with ring and pinion gears, bevel gear differential, housing, and input and output flanges. Bevel gear differentials have become the most widespread type. The side pinions ("spider gears") of a differential act as "balance beams" to bring the torque on both driven wheels into equilibrium.

On road surfaces with dissimilar traction for the driven wheels, the wheel with the lowest friction coefficient (modified by the balance beam effect) deter-

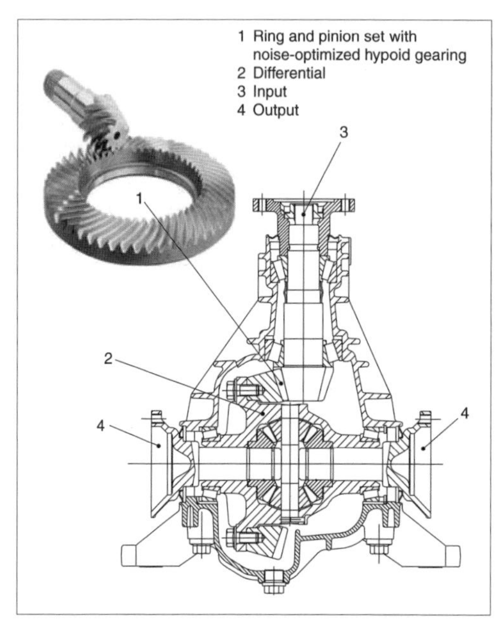

Fig. 5.4-3 Passenger car final drive unit (rear-wheel drive; ZF HAG 210).

mines the maximum forward thrust available to the vehicle. In the event of excessive drive torque, this wheel will spin. This loss of traction may be reduced by means of a limited-slip differential. Distinction is made between locking and friction limited slip. Locking differentials rigidly connect both halfshafts, resulting in forced slip and binding in cornering.

Friction limited slip differentials permit speed differences between driven wheels. Their locking moment is either proportional to drivetrain torque (friction generated by clutch packs, by worm gears, or by helical gears) or dependent on wheel speed differences (viscous limited slip).

Electronically controlled locking differentials were introduced in the past but have not enjoyed much market success.

Increasingly, limited slip differentials are being replaced by electronically controlled brake application. Brake activation on a spinning wheel can generate torque which ensures traction of the nonspinning wheel.

5.4.1.6 All-wheel-drive transfer case

All-wheel drive improves traction and vehicle safety, especially on wet or slippery surfaces. Transfer cases are either manually selectable or permanently engaged (see Section 5.5, Fig. 5.5-2).

The vehicle's front and rear axles are driven from a centrally located differential, with 50/50 or asymmetrical front/rear torque split.

The transfer case is often configured with planetary spur gears and with locking differential (in the case of some off-road vehicles), friction limited slip, or open differential action. In the latter case, intelligent control of individual wheel braking may be used to achieve limited slip effects on axle differentials as well as on the center differential. Some all-wheel-drive passenger cars dispense with the central differential entirely, relying instead on a viscous coupling or an electronically controlled laminar clutch.

5.4.1.7 Articulated shafts

Articulated shafts bridge the spatial gaps between transmission and final drive (also called driveshafts or propshafts) and between transaxle and driven wheels (halfshafts). If axles are not kept in precise alignment, they must be connected by means of articulating joints. In addition, provision must be made for length changes, to compensate for manufacturing tolerances, elasticity, and kinematic effects. Compensation for angular movement is accomplished by means of Hooke joints (Cardan joints), constant velocity (CV) joints, or flexible disc joints.

Halfshafts of front-wheel-drive vehicles are subjected to especially demanding service. They must not only compensate for engine and wheel movement but also be capable of extreme angularity imposed by steering movements. The predominant constant velocity joint configuration, based on design work by Hans Rzeppa, is in the form of fixed-length universal joints or, alternatively, in the form of tripod (tripot), Rzeppa, or cross groove universal joints with provision for axial movement (see Fig. 5.4-4).

Driveshafts for rear-wheel-drive vehicles are either a single piece or, for longer drivelines, composed of several sections located by intermediate bearings.

Details regarding drivetrain components and their mission may be found in the technical literature [1, 2, 3, 4].

5.4.1.8 Vibratory system

A vehicle drivetrain represents a vibratory system. If its characteristic frequencies are resonantly excited, vibratory amplitudes may, depending on damping (or lack

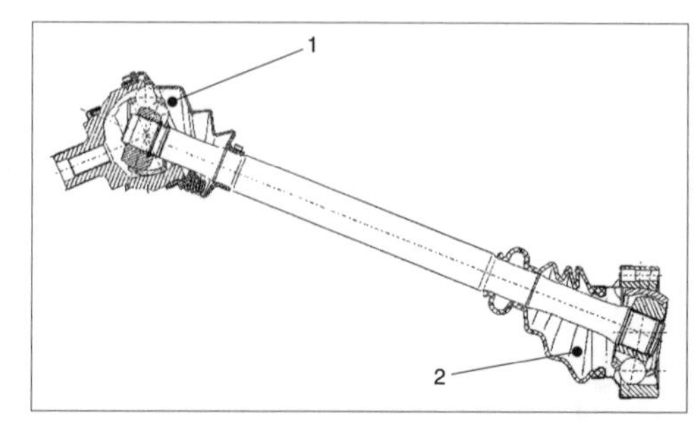

Fig. 5.4-4 Constant-velocity halfshaft for front-drive passenger car. 1, outboard (wheel side) constant velocity joint; 2, inboard (differential side) constant velocity joint.

thereof), assume extreme values. The consequences are service life problems and undesirable noise (e.g., body rumble, rattles).

During the design process, the entire drivetrain should always be regarded as a system.

Drivetrain vibration may appear in the form of torsional vibration or bending vibration of articulated shafts as well as vibration of the engine-transmission unit (engine crankcase, transmission case) in bending.

The determining factors for resonant frequencies are moments of inertia, masses, and torsional and bending stiffness of individual drivetrain components.

Excitation is typically caused by the combustion engine (number of cylinders, configuration) and the driveline.

Measures to avoid torsional vibration problems include:

- Damping of the resonance path as well as decoupling of combustion engine and drivetrain in the supercritical region by means of torsional dampers (clutch disc, in the case of conventional manual transmissions), dual-mass flywheel, hydraulic torque converter (in some cases, lockup torque converter with slip control)
- System tuning by means of torsional dampers
- Attempts to limit angularity of articulated shafts (bearing in mind hardware suppliers' recommendations)

Measures to avoid bending vibration problems include:

- Short articulated shafts (if necessary, two shafts with intermediate bearing)
- Stiff engine and transmission housings

Details regarding drivetrain vibration may be found in the technical literature [5, 6, 7].

5.4.2 Starting components

5.4.2.1 Clutches

Manually or automatically actuated dry plate friction clutches, multiplate oil bath clutches, and electric powdered metal clutches are suitable for achieving smooth takeup and interruption of transmitted power. Manual transmissions require very rapid and complete interruption of power flow with the smallest possible rotational inertia. This can only be achieved by the dry-plate clutch. Continuously variable transmissions permit other, alternative clutch configurations.

The dry-plate friction clutch (Fig. 5.4-5, hereafter simply the clutch) consists of a pressure plate and diaphragm spring attached to the engine flywheel, a clutch disc free to slide on an axially splined transmission input shaft, and a release bearing (also known as throwout bearing) that transfers clutch release movement from nonrotating actuating elements (release fork, cable) to the pressure plate [8]. Dual-disc clutches no longer enjoy any market significance. A torsional damper is necessary between the engine and transmission, integrated either in the clutch disc or the engine flywheel (dual-mass flywheel).

Transmitted torque

The clutch must be capable of withstanding maximum engine torque plus dynamic deviations (spikes) beyond the nominal maximum. Transmitted torque is the product of the clutch pressure (diaphragm spring force minus spring "set," return spring force, centrifugal force of diaphragm spring fingers, friction including friction of the clutch disc on the transmission input shaft, and losses in the clutch actuation system), clutch disc lining coefficient of friction, mean radius of the clutch friction disc, and the number of friction surfaces (single disc clutch = two surfaces) [2]. By modulating pressure plate pressure, the clutch becomes a torque converter in which torque is transmitted without modification. Torque transmission with slippage warms the clutch disc and results in wear of the clutch lining. German vehicle type certification requires maximum clutch loading consisting of five starts from standstill, within five minutes, on a 12% grade, at rated maximum gross vehicle weight (including trailer). This will result in a reduction in friction coefficient (fading) and, because of the temperature differential between friction side and back side of the pressure plate and flywheel, conical deformation of these components, which reduces both effective contact area and effective radius. Spring action of the clutch disc itself partially compensates for this deformation. Furthermore, this spring action also reduces clutch lining wear, reduces self-induced frictional oscillations (chatter, perceptible as vehicle oscillations at about 10 Hz), and contributes toward jerk-free starts. In the clutch, in contrast to the brakes, heat storage capacity is more important than cooling. The bandwidth of practical designs is shown in Fig. 5.4-6.

In the standard design, release pressure is composed of clutch pressure divided by the mechanical advantage of the diaphragm spring (i = 3–4). Alternatively, an auxiliary spring, working against the diaphragm spring, can be used to reduce clutch release effort. A prerequisite for this is a diaphragm spring location independent of lining wear, which in turn necessitates a mechanism for automatic lining wear compensation.

The clutch disc must be capable of withstanding angular and axial misalignment of crankshaft and transmission input shaft, as well as flywheel wobble caused by crankshaft bending. Failure to compensate for these will result in wear of the clutch disc and input shaft splines.

The most important aspects of friction linings are friction coefficient, bursting speed, wear of lining and mating surface, distortion due to thermal loading, weight and rotational inertia, ability to modulate torque rise, and tendency to chatter or squeal. Coefficients of dynamic friction (during clutch slip) and static friction (clutch fully engaged) are practically identical, between 0.3 and 0.4 under normal circumstances, dropping to 0.2 in fading conditions. Linings consist of yarns (primarily

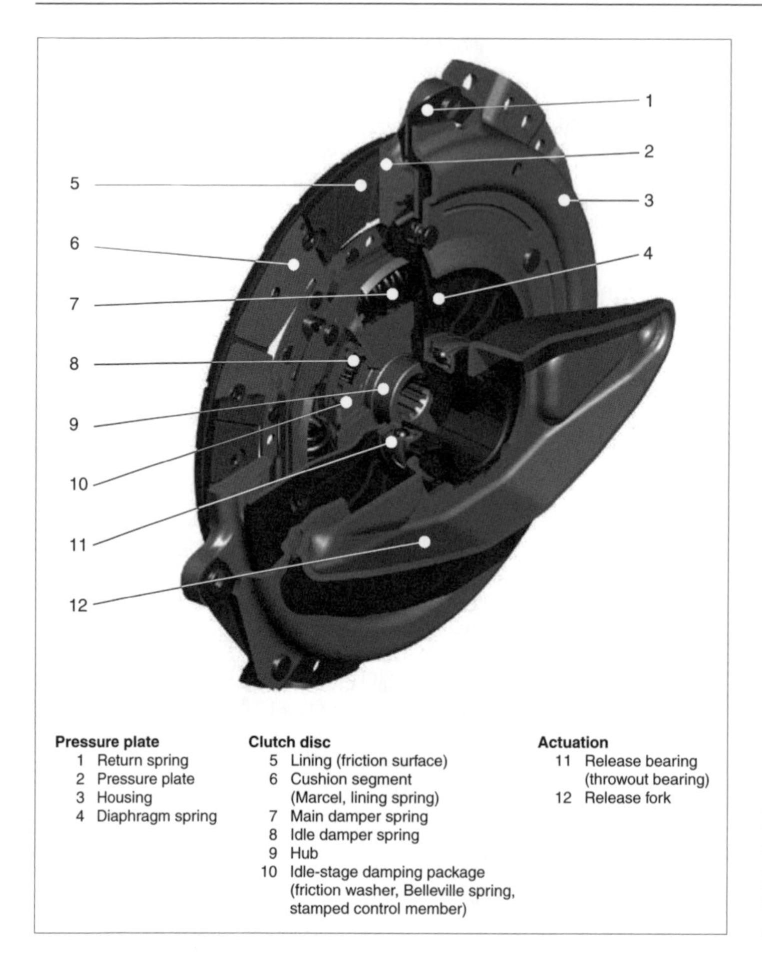

Fig. 5.4-5 Passenger car clutch, consisting of pressure plate with diaphragm spring, clutch disc with torsion damper, and release bearing.

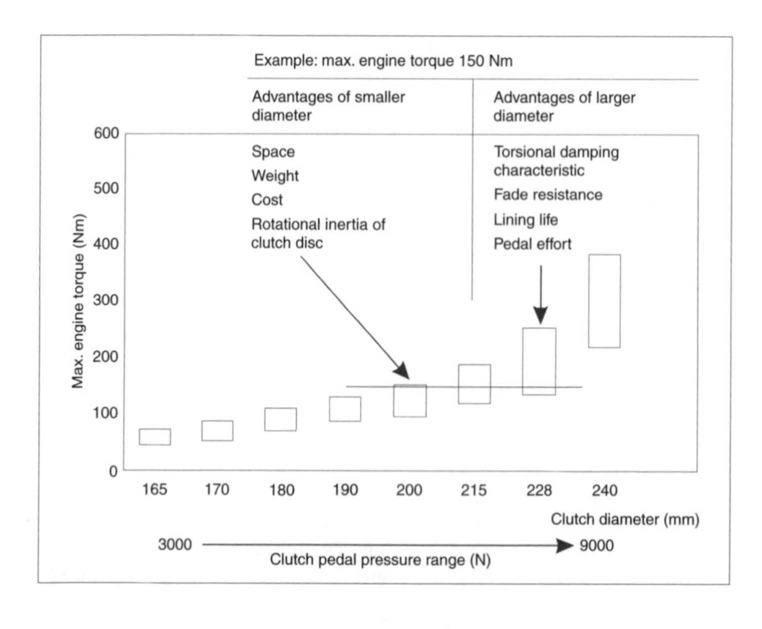

Fig. 5.4-6 Correlation between maximum engine torque and clutch diameter.

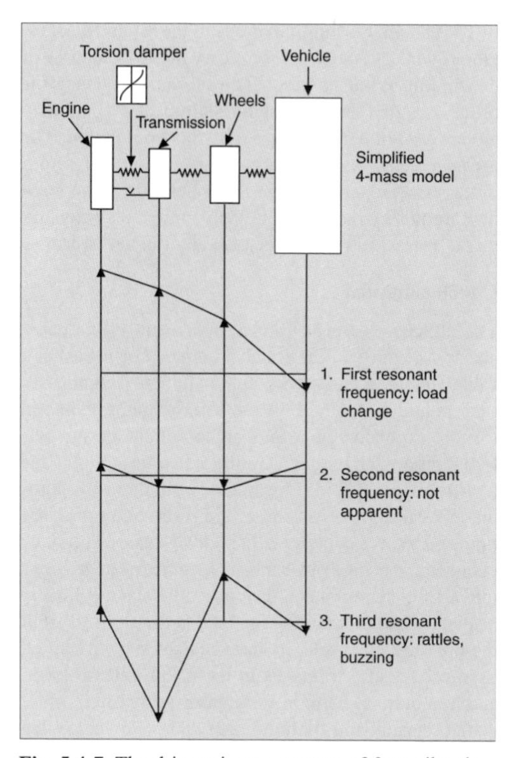

Fig. 5.4-7 The drivetrain as a system of four vibrating masses.

glass fiber bundles with brass or copper wires) embedded in a "friction cement" of resins, various rubbers, and filler materials. Sintered metal linings are unsuitable due to chatter.

Separation

After the clutch is disengaged, drag torque should be no more than 0.2–0.5 Nm. Coastdown time for the clutch disc to come to a stop depends on its own rotational inertia and transmission drag torque.

Torsional vibration damping

For weight and comfort considerations, the driveline is relatively soft in torsion. In first gear at maximum engine torque, up to about 90° of twist may be encountered. Firing-induced nonuniform engine rotation and load changes in response to rapid throttle or clutch pedal movements also excite torsional vibrations (Fig. 5.4-7). The first case may result in transmission rattle and body noises, while the second case may give rise to vehicle longitudinal oscillation and impact noises ("clunks"). During the clutch slip phase, the engine is decoupled from the remainder of the drivetrain with respect to torsional vibrations.

As a result of the torsion damper in the clutch disc, the resonant frequency due to engine firing frequency and, therefore, the frequency of rattle and buzzing range from 40 to 70 Hz, depending on vehicle type and selected gear. With a four-cylinder engine, for example, this represents an engine speed of 1,200 to 2,100 rpm. Fig. 5.4-8 shows a typical torsion damper characteristic with both driving and motoring loads and stops to limit torque peaks. The main damper is designed so that given the weakest possible spring stiffness, the exact amount of friction is found to give the best compromise between behavior in resonant frequency sweep and deterioration of supercritical uncoupling. Rattle also appears at engine idle with no gear selected. This is countered by the idle damper, which may be designed supercritically. Its spring rate is about 1% that of the main damper. An angular deflection of ±2–±4° is superimposed on the static angular deflection resulting from the transmission's temperature-dependent drag torque.

The dual-mass flywheel (Fig. 5.4-9) employs a grease-lubricated torsion damper between the primary and secondary flywheel sections. As in the conventional

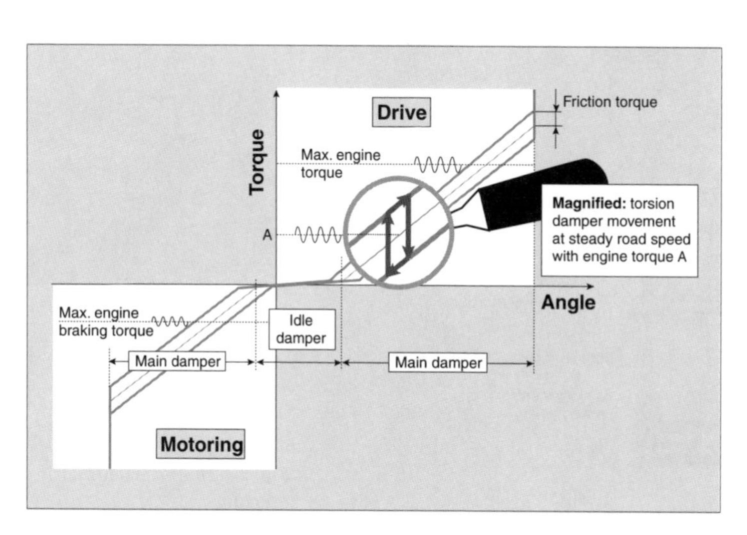

Fig. 5.4-8 Torsion damper characteristic.

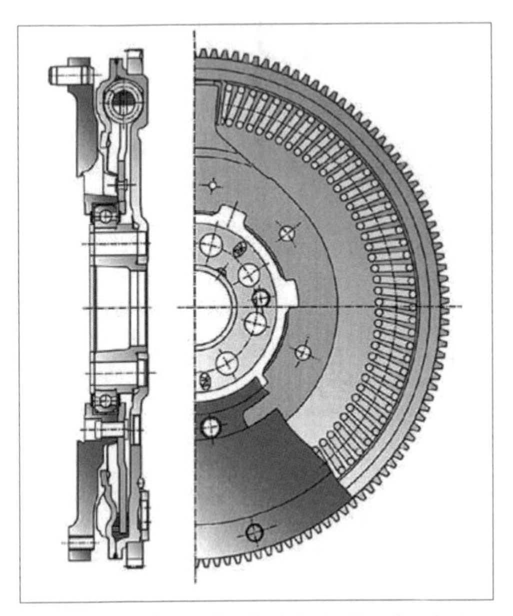

Fig. 5.4-9 Dual-mass flywheel (arched spring design, greased).

flywheel, the primary flywheel is bolted to the crankshaft and carries the starter ring gear and ignition timing marks on its periphery. The secondary flywheel, which carries the pressure plate, is mounted on the primary flywheel and forms one of the clutch disc's two mating friction surfaces. Resonant frequency drops to 8–12 Hz—that is, engine idle is in the supercritical region. Thanks to the large secondary mass, the degree of decoupling is outstanding. The resonant frequency is briefly excited during engine startup and shutdown. Noises are not permissible during this brief period. The greatest loads on the dual-mass flywheel result from driver error: extended periods of operation in the resonant frequency range or extremely rapid clutch engagement. Special idle tuning is generally not necessary.

Clutch actuation

The clutch release bearing is permanently lubricated and self-centering. The clutch is actuated by means of a cable (in some cases with automatic wear compensation) or hydraulically (Fig. 5.4-10). Hydraulic actuation is more expensive but offers higher efficiency and better vibration damping. Hydraulic actuation components consist of a self-bleeding master cylinder with automatic wear compensation, a fluid reservoir, a pressure line, and a slave cylinder with preload spring. The slave cylinder and release mechanism are sometimes combined into a single component, the concentric slave cylinder.

Pedal characteristics (Fig. 5.4-11) are derived from the pressure plate release force characteristic modified by mechanical advantages in the pedal, release lever, and hydraulic system; in some cases an auxiliary (over center) spring; and friction losses and travel losses due to elastic deformation. The latter are, however, required for good clutch modulation.

For automatic clutch actuation (Fig. 5.4-22), discussed in Section 5.4.3, an electrohydraulic or electro-

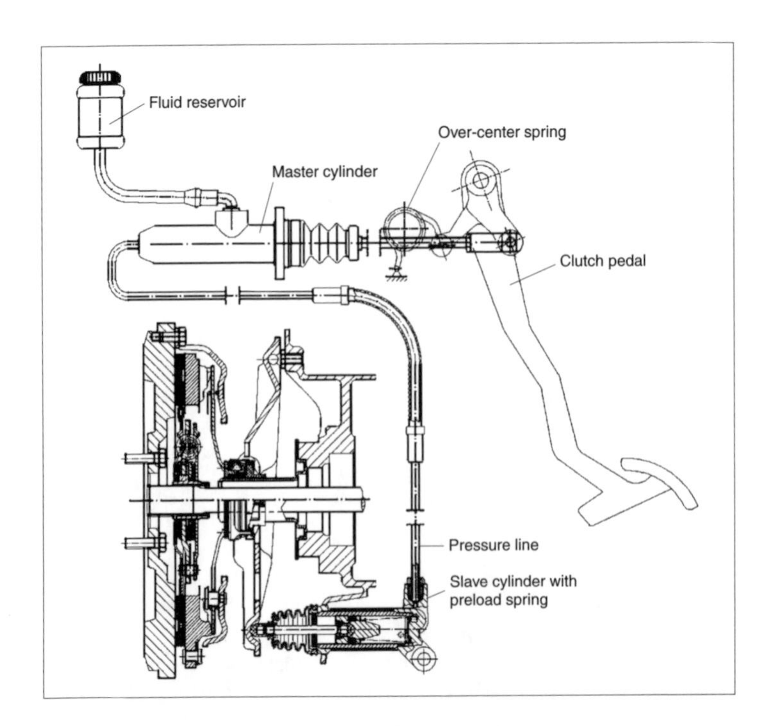

Fig. 5.4-10 Hydraulic clutch actuation.

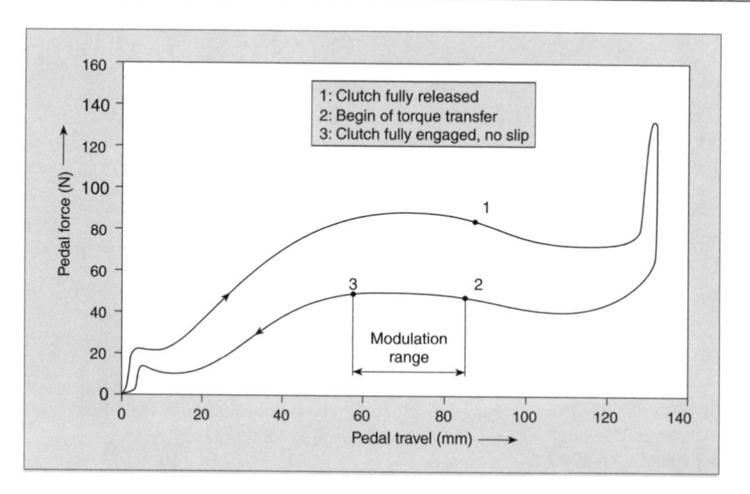

Fig. 5.4-11 Clutch pedal characteristic.

mechanical actuator does the work of the driver's clutch foot. This actuator is controlled electronically by a unit which employs sensors to discern driver intentions and specific details of vehicle motion. Driver demands are derived from throttle pedal angle and pedal acceleration. In order to achieve rapid gear changes, clutch release, gear-shifting, and clutch re-engagement must overlap in time. This in turn requires a gear sensor. As a component for automatically shifted gearboxes (ASG), automatic clutch actuation will acquire major significance.

5.4.2.2 Hydrodynamic torque converter

At present, hydrodynamic torque converters find application in passenger car drivetrains in conjunction with discrete-gear automatic transmissions employing up to six forward speeds, as a driveaway (startup) element with continuously variable transmissions (CVTs), and for torque multiplication.

Hydrodynamic torque converters have completely displaced fluid couplings from passenger car practice. The added complexity of the torque converter (which requires an additional reaction element, the stator) is justified by the added benefit of torque multiplication.

Construction

Torque converters for passenger car applications are generally based on the "TriLok" principle, with opposed impeller and turbine, between which is a stator at their inner periphery. In addition, the torque converter housing often includes lockup clutches and torsion dampers (see Fig. 5.4-12).

The impeller is connected directly to the engine crankshaft, while the turbine is coupled to the transmission input shaft. A stator shaft joins the stator to the transmission housing. The torque converter is connected to the transmission oil system and completely filled with oil during operation. To eliminate waste heat from the torque converter, oil circulates continuously.

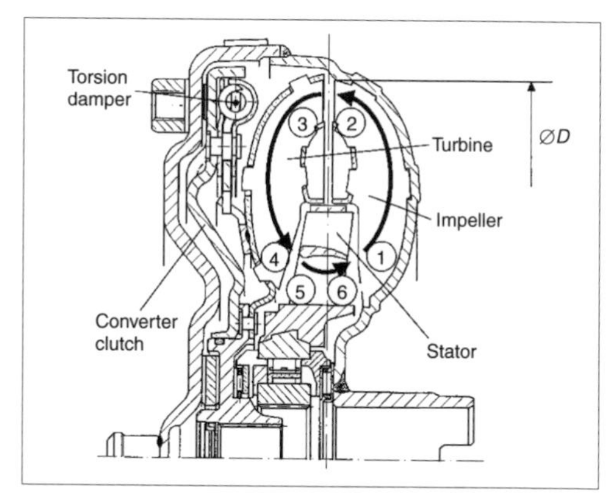

Fig. 5.4-12 Torque converter with lockup clutch and torsion damper.

Operating principle

The present state of torque converter development encompasses a total of four different operating modes:

1. Hydrodynamic torque converter operation

Hydrodynamic operation is characterized by the fact that torque transfer is only possible if there is slip between impeller and turbine. This slip is represented by the speed ratio

$$v = \frac{\omega_T}{\omega_P}.$$

Torque absorbed by the impeller is given by the Euler turbine equation:

$$T_P - \rho \cdot \dot{V} \cdot \Delta(r \cdot c_p)$$

which may be approximated by

$$T_P \approx \rho \cdot \dot{V} \cdot \omega^2 \cdot D^5.$$

By introducing a proportional power factor λ, we have

$$T_P = \lambda \cdot \rho \cdot \dot{V} \cdot \omega^2 \cdot D^5.$$

Power factor λ represents a dimensionless quantity, which permits comparing torque converters independent of their turbine wheel diameter D, impeller speed ω_P, or density ρ of their working medium.

Converter torque ratio is defined as

$$\mu = \frac{T_T}{T_P} \geq 1 .$$

Another important quantity is efficiency,

$$\eta = \frac{T_T \cdot \omega_T}{T_P \cdot \omega_P} = \mu \cdot \nu .$$

In practice, torque transfer behavior of a torque converter in hydrodynamic operation is evaluated by plotting instantaneous impeller torque at a speed $n_P = 2000$ rpm (M_{P2000}), torque ratio μ and efficiency η as functions of speed ratio ν. Fig. 5.4-13 shows typical examples of these characteristic curves.

The quantity M_{P2000}, however, does not represent a direct relationship between the maximum engine torque and the size of the torque converter necessary for its transmission. Fig. 5.4-14 shows maximum engine torque against corresponding torque converter size in the form of a scatter diagram. The scatter band arises from the different constructive configurations of hydrodynamic circuits (converter aspect ratio, etc.) as well as from lockup clutch design details (single or multiplate clutch).

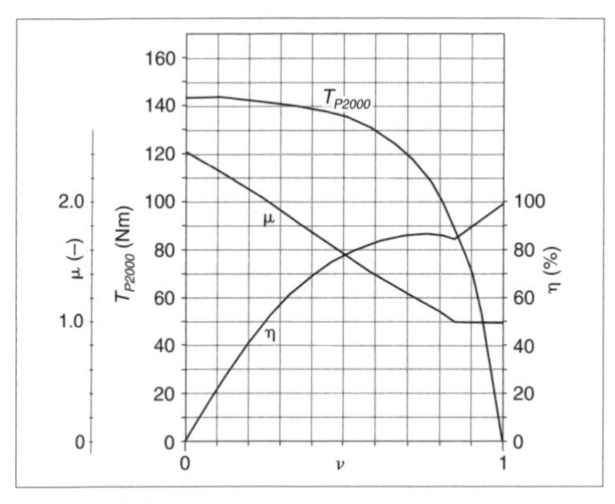

Fig. 5.4-13 Torque converter characteristic curves.

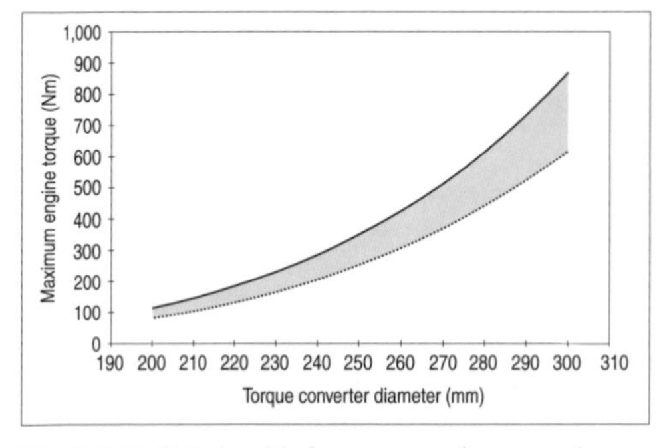

Fig. 5.4-14 Relationship between maximum engine torque and torque converter size.

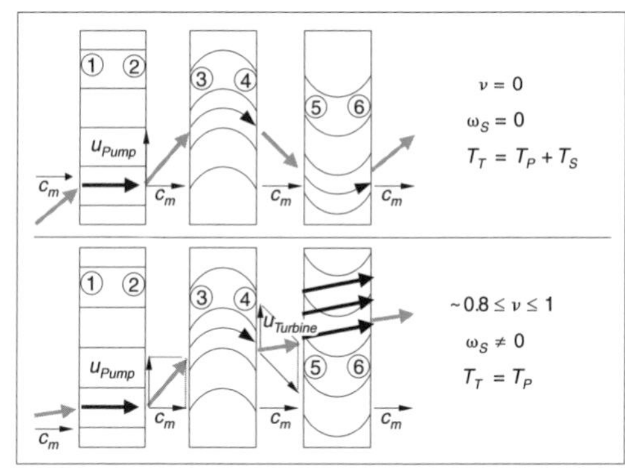

Fig. 5.4-15 Flow parameters at various operating points. c_m, fluid speed at centerline.

To better understand flow in a torque converter, a representation was chosen to clarify the internal relationships between the blade rows and individual rotors of the hydrodynamic circuit. Fig. 5.4-12 indicates the rotor inlet and outlet zones with numerals from 1 to 6. These are repeated in the modified representation of Fig. 5.4-15.

In the figure, fluid particles are accelerated by the impeller between 1 and 2, thereby gaining energy. This energy-laden fluid stream enters the turbine at 3. On its way through the turbine toward 4, the fluid stream gives up some of its energy to the turbine. This energy transfer establishes turbine torque and therefore the transmission input torque. Between 5 and 6, the stator causes an additional change in flow direction, resulting in stator torque, counteracted by the transmission housing. For small speed ratios (large speed differential) between impeller and turbine, torques at the impeller and stator have the same sign, so that $T_T = T_P + T_S$.

2. Hydrodynamic clutch operation

If external operating conditions result in greater speed ratios—that is, the speed differential between impeller and turbine becomes smaller—the direction of flow impacting the turbine changes to reverse the moment direction of the stator. This means that with $T_T = T_I - T_S$, the resulting turbine torque is smaller with respect to the impeller torque. Torque multiplication is no longer present. To avoid this situation, an overrunning (one-way) clutch is inserted between the stator and its connection to the housing. This permits one-way torque transfer to the stator mount. As a result, stator torque is zero. With $T_T = T_I (+ T_S = 0)$, the torque converter becomes a hydrodynamic clutch. Torque multiplication $\mu = 1$.

3. Lockup torque converter operation

Increasing demands for reduced energy consumption by passenger cars have forced the introduction of lockup torque converters to eliminate power-wasting slip within the torque converter. This clutch represents a mechanical connection between impeller and turbine, eliminating hydrodynamic power transfer between these components. Torque transfer is then achieved with zero slip.

Axially mounted within the torque converter housing is a piston with a large surface area carrying a friction lining. This piston is rotationally locked to the turbine. Engaging or disengaging the lockup clutch is accomplished via the transmission fluid; the direction of fluid volume flow to the converter is varied. Depending on the transmission fluid flow direction, fluid pressure on the right and left sides of the piston will differ, which moves the piston in the "engaged" or disengaged" direction. Along with the advantages of slip-free power transfer, a lockup torque converter also carries the disadvantage of a rigid connection between engine and transmission. The outstanding vibration damping properties offered by hydrodynamic clutch operation are no longer available. As a result, an additional torsional vibration damper, similar to those found in conventional clutches for manual transmissions, is integrated in the torque converter housing. This significantly expands the operating range, permitting active lockup torque converter operation with adequate comfort.

4. Controlled-slip torque converter operation

Expanded functionality is achieved through application of controlled-slip clutches. In this form of operation, the torque converter clutch is not completely locked. There remains a definite amount of slip between input and output sides of the converter. Slip prevents full transfer of engine-generated torsional vibrations to the transmission and therefore to the driveline. Because torsional vibrations are dependent on engine type and operating conditions, a map of required slip speeds for any operating condition may be obtained from theoretical considerations and actual vehicle testing. Using this map, transmission slip speed is established by the electronic transmission controls acting through a pressure control system. Where comfort considerations once dictated an "open" converter clutch in certain operating ranges, the controlled-slip converter permits operation with considerably less mechanical slip, resulting in an appreciable increase in efficiency. Operating regimes which previously permitted complete lockup are retained.

Because power lost through clutch slippage is converted to heat, suitable measures must be taken to prevent unacceptable temperature increases in components engaged in the friction process. This is achievable because even with an activated converter clutch, oil flow is still directed through the converter. The oil flow is arranged to provide direct cooling of friction surfaces.

Application examples

As already indicated, modern passenger car torque converters are almost exclusively fitted with lockup clutches. Depending on engine torque, these are configured as single or multiplate clutches. In addition to their diameter D, hydrodynamic circuits are also differentiated by their cross-sectional aspect ratio (see Fig. 5.4-16).

Prognosis

Ever more restrictive fuel economy and emissions regulations are forcing more efficient passenger car drivetrain designs. For the torque converter, this means that slip endemic to hydrodynamic operation must be avoided by converter lockup. The resulting sacrifices in comfort demand a capable vibration damping system to ensure desired comfort levels even with expanded lockup clutch application. In addition, operating regimes which must, as before, be negotiated in hydrodynamic operation mode demand a high level of hydrodynamic efficiency. Simultaneously, torque converter space allocation continues to be reduced.

Accordingly, current and future development must be directed toward optimizing individual torque converter, hydrodynamics, and vibration damping functions

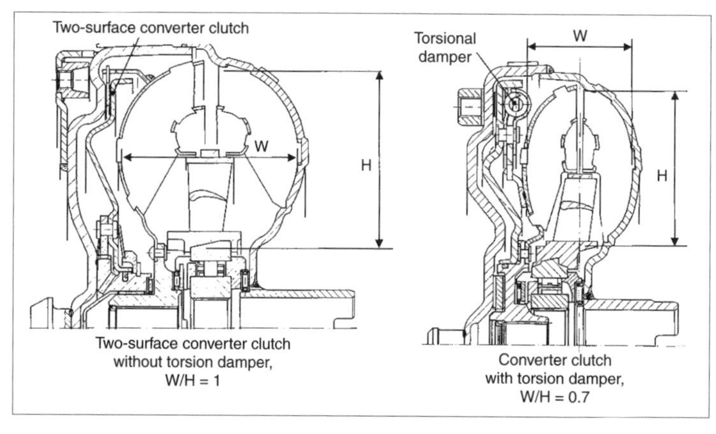

Fig. 5.4-16 Application examples.

within the available space and application-specific requirements. Special attention must be paid to optimizing the aforementioned functions not as individual components but rather as parts of an overall system.

5.4.3 Manual transmission systems

5.4.3.1 Construction and operation

The essential elements of manually shifted transmissions are:

- Foot-operated clutch for driveaway and driveline separation
- Synchromesh transmission with four to six individual gear pairs
- Transmission actuation and transfer of shifter motion and force from shift lever to transmission

Individual gears are positively locked during the shift process. It is therefore necessary to interrupt power flow from the engine by means of a clutch. After shifting out of the previously selected gear and, by means of synchromesh, matching speeds for the shift element to be selected, the next desired gear may be selected and locked. The shift process is completed when the clutch is re-engaged.

To achieve multiple gear ratios, gear pairs with differing numbers of teeth are assembled in a countershaft gearbox. Transmissions typically have a single countershaft, sometimes two or three. With the exception of reverse gear, which may consist of a sliding gear, all gear pairs are in constant mesh.

Manually shifted transmissions fall into one of two basic types, with either coaxial input and output shafts or with staggered shaft axes (Fig. 5.4-17). The coaxial configuration permits a direct drive without any gears and is particularly suited to vehicles of standard drive (front engine/rear drive) configuration. Transmissions with staggered axes do not have a direct gear. They are installed in vehicles with front or rear drive and with designs employing separate transmission and final drive.

Manually shifted transmissions are characterized by highly efficient power transfer. Minimal losses are due to meshing gears, friction between shift components

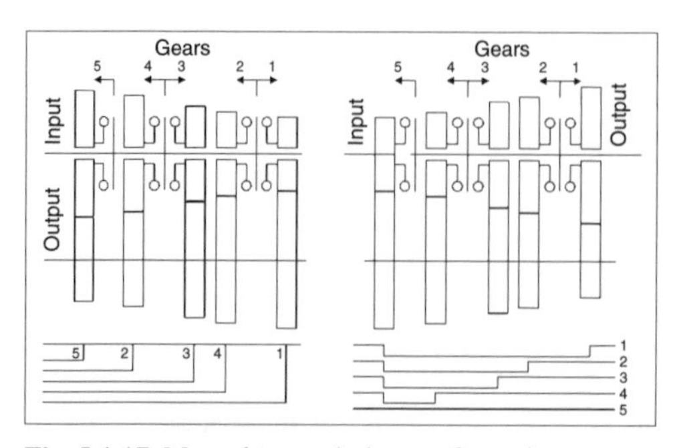

Fig. 5.4-17 Manual transmission configurations.

with differing rotational speeds, idling gears, and dynamic seals for shafts and bearings, as well as oil splash. Transmissions with direct power transfer in one specific gear achieve efficiencies of up to 99%.

5.4.3.2 Gearing

Vehicle transmissions use involute gearing exclusively. Involute gears are easy to manufacture and relatively insensitive to changes in shaft spacing. In passenger car applications, helical-cut gears are generally employed (with the exception of reverse gear).

Gearing is designed in keeping with collective loading parameters and must provide assurances against gear tooth breakage, flank fatigue, galling, pitting, and spalling.

Elimination of gear noise takes on special importance in passenger car applications, particularly in view of steadily increasing acoustic demands. In addition to helical gearing and precision manufacturing, important measures to reduce gear noise include high tooth contact ratios, deep tooth forms, and suitable gear design corrections (helix angle, radial or axial crowning).

Modern finite element-based calculation methods now make it possible to apply gearing corrections in the transmission design phase for shaft bending and housing deformation under load, thereby improving transmission load carrying capacity and noise emissions.

5.4.3.3 Synchromesh

To match rotational speeds of meshing parts, double clutching with throttle application was once customary even in passenger cars. Today, this is accomplished by a speed conversion device interposed ahead of a dog clutch ("jaw clutch"). Blocking synchromesh systems have become the predominant form of gear speed synchronization. These consist of friction clutches to achieve matched rotational speeds. Synchromesh permits positive gear engagement only after speeds are matched and assures good shift feel, particularly in terms of shift smoothness.

There are several different types of blocking synchromesh. The most common is the Borg-Warner system, available with single, double, or triple-cone synchromesh (Fig. 5.4-18). Multiple-cone synchromesh components are usually only applied to lower gears, as they are capable of withstanding higher thermal loads. They also reduce shift effort, thereby equalizing shift efforts across various gears.

Possible friction pairings include brass, sintered powder metal, and molybdenum, all acting against steel.

Because each transmission includes multiple synchromesh units, development has of course paid close attention to cost reduction. Preferred manufacturing methods include modular synchromesh packages with a maximum number of common parts and application of sheet metal forming and sinter technology.

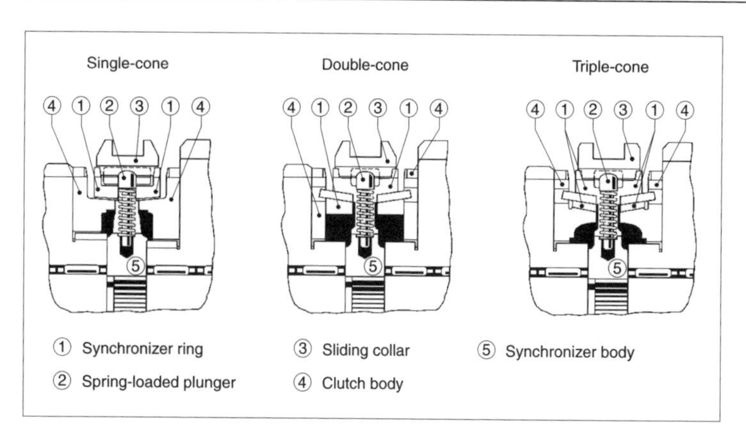

Fig. 5.4-18 Synchromesh systems.

5.4.3.4 Additional transmission components

In addition to low manufacturing cost and high reliability, lightweight design and low noise emissions are primary considerations in modern manual transmissions.

Preferred housing materials are aluminum and, increasingly, magnesium. To achieve the greatest possible rigidity and low noise radiation, housings are designed with the aid of finite element methods to optimize shape, wall thickness, and reinforcing ribs.

Additional weight reduction may be achieved by using hollow shafts.

Shafts and gears are almost exclusively located by rolling-contact bearings: needle bearings for idler gears and ball, roller, or tapered roller bearings for housing-mounted shafts.

Dynamic sealing is accomplished by means of elastomeric radial shaft seals, in some cases with helix seal ribs for better sealing. Solid flat gaskets as well as silicon or anaerobic liquid sealing materials are employed for static sealing of housing parting surfaces and covers.

Modern manual transmissions are lubricated for life. The most commonly used lubricant is automatic transmission fluid (ATF). The lubricant must exhibit favorable

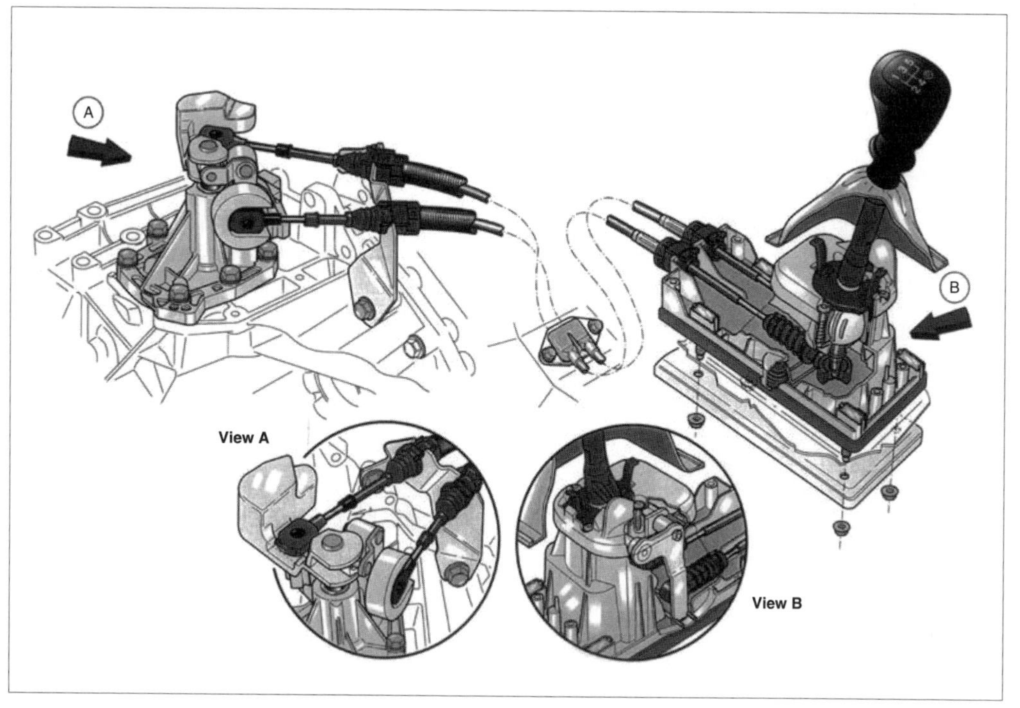

Fig. 5.4-19 External cable shift mechanism.

viscosity/temperature properties to prevent difficult shifting at low temperatures, and suitable additives must be used to achieve uniform synchronizer friction behavior under all operating conditions. At the same time, the lubricant should provide insurance against gear galling and pitting.

Depending on transmission size and configuration, oil fill quantity may be between 1.5 and 3 L. Transmission lubrication is achieved by oil splash and specific oil control by means of sheet metal deflectors. Heavily loaded transmission applications (e.g., off-road vehicles) may include a gear oil pump to ensure adequate oil supply.

5.4.3.5 Transmission shift mechanism

The shift mechanism transmits the driver's shift movements to the corresponding transmission shift component. Distinction is made between external and internal shift mechanisms.

The external shift mechanism extends from the shift lever to the transmission. It may consist of shift rods with solid connections or a cable shifter with flexible connections. The present development trend is toward cable shifters with their attendant advantages in terms of space requirements, vibration decoupling between the engine/transmission package and the vehicle cabin, weight, and assembly, particularly in the tight confines of front-drive transverse-engined vehicles [9].

The internal shift mechanism conveys shift movements by means of shift rails and forks or arms to individual shift sleeves.

Fig. 5.4-19 illustrates an external cable shift mechanism for a five-speed manual transmission.

5.4.3.6 Application examples

The Mercedes-Benz five-speed Type SG 150 transmission (Fig. 5.4-20) is designed for compact transverse front-engined vehicles. It uses a two-shaft configuration employing offset input and output shafts with a maximum input torque capacity of 180 Nm [10]. Its transmission ratio range is 4.7. First and second gears are fitted with double-cone synchromesh; the remaining gears, including reverse, have single-cone synchros.

For weight reduction, its transmission shafts are hollow. The transmission weighs 32 kg, including 1.8 L of oil.

The ZF Type S 5-39 transmission (Fig. 5.4-21) is designed for standard configuration (front engine/rear drive) mid-range and luxury passenger cars equipped with powerful, high-torque diesel engines. Its maximum input torque is 400 Nm. The transmission ratio range is 5.24, with a direct fifth gear. The high proportion of fifth gear operation therefore results in especially good efficiency. Transmission dry weight is 45 kg. The 1.6 L of oil in this transmission provide lifetime lubrication.

The external shift linkage is connected to a central shift rail. This is designed for torsional stiffness and, in conjunction with deliberately reduced shift effort for fifth gear, provides a high degree of shift precision and confidence.

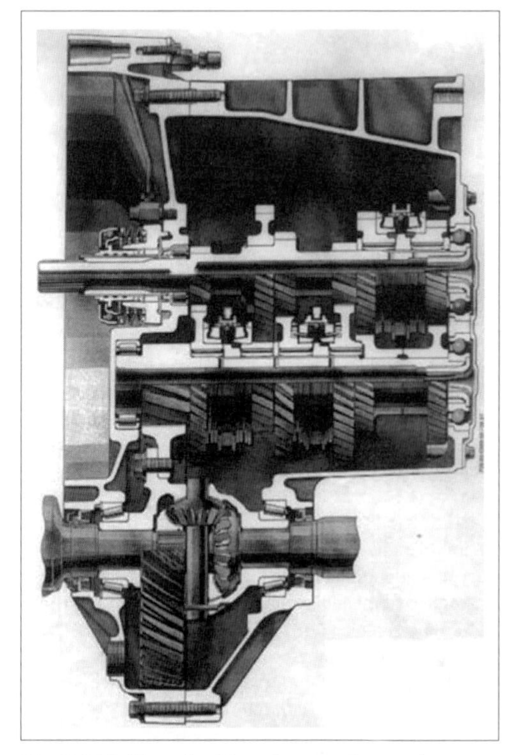

Fig. 5.4-20 Mercedes-Benz five-speed manual transmission, Type SG 150.

5.4.3.7 Automatically shifted manual transmissions

Automatically shifted manual transmissions have been introduced in series production for some passenger cars in the lowest performance class as well as in the very high-performance vehicles. These systems are based on manually shifted synchromesh transmissions fitted with electrohydraulic or electromechanical actuators for clutch and transmission operation (Fig. 5.4-22). For reasons of comfort and to provide insurance against over-revving, the engine is controlled by the transmission electronics during the shift process. Corresponding engine speed after completion of the shift process is calculated and set by the electronic control system.

Automatically shifted manual transmissions combine the high efficiency of manual transmissions with shift programs emphasizing fuel economy. One disadvantage is the clearly perceptible thrust interruption during shifts, particularly while accelerating at high loads in lower gears.

5.4.4 Geared automatic transmissions

5.4.4.1 Operation

There are three primary differences between automatic and manual transmissions:

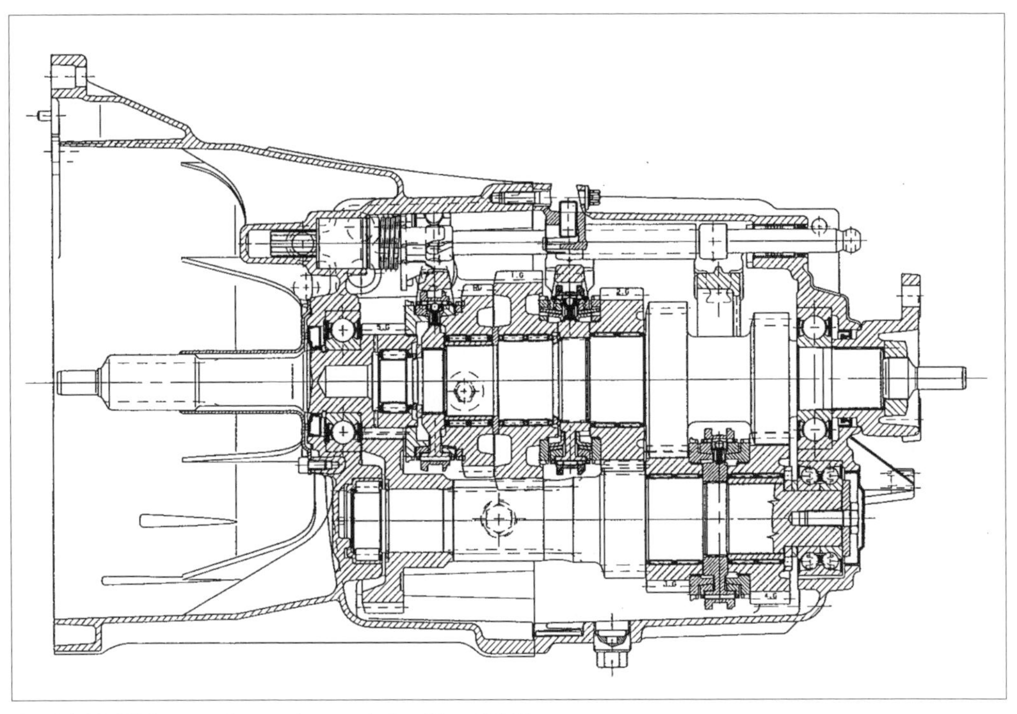

Fig. 5.4-21 ZF five-speed manual transmission, Type S 5-39.

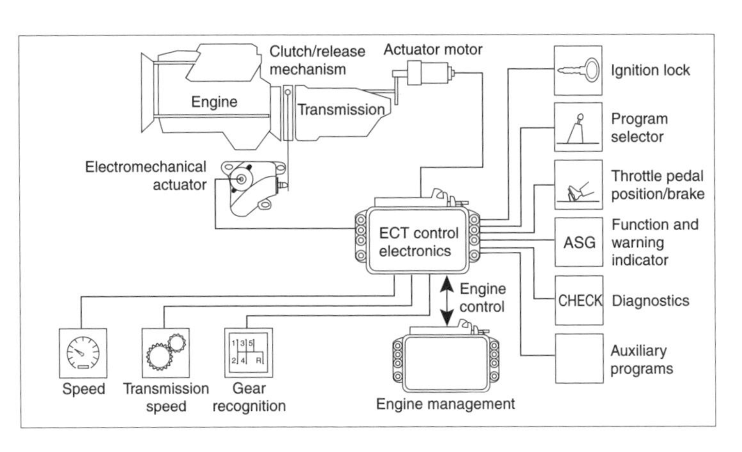

Fig. 5.4-22 System schematic for automatically shifted manual transmission.

1. Driveaway without clutch operation

Automatic transmissions generally employ a hydrodynamic torque converter as the driveaway (startup) element (compare to Section 5.4.2). The minimal torque transfer capability of the converter at engine idle speed permits bringing the vehicle to a stop using low braking effort. If the brake is released, the vehicle will creep. This permits very precisely controlled movement and parking maneuvers. Furthermore, creeping torque can hold a vehicle against the pull of slight grades. In addition to decoupling the drivetrain while the vehicle is stationary, the torque converter operating principle also permits very comfortable driveaway and, thanks to torque multiplication, increased thrust with more rapid acceleration for a given overall gear ratio.

2. Shifts take place under load, without power flow interruption

During manual transmission gear changes, the clutch is disengaged, separating engine and transmission. This interrupts power flow and, therefore, vehicle acceleration (see Section 5.4.3). By contrast, gear changes in an automatic transmission are accomplished under load, without interruption in power flow. This requires several shift elements (clutches, brakes) adequately dimensioned to withstand and shift the engine's full power. Fig. 5.4-23 illustrates the principle of an automatic transmission using the example of a two-speed full-load-shifting transmission. First gear is transmitted by clutch C1, while clutch C2 is engaged for second gear. Fig. 5.4-24 shows torque history at both clutches, transmission output torque history, and engine speed history for an upshift under load. Gear changes are carried out by means of an "overlap shift." During this overlap phase, the torque transmitted through the clutch to be disengaged is decreased. Simultaneously, torque transmitted through the clutch to be engaged is increased. At the end of the overlap phase, the newly engaged clutch transmits the torque of the newly selected gear, even while the rotational engine speed associated with the previous gear is still present. Further increase of torque transmitted by the engaged clutch synchronizes engine-side rotating masses to the speed dictated by the newly engaged gear (torque increase phase). From the engine output torque trace, it is apparent that during the upshift process, torque drops to the level of the newly selected gear immediately after the overlap phase. Next, in the

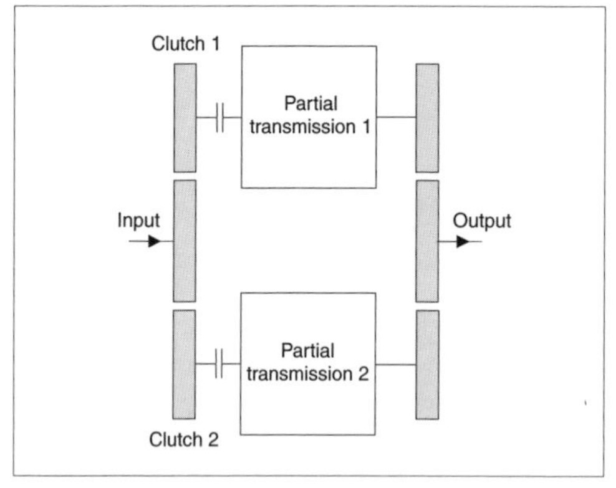

Fig. 5.4-23 Principle of a two-speed transmission with power-shifting capability.

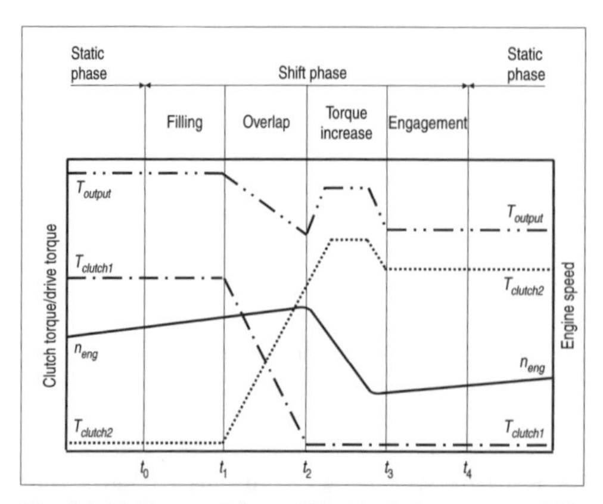

Fig. 5.4-24 Torque and speed history during a power shift.

torque increase phase, drive torque increases as energy stored in engine-side rotating masses is transferred to the driveline. Once rotating speeds are synchronized, the working engine speed and torque are matched to the new gear. The engine torque trace during the torque increase phase may be controlled by modulation of the newly engaged clutch and by means of the engine control system (see Section 5.4.6.4). Other shifts are carried out similarly. Detailed descriptions of the process may be found in the literature [1].

3. Shifts take place automatically

In automatic transmissions, gear changes are not accomplished by a conscious effort on the part of the driver; instead they take place automatically. This is accomplished by a shift program, largely determined by the two parameters of vehicle speed and throttle position (see Sections 5.4.4.5 and 5.4.6.4).

5.4.4.2 Construction

Planetary automatic transmissions

Automatic transmissions are generally configured as systems of planetary gears. Reasons include:

- Ability to achieve multiple gear ratios with a single gearset. Depending on whether input is via sun gear, planet carrier, or ring gear, various output ratios may be achieved by choosing one of these three elements. Coupling multiple sets of planetary gears offers a variety of possible gear ratios.
- Coaxial mounting of core components. Input and output are located on a common axis, simplifying mounting of rotationally symmetric shift and coupling elements.
- Compact dimensions. Engine output is transferred by multiple gear engagements (power split). This leads to more compact transmission dimensions. Furthermore, planetary gears do not generate external radial forces.
- High efficiency. Planetary transmissions exhibit high efficiency, as only a portion of the transmitted power is subject to rolling contact losses in meshing gears; the remainder is transmitted mechanically without losses by rotary motion.

Fig. 5.4-25 is a schematic representation of a ZF-5 HP 19 five-speed automatic transmission, a typical representative modern discrete-gear automatic transmission for rear-wheel-drive applications. The transmission schematic illustrates relationships between the major power-transmitting components:

- Controlled-slip torque converter
- Clutches (A, B, E, F)
- Brakes (C, D, G)
- Freewheel in torque converter and first gear
- Gear trains (Ravigneaux gear train and simple planetary gear train)
- Mechanical components (shafts, carriers, etc.)

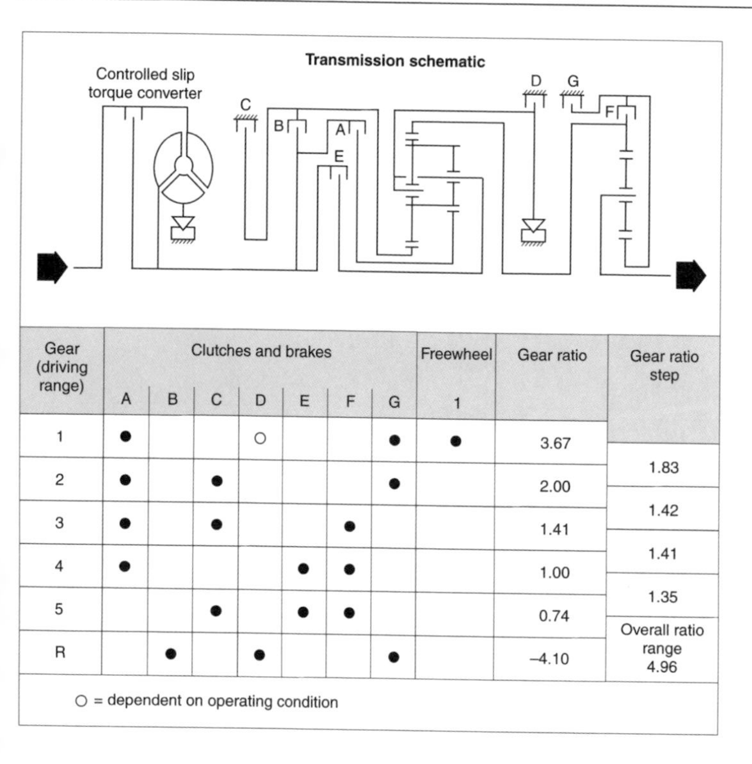

Gear (driving range)	Clutches and brakes A	B	C	D	E	F	G	Freewheel 1	Gear ratio	Gear ratio step
1	●			○			●	●	3.67	
2	●		●				●		2.00	1.83
3	●		●			●			1.41	1.42
4	●				●	●			1.00	1.41
5			●		●	●			0.74	1.35
R		●		●			●		–4.10	Overall ratio range 4.96

○ = dependent on operating condition

Fig. 5.4-25 Schematic of ZF5 HP 19 transmission.

The associated shift diagram indicates which elements are engaged in which driving range ("gear"). To transmit power, three shift elements are engaged in each gear. Gear shifts are accomplished by engaging or disengaging a single shift element. Gear ratios of individual driving ranges are achieved by different couplings of planetary gear train components.

In developing an automotive transmission planetary gear train system, it is important to find a suitable set of gear ratios for the desired range of service. In addition to simple planetary gear trains, special gear trains such as Ravigneaux, Simpson, or Wilson gear trains may find application [1, 11, 12].

Automatic preselector transmissions

In addition to automatic transmissions based on planetary gear trains, modern applications include preselector transmissions. These also employ a hydrodynamic torque converter as a driveaway element. Gearsets are formed by spur gears as in a manual transmission, but in contrast to conventional manual transmissions, these may be power shifted. Fig. 5.4-26 schematically illustrates one example of this transmission type, the five-speed W5A180 by DaimlerChrysler, installed in the front-wheel-drive A Class [13].

Double clutch transmission

Double clutch transmissions are similarly derived from conventional manual transmissions. These consist of two interlocked preselector transmissions, one of which serves even-numbered gears, the other for odd-numbered gears (compare to Fig. 5.4-23). On the input side, each partial transmission is equipped with its own clutch; on the output side, both partial transmissions merge on the output shaft. Gears are shifted by the synchronizing mechanism in the partial transmission which is not transmitting power at the moment [14]. To date, this transmission principle has not been applied to production passenger cars.

5.4.4.3 Planetary gear train components

A planetary gear train (Fig. 5.4-27) consists of three core components: sun gear (1), planet gear carrier (3), and ring gear (4). Planet gears (2) are attached to the planet gear carrier. A planetary gear train permits the following three gear ratios:

$$i_3 = \frac{z_4}{z_1} \text{ ("stationary" gear ratio)}$$

The planet gear carrier is held stationary. Input is by means of the sun gear, output through the ring gear. The number of teeth on the ring gear is z_4 (a negative number, since the ring gear has internal teeth), while z_1 is the number of teeth on the sun gear.

$$i_{13} = 1 - i_3$$

The ring gear is held stationary. Input is by means of the sun gear, output through the planet gear carrier.

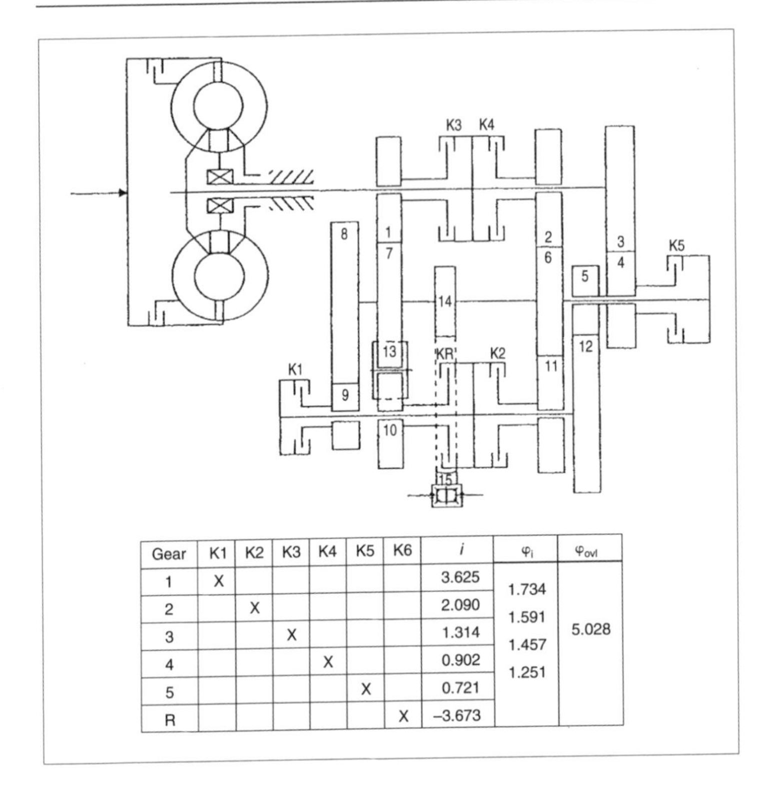

Gear	K1	K2	K3	K4	K5	K6	i	φ_i	φ_{ovl}
1	X						3.625	1.734	
2		X					2.090	1.591	
3			X				1.314	1.457	5.028
4				X			0.902	1.251	
5					X		0.721		
R						X	–3.673		

Fig. 5.4-26 Schematic of DC-W5A180 transmission.

$$i_{43} = 1 - \frac{1}{i_3}$$

The sun gear is held stationary, input is by means of the ring gear, output through the planet gear carrier.

By exchanging input and output, reciprocals of the stated gear ratios may be obtained. Depending on realized gear ratio and power demands, planetary gear trains are built with three to five planetary gears. Sun and planetary gears are made of case-hardened steel. Ring gears are broached in heat-treated steel blanks and nitrided for highly loaded applications. Planet gear carriers are built up of welded or riveted steel stampings, or they consist of aluminum die castings [15]. Construction of Ravigneaux, Simpson, and Wilson gear trains is described in the literature [1, 11, 12].

Shift components

Automatic transmission power shifting components are composed of multiplate clutches and multiplate brakes. Construction of a multiplate clutch is shown in Fig. 5.4-28. This consists of an outer plate carrier, which is splined internally to allow axial motion of the steel clutch plates; an inner plate carrier with lined clutch plates; a cylinder containing a passage for pressurized oil; a

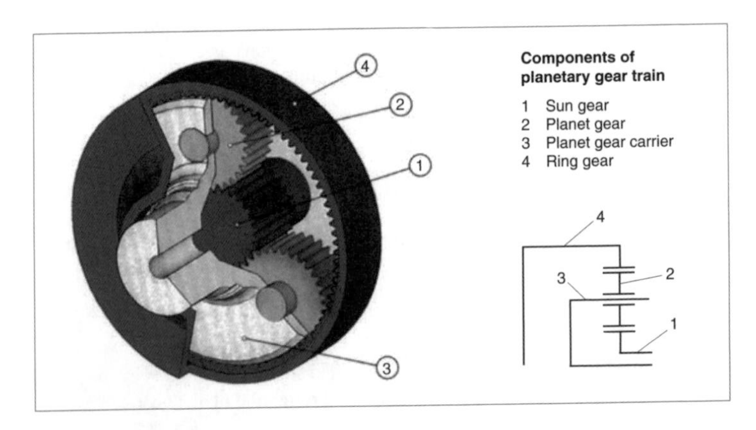

Fig. 5.4-27 Schematic and cutaway of planetary gear train.

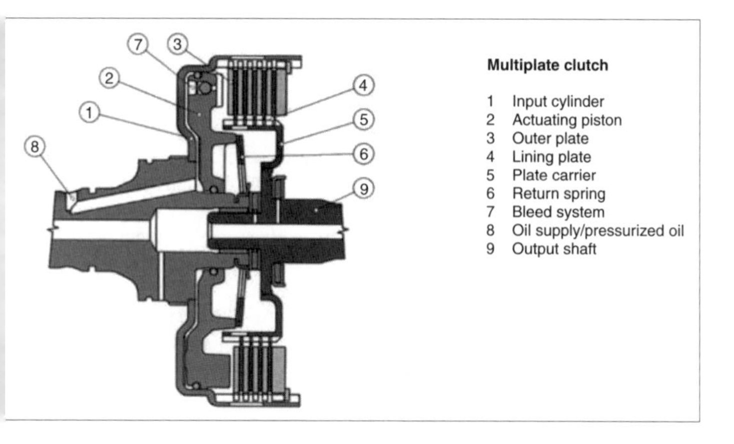

Fig. 5.4-28 Multiplate clutch.

piston which presses against the steel plates under the action of oil pressure; and a return spring which retracts the piston when oil pressure is cut off. Configured as a clutch, both plate carriers are designed to rotate; as a brake, the outer plate carrier is fixed to the housing.

Under static conditions, the shift elements are unpressurized and therefore open, or they are hydraulically closed by main system pressure and are positively engaged to transmit the applied torque. For a given torque T_{clutch} to be transmitted, the necessary main system pressure may be calculated from

$$P_{\text{stat}} = \frac{4T_{clutch}}{n \cdot \mu_{\text{stat}} \cdot r_m (D_O^2 - D_I^2)\pi} \quad \text{for } \omega = 0$$

where n is the number of friction surfaces, μ_{stat} the static coefficient of friction, r_m the mean friction radius of the clutch plates, D_O the outer and D_I the inner diameters of the actuating piston. Shift pressure layout follows from the same equation, but with T_{clutch} taken as the maximum torque during the shift (engine torque plus torque due to rotational inertia) and with μ_{dyn} replacing μ_{stat}.

Steel is used for outer plates. Plate thickness dimensions follow from consideration of heat generated during a shift, which must be absorbed by the steel plate given an acceptable temperature rise. Inner plates, consisting of a carrier plate (~0.8 mm thick), are lined with a friction lining applied to both sides. The friction material is a so-called "paper lining" with a carrier consisting of cellulose, aramid fibers, plastic components, and minerals, all soaked in phenolic resin. The lining friction coefficient and its response over the range of rotational speed may be modified by changing details of the lining composition. This not only has a marked effect on clutch capacity, but also on transmission shift comfort.

Friction linings are grooved to permit oil flow even when the clutch is engaged. This permits heat stored in the steel plate to be removed after completion of the shift process. Outer plate carriers, inner plate carriers, piston, and cylinder are made of steel stampings or die-cast aluminum. They are usually welded or riveted to shafts, planet carriers, or other elements or are integral with adjoining functional components [15]. Pistons are sealed by O-rings.

Oil supply

The oil supply of an automatic transmission fulfills the following functions:

1. Cooling of the hydrodynamic torque converter
2. Lubrication and cooling of the transmission's mechanical components (shift elements, gears, bearings)
3. Pressure supply to the hydraulic control system
4. Pressure supply to hydraulic actuators

A transmission contains 6–8 L of ATF (automatic transmission fluid). Modern ATFs consist of a mineral base oil plus specific chemical additives. These additives permit ATF to meet the various demands imposed on automatic transmissions:

- Thermal stability between −40 and +150°C
- Resistance to thermal aging
- High coefficient of static friction and increasing friction coefficient with increasing slip
- Minimal viscosity change over temperature
- Resistance to foaming
- Resistance to deposit formation
- Corrosion resistance
- Compatibility with sealing materials

Oil is supplied by a gerotor pump driven at engine speed by the torque converter neck. The gerotor design has become dominant because it fits easily into the available space between torque converter and transmission mechanical section and because of its simplicity and durability. Depending on transmission requirements, its theoretical oil supply rate is between 14 and 23 cc/revolution; the speed range is 600–7,000 rpm; and the pressure range is ~3–24 bar. Because the pump must be designed for operation at engine idle speed and elevated oil temperatures, its supply rate at high speeds is much too great. This excess flow must be returned to the oil sump or to the suction side of the pump with the minimum possible of flow losses. To avoid this disadvantage, some transmissions use controlled vane pumps. These, however, have more complex construction and lower efficiency and require more complex controls.

The oil supply system includes an oil filter located in the sump at the pump suction side. The filter prevents transmission manufacturing contaminants and transmission wear particles from entering the oil pump and hydraulic control system.

Hydraulic control system

The importance of hydraulic transmission controls has been diminished by the introduction of electronically controlled transmissions (ECTs; see Section 5.4.6). However, in order to translate ECT functions to power-transmitting drivetrain components, hydraulic components are as important as ever. Along with translation of analog and digital pressure signals, in which the system's dynamic behavior plays a major role, the hydraulic system retains its basic functionality including pressure boost or reduction and safety and limp-home functions.

The hydraulic control system consists of the following subsystems:

- Main pressure supply
- Shift pressure control
- Gear change control
- Torque converter pressure system with converter clutch control
- Lubricating oil system
- Limp-home system

The complete scope of hydraulic control functions is described by a schematic diagram and realized by a hydraulic control unit. This consists of multiple die-cast aluminum housings containing passages covered by an intermediate plate and bores for valves which control oil flow or pressure. Fig. 5.4-29 illustrates a cross section through several control unit valve bores. This also shows the design details of individual valves.

Forward (D) or reverse (R) gears are activated by the selector valve, which is mechanically connected to the driver's gear selector lever. The neutral (N) position relieves pressure to at least one shift element, thereby splitting the drivetrain. The selector valve also ensures that in the event of an electronic malfunction or loss of electrical continuity between the ECT unit and solenoids, a limp-home mode with one forward gear and reverse is available. Spool valves are activated externally by solenoids controlled by the ECT unit or by internal pressures. These spool valves control shifts between gears. Control valves are similarly modulated by the ECT unit by means of analog pressure signals. These control valves govern pressure history to the engaging and disengaging clutches during a shift process, set main system pressure proportional to engine torque, and govern the torque converter clutch to a specific predetermined slip.

The hydraulic control unit is usually bolted to the underside of the transmission. Oil flows to the oil pump, converter, and shift elements and, for general lubrication, by way of connecting passages in the control unit. The hydraulic control unit usually also contains sensors for rotational speed and oil temperature, wiring for the solenoids, and the transmission connector plug. By suitable arrangement of hydraulics and electrical package, the hydraulic and electrical control components may be integrated into a single electrohydraulic unit [16].

5.4.4.4 Actuation

Selector valve actuation

Automatic transmissions are fitted with a selector lever as an external shift mechanism. This is connected to the

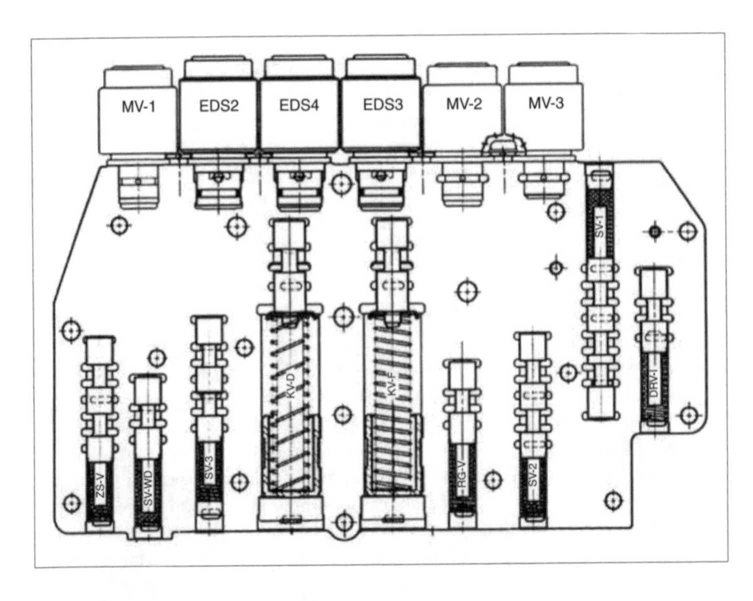

Fig. 5.4-29 Hydraulic control unit.

transmission gear selector shaft by means of a linkage or Bowden cable. The selector lever controls basic transmission modes P (park), R (reverse), N (neutral), and D (drive). A mechanical linkage acts directly on the selector valve and parking pawl. The gear selector shaft is locked in position by a spring-loaded detent plate. Usually a transmission mode indicator switch is located external to the transmission on the selector shaft. This recognizes the selector lever position, which is displayed on the instrument panel, and controls the backup lights and ignition interlock to prevent starting the engine with any gear engaged.

Parking mechanism

Vehicles equipped with manual transmissions may be secured against rolling away by simply leaving a gear engaged after the engine is shut down. This is not possible in the case of an automatic transmission; because there is no supply of pressurized oil available after engine shutdown, all shift elements are open and the transmission is effectively in neutral. To prevent the vehicle from rolling away, without applying the parking brake, automatic transmissions are equipped with a parking mechanism. A park pawl engages a park gear, locking the transmission output shaft against the housing. In selector positions R, N, and D, a return spring keeps the park pawl from engaging. When P is selected, a wedge forces the pawl against the park gear. If the pawl meets a corresponding space between teeth, it engages and locks the drivetrain. If it meets a tooth, it is preloaded by a spring mechanism so that as the drivetrain turns, spring preload engages the pawl in the next available park gear tooth gap. Spring forces, park pawl, and gear geometry are chosen so that the pawl can be disengaged from forward or reverse even on steep grades and so that the pawl cannot accidentally engage if P is selected while the vehicle is in motion.

Shift lever

In addition to the main selector positions P, R, N, and D, the classic automatic transmission shift lever also offers various specific driving ranges ("gears"), such as 4, 3, and 2, which can be selected to block upshifts. The primary purpose of this feature is to provide engine braking when descending grades, but it may also be used by performance-minded drivers to keep the engine operating at higher speeds. Gear selector levers normally have a locking function to prevent unintentional operation; to move out of P, to move past N to R, or to select lower gears, a latch must be released. Other variations permit selecting certain positions only through an additional transverse movement of the lever, thereby making more difficult any unintended lever movement to a prohibited gear. Newer shift lever designs add a "tip" gate alongside the usual automatic shift gate [17]. When the lever is moved to this parallel gate, the ECT switches from automatic mode to manually shifted mode. Moving the lever forward (+) shifts up; moving the lever to the rear (−) shifts down. This feature was introduced in order to offer the driver of an automatic-equipped car the option of driving a manually shifted transmission if desired. Fig. 5.4-30 illustrates such a shift lever.

Fig. 5.4-30 Gear selector lever with "tip" gate.

5.4.4.5 Operational behavior

Customer-relevant operational behavior of a vehicle equipped with an automatic transmission is evident in its shift quality—that is, the way and manner in which gear shifts are executed, whether and how shifts are perceived, and, in terms of its shift schedule, which gear is engaged and when gears are shifted.

Shift quality

Because, by definition, shifts in an automatic transmission take place automatically—that is, are not consciously initiated by the driver—they must take place largely imperceptibly; otherwise, they will be perceived as disturbing. Furthermore, the driver, in operating the throttle pedal, expects a direct reaction in the form of vehicle acceleration which as a rule is achieved by an immediate downshift. Therefore, the objective of shift process control is to react spontaneously to driver commands and to execute shifts as free of jerks as possible. The basic prerequisites for good shift quality are found in the construction of the mechanical transmission and in the hydraulic control system. Good shifting is easier to achieve with small ratio differences between gears than with large gear steps. Good system behavior within the hydraulic control system is a determinant for good shift quality: in other words, the overall strongly nonlinear signal path must exhibit tight tolerances and remain stable under all operating conditions. If these prerequisites are met, appropriate control algorithms within the ECT can be used to optimize shift processes. Precise determination of transmission input and output speeds permits rotational speed control of the engaging and disengaging clutches during the shift process [18]. This speed control leads to continuous, always consistent shifts independent of operating conditions. In shift regimes where speeds do not change, adaptive strategies are employed in an attempt to adjust clutch control pressures to appropriate levels. Adaptive algorithms ensure that variations during the transmission's operating life—clutch wear, for example—may be compensated for and do not exert a negative influence on transmission shift behavior or shift quality.

Shift program

Automatic transmissions are shifted in response to vehicle speed and throttle opening. A transmission shift schedule consists of up- and downshift lines for all possible gear shifts as functions of these two parameters (Fig. 5.4-31). If, for example, the vehicle is accelerated under constant load and crosses an upshift line, a shift to the next higher gear takes place. If, on the other hand, actuation of the throttle pedal signals a demand for vehicle acceleration, thereby crossing a downshift line, a downshift of one or more gears takes place. Electronic transmission controls offer a great degree of freedom in configuring such shift maps. Several maps may be stored in the ECT system for different driving characteristics. It has become customary to offer "E" programs (economy mode), providing fuel consumption oriented behavior, with early upshifts and delayed downshifts, and "SP" programs (sport mode) for performance-oriented driving. These programs may be selected by the driver by means of a program selector switch on the shift lever, or they may be chosen automatically by the transmission electronics. This requires a shift strategy which undertakes classification and evaluation of driving conditions and driving behavior as functions of various parameters—for instance, throttle pedal position, throttle pedal movement, vehicle speed, longitudinal and lateral acceleration, brake actuation, etc.—and selects a driving program appropriate for the situation at hand [17]. Fig. 5.4-32 illustrates such a system in the form of a flowchart. Another means of matching the shift program to conditions is use of a transmission shift schedule in which variable shift lines are matched to driving situations as functions of various external parameters [19]. An overview of all current means of achieving desired shift programs may be found in the literature [20].

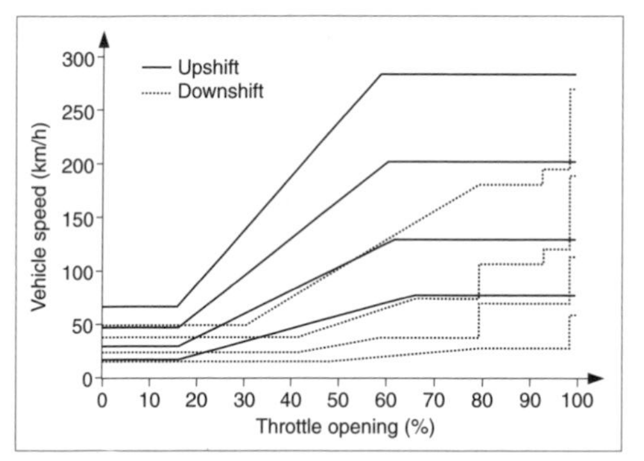

Fig. 5.4-31 Shift schedule.

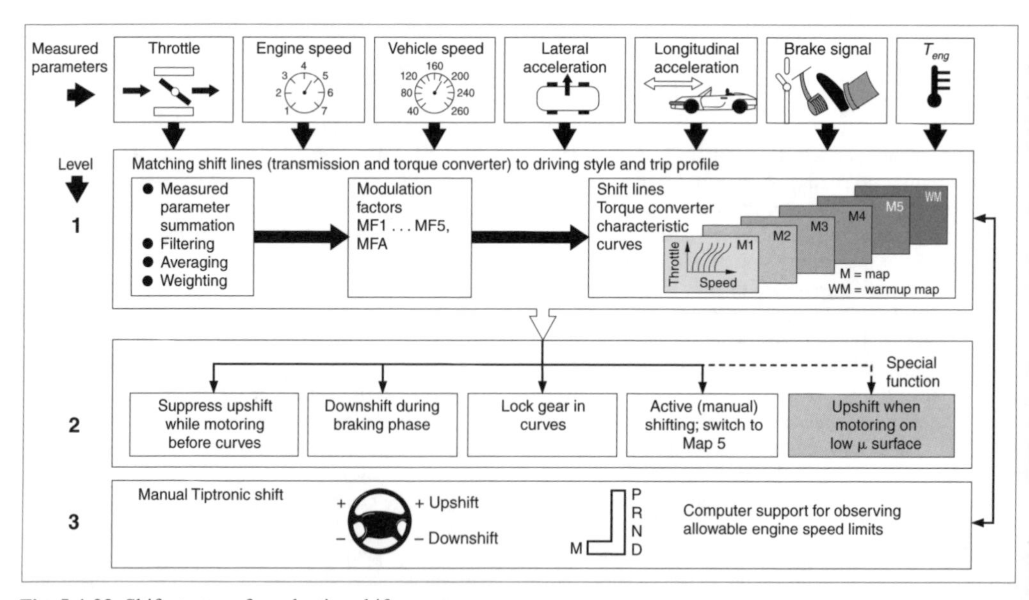

Fig. 5.4-32 Shift strategy for selecting shift programs.

5.4.4.6 Application examples: transverse front drive

Passenger cars with front-wheel drive and transversely mounted powerplants offer relatively little space for the transmission. Consequently, the transmission must occupy a very short package. Due to this limited installation space as well as the orientation of engine and halfshafts, the location of other assemblies at the front of the car, and high demands for crash safety, the transmission design must meet requirements which are strongly dependent on the overall vehicle concept at hand. Along with three-speed transmissions, still in common use, these applications invite consideration of four-speed solutions. One example for high-torque engines is the ZF-4 HP 20, which will handle up to 330 Nm of torque and power output of up to 160 kW in passenger cars, vans, and light trucks. Fig. 5.4-33 shows a cross section of this particular transmission. The overall configuration and major components are readily visible.

In this transmission, the four forward gears and reverse are realized by two planetary gear trains and five shift elements. A slim torque converter and "interlocking" arrangement of gearsets and shift elements yield a compact design with an overall transmission length of 368 mm. The transmission does not have any freewheels. All shifts are carried out as controlled power shifts. One characteristic feature of front-drive transmissions is the integral final drive. By selecting an appropriate spur gear package, the final drive ratio may be matched to vehicle requirements in terms of hillclimbing ability and top speed.

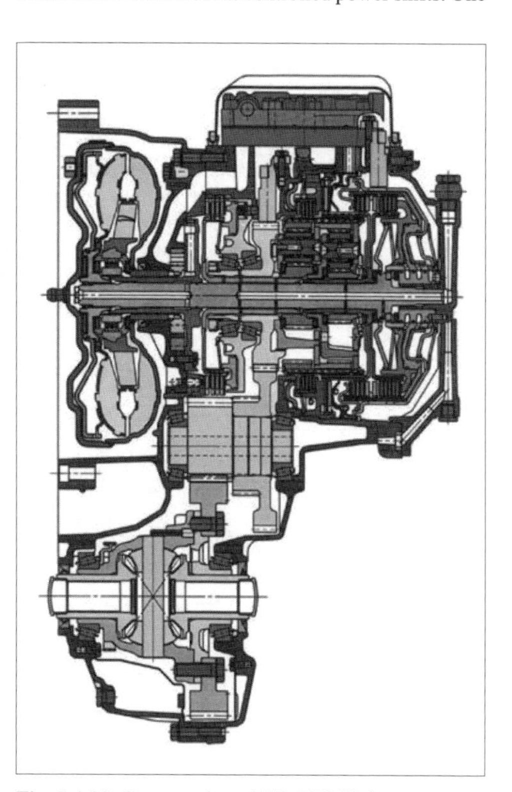

Fig. 5.4-33 Cross section of ZF-4 HP 20 for transverse front-drive installations.

Standard drive

Among passenger cars with standard drive (front engine/rear wheel drive), European manufacturers have largely adopted five-speed transmissions. In comparison to four-speed transmissions, these have up to 25% higher overall gear ratios, which has positive effects on performance, fuel economy, exhaust, and noise emissions. The ZF-5 HP 19 may be used in applications developing up to 300 Nm and power output up to 150 kW, shown in the cutaway in Fig. 5.4-34 (see also Fig. 5.4-25). In addition to a mechanical package which has been proven in many years of service, this design exhibits a number of detail solutions which serve to optimize its operational behavior, improve production quality, increase maintenance friendliness, and improve environmental compatibility and recyclability [18].

Longitudinal front engine/front-wheel drive and all-wheel drive

A front-wheel-drive variant may be derived from vehicle designs with a front longitudinal engine/standard transmission configuration. Such a front-drive variant would use the same basic transmission but with drive taken out via chain, forward through a shaft running parallel to the transmission, to a front-mounted final drive unit powering the front-drive halfshafts. Adding an integral transfer case permits realization of a very compact all-wheel-drive version for passenger car AWD applications. Bolting an external transfer case to the transmission output side also makes possible an add-on all-wheel drive, primarily used for off-road vehicles.

5.4.5 Continuously variable transmissions

5.4.5.1 Operating principles

In contrast to discrete-gear transmissions, continuously variable transmissions offer an advantage in that they eliminate the fixed relationship between vehicle speed and engine speed for individual driving ranges. This results in two separate effects. First, overall ratios can be matched exactly to the thrust hyperbola (Fig. 5.4-35). This in turn results in increased thrust and therefore improved vehicle performance. Second, overall ratios in part-load operation, which currently makes up the major portion of a vehicle's operating cycle, may be set to allow the engine to operate in its best fuel consumption regime (Fig. 5.4-36), contributing to fuel savings. While present-day discrete-gear transmissions employ gearsets in countershaft or planetary configurations, continuously variable power transmission may be realized through a variety of means. Detailed representation of

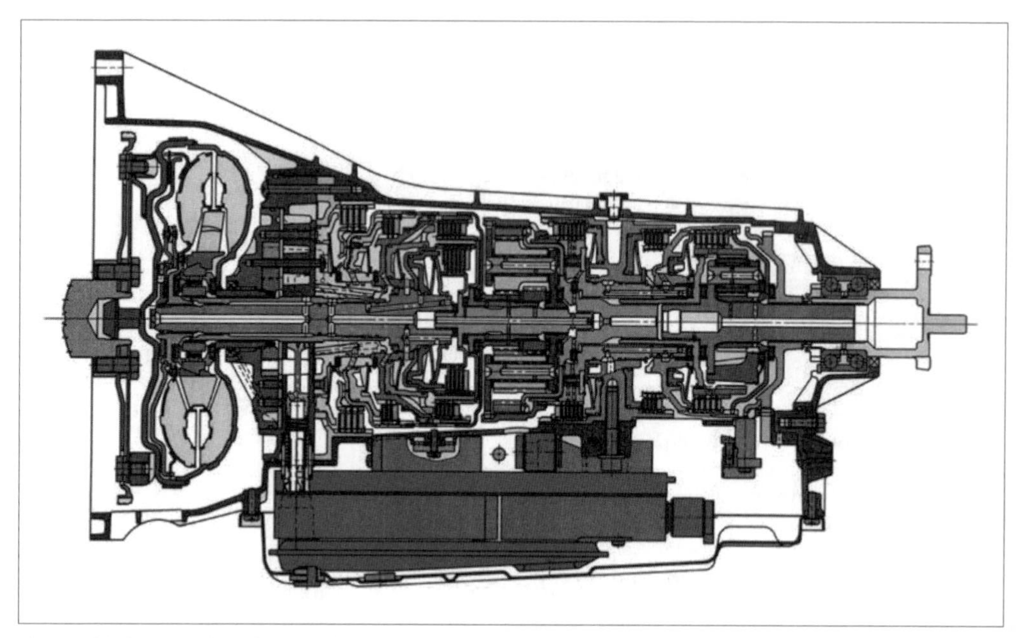

Fig. 5.4-34 Cross section of ZF-5 HP 19 for standard-drive applications.

mechanical, hydrodynamic, hydrostatic, and electric transmissions as well as possibilities for power branching may be found in the literature [21].

While electric drive thus far has found only limited application in motor vehicles, continuously variable hydrostatic drive may be found in vehicles used in construction, in tractors and agricultural equipment, and in special-purpose commercial vehicles and buses. A hydrodynamic continuously variable transmission—in other words, the torque converter—is currently coupled to virtually every discrete-gear automatic transmission as its driveaway element. In mechanical continuously variable transmissions, a principal distinction is made between power transmission via traction (rollers) and belts. Toroidal transmissions are the most highly developed subset of the roller transmission class. Power is transferred between an input and an output disc, both of which are toroidal in shape, via movable rollers. By changing the friction roller inclination angle, the speed ratio between input and output may be varied continuously over its operating range. Power is transferred by friction. This leads to very high Hertzian contact stresses on the components involved and necessitates a special traction fluid capable of withstanding high contact

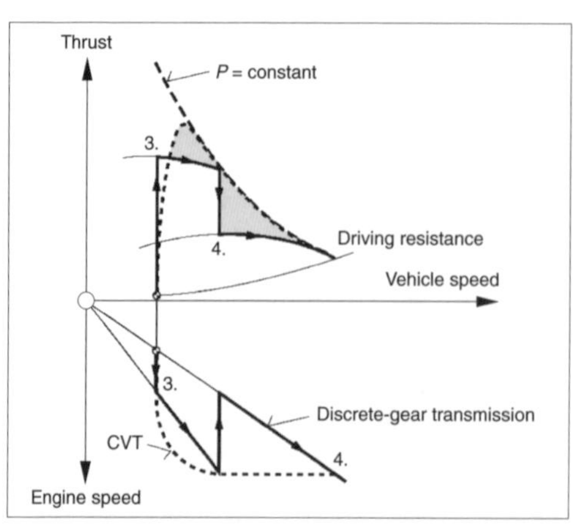

Fig. 5.4-35 Matching overall ratios to the thrust hyperbola.

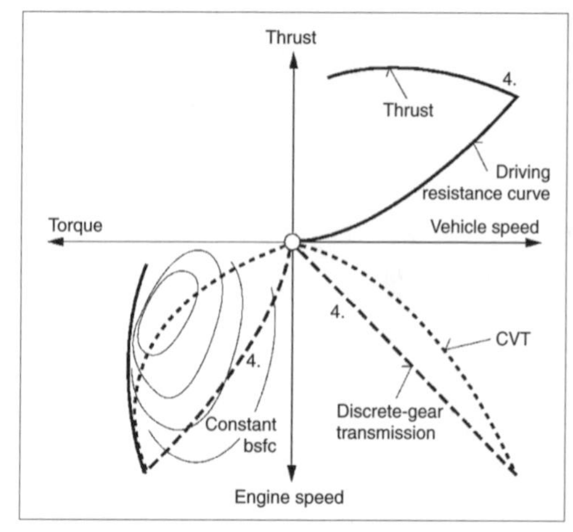

Fig. 5.4-36 Operation in an engine's optimum consumption range.

pressure and transmitting the circumferential forces. Toroidal transmissions have been used since the 1930s [22] but to date have not been able to prevail against discrete-gear automatic transmissions.

Present-day passenger car drivetrain applications employ CVTs (continuously variable transmissions) as nondiscrete mechanical transmissions in the form of a belt transmission. The core element is known as the variator, consisting of two pairs of conical discs. Power is transferred by a form of belt. The pitch diameter of the belt, and therefore the speed ratio, may be altered by means of axial variation of the distance between disc halves. Power transmission between disc pairs and the belt is accomplished by friction. Axial pressure between the discs is applied by pistons and hydraulic pressure. Changing the speed ratio is also accomplished hydraulically.

5.4.5.2 Construction

Fig. 5.4-37 shows in schematic form a CVT transmission for transverse front-engined applications [23]. Component assemblies typical of CVTs are identified. Instead of a hydrodynamic torque converter, which is locked once the vehicle is in motion, a dry or wet multiplate clutch or magnetic powder clutch may be used as the startup (driveaway) element. Switching between forward and reverse is realized by a wet multiplate clutch. Similarly, reverse may be realized by means of a countershaft gearset engaged by a dog clutch. In general, a gear pump is available to supply pressure. A chain or push belt serves as the transmission element. Final drive consists of helical-cut spur gears. Differential action is via bevel gears. Hydraulic and electronic controls are conceptually similar to those found in discrete-gear transmission controls, with the difference that required clamping pressures are as much as five times those found in discrete-gear automatics.

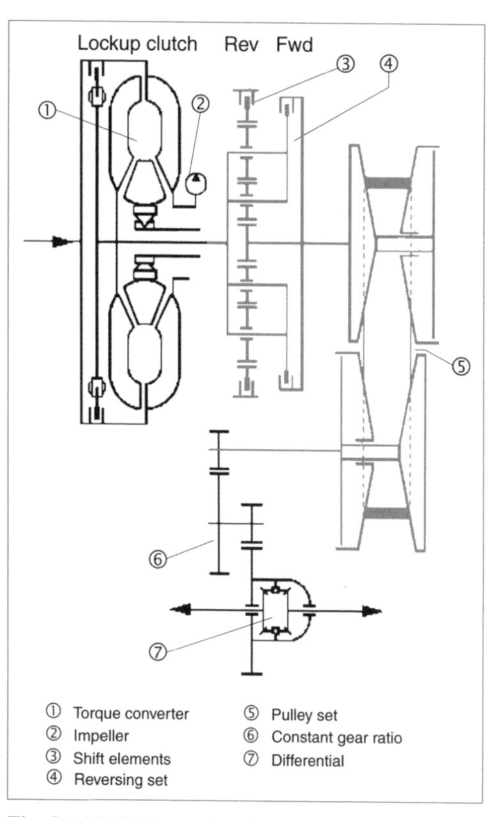

Fig. 5.4-37 ZF Ecotronic schematic.

5.4.5.3 Component groups

In addition to the components and assemblies familiar from discrete-gear transmission practice (see Section 5.4.4.3), certain CVT-specific parts, including the (belt) transmission element, variator, and control system, deserve special attention.

Transmission element

Continuously variable industrial transmissions customarily employ tension chains of various types as transmission elements. Modifications of these chains to forms suitable for automotive applications are currently under development. The "push belt" developed by Van Doorne's Transmissions (VDT) is a well-proven transmission element for production automotive applications (Fig. 5.4-38). This consists of two multilayer bands, each composed of continuous high-strength steel rings about 0.2 mm thick which hold individual stamped-steel elements. In contrast to a conventional chain drive, power transmission from the primary to the secondary pulley is not by means of tension, but rather by compression of the elements—the belt "pushes." Belt widths of 24 and 30 mm with 6, 9, 10, and 12 steel rings are available for different performance classes and up to 150 kW of transmitted power. For lower power levels (engine output torque <65 Nm), rubber bands may be used.

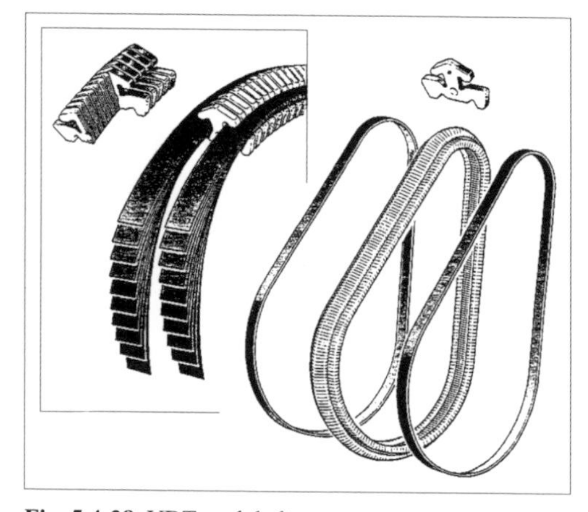

Fig. 5.4-38 VDT push belt.

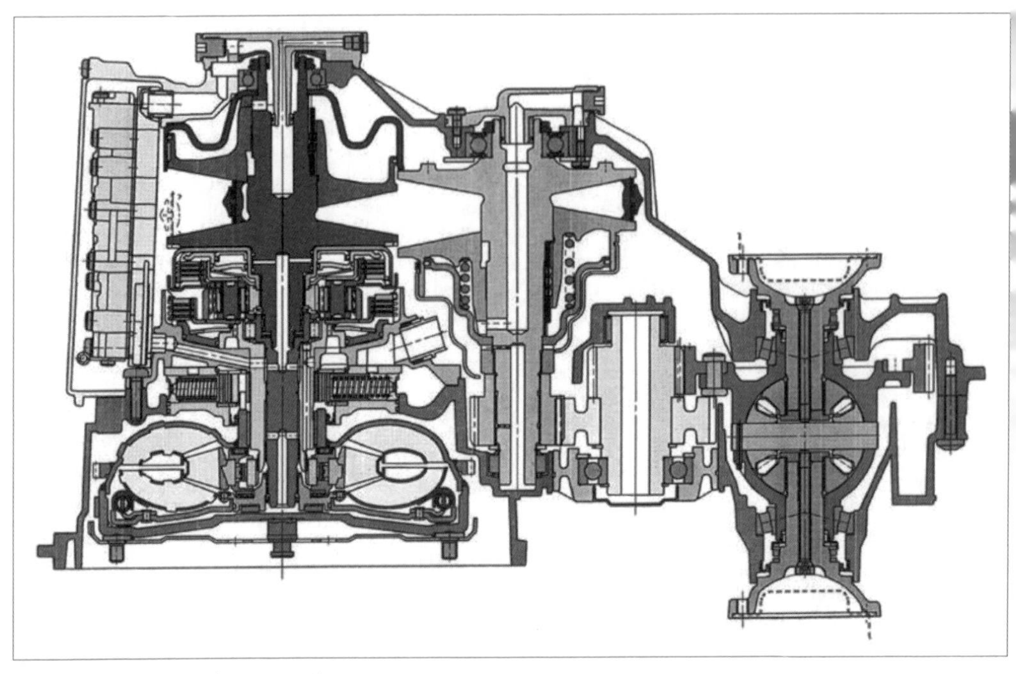

Fig. 5.4-39 ZF Ecotronic cross section.

Variator

Fig. 5.4-39 shows a ZF Ecotronic transmission in cross section. The conical primary and secondary pulley sets are visible. A pulley set consists of a fixed half and a sliding half that may be moved axially along its shaft. The sliding pulley half moves on roller splines, minimizing friction losses between pressure cylinder and the point of contact between band and pulley. Low friction also manifests itself in rapid pulley adjustments,

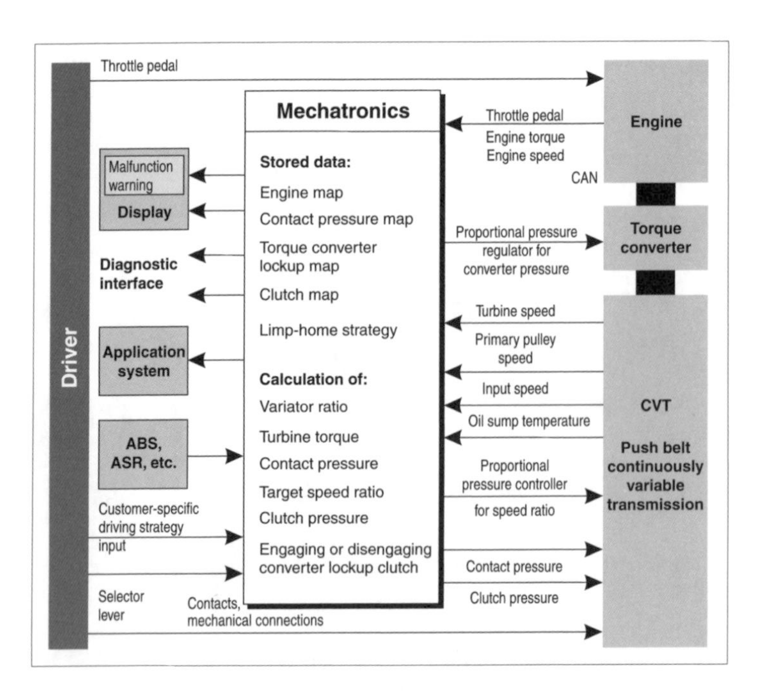

Fig. 5.4-40 Block diagram of a CVT transmission control system.

resulting in good variator dynamic response. The pulley halves are made of hardened steel with ground conical surfaces. The pulley shape is optimized to keep component weight within acceptable limits while assuring minimal deformation. Primary and secondary pressure cylinders consist of stamped steel components. The secondary pulley set is preloaded by a spring. The cylinder space is fitted with an additional sheet metal cover to compensate for lubricating oil centrifugal effects.

Control

Continuously variable transmissions have no gears to be shifted. However, the transmission control system must fulfill several special tasks, including control of variator pressure for belt contact pressure and variator adjustments for speed ratio control. Belt pressure should be proportional to transmitted torque; insufficient belt pressure will result in belt slippage and failure.

Excessive belt contact pressure requires high oil pressure and therefore high pump output. This in turn leads to low transmission efficiency. It is therefore extremely important to control torque-dependent contact pressure as precisely as possible. Speed ratio control is usually achieved by modifying secondary pressure. Belt contact pressure as well as speed ratio control are calculated by transmission electronics in response to vehicle and transmission sensor signals. Electromagnetic pressure regulators convert these signals to hydraulic pressures, which in turn control the appropriate valves in the hydraulic system. Fig. 5.4-40 shows a system schematic of the transmission control system, including signal inputs and functions.

5.4.5.4 Actuation

As far as external shift components are concerned, the selector lever, position indicator, instrument panel display, and parking device as described in section 5.4.4.4 for discrete-gear automatic transmissions are equally applicable to CVTs.

5.4.5.5 Operational behavior

In terms of operational behavior, a continuously variable transmission offers a higher degree of freedom than a discrete-gear transmission. The forced relationship between vehicle speed, selected gear, and engine speed is removed. This relaxation of constraints permits significant intervention in a motor vehicle's driving characteristics.

Driving strategy

Operational behavior of a continuously variable transmission is affected by the variator regulation strategy. A complete range of possible strategies is available, from "extremely economical" to "very sporty." Fig. 5.4-41 illustrates driving program controls for various strategies superimposed on an engine map. It is possible to select an operational characteristic by means of a predetermined program or to employ adaptive operating

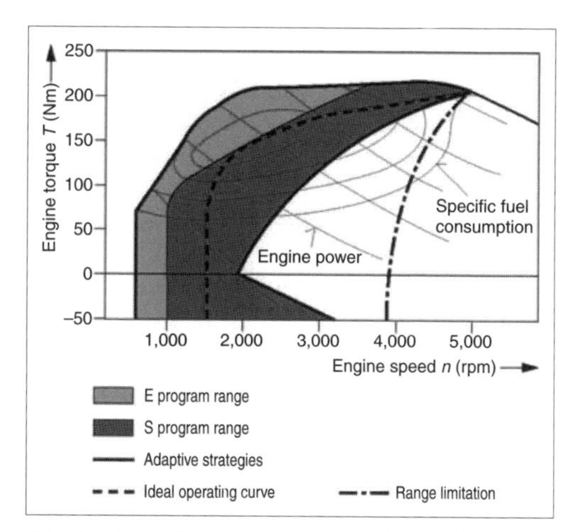

Fig. 5.4-41 Driving program control for a CVT transmission.

point control to achieve the most suitable point on the map as a function of the operating situation at hand.

Fuel consumption

A major motive for selecting a continuously variable transmission is the possibility of reducing fuel consumption. Compared to a four- or five-speed transmission (I = 4.0 to 5.0), a CVT offers a wider ratio range (I = 5.5 to 6.0) and the ability to operate the engine in its most fuel-efficient operating range during a large portion of its driving cycle. Given a driving strategy optimized for economy, a CVT can achieve fuel economy comparable to a five-speed manual transmission. Compared to a four-speed automatic, fuel economy gains of up to 10% may be realized.

Acceleration behavior

The ability to follow exactly the thrust hyperbola during acceleration allows a CVT to "fill out" the gaps in tractive effort exhibited by a discrete-gear transmission. This results in a performance gain and improved 0–100 km/h acceleration. Tests have shown up to 8% better acceleration than with a four-speed automatic transmission.

Comfort

Because, by definition, continuously variable transmissions do not perform any gear shifts, there are no shift jerks. In terms of shift comfort, the CVT is an ideal transmission. Depending on control strategy, engine speed behavior may require some acclimatization. With electronically controlled transmissions, however, driving behavior can largely be made to match that of a discrete-gear transmission.

Manual shift mode

For die-hard fans of human-controlled transmissions, CVTs may be fitted with a manual shift mode. Shifts are carried out by means of a selector lever in a separate gate with three positions: + for upshifts, a central position, and − for downshifts. The CVT simulates discrete gears, giving the option of driving, for example, a six-speed "Tiptronic"-like transmission.

5.4.5.6 Application examples

A cross section of the ZF Ecotronic transmission is shown in Fig. 5.4-39. The CVT is fitted with a torque converter to provide startup behavior and parking and maneuvering comfort like that of a discrete-gear transmission. One particular detail is the rotary piston pump, located between torque converter and pulley sets. Volume flow is restricted to a maximum of 22 L/min by suction-side throttling. This limits pump power consumption and provides good transmission efficiency at engine speeds above 2,200 rpm. The variator has been described in section 5.4.5.3; power is transmitted by a VDT push belt. The double spur gear final drive reduction allows installation of various gearsets, giving overall reduction ratios at startup ranging from 12.5 to 17.2:1.

The cutaway in Fig. 5.4-42 shows a continuously variable transmission variant with wet multiplate clutches as startup elements. The planetary gearset forward drive clutch and reverse gear brake are configured to serve as startup elements for the corresponding direction of travel. Oil is supplied by an external gear pump mounted at the back of the transmission and driven by a quill shaft passing through the primary set. This transmission is characterized by compact dimensions and low weight.

In addition to the preferred transverse front engine vehicle applications, continuously variable transmissions may of course also be fitted to vehicles with front-wheel drive and longitudinal engines or to standard-drive vehicles. Modular transmission components make it possible to satisfy these different applications with a maximum of identical or similar components [24].

Fig. 5.4.42 ZF VT1 cutaway.

5.4.6 Electronic transmission controls

After electronics began to pervade the automotive world, electronic transmission controls also were developed. The first European electronically controlled passenger car transmission entered the market in 1983. By 1990, electronically controlled transmissions had achieved a worldwide market share of 27%. Five years later, 83% of all automatic transmissions were electronically controlled [25]. Today, it is taken for granted that automatic discrete-gear and continuously variable transmissions as well as automated manual transmissions are equipped with electronic controls. As a result, an increasing share of control functions has been transferred from hydraulic to electronic control. While the first generation of electronic transmission controls only manipulated the shift program, modern transmissions use the hydraulic system only to translate electronic systems into hydraulic pressures and to perform a few basic functions such as safety and limp-home features. The most important functions are realized by the electronic transmission control unit (TCU).

5.4.6.1 Overall system

The TCU consists of an electronic hardware module, which receives sensor signals from the transmission and vehicle plus information from other control units, processes this input, and generates output signals to drive actuators in the transmission and elsewhere in the vehicle. Fig. 5.4-43 illustrates how the ECT is integrated in the vehicle communications system [19]. Data exchange with other control units (engine control system, vehicle dynamics systems, instrument panel displays, selector lever, etc.) is achieved by means of a data bus system (CAN, or controller area network). Vital input parameters supplied by the engine control system include engine torque, engine speed, engine temperature, throttle opening, and throttle pedal position. Primary inputs from the vehicle dynamics control units are wheel speeds. Selector lever signals for lever position and the desired shift program are usually transmitted by separate wires. Sensor information provided by the transmission (rotational speeds, position recognition, transmission temperature, etc.) are fed to the control unit by means of a wiring harness.

The valves inside the transmission are actuated by the control unit. Pressure regulators exhibit a specific current/pressure relationship, enabling control of pressure by means of current, while digital control suffices for switching valves (on/off) and pulse-width–modulated valves. Electrical supply for sensors and grounding of transmission internal electrical devices also requires two connecting cables. An additional output is the diagnostic

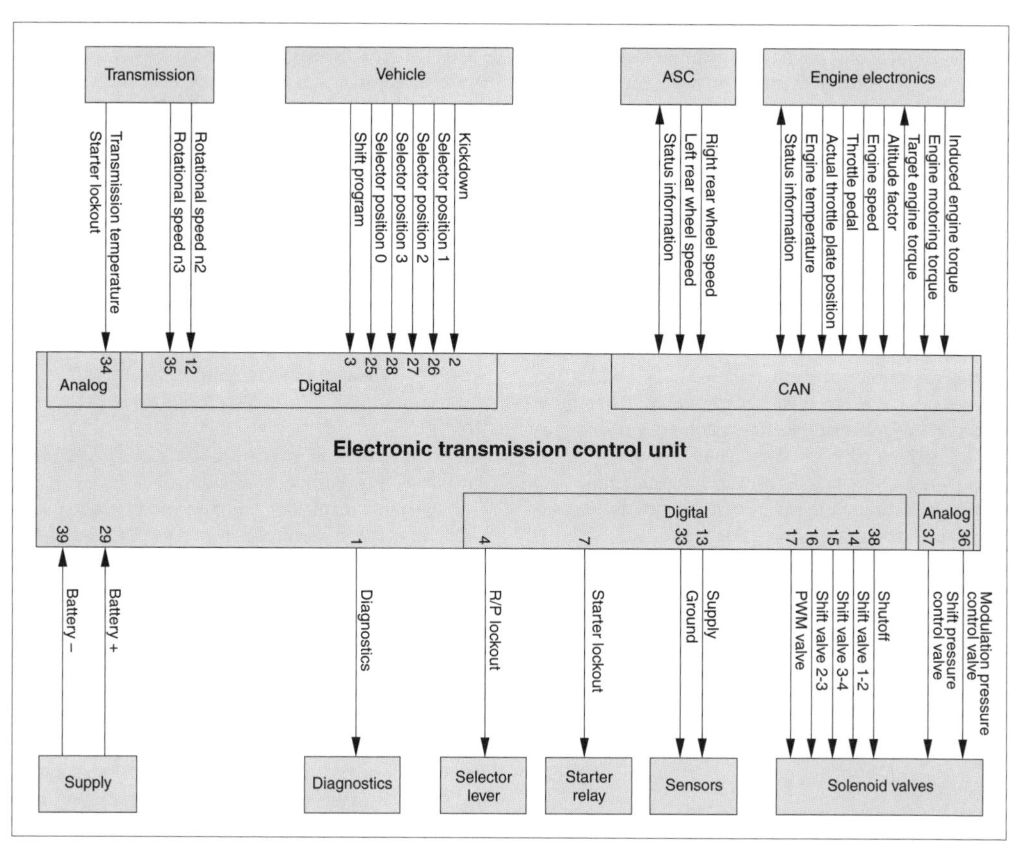

Fig. 5.4-43 Overall TCU system.

line, which not only serves diagnostic functions but also permits programming the control unit. If dashboard instrumentation is not CAN-capable, gear indication in the cockpit is also fed by separate wires from the transmission control unit. Special requirements may include activation of the backup lights, shift lock, and key lock functions.

5.4.6.2 Control unit

In general, transmission control units are supplied as stand-alone devices employing printed-circuit technology mounted either inside the vehicle cabin or in the engine compartment. Fig. 5.4-44 shows a printed-circuit transmission control unit. Major components include microprocessor, working memory (RAM), parameter memory (EPROM and EEPROM), timer, watchdog, CAN component, input signal processors, output elements including power amplifiers, capacitors, diodes, transistors, and power supply as well as mechanical components such as heat sinks, plugs, and housing. Modern control units employ 16- or 32-bit processors. Storage capacity may range up to 512K of ROM and 64K of RAM. Flash memory is often used for possible data modification in production units, permitting reprogramming of installed units. Clock cycles are typically 10–20 msec.

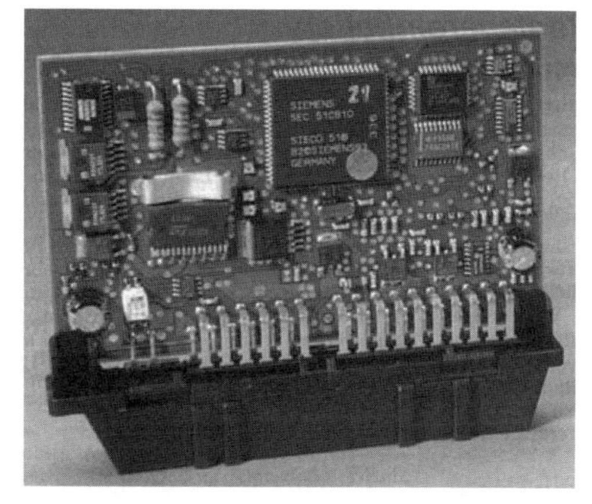

Fig. 5.4-44 Electronic transmission control unit.

In addition to stand-alone versions, combined engine and transmission controls in a single unit (powertrain controllers) are also employed, especially in the United States. A new direction is represented by transmission controls integrated in the transmission itself. These combine electronics, rotational speed sensors, temperature sensors, pressure sensors if required, internal transmission connections, plugs, and position switches in a single mechatronic unit mounted on the hydraulic transmission control unit [13]. The environment within a transmission places especially high demands on electronic components with regard to thermal and vibration loads and on the housing with regard to sealing integrity. This configuration, however, has advantages with regard to space, weight, reliability, tolerances, transmission inspection, and system costs.

Modern electronic transmission control programs are almost exclusively written in high-level languages (usually C or C++). Due to the extensive and highly complex functions involved, function development and programming is accomplished with the aid of appropriate software tools that generate a comprehensible program structure, simplify documentation, and, in conjunction with simulation software, permit immediate functional testing on the computer. Programs are usually divided into a program section and a data section in which the data component is subdivided into nonvariant parameters, variable-dependent data, and application parameters. Often a distinction is still made between control unit, transmission, and vehicle-specific program segments and data sets, which may be developed under various areas of responsibility.

5.4.6.3 Components

Sensors

Inductive pickups or Hall effect sensors are used to generate rotational speed signals. Commonly measured parameters are transmission input and output speeds, and, depending on the transmission concept, specific internal speeds. An output speed sensor may be dispensed with if individual gear speeds are available as sufficiently good signals with adequate dynamic response. Hall effect sensors involve more complex technology, but permit capturing lower rotational speeds than inductive pickups. Transmission oil temperature is sensed by a thermistor soldered directly to the transmission internal wiring. Gear selector positions are usually sensed by a position switch mounted externally on the transmission selector shaft. Depending on service requirements, in addition to the customary P, R, N, and D positions, the electronic control unit may also receive information for gears 4, 3, 2, and 1. Backup lights are activated when the selector is in the R position, and positions P and N are used to engage shift and ignition key locks.

Actuators

Electrical current signals from the electronic control unit are converted into hydraulic pressures by actuators—pressure regulators, on/off (switching) and pulse-width–modulated solenoids. Fig. 5.4-45 shows a sectioned pressure regulator with its associated electrical current/pressure characteristic. This particular example illustrates a flat-seat regulator with falling characteristic: that is, in its inactivated condition, it is opened by applied pressure, and it is closed by increasing control current. Depending on requirements, pressure regulators—closed by spring pressure and exhibiting rising characteristic curves—may be used. Modern pressure regulators work with pressures of 0.5–7.5 bar and control currents of up to about 1 amp. Fig. 5.4-45 also shows a 3/2-way solenoid as used for switching during gear shifts. This is a plate armature solenoid valve with ball seat, normally closed (no applied current). Switching (on/off) solenoids are customarily controlled by pulse-width modulation. Higher pull-in current is ap-

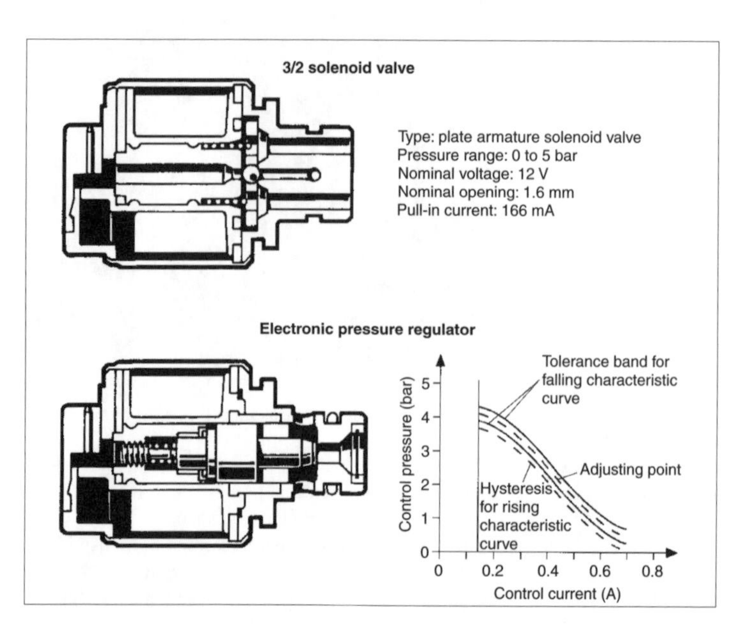

Fig. 5.4-45 Pressure regulator and switching solenoid valve.

plied during the switching phase in order to quickly and positively switch the solenoid. In the holding phase, current is cut back in order to reduce current draw and power losses. Hydraulically, however, the solenoid stays at the limit of its travel. The situation for PWM valves is different: the modulated electrical signal results in corresponding hydraulic modulation. Continuous opening and closing of PWM valves also permits realization of pressure characteristic curves, but with precision not comparable to that of a pressure regulator. PWM valves are used primarily to control lockup torque converter clutches. With rotational speed sensing, closed loop control is possible, resulting in less stringent demands for valve precision than in the case of clutch control.

Wiring and connectors

Modern transmission designs attempt to place all electrical components within the transmission itself, connect them using a single wiring harness, and feed all lines to the external control unit through a single connector. Internal transmission wiring consists of a temperature- and oil-resistant harness. Sensors and actuators are either connected directly (integral wire leads) or by means of plug and socket connections. Terminations are soldered or crimped to the transmission connector. The number of pins in the connector is strongly dependent on the overall system configuration. Stand-alone control units may have 11 to 21 pins. A compact and economical alternative may be offered by an electronic control module integrating all electronic components, with electrical leads formed by stamped conductors [16]. Further integration of the electronic control unit into this module permits additional reduction in wiring complexity and effort. In one example of integrated transmission control, the transmission connector needs only five pins [13].

5.4.6.4 Functions

Pressure control

In electronically controlled transmissions, pressure control functions are realized in ECU software. Translation of electronic signals into clutch pressures is achieved by pressure regulators in the hydraulic transmission control system. The following functions are required to apply the necessary pressure to a transmission's shift elements:

- Setting main pressure as a function of engine torque and torque ratio to a level capable of transmitting the transmission input torque.
- Control of shift pressures to modulate power shifts. The principal events of a power shift event are described in Section 5.4.4.1 and illustrated in Fig. 5.4-24. Possibilities for pressure adaptation and pressure control are described in Section 5.4.4.5.
- Gear selection while stationary and while in motion. Filling pressure and fill time adaptation also find application in the gear selection process.
- Torque converter clutch control. In addition to open and locked clutch, slip control and transitions between modes must be modulated. The multitude of input parameters required for pressure control of a controlled-slip converter clutch are shown in Fig. 5.4-46 [19].

Pressure control determines a transmission's shift quality and, therefore, directly affects driving comfort. Thus, functional development and application of pressure control take on considerable importance.

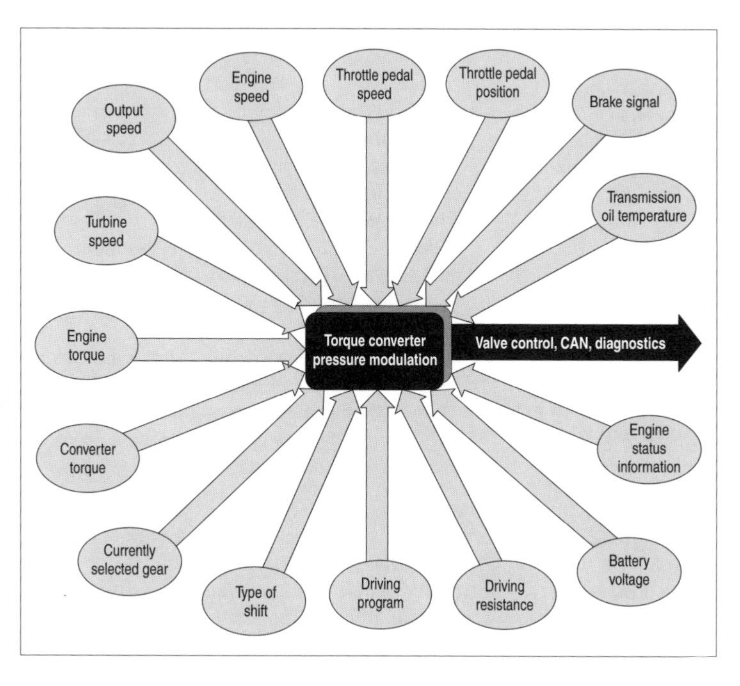

Fig. 5.4-46 Factors influencing torque converter pressure control.

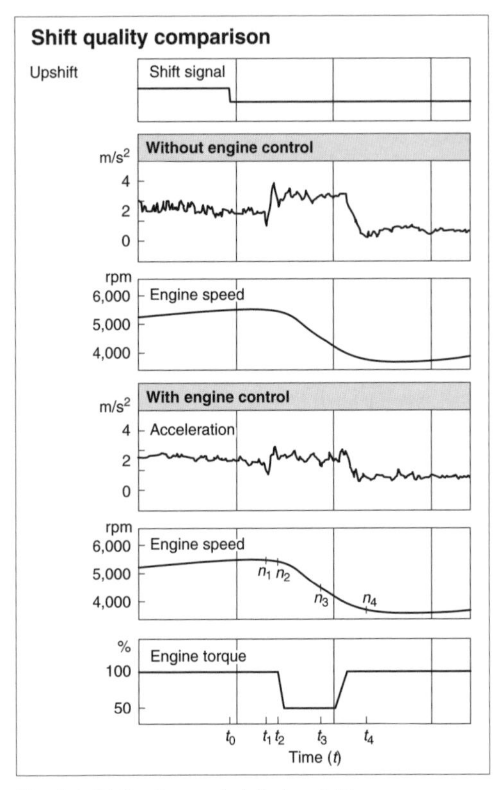

Fig. 5.4-47 Engine control during shift process.

Powertrain management

Transmission shift quality through engine torque modulation may be improved by means of communication between engine and transmission controls. Fig. 5.4-47 illustrates this using an example of a power upshift. During the overspeed phase, synchronization of engine rotating masses is achieved not only through increased pressure to the newly engaged clutch but also by engine torque reduction. Because this process must take place very dynamically and in close relation to transmission-side pressure control, it is achieved by modifying ignition timing [26].

Additional functions are limitation of engine startup speed and engine startup torque. Upon moving the selector lever into gear, engine speed can be limited until the transmission is locked up. This prevents subjecting transmission and driveline to increased shock loads. Similarly, in this operating condition, engine torque can be limited to prevent unacceptable heat loading of friction elements in the shift mechanism. When starting in first gear or reverse with full torque multiplication, input torque can be reduced by directly controlling the engine so that the transmission is not required to transmit the full converter torque multiplication. This permits more appropriate transmission and driveline layout and sizing.

Shift program

Introduction of electronic transmission controls has opened many new shift program possibilities [17, 19, 20]. Along with software representations of various shift programs and consideration of driving behavior through driving condition identification (see configurations in Section 5.4.4.5 and Fig. 5.4-32), the following major functions have been realized:

- Compensation for driving resistance changes. This encompasses not only grades but also varying vehicle load conditions including towing. Suitable shift program modification matches vehicle performance to actual demands and prevents transmission "hunting" between gears.
- Altitude compensation. Combustion engines lose power at high altitudes, such as when negotiating mountain passes. Because shift programs are established at normal geodetic altitudes, they must be modified for such operational conditions.
- Shift prevention in curves. When entering a curve, throttle closure induces an upshift. This may be prevented by sensing throttle pedal movement and vehicle transverse acceleration. The current gear is locked in while negotiating the curve. This leads to better shift harmonization.
- Engine braking downshifts while descending grades. By sensing road grade, the vehicle's service brakes can be augmented by downshifts to lower gears.
- Dynamic full-throttle downshifts. Downshifts do not take place at rigidly defined shift points, but rather as a function of throttle pedal gradients. As a result, a leisurely driving style results in late downshifts. The shift program seems well balanced and does not convey a "nervous" impression, yet spontaneous downshifts and therefore sporty shift behavior are readily available in response to rapid pedal movements.
- Warm-up program. In a cold condition, elevated shift points cause the engine to operate at higher speeds, thereby bringing the catalytic converter up to its operating temperature more rapidly.
- Traction demands. Traction control systems are supported by special gear selection. For example, second gear may be selected when starting up on low-coefficient surfaces.
- Speed control ("cruise control") demands. Control quality in speed control mode may be improved by altering shift schedules. Downshifts to enable engine braking on downgrades permit maintaining constant speed even while descending hills.

Additional vehicle- and transmission-specific functions are being developed to improve driving behavior of passenger cars equipped with automatic transmissions. Increased vehicle electronification and system networking will offer new possibilities which are not currently available in production vehicles. Examples include modifying shift programs in response to telematic information systems such as GPS and optical recognition of the road surface and immediate traffic environment.

Safety concept and diagnostics

The safety concept of electronically controlled transmissions is configured so that in the event of electrical failure or loss of electronic signals, the electronic transmission control system attempts to fall back on substitute parameters. While this may affect transmission shift quality or restrict shift program functions, transmission functionality is fully and completely retained. Only the failure of basic information, without which safe operation may no longer be assured, or the failure of hardware such as disconnected wiring will cause the transmission control system to enter an emergency shutdown mode, while the transmission continues nevertheless to function in a hydraulic limp-home mode.

Operational malfunctions are detected by diagnostic functions and stored in diagnostic memory. A service facility can read this memory with the aid of a diagnostic computer and a diagnostic connection. In the event of a malfunction, this permits rapid and accurate fault determination and repair of the defective component. Diagnostic data permit not only fault identification in electrical and electronic components, but they also facilitate drawing conclusions regarding faults in transmission mechanical components.

5.4.7 Prognosis

The passenger car sector has long been dominated by manual transmissions and power-shifting automatics with hydraulic torque converters. Distribution of these two transmission concepts varies by geographic region. Compared to the past, the present picture shows considerably greater variety. If we regard the conventional-layout (front engine/rear drive) passenger car, we now find, in addition to these two conventional transmission types, manually shifted transmissions with automatic clutches (so-called semiautomatic transmissions), automated manual transmissions, and, in front-drive vehicles, belt-type continuously variable transmissions (CVTs). Moreover, continuously variable traction drives are under consideration for the highest-performance standard-drive passenger cars.

Electrified drivelines appear attractive for passenger cars with alternative propulsion systems (such as battery-powered electric vehicles, hybrids). This also applies to fuel-cell propulsion systems, which are objects of intensive research.

In view of their limited storage capacity, battery-powered electric vehicles will not enjoy any chance of widespread market penetration. Instead, their use will remain limited to special applications.

Hybrid drives—that is, vehicles which combine a combustion engine with an additional propulsion system (electric, hydraulic, or mechanical with flywheel)—require demanding, complex driveline technology. Their cost/benefit relationship must therefore be examined critically, leading to the conclusion that in these cases, too, only certain limited applications may be accorded any chance of future success.

If fuel cells should achieve a breakthrough, the result will be long-lasting major effects on mechanical drivetrain technology, as it moves electric propulsion combined with mechanical transmission solutions to the forefront.

At present, the future of transmission technology for conventional-drive passenger cars appears as follows:

1. Manual transmissions will maintain their position as cost-effective solutions.
2. The automatic clutch will enjoy only limited market penetration, as this represents a halfway measure compared to automatic transmissions.
3. Automated manual transmission applications will expand, primarily in small, economical vehicles capable of only moderate acceleration performance. They will, however, prepare the ground for future market penetration by comfort-oriented power-shifting automatics and continuously variable transmissions.
4. Continuously variable transmissions will continue to gain in importance, particularly in transverse front-drive applications.

Symbols appearing in Section 5.4

Units and dimensions		
a	m/s^2	Vehicle acceleration
c_p	m/s^2	Circumferential flow velocity
c_d	—	Aerodynamic drag coefficient
f	—	Coefficient of rolling resistance
g	m/s^2	Acceleration of gravity
i	—	Ratio, mechanical advantage
i_f	—	Final drive ratio
i_g	—	Discrete gear ratio
m	kg	Vehicle mass
n	—	Number
p	Pa	System pressure
r	m	Radius, tire dynamic radius
r_m	m	Mean friction radius
v	m/s	Vehicle speed
z	—	Number of gear teeth
A	m^2	Cross-sectional area
D	m	Pitch diameter
D_o	m	Outside diameter
D_i	m	Inside diameter
F_B	N	Brake force
F_d	N	Driving resistance force
I	—	Transmission ratio range
J	kg m^2	Moment of inertia
M_g	Nm	Clutch torque
T	Nm	Torque
$\dot{V}$	m^3/s	Volume flow rate
α	°	Grade
Δ	—	Difference
η	—	Efficiency
κ	—	Factor for rotationally accelerated vehicle components

λ	—	Power factor
μ	—	Torque ratio
μ_{dyn}	—	Coefficient of dynamic friction
μ_{stat}	—	Coefficient of static friction
ν	—	Speed ratio
ρ	kg/m^2	Density
φ	—	Overdrive factor
ω	rad/s	Angular velocity
$\dot{\omega}$	rad/s^2	Angular acceleration
Subscripts		
ovl	Overall	
max	Maximum	
min	Minimum	
eng	Engine	
0	Relative to maximum power	
L	Stator	
P	Impeller ("pump")	
T	Turbine	
1	Sun gear	
2	Planet gear	
3	Planet gear carrier	
4	Ring gear	

References

[1] Förster, H.-J. *Automatische Fahrzeuggetriebe*. Springer-Verlag, Berlin, Heidelberg, New York, 1990.
[2] Förster, H.-J. *Die Kraftübertragung im Fahrzeug vom Motor bis zu den Rädern, Handgeschaltete Getriebe*. Verlag TÜV Rheinland GmbH, Cologne, 1987.
[3] Lechner, G., and H. Naunheimer. *Fahrzeuggetriebe-Grundlagen, Auswahl, Auslegung und Konstruktion*. Springer-Verlag, Berlin, Heidelberg, New York, 1994.
[4] Pierburg, B., and P. Arnborn. *Gleichlaufgelenke für Personenkraftfahrzeuge*. Verlag Moderne Industrie, Landsberg/Lech, 1998.
[5] Schmidt, G. "Schwingungen in PKW-Antriebsträngen." VDI-Berichte 1220, VDI-Verlag, Düsseldorf, 1995.
[6] Hafner, K.E., and H. Maass. *Die Verbrennungskraftmaschine*, Volumes 1 through 4. Springer-Verlag, Berlin, Heidelberg, New York, 1981/84.
[7] Duditza, F. *Kardangelenkgetriebe und ihre Anwendungen*. VDI-Verlag, Düsseldorf, 1973.
[8] Drexl, H.-J. *Kraftfahrzeugkupplungen*. Verlag Moderne Industrie, Landsberg/Lech, 1997.
[9] Ersoy, M. "Entwicklungstendenzen für Getriebe-Außenschaltungen." VDI-Berichte 1393, VDI-Verlag, Düsseldorf, 1998, pp. 273–286.
[10] Eberspächer, R., T. Göddel, and C. Wefers. "Das Schaltgetriebe und Schaltungskonzept der Mercedes-Benz A-Klasse." VDI-Berichte 1393, VDI-Verlag, Düsseldorf, 1998, pp. 491–511.
[11] Looman, J. *Zahnradgetriebe*. Springer-Verlag, Berlin, Heidelberg, New York, 1970.
[12] Dach, H., and P. Köpf. *PKW-Automatgetriebe*. Verlag Moderne Industrie, Landsberg/Lech, 1994.
[13] Göddel, T., H. Hillenbrand, C. Hopff, and M. Jud. "Das neue Fünfgang-Automatikgetriebe W5A180." Sonderausgabe ATZ und MTZ, Mercedes A-Klasse, 1997, pp. 96–101.
[14] Flegl, H., R. Wüst, N. Stelter, and I. Szodfridt. "Das Porsche-Doppelkupplungs-(PDK)-Getriebe." ATZ 89, 1987, 9, pp. 439–452.
[15] Wagner, G. "Gestaltung und Optimierung von Bauteilen für automatische Fahrzeuggetriebe." Konstruktion 49, 1997, 6, pp. 31–35.
[16] Rösch, R., and G. Wagner. "Elektrohydraulische Steuerung und äußere Schaltung des automatischen Getriebes W5A330/580 von Mercedes-Benz." ATZ 97, 1995, 10, pp. 698–706.
[17] Maier, U., J. Petersmann, W. Seidel, A. Strohwasser, and T. Wehr. "Porsche Tiptronic." ATZ 92, 1990, 6, pp. 308–319.
[18] Burger, A., G. Gierer, J. Haupt, and J. Völkl. "Das neue Fünfgang-Automatikgetriebe für die neue BMW 3er-Baureihe." Sonderausgabe ATZ und MTZ: BMW 3er, 1998, pp. 134–140.
[19] Rösch, R., and G. Wagner. "Die elektronische Steuerung des automatischen Getriebes W5A330/580 von Mercedes-Benz." ATZ 97, 1995, 11, pp. 736–748.
[20] Tinschert, F., G. Wagner, and R. Wüst. "Arbeitsweise und Beeinflussungsmöglichkeiten von Schaltprogrammen automatischer Fahrzeuggetriebe." VDI-Berichte No. 1175, VDI-Verlag, Düsseldorf, 1995, pp. 185–203.
[21] Förster, H.-J. *Stufenlose Fahrzeuggetriebe*. Verlag TÜV Rheinland, Cologne, 1996.
[22] Gott, P.G. *Changing Gears: The Development of the Automotive Transmission*. SAE historial series 90-21369, Warrendale, 1991.
[23] Boos, M., and W.-E. Krieg. "Stufenloses Automatikgetriebe Ecotronic von ZF." ATZ 96, 1994, 6, pp. 378–384.
[24] Boos, M., and H. Mozer. "ECOTRONIC—The Continuously Variable ZF Transmission (CVT)." SAE 970685.
[25] Pieper, D. "Automatic Transmissions—An American Perspective." VDI-Berichte 1175, VDI-Verlag, Düsseldorf, 1995, pp. 25–39.
[26] Neuffer K. "Elektronische Getriebesteuerung von Bosch." ATZ 94, 1992, 9, pp. 442–449.

5.5 All-wheel drive, braking, and traction control

5.5.1 All-wheel-drive concepts

5.5.1.1 Application of all-wheel drive

All-wheel drive or four-wheel drive has a long tradition in automotive engineering and maintains an important position in modern powertrain technology. From a technical standpoint, two main categories of all-wheel-drive vehicles may be defined:

- The segment containing off-road-capable vehicles, which require four driven wheels for traction reasons
- The sports car and sedan segment, in which all-wheel drive provides, in addition to improved traction, benefits in vehicle dynamics

For pure all-terrain vehicles that see frequent off-road use, good traction means optimum utilization of the available friction coefficient at all wheels. This is best achieved by locking axles as needed (eliminating any differential effects).

All-wheel drive assures optimum acceleration for high-powered passenger and sports cars, regardless of road conditions. In addition, targeted torque distribution within the drivetrain permits specific tuning of vehicle behavior.

For this vehicle category, rigid coupling of axles is ruled out for reasons of vehicle dynamics, wear, and comfort (excessive driveline windup). To avoid these stresses and achieve the desired steering tendency (oversteer, understeer), a central differential or appropriate coupling units provide the necessary speed equalization. In addition, locking differentials have proven very useful for taking full advantage of available traction at all wheels. It should be noted, however, that for braking, completely independent wheels present the best solution. Each coupling affects braking stability to a greater or lesser degree and requires corresponding additional measures.

5.5.1.2 All-wheel-drive characteristic curves

The traction diagram (Fig. 5.5-1) has proven highly useful in evaluating various systems. This plots drive torque distribution against vehicle dynamic weight distribution, thereby enabling basic determination of vehicle behavior (over- or understeer). Acceleration or grade is plotted on the vertical axis and drive torques or weight distribution on the horizontal axis.

Rigidly coupled axles result in equal slip and therefore equal friction coefficient utilization. From the definition of friction coefficient,

$$F_f/F_r = W_f/W_r$$

That is, for rigid all-wheel drive, the circumferential forces and therefore torques are distributed in proportion to the corresponding dynamic axle loads (straight line "A" in Fig. 5.5-1, where all friction coefficients intersect). Any weight distribution to the right of this emphasizes the front axle and therefore a tendency to understeer, while the left side of the diagram represents a tendency toward oversteer.

Open center differentials distribute drive torque in a constant ratio (B). Viscous couplings (D, E) and locking differentials (C) distribute torque as a function of slip, again dependent on grade or acceleration.

In addition to the nominal design condition, extreme loads must also be considered. In some cases, all-wheel drive system characteristics are represented not by curves

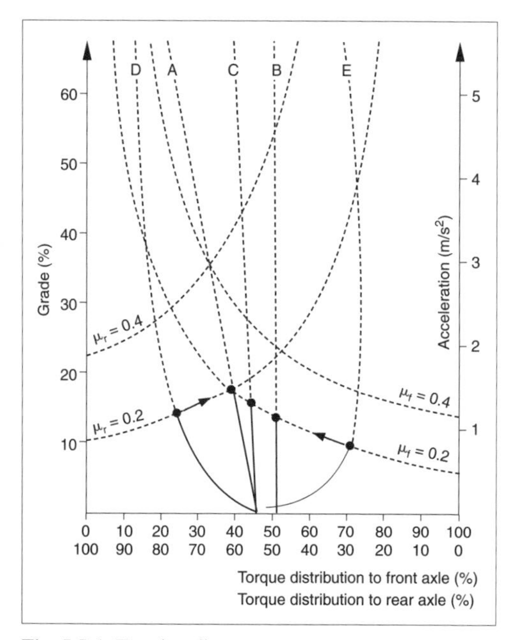

Fig. 5.5-1 Traction diagram.

but rather by surfaces. The extremes of these surfaces must be taken into consideration. Evaluation of a vehicle is only possible upon examination of all possible conditions. An ideally designed system conveys a nonvarying driving feel, regardless of load and operating conditions. Whether the design seeks to achieve oversteer, understeer, or neutral steer is driven by the manufacturer's philosophy, which usually seeks to have derivative 4×4 models maintain a basic similarity to their two-wheel-drive base vehicles.

5.5.1.3 Drivetrain taxonomy

Any attempt to classify drivetrains in terms of design characteristics would quickly result in a confusing picture. Classification according to torque distribution options yields a much more clearly understood system. Assuming homogeneous road conditions and linear vehicle motion, speed and acceleration or grade are varied. Based on these criteria, four groups or generations may be established, which, however, should not in any way be taken to represent a performance evaluation of individual systems.

An appropriate system must be chosen to meet individual application goals (traction or vehicle dynamics), vehicle type, and package. It must also be remembered that along with engineering considerations, target costs exert a decisive influence on system selection.

The first group includes selectable four-wheel drive and permanent four-wheel drive with a center differential (with or without mechanical lockup).

The second group is largely based on the differential power split exemplified by the first group but overlays this basic configuration with a system-internal limited slip system (e.g., viscous coupling or Torsen differential). This group also includes driving the second axle by means of a speed difference sensing coupling which replaces the center differential and which transmits torque as a function of wheel slip. In addition to the familiar viscous couplings, this group may include multiplate clutches which respond to pressure generated by speed differential sensing pumps (e.g., Geromatic, Honda Dual Pump, Viscolok, and the Toyota RBC system).

Typical of the third group is application of electronics for external control of clutches and limited-slip devices for center differentials. This group is represented by controllable viscous couplings (Viscomatic) and multiplate clutches made by Haldex (VW 4motion). Both systems superpose external controls on the intrinsic characteristics of the base system. Multiplate clutches or limited-slip units are strictly externally controlled, as will be described below for the Nissan ETS.

A fourth group consists of all-wheel drive with freely selectable power distribution. Technically, this may be achieved by means of suitable controllable multiplate clutches or speed modulating gearboxes. Due to higher costs and the resulting negative impact on the cost/benefit ratio, such solutions are at present still limited to niche vehicles.

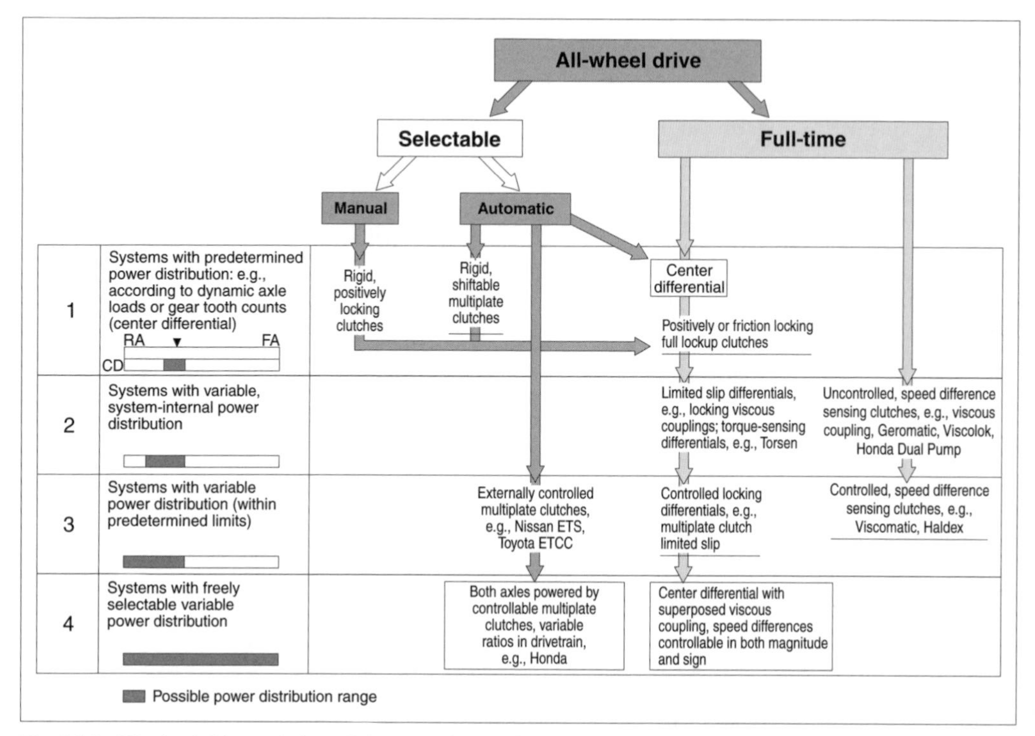

Fig. 5.5-2 All-wheel drive variations. RA, rear axle; FA, front axle; CD, center differential.

5.5.1.4 Components

To stay within the scope of this book, the following discussion will limit itself to widely established systems:

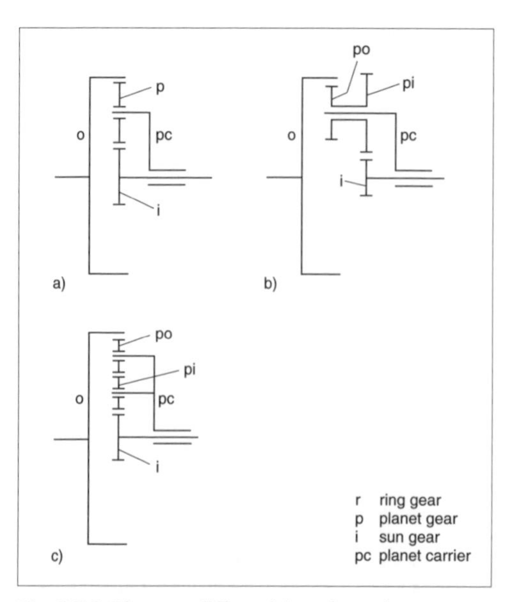

Fig. 5.5-3 Planetary differential configurations.

Center differential with/without limited slip

Alongside conventional bevel gear differentials, planetary gearboxes are also suitable as center differentials, as appropriate geometric design of planetary components (sun, planet, and ring gears) and choice of torque input and output modes may be used to achieve a specific drivetrain torque distribution.

In the most common arrangement (Fig. 5.5-3a), with input through the planet carrier, torque distribution may be calculated from the gear pitch diameters

$$T_{\text{ring gear}} / T_{\text{sun gear}} = d_{\text{ring gear}} / d_{\text{sun gear}}$$

Possible torque distributions are:

Configuration a: ring gear/sun gear = 65:35 ($\pm 5\%$)

Configuration c: carrier/sun gear = 65:35 to 50:50

An example of an engineering solution for configuration c is shown in Fig. 5.5-4 (Mercedes-Benz 4matic). Gearing may be straight-cut or helical, depending on noise requirements.

Limited-slip units

System-integral, fixed torque distribution provided by so-called "open" center differentials can only be matched to the maximum traction value at one specific operating point. In order to provide the greatest possible tractive effect under conditions deviating from this single

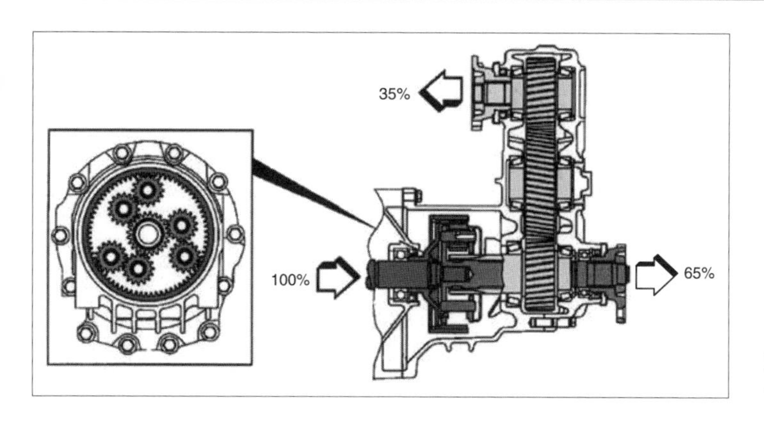

Fig. 5.5-4 Mercedes-Benz 4matic.

operating point, it is necessary to employ some form of limited slip.

A special case of "limited slip" is represented by active brake actuation. This is discussed in Section 5.5.2.

Possible configurations for limited slip are:

- Automatic limited slip
  - Speed-sensing limited slip (viscous couplings, etc.)
  - Torque-sensing limited slip (Torsen, GKN Powrlock®)
- Externally switched/controlled limited slip
  - Manually actuated by driver
  - Electronically controlled (Steyr ADM, multidisc clutches, etc.)

In comparing limited-slip systems, the so-called limited-slip percentage or locking factor is often used:

$$S = (T_{high} - T_{low}) \times (T_{high} + T_{low})\ [\%]$$

Usually, the torque bias ratio is expressed as:

$$T_{low} : T_{high} = 1 : x \text{ or } T_{high} : T_{low} = x : 1$$

Torsen differential, GKN Powrlock®

These differential units have torque-dependent locking. The effect is preset by appropriate friction elements acting on the faces of helical gears.

With very low traction or with one extremely unloaded wheel, however, locking torque approaches zero, with a corresponding collapse of forward progress.

With a torque-sensing locking differential (Torsen differential, GKN Powrlock®) as a center differential, it is not possible to obtain torque distribution differing from axle load distribution, since internal friction results in a locking effect similar to a rigid coupling. The vehicle therefore reacts in a more or less neutral manner.

The great advantage of the Torsen unit is its freedom from windup during cornering, as speed differences "desired" by the vehicle are permitted. This also applies to braking, since it provides braking stability similar to that of an open differential.

Couplings with self-acting torque adaptation

Viscous coupling

Viscous couplings are members of the class of power transmission elements which have system-intrinsic behavior in terms of automatic torque distribution as well as automatic, longitudinal or transverse limited slip dependent on rotational speed difference. Torque is transmitted by means of fluid friction between clutch elements as a result of a forced rotational speed difference between inner and outer sections.

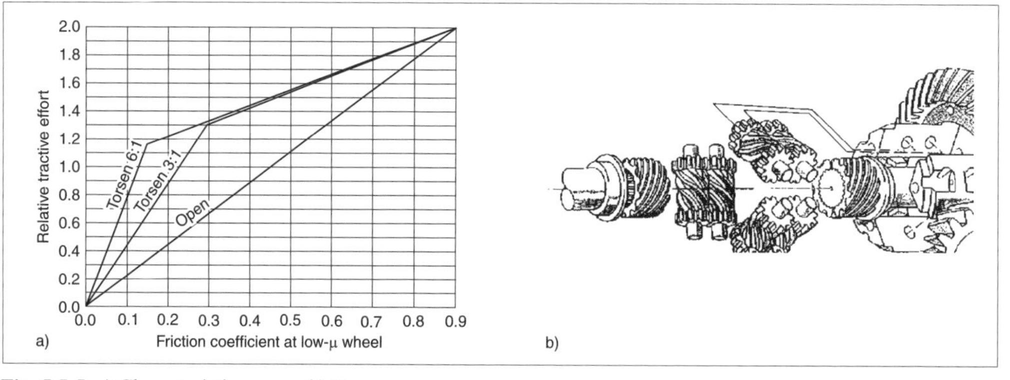

Fig. 5.5-5 a) Characteristic curves; b) Torsen.

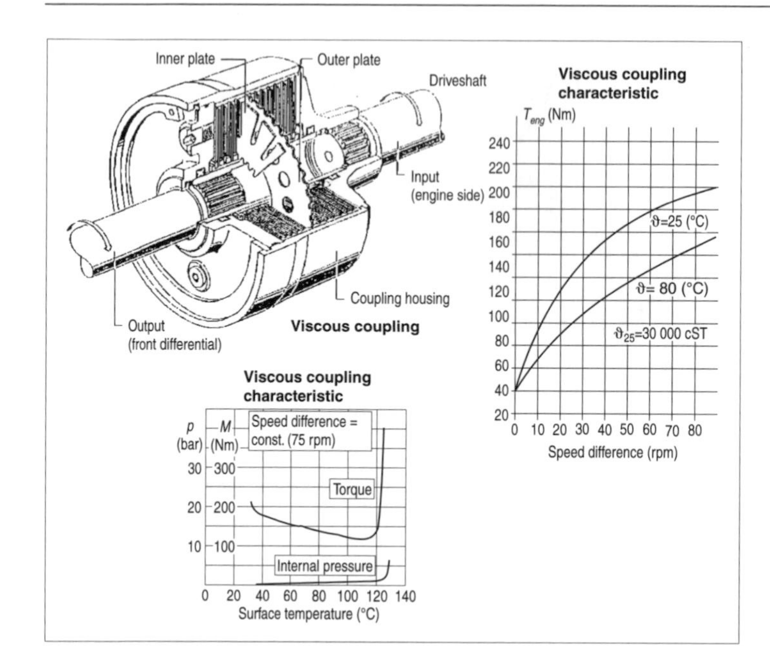

Fig. 5.5-6 Characteristic curves and cutaway of viscous coupling.

Inner and outer clutch discs, installed in pairs, contain slots or holes with geometric protrusions which, in the event of speed differences, cause increased shear forces in the surrounding silicone fluid.

In addition to disc geometry, the characteristic curve of a viscous coupling is also dependent on the following guidelines:

- Plate diameter (90–125 mm)
- Plate gap (0.20–0.35 mm)
- Number of plates
- Fluid fill (viscosity ~25,000 cSt, 90%)
- Temperature
- Absolute rotation speed

A special characteristic of viscous couplings is their intrinsic protection against excessive temperatures, often referred to as the "torque hump phenomenon" (due to the prominent kink in the characteristic curve). Silicone oil expands as temperature rises. Because of solubility of air in this oil, the pressure gradient is initially flat; only at 100% filling does the pressure gradient rise steeply. This also alters flow conditions in the coupling; the plates come into intimate contact with one another, a process which is assisted by their ability to slide axially and by openings in the discs. The particular shape of the plate slot edges causes a "windshield wiper effect." Rising absolute pressure promotes a pressure differential at any given pair of plates, turning these into a true friction coupling (excessive wear is prevented by hardened surfaces).

In hump mode, the clutch automatically reaches thermal equilibrium; that is, the power loss ($P_{loss} = (\pi/30) \cdot T \cdot \Delta n$) equals rejected heat. The coupling is essentially rigidly locked.

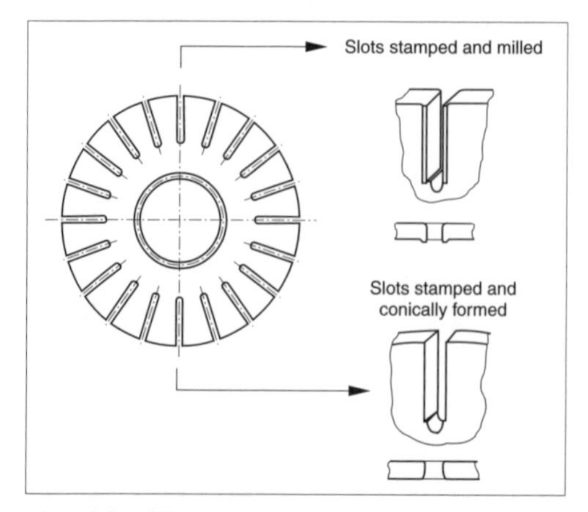

Fig. 5.5-7 Differences in clutch plate geometry.

Gerodisc®, Geromatic® and Twin Geromatic®, Viscolok

In terms of design, these speed-sensing couplings are based on a mult-disc wet clutch, with self-generated pressure applied by a pumping element and piston.

In detail, the Geromatic® consists of a rotating housing containing a hydrostatic displacer unit, an output shaft, and a multidisc clutch to join output shaft and driven housing. In the event of a rotational speed difference, the clutch plates are pressed together by means of a piston actuated by pressure from a gerotor pump. Two inlet and two outlet valves control oil flow for both relative rotation directions. A temperature-compensating throttle valve within the piston gives progressively rising pressure with increasing speed differ-

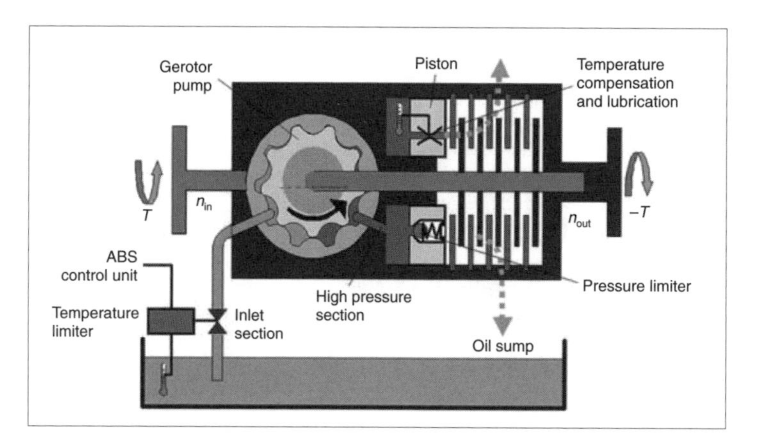

Fig. 5.5-8 Geromatic®.

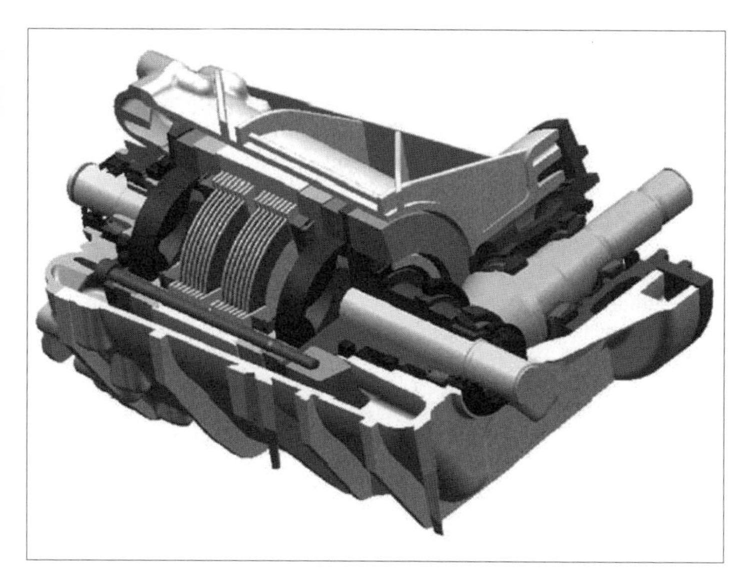

Fig. 5.5-9 Twin Geromatic®.

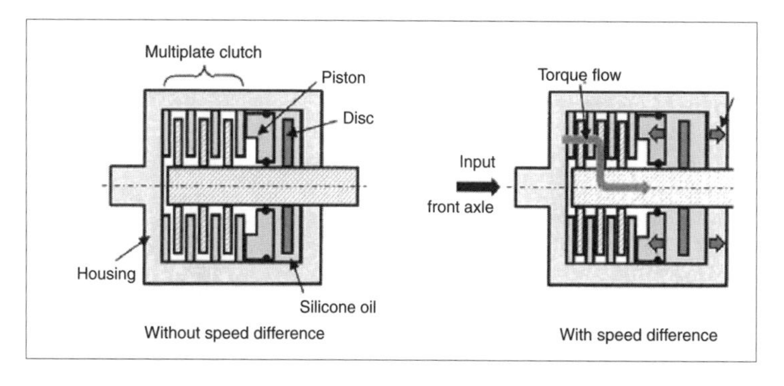

Fig. 5.5-10 Schematic of Toyota RBC©.

ence and thereby increased torque. The outflowing oil provides disc cooling and lubrication.

If necessary, a pressure limiting valve can limit torque transmitted to the second axle.

The Twin Geromatic drives each rear wheel of an all-wheel-drive vehicle by means of its own dedicated unit, eliminating the need for a mechanical differential. In addition to their function in coupling front and rear axles, these units also produce the effect of a limited slip differential across the rear axle.

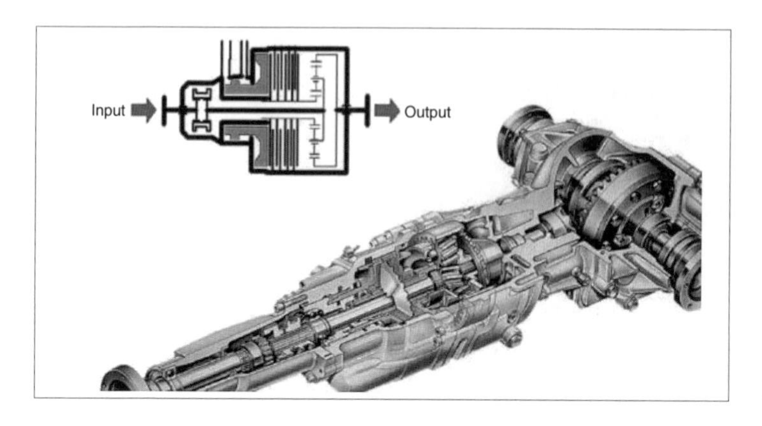

Fig. 5.5-11 Viscomatic.

Toyota RBC© (Rotary Blade Coupling)

Similar to the GKN Viscolok, a single-plate viscous coupling is employed as a speed difference sensing pump element. Torque is transmitted by a friction disc clutch. The curve of torque as a function of speed difference is slightly degressive.

Externally controlled couplings

The characteristic curve of an uncontrolled coupling is always a compromise between traction, vehicle dynamics, and comfort. Often, braking stability (see Section 5.5.7) can only be achieved by means of an inline coupling to separate the driveline or of an overrunning clutch. External control permits a less compromising solution to these challenges. In such systems, the following effects can be sensed electronically and applied for drivetrain control:

1. Wheel rotational speed
2. Engine torque, throttle movement
3. Braking situation
4. ESP signal
5. Steering angle, driving in reverse, clutch actuation
6. Special situations (towing, excessive temperatures, dynamometer testing)

The desired self-steering tendency is determined by drive torque distribution as a function of the above parameters. Control is exerted in the form of the rotational speed difference Δn, as this also takes dynamic axle loads into account.

It is therefore necessary to network control units for engine, transmission, and brakes by means of a BUS system.

Demands on such systems are, in particular:

- High torque transfer at very small differences in rotational speeds, which translates into optimum traction and good efficiency combined with low thermal loading
- Minimum torque virtually independent of rotational speed differences
- Rapid torque ramp-up and ramp-down to satisfy vehicle dynamic and braking stability demands ($T_{max} \rightarrow T_{min} \leq 100$ ms)

Principal examples of this group are:

Viscomatic©

This is based on a controlled viscous brake, which governs the reaction torque on the sun gear of a planetary gear train.

The assembly consists of three main component groups:

1. Planetary transmission and viscous brake
2. Hydraulics
3. Electronics and control logic

A piston is used to vary clutch plate clearance (between 0.50 and 0.15 mm) and clutch volume (fill factor changes, between 55 and 92% silicone oil); both parameters are determinant for torque level. Intensive mixing of silicone oil and air (no centrifugal force, otherwise the system would cease operation) is of critical importance.

Variable braking of the sun gear by means of the viscous unit permits altering drive distribution between "single axle operation" and "locked all-wheel drive" (with nearly locked sun gear).

Haldex©

Based on a combination of a speed difference sensing pump system (swash plate and annular piston) and multiplate clutch, the internal torque transmission bandwidth is extended by external controls. A stepper motor

Fig. 5.5-12 Haldex© coupling.

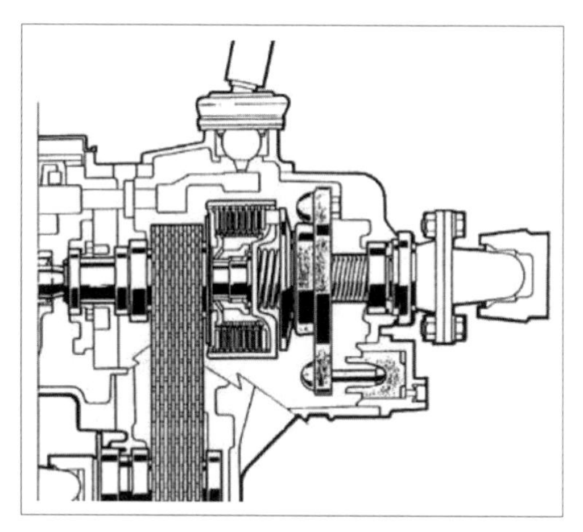

Fig. 5.5-13 Schematic of Nissan ETS.

driven throttle valve permits adjusting hydraulic circuit pressure and thereby the actual operating condition over a very wide control range.

Nissan ETS

This system, developed to drive front axles, also employs a multiplate clutch and pressure modulation. In contrast to the Haldex© coupling, pressure in the Nissan ETS is supplied by an engine-driven pump; the system therefore has no means of self-regulation.

This lack of a system-internal reaction places special demands on the control unit, as clutch lockup is usually undesirable for reasons of vehicle dynamics.

5.5.1.5 System selection

As already described under the heading of all-wheel-drive system taxonomy (5.5.1.3), the four classification groups do not in any way represent a relative evaluation of these systems.

Decisive factors in choosing a concept are all items specific to all-wheel drive contained in the vehicle performance specifications—first and foremost, the intended nature of service in the hands of the customer, market positioning, basic vehicle concept, etc.

Seen worldwide, increased traction is still the predominant customer expectation of AWD systems; however, for Europe, vehicle dynamics properties are at least as important.

Common to all are demands for low costs, low weight, and good efficiency. There is a distinct trend away from operational elements specific to all-wheel drive and toward requirements for absolute compatibility of all-wheel drive with other control systems (ABS, ESP).

5.5.1.6 Effect on crash behavior

All-wheel drive also has a significant effect on crash behavior. In principle, the added drivetrain mass as well as altered stiffness due to the extended driveline play a role. With transverse engines, the drivetrain usually increases crash stiffness, thereby reducing passenger cell deformation and increasing the crash impulse.

5.5.1.7 Noise, vibration, and harshness (NVH)

Fundamentally, every increase in the number of moving parts leads to an increase in NVH issues.

Additional NVH sources include not only stick-slip effects in multiplate units and Torsen systems, but also differential windup. Relief may be available in the form of special oils or plate configurations with improved surface quality. Increases in alternating-load "clunks" due to possible motion of bracing elements are countered by careful tuning of engine and transmission units and suspension mounts and through engine management design to smooth abrupt torque changes.

5.5.1.8 Dimensioning

Two load conditions represent the primary determinants in dimensioning transmissions within the drivetrain:

- Continuous load according to a load spectrum, determined experimentally or synthetically. All components subject to alternating or varying loads such as gears, undercut shafts, and rolling-element bearings are designed accordingly. In addition, calculations are carried out for such components as are subject to operation under temporary loads sufficiently high to affect fatigue life (e.g., rapid acceleration).
- Protection against failure under extreme load conditions—for example, rapid clutch engagement, loss of traction at one axle (or one wheel) while accelerating, or off-road conditions. Depending on engine choices, boundary conditions are either maximum possible drive torque or breakaway torque. Given these conditions, any permanent deformation or cracks which may lead to later fatigue failure are not permissible.
- In order to achieve lighter drivetrains with smaller dimensions, both of the above loads must be reduced. The load spectrum may be lowered by adapting the all-wheel-drive coupling characteristic curve, while the maximum torque can be limited by suitable controllers.

5.5.1.9 Braking behavior of all-wheel-drive vehicles, all-wheel-drive systems, and control systems

Brake force distribution as determined by braking system design is heavily biased toward the front axle (see Fig. 5.5-14), which is therefore able to utilize more of its available traction under braking and with greater brake slip, turning at a lower speed than the rear axle. If this rotational speed difference is reduced or nullified by the all-wheel-drive system, torque is transferred from the rear to the front brakes, altering the actual brake force distribution. The intended wheel lockup sequence is compromised; front and rear wheels may lock nearly simultaneously, or even simultaneously.

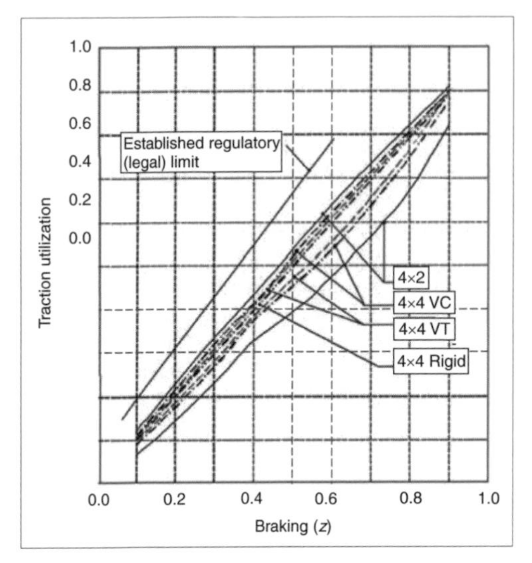

Fig. 5.5-14 Braking of 4WD vehicles.

Particularly under "split μ" conditions, the limited buffer or "safety margin" for wheel lockup collapses; as a result, vehicles with open center differentials exhibit the least—and vehicles with locked drivetrains the greatest—yaw tendency.

Along with ABS (see Section 5.5.2), there are several all-wheel-drive solutions to this problem:

- Freewheel to rear axle drive
- Controlled fully disengaging couplings
- Open center differential or Torsen
- Externally controllable all-wheel couplings (see Haldex, Viscomatic, etc.)

ABS and vehicle dynamics control systems (ESP) will soon be part of the standard equipment list for passenger cars as well as light utility vehicles.

Here, too, there is a reaction, depending on the degree of coupling between axles or wheels. Problems arise with ABS in establishing reference speed (which may be countered by use of a deceleration sensor). With ESP, brake application at an individual wheel results in braking or drive moments at the other wheels which, although considerably smaller in magnitude, nevertheless reduce system effectiveness. Full compatibility with ESP requires complete disconnection of the wheels.

Bibliography

[1] "SAE Innovations in Four Wheel Drive/All Wheel Drive Systems." TOPTEC, Proceedings, April 12–13, 1999.

[2] Dick, Wesley M. *All-Wheel and Four-Wheel-Drive Vehicle Systems*. SAE SP-1063.

[3] European ISTVS Conference, "Off Road Vehicles in Theory and Practice," Proceedings, Vienna, Austria, September 28–30, 1994.

[4] Richter, B. (editor). *Allradantriebe. Fortschritte der Fahrzeugtechnik 11*, Vieweg, 1992.

[5] Preukschat, A. *Fahrwerktechnik: Antriebsarten*. Vogel-Buchverlag, Würzburg, 1985.

[6] "Allradantriebe beim PKW." Fortschritt-Berichte VDI Reihe 12 No. 81, 1986.

[7] Pfyl, J. *Moderne Allradantriebstechnik*. AT Verlag, Aargauer Tagblatt AG, 5001 Aargau, 1988.

5.5.2 Traction and braking control

5.5.2.1 Active safety

A carefully designed chassis, with its axles, springs, damping, steering, and braking systems, represents a significant contribution toward the active safety attributes of a vehicle. In conjunction with vehicle weight distribution, these items play a significant role in determining vehicle traction, stability, and steering properties.

Traction is dependent on axle loading and the available coefficient of friction between tires and road surface. Maximum power transmission is achieved when the tires are not spinning and are operating in the optimum drive slip regime. In the case of all-wheel drive, the first drive axle to exceed optimum slip determines the maximum acceleration or grade climbing ability of the vehicle.

The two driven axles of an all-wheel-drive vehicle permit greater traction than is available for single-driven-axle vehicles (rear- or front-wheel drive). By means of special technical means (traction control systems), however, it is possible to achieve nearly complete utilization of the available traction for both single-driven-axle and all-wheel-drive vehicles.

5.5.2.2 Traction control systems

Along with mechanical limited-slip differentials, single-driven-axle as well as all-wheel-drive vehicles may employ drive components which make use of electronic control and external energy sources (hydraulic or electrical) to achieve differential locking or controlled limited slip (Fig. 5.5-15).

In addition to controlled limited slip, all-wheel-drive systems may use controlled couplings that activate a second drive axle as needed. Combination of these systems with additional transverse and longitudinal limited slip units has also been implemented [1]. Configurations with electromagnetic longitudinal and electrohydraulic transverse limited-slip differentials are in use [2], as are fully hydraulic solutions. Electronic limited-slip control uses wheel rotational speed signals from the ABS system. Using individual axle speed differences or side-to-side differences at front or rear axles, a control unit determines the necessary interventions to individual limited-slip units.

For reasons of simplicity, controlled limited-slip function may be achieved by brake actuation at the drive wheels (Audi "Electronic Differential Traction Control" or EDL; Nissan "Active Brake" limited slip; Rover "Electronic Traction Control" or ETC; Porsche and Mercedes "Automatic Brake Differential" or ABD, etc.) [3]. This, however, assumes that the brakes are adequately dimensioned to withstand activation over longer periods of time. If control conditions for brake actuation are met,

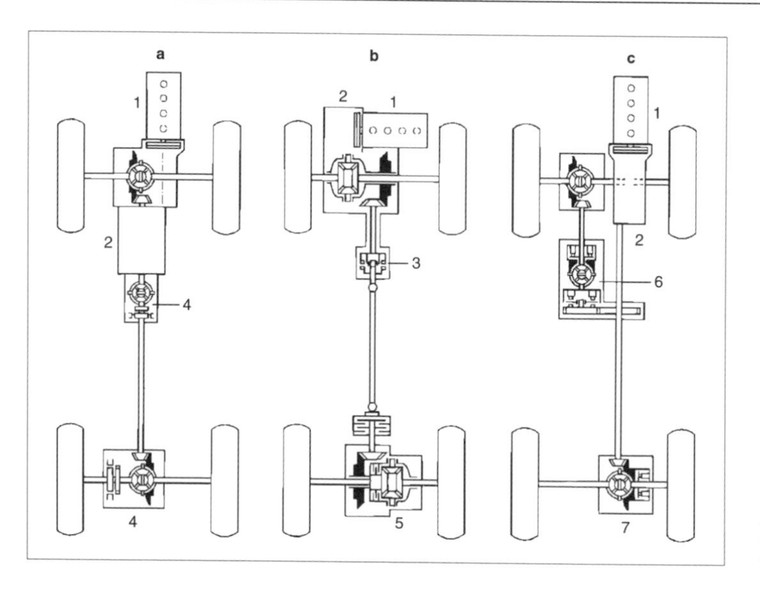

Fig. 5.5-15 All-wheel-drive concepts.

the ABS pump meters brake fluid volume to the appropriate brake. By means of stepped pulse modulation, solenoid valves ensure that brake pressure maintains the wheel at optimum slip. As a result of such single-wheel brake application, the opposite drive wheel can transmit increased torque, the added torque being equal to the brake torque set by the control system (Fig. 5.5-16).

The principle of brake application on the drive axle may be found among all drivetrain configurations. Expanded versions of brake application systems include engine intervention, which in addition to pure traction control also permits influencing vehicle stability (vehicle stability systems).

5.5.2.3 Vehicle stability systems

Systems which are not solely limited to traction optimization but which also emphasize vehicle stability require additional engine intervention options and, in some

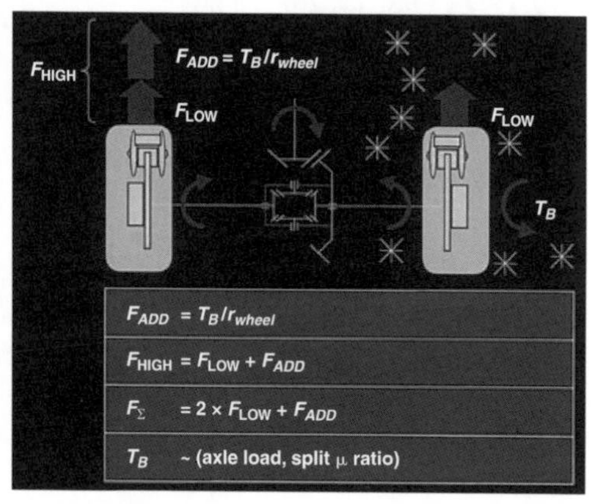

Fig. 5.5-16 Operating principle of Automatic Brake Differential (ABD).

cases, additional sensors to capture the vehicle's instantaneous dynamic condition. Along with passive systems that react to vehicle instability by reducing tractive demands on the drive wheels, some active systems achieve vehicle stabilization by directed corrective moments (brake actuation at individual wheels).

5.5.2.3.1 Passive systems (ASC, ASR)

All passive vehicle stability systems are based on ABS braking systems, which, by means of appropriate changes to actuators and electronics, are modified for an expanded mission.

Passive stability systems, like ABS, are generally limited to input signals from wheel speed sensors; only in special cases do they employ additional signals. Stability control systems such as ASC (Automatic Stability Control) [4] or ASR (Acceleration Slip Regulation) [5] are built upon a foundation of basic ABS hardware. An interface to the drivetrain permits drive torque modification in critical situations.

Fig. 5.5-17 illustrates the essential hydraulic elements of an ASC system for vehicles with diagonally split brake systems. To control acceleration slip at the drive axles, additional solenoid valves to separate the master cylinder from individual brake circuits are installed, permitting individual pressurization for ABD (automatic brake differential) action.

Interfaces to the engine management system permit incorporation of diesel as well as gasoline engines in the control loop. In the event critical acceleration slip levels are exceeded, drive torque is reduced by means of the torque interface to the engine control system. Applicable control strategies include combined ignition/fuel injection shutoff, ignition timing modification, and throttle position control for gasoline engines or fuel injection quantity control for diesel engines.

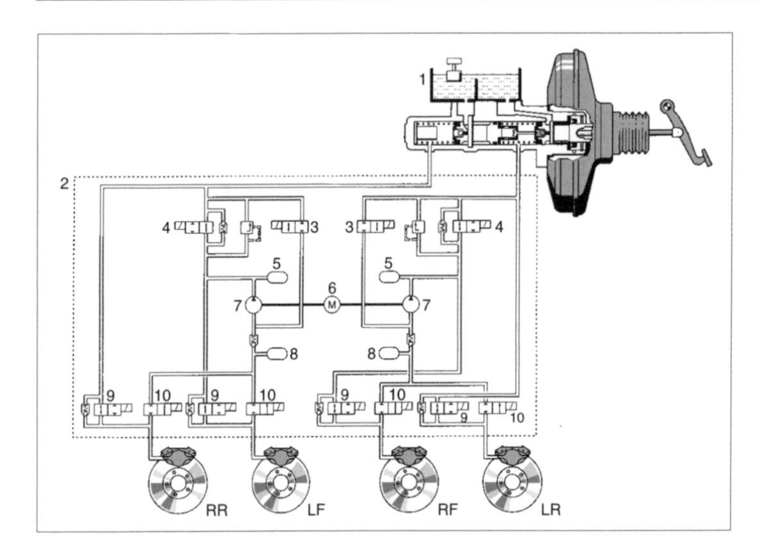

Fig. 5.5-17 ABS/ASC hydraulic schematic.

Actuators for throttle position control may consist of an additional inline-mounted prethrottle, with an electric positioning motor or an electronic engine output control (E-gas, drive-by-wire). Passenger car diesels employ electronic injection pump control.

In some cases, gasoline engines dispense with the throttle as an element in the control loop, instead operating only with ignition and injection control loops.

To avoid undesirable transmission "hunting" during an ASC control event, ASC control information is also sent to the automatic transmission control unit.

Negative effects on vehicle stability of excessive braking slip at the drive wheels during aggressive clutch engagement on downshifts may be eliminated by control loops which apply engine intervention to reduce engine drag torque (MSR, engine drag torque control).

Rear axle brake pressure proportioning valves may be replaced by electronic controls which ensure that braking slip at the rear wheels remains below the ABS activation threshold in noncritical stability regimes, independent of vehicle load and driving situation and under all braking processes (EBD, electronic brake force distribution) [6].

In brake application during special driving situations—preferentially in curves or lane changes—instabilities may be countered by Cornering Brake Control (CBC) [7]. By means of lateral differential braking, CBC introduces compensating moments to generate vehicle stabilizing effects.

Combining the control unit and hydraulic unit into a single assembly (actuator with pump and solenoid valves) results in compact devices with a reduced number of external electrical interfaces. This minimizes wiring complexity, as solenoid valve and electrohydraulic pump activation are accomplished by internal connections. The control unit and coil carrier are connected to the hydraulic unit by means of a "magnetic connector" (Fig. 5.5-18).

Configurations with slip control hydraulics integrated in the actuation system have not been able to establish themselves because of the associated proliferation of variations. At present, among conventional booster-equipped brake systems, the field is dominated by add-on technology integrated with existing brake systems.

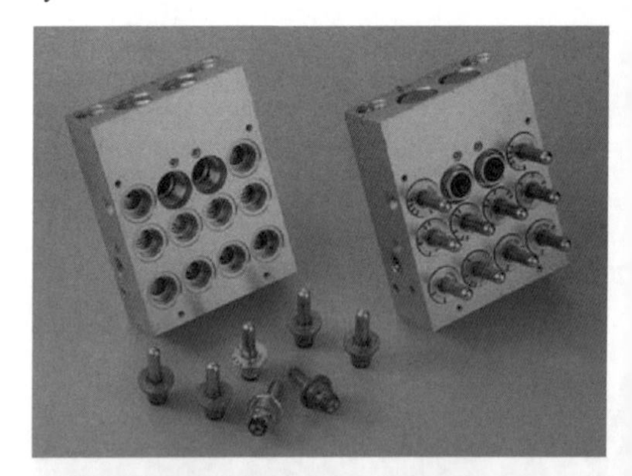

Fig. 5.5-18 Hydraulic slip control hydraulic unit and solenoid valves.

5.5.2.3.2 Active systems (DSC, ESP)

Stability and traction control systems ease the task of driving on critical-coefficient road surfaces. Passive systems, which largely only counteract longitudinal slip as determined by wheel rotational speeds, permit different brake or engine interventions as a function of vehicle speed. They operate either by reducing engine torque or redistributing engine torque among the drive wheels.

Application of additional vehicle dynamic sensors (lateral acceleration, yaw rate, steering angle) permit determination of driving condition as well as driver commands with respect to the intended vehicle path. To counteract deviations from the intended course of action, various control elements, interfaced to the control unit, may be activated. Active vehicle stabilization is achieved by countering undesirable yaw moments by means of compensating moments through brake pressure application to individual wheels (Fig. 5.5-19).

As with ASC, such systems—which include (among others) DSC (Dynamic Stability Control) [8] and ESP (Electronic Stability Program) [9]—have, in addition to their pure DSC function for yaw moment compensation, CBC (Cornering Brake Control) as their underlying control loop. CBC smoothes the transition to DSC activation and permits comfortable control system intervention in any driving situation. The principle of CBC and DSC operation is shown in Fig. 5.5-20.

For low-temperature operation of DSC, preload or other special measures in the region of the master cylinder and DSC pump are necessary to counteract the negative effects on brake pressurization of increased viscosity at low temperatures.

Preload may be achieved either by means of an electric preload pump or an actively controlled booster (Fig. 5.5-21). Some more unusual systems intervene only on the front wheels; the rear wheels are not pressurized. DSC systems may be integrated in conventional vacuum-assisted brake systems or realized with the aid of hydraulic boosters. To date, the higher costs of hydraulically boosted brake systems have limited their market penetration.

5.5.2.4 DSC and ESP with power brakes

Power brakes achieve brake torque by means of an external energy source, controlled directly by the driver or by an electronic control system. In addition to hydraulic energy, such systems may employ electrical energy. For design reasons, purely hydraulic power brakes provide a good basis for active stability controls but have to date achieved only limited market penetration.

Significant systems which must be considered in this context are electrohydraulic brakes (EHB) and electromechanical brakes (EMB). Because signals are carried electronically, both systems are collectively classified as brake-by-wire systems.

5.5.2.4.1 Brake by wire

The transition from add-on systems to integrated installations, which incorporate service brake functionality as

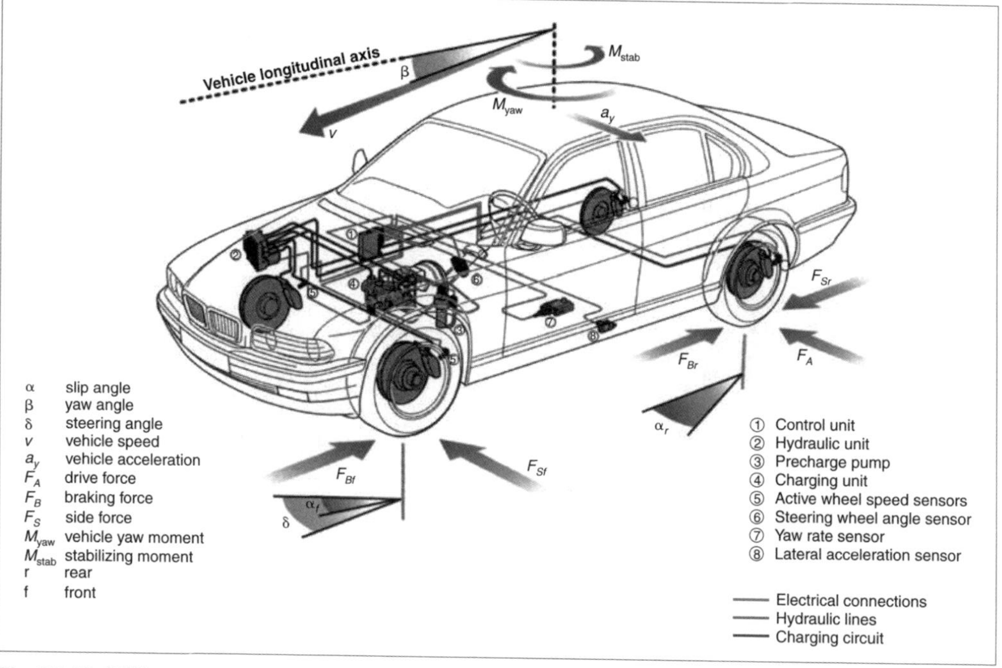

Fig. 5.5-19 DSC components and function.

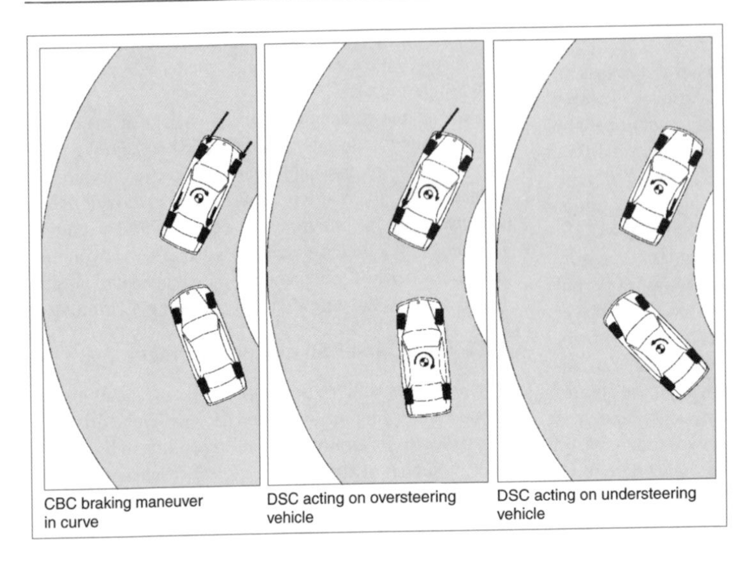

Fig. 5.5-20 CBC and DSC in cornering.

well as stability control, may be achieved by brake-by-wire systems. Brake-by-wire solutions may be implemented as electrohydraulic brakes (EHB) [10] or electromechanical brakes (EMB) [11]. In both cases, driver brake inputs are detected, processed by a control unit, and transmitted as commands to the appropriate actuators. Feedback of the established brake torque is achieved via pressure sensors or direct measurement of brake clamping force.

For EHB systems, in the event of a loss of external force, direct application of pressure proportional to brake pedal force is provided to one or both brake circuits by a standby master cylinder (Fig. 5.5-22).

EMB systems employ a dual-circuit basic system layout, including energy supply, and therefore dispense with any possibility of human-muscle–actuated brake application. Brake clamping force is generated by an electric motor located at the brake caliper (Fig. 5.5-23).

EH systems can be combined with an accumulator and corresponding valves, or electrohydraulic plungers can be used.

Electromechanical designs as well as electrohydraulic plunger solutions present increased demands on the electrical supply system. Transition to a 42 V onboard electrical system (36 V battery voltage) is desirable for such systems (see Section 5.6).

Fig. 5.5-21 Active brake booster for ESP preloading.

5.5.2.5 Sensors

Stability control systems are dependent on sensor signals which precisely describe wheel behavior as well as driving condition. The most significant configurations are wheel speed sensors and vehicle dynamics sensors.

5.5.2.5.1 Wheel speed sensors

Distinction is made between passive and active wheel speed sensors. Passive rotational speed sensors operate on the induction principle and provide a voltage signal proportional to rotational speed, which is in turn processed by appropriate input signal processing devices in the control unit.

The speed sensor "reads" an impulse disc, located as an independent element somewhere on the axle or integrated in a wheel bearing. As motion of the impulse disc is a prerequisite for sensor voltage generation, passive sensors do not permit measurement under conditions of zero speed.

Active systems are based on Hall effect or magnetoresistive sensors. These sensors are capable of providing speed signals even while stationary. Their voltage supply permits a larger usable air gap between the sensor and trigger disc. A further increase in usable air gap may be achieved by application of magnetically coded trigger discs.

5.5.2.5.2 Vehicle dynamics sensors

In addition to wheel speed information, vehicle dynamics control systems require sensors to capture dynamic information (Fig. 5.5-24). Sensor signals for yaw rate [12], lateral acceleration, and steering angle are pro-

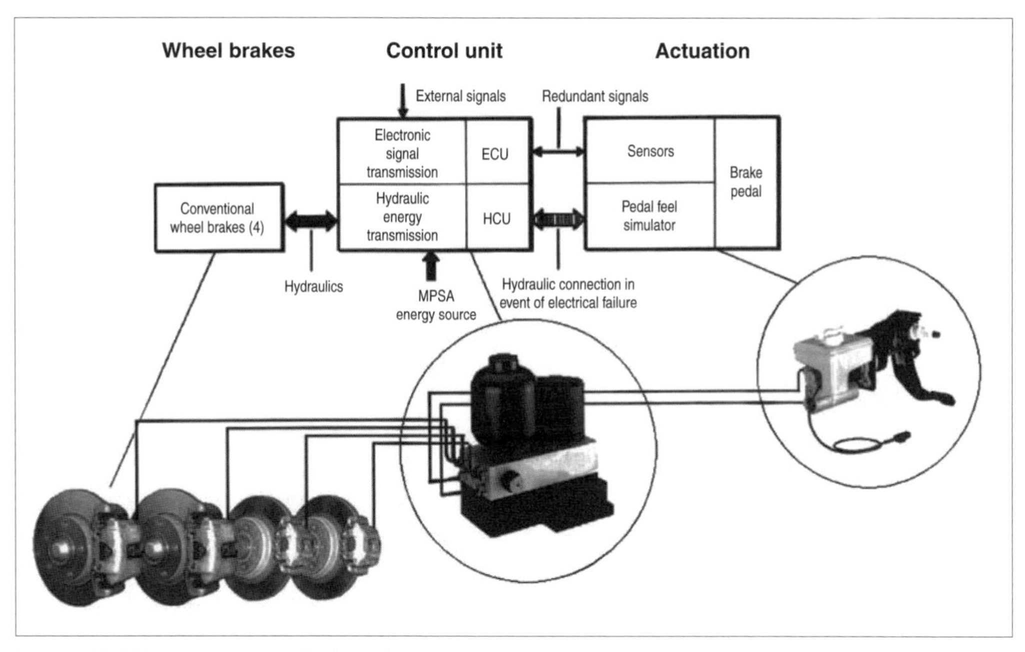

Fig. 5.5-22 EHB components and schematic.

cessed to determine intended and actual vehicle path. In order to better recognize transitional situations, brake pressure is also sensed.

Steering angle sensors to determine driver intentions in DSC systems must determine not only actual steering wheel position but also steering angle. Along with potentiometer-based sensors, digital sensors based on optoelectronics or Hall effect are used. Pressure sensors are required for brake-by-wire solutions as well as in conventional DSC systems employing amplifier technology.

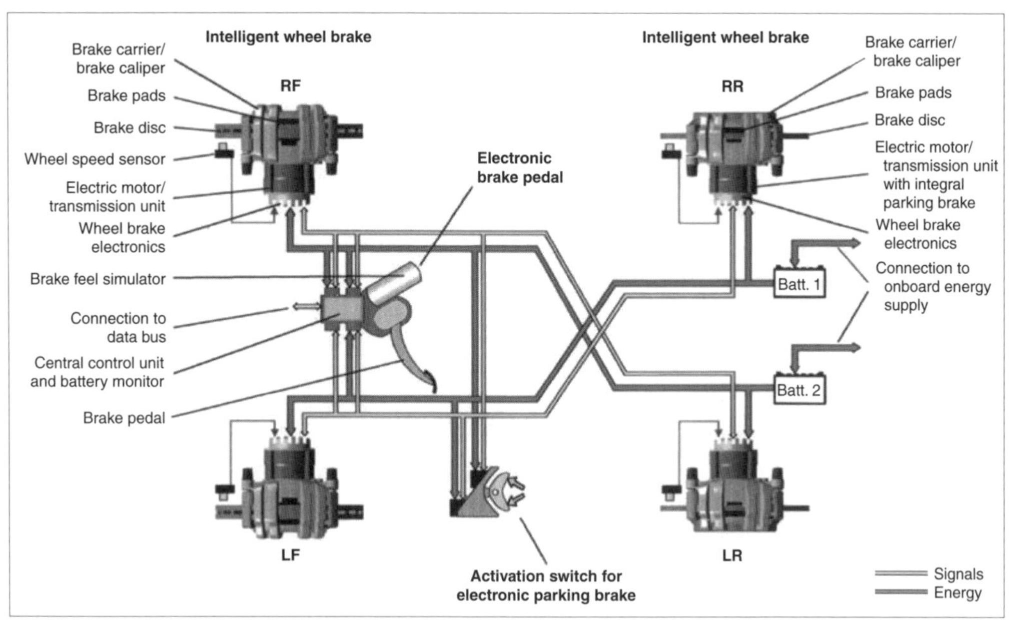

Fig. 5.5-23 EMB components and schematic.

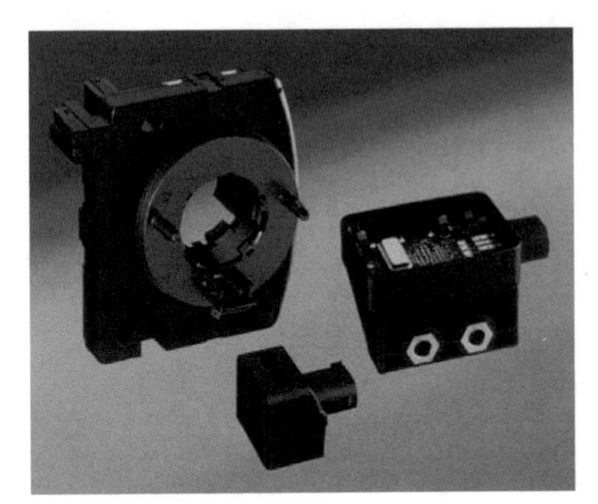

Fig. 5.5-24 Vehicle dynamics sensor package (steering angle, lateral acceleration, vehicle dynamics).

5.5.2.6 Adaptive cruise control (ACC)

Adaptive cruise control (ACC) is the first control system to utilize information from the vehicle's immediate surroundings in order to implement a driver assistance function. The first-generation ACC systems described below expand upon the familiar speed control function ("cruise control") to add vehicle following distance ("headway") control for increased comfort and driver workload reduction in flowing traffic.

Object sensing

To determine vehicle following distances, relative speeds, and transverse position, the core component of an ACC system is its 76 GHz multibeam radar array. These have largely weather-independent detection properties. A typical detection range of 120–180 m may be achieved with a lateral spread of ±3–5°. Multiple objects in the field of view must be resolvable by the sensors and individually measurable (multitarget capability). Precise sensor azimuth and elevation adjustment within the vehicle is necessary to ensure accurate transverse position determination and high detection assurance, particularly at greater ranges.

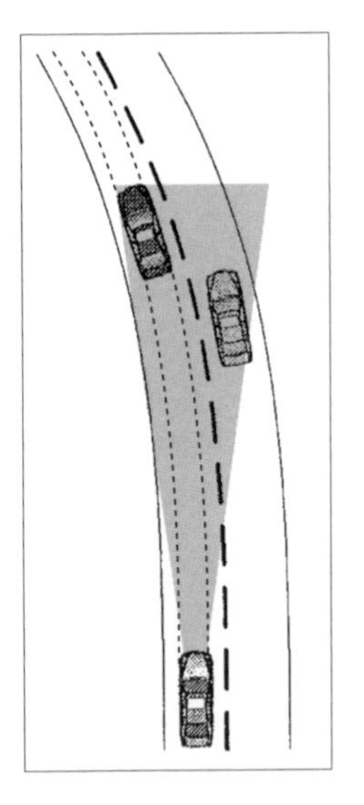

Fig. 5.5-25 Object selection based on lane determination.

Lane determination

Without reference to the surrounding traffic situation, object data alone is largely useless for any control application. In order to determine its relevance for vehicle longitudinal control, this data must be associated with the vehicle's own anticipated traffic lane. The present state of technology does not yet permit predictive lane determination. Instead, systems depend upon estimation of future conditions (lane prediction), based on actual vehicle dynamic conditions. The radius of the currently negotiated curve, preferably as determined by measured yaw rate, continues to be taken as valid for path determination, which is largely true for steady-state motion on straight roads or in curves on freeways, highways, and well developed secondary roads—the pri-

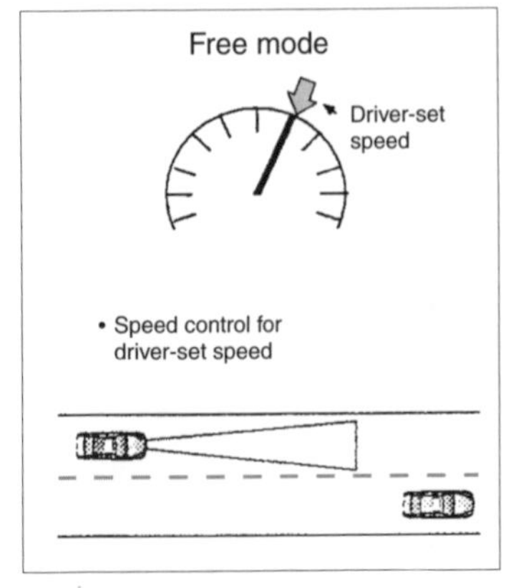

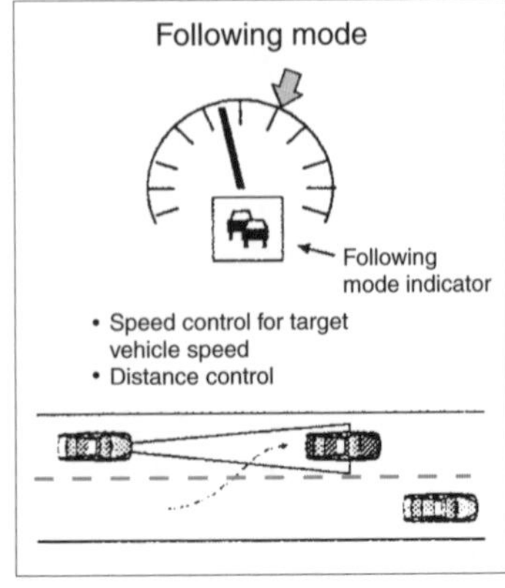

Fig. 5.5-26 Operating conditions for ACC: free and following modes.

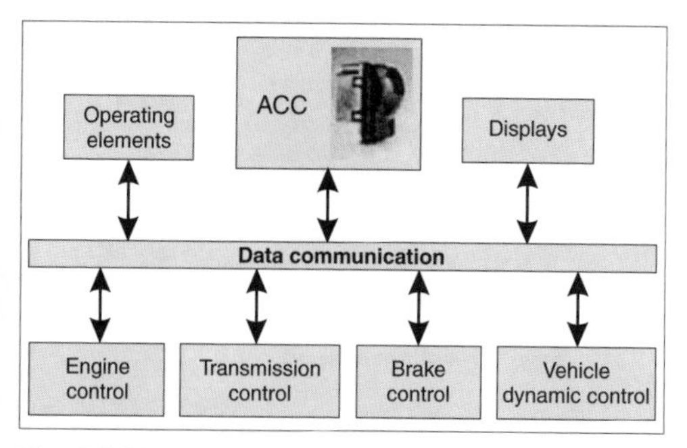

Fig. 5.5-27 ACC system integration.

mary environments for ACC application. Corrections derived from observations of relative lane movement of moving or stationary objects may be utilized in support of lane prediction. In principle, however, no true predictive knowledge of lane alignment is available, so that absolutely certain object/lane correlation is not possible for first-generation ACC systems. It must therefore be assumed that erroneous traffic situation interpretation is possible. This in turn requires limits on system interventions in vehicle longitudinal dynamics (to ~ -2.5 m/sec^2 to $+1$ m/sec^2). A further important consequence is that ACC systems clearly must be regarded as comfort systems which require responsible application on the part of the driver and in which the driver always retains full responsibility for the driving task.

System limitations

ACC systems exhibit various technical limitations—for example, limited fields of view and deceleration capability, uncertainty in object/lane correlation, lack of object classification, and, as a consequence of the last point, the necessity of ignoring all stationary objects. This in turn necessitates driver intervention in various situations. It must therefore be assured that the driver is able to override the system by means of accelerating or braking in any situation and without conflict. The limits of system implementation must be learnable by the driver.

Speed and headway control

The basic function of ACC is speed control. The system maintains the target speed selected by the driver under unrestrained driving conditions as long as no slower vehicle is detected in the lane of the vehicle. If a relevant vehicle is detected, ACC switches to distance-following mode: that is, speed is matched to that of the slower vehicle and a following distance appropriate for the situation is maintained. In accordance with the driver's comfort expectations, following-distance control is intended to match the speed variations of the leading vehicle, thereby maintaining a constant time headway τ, which typically may be set by the driver in the range 1–2 s. The following distance in meters is therefore proportional to speed: $\tau \cdot v$.

Human-machine interface

ACC employs controls similar to those found in speed control (cruise control, or CC) systems, augmented by the ability to select headway. Functional transparency requires indication of system activation and the current operating (following or free) mode in order to inform the driver of system reactions when switching modes (e.g., acceleration to the target speed, which should also be displayed). It may also be advantageous to indicate maximum system deceleration events to the driver as a warning that intervention in the form of more forceful manual brake application is necessary. As in cruise control, manual brake application deactivates the system.

System integration

Complex system integration is necessary to achieve ACC functionality within the vehicle. The ACC unit, preferably in the form of a single combined sensor/control unit, assumes its longitudinal control function in cooperation with other systems. If acceleration is required to meet operational targets (previously set by means of driver-actuated control elements), the ACC translates its commands by means of an E-gas (drive-by-wire) system acting as an intelligent engine torque control. Ideally applied in conjunction with an automatic transmission, ACC can also influence shift behavior. If deceleration is called for and engine drag torque cannot provide sufficient retardation, ACC systems with brake control unit intervention capability may utilize a power brake system. Brake pressure may be generated hydraulically by a slip control system hydraulic pump or by a controllable vacuum brake booster. A vehicle dynamic control or slip control system provides vehicle dynamics data for ACC path prediction.

System safety

Because an integrated system may demand safety-relevant control interventions, both ACC and its associated actuator systems require intrinsically safe behavior. Every partial system must be capable of autodiagnosing its own functions. In the event of malfunction, the overall integrated ACC system must revert to a safe shutdown condition.

The complex functional interaction of partial systems requires comprehensive communication by means of a suitable data channel (e.g., CAN bus). From both hardware and protocol standpoints, this data channel must meet the most stringent safety demands with regard to data integrity, fail-safe operation, and real time behavior.

References

[1] Schöpf, H.U. "Mercedes-Benz Fahrdynamik-Konzept, ASR, ASD und 4MATIC." Autec, 1986.
[2] Sagan, E., and T. Stickel. "Der neue BMW 525iX—Permanentallradantrieb mit elektronisch geregelten Sperrdifferentialen." ATZ 94, 4, 1992.
[3] Müller, A., and B. Heissing. "Das Fahrwerk des Audi A4. 5." Aachener Kolloquium Fahrzeug- und Motorentechnik, Aachen, October 1995.

[4] Leffler, H. "Entwicklungsstand der ABS-integrierten BMW-Schlupfregelsysteme ASC und DSC." ATZ 96, 2, 1994.
[5] Gaus, H., and H.-J. Schöpf. "ASD, ASR und 4MATIC: Drei Systeme im "Konzept Aktive Sicherheit" von Daimler-Benz." ATZ 88, 5, 6, 1986.
[6] Kohl, G., and R. Müller. "Bremsanlage und Schlupfregelsysteme der neuen 3er-Baureihe von BMW - Teil l." ATZ 100, 9, 1998.
[7] Fischer, G., R. Müller, and G. Kurz. "Bremsanlage und Schlupfregelsysteme der neuen 5er-Reihe von BMW." ATZ 98, 4, 1996.
[8] Debes, M., E. Herb, R. Müller, G. Sokoll, and A. Straub. Dynamische Stabilitäts Control DSC der Baureihe 7 von BMW." ATZ 99, 3, 4, 1997.
[9] Fennel, H. "ABS plus und ESP—Ein Konzept zur Beherrschung der Fahrdynamik." ATZ 100, 4, 1998.
[10] Jonner, W-D., H. Winner, L. Dreilich, and E. Schunck. "Electrohydraulic Brake System—The First Approach to Brake-by-Wire Technology." SAE 960991.
[11] Bill, K.-H., and M. Semsch. "Translationsgetriebe für elektrisch betätigte Fahrzeugbremsen." ATZ 100, 1, 1998.
[12] Rittmannsberger, N. "Der Drehratensensor für die Fahrdynamikregelung." ATZ/MTZ Sonderausgabe System Partners 97.
[13] Naab, K., and G. Reichart. "Driver Assistance Systems for Lateral and Longitudinal Vehicle Guidance—Heading Control and Active Cruise Support." *Proceedings* of AVEC '94, 1994, pp. 449–454.
[14] Naab, K., and R. Hoppstock. "Sensor Systems and Signal Processing for Advanced Driver Assistance." Seminar on Smart Vehicles, Delft, Netherlands, February 13–16, 1995, Swets & Zeitlinger, 1995.
[15] Winner, H., S. Witte, W. Uhler, and B. Lichtenberg. "Adaptive Cruise Control System, Aspects and Development Trends." SAE 961010.
[16] Winner, H. "Adaptive Cruise Control." Contribution to *Automotive Electronics Handbook*, 2nd edition, Jurgen, R. (Editor). McGraw Hill Inc., 1999.
[17] Konik, D., R. Müller, W. Prestl, T. Toelge, and H. Leffler. "Elektronisches Bremsenmanagement als erster Schritt zu einem integrierten Chassis-Management." ATZ 101, 4, 1999, and 102, 5, 1999.

5.6 Optimization of auxiliary devices

5.6.1 Rising motor vehicle energy consumption due to the growing number of electrical consumers

There are growing demands for reduced consumption of fossil energy, with a concomitant reduction in CO_2 emissions. Until very recently, the main burden for energy conservation was borne by the powerplant itself—for example, through improved efficiency of gasoline or diesel engines along with the transmissions to which they are coupled. The affirmation of the European automobile industry (ACEA) to reduce CO_2 emissions of vehicle models certified as of 2008 to less than 140 g CO_2 per km imposes restrictions in all sectors of the vehicle: efficiency improvements, lower rolling resistance and aerodynamic drag, lower vehicle weight. In the process, all auxiliary drive components, whether mechanical, hydraulic, or electrical in nature, take on greater importance—for instance, installation of roller followers for valve actuation, electrically driven water pumps, and improved engine oil pump efficiency. One special topic is the great variety of electrical consumers in a vehicle. Fig. 5.6-1 illustrates that even today, electrical

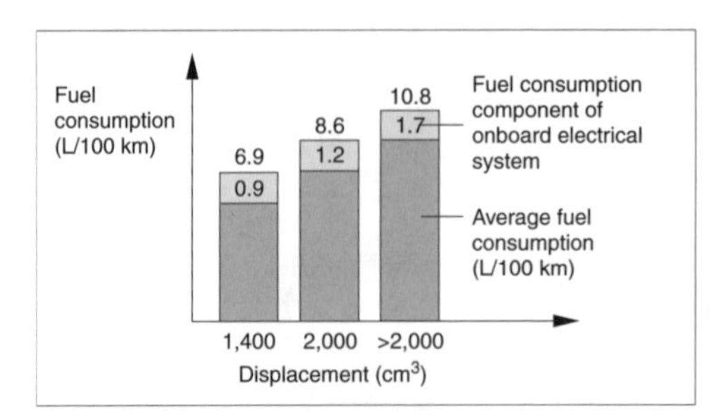

Fig. 5.6-1 Fuel consumption component of onboard electrical system.

energy on average accounts for more than one liter per 100 km of a vehicle's energy consumption. It is already apparent that the number of electrical consumers will increase in the future. The following overview shows a few of the systems currently in use or expected to enjoy widespread use in the future [1]:

Present-day electrical systems

Starter
Ignition
Fuel pump
Engine and transmission control units
Fuel injectors
Throttle actuation (drive-by-wire)
Air bypass valve positioner
Heated oxygen sensor
Speed control system
Antilock brake system
Traction control
Electrohydraulic steering system
Climate controls
Heated windshield washer nozzles
Heated seats
Front and rear wipers
Headlight cleaning system
Headlights and illumination
Heated backlight
Instruments
Navigation system
Radio/CD player/cassette tape/amplifier
Power windows (front and rear)
Power sunroof
Power door and trunk locks
Power headrest and seat adjustments
Power outside mirrors
Power antenna
Antitheft system
Infrared locking system
Airbags and seat belt pretensioners

Future or increasingly popular electrical systems

Electrically assisted power steering
Electric engine cooling
Electric water pump

Air pump for emission control concepts
Active suspension
Brake-by-wire
Electric A/C compressor
Electronically controlled five- or six-speed mechanical transmission
Heated catalytic converter
Adaptive cruise control (following-distance control)
Braking assistant
Voice-activated controls
Heated windshield
Assisted night vision
Blind-spot monitor
Parking assistant
Tire pressure monitor
Active engine mounts
Telephone/fax
Intelligent restraint systems

Individual consumers may well draw several hundred watts or even kilowatts. Moreover, it has been shown that once such advances are attained, consumers are loath to relinquish them. Consequently, this means that very fuel-efficient vehicles equipped with the expected comfort and convenience features, including air conditioning, can hardly be expected to achieve the necessary fuel economy improvements unless the onboard electrical system is improved as well. This is equally valid for future propulsion systems such as hybrid vehicles or a special variant, fuel cells.

5.6.2 Strategies for reduced fuel consumption

The following strategies may be applied to minimize electrical consumption:

- Reducing the number of electrical consumers
- Improving efficiency of onboard electrical generation
- Improving efficiency of accessory consumers in the vehicle
- Application of an energy management system

It seems apparent that due to growing demands for improved safety, reduced fuel consumption and exhaust emissions, and increased comfort, a reduction in the number of electrical consumers is not a viable solution. There remain three alternatives.

5.6.2.1 More efficient onboard electrical generation

Several examples are given to illustrate this composite network.

5.6.2.1.1 Generator improvements

Fig. 5.6-2 shows necessary and potential generator improvements (after Pasemann [2]) in association with an increase in the onboard voltage from 12 to 42 V as discussed in that source.

Water-cooled generators are even simpler to realize than speed-controlled (variable drive ratio) generators. Improved heat rejection leads to greater efficiency, particularly at high rotational speeds. Fig. 5.6-3 illustrates the resulting changes in comparison to a conventional generator.

5.6.2.1.2 "Smart power switch"

Power losses in modern relays are far too high. Their replacement by compact switching modules with integrated semiconductor switches is under way and has already gained a foothold in production vehicles and in hybrid vehicle controls.

5.6.2.1.3 The ISAD system

The ISAD system (Integrated Starter Alternator Damper [3]), shown in Fig. 5.6-4, replaces both starter and generator with an appropriate asynchronous motor in an effort to achieve better efficiency, particularly at higher rotational speeds. The intended increase in onboard voltage from 14 V to 42 V improves the efficiency of the electrical machine. The intended crankshaft damping function, however, appears to be more difficult to achieve. This is attributable to the relatively large amount of energy that must be absorbed to achieve the necessary damping of crankshaft torsional vibrations.

5.6.2.1.4 Adaptronics

Because adaptronics are themselves consumers of electrical energy (depending on application), they cannot

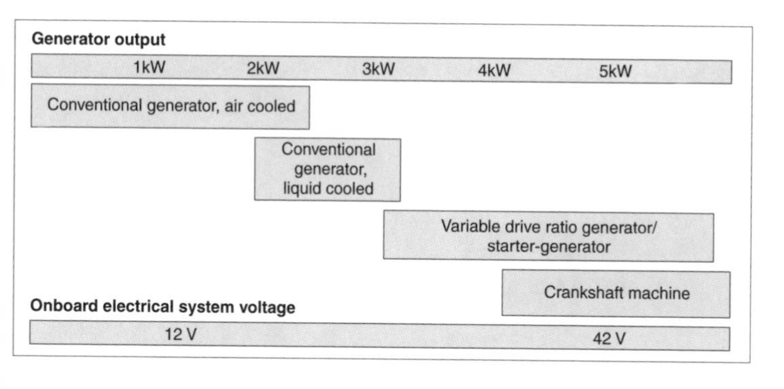

Fig. 5.6-2 Generator type as a function of power demands.

Special features	Customer advantages
• Hybrid rotor design	• Improved performance • Compact dimensions, low specific weight (mass/output) • Low rotational inertia
• Liquid cooling of generator stand and electronics	• Rejection of thermal losses to coolant
• No fan	• Increased overall efficiency • Reduced noise
• All coolant passages cast into a single component	• Sealless concept • Improved mechanical rigidity
• Improved thermal design • Pressed-in diodes	• Higher performance and durability compared to air-cooled generators at high operating temperatures
• Single-chip voltage regulator with additional functions (ASVR, all-silicone voltage regulator)	• Increased reliability, durability, and improved thermal performance capabilities
• Axial or radial battery and coolant connections	• Greater flexibility for customer-specific demands

Fig. 5.6-3 High-performance liquid-cooled generator (source: Delphi Automotive).

replace electrical consumers. Adaptronics, however, offer the possibility of replacing the weight used for "absorption and counterweighting" damping measures and so on. Fig. 5.6-5 shows their basic function [4]. Two examples are currently under investigation: the adaptronic vehicle headliner and adaptronic engine mounts. The headliner is intended to improve interior acoustics by means of piezoelectrically driven synthetic fibers; adaptronic engine mounts seek to achieve better damping characteristics.

5.6.2.1.5 Increased onboard voltage

Increased onboard voltage requires a new onboard electrical system [2, 5]. As a first step, shown in Fig. 5.6-6, two onboard voltages will be used, namely 14 and 42 V. The illustration shows a centralized concept; decentralized solutions are also possible. Implementation of such systems is imminent. In a second step, 192-V systems are anticipated for high-performance applications such as active suspension or electromagnetic valve actuation.

5.6.2.1.6 The 5 kW fuel cell

Among other projects, research efforts by BMW (Fig. 5.6-7) [6] provide an indication of how much attention is directed toward improvements in electrical energy consumption. It is apparent that a fuel cell has been installed at the rear of the vehicle. This 5 kW PEM (proton exchange membrane) operates as an APU (auxiliary power unit), serving all of the vehicle's conventional electrical consumers. This system is conceived by International Fuel Cell Corp. (IFC) as a closed system; it operates on stored hydrogen, without any other energy sources.

5.6.2.2 Improved efficiency and reduced consumption of auxiliary devices

5.6.2.2.1 Electric power steering

Hydraulic power steering, which operates at all times during vehicle operation, consumes energy continuously. New solutions replace hydraulics with electrohydraulic or purely electrical units. The energy savings attributable to such solutions are on the order of 0.2 L/100 km. With proper design, the safety question becomes noncritical—that is, no steering lockup in the event of system malfunction [7].

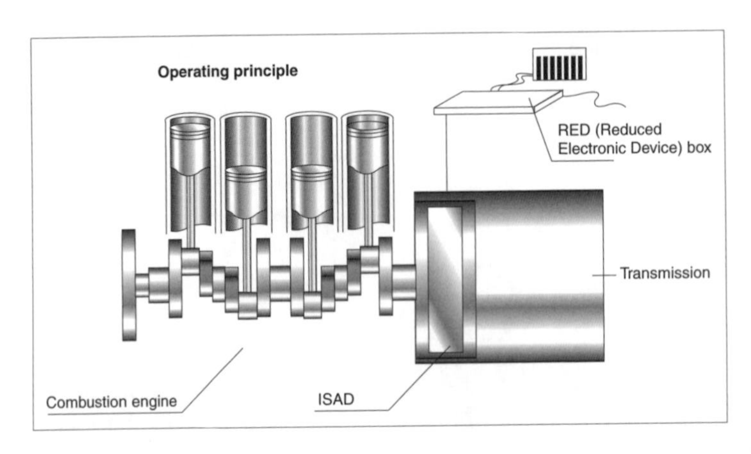

Fig. 5.6-4 Operational schematic of ISAD (source: Conti Tech).

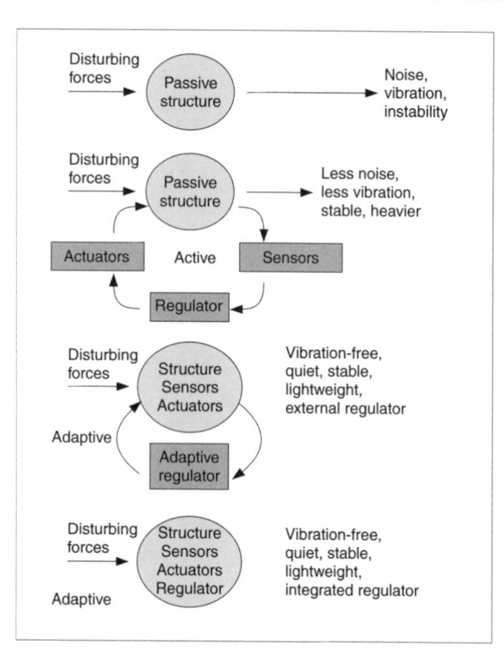

Fig. 5.6-5 Comparison of various types of structural systems (source: [4]).

5.6.2.2.2 Demand-driven fuel supply pump

Supply of a constant fuel flow to the engine, with excess fuel returned to the tank, is currently being replaced by demand-driven fuel pump concepts, which contribute to reduced average electrical power demands.

5.6.2.2.3 Electrohydraulic/electromechanical brake system

After ABS (antilock brakes) and traction control made inroads into vehicle brake systems, a further step was realized in the form of ESP—electronic stability program. Additional future solutions include electrohydraulic and electromechanical braking systems, which

Fig. 5.6-7 Fuel cell batteries, here shown in a BMW 7-series (source: BMW).

are currently under development. Electrohydraulic systems are likely to appear first. The electromechanical solution [8], shown in Fig. 5.6-8, demands highly refined security systems to ensure that vehicle stability is not endangered by unintentional single-wheel lockup.

5.6.2.3 Energy management systems

A recent dissertation (Energie- und Informationsmanagement für zukünftige Kfz-Bordnetze – "Energy and information management for future automotive electrical systems" [9]) outlined the following four possibilities for implementation of motor vehicle energy management systems:

- Implementation of optimum operation of electrical consumers, which are designed with sufficiently high power to meet extreme demands, but whose maximum performance is not required for certain driving situations and applications (for example, by means of pulse-width-modulation–control consumers)
- Intelligent demand- and condition-dependent switching of electrical consumers on the basis of plausibility tests and evaluation schemes, with simultaneous monitoring of switching processes (for example, momentary deactivation of windshield heating while brake lights are activated)

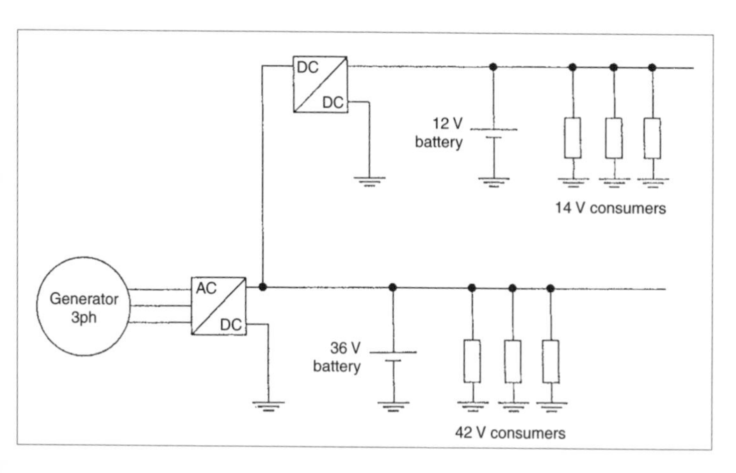

Fig. 5.6-6 Central concept of a 42-V onboard electrical system within passenger car electrical system structures (source: [5]).

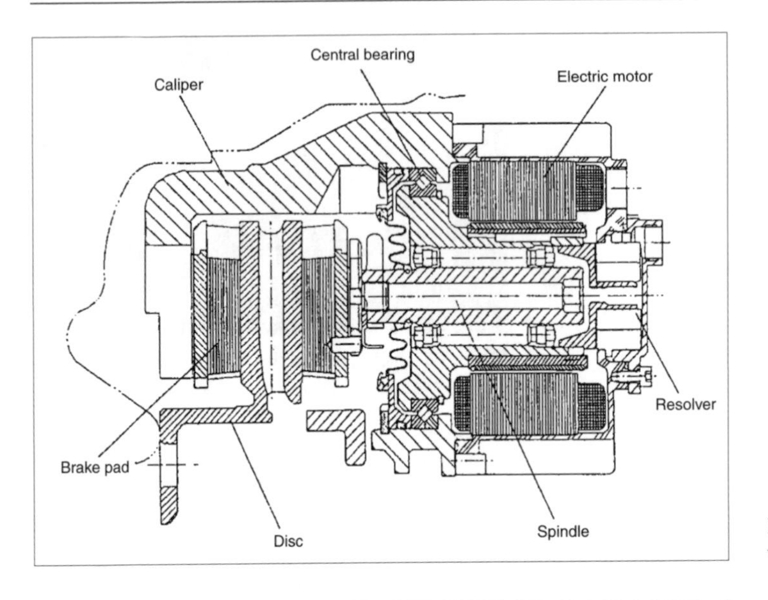

Fig. 5.6-8 Electromechanical wheel brake (source: [8]).

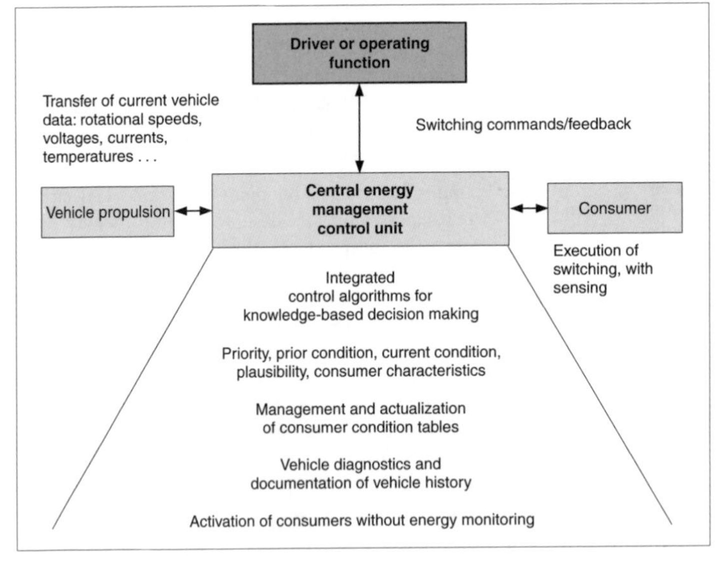

Fig. 5.6-9 Principle of motor vehicle electrical energy management [9].

- Determination of maximum achievable savings potential of a motor vehicle energy management system through possible complete load elimination, with the maximum comfort sacrifices tolerable to the driver
- Avoidance of false driver interpretation based on automated control inputs and thereby improvement in system acceptance

Intervention in the vehicular energy balance may be achieved by:

1. Manual activation or deactivation of electrical consumers (e.g., by the driver)
2. Semiautomatic control (e.g., by timer switches for manual activation of rear window defrost and au-

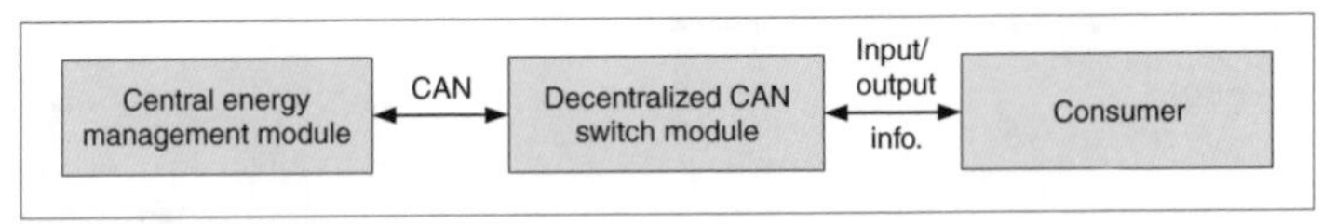

Fig. 5.6-10 Principle of decentralized consumer control by means of an energy management control module (EM module) via a CAN bus for a single consumer switching branch [9].

tomatic deactivation after a predetermined heating period)
3. Fully automatic activation and deactivation in accordance with driver or vehicular commands
4. Fully automatic intelligent switching of electrical consumers with condition sensing, evaluation, and integrated energy and information management

Fig. 5.6-9 shows the electrical energy management principle as applied to a motor vehicle.

In order to undertake more efficient energy management, electrical consumers must be capable of networked communication with one another.

The network structure of a decentralized consumer structure may be achieved by means of the CAN bus. Fig. 5.6-10 shows a corresponding design.

In addition to networking, a corresponding classification of electrical consumers is part of the problem solution; for example, classification may be prioritized as:

1. Safety functions (light functions, brake systems such as ABS, airbag, etc.)
2. Limp-home functions (engine control module, other vehicle control units including electrical supply, etc.)
3. Basic vehicular functions (starter, engine electrical auxiliary devices, radiator fan, fuel injection system, etc.)
4. Standard comfort functions (interior ventilation fan, climate control system, window heating, motors for mirror adjustment, etc.)
5. Expanded luxury and comfort functions (four power windows, four heated seats, navigation, high-end sound system, auxiliary heaters, etc.)
6. Service, diagnostic, and vehicle operation functions

These few examples illustrate the importance of the onboard electrical system and the associated optimization of energy generation and consumption. Ambitious fuel economy goals cannot be realized without improvements in this vital area.

References

[1] Seiffert, U. "The changes in automobile electrical systems." WiTech Engineering GmbH, Brunswick, 1997.
[2] Pasemann, K. "Das 42-V-Bordnetz im PKW." IIR-Tagung, Stuttgart, April 13, 1999.
[3] Pels, T. "ISAD—das integrierte Starter-Alternator-Dämpfer-System." VDI-Berichte 1418, Düsseldorf, VDI-Verlag, 1998.
[4] Melcher, J. "Adaptronik im Automobilbau." ATZ 100. Jahrgang/No. 4, Wiesbaden, April 1998.
[5] Ehlers, K. "Positionspapier des Forums Bordnetzarchitektur." Haus der Technik, 16. Tagung "Elektronik im Kfz," Essen, 1996.
[6] BMW press release, Geneva auto show, 1999.
[7] Connor, B. "Elektrische Lenkhilfen für PKW als Alternative zum hydraulischen und elektrohydraulischen System." ATZ 98. Jahrgang No. 7/8, Wiesbaden, August 1996.
[8] Belschner, R., et al. "Trockenes Brake-by-wire mit fehlertolerantem TTP/C-Kommunikationssystem." VDI-Bericht No. 1415, 1999.
[9] Bäker, B. "Energie- und Informationsmanagement für zukünftige Kfz-Bordnetze." Dissertation at the Institute for Electronic Measurement Technology and Fundamentals of Electrotechnology, Brunswick Technical University, 1999.

5.7 The two-stroke engine: opportunities and risks

Thanks to favorable boundary conditions, the two-stroke engine as a passenger car powerplant enjoyed renewed attention in the automobile manufacturers' development departments in the early 1990s. Most manufacturers viewed this engine concept as an alternative to existing four-stroke gasoline engines in terms of fuel economy, weight, and cost.

5.7.1 The two-stroke process

The principle of the two-stroke process is shown in Fig. 5.7-1, in comparison to the four-stroke cycle. The basic difference is that the induction, compression, expansion, and exhaust processes take place over 360 instead of 720 degrees of crankshaft rotation. This is also shown in Fig. 5.7-2, which indicates combustion chamber pressures encountered in the course of a single working cycle.

In contrast to the four-stroke engine, whose gas exchange processes are governed by the valve train, two-stroke practice primarily uses ports in the cylinder bore, which admit precompressed mixture (or, in the case of direct-injection engines, air only) from the crankcase into the combustion chamber. There, the fresh charge displaces exhaust gases generated during the previous cycle, forcing these out through simultaneously open exhaust ports. In order to optimize the gas exchange process with respect to load and rotational speed, a rotary valve is used on the exhaust side. More rarely, poppet exhaust valves are used to reduce loss of fresh charge through the exhaust ports. Fig. 5.7-3 shows examples of different scavenging concepts.

The concept of cross-scavenged engines with precompression and variable exhaust timing has been examined in greater detail. Potential advantages and well-known issues, as outlined in the following lists, represent the main points of two-stroke development.

Advantages:

- Better packaging
- Lower weight
- Better part-load fuel economy
- Higher specific output
- Improved comfort
- Lower service costs

Issues:

- Increased oil consumption
- High fuel consumption, exhaust emissions
- Poor idle quality
- Unbalanced engine characteristics
- Low engine braking effect
- Inadequate durability
- Unfavorable noise behavior
- Manufacturing difficulty

5.7.2 The applied concept

A modern crankcase-scavenged two-stroke engine with electronically controlled, air-assisted direct fuel injection,

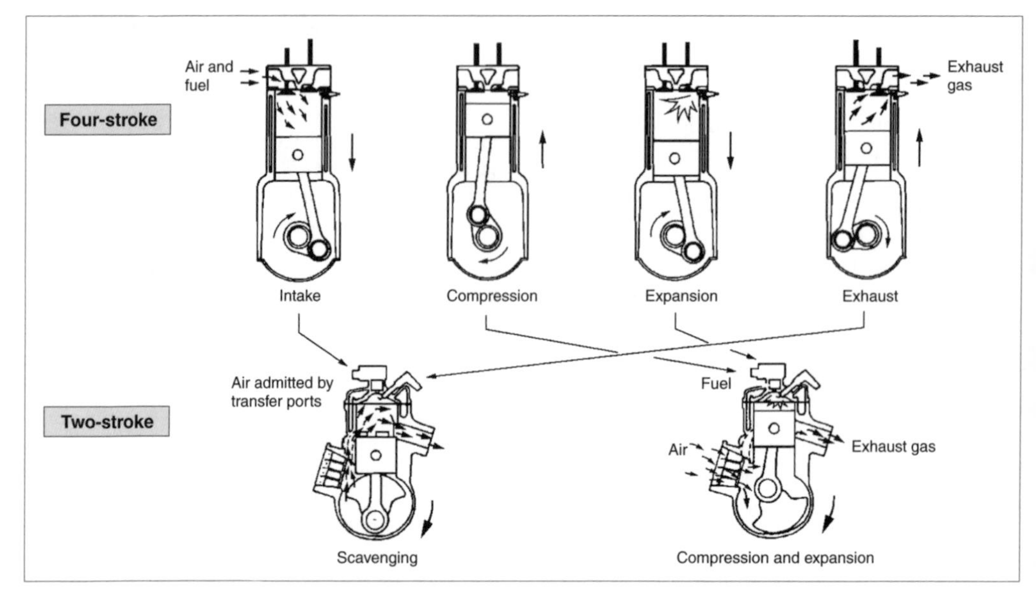

Fig. 5.7-1 Operating principle.

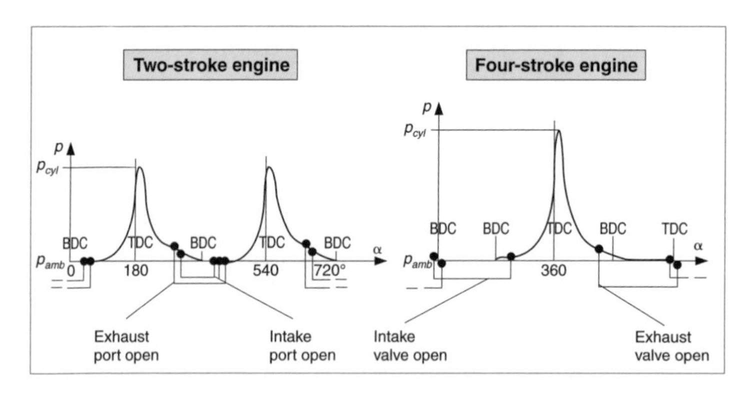

Fig. 5.7-2 Cylinder pressure history of two- and four-stroke engines.

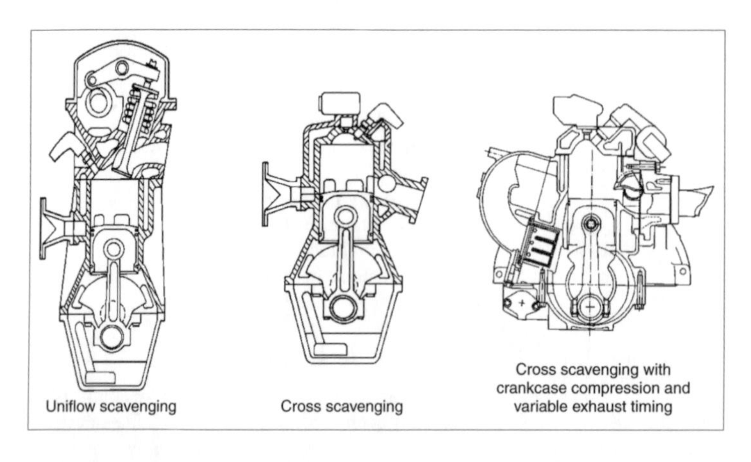

Fig. 5.7-3 Scavenging concepts.

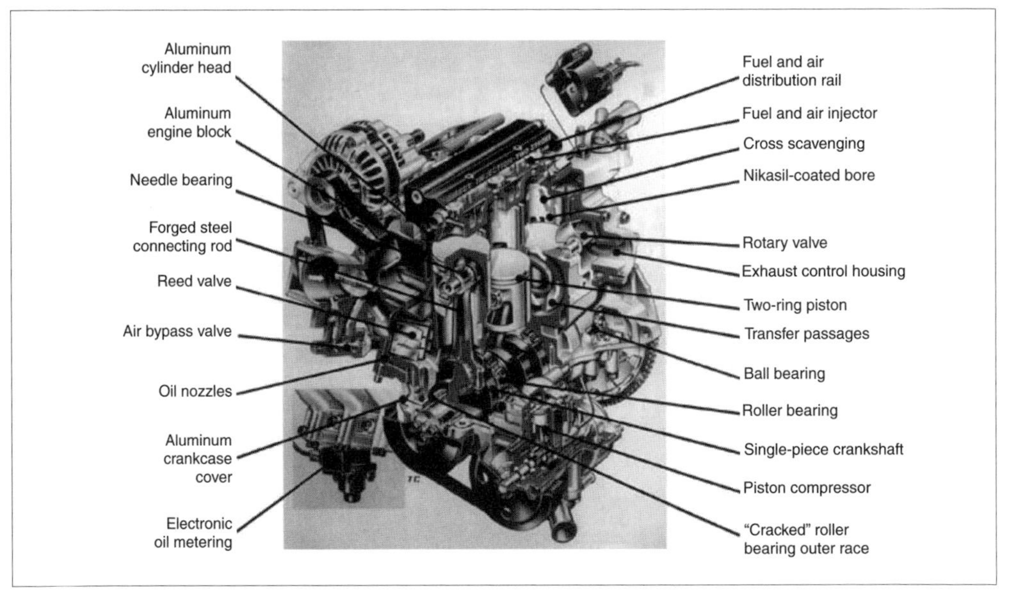

Fig. 5.7-4 Modern crankcase-scavenged two-stroke engine.

oil metering, and variable exhaust timing (rotary valve) was chosen as a test bed. Engine details are shown in Fig. 5.7-4.

Other important engine details are:

- Scavenging concept
  - Crossflow scavenging with crankcase compression and reed valves to prevent backflow of fresh charge air into the intake system
- Compression ratio:
  - Geometric: 10.5:1
    Variable exhaust timing yields the following effective ratios:
  - Exhaust rotary valve open: 6.4:1
  - Exhaust rotary valve closed: 9.3:1

This concept seemed an attractive test as it combined the constructive simplicity of a two-stroke engine, without a valve train, with a potentially high degree of gas exchange variability. The choice of a direct-injection system reduces the danger of fresh mixture loss into the exhaust passage; at worst, only excess air will be forced into the exhaust. This is unavoidable if any realistic possibility of meeting exhaust standards is to be retained.

5.7.3 Development focus

Development focus was guided by the expected advantages and well-known weaknesses of two-stroke engines: emissions and NVH, fuel consumption, mechanical durability, package, weight, and costs.

5.7.3.1 Emissions

Emissions of the two-stroke engine under consideration are marked by the specific requirements of its gas exchange process and mixture formation by means of air-assisted direct injection. Exhaust gas aftertreatment is made more difficult by a nonstoichiometric mixture and a total-loss lubrication system, with the resulting catalytic converter contamination by oil combustion products.

Thanks to lean operation, NO_x feedgas emissions should already be so low that emission limits can be met without exhaust gas aftertreatment. However, a target conflict is created between a fuel-efficient shift of operating range to low engine speeds and high loads and the associated high oxides of nitrogen emissions. Moreover, formation of nonhomogeneous mixtures (exhaust gas stratification, scavenging process deficiencies) restricts the degree of possible lean operation.

Aftertreatment of hydrocarbon emissions is problematical due to the lean-running engine's low exhaust gas temperatures. Similarly, feedgas HC emissions are higher as a result of incomplete stratified charge combustion (higher residual exhaust gas proportion and nonhomogeneous stratification at low loads/speeds, incomplete mixture cloud integrity). Long-term tests showed that over an operational life of 80,000 km, NO_x emissions remained at a nearly constant level but HC emissions rose with increasing mileage. Stage IV exhaust-emissions limits mandate exhaust gas aftertreatment by means of a deNO_x or NO_x storage catalyst. Furthermore, it will be necessary to develop lubricating oil that prevents catalyst contamination and concomitant exhaust gas aftertreatment deterioration over the service life of the engine.

5.7.3.2 Noise, vibration, and harshness

Analysis of baseline conditions showed that the two-stroke engine was at the upper range of the scatter band

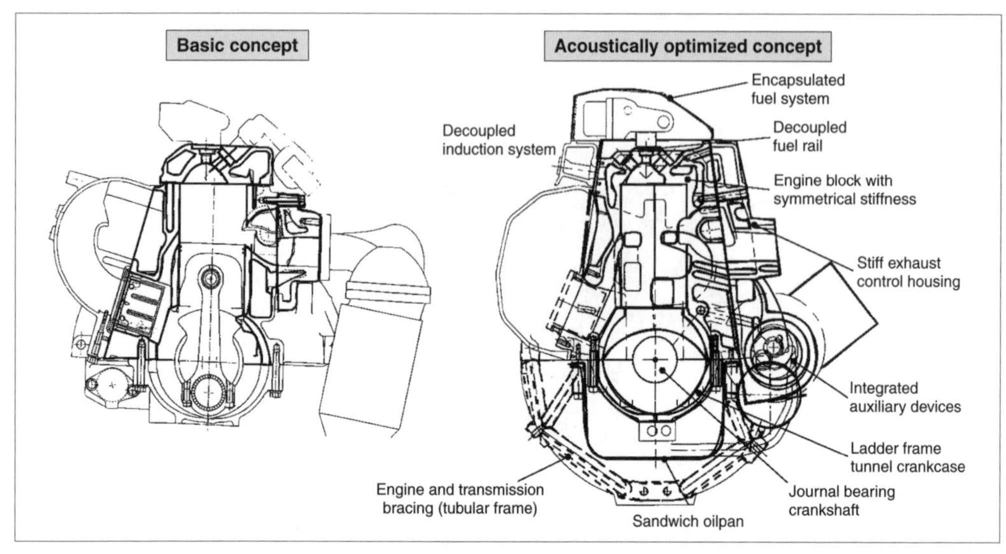

Fig. 5.7-5 Acoustic optimization of four-stroke engine.

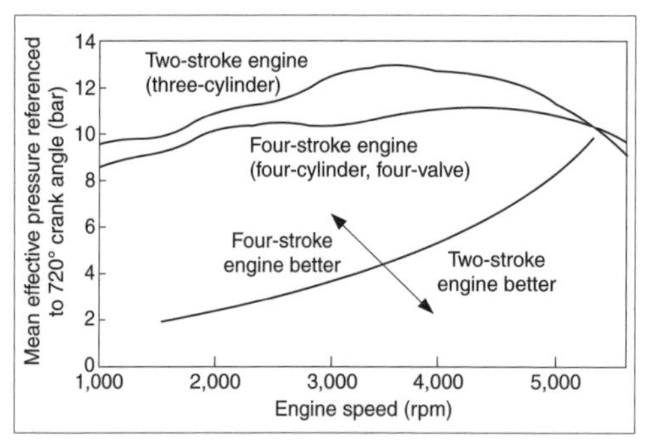

Fig. 5.7-6 Differences in specific fuel consumption.

of comparable eight-valve and 16-valve four-cylinder four-stroke gasoline engines. Furthermore, it was determined that mechanical noise was dominated by injection system noise at low engine speeds, by direct and indirect combustion noise at medium engine speeds, and by mechanical noise at high speeds. Fig. 5.7-5 shows countermeasures undertaken to combat this noise behavior, comparing the basic concept with an acoustically optimized concept.

Additional measures include application of graphite-coated pistons, increased rotational inertia, and reduction of the pressure rise d_p/d_α by retarded spark timing. This latter measure, however, must be weighed against the resulting loss of engine torque. At low speeds and loads, it is apparent that due to its doubled firing frequency, the two-stroke's rotational force is more uniform compared to a four-stroke engine, but cyclic variations are greater because of higher and less well defined residual gas content.

It must also be recognized, however, that an acoustically optimized two-stroke engine no longer represents the simplest and most cost-effective engine concept.

5.7.3.3 Fuel consumption

One of the main reasons for resumed interest in two-stroke engine development was the expectation of improved fuel economy based on reduced throttling losses and friction as well as on lean combustion. However, a comparison of fuel consumption maps for two- and four-stroke engines showed the differences illustrated in Fig. 5.7-6.

With reference to a four-stroke cycle, the two-stroke engine achieves higher mean effective pressures at full load. Given transmission layouts for equal vehicle performance, this leads to improved fuel consumption through a shift in operating points. Throughout the entire engine speed range, the four-stroke engine showed itself more economical in its upper load range. Consequently, for the extraurban portion of the new European driving cycle, the two-stroke engine was shown to have a fuel consumption disadvantage.

Analysis of friction and thermodynamics should provide insight into this behavior. As the total friction losses of two- and four-stroke engines lie at about the same level, a detailed analysis was carried out.

Fig. 5.7-7 shows a comparison of friction components as functions of engine speed, attributable to crankshaft, piston and connecting rod, and auxiliary drives (water pump, alternator) for two- and four-stroke engines and to the air compressor of the two-stroke and the valve train and oil pump of the four-stroke. It is apparent that the lack of valve train is offset by the addition of a compressor (~20% of total friction). The high friction component attributable to the piston and connecting rod

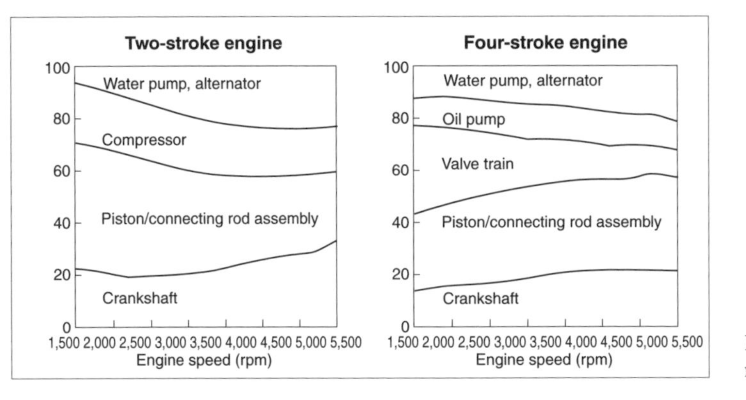

Fig. 5.7-7 Contributions to mechanical friction.

assembly of the two-stroke engine is especially noteworthy. The grounds for this are boundary lubrication in the vicinity of gas exchange ports, cylinder bore distortion, and oil deposits in the ring grooves. Experiments with monoblock engines and alternative oil formulations indicate that these problems may be manageable.

Thermodynamic analysis demonstrates the disadvantages of the two-stroke, and they are a result not only of its low effective compression ratio. Similarly, thermal losses are greater due to an overall higher basic temperature caused by its doubled firing frequency.

Compared to a modern four-stroke engine operating stoichiometrically, fuel economy advantages of 5–10% may be achieved in the two-stroke engine. Meanwhile, however, this has already been surpassed by current direct-injected four-stroke engines.

5.7.3.4 Mechanical durability

Two-stroke engines are expected to satisfy the same demands for durability as any four-stroke gasoline or diesel engine. In the past, two-stroke engines have generally not been able to achieve such high mileages. Failure modes are usually in the areas of piston, piston rings, and cylinder bore, which for design reasons experience conditions different from those in a four-stroke engine. High specific output, doubled firing frequency, and large piston crown areas on short-stroke engines result in greater heat input to the piston. Piston thermal issues must be carefully considered to avoid premature failure. Cylinder walls interrupted by scavenging ports exhibit insufficient heat rejection. This situation is aggravated by extreme bore distortion due to inadequate cylinder wall support attributable to scavenging ports. Fig. 5.7-8 illustrates these relationships.

In operation, cylinder bore distortion is aggravated by nonuniform thermal loading of inlet and exhaust ports. Necessarily larger piston clearances lead to carbon deposits (coking) in the ring grooves and vaporization of lubricating oil in the ring area.

A further weak spot is represented by the lubrication system itself. Oil is mixed with the airstream in the reed valve area by means of a map-controlled metering pump. Excess oil accumulates in the crankcase and is drawn off to a collection tank. Targeted lubrication of critical areas is therefore not possible. Furthermore, oil quality requires further optimization.

Functionality of reed valves and crankcase bearings can be compromised by free carbon particles which enter the induction system, and therefore the crankcase, by way of exhaust gas recirculation.

Suitable means of addressing the aforementioned problems include manufacturing the engine as a monoblock design to minimize distortion as well as using sealed bearings and pressure lubrication. These measures,

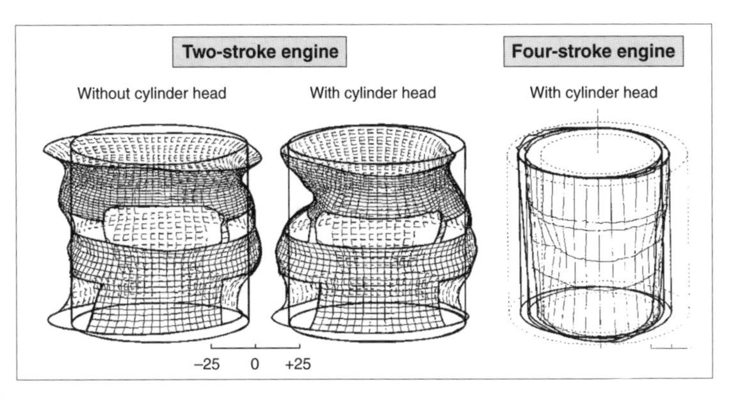

Fig. 5.7-8 Static cylinder bore deformation of two- and four-stroke engines.

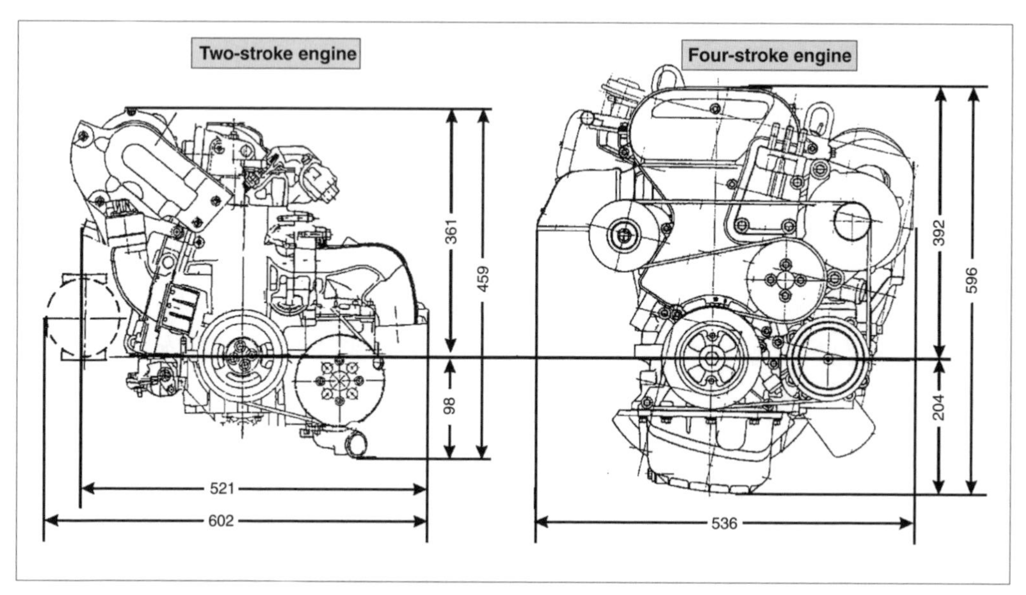

Fig. 5.7-9 Package comparison: width and height.

however, are in opposition to the desire for a simple, low-cost two-stroke engine.

5.7.3.5 Package and weight

Figs. 5.7-9 and 5.7-10 illustrate the differences in exterior dimensions for two- and four-stroke engine concepts.

In both cases, engine width is determined by intake and exhaust manifolding as well as configuration of auxiliary devices. The two-stroke engine is not as tall as the four-stroke. This advantage, however, largely manifests itself below the crankshaft centerline and is therefore only marginally useful, as crankshaft location is determined by the transmission and driveshaft if a given vehicle is designed to accept both two- and four-stroke powerplant options. Only a transmission concept specifically tailored to a two-stroke engine can benefit from this potential package advantage. In terms of overall length, the two-stroke engine under consideration has

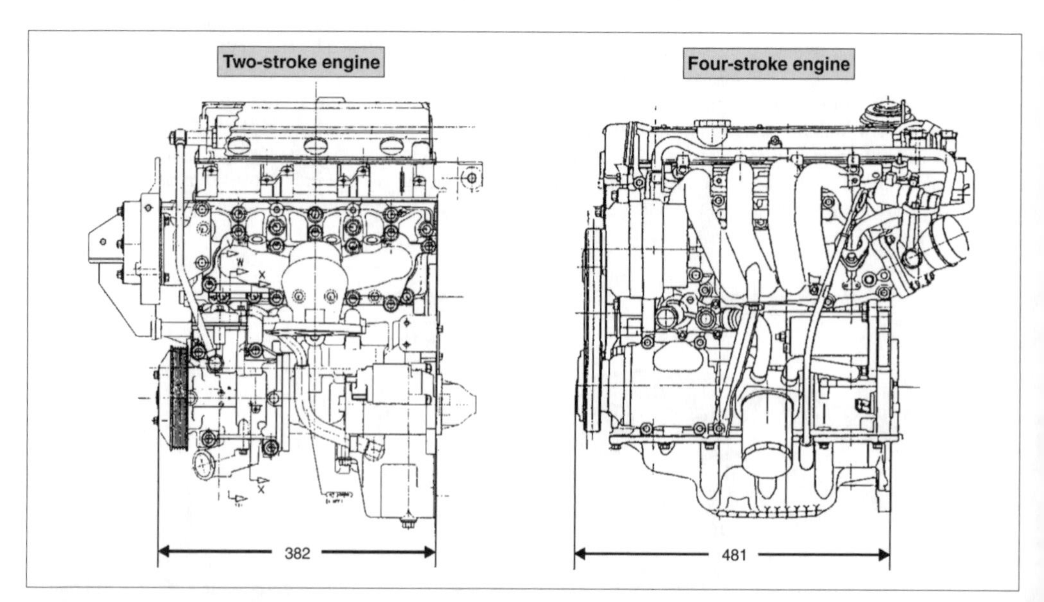

Fig. 5.7-10 Package comparison: length.

the advantage of three-cylinder design compared to a four-cylinder four-stroke engine. If the four-stroke were configured as a three-cylinder, it too would gain this benefit. Optimization of the two-stroke concept in terms of acoustic behavior and durability considerations (encapsulation, journal-bearing crankshaft, pressure lubrication) eliminates the option of retaining any possible two-stroke space advantages.

The same reasons that make any two-stroke package advantages unlikely can also be applied to weight considerations, demonstrating that it would be difficult to match weight benchmarks already established by current four-stroke engines.

5.7.3.6 Costs

The simple construction of two-stroke engines leads to an expectation of considerable cost savings. If this problem is examined more closely, the cost savings are primarily attributable to elimination of the valve train and complex four-stroke cylinder head. However, two-stroke cylinder block manufacturing becomes more complex, as it must incorporate charge transfer elements and air handling for the subsequent working stroke. Similarly, crankshaft manufacturing and assembly, with split roller bearings on main and connecting rod bearings, also becomes more involved. Two-stroke-specific components include reed intake valves, which permit precompression of charge air in the crankcase, and the rotary valve which controls exhaust gas flow. Moreover, the mixture injection system must be taken into account when considering costs. Overall, the cost advantage of two-stroke engines, given identical boundary conditions, is marginal.

5.7.4 Summary and evaluation

At present, two-stroke engines may be summarized as follows:

- Emissions standards may only be met by means of complex exhaust gas aftertreatment ($DeNO_x$ "catalyst," NO_x storage catalyst) analogous to four-stroke lean-burn engines or direct-injection gasoline engines.
- Complex design measures are required to achieve satisfactory NVH (noise, vibration, harshness) behavior.
- There is a 5–10% potential fuel economy advantage compared to stoichiometrically operating four-stroke engines (but these may also be operated with homogeneously lean mixtures or as direct injection engines).
- Complex measures are needed to achieve adequate mechanical durability.
- Package advantages are only in terms of engine length, due to the smaller number of cylinders with the same specific output.
- The weight advantage is limited.
- There is limited or no cost advantage.

In the brief development phase of the early 1990s, two-stroke engines could not be developed to a level where they would offer a convincing alternative for widespread application as automotive powerplants. However, thanks to their specific advantages, two-stroke engines present an attractive concept for other applications. This is especially true if demands for compact dimensions and limited operating conditions are only marginally impacted by other demands. Consequently, applications include small-displacement engines for motorcycle applications and the large field of primarily stationary engines for generators, watercraft, machinery (chain saws, etc.), combustion engines in hybrid vehicles, and so forth.

Bibliography

[1] Blair, G.P. *The Basic Design of Two-Stroke Engines*. Queen's University, Belfast, and Society of Automotive Engineers, ISBN 1560910089, 1990.
[2] Krickelberg, T. "Zukünftige Chancen des 2-Takt-Motors als PKW-Antrieb." Unpublished presentation, TU Vienna/TU Graz, November 1995.
[3] Königs, M. "Der Zweitakt-Motor." Ford Technology Presentation. Cologne, 1991.
[4] SAE SP-988. *Elements of two-stroke engine development*. Collected papers, 1993.
[5] SAE SP-1019. *Two-stroke engines: theoretical and experimental investigations*. Collected papers, 1994.
[6] SAE SP-1049. *Two-stroke engine design and emissions*. Collected papers, 1994.
[7] SAE SP-1131. *Progress in two-stroke engines and emission control*. Collected papers, 1995.
[8] SAE SP-1195. *Design, modelling and emission control for small two-stroke engines*. Collected papers, 1996.
[9] SAE SP-1254. *Design and application of two-stroke engines*. Collected papers 1997.

5.8 Alternative fuels in comparison

5.8.1 Possible sources of propulsion energy for mobile consumers

To propel a road vehicle, mechanical energy—that is, available energy (exergy)*—must be supplied to the wheels. This may be provided by two fundamentally different technical means: by extracting the ability to do work (exergy) within the vehicle itself or by supplying nearly pure exergy to the vehicle.

Extracting available energy from an energy carrier within the vehicle itself requires an engine (motor) mounted within the vehicle which obtains the necessary energy by means of a physical/chemical process. Possible energy carriers are:

- Chemically bound energy. Mechanical energy may be extracted by means of internal or external combustion in Otto, Diesel, Rankine, Stirling, or other processes.
- Physically bound energy in the form of:
  - Sensible heat
  - Latent heat
  - Mixed systems

*energy = exergy (available energy) + anergy (unavailable energy)

Mechanical energy may, for example, be extracted from these carriers by means of Stirling or Rankine processes. Furthermore, mechanical forms of energy such as
- Pneumatic energy
- Hydrostatic energy
- Kinetic energy

may in principle also be considered.

If the equipment required by the energy conversion process is not to be carried aboard the vehicle, nearly pure exergy must be supplied to the vehicle. Losses and emissions associated with the energy conversion process are then relocated elsewhere—in other words, to a generating station. Storage of nearly pure exergy, however, imposes special automotive engineering challenges.

Possible forms of nearly pure exergy are:

- Directly applied mechanical energy (e.g., cable car, flywheel storage)
- Electrical energy
  - With intermediate storage in the vehicle:
    Battery; i.e., conversion to electrochemical energy
    Capacitor; i.e., conversion to electrostatic energy
    Inductor; i.e., conversion to magnetic energy
    Flywheel; i.e., conversion to mechanical energy
  - Delivered on demand:
    Via catenary (overhead wire)
    Inductively

Essentially, for road traffic, only systems employing intermediate storage enter consideration. Buses powered by overhead wires no longer play a significant role in vehicular traffic. The concept of an inductive energy supply for road vehicles by means of suitable systems within the roadway is primarily being investigated in the United States but has hitherto not achieved any practical relevance.

5.8.2 Criteria for selection of energy carriers for vehicular traffic

Energy carriers for automotive propulsion must satisfy a multitude of requirements. Therefore, the following will outline some of the criteria (in no particular order) which must be met by an ideal vehicular energy carrier.

With respect to the vehicle:

- High fuel system energy density (energy carrier, energy reservoir, peripherals)
- Least possible energy losses during nonoperation of vehicle
- Low aggressivity toward fuel system materials
- Closely maintained tolerances for physical and chemical properties

With respect to the engine:

- For internal combustion engines, ignitability or knock resistance within the required range
- For spark ignition (Otto cycle) fuels: sufficiently high vapor pressure for cold starts at low temperatures
- No harmful deposits of fuel constituents or combustion products in the mixture formation system, combustion chamber, or exhaust system
- No substances which may poison catalytic converters
- Low demands on engine technology (e.g., avoidance of need for extreme injection pressure in diesel engines)
- Clean combustion, in particular minimization of raw NO_x, HC, and soot emissions; consequently, adherence to emissions limits by means of simple exhaust gas aftertreatment
- Compatibility with lubricating oils

With respect to fuel handling and safety:

- Storable and transportable under all ambient conditions
- Safe handling by laypersons
- Low vapor pressure to minimize evaporative losses
- Low tendency to form explosive mixtures during handling and in accidents, both in enclosed spaces and open air
- Ability to park vehicle in garages without restrictions

With respect to economic operation:

- Acceptable costs, long-term cost predictability
- Worldwide availability, or at least availability in significant vehicle markets via dense infrastructures
- Primary energy and raw materials available in large quantities from various world regions

With respect to human impact:

- No danger from physical contact
- Nontoxic
- Non-narcotic
- Noncorrosive
- Lowest possible odor impact, lowest possible allergenic effect
- Easily washed or rinsed off

With respect to environmental impact:

- Low emissions in obtaining raw materials and primary energy
- Low emissions in manufacture
- Low losses and emissions during transport, storage, and filling
- No release of chemical elements and compounds which are not naturally present in air; low photochemical potential for exhaust gases (low contribution to smog and ozone formation)
- Not a contributor to greenhouse gases
- Lowest possible damage to environment in the event of accidental release to soil, water, or air; easily broken down biologically

All of the aforementioned possible means of supplying energy to vehicles have been investigated. Because

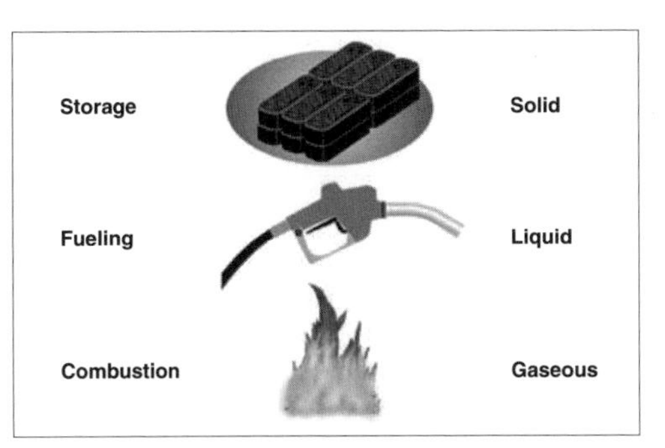

Fig. 5.8-1 Ideal bulk conditions for energy carriers.

of frequently conflicting demands, only a few have established themselves. It is obvious that no single energy form meets all of these demands to an equal degree.

Above all, for reasons of energy density and ease of handling, chemically bound energy, released by combustion processes, exhibits significant advantages. For reasons of constructive effort (weight, costs) and response, internal combustion processes employing spark ignition and diesel engines have completely displaced external combustion processes such as steam or Stirling engines. Solid fuels have therefore also been eliminated.

Gaseous fuels have drawbacks in terms of storage and therefore play only a minor role. Fig. 5.8-2 illustrates energy density of fuels and containers for significant gases in comparison to gasoline.

By far the widest distribution is enjoyed by chemical energy stored in mixtures of liquid hydrocarbons, commonly known as gasoline and diesel fuel.

Fig. 5.8-3 shows the mean calorific value, as a significant characteristic for evaluating suitability of various energy carriers as a motor fuel, compared to gasoline and diesel fuel. The mean calorific value is plotted against stoichiometric air/fuel ratio. This is directly related to thermal condition by fuel density.

Despite their different calorific values, the mean calorific values of various fuels are of about the same order of magnitude. It follows that the achievable mean effective pressures of different fuels in any given engine are about the same. The high stoichiometric air/fuel ratio of hydrogen does not constitute a basic disadvantage but instead shows that in comparison to other fuels, appreciably less hydrogen is needed to burn with a given

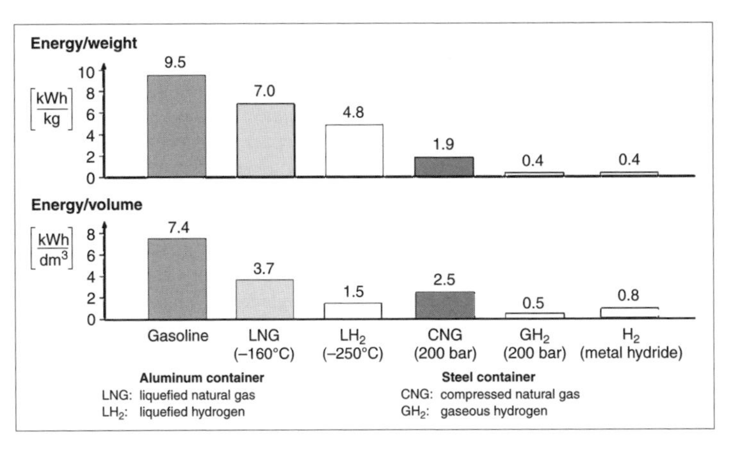

Fig. 5.8-2 Energy storage density of significant gases in comparison to gasoline.

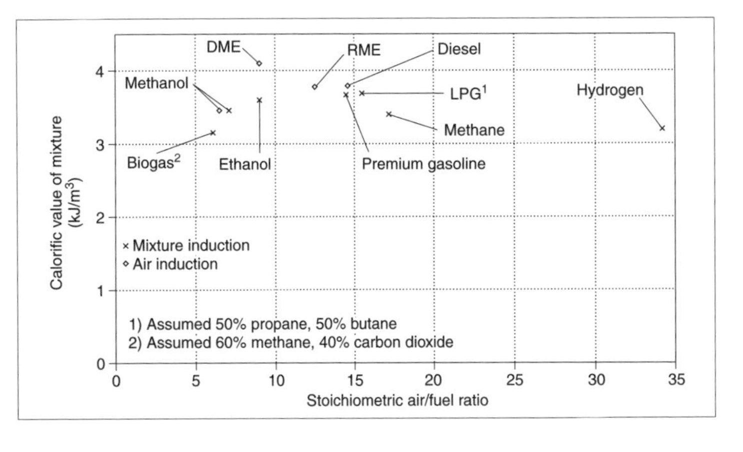

Fig. 5.8-3 Mixture calorific values of various conventional and alternative fuels at $\lambda = 1.0$, $T = 0°C$, and $p = 760$ Torr.

mass of air. This is attributable to the high gravimetric energy content of hydrogen.

If all criteria are considered, it is apparent that, to date, gasoline and diesel fuel have achieved dominant roles as mobile energy carriers to the exclusion of almost all other sources—and for good reason. From a global standpoint, they represent a highly attractive and, in the summation of their properties, acceptable solution to the aforementioned requirements. However, new engineering developments may result in future viability and widespread application of options which have hitherto appeared unattractive.

5.8.3 Conventional fuels and possibilities for their continued development

Gasoline and diesel fuels may be greatly improved by means of targeted development [1]. One such development, currently under way, is variously known as "desulfurization," "reduced aromatic content," or more generally "designed fuels." Heretofore, gasoline and diesel fuels have almost exclusively been obtained by cracking that naturally occurring mixture of materials collectively known as "crude oil." Chemical transformations of this feed stock are only applied to a very limited degree. The accompanying refinery processes will not be described in greater detail; literature describing these may be found in [1].

There are many alternatives to conventional refining processes. Of greatest interest is the Fischer-Tropsch process for production of diesel fuel and the Mobil Oil process for production of gasoline [1]. In both cases, synthesis gas serves as the point of origin for subsequent steps in the process. With the addition of external energy, synthesis gas may in principle be produced from any available raw material containing hydrocarbons, oxygen, and hydrogen. As a result, this process permits complete decoupling of fuel production from crude oil. Furthermore, any available primary energy source may be engaged in synthetic fuel production processes. In terms of process energy as well, crude oil as an energy source may be completely abandoned. Natural gas or oil may constitute the material basis for fuel just as readily as biomass; even extraction of carbon dioxide from air and water is theoretically possible, if the necessary process energy is available at acceptable cost.

Synthesis gas as the point of origin has one additional advantage: in contrast to fuel production based on crude oil, in which compounds are separated from the available mixture and subjected to minor chemical alterations, in synthesis gas the desired chemical compounds are deliberately built up from very small basic molecules. This process can be controlled precisely, providing gasoline and diesel fuel of quality well superior to that available through classic refining processes. Such "designed fuels" permit achieving appreciable improvements in vehicular emissions behavior.

Synthesis gas may be used to produce numerous other compounds suitable as motor fuels: methanol, ethanol, DME, etc. In each case, it must be determined whether these indeed have any advantages with respect to the requirements outlined above compared to the qualities exhibited by current or improved diesel fuel and gasoline.

5.8.4 Significant reasons for alternative energy carriers in motor vehicles

In addition to gasoline and diesel fuel, there are many other substances which meet the above requirements to a similar degree but with different emphasis. Foremost among these are the hydrocarbon compounds methanol, ethanol, methane (CNG—compressed natural gas or LNG—liquefied natural gas), and LPG (liquefied petroleum gas).

Among the many demands imposed on motor vehicle energy carriers, ecological properties in recent years have assumed special relevance. On the one hand, emissions of substances which normally have little or no presence in the atmosphere—CO, oxides of nitrogen, hydrocarbons, particulates, benzene, etc.—must be limited to a level which can confidently be considered safe. On the other hand, release of carbon dioxide should be minimized in order to avoid global warming as a result of the greenhouse effect [2, 3].

In the first case, questions regarding fuel origins and processes required to refine them are of only limited interest. It is "only" necessary to reduce emissions (i.e., those emanating from tailpipes as a result of vehicle operation). To this end, combustion and exhaust gas aftertreatment must be matched to each other; this is best achieved by means of close tolerance fuels—that is, fuels whose engine-specific chemical and physical properties are known to a high degree of precision. Hydrogen exhibits an especially high potential for improved environmental characteristics.

Electric vehicles exhibit their greatest advantages in terms of local ecological effects. These vehicles primarily employ batteries as storage media for electrical energy. In the longer term, our current state of knowledge indicates that fuel cells may prove useful for generating electricity from suitable chemical energy carriers. At present, all other variations appear to be completely unsuitable or only suitable for special applications.

In considering climatic effects, the entire chain—from obtaining raw materials (e.g., crude oil) to use in the automobile itself—must be taken into account. After all, as far as the greenhouse effect is concerned, the actual source of greenhouse gases is immaterial. The choice of vehicular energy carrier is not the sole determining factor; rather, the raw materials, primary energy, and manufacturing processes enter into the picture. A significant goal is to avoid use of fossil fuels for primary energy.

With this as background, the following fuels will be examined here:

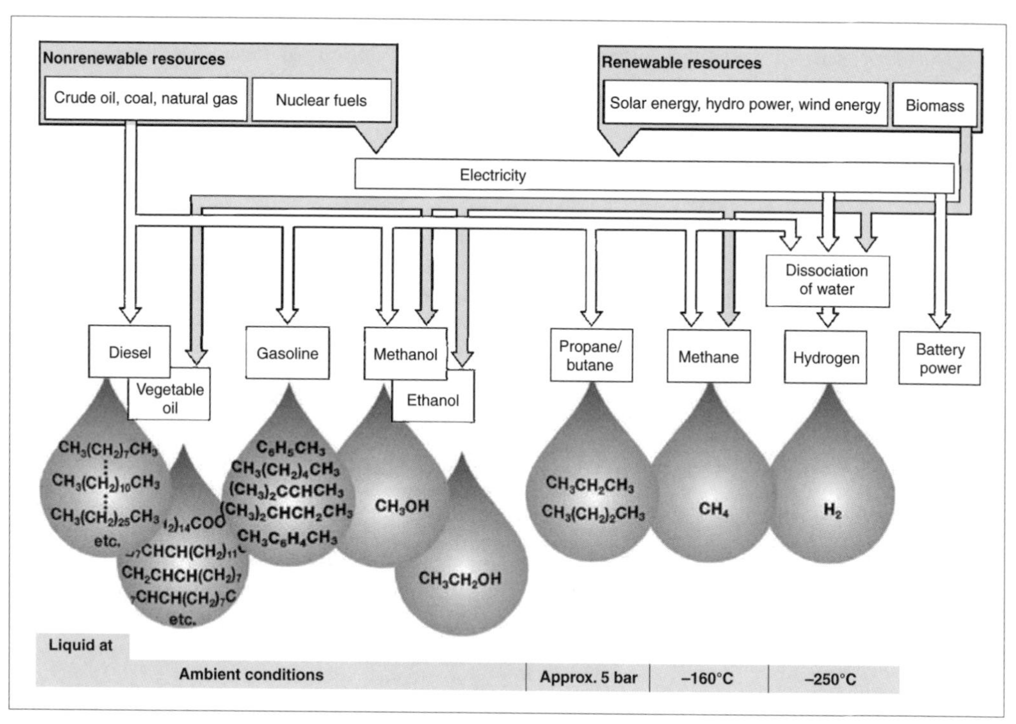

Fig. 5.8-4 Possible means of production for the most significant energy forms.

- Biofuels
- Methanol
- Natural gas
- Hydrogen

5.8.5 Biofuels

In any discussion of alternative fuels, agricultural products take on a special role. Several species of plants produce oils which, after relatively minor chemical transformation, are suitable as diesel fuels. In Europe, the most significant of these is rapeseed oil; in tropical regions, palm oil enters consideration. In terms of local emissions effects, these biofuels have no special advantages or disadvantages relative to present-day diesel fuel. When the entire process chain is examined, including soil preparation, harvesting, production, and consumption, it is found that slightly less carbon dioxide is released from biofuels than in the case of diesel fuel produced from crude oil. This advantage, however, is obtained at appreciably higher cost (not including any possible tax benefits). At present, from an economic standpoint, vegetable oils do not represent a viable alternative to conventional fuels in developed countries [1].

Ethanol, well suited as a fuel for spark ignition engines, may be produced from sugar or starch using current, familiar processes [4]. Brazil most notably, beginning in the 1970s, has promoted ethanol fuel on a massive scale by means of government programs. However, in terms of local emissions, ethanol does not exhibit any significant advantages in comparison to gasoline. If the entire production chain is examined, from fertilizer production to farming and ethanol generation to consumption in motor vehicles, ethanol does not offer any significant benefit in terms of the greenhouse effect. Here, too, economic costs must be compared to limited ecological utility. Additional rationalizations—employment policy, agrarian structures, or other issues—must provide justification for this approach [1].

5.8.6 Methanol

The advantages of methanol (CH_3OH) as a motor fuel lie in its potential for reduced ozone formation compared to gasoline and in its relatively simple production from synthesis gas. Against this backdrop, extensive research and large-scale experiments investigating the suitability of methanol as a motor fuel were carried out in the 1970s and 1980s [5].

Methanol's disadvantages are above all its high toxicity and, in comparison to gasoline, approximately 50% lower energy content. Establishment of an additional methanol infrastructure is universally regarded as ill-advised.

Methanol production from coal exhibits a maximum production efficiency of about 50%, while production from natural gas has a maximum efficiency of about two-thirds. It is therefore more sensible to apply these very plentiful natural gas resources directly as energy carriers for motor vehicle operation, thereby also taking advantage of the appreciably better emissions potential of natural gas. The resulting infrastructure for liquefied natural gas could, with certain modifications, later also be used for hydrogen fuel.

5.8.7 Natural gas

Natural gas is the most environmentally friendly fossil fuel. It consists primarily of methane (CH_4), which has the highest hydrogen content of any hydrocarbon. The lower the carbon content of a fuel, the lower its carbon-based emissions (carbon monoxide, carbon dioxide, soot, and hydrocarbons). Natural gas resources are even more readily available than crude oil, making it possible to diversify from sole dependence on oil [6].

Natural gas may be used in combustion engines without any chemical modification. However, depending on its geographical source, cleaning (primarily, dehydration) is necessary.

For transport in a vehicle, the volumetric energy density of natural gas must be increased. This is accomplished either by storage at high pressure or at very low temperature. Pressurized storage is usually done at pressures up to 200 bar. Vehicles meeting specially developed Euro-

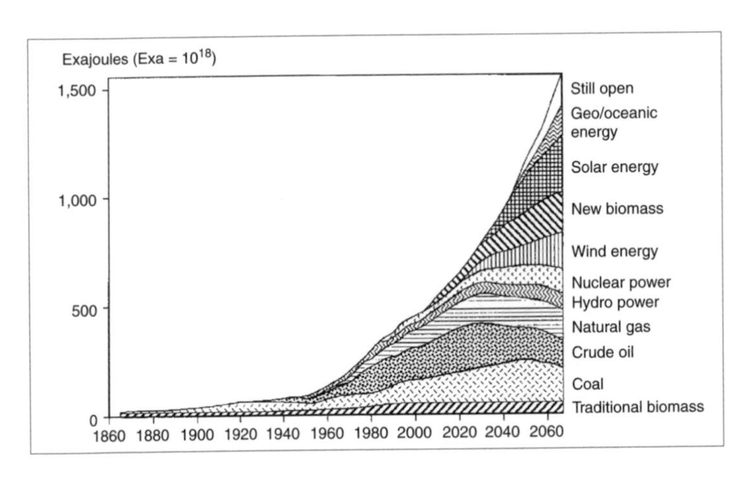

Fig. 5.8-5 Scenario for worldwide energy demand development.

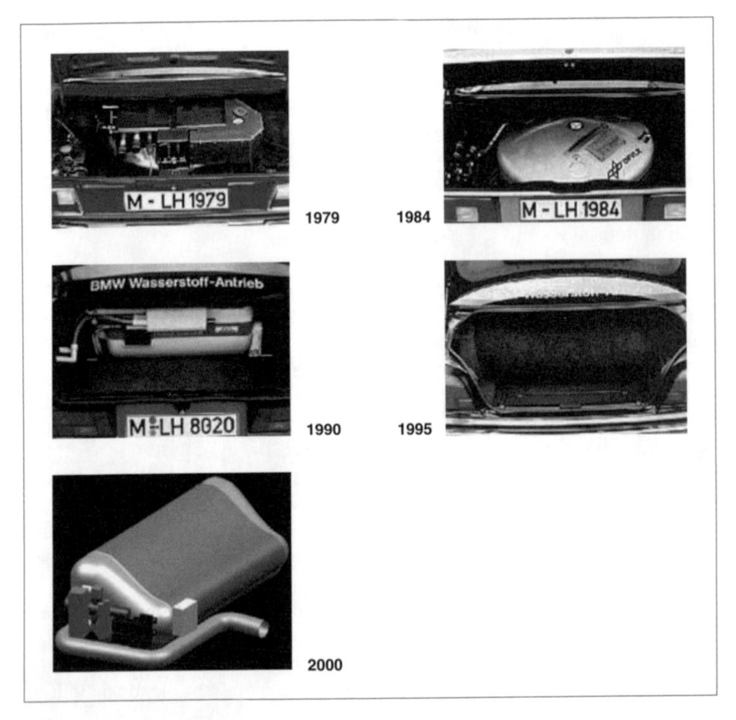

Fig. 5.8-6 Liquid hydrogen tank system development stages.

pean Union guidelines (see Section 7.6) have already been granted type approval. Within the gas industry itself, a general trend toward ultracold liquefied gases may be discerned. The highest energy density is achieved in the liquid phase at −160°C, even when effective volume including thermal insulation is taken into account. Storage technology is described in greater detail in Section 7.6.

Overall efficiency (from source to tank) for compressed or liquefied natural gas may reach 90%, comparable to the favorable levels obtained with modern conventional fuels [7]. Since natural gas has largely displaced oil in the heating energy market, it is quite conceivable that this "cleanest" of all fossil fuels may also play a role in the transportation sector.

5.8.8 Hydrogen

In terms of avoiding local emissions, hydrogen-fueled spark ignition engines present a nearly ideal situation. Suitable process control results in only minimal emissions of nitrogen oxides; the most significant oxidation product is water. In this situation, as in others, production of the energy carrier itself is decisive in evaluating the overall process. At present, hydrogen is generated almost exclusively by steam reforming of natural gas. In the process, the entire energy content of the natural gas feed stock is released as carbon in the form of CO_2. A much more attractive perspective is offered by utilization of electrical energy generated by renewable resources—water power, solar energy, wind energy, or nuclear primary energy. In this way, hydrogen may be produced from water with almost no emissions.

To supply worldwide energy demand, new large-scale applications for nonfossil fuels are under development. It is estimated that the coming decades will see consumption of fossil energy sources peak and decline, with most of the additional energy demand met by renewable resources [8]. Large-scale solar energy generation in solar thermal powerplants may assume a key role.

Development of components and their interactions to create a functional hydrogen energy economy was advanced most notably in a pilot project by Solar Wasserstoff Bayern GmbH [9] and the Euro-Québec Hydro-Hydrogen Pilot Project [10].

Use of hydrogen for mobile applications began in the mid-1970s. Hydrogen storage technology is very similar to that used for natural gas storage, although the energy density of hydrogen is lower. Pressurized gas systems (200 bar) have a long service history. They are primarily suited to commercial vehicles with minimal range requirements. Technology for deep-cooled (−253°C) liquid hydrogen was developed for space travel and then gradually refined in stages for automotive applications [11, 12]. Fig. 5.8-6 illustrates development of liquid hydrogen tank systems since 1979.

Technical details may also be found in Section 7.6.

In the early 1980s, extensive research was conducted on chemical storage of hydrogen in metal hydrides [13]. Storage densities similar to those of pressurized hydrogen systems were achieved (see Fig. 5.8-2).

For passenger cars, research has concentrated primarily on liquid hydrogen technology due to its relatively high energy density, although its overall efficiency of at most 70% is less favorable than that of pressurized hydrogen systems.

Long-term success of hydrogen as an energy carrier will probably depend primarily on:

- How efficiently hydrogen can be produced
- To what degree additional stationary consumers may utilize hydrogen
- What safety measures will lead to widespread public acceptance

References

[1] Schindler, V. *Kraftstoffe für morgen, eine Analyse von Zusammenhängen und Handlungsoptionen.* Springer-Verlag, Berlin, Heidelberg, New York, 1997.

[2] "Klimaänderungen 1995: Zweiter Sachstandsbericht von IPCC (Intergovernmental Panel on Climate Change)." ProClim, Bern, 1996.

[3] Fahl, U., et al. "Kostenvergleich verschiedener CO_2-Minderungsmaßnahmen in der Bundesrepublik Deutschland." Forschungsbericht No. 40 des Instituts für Energiewirtschaft und Rationelle Energieanwendung, Stuttgart University, 1997.

[4] Menrad, H. "Äthanol als Kraftstoff für Ottomotoren." ATZ 81, 6, 1979.

[5] Ingamells, J.C., and R.H. Lindquist. "Methanol as a Motor Fuel or a Gasoline Blending Component." SAE 750123, 1975.

[6] World Energy Council. *Survey of Energy Resources.* 18th edition, London, 1998.

[7] Mauch, W., and R. Fröchtenicht. "Energieaufwand und Kosten für die Bereitstellung von Erdgas als CNG und LNG." ETG-Fachbericht 65, VDE-Verlag, Berlin Offenbach, 1997.

[8] Shell. "Aktuelle Wirtschaftsanalysen. Energie im 21. Jahrhundert, Betrachtungen zur Entwicklung des Welt-Energieverbrauchs." Hamburg, 1995.

[9] Szyszka, A. "Ten Years of Solar Hydrogen Demonstration Project at Neunburg vorm Wald, Germany." International Journal of Hydrogen Energy, Vol. 23, 10, 1998.

[10] Dorlet, B., J. Gretz, D. Kluyskens, F. Sandmann, and R. Wurster. "The Euro- Québec Hydro-Hydrogen Pilot Project [EQHHPP]: Demonstration Phase, Hydrogen Energy Progress X." Proceedings of the 10th World Hydrogen Energy Conference, Cocoa Beach, June 20–24, 1994.

[11] Peschka, W. *Liquid Hydrogen: Fuel of the Future.* Springer-Verlag, Vienna, New York, 1992.

[12] Braess, H.-H., and W. Strobl. "Hydrogen as a Fuel for Road Transport of the Future: Possibilities and Prerequisites." In: Hydrogen Energy Progress XI, Proceedings of the 11th World Hydrogen Energy Conference, Stuttgart, June 23–28, 1996.

[13] Förster, H.J. "Alternative Antriebskonzepte Elektro-, Hybrid- und Wasserstoffantrieb." VDI-Berichte No. 817, VDI-Verlag, Düsseldorf, 1990.

[14] Nierhauve, B. "Neue Herausforderungen an die heutige und zukünftige Kraftstoffversorgung des Straßenverkehrs." ATZ, 1999, pp. 154–160.

[15] Beutler, M., and M. Schmidt. "Erdgas - Ein alternativer Kraftstoff für den Verkehrssektor." ATZ, 2000, pp. 176–182.

[16] Osborne, K.D., and M.S. Sulek. "Ford's Zero Emission P2000 Fuel Cell Vehicle." SAE 2000-01-C046.

[17] Forum für Zukunftsenergien (publisher). "Hyforum 2000, Proceedings of the International Hydrogen Energy Forum 2000." Munich, Sept. 11–15, 2000.

6 Body

6.1 Body structural practice

6.1.1 Unit body

In the early years of the automobile, bodywork—in keeping with horse-drawn-coach-building practice—was attached to a frame structure. Today, this structural practice is only found in commercial and off-road vehicles. Within passenger car practice, unit body construction—also called self-supporting bodywork—has become predominant. This very complex structural practice must satisfy many diverse requirements, ranging from systematic application of lightweight design principles through effective occupant protection to attractive appearance. One more criterion must be added to these: because all components of a vehicle must be attached to the bodywork, the bodywork represents the single most important component carrier.

6.1.1.1 Development demands

The catalog of requirements for an automobile body is extensive. Form and color (design) and spatial relationships as well as occupant protection and collision behavior represent purchase-decisive criteria for customers. To these are added company-internal development requirements self-imposed by individual auto manufacturers (Table 6.1-1; see also Section 4.2).

As a result of this complexity, which we have touched on only briefly (in practice, the catalog of requirements imposed on a body structure includes many hundreds of demands and parameters), development is increasingly being carried out by means of computational methods. The primary task is to assign a realistic target for each demand and then subordinate these partial targets to the desired overall target. In view of ever shorter development times, these target values may only be met if the entire development process is largely supported and guided by computational methods. Today, the primary function of test activities is to validate results obtained through analytical methods or to refine and tune the computational method.

Table 6.1-1 Demands on body development

Customer-relevant criteria	Production-relevant criteria
Successful design	Simple assembly sequence
Maximum safety	Utilization of existing manufacturing facilities
Low fuel consumption	Low parts count
High comfort	Simple joining technology
High functionality	Ease of manufacturing
High quality and longevity	Good welding accessibility
Attractive price	High process quality
Low repair costs	Shared parts and platform solutions
Low noise level	Optimized application of materials
Dimensioned for everyday utility	Low manufacturing costs

6.1.1.2 Exterior skin

The exterior contours of a vehicle are determined by three main factors: the basic requirements established by the package, the design, and aerodynamics. Market research has repeatedly shown that of these, design is by far the most significant factor affecting purchase decisions.

6.1.1.2.1 Design

Because design requirements as a rule affect body dimensions and characteristics, design (Section 4.1) and body development take place simultaneously in a networked process. In this process, designers are given the vital role of coordinators who, even in the conceptual stage, must take into account physical properties of materials, engineering requirements, and production considerations. Consequently, from the very beginning, they must secure the specialized services of design and manufacturing engineers.

In short, design exerts a considerable influence on the basic shape of a body shell. In modern practice, all desired body variations are taken into consideration from the very outset of vehicle development. A good example is provided by sedan and coupe versions of a given model line: in planning the sedan, possible lack of a B-pillar on the coupe model must be taken into account. In dimensioning pillars and struts, designers must consider their future visibility in the later vehicle. Installation of interior components should be kept as flexible as possible in order to permit various trim levels. Visible joints and gaps (shutlines)—for example, between side panels, doors, and decklid—must be executed to a high degree of precision in order to meet rising customer expectations for vehicle visual quality.

6.1.1.2.2 Aerodynamics and aeroacoustics

Preliminary aerodynamic body optimization in the wind tunnel is conducted as early as the first 1/5 scale clay models. Time and again, in the further course of development, any and all changes in design and package requirements must be taken into consideration. The final shaping and optimization is carried out on a full-scale plastic model. In particular, front and rear bumper areas,

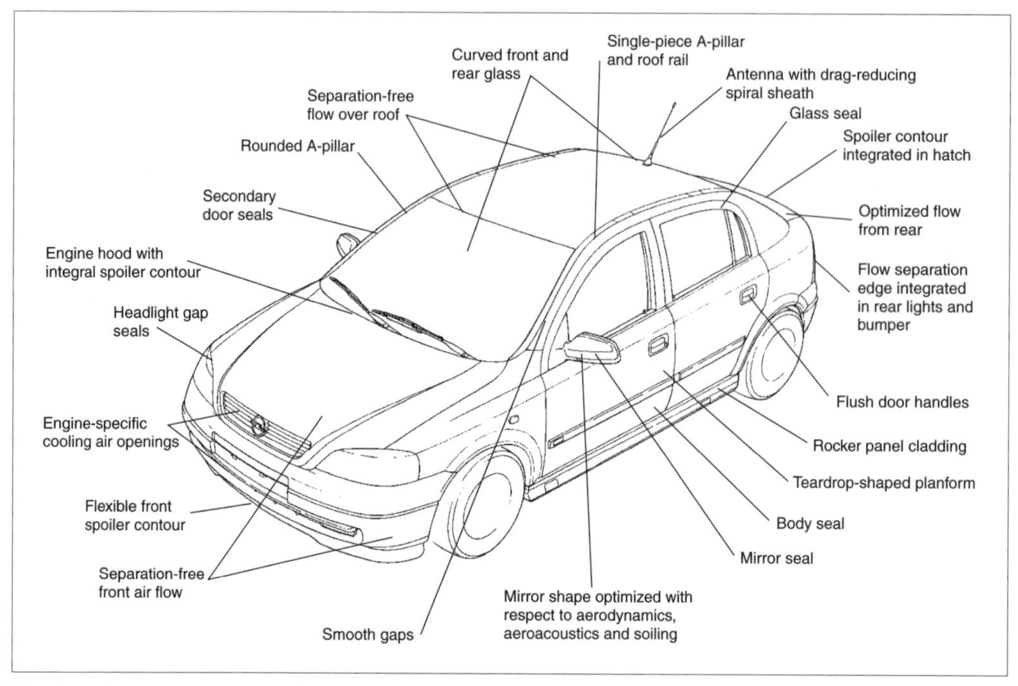

Fig. 6.1-1 Measures toward aerodynamic optimization of exterior skin.

outside mirrors, A-pillars, roof gutters, rocker panels, and underbody are subjected to detailed scrutiny.

Aerodynamic optimization encompasses a multitude of individual measures (Fig. 6.1-1; see also Section 3.1). On the usual basic body shapes, the following design measures [1] have proven effective in achieving the lowest possible aerodynamic drag coefficient:

Flow around the front of the vehicle should take place with as little flow separation as possible. It is therefore necessary to round off any transitions. A steeply raked windshield (in keeping with outward visibility requirements) combined with rounded A-pillars is desirable. Flow is best directed over a crowned roof, in which the curvature is kept constant. At the rear of the vehicle, a defined aerodynamic flow separation is required. A drawn-in tail ("boat tailing") is helpful in keeping the stagnation area as small as possible, which also has a beneficial effect on the body's tendency to resist soiling. In notchback sedans, such boat tailing may be achieved by drawing the roofline downward. Suitable angling of the tail may minimize induced drag created by turbulent flow, especially in notchback designs.

Shaping of the underbody has a considerable effect on aerodynamic drag of a vehicle, as well as on aerodynamic lift forces (which are to be minimized). The objective is to achieve as smooth an underbody as possible although, of course, compromises must be reached—for example, in assuring cooling airflow for brakes and catalytic converter. A large contribution to overall aerodynamic drag is attributable to the vehicle's wheels (as much as $c_d = 0.06$). Attempts to effect a significant reduction of this contribution by means of design measures—for example, with rear wheels mounted flush with the bodywork—have to date not yielded the desired results.

Further development stages seek to reduce wind noise and soiling of side and rear glass in rainy conditions induced by airflow over the bodywork. To this end, sharp edges and door gaps should be given special attention, as these progressively increase wind noise with increasing vehicle speed (see also Sections 3.1 and 3.4).

6.1.1.3 Package

Establishing body shapes and dimensions also defines the space available for all installed vehicle systems and components. The product of this defining process, which takes place in parallel to design and aerodynamic development, is termed the "package" (Section 4.2).

As a rule, key dimensions have already been established before vehicle development is begun (Table 6.1-2 and Fig. 6.1-2). These include not only overall length and width, but also an entire series of other notable vehicle coordinates which describe the body: the transition between engine hood and windshield, and radiator grille, ramp angles, ground clearance, location and shape of bumpers, front and rear H-points (occupant hip locations), distance between driver and steering wheel, and luggage volume, to name just a few examples.

A vehicle's package plays a significant part in determining the spatial impression experienced by its

Table 6.1-2 Key package dimensions

Exterior dimensions	Interior dimensions
Vehicle length (L103)	Effective Seating Reference Point (SgRP) leg room accelerator/second (L34/L51)
Vehicle width (W103)	Minimum knee clearance second (L48)
Height (H101)	SgRP couple distance (L50)
Wheelbase (L101)	Effective T-point head room front/rear (H75/H76), head clearance diagonal driver (W27)
Tread front/rear (W101/W102)	SgRP front to heel/second to heel (H30/H31)
Enclosed luggage compartment volume (ISO V210)	Vision angle to windshield upper DLO (H124)/lower, vision angle to backlight upper/lower
Fuel tank volume	Shoulder room front/second (W3/W4), width between armrests
	SgRP—front/y-coordinate (W20)

occupants. The following criteria exert a positive effect on this spatial perception:

- Tall roofline
- Engine hood/windshield transition pulled far forward
- Long wheelbase
- Wide track

Naturally, these parameters can only be implemented to the extent permitted by design, aerodynamics, and the major criteria dependent on the established vehicle class.

Driver seating position forms the basis of interior layout. First, seat longitudinal and vertical adjustment ranges are established. The longitudinal adjustment must provide adequate space for the driver; he or she must be able to reach the steering wheel and pedals comfortably, without crowding the rear-seat passengers. For seat height adjustment, outward visibility over the instrument panel, headroom (H75 in Fig. 6.1-2), pedal reachability, entry and exit, and all sight lines must be considered.

Next, a reference point for seat adjustment is defined, the so-called hip point (H point). This represents the seat adjustment for a 95th-percentile male (only 5% of the target market population is taller). This seating position is determined by seat height and the heel point. The sole of the foot must be able to touch the throttle pedal (see also Section 6.4.1).

The H-point location may be calculated or measured. Based on this, remaining interior dimensions such as headroom and legroom are determined. Rear seat knee- and legroom are determined largely by the distance between driver and rear occupant H points.

6.1.1.4 Body structure

The body structure of a vehicle must perform the following functions:

- Absorb all forces and moments
- Define the interior space
- Provide peripheral energy conversion zones
- Provide mounting points for all driveline components and suspension modules

The following sections provide an example illustrating the process of body shell design. The case at hand involves a five-door compact vehicle introduced in 1998 [2]. Although individual design elements differ from manufacturer to manufacturer, dependent upon their

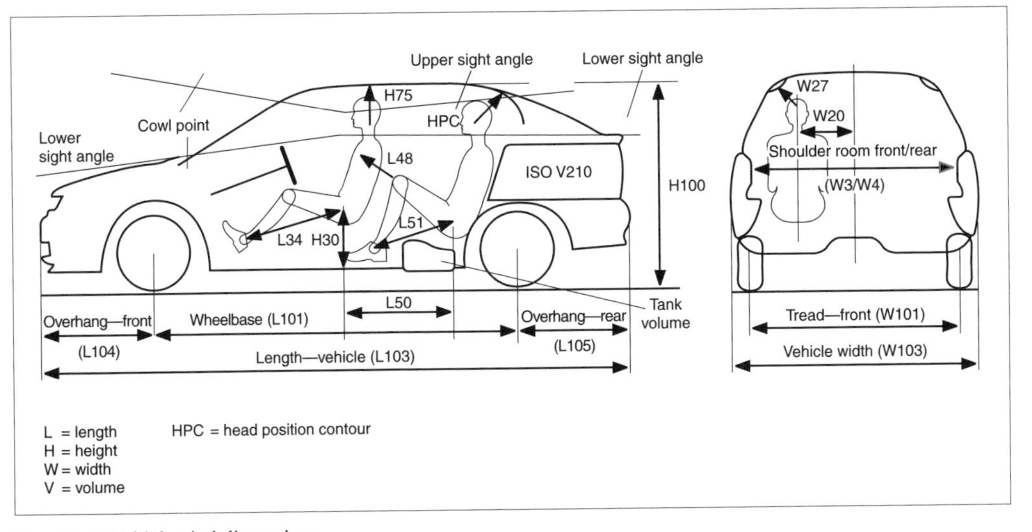

Fig. 6.1-2 Vehicle vital dimensions.

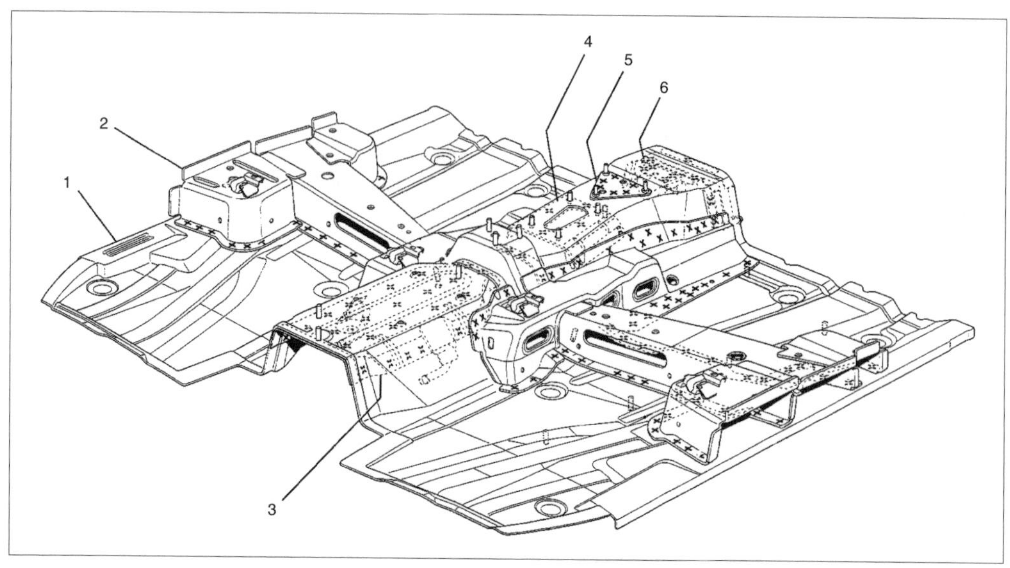

Fig. 6.1-3 Example of front floorpan. 1, front floor; 2, front seat crossmember; 3, front tunnel reinforcement; 4, air bag control unit mount; 5, parking brake handle reinforcement; 6, exhaust system bracket.

individual design philosophies, vehicle class, and model generation, this description nevertheless covers a variety of applications. The sequence of individual segments corresponds to the production sequence in the body shell assembly facility.

6.1.1.4.1 Floorpan

Longitudinal and transverse members form the foundation of the floorpan. These are closed, front and rear, by floorpan sheet metal. See Fig. 6.1-3.

The two front longitudinal members, composed of the forward frame rails and their extensions, are fitted with blanking plates at their front extremities to which the bumper is bolted. On the right side, a welded-in reinforcement is fitted with a threaded plate to receive the removable towing eye. Farther along the frame rails, front suspension pickup points are welded to their undersides. Engine mounts are also attached to these rails.

The forward frame not only is required to carry the engine and front suspension but also plays a central role as an energy-absorbing component in a frontal collision. Specific areas subjected to the greatest loads are given added reinforcement. In the example shown, the forward frame extension ends at the connection between the seat crossmember and floorpan. In other floorpan concepts, these frame extensions may reach as far as the rear-seat kick panel. Above all, such extensions prevent bending of the floorpan in the event of an accident as well as in normal operation. The seat crossmember in conjunction with seat mounts attached to the floorpan provides rigid attachment for the front seats. Seat-belt retractors are attached to anchorages at the tops of the seat mounts. In the event of a frontal collision, these retractors are subjected to very high forces, which are best absorbed through the seat mounts.

The structure for the rear floorpan is formed by two rear longitudinal frame rails and three crossmembers for the rear floorpan, rear suspension, and transom (Fig. 6.1-4). In order to maintain better load paths, the rear frame is joined directly to the rocker panels. This provides advantages in rear collisions and increases the overall body stiffness. Rear suspension mounts are also located in this area. The center of the frame serves to support rear suspension springs. The towing eye is attached at the left rear; a muffler support is located at the right rear. Application of tailored blanks permits dispensing with any additional reinforcements at the rear of the floorpan (in contrast to the front frame structure).

In operation, the crossmember at the rear floor prevents excessive bending in the rear seat area. Correspondingly, it resists deformation of the floorpan in the event of a rear collision. This member also provides pickup points for the front fuel tank mounting straps.

The rear suspension crossmember is decisive in rear collisions as well as in providing torsional rigidity. In a rear collision, this crossmember stabilizes the rear frame and prevents uncontrolled buckling. As a bending moment carrier, it opposes extreme twisting of the rear frame that might result from opposite deformation of the rear frame profile. In order to optimize rear suspension crossmember behavior, its connection to the rear frame must be designed accordingly. It is therefore necessary to join the crossmember to the rear frame, along its entire depth, in addition to welding a tab to the frame underside. The width of the connection must be chosen so that the flanks of the crossmember support the rear

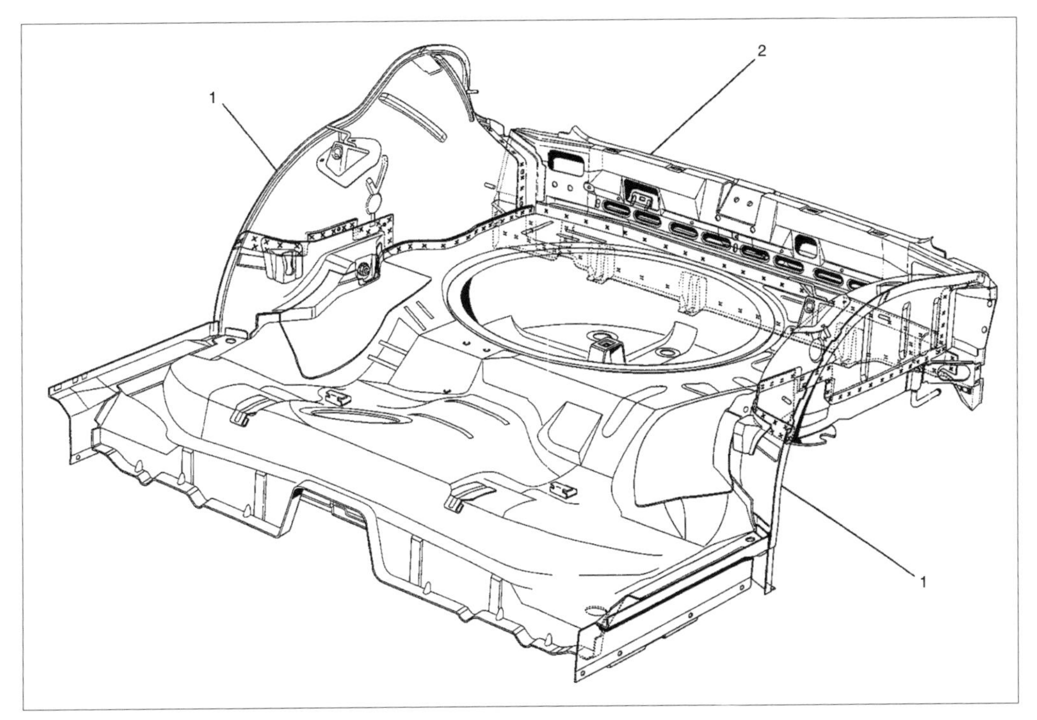

Fig. 6.1-4 Rear floorpan. 1, inner fender; 2, lower transom.

frame exactly at those points where it is most likely to buckle or twist in the event of an accident. Furthermore, this crossmember is a vital mounting element, with attachment points for the rear tank mounting straps, rear seat bench, and center rear seat belt.

The transom (rear crossmember) is intended to prevent bending in the spare tire well area. The rear substructure is joined to the floor. In the example shown, the spare tire well is welded directly into the rear floor panel. Other designs have the well formed into the floor panel itself.

The wheel wells (inner fenders) are welded to vertical flanges of the rear frame. Stiffening of the upper rear crossmember is accomplished by extensions from the wheel wells. These extensions are vital for overall torsional stiffness of the body and local stiffness of the trunk opening. They also provide an additional load path in the event of a rear collision.

6.1.1.4.2 Bodywork

Front components are first attached to the floorpan (Fig. 6.1-5). The upper front section provides lock and latch for the engine compartment hood. This component also mounts bumpers or pads to adjust engine hood gaps as well as radiator mounting points. The front side sections provide headlight mounting points. To these sections are joined the front inner fenders, each of which may consist of one or two pieces. Suspension strut mounting points, so-called "strut towers," are located at the tops of the inner fenders. These absorb all suspension strut reaction forces. To provide adequate stiffness, the inner fender must be joined to the forward frame by means of a member acting as a tension brace, with as straight a run as possible.

A frame horn is located above each inner fender, with the task of directing forces to the A-pillar and doorpost in the event of a frontal collision before it also absorbs energy by means of controlled deformation. Behind the frame horn, each inner fender is joined to the side panel bulkhead. The cowl runs transverse to this; together with the firewall and floorpan extension, it separates the engine compartment from the passenger compartment. Firewall and cowl serve as mounting points for various components and assemblies. The firewall shields the passenger compartment from noise and dirt and prevents entry of the engine or other components into the compartment in the event of a collision. Firewall deformation after a frontal collision is therefore one measure of a body design's structural quality and is often used for comparison purposes. A reinforced panel is located above the cowl; together with the A-pillars and the front roof frame, this forms the windshield opening.

6.1.1.4.3 Side panel assembly

The side panel assembly consists primarily of an inner and an outer component (Fig. 6.1-6). In the vicinity of door openings, these shells together with the inner A-pillar also form the pillar cross sections. On many cars,

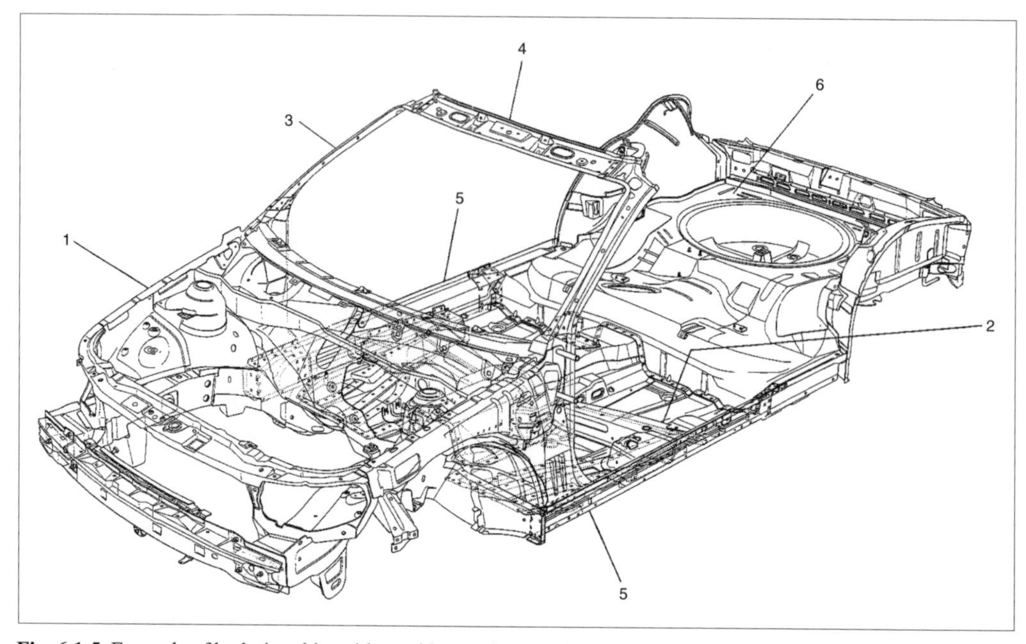

Fig. 6.1-5 Example of body-in-white without side panels or roof. 1, front section; 2, floorpan front; 3, inner A-pillar; 4, front roof frame; 5, front floorpan side; 6, rear floorpan.

the A-pillar cross section, Fig. 6.1-6a, may be seen to have an additional reinforcement. On convertibles, this may consist of high-strength tubular sections. While the windshield is attached to the inner flange of the A-pillar, the door seal is attached to the outer flange. The B-pillar cross section, Fig. 6.1-6b, has door seals on both sides. In addition, the B-pillar carries the rear door hinges. This particular example shows welded hinges. In many vehicles, these components are still being bolted in place. As both mounting variations exhibit advantages and disadvantages, their choice is dependent upon manufacturer philosophy. Additional local reinforcement is applied in the vicinity of the door hinge attachment points. In the example shown, these reinforcements are welded to the outside, although many other vehicle manufacturers locate these between the inner and outer side panel members. During vehicle operation, these reinforcements also serve to reduce diagonal deformation of door openings, an acoustically relevant consideration. Furthermore, the B-pillar reinforcement is very important in side impacts, as it reduces passenger compartment intrusion.

At the rear, an extension is added to the side panel. An attached panel encompasses the taillight openings. This is an important contributor, as are the wheel well extensions, to increased torsional and rear body diagonal stiffness.

6.1.1.4.4 Roof

The last component in the production sequence for the body shell under consideration here is the roof. This consists of an outer skin reinforced by glued-in ribs and

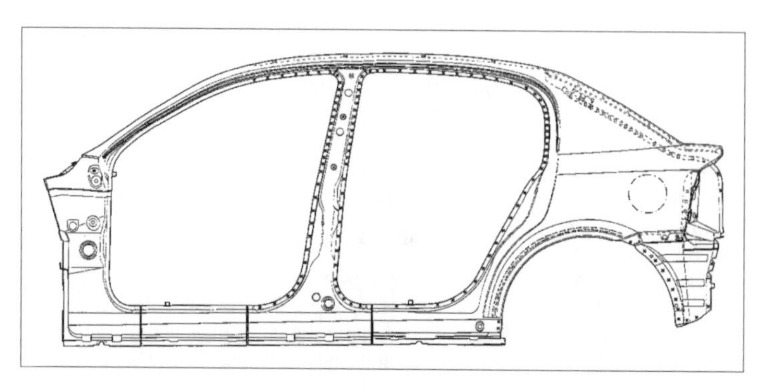

Fig. 6.1-6 Side panel assembly.

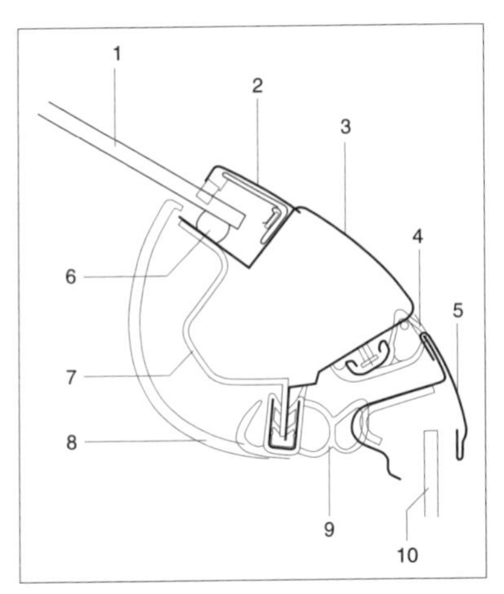

Fig. 6.1-6a Cross section of A-pillar. 1, windshield; 2, windshield bezel; 3, outer A-pillar; 4, secondary seal; 5, front door window frame; 6, windshield cement; 7, inner A-pillar; 8, A-pillar trim; 9, door seal; 10, front door glass.

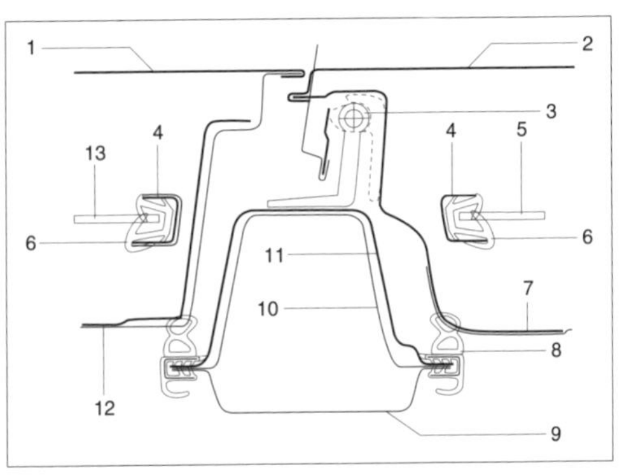

Fig. 6.1-6b Cross section of B-pillar. 1, front door exterior skin; 2, rear door exterior skin; 3, rear door hinge; 4, guide rail; 5, rear side glass; 6, side glass seal; 7, rear door frame; 8, door seal; 9, inside side panel; 10, outside side panel; 11, B-pillar reinforcement; 12, front door frame; 13, front side glass.

the rear roof frame. While the ribs have no effect on overall body torsional stiffness, the rear roof frame plays a decisive role. Optionally, the roof may include a sliding or tilting/sliding sunroof. The latter (Fig. 6.1-6c) is more popular today. Simple sliding sunroofs are no longer of any great importance, and tilt-only roofs are available solely as aftermarket accessories. In order to limit proliferation of sliding roof variations, consideration must be given during the design phase to ensure uniformity in defining those surfaces adjoining the sunroof. Given these preconditions, the tilting/sliding sunroof becomes a predetermined design element. The complex problem which must be solved is to integrate the entire sunroof mechanism, including its electronic drive, in the available space of no more than 6 cm. This dimension is created by two contradictory requirements: first, the laws of aerodynamics demand low overall vehicle height; second, occupant space requirements must be satisfied, given that the modern "95th percentile male" is 1.89 m (6 ft 2.4 in.) tall. Seats may not be lowered indefinitely, and adequate seat cushion spring travel is required for comfort reasons. One important detail is sunroof water drainage. Rainwater from the sunroof is carried by a hose within the roof leading to the wheel well by way of the A-pillar. This detail requires that the A-pillar not be blocked by closing panels.

6.1.1.4.5 Attached parts

The term "attached parts" is understood to mean all body components that are not welded to the body shell. These include doors, lids, fenders, and bumpers. Development of attached parts is characterized by two trends:

1. Modularization

Due to outsourcing of development and production to outside firms, attached parts are increasingly being delivered to the car manufacturer's assembly line as

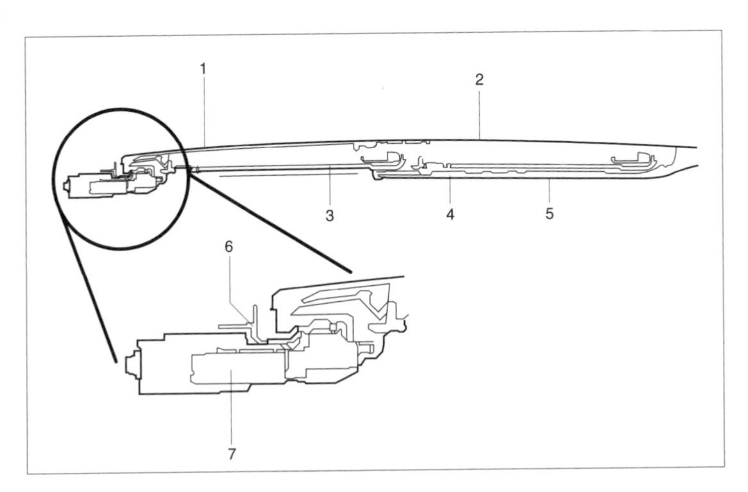

Fig. 6.1-6c Sliding/tilting sunroof. 1, glass cover; 2, roof; 3, sunshade; 4, lower frame; 5, formed headliner; 6, frame; 7, motor.

complete modules. One example is the front end module, which, in addition to the front bumper and headlights, might in future also include the radiator. Door modules now consist not only of exterior skin, windows, and inner panels but also the entire suite of mechanical and electronic equipment required for door locking and window operation.

2. Application of lightweight materials

Large-scale application of new steel body materials with lower specific weight (see Section 6.1.4) has already begun, primarily in attached parts. Most bumpers and an increasing number of fenders consist of plastic components. Increasingly, engine and trunk lids are being made of aluminum. For extremely fuel-efficient vehicles (e.g., the "three-liter car"), large-area magnesium lids are under development.

6.1.1.4.6 Joining technology

Choice of the optimum joining technology for any given application is decisive in establishing body quality; for example, these choices influence torsional stiffness and corrosion resistance [3]. For the all-steel body under consideration here, spot welding is by far the most significant joining technology. Depending on vehicle type, a body may contain 3,000 to 5,000 spot welds. Today, these are almost exclusively made by automated industrial robots or in multipoint welding facilities. Establishing weld point locations by means of CAE tools (Section 10.2) is an integral part of body development.

Of the various gas metal arc welding (GMAW) techniques, body assembly primarily uses metal active gas (MAG) welding. This is employed for highly stressed parts, such as nut plates for mounting suspension components, or in areas where restricted access does not permit spot welding.

Arc welded studs provide strong joints in areas where, in part due to the manufacturing sequence, only one side of a panel is accessible. On average, a body shell contains 150 welded-on studs. Most of these are in the floorpan area.

In the last few years, laser beam welding has made its first inroads in mass-production applications [4]. In addition to the high weld speeds made possible by this technology, its primary advantage is in improved surface quality. A laser-beam weld seam at the body surface can be removed prior to painting much more quickly than can a conventional weld seam. In future, mechanical joining processes such as clinching and pierce riveting, as well as these technologies in combination with adhesive joining, will gain importance [5].

6.1.1.5 Body characteristics

6.1.1.5.1 Assembly tolerances

In recent times, the topic of "gap dimensions" has garnered greater significance in discussion of automotive visual quality. The overall goal of many automobile manufacturers is to keep gaps, especially in visible areas, as small as possible. At the same time, the gaps must be as parallel as possible. A program of tolerance management to accompany the development process establishes the cornerstone for a high-quality body shell.

In order to assure constant repeatability in body manufacturing, assembly studies for any new body design are conducted at the beginning of development. These establish all measuring, clamping, and mounting locations for the manufacturing tooling, not only for the body as a whole but also for all individual components. In this process, gap tolerances are established before component tolerances. Care must be taken that clamping points are always at the same locations so that the manufacturing facility can be given specific targets for training and, later, for quality assurance. Design of tooling and inspection facilities invariably follows these assembly studies and product drawings.

The joining station of Fig. 6.1-7 is an example showing implementation of high quality demands under production conditions. This station permits welding the floorpan group to the side panels and roof while maintaining invariant precision. This is made possible by the station's clamping frame, which forms a closed system, immune to outside influences, and which ensures that body components are always clamped according to the specifications determined during the assembly study.

6.1.1.5.2 Body stiffness

It is always a development objective to achieve a stiff body with a homogeneous structure which forms the basis of a vehicle with good dynamic noise behavior and good driving characteristics. Stiffness is largely designed into a body by means of computational methods (see Section 10.2).

Torsional stiffness is understood to mean deformation of a body around an axis in response to an applied moment. Under actual driving conditions, such moments are often applied by road irregularities, for example, leading to relative motion between the body and components attached to the body. These irregularities can result in undesirable vibration and noise, which is why every effort is made to achieve high torsional stiffness (see also Section 3.4).

Means to achieving high torsional stiffness include:

- Optimized transitions and connections between sections (structural nodes)
- Section design of longitudinal and transverse members
- Avoidance of articulated joints and restricted sections
- Multifunctional reinforcements
- Optimized weld locations
- Application of high-strength steels or tailored blanks
- Additional transverse members

The following torsional stiffnesses are achievable in modern automobile bodies:

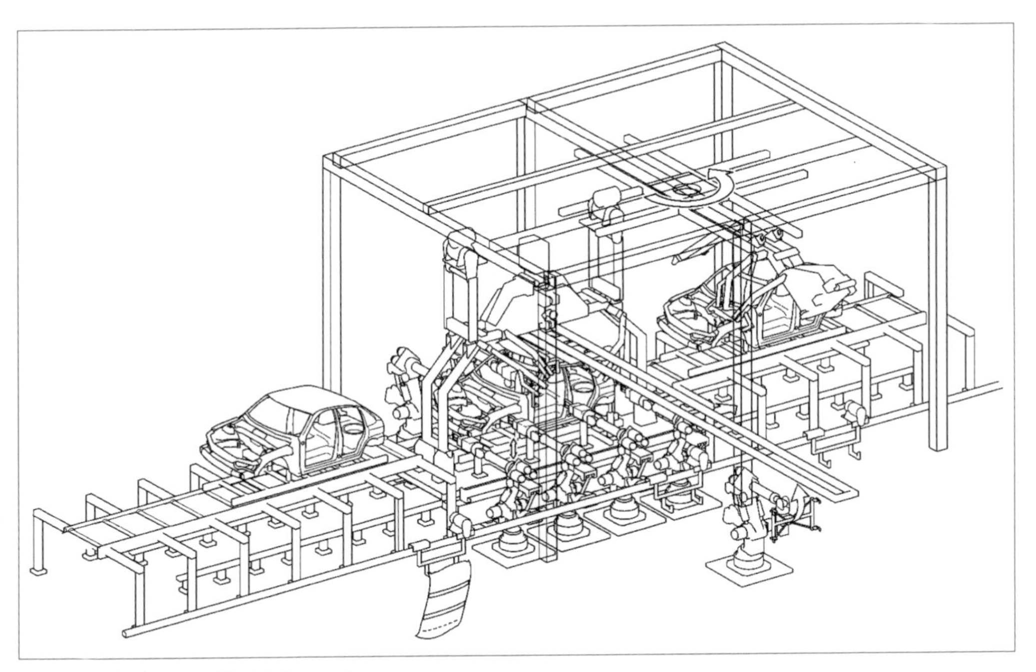

Fig. 6.1-7 Body assembly joining station.

— Sedans: 12–18 kNm/rad
— Fastbacks: 10–14 kNm/rad
— Station wagons: 10–14 kNm/rad
— Vans: 10–12 kNm/rad

In any case, overall stiffness cannot be regarded as the sole objective. The designer should optimize local stiffness for every component. Accordingly, torsional stiffness c_T taken by itself has only limited indicative value. It must be regarded in combination with mass m and footprint area A of a vehicle, as there is a goal conflict between lightweight design and stiffness. A typical value for evaluating "lightweight design quality" (Fig. 6.1-8) may be defined as follows:

$$L = \frac{c_T A}{m}$$

It must be assumed that future development will see more extensive application of safety measures, such as reinforcements, to meet even higher crash standards (e.g., Euro-NCAP) [2]. The resulting weight increases partially offset the positive development trend of lightweight design quality.

Bending stiffness indicates what force (kN) must be applied to deflect the body by a specified amount (for example, 1 mm) between the axles. Under actual driving conditions, such forces are encountered in driving over undulating pavement. High stiffness is achieved either by increased section depth of longitudinal members, front and rear frames, and rockers or by applying reinforcements.

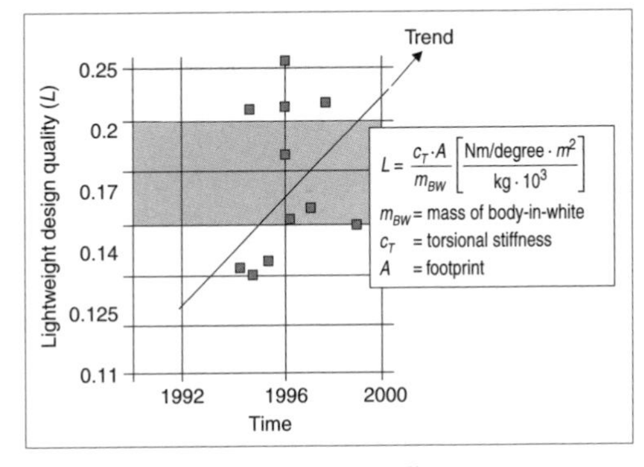

Fig. 6.1-8 Lightweight design quality.

Pickup points for chassis components such as suspension, shock absorbers, and springs require locally stiff body input points—so-called coupling points. This requirement results from the desire to eliminate body compliance in chassis tuning. In order to reach such high stiffness, connections must be made to solid frame members, and, in part, designers must employ local reinforcements.

The torsional resonant frequency of a complete vehicle should fall between 23 and 24 Hz (see also Section 3.4). To achieve this, the target value for the body-in-white is 35–45 Hz for the first symmetric resonant frequency (bending) and 30–40 Hz for the first asymmetric resonant frequency. The higher these values are, the stiffer and correspondingly lighter the body. Also important is a general stiffness extending throughout the body in order to keep vibration amplitudes low.

6.1.1.5.3 Collision behavior

One vital body development goal is, of course, optimum behavior in every type of collision. The primary consideration is always to keep the passenger cell intact. Intrusion of body parts or components must be avoided under all conditions. The energy dissipation required to limit occupant acceleration in response to collision impulse should be taken up by the front structure and longitudinal members, especially in the event of a frontal collision. Individual measures in terms of body structure are discussed in Section 6.1.1.4; details are also found in Section 6.5.

6.1.1.6 Prognosis

Future automotive body construction will, above all, be characterized by the drive toward greater weight reduction for the purpose of improved fuel economy. Reducing body weight by 100 kg results in secondary weight savings on other components, such as brakes and transmission, on the order of 16 kg [6].

Thanks to its proven manufacturability and attractive material costs, the all-steel body will continue to enjoy a high market share. While the proportion of alloy steel will be considerably reduced, high-strength steel varieties and bake-hardening steels, whose strength is increased by additional heat treatment, will enjoy more widespread application.

References

[1] Hucho, W.-H. *Aerodynamik des Automobils*. 3rd ed. VDI-Verlag, Düsseldorf, 1994.
[2] Hiemeni, R., et al. "Optimierung von Karosserie und Fahrwerk des neuen Opel Astra als Beispiel für erfolgreiches Frontloading." In: VDI-Bericht 1398, 1998, pp. 101–122.
[3] Leuschen, B., and B. Hopf. "Fügen von Stahl, Aluminium und deren Kombination." In: VDI-Bericht 1264, 1996.
[4] Geiger, M., et al. "Laserschweißgerechte Konstruktion und Fertigung räumlicher Karosseriebauteile." FAT-Bericht 118, 1995.
[5] Hahn, O., and D. Gieske. "Ermittlung fertigungstechnischer und konstruktiver Einflüsse auf die ertragbaren Schnittkräfte an Durchsetzfügeelementen." FAT-Bericht 116, 1995.
[6] Braess, H.-H. "Negative Gewichtsspirale." In: ATZ 101, No. 1, 1999.
[7] Further information may be found in VDI-Berichte, particularly 665 (1988), 818 (1990), 968 (1992), 1134 (1994) and 1398 (1998) as well as in ATZ.

6.1.2 Space frames and cladding

6.1.2.1 Introduction

Over the past few decades, a marked worldwide increase in vehicle weight has become apparent within the auto industry. For example, compared to the situation of the 1970s, weight increases within the lower middle class (e.g., VW Golf, Audi A3) amount to approximately 40% (Fig. 6.1-9). This is attributable to more demanding safety requirements such as passenger cell rigidity, deliberate body deformation properties, increased torsional stiffness, airbags, and more stringent exhaust standards, as well as greater comfort demands such as power steering and brakes, air conditioning, electric windows, and the universality of these vehicles. Given identical vehicle performance, this weight increase requires adaptation of engines and transmissions, more robust suspensions and brakes, and greater fuel tank volume. In the upper class, these factors result in vehicle curb weights of up to 2,000 kg.

During the service life of a vehicle, energy consumption is determined by overall driving resistance and onboard energy consumption (Section 5.6). Consequently, mass is a significant factor influencing driving resistance. A weight reduction will manifest itself in decreased fuel consumption [1, 2].

These effects conspire to create a "weight spiral," which leads to increased energy consumption and greater environmental impact. Generally, depending on vehicle type and engine configuration, a 100-kg weight increase is associated with a consumption increase of 0.41–0.61 L per 100 km [1].

Modern vehicles must meet demands for lower fuel consumption, lower emissions, and improved recyclability—and not just for ecological reasons. These goals, and therefore a reversal of the weight spiral, can only be achieved by systematic utilization of the potential offered by lightweight design. Because safety requirements, exhaust regulations, and comfort expectations will only become more demanding in the future, this reversal is not achievable through combining all currently known lightweight design possibilities for individual components; such a reversal rather can be attained only by means of a new overall engineering concept.

6.1.2.2 Audi space frame concept

The greatest weight-saving potential is represented by the body itself (including interior equipment). In conjunction with systematic application of lightweight materials throughout the vehicle, lightweight body design as a primary measure enables additional secondary weight reductions (Fig. 6.1-10). Therefore, smaller engines may be installed without incurring any significant sacrifice in vehicle performance. These weight reductions result in lower chassis and transmission loads, thereby producing secondary weight reductions in these departments as well as a reduction in the required fuel tank volume [3].

Application of lightweight materials offers one advantageous possibility for weight reduction. Analysis of possible lightweight materials in automotive body shell construction with respect to specific weight, strength, stiffness, crash behavior, availability, and energy requirements indicates clear advantages for aluminum.

If one merely substitutes aluminum for steel in existing bodies without any additional optimization (e.g.,

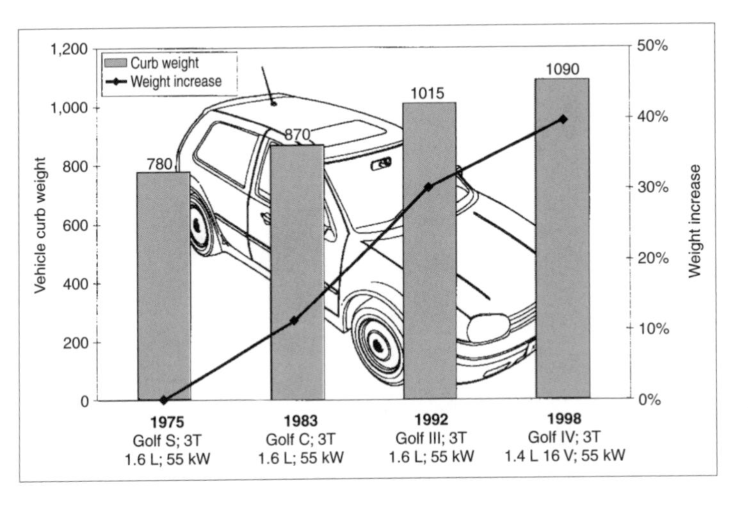

Fig. 6.1-9 Development of vehicle curb weight since 1975, as illustrated by the VW Golf.

ULSAB [4]), the result will be a maximum theoretical weight savings of 66%. However, due to differences in material properties, this leads to unacceptably lower body rigidity and strength. The problem may be alleviated by local reinforcement (and therefore application of additional weight), but conceptually, aluminum's intrinsic advantages are not fully utilized.

As a lightweight material, aluminum in its various semifinished forms exhibits clear advantages over steel. In addition to sheet metal, the primary form for steel, cost-effective semifinished aluminum components are available in the form of extrusions and castings, which may be joined in an aluminum-specific design.

Application of all these semifinished forms to automotive body design leads to a new concept, the Audi Space Frame (ASF®). With optimum design, this cellular structure is characterized by high static and dynamic stiffness—a yardstick for safety and comfort—with high strength. Compared to a similar modern steel-unit body, the achievable weight savings amounts to about 40% [5].

Development of the space frame concept is based on "skeleton" construction, which has been applied from the earliest days of automotive industrial production. A steel or wooden skeleton forms a solid framework to which secondary and tertiary body parts are attached as nonstressed members by means of various joining techniques [6]. In ultrahigh-performance sports cars (e.g., Lamborghini), parts of the body are fabricated in steel using this skeleton technique. In the 1960s, for example, the Touring coachbuilding firm patented its "Superleggera" concept in which a steel space frame was clad with aluminum skins. Various limited-production sports cars such as Aston Martin DB4 GT (1960), Lamborghini 350 GT (1964), Sunbeam Venezia (1963), and Maserati 3500 GT (1957) were produced using this concept. Even today, some extremely limited production cars employ wooden frames, such as the British Morgan marque [7].

The ASF® structure consists of extrusions, castings as nodal elements, and large castings with integral functional surfaces. These components are connected to each other by means of various joining operations. Sheet

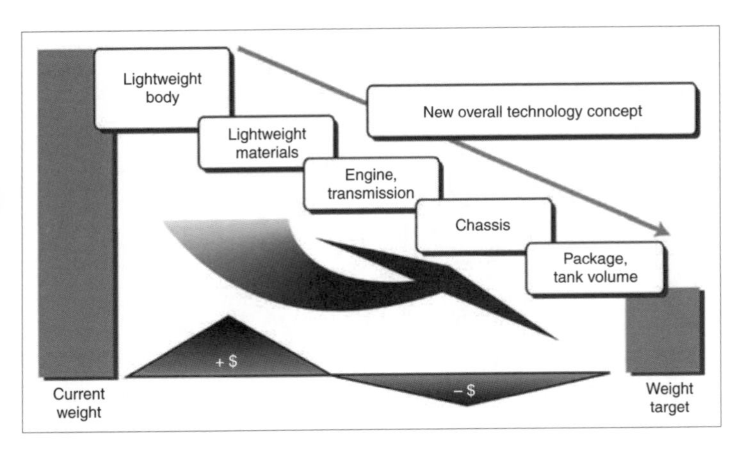

Fig. 6.1-10 Ecologically effective weight reduction by means of lightweight design.

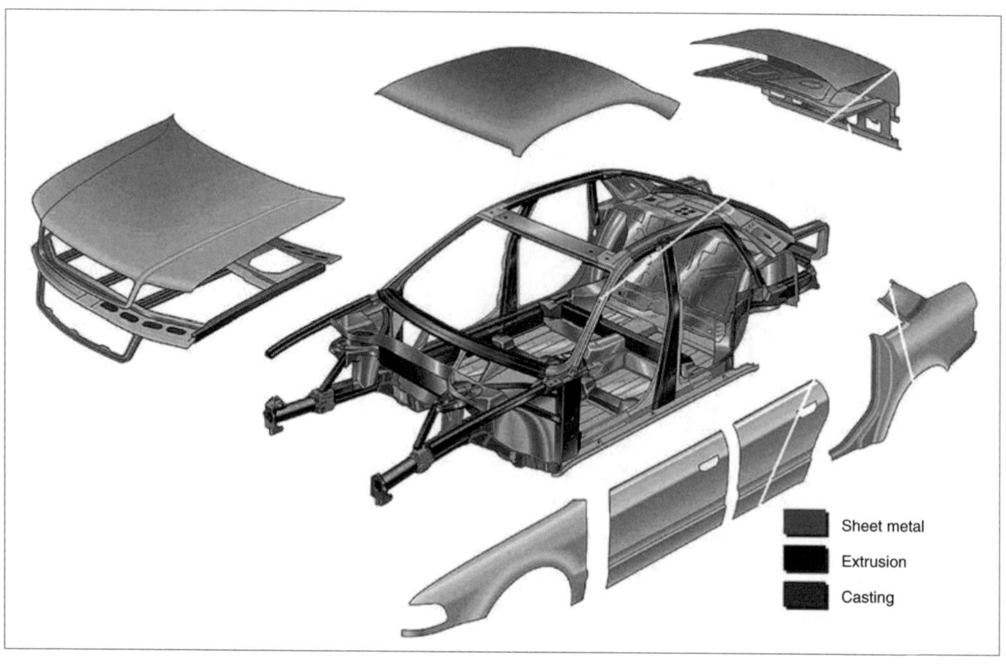

Fig. 6.1-11 Structure and components of the Audi A8 ASF®.

metal panels are incorporated in the resulting cellular structure to form closed surfaces, in particular acting as shear webs [8]. Fig. 6.1-11 shows the ASF® structure of the Audi A8 with its associated sheet metal components.

6.1.2.3 Materials

The ASF® employs various aluminum alloys. These may be divided into hardenable (AlMgSi) and nonhardenable (AlMgMn) alloys. After cold rolling, hardenable alloys are annealed and quenched, giving them good forming properties (T4 condition). Extruded shapes are quenched from their hot forming temperature. After being formed into the required component shape, tensile strength may be further increased by heat-treating (T6 condition) [9].

For nonhardenable alloys, which are only used as sheet metal components, increases in tensile strength are only realizable by means of cold working. Subsequent thermal influences may cause a partial reduction of this effect. These alloys are employed for their outstanding formability into complex shapes, which are not possible with hardenable sheet metal.

All alloys are characterized by very good corrosion resistance. For corrosion reasons, German applications do not employ AlCuMg alloys in automotive body applications. If different metals come into mutual contact (e.g., steel bolts and aluminum panels), it is necessary to isolate these from one another or apply a protective coating. For this, a Dacromet® coating has been found to be outstandingly effective.

Typical mechanical properties of the materials used are listed in Fig. 6.1-12.

Heat treatment of sheet metal is accomplished in the assembled condition at 205°C component temperature for 30 min. This heat treatment is necessary above all for achieving high dent resistance for the exterior skin

	Alloy	Yield strength $R_{p0.2}$ (MPa)	Tensile strength R_m (MPa)	Elongation A_5 (%)
1) Sheet metal	AA6009 T6	230	280	10
	AA6016 T6	200	250	14
	AA5182	135	270	25
2) Vacural casting	AlSi10Mg T6	120–150	180	15
3) Permanent mold casting	AlSi7Mg T6	200	230–250	5
4) Extrusion	AlMgSi0,5 T6	210–245	225–265	11

Fig. 6.1-12 Aluminum materials in the Audi A8.

(hail impact, etc.) Tensile strengths meet or exceed those of conventional deep-drawing steels [2, 5, 8].

For castings, heat-treatable aluminum casting alloys of the G-AlSiMg family are employed. Vacuum die casting methods are used to achieve high ductility and good weldability. These processes evacuate the casting space, resulting in less gas content in the finished casting. Subsequent heat treatment increases tensile and yield strength. Wall thicknesses are between 2 and 5 mm; greater wall thicknesses are produced using permanent mold-casting methods.

6.1.2.4 ASF structural elements

The ASF® structural system is a pure frame structure, in which each surface element is an integral load-bearing member. Castings, generally of complex configuration, serve as nodal elements with high local stiffness, for tolerance adjustment, and as load paths—for example, for suspension pickups. By means of reinforcing ribs and load-appropriate wall thicknesses, a weight-minimized construction suitable for the intended service may be realized. The manifold possible configurations of die-cast components permit functional integration and thereby a reduction in the number of parts and joining operations.

Extrusion is a process in which aluminum excels over steel. Extrusion results in semifinished shapes with high intrinsic stiffness. The ASF concept uses primarily closed-section extrusions with up to 12 hollow chambers in each section. Changes in wall thickness in cross section or at flanges to achieve the required strength or rigidity may be chosen almost at will. Bending of extruded profiles is typically done by stretching. Additional processing steps such as trimming ends, holes, and punched tabs are made by conventional processes or in some cases by means of high-speed machining (HSM) [8].

Extrusions, which are individually heat treated and finish machined, are joined to the body structure by means of MIG welding and cast nodal components. To adjust tolerances, castings and extrusions are often telescoped into one another. In some cases sheet metal components, not yet hardened, are attached to this structure by means of pierce rivets.

To increase the strength of sheet metal parts, the entire body shell is heat-treated in a separate tunnel oven at 205°C metal temperature for 30 minutes. The resulting increase in strength meets hail impact criteria with reduced wall thicknesses, thereby permitting additional weight reduction. Further development of these alloys seeks to integrate this separate heat treatment with the painting process [5, 8].

The dimensional quality of a body-in-white is in part determined by the precision of individual parts and in part by the chosen joining process combined with the type of assembly fixtures. In the first-generation ASF® bodies (Audi A8, in production from 1993 onward), cast and extruded components have tolerances of ±1.0 mm. Due to greater "springback," tolerances of aluminum sheet metal parts are somewhat higher than similar steel components. By means of a specific assembly sequence from subassemblies to finished ASF body, functionally determinant dimensions are held to ±1 mm.

Cast and extruded components are provided with fixturing holes for fitting, bending, and cutting operations. These so-called reference holes assure correct mounting of individual parts in clamping and welding fixtures. Dimensional accuracy of subassemblies after welding and cooling is significantly influenced by the type and sequence of the joining process. This is represented by the effort expended in developing optimum welding sequences. Surface-defining sheet metal is attached by various joining methods to the frame structure, which is composed of joined subassemblies. This results in a rigid structure with a high degree of dimensional accuracy. The body shell is completed by joining subassemblies in rigid fixtures. In the first-generation space frame, most joining operations were carried out manually.

For the second-generation space frame, the A2 (shown in 1997 as the A12 study), tighter tolerances of individual parts were achieved by continued development of old manufacturing processes as well as introduction of new manufacturing methods. Only by these means is transition to fully mechanized body assembly assured. The high degree of manufacturing mechanization and the joining methods employed demand individual extrusion tolerances on the order of ±0.2 mm and casting tolerances of ±0.5 mm. New methods such as hydroforming and calibration permit maintaining tight tolerances in extruded sections. For castings, modified heat treatment conditions assure that these tolerances can be maintained.

6.1.2.5 Manufacturing technology

For steel unit bodies in monocoque sheet metal, the primary joining process is resistance spot welding. In some areas of the body, joining may be accomplished by adhesives, gas metal arc welding (in particular, metal active gas—MAG—welding), stud welding, brazing, and laser welding.

These joining processes cannot simply be carried over to the space frame concept and in general are not suitable for aluminum-specific conditions, due in part to the specific properties of aluminum; also, different semifinished parts, compared to steel parts, permit a different structural concept, and therefore, in some cases, require a one-sided joining method.

While resistance spot welding is the most significant joining method for steel bodies, it plays only a subordinate role in the aluminum body structure of the space frame concept. One reason is the appreciably lower process assurance compared to the competing pierce rivet method (see below), such as low electrode idle time and "tacked" spots with low strength. The other reason is the higher welding current required for aluminum as compared to steel due to the surface oxide layer. These factors lead to higher facility investment costs.

Metal inert gas (MIG) welding is well suited to joining castings, extrusions, and sheet metal parts. Welding speeds of up to 0.7 m/min are possible. Thermal distortion due to high heat input is reduced by suitable weld bead sequences, and heat rejection is reduced by cooling elements integrated in the clamping fixture. Due to the hot crack sensitivity of AlMgSi alloys, welding is generally done with a suitable filler material. As a single filler material for semifinished sheet metal and extruded and cast parts, AlSi12 in the form of wire has proven an outstanding choice [10].

Laser welding is a relatively new joining process for aluminum components; it has not yet been applied to series production of body shells. High beam intensity and the resulting low heat input—and thus higher welding speeds—permit distortion-free joining and will therefore account for a significant share of welds in the next generation of ASF® vehicles. Welding speeds of up to 7 m/min for body-relevant material thicknesses also guarantee economical manufacturing. This process imposes high demands on the fit of parts to be joined as well as the precision of fixtures and tooling. Depending on the alloy in question, laser welding may also involve filler material. In contrast to steel, the desirable coupling characteristics resulting from the wavelength of Nd:YAG lasers makes these the preferred choice for welding of aluminum bodywork. In addition, these lasers permit the use of fiber optic light pipes, which permit much greater flexibility compared to the mirrors used with CO_2 lasers [3, 8, 10].

Pierce riveting is a new manufacturing process pioneered by Audi's all-aluminum bodywork. This is a mechanical process without any heat input to the joined parts. Pierce riveting joins parts using so-called pierce rivets, which create a force- and fit-controlled joint. Although this combined riveting and cutting process is accomplished without any predrilling (Fig. 6.1-13), access to both sides of the joint is required. The ASF® concept uses semihollow rivets. The joint is achieved by a stamping and forming process: The parts to be joined are laid on a die; the stamping tool holds the rivet and stamps it into the upper component, while plastically forming the lower part into a locking head. The shape of the locking head is determined by the die contour. During the setting process, the rivet shaft spreads open to draw and pin the parts together. The material punched out of the upper part is enclosed by the hollow rivet shaft.

Pierce rivets must have higher strength than the parts

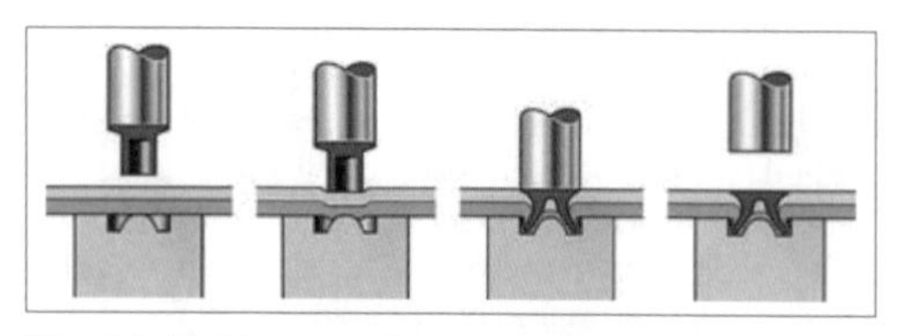

Fig. 6.1-13 Pierce riveting sequence.

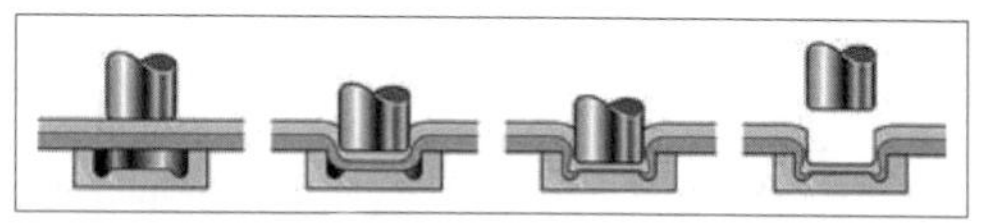

Fig. 6.1-14 Clinching sequence.

to be joined. In applying steel rivets to aluminum components, an appropriate surface treatment of the pierce rivet is required for corrosion considerations. For properly applied pierce rivets, strength in shear as well as tension is appreciably greater than that of a corresponding spot welded joint. Given access to both sides, all possible combinations of semifinished components may be joined with a high degree of process assurance [8, 10].

Clinching involves a joining technology similar to that of pierce riveting, but without the use of any additional joining components. It is a purely deformation-based technology and is differentiated into processes with and without a cutting phase (Fig. 6.1-14). Here, too, a force- and form-fit joint is created; however, due to the joining concept, this joint achieves comparatively only limited strength, especially in tension. In aluminum joining applications, clinch tooling tends to pick up aluminum deposits, necessitating more tool maintenance. Suitable tool coatings can greatly extend service intervals for such tools.

Stud welding, usually accomplished in body shell assembly by means of the drawn arc process, is a welding method for attaching threaded studs, for example, to the body. These studs typically serve to attach lines, hoses, and heat shields. For strength reasons, these studs are not suitable for attaching highly-stressed, safety-relevant components.

Bonded joints in the ASF® concept are based on thermal curing epoxy systems and may be classified as bond-controlled (as opposed to force- or fit-controlled) joints. Curing of epoxy adhesives normally takes place during the heat treatment of aluminum bodies. In contrast to welded joints, bonded joints are "cold joints" and therefore free of thermal distortion. A key advantage of properly designed bonded joints compared to spot welding processes is their uniform stress distribution between adhesive layer and bonded components. It is necessary in a thermal curing adhesive system to use appropriate fixturing to hold components until the adhesive has set. In practice, adhesives are applied as part of compound joining methods—for example, in conjunction with pierce riveting. The rivets then serve as fixturing elements. If process-specific requirements are observed, the strength behavior of bonded joints may be described as "very good."

To assure long-term behavior of bonded joints as well as to improve process assurance of thermal joining methods, aluminum components are subjected to a surface pretreatment process. In the Audi space frame, application of a conversion coating has proven optimal. This coating produces a defined, durable surface which prevents creep into the bonded joint.

6.1.2.6 ASF properties

The ASF® concept is distinguished by high stiffness combined with low weight. Weight savings in the first ASF generation, compared to a conventional steel monocoque body, amount to nearly 40%. Future safety requirements—either regulatory or company-specific—lead to the expectation that this weight advantage may be expanded, thanks to the higher specific energy absorption capability of aluminum compared to steels used in automobile bodies. Additionally, new higher-strength alloys currently under development, as well as new manufacturing technologies, will reinforce this trend. The ASF® concept has convincing advantages in terms of:

- High strength, therefore a high level of safety
- High stiffness, therefore good road behavior and comfort
- Low weight, therefore better fuel economy

Operational durability over the life of the vehicle is assured by wall thicknesses chosen for high stiffness and by the possibility of shaping aluminum components appropriate for stresses in load path areas. In particular, replacement of spot welds by pierce riveting assures unchanging high stiffness over the entire life of the vehicle.

The ASF body structure forms a stable compound frame around the passenger cell. In the event of an accident, this safety cage provides sufficient resistance to deformation. At the front, the ASF structure is configured as a forward section of defined deformability. The system of longitudinal members, composed of extrusions and cast connecting nodes, converts collision energy into deformation work. The sequential deformation force levels of this system of longitudinal members achieves optimum crash behavior under varying collision speeds. Fig. 6.1-15 illustrates that the tubular front longitudinal members exhibit especially high energy absorbing capacity by virtue of optimum collapsing behavior. At the rear, too, the system of longitudinal members achieves defined deformation with high energy absorption. As a result, the fuel tank area and passenger cell are free of any significant deformation. The ASF® body therefore combines two survival-relevant passive safety features: a rigid safety cage forming a survival space for its occupants and energy-absorbing body sections to achieve very good deceleration rates.

In the event of accident damage, the design of the ASF® body structure with its defined deformation zones leads to sectioning repairs predefined by the design. In the event of damage, cast parts and extrusions must generally be replaced; extrusions can be repaired by sectioning and inserting bushings or sleeves [11], as shown in Fig. 6.1-17. The load-carrying structure is then MIG welded. The front longitudinal member, statistically more likely to be subjected to collision damage, can be quickly and economically unbolted and replaced (Fig. 6.1-18). Repair of exterior panels is accomplished by riveting and bonding. The subsequent paint process is no different from that of a steel-bodied vehicle. The ASF® construction method has proven to afford a high degree of reparability; with bolted, riveted, and bonded joints combined with modular design of individual body elements, it is not at a disadvantage with respect to repair costs in comparison to a steel body [12]. This results in lower repair costs and therefore more attractive insurance classification.

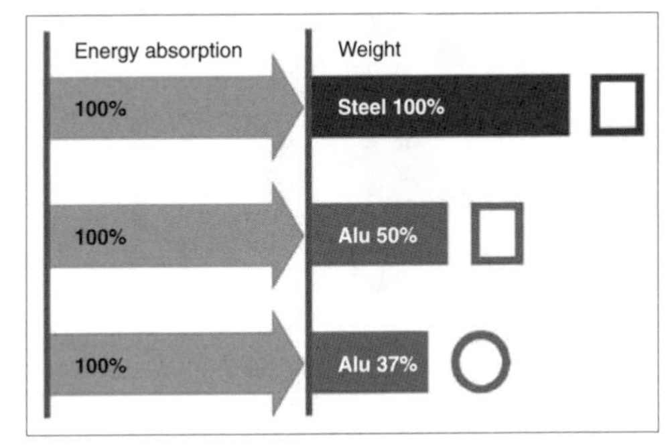

Fig. 6.1-16 Energy absorption of various longitudinal structures as a function of weight. Alu, aluminum.

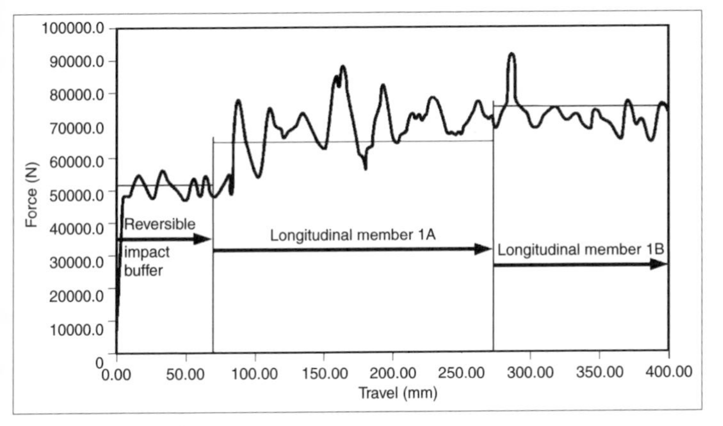

Fig. 6.1-15 Force as a function of deformation travel of Audi A8 longitudinal member.

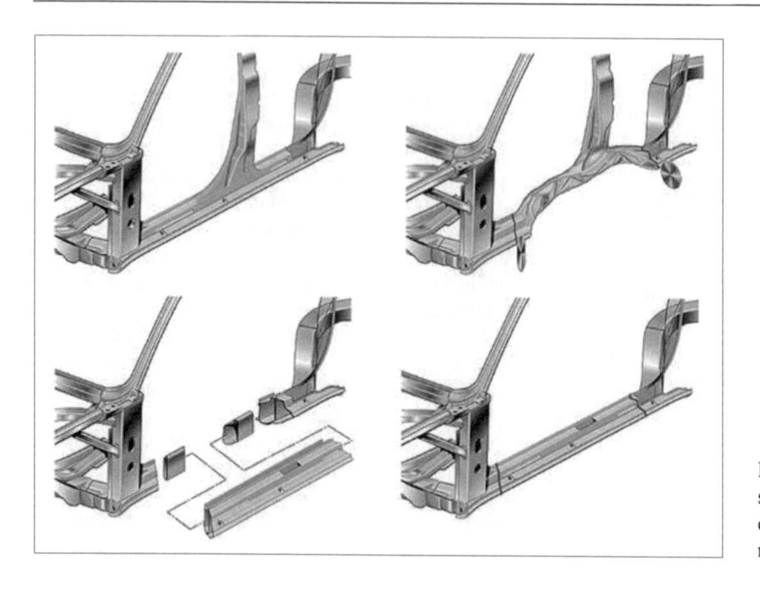

Fig. 6.1-17 Audi A8 door sill repair concept as an example of extruded member repair procedures.

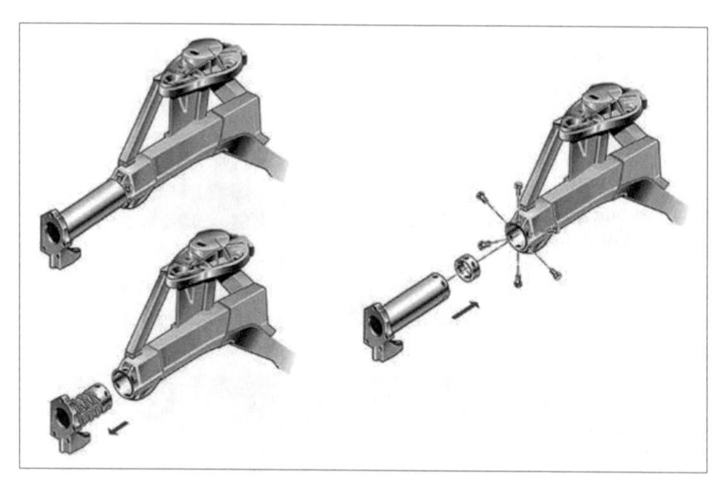

Fig. 6.1-18 Audi A8 front longitudinal member repair sequence after deformation from a 15 km/h impact.

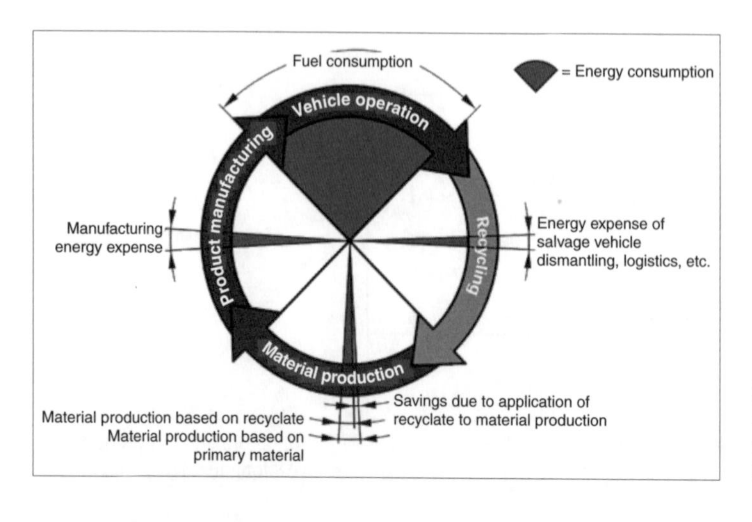

Fig. 6.1-19 Proportion of energy consumed by various stages of a passenger car material cycle.

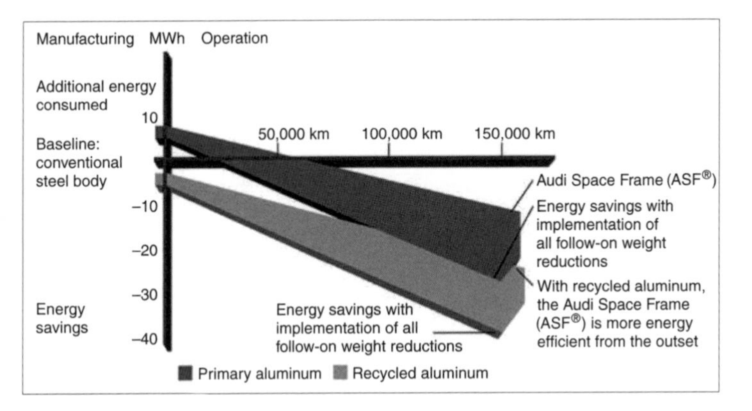

Fig. 6.1-20 Energy budget for manufacturing and operation of an aluminum vehicle, as illustrated by the Audi A8.

6.1.2.7 Energy budget

Today, improved environmental compatibility is of central interest, especially for motor vehicles. It is therefore necessary to examine and optimize the entire product life cycle. Low pollutant emissions and efficient use of nonrenewable resources help to minimize environmental impact. This is true for the entire product cycle, from mining and production of materials through product use to recycling for the next cycle or for environmentally friendly disposal. These environmental considerations imply an exercise in optimization, which may require compromises in individual phases to the benefit of the overall picture.

Fig. 6.1-19 shows the results of a study examining the overall energy budget of an aluminum vehicle [1]. It is apparent that the overwhelming proportion of energy is consumed during vehicle operation. It is therefore ecologically advantageous to accept greater complexity in materials production, vehicle manufacturing, or recycling if these may lead to lower environmental impact during operation.

More energy is required to manufacture primary aluminum than for a corresponding amount of steel. The energy demand in manufacturing a passenger car of primary aluminum is about twice that of a comparable steel body. In the course of vehicle operation, however, the total energy consumed is less than that of a similar steel-bodied car, thanks to improved fuel economy. To date, only a few casting alloys have been made of secondary (recycled) aluminum, but ongoing developments suggest that we may expect application of recycled materials in the field of alloy sheet metals as well. The aforementioned study indicates that even with secondary aluminum accounting for only 2/3 of the body weight, manufacturing energy consumption is lower than that of a similar steel-bodied vehicle. Fig. 6.1-20 illustrates these relationships using the Audi A8 as an example [2].

6.1.2.8 Prognosis

Production of A8 vehicles using the ASF concept, which has continued for ten years at this writing, has proven the series production viability of this concept. In only a short period of time, a quality level was achieved which makes no excuses to that of a corresponding steel-bodied car. Manufacturing and assembly technology, such as ductile die casting and fully automated pierce riveting, are now almost taken for granted. At the same time, experience has imparted three lessons:

Involvement with new materials and with corresponding, materials-appropriate vehicle concepts requires additional, new manufacturing technologies which may be applied to the first new generation. Continued intensive development shows the way to completely new concepts and technologies, which improve economy and may be applied to a second generation.

Application of high technology is not to be equated with high component costs. Moreover, sensible combination of manufacturing steps to achieve integrated manufacturing processes may result in considerable cost reduction.

The future of ASF technology has only just begun.

References

[1] Rink, C. "Aluminium als Karosseriewerkstoff, Recycling und energetische Betrachtungen." Dissertation, Hannover, 1996.

[2] Haldenwanger, H.G. "Zum Einsatz alternativer Werkstoffe und Verfahren im konzeptionellen Leichtbau von PKW-Rohkarosserien." Dissertation, TU Dresden, 1997.

[3] Mathias, K., and D. Dieckmann. "Das Leichtbaupotential bei der Entwicklung von Personenwagen-Aufbauten und der Einfluss auf den Kraftstoffverbrauch." VDI-Berichte No. 968, Entwicklungen im Karosseriebau, 1992.

[4] Winkelgrund, R. "Ultraleichte Stahlkarosserie fertiggestellt." Werkstoffe im Automobilbau 98/99. Sonderausgabe ATZ(MTZ), pp. 76–80.

[5] Paefgen, F.J., and W. Leitermann. "Audi Space Frame—ASF, ein neues PKW-Rohbaukonzept in Aluminium." VDI Berichte No. 1134, Entwicklungen im Karosseriebau, May 1994.

[6] Anselm, D. *Die PKW-Karosserie*. Vogel Verlag, Würzburg, 1997. English language edition: *The Passenger Car Body*. SAE International, Warrendale, PA, 2000.

[7] *Die Chronik des Automobils*. Chronik Verlag, Gütersloh, 1994.

[8] von Zengen, K.H. "Space Frame Quo vadis?" IBEC *Proceedings* 1999, Detroit, January 1999.
[9] Klein, B. *Leichtbaukonstruktion*. Vieweg, 1994.
[10] Müller, S. "Robotereinsatz beim Fügen von Aluminium-Leichtbaustrukturen." Fügeverfahren zur Realisierung von innovativen Leichtbaukonzepten, Erding, April 1999.
[11] Ullrich, W. "Das Kundendienstkonzept zur Aluminium-Karosserie." Auditorium, Aluminium-Technologie im Karosseriebau, October 1993.
[12] Stümke, A., H. Bayerlein, and F. Eckl. "Laseranwendungen bei AUDI." Lasermaterialbearbeitung im Transportwesen, BLAS Verlag, Bremen, 1997.
[13] Rottländer, H.P. "Laserverbindungstechnik im Automobilkarosseriebau." Aachener Kolloquium Lasertechnik, Aachen, 1998.

6.1.3 Convertibles

6.1.3.1 Introduction

Convertibles (or, in some markets, "cabriolets") are understood to be vehicles whose roof may be stowed or removed and whose side windows may be completely lowered. Convertibles are intended to achieve the ride comfort of a corresponding sedan.

Fig. 6.1-21 Side view of a Volkswagen Golf Cabriolet.

Convertibles are differentiated from roadsters and spyders. Both of these are two-seat vehicles. Roadsters have a rudimentary top, while spyders are vehicles with no top at all. In modern usage, the "roadster" label is also applied to two-seat convertibles.

Convertibles have enjoyed a recent renaissance. In 1987, German new vehicle registrations included 30,400 new convertibles. In 1997, this number more than quadrupled, to 122,000. In that year, convertibles represented 3.5% of German new car registrations. In 1998, nearly 50 different convertible models were offered by 34 manufacturers.

Because convertibles remain niche vehicles despite their increased market share, they are often developed due to cost considerations as derivatives of large-volume production vehicles (e.g., Fig. 6.1-21).

6.1.3.2 Body shell

If the roof of a sedan is to be replaced by a movable top system, it is necessary to carry out structural changes to the body shell in comparison to a sedan. Fig. 6.1-22 shows a convertible body shell. Convertible-specific body shell structural changes are marked as shaded areas. The inset illustrations will be explained in the course of this section. These illustrations are referred to by the number at the upper left of each.

Design of convertibles must provide space for

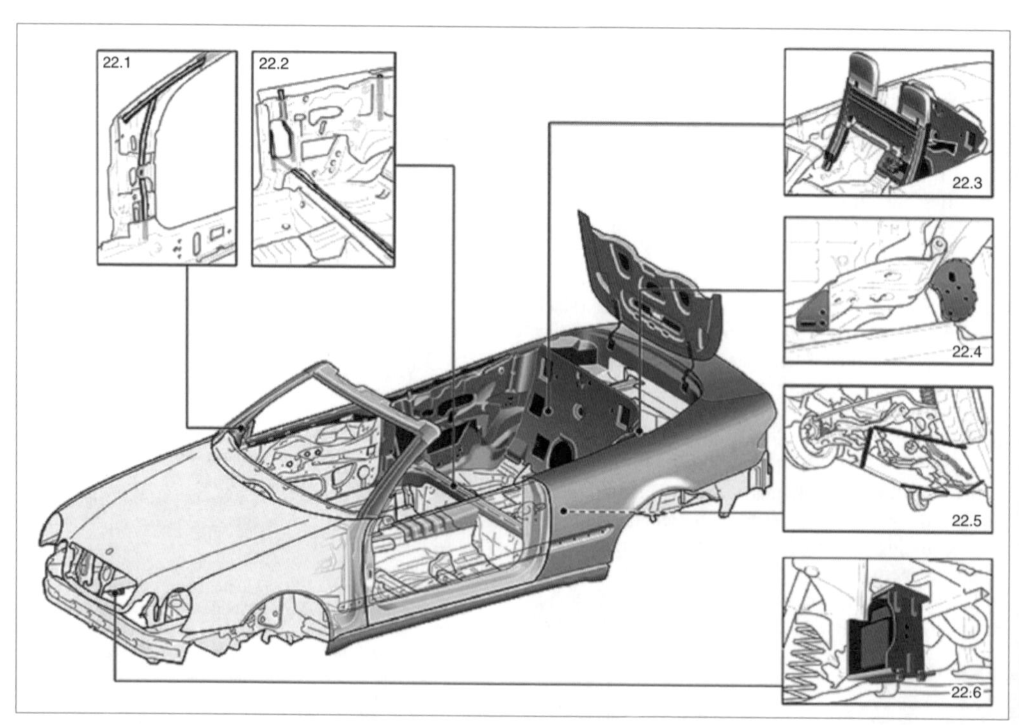

Fig. 6.1-22 Convertible body shell.

stowing the top. Most convertible tops may be stowed completely within the bodywork, for which a well or "tub" is provided. This well either is configured as a rigid structure for reasons of body stiffness or is variable—for example, as a rollup unit—permitting larger luggage capacity with the top up. A lid or hard tonneau cover protects the convertible top material in its stowed condition. If the top is not fully stowed within the bodywork, it may be protected by a flexible tonneau cover.

Points for attaching the top are shown in inset 22.4.

If a tub is part of the convertible design, the trunk lid of the parent large-scale production model must be altered or redesigned.

Because most automobile bodies are designed as unit bodies, removal of the roof leads to a significant loss of body rigidity. This would have detrimental effects on vibration behavior as well as on the convertible's crash concept, and must therefore be countered by appropriate reinforcing measures.

Often, the only body items carried over from the parent sedan to a derivative convertible model are the front section and a portion of the floorpan.

In newly designed convertibles, roadsters, or spyders, unit body construction or space frame technology (Section 6.1.2) may be employed. This latter system, resembling a tube frame design, offers an advantage in that, for example, the floorpan area can be given greater stiffness. Moreover, this technology offers advantages in terms of weight and deformation behavior in the event of a collision. An exterior skin of composite material or sheet metal panels is attached to the frame structure.

6.1.3.2.1 Body stiffness

One goal of body shell development is to achieve high static and dynamic stiffness.

Insufficient static stiffness may result in doors, lids, and, above all, a convertible top which do not open or close freely if, for example, the vehicle is parked at a curb.

Dynamic stiffness of a body shell significantly determines a vehicle's ride comfort. The body's resonant frequencies should not match those of components (e.g., suspensions, engine, exhaust system, steering system), which are themselves capable of vibration. The goal of dynamic body design is to achieve the highest possible stiffness and to provide resonant frequency ranges for components which do not coincide with those of the body so that mutual resonance or coupling effects are avoided. Similarly, the primary torsion and primary bending frequencies should be well separated (by at least 3 Hz) to avoid resonant coupling.

The primary torsional frequency must be regarded as especially critical for ride comfort, as this may be easily excited by engine vibrations.

Removing the roof of a sedan reduces torsional stiffness to about one-sixth of its former value. The torsional resonant frequency also drops along with stiffness. If a convertible is developed as a sedan derivative, overlaps between body and component resonant frequencies are to be expected. The result is irritating vibrations, perceptible, for example, through the steering wheel or the seat or through visible transverse movement of the windshield (A-pillars).

A further drawback of low body resonant frequencies is that given the same excitation level (e.g. road speed), larger vibrational amplitudes than would accompany a higher resonant frequency can be expected.

During convertible body development, the following measures may be applied effectively to combat loss of rigidity:

- Larger longitudinal member cross sections (e.g., door sills)
- Bulkheads
- Improved joints at sill and pillars
- Front and rear diagonal braces (see inset 22.5)
- Closed tunnel cross section

Many stiffening measures increase body weight without any observable rise in resonant frequency. Despite lacking the mass of a roof, convertibles often weigh 100 kg more than the corresponding sedan.

The measures listed above cannot compensate for the convertible's loss of rigidity. The lowest body shell resonant frequency of a modern sedan is at least 40 Hz. Depending on model, resonant frequencies of four-seat convertibles fall between 17 and 26 Hz.

6.1.3.2.2 Mass damper

If during the convertible development process it is determined that ride comfort is inadequate, the body's first torsional mode is damped with the aid of vibration dampers (see inset 22.6). Battery, engine, and, in the case of automatic convertible tops, the hydraulic pump or additional masses are employed as damping masses.

Fig. 6.1-23 illustrates vibration damper operation. Shown is the measured transverse displacement amplitude at the windshield frame as a function of the excitation frequency of a vehicle with and without a damper. In the process, the body is excited to undergo torsional vibration by a four-poster hydropulse rig. Without the damper, a maximum deflection at about 15 Hz is apparent. Modal analysis permits attribution of this maximum to the primary torsional resonant frequency. The graph shows that application of a tuned vibration damper permits appreciable reduction of the resonance amplitude.

Locations with the largest vibration amplitudes in first-order torsion are well suited for damper placement. These will always be at the outer corners of a body. Because additional masses are most easily installed in these areas, this is also where they are most commonly found in convertibles. Many convertibles employ dampers with masses of 20 kg or more.

It is more difficult to use the engine as a damper, because changes to the engine mounts generally have a considerable effect on overall vehicle acoustics.

It is advisable to use not just a single damper, but instead to distribute the necessary damping mass through

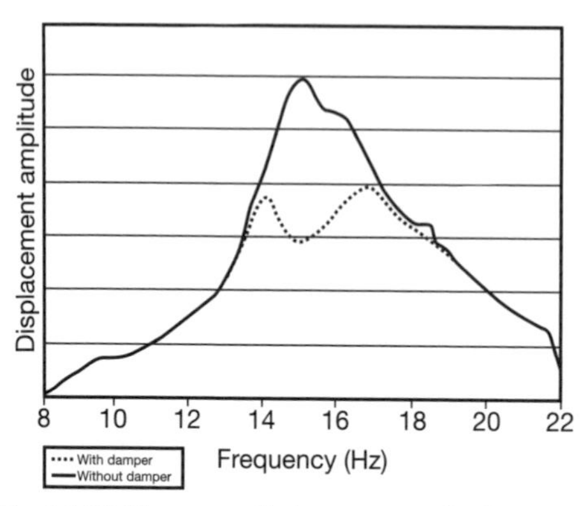

Fig. 6.1-23 Transverse displacement amplitude at windshield frame as a function of exciter frequency for a vehicle with and without vibration damper.

several dampers. For example, installation of a large damping mass at the rear of the vehicle can increase vibration amplitudes at the front of the car.

6.1.3.3 Reducing accident consequences

Modern convertibles must fulfill the exact same safety requirements as sedans. The absence of a roof leads to a convertible safety concept which differs from that of a sedan.

In a frontal collision, the doorsills of a convertible are subjected to greater bending loads than those of a corresponding sedan. In a rollover, sufficient survival space must be provided. Furthermore, stiffening measures must be undertaken on convertibles to counter side impacts.

To increase bending stiffness of doorsills, double sheet metal thicknesses, tailored blanks, or reinforcing tubes may be applied. In addition, load paths are designed to go through the doors.

Side impact protection is provided in part by the reinforced doorsills and by installing reinforcing elements such as tubes in the B-pillar areas (see inset 22.2). Also, many convertibles offer occupant protection in the form of side airbags, and, most recently some provide head airbags.

In a convertible rollover, the A-pillar is subjected to especially high loads. Buckling of the pillar must be avoided in order to guarantee adequate survival space. To this end, A-pillars are reinforced with integral tubes of high-strength steel, and the cowl is reinforced by thicker gauge sheet metal and deeper sections (see inset 22.1).

Despite these reinforcements, it is often not entirely possible to prevent buckling of the A-pillar. Therefore, many convertibles employ a roll bar to assist the A-pillar in its task and, in four-seat convertibles, to provide added safety for rear-seat passengers.

Roll bars may be mounted near the B-pillars, or automatically deploying roll bars may be mounted behind the rear seats. Automatic roll bars are used for reasons of appearance.

Level sensors recognize the danger of an impending rollover. These devices monitor vehicle roll and acceleration around the longitudinal and transverse axes and, in the event certain limits are exceeded, activate the roll bar. In addition, a g-sensor checks to see if the vehicle is in a weightless condition (airborne), which equates to loss of contact with the roadway. As an alternative to the g-sensor, some systems check whether the rear wheels have become airborne.

Constructively, rollover protection is achieved by a rollover cassette, which uses springs to rapidly deploy the rear seat headrests in the event of an emergency (see inset 22.3) or as a flip-up roll bar.

In place of a fixed roll bar, special integral-seatbelt seats, whose headrests serve as roll bars, may be used.

Convertibles generally cannot be fitted with height-adjustable front and rear seat belts—the exception being vehicles with fixed roll bars. Seat belt anchorages are usually a compromise between appearance, regulations, and comfort. Usually, the rear seating area of convertibles is so narrow due to placement of the top mechanism that a third rear seating position is not possible.

6.1.3.4 Doors

Convertibles require frameless doors, so as a rule the doors of the parent large-volume sedan must be altered. In a frameless door, the side glass is subjected to pressure, torsion, and bending forces which would otherwise be taken by the door frame. These forces result from the closing pressure of the door against the A-pillar and convertible top and, while under way, by aerodynamic forces caused by lower pressure against the side glass.

To address these challenges, side glass thickness must be increased. Forces from the side glass are transmitted to the door post and window mechanism. Therefore, in a frameless door, the door post area must be reinforced. Greater loads on the window mechanism demand that this, too, be modified for service in convertibles.

6.1.3.5 Convertible top

A convertible top is expected to be usable year-round, meet demanding visual expectations, and, if possible, match the longevity of the vehicle itself. The ultimate goal is a pleasing, if possible coupelike, appearance.

Except for the Mercedes SLK and Ford Thunderbird, which are fitted with retractable hardtops, convertible tops consist of a top framework covered with cloth.

6.1.3.5.1 Convertible top construction

A convertible top may encompass more than 300 parts and nearly every material commonly used in automotive engineering.

For weight reasons, top frames are increasingly taking advantage of lightweight design. Aluminum or magnesium extrusions or die castings are being used. Today, only highly stressed parts are still made of steel.

Fig. 6.1-24 shows the components of a convertible top mechanism. Distinction must be made between tops with folding top supports and those with fixed body attachment points.

Convertible top bows give the top mechanism high transverse stiffness and simultaneously serve to support the top fabric.

Convertible tops may operate manually or, to an ever increasing degree, automatically or semiautomatically. In contrast to fully automatic tops, semiautomatic tops must be latched manually. For both variations, opening is automatic.

The top header is locked to the windshield frame by two latches, located near the left and right ends, or by a central lock. New designs seek to implement central locks. Tops with folding bows and without over-center action require an additional one or two latches to lock the bows to the (stowed) convertible top lid.

Electromechanical or electrohydraulic drives are used for semi- or fully automatic tops. In the event of system malfunction, emergency manual operation is possible. Electrohydraulic drives have an advantage in that their layout is more flexible and their multitude of possible cylinder sizes makes it possible to fit them even in restricted spaces. If a hydraulic system is installed in a vehicle, other elements such as lids and door locks may be actuated hydraulically.

A further differentiating feature of convertible tops is the backlight, either in the form of rigid glass or flexible plastic. A rigid glass panel imparts higher stresses on the top during opening or closing, and requires more package height. In contrast to the plastic backlight, glass can be heated and is less prone to scratching.

6.1.3.5.2 Textile package

The textile package contained in a convertible top system consists of a top cover, padding, and headliner. A roadster top is not padded and, in some cases, dispenses with an inner headliner.

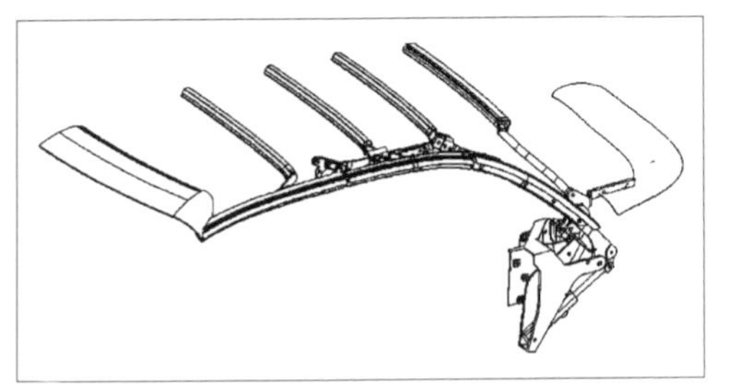

Fig. 6.1-24 Representation of convertible top mechanism components.

In selecting top materials, care must be given to ensure good crease resistance and to thermal and environmental effects.

The convertible top provides a weather seal against rain. The current state of the art for top materials is a three-layer product. The top layer consists either of a textured PVC (vinyl) material or a woven, usually polyacrylic, fabric. A rubber coating between the upper and bottom layers provides the top's resistance to water intrusion. The bottom layer, again fabric (usually polyester), is for appearance and durability.

For thermal and acoustic insulation, the top is often padded. Furthermore, padding can serve to hide the top bows. The core of a padding mat generally consists of a felt pad, which compresses easily when the top is stowed and demonstrates good recoverability when the top is erected. In order to allow the padding to withstand tensile forces, its upper and lower sides are enclosed by woven mesh.

The interior headliner is for appearance and consists of dyed, woven pattern fabric.

One objective of top design is to provide completely adhesive-free joining technology in order to minimize the rejection rate during assembly due to accidental glue soiling.

6.1.3.5.3 Seals

A top must seal against the body and glazing in such a manner that the convertible may be run through a car wash without leaks. Convertibles demand their own unique sealing concepts.

Large-volume seals are used in order to compensate for tolerances. Because many seals are readily visible with the top open, they must also satisfy appearance requirements.

To seal the transition between side glazing and top, windows are often cycled automatically. When a door is opened or closed, the windows lower automatically for a short distance. Once the door is closed, the window runs up to its closed position, sealing the transition.

Despite all these measures, water leaks cannot be eliminated completely. In order to avoid inconveniencing the vehicle's occupants, water entering the cabin must be removed by carefully placed channels near the A-pillars, top header, and main top mechanism bearings.

Minimization of wind noise and sound insulation are important secondary goals of seal development.

References

Emmelmann, H.-J., and W. Wilhelm. "Karosserie- und Verdeckentwicklung des Golf III-Cabriolets." *Cabrio-Systeme,* Haus der Technik e.V., Essen, 1996.

Emmelmann, H.-J., and B. Schröder. "Zukünftige Cabriolet-Technologien." *Cabrio-Systeme,* Haus der Technik e.V., Essen, 1996.

Freymann, R. "Strukturdynamische Auslegung von Fahrzeugkarosserien." VDI Berichte No. 968, 1992.

Grunau, R., M. Heidrich, M. Müller, and A. Paul. "Auslegung von

Karosserie-Schwingungstilgern." ATZ Automobiltechnische Zeitschrift 99, 1997.

Hanus, K.H., and A. Paul. "Reduzierung von Karosserieschwingungen und Innengeräusch mit Hilfe von Tilgern." Stuttgarter Symposium Kraftfahrwesen u. Verbrennungsmotoren, 1997.

Reuter, D., and J. Just. "Das neue 911 Carrera Cabriolet." ATZ Automobiltechnische Zeitschrift 96, 1994.

Rohardt, H. "Die Karosserie des neuen Porsche 911 Carrera Cabriolet." mobiles 24, 1998/99.

Wohlgemuth, J., and R. Nordhoff. "Verdecktextilien von der Entwicklung bis zur Fertigung, dargestellt am Beispiel des MB CLK Cabriolets." *Cabrio-Systeme,* Haus der Technik e.V., Essen, 1998.

Wohlgemuth, J., and R. Nordhoff. "Anforderungen an den Verdeckstoff durch moderne Verdecksysteme am Beispiel des Mercedes-Benz CLK-Cabriolets." VDI-Gesellschaft Kunststofftechnik: Kunststoffe im Automobilbau—Zukunft durch neue Anwendungen mit Fachtagung Textil und Oberflächenmaterialien, March 1998.

6.1.4 Hybrid and mixed-material body construction

Innovative concepts and processes, based on development of new materials and methods, form the basis for lightweight construction-oriented improvements necessary to satisfy standards for all-encompassing considerations of energy availability, energy prices, raw materials availability (including raw material prices), the environment, and effects on immediate surroundings, as well as diverse vehicle-specific regulatory issues. In the conceptual product development of automobile bodies, the entire life cycle—from the launch of a new development through service life to scrapping—must be examined with respect to resource conservation and environmental efficiency evaluation of the applied materials.

In view of lightweight construction criteria for large-volume body production, multifunctionality in the sense of component and functional integration—so-called integral construction—leads to more weight- and cost-effective hybrid structures. In particular, this opens new avenues for materials with high weight-specific stiffness or strength which are available for use with mass production manufacturing technologies using precision chipless metalforming methods. In the future, new system-appropriate material combinations with high recycling potential will establish themselves. Lightweight criteria will also be applied to the construction methods commonly found in commercial vehicles, buses, and railway vehicles—differential construction, for example—with separate load-bearing frames and superstructures.

Implementation of lightweight principles via hybrid construction requires new design strategies, whose hallmarks may in brief be classified as follows:

- Construction of lightweight shapes in which effort is made by means of improved force distribution and shaping to achieve high load-carrying capacity of a structure while minimizing material usage
- Lightweight materials construction, in which a specifically heavier material is replaced by lighter, stronger materials which are suitable for high volume shaping and joining methods
- Composite lightweight construction, with case-specific optimum combinations of different materials which together provide high stiffness and load-carrying ability with minimum weight
- Conceptual lightweight construction, characterized by systematic selection of individual components optimally matched to the total system, including component location (vehicle package) and design
- Modular lightweight construction, in which chaining and networking of individual component and assembly functions (which may be regarded as partial integration of vehicle functions) leads through hybrid construction to mixed-material construction
- Environmental, local, conditional lightweight construction, which suggests reexamination of overemphasized safety and strength requirements as well as a more advantageous geometric setting for the lightweight component to be integrated

Always hidden behind this problem structure of lightweight construction is an individual technological effort whose objective is achieving the optimum degree of lightweight construction as a measure of the most advantageous relationship between weight savings and cost. Materials choice—and therefore component design—assisted by computational simulation of components and assemblies under the most realistic boundary conditions of operating profile and manufacturing technology lead to the desired development goal represented by the complex structure of an automobile body. Fig. 6.1-25 illustrates the stages of component design.

After establishment of the vehicle package on the basis of a multifunctional requirement profile imposed on the body, the potential for lightweight construction is almost entirely represented by material- and technology-specific conceptual lightweight design using mixed materials. This encompasses a choice of suitable lightweight materials such as

- High-strength steel sheet metal
- Aluminum (cast, extruded, and alloy forms)
- Cast magnesium alloys
- Fiber-reinforced thermoplastics and thermosetting plastics
- Fiber composites (directional fiber reinforcement)

as well as material-specific shapes and joining technology, including corrosion protection using material-specific sealing and painting technology. In the process, the aforementioned lightweight materials, in new or modified forms, often result in new, more attractive vehicle technology.

With the introduction of volume production steel unit bodies employing differential construction methods

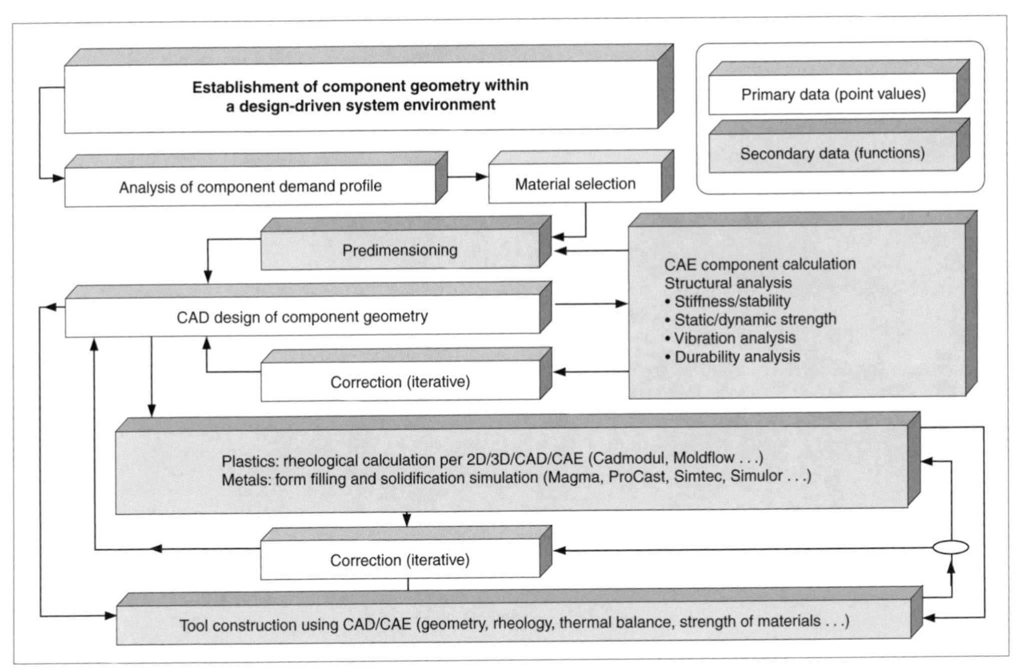

Fig. 6.1-25 Stages of component design.

combined with increased demands for vehicle safety measures to reduce accident consequences and demands on engineering functions of the body itself, substitution of rubber-mounted glazing by glazing that is elastically joined to the body structure by means of polyurethane cement represents a significant contribution to structural stiffening of the body. Cemented glazing acts not only as a shear reinforcement, capable of adding 25–34% to body-in-white torsional stiffness, but also reduces loading of window openings in nodal regions by 25–60% (depending on vehicle condition). The resulting achievable primary weight reduction, as a portion of body shell weight without lids or doors, approaches 1%. However, secondary weight advantages, combined with the following vital engineering functional necessities, must be regarded as positive steps toward optimum multifunctional component lightweight construction using mixed materials:

- High bond strength between glass and bodywork within the plastic deformation range of the body (satisfying crash requirements in the windshield area)
- Independent styling, particularly in bus and rail vehicle construction, through application of any desired window shape and realization of extremely large window dimensions
- Acceptance of large tolerances between window and body by means of automatic adaptation of the cement bead as well as leak reduction (elimination of supplementary sealing compound customarily employed with rubber seals)
- Simplified window installation, with great potential for automation

If attached parts such as

- Doors and lids
- Sliding roof and roof hatches
- Fenders
- Front end module

on unit bodywork or body-on-frame designs are simply replaced by sheet aluminum components without any conceptual changes—for example in AlMg5 clinching/pierce riveting/adhesive technology or with MIG or laser welding with or without AlSi12 as a supplementary welding (filler) material—a weight reduction potential of about 15% may be achieved. Fig. 6.1-26 illustrates a steel sheet metal body structure with attached ("hang-on") parts in aluminum, magnesium, or fiber-reinforced composite. The greatest potential for mixed-material construction lies in the front-end module as a single unit encompassing bumper system, radiator assembly, headlights, and hood latch receiver. In other words, the modular front end may be made conceived in any of the following materials:

- Die-cast aluminum (AlSi10MgMn)
- Die-cast magnesium (MgAl9Zn1/AZ91 alloy)
- Glass-fiber-mat–reinforced thermoplastic (GMT/PPGF30)
    - Pressed technology

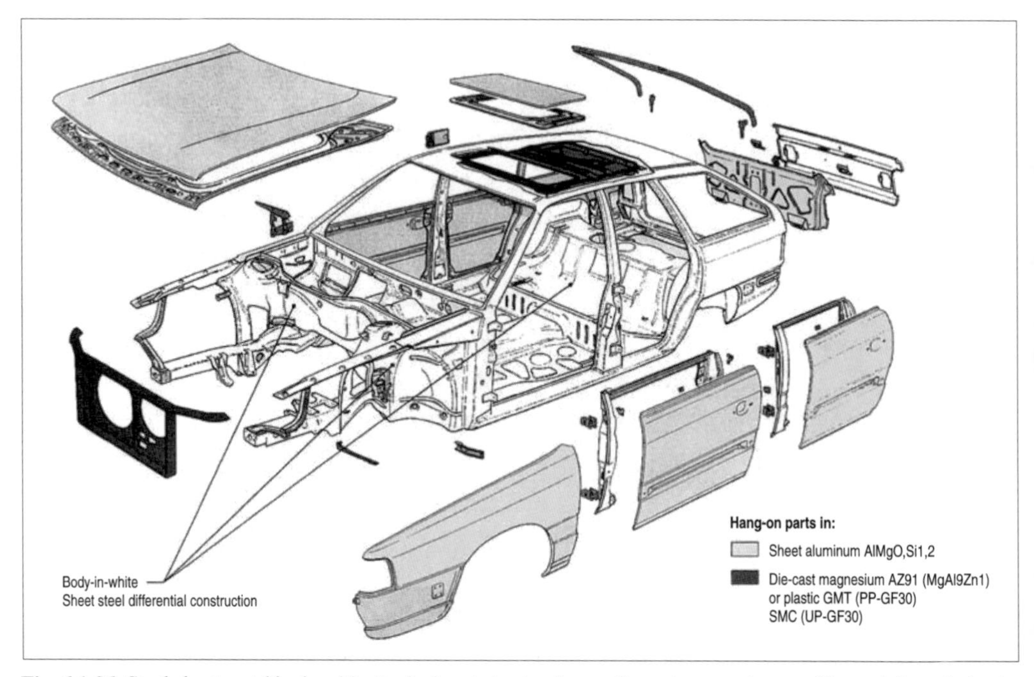

Fig. 6.1-26 Steel sheet metal body with attached parts in aluminum, die-cast magnesium, or fiber-reinforced plastic.

— Glass-fiber–reinforced thermosetting plastic (SMC/UP-GF30)
  — Pressed technology
— Metal-insert–reinforced injection-molded thermoplastic (PA6.6.-GF30) (see Fig. 6.1-27).

An additional weight-reducing measure is simple substitution of front and rear lids with fiber-reinforced composite, most commonly SMC/UP-GF30, BMC/UP-GF25, or GMT/PA6.6-GF25 (PP-GF30). For example, using SMC composite to mold an integral trunklid and integrating components and functions such as

— Aerodynamic spoiler as an exterior skin element
— Hinges integrated in the frame
— Attachments for electrical wiring inputs and functional parts molded into the interior panel

may yield up to 25% weight savings compared to the steel version using differential construction (Fig. 6.1-28). Weight savings of similar magnitude may be achieved by application of fiber-reinforced thermoplastics (PPO/PA) in the rear underbody area, in the integral spare tire well, and at the front fenders.

Application of high-strength sheet metal, in practice with tensile strength exceeding that of St13 and St14 by 40% or more combined with sheet metal of varying gauges joined together as welded plates and then drawn or hydroformed to form components can only reduce the weight of a mono-steel sheet metal body by 6–8%. Fig. 6.1-29 shows application of high-strength sheet metal on a volume production body, while Fig. 6.1-30 illustrates the possibilities for such welded plates (tailored blanks) in combination with various sheet metal gauges and steel grades. As stiffness of the body-in-white is one of the most important design criteria for an automobile body, the potential for lightweight construction—even with the use of high-strength sheet steel of various qualities, which does not offer any advantages in terms of stiffness but does provide better service durability and energy absorbing characteristics—must be regarded as limited.

In the face of multifaceted requirements imposed on the body shell, the broad spectrum of available materials offers possibilities for applying different basic concepts to body construction. However, despite a high degree of lightweight construction achieved by mixed-material design, examination of unconventional material combinations and their associated construction methods (in the form of concept cars and experimental vehicles), compared to classic monomaterial body structures (steel or aluminum), demonstrates that hybrid construction does not satisfy vehicle-specific manufacturing technology demands particularly well when viewed as part of the bigger picture, including suitability for volume production. This is especially true for manufacturing requirements imposed on individual materials (steel, aluminum, magnesium, plastics) by the associated painting technology: for a mixed-material body, the paint structure must be designed in keeping

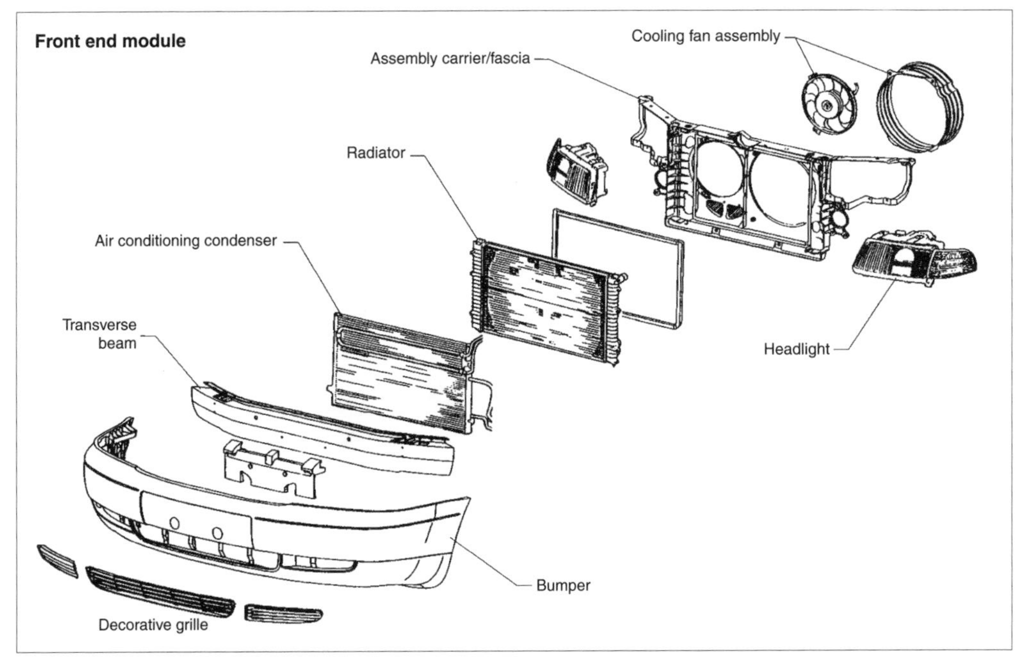

Fig. 6.1-27 Modular front end unit.

with these materials' respective off- or online painting requirements.

As various body substructures must often satisfy entirely different demands, it makes sense that these should be made of different materials. Fig. 6.1-31 indicates individual body weight savings potentials. In mixed-material construction, metal is the preferred material for the frame structure as its properties, suitability for

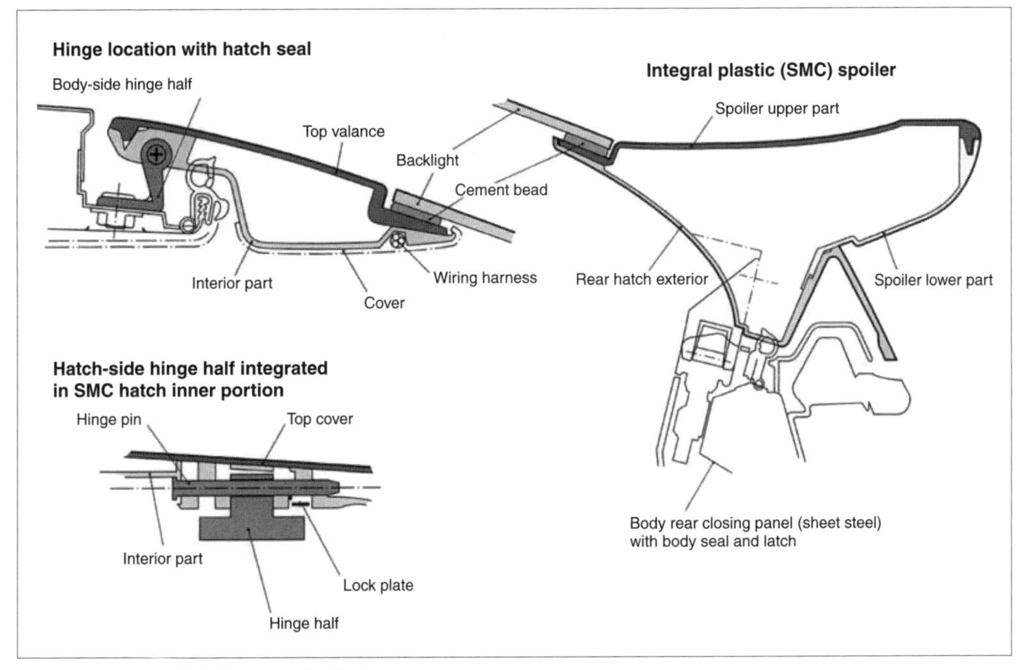

Fig. 6.1-28 Integral SMC hatch (Audi Coupe).

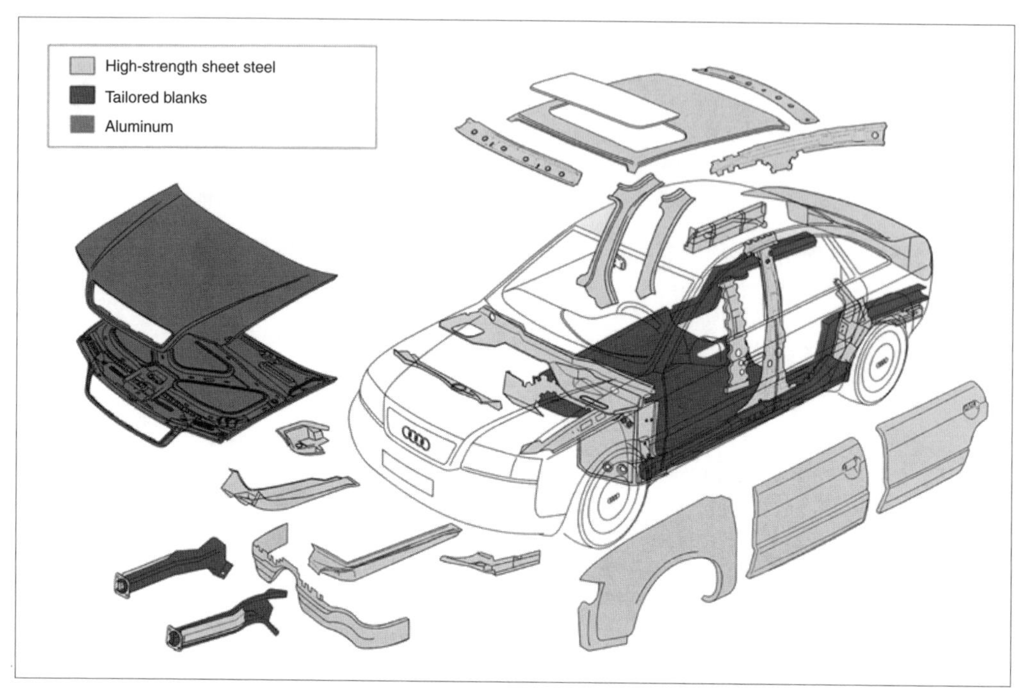

Fig. 6.1-29 Application of high-strength sheet steel to volume production bodywork (Audi A4).

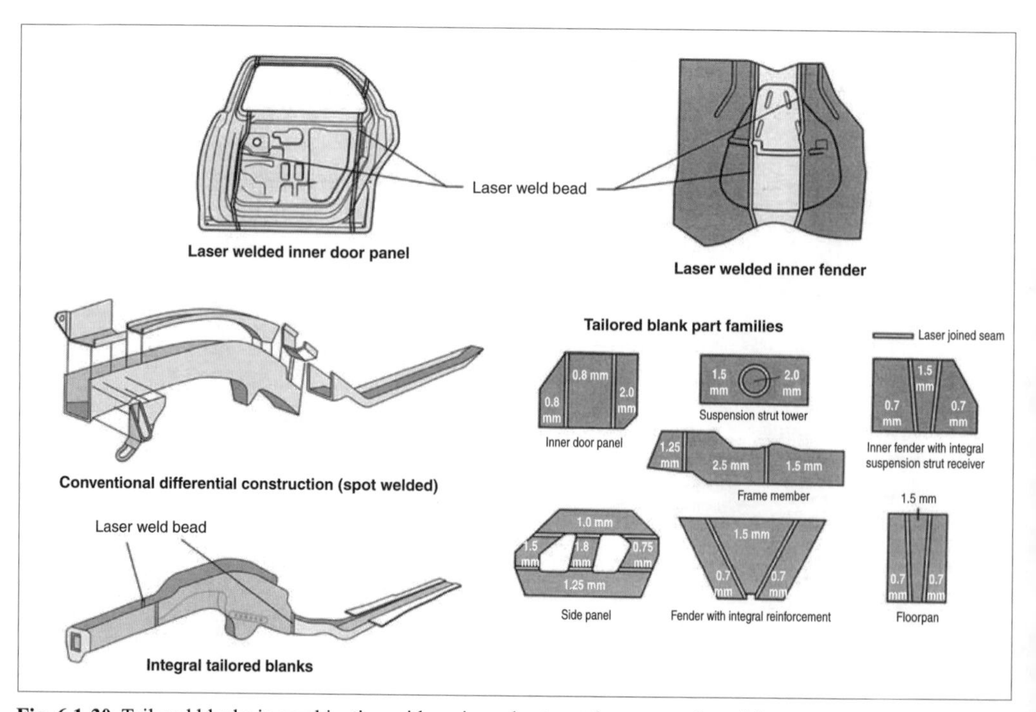

Fig. 6.1-30 Tailored blanks in combination with various sheet metal gauges and qualities.

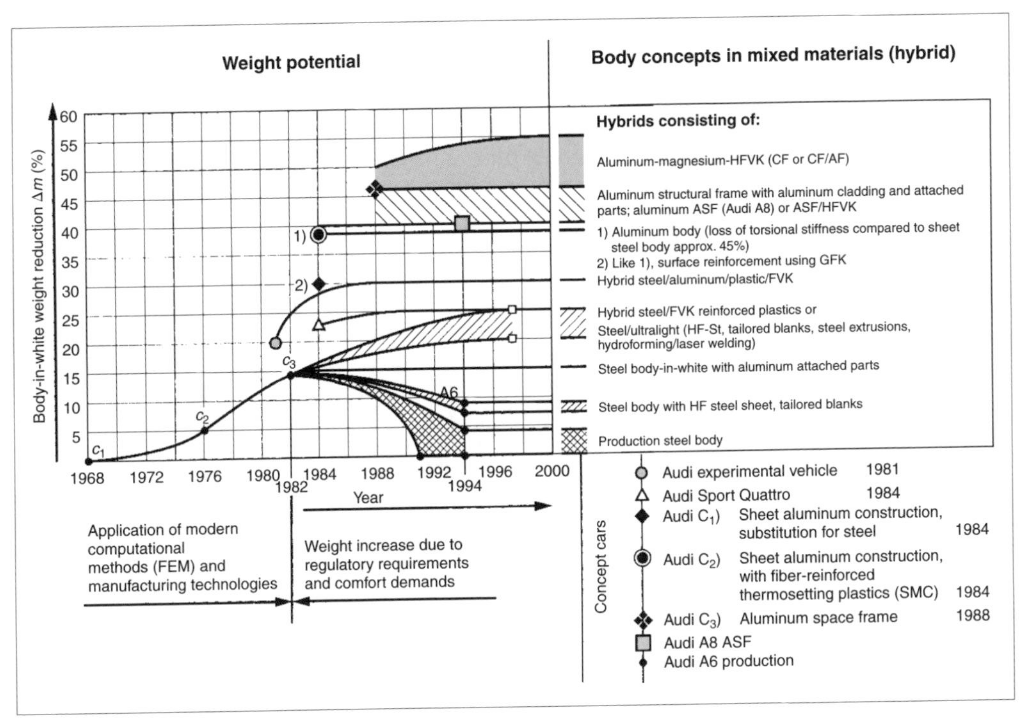

Fig. 6.1-31 Body weight potential using mixed-material (hybrid) construction.

volume production, and costs are unmatched by alternative materials such as sandwiches or fiber composites. The framework is completed by attached parts and cladding, which may themselves consist of various materials (Fig. 6.1-32 and 6.1-33).

If thermoplastic materials are chosen as cladding, either unreinforced or reinforced with chopped glass fibers, glass spheres, or flakes, consideration must be given not only to properties such as

— Thermal stability
— Cold impact ductility
— Longitudinal and transverse shrinkage
— Thermal expansion behavior near attachment points and body gaps
— Paint technology

but also to vibration behavior with respect to body first-order bending and torsional frequencies (reduction in resonant frequency of 8–10% may be expected for hybrid body concepts) in order to keep vibration comfort sacrifices within reasonable limits during vehicle operation.

In modern automobile body construction, individual joining techniques may be selected from

— Material bonding (welding)
— Mechanical
— Adhesive
— Detachable

joining methods, according to the nature and properties of the basic materials, semi-finished parts, and component geometries. For the mixed-materials concept, optimized lightweight component applications increasingly employ a combination of material and mechanical joining methods, such as the following examples, with adhesive bonding technology:

— Weld bonding
— Clinch bonding
— Pierce rivet bonding
— Solid rivet bonding

Without exception, this approach applies to mixed-material construction, where steel, aluminum alloy, magnesium alloy, and fiber-reinforced composites must be joined. Adhesive selection must concentrate on strength-increasing as well as corrosion-resisting properties of the bonding zone. Given a basis of mixed-material joints, mechanical joining methods and their combination with various adhesive joining methods offer an advantage not only in creating force- and form-fit joints (Fig. 6.1-34) but also in matching the level of material bonds (e.g., welding; Fig. 6.1-35). The major mechanical joining methods include pierce riveting, in which multiple layers of different materials may be joined with high-strength, fluid-tight bonds in a single low-noise, emissions-free, and low-energy manufacturing step. Because proof of joint strength as a function of tool travel

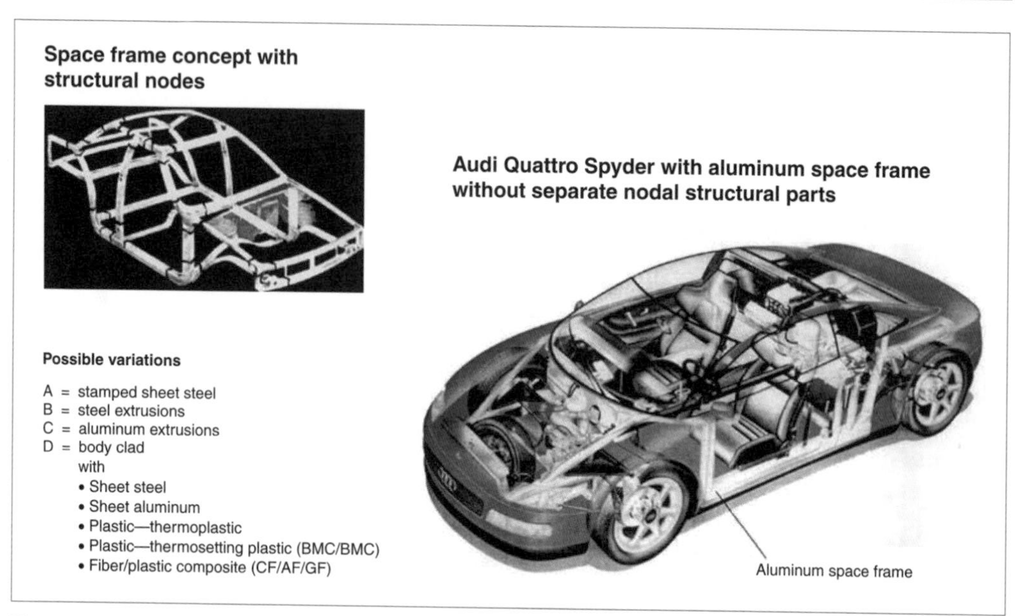

Fig. 6.1-32 Clad body/structural frame concept.

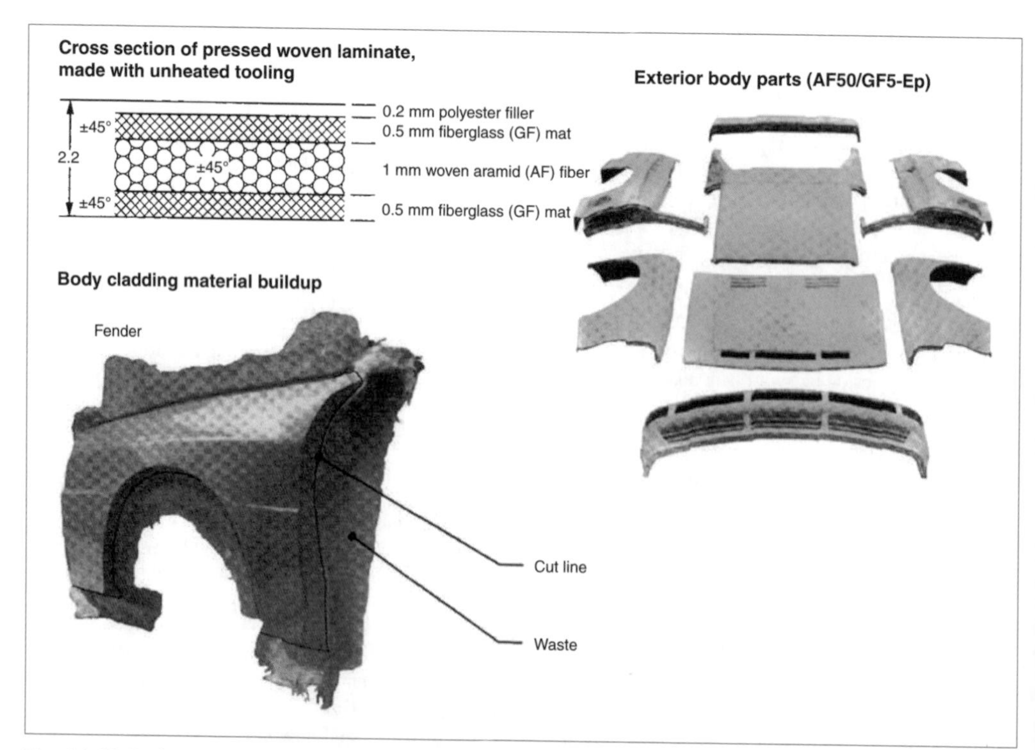

Fig. 6.1-33 Body exterior panels in high-performance fiber composite.

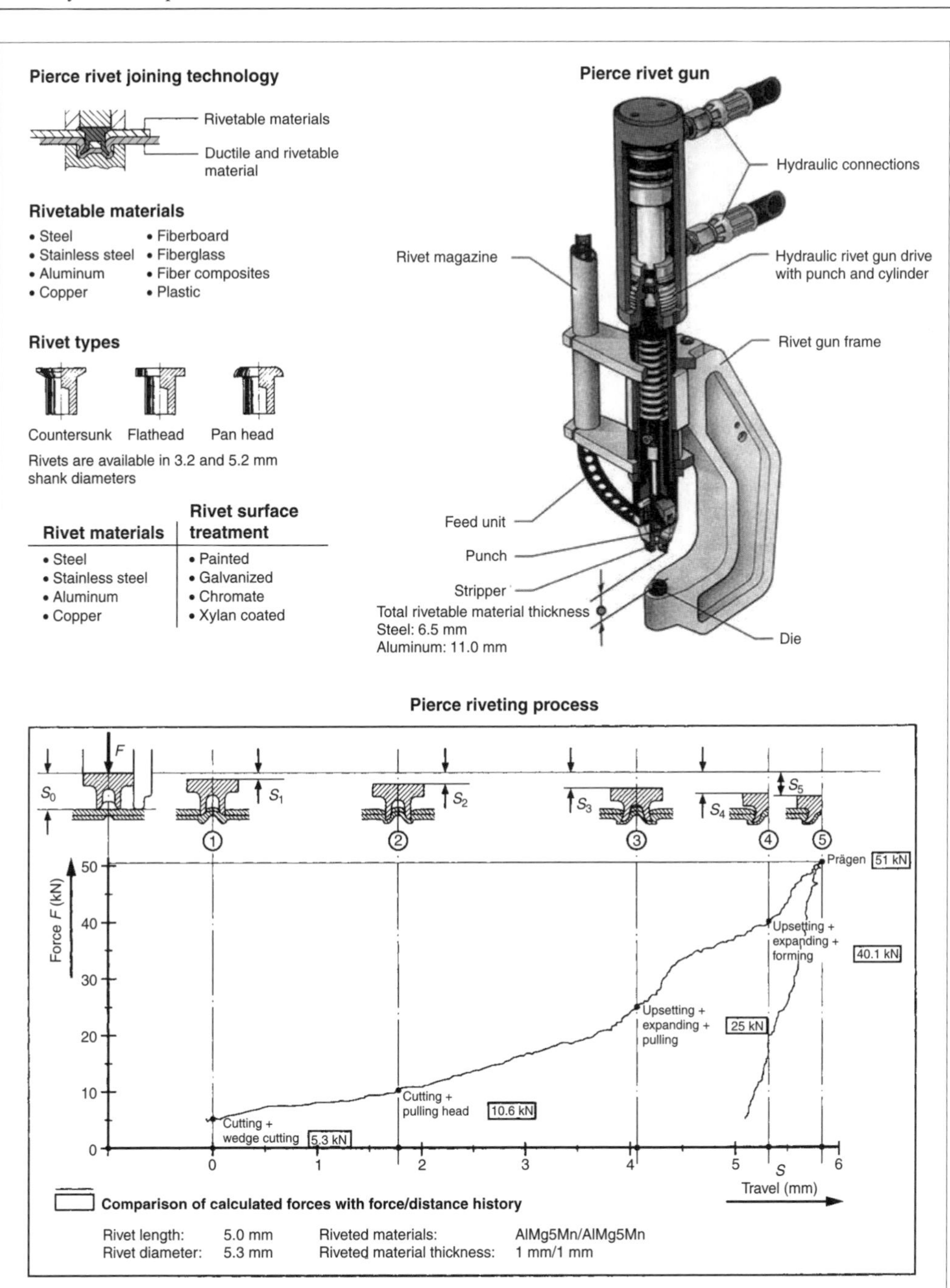

Fig. 6.1-34 Pierce rivet technology.

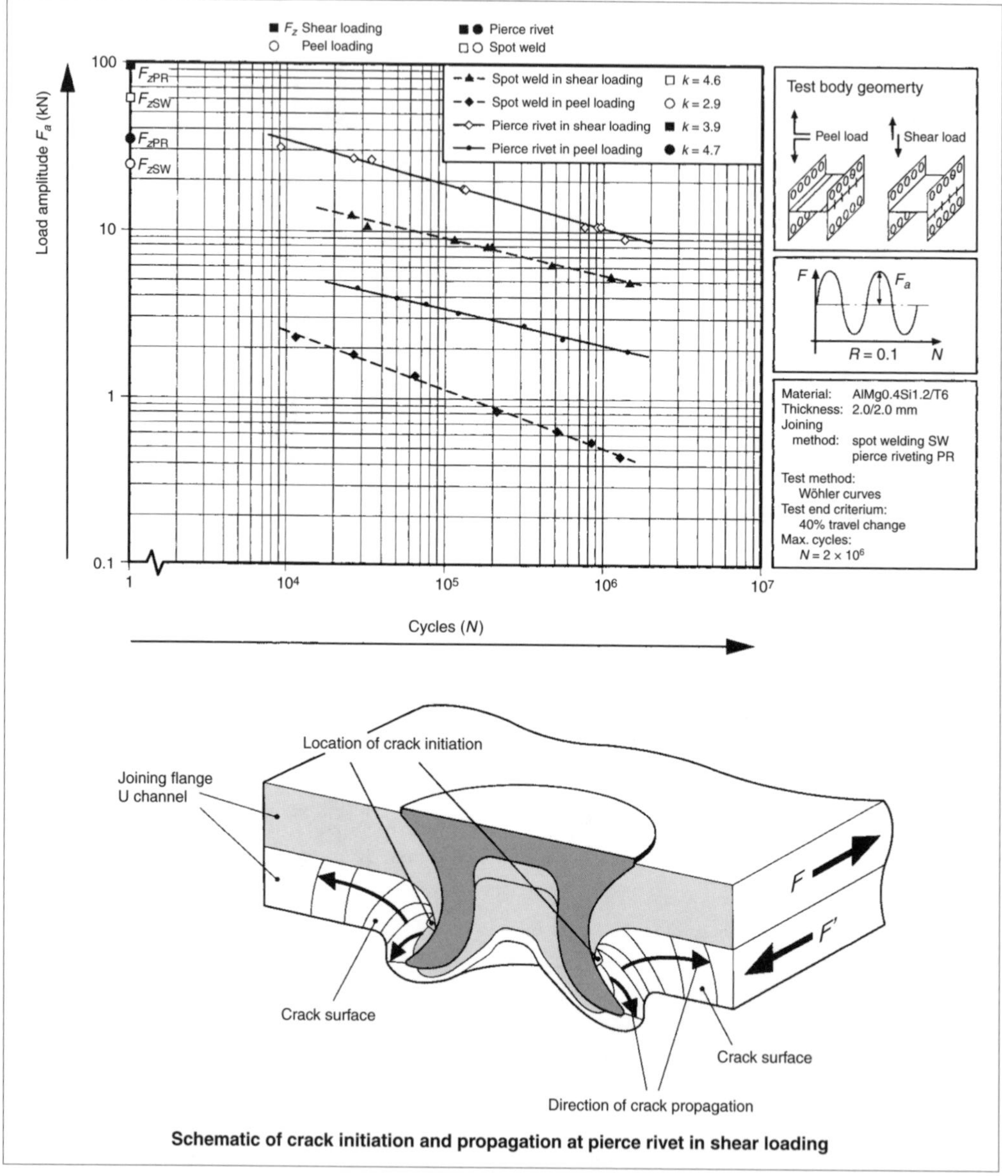

Fig. 6.1-35 Comparison of fatigue resistance of spot welded and pierce riveted shear and peel test samples.

may be obtained at every phase of the joining process for both clinching and pierce riveting, their suitability for large series production may be regarded as assured. Fig. 6.1-36 compares durability of various joining methods when subjected to vibration.

Multimaterial joining as a body concept strategy combined with application of lightweight materials, components, and assemblies as well as joining techniques appropriate for lightweight construction represent a significant step on the systematic path toward lighter automobile bodies and therefore toward a weight-oriented vehicle. Of necessity, synergies will be transferred to related operational vehicle design fields such as

— Bus construction
— Self-supporting internal-frame-based commercial vehicle structures
— Railway vehicle structures

To this end, integration of new lightweight materials in manufacturing processes suitable for volume produc-

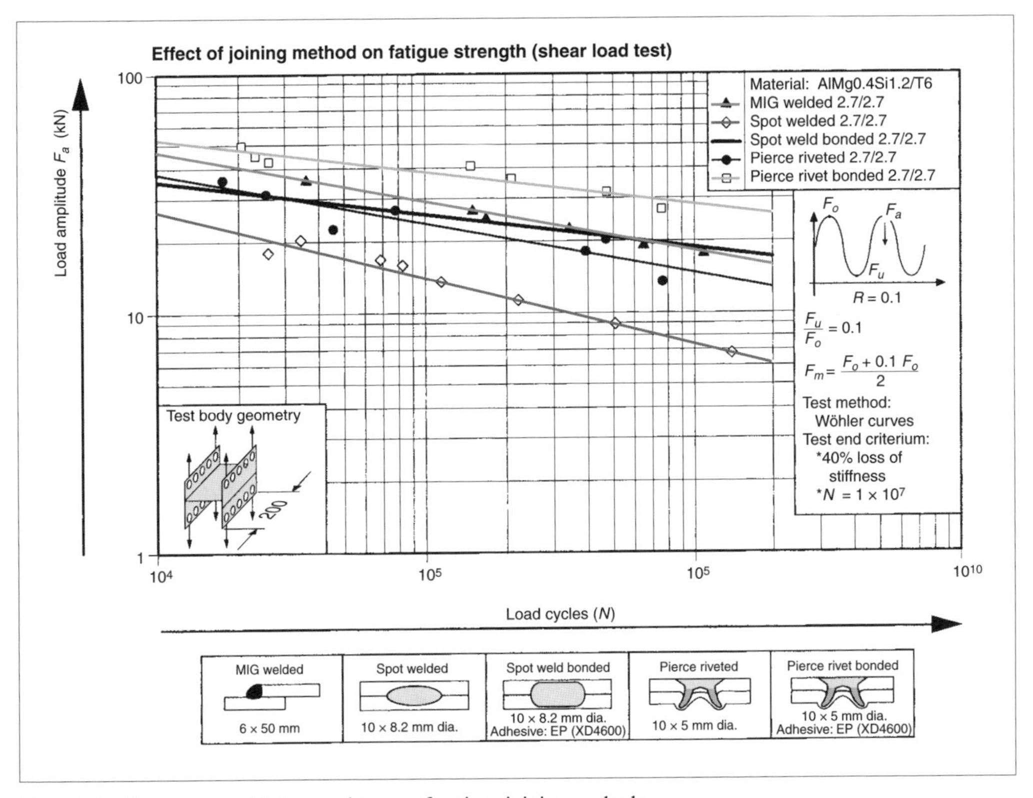

Fig. 6.1-36 Comparison of fatigue resistance of various joining methods.

tion with process-assured quality will gain increasing importance in modern automobile body construction.

Reference

[1] Haldenwanger, H.-G. "Zum Einsatz alternativer Werkstoffe und Verfahren im konzeptionellen Leichtbau von PKW-Rohkarosserien." Dissertation, TU Dresden, 1997.

6.2 Automobile body materials

Modern automobile body construction draws on an almost bewildering variety of materials and material combinations. In auto body construction, evolutionary substitution of traditional materials is driven by constant transformation within the

human / technological / economic / traffic / regulatory

environment and by worldwide competitive pressure among automobile manufacturers to match the auto body to the demands of its environment. In individual cases, this often forms the basis of a new orientation in product technology and, therefore, a renewal of the overall body/superstructure concept with respect to changed human needs, altered economic pressures, and new technological possibilities.

In choosing new or modified materials or material combinations, the following development goals must be considered:

— Application of energy-saving materials
— Energy-saving manufacturing methods
— Recyclability of applied materials
— Reduced waste
— Improved long-term behavior
— Improved reliability
— Optimized overall comfort

With goal-oriented application of new materials or technologies, the developer is concerned with

assuring reliable function of a component or assembly, in the required quality, with justifiable economic effort, in keeping with logistical preconditions (requirements) *concomitant with volume production.*

New technology is suitable for a component or assembly if—in comparison to the prior state of scientifically proven, high-grade materials technology or customary practice—it is not discovered to be disadvantaged in any of the above four considerations and at least one of these goals is more easily achieved with the new technology than with the old. Systematic, evolutionary development of traditional inorganic and organic monomaterials as well as fiber-reinforced material variants combined with material-appropriate manufacturing technologies offer new possibilities to satisfy the manifold functional demands of an automobile body.

Individually, these are:

- High strength steel alloys
- Aluminum alloys for sheets, extrusions, and castings
- Magnesium alloys for castings, extrusions, and, to a limited extent, sheet metal components (forming, with limited formability, possible below 200°C).
- Fiber-reinforced aluminum and magnesium casting alloys
- Temperature-resistant plastics (>200°C)
- Nondirectional fiber-reinforced plastics
- Directional fiber-reinforced plastics
- Hollow titanium extrusions
- Powdered metals (Metal Injection Molding—MIM).

The greatest usefulness of a new development may manifest itself in such diverse details as

- Increased design freedom for a function-oriented, multifunctional component concept; i.e., integration of components and functions (aesthetics)
- Weight-specific higher mechanical load-carrying ability, permitting lighter designs
- Improved aging mechanisms, such as reduced corrosion
- Improved failure behavior, with longer service life or "benign" failure
- Wider service temperature range for high and low temperatures
- Increased damping
- Improved manufacturing with cost and quality advantages
- Multiple recycling possibilities

Choice of manufacturing technologies appropriate for the given materials and material properties has a major effect on resource loads imposed by operation of a vehicle. Fig. 6.2-1 shows the specific energy consumption for typical parts made of various materials. For metals, including steel but especially for lightweight alloys, the volume-specific energy cost is many times that of reinforced or unreinforced plastics. Not reflected in this representation are the different volumes required to provide the needed tensile strength and stiffness. Consequently, the relative energy consumption needed to produce a given part using various materials may be compared with the aid of lightweight construction numbers (Fig. 6.2-2) and their derivation (Fig. 6.2-3) as functions of material density for stiffness as well as tensile strength (Fig. 6.2-4). The effects of vehicle operation on energy demand—that is, on fuel consumption—are reflected by driving resistance. As individual resistances (with the exception of aerodynamic drag) are determined by translational and rotational (oscillating) masses, application of weight-reducing materials to realize a high degree of lightweight construction potential is a worthwhile development goal in terms of resource conservation (Fig. 6.2-5).

6.2.1 Metals in automobile body construction

Generally, in translating lightweight auto body construction principles into reality, thin-walled, delicately

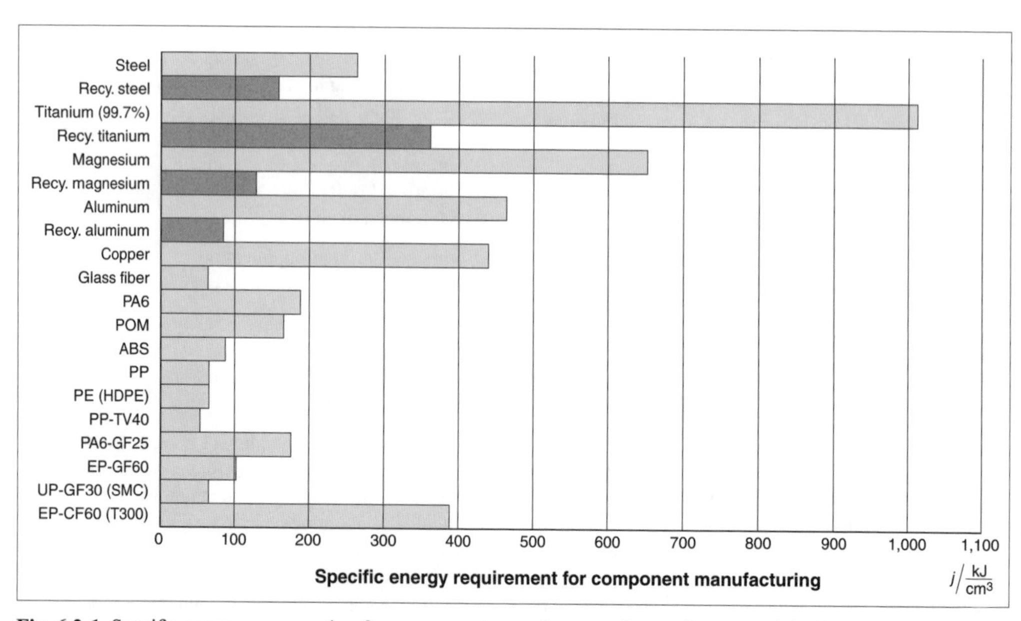

Fig. 6.2-1 Specific energy consumption for component manufacture using various materials. Recy., recycled.

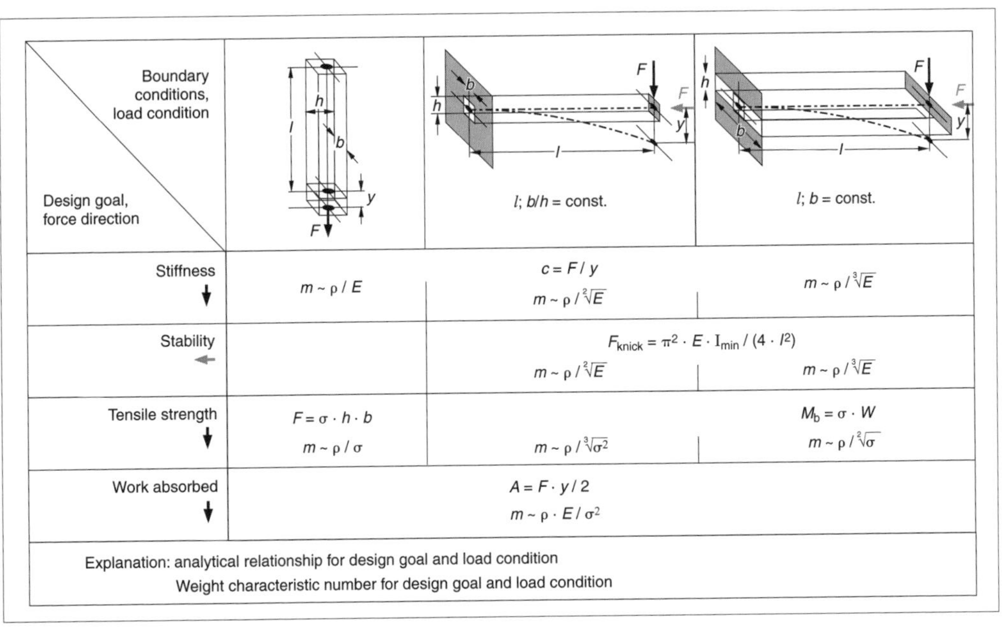

Boundary conditions, load condition / Design goal, force direction	(tension member)	l; b/h = const.	l; b = const.
Stiffness ↓	$m \sim \rho / E$	$c = F / y$ $m \sim \rho / \sqrt[2]{E}$	$m \sim \rho / \sqrt[3]{E}$
Stability ←		$F_{knick} = \pi^2 \cdot E \cdot I_{min} / (4 \cdot l^2)$ $m \sim \rho / \sqrt[2]{E}$	$m \sim \rho / \sqrt[3]{E}$
Tensile strength ↓	$F = \sigma \cdot h \cdot b$ $m \sim \rho / \sigma$	$m \sim \rho / \sqrt[3]{\sigma^2}$	$M_b = \sigma \cdot W$ $m \sim \rho / \sqrt[2]{\sigma}$
Work absorbed ↓	$A = F \cdot y / 2$ $m \sim \rho \cdot E / \sigma^2$		

Explanation: analytical relationship for design goal and load condition
Weight characteristic number for design goal and load condition

Fig. 6.2-2 Lightweight construction characteristic numbers.

shaped hollow sections integral to the body structure must be used. A lightweight design optimally matched in terms of strength and stiffness to the functional requirement profile generally offers advantages compared to a conventional "massive" design. Often, this also encompasses damage behavior as well. In order to satisfy defined functionality and quality requirements, great significance must be given to long-term proof of the

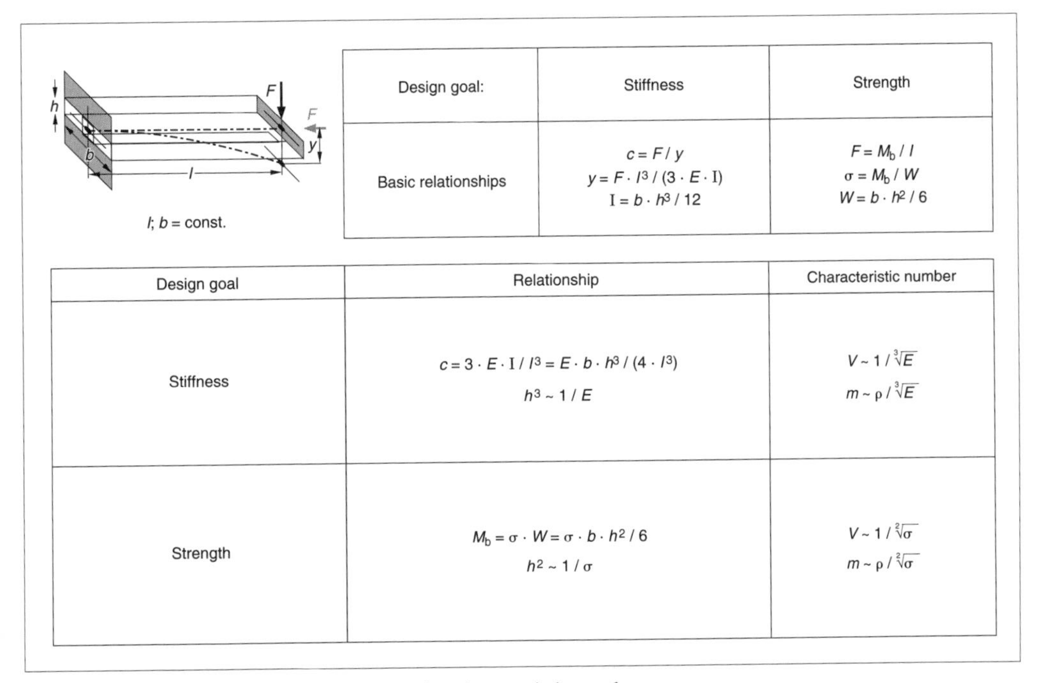

l; b = const.

Design goal:	Stiffness	Strength
Basic relationships	$c = F / y$ $y = F \cdot l^3 / (3 \cdot E \cdot I)$ $I = b \cdot h^3 / 12$	$F = M_b / l$ $\sigma = M_b / W$ $W = b \cdot h^2 / 6$

Design goal	Relationship	Characteristic number
Stiffness	$c = 3 \cdot E \cdot I / l^3 = E \cdot b \cdot h^3 / (4 \cdot l^3)$ $h^3 \sim 1 / E$	$V \sim 1 / \sqrt[3]{E}$ $m \sim \rho / \sqrt[3]{E}$
Strength	$M_b = \sigma \cdot W = \sigma \cdot b \cdot h^2 / 6$ $h^2 \sim 1 / \sigma$	$V \sim 1 / \sqrt[2]{\sigma}$ $m \sim \rho / \sqrt[2]{\sigma}$

Fig. 6.2-3 Derivation of lightweight construction characteristic numbers.

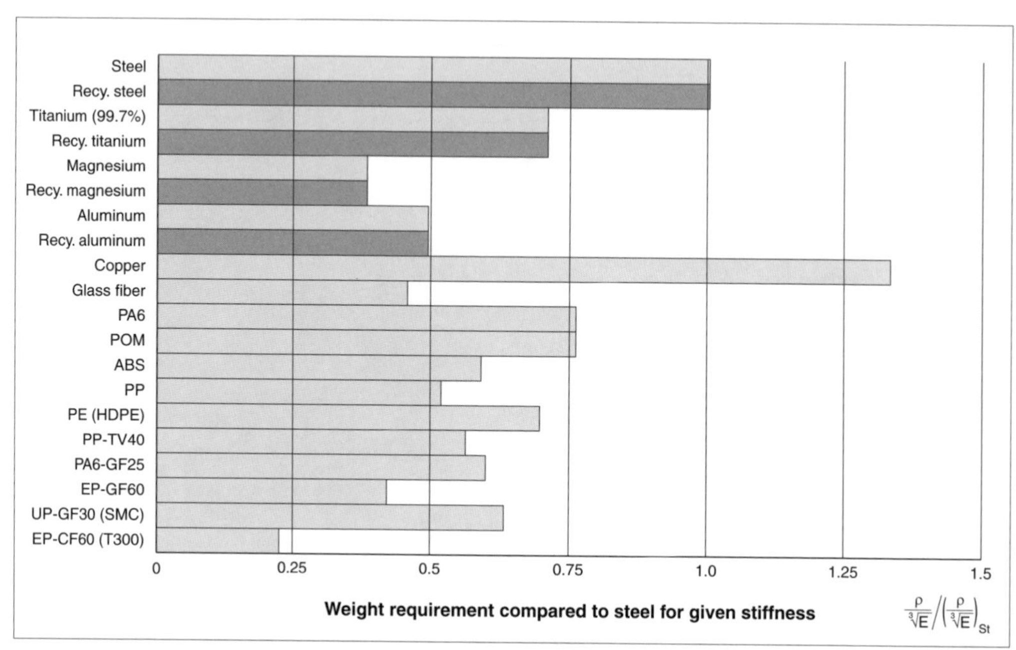

Fig. 6.2-4 Relative weight requirement relative to steel for various materials with given bending stiffness target. Recy., recycled.

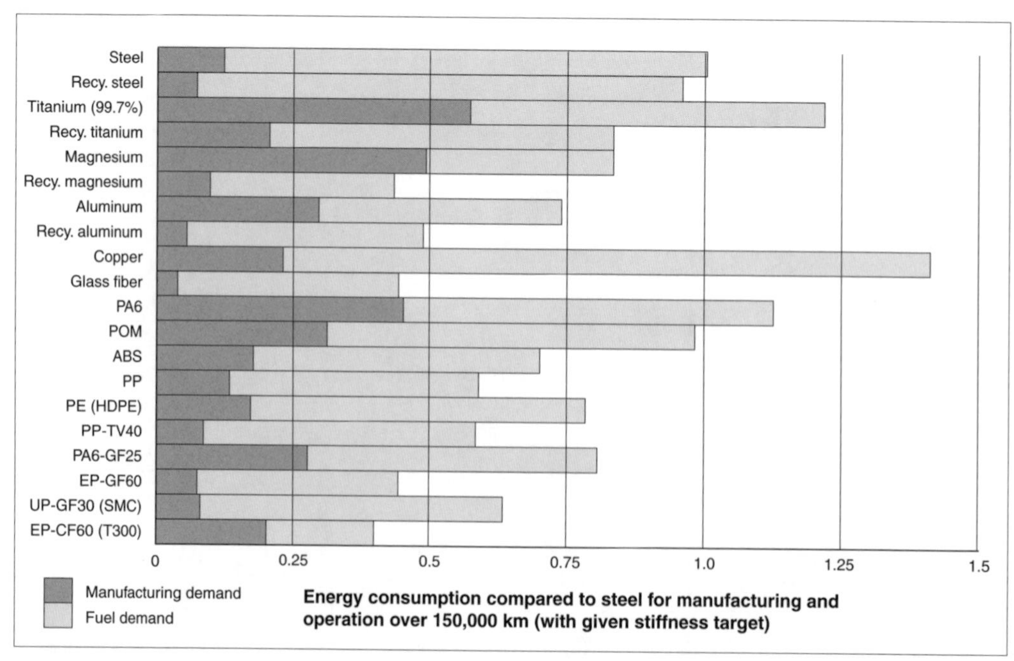

Fig. 6.2-5 Various materials for lightweight construction potential of weight-specific stiffness and energy consumption in comparison to steel. Recy., recycled.

body structure's load-carrying ability, safety, and utility (Fig. 6.2-6). Proof of load-carrying ability and safety margins as determined by a service-life–oriented component analysis is customarily employed for safe dimensioning (Fig. 6.2-7). Depending on application, two basic concepts are applied for materials selection and dimensioning:

— Safe life quality: a requirement for absolute freedom from failure over the entire life cycle
— Fail-safe quality, which predicates basic failure tolerance and adequate safety margins required of the body or superstructure

Increasingly, however, given sufficient complexity, optimization strategies must be applied, for which numerical simulation methods are ideally suited.

For body manufacturing under high-volume conditions, thin-gauge sheet steel stampings offer great potential for lightweight construction, saving weight while meeting high strength and safety demands at safety-relevant sections of the body structure (passenger cell). Forming by means, for example, of the following techniques requires a high degree of plastic deformability:

— Deep drawing
— Hydroforming
— Extruding
— Stamping
— Bending

Along with *ultimate strain*, a material's r_m value is significant. This indicates the ratio of width to thickness change of sheet metal. Low *strain limits* make forming easier. However, in finished condition, a high strain limit is desirable. Accordingly, forming processes which achieve work hardening (n_m value) offer functional advantages (Fig. 6.2-8). These advantages may also be achieved by means of paint baking temperatures (bake hardening, or BH, thin steel sheet). Compared to unalloyed deep-drawing steel (St14), the n_m value may be raised by reducing the carbon content to 0.02% while raising titanium and niobium content, which results in complete precipitation of free carbon and nitrogen (interstitial free, or IF, thin steel sheet).

Attractive possibilities are offered by phosphorus-based microalloys (StE, for example). Instead of precipitation hardening by vanadium, niobium, or titanium, this material increases strength by solid solution hardening with phosphorus (~0.1% P). Because carbon atoms are not firmly bound in this case, paint baking (170°C, 20 min) produces hardening similar to aging (ZstE 300 BH). In dual-phase steels (DP), fine-grained structure consisting of ferrite with, for example, 15% austenite is quenched from the dual phase field, transforming austenite to martensite grains. This harder phase is scattered through softer ferrite. These thin-gauge sheet steels do not exhibit a definite yield point and therefore may be formed without skin passing. Initial work hardening is relatively high, as is tensile strength, but the r_m value is low. Transitional materials with properties between those of the described thin-gauge steels are possible. For example, DP steels also exhibit the BH effect and may additionally be strengthened by the addition of phosphorus. If the BH effect is not required, phosphorus steels can be combined with IF properties. Principally, in view of their cold-working properties, P and DP steels may be regarded as having the greatest lightweight potential in "steel intensive" auto body construction.

An important selection criterion for application of thin-gauge steel, particularly higher strength varieties, is surface coating for improved corrosion behavior. An electrolytically deposited layer of zinc, about 10 μm thick, provides effective corrosion protection; single-sided application provides advantages in spot welding.

In describing the most important forming processes

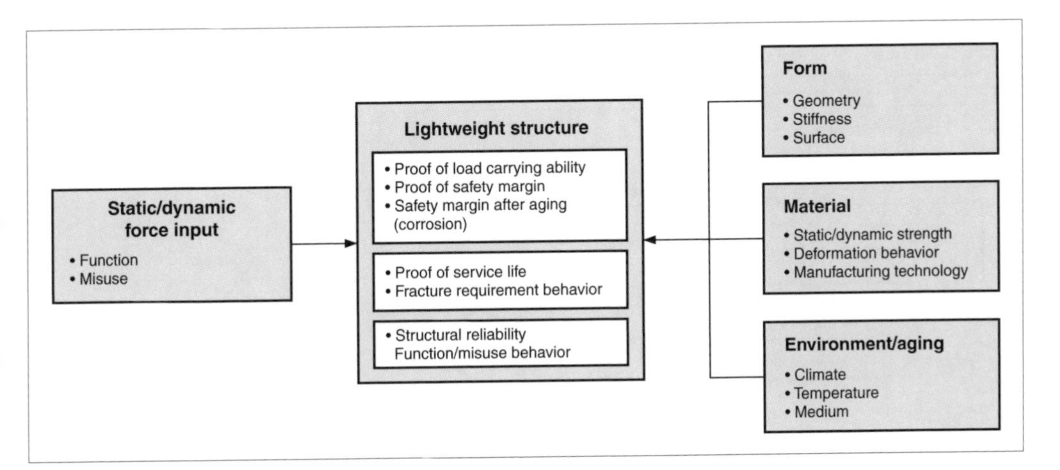

Fig. 6.2-6 Proof parameters for lightweight structures.

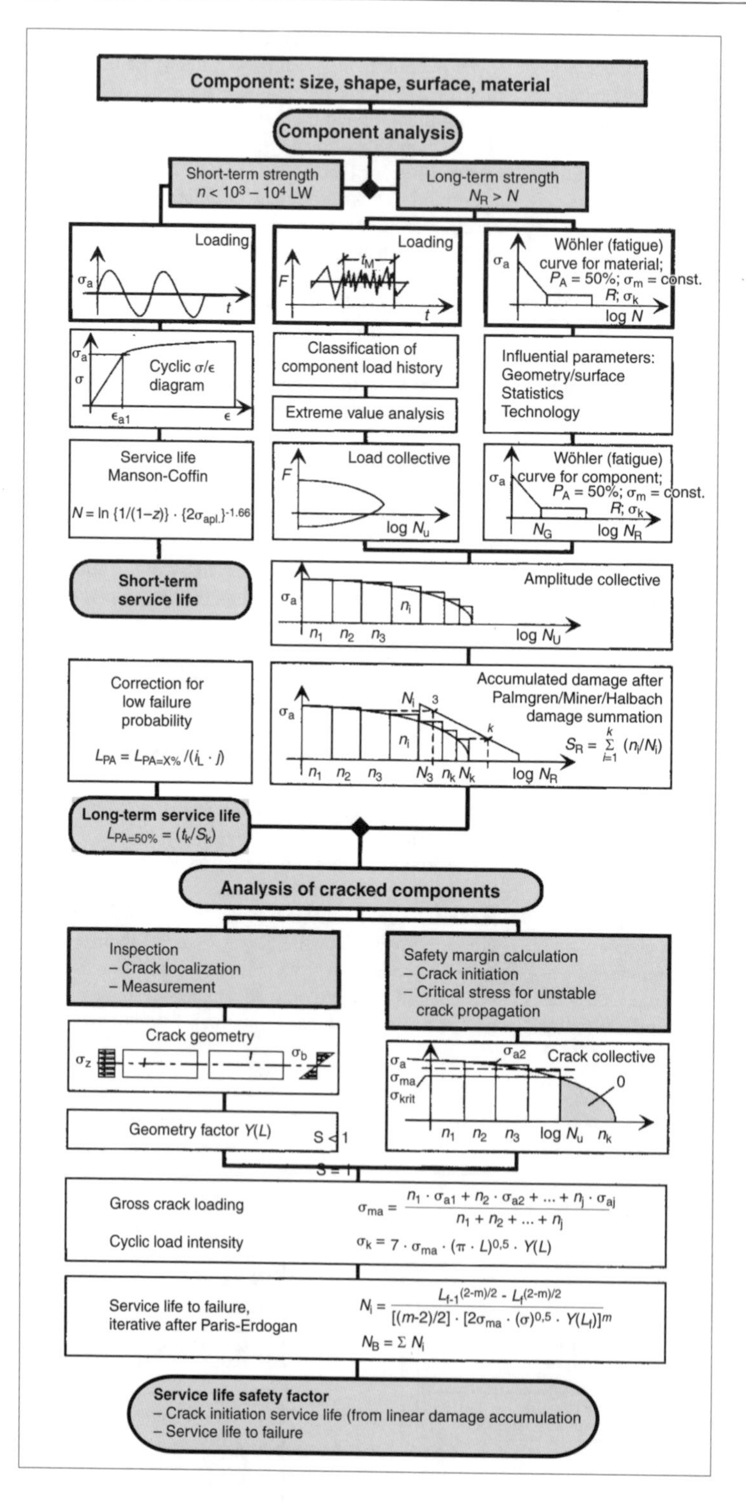

Fig. 6.2-7 Service life oriented component analysis.

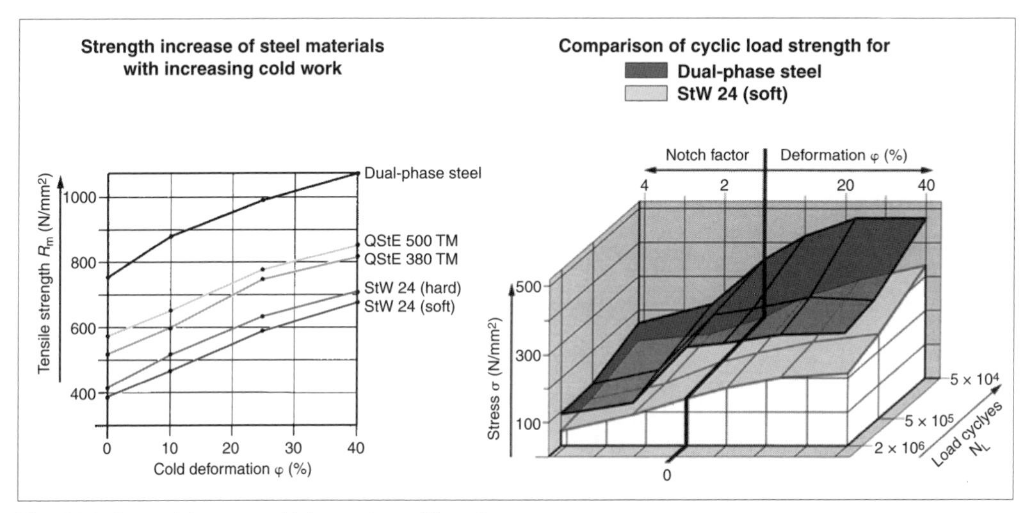

Fig. 6.2-8 Strength increase with increasing cold work.

for thin-gauge steel, it is apparent that deep-draw forming by means of internal hydroforming (IHF) results in weight and manufacturing time savings by eliminating joining flanges (as used for spot welding, adhesive joining, or pierce riveting). Compared to deep-drawing sheet metal operations, hydroforming is a more complex and time intensive but dimensionally a more accurate technology. Main stages consist of:

— Expansion by fluid pressure and pre-bending
— Swaging against inside pressure (cold swaging)
— Expansion with fluid pressure ($p_i \gg 1{,}000$ bar) and calibration

In view of the relationship between lightweight construction costs and lightweight utility, hydroforming can only be justified if it results in additional functional value and therefore functional integration. Higher tensile strength resulting from cold working during the hydroforming process and thus higher energy absorption in the event of a collision may be applied toward a lightweight design strategy in that the wall thickness of a structural member is correspondingly reduced. Furthermore, internal hydroforming permits manufacturing hollow sections with varying wall thickness, starting with laser-welded blanks, in keeping with the concept of functionally integrated lightweight construction (Fig. 6.2-9). Because internal hydroforming may also be used to create connector geometries, various conventional cast connectors used in the aluminum space frame concept, for example (see Section 6.1.2), may be substituted or simplified (Fig. 6.2-10). Consequently, a welded lap joint is no longer necessary, and, with IHF, the process-specific wall thickness of conventional cast connectors can be reduced to the functionally required thickness,

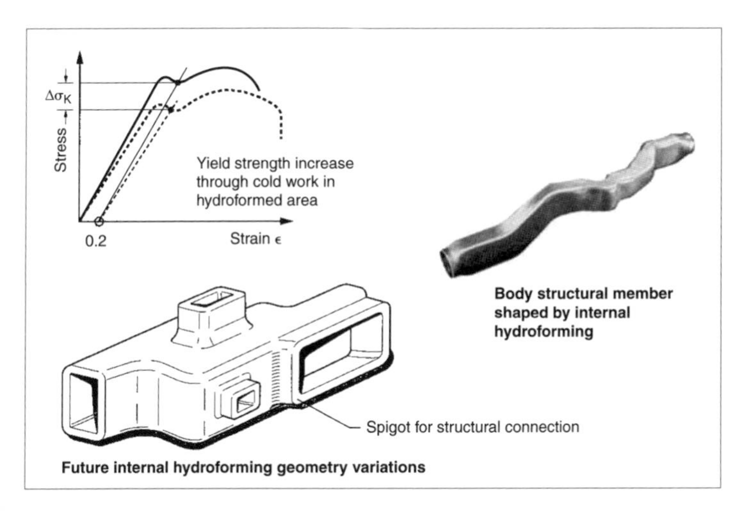

Fig. 6.2-9 Structural member geometry of flangeless hollow profile produced by internal hydroforming.

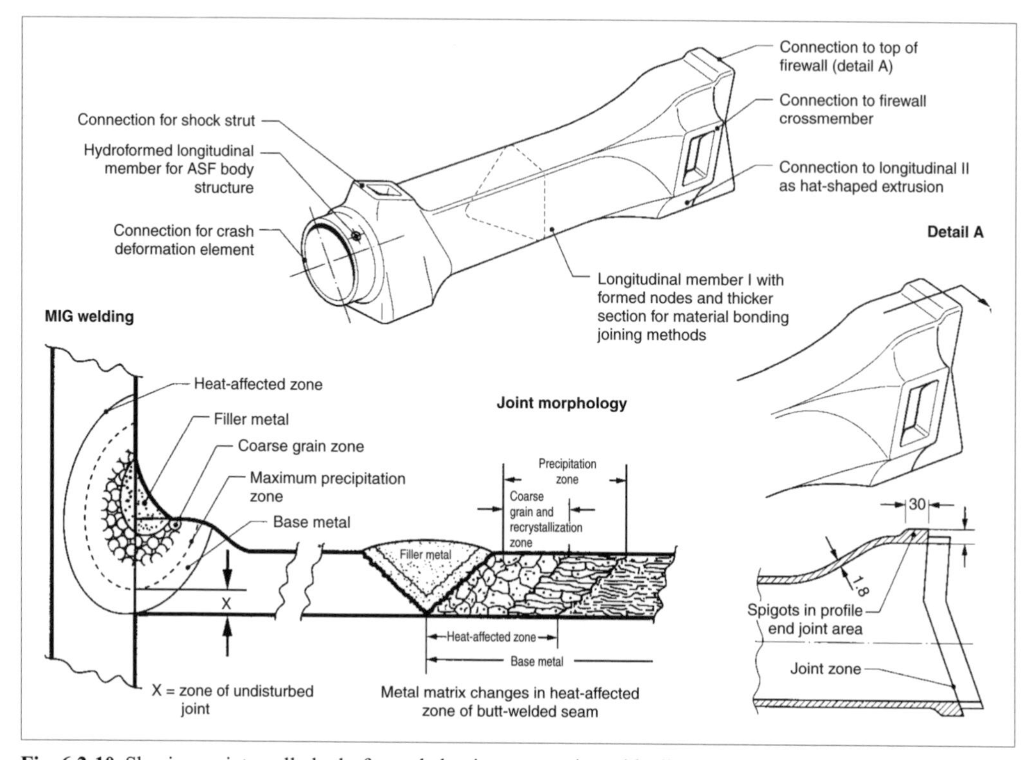

Fig. 6.2-10 Shaping an internally hydroformed aluminum extrusion with elimination of cast connecting nodes.

thereby reducing weight and cost. In principle, extruded shapes, which are formed or bent by the IHF process and shaped into connectors at the joint locations, then swaged at their ends and joints (Fig. 6.2-11), are especially suited for space frame structures, for example in:

— Bus construction
— Box structures for trailers and semitrailers
— Wagon structures for railway vehicles

Corresponding design examples are shown in Fig. 6.2-12.

Of the lightweight metals aluminum and magnesium, the former—in the form of sheet, extruded semi-finished parts, or a wide variety of castings—offers high technological potential for lightweight body construction. The user is faced with an extensive palette of proven wrought aluminum and aluminum casting alloys whose material properties, depending on manufacturing process, provide great opportunities for optimization. For wrought aluminum alloys, the highest tensile strength values are achieved by precipitation hardening. This is heat treatment the objective of which is to generate coherent or semicoherent particles of a second phase, which act to create stresses in the crystal lattice and therefore give rise to increased tensile strength. Typically, heat treatment—so-called artificial aging—is done at a temperature of 160 to 210°C. In precipitation hardening aluminum alloys, the so-called aging temperature represents the upper limit for hot tensile strength. Another possible way to increase the resistance of aluminum alloys to shape changes is dispersion hardening, in which, analogous to precipitation hardening, specific phases are finely distributed throughout the metal matrix. Hardenable aluminum alloys are subjected before heat treament to joining processes such as:

— Shielded gas welding (TIG or MIG)
— Laser welding
— Pierce riveting or clinching

In automobile body construction, the following alloys represent the most promising compromise for functionally appropriate lightweight construction (space frame concept):

— AlMg0.4Si1.2 (AA 6016) for sheet metal
— AlMgSi0.5 for extrusions
— AlSi10MgMn ($\varepsilon \geq 0.5$) for cast connectors

Compared to Al alloys, magnesium alloys exhibit less tensile strength at both room and elevated temperatures. The small difference in dynamic strength, Fig. 6.2-13, is attributable to the greater notch sensitivity of cast Mg alloys in practice. Smooth surfaces and avoidance of notches are imperative. Their lower modulus of elasticity makes Mg alloys less sensitive to impact and

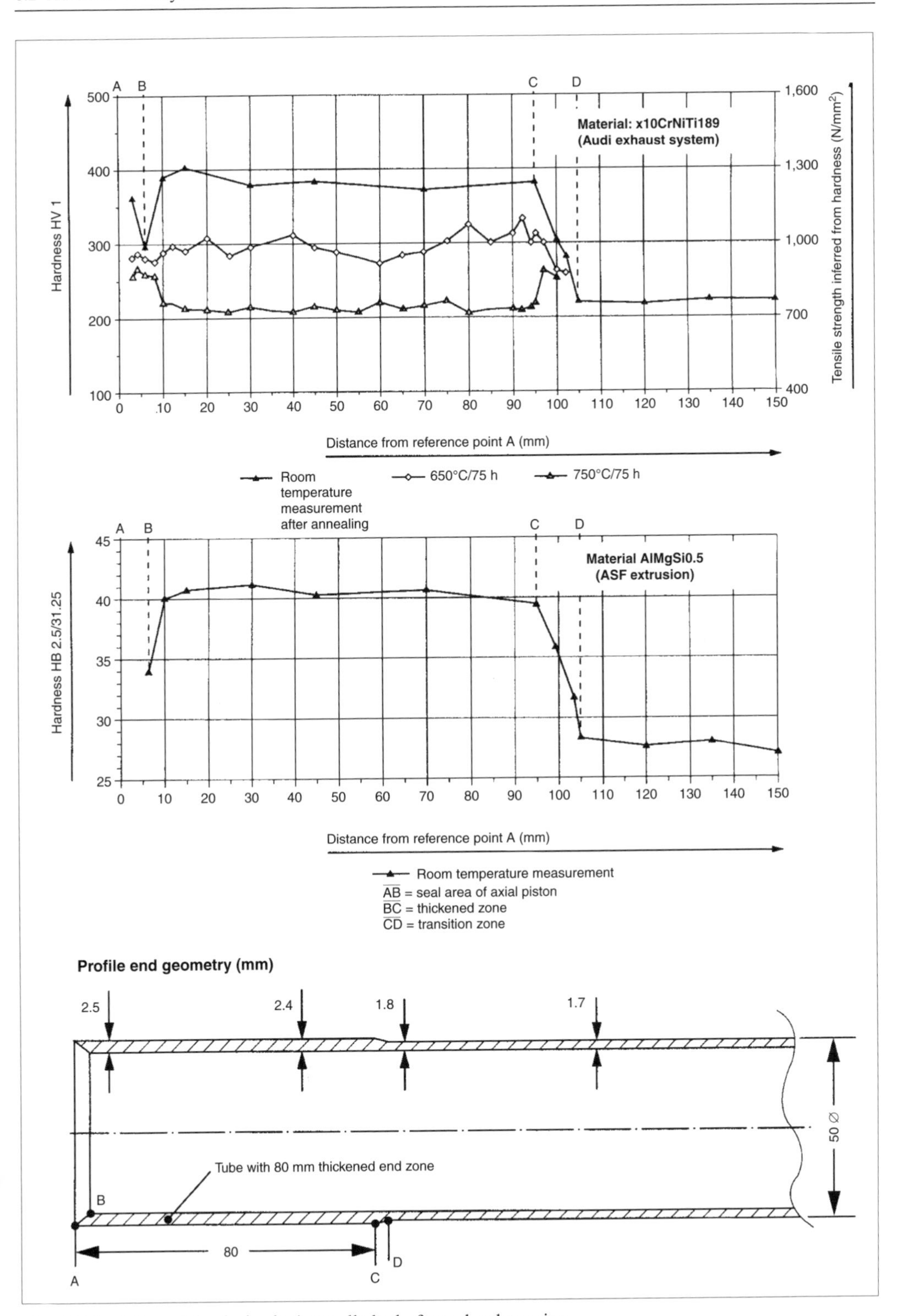

Fig. 6.2-11 Cold work hardening by internally hydroformed end swaging.

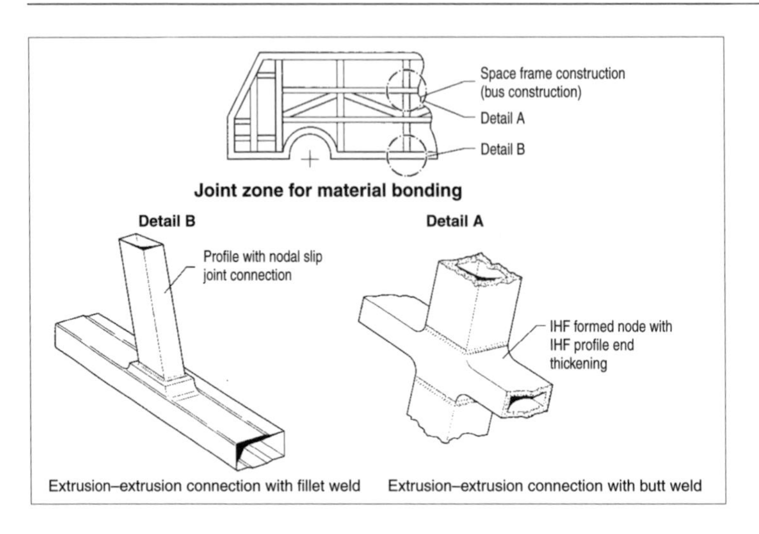

Fig. 6.2-12 Joint geometry for extruded profiles in space frame construction.

shock loading and provides them with a capacity for noise damping. Forming of wrought Mg alloys for automobile body construction is accomplished by extrusion, hot pressing, rolling, and drawing above 200°C. Because of the hexagonal lattice structure and resulting danger of stress crack corrosion, cold forming should be avoided (stress relief annealing at 280–300°C). It should be noted that the oxidation propensity of molten magnesium requires special precautions during casting and welding (gas shielding, e.g., SO_2 or SF_6). Furthermore, Mg alloys, due to their extreme electronegative potential, demand special corrosion protection. The tensile strength of permanent mold or Vacural™ die cast aluminum connectors of alloy AlSi10Mg0.3, employed in modern light-alloy body construction, may be improved in comparison to conventional castings by new processing methods (Figs. 6.2-14 and 6.2-15) such as:

— Thixoforming/thixocasting
— Squeeze casting

The degree to which magnesium alloys are suitable for cast body connectors is dependent on thermomedial

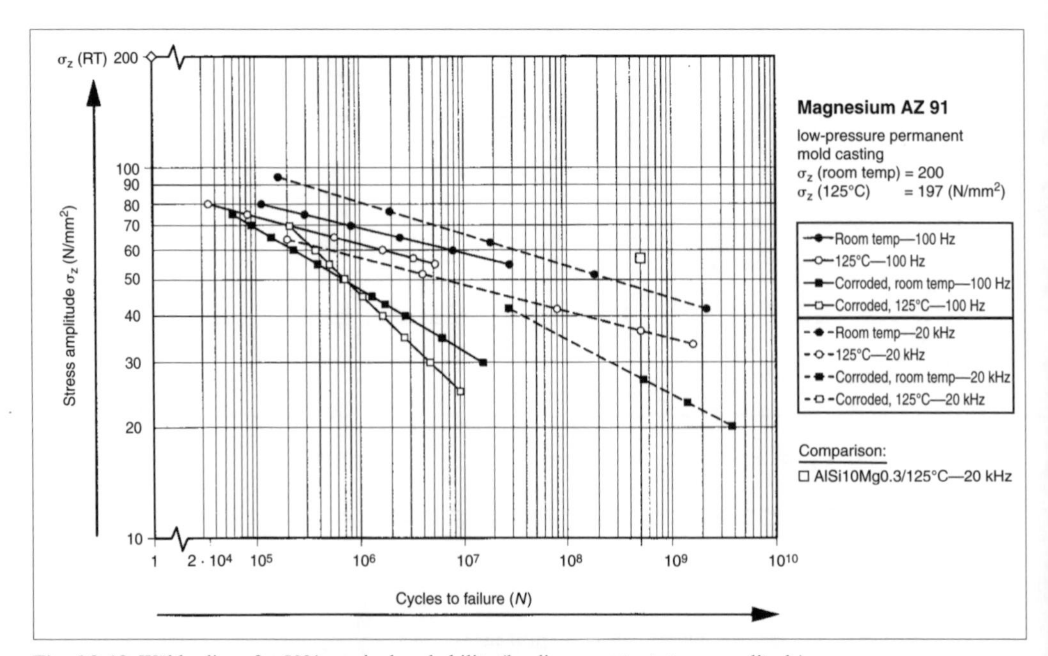

Fig. 6.2-13 Wöhler lines for 50% survival probability (loading: constant stress amplitude).

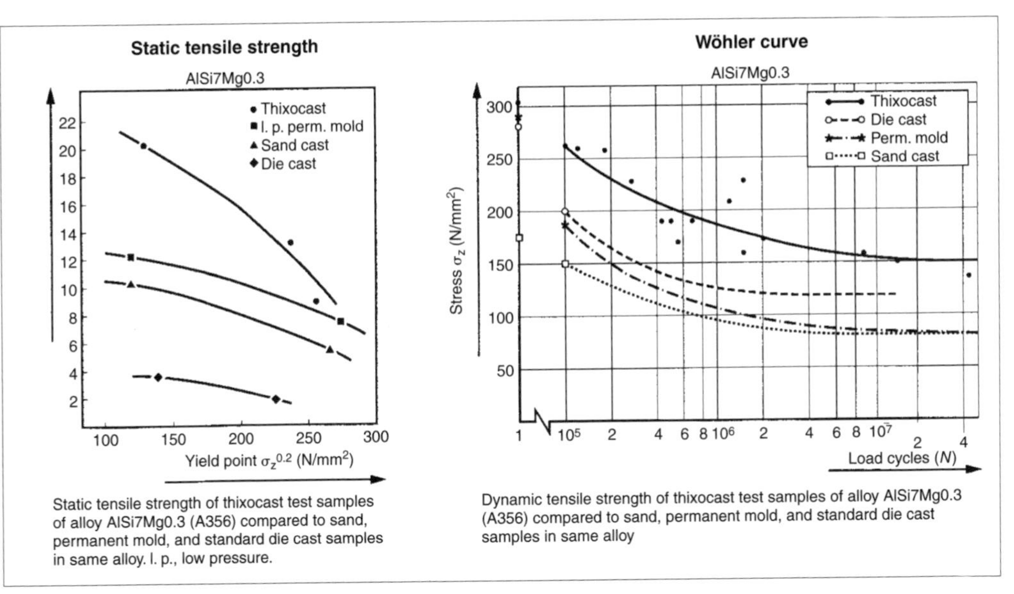

Fig. 6.2-14 Static and dynamic tensile strength of thixocast test samples (AlSi7Mg0.3).

loading (especially corrosive media combined with heat effects).

Another materials group currently under development for lightweight construction includes aluminum foams, made by melting or powder metallurgy. The low density of 0.3 g/cm³ with a porosity of 90% in aluminum foams offers great lightweight construction potential for thermal and sound insulation functions as well as absorption of impact energy, while still providing adequate strength levels.

New body construction options are offered by aluminum matrix composite materials, with ceramic silicon carbide inserts and aluminum oxide fibers (particles), as well as carbon fiber or aluminum sandwich structural elements (Fig. 6.2-16 and 6.2-17). In the constructive design of body components, depending on type and position of reinforcing components, attention must be directed to their degree of anisotropy. Improved mechanical properties are especially marked in long-fiber applications. Here, however, care must be taken that plastic deformation with respect to the composite material will no longer be possible after long fibers are introduced

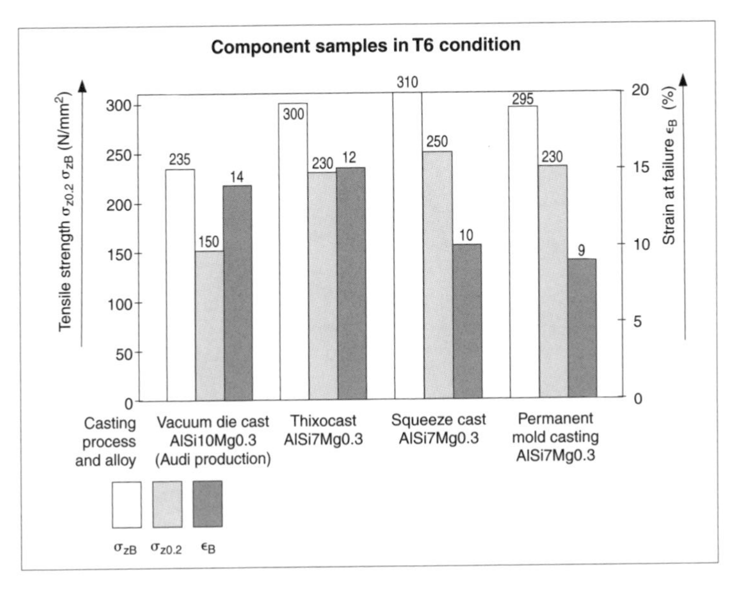

Fig. 6.2-15 Comparison of aluminum alloy material properties as a function of various casting methods.

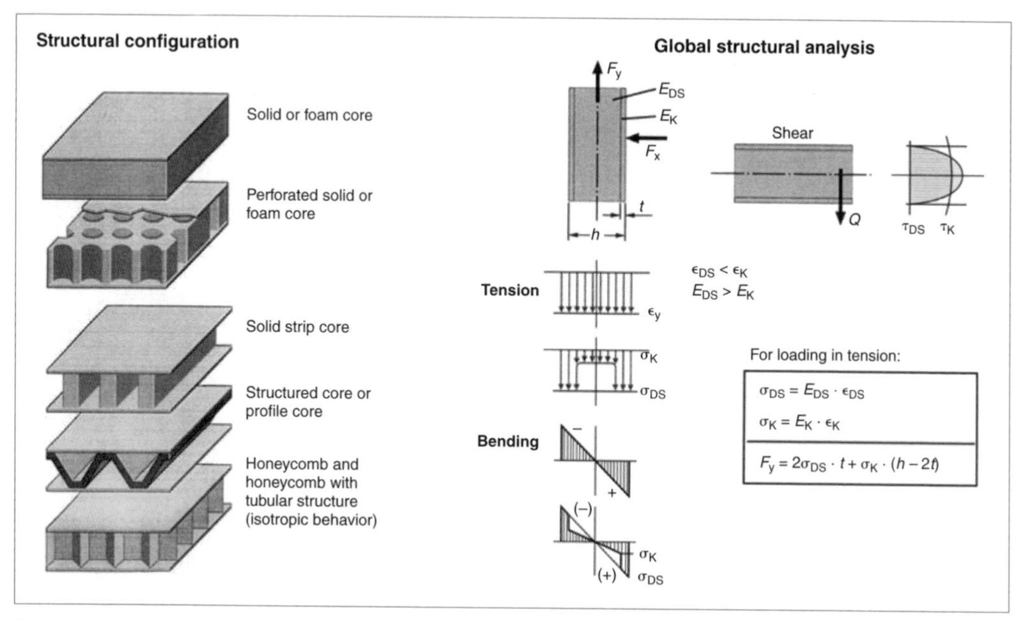

Fig. 6.2-16 Sandwich structures.

into the metal matrix (fiber breakage). As even sandwich composite panels (Fig. 6.2-16) permit only limited three-dimensional deformation, application opportunities for the described composite materials with metal, metal matrix, fiber composite, or covering layers must be regarded as limited. Nevertheless, sandwich composite panels offer possible solutions for bus, commercial vehicle, and railway vehicle construction.

6.2.2 Nonmetallic materials in auto body construction

Unreinforced and reinforced thermoplastics and thermosetting plastics, as well as fiber composite materials—directional fiber reinforced weaves, cloth, and knits—offer new materials technology application opportunities for body construction and therefore, in many cases, at-

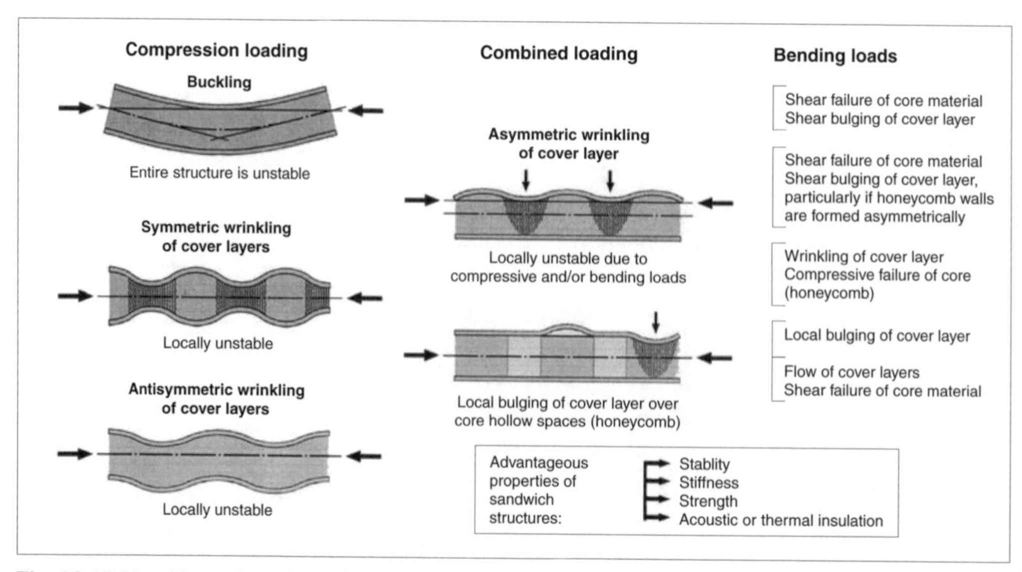

Fig. 6.2-17 Unstable configurations of sandwich structures.

tractive technology. Especially in the case of body structure cladding, these lead to products which are mainly distinguished by better multifunctional properties than are possible with conventional materials and manufacturing methods.

In principle, however, these new materials, based on chemical substances and their associated new technologies, must be integrated into existing structures within the automobile industry. This applies in particular to body assembly and finishing. Both of these body manufacturing steps are dominated by the needs of metal technology, so that thermoplastics or thermosetting plastics are difficult to integrate due to their different technological processes. Therefore, the problem of materials-appropriate manufacturing technology must be solved by mixed construction strategies such as:

— Offline painting of plastic body parts
— Offline preassembly of plastic body parts

Plastics are distinguished not only by their relatively great design freedom—with integration of components and functions during development of body parts—but also, above all, by the great variety of available synthetic polymers. Even at the material manufacturing stage, the basic building blocks of polymers—molecular chains—may be matched to the engineering requirements of the final product. In the next stage as well—material processing—polymers may be custom-tailored to engineering requirements. Various constraints may be imposed on polymers, as on metals or composite materials, so that, for example, all molecular chains are oriented in the same direction, resulting in greater strength in the direction of the chains. A further property improvement may be achieved through polymer alloys or mixtures and the addition of reinforcing materials such as:

— Short fibers, usually based on glass
— Glass powder
— Glass flakes
— Glass microspheres
— Talcum

Among the most important properties of thermoplastics are their manifold manufacturing possibilities, such as:

— Injection molding
— Transfer molding
— Gas-assisted injection molding
— Extruding/pultruding
— Thermo-molding
— Thermo-pressing

Also important are the manufacturing properties of thermoplastics, which permit the forming of complex shapes at temperatures above 200°C.

The material strength of plastics only rarely (as in the case of the tensile strength of fiber-reinforced plastics) approaches the levels of light alloys. However, this may often be compensated for by comparatively low weight (densities of 0.9–1.8 g/cm³). One advantage of attached body parts is their resistance to chemicals and corrosive environmental conditions. In evaluating mechanical properties of thermoplastics and thermo-setting plastics, it should be noted that these are in large part temperature-dependent, so that along with strength properties, deformation behavior as a function of temperature and relative humidity must be considered. A measure of thermal form stability is provided by the Martens or Vikat number or shear modulus as a function of temperature and material damping.

Thermoplastics offer basic advantages in volume-production injection molding technology applications for exterior body component applications in body color, such as:

— Bumpers
— Fenders
— Rocker panels
— Radiator grille
— Aerodynamic aids (such as spoilers)

Additional potential applications without significant demands for component surface quality include:

— Inner fender liners
— Engine compartment noise shrouding
— Underbody shrouding as an aerodynamic aid and corrosion protection

Polyurethane materials have similar importance for the above applications and, under consideration of production volume demands, for exterior body applications.

Thermoplastics compete with certain duromers, which have similar or better useful properties in terms of

— Thermal stability
— Thermal expansion similar to steel, aluminum, or magnesium
— Elastic modulus as a function of temperature

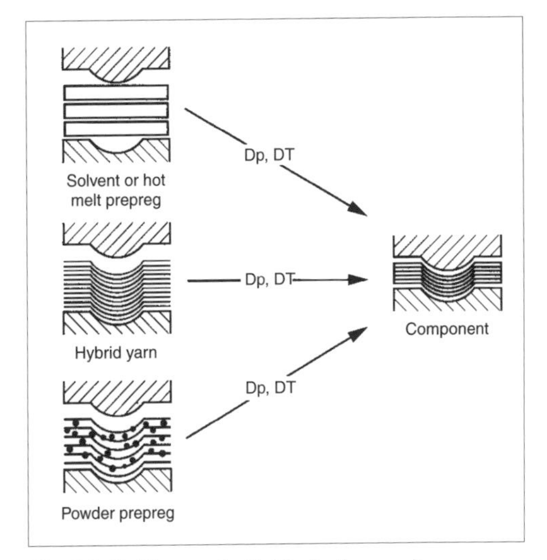

Fig. 6.2-18 Thermoplastic blanks (prepreg).

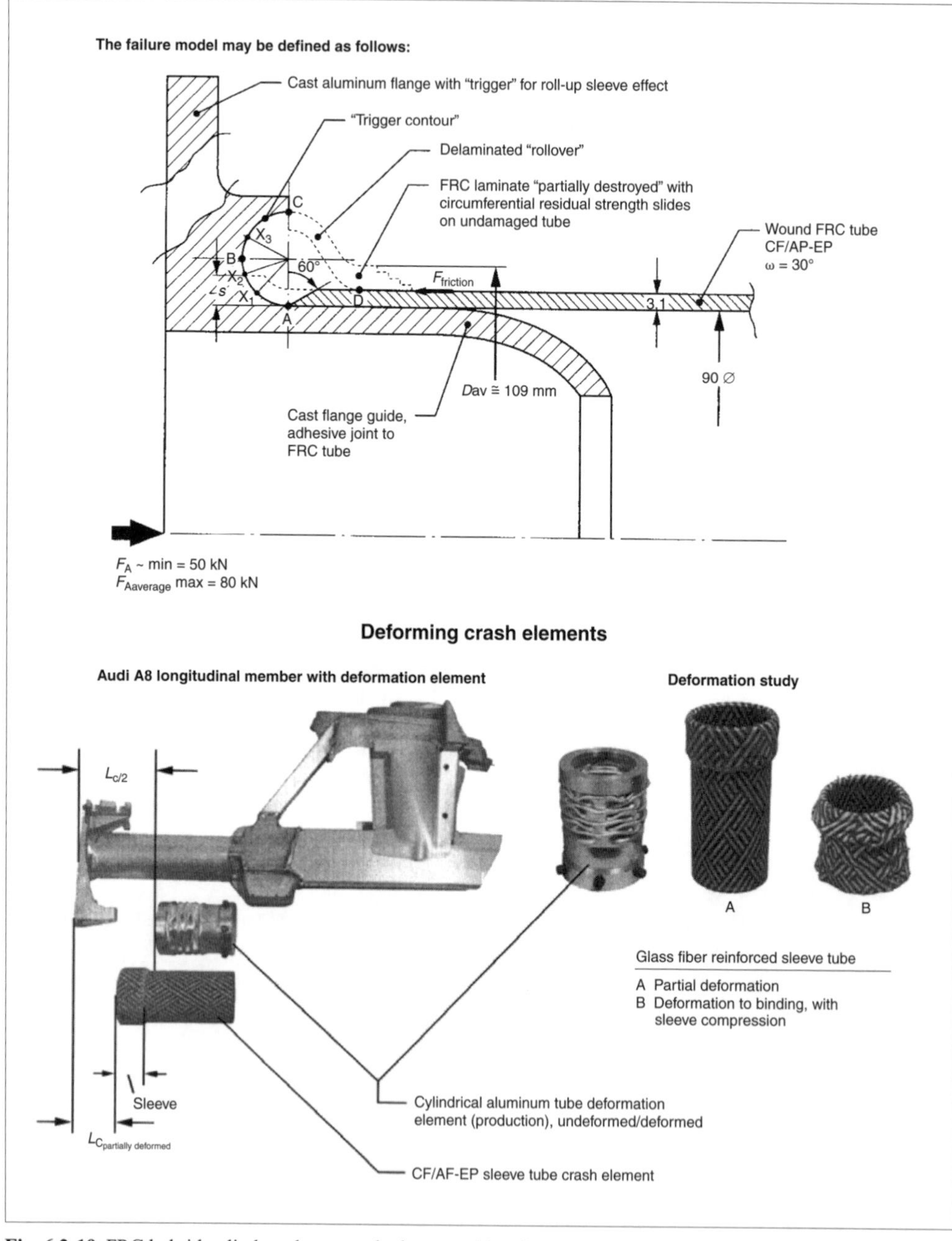

Fig. 6.2-19 FRC hybrid cylinder tube as crash element with roll-up sleeve effect.

Consequently, in small volume (<250 per day) auto body construction, unsaturated glass-fiber–reinforced polyester resins in the form of so-called sheet molding compound (SMC) or bulk molding compound (BMC) have proven to be a cost-effective alternative to metals for structural body parts (spare tire well with integral rear floorpan) as well as automobile body parts such as:

- Front hood (single piece) with reinforcing ribs
- Front hood (two-piece) with bonded halfshell inner part
- Rear lid (two-piece) with integral aerodynamic aids (spoiler)
- Fenders
- Sunroof cover (single piece) with reinforcing ribs

Moreover, SMC and BMC have also been used cost-effectively in bus and commercial vehicle parts including:

— Cab front components
— Side cladding
— Cargo compartment doors
— Rear decklid and engine lids
— Bumper beams

Application of SMC or BMC with high-grade exterior skins (class A surface) and the opportunity to apply in-mold or powder mold coat (IMC/PMC) on the visible surface offer advantages for quality improvement. Along with the described SMC varieties with about 25 weight percent fiber content, special SMC types with greater glass fiber content of 50 to 70 mass percent or with partially oriented fiber strands are also employed. With targeted inserts applied during the molding process, these permit production of parts with higher strength and stiffness, either throughout the part or in localized areas.

Thermoplastic prepregs offer an alternative material for the described body component applications and extend beyond these (e.g., engine encapsulation, underfloor protection, bumper beam, etc.). For glass-fiber-mat–reinforced thermoplastics (GMT), distinction is made between:

— Solvent or hot melt prepreg
— Hybrid yarn prepreg (commingled yarn)
— Powder prepreg

The different types of semifinished blanks used to manufacture high-performance thermoplastic matrix composite components are shown in Fig. 6.2-18.

On the other hand, continuous-fiber–reinforced high performance composites (FRC, or fiber-reinforced composites), commonly employed in aerospace and auto racing applications, may be regarded as less problematic in primary body structures (Fig. 6.2-19 and 6.2-20). This also applies to shear-reinforcing body components, which, as layered composites of very stiff carbon fibers and impact-resistant aramid fibers, offer new possibilities for functionally integrative lightweight construction. Shear-reinforcing body components include:

— Firewall
— Front floorpan
— Rear floorpan
— Rear bulkhead

Due to the anisotropic, nonlinear behavior of these materials, fiber-composite–appropriate computational methods must be employed. Deformation behavior of layered fiber composites is shown in Fig. 6.2-21. A precondition for application of computational methods to layered composites is knowledge of primary and secondary material characteristics as a function of environmental conditions.

Resin transfer molding (RTM) also presents application opportunities for visually high-quality exterior body panels in high-performance fiber composites.

In order to improve instability behavior of hollow sections or local deformation behavior of body structural components, plastic foams with high potential for lightweight construction offer good preconditions for design solutions (Fig. 6.2-22).

However, polyurethane sandwich systems with glass mat reinforcement and metal inserts for local force input or transmission offer solution possibilities for body components with high integration potential and light weight (Fig. 6.2-23).

Because the sandwich must be regarded as a multi-

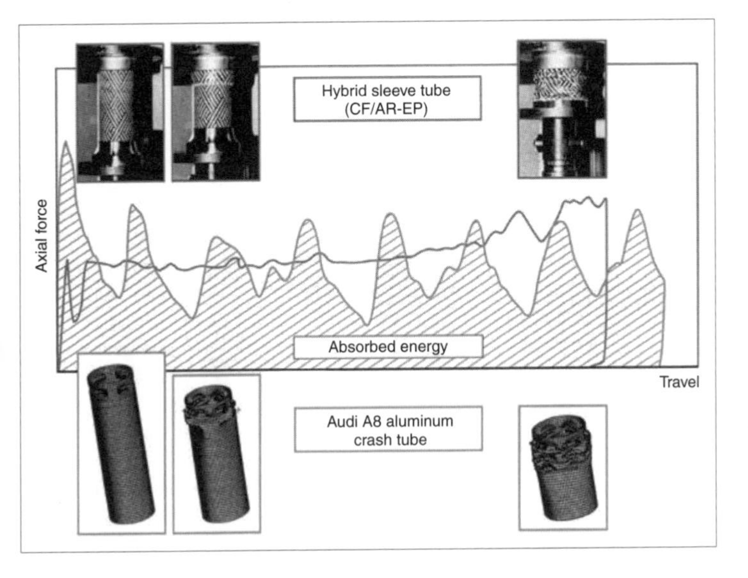

Fig. 6.2-20 Energy absorption comparison: hybrid sleeve tube CF/AR-EP against aluminum tube (identical outside diameter).

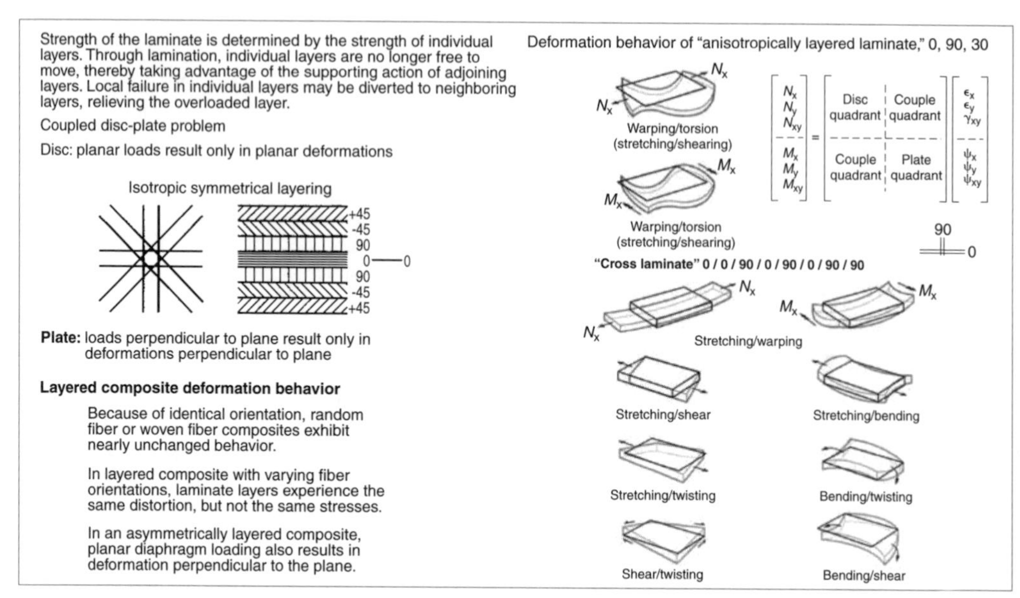

Fig. 6.2-21 Laminated fiber composite deformation behavior.

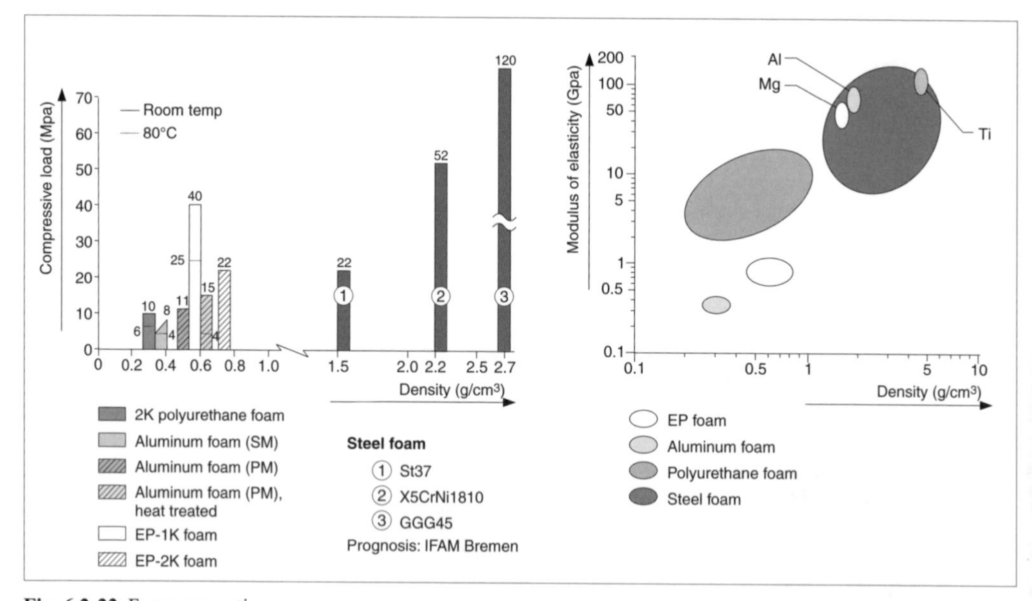

Fig. 6.2-22 Foam properties.

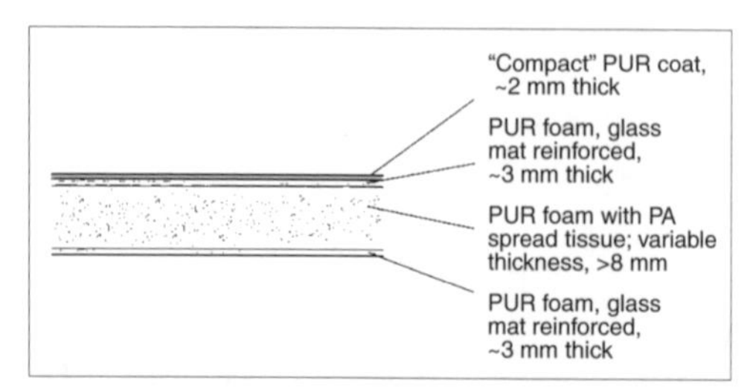

Fig. 6.2-23 Construction of a polyurethane resin (PUR) sandwich system to meet body exterior quality requirements.

layer composite, mechanical behavior of the composite is mainly dependent on properties of the individual layers, the sequence of layers, and the quality of the bond at the boundary layers of the composite. If the PUR (polyurethane resin) sandwich is not built up symmetrically, the so-called "bimetal effect" must be considered (although its effects may be calculated). Polyurethane/glass mat sandwich systems are not only suitable for auto body floor structural elements, but also for convertible hardtops (Fig. 6.2-24), passenger cabins in bus construction and box structures for low-volume commercial vehicles at limited investment cost.

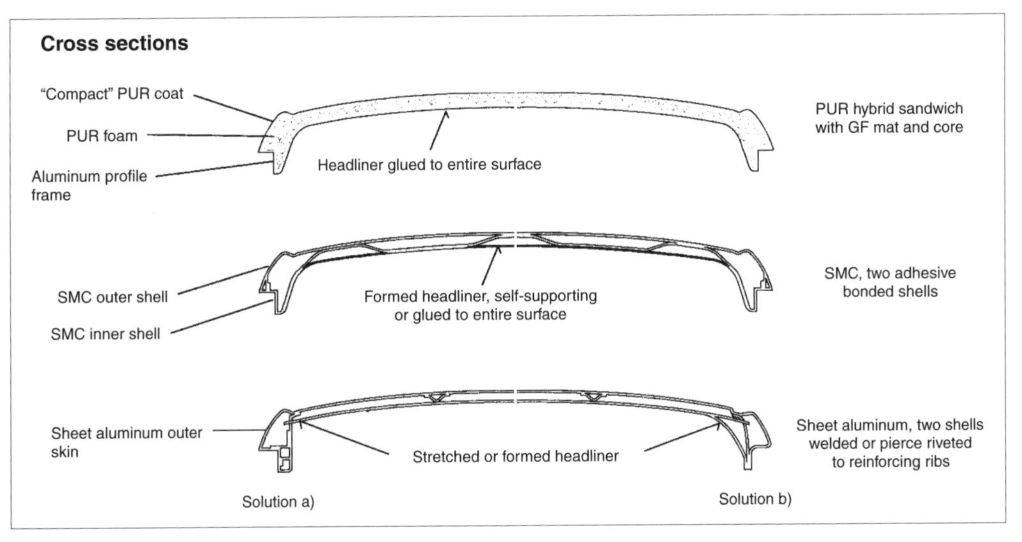

Fig. 6.2-24 Material-specific hardtop concepts.

6.2.3 Multifunctional materials (smart materials) in auto body construction: The future

Multifunctional materials—"smart" materials—possess the ability to carry out sensor well as actuator functions. A distinction is made between active and passive material systems: passive systems react to changes in their environment without assistance, while active systems utilize a feedback system which permits analysis of the sensor signal, recognition of changes, and, by means of a control loop, initiation of an appropriate reaction. Sensor as well as actuator functions within multifunctional materials are made possible by means of changes in macroscopic material properties during phase transformations. These phase transformations may be described by changes in order parameters. In the future, composite materials and combined materials with sensor

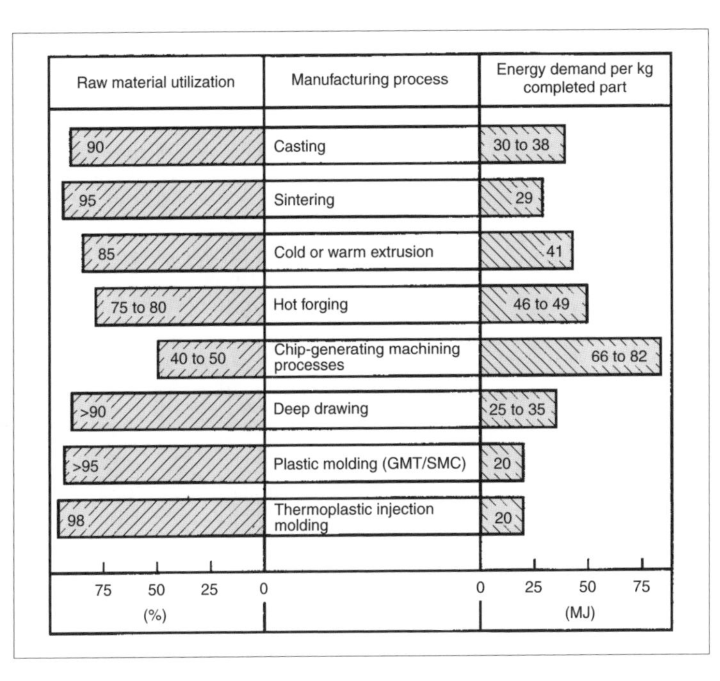

Fig. 6.2-25 Raw material and energy utilization for various manufacturing processes.

and actuator functions will be suitable as structural components for adaptive vibration damping, micropositioning, and geometrical adaptation. Conceivable future applications may include the body structure's firewall.

In principle, materials engineering must be regarded as the key technology for all materials-related engineering development. Moreover, functional materials are available in all material categories, enabling mutual implementation of various physical quantities in such a manner as to permit system-appropriate, vehicle-specific lightweight construction with high functionality for automotive body development (Fig. 6.2-25).

References

[1] Haldenwanger, H.-G. *Hochleistungs-Faserverbund-Werkstoff im Automobilbau, Entwicklung, Berechnung, Prüfung, Einsatz von Bauteilen*. VDI-Verlag, Düsseldorf, 1993.

[2] Haldenwanger, H.-G. "Zum Einsatz alternativer Werkstoffe und Verfahren im konzeptionellen Leichtbau von PKW-Karosserien." Dissertation, TU Dresden, 1997.

[3] Bleicher, W. *Konstruieren mit Aluminium*. Düsseldorf, 1974.

[4] Klein, B. *Leichtbau-Konstruktion, Berechnungsgrundlagen und Gestaltung*, 2nd edition. Vieweg-Verlag, Brunswick, 1994.

[5] Haldenwanger, H.-G. "Neue Werkstoffe und Verbundstrukturen im Automobilbau." VDI-Bericht No. 734, 1989.

[6] Haldenwanger, H.-G., J. Heiss, and H. Reim. "PUR-Hardtop für das AUDI Cabrio." VDI-K Kunststoffe im Automobilbau, VDI-Verlag, 1998.

[7] Haldenwanger, H.-G. "Numerische Simulation des Gießprozesses als Funktion der erzielbaren Festigkeiten." Tagungshandbuch Kolloquium–Westsächsische Hochschule Zwickau, 1997.

[8] Haldenwanger, H.-G., H. Bauer, and W. Schneider. "Recycling von Altfahrzeugen: Kreislaufeignung von Werkstoffen als Funktion der Demontagewirtschaftlichkeit." VDI-K Kunststoffe im Automohilbau, VDI-Verlag. 1997.

6.3 Surface protection

Protection of all automobile surfaces is of vital importance. Occasional encounters with aggressive media found in the environment (acid rain, road salt) as well as intense competition within the automobile industry, conducted in part by means of long-term warranties against rust perforation, have led to development of sophisticated paint and corrosion protection systems. With increased environmental awareness on the part of manufacturers and customers alike, paint processes featuring reduced emissions—for example, water-based finishes—found their way to volume production in about 1990.

The following section is a guide to current techniques for achieving good (which is to say long-term stable) surface protection. Unless explicitly noted otherwise, all of the described systems are based on conventional steel unit bodies. These processes must be adapted and modified for alternative body materials such as aluminum and plastic.

6.3.1 Benefits of surface protection

Surface protection treatments applied to motor vehicles may be divided into three functional groups:

- Protection of the steel body material against corrosion
- Protection of the paint, which represents high value as a design element for customers
- Temporary measures to protect the vehicle during transport from the manufacturer to the customer

6.3.1.1 Corrosion protection

Effective corrosion protection has several positive effects:

- Constant structural integrity throughout the vehicle life
- Long-term value retention and therefore improved economy
- Resource preservation through longer operating life

What causes automobile body corrosion? Most metals occur in nature as oxides, sulfides, or carbonates—in other words, as ores—because these chemical compounds represent the most stable form in terms of energy level. Only through addition of energy, in the smelting process, is metallic iron derived from iron ore. If steel produced from this iron is subjected to environmental influences or aggressive media, the material returns to its lower energy state: it reverts to oxide. In the case of metals, this is known as corrosion.

DIN 50900 defines the concept of corrosion as follows: "Corrosion is the reaction of a metallic material with its environment, causing a measurable change in the material, and may result in compromised function of a metallic component or an entire system. In most cases, the reaction is electrochemical in nature. In some cases, however, it may be chemical or metallophysical in nature."

If a design consisting of metallic components is to remain stable over a long time span, the material-based behavior must be taken into consideration in design and production planning.

Corrosion occurs when electrons released in an anodic reaction (metal dissolution, oxidation) are transported by an electrical conductor (the metal itself) to a cathodic reaction (reduction). An ionic conductor (electrolyte) completes the electrical circuit. In corrosion damage to automobile bodies, the most important electron-consuming reaction is reduction from atmospheric oxygen (Fig. 6.3-1). Along with the material itself, the degree of corrosive attack depends to a great extent on composition of the electrolyte (for example, its pH) and ambient temperature.

As a rule, three types of corrosion can occur on an automobile body: surface corrosion, crevice corrosion, and galvanic corrosion. If the described process takes place on an exposed metallic surface, it is termed surface corrosion.

Crevice corrosion on an auto body can occur, for example, in flanges, crimped joints, and beneath gaskets. It is primarily attributable to the fact that moisture cannot easily escape from crevices. At the same time, introduced salt residues behave hygroscopically: that is, they keep the crevice moist. Because surfaces in these areas are often inaccessible and not always properly coated

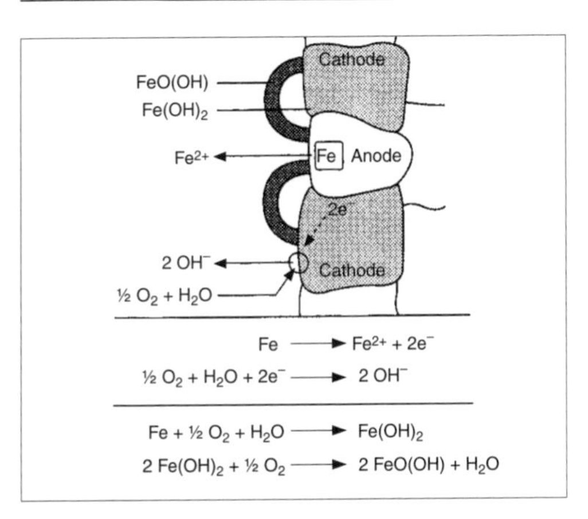

Fig. 6.3-1 Electrochemical corrosion processes.

and protected, any design-mandated crevices must be either relocated to a dry area or sealed.

If two different metals are in electrical contact with one another in the presence of an electrolyte, galvanic corrosion may result. The greater the electrochemical potential difference between the two materials and the smaller the anode in comparison to the cathode, the more rapid the rate of corrosion and resulting damage. To avoid galvanic corrosion, one must either avoid materials with dissimilar electrochemical potentials or galvanically separate the components in question.

6.3.1.2 Surface protection

Automobile paint not only protects the underlying material from environmental effects but also represents a vital design element. Therefore, the exterior paint system, generally consisting of two layers—a color coat and clear coat—must resist the following effects:

- Chemicals (fuels, bird droppings, etc.)
- UV radiation
- Mechanical effects (stone impacts, car washes)

6.3.1.3 Transport protection

Transport protection is designed to protect a new vehicle on its way from manufacturer to dealer or customer from mechanical effects and aggressive environmental influences. For example, wear particles from overhead electrical catenaries or railcar brakes might attack auto bodies during railway transport. In general, waxes and/or adhesive films are employed. Transport protection is primarily differentiated from all other surface protection measures by its limited duration.

6.3.2 Surface protection development and production

The design target catalog requirements for surface protection of the automobile body and its components are, in general, laid down in company internal guidelines. Supplementary to these is a comprehensive structure of DIN/ISO norms covering corrosion protection. Validation of individual vehicle components takes place in experimental work using tests which are also largely standardized. These include, among others, stone impact, adhesion, moist climate, UV, salt spray, and abrasion resistance tests. These are conducted in the laboratory as well as in open-air weathering facilities. Moreover, the newly developed complete vehicle is tested on various test tracks and subjected to corrosion tests.

Auto body surface protection systems generally follow the outline presented in Table 6.3-1.

6.3.2.1 Metal surface precoating

Today, galvanized or zinc-coated steel sheetmetal is widely used in auto body construction. The zinc layer is a highly effective corrosion barrier. Above all, its effectiveness is developed if the paint layer is damaged. At that point, zinc, by virtue of its low electrochemical potential, acts as a galvanic element. In the presence of an electrolyte, at first only the zinc layer corrodes. As a rule, auto body sheetmetal is galvanized by the steel manufacturer. Various types of precoating may be applied. Along with electrolytic (electroplated) galvanizing and hot dip galvanizing, "galvannealed" sheets have their zinc layer converted to a zinc-iron alloy by thermal treatment after hot dip galvanizing. In the case of duplex coatings, a galvanized sheet is additionally coated with organic materials. In addition to electrolytically deposited pure zinc, some auto manufacturers employ zinc-nickel alloys with 11% nickel content, which exhibit better barrier action.

Table 6.3-1 Buildup of auto body surface protection coatings

Layer	Function	Thickness in μm
Paint color coat	Color and protection	40–70
Primer coat	Provides a smooth surface	30–35
Cataphoretic dip primer	Corrosion protection, especially in body cavities	18–35 (exterior), at least 10 (in cavities)
Phosphating	Corrosion protection and paint adhesion	1.5
Zinc (electrolytically or hot dip applied; zinc alloys (electrolytically applied)	Corrosion protection	8
Steel sheet metal	Mechanical strength	

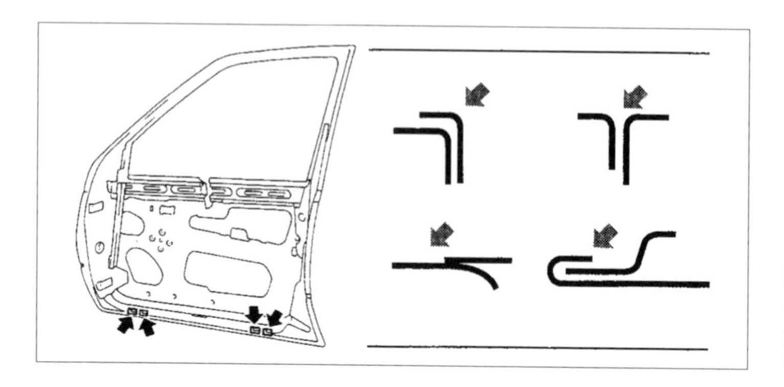

Fig. 6.3-2 Location of vent openings and design possibilities for corrosion-optimized joints.

6.3.2.2 Anticorrosion measures in auto body construction

Surface protection and anticorrosion measures may be positively influenced during the body design phase. This begins with material selection. Additionally, suitable material pairings or avoidance of corresponding points of contact may preclude galvanic corrosion from the outset. As another example, body cavities, subjected to moisture, may be deliberately designed to provide good ventilation. Such cavities may be found in doors, door sills, and structural members. Ventilation openings also provide good access for paint during application of cataphoretic dip primer.

From a design standpoint, all components should provide good water drainage. In joining two sheetmetal parts, sharp edges or transitions should be avoided (Fig. 6.3-2). In addition, the designer must determine whether cavities should, for example, be protected by wax coatings or whether joints are to be sealed by caulking and adhesives.

The watchword in any case is process-appropriate vehicle design. In this, certain guidelines must be observed, which affect, for example, handling of the body-in-white during the production process (fixturing holes for the product conveying system). Depending on the choice of manufacturing method, openings must be provided in the body to permit entry and drainage of process materials (Fig. 6.3-3). Flanges, crimps, and lap joints must be configured to permit easy and process-assured application of adhesives and sealants.

6.3.2.3 Production measures

Fig. 6.3-4 illustrates typical stations in the automobile assembly process. These may vary in detail from manufacturer to manufacturer as a result of differing product lines and differing manufacturing processes which evolved over time.

6.3.2.3.1 Bonding and sealing

Welded and crimped flanges, unavoidable from a construction standpoint, must be sealed against entry of water, corrosive media, and atmospheric oxygen by means

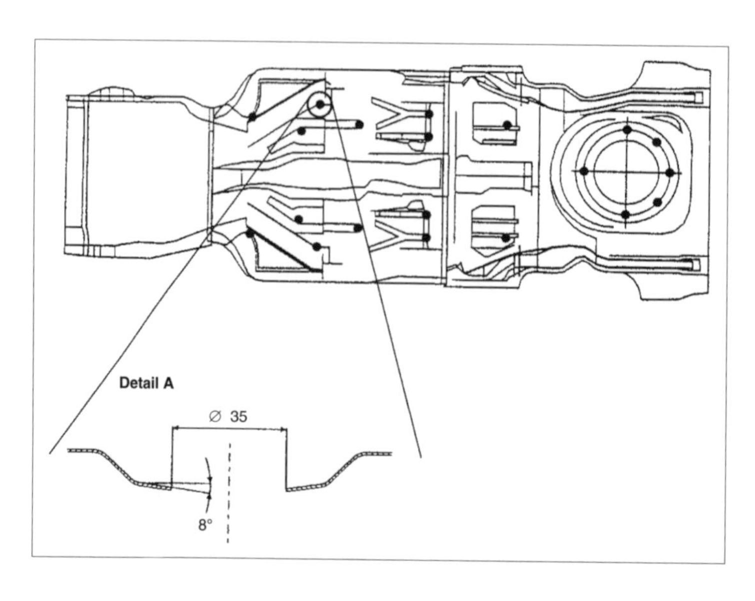

Fig. 6.3-3 Location and design of floorpan drain holes.

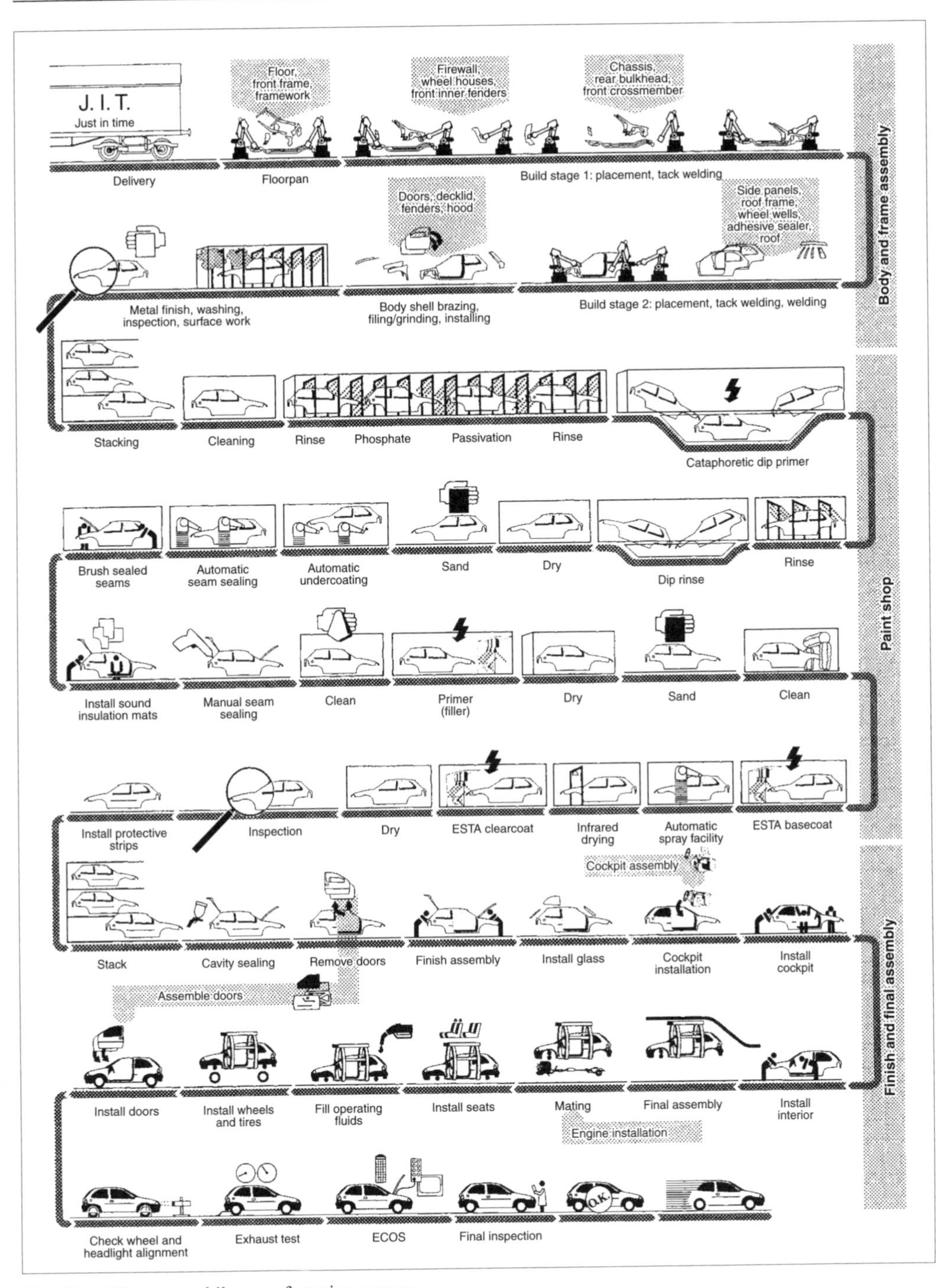

Fig. 6.3-4 The automobile manufacturing process.

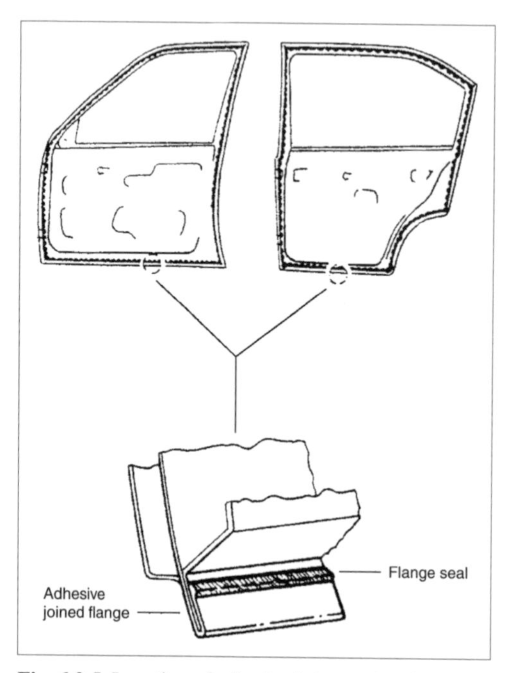

Fig. 6.3-5 Location of adhesive joints and seals.

of adhesive or sealant (Fig. 6.3-5). Application of adhesive, a bonding process, is enjoying increasing significance within the automotive industry. Its advantage over a sealant is that adhesive bonding both bonds and seals the joint. Given process-appropriate design, adhesive bonds may meet or even exceed the strength of spot welds.

The necessary sealing operations are carried out either in body assembly or in the paint shop. All areas of the body which will be inaccessible after assembly must be sealed during the body assembly operation (e.g., hidden crimped flanges or reinforcing ribs required under exterior panels). In this processing sequence, establishment of body assembly cycle times must take into consideration the necessary pre- and post-treatment times for adhesive bonding. If sealing operations are conducted in the paint shop, material application normally takes place on degreased, cataphoretically dip-primed surface. As curing for adhesive and paint takes place simultaneously, no additional hardening process is required.

Table 6.3-2 lists various sealing materials. A decision for application of a specific product should always take into consideration the desired rigidity or elasticity of the finished assembly.

6.3.2.3.2 Pretreatment

Prior to the actual paint process, exterior body surfaces are subjected to pretreatment. This serves as corrosion protection and promotes improved paint adhesion. The standard process for all-steel bodywork is cleaning/degreasing, rinsing, activation, phosphating, rinsing, passivation, and multiple rinses with deionized water.

The body shell delivered by the body assembly department retains traces of anticorrosion oils, lubricants, grinding dust, and other contaminants, which must be removed to assure paint process quality. Cleaning and degreasing of the body is carried out by means of aqueous alkali and tenside solutions. Subsequent rinsing removes any remaining traces of these cleaning solutions. Activation of the surface by means of titanium phosphate prior to application of the actual phosphate coating is intended to speed up the phosphating process, reduce the weight of the layer, and therefore provide especially fine crystalline layers.

Phosphating is carried out with a solution consisting mainly of zinc, nickel, and manganese ions as well as phosphoric acid and an accelerator (an oxidizer such as nitrite, peroxide, or hydroxyl amine). Additionally, for treatment of steel/aluminum bodies, free fluoride is required. The acidic phosphate solution works as an etchant. During the etching reaction, iron (II) and zinc ions as well as atomic hydrogen are formed. The accelerant depolarizes the hydrogen, thereby freeing the metallic surface of hydrogen bubbles which might disturb the phosphating reaction. As a consequence of the etching reaction, acid concentration at the metallic surface decreases sharply; therefore, a layer of less soluble zinc or zinc/iron phosphate is formed. Its layer weight lies between 1.5 and 5 g/m^2. The resulting iron (II) phosphate is converted to less soluble iron (III) phosphate by the oxidizing agent and removed as sludge. Any sludge particles remaining on the body are removed by a subsequent rinse.

Any remaining pores in the phosphate layer are sealed by passivation with an aqueous solution of hexafluorozirconic acid or a postrinse solution containing chromium. Finally, any remaining water-soluble salts are rinsed away with deionized water.

Table 6.3-2 Auto body sealing materials and adhesives

Material	Main application	Strength	Elasticity
Epoxy resin	Structural adhesive joints on body-in-white, bonding crimped joints	++++	+
Rubber-based products: hot butyl	Padding adhesive	+++	++
PVC plastisols	Sealing in paint shop	++	+++
Acrylic plastisols	Sealing body-in-white	+	++++

Phosphate application is carried out either as a spray or dip process or as a combination of both.

6.3.2.3.3 Electrophoretic dip primer

By virtue of its great economic benefits and very consistent layer formation, electrophoretic ("electrocoat" or "E-coat") dip priming has become an established means of corrosion protection within the auto industry. In this process, the object to be primed is immersed in a paint/water bath containing between 15 and 25% dispersed paint particles. Application of direct current between the object and bath causes paint particles to wander toward the object surface and be deposited there. Distinction is made between anodic and cathodic electrocoat dip priming. Due to its better anticorrosion properties, cataphoretic dip priming, in which the auto body serves as the cathode (Fig. 6.3-6), has become the more widely used process.

Electrochemical deposition typically takes place within 2.5–4 min. For larger production volumes, the dip tank is usually configured as a continuous-flow facility, with between five and seven rectifier fields and various voltages. Depending on paint type and process, coating thicknesses on exterior parts of 18–25 μm are targeted, while cavities must have coating layers at least 10 μm thick.

After the cataphoretic primer bath, a cascade rinse process removes any paint particles which are not electrochemically bound. To reduce water consumption and recover unused primer, an ultrafiltrate, made from dip primer by an ultrafiltration unit, is used as a rinsing medium. The auto body cascade rinse counterflows back "upstream" along the process direction to conserve deionized water. Rinse and paint remnants are returned to the dip basin. After a final rinse with deionized water, the primer adhering to the body is baked on for about 20 minutes at 170–180°C. Simultaneously, this drying process may be used to harden bake-hardening steel, which reaches its ultimate strength only after heat treatment, as well as for curing of adhesives and sealants.

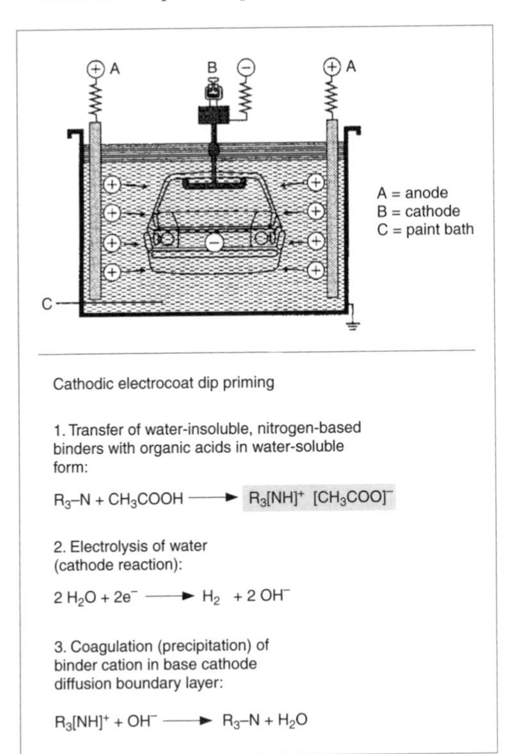

Fig. 6.3-6 Cataphoretic dip primer: schematic and reaction equations.

6.3.2.3.4 Primer and color top coat

Exterior surface protection and body color are provided by wet sprayed paint coats or in individual cases by powder paint. Primer, also known as primer/filler or primer/surfacer, provides a smooth substrate for the color coat, covering any rough or uneven spots remaining from previous process steps. Additionally, it provides grip for the color coat. Along with single-color—usually gray—primers, colored or even body-color-specific primers are used. Colored primers have an advantage in that the thinner color coats can be applied (cost advantage), and minor mechanical damage incurred during vehicle service is less apparent.

As a rule, color coats employ two-coat systems: a base coat for color and a clear coat. Base coats are differentiated into solid and effect colors (metallic, pearlescent). In addition to color pigments, effect colors also contain aluminum or other reflective particles which provide different color or glitter effects depending on illumination. In developing colors, it is especially important to ensure that the chosen pigments are stable under long-term light exposure and that they are completely embedded in the binder; otherwise, color shifts may become apparent or pigments may collect at the paint/primer boundary and lead to adhesion problems—for example, as a result of continuous exposure to sunlight.

The clear coat is applied as a protective layer to smooth the finish and achieve a gloss effect. Clear coat chemical makeup must be such that it resists yellowing even under extreme UV exposure. Furthermore, it must protect against scratching, as may be encountered in automatic car washes, and it must withstand organic and inorganic substances such as fuel, bird droppings, tree sap, and acid rain.

The complete system, consisting of primer, color coat, and clear coat, must also protect underlying layers against penetration by moisture and UV radiation. The cataphoretic dip primer in particular contains epoxy resins which could be destroyed by ultraviolet light. The result would be loss of primer adhesion followed by peeling of the entire paint structure. An additional requirement is resistance against mechanical factors such as stone impact; this may be achieved by good adhesion of individual layers and corresponding flexibility.

Wet paints consist of volatile and nonvolatile components. The volatile components, organic or aqueous-organic solvents, serve to facilitate the paint application process. After curing (drying), the nonvolatile components form the actual paint film. The included binder, usually consisting of alkyd, acrylic, or melamine resins, forms the paint layer. Color pigments are embedded in the binder, as are various additives including wetting, flowing, matting, and antiseparation agents; plasticizers; film formation agents; and fillers.

Paint is applied in spray booths. Whenever possible, these are located in a separate area of the building (clean room area). This separation addresses the process sensitivity of paint materials to dirt or other contaminants (such as silicon). During application, close tolerances for ambient temperature and humidity must be maintained in order to prevent shifts in paint shade. For water-based paints, for example, these may be 23°C air temperature (±3°C) and 65% relative humidity (±5%).

Although interior painting is generally carried out by robots or manual (human) painters, exterior painting is done by robots and, above all, by automatic spraying machines. Paint is usually applied by means of air atomized spray guns or high speed electrostatic rotary atomizers (ESTA). The painted body is dried in a tunnel oven by radiant heat, ambient air, or a combination of both. Drying time is generally about 30 minutes at temperatures of about 165°C for primers and 140°C for color coats. The paint system cures during this drying process; that is, molecular chains of the binding agent react and intertwine under the influence of elevated temperatures at previously masked areas. The resulting macromolecules then form a stable union. Pigments and additive agents mentioned earlier are trapped in this molecular network and influence its properties—for example, in terms of hardness, sandability, flow-out, and gloss.

In addition to conventional wet paints, powder paints or powder slurry may be applied. At present, however, these do not enjoy a significant market share in the automobile industry. Powder paints lack any solvent; paint particles not adhering to the workpiece (overspray) may be returned directly to the process. Frequent color changes and low production volumes burden the economic viability of this process. Retouching is generally only possible with wet paints, which requires additional painting facilities. For this reason, powder paints are primarily applied as single-color primers or clear coats. Powder slurry paints represent a combination of both processes. In these, the paint manufacturer provides a powder paint combined with an aqueous solvent. The paint can be applied in existing, modified wet paint facilities. Like powder paint, this technology uses virtually no organic solvents.

6.3.2.4 Body cavity sealing and undercoating

Due to severe conditions in the affected areas, effective corrosion protection of cavities and underbody requires not only special sealing operations but also additional measures such as cavity sealing by means of waxes. This generally takes place immediately after painting. Undercoating is applied in part in the paint shop and in part at the end of the production process.

Table 6.3-3 Body cavity sealing waxes

Material	Solid body content, %
Hot flooding wax	100
Hot spray wax	100
Solvent-free spray wax	>99
Solvent-based spray wax	>70
Water-based spray wax	>40

6.3.2.4.1 Body cavity sealing

Cavity sealing by means of wax serves to seal crevices and flanges, thereby combating crack corrosion. Table 6.3-3 outlines the types of wax employed for this operation.

Penetration behavior of waxes is decisive in determining sealing effectiveness in narrow crevices. This may be tested by the following simple method: two rectangular test plates, separated by a specified distance, are stacked horizontally at room temperature and loaded by a specified weight. One edge is sealed by application of excess wax. After the specified test period, the plates are separated and the wax penetration distance measured.

Wax application methods must be taken into account as early as the vehicle design phase. Cavity sealing may be accomplished by hot flooding or spray processes. In the hot flooding process, the area of the body to be sealed is heated to temperatures between 60 and 70°C, and the cavities filled with wax at a temperature of about 120°C. Some of the wax solidifies and adheres to the cooler auto body sheetmetal while the remainder drains away. For this process, cavities must be designed to allow assured, defined drainage. Air pockets at the top of profile sections must be avoided.

In contrast to the hot flooding method, spray wax application takes place at room temperature, in which both object and material temperature should be within 10°C of each other. Wax is sprayed at high pressure (between 2,000 and 12,000 kPa), with or without air mixing, into body cavities. From a design standpoint, it is vital that sufficient openings be provided for material application.

6.3.2.4.2 Undercoating

The vehicle underbody is subjected to severe mechanical loading in the form of stone impacts. Furthermore, the underbody is highly susceptible to corrosion from moisture and road salt. To protect against these, materials based on asphalt, PVC, and polyurethane have proven themselves. Areas not to be protected by these materials are sealed with wax (Fig. 6.3-7). This treatment simultaneously protects suspension components in these areas.

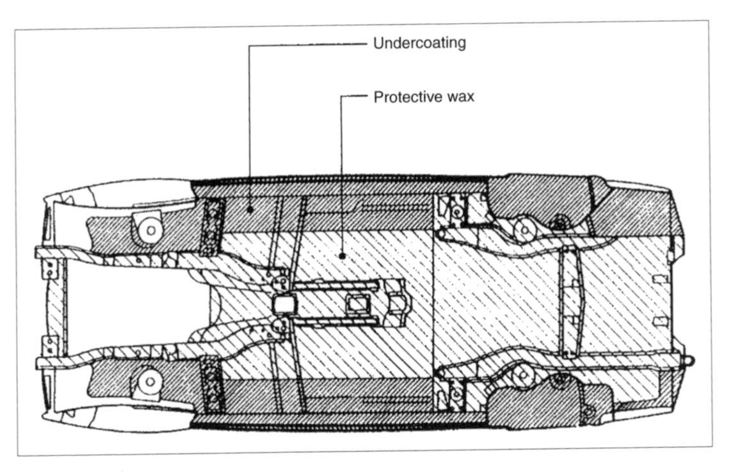

Fig. 6.3-7 Application example of undercoating materials and protective wax.

With the exception of hot wax, the same materials may be applied as are used for cavity sealing. In the underbody area, the waxes used must exhibit high resistance to being washed off rather than penetrating ability.

For this application, any of the usual spray methods may be employed. Automated facilities must have model-specific controls, as not all areas of the underbody should be coated. For example, zones near the exhaust system, including the catalytic converter, should be kept free of coatings.

Table 6.3-4 Transport protection

System	Implementation
Waxes (solvent- or water-based)	Currently in use
Acrylic systems	Currently in use
Films	Currently in use
Sprayable films	Experimental
Transport hoods	Experimental

6.3.2.5 Transport protection

In taking delivery of a new vehicle, the customer demands a qualitatively and visually faultless vehicle. To this end, additional transport protection is applied in manufacturing. This protects the fresh paint against harmful effects for a period of up to six months. Table 6.3-4 lists available transport protection methods and materials.

6.3.3 Prognosis

Auto body surface protection has achieved a high technical standard, as evidenced by the ever more comprehensive rust-through warranties offered by the industry. In further development of these processes, manufacturers of automobiles, production facilities, and paints have in recent years introduced increasingly environmentally compatible products and processes into production. These mutual efforts have shown dramatic results. Above all, application of water-based paints has

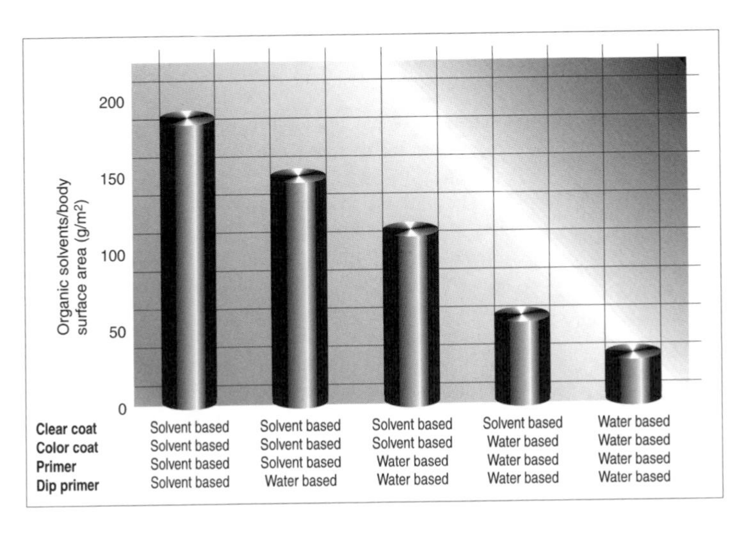

Fig. 6.3-8 Reduction of organic solvent emissions.

resulted in a considerable reduction of surface-related solvent emissions (Fig. 6.3-8). If the application spectrum of powder paints should improve, further emissions reductions are possible.

Other examples of steps toward environmentally compatible processes are conversion to lead-free electrocoating of primers as well as chromate-free rinse solutions. To reduce water consumption, paint processes are increasingly using closed-loop systems.

The drive toward lightweight body construction will see increased use of hybrid (mixed material) metal automobile bodies, particularly with attached parts of aluminum or plastic. Such a materials mix demands adaptation of all processes.

References

DIN norm handbooks for paint and related coating materials, pigments, and fillers (Nos. 30, 49, 117, 143, 168, 195, 201, 232, 278). Beuth, Berlin.

Jahrbuch für Lackierbetriebe, Vinzentz. Appears annually.

Müller, Klaus-Peter. *Praktische Oberflächentechnik: Vorbehandeln—Beschichten—Prüfen*. Vieweg, Wiesbaden, 1998.

Prüftechnik bei Lackherstellung und Lackverarbeitung. Vinzentz, 1996.

Trade Journals

Adhäsion-kleben und dichten, Verlag Vieweg.

Farbe & Lack, Vinzentz Verlag.

Galvanotechnik, Eugen G. Leutze Verlag.

Industrie Lackierbetrieb, Vinzentz Verlag.

JOT—Journal für Oberflächentechnik, Verlag Vieweg (September issues are dedicated to the subject of automobile paints).

mo—Metalloberfläche, Beschichten von Metall und Kunststoff, Carl Hanser Verlag.

6.4 Vehicle interior

6.4.1 Ergonomics and comfort

If one had to select an automotive term which has enjoyed frequent exposure in both general usage and specialist media over the past decade, one could hardly overlook the ever-popular "ergonomics."

In our present civilized world, ergonomics appear in many different forms. Following are only the most important fields of application:

- General ergonomics in technical systems such as mechanical engineering and road vehicle, aircraft, and marine vessel engineering.
- So-called micro-ergonomics: in other words, proper arrangement and use of a workplace and its environs, including operation of doors and lids, robots, etc.
- The multitude of different consumer goods which we encounter in our daily lives: their correct, safe, and comprehensible operation is also included under "ergonomics."

In principle, ergonomics is the study of people and their work. The objective of ergonomics is to optimally match labor and the work environment to human beings. This chapter will concern itself primarily with the ergonomics of motor vehicles.

6.4.1.1 Ergonomic demands on the total vehicle

Today, the automotive industry incorporates wide-ranging application of ergonomic methodologies as well as the ensuing demands (Fig. 6.4-1). In automobile construction, primary applications of ergonomics are spatial geometry and the man/machine interface.

Fig 6.4-2 illustrates the *circle of optimum adaptation*. This concept can be described as follows:

It is the objective of spatial geometry to meet the objective and subjective expectations of the customer. The design goal of man/machine interaction is optimum workload relief for the driver (neither excessive nor insufficient demands).

In automobile development, the package phase is decisive in establishing "proper" spatial ambiance for both driver and passengers. The "package" (Section 4.2) encompasses all relevant vehicle components, which are "sketched in" to fit within the given exterior dimensions. The vehicle occupants themselves represent critical components. It is fundamentally important to establish from the outset how many and which persons will "fit into" the new vehicle. It is obvious that a five-seat sedan must accommodate five persons plus luggage. In a sports coupe, however, it must be decided whether fully usable rear seats should be included or whether a 2+2 concept applies (which accommodates children, or, for example, 50th percentile persons, in rear seats).

During this phase of establishing "dimensional chains," it is vitally important to ergonomics that nearly the entire life cycle of the automobile be taken into consideration. As a rule, the following breakdown applies to all volume-production vehicles:

Development time (present state of the art)	3–4 years
Production life	5–8 years
Service life (first and second owner)	5–10 years

In other words, one could say that including development time, the life of an automobile covers from about 15 to nearly 25 years.

The increase in average size of central Europeans has, at least historically, been about 1.5 cm per decade. This means that in their spatial concept, ergonomic engineers must take into consideration the fact that in 15 years, occupants will probably be taller than at present.

Into the 1980s, interior space layout was accomplished with two-dimensional templates (in side view, not in width). Since the 1950s, an SAE template has been available which even today, in conjunction with the H-point machine in significantly revised form, establishes the basis for a multitude of homologation validation requirements. The SAE template consists of a

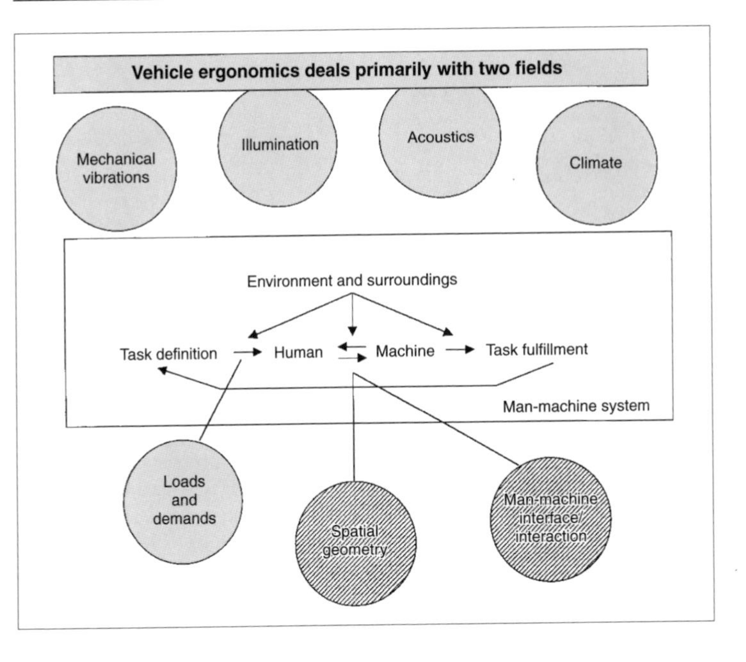

Fig. 6.4-1 Ergonomic methods and challenges.

torso in side view with movable legs (Fig. 6.4-3). It may be supplemented by head contour locator line and eye ellipse templates (the latter are included in side and plan views).

The so-called Kiel dummy (DIN 33408 body template) has been available since the early 1970s. This template shows a human model in side view, with movable arm, leg, head, and torso. Joint movement is modeled on human movement ranges: that is, connections on the model equipped with scales permit only those motions corresponding to actual human joint movement.

Templates are available for different percentiles (body size) as well as for male and female bodies. They are available in full scale as well as in smaller scales.

All of these templates suffer from three drawbacks:

— These models (with the exception of the eye ellipse) exist only in side view form.
— In contrast to the actual human population, these templates do not grow.
— They do not include any information as to which settings represent a natural human posture—which, however, does not always represent a comfortable position.

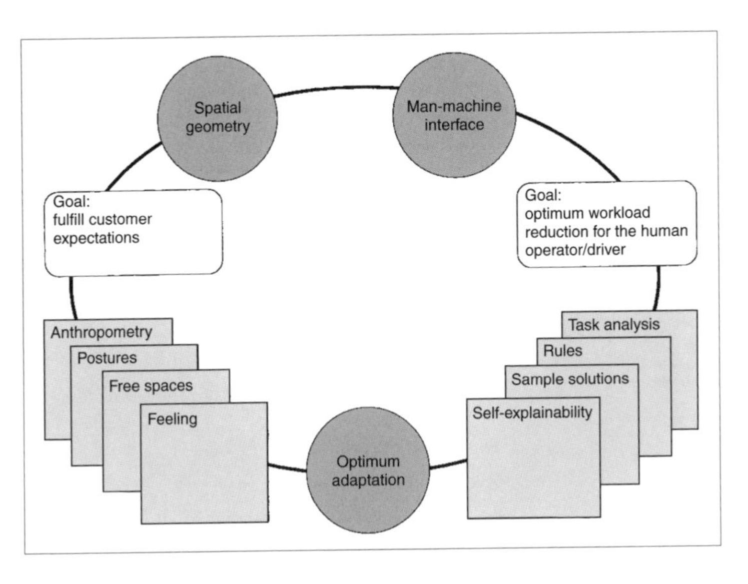

Fig. 6.4-2 The ergonomic process.

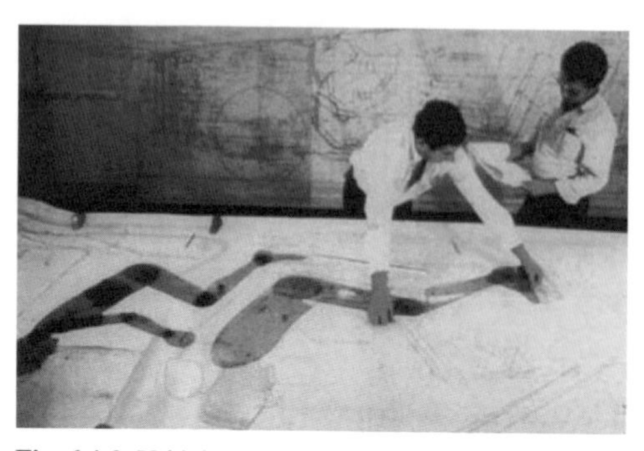

Fig. 6.4-3 Vehicle layout using SAE H-point template.

With the introduction of CAD techniques, disadvantages of two-dimensional templates became especially apparent.

The driver workplace

The most important interface between driver and vehicle is the seat, followed by components which impart information (instruments and indicators) and vehicle controls (Fig. 6.4-4).

Above all, the seat is of vital importance in performing its functions of supporting, holding, and positioning the driver. As a rule, seat adjustment is determined individually by the driver himself, but often the spatial arrangement of indicators and controls mandates a different seat position, which may result in an uncomfortable, forced pose and therefore lead to muscular strain. This must be avoided if at all possible.

The driver should be able to sit in a comfortable basic position with sufficient freedom of movement by means of seat adjustment to achieve optimum (orthopedic) body posture. In the technical jargon, this is referred to as the "seat adjustment envelope."

This range of seat adjustment, established by vehicle engineers in the package, shows possible seat settings for reach and height. The so-called seating reference point (SgRP, also known as H-point or hip point) of a small-stature so-called fifth percentile female would occupy the upper left of the envelope, while the pelvis of a large male (95th percentile) would fall at the lower right (Fig. 6.4-5).

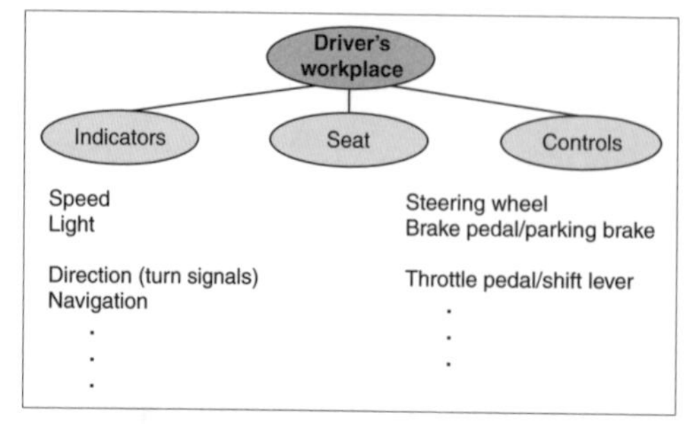

Fig. 6.4-4 The driver's workplace.

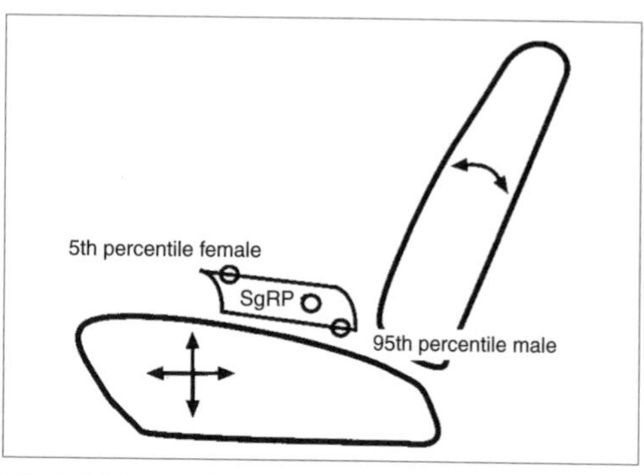

Fig. 6.4-5 Seat adjustment range (SgRP = seating reference point).

Throughout this range of adjustment, it is important that every possible setting still provide optimum view of instruments, indicators, and controls as well as accessibility of storage compartments. Therefore, most modern midrange vehicles offer adjustable steering wheels and/or steering columns to achieve optimum driver positioning.

The seat itself is provided with a wide range of adjustment functions:

SLV	Seat reach
SHV	Seat cushion height
SNV	Seat cushion angle
STV	Seat cushion length adjustment
SBV	Seat cushion width adjustment
LNV	Backrest rake adjustment
LTV	Lumbar support thickness adjustment
LHV	Lumbar support height adjustment
LKV	Backrest shoulder support
KHV	Head restraint height adjustment
KNV	Head restraint angle adjustment

Even with all of these available adjustments, improvements are being made on a continuous basis. Today, for example, special "climate" or "vibration" seats are available.

In shaping the "seating surface," it is imperative that ergonomic aspects be considered, such as, for example:

Backrest design:

- Lateral positioning of pelvis
- Lateral positioning of torso
- Freedom of motion for elbow (gear shifting, parking brake actuation for small and large drivers)

Seating surface:

- Improved egress by means of relieved padding in midthigh area
- Seat width adjustment

Overall seat adjustability for:

- Good access to and visibility of all indicators and controls

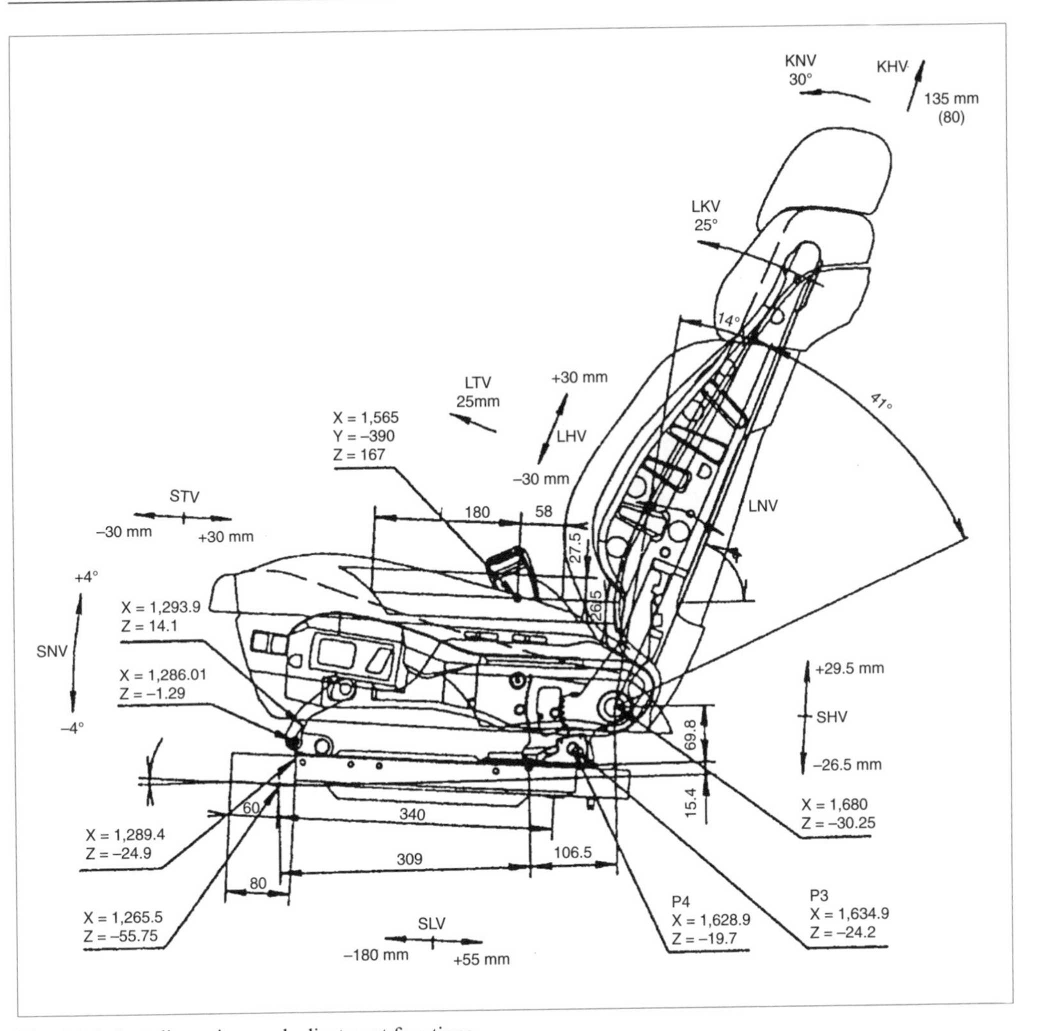

Fig. 6.4-6 Seat dimensions and adjustment functions.

— Good differentiability and clarity of operating elements

6.4.1.2 Basic ergonomic layout

Introduction of CAD technology—and therefore three-dimensional representation—to the automobile development process has not only accelerated the early definition phases of a package layout but also made possible early and accurate predictions of design and layout implications.

In our modern industrial society, time has become a decisive competitive factor. Increased complexity within the automobile (and motorcycle) development process means that more simultaneous design targets need to be met sooner and faster during the layout phase.

With introduction and proliferation of CAD programs in design departments, it was therefore logical to represent human models in CAD form and offer these to development engineers. There are a wide variety of offerings in this sector, with added themes such as crash simulation (e.g., Madymo) and climate perception.

Parallel to these, an even greater number of general human models was developed for workplace layout. All of these computational models are intended to more or less accurately represent the human body in specific application cases; they usually provide a three-dimensional representation.

Within the automobile industry, the RAMSIS human model, developed by German auto manufacturers, has established itself as a standard. The current state of the art of the RAMSIS model permits static position and motion investigations of the legs and hand-arm system inside vehicle concepts. Furthermore, it enables prediction of outward visibility and reachability of control el-

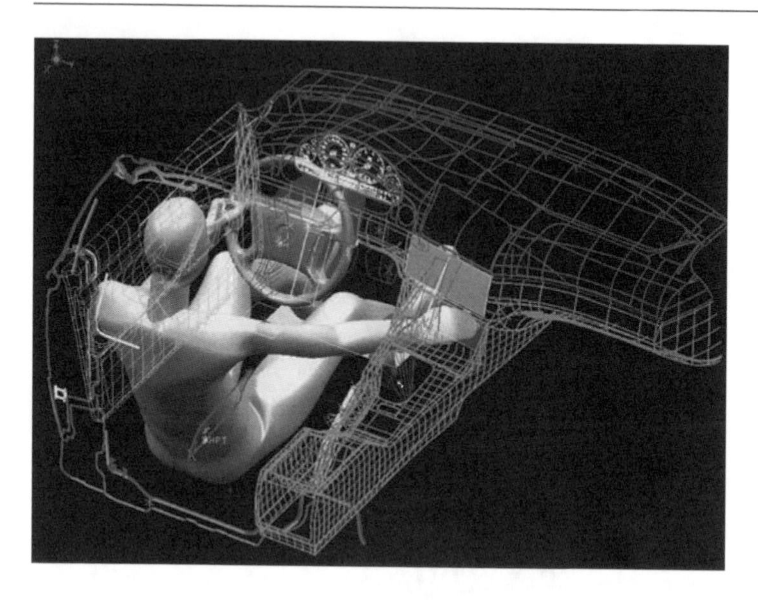

Fig. 6.4-7 RAMSIS cockpit layout.

ements. RAMSIS is even able to "see" through the driver's eyes and recognize obstructions caused by the steering wheel or instrument hoods.

The computationally derived posture is validated using an actual human and vehicle simulator experiments and special test drives in an actual vehicle (Fig. 6.4-7 and 6.4-8).

One advantage is the possibility of using the comfort model at least to discern tendencies in comfort improvement. For checking static headroom, for example, the head contour locator line is a useful aid which represents behavior of an actual person.

The deciding factor in applying RAMSIS is appropriate choice of geometric constraints. In this, values drawn from experience must be available; these may vary from manufacturer to manufacturer.

In a modern development process, layout positions of relevant persons are determined and their hand reach envelopes established. These surfaces are transferred to the ALIAS styling program. A CAS (computer-assisted styling) designer creates the interior, taking into consideration these reach envelopes. The designer positions components such as heating and radio controls in cooperation with ergonomic engineers (see also Section 4.1).

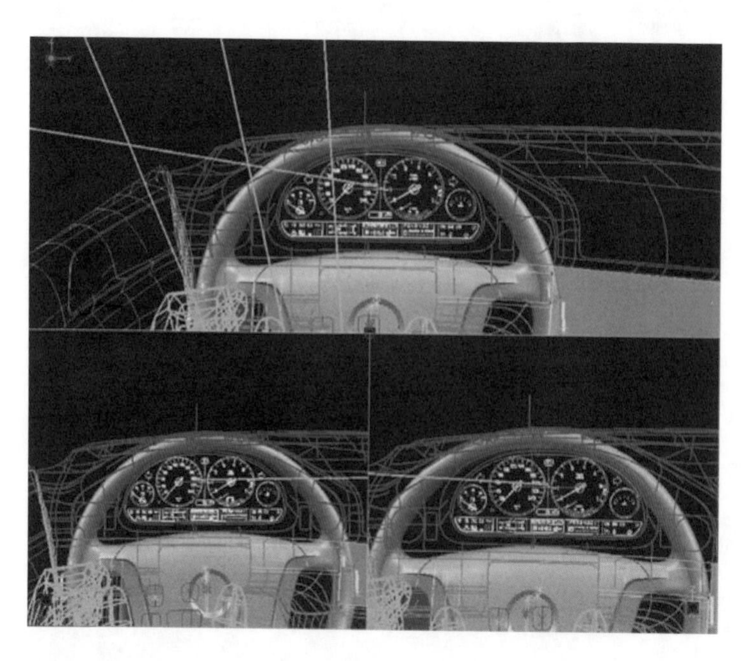

Fig. 6.4-8 Visual obstructions recognizable by the "RAMSIS eye."

The completed ALIAS model is then milled in foam and installed in a seating buck, where it can be evaluated by test personnel and decision makers. Using this process, a large number of concepts can be "played out" before a concept indicative of the final target is worked into the time- and cost-intensive construction of a seating buck.

Construction of a traditional seating buck is usually divided into mock-ups and operating element design. Operating element design mainly involves building an actual functional prototype, which serves the user interface team in translating operating concepts into hardware and conveys information regarding haptics, materials, operating acoustics, etc. Mock-ups are called upon for anthropometric and perception test purposes.

The goal of built-up models is always validation of package and design proposals. In the past, anthropometric examinations of vehicle geometry, reach envelopes, and fields of view were conducted in this way (Fig. 6.4-9). Today, as outlined above, many of these examinations may be conducted during the earliest design phases.

Evaluation of spatial impressions—that is, the vehicle character—is becoming an increasingly important design factor. For example, at what height does a beltline impart a feeling of safety and security, and at what height does it interfere with outward visibility?

Variable models are constructed to enable comparative representation of various geometric variations. These models permit occupants to set different spatial geometries using electric motors and to experience these in direct comparison. This may cover a wide range of assignments: for example, comparison of old and new vehicle generations, comparison against comparison vehicles, evaluation of alternative seat locations, or experiencing various exterior variations from an occupant viewpoint (Figs. 6.4-10 and 6.4-11). One great advantage is the combination of exterior and interior; the chosen interior may easily be reflected by the vehicle's exterior proportions and vice versa (Fig. 6.4-12). More complex variations can be precisely and comfortably realized with the aid of programmable controls.

In the past, models were built in sectional form; surfaces were represented by a number of stations or lofts whose edges were sometimes connected to one another. Today, the usual method is to mill foam components using CA data (Fig. 6.4-13).

Customer-appropriate and safe vehicular interface design sets entirely new challenges for designers, development engineers, and ergonomicists. The software industry can provide only limited assistance, because desktop computers are intrinsically different from mobile applications—that is, from actual driving tasks.

In the past few years, the number of available operating functions has risen dramatically. While in the 1960s 20 or 30 functions were available (in addition to pure driving functions), a well-equipped modern luxury car has more than 500 functions. Fig. 6.4-14 shows the numerical proliferation of control elements and indicators in motor vehicles.

In the late 1980s, a reversal in this trend was noted. This was achieved by installation of "multifunction displays" or text displays. This eliminated the need for many vehicle condition warning lights (e.g., windshield wiper fluid level).

Another trend reversal became apparent in the early 1990s. For the first time, the number of control elements actually decreased. This again was achieved using multifunctionality, in this case by freely assignable and selectable buttons. Optional display screens were installed with new representational possibilities and configuration of functions. Navigation systems became available for the first time, even including moving map displays on which a driver could plan his route and trace his progress (Section 8.3).

In the future, networked information and the fact that information will become ubiquitous—available nearly

Fig. 6.4-9 Model structure for determining reach envelope.

Fig. 6.4-10 Long version of an adjustable seating buck.

Fig. 6.4-11 Short version of an adjustable seating buck.

Fig. 6.4-12 Determining lateral head contour locator line.

Fig. 6.4-13 CA-milled foam components for early determination of spatial feel.

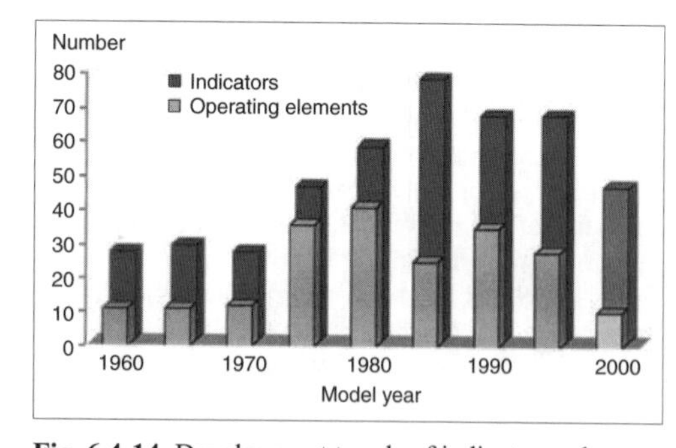

Fig. 6.4-14 Development trends of indicator and operating elements.

anywhere, at any time—will result in a further increase in so-called secondary functions—functions, that is, which are not immediately related to driving.

6.4.1.3 Development methods and integration of ergonomics in the creative process

Human modeling in the vehicle development process

As already described, computer simulation during the definition and design phase of a new product may help reduce the need for lengthy and costly testing. From an ergonomic viewpoint, it is remarkable that until recently, the human being, in terms of his various characteristics and needs, was only represented in limited segments of the CAx chain. Consequently, not all goals were always given human-appropriate weighting. These

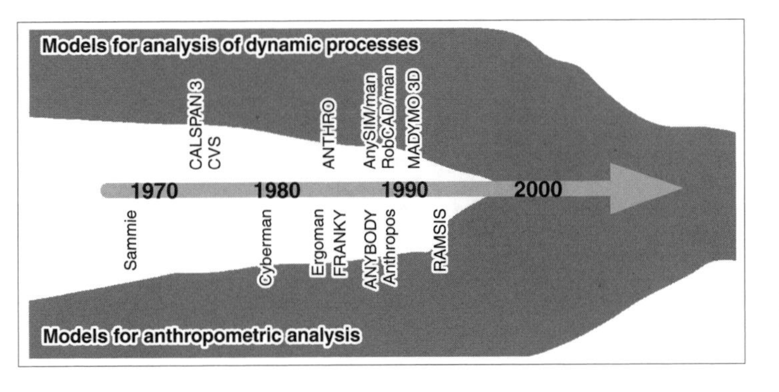

Fig. 6.4-15 Implementation timeline for new ergonomic development methods.

deficits manifested themselves in ergonomic quality of the product as well as workplace design in the vehicle production facility.

Since the 1960s, about 150 different human models have been developed worldwide (Fig. 6.4-15) with different objectives (including layout/ergonomic evaluation of seated workplaces or manual work stations, climate dummies, crash dummies, animation tools). Most of these were developed in-house by industrial concerns to address their own specific needs.

Some of the tools used to simulate human behavior should be explained in the context of the product creation process, using the automobile (and motorcycle) as an example. Here the focus is on a process-oriented examination. If one evaluates the entire process (Fig. 6.4-16), it becomes obvious that later changes to the product incur higher costs; the level of detail in the product and the associated cost effects increase roughly exponentially with development time.

Understandably, simulation tools are most effective at the beginning of the development process (Section 10.2). As a result, all conceptual product parameters must be laid down during the target definition phase ("product vision") and the concept phase. During these phases, work proceeds on a large number of possible alternatives, which are gradually weeded out as the project status is advanced. In addition, nearly all specialist departments involved in the process input their requirements during the predevelopment phase. This results in two main requirements to be met by any simulation tool:

— Speed. To do justice to the growing number of concept ideas and product variations, ergonomic detail investigations must be processed in ever shorter time spans and with ever greater information content.
— Transparency. To provide all project partners with efficient decision-making tools, in terms of representation, content, and graphics, investigative results must be easily, quickly, and precisely translatable.

The human models presented here are at present predestined for application in automotive development. Their objectives, however, are widely different. As a

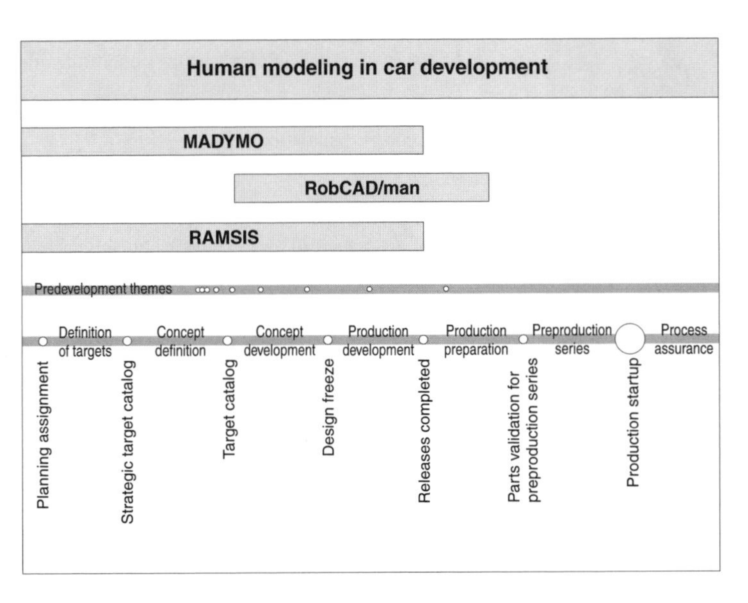

Fig. 6.4-16 The product creation process.

Fig. 6.4-17 Placement of human models in vehicle.

rule, three simulation tools for human properties are currently employed by the automobile industry:

— RAMSIS ("**R**echnergestütztes **a**nthropometrisch-**m**athematisches **S**ystem zur **I**nsassen-**S**imulation"—"Computer-Aided Anthropometric-Mathematical System for Occupant Simulation") is a three-dimensional human model for layout and analysis of driver workplaces (Fig. 6.4-17).

The most prominent feature of RAMSIS is its posture models. In tests using videogrammetry or marker systems, subjects in various scenarios (automobile, commercial vehicle, motorcycle, etc.) are evaluated by RAMSIS itself to provide statistical models for joint angle distribution functions. RAMSIS is then in a position to simulate exact real postures using restrictions within the job definition.

Along with the classical analysis tools such as sight line simulation and reach envelopes, RAMSIS offers numerous analysis tools such as mirror vision, comfort evaluation of postures, and safety belt analyses. RAMSIS is implemented during the product creation process.

Meanwhile, mock-ups have been completely eliminated as tools for analyzing static questions (classical anthropometric problems) during the definition phase.

Since 1995, industry, vendors of measuring technology, and universities have been working on the RAMSIS Dynamic Project (Fig. 6.4-18). In this, the human model is successively converted into a dynamic model. In time, the current, previously described test rigs (seating bucks) might be eliminated entirely. In actuality, however, for the near future, validated concepts will still be presented to decision-makers in the form of an overall model, a seating buck.

Fig. 6.4-18 Further development of RAMSIS for dynamic entry and egress.

— RobCAD/man is an application developed by the Institute for Machine Tools and Industrial Engineering, Munich Technical University, which permits worker simulation within an examination cell of the RobCAD (by Tecnomatrix) assembly planning system. In this, all relevant worker activities may be "virtually validated" in terms of ergonomic and time considerations. Specifically, accessibility (grasp and reach envelopes), perception, reachability, cycle times, etc. are investigated.
— The third tool currently being used includes climate and acoustic simulation tools.

Prognosis

The trend for the future is toward fewer 3-D "software dummies" accompanied by expanded application spectra. Cooperative efforts among manufacturers of modern human models will result in jointly developed functionalities. The greatest challenge within the framework of developing dynamic simulation tools lies in the task of realistically simulating human properties beyond the current capabilities of individual models.

Possible future solutions

Vehicle display and control principles will play a decisive role in future marketplace competition. All vehicle manufacturers and system vendors are currently developing solutions which, although differing in detail, point in the same direction. New technologies such as the Internet, provider services, and telematics (a combination of the concepts "telecommunications" and "informatics") will in coming years set new challenges for research and development and attain a hitherto unknown dimension in comfort, safety, and service within the automobile (Sections 6.4.2 and 8.3).

Compared to the present situation, many more functions will be available to rear-seat passengers. Display screens, telephone and fax connections, and new means of data entry such as speech and gesture recognition will achieve major significance in many future vehicles. Development of communications technology, proceeding as it does by leaps and bounds, must be taken into consideration. New job descriptions must be introduced into the design process to meet the demands imposed by new media applications. Another avenue of development, until now largely ignored, involves vehicle operation and design that address the needs of senior citizens. This involves not only improved entry and egress and easily legible instrumentation but also a fast, appropriate—indeed a "thinking"—car.

It is not yet possible to say where this development will lead in the next decade or two. Even though humans now operate their communications centers manually, with their fingers, these will soon employ speech recognition, as is already the case for in-car telephones.

This rapid development pace requires very careful guidance on the part of developers and especially ergonomicists. The actual task of driving will be with us for the next few decades. Many auto manufacturers are already planning to enter the field of driver assistance systems with future vehicle generations. Yet with all the technological possibilities opening before us, the human being as driver and passenger must be able to comprehend this new world and to function within it. This is a new challenge for ergonomicists, designers, planners, and developers, as yet unimaginable in all its consequences.

References

Bullinger, H.-J., et al. "Anthropometrische und kognitve Evaluierung der Fahrer/Fahrzeug-Schnittstelle im PKW." ATZ 98, 7/8, 1996.

Bubb, H. "Ergonomie in Mensch-Maschine-Systemen." Conference "Komfort und Ergonomie in Kraftfahrzeugen," Haus der Technik, Essen, 1997.

Krist, R. "Modellierung des Sitzkomforts—Eine experimentelle Studie." Dissertation at Catholic University of Eichstätt, in FAT-Bericht 123, 1995.

Seidl, A. "RAMSIS: das führende Ergonomiewerkzeug für Design und Entwicklung von Kraftfahrzeugen." Conference "Fahrzeugkomfort," Haus der Technik, Essen, 1999

SAE J1100. "Motor Vehicle Dimensions." Society of Automotive Engineers Inc. Warrendale, PA, June 1998.

SAE J826. "Devices for Use in Defining and Measuring Vehicle Seating Accommodation." Warrendale, PA, July 1995.

Elsholz, J. "Die Gestaltung des Arbeitsplatzes nach ergonomischen und physiologischen Gesichtspunkten." Fourth BMFT Status Seminar, 1976, Verlag TÜV Rheinland 1977, pp. 473–485.

Elsholz, J., and M. Bortfeld. "Anwendungsorientierte Untersuchungen an Bedienelementen nach ergonomischen und physiologischen Gesichtspunkten." 18th FISITA Congress, Budapest, 1978. Vol. 1, pp. 231–244.

Färber, B. "Mehr Instrumente, mehr Sicherheit?" VDI-Bericht 819, 1990, pp. 1–18.

FAT-Bericht 123. "RAMSIS—Ein System zur Erhebung und Vermessung dreidimensionaler Körperhaltungen von Menschen zur ergonomischen Auslegung von Bedien- und Sitzplätzen im Auto."

FAT-Bericht 135. "Mathematische Nachbildung des Menschen—RAMSIS 3 D Softdummy." 1997.

Landau, K. (editor). "Mensch-Maschine-Schnittstellen." Fall conference of the Gesellschaft für Arbeitswissenschaft, October 1998. Verlag Institut für Arbeitsorganisation, Stuttgart, 1998.

FAT-Bericht 64. "Sicherheitsorientierte Bewertung von Anzeige- und Bedienelementen in Kraftfahrzeug—Grundlagen." 1987.

FAT-Bericht 74. "Sicherheitsorientierte Bewertung von Anzeige- und Bedienelementen in Kraftfahrzeugen"—empirical results, 1988.

6.4.2 Communications and navigation systems

6.4.2.1 Goals and solutions

Car-mounted communications and navigation systems essentially address three topics: they serve to optimize the driving route—for example, by helping avoid unnecessary searching or waiting in traffic jams; they offer entertainment during the journey—for example, in the form of music delivered by radio or descriptions of sights along the way; and they increase safety and comfort—for example, by enabling transmission of information from the car in the event of a breakdown or by finding just the right hotel and making reservations.

Realization of these systems is founded on three basic devices: the car radio, the mobile phone, and navigation systems. By combining their basic functions, new functions may be derived. Because of the amount of software involved, it is to be expected that technology from the personal computer (PC) sector will enjoy increasing influence (Fig. 6.4-19).

The most commonly used standards and systems are

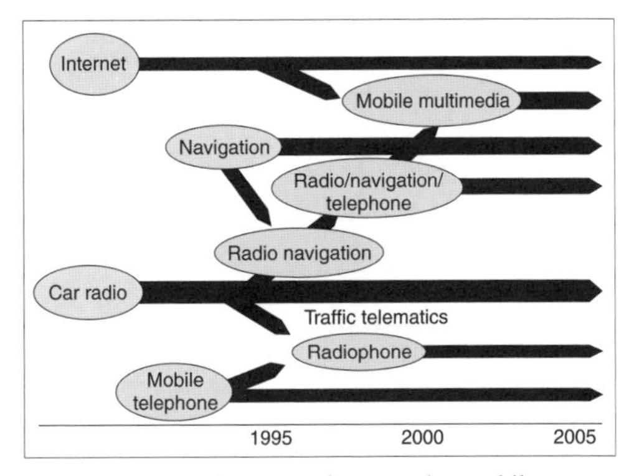

Fig. 6.4-19 Development of automotive mobile communications.

FM radio broadcasts, GSM (Global System for Mobile Communications), and GPS (Global Positioning System). However, other broadcast bands, telephone standards, and satellite navigation systems offer similar services. Characteristically, most of these standards were not developed primarily for use in automobiles; their adaptation to this field requires added effort, in terms of labor as well as hardware.

The following will examine in greater detail these basic devices and several of their combinations.

6.4.2.2 Car radio

The car radio [1] is a general concept covering broadcast receiver, cassette drive, CD player or changer, and similar primarily entertainment-oriented devices. The current trend is toward a standard unit with equipment features: CD player, AM and FM receiver, radio data system or RDS (see below), and four 20-watt amplifiers. The number of devices incorporating a cassette player is declining; devices with other drives such as minidisc are as yet still rarities.

In Europe, FM is the radio band most listened to. Technical solutions applied to FM receivers are indicated in the block diagram of Fig. 6.4-20.

Primary selection of a desired station is accomplished by tuned bandpass filters in a regulated preamplifer section. A mixer converts the signal to an intermediate frequency of 10.7 MHz with the aid of a PLL-controlled oscillator. It is here that the actual station selection takes place. The intermediate signal is demodulated to yield a multiplex signal (MPX-S). Fig. 6.4-21 illustrates its frequency spectrum.

The multiplex signal contains the audio signal in the form of a mono and a stereo signal, the added RDS signal, and, eventually, data radio channel (DARC). Interference generated by the automobile ignition system is recognized as distortion in the MPX-S and eliminated.

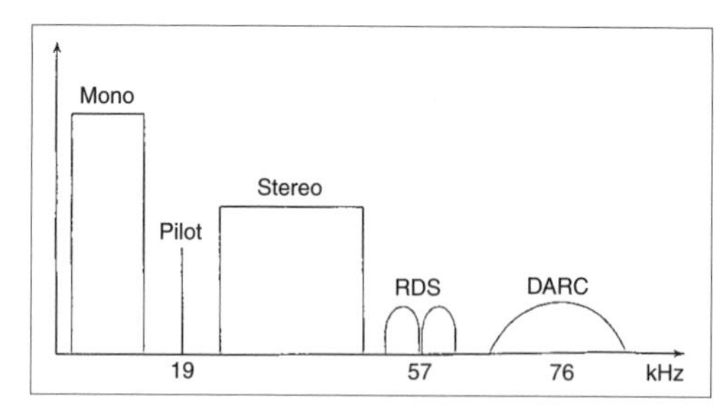

Fig. 6.4-21 Spectral distribution of audio and auxiliary signals within an MPX signal.

After stereo decoding and equalization of transmitter-generated treble boost (de-emphasis), the original audio signals are again available.

6.4.2.2.1 RDS (Radio Data System)

RDS [2] carries inaudible control and information signals within the broadcast signal (see Table 6.4.1). Markers for traffic bulletins and for stations carrying traffic information are the most important information carried by RDS.

In countries in which the same program is transmitted by aerials in different geographic regions, information regarding alternative frequencies (AF) for the same broadcast is also useful. Because the region in which a broadcast program may be received is usually greater than the range of any individual station, cross-country trips involve frequent frequency changes. Conversely, a program may be received locally on several frequen-

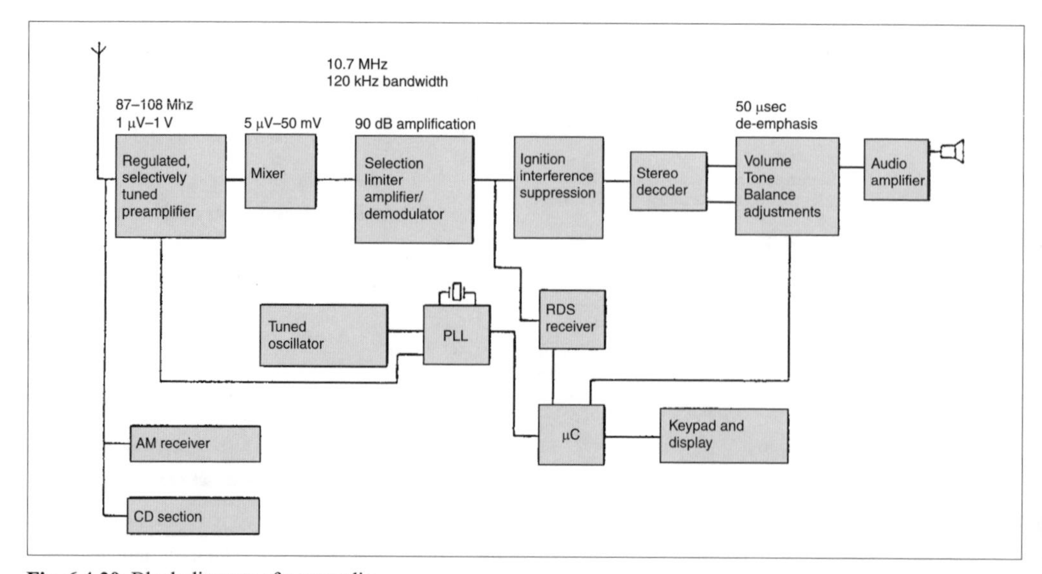

Fig. 6.4-20 Block diagram of a car radio.

Table 6.4-1 Radio Data System (RDS) information content

PS	Program Service name (max. 8 ASCII characters)
PTY	Program Type code (e.g., news, sports, talk, etc.)
TP	Traffic Program flag (channel carries traffic program)
AF	Alternative Frequencies lists (other channels carrying same program)
EON	Enhanced Other Networks info (maximum 8 other program codes; e.g., PTY, TP, TA, AF, etc.)
TA	Traffic Announcement flag
TDC	Transparent Data Channel
CT	Clock Time and date

cies, with reception quality varying within the space of only a few meters.

AF information permits automating the frequency switching function within the radio itself.

RDS properties

Transmission method: amplitude modulated with suppressed carrier frequency of 57 ± 2.4 KHz, synchronized to pilot subcarrier, data rate 1.2 kbit/s.

6.4.2.2.2 TMC

One application of the RDS TDC feature is TMC (Traffic Message Channel). Highly compressed traffic information can be transmitted on this channel. For example, about 100 bits/s are required to convey all German traffic information, covering all incidents, nationwide. This information may be used to synthesize radio traffic announcements or to add a dynamic dimension to navigation systems.

In TMC, all possible traffic announcements are conveyed schematically. To cover every possible situation, standard phrases are stored in the car radio; for example "Exit at . . . and take detour . . ." TMC conveys information regarding incidents and their locations. Based on this encoded information, suitable standard phrases are selected by the radio and combined with current information generated by a speech synthesizer.

Advantages of this process include the following: Traffic announcements are transmitted silently, parallel to the audio signal; they can be captured using limited storage capacity in the car radio and called up by the user specifically for his planned route: and announcements can be synthesized in the user's native language, even while traveling in foreign countries.

6.4.2.2.3 DARC

DARC (Data Radio Channel) is a new system that has been standardized worldwide. In Europe, it is known as SWIFT (System for Wireless Infotainment Forwarding and Teledistribution) [3]. Compared to RDS, DARC permits a tenfold increase in the data transmission rate (see Table 6.4-2).

Table 6.4-2 Possible DARC services

Information as in RDS Telegram incl. TMC
Expanded traffic alerts
Timetables (bus, rail) incl. delay advisories
Traffic guidance advisories (detours, etc.)
Subscription services such as paging, stock quotes

DARC properties

Transmission method: LCMSK (Level-Controlled Minimum Shift Keying) at 76 ± 17.5 KHz, synchronized to pilot subcarrier, data rate 16 kbits/s. Every receiving device may be individually addressed.

6.4.2.3 DAB

DAB [4] is a new digital transmission system. It represents a synthesis of broadcast and data services. Transmissions are no longer differentiated between audio, video, and data services of other applications. Data streams may be freely combined from different partial data streams to form a single multiplex signal.

The transmission method is conceived for mobile reception, is tolerant of multipath reception, and has extensive error correction capabilities. Its high data transmission rate makes DAB particularly suitable for added services with graphic representations.

DAB properties

Transmission method: COFDM (Coded Orthogonal Frequency Division Multiplex). Transmission frequency: terrestrially by means of regionally differentiated portions of VHF band III (174–230 MHz), terrestrially and via satellite in portions of the L band (1,452–1,492 MHz), data reduction, and simultaneous broadcasting. Another related system is DVB (Digital Video Broadcasting) [5].

6.4.2.4 Car telephony

The need to place calls from within an automobile and to be accessible by telephone while in a car is not new. This medium offers the ability to exchange information with one's surroundings, individually and bidirectionally. Consequently, technical solutions for this need have long been available. Today, Europe and a large part of the world have adopted GSM (Global System for Mobile Communications) as their telephony standard. In Germany this is implemented in the D and E networks, operating at different frequencies [6, 7, 8].

The primary function of a mobile telephony network is to ensure that the user can be reached or can place calls anywhere at any time. To enable this, the network consists of a comprehensive hierarchical structure. Fig. 6.4-22 outlines its configuration.

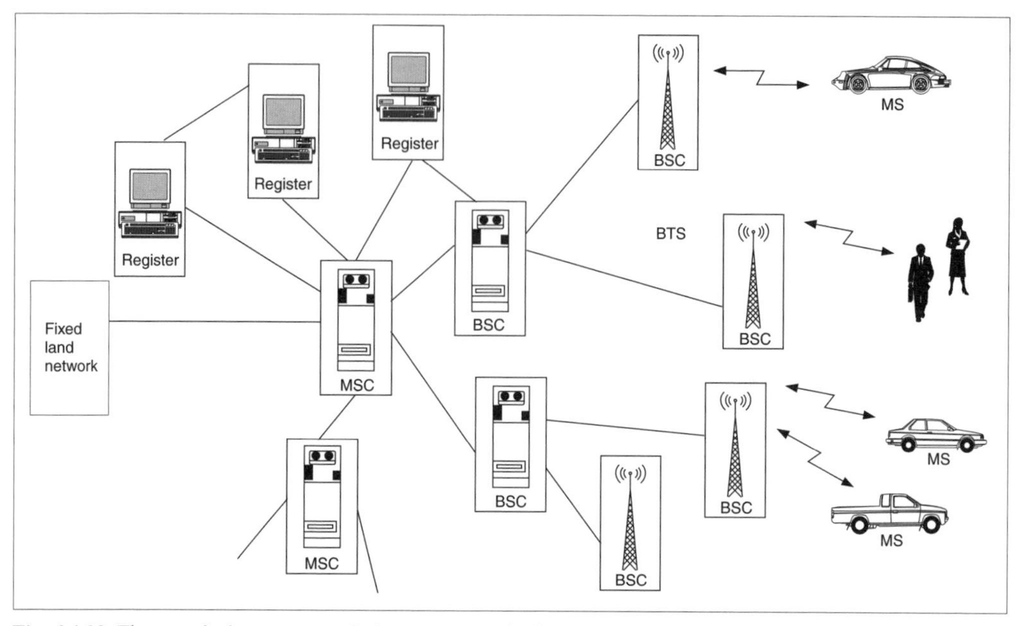

Fig. 6.4-22 The car telephone as part of a larger communication network.

Mobile stations (MS) communicate with mobile switching centers (MSC) via base transceiver stations (BTS) and base station controllers (BSC). Within this composite structure, physical properties of any individual communication such as frequency channel, field strength, time slot, and so on are established for the mobile station. On the basis of field strength, error rate, and other measurements reported by the mobile station, a base station is assigned or a change of base station is initiated. Authorization is checked by a login sequence. Presence of every logged-in user is tracked anonymously, and data received from the mobile unit is processed for the fixed network, or conversely data originating from the fixed net is processed for the mobile net.

Various registers are assigned to certain MSCs; these registers contain stored data regarding user, devices, etc.

One special feature of GSM is the SIM (Subscriber Identity Module) card. This authorization card, inserted into the end unit, carries among other things user-specific data which identify the user to the network.

Usable data in GSM include speech, general, or fax data. Short messages (SMS, Short Message Service) may be transmitted in parallel to an ongoing telephone conversation.

Properties of GSM

Transmission frequency from mobile station to base station, D Network (E Network in parenthesis): 890–915 MHz (1,710–1,785 MHz), in reverse direction 935–960 MHz (1,805–1,880 MHz). Modulation method: GMSK (Gaussian Minimum Shift Keying), time multiplexed, training sequence to correct for transmission range, channel coding, framework structure for logical channels.

6.4.2.4.1 Automobile installation of cell phones

Most mobile telephones are cell phones, portable devices which were not primarily developed for use in automobiles. The main components of a cell phone are a 2-W transmitter/receiver unit, a digital signal processing unit for channel and speech encoding, and a control unit which coordinates interaction with the rest of the network. Added to these are microphone, speaker, antenna, keypad, display, battery, and SIM card reader.

Installation kits are available to adapt cell phones for automotive service. These provide a car-mounted bracket, which normally incorporates power connections to charge the cell phone batteries, and an antenna connection.

This last is especially important as it carries electromagnetic radiation generated by the cell phone by means of a shielded cable to an antenna attached outside the car body, which acts as a Faraday cage. Transmission within the car might otherwise result in overlapping reflected signals and locally increase field strengths.

Telephony while driving could handicap the driver. Therefore car installation kits often include a hands-free feature, with which speech by the occupants is picked up by a microphone—for example, mounted near the inside rear-view mirror; the telephone audio signal drives a separately installed speaker. Hands-free installations must suppress echoes of incoming speech to avoid feedback. In the future, complex echo-compensating solutions based on digital signal processing will permit simultaneous speech while listening to near (in car) and far (on network) participants. Presumably, this will employ algorithms which also suppress vehicle driving noise in the microphone signal.

6.4.2.4.2 Permanent telephone installations

Permanently installed telephones usually consist of a remotely mounted transceiver unit with a maximum 8 W of transmitter power and a handset with integrated hands-free operation.

Compared to portable cell phones, permanently installed units satisfy somewhat more demanding specifications in terms of reception and transmission properties.

6.4.2.4.3 Radiophone

Miniaturization resulting from a high degree of component integration within telephone modules permits their integration within a standard car radio chassis. Advantages of this solution are cost savings through common use of many hardware components for both applications, simplified installation, unified operation and improved hands-free operation through use of car radio speakers.

6.4.2.5 Beacon communications

Beacons [9, 10] communicate with an OBU (On Board Unit) installed behind the vehicle windshield over distances of a few meters. These are generally installed on bridges directly over the roadway or at the edge of the road. Their main application is toll collection—for example, at toll plazas with rapidly moving traffic—or with automatic billing systems.

Microwaves have crystallized as the transmission medium of choice. The objective of the technical concept is to produce OBUs as cheaply as possible. Up- and downlinks are therefore achieved by different means. The transponder principle is used for uplink; a 5.8-GHz carrier broadcast by the beacon is modulated by a phase-modulated 1.5- or 2-MHz subcarrier within the OBU and retransmitted. Data rates of 250 kbits/s are achieved. For downlink, the beacon modulates the 5.8-GHz carrier amplitude for a data rate of 500 kbits/s.

For security reasons, the OBU often includes an anonymous chip card containing an account balance from which tolls are deducted.

6.4.2.6 Navigation

Navigation systems [11] assist the driver in traveling to and through unfamiliar territory. Most solutions are guidance systems: that is, the driver is guided to his destination by means of direction recommendations. These devices calculate the best route to the destination based on user-selected optimization criteria. If the driver deviates from these recommendations, a new route is calculated automatically.

The key components of a guidance system are devices to determine actual location, a digitized map, a data entry unit for destination input, output devices to convey driving directions, and comprehensive software to control the associated processes.

A gross coarse position determination is accomplished by GPS (Global Positioning System). The navigation system calculates the driver's exact location by correlating range, distance covered, and direction of travel with possible route profiles from a map. Speedometer signals, wheel sensors, the earth's magnetic field, or yaw rate sensors are employed to determine length and direction of the path already covered. The achieved accuracy is approximately ± 5 m.

Digital maps are usually stored on CD-ROMs. A CD with about 650 MB storage capacity can, for example, store the entire German road network. A disadvantage of this storage medium is long access times, measured in seconds.

More comprehensive information such as descriptions and images of sights along the way require considerably more storage capacity. It is anticipated that the more densely packed DVD (Digital Versatile Disc) will become the predominant storage medium for future applications.

The algorithm applied to route calculations sequences appropriate road segments contained in the digital map until it establishes an unbroken connection between origin and destination. Along with its length, each road segment has an assigned average speed, permitting calculation of estimated travel time. Depending on chosen optimization criteria, a route optimized for distance or time is selected from the possible road connections.

Route calculation should be accomplished so quickly that the driver, if he misses a turn, is given updated directions before reaching the next intersection. Route suggestion should present as little distraction as possible from the driving task. Audible instructions are especially suited for this. Parallel to audio instructions, visual displays have proven useful as more persistent and more detailed representations.

6.4.2.6.1 Dynamic navigation

By combining radio traffic information regarding traffic jams with route calculation by means of a navigation system, a uniquely optimized route may be calculated taking into account actual traffic conditions [12].

A precondition for this is traffic reports which provide an exact location of the problem, displayable on the navigation CD map, with information on length and anticipated duration of the situation and classification of the degree of obstruction, such as "full closure" or "stop-and-go traffic."

Traffic Message Channel (TMC) and queryable traffic information delivered by service providers via mobile telephone supply these data. Both DARC and DAB are expected to provide these as well.

For road segments with traffic obstructions, the navigation system establishes a new average speed using traffic reports and uses this to calculate a new route. A similar process is carried out if the traffic alert is lifted.

By user choice, route calculations may be carried out automatically. To inform the driver, traffic reports can be conveyed verbally or in written form (visual display); or maps can be displayed, for example, showing the affected section of road and the revised route recommendation.

6.4.2.7 Traffic telematics

The combination of mobile telephone, position determination, and possibly other sensors or components of a networked automobile, in conjunction with a service center, permit realization of comprehensive services. "Traffic telematics" draws on these resources to encompass a variety of features already familiar, currently in development, or envisioned for the future [13, 14]. Telematic functions are currently highly country-specific and in large degree dependent on service offerings. Following are a few of the services presently available or soon to be introduced:

- Dynamic route guidance: In addition to the above-mentioned dynamic navigation, solutions are in service in which a central office does not report traffic jams but rather preselects the route for the navigation system, thereby actively and deliberately redirecting traffic streams.
- Route guidance on demand: Drivers of cars without built-in navigation systems may, if desired, allow themselves to be guided by a service provider. The vehicle reports its position to a central office and obtains driving directions based on real-time traffic conditions.
- General travel services: Real-time information can be accessed about sights along the way, events, hotels, museums, emergency services, and so on in the vicinity of the driving route. This function could also be used to book hotel reservations and the like.
- Emergency service: This is a preprogrammed outgoing emergency call with location information, actuated by a very simple manual procedure (emergency button); or, in the case of an accident, it could automatically summon police and/or emergency services.
- Breakdown services: This is a preprogrammed sequence with which the driver can establish a voice connection with a central office in the event of a mechanical breakdown. On the basis of verbal fault description, possibly supported by remote diagnostics, and vehicle location report, the office could dispatch assistance or provide advice for simple repair procedures.
- Antitheft protection: A combination of mobile telephone, vehicle positioning, and possibly engine control system or vehicle immobilizer permits development of several systems which provide police with information on the actual location of a stolen vehicle or which restrict mobility options for a vehicle.
- Floating car data: Knowledge of actual traffic conditions is the most important prerequisite for traffic advisory services. In the currently implemented system, a central office employs installed mobile telephones to query a large number of vehicles regarding their current position (e.g., as determined by GPS) and road speed (e.g., from speedometer signals). A mathematical model of traffic streams combines results of this sampling with information from other sources, permitting description of the real-time traffic situation. The objective is to derive traffic prognoses, permitting wide-area detours of traffic jams that develop during trips already in progress.

6.4.2.8 Driver information systems

Despite all individual optimization efforts, the multiple display and operation elements of communication systems may overload the driver. Information systems [10] therefore also bear a responsibility of providing the driver with a unified user interface.

Two technical solutions and their hybrid form suggest themselves.

First, the "all in one" solution, in which for space reasons the smallest possible solutions are sought for individual components. These are then combined with operation and display units in a single housing fitting a standard bay and installed in place of the car radio.

The other is the decentralized solution, in which individual components are installed at various locations throughout the vehicle. These components exchange control and information display data with a central control/display unit by means of a bus system. Input of the most important control functions is preferably accomplished from the driver's immediate reach space, if possible by means of elements integrated in the steering wheel. Graphics-capable color liquid crystal displays are often employed to display information. Their incorporation into the combination instrument is advantageous (Fig. 6.4-23).

In their structural configurations, these solutions differ in their incorporation of display and keypad within an overall system. Some implementations have operating software for all applications concentrated in a central service unit; others allow all components access to these resources. Advantages of central operating software are apparent as functions which combine contributions from several components; advantages of distributed software include the possibility of independent continued development of individual components' user interfaces.

An example of steering wheel actuation is the accessory equipment for a radiophone, with which telephone conversations may be initiated or accepted; telephone numbers, radio stations, or CD titles called up sequentially; sources selected; and volume adjusted. Data transmission is by means of an infrared link. In other solutions,

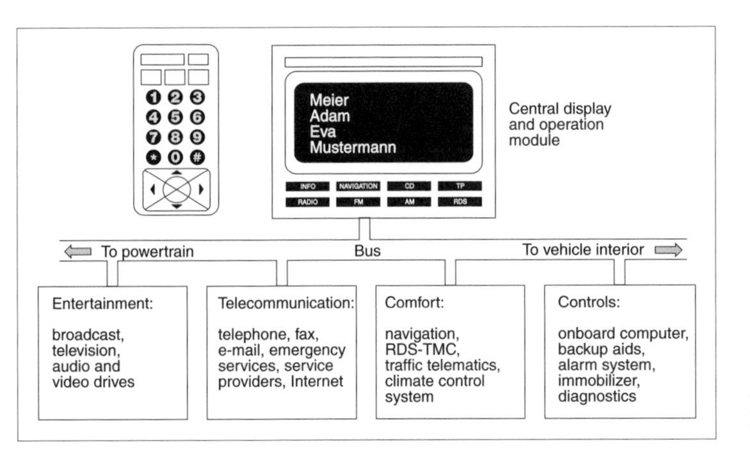

Fig. 6.4-23 Structure of a vehicle information system.

data is transmitted from buttons integrated in the steering wheel by means of wiring secured against twisting.

In addition to keypad and display, voice is enjoying increasing application as an information and control element for driver information systems, as this human sensory channel is least involved in the driving task.

Depending on their intended user base, voice recognition systems are differentiated into speaker-dependent or speaker-independent systems; discrete speech input, continuous speech input (e.g., digits of a telephone number), or natural speech input (capable of filtering commands from context); and according to vocabulary size. For vehicular applications, algorithms to suppress driving noise from speech samples are significant.

At present, devices with speaker-independent discrete and continuous speech recognition of some words (e.g., a phone number) and speaker-dependent individual vocabularies of several hundred words are in widespread service. Larger vocabularies include a spell checker.

For navigation system destination entry in particular, speech recognition systems with very comprehensive vocabularies of several thousand words are currently being tested.

References

[1] *Elektronik im Kraftfahrzeugwesen.* expert Verlag, Vol. 437, 1994, ISBN 3-8169-1024-6.
[2] RDS-Standard. CENELEC EN 50067.
[3] DARC (SWIFT)–Standard. ETS 300 751. ITU-R Recommendation BS 1194 "Data Radio Channel (DARC)."
[4] From the publication series issued by the DAB Plattform e.V., Am Moosfeld 31, D-81829 München, Germany. *Digital Audio Broadcasting (DAB)—Der Radiohighway*, e.g., Vol. 14, "Digitale Systeme für Hörfunk und Fernsehen," Frank Müller Römer, 1996.
[5] Reimers, Ulrich. *Digitale Fernsehtechnik—Datenkompression und Übertragungstechnik für DVD.* Springer-Verlag, Berlin, Heidelberg, New York, 2nd edition, 1997. ISBN 3-540-60945-8.
[6] "GSM recommendations." Special Mobile Group (Technical Committee SMG), ETSI, European Telecommunication Standards Institute, F-06921 Sophia Antipolis Cedex. (Very comprehensive GSM specifications).
[7] Mouly, M., and M. Pautet, 4, rue Elisee Reclus, F-91120 Palaiseau. "The GSM System for Mobile Communications," self-published, 1992. ISBN 2-9507190-0-7.
[8] Redl, S., and M. Weber. "D-Netz-Technik und Messpraxis," Franzis (Funkschau: Technik), Munich, 1993. ISBN 3-7723-4851-3.
[9] CEN, ENV 12253, CEN. ENV 278/9/#64 and 65, "Road Transport and Traffic Telematics."
[10] Detlefsen, W., and W. Grabow. "Interoperable 5.8 GHz DSRC Systems as Basis for Europeanwide ETC Implementation," 27th European Microwave Conference EMC, Jerusalem, September 1997.
[11] "Elektrik und Elektronik für Kraftfahrzeuge, Sicherheits- und Komfortsysteme." Volume 8/99. Published by Robert Bosch GmbH, Stuttgart, order no. 1987 722 037.
[12] Neukirchner, E. "Einfluss von Verkehrsmeldungen auf die Routenempfehlungen fahrzeugautonomer Zielführungssysteme." ITG-Fachbericht "Informatik im Verkehr," presentations at the 1996 VDE-Kongess in Braunschweig. Copyright VDE-Verlag GmbH, Berlin and Offenbach, October 1996. ISBN 3-8007-2205-4.
[13] Evers/Kasties. *Kompendium der Verkehrstelematik, Technologien, Applikationen, Perspektiven.* TÜV-Verlag GmbH. ISBN 3-8249-0421-7.
[14] DGON-SATNAV 98—Satellitennavigationssysteme—Grundlagen und Anwendungen. October 5–9, 1998, Dresden.

6.4.3 Interior comfort/thermal comfort

Whether a vehicle occupant perceives the interior as comfortable depends on many factors: for example, on seats (contours, materials), on materials used for the dashboard and inner door panels, on ride and operating comfort. Not the least of these is thermal comfort, which plays an extraordinarily important role, as freezing or perspiring occupants would not feel at home in even the most comfortable seats.

As may be seen in Fig. 6.4-24, the climate control system consists of several subsystems: a heating circuit, a cooling circuit, the climate control unit, air ducts to the climate control unit and to vents, and the control panel for the climate control system (including the associated controls and sensors). Heating and cooling circuits have already been discussed in Section 3.3; the emphasis of this section will be on the climate control

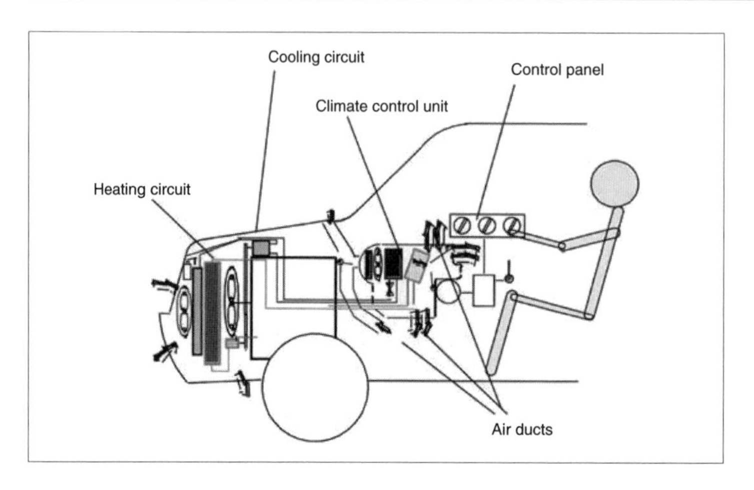

Fig. 6.4-24 Climate control system.

unit—its air ducting/air distribution and operation/control systems.

6.4.3.1 Driver comfort requirements

Climate control addresses the comfort requirements of vehicle occupants and provides traffic safety benefits.

Reference [1] contains interesting recommendations for evaluating comfort. For example, it is shown that perception of discomfort forms the basis of the hierarchy shown in Fig. 6.4-25. This is analogous to Maslow's well-known "hierarchy of needs," in that human beings first seek to satisfy physiological needs such as food, drink, and sleep. Once these are satisfied, a need for security is perceived. After this come social relationships and so on. The comfort hierarchy may be understood in the same way. Here, too, shortcomings are first perceived when all needs below it in the hierarchy are satisfied.

Conditions for thermal comfort were first described by Fanger in 1970 [2]. This was followed by a series of investigations which described in greater detail thermal condition demands as applied to vehicle interiors. Thermal conditions in motor vehicles, for example, are completely different from those in buildings. Because of the arrangement of glazed surfaces in the upper half of the vehicle, incoming (solar) thermal radiation varies drastically from head to feet. As a result, different airflow velocities and temperatures are necessary to ensure thermal comfort. Fig. 6.4-26 shows experimental results by Wyon et al [3]. Additional references may be found in sources [4] through [8].

In addition to the automobile driver's desire for a comfortable environment, it must be remembered that climate conditioning of the passenger cabin also benefits traffic safety [9]. Numerous recommendations and regulatory requirements ensure that the environs of stationary workplaces induce as little stress as possible.

For example, one recommendation governs temperatures not to be exceeded while the human body is subjected to certain specific physical stresses. Such regulations are nonexistent as far as automobile drivers are concerned, although summertime thermal stress in the cabin may be very high indeed.

High temperatures in particular are critical with respect to driver condition. For example, in tests of reaction time, sensory perception, and associative processes, a temperature increase from 25–35°C resulted in deterioration of these abilities by more than 20% [10]. Tests under actual road conditions show that temperature inside the vehicle has a significant effect on driving safety. A temperature rise from 21–27°C in actual road

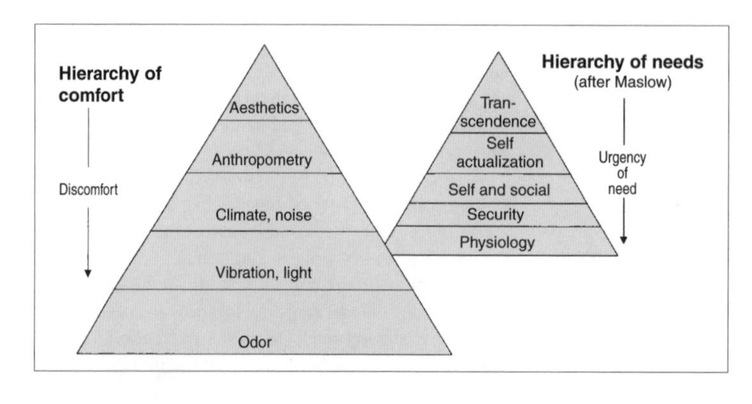

Fig. 6.4-25 Hierarchy of comfort.

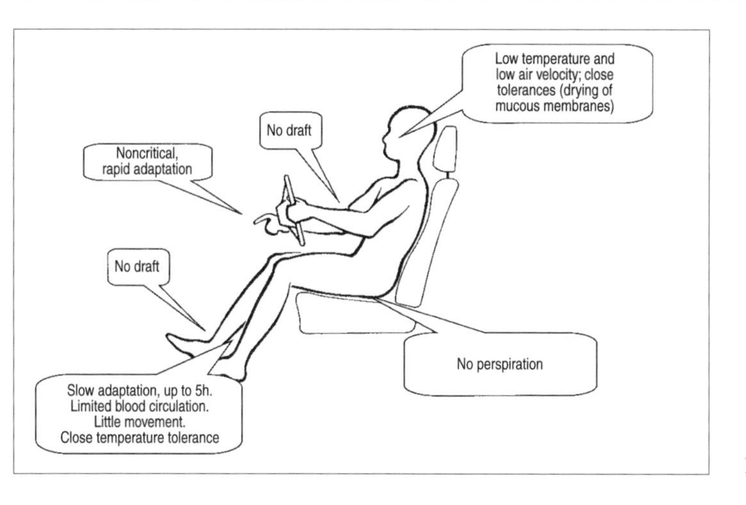

Fig. 6.4-26 Thermal comfort.

tests increased driver reaction time by 20% or more. At the same time, attention span diminished. At 27°C, twice as many signals (car horns, flashing lights of emergency vehicles, or warning lights inside the vehicle, activated by a computer) were missed compared to those perceived at 21°C [11].

6.4.3.2 Climate control unit functions and construction

The climate control system consists of a heating circuit, a cooling circuit, the climate control unit, air ducts to the climate control unit and vents, and the climate control system control panel (including the associated control unit and sensors). "Climate control unit" is understood to mean a modular unit, normally installed in the vehicle interior below the instrument panel. Its main components (see also Fig. 6.4-27) are:

- Blower
- Air filter
- Evaporator (as part of the cooling circuit; see Section 3.3)

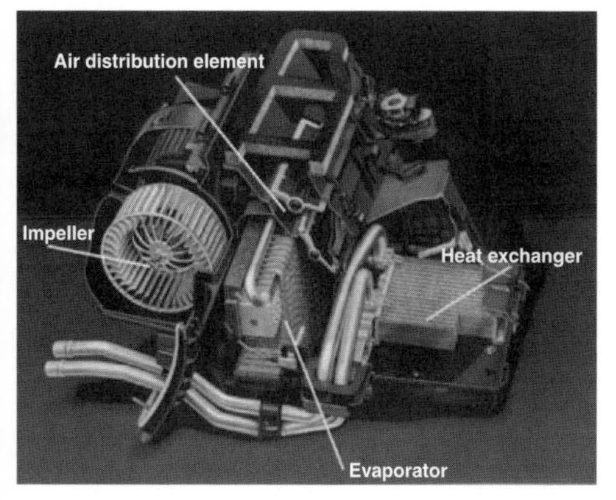

Fig. 6.4-27 Main climate control unit components.

- Heater core (as part of the heating circuit; see Section 3.3)
- Temperature control system
- Air distribution flaps

As layout and design of heating and cooling circuits has already been described in Section 3.3, we may assume that adequate heating or cooling capacity for treating the air volume in question is available. Therefore, thermal comfort within the cabin is only dependent on how well the climate control system fulfills its functions. These functions are:

- Supply air
- Cleanse air
- Control air temperature and humidity
- Distribute air

The configurations to be described below are principally independent of the climate system's degree of (human) control automation and climate control automation. Similarly, these configurations are as valid for simple climate systems as for systems with individual left/right temperature separation and air distribution. In the latter case, all described elements must of course be available symmetrically and controllable individually. The following descriptions also apply to pure heating devices, in which case the evaporator for air cooling and dehumidification is absent.

6.4.3.2.1 Climate control unit functions: supply air

Drawn in by the blower mounted in the climate control unit, air normally enters the vehicle through openings along the engine hood, through a water separator below the hood, and into the climate control unit. From there, air is supplied to the vehicle cabin via air ducts. Finally, air leaves the vehicle through extractor vents (Fig. 6.4-28). Due to pressure losses in these exhaust vents, a closed vehicle (windows and sunroof closed) has a slight overpressure of about 80 Pa. This also prevents air from

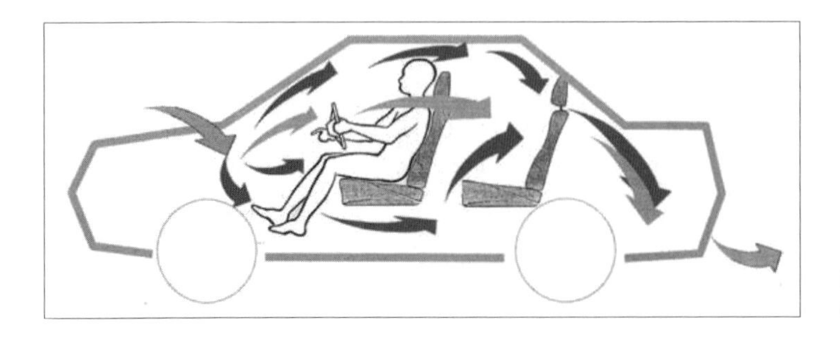

Fig. 6.4-28 Ventilation openings.

entering the vehicle by way of (unintended and undesirable) leaks at doors or windows.

The water separator is normally mounted in the engine or powerplant compartment. Ambient air is usually drawn in from just ahead of the windshield, where only negligible stagnation pressure is available at higher speeds. By means of appropriate geometric design of the water separator, larger particles as well as rain and windshield washer fluid are removed from the airstream before it passes through a short duct through the firewall and into the climate control unit.

An important part of the climate control unit is its so-called blower tract. The suction side of the blower includes a suitable flap mechanism which permits switching the air intake between outside air and cabin air. In the latter case, air from inside the passenger compartment is drawn from the footwell area below the instrument panel and recirculated.

The blower itself consists of an impeller wheel, a scroll housing, and a blower motor. The motor includes a control unit which provides various voltages to permit different blower speeds and therefore different airflow rates. This is usually achieved by means of a simple network of dropping resistors with four or five different possible settings. Alternatively, electronic controls may be used which permit smaller speed steps or infinitely variable blower speed control. Modern blower motors are almost exclusively permanent magnet motors with internal rotating armatures. Alternatively, electronically commutated motors may be employed.

Blower aerodynamics are primarily determined by the combination of impeller and scroll size. The exact configuration strongly depends on individual vehicle requirements. Significant factors include cockpit packaging as well as maximum blower power. A typical value for maximum airflow with outside air is 8 kg of air per minute. Impeller diameter is typically 100–170 mm with a width of 50–90 mm. For space reasons, ideal impellers are seldom used [12]; instead, the motor encroaches into the impeller wheel from one side.

6.4.3.2.2 Climate control unit functions: control temperature and dehumidify

The blower tract described in the previous section is joined to an air distribution box. Its job is to control air temperature and distribution to various points within the passenger cabin.

From the blower, air flows through the evaporator (Section 3.3), where it is cooled from its ambient temperature. Because cooled air cannot contain as much water vapor as warm air, part of the available cooling capacity is used only to dehumidify air. Water condenses on the evaporator surface. Drains at the bottom of the climate control unit permit discharge of condensate from the unit and vehicle interior. At maximum cooling capacity and under normal operating conditions, air exiting the evaporator is typically at a temperature of 3–5°C and a relative humidity of about 90%.

The cooled airstream is at least in part directed through the heater core (Section 3.3). There are two air temperature control system variations.

Water-side temperature control

In this case, differing heating demands on the heater core are met by varying the engine coolant flow rate through the core. To maintain the desired air temperature downstream of the heater core, continuously variable (manually or electrically controlled) valves or quasi-continuous pulse-width–modulated valves control coolant flow between zero and the maximum possible volume flow rate. One important consideration is that at part throttle, volume flow rates of much less than 100 L/hour are often sufficient. Fig. 6.4-27 illustrates a climate control unit with water-side temperature control.

Air-side temperature control

In this configuration, coolant flows through the heater core continuously under all operating conditions. Temperature control is achieved entirely by means of one or more mixing flaps that divert a larger or smaller portion of the airflow through the heater core. Fig. 6.4-29 shows a cross section of an air-side climate control unit. The remaining airflow is diverted through a bypass duct and past the heater core, unheated. Downstream of the heater core, cold and warm air are combined in the mixing chamber in response to mixing-flap settings to achieve the desired air temperature. The ratio of the two airflow rates, established by the mixing flaps, determines temperature.

The advantages of air-side temperature control are

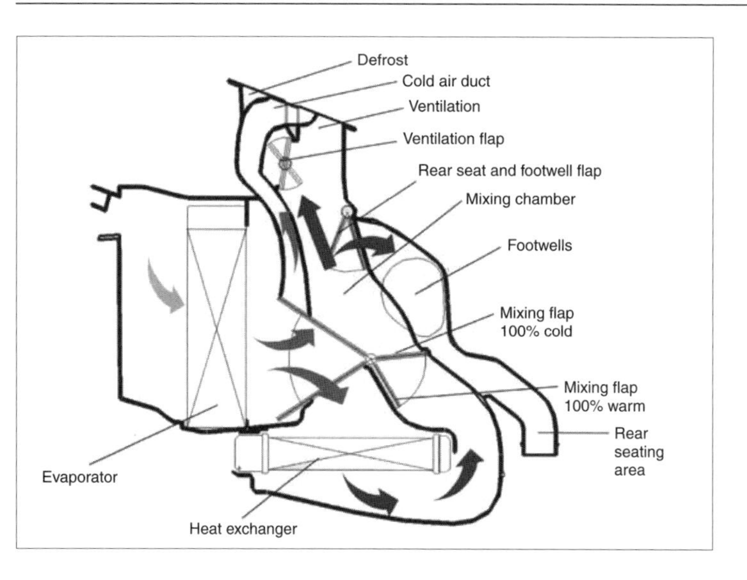

Fig. 6.4-29 Cross section through a climate control unit.

its rapid response to changes in target temperature settings (while in the case of water-side control, thermal inertia of the heater core results in delays); its excellent manual controllability (manual coolant valves have unfavorable control characteristics, especially at low flow rates); and its insensitivity to changing coolant volume flow rates caused by engine speed changes. (In water-side control, changing coolant flow rates have an immediate effect on air temperature.) On the other hand, installation volume requirements for air-side control are somewhat greater than for water-side control.

6.4.3.2.3 Climate control unit functions: air distribution

Outlets to individual air ducts serving various zones of the vehicle interior are located downstream of the heater core or mixing chamber. At the least, these consist of outlets for ventilation (midlevel), defrost, and footwells (Figs. 6.4-29 and 6.4-30). More sophisticated systems may add outlets for rear passenger footwells, rear seating area, or B-pillars (Fig. 6.4-31). Appropriate flap mechanisms permit different airflow rate proportions to individual vents.

Even at fixed temperatures, airstreams to various ducts are not at the same temperatures. To address the vehicle occupants' thermal comfort perceptions, air directed to the dash vents is generally somewhat cooler, while that going to the footwells is somewhat warmer (temperature spread, stratification). One exception to this rule is the "defrost" setting, in which maximum heated air is sent directly to the windshield. The objective is to remove condensation ("fogging") from glass areas. For this reason, a certain amount of air leakage is often directed to the defrost vents.

Air ducts are connected to the outlets of the climate control unit. These ducts carry air from the climate control unit to adjustable, visible vents located in the instrument panel, center console, or sometimes in the rear seating area. Vents in footwell areas are usually not visible and not adjustable. The task of ductwork is to transport air to specific locations in the vehicle interior silently and with the lowest possible pressure drop. Depending on the desired temperature, air distribution is handled in different ways. For heating, defrost and footwell vents are given priority; for cooling, it is likely that so-called midlevel vents will carry most of the airflow. Ideally, cooled air will descend, while heated air will rise from the footwells.

Air is exhausted from the cabin by specifically designed extraction vents. These are usually found in the vicinity of the rear seats or are realized by perforations in the rear package shelf. Air leaves the rear of the car at bumper height.

6.4.3.2.4 Climate control unit functions: air filtering

Use of filters (see also [13]) with approximately 600 to 1,000 cm^2 frontal area and 20–55 mm thickness has become standard in newer vehicles (Fig. 6.4-32).

These so-called particulate filters remove impurities ranging from dust particles, at about 100 μm, to microscopic particles with diameters of only 0.01 μm. This bandwidth contains particulates which are critical for humans, especially respirable particles of <2 μm. The filter material usually consists of synthetic microfibers or a mixture of microfibers and impregnated cellulose fibers.

To meet growing comfort demands, so-called activated charcoal filters are finding increasing application. These remove gaseous contaminants and objectionable odors. Such filters employ open-pored substrate materials to which charcoal particles are attached. Along with

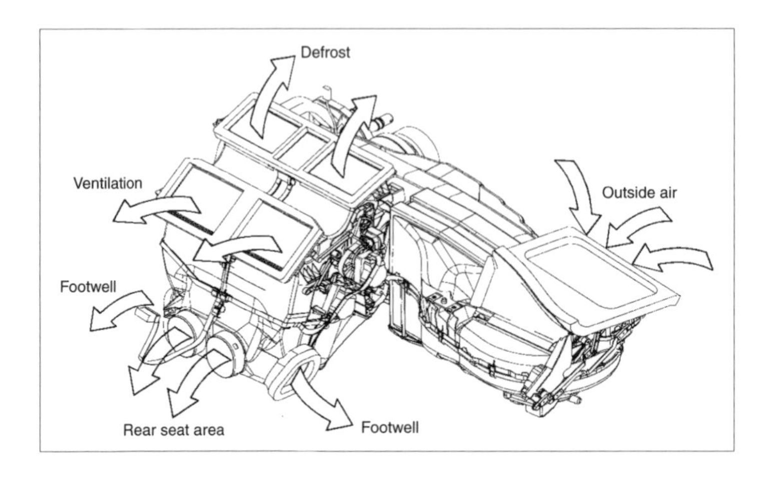

Fig. 6.4-30 Air vents.

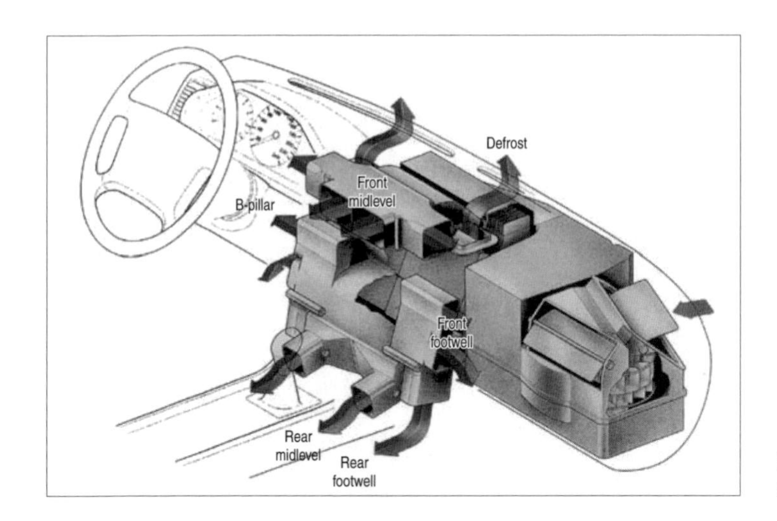

Fig. 6.4-31 Air vents in sophisticated systems.

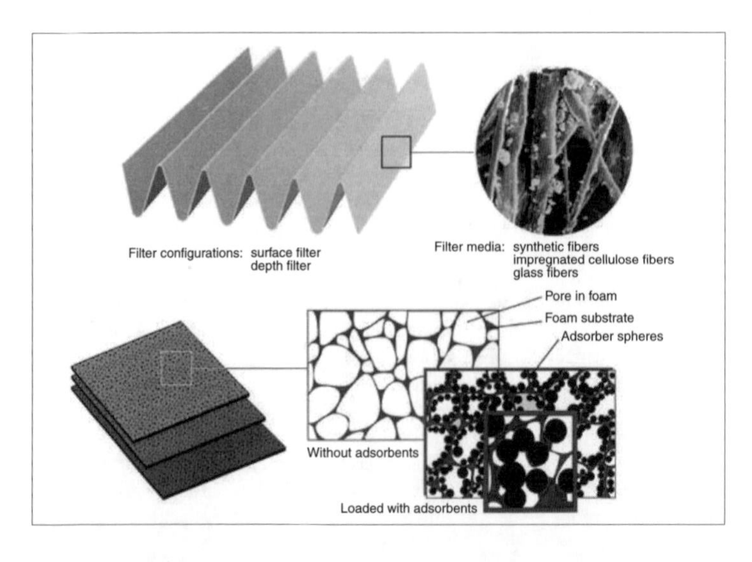

Fig. 6.4-32 Filters.

particulate filters and activated charcoal filters, so-called hybrid filters are often employed. During pleating of the filter material, particulate filters are joined to several layers of activated charcoal filter. The useful life of a hybrid filter, at 30,000 to 50,000 km, is less, however, than that usually obtained with a large activated charcoal filter package.

In principle, particulate filters may support growth of bacteria and funguses (but not viruses). This requires three things: temperature, nutrients (dirt), and moisture. If filters remain idle for longer periods, there is an increased risk of microorganisms colonizing the filter and penetrating to the clean side.

This possibility exists because funguses require a water activity of ≥0.7 and bacteria a water activity of about 1 (i.e., presence of liquid water). Such conditions cannot be positively eliminated from filters. To minimize the risk, filters must be replaced at recommended intervals.

Two main filter mounting variations are commonly encountered. In the first, the filter is located in the engine or powerplant compartment on the suction side of the climate control unit. This variation permits easy filter replacement at the end of its service life. The second variation is mounting between the blower and evaporator. The filter is therefore located inside the climate control unit and cleans air drawn from outside as well as recirculated air. Filter replacement is done from inside the passenger compartment. This design solution offers several advantages, including reduced noise generation from the overall climate control unit, as the filter works as a sound absorber and may have a beneficial effect on airflow.

6.4.3.2.5 Climate control unit configurations

Auto manufacturers' individual space requirements have resulted in basically two different climate control unit configurations.

Symmetrical configuration

In this arrangement, components such as blower, evaporator, and heater core are arranged along the vehicle longitudinal axis (x direction) sequentially and symmetrical to the vehicle centerline (see, e.g., Fig. 6.4-27). This results in good flow conditions, short airflow paths, and correspondingly low pressure drop in the climate control unit. Another advantage of this arrangement is that due to symmetry around the longitudinal axis, left- and right-hand-drive vehicles may be equipped with the same climate control unit. This configuration requires sufficient space in the x direction. In return, dimensions in the y (transverse) direction are relatively compact.

Asymmetric configuration

In this arrangement, the blower and intake tract are displaced left or right of the vehicle centerline along with the evaporator, heater core, and distribution box. Additional filter elements are often installed between the blower and distribution unit. Considerably more space in the y-direction is needed for this configuration, but compared to symmetrical designs, dimensions along the longitudinal axis are more compact (Fig. 6.4-30). Side induction may lead to nonuniform air distribution to the evaporator. Appropriate housing design and possibly air guide elements may be employed to compensate for this disadvantage.

6.4.3.2.6 Climate control unit layout

The basic layout is determined by auto manufacturer requirements in terms of installation space and functionality (e.g., right/left separation, number and location of air ducts, location of outside air connections, refrigerant and coolant lines). In principle, given these requirements, a climate control unit concept may be established; see also [14] and [15].

As described in Section 3.3, layout of heating and cooling circuits employs thermodynamic simulation models. This establishes the shape and size of heat exchangers.

In the concept phase, aerodynamic performance data must be matched to the space available for the blower. An airflow rate of 0.5 m^3/min per person is required—in other words, about 2.5 m^3/min for a typical passenger car. This airflow rate is normally achieved at the lowest blower setting; for comfortable climate control, however—especially during warm-up and cooldown phases—appreciably higher airflow rates are needed. Blower layout is accomplished with the aid of one-dimensional simulation methods. First, pressure reductions across individual components or the entire climate control unit are applied along with information regarding water separation rate, air ducts, and air exhaust. The resulting air consumption curve is compared to blower characteristic curves, as shown in Fig. 6.4-33, to select a blower matching the known requirements.

Today, computational fluid dynamics (CFD) is a well-entrenched component of climate control system development. Based on 3-D CAD models of the heating/air conditioning unit, flow calculations are performed to

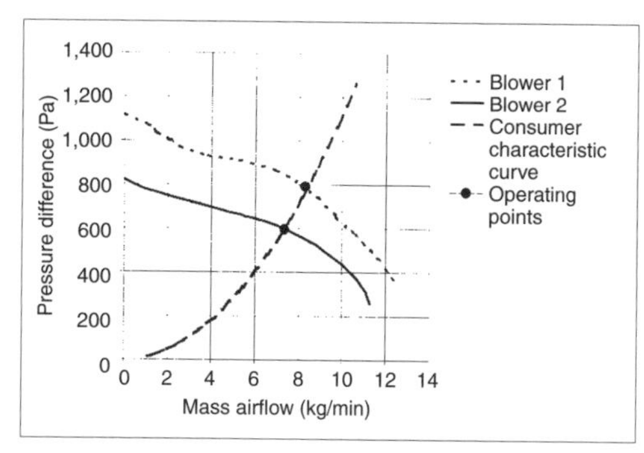

Fig. 6.4-33 Air consumer characteristic curve.

provide insights into aerothermodynamic processes inside the unit. On the basis of these computational results, targeted measures may be applied to improve velocity and temperature distribution. The main application is flow optimization in the distribution box as well as its associated air ducts. Another application for CFD is calculation of flow through filters or heat exchangers.

6.4.3.3 Climate system controls

Seen as a subsystem of the climate control system, the control subsystem consists of a user control panel, sensors for determining conditions inside and outside the passenger cabin, as well as actuators for control elements. These control elements affect thermal comfort inside the passenger cabin.

The most important parameters affecting passenger cabin comfort are interior temperature, interior humidity, insolation (solar radiation input), and pollutants drawn in with ambient air, as well as ambient temperature and humidity.

In high-comfort systems, all of these parameters are captured and compared to target values by a control unit. The system reacts to counter deviations and attempts to reach the defined comfortable condition as quickly as possible. The most important actuating elements are the air conditioner compressor clutch and/or control, water (engine coolant) pump, coolant valves or mixing flaps, auxiliary heating, the climate control system blower output, air distribution flaps, and flaps to control fresh or recirculated airflow.

It is very difficult, if not impossible, to define a general perception of comfort that can be used as a control system target and that is identical for all vehicle occupants. All controls are therefore well-balanced compromises between ever-changing passenger needs and ambient conditions. Needs also change as a function of clothing, mood, and physiology. These also change the currently valid subjective "comfortable temperature." Therefore, even a highly automated system still requires the ability to set target values by means of a user control panel.

6.4.3.3.1 Degrees of automation

According to equipment variations, climate control systems may be classified into three main groups:

Manual climate control systems
Semiautomatic climate control systems
Fully automatic climate control systems.

Control of manual systems

The simplest manual climate control systems offer only a simple (automatic) control to prevent evaporator icing (Section 3.3). All other functions such as temperature preselection, air distribution and blower speed are set manually/mechanically. These must be explicitly set by the user in response to changing conditions.

Control of semiautomatic systems

In the next development stage, semiautomatic climate control systems add cabin temperature control. The user merely inputs a desired temperature by means of the operating unit; the electronic control system uses this preset and boundary conditions (e.g., coolant and ambient temperature) along with the climate system characteristic curve to calculate an output air temperature. The mixing ratio of warm air from the heater core and cold air from the evaporator is adjusted accordingly, or the coolant control valve is actuated. For water-side-controlled systems, a temperature sensor is usually installed downstream of the heater core in a secondary control loop. Air-side-controlled systems, on the other hand, usually do without such a sensor. The control loop is closed by sensing actual cabin temperature.

Control of fully automatic systems

The highest development stage of climate control systems also automatically controls, in addition to temperature, blower output and air distribution. Various electronic blower controllers are installed that set blower output by means of a control input. These blower controllers are actuated by the main interior climate controller. Often, additional functions are included for special operational or boundary conditions.

Automatic air distribution requires actuators that close individual climate system vents or nozzles by means of flaps. A block diagram of a fully automatic climate control system is shown in Fig. 6.4-34.

Many fully automatic climate control systems have separate left and right channels which can be set independently, doubling the number of some necessary flaps. Increasingly, switching between fresh and recirculated air is also controlled by automatic systems that sense ambient air quality and, if necessary, block pollutant-laden outside air and switch to recirculated cabin air.

6.4.3.3.2 Operation of climate controls, ergonomics

It is necessary to input climate control system target values independently of the degree of automation. In the simplest case of a manual climate control system, operation consists of changing basic climate system functions. In other words, the following functions must be exercised by the operator: increase or decrease air volume, change temperature setting or if necessary turn on the climate control system, change air distribution (e.g., select the defrost setting), and operate the fresh air selector. As climate control systems become more complex, other functions are added, such as right/left separation and corresponding control functions, or a temperature-adjustable center vent.

User control elements and symbols must be recognizable; that is, they must be sufficiently bright and large and, of course, understandable; conceptually, any given symbol must be clearly associated with a certain function. Graphic symbols are more readily understood

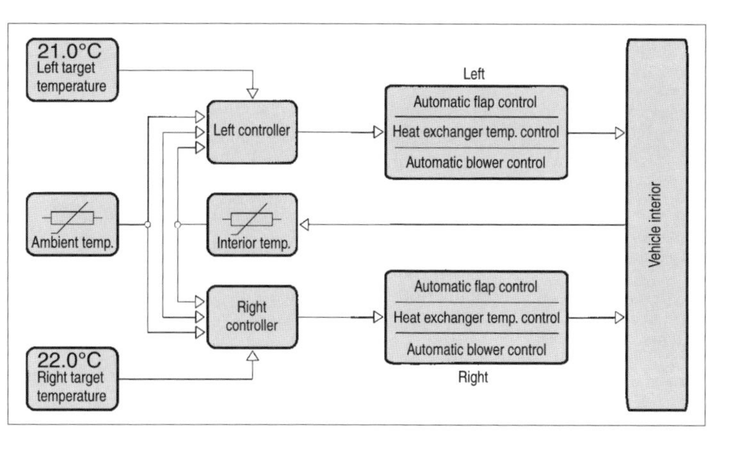

Fig. 6.4-34 Fully automatic climate system control.

than lettering. Naturally, the shape of control elements must clearly indicate their mode of actuation, such as pressing, rocking, turning, or sliding. Furthermore, control elements must be secondarily compatible; this means that operator expectations match reality. For example, an upward movement is seen as "from less to more." Color coding—blue for cold, red for warm—also satisfies this requirement.

A typical control panel with examples of various control elements is shown in Fig. 6.4-35.

6.4.3.3.3 Actuator drives and sensors

Adjustable airflow control flaps are installed in all climate control systems to divert air to various outlets. Climate control systems with a higher level of automation replace purely mechanical actuation systems—such as, for instance, Bowden cables, with electromechanical actuators.

Vacuum solenoids are the most common form of mechanical actuator for climate systems. In the most basic case, the solenoid can take either of two positions (two-chamber servos allow three positions). An electrically actuated valve switches a vacuum hose between a central vacuum reservoir and the vacuum solenoid working chamber. When actuated, the solenoid performs a linear or rotary motion. When the air chamber is again vented to atmosphere via the valve body, a spring returns the solenoid to its rest position.

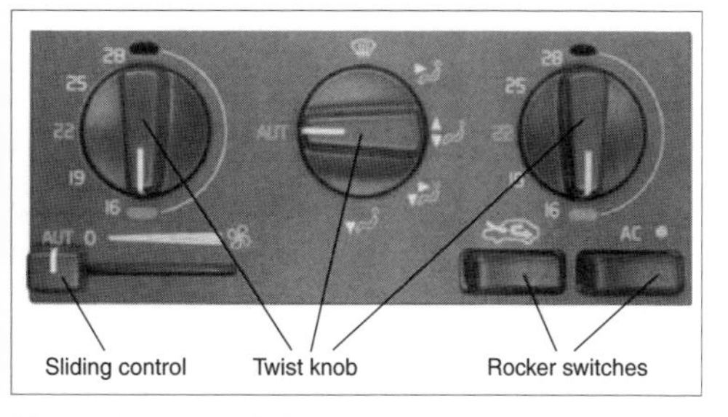

Fig. 6.4-35 Control elements.

Electric actuators are becoming increasingly popular for exact positioning of air control flaps. For climate control system applications, a variety of actuator configurations is known, all differing according to application and purpose. Motors range from simple DC motors to AC motors with reporting potentiometers that permit precise determination of flap position to stepper motors. In recent years, stepper motors [16] have gained importance and widespread acceptance for climate control system applications; reasons include continued development of electronics, miniaturization and, not least, the fact that stepper motor positioning is reproducible without resorting to a separate means of position feedback. Another stepper motor advantage is low wear, which makes sense in climate systems; unlike DC motors, steppers do not contain any sliding elements. With a growing number of similar actuator drives in a climate control system, the wiring harness and the number of connectors on the manual control unit rapidly grow. Bus systems offer a familiar solution for reducing wiring harness complexity. This would require a certain degree of intelligence in the actuator system to carry data transmissions between control unit and actuator and to drive the final stage directly.

A multitude of sensors controlling a variety of different physical properties is used in modern climate control systems which make a significant contribution to passengers' comfort perceptions. Due to many different requirements for installation space and boundary conditions, climate control systems exist in many different forms. References [17] to [21] include information on climate control system construction, measurement principles, and evaluation.

References

[1] Bubb, H. "Ergonomie in Mensch-Maschine-Systemen," lecture notes, "Komfort und Ergonomie im Kraftfahrzeug," Haus der Technik, Essen, 1995.

[2] Fanger, P.O. *Thermal Comfort, Analysis and Applications in Environmental Engineering.* Danish Technical Press, Copenhagen, 1970.

[3] Wyon, D.P., S. Larsson, B. Forsgren, and I. Lundgren. "Standard Procedures for Assessing Vehicle Climate with a Thermal Manikin." SAE Paper 890049, 1989.
[4] Bohm, H., A. Browen, O. Noren, I. Hohner, and H. Nilson. "Thermal Environment in Cab." The American Society of Agricultural Engineers.
[5] Hucho, W.-H. *Aerodynamik des Automobils*. 3rd ed., VDI-Verlag, Düsseldorf, 1994.
[6] *ASHRAE Handbook Fundamentals: Physiological Principles and Thermal Comfort*, Atlanta, 1993.
[7] "Grundlagenuntersuchung zum Einfluss der Sonneneinstrahlung auf die thermische Behaglichkeit in Kraftfahrzeugen." FAT No. 81, December 1989.
[8] "Einfluss der Sonneneinstrahlung auf die thermische Behaglichkeit in Kraftfahrzeugen." FAT No. 109, May 1994.
[9] Arminger, G. "Einfluss der Witterung auf das Unfallgeschehen im Straßenverkehr." ATZ August 1999.
[10] Mackworth. *Researches in the Measurement of Human Performance*. Cited in Wenzel et al., *Klima und Arbeit*, 3rd edition, Bayrisches Staatsministerium für Arbeit und Soziales, 1984.
[11] Taxis-Reischl, B. "Wärmebelastung und Fahrverhalten." ATZ August 1999.
[12] Eck, B. *Ventilatoren*, 5th edition, Springer-Verlag, 1991.
[13] Hipp-Kalthoff, C., and T. Rinckleb. "Gesundheitsschutz und Erhöhung des Fahrkomforts durch Filterung der Fahrzeug-Innenluft." ATZ September 1995.
[14] Colinet. *La Climatisation*, Boulogne, 1987.
[15] Wertenbach, J., and D. Wahl. *Thermal Comfort of Occupants*.
[16] Bollinger, H., and D. Hrischi. "Der Schrittmotor—das Antriebselement mit Zukunft." Special edition of Landis & Gyr-Mitteilungen No. 1-84.
[17] Anon. "Exhaust sensors for auto-dumper system." Figaro Engineering Inc., Japan.
[18] Käfer, O. "Umluftautomatik mit Feuchteregelung im Fahrzeuginnenraum." ATZ July 1998.
[19] Anon. "Optical Sensors." Control Devices, Inc., USA.
[20] Knittel, O., and C. Ruf. "Von der Erfassung der Luftfeuchtigkeit zum komfortoptimierten Klimabetrieb." VDI Berichte 1415.
[21] Anon. "HyCal Sensing Products, Temperature & Moisture Sensors." Honeywell Inc., USA.

6.5 Vehicle safety

6.5.1 General

The high growth rate of passenger and freight transport by means of passenger cars and commercial vehicles has its advantages but also its disadvantages: up until the early 1970s, this growth rate has been partly responsible for a rise in serious injuries and fatalities per passenger-kilometer. Statistics show that in the intervening years, the number of fatal injuries per transport kilometer, over time, has dropped thanks to numerous road improvements (roadway, signaling), traffic education, and vehicle performance. The topic of vehicle safety also gained significance in the scientific community. As early as the mid-1950s, Prof. Koessler, director of the Institute für Fahrzeugtechnik [Institute for Automotive Technology] at the University of Brunswick, Germany, defined the tasks of automobile developers as follows: "Vehicles must be designed and constructed so they may transport a commodity as quickly, safely, and comfortably as possible from Point A to Point B." Automobile manufacturers intensified their activities in the fields of accident prevention as well as reducing the effects of accidents, popularly known as "active and passive safety" [1], as shown, for example, by Béla Barényi's seminal 1952 patent for "the principle of the rigid safety cell" [2]. Although, in the United States, occupant protection in the event of a collision was and remains at the forefront, European regulations have conferred greater responsibility on the vehicle driver. Automotive safety in the United States was given a remarkable boost in the mid-1960s by consumer advocate Ralph Nader and the creation of NHTSA (National Highway Traffic Safety Administration). Meanwhile, we have worldwide safety regulations supplemented by consumer information and product liability. Consumer information is disseminated by large insurance firms, regulatory agencies, consumer protection organizations, and automotive magazines. Product liability has also provided vital impetus for the engineering of safety-related systems. Vehicle manufacturers assume responsibility for the technical execution of their products, ensuring that the vehicles they offer for sale at least represent the current state of the art. Among customers as well, vehicle safety enjoys high priority; a car with notable safety features enjoys certain market advantages.

Tables 6.5-1 and 6.5-2 provide an overview of safety-relevant regulations in the United States and the European Community, respectively. These regulations themselves are in a constant state of flux, thanks to changing technology and new engineering insights.

Table 6.5-1 FMVSS (Federal Motor Vehicle Safety Standards, USA)

FMVSS	Content
101	Control and displays
102	Transmission shift lever sequence, starter interlock and transmission braking effect
103	Windshield defrosting and defogging systems
104	Windshield wiping and washing systems
105	Hydraulic brake systems
106	Brake hoses
108	Lamps, reflective devices, and associated equipment
109	New pneumatic tires
110	Tire selection and rims
111	Rearview mirrors
113	Hood latch system
114	Theft protection
116	Motor vehicle brake fluids
118	Power-operated window, partition, and roof panel systems
119	New pneumatic tires for vehicles other than passenger cars
120	Tire selection and rims for motor vehicles other than passenger cars
121	Air brake systems
124	Accelerator control systems
126	Truck-camper loading (superseded by TP575.103)

Table 6.5-1 (continued)

FMVSS	Content
129	New nonpneumatic tires for passenger cars
135	Passenger car brake systems
201	Occupant protection in interior impact
202	Head restraints
203	Impact protection for the driver from the steering control system
204	Steering control rearward displacement
205	Glazing materials
206	Door locks and door retention components
207	Seating systems
208	Occupant crash protection
209	Seat belt assemblies
210	Seat belt assembly anchorages
211	Wheel nuts, wheel discs, and hub caps (reserved)
212	Windshield mounting
213	Child restraint systems
214	Side impact protection
216	Roof crush resistance
219	Windshield zone intrusion
301	Fuel system integrity
302	Flammability of interior materials
303	Fuel system integrity of compressed natural gas vehicles
304	Compressed natural gas fuel container integrity

6.5.2 Automobile safety domains

Automobile safety may be divided into the categories shown in Fig. 6.5-1. An early thorough representation was developed by Wilfert [3]. As a result of numerous expansions, the general concepts "passive" and "active" safety no longer adequately describe the field [4]. Additionally, the following concepts are defined:

— Accident prevention (colloquially, "active safety"): all measures which serve to avoid accidents
— Accident mitigation (colloquially, "passive safety"): all measures which serve to minimize the effects of accidents
— Exterior safety: design of a vehicle's exterior contours with the goal of minimizing injuries to "collision partners"
— Interior safety: design of vehicle interiors and components with the goal of minimizing injuries to occupants
— Occupant restraint systems: vehicle components which affect motion of occupants relative to the vehicle
— Primary collision: collision of a vehicle with an obstacle
— Secondary collision: impact of vehicle occupants against vehicle interior components
— Active safety systems: safety and restraint systems which are activated manually (e.g., safety belts)
— Passive safety systems: safety and restraint systems which, in the event of a serious accident, are activated automatically by sensors (e.g., airbags) or safety belts which are automatically applied (e.g., by closing the vehicle door).

Table 6.5-2 Safety-relevant European Community directives[a]

70/156	Type approval of motor vehicles and their trailers
70/221	Fuel tanks and rear protective devices
70/311	Steering equipment
70/387	Doors
71/127	Rearview mirrors
71/320	Braking devices
72/245	Suppression of radio interference
74/60	Certain interior fittings
74/61	Antitheft devices
74/297	Behavior of steering mechanism in event of impact
74/408	Strength of seats and their anchorages
74/483	External projections
75/443	Reverse and speedometer equipment
76/114	Location and attachment of statutory plates and inscriptions
76/115	Anchorages for safety belts
76/756	Lighting and light signaling equipment
76/757	Reflex reflectors
76/758	End outline marker lamps, side lamps, rear lamps, and stop lamps
76/759	Direction indicator lamps
76/760	Rear registration plate lamps
76/761	Headlamps and filament lamps for them
76/762	Front fog lamps and filament lamps for them
76/769	Dangerous substances
77/389	Towing devices
77/538	Rear fog lamps
77/539	Reversing lamps
77/540	Parking lamps
77/541	Safety belts and restraint systems
77/649	Field of vision
78/316	Identification of controls, tell tales, and indicators
78/317	Windscreen defrosting and demisting systems
78/318	Windscreen wiper and washer systems
78/548	Heating systems
78/549	Wheelguards
78/932	Headrests
92/22	Safety glass
92/23	Tires
92/114	External projections forward of the cab's rear panel (commercial vehicles)
94/20	Mechanical couplings
96/27	Occupant protection, side impact
96/79	Occupant protection, frontal impact

[a] Only basic directives are listed. Other regulations also apply (e.g., trip recorder).

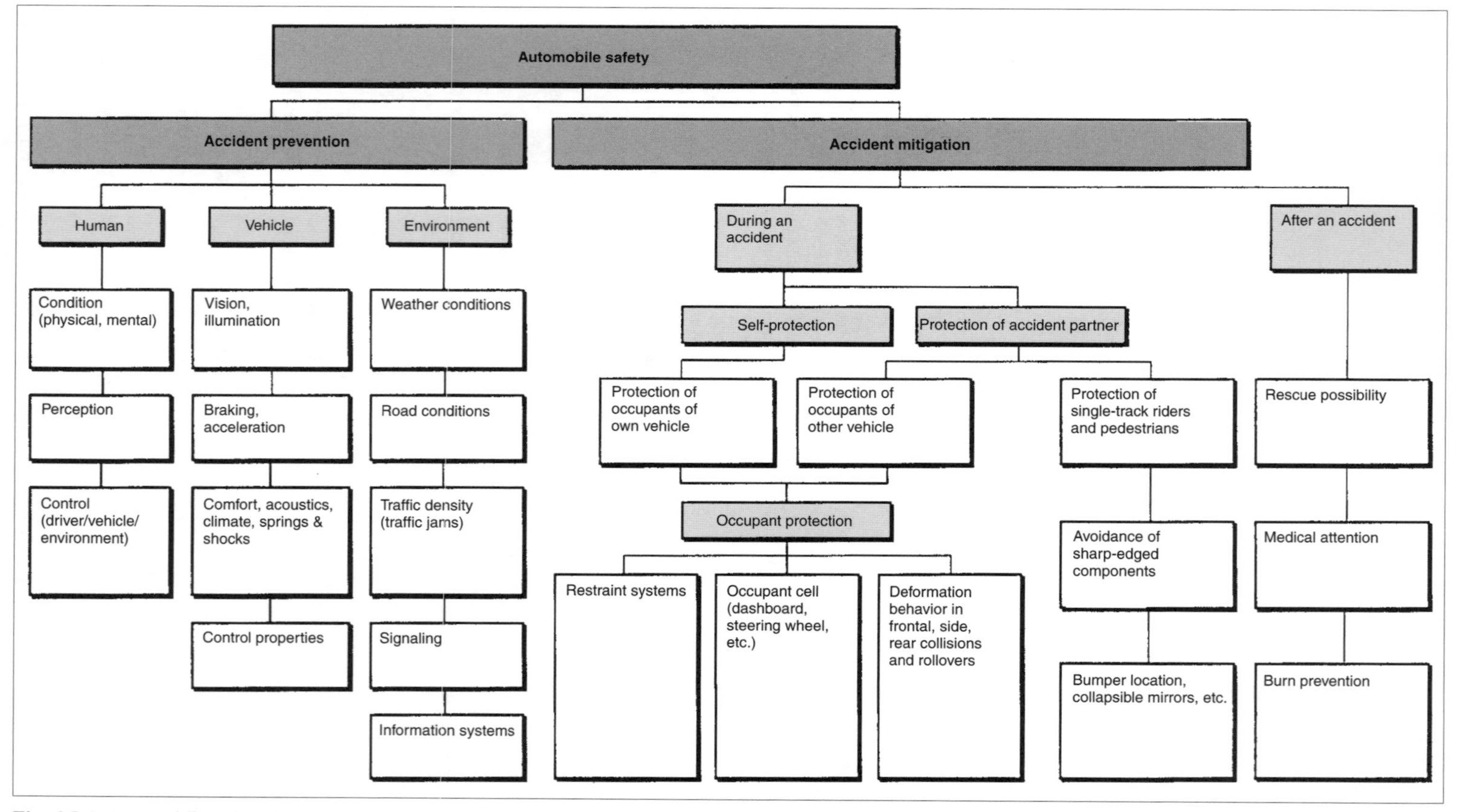

Fig. 6.5-1 Automobile safety domains.

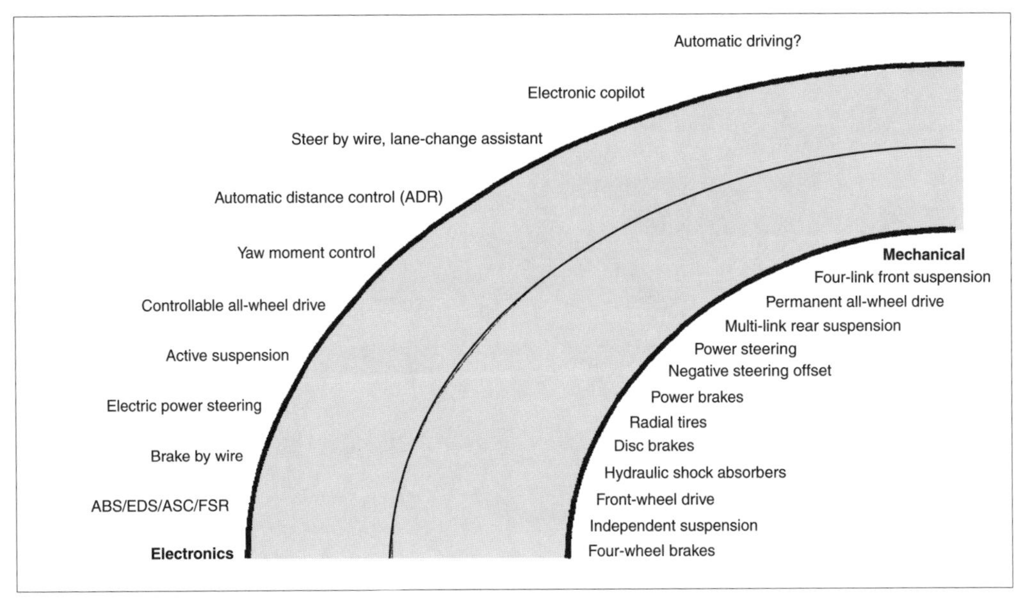

Fig. 6.5.2 Accident prevention innovations.

Along with driver behavior, accident-preventive measures include "benign" vehicle behavior in extreme handling situations, environmental impact, information presented to the driver, road and traffic conditions, and, of course, the technically faultless condition of the vehicle.

Human error has been and remains the main cause of accidents. Accident rates will continue to decline as a result of alcohol-free driving, good comfort and ergonomics, adequate illumination and good visibility, numerous assistance systems—especially in terms of vehicle behavior (brakes, powerplant, ABS, yaw control, brake assistant; see also Chapter 7), highly stable handling, vehicle insensitivity to crosswinds, as well as use of information and communication systems. Fig. 6.5-2 illustrates the path of change as a portion of the accident prevention field [5].

6.5.3 Statistical accident data

The historical development over time of accident statistics is recorded and disseminated worldwide. Prof. Max Danner of Munich and Berlin has garnered especially well deserved recognition for his systematic capture and evaluation of accident data (e.g., [6]). Fig. 6.5-3 [7] illustrates development of the traffic fatality rate as a function of billions of vehicle kilometers per year for various countries. Comparing the years 1996 and 1970 shows a reduction in the fatality rate of more than 70% on average, with Great Britain leading, in a positive sense. Comparing the same years, public transportation and railways are safer by a factor of ten.

The main group of accident participants is passenger car occupants, followed by pedestrians, bicyclists, and riders of single-track motor vehicles. Fig. 6.5-4 examines accident participants in greater detail. Most apparent is the large proportion of single-car fatal accidents and personal injuries in car-to-car collisions. Overall, the proportion of vehicle occupant fatalities relative to total deaths and injuries has decreased markedly over the last few years. Thanks to a larger database in the Federal Republic of Germany [8], fatal passenger car accidents are broken down as follows:

- Side impact in front door area, 25.9–30.5%
- Frontal impact with zero offset, 17.7–22.1%
- Pure roof impact after a rollover, 14.8–17.5%
- Offset frontal impact, 13.2–14.2%
- Large-area side impact, 9.4–11.9%
- Car-to-car side impact, oblique from front, 5.7–7.4%

There are numerous other parameters such as impact speed, speed changes, seat occupancy, occupant type, collision object, severity of injuries to various body parts, and so forth. Consequently, one may recognize very precisely in the case of car-to-pedestrian collisions, for example, which pedestrian body areas are more likely to be injured [9].

Fig. 6.5-5 illustrates car-to-passenger collisions with accident injury severity on the Abbreviated Injury Scale (AIS) >2 (medium severe injuries such as deep flesh wounds and concussion with unconsciousness lasting less than 15 min) as a function of collision speed with pedestrian.

Accident analysis forms the basis for numerous accident simulation tests and crash dummy criteria.

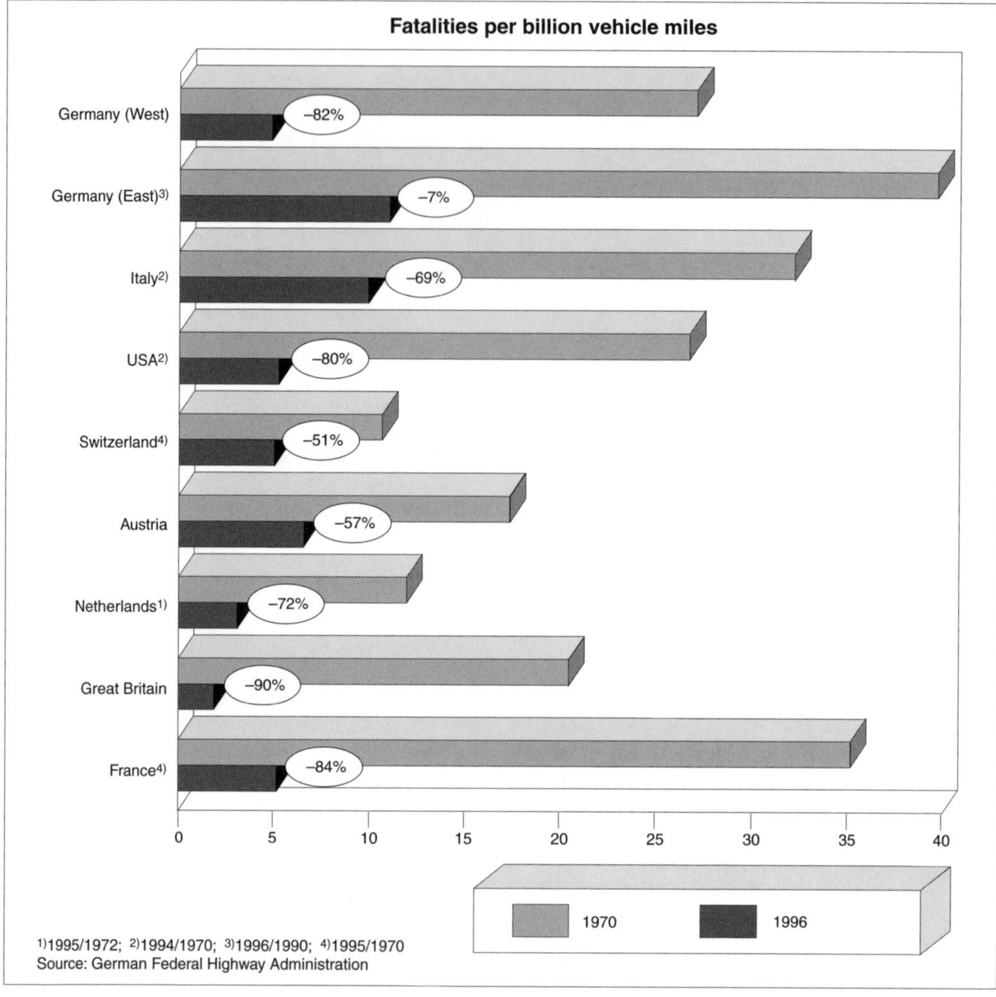

Fig. 6.5-3 Development of traffic fatality rates in western Europe and USA.

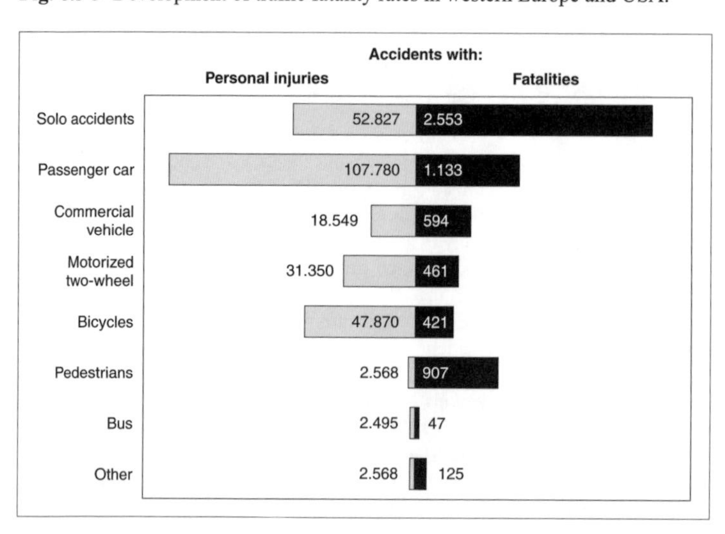

Fig. 6.5-4 Opponents in passenger car accidents involving personal injuries, and fatal accidents; 1994.

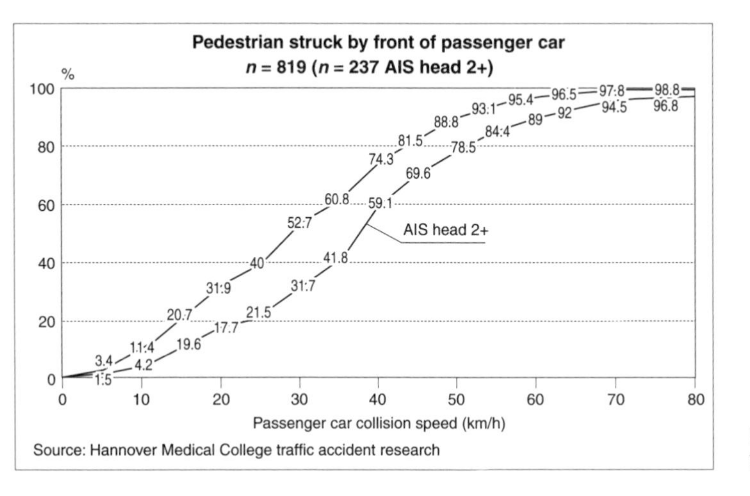

Fig. 6.5-5 Pedestrians struck by front of passenger car.

6.5.4 Biomechanics and protection criteria

6.5.4.1 Biomechanics

6.5.4.1.1 Basics

First in the United States and later in Europe, intensive research has been and continues to be conducted in the field of biomechanics. Significant contributions to the investigation of humans' ability to withstand shock loads were made by United States Air Force Col. John Stapp, who, as his own test subject, became the first human being to decelerate from 632 mph (approximately 1,000 km/h) to a full stop in 1.4 s. Assuming constant deceleration (step impulse), this implies a constant 20 g deceleration. An ongoing series of annual conferences on this topic, held in the United States, have been named for Colonel Stapp, who has been bestowed with the highest awards and honors in recognition of his activities. The foreword and introductory dedication to the *Proceedings* of the 8th Stapp Conference [10] reads in part:

The Stapp Car Crash Conferences are named in honor of Colonel John Stapp, USAF (MC), who pioneered (and is still pioneering) in establishing human impact tolerance levels. His historic rocket sled rides at Holloman Air Force Base, New Mexico, in 1954, in which he voluntarily subjected himself to up to 40 G accelerations while stopping from a speed of 632 miles per hour in 1.4 seconds still represent the best basis for quantitating human tolerance to acceleration. In addition to his own dangerous volunteer work, he has directed countless other safety research programs involving human volunteers, animals and cadavers. The equipment and techniques developed under his guidance have become standard in this research area and have contributed much to the advancement of safety. The naming of these conferences after Colonel STAPP is a fitting tribute to a man who has dedicated his life—even to the point of risking it—to research aimed at increasing man's chances of survival in adverse crash environments.

Along with the Stapp Conferences, another important institution for exchange of biomechanical research findings is IRCOBI, the International Research Council on the Biomechanics of Impacts [11]. For vehicle safety, biomechanics is a means of describing injury mechanisms and an instrument for determining mechanical load-bearing capacity of the human body. Results of biomechanical research lead to establishment of load limits. Protection criteria, derived from these limits, are physical quantities which can be measured by experimental equipment and whose raw values or derived limits may not be exceeded.

6.5.4.1.2 Human load-bearing capacity

Load-bearing capacities describe, among other things, bone fractures, organ damage, and other injuries. Classification is by means of the "Abbreviated Injury Scale" (AIS) or the "Overall Abbreviated Injury Scale" (OAIS). The AIS and OAIS evaluate, respectively, individual and overall injuries and assign values from 0 to 6 [12]. Table 6.5-3 outlines classification of various injuries.

Loading limits are affected by age, gender, anthropometry, body mass, and mass distribution. In accident simulation tests, it is therefore difficult to represent all persons who might be affected by an accident—vehicle occupants, pedestrians, etc. Researchers try to capture as wide a spectrum as possible with the aid of test dummies.

The following will describe some human loading values in greater detail.

External injuries

In older vehicle designs, impact with the windshield could result in facial and neck lacerations. Evaluation of these injuries was conducted by Prof. Lawrence M. Patrick of Wayne State University. Bonded-in laminated windshields have been a major contributor to a reduction in lacerations, in addition to containing driver and passenger airbags.

Table 6.5-3 Abbreviated Injury Scale (AIS)

AIS stage	Injury severity
1	Minor injury, e.g.: – laceration and abrasions – contusions
2	Moderate injury, e.g.: – large lacerations – cerebral injury with unconsciousness (<15 min)
3	Serious but not life-threatening injury, e.g.: – cerebral injury with unconsciousness (<1 h) – rupture of diaphragm – loss of eye
4	Severe injury, life-threatening but survival likely, e.g.: – cerebral injury with unconsciousness (<24 h) – abdominal rupture – amputation of leg, above knee
5	Critical injury with uncertain survival, e.g.: – cerebral injury with unconsciousness (>24 h) – myocardial rupture – cervical spine injury with quadriplegia
6	Very severe injury, survival unlikely, not treatable, e.g.: – skull fracture – crushed thorax – cervical rupture at or above third vertebra

Cranial fractures as a result of impact with vehicle interior components were summarized by John J. Swearingen [13] as follows: skull deceleration multiplied by head mass results in a force resulting in skull fracture in the event of distributed impact load. Deceleration values for the skull are 200 g, 30 g for the nose, and 40 g for the chin. Chest injuries may result from impact with the steering wheel and instrument panel. Therefore, impact reaction force should be less than 8,000 N if possible. Chest deformation may not exceed 6 cm, with 5 cm as a stated limit.

Femur fracture may be assumed in the event of longitudinal forces in excess of 11,000 N. Recently, injuries to lower extremities (especially ankles) have become mor prominent now that head and upper body are well protected.

Internal injuries

Human internal injuries are much more difficult to record. The biggest problem is unquestionably loading of the brain and neck. As a guideline for head injuries, in the anterior-posterior direction, acceleration of 80 g over a period of 3 ms should not be exceeded. Unconsciousness and severe brain damage are evaluated according to the so-called Patrick curve (Wayne State University Concussion Tolerance Curve [14]; Fig. 6.5-6). The time-deceleration function indicates that the level of loading must decrease over time. The Head Injury Criterion (HIC) is derived from the Patrick curve.

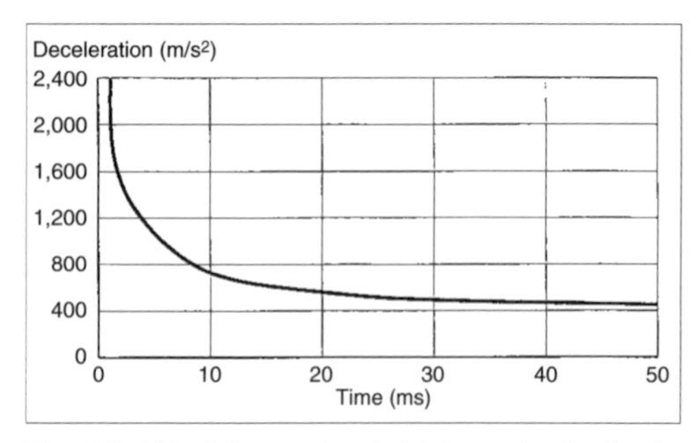

Fig. 6.5-6 Patrick curve (yardstick for evaluating loads on human brain).

Dr. Ernst Fiala [15] provides insight into load limits for angular acceleration. Fiala determined that with a brain mass of 1,300 g, angular acceleration of less than 7,500 rad/sec^2 are permissible.

The neck is just as critical as the head. As a connecting element between torso and head, it is subjected to more or less severe loads in any accident. This especially affects the seven neck vertebrae, which are highly loaded during a forward movement of the head relative to the torso (flexure) and relative rearward movement (extension). These movements result in tension, compression, and inertia forces as well as torques. Depending on occupant musculature and behavior, critical injuries might result. Excessive torque at the occipital condyles (rounded prominences on each side of the base of the skull) may be especially critical.

Neck injuries are becoming increasingly prominent since safety belts and airbags provide such good head and body protection.

6.5.4.2 Protection criteria

Numerous criteria exist for use with installations using test dummies or individual body parts. Major criteria are listed below. Country-specific influences must be considered. In the United States, for frontal impact, the following apply:

— *Head Injury Criterion (HIC)*

$$HIC = \left[(t_2 - t_1) \left\{ \frac{1}{t_2 - t_1} \int_{t_1}^{t_2} a(t)dt \right\}^{2.5} \right]_{max} \leq 1{,}000$$

— Resulting chest acceleration: <60 g over a period of more than 3 ms
— Sternum deformation: <3 in. (76.3 mm)
— Femur force: <10 kN per femur

European criteria are as follows:

— Head: HPC, *Head Protection Criterion*, <1,000 and resulting head acceleration <80 g, with no spikes <3 ms
— Thorax: thorax deformation (TCC, *Thorax Compression Criterion*) <50 mm
— *Viscous criterion* VC <1.0 m/s
— Legs: femur force <9.07 kN at 0 ms and <7.56 kN after 10 ms
— TCFC, *Tibia Compressive Force Criterion* <8 kN
— TI, *Tibia Index*, measured at top and bottom of each tibia, must not exceed 1.3
— The Tibia Index is $|M_R/(M_C)_R| + |F_Z/(F_C)_Z|$
Where $M_R = \sqrt{(M_X)^2 + (M_Y)^2}$ and $|M_X|$ and $|M_Y|$ are bending moment around x- or y-axis respectively; $(M_C)_R < 225$ Nm, critical bending moment;
$F = |F_Z|$, absolute value of the measured axial force; and
$(F_C)_Z < 35.9$ kN, critical compression force in z direction
— Knee displacement <15 mm
— NIC, *Neck Injury Criterion*, may not exceed the tensile and shear forces specified in kN as functions over time. The bending moment around the y axis in rearward motion should be <57 Nm. Forward motion is at present only documented (limiting value <190 Nm).
— Steering wheel movement: the residual steering wheel displacement at the center of the steering wheel shall be <80 mm vertically and <100 mm horizontally.

Fig. 6.5-7 provides an overview of some of these requirements.

For side impact, the following criteria apply in the United States:

— Thorax: *Thoracic Trauma Index* (TTI), passenger cars with four doors <85 g; with two doors <90 g.
— $TI = ½(G_R + G_{LS})$ where G_R is the maximum acceleration of the upper and lower ribs and G_{LS} is the maximum acceleration of the lower spine in g
— Pelvis: the resulting accelerations must not exceed 130 g

The following criteria apply in Europe:

— Head: *Head Protection Criterion* (HPC) <1,000 (−), in the event the head touches a vehicle component
— Thorax: thorax compression <42 mm, *viscous criterion* VC <1.0 m/s
— Pelvis: *Pubic Symphysis Peak Force* (PSPF) <6 kN
— Abdomen: maximum abdominal internal loading (*Abdomen Performance Criterion*, APF) <2.5 kN ≈ external loading <4.5 kN

Fig. 6.5-8 shows some requirements for evaluation of side impact protection performance in the United States and Europe.

In designing vehicles and restraint systems, tolerances of the overall system must be considered. In other words, to ensure that, for example, no vehicles exceed the HIC limit of 1,000, the development target must be considerably lower.

6.5.4.3 Simulation apparatus

6.5.4.3.1 Head

To simulate head impact with the instrument panel or within the vehicle interior, a reduced pendulum mass of

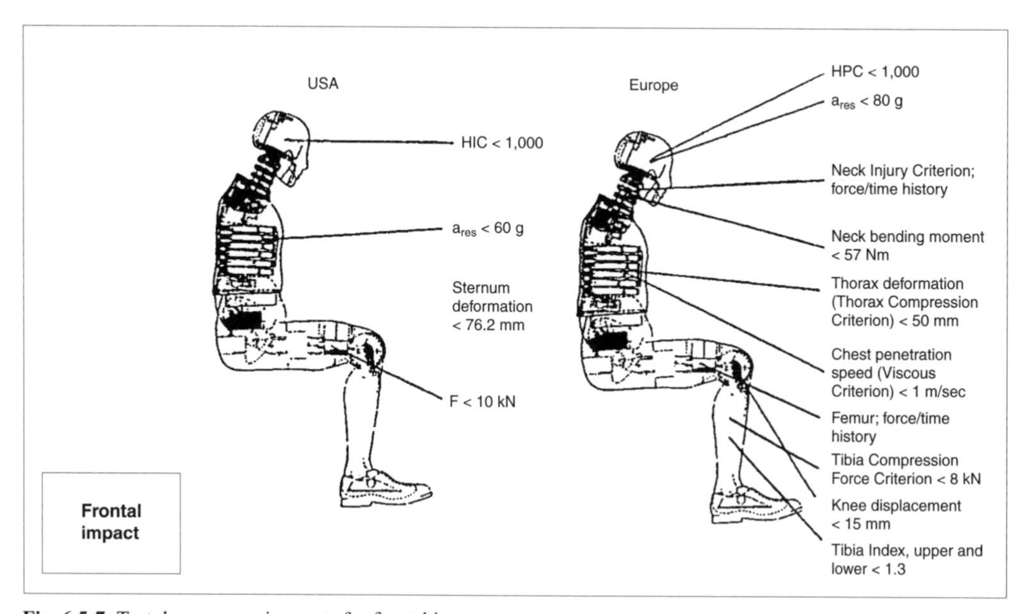

Fig. 6.5-7 Test dummy requirements for frontal impact.

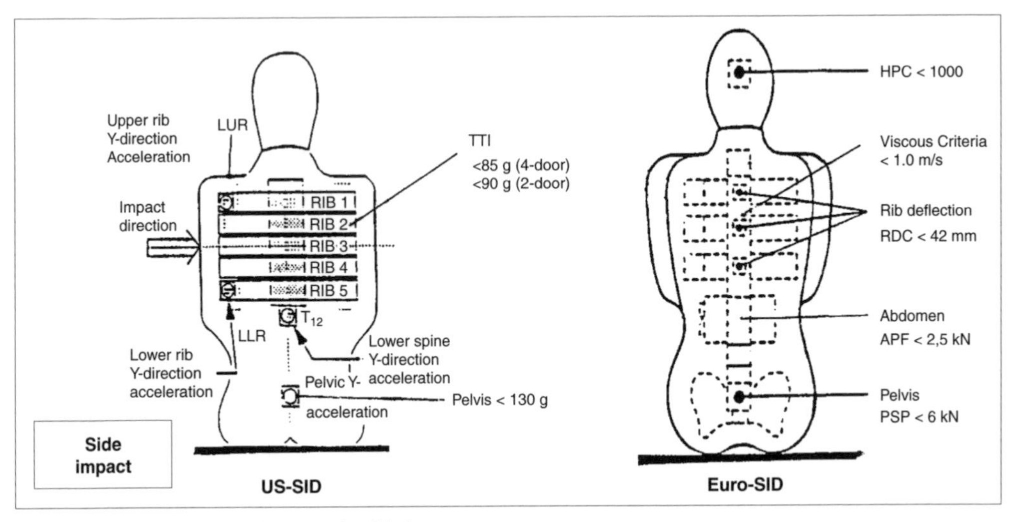

Fig. 6.5-8 Test dummy requirements for side impact.

6.8 kg as specified in SAE J921 [16] is fitted with accelerometers. The head may be covered with a film to evaluate maximum surface pressure during impact.

6.5.4.3.2 Thorax

To measure horizontal force during an impact with the steering system, a body as described in SAE J944a is used. This body segment represents the thorax of a 50th percentile male and weighs 36 kg.

6.5.4.3.3 Overall body

Numerous test dummies are available for simulation of the overall body. These range from child test dummies (simulating various ages) to 5th percentile female, 50th percentile male, and 95th percentile male. (The percentile number always includes the corresponding range).

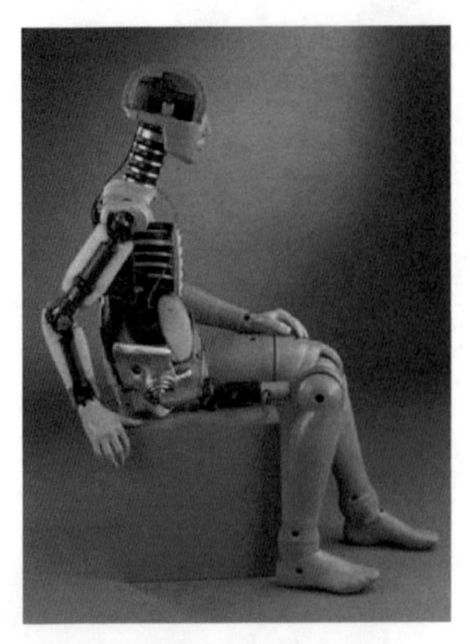

Fig. 6.5-9 Hybrid III test dummy.

For actual testing to meet regulatory requirements, the Hybrid III dummy (Fig. 6.5-9) is used for frontal impact, and for side impact the U.S. SID (Side Impact Dummy) and EuroSID are used. Fig. 6.5-10 illustrates the construction differences between the two test dummies. All dummies have sensors installed in the head, neck, thorax, pelvis, and femurs. Measured parameters are acceleration, force, and deformation.

As insight into injury mechanisms and loading limits increases, improved test equipment and methods are part of a continuous optimization process. For example, in Sweden, the firms Saab, Volvo, and Autoliv, in cooperation with Chalmers University of Technology, developed BioRID and BioRID-II (Biofidelic Rear Impact Dummy) to permit evaluation of whiplash injuries. Development is continuing on mathematical models as well as actual test dummies.

6.5.5 Quasistatic demands on vehicle bodies

6.5.5.1 Seat and seat belt anchorages—tests

In the event the inner seat belt latch is attached to the seat, both seat and anchorages are tested simultaneously. Force is uniformly applied to the body blocks by means of tension bands. Each seat must withstand at least 14,000 N. Reinforcement at the upper B-pillar mounting points must be configured to avoid tearing the B-pillar as a result of excessive local stiffness.

In general, the seats themselves cannot absorb the forces applied to the anchorages, as the seat only needs to withstand a force of twenty times its own weight over a period of 30 ms. The inner anchorage makes use of the rigid central tunnel, to which a serrated seat rail is attached. When loaded by a sufficiently severe accident, the anchorage latches firmly. Seat backrests, which also contain the upper seat belt anchorage, must withstand

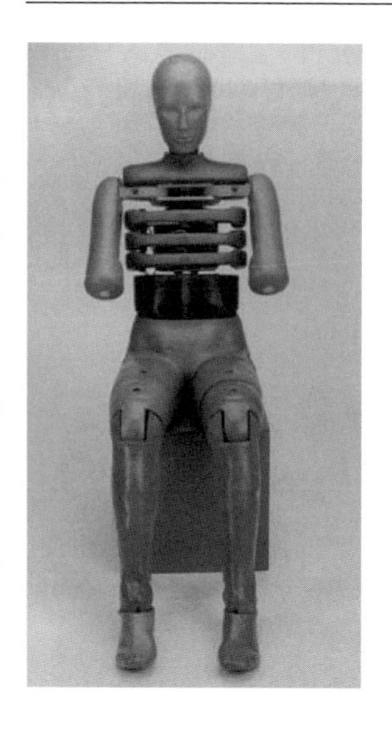

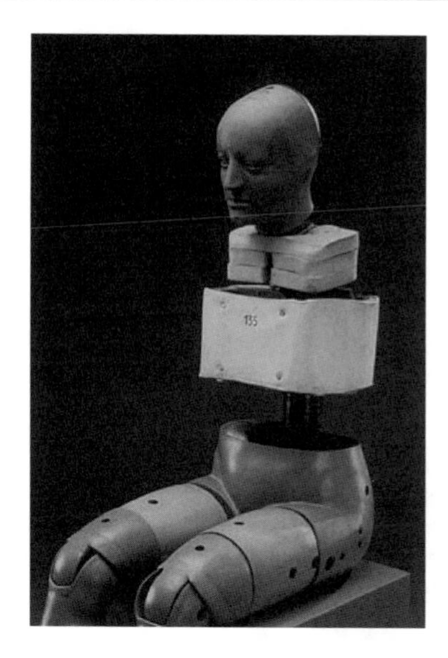

Fig. 6.5-10 Side impact test dummies.

forces and moments and, because of the required rigidity, are of relatively heavy construction.

6.5.5.2 Roof strength

Roof strength (crush resistance) in accordance with FMVSS 216 is tested using a plate angled 25° toward the outside (with respect to the vehicle horizontal longitudinal plane) and 5° forward. With a load of 1.5 times the vehicle curb weight (but no more than 5,000 lb or 2,267 kg), crush (measured perpendicular to the plate surface) must not exceed 5 in. (12.7 cm).

6.5.5.3 Side structure

In addition to dynamic tests, a static test to evaluate capability of passenger car door systems to resist a concentrated lateral inward load is available in accordance with FMVSS 214. In this test, a cylinder is forced into the door, perpendicular to the vehicle longitudinal axis. The lower edge of the cylinder is 5 in. (12.7 cm) above the lowest point of the door. The loading device has a diameter of 12 in. (~30.5 cm). Its height must be such that it rises at least 1/2 in. (1.27 cm) above the lower edge of the window opening. Fig. 6.5-11 shows the side impact resistance of a vehicle with and without a door intrusion beam (side impact beam). In such tests, the attainable maximum force is limited by the ability of doors and locks to transmit forces to the A- and B-pillars, so that in both cases (with and without door intrusion beam) roughly equal maximum forces are achieved. Clear differences in force level are discernible at deflections of 6 inches (15.2 cm) and 12 inches (30.5 cm). Especially at the beginning of deformation, a door equipped with an intrusion beam provides appreciably more resistance to intrusion of another body (obstacle or accident partner).

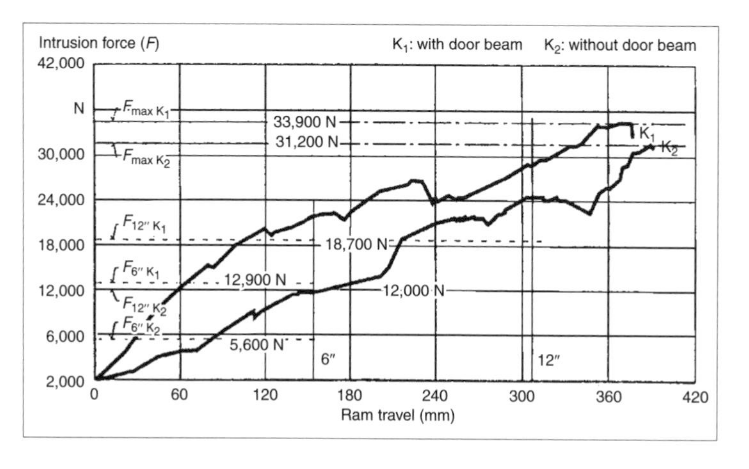

Fig. 6.5-11 Door intrusion test results (for door with and without door intrusion beam).

6.5.6 Dynamic vehicle collisions

6.5.6.1 Frontal collision

Impact against a fixed barrier and pendulum impact tests against the bumper are conducted to ascertain possible damage to headlights, turn-signal lights, etc., and to safety-relevant components (e.g., movable suspension components) in low-speed collisions. While the United States has legislation covering these tests, in Europe these are only mandatory in certain countries. The extent of vehicle damage in low-speed collisions is indirectly reflected by the vehicle's insurance classification, which emphasizes repair costs rather than safety aspects. If possible, no permanent damage should be incurred in any impacts below 8 km/h. For evaluation of repair damage, tests are conducted at impact speeds of up to 16 km/h.

At higher impact speeds, the kinetic energy of the collision partner must largely be dissipated by deformation of vehicle components. In a frontal collision against a fixed barrier perpendicular to the vehicle's longitudinal axis, changes in deceleration and deformation over time, as shown in Fig. 6.5-12, are obtained. Rebound, recognizable as a negative velocity, indicates that in this particular case, a vehicle elastic component of about 10% remains, so that with impact against a fixed barrier at 50 km/h, the total speed change amounts to approximately 55 km/h. Forces transmitted by the impacting vehicle to the barrier are shown in Fig. 6.5-13. Deformation forces are measured by 325 × 325 mm test segments attached to the face of the barrier. A 50 km/h fixed barrier impact of a passenger car weighing about 1,000 kg resulted in a maximum individual force of 128 kN; addition of the individual forces resulted in ~300 kN as maximum value. Impact against a fixed barrier may be described by the following quantities (assuming a plastic collision):

$\ddot{s}_V$ deceleration of vehicle during collision as $f(t)$
$\dot{s}_V$ velocity of vehicle during collision as $f(t)$
s_V distance traveled by vehicle during collision as $f(t)$
$F, \bar{F}$ deformation force, average deformation force
v_i impact velocity
Δv velocity change, from $v_i^2 = 2\alpha \cdot s_v$

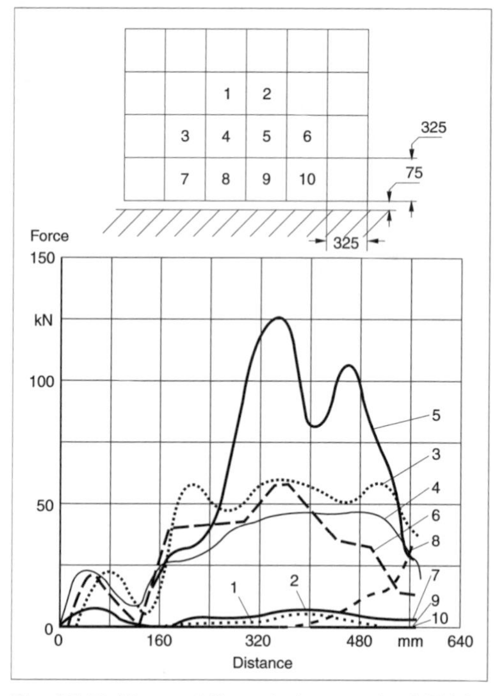

Fig. 6.5-13 Measured forces in impact of ~1,000-kg passenger car against fixed barrier at 50 km/h.

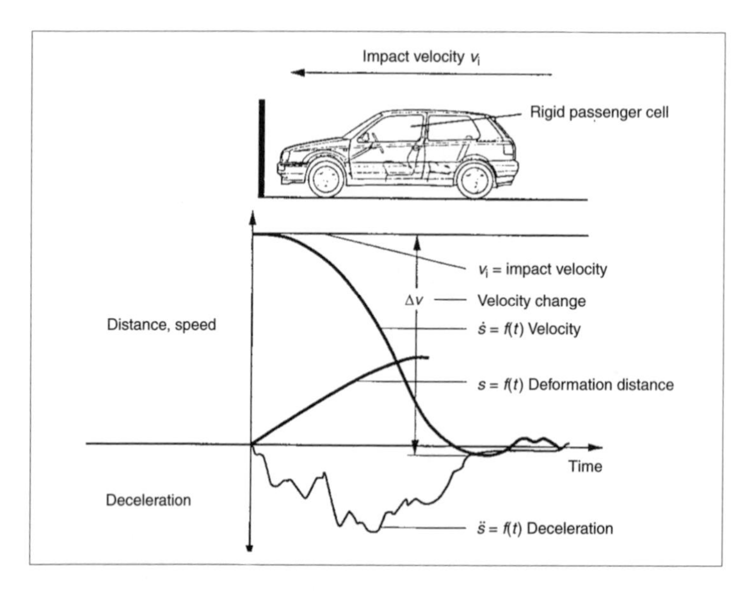

Fig. 6.5-12 Principal deceleration, speed, and deformation histories in a frontal impact.

It follows that deformation forces in a frontal impact may vary. The greater the vehicle mass, the greater the value of $F = ma$. It is also true that because of occupant loads (given identical deceleration), deformation travel in impact against a fixed barrier is impact speed dependent and not mass dependent. With an impact speed of about 50 km/h, most vehicles exhibit deformation of 450–750 mm. With speeds of 35 mph (~56 km/h) these increase by ~100–150 mm.

Behavior in a 30° angled barrier impact is very similar to a frontal car-to-car impact. The "glancing blow" phase in comparison to a frontal impact is likely to result in reduced occupant loading. On the other hand, an offset impact is an extremely severe test of vehicle structure. Most of the absorbed energy must be dissipated by the vehicle side. The corresponding velocity changes (at the center of mass) Δv_x and Δv_y are, relative to the center plane of the vehicle,

30° impact
$v_i = 50$ km/h $\Delta v_x \approx 57.6$ $\Delta v_y \approx 14.4$ km/h

40% offset impact
$v_i = 50$ km/h $\Delta v_x \approx 52.2$ $\Delta v_y \approx 13.7$ km/h

In a frontal impact between two vehicles of masses m_1 and m_2, the following relationships hold:

$$\frac{m_1 \cdot v_{i1}^2}{2} + \frac{m_{21} \cdot v_{i2}^2}{2} = \frac{1}{2}(m_1 + m_2) \cdot u^2 + F \cdot (\Delta s_1 + \Delta s_2) \quad (1)$$

where u is the common velocity after collision:

$$u = \frac{m_1 \cdot v_{i1} \cdot m_2 \cdot v_{i2}}{m_1 + m_2} \quad (2)$$

Inserting the initial relative speed

$$v_r = v_{i1} - v_{i2} \quad (3)$$

of both vehicles gives us (from Eqs. 1 and 2)

$$v_r^2 = 2F \cdot \Delta s_1 + \Delta s_2 \cdot \frac{m_1 + m_2}{m_1 \cdot m_2} \quad (4)$$

Therefore the velocity changes for vehicles 1 and 2 are:

$$\Delta v_1 = v_{i1} - u = v_{i1} - v_{i2} \cdot \frac{m_2}{m_1 + m_2} = v_r \cdot \frac{m_1}{m_1 + m_2} \quad (5)$$

$$\Delta v_2 = v_{i2} - u = v_{i2} - v_{i1} \cdot \frac{m_2}{m_1 + m_2} = v_r \cdot \frac{m_1}{m_1 + m_2} \quad (6)$$

For the often-raised question of what speed should be chosen for impact against a fixed barrier in order to simulate a frontal collision between two identical vehicles with identical initial velocities $v_{i1} = v_{i2,}$ it is necessary that the velocity changes of the two vehicles must be identical; that is, Δv_{iB} in an impact against a fixed barrier is equal to Δv_{iV} in a vehicle-to-vehicle collision. For a plastic impact, the following relationships hold:

Case I. Impact against a fixed barrier

$$\Delta v_{1B} = v_{i1} - v_{i2} \cdot \frac{m_2}{m_1 + m_2}$$

with $m_2 =$ barrier $= \infty$ and $v_{i2} =$ barrier $= 0$ (7)

$$\Delta v_{1B} = v_{i1} \quad (8)$$

Case II. Impact against another vehicle

$$\Delta v_{1V} = v_{i1} - v_{i2} \cdot \frac{m_2}{m_1 + m_2}$$

with $m_2 = m_1 = m$ and $v_{i2} = -v_{i1}$ (9)

$$\Delta v_{1V} = v_{i1}$$

Therefore a barrier impact of a vehicle with v_{i1} is equivalent to a car-to-car impact in which each vehicle is traveling at v_{i1} (in opposite directions, of course), or impact of a vehicle at $2 \cdot v_{i1}$ against a stationary vehicle. A 50-km/h impact against a fixed barrier has (assuming plastic impact) the same consequences as impact of two vehicles of the same mass with a relative velocity of 100 km/h. If, however, we take into account the fact that elastic behavior is more pronounced in a barrier crash, then the barrier impact represents a vehicle-to-vehicle speed of about 110 km/h. With inclusion of the usual rebound phase due to partial elasticity, the equivalent test speed for impact against a fixed barrier is about 10% lower than in an actual collision event.

Fig. 6.5-14 illustrates the requirements for frontal impact investigations.

6.5.6.2 Side impact

Simulation techniques are also available for side impacts:

- For example, the technique described in ECE regulations, in which a deformable barrier (mass 950 kg) is driven perpendicularly into the side of the test vehicle. This test is required for all vehicles newly homologated in Europe as of 1998.
- In the United States, impact against a fixed 1,818 kg barrier perpendicular to the side of the test vehicle forms the basis for vehicle tests including fuel system and tank integrity (Section 7.6). The relatively heavy barrier has a noncontoured face. The test is conducted on both sides of the vehicle; impact speed is at least 20 mph (32 km/h) [17].
- Since 1993, the United States has required an additional side impact test according to FMVSS 214. In this case, the center of a movable barrier of 1,365 kg is driven at a speed of 54 km/h in a crabbed condition against the side of the test vehicle.

6.5.6.3 Rear collision and vehicle rollover

Rear collision tests, like side impacts, employ a fixed 1,800-kg barrier driven against the rear of the test vehicle with a speed of 48.3 km/h (30 mph) in the vehicle

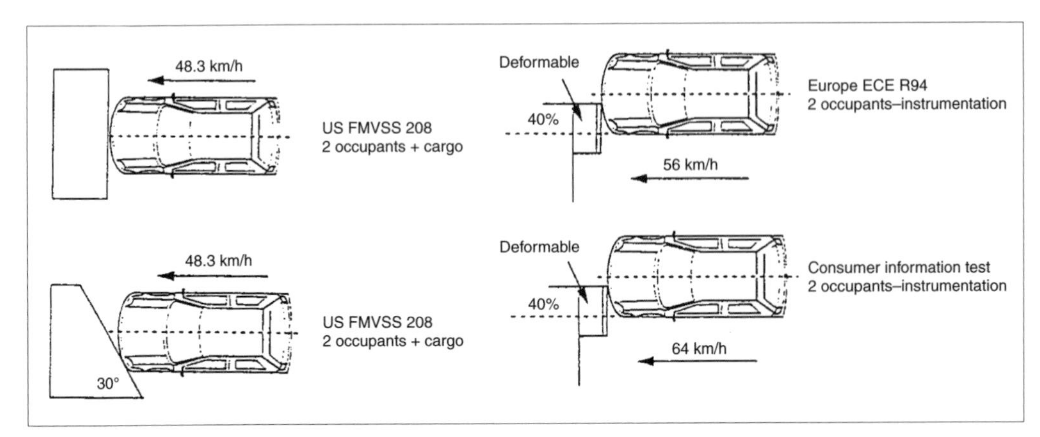

Fig. 6.5-14 Different frontal impact test methodologies.

longitudinal direction. The test criterion is overall fuel system integrity.

Two techniques are used to simulate vehicle rollover. Fuel system integrity is tested before and after frontal, side, and rear impact tests by mounting the vehicle on a rotisserie and rotating it in 90 degree increments around its longitudinal axis. The vehicle is held in each position for five minutes. Gravity check valves are installed in the lines between fuel tank and active charcoal filter; these close if the vehicle is tilted (Section 7.6).

In conjunction with tests of occupant behavior, the entire vehicle is subjected to a dynamic rollover test. The vehicle is mounted on a dolly at an angle of 23° to the horizontal. The dolly is subjected to a precisely defined deceleration from a speed of 30 mph (48.3 km/h). The test vehicle then undergoes multiple rotations, as shown in Fig. 6.5-15. Of interest is the much longer time span of several seconds, compared to frontal and side impacts. These guarantee occupants a high probability of survival without major injuries, if the survival space remains intact.

These described test procedures raise a series of engineering questions which impose especially high demands on body engineering and detail solutions.

6.5.6.4 Body

The limited deformation possibilities in the deformation zones of modern compact vehicles imposed by design constraints demand deformation elements with the highest possible volume-specific energy capacities. Their force/time history should be as constant as possible. Solutions include accordioning, bulging, relocation of powertrain components, and stable crossmembers (resulting in downward displacement of the powerplant in a collision).

To keep it in place during an accident, bonding the laminated glass windshield to the body has become the predominant mounting method.

Door lock design is an important factor in meeting the nonregulatory criterion that vehicle doors be openable without excessive effort even after a severe collision. The door lock design must guarantee that even with limited relative motion in the vehicle longitudinal direction, the locks remain intact and the opening mechanism not be rendered inoperable.

For a side impact, in addition to solid design of the A- and B-pillars and side beam reinforcement, transverse bracing of the pillars must be rigid. To achieve this, the seat crossmember and other transverse reinforcements—for example, under the dashboard or at the rear of the car—must provide the required strength (Section 6.1.1).

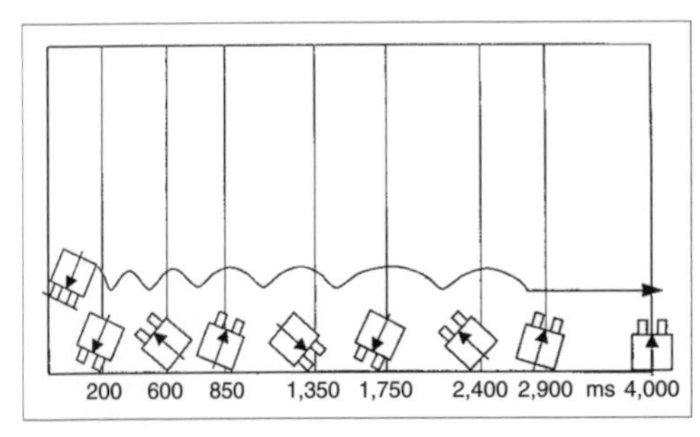

Fig. 6.5-15 Vehicle motion sequence in rollover test.

6.5.7 Occupant protection

6.5.7.1 Vehicle interior

In addition to classic restraint systems (safety belts), the entire vehicle interior is subject to special requirements. For example, component radii within the possible head impact zones (tested with a 165-mm diameter sphere) must be greater than 2.5 to 3.2 mm (depending on location).

The new version of FMVSS 201 requires that specified maximum acceleration levels not be exceeded in tests with a mechanical head form against numerous impact areas within the vehicle.

All remaining vehicle interior components have a

direct or indirect effect on vehicle safety. Design of the interior door panels is of special significance for side impact protection, as these may help dissipate loads on vehicle occupants.

6.5.7.2 Restraint systems

Restraint systems are divided into active systems whose protective function is achieved by actions on the part of vehicle occupants—fastening seat belts or child seats are good examples—and those systems whose protective function is achieved in the event of an accident without any action by the occupants, such as seat belt tensioners, seat belt force limiters, and airbags.

Modern belt systems, particularly when combined with airbags, exhibit extremely good performance and guarantee vehicle occupants outstanding protection in the event of a collision.

A deciding criterion for quality and effectiveness of a belt restraint system is optimum matching of individual components to each other; in other words, only perfect interplay of vehicle structure, steering wheel movement, seat behavior, interior design, belt characteristics, and airbags can guarantee optimum occupant protection.

6.5.7.2.1 Safety belts

Worldwide, the three-point belt with automatic retractor has become predominant. Three-point belts have become mandated equipment for all outer seats and are becoming increasingly common on rear center seats. For adjustable seats, the lock is mounted directly to the seat and the outer part of the belt to the B- or C-pillar.

The zone of permissible seat belt anchorage location is precisely described by regulatory standards (Fig. 6.5-16). The outer top point in most vehicles is adjustable in height. Passenger perception of a comfortable seat belt height is directly related to effectiveness of the belt as a safety device.

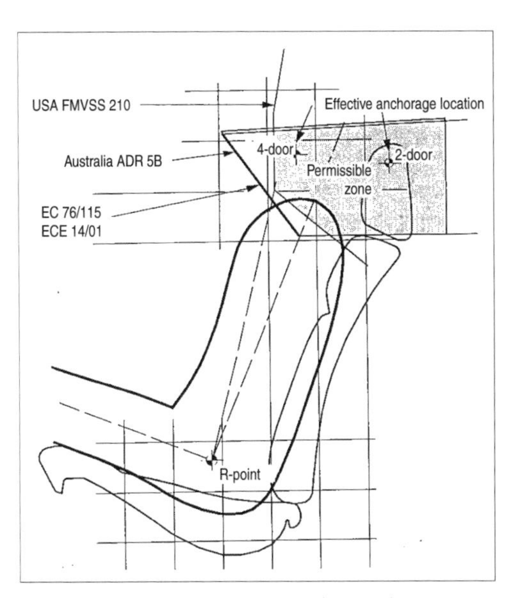

Fig. 6.5-16 Legislatively mandated zones for geometrical placement of upper seat belt anchorage.

Belt locking in the event of an accident is accomplished by two mutually independent mechanical Emergency Locking Retractor (ELR) systems. One system (Vehicle Sensitive Retractor, VSR) uses vehicle acceleration or deceleration (pendulum principle). The locking function is engaged by an impulse of more than 0.4 g. A second system (Webbing Sensitive Retractor) reacts to belt extension acceleration, where acceleration beyond a predetermined level causes locking: that is, given extension acceleration of more than 1 g, locking is engaged with no more than 50 mm of payout.

A variety of subsystems are available to ensure optimum interaction between safety belts and vehicle structure. These fall into the two general categories of webbing clamping devices ("webbing grabbers") and pretensioners (mechanical or pyrotechnic). In detail, these systems work as follows.

Webbing clamping devices clamp the belt above the retractor after the automatic locking function has been activated. This prevents the remaining portion of the belt webbing stored on the reel from paying out in response to restraint loads; depending on seat position, this could amount to as much as 10 cm of payout (without clamping).

Mechanical belt pretensioners release a spring if a deceleration threshold is exceeded. Within 10 ms, the belt is pretensioned with a force of 2,000 N. Mechanical belt tensioners are often arranged as lock extensions.

Pyrotechnic belt pretensioners are sensor-activated and sensor-controlled as a function of accident severity. The belt webbing is tensioned after a deceleration threshold is exceeded. For a 50 km/h frontal impact, the belt is pretensioned to 1,000 N within 20 ms. After the activation impulse, a cable attached to the piston is retracted so that the retractor spool is pulled back opposite to the extension direction. Other locking arrangements employ a rotation principle.

Choice of subsystems is dependent on interaction of all components relevant to the accident process: vehicle deformation characteristics, seat, restraint systems, and geometrical arrangement.

Comfort is an equally important factor in seat belt acceptance: location of the latch on the seat and belt height adjustment have made significant positive contributions. Another measure to minimize webbing payout in three-point belts is optimization of retraction force by means of automatic belt retractors which guarantee problem-free belt storage without applying excessive force to the belt as worn.

6.5.7.2.2 Child-restraint systems

There is no single child-restraint system that can satisfy the demands of all ages and stages of child development. In designing child-restraint systems, particular

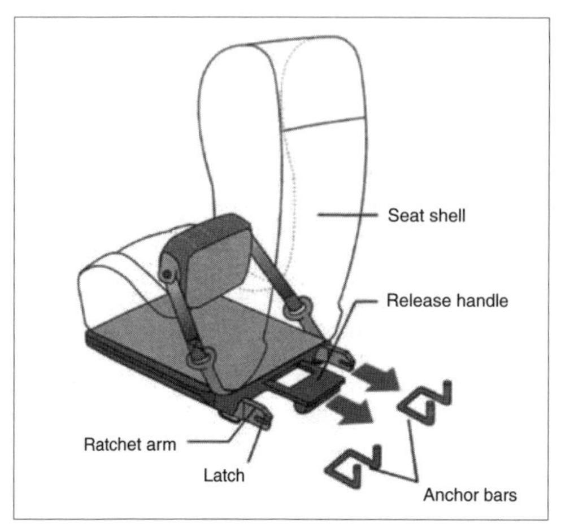

Fig. 6.5-17 ISOFIX attaching points for Volkswagen child seats.

attention must be given to differences in body mass distribution of developing people compared to fully developed adults.

Some manufacturers offer child seats which are optimally designed for attachment inside vehicles.

Fig. 6.5-17 shows the ISOFIX attachment system of Volkswagen child seats [18]. It would be most beneficial if international norms for such systems became regulatory requirements.

There are two basically different installation possibilities: rearward-facing child seat systems and forward-facing systems. In the rebound system, the child faces opposite to the direction of vehicle motion. This arrangement is especially recommended for infants and small children, because it minimizes the danger of severe neck injuries. In this age group, the head is too heavy relative to neck muscle development. Vehicles with passenger airbags have these deactivated with installation of the child seat. Older children with correspondingly developed neck musculature may safely use forward-facing child seats.

6.5.7.2.3 Airbag systems

Airbags also exhibit different configurations, attributable to American regulatory requirements.

American requirements for "passive safety," which are not always well-advised, have resulted in decreasing willingness on the part of airbag customers to wear safety belts. This, however, is a perversion of the original safety logic. Airbag systems should work and should only be allowed to work as extensions of belt systems. Different occupant kinematics, belted or unbelted, require different airbag systems, which must be matched to the installed restraint system.

In three-point belt systems, the occupant moves forward relative to the vehicle before the belt tightens. Thereupon forward motion continues but at a reduced rate. This is affected by the lap belt, with its downward vertical force. A pronounced head rotation takes place in the final impact phase, which may be considerably reduced by the addition of driver and passenger airbag systems.

Such airbag systems consist of compact units. A crash sensor evaluates the time/deceleration history—in other words, accident severity. If the sensor is triggered, an initiator inside an igniter tube ignites a gas generant pellet. The resulting gas fills the bag. Sensor firing time in a 50 km/h barrier impact may be up to 30 ms; the time for subsequent bag inflation about 25 ms. The initial prevailing bag pressure is 1.8–2.2 bar. The main sequence of airbag deployment for driver and passenger airbags is shown in Fig. 6.5-18.

In countries with lower seat-belt usage rates—for example, the United States—permissible occupant loading must also be provided by systems without seat belts. This requires appropriate design of the area below the dashboard—for example, by installation of an energy-absorbing knee bolster similar to that of vehicles with passive belt systems. With this layout (airbag as sole restraint system), the sensor system must work much more precisely and still assure firing even in oblique impacts (-30–$+30°$ to vehicle longitudinal axis). The system is designed so that it alone can carry out all occupant restraining functions. The driver bag volume of such systems is about 80 L, passenger bag volume about 150 L. Bag layout may not be conducted assuming a 50th percentile male; it must also accommodate smaller persons (in extreme cases, children standing in front of the dashboard) and very large, heavy occupants. Openings in the bag provide a controlled force/distance response. Rapid airbag inflation has also resulted in a small number of fatal injuries. For this reason, future airbags will be ignited in response to the actual accident situation, the size of the occupants, and their seating position, with varied inflation pressure and timing. To date, about 100 persons have sustained fatal injuries due to existing US airbag regulations; in response, intense effort is currently under way to modify the law [19].

Airbag systems are now also used for other seating positions and accident types—for example, rear passengers and side impact. Recognition of a severe side impact accident is an especially challenging task. The time used to recognize the accident and inflate an airbag must necessarily be very short; in a 50-km/h side impact, the occupant may contact the inside door panel in about 25 ms.

6.5.7.2.4 Seats, backrests, and head restraints

Increasingly, improvements in frontal and side impact collision behavior have directed more attention to rear-end collisions and perfect head restraint function. Cervical injuries often result in prolonged recovery processes. At present, new solutions to improve this situation are in development. One example is offered by

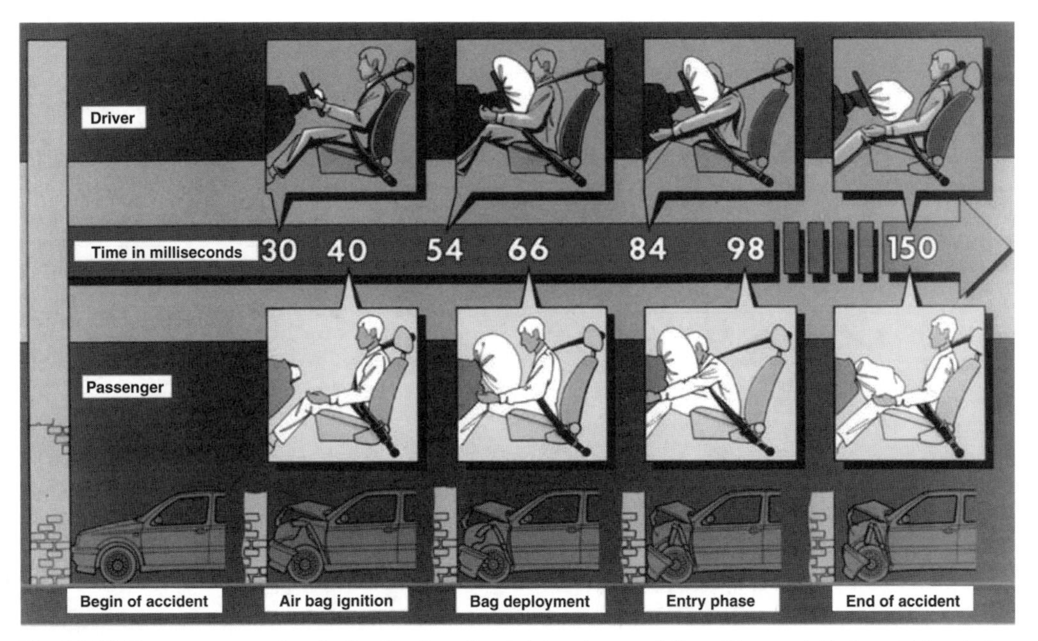

Fig. 6.5-18 Time sequence and protective function of restraint system consisting of three-point belt and airbag during frontal impact (driver and passenger).

Fig. 6.5-19 Head restraint (Saab) which moves up and forward in a rear-impact collison.

Volvo, in which the backrest moves rearward parallel to the upper body. Another is Saab (Fig. 6.5-19), in which the head restraint moves upward and simultaneously toward the head.

6.5.8 Interaction between restraint systems and vehicle

6.5.8.1 Unbelted occupant

In a frontal impact of a vehicle against a fixed barrier or another vehicle, an occupant, owing to his inertia, will move relative to the vehicle in the direction of the dashboard and/or steering wheel. As a first approximation,

Vehicle	Occupant, to impact	
$\ddot{s}_V = -a$	$\ddot{s}_{occupant} = 0$	(10)
$\dot{s}_V = -a \cdot t + v_1$	$\dot{s}_{occupant} = v_1$	(11)
$s_V = \dfrac{a \cdot t^2}{2} + v_1 \cdot t + s_0$	$s_{\text{occupant}} = v_1 \cdot t + s_0$	(12)

The relative distance between occupant and vehicle is

$$\Delta s = (s_{occupant} - s_v) = \frac{a \cdot t^2}{2} \qquad (13)$$

In a 50-km/h impact, given an average vehicle deceleration of 15 g and a distance between occupant and steering wheel of 0.30 m, the occupant will strike the steering wheel after ~64 ms. The speed differential at that time will still be ~33.4 km/h. Without a restraint system, the kinetic energy of the occupant must be absorbed by the steering wheel, instrument panel, and windshield. If we assume a possible deceleration distance of ~0.1 m, this gives an average occupant deceleration of 44 g over a period of ~21.5 ms. Fig. 6.5-20 illustrates this theoretical process.

6.5.8.2 Three-point belt

In the case of a three-point belt without belt clamping or pretensioning devices, after locking of the belt mechanism, the occupant moves forward until the webbing on the spool is pulled tight. The lap and shoulder belt

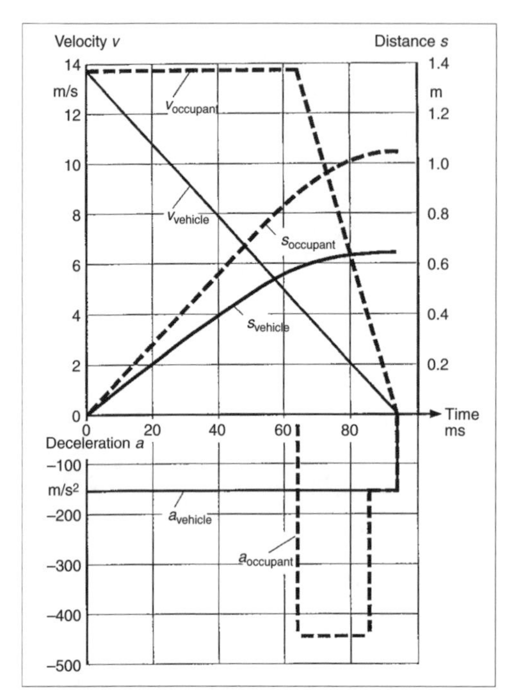

Fig. 6.5-20 Deceleration/time, velocity/time, and distance/time histories for vehicle and occupant in a 50-km/h impact against fixed barrier.

restrain the occupant relative to the vehicle. Under increasing loading, the lap belt also generates a downward vertical force so that the seat cushion shell design and its support must be included in the occupant protection system. Fig. 6.5-21 shows typical head deceleration/time and thorax deceleration/time functions and the corresponding belt tension history for a test dummy secured by a three-point belt in a 50-km/h impact.

Belt clamping or pretensioning devices are applied in order to limit forward displacement of occupants, particularly on the driver side. After a vehicle-specific impact threshold is reached, mechanical tensioners are applied directly and pyrotechnic tensioners are activated by a sensor.

Two effects are apparent: the increase in belt tension begins sooner and the force level is somewhat lower than for a normal three-point belt with the same deceleration/time history. Of course, a similar effect can also be achieved by reducing passenger cabin deceleration in a barrier impact.

6.5.8.3 Airbag systems

The two different configurations—"airbag plus three-way belt" and "airbag without belt system"—lead to different occupant motions. First, consider the system with a three-way belt. Initially, the occupant undergoes a motion similar to that in a conventional restraint system. After airbag inflation, head rotation is dramatically reduced and thorax deceleration values are improved. Fig. 6.5-22 shows head and thorax decelerations on test dummies for sled simulation of a 50-km/h barrier test, as well as knee forces. Special tests also measure contact pressure resulting from impact with an energy-absorbing steering wheel compared to an airbag steering wheel. In the case of the airbag wheel, surface contact pressure is appreciably less.

In the airbag-only scenario, in which the airbag is to a large extent designed for the case of an unrestrained occupant, deceleration of the driver-side occupant is applied by the airbag, steering column, and energy-absorbing knee bolster. On the passenger side, energy must be absorbed by the airbag and knee bolster. Without a three-point belt, the problems of an "out-of-position occupant"—for example, not sitting in line with the airbag—are especially critical. It is therefore imperative that the three-way belt be worn.

Future changes encompassed by the "smart restraint" concept may be seen in Table 6.5-4.

In the longer term, the PRE-CRASH solution may become feasible: that is, accident severity is evaluated before the event actually begins.

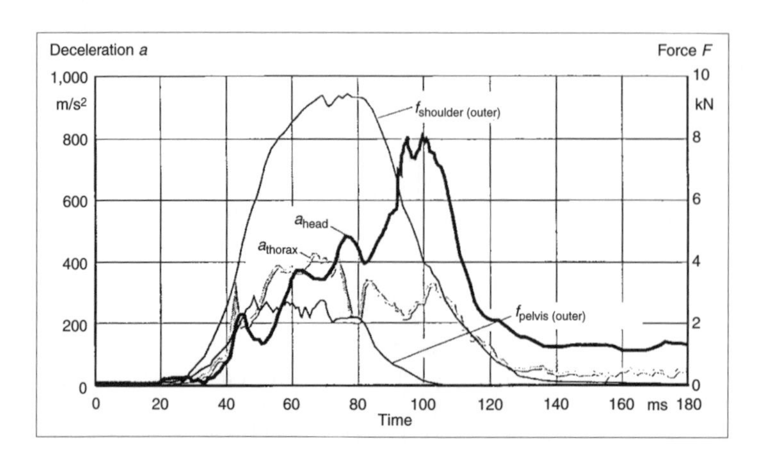

Fig. 6.5-21 Deceleration/time and belt force/time functions for a test dummy restrained by three-point belt during a 50-km/h impact.

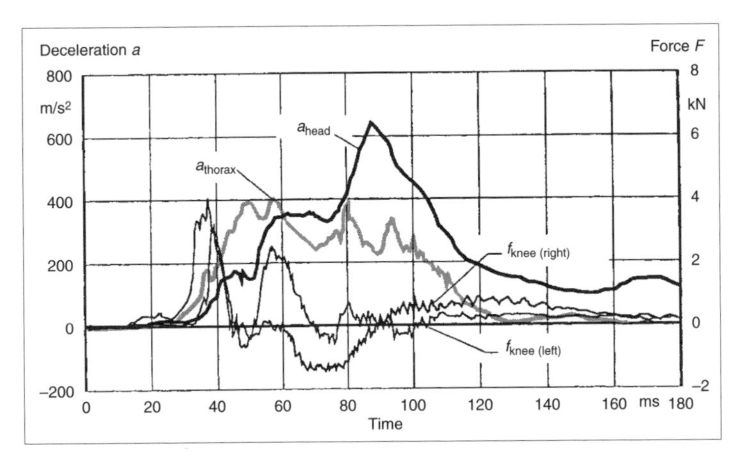

Fig. 6.5-22 Head and thorax deceleration values as well as knee force (sled test, 50 km/h, restraint system with airbag and three-point belt).

6.5.8.4 Effects on restraint system capability

The complexity of actual accident events and accident simulation methods emphasizes that numerous parameters must be considered: deformation behavior of the vehicle structure, interior integrity, steering wheel and steering column, seats, restraint systems, geometric arrangement of seat belt anchorages, etc. It is not optimization of individual components but rather the functioning of the entire system which determines accident-mitigating vehicle safety.

6.5.9 Side collisions

Proximity of occupants to vehicle doors limits opportunities for application of special side-impact restraint systems. Factors which can be modified include body resistance to penetration by the opposing vehicle—for example, by a collision-appropriate rocker panel design or by side intrusion beams and door reinforcements at shoulder height. Transverse reinforcements inside the vehicle—such as under the dashboard, the seat cross-member, the seat design, and transverse reinforcements at the rear of the vehicle—help to improve occupant protection.

A second factor in side collisions is design of the vehicle interior. Door inner-panel shape and material carry special significance. Side airbag systems, now being introduced, are dependent on a functioning sensor system which must be capable of recognizing the onset of a severe accident at an early stage in order to deploy the airbag before the head of the occupant can touch the side window.

6.5.9.1 Theoretical considerations

Compared to frontal impacts, side impacts have an even greater range of variations for the possible accident process (e.g., impact direction, impact point). Furthermore, limited deformation space is available for the occupants. Side impact is governed by a larger number of factors, which will be examined in this section in greater detail.

Significant factors are:

— Involved collision partners (mass, structural stiffness, structural geometry)

Table 6.5-4 Further airbag development

Current state of the art	Future
Crash signal, evaluation of deceleration/ time function.	Crash signal is evaluated, first or second stage is ignited (or not) depending on crash severity. Belt system is checked. Driver monitored by means of ultrasound and infrared sensors: Seat occupancy Out of position Child seat
↓	↓
Control unit provides ignition signal dependent on situation.	Control unit provides ignition signal dependent on situation.
↓	↓
Generator inflates airbag.	Two-stage generator is ignited. Depending on situation, only driver side or both driver and passenger airbags are inflated.

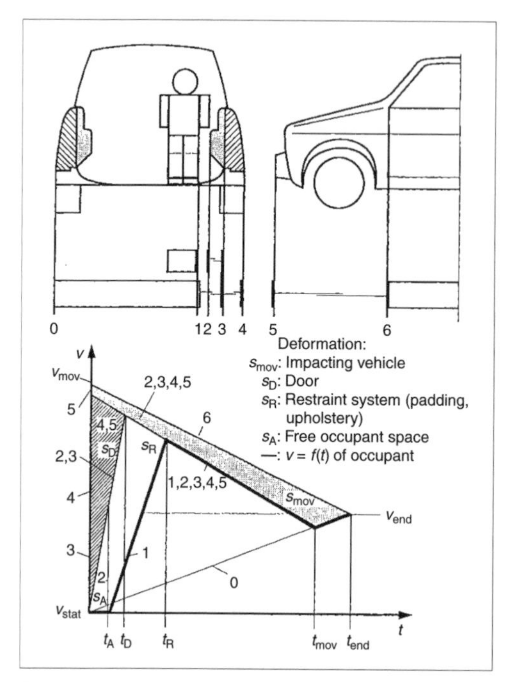

Fig. 6.5-23 Schematic speed/time representation for vehicle structure and occupants in perpendicular side impact.

— Impact point and angle, impact speed, and occupant placement
— Vehicle interior design and usage of restraint systems

Fig. 6.5-23 explains in principle the car-to-car side impact process. Shown are both the impacted and impacting vehicles, with an occupant on the impacted side, the derived mechanical model, and simplified motion of the impact process using a velocity/time diagram as an example.

In the figure, numbers 1 through 6 mark the velocity/time histories of vehicle-based indicated parts and the occupant. The surfaces bounded by these "speed bands" represent the relative spaces (deformation spaces) between the indicated components or between components and the vehicle occupant.

For simplification, force/distance characteristics of the involved structures and the lateral restraint system (padding, upholstery) are assumed to be rectilinear. Vibration of the mechanical model is not considered. The resulting functions $a = f(t)$ and $s = f(t)$ therefore deviate from those of an actual accident. As a result of these boundary conditions and the single-mass occupant system, only a qualitative comparison of the collision model (discussed below) with actual conditions is possible. The striking vehicle is driving with velocity v_{mov} perpendicular to the side of the stationary target vehicle. After only a short period of time, door exterior skin 4 has the same velocity as bumper 5. At time t_D, door inner panel 3 has the same velocity as bumper 5. Contact between the occupant and restraint system 2 begins at time t_A. The relative velocity between occupant and door is greatest at time t_D. It is greater than the velocity change of the impacting vehicle. The occupant is decelerated by padding. The deformation distance is s_R. At time t_R, occupant 1 and bumper 5 have the same velocity. Bumper and occupant are simultaneously decelerated until time t_{end}. Deformation of the side structure ends at t_{mov} (seats are fully compressed; i.e., no additional deformation space is available). The front structure of the impacting vehicle has deformed by a distance s_{mov}. Deformation of the side structure is determined by relative motion of door 3 to an undeformed point on the vehicle, 6.

6.5.9.2 Side impact test (as defined in USA)

Tests with movable barriers give the deceleration/time plots, shown in Fig. 6.5-24, against the impact barrier and against the tested vehicle. The following conditions apply to this test:

— Test vehicle $m_2 = 1{,}435$ kg
— Initial velocity $v_2 = 0$ km/h
— Barrier $m_1 = 1{,}385$ kg
— Initial velocity $v_1 = 54.5$ km/h

It is apparent that the impact ends after ~100 ms. Criteria are thorax and pelvis acceleration.

Since 1998, a test procedure has been implemented in Europe using a dedicated barrier and a special test dummy (EuroSID). EuroSID is also intended to evaluate head acceleration, thorax deformation, and thorax velocity.

Unfortunately, differences between European and US test dummies also manifest themselves in side impact protection differences. Initially, protection in side impacts was sought from the vehicle itself—rigid body and energy-absorbing interior. After introduction of an airbag deploying from the seat or windowsill, an increasing number of upper body and head airbags were implemented, as shown in Fig. 6.5-25 [20]. The illustration shows a "window airbag," which in this case is 2 m long, 250 mm wide and 60 mm thick. With nine chambers, it has a volume of 12 liters. It inflates within 25 milliseconds.

6.5.10 Compatibility

6.5.10.1 General observations

Because considerable progress has been made in the field of accident mitigation, compatibility of traffic participants (vehicles) to each another will acquire increasing importance with respect to safety-relevant measures. The first published considerations on this topic date back to the early 1970s [21, 22, 23].

In traffic accidents, the following participants deserve special attention: pedestrians; single-track riders;

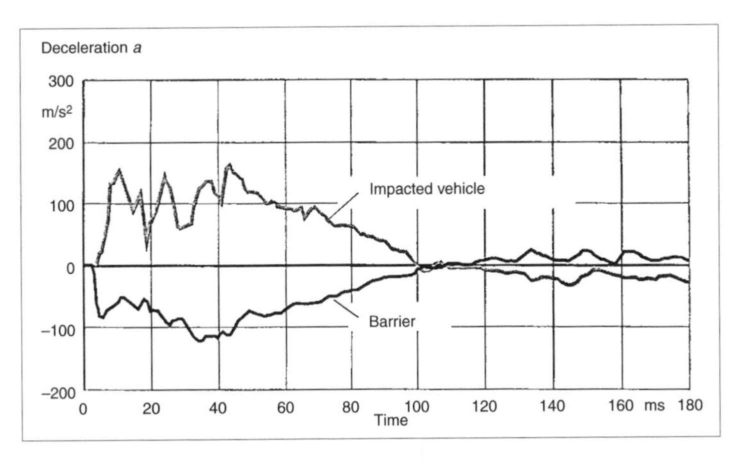

Fig. 6.5-24 Deceleration/time histories for impact barrier and test vehicle (test according to US side impact standard FMVSS 214).

and passenger car, commercial vehicle, and bus occupants. Injuries may be incurred in accidents with other collision partners or in solitary accidents. Compatibility investigations have shown that a global effort encompassing all collision groups is extraordinarily complex and possible solutions very difficult. For vehicle layout in view of compatibility considerations, the following criteria enter into the collision process:

— Mass ratio, vehicle structural geometry, force/distance relationship of the vehicle structure, locations and masses of powerplant components
— Restraint systems, geometry of passenger cell (survival space)
— Behavior of steering, dashboard, upholstery, etc.

In compatibility considerations, reduction of accident consequence costs and all resulting relevant side effects, such as reduced comfort or increased energy and raw materials consumption, must be weighed against each other. Of course, today's vehicle owners demand that self-protection—that is, protection of the occupant in his own vehicle—should not be diminished. Along with description of the accident type, information regarding the injury type and severity must be available. To evaluate any given safety measure, a relationship must be found between injury severity, as described by the AIS scale, and the injury criterion, as measured and calculated by the use of test dummies.

For different mass categories and restraint systems, force/deformation characteristics may be modified to the point where a minimum in overall costs (accident consequence costs, manufacturing and operating costs) is achieved. It is possible to optimize vehicles and restraint systems with respect to minimal overall costs if one knows:

— The events of the accident in correlation to the severity of injury according to AIS
— The correlation between AIS occupant injury to test dummy data

Fig. 6.5-25 Head airbag in side impact [20].

— Financial evaluation of occupant injury and the costs incurred from additional vehicle construction costs

In a comprehensive study, such as one promulgated by the German Federal Ministry of Research and Technology, a special optimization process was developed using frontal and side-impact models in order to find compatible vehicles—that is, vehicles which "tolerate" one another. The study [24] provided the following conclusions:

— Greater conversion of deformation energy in the space between the vehicle front section and engine decreases aggressivity of the vehicle; in impact against another vehicle, reducing deformation energy in the upper part of the front structure improves side impact protection. Conversion of deformation energy between engine and occupant protection system increases occupant protection. In principle, a light vehicle should have a stiffer structure, and occupant restraint systems must be optimally implemented (belt tensioner, airbag, deployment time).
— Side impact protection can be considerably improved by adequate layout of the vehicle front section. Side airbags are to be regarded positively.
— With applied seat belt, the safety level of modern vehicle designs is adequate; however, the airbag system helps reduce head injuries.
— Occupant protection is improved by matching vehicle structural characteristics to the safety belt system.

Recently, numerous investigations by scientists and automobile manufacturers [25] have focused greater attention on this topic.

Recent investigations [26] affirm the necessity for three-dimensional models to more precisely define information on vehicle compatibility behavior. Local force spikes automatically imply reduced compatibility behavior. Therefore, the degree of force distribution determined by application of FEM (Finite Element Methods) models and/or force measurement in frontal impacts may be applied as a measure of compatibility.

6.5.10.2 Car/truck collision

A special compatibility problem is collisions between trucks and cars, because not only the mass distribution, but also, above all, the geometrical relationships are critical. In the United States, incompatibility between light trucks and passenger cars is regarded as very worrisome.

For the truck sector, Schimmelpfennig [27] has indicated a pathfinding solution: a guard rail frame surrounding the entire truck and trailer, thereby preventing underride of an impacting car. Fig. 6.5-26 illustrates the principal arrangement in plan view.

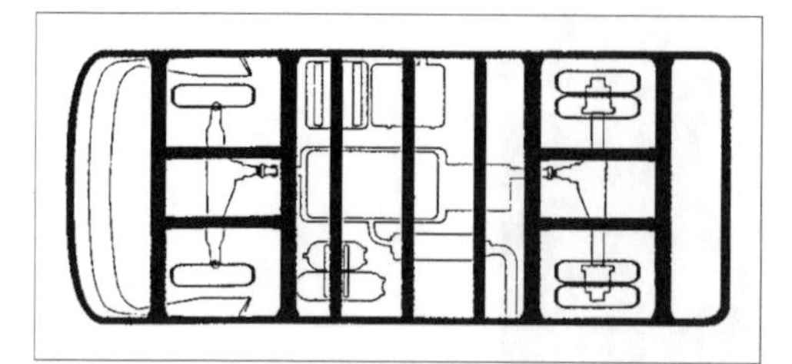

Fig. 6.5-26 Guard rail frame for improved underride protection.

6.5.10.3 Pedestrian collision

A further incompatibility is found in collisions between pedestrians and passenger cars. Mass ratios as well as differing body sizes and low injury resistance of pedestrians permit only limited opportunities to improve the situation. In general, these measures can be implemented to defuse the situation:

- Reduced waiting time at light signals (<40 s)
- Separate signaling for turning traffic and pedestrians
- Increased driver awareness of speed effects on braking
- Traffic education, particularly in kindergarten/schools, about the value of bright-colored clothing and reflectors in making pedestrians more visible, calling attention to the danger posed by turning traffic

Additionally, in Europe, intensive discussion is under way with regard to future regulations for vehicle-based measures [28]. Fig. 6.5-27 shows one of the proposed requirements which would have considerable consequences for present-day vehicle designs.

6.5.11 Computational methods in development of safety components

6.5.11.1 Basic principles

Even in the predevelopment phase of a vehicle, an optimization process based on geometric preconditions must be found to relate accident behavior, mass, interior acoustics, and vibration behavior. A breakthrough in the field of computational methods was realized with development of stable software programs and continued development of computer systems. Even though many parameters affect the results of accident simulations, computational methods nevertheless are an important contribution toward improving the overall development process. In the product creation process, numerical methods play an ever growing role in the interplay between development, production, marketing, and finances, as well as quality assurance and quality planning (Section 10.2).

6.5.11.2 Description of numerical tools

The most important assignment in any computationally supported engineering task is development and application of physical/mathematical models [29].

Even in the model creation phase, physical input parameters and the component in question must be known. Depending on the project phase, investigative results

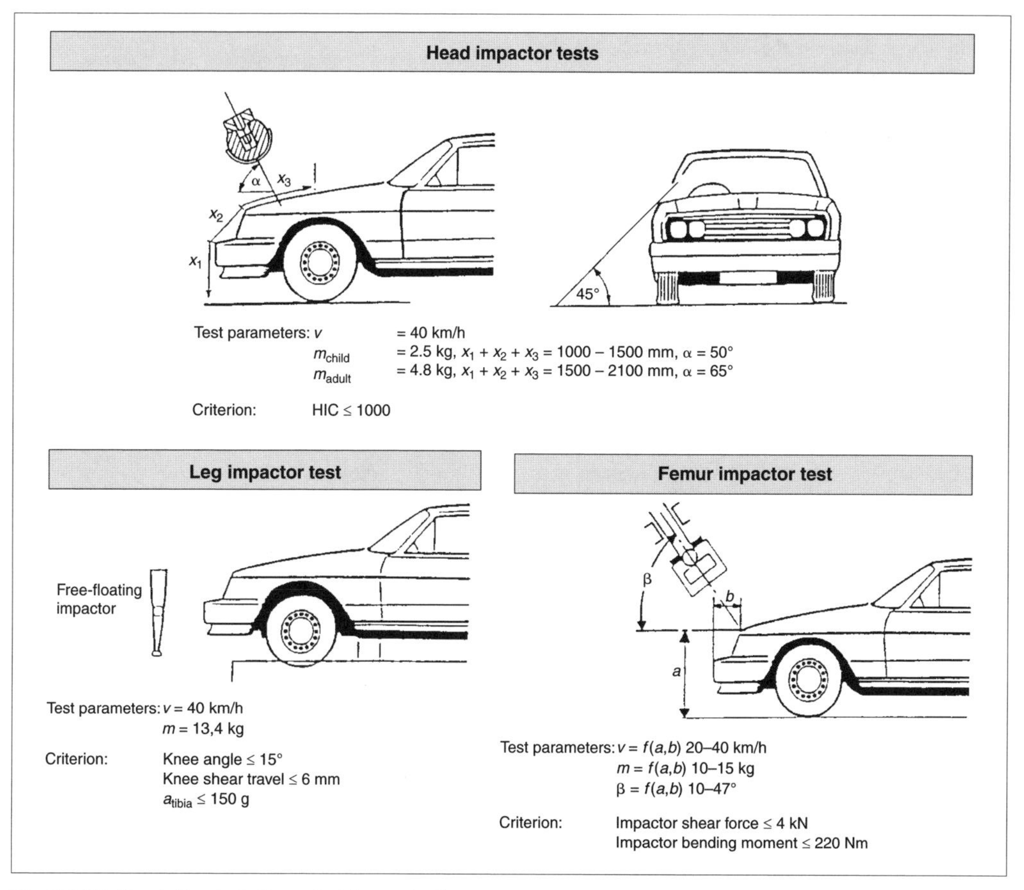

Fig. 6.5-27 Simulation tests for pedestrian protection.

may range from trend predictions to detail solutions all the way to full predictability.

Compared to the past, we have available today a wide range of aids to transform the physical model into a corresponding mathematical model.

The most important aspect for an automotive engineer is not so much the actual development of new software, but rather the creative application of programs. For example, numerous program packages are applied to crash calculations (Fig. 6.5-28). These simulation

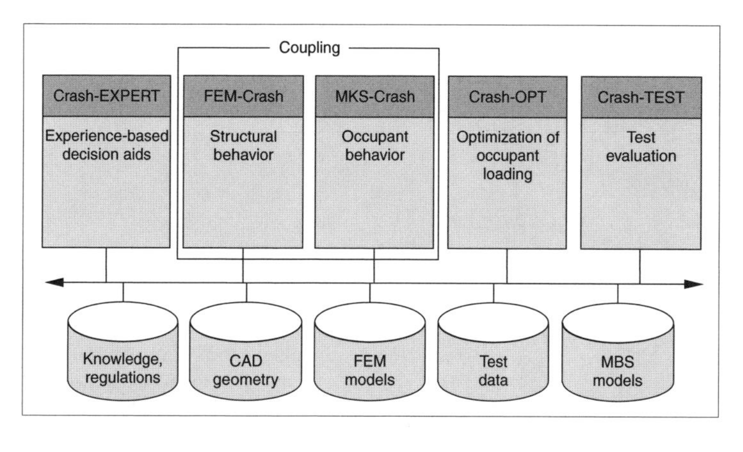

Fig. 6.5-28 Computational modules for computer-aided safety calculations.

programs notwithstanding, many actual tests will still have to be conducted in the future. Accordingly, the crash test module includes test analysis software which permits comparison of sled and crash tests.

CAD (Computer Aided Design) data forms the basis for FEM models (Finite Element Methods). These are employed to examine vehicle structural behavior from individual components all the way to the entire vehicle structure. Software installed on most auto manufacturers' supercomputers represents an important element in the computer simulation process. For example, the MBS (Multi-Body Systems) module contains the computational program to simulate occupant motion, including interaction with restraint systems and vehicle interiors.

One approach is software-side coupling. This permits not only replacement of specific MBS modules with FEM building blocks but also integrated computation of a complete vehicle with restraint systems and occupants.

Building blocks named Crash-OPT ("Optimale Struktur für bestimmte Rückhaltesysteme"—Optimal Structures for Specific Restraint Systems) and Crash-Expert (a knowledge- and experience-based decision system) are still in the development stage. Both modules, FEM and MBS-Crash, employ numerous pre- and postprocessors including connections between the modules themselves.

6.5.11.3 Component calculation

In many aspects of accident research, the most important components are the front longitudinal members. Many theories have grappled with the phenomenon of controlled collapse as a means toward optimum energy conversion. The following relationship holds for the average controlled collapse force [30]:

$$F = \sigma_F \cdot \delta_F \cdot \alpha_F \frac{s_x^2 \cdot Ux}{Ua}$$

where σ_F is the form factor, δ_F the yield strength of the material, α_F a speed-dependent factor ranging from 1.0 to 1.5, s_x the sheet metal thickness, U_x the circumference of the profile section with sheet metal thickness s_x, and U_a the total circumference of the profile.

Fig. 6.5-29 shows a computer simulation of a longitudinal member [30].

The programs DYNA3D and PAMCrash may already be used to solve such problems using finite element methods. Longitudinal and transverse bracing is especially important, such as the front cross-members which introduce forces into the body. The onset of controlled collapse is clearly visible. Most energy is converted in the collapse process. The least energy is absorbed in buckling due to high bending moments.

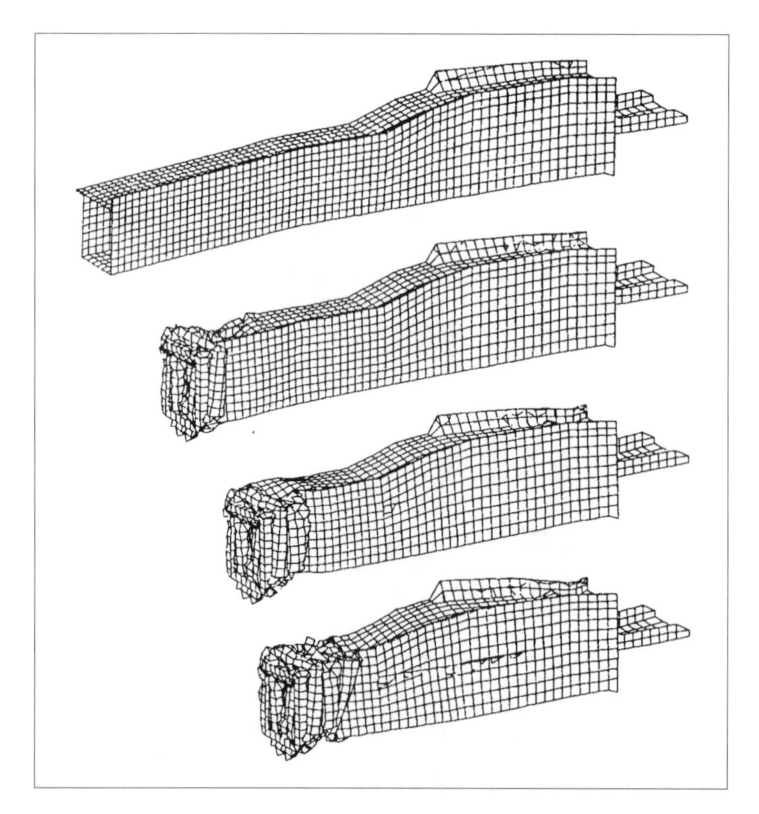

Fig. 6.5-29 Finite element model of longitudinal member (behavior during deformation).

6.5.11.4 Overall vehicle layout

As early as the concept phase, design engineers may conduct initial calculation runs using known boundary conditions such as wheelbase, track, powerplant location, powerplant size, and other dimensional data. As concept studies, however, these do not offer any quantifiable results. They do advance construction of prototypes whose safety behavior may be reasonably well predicted. In later development phases, much more precise results may be expected from computational methods as well as actual testing.

Thanks to increasing computer power (parallel processors), detailed FEM models can be used so that vehicle-to-vehicle collisions can be simulated within an acceptable expenditure of computer time. Fig. 6.5-30 shows an example of a car-to-car collision [26].

6.5.11.5 Overall model

It is necessary to evaluate the vehicle as well as its occupant behavior in an overall computation. The model used to conduct these simulations is realized by coupling a FEM program (e.g., PAMCrash) with an occupant simulation program (e.g., MADYMO). Both coupling models are described below.

6.5.11.5.1 Vehicle model

The vehicle model defined in the FEM program may, for example, consist of:

- A front structure with 72 beam elements
- An engine block
- A flat shell structure (220 shell elements) as contact surface
- A rigid passenger cell

The front structure of beam elements permits variable stiffness distribution in longitudinal, transverse, and vertical directions, so that model variations can be examined with less effort and computer time than in more complex FEM models. Deformation of beam elements is defined by the user. The force/distance relationships of these beams can be defined at will and also used to “illustrate” tests using actual vehicles or conventional FEM computations.

6.5.11.5.2 Occupant simulation

Occupant simulation by means of MBS programs permits optimization of occupant restraint systems. Exact input data, particularly for test dummies, is required for model development. Auto manufacturers work closely together in simulating test dummies for model development purposes. MBS simulation permits conducting safety development in three main areas:

- Concept and trend analysis
- System plausibility checks
- Component data development

The MBS method can be used to for optimization of safety belts (e.g., analysis of effects due to belt clamping devices, belt tensioner, seat behavior, location of seat belt anchorages) and airbags.

Relative motion between test dummy and vehicle body or individual components can be represented as well

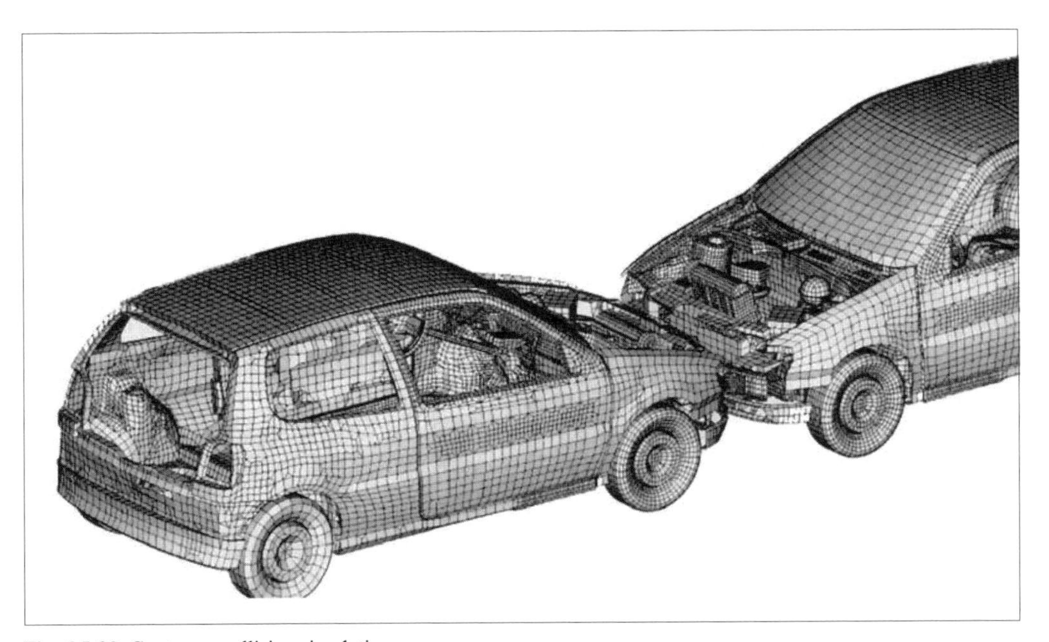

Fig. 6.5-30 Car-to-car collision simulation.

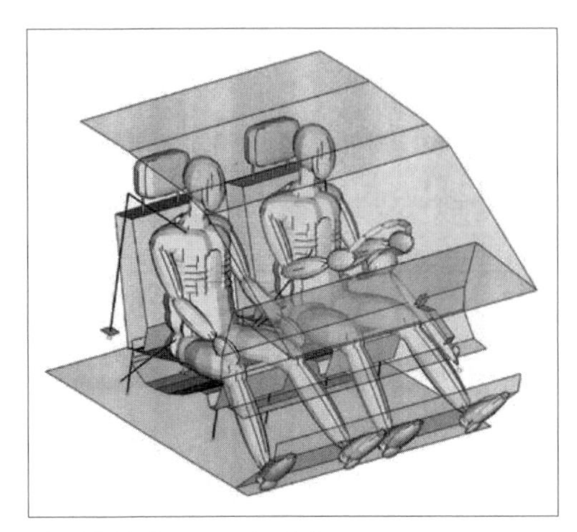

Fig. 6.5-31 Occupant simulation.

as deceleration/time, force/time, and distance/time relationships. Coupling of structure and occupant behavior offers the possibility of computationally evaluating the entire system (vehicle and occupants).

Fig. 6.5-31 shows two Hybrid III dummies, a vehicle interior, a steering system, and a driver airbag. In contrast to most occupant models, the dummies of this model are not subjected to an acceleration field. Moreover, the interior—in other words, the occupants' surroundings—is coupled to the model's passenger cell.

This coupling is interactive; therefore, reaction of the occupants with the vehicle is possible. Without the added effort of a subsequent computational process, coupling permits subjecting occupants to the three-dimensional motion of the vehicle itself.

This simulation of impact behavior including the restraint system, taking into account test dummies, is an important aid in the development process. The effort expended on model development may be considered justified if the model forms the basis for an entire model line. On the basis of manifold engine and transmission combinations, the computation is also helpful in a "worst case" contemplation—that is, consideration of the worst possible configuration. Nevertheless, even with these tools, it will not be possible to entirely dispense with actual development and validation tests.

In the future, advances in the computational field will permit a marked decrease in the experimental effort expended per vehicle model [31] [32]. The freed capacity is urgently needed for a greater number of vehicle model variants and increased number of experimental tests.

6.5.12 Summary

Vehicle safety is no longer regarded as an isolated aspect of vehicle development, but rather forms an integral component of the product creation process. This transformation occurred analogously to the metamorphosis of the automobile from an independent means of transportation to an integral component of a transportation system. Without this conversion, the automobile in the medium term would find itself in a difficult position in public debate. The possibilities for making the vehicle an even safer means of transportation have been assisted by introduction of modern technology. Independent of unquestionably necessary regulatory requirements, initiatives on the part of individual auto manufacturers have brought new movement into this important field. The task before us is to realize the vision of the automobile matching railways in terms of safety.

References

[1] Braess, H.-H. "Aktive und passive Sicherheit im Straßenverkehr." *Zeitschrift für Verkehrssicherheit* 42, 1996.
[2] Barényi, Béla. "Das Prinzip des gestaltfesten Fahrerraums." German Federal patent 854.157, 1952.
[3] Wilfert, K. "Entwicklungsmöglichkeiten im Automobilbau." *ATZ* 1973, S. 273-278.
[4] Seiffert, U. *Fahrzeugsicherheit Personenwagen*, VDI-Verlag Düsseldorf 1992, ISBN 3-18-401264-6.
[5] Seiffert, U. "Die Automobiltechnik nach der Jahrtausendwende." Euroforum Munich, May 25 and 26, 1998.
[6] Danner, M. "New Investigations of HUK Accident Research, Interior Safety of Automobiles." Report, 4th International Technical Conference on Experimental Safety Vehicles, Kyoto, 1973.
[7] VDA-Jahresbericht, Verband der Automobilindustrie, Auto 1998, Frankfurt.
[8] Langwieder, K. "Realitätsbezogene Crashtests—eine Grundlage für mehr Sicherheit von PKW." Crash-Test 1996 conference, TÜV-Akademie, Munich, October 15 and 16, 1996.
[9] Otte, B. "Verletzungsmuster und Verletzungsschwere bei Crashkonfigurationen im realen Unfallgeschehen." Crash-Test 1996 conference, TÜV-Akademie, Munich, March 9 and 10, 1998.
[10] Eighth Stapp Car Crash Conference, Wayne State University Press, Detroit, 1966.
[11] IRCOBI *Proceedings* 1998, International IRCOBI conference on Biomechanics of Impact. Göteborg, Sweden, September 16–18, 1998.
[12] AAAM (American Association for Automotive Medicine), *The Abbreviated Injury Scale*, revised 1980.
[13] Swearingen, J.J. "Tolerances of Human Face to Crash Impact." Report no. AM 65-20, Federal Aviation Agency, Oklahoma City, July 1965.
[14] Patrick, L.M., et al. "Survival by Design—Head-Protection." Seventh Stapp Car Crash Conference, Oklahoma City.
[15] Fiala, E., et al. "Verletzungsmechanik der Halswirbelsäule." Experimental report, Technische Universität Berlin, March 1970.
[16] Appendix to SAE J921, "Motor Vehicle Instrument Panel Laboratory Impact Test Procedure—Head Area." November 1971.
[17] US FMVSS No. 301. "Fuel System Integrity." National Highway Traffic Safety Administration.
[18] Degener, M. "ISOFIX, das neue Befestigungssystem für Kindersitze." Haus der Technik, Essen, Sept. 29, 1998.
[19] Status Report, Insurance Institute for Highway Safety, Vol. 33, No. 9, October 1998.
[20] Special issue of *ATZ/MTZ*, "Die neue S-Klasse." Wiesbaden, 1998.
[21] Ventre, Phillippe. "Homogeneous Safety amid Heterogeneous Car Population?" Third International Technical Conference on Experimental Safety Vehicles, Washington, D.C., June 1972.

[22] Seiffert, U. *Probleme der Fahrzeugsicherheit.* Dissertation, Technische Universität Berlin, 1974.
[23] Appel, H. "Sind kleine Wagen unsicherer als große?" *VDI Nachrichten*, No. 7, Düsseldorf, February 2, 1975.
[24] Richter, B., et al. "Entwicklung von PKW im Hinblick auf einen volkswirtschaftlich optimalen Insassenschutz." Final report, BMFT [German Federal Ministry of Research and Technology], requested by Federal Minister of Research and Technology, 1984, p. 116.
[25] Huber, G. *Passive Safety of Vehicles including Partner Protection.* Fisita, Prague, 1996.
[26] Relou, et al. "Entwicklung kompatibler Fahrzeuge mittels kompatibilitätsbewertender Crashsimulation." VDI, Berechnungstagung [Computational Conference], Würzburg 1998.
[27] Schimmelpfennig, K.-H. "Bord Frame, a possible contribution to improve passive safety." 15. International Technical Conference on the Enhanced Safety of Vehicles." Melbourne, Australia, May 1996.
[28] Dickison, M. "Development of passenger cars to minimise pedestrian injuries." SAE Technical Paper 960098. Detroit, February 1996.
[29] Seiffert, U., and Z. Scharnhorst. "Die Bedeutung von Berechnungen und Simulationen für den Automobilbau." VDI-Bericht 699, VDI-Verlag, Düsseldorf, 1988.
[30] Beermann, H.-J, et al. "Aufpralluntersuchungen mit vereinfachten Strukturmodellen." Text of presentation at 4th IfF Conference, Brunswick, June 1982.
[31] VDI-Berichte 1354, *Innovativer Insassenschutz im PKW.* VDI-Verlag Düsseldorf, 1997, ISBN-3-18-091354-1.
[32] VDI-Berichte 1411, *Berechnung und Simulation im Fahrzeugbau.* VDI-Verlag Düsseldorf 1998, ISBN 3-18-091411-4.
[33] Siebertz, K., et al. "Beurteilung des Insassenschutzes mit Out-of-Position Models." In VDI-Bericht 1411, *Berechnung und Simulation im Fahrzeugbau.* VDI-Verlag, Düsseldorf, 1998.

General

Kramer, F. *Passive Sicherheit von Kraftfahrzeugen. Grundlagen—Komponenten—Systeme.* Brunswick/Wiesbaden, Verlag Vieweg, 1998, ISBN 3-528-06915-5.
VDI-Berichte 1471, *Innovativer KFZ-Insassen- und Partnerschutz.* VDI-Verlag, Düsseldorf, 1999, ISBN 3-18-091471-8.

7 Suspension

7.1 General

Automobiles are vehicles whose movement on a dedicated surface (generally a road, which is part of a road network) may be controlled by a driver in longitudinal and transverse directions as well as around the vertical axis (yaw axis) within the limits imposed by the roadway or by physical laws. Transverse and yaw movements are closely coupled to one another.

In the vertical direction, the automobile follows the roadway without any action on the part of the driver (ascending or descending hills). Transmission of road irregularities to the vehicle, however, should be minimized to ensure ride comfort and safety.

7.1.1 Definition of the suspension concept

With the exception of gravity and aerodynamic forces and moments, all external forces and moments acting on a vehicle are transmitted through the areas of tire contact with the roadway—the tire contact patches.

In a more restricted sense, the suspension may be understood as a composite of vehicle systems which generate forces at the tire contact patches and serve to transmit these to the vehicle: wheels and tires, brakes, hubs/suspension arms/steering systems, springs/shock absorbers. In a larger sense, all systems necessary for guiding the vehicle are included: brake, clutch, and throttle actuation, steering wheel, steering column, pedal cluster, control systems to support suspension action, as well as driver assist systems.

This chapter provides an overview of the functions of wheel location, steering, and springing as well as vehicle dynamic control systems. Control systems that govern braking and acceleration slip are treated in Section 5.5.

7.1.2 The suspension task

In its role as the connecting link between road and vehicle, the suspension is a significant contributor to handling dynamics and ride comfort. Moreover, the suspension affects space utilization, aerodynamics, and costs.

The high priority given the suspension system within the automobile as an aggregation of systems may be discerned from the following:

1. An automobile with well-tuned handling dynamics may be driven with little effort. It translates driver inputs immediately, predictably, and precisely and imparts a feeling of security and even satisfaction. For many customers, this very tangible impression as well as evaluations of handling dynamics in the specialist press represent an important factor in the purchase decision.
2. The handling dynamics of an automobile very significantly determine the driver's opportunities to avoid or master critical situations. In Germany, about one-third of traffic fatalities are attributable to leaving the roadway and about one-fifth to collisions with oncoming vehicles (Fig. 7.1-1). Barring analysis of individual accidents, an exact statement regarding the potential presented by further increases in active safety is not possible. Nevertheless, even if significant influence is attributed to the driver, it appears that an appreciable reduction in the number of accident victims is possible.
3. A high degree of ride comfort is not only perceived subjectively as a pleasant experience, providing a market-competitive criterion; ride comfort also has a demonstrated influence on driver physical and psychological performance and therefore on safety [1].

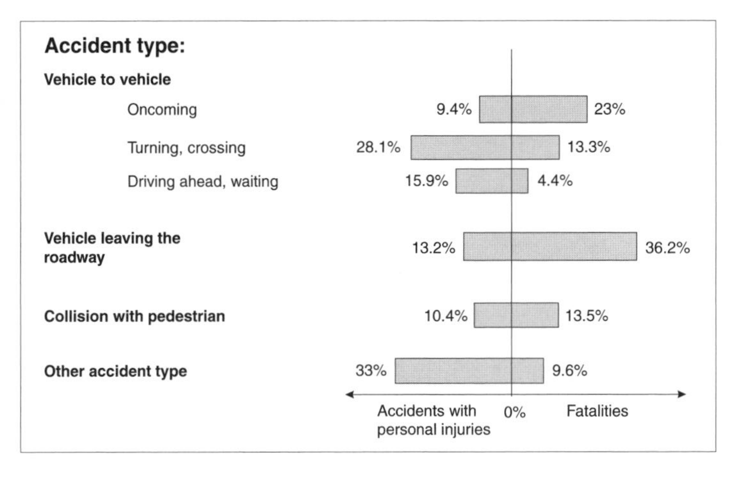

Fig. 7.1-1 Distribution of injuries and fatalities as a function of accident type (source: German Federal Statistics Office, 1995).

The suspension alone is not decisive in establishing handling dynamics and ride comfort; factors given equal weight are overall vehicle size, center of gravity height, weight distribution, radii of gyration, wheelbase and track, and, at higher speeds, aerodynamic properties.

A compilation of frequently encountered symbols and concepts is beyond the scope of this introductory chapter. Reference is therefore made to the literature, for example [2].

7.1.3 Vehicle dynamics and suspension forces

Building upon the previously outlined concept—that driving an automobile consists of moving a body within a gaseous atmosphere and a gravity field, and in the process, vertical forces, propulsion and braking forces, and lateral guiding forces are transmitted to the roadway by tire contact patches—it appears logical and sensible to analyze vehicle dynamics on the basis of these forces.

This approach conforms to the usual division of vehicle dynamics into longitudinal, transverse, and vertical dynamics. The six degrees of freedom of the vehicle, which for simplicity will here be regarded as a single-mass system, take part in longitudinal, transverse, and vertical dynamics as outlined in Table 7.1-1.

Longitudinal, transverse, and vertical dynamics are mutually interdependent in both the primary and secondary degrees of freedom. Moreover, there are opposing, complex interdependencies among tire force components (Section 7.3).

7.1.3.1 Transverse dynamics: transverse suspension forces

In the transverse direction, the primary forces are tire side forces transmitted from the roadway to the vehicle body. For steered wheels, lateral force components also result from tangential tire forces, which cannot be ignored at large steering angles.

By definition, the primary task of tire side forces is controlling vehicle transverse dynamics in terms of maintaining the course desired by the driver and also controlling rotation around the vertical axis. The center of gravity height with respect to side force components acting on the body usually results in an undesirable roll motion of the vehicle, decoupling the transverse motion of the vehicle center of gravity from that of the tire contact patches. The vehicle and its surroundings represent the control path, with the driver as controller. Concepts and processes familiar from control and feedback technology are therefore also used in vehicle dynamics (see, e.g., [3]).

Table 7.1-1 Contributions of vehicle degrees of freedom to longitudinal, transverse, and vertical dynamics

Designation	Primary degrees of freedom	Secondary degrees of freedom
Longitudinal dynamics	Longitudinal motion, pitch	Vertical motion
Transverse dynamics	Transverse motion, yaw, roll	Pitch, lift motion
Vertical dynamics	Vertical motion, pitch, roll	Vertical motion

Tire side forces are largely a function of slip angle, with (vertical) wheel load and tangential force as the most important parameters (Section 7.3). Tire side force components resulting from suspension camber or tire asymmetry are of secondary importance and will not be considered here.

Slip angles have various causes, which may be classified by their generating mechanisms.

7.1.3.1.1 Steering the wheels

By imposing a steering angle on the wheels, usually at the front axle, the driver causes an immediate change in slip angle as the rolling direction of the wheels, free of lateral forces, changes with respect to the instantaneous velocity vector of the tire contact patch. The resulting side force or side force change represents the first phase of tire side force generation. This causes a change in the lateral acceleration and, as it acts eccentrically to the vehicle center of gravity, a rotational acceleration. In conventional motor vehicles, rotational acceleration includes a moment around the yaw axis as well as a component around the roll axis [4]. As a result, velocity components at right angles to the tires are built up in the contact patches, at both the front and rear axles, which reduce the original slip angle of the front axle or delay its continued increase, while at the rear axle, increasing transverse velocity marks the beginning of side force generation. The original direction of this side force depends on the center of gravity location, radii of gyration in yaw and roll, and wheelbase and roll center height at the front and rear axles. It significantly defines the turn-in behavior of the vehicle. Thereafter tire slip angles and side forces are generated as a result of the increasing vehicle yaw angle and yaw rate, as may be readily derived for steady-state cornering.

In addition to driver inputs, steering movements of the wheels are also caused by suspension design measures. These steering movements, which may also be termed toe changes, are caused by the following:

- Vertical spring motion, in part as a result of suspension kinematics and in part due to changes in lateral force from spring and shock absorber forces and the resulting elastic deformations
- Tire longitudinal and transverse forces as well as tire aligning moment resulting from elastokinematic properties of the suspension
- Elastic deformation in the steering train
- Active steering systems (Section 7.4.4)

Targeted application of these possible means of steering angle change is used to achieve the desired degree of vehicle stability; to improve phase shift between steering angle, yaw rate, and lateral acceleration during transient maneuvers; and to optimize directional stability during throttle changes and braking in turns.

Conventional passive suspensions may be equipped with a type of "mechanical control" which automatically executes corrective steering movements at the front and rear wheels. To improve turn-in behavior and stability, many rear suspension designs are designed to build up a steering angle in turns that increases the axle lateral force gradients as a function of slip angle while the opposite effect is sought at the front axle.

The disadvantage of these passive mechanical controls is that they can only achieve very simple dependence of induced steering angle components as a function of wheel travel and acting forces. Furthermore, movement of the wheel in jounce as well as tire longitudinal and tangential forces do not represent ideal parameters for describing instantaneous vehicle dynamic conditions. For example, it would be impossible to determine whether a lateral force acting on a tire is introduced by the driver in order to achieve a change in direction or whether it represents a disturbing force. Consequently, some manufacturers, particular Japanese automakers, install active steering systems at the rear axle to affect side force generation by means of computer-controlled rear axle steering. Research and advanced development programs are also investigating systems which permit intelligent interpretation of driver inputs at the front axle in order to optimize the vehicle/roadway control path. The analogy of computer-controlled front or rear axle steering to "fly by wire" as used in aviation is apparent. Therefore, the term "steer by wire" is often used if the obvious mechanical connection between the steering mechanism and wheels is interrupted by a controller.

7.1.3.1.2 Stabilizing the vehicle on a predetermined path

In this case, slip angles and the resulting tire side forces are the result of deviation from the vehicle path due to external disturbances such as crosswind, road camber, and road irregularities, as well as disturbances to the front/rear axle side force ratio as the result, for example, of longitudinal acceleration initiated by the driver. If we begin with the simplifying assumption that the vehicle is in a stable, steady-state condition, the disturbing force gives rise to vehicle transverse, pitch, and yaw motion components, which will alter slip angles at the tires. In a dynamically correct suspension layout, force changes at the wheels will result in a small unavoidable deviation from the driver's intentional path to a new, stable, steady-state condition. The magnitude of the deviation from the original path is a measure of the vehicle's dynamic sensitivity to transverse disturbances.

Sensitivity to disturbances increases with increasing speed because:

- The tire side force is dependent on slip angle; that is, a doubled vehicle speed implies doubled transverse velocity with the same lateral disturbing force.
- Yaw damping is roughly inversely proportional to vehicle speed.
- Given identical crosswind speeds, aerodynamic side forces and yaw moments increase proportionally with vehicle speed and aerodynamic lift increases as the square of vehicle speed.

Active steering systems offer the potential of "controlling out" deviations from target parameters. A precondition, however, is earlier recognition of deviations than could be achieved by the driver, which, given the technology currently available, is not yet sufficiently assured for small deviations.

7.1.3.2 Longitudinal dynamics: longitudinal suspension forces

The major longitudinal forces transmitted from the roadway to the vehicle body are propulsion and braking forces. In the case of steered wheels, components of tire side force also act in the longitudinal direction; at larger steering angles, these cannot be ignored.

By definition, the primary task of propulsion and braking forces is to control longitudinal dynamics of the vehicle in terms of acceleration and speed. Longitudinal acceleration also gives rise to pitching and vertical motions. Coupling depends on center of gravity location, wheelbase, and suspension kinematics (antisquat and antidive geometry; see Section 7.4). Stabilizing factors are pitch and lift stiffness, as well as shock absorber damping. In the case of live axles, driveshaft torque reaction on the body, acting as a roll moment, must also be considered.

The following physical relationships affect lateral dynamics:

Dynamic weight transfer as a result of vehicle acceleration or braking causes an opposite change in instantaneous normal force at the front and rear axles. Because of tire side force dependence on normal force, this changes the lateral force distribution between front and rear axles; braking in a turn causes a turning-in moment, while accelerating out of a turn causes a turning-out moment.

Elastokinematic toe changes, directly caused by changes in tire tangential force and indirectly by changes in wheel travel and the resulting change in side forces, generate side force components which affect vehicle handling (Section 7.4). Toe changes, briefly outlined here, allow tuning of handling response to load changes or braking in turns.

Tangential force differences between left and right sides directly produce a moment about the vehicle vertical axis. As a rule, attempts are made to keep differences small by appropriate choice of brake design concepts and, in the case of driven axles, by use of a differential. Brake and slip control systems are similarly tuned (Section 5.5.2).

The yaw moment caused by unequal tangential forces may also be employed positively. Limited-slip and locking differentials generate a locking effect which tends to produce a correcting yaw moment, tending to bring the car back to a straight path. In order to prevent exaggerated understeer in acceleration, the differential locking effect is limited. On the other hand, in decelerating using engine braking, differential locking torque helps to compensate for oversteering moments caused by dynamic weight transfer. To better utilize this effect, some vehicles have differentials with a higher limited slip factor while under engine braking than in acceleration.

Vehicle stability control may be realized by means of differentials with controllable locking torque, which permit introduction of a correcting yaw moment. By means of special, complex final drive units, turning-in as well as turning-out yaw moments may be generated to affect vehicle handling [5].

Modern brake control systems use deliberate, controlled, asymmetric, driver-independent brake intervention to stabilize the vehicle at the handling limit [6]. The possibilities of deliberate asymmetrical brake force distribution are even applied in situations well removed from the handling limit in order to improve handling while braking in turns (Sections 5.5 and 7.2).

To correct course deviations encountered in braking, future electrohydraulic and electromechanical braking systems will make use of individual wheel brake force distribution and precisely controlled introduction of asymmetrical braking force.

Dependence of tire lateral force on tangential force permits altering brake force distribution to decrease the lateral force of one axle compared to the other, thereby generating a moment about the vertical axis. Analogously, tire side force capability is reduced under acceleration or engine braking. For all-wheel-drive configurations, tractive effort distribution between front and rear axles naturally affects vehicle handling. However, the effect is minimal in the usual range of braking or acceleration slip and is masked by the effect of weight transfer. Still, within the range of maximum transmittable forces for any given axle, targeted distribution of braking forces and, in the case of all-wheel drive, of acceleration forces represents a significant influence on handling [7]. In this situation, instantaneous wheel load and the available coefficient of friction are important parameters in optimizing lateral dynamics (Fig. 7.1-2).

7.1.3.3 Vertical dynamics: vertical suspension forces

Vertical forces, acting between suspension and body, consist mainly of spring and shock absorber (damping) forces. Their task is to support the vehicle mass on the wheels and impose tight limits on vehicle lift, roll, and pitch motion relative to the roadway. Roll represents an important, albeit usually undesirable, degree of freedom for vehicle lateral dynamics. To the extent permitted by spring travel, the body should be isolated against roadway irregularities and dynamic changes in wheel load minimized. Tire normal forces differ from the corresponding spring and shock absorber forces introduced to the body by the inertia of unsprung masses.

Because of the degressive dependence of tire side force on tire vertical load, from moderate lateral acceleration levels upward, the degree of understeer may be influenced by means of the front to rear roll stiffness distribution. The roll moment of individual axles is not only dependent on roll angle but also on lift and pitch angle, so that roll stiffness distribution provides an effective tool for tuning cornering behavior, even under braking or acceleration.

Along with spring and shock absorber forces, vertical reaction forces from tire side and tangential forces also affect the body, as do unsprung masses. These forces result from coupling of the tire contact patch longitudinal and transverse movement in response to vertical suspension motion (antidive, antisquat, sideforce reaction angle). These effects are used to reduce roll and pitch angles. For rapidly initiated steering maneuvers, the fact that roll acceleration and speed are lower is particularly advantageous for roll stabilization, compared to stabilization by means of roll stiffness. In this way,

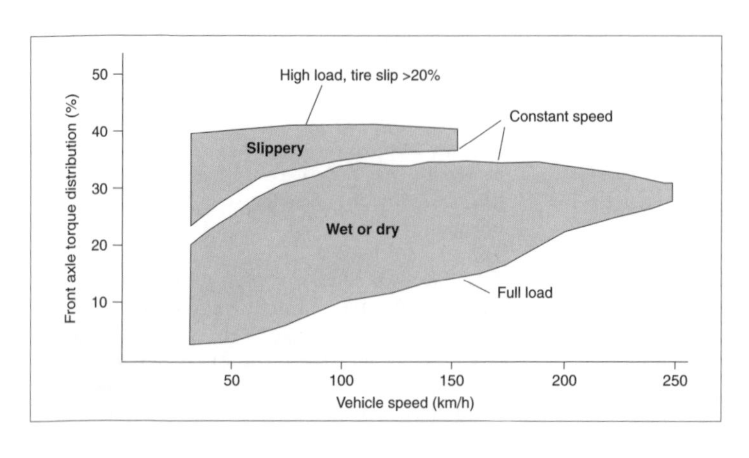

Fig. 7.1-2 Tractive force distribution as a function of vehicle speed (Porsche Carrera 4) [7].

turn-in and roll overshoot are positively affected. However, suspension kinematics must be designed to avoid adverse effects, especially jacking in cornering.

Goal conflicts between the demands of comfort, tire traction, appropriate wheel load distribution for good handling, and stiff, highly damped coupling of the vehicle to the roadway during transient maneuvers may only be solved to a limited degree using conventional spring systems. The problem is compounded by the fact that vehicle weight and weight distribution vary in response to often not insignificant cargo loading [3]. Load-leveling semiactive suspension with situationally dependent tuning of system parameters and, above all, actively controlled springing and damping systems offer great potential for reducing these goal conflicts (Section 7.4).

In vertical dynamics, aerodynamic lift and aerodynamic pitching moment also act as external forces and moments, along with weight and tire normal forces. These aerodynamic effects are usually described with the aid of front and rear axle lift (Section 3.2). At higher speeds, magnitude and distribution of aerodynamic lift forces have especially important effects on lateral dynamic behavior, as they alter tire normal-force suspension parameters including toe, caster, camber, and spring rate.

7.1.4 Basic goal conflicts

Goal conflicts arise in optimizing vehicle handling. Their magnitude serves as an indicator for the quality of the chosen suspension as well as the overall vehicle concept:

- Depending on the vehicle mass package, microsteering movements as functions of wheel travel, tire side force, and aligning moment are needed to achieve good turn-in and cornering behavior. However, the suspension should not generate any tire forces or moments to disturb straight-line stability on uneven roadways. Vehicle center-of-gravity location and yaw moment of inertia should therefore be considered during vehicle concept and package planning.
- In some suspension designs, the immediate dependence of elastokinematic steering behavior on hub carrier transverse stiffness requires high transverse stiffness of the suspension arm bushings. However, to isolate high-frequency transverse vibrations of the tire/wheel system, sufficient transverse elasticity is desirable. Therefore, suspension concepts which exhibit good elastokinematic steering behavior while permitting sufficient transverse elasticity are to be preferred.
- Traction and braking forces should be transmitted as directly as possible—that is, without vibration-causing phase differences between braking and traction moments on the body. However, this requirement is in conflict with the goal of reducing transmission of road impacts by means of targeted longitudinal compliance and damping. The goal conflict may be defused by selecting suspension concepts with greater transverse torsional stiffness and simultaneously greater longitudinal elasticity of the hub carrier with respect to the body. Longitudinal compliance and damping of steered wheels also has a significant effect on excitation of steering wheel rotary oscilla-

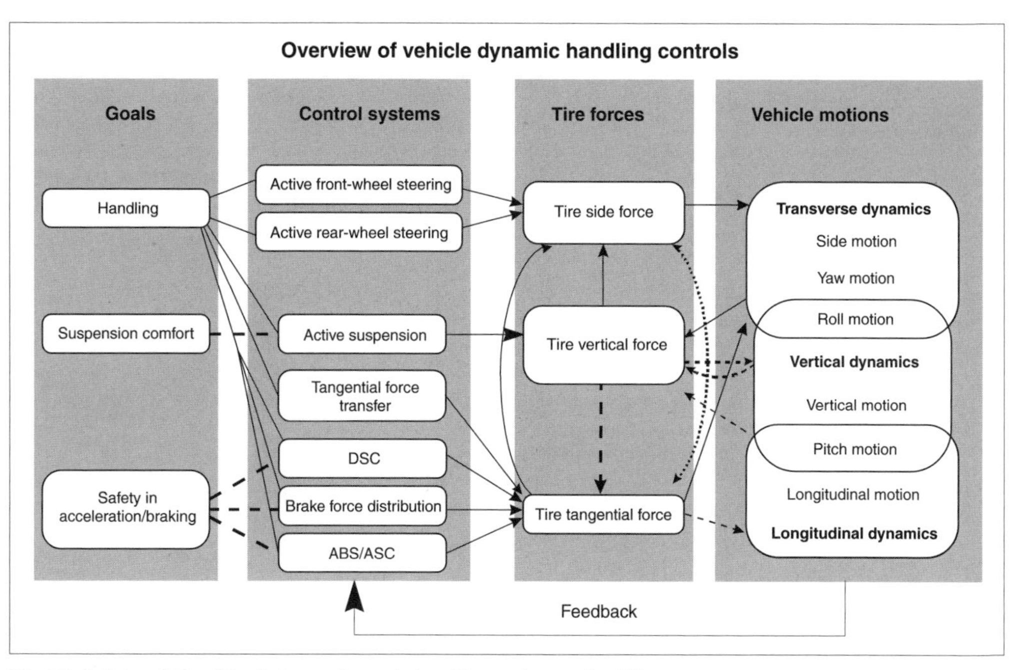

Fig. 7.1-3 Interrelationships between dynamic handling systems, after [5].

tions. In this situation, resonant frequencies of longitudinal and steering vibration modes for unsprung masses as well as the degree of coupling of these modes are of great significance [6].

- A major aspect in the design of the spring/damper system is achieving a good handling/ride comfort compromise appropriate to the character of the vehicle. This is made more difficult because the most suitable chassis tuning is dependent upon road surface, instantaneous vehicle speed, and maneuvers initiated by the driver. In order to resolve goal conflicts in suspension tuning, interactions between center-of-gravity height, pitch, and roll radius of gyration, wheelbase, and track should be considered in establishing the overall vehicle concept.

As may be seen from these conflicting goals, the vehicle suspension must fulfill a complex task, especially because subjective human impressions enter into its evaluation (Section 7.5). A suspension system should satisfy the stated multiplicity of requirements with minimum expenditure in terms of weight, space, and cost with as little impact as possible from environmental conditions, and it should do so over the life of the vehicle as constantly as possible.

In satisfying this complex assignment, the possibilities of controlled dynamic suspension systems (Fig. 7.1-3) may provide an important contribution, both in solving functional goal conflicts and in achieving a new quality in handling dynamics [8].

References

[1] Wierwille, W.W., J.C. Gutmann, T.G. Hicks, and W.H. Muto. "Secondary task measurement of work load as a function of simulated vehicle dynamics and driving conditions." *Human Factors*, 1977, pp. 557–575.

[2] Mitschke, M. *Dynamik der Kraftfahrzeuge*, Band C, Fahrverhalten. Springer Verlag, 1990.

[3] Zomotor, A. *Fahrwerktechnik: Fahrverhalten*. Vogel Fachbuch, 1987.

[4] Pauly, A. "Dynamique transitoire, contribution à l'étude du comportement de guidage non stationnaire." Presentation S.I.A., Ecole Centrale de Lyon, 1979.

[5] Furukawa, Y., and M. Abe. "Advanced Chassis Control Systems for Vehicle Handling and Active Safety." *Vehicle Systems Dynamics*, 28, 1997, pp. 59–86.

[6] van Zanten, A.T., R. Erhardt, and G. Pfaff. "Die Fahrdynamikregelung von Bosch." *Automatisierungstechnik* 44, 1996, p. 7.

[7] Birch, S. "New Stability from the Porsche Stable." *Automotive Engineering International*, February 1999, p. 16.

[8] Beiker, S. "Verbesserungsmöglichkeiten des Fahrverhaltens von PKW durch zusammenwirkende Regelsysteme." Dissertation, TU Brunswick, 1999.

7.2 Brakes

7.2.1 Introduction

Brakes are among the most important components of any motor vehicle. The safety of every vehicle and therefore of its occupants as well as other traffic participants and environs depends on the ability of breaks to function. Brakes are therefore safety-related components and are subject to strict regulatory demands.

Within the realm of passenger cars, the term "brake" is always used in reference to friction brakes. Continuously acting brakes (e.g., retarders) are used only on heavy commercial vehicles and will not be considered in this volume. In addition to basic requirements, further demands on brakes include:

— Speed reduction, if necessary to bring the vehicle to a stop at a specific location (deceleration brake)
— Prevention of undesired acceleration while descending grades
— Prevention of undesired motion of a parked vehicle (parking brake)

Wheel slip control systems were created with the aid of expanded and highly capable hydraulic control units and "intelligent" electronics. These systems include:

— ABS (anti-lock braking systems)
— EBD (electronic brake force distribution)
— ASR (antislip control)
— ESP (electronic stability program)

7.2.2 Brake system basics

7.2.2.1 Physical principles

Whether and how fast a vehicle moves, whether it changes direction as commanded by the driver, and whether it slides or spins are all largely determined by forces transmitted between the tires and roadway (Section 7.1).

(Tire) normal force is that portion of the vehicle weight borne by each individual tire (wheel) acting vertically on the road surface.

Braking force is generated by an *actuating force* (foot pressure on the brake pedal) supported by a brake booster. Actuating force is converted to hydraulic pressure in a master cylinder. This in turn acts on a wheel cylinder, which imparts clamping or pressure forces to

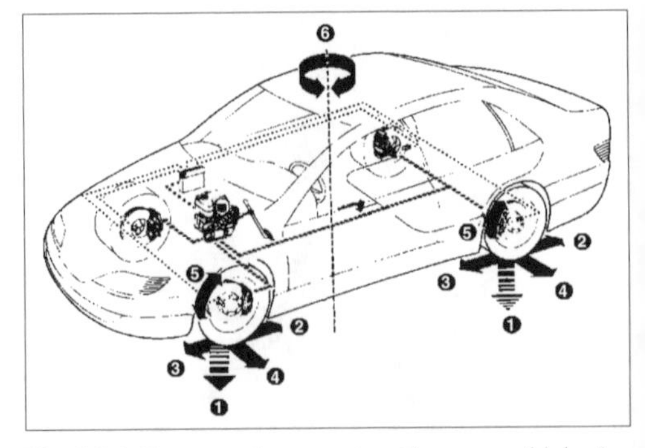

Fig. 7.2-1 Forces and moments acting on a vehicle. 1, Normal force; 2, braking force; 3, drive force; 4, lateral force; 5, inertia force; 6, yaw moment and polar moment of inertia.

the wheel brakes. These generate braking torque at the wheels, which acts as braking force on the roadway.

Cornering force is that force which imparts the desired directional movement to the vehicle and maintains its course in the presence of external disturbing forces (e.g., crosswinds).

Yaw moment arises from different longitudinal and/or lateral forces at the wheels. It results in rotation of the vehicle about its vertical axis.

7.2.2.2 Types of brake systems

All motor vehicles must be equipped with at least two mutually independent brake systems. Distinction is made between:

1. **Service brake system**: Service brakes are used in normal vehicle operation by means of the brake pedal and primarily serve to reduce vehicle speed.
2. **Emergency brake system**: Emergency brakes are used to stop a vehicle in the event of a malfunction in the means of operation and control of the service brake system. In general, the second service brake circuit serves as the emergency brake system. In the past, a modulatable parking brake served as the emergency brake system.
3. **Parking brake system**: To prevent unintentional rolling of a stationary vehicle, parking brakes may be applied and locked in their braking position (usually self-locking), with manual unlocking. Actuation is generally by means of the "hand brake."
4. **Supplemental brake system**: This serves to relieve the service brakes on longer downgrades and in frequent brake applications. Passenger cars do not use such systems.

General requirements

Brake systems must have certain basic features: The action of service and emergency brake systems must be modulatable (i.e., degree of brake action is controllable); the condition of brake components must be readily examinable; and the wear components must be designed to be adjusted or replaced at appropriate service intervals.

Uncontrolled brake systems consist of the following major components (Fig. 7.2-2):

- Brake pedal assembly
- Brake booster
- Dual master cylinder with fluid reservoir
- Wheel brakes (disc or drum)
- Brake pressure proportioning valve for the rear axle (in some cases)

7.2.2.3 Regulatory requirements

Most countries have their own individual requirements for brake systems. In Germany, these are described by §41 of the StVZO (Strassenverkehrszulassungsordnung—the German motor vehicle certification code). In the European Community, brake systems must meet the re-

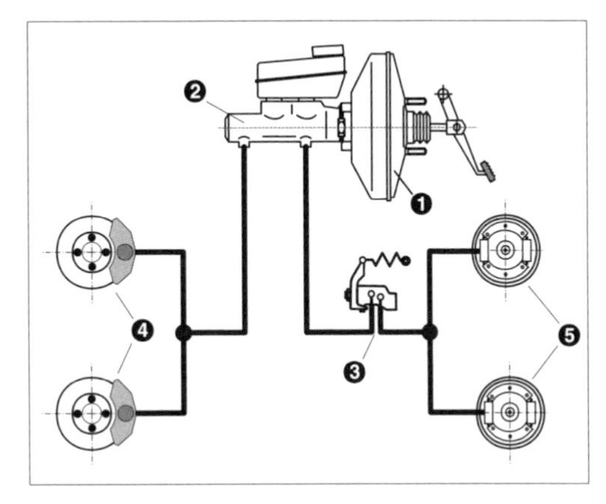

Fig. 7.2-2 Uncontrolled brake system. 1, Vacuum brake booster; 2, tandem brake master cylinder with fluid reservoir; 3, proportioning valve; 4, front axle wheel brakes (here shown as disc brakes); 5, rear axle wheel brakes (here shown as drum brakes).

quirements of EC directive 71/320. Alternatively, type certifications meeting ECE Regulation R 13, issued by the United Nations in Geneva, are acceptable. The United States has its own regulations. For passenger car brakes, these are contained in FMVSS (Federal Motor Vehicle Safety Standard) 135, whose requirements differ somewhat from those of ECE R 13 and 71/320.

7.2.2.4 Split brake systems

Splitting the brake system into separate circuits provides added insurance against total brake failure in the event of fluid leakage in one part of the system. The vehicle can still be adequately braked with the remaining intact brake circuit.

The most common dual circuit brake system configurations are:

Front/rear brake split

Here, the front and rear wheels each represent a separate braking circuit. The chief advantages of this configuration are:

- No asymmetrical pull in the event one circuit fails
- The possibility of employing a staged dual master cylinder
- No need to provide a separate brake line to the rear axle
- No danger of simultaneous compromise of braking ability at all wheels due to overheating (brake fluid boiling) at both brakes of a single axle

Diagonal brake split

In this configuration, diagonally opposed brakes are joined in a single brake circuit. The higher braking ability of at least one front wheel is used even if one circuit fails. Brake pulling in the event of one circuit failing may be compensated for by appropriate design, such as a negative steering offset.

7.2.2.5 The braking process

The shortest possible braking distances demand, among other things, the following constraints:

- High coefficient of friction between tires and road surface
- Rapid rise of actuating force
- High actuating force
- Rapid brake system response (actuation, transfer apparatus, wheel brakes)
- Ideal brake force distribution
- Optimum slip control to maximize utilization of available tire/roadway friction coefficient

In the course of the entire braking process, from the moment of objective necessity to deceleration to the moment the vehicle comes to a stop, the following phases may be defined:

- Unbraked motion
- Objective necessity to brake, t_o
- Perception time, t_p
- Decision time, t_d
- Reaction time, t_r
- Actuation time, t_a
- Foot pressure rise time, t_f
- Brake response time, t_b (from touching of pedal to 10% of final brake system pressure; this point is conveniently termed the "brake pressure point")
- Brake pressure rise time, t_s (from t_b to 75% of final brake system pressure)
- Total braking time t_{tot}

Initial speed and braking distance are used in calculation as well as measurement of the achieved deceleration rate (often termed "braking power," which, however, is physically incorrect terminology). In the case of nonconstant deceleration throughout the entire braking distance, calculation of average deceleration will provide differing results depending on which values are used in the calculation. Because distance is the deciding criterion for braking performance, this and the initial speed will be used as input parameters. Braking distance is customarily measured between the point where the brake is applied, and the point where the vehicle comes to a full stop. Initial speed is vehicle speed at the point of application of the brake light switch. Therefore, deceleration over the braking distance is

$$b = \frac{v_0^2}{2s}$$

where

b = average deceleration

v_0 = speed

s = braking distance

In releasing the brakes, the so-called "release time" or "release duration" is significant. This represents the time between the beginning of pressure release on the actuating mechanism (withdrawal of foot pedal pressure) and complete disappearance of braking force.

7.2.2.6 Brake system design

7.2.2.6.1 Brake force distribution

Ideally, in order to achieve neutral handling on a homogeneous road surface in a partially braked (i.e., not a maximum braking effort) condition, it is desirable to make the same use of the available coefficient of friction between tires and roadway at all wheels. Relative to the axles, this means that braking force at the axles should be proportional to the respective axle loads.

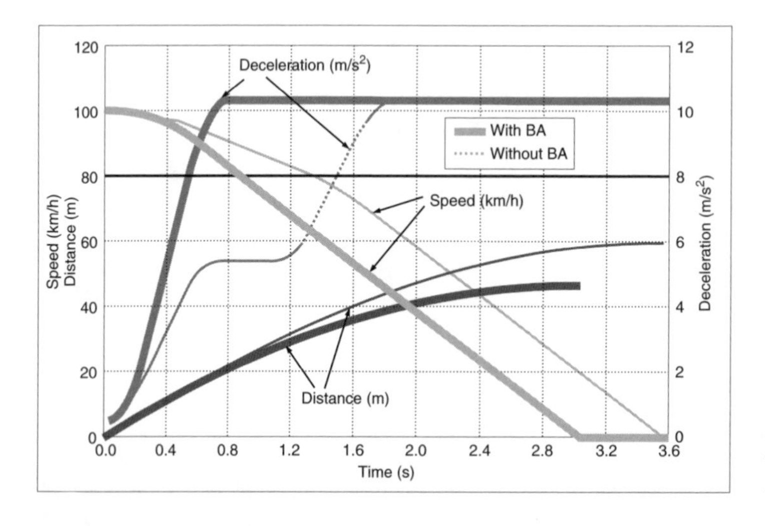

Fig. 7.2-3 Speed and braking distance as functions of time, with and without braking assistant.

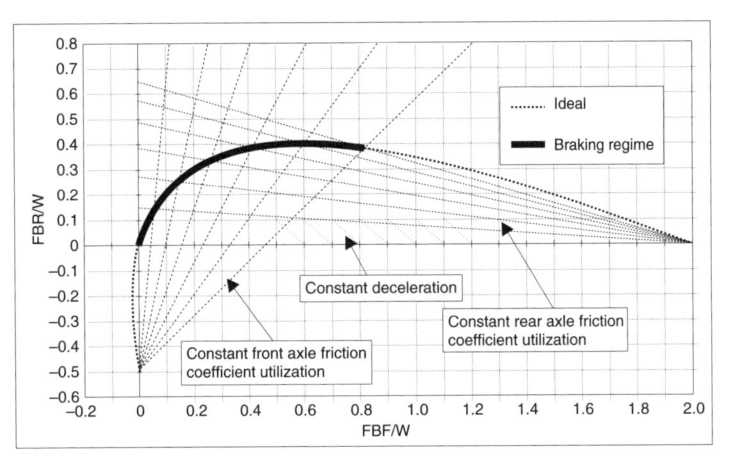

Fig. 7.2-4 Brake force distribution diagram (here shown as a complete circumferential force diagram): FBF, front axle braking force; FBR, rear axle braking force; W, weight of vehicle.

An ideal brake force distribution (Fig. 7.2-4) describes a parabolic curve. However, the brake force distribution built into the vehicle describes a linear response.

The installed brake force distribution is chosen so that it is always below the ideal curve: that is, it only intersects this at a deceleration rate of 0.8 g. This brake force distribution applies to vehicle motion in a straight line. In curves, the ideal brake forces are unevenly distributed between the brakes on any one axle, as wheel loads change in response to lateral acceleration (similar to axle loads in a longitudinal direction).

Application of electronic slip control systems (EBD: electronic brake force distribution) permits brake force distribution to achieve the ideal distribution even in cornering without the addition of any other hydraulic components.

7.2.2.6.2 Brake force distribution diagram

7.2.2.6.3 Dimensioning

Wheel brake size is primarily determined by vehicle weight and achievable maximum vehicle speed. Vehicle weight and targeted deceleration rate imply a brake torque that must be withstood by the effective radius of the brake disc and the clamping force of the brake caliper. The required clamping force and chosen brake caliper piston area imply a system pressure, which must be provided by the brake actuation mechanism. The volume of the brake system is determined by the volume of the actuation mechanism. Suspension geometry and wheel rim contours determine which combination of caliper, brake pad, and brake disc can be selected.

7.2.2.6.4 Thermal layout

Brakes convert kinetic energy to heat by means of friction. This heat is in part stored in appropriately dimensioned brake components and in part rejected to the environment through air cooling.

Thermal loading of brakes is therefore determined primarily by the maximum attainable vehicle speed. The faster the car, the more kinetic energy there is that must be converted to heat.

The second major criterion for thermal layout is long downgrades, which result in extreme brake heating. In this situation, potential energy is converted to heat over an extended period of time without permitting an adequate cooling air supply to reject this heat to the surroundings. Because of low speeds encountered in descents, air cooling of wheel brakes is comparatively ineffective. Furthermore, a conflict arises in that improved cooling brings with it easier access for water and dirt. In winter driving especially, contamination of brake discs and pads with road salt can lead to diminished brake effectiveness due to reduced pad friction coefficient.

Brakes are designed so that the manufacturer's established limiting rotor temperatures are not exceeded under realistic extreme conditions; the consequence would be reduced brake performance (fading) or even destruction of the brake itself.

7.2.2.6.5 Pedal characteristics (ergonomics)

Pedal characteristics, or so-called "pedal feel," provide the driver with feedback regarding the braking process and condition of the brake system. These characteristics vary as part of a vehicle's brand- and model-specific identity. The major parameters affecting pedal feel are:

- Activation force
- Pedal freeplay
- Threshold pressure*
- Amplification (deceleration/pedal force)
- Hysteresis
- Pedal travel
- Pedal travel and pedal force at booster vacuum runout point
- Pedal travel increase and pedal force increase during fading
- Response time
- Release time

* Threshold pressure is that pressure which the brake booster applies to "preload" the brake system, even at

very low brake pedal pressure, at the beginning of brake actuation. This eliminates any so-called "dead pedal" feel resulting from seal friction in the master cylinder and wheel cylinders. Brake pressure modulation in this regime is almost exclusively a function of pedal travel (rather than pressure).

Development of brake actuation systems is tending toward reduced pedal freeplay, lower activation force, and higher threshold pressures to achieve the most direct brake system response possible; and toward high amplification (boost, power assist) and short pedal travel for increased comfort.

7.2.3 Wheel brakes

Wheel brakes are configured as friction brakes. During the braking event, wheel brakes generate torques and convert kinetic energy into heat.

The defining characteristic of brakes is the ratio of generated tangential force at the rotor or drum to applied clamping force, the so-called *C** value. It simplifies brake system calculations.

Until 1960, brakes were almost exclusively configured as drum brakes.

Considerably reduced *C** variation as a function of varying friction coefficients (Fig. 7.2-5) and ability to withstand higher thermal loading have made disc brakes the predominant brake configuration in use today.

7.2.3.1 Drum brakes

Drum brakes are radial brakes. They are fitted with two inner brake shoes which are pressed against the friction surface of the drum during a braking event by means of hydraulically actuated cylinders. Springs draw the brake shoes back to their rest position, resulting in play between the drum friction surface and brake linings.

Adjustment

Brake lining wear may be compensated by manual adjustment using simple tools. However, as service intervals are being set at ever-longer durations and increased brake pedal travel between servicing is undesirable, automatic adjustment mechanisms are installed.

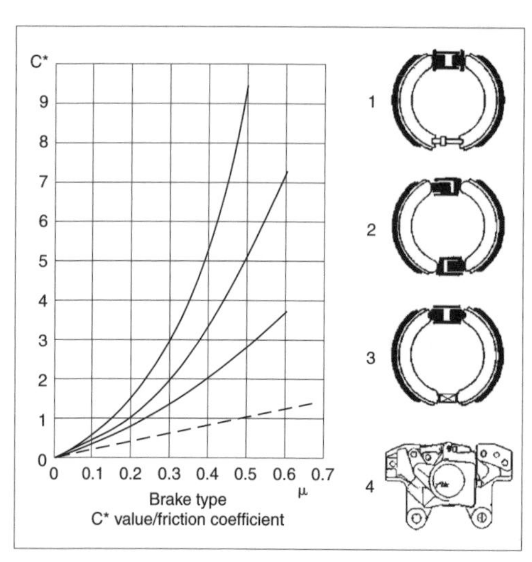

Fig. 7.2-5 Brake C* values. 1, simplex drum brake; 2, duplex drum brake; 3, duo servo drum brake; 4, disc brake.

7.2.3.1.1 Parking brake

Parking brake function is easily achieved using a drum brake. The parking brake mechanism, actuated from the driver seating position, applies cable force against drum brake levers. Transfer and distribution of brake forces are at lower efficiency than in hydraulic application of the service brakes. At present, parking brake function is achieved almost exclusively through mechanical actuation by means of a parking brake handle (or dedicated foot pedal). In future, this will be replaced by electric motor actuation, the so-called "electric parking brake" (EPB; Section 7.2.7.1).

7.2.3.1.2 Simplex drum brake

This brake configuration is commonly employed on the rear axle of passenger cars with maximum speeds no greater than about 170 km/h (106 mph). Low generated brake torque sensitivity to friction coefficient variations (*C** = 2.0 to 2.3) provides adequate handling stability under braking. The forward (in the direction of vehicle travel) brake shoe (the primary brake shoe) provides about 65% of the braking torque, while the rear (secondary) shoe provides the remaining 35%. Accordingly, either the primary lining is thicker or differing arc lengths* are chosen for the two shoes.

* Circumferential arc length over which the lining contacts the drum.

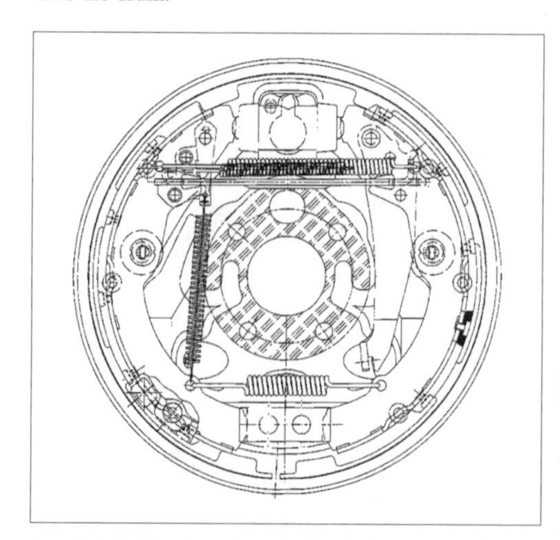

Fig. 7.2-6 Simplex drum brake (with automatic adjustment).

7.2.3.1.3 Duplex drum brake

Each of the identically sized brake shoes pivots against its own fixed point on the brake backing plate and is pressed against the drum by a single-acting blind hole cylinder. Both shoes are primary shoes with increased self-energizing effect. Because the higher C^* value of about 2.5–3.5 results in more difficult modulation and because integration of a parking brake is more complicated, this configuration is of only limited significance.

7.2.3.1.4 Duo-servo drum brakes

This configuration generates very high brake torques, as the self-energizing effect of two sequential brake shoes is especially pronounced ($C^* = 3.5–6.5$). Duo-servo brakes are therefore preferred for vehicles with high payload—for example, small to medium-sized commercial vehicles.

A parking brake arrangement can easily be achieved by means of a lever attached to the secondary shoe and an actuating rod to transmit reaction force to the primary shoe.

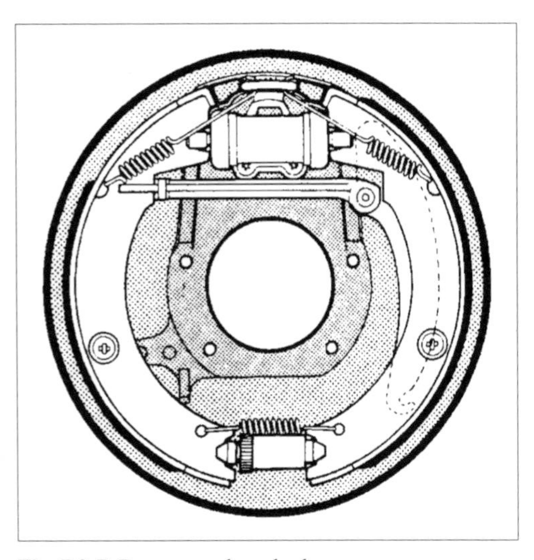

Fig. 7.2-7 Duo-servo drum brake.

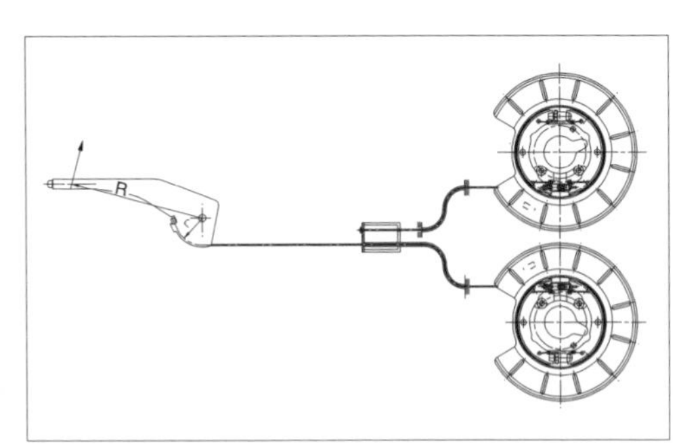

Fig. 7.2-8 Parking brake actuating mechanism with combined disc/duo-servo drum brake.

A very functional duo-servo application works in combination with disc brakes. Here, a purely mechanically actuated duo-servo drum brake provides the parking brake function, while the disc brake provides all service brake functions on the rear axle. One advantage here is optimum lining selection independently for parking and service brakes.

7.2.3.1.5 Brake drums

The version A drum (Fig. 7.2-9) is generally used today because it has the lowest manufacturing cost. Drum B is a composite casting; the other component is an aluminum alloy, while the inner surface is a gray iron alloy, chosen for its more suitable friction pairing against the brake lining material. A further stage in weight-optimized drum design is shown in configuration C. Here, the aluminum consists of a cast-in matrix of ceramic or aluminum oxide. Aluminum brake drums are difficult to manufacture, and their performance range is limited due to their lower melting point.

7.2.3.1.6 Calculation of drum-brake design parameters

Based on the geometrical situation of Fig. 7.2-10, the circumferential force of a brake shoe may be described by

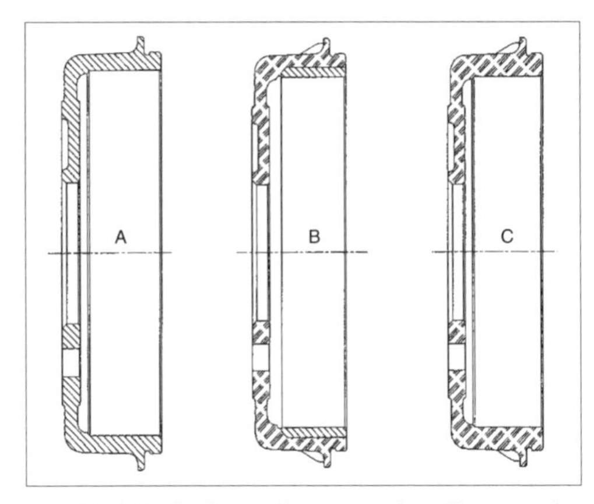

Fig. 7.2-9 Brake drums. A, gray cast iron; B, composite casting; C, aluminum/ceramic composite casting.

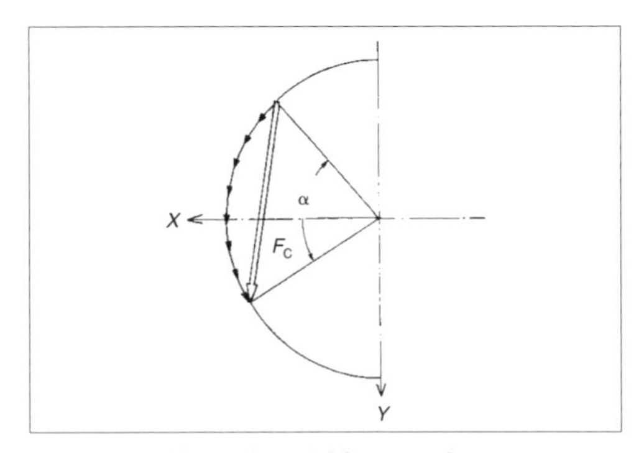

Fig. 7.2-10 Circumferential force resultant.

$$F_C = \int^{\infty} dR_x + \int^{\infty} dR_y$$

For a simplex shoe, the internal self-energizing effect is the ratio of ideal total circumferential force F_C to the applied spreading force F_S.

$$C^* = \frac{F_C}{F_S}$$

This calculation must be performed separately for each brake shoe; addition of the results provides the total shoe factor (C^*) for the brake.

(Other criteria and formulae for brake design calculations are described by H. Strien in *Auslegung und Berechnung von PKW-Bremsanlagen*, Alfred Teves GmbH, 1980).

7.2.3.2 Disc brakes

7.2.3.2.1 Disc brake configurations

Disc brakes are axial brakes. Each brake lining area covers a portion of a flat ring surface (partial-lining disc brake). If not expressly stated otherwise, the concept "disc brake" always implies a partial-lining disc brake.

Fully annular disc brakes, in which a toroidal lining contacts the entire disc, are not used in passenger car construction. Vehicles with high performance potential use disc brakes almost exclusively.

One important advantage of disc brakes is that tangential force changes linearly with friction coefficient; that is, C^* increases linearly with increasing friction coefficient (Fig. 7.2-5).

Other important advantages of disc brakes, as compared to drums, are:

- Ability to withstand greater thermal loads
- Less sensitivity to changes in pad friction coefficient (see above)
- More consistent response (reproducibility)
- More uniform pad wear
- Simple (self-acting) adjustment
- Simple pad replacement

Caliper designs

Disc brake configurations are dominated by two different caliper designs—fixed and floating calipers. Fixed calipers contain pistons on both sides of the brake rotor (Fig. 7.2-11), while floating calipers have pistons on one side only and are mounted in such a way as to permit lateral motion (Fig. 7.2-12).

Clamping forces in the brake caliper are applied in an axial direction to brake pads by means of hydraulic cylinders. Brake pads act on both sides of the flat friction surfaces of the brake rotor. Pistons and pads are contained in a caliper which arches over the outside diameter of the rotor. The pads are braced against rotation by a statically mounted component attached to the wheel carrier.

Sealing the pistons within the caliper is achieved by rectangular-Section seal rings fitted to a profiled housing groove. These have proven themselves in their ability to withdraw brake pads after the braking event (piston seal "rollback") and conversely to return the piston to its nominal position in response to the piston being forced back into the cylinder bore by axial deformation and displacement of the brake rotor ("knockback").

A dust seal protects the area between piston and housing bore against dirt and moisture. This seal is in the form of a bellows, as it must compensate for pad and rotor wear as well as axial tolerances.

7.2.3.2.2 Brake discs

Solid discs/ventilated discs

About 90% of the energy converted by brakes first enters the disc, from which it is transferred to the surrounding air. The friction ring may reach temperatures up to 700°C (red hot), for example in descending a long downgrade. Because of improved cooling action, internally *ventilated disc brakes* are enjoying increased application. A further measure for improved cooling action and additionally for reducing sensitivity to water is cross-"drilled" or slotted brake rotors. Disadvantages of these include higher cost and sometimes increased noise generation.

The so-called "ATE Power Disc" was developed to avoid such drawbacks. In this design, the ring surfaces contain a continuous groove. Other major advantages of this multifunction groove are:

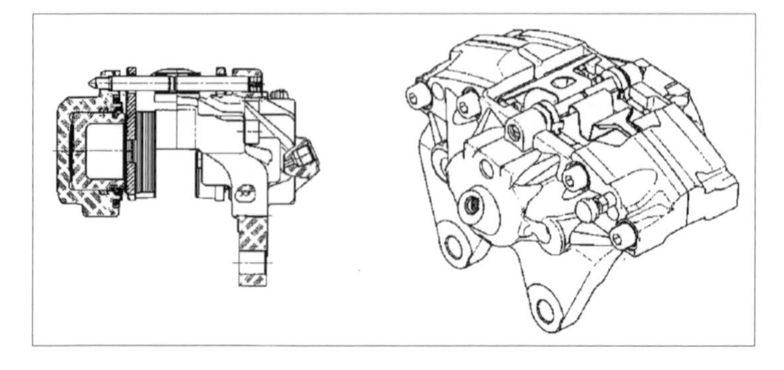

Fig. 7.2-11 Fixed caliper.

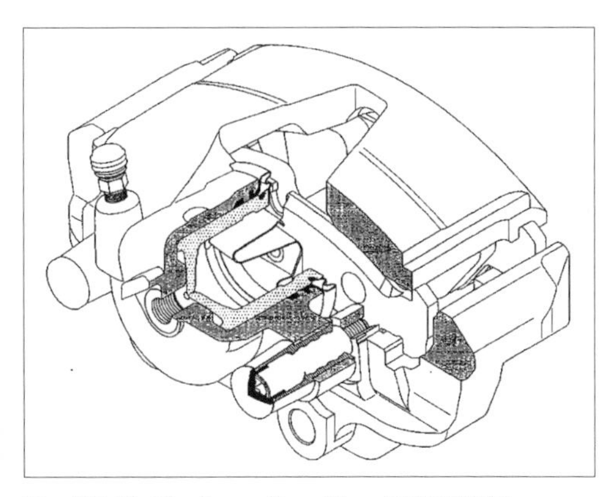

Fig. 7.2-12 Floating caliper (Type TEVES FN).

- Visual wear limit indication
- Improved wet braking
- Reduced fading
- Groove-free wear of pad and rotor

As the rotor, depending on installation, must have an attaching flange ("disc brake hat") on one side, the rotor tends to "cone" as it heats up; that is, the friction surfaces distort, changing from flat to conical. This may result in uneven pad wear due to point contact of the pads on the friction ring. The result is development of undesirable noise. Design measures may keep this distortion within acceptable limits (Fig. 7.2-13). However, completely neutral coning behavior should be avoided; otherwise, vibration (so-called judder) might arise during brake application due to the rotor assuming a wavy profile.

The term "judder" is typically applied to a combination of steering wheel vibrations, brake pedal pulsation, and low-frequency noise generation. Other causes may include pad deposits on the rotor—for example, as a result of extreme thermal loading—or rotor thickness differences (washouts) due to uneven rotor wear.

7.2.3.2.3 Fixed calipers

Fixed calipers are recognizable by brake cylinders mounted on both sides of the rotor surface and by a rigidly mounted housing (Fig. 7.2-11). The pistons on either side of the rotor are hydraulically joined by passages drilled in the caliper halves. The point where these passages cross the rotor outside diameter is thermally sensitive. Special measures ensure adequate cooling air supply, and generous disc dimensioning prevents brake fluid vapor formation, which might lead to excessive fluid takeup ("pedal to the floor") and resulting brake failure. Brake pads are tangentially supported on either side of the pistons.

Fixed front axle calipers are the most common form encountered on heavy rear-drive passenger cars because these vehicles provide ample installation space. Furthermore, these vehicles usually have a strongly positive steering offset, so that the caliper does not need to be located very far into the wheel.

7.2.3.2.4 Floating frame calipers

In order to take full advantage of the benefits of negative steering offset (Section 7.4.5), it is desirable to install brake discs deeper (i.e., farther outboard) into the

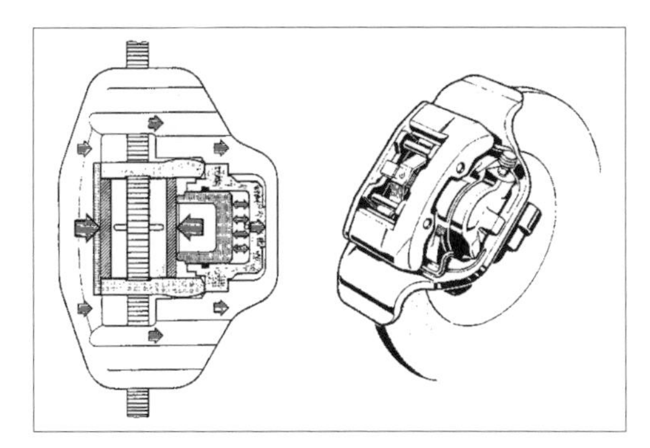

Fig. 7.2-14 Floating frame caliper.

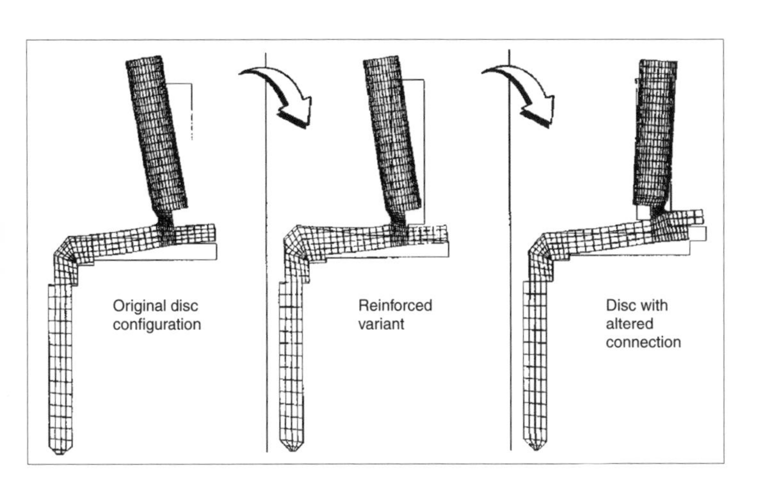

Fig. 7.2-13 Finite element analysis to reduce disc "coning."

wheel. The first design solution to be applied to the disc layout was the floating frame caliper.

In this design, a single hydraulic brake cylinder is used, located toward the inside of the disc. Its reaction is transmitted by means of a frame to the pad on the wheel side of the rotor. The cylinder housing is located within the frame. Tangential torque of both brake pads is taken by two arms extending from the anchor plate (pad holder), which is rigidly bolted to the hub or spindle. The advantage of floating frame calipers is low fluid temperatures in the cylinder, as the large, open pad slots permit good cooling airflow to the pads.

7.2.3.2.5 Sliding calipers

Sliding calipers also permit negative steering offset, because here too the piston is located on the inboard side of the caliper and therefore wheel-side space requirements are minimal. Front-wheel-drive layouts (usually with diagonally split brake circuits and negative steering offset) require added space inside the front wheel, which is provided by sliding calipers.

Other advantages of sliding calipers are:

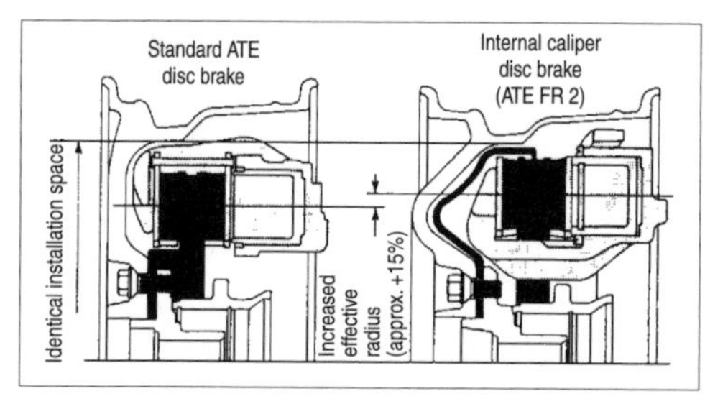

Fig. 7.2-15 Internal caliper brakes.

— Large pad area
— Optimum pad shape
— Low weight
— Small form factor

The usually single-piece housing slides on two guideways located by a firmly attached bracket or on the wheel carrier itself. These two guideways may be connected on the wheel side by a cast brace or, as in the TEVES FN principle (Fig. 7.2-12), by hook-shaped ends on the pad carrier plate. In this way, it is possible to at least partially have the brake pads applied by a pulling force ("push–pull principle"). Axial friction forces in the pad carrier guides are on the "leading" side, which achieves two functional advantages: first, the pads conform themselves evenly to the rotor surface, an important contribution toward the goal of parallel pad wear; second, noises, especially squealing, are reduced.

7.2.3.2.6 Optimized sliding caliper designs

Internal caliper brakes

The caliper design solution shown in Fig. 7.2-15 permits a very large brake disc within a given wheel size. Suitable for vehicles of <250 kW output, this is a complex design, particularly if a special version with a corrosion-resistant stainless steel brake disc carrier is chosen.

Nearly the same rotor diameter may be achieved using the FN3 configuration (Fig. 7.2-16), in which the caliper bridge across the rotor is very long, as seen along the rotor circumference, and therefore thin without compromising caliper stiffness (and therefore fluid volume takeup).

Even larger rotor diameters are possible with the FNR sliding frame caliper (Fig. 7.2-17). Here, the caliper bridge runs outward around the two guideways and

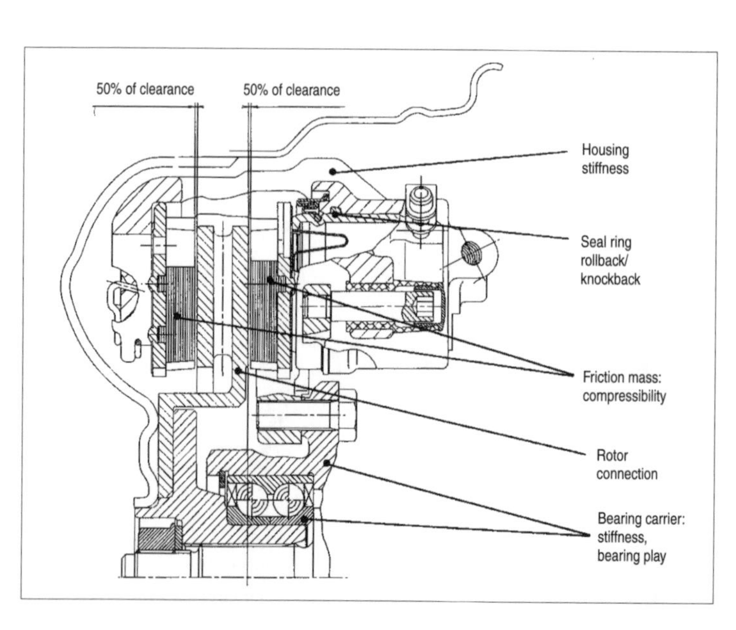

Fig. 7.2-16 Sliding caliper installation.

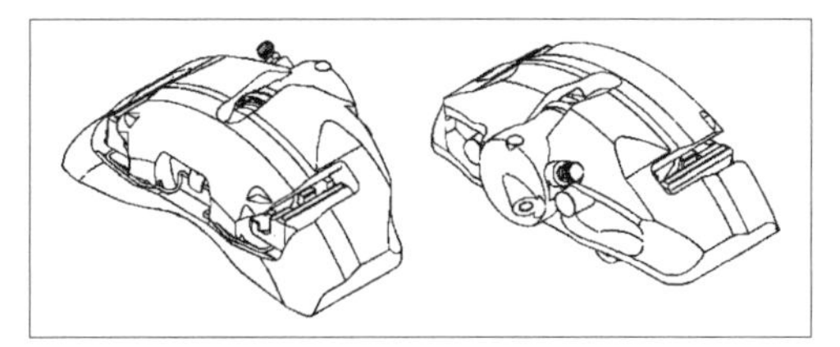

Fig. 7.2-17 Sliding frame caliper.

joins the central caliper tang on the wheel side as a single casting.

7.2.3.2.7 Disc brake design calculations

In disc brake designs, the shoe factor C^* is employed, which in this case results from the ratio of twice the friction force F to the caliper clamping force F_C (A_K is piston area, p is hydraulic pressure, μ is coefficient of friction):

$$C^* = 2\frac{F_T}{F_C}\text{; if}$$

$$F_T = A_K \cdot p \cdot \mu \quad \text{and} \quad F_C = A_K \cdot p,$$

then $C^* = 2\mu$

The clamping force is assumed to be acting at the center of the piston. Typical friction coefficient values for disc brakes fall between $\mu = 0.35$ and 0.50 (i.e., $C^* = 0.7$ to 1.0), where μ is defined as the average operating friction coefficient per any given pad material type. It may vary by $\pm 10\%$ in response to rotor temperature, vehicle speed, contact pressure, etc.

7.2.3.2.8 Combined sliding caliper

This design integrates the parking brake function within the brake caliper. As the C^* factor is only that of a disc brake, equal to 2μ, extra piston force is needed for the parking brake function (Fig. 7.2-8). Actuation is accomplished by means of a hand lever.

For the parking brake function, the simple wear adjustment feature of (disc) service brakes must be reproduced by means of a more complex adjustment mechanism.

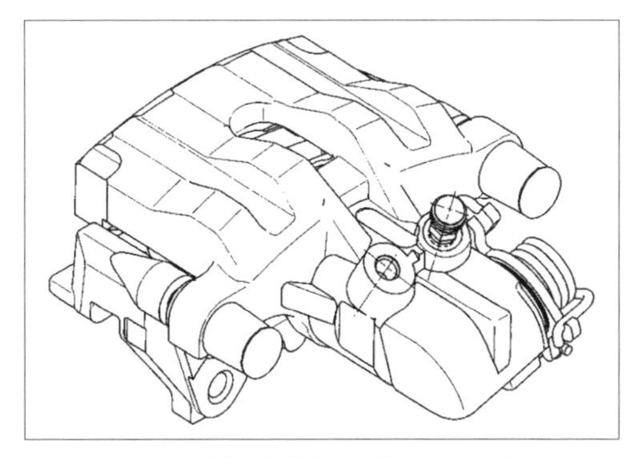

Fig. 7.2-18 Combined sliding caliper, Type FNc.

7.2.3.3 Materials

Sliding calipers are generally cast of ductile iron, of grade GGG50 to GGG60. For special lightweight requirements, bolted calipers are used, in which the cylinder side is made of high-strength cast aluminum and the caliper tang that arches over the rotor to the wheel-side pad is likewise cast in ductile iron. For some applications, the entire caliper consists of a single aluminum piece.

Brake caliper stiffness is indirectly defined by its fluid takeup as a function of cylinder hydraulic pressure, the so-called *volume takeup*. This is calculated by means of 3-D computer models, at pressures between 5 and 160 bar, and checked by actual experimental measurement.

Fatigue strength of a caliper is assured by hydraulic testing with pulsating pressure, between 0 and 100 bar, for 500,000 cycles. This test is conducted in functional testing with precisely defined parameters, as design influences cannot yet be completely accounted for using numerical methods.

However, *stress distribution* throughout a component may be determined without great difficulty and represented in color, as in Fig. 7.2-19.

Brake pistons may be made of steel, cast gray iron, aluminum alloys, or injection molded plastics (Fig. 7.2-20). In the case of steel pistons, deep drawing and extrusion are customary manufacturing methods. In order to assure the required surface finish and above all the diametral tolerance, grinding the outside diameter is imperative.

7.2.3.4 Brake linings and pads

Core requirements for brake linings and pads include:

- Friction coefficient, μ
- Good friction coefficient consistency (e.g., at different temperatures, in wet or dry conditions, in the presence of dirt or road salt)

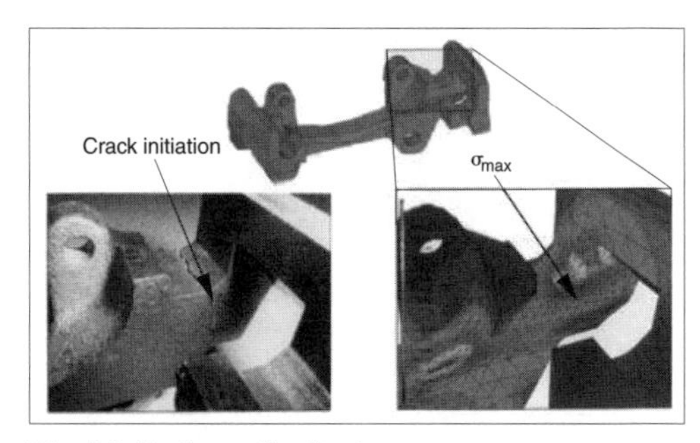

Fig. 7.2-19 Stress distribution.

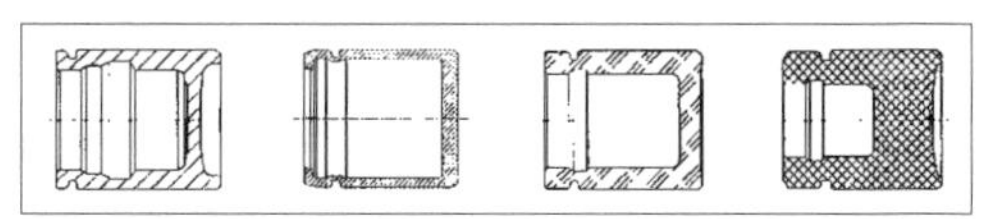

Fig. 7.2-20 Pistons of various materials.

— Low pad wear as well as low wear of the mating friction surface (drum or rotor)
— Low noise generation or good noise damping (e.g., against high frequency squealing and low frequency rumbling)
— Low compressibility

Beyond these requirements, certain standards must be maintained to continuously meet the above core requirements. These standards concern:

— Resistance to shear and compressive forces
— Resistance to fracture and tearing
— Fatigue
— Thermal expansion
— Thermal conductivity

Friction materials used in drum and disc brakes are composed primarily of:

— Metals, in fiber or powder form (~14% by volume)
— Fillers (e.g., inorganic fibers, ~23% by volume)
— Polymers (e.g., resins, rubbers, organic fibers and fillers, ~35% by volume)
— Solid lubricants (~28% by volume)

7.2.4 Actuation system

"Actuation system" refers to the combination of brake booster and master cylinder. Actuating force (pedal force) is transmitted to the wheel brakes as hydraulic pressure at high efficiency. With various diameters for master cylinder and wheel cylinders or pistons, so-called "external ratios"* and therefore increased force are obtained.
* external ratio = (pedal travel)/(2 × actuation distance inside a wheel brake)

7.2.4.1 Tandem master cylinder

The tandem master cylinder (also called a dual master cylinder) consists of two sequential master cylinders within a single housing. This arrangement permits pressure buildup and dropoff within the brake system, and, in the event of volumetric changes in the brake system (e.g., due to temperature changes or brake pad/lining wear), it provides the necessary volume compensation by means of a compensating port.

The tandem cylinder design satisfies the dual-circuit braking system regulatory requirement.

The pistons close two chambers for the primary and secondary circuits. If one circuit fails— for example, as a result of a failed seal at a wheel cylinder— this becomes apparent due to increased pedal travel, as the corresponding piston must be pushed to its stop before the remaining, intact brake circuit can build up pressure.

Depending on the configuration of the valve separating the brake fluid reservoir and the master cylinder, a distinction is made among:

— Vent port tandem master cylinder
— Central valve tandem master cylinder
— Plunger tandem master cylinder

7.2.4.1.1 Vent port tandem master cylinder

In a conventional vent port master cylinder, connections to the brake fluid reservoir consist of small holes (variously called vent holes, compensating ports, and replenishing ports) in the cylinder bores. As the piston is stroked, cup seals pass over these holes and interrupt connections between the main cylinder chambers and reservoir chambers. Pressure buildup in the brake system is possible from this point onward. As the brakes are released, the pistons return to their retracted positions and brake fluid flows from the wheel brakes back into the master cylinder.

7.2.4.1.2 Central valve tandem master cylinder

In the central valve tandem master cylinder, the chambers are connected to the brake fluid reservoir by means of so-called central valves located in the pedal pushrod (primary) piston and the floating (secondary) piston. In their inactive condition, these valves are held open by stop pins.

When the brake pedal is actuated, the primary piston is forced forward by the brake booster pushrod. Simultaneously, with opened central valves, the secondary (floating) piston moves forward in response to pressure from a preloaded spring between the primary and secondary pistons. The captive spring permits nearly simultaneous closing of both central valves, which may reduce the total closing travel of the tandem master cylinder to half that of sequential closing designs. This has

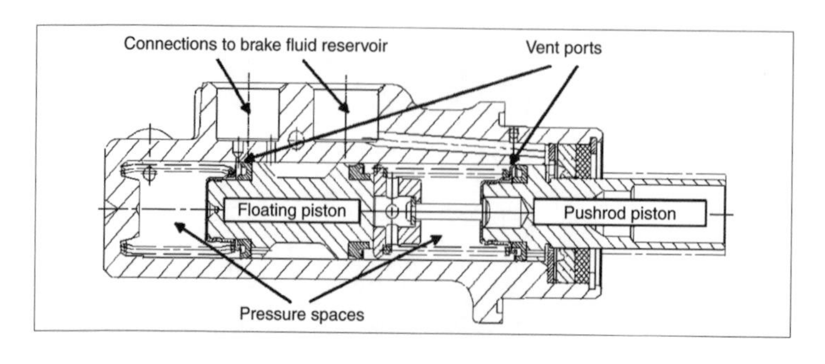

Fig. 7.2-21 Vent port tandem master cylinder.

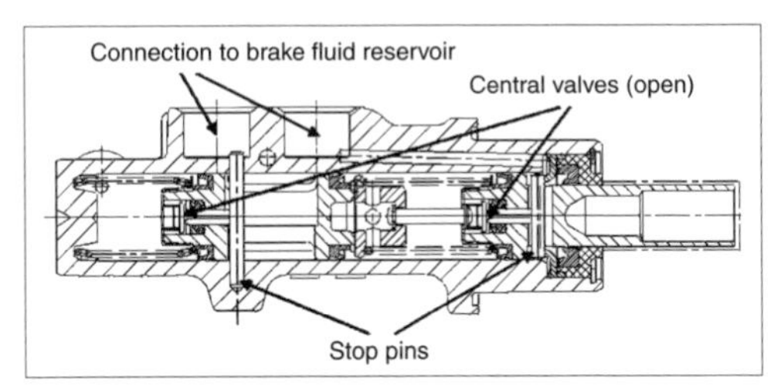

Fig. 7.2-22 Central valve tandem master cylinder.

a positive effect on pedal dead travel, which should be as small as possible to achieve immediate brake system response.

With the central valves closed, return flow from the tandem master cylinder to the reservoir is blocked. Fluid volume displaced by additional piston movement flows through the hydraulic control unit to the wheel brakes.

When combined with ABS, ASR, or ESP, tandem master cylinder pistons could be pushed back to their retracted position. While under pressure, seal travel over the vent holes could result in damage to the cup seals. Therefore, in this case, brake fluid under high pressure must be able to flow back to the reservoir via the central valves.

7.2.4.1.3 Plunger tandem master cylinder

Tandem master cylinders in plunger configuration are preferred in applications where a short installed length for the actuating mechanism is required. Pressure springs, sealing, and guide elements are in part arranged concentrically, which has advantages in reducing the required master cylinder overall length.

The valve mechanism is also suitable for ASR and ESP applications. In the retracted position, valve ports in the pistons connect each tandem master cylinder chamber with the corresponding reservoir chamber. As the cylinder is actuated by the pushrod, the passages, located ahead of the cup seals as seen in the direction of actuation, are pushed under the seal lips, thereby closing these passages. As actuation continues, these passages are again opened by the cup seal on the chamber side, which takes place without seal damage as, in this condition, the same hydraulic pressure is found on both sides of the sealing lips. This is especially important on brake release because, depending on possible prefilling of the brake system, for example by ASR or ESP controls, a high pressure condition may exist in the master cylinder in this position.

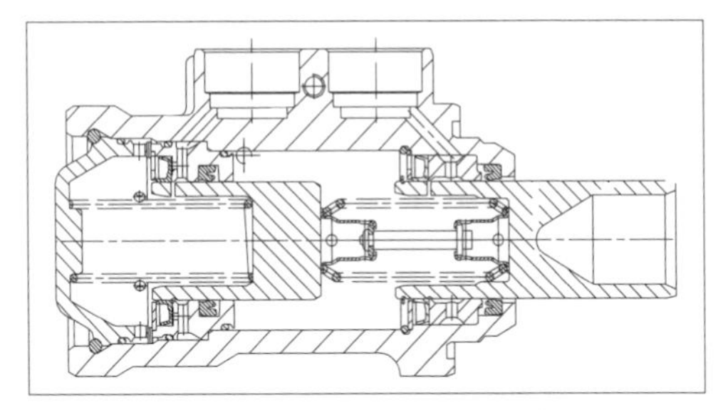

Fig. 7.2-23 Plunger tandem master cylinder.

7.2.4.2 Brake fluid reservoir

The brake fluid reservoir is attached to the top of the tandem master cylinder by means of so-called "reservoir plugs" and is customarily joined to the master cylinder by an additional fastening arrangement which permits high-pressure filling during vehicle assembly and ensures that in the event of an accident, flammable brake fluid does not escape. The reservoir:

- Serves to contain fluid to compensate for brake pad or lining wear
- Ensures volume compensation within the brake system for various environmental conditions
- Separates the two main cylinder circuits in the event of dropping fluid level
- May serve as a volume reservoir for a hydraulically actuated clutch or for an ESP preloading pump
- If needed, contains brake fluid for charging a hydraulic pressure accumulator

To ensure that the brake system is pressureless in its released state, the reservoir interior is connected to atmosphere through the reservoir cap. This is accomplished by a labyrinth seal in the cap or by a slotted membrane in the cap.

In the event of brake fluid loss, a reservoir warning system illuminates a warning light on the instrument panel.

7.2.4.3 Brake boosters

Brake boosters amplify foot pressure applied to the brake pedal, thereby increasing operating comfort and vehicle safety. Two design configurations are employed:

- Vacuum power brake boosters
- Hydraulic power brake boosters

7.2.4.3.1 Vacuum power brake boosters

Vacuum power brake boosters, also called vacuum boosters, have been favored over hydraulic power brake boosters despite their appreciably larger size. The main reason for this is the no-cost availability of a vacuum source on most gasoline engines.

A vacuum line connects the vacuum chamber to the gasoline engine manifold or, in the case of gasoline engines with limited manifold vacuum or diesel engines, to a separate vacuum pump.

Starting with the brake released condition (also called the ready condition), vacuum boosters function as follows.

Released condition (Fig. 7.2-24)

In this condition, the same pressure is applied to both sides of the diaphragm. The power piston return spring holds the diaphragm support plate in its retracted posi-

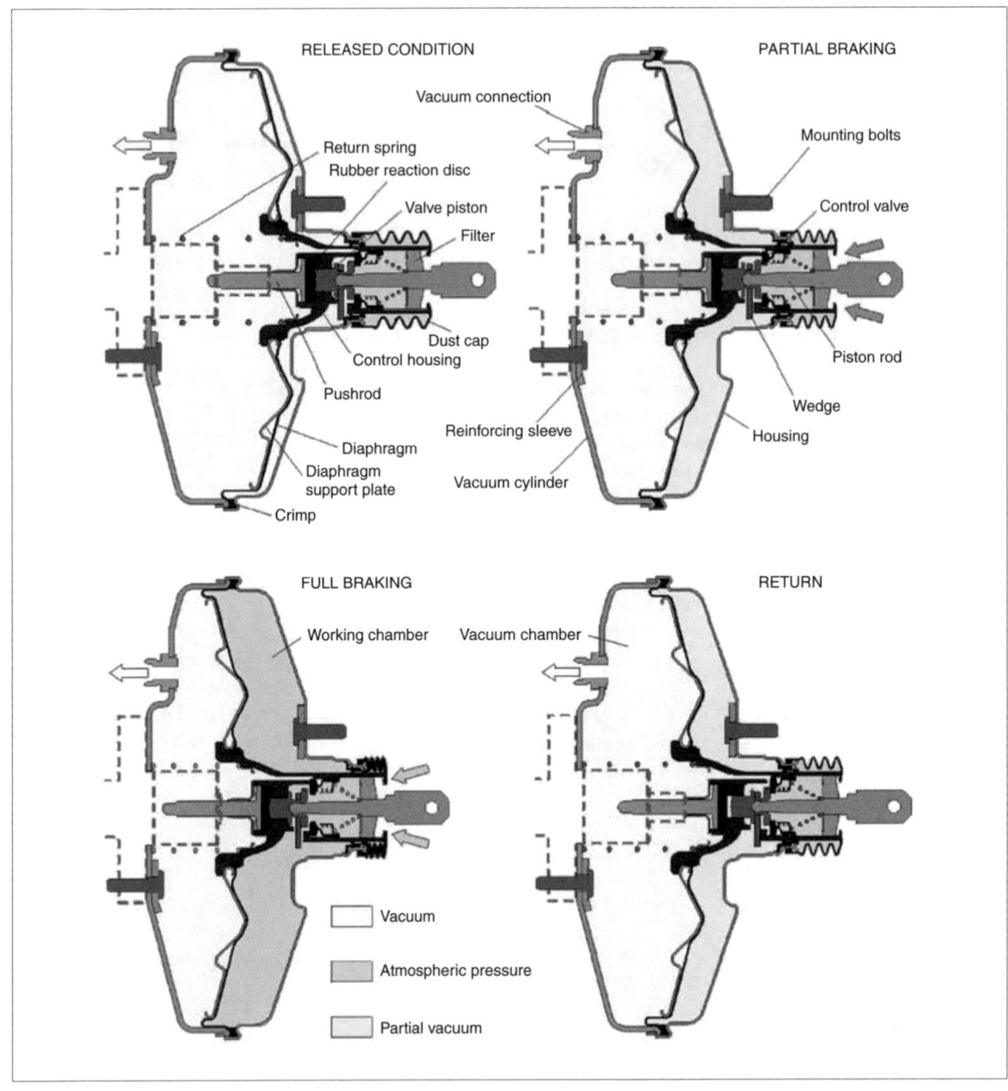

Fig. 7.2-24 Operating modes of a conventional single vacuum brake power booster (configuration: wedge horn).

tion against atmospheric pressure acting on the control port in the booster housing.

Partial braking (Fig. 7.2-24)

As the input pushrod is actuated by the brake pedal, the control valve closes the vacuum control port on the pedal-side chamber. As the pushrod continues its motion, the atmosphere control port is opened, thereby permitting atmospheric pressure to build ahead of the diaphragm support plate. As a result, a pressure differential exists between the forward and rear booster chambers, which moves the membrane support plate toward the tandem master cylinder, thereby amplifying the driver's foot pressure.

In the tandem master cylinder, hydraulic pressure builds in response to the forward movement of the piston. Given constant foot pressure, the pistons in the tandem master cylinder, the input pushrod, and the valve plunger come to rest at a displaced position determined by the foot pressure level. Now the reaction disc causes the valve plunger to rest against the control valve, cutting off the supply of outside air.

This is the "holding" position; any change in pedal pressure will cause an increase or decrease in the pressure difference between the two sides of the diaphragm support plate. Analogously to pedal pressure, hydraulic pressure in the brake system is raised or lowered, thereby setting the desired vehicle deceleration. The amplification factor (output force to input force) is determined by the ratio of valve plunger area to that of the rubber reaction disc.

Full braking (Fig. 7.2-24)

In the full braking position, the connecting passage between the pedal-side working chamber and vacuum chamber is completely closed, and the atmosphere control port is constantly open. Full atmospheric pressure is therefore applied to the diaphragm support plate, and the maximum possible boost in brake force is achieved. This condition is termed the booster vacuum runout point. Any additional increase in pressure on the tandem master cylinder piston can only be achieved by increased pedal pressure.

In addition to this conventional vacuum booster design, *tie-rod–actuated brake boosters* are currently in use. In this design, tensile forces in actuation are not transmitted by the booster housing (power chamber halves) but by a tie-rod which completely penetrates the device, including membrane and membrane support plate. This design permits appreciably thinner-walled or aluminum booster halves, with corresponding weight savings, without encountering the expansion found in conventional designs.

For larger vehicles, the working capacity of single-diaphragm boosters is inadequate. In such applications, tandem brake boosters are employed; these consist essentially of two single-diaphragm boosters, back to back, in a single unit (Fig. 7.2-25).

In some cases, installation of brake equipment from inside the engine compartment may be advantageous. So-called "front bolt devices" were developed for these applications. In these, the tie-rod is hollow, so that bolts may be inserted from the engine compartment side to bolt the unit to the firewall.

7.2.4.3.2 Vacuum pump

The available vacuum found in the intake manifold of gasoline-engined vehicles represents an economical energy source for power brake boosters. However, for injected engines (diesels and, in the near future, gasoline-injected engines), vacuum pumps will be necessary. These are usually rotary vane pumps attached to the engine and driven by the camshaft. Because they are permanently driven, they result in increased fuel consumption. Consequently, demand-driven, controlled, electrically powered vacuum pumps are under development.

7.2.4.3.3 Expanded-function vacuum brake boosters

For many expanded functions found in modern brake systems, electrically controllable so-called "active boosters" are employed.

To ensure good pressure rise dynamics, particularly at low ambient temperatures, active boosters are employed to preload ESP system pumps. In "Electronic Brake Assist" systems, the active booster serves as an aid to full braking in panic situations. With Adaptive Cruise Control (ACC, Section 5.5.2), it assumes the task of maintaining following distance, comfortably and independent of brake pedal actuation, with the aid of radar sensors by means of partial braking control.

These applications employ a brake booster with an electromagnetic solenoid integrated within the control housing. By means of a sliding sleeve, it is possible to actuate the control valve. This is done so that initially, the flow passage between the vacuum chamber and working chamber is closed. With more current to the solenoid, the flow passage between the working chamber and ambient air is opened and the booster activated.

To accurately ascertain driver intentions, a so-called "activation switch" is integrated in the control housing. In the case of brake assist, the primary criterion is brake pedal release; for ESP preloading, it is pedal actuation.

7.2.4.3.4 Hydraulic brake booster

Compared to vacuum brake boosters, hydraulic boosters offer the advantage of reduced installation space requirements and generally much higher booster runout

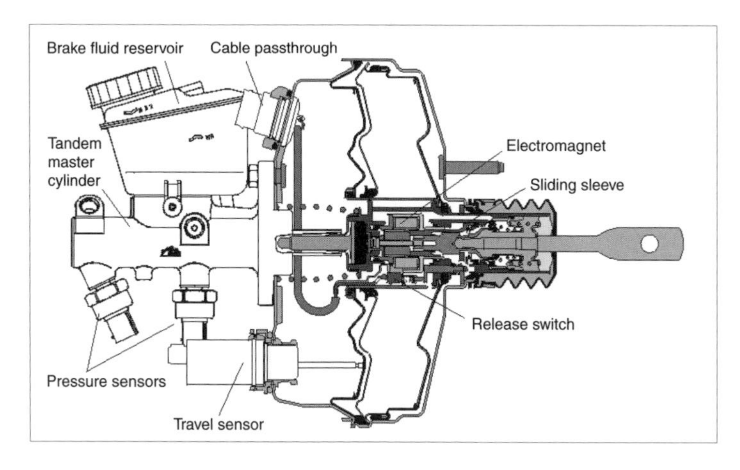

Fig. 7.2-25 Active tandem brake booster.

point. Disadvantages include higher cost and the "dead pedal" feel (no threshold pressure).

7.2.4.4 Brake proportioning valve

In order to achieve short braking distances, it is necessary to make the best possible use of rear brake power; however, rear wheels should never be permitted to lock up before the front wheels. This regulatory requirement is met by smaller rear brake dimensions than those of the front wheels. In general, though, because of the nonlinear nature of the ideal brake force distribution curve, smaller dimensioning alone is insufficient, so that a so-called brake proportioning valve (or electronic brake force distribution, or EBD; see Section 7.2.2.6.1) is employed.

Because of variations in pad or lining friction coefficient (which may undergo extreme changes), the effect of engine motoring torque, and manufacturing tolerances and friction within the pressure-limiting valve, the layout of the brake proportioning valve must be well below the ideal brake force distribution curve.

7.2.4.4.1 Brake pressure limiters

Brake pressure limiters limit the output-side pressure transmitted to the wheel brakes to a predetermined cutoff pressure.

7.2.4.4.2 Brake proportioning valve (preset)

Brake proportioning valves (also known as brake pressure regulators) are applied to vehicles in which only minimal axle load changes are expected. They have a fixed inflection or "split" point, above which the rear brake line pressure is reduced by a fixed ratio compared to front line pressure.

7.2.4.4.3 Load-sensing proportioning valve

Load-sensing (also known as height-sensing) proportioning valves are often used in vehicles that may encounter extreme changes in axle load due to cargo load. Installation of load-sensing proportioning valves is also indicated for small cars with a short wheelbase and high center of gravity that exhibit strongly deceleration-dependent weight transfer effects.

Vehicle loading is indirectly sensed by means of the vehicle ride height (compression of the suspension springs). When laden, the reduced space between body and axle results in increased spring force and therefore a raised inflection point for the split point of the proportioning curve.

7.2.4.5 Brake connections

High-pressure metal brake tubing, brake hoses, and flexible lines are used to connect brake system hydraulic components.

Brake tubing is used between rigid, nonmoving points on the body. This consists of double-walled, brazed steel tube. As protection against environmental influences, tube surfaces are galvanized and additionally covered with a plastic coating.

Brake hoses are used at transitions to dynamically loaded components such as wheel carriers or brake calipers. They ensure unimpeded transfer of fluid pressure to the brakes, even under extreme conditions. Along with mechanical strength, the most important characteristics include pressure capacity, minimal volume change under pressure, chemical resistance to such substances as oil, fuels, and salt water, and good thermal stability.

Brake hose construction consists of an inner hose, a two-ply woven layer to resist pressure, and an outer rubber layer to protect the pressure layer against external influences.

Flex lines, like brake hoses, are used at transitions to dynamically loaded components.

Due to limited flexibility resulting from their construction, PTFE (polytetrafluorethylene—also known as Teflon) lines, with a stainless steel braided cover as a pressure carrier and another thermoplastic elastomer as

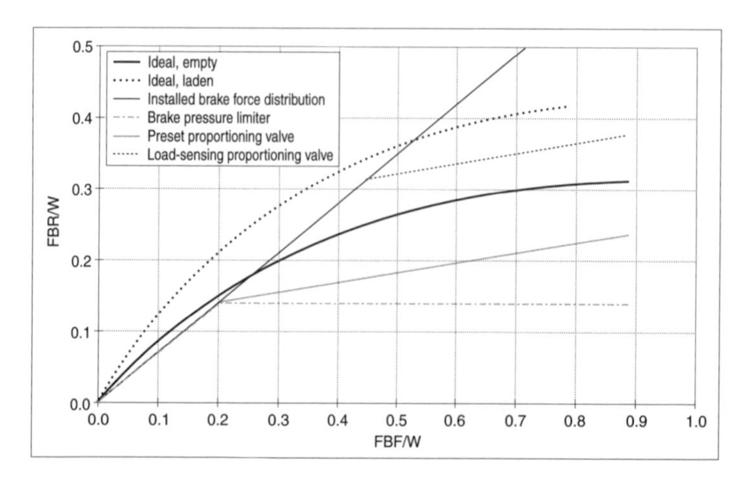

Fig. 7.2-26 Brake proportioning valve characteristic curves.

outer protective layer, are limited to applications with minimal motion, for example at (floating) calipers as a result of pad wear.

7.2.4.6 Brake fluid

Within the hydraulic section of the brake system, brake fluid serves as the energy transmission medium between tandem master cylinder or hydraulic control unit and wheel brakes. Additionally, the fluid is tasked with lubricating moving parts such as seals, pistons, and valves and providing these with corrosion protection.

Brake fluid should have as low viscosity as possible, even at extremely low temperatures (to −40°C), to provide good brake actuation and release behavior as well as to permit good electronic control system functions. Moreover, brake fluid should have as high a boiling point as possible, so that even the most extreme thermal loading of the brake system does not result in formation of vapor bubbles. As a result of the limited volumetric capacity of the tandem master cylinder, compressibility of vapor bubbles would result in inability to build up sufficient brake pressure.

Brake fluid is differentiated into conventional and silicone-based fluids.

Conventional fluids, based on polyglycol and polyglycol ether, are hygroscopic: they absorb and bind water. In this way, water entering the system is not separated and cannot boil to form vapor bubbles. Numerous international standards, such as DOT 3, DOT 4, and DOT 5.1, require that brake fluid, when exposed to water according to a specific test procedure, exhibit a certain minimum so-called "wet boiling point." Brake fluid replacement at regular intervals is necessary to assure sufficiently low water content under operational conditions.

Silicone brake fluids meeting the DOT 5 standard are based on hydrophobic silicone oil, which can absorb only small traces of water. Any undissolved water present may boil (formation of vapor bubbles, see above) or lead to component corrosion. Compressibility and quantity of dissolved, sometimes outgassing, air is greater in silicone brake fluids than in conventional fluids. Because of their particular properties, these fluids may only be used after specific approval by the vehicle manufacturer. These are primarily of interest for vehicles which must be stored for several years (e.g., military vehicles).

7.2.5 Electronic control systems

7.2.5.1 General

Electronic control systems are among the most important technical advances in automotive engineering in recent years, and they will continue to provide revolutionary expansion of individual mobility in the foreseeable future (see Section 7.2.7, "Development of future brake systems").

In order to ensure that neither safety nor reliability of brake systems is compromised in the face of increasing complexity, electronic control systems include sophisticated safety circuits and modular safety algorithms. These ensure that in the event of failure in a subsystem, the remaining lower-level functions, down to purely mechanical/hydraulic functions, are retained. Even in the event of total failure (e.g., loss of electrical power), conventional brake operation remains intact.

7.2.5.2 Physical foundations

7.2.5.2.1 Dynamics of a braked or powered wheel

The laws governing forces between tire and roadway are more complex than the laws of friction between two solid bodies, familiar from classical physics. The reason for this is the characteristic of elasticity of the rubber tire (see Section 7.3).

7.2.5.2.2 Longitudinal tractive forces

Braking slip/drive slip

A tire can transmit forces only if it deforms due to its elasticity or if it slides. The tangential velocity of a braked wheel is therefore less than that of the wheel center.

To simplify these relationships, special ratio identities are introduced: percent (braking) slip, λ_B, and percent drive slip, λ_T.

$$\lambda_B = \frac{(v_V - v_C)}{v_V} \cdot 100\%$$

where

λ_B = (brake) slip in %

v_V = vehicle velocity (speed of wheel center)

v_C = circumferential velocity of wheel

$$\lambda_T = \frac{(v_C - v_V)}{v_C} \cdot 100\%$$

where

λ_T = drive slip in %

Braking force coefficient and other influential quantities

In order to obtain a characteristic number independent of weight and loading condition of a vehicle, the so-called "traction coefficient" or "braking force coefficient" μ_B is defined as:

$$\mu_B = \frac{F_B}{F_A}$$

where

μ_B = braking force coefficient

F_B = braking force

F_A = tire normal force

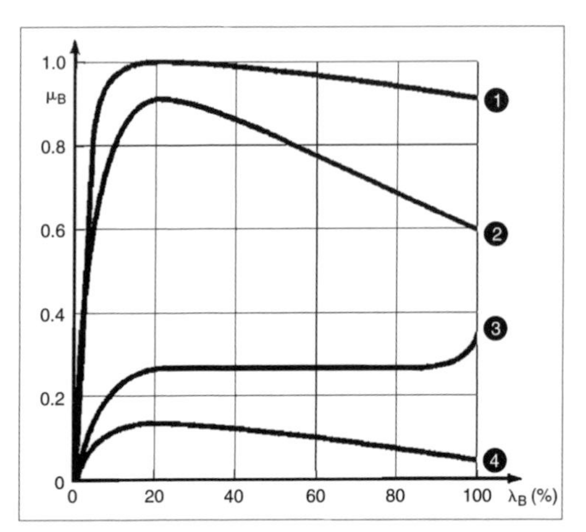

Fig. 7.2-27 Braking force coefficient vs. slip. λ_B, brake slip, μ_B, braking force coefficient; 1, dry asphalt; 2, wet asphalt; 3, unpacked snow; 4, ice.

For each combination of tire and road surface, there is a specific functional relationship between braking force coefficient and slip (Fig. 7.2-27).

The braking force coefficient shows an almost identical response as a function of slip under nearly all conditions. Even at low slip values, braking force coefficient rises steeply up to the point of critical slip, λ_{BK}, where the braking force coefficient generally reaches its maximum value somewhere between 8 and 30% braking slip. With increased slip beyond critical slip, braking force coefficient drops more or less steeply. The rising branch of the μ – slip curve (braking force coefficient curve) is termed the stable regime. The unstable region begins on either side of critical slip. If brake pressure is not reduced quickly enough in this region, wheel lock is imminent.

The road surface is a major determining factor of traction coefficient. Furthermore, different tires will have different braking force coefficient curves; contributing factors include tire type and configuration (e.g., summer or winter tires, tread pattern, amount of wear), dimensions, inflation pressure, temperature, and vertical load.

The braking force coefficient curve shows one special characteristic for loose snow and gravel: the maximum value is obtained at 100% slip. This is caused by a wedge of snow or gravel pushed ahead of the locked tire as slip increases. Yet here, too, there is a point of critical slip. It is reached not when braking performance diminishes, but when the vehicle loses stability and the ability to steer.

7.2.5.2.3 Transverse tractive forces

A tire can only generate sufficient lateral force if it deforms laterally and slips ("sideslip").

This phenomenon is only possible if the direction of motion of the wheel center differs from the tire longitudinal direction. The angle between these two directions is known as the slip angle (Fig. 7.2-29).

To quantify lateral force reserves, the so-called "lateral force coefficient" μ_C is defined as

$$\mu_C = \frac{F_C}{F_N}$$

where

μ_C = lateral force coefficient

F_C = lateral force ("cornering force")

F_N = tire normal force

An example of the functional relationship between lateral force coefficient and tire slip angle is shown in Fig. 7.2-28.

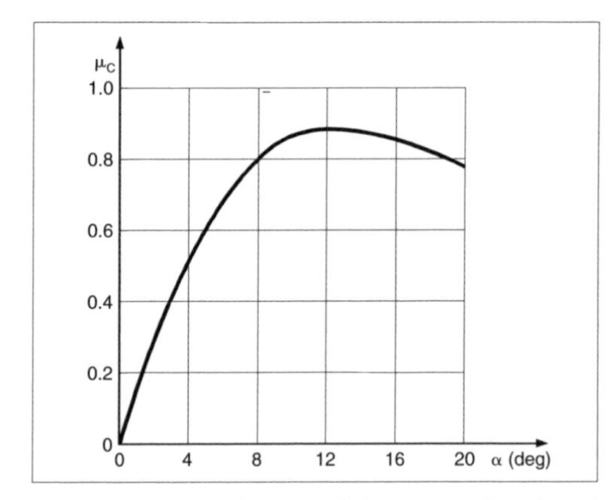

Fig. 7.2-28 Lateral force coefficient μ_C as a function of slip angle α.

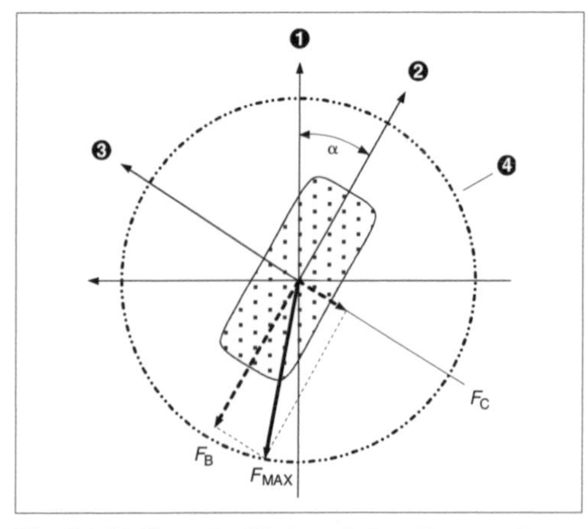

Fig. 7.2-29 Kamm's friction circle. F_{max}, maximum transmittable total force; F_B, braking force; F_C, lateral force; α, slip angle; 1, direction of wheel center motion; 2, wheel longitudinal direction; 3, wheel transverse direction; 4, Kamm's friction circle.

7.2.5.2.4 Combined longitudinal and transverse tractive forces

Under any given set of tire, road surface, and tire normal force conditions, a tire can only transmit a limited total force. The resolution of this total force into two components may be approximated graphically by Kamm's friction circle (Fig. 7.2-29).

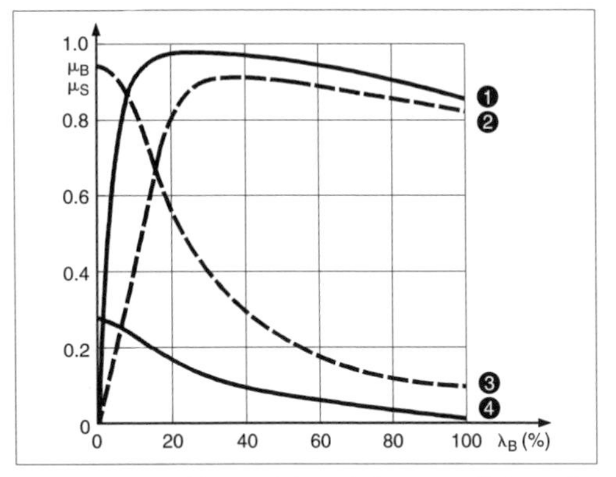

Fig. 7.2-30 Braking and lateral force coefficients for various slip angles α on dry pavement. λ_B, brake slip; μ_B, braking force coefficient; μ_C, lateral force coefficient; 1, braking force coefficient at $\alpha = 2°$; 2, braking force coefficient at $\alpha = 5°$; 3, lateral force coefficient at $\alpha = 5°$; 4, braking force coefficient at $\alpha = 2°$.

The geometric sum of both force components may not exceed the total force transmissible by the tire. This is illustrated by two practical examples:

— In extreme cornering, the transmissible braking force is considerably less than in straight-line motion.
— During extreme deceleration, steerability is restricted, compared to an unbraked condition.

The extreme case, full braking with locked wheels, leads to total loss of lateral forces and therefore loss of steerability and stability.

Fig. 7.2-30 illustrates how lateral force coefficient decreases in response to increased slip drawing upon the (longitudinal) braking force coefficient.

7.2.5.3 Sensors for electronic (braking) control systems

Reliable operation of electronically controlled system functions is only possible through application of multiple special sensors.

7.2.5.3.1 Wheel sensors

Wheel sensors capture the instantaneous rotational speed of each wheel.

The sensor unit consists of a sensor and an excitor ring. The ring is rigidly attached to the wheel while the sensor is attached to the wheel carrier.

For rear-wheel-drive vehicles, capturing rear axle

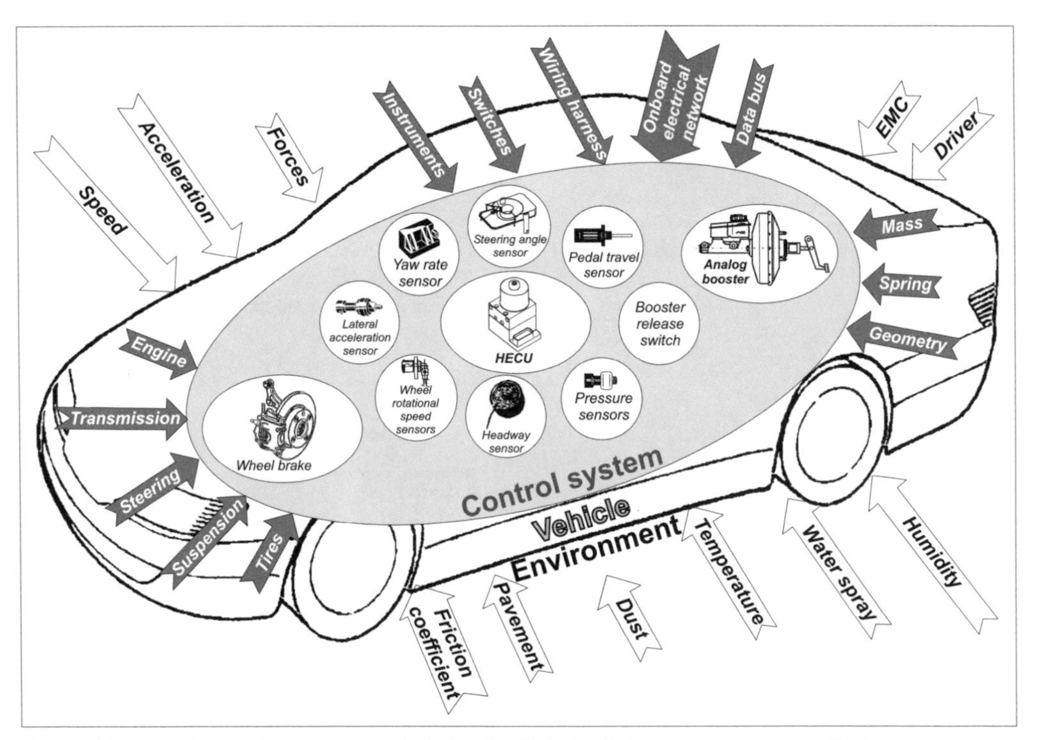

Fig. 7.2-31 Control system interfaces. HECU, hydraulic ECU; EMC, electromagnetic compatibility.

wheel speeds is possible using only a single sensor. This is mounted on the input side of the differential. In this case, the sensor signal represents the arithmetic mean of rear wheel speeds.

Two primary sensor types, employing different operating principles, are in current use. So-called "passive sensors" are based on magnetic induction, while "active sensors" operate magnetoresistively.

Inductive (passive) wheel speed sensors

Passive wheel speed sensors operate on the principle of induction.

The sensor head, as shown in Fig. 7.2-33, contains a permanent magnet (1), a coil (2), and an external cable connection (3).

The magnetic flux changes as the teeth of the ferromagnetic excitor ring pass by. Magnetic induction causes a changing voltage whose frequency is proportional to wheel speed.

Magnetoresistive (active) wheel speed sensors

The sensor assembly consists of thin magnetoresistive layers (thin film sensors) in a Wheatstone bridge arrangement combined with an electronic signal conditioning circuit. The sensor operating principle is based on electrical resistance changes of the magnetoresistive layers in response to a changing magnetic field running parallel to these layers.

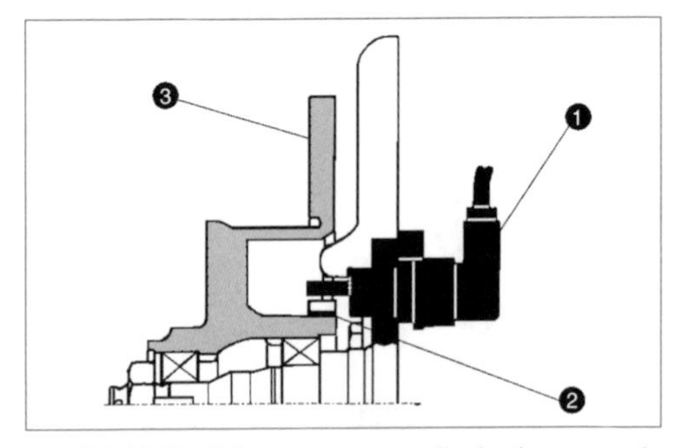

Fig. 7.2-32 Radial arrangement of wheel sensor. 1, Sensor; 2, excitor ring; 3, brake rotor.

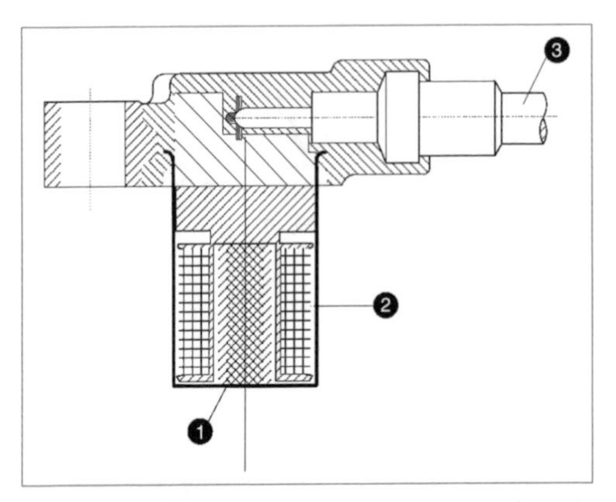

Fig. 7.2-33 Inductive wheel sensor construction. 1, Permanent magnet; 2, coil; 3, cable connection.

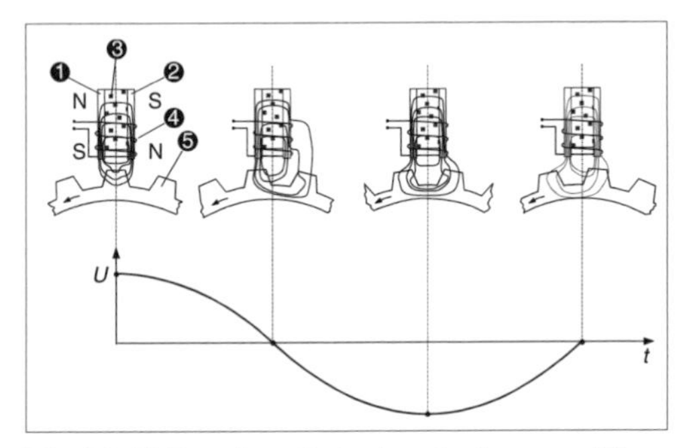

Fig. 7.2-34 Function of inductive wheel sensor. *t*, Time; *U*, voltage; *N*, north pole; *S*, south pole; 1, 2, permanent magnets; 3, soft iron core; 4, coil; 5, excitor ring.

There are two basic variations. In the first, a permanent magnet is mounted behind the thin film sensor. A rotating, ferromagnetic impulse ring (e.g., a gear) in front of the sensor produces magnetic field strength changes in the sensor substrate. In the second variation, a magnetic encoder rotates in front of the sensor, which contains only the sensor element. This sensor element is equipped with a small magnet to generate a "stabilizing field." The stabilizing field eliminates the possibility of a frequency doubling effect in the sensor element by means of small air gaps.

The encoder track of the magnetic encoder consists of a whole number of identical, alternating north and south pole areas which form a closed circle. Two successive north and south poles represent one increment and are equivalent to a "tooth" on a ferromagnetic excitor ring. In operation, these sensors are supplied with electrical energy by an electronic control unit, and they return a signal stream (in this case, a square wave signal).

Together, the magnetic field sensor, magnet, and evaluation circuit form a single ready-to-install unit (Fig. 7.2-35).

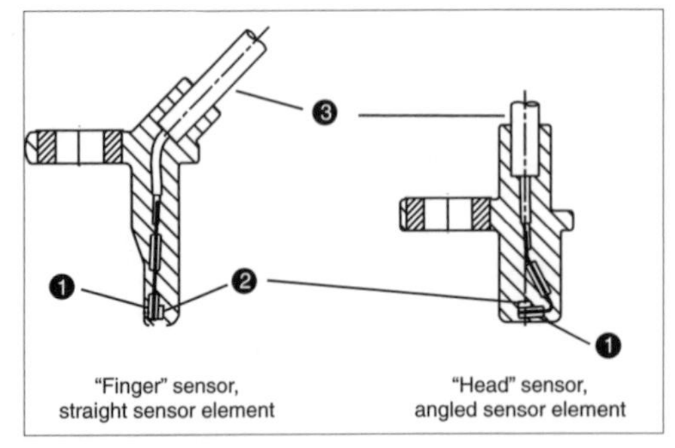

Fig. 7.2-35 Cross section of magnetoresistive sensor. 1, Sensor element; 2, stabilizing magnet; 3, cable connection.

Compared to inductive sensors, magnetoresistive wheel speed sensors offer a series of advantages. These include:

— Sensing of low speeds (down to $v = 0$)
— Improved signal quality (high-resolution digital signal, large air gap possible, etc.)
— Signal largely insensitive to temperature differences and vibration
— Reduced weight and space requirements

7.2.5.3.2 Brake actuation travel sensor

The travel sensor (Fig. 7.2-25) is employed in TEVES electronic brake assist (see Section 7.2.6.1). The task of the travel sensor in brake assist applications is to provide an electrical signal to enable the electronic control unit to recognize pedal position and actuation speed. The travel sensor plunger is connected to a wiper which slides across a linear-response potentiometer winding. With the potentiometer acting as a voltage divider, the electronic control unit is provided with a defined voltage dependent on wiper position to evaluate pedal position and speed.

7.2.5.3.3 Acceleration switch (G switch)/ accelerometer G sensor

The acceleration switch or sensor provides information on vehicle deceleration or acceleration. In vehicles with only two wheel sensors and in all-wheel-drive vehicles with locked drive between front and rear axles, this information is necessary (in addition to measured wheel speeds) for optimum ABS control.

The acceleration switch contains circuit elements with defined inclination angles representing the desired acceleration stages built into the switch housing.

Mercury switches, previously used in acceleration switches, are no longer used for environmental reasons. They have been replaced by mechanical pendulum elements which work similarly: that is, as the pendulum moves, electrical contacts are opened or closed.

Modern electronic brake control systems often use a longitudinal acceleration sensor in place of an acceleration switch. In configuration and operation, this is identical to a lateral acceleration sensor but is installed rotated 90° (horizontally) to a lateral sensor (See Section 7.2.5.3.5, Lateral acceleration sensor).

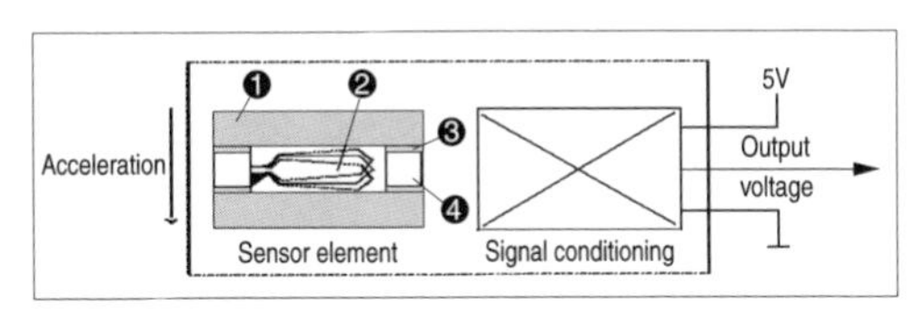

Fig. 7.2-36 Lateral acceleration sensor. 1, External electrode; 2, cantilevered beam (seismic mass); 3, insulator and electrode connection element; 4, center electrode.

7.2.5.3.4 Steering wheel angle sensor

The steering wheel angle sensor provides the ESP control unit with information regarding instantaneous steering wheel angle and therefore the desired direction of vehicle motion.

The steering wheel angle is optically determined by suitable arrangement of multiple photocells and light barriers which are converted to data words. For safety reasons, conversion is done by two microprocessors.

7.2.5.3.5 Lateral acceleration sensor

The lateral acceleration sensor provides a signal proportional to vehicle lateral acceleration and, in conjunction with the yaw rate sensor, provides vehicle lateral dynamics information necessary for yaw control.

The acceleration sensor consists of a small cantilevered beam which changes position, and therefore capacitance of a capacitor, in response to lateral acceleration. Capacitance changes are electronically evaluated and the resulting signal passed to an electronic control unit.

7.2.5.3.6 Yaw rate sensor

Yaw rate is understood to mean the angular velocity of a vehicle about a vertical axis. The yaw rate sensor provides a signal proportional to the vehicle yaw rate and, in conjunction with the lateral acceleration sensor, provides vehicle lateral dynamics information necessary for yaw control.

The sensor element consists of two conjoined quartz tuning forks mounted parallel to the vehicle vertical axis (Fig. 7.2-37). An electronic circuit excites the upper tuning fork into sinusoidal oscillation. As the vehicle rotates about its vertical axis, a Coriolis force dependent on the rotation rate is exerted on the tines of the tuning fork. This is transmitted to the lower tuning fork where it excites sinusoidal oscillation, which is converted to a signal proportional to rotation rate by a suitable amplifier circuit.

In addition to vibration excitation of the upper tuning fork, the circuit concept includes signal processing and security elements capable of recognizing internal sensor faults.

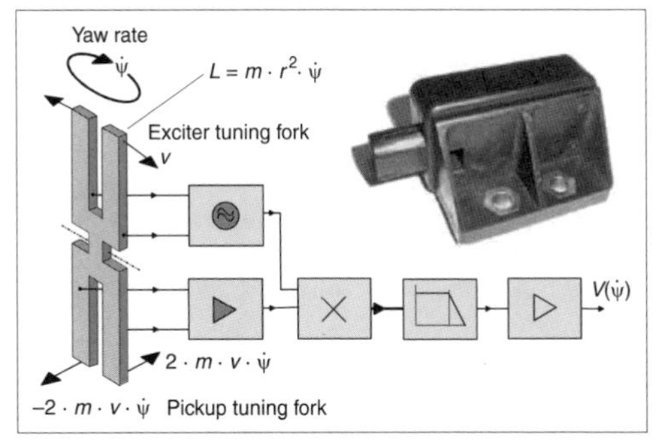

Fig. 7.2-37 Yaw rate sensor.

7.2.5.3.7 Pressure sensor

The task of the pressure sensor is to measure brake pressure as applied by the brake pedal.

For yaw control, ESP control units must provide wheel-specific brake pressures independent of pedal pressure. To achieve course correction, the tandem master cylinder is decoupled from the brake system. During brake pedal activation, a "deceleration command" associated with master cylinder pressure is used to calculate target pressures for the individual wheel cylinders. For safety reasons, ESP units employ two tandem master cylinder pressure sensors (redundancy). Continued development of intelligent safety algorithms seeks to eliminate the need for a second pressure sensor.

The pressure sensor is based on a ceramic sensor element which changes capacitance in response to pressure. The electronic signal processing circuit and ceramic element are contained within a metal housing.

7.2.5.4 Hydraulic/electronic control unit for electronic (brake) control systems

On current ABS/ASR/ESP systems (e.g. TEVES MK20), the hydraulic control unit, as shown in Fig. 7.2-38, consists of:

- A central hydraulic housing containing valves (1)
- An integral pump and attached electric motor (2)
- A solenoid carrier (3) containing electronic elements, attached by means of a so-called "magnetic plug"

Valves

The inlet and outlet valves incorporated in the hydraulic housing permit modulation of wheel brake pressures. Each controlled braking circuit is served by one inlet valve and parallel check valve and one outlet valve. The inlet valve is normally open (NO), and the outlet valve normally closed (NC); see Fig. 7.2-41.

Magnetic plug concept

The HCU (hydraulic control unit) solenoids consist of a hydraulic/mechanical component pressed into the valve block and the solenoid coil, contained in the electronic housing. When the ECU (electronic control unit) and HCU are assembled, the coils are guided by the valve domes, effectively joining the "magnetic plugs" and creating functional solenoid valves.

Pump

The integral dual-circuit hydraulic pump supplies brake fluid from the low pressure reservoir to the appropriate brake circuits of the tandem master cylinder, replacing the fluid volume drawn off by ABS control action.

In active brake control processes—that is, those in which the brake pedal is not actuated (e.g., in ASR or ESP action)— the pump unit provides the fluid volume required in the pressure rise phase of a control cycle.

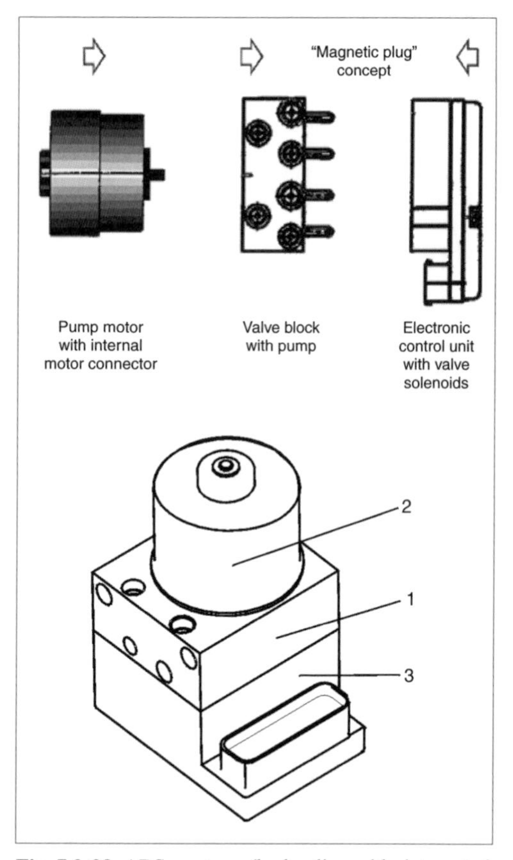

Fig. 7.2-38 ABS system (hydraulic, with integrated electronics).

7.2.5.5 Electronic control unit for electronic (brake) control systems

Given input data in the form of wheel speeds, yaw rate, steering wheel angle, etc., the electronic control unit (ECU) employs complex control logic to calculate inputs for brake actuators and engine controls in order to keep wheels turning with the desired slip and appropriate for the situation at hand. Other important tasks for the ECU are:

- Signal level conditioning and conversion of input and output signals
- Safety monitoring of the electronic control system
- Fault diagnosis

In the current state of the art, the ECU takes the form of a microprocessor system. Input parameters to an electronic control unit include, among others:

- Wheel sensor signals
- Additional signals (e.g., steering wheel angle, yaw rate)
- Signals from various switches (e.g., brake light switch)

— Engine information for control systems acting on the engine
— Operating voltage

Output parameters include, for example:

— Switching signals for solenoid valves
— Switching signals for the hydraulic unit electric motor
— Signals to match engine drag or drive torque
— Signals to monitor safety-relevant assemblies
— Switching signals for warning and functional lights
— Fault condition information

The structure realized within the control logic may be termed an "adaptive controller"; that is, by means of a search routine, the working point is continuously matched to the optimum control path.

The control logic includes: (a) basic algorithms, independent of any particular vehicle; (b) algorithms which are matched to various vehicle models by suitable adaptation of parameters; and (c) special features developed for a particular vehicle manufacturer or specific vehicle model.

Written in the high-level language "C," control algorithms are spread across numerous modules. This permits rapid software adaptation, software maintainability, a justifiable level of complexity, and combinability of various modules (ABS, ASR, ESP, etc.)

7.2.5.6 ABS braking

7.2.5.6.1 ABS functionality

Without ABS, excessively hard braking (with respect to road surface condition or driving situation) leads to unfavorable slip values or even locked wheels. This not only may lead to the vehicle leaving its intended lane and compromised steering ability, but in most cases may also increase braking distances. ABS systems permit maximum-effort braking without locked wheels, an impossible feat even for above-average drivers in the face of critical road surface conditions or in dangerous situations. The required individual modulation of optimum braking force at each wheel is not possible by means of the brake pedal alone. This is especially apparent on split-coefficient surfaces, where individual wheels demand different optimum brake line pressures.

ABS improves the following vehicle performance parameters

— **Vehicle stability**
If brake pressure during a maximum-effort stop rises and exceeds the limit for locked wheels, ABS prevents wheel locking and therefore a breakout of the vehicle from its intended course and rotation about the vehicle vertical axis (spinning).
— **Steerability**
The vehicle remains steerable even in maximum-effort braking under the most varied road conditions. The vehicle can negotiate curves or avoid obstacles despite fully applied brakes.
— **Optimum braking distance**
ABS reacts to changes in road surface grip, such as from dry to wet asphalt. The available coefficient of friction between tire and roadway is always used to the maximum degree possible.

Except for a few exceptions caused by physical conditions (e.g. unpacked snow or gravel), braking distances with ABS are shorter than distances without ABS.

Other advantages

— Under maximum-effort braking, ABS prevents flat spotting of tires.
— In dangerous situations, drivers can concentrate on traffic conditions around them even during extreme braking.

The limits of ABS

ABS cannot repeal the laws of physics:

— On slippery road surfaces, stopping distances will be greater than on dry, grippy pavement, ABS notwithstanding.
— Given excessive speed through a turn, braking with ABS cannot increase tire lateral force. Even with ABS, the vehicle may depart from its lane – or leave the road entirely.

7.2.5.6.2 ABS operating range

The best possible brake action is not achieved with maximum brake pressure, but rather when modulated pressure results in individual wheel braking in the regime of optimum slip; that is, critical slip is not exceeded. The optimum value depends on actual tire and roadway conditions.

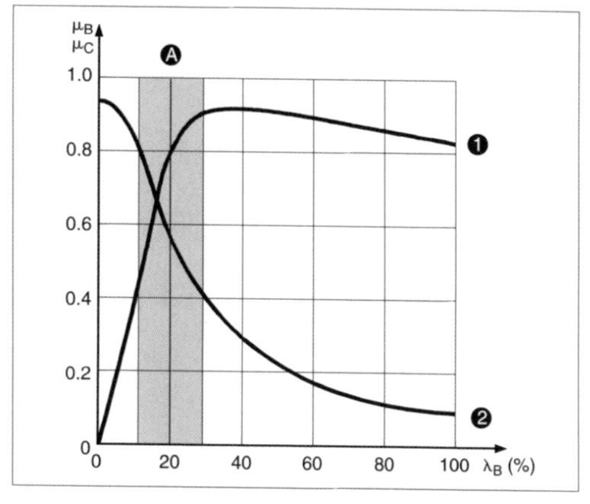

Fig. 7.2-39 ABS control regime (example). λ_B, Braking slip; 1, braking force coefficient; μ_B, braking force coefficient; 2, lateral force coefficient, μ_C, lateral force coefficient; A, ABS control regime.

The operating range of ABS control is always selected to provide simultaneously the best possible vehicle stability and steerability.

ABS control is activated if and when a wheel is decelerated to the point where the optimum slip regime is exceeded.

Fig. 7.2-40 illustrates a braking event without ABS control. Section I shows unbraked motion. Brake line pressure is zero, tire rotational speed represents vehicle speed (no slip), and vehicle speed is approximately constant.

In Section II, the brakes are applied gently. Slight wheel brake pressure is present. Wheel rotational speed is somewhat less than vehicle speed (slip within the stable regime), and vehicle speed drops steadily.

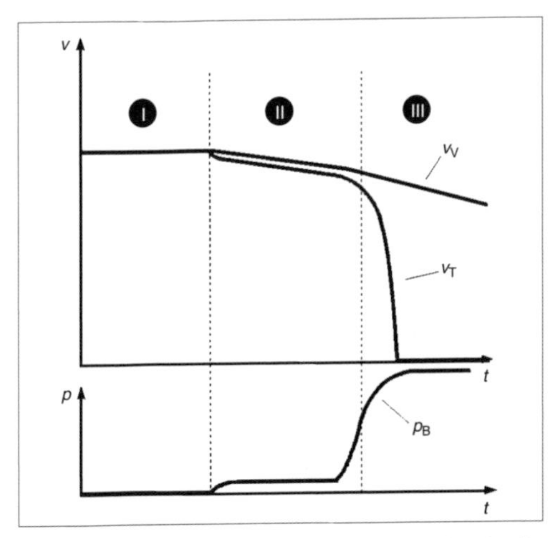

Fig. 7.2-40 Braking event without ABS (single wheel). t, Time; v, speed; p, pressure; I, unbraked motion; II, partial braking; III, maximum-effort braking without ABS; v_V, vehicle speed; v_T, tire circumferential speed; p_B, brake pedal pressure; p_R, wheel brake pressure.

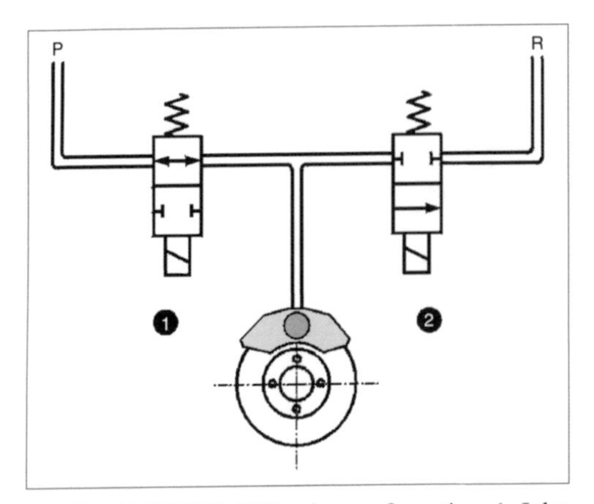

Fig. 7.2-41 TEVES ABS valve configuration. 1, Inlet valve (NO); P, connection to actuation; 2, outlet valve (NC); R, return.

Section III represents maximum-effort braking. Wheel brake pressure exceeds wheel lockup pressure, wheel rotational speed decreases to zero, and vehicle speed decreases, with deceleration rate set by the coefficient of sliding friction.

If the electronic control unit detects very rapid wheel speed changes characteristic of a tendency to lock wheels (wheel deceleration exceeds the maximum possible vehicle deceleration), appropriate brake pressure modulation commands are sent to the control valves.

In principle, an ABS control cycle involves three phases.

Phase 1: Pressure hold

As the pedal is applied, wheel brake pressure rises, and wheel rotational speed progressively decreases. If wheel speed indicates a tendency toward locking, an inlet valve is closed. Brake line pressure in that circuit can no longer be increased, even if pedal pressure is increased.

Phase 2: Pressure decay

If wheel rotational speed drops despite constant brake line pressure—that is, slip increases—brake line pressure to this wheel is reduced. To accomplish this, the electronic control unit holds the inlet valve closed and briefly opens the outlet valve. As a result of the resulting pressure drop, brake torque at the wheel is reduced.

The duration of this pressure reduction pulse may be determined on the basis of wheel deceleration, allowing the wheel to accelerate briefly (so-called "predictive control"). If the wheel does not exhibit the expected behavior at the end of this period, further pressure reduction may be applied in extreme cases, such as a friction coefficient jump from asphalt to ice, until the wheel exhibits the desired re-acceleration.

Phase 3: Pressure build

As soon as wheel rotational speed has increased to just below the point of optimum slip, brake line pressure is again built up in stages. The outlet valve remains closed, while the inlet valve is repeatedly opened briefly.

This three-phase control cycle of pressure holding, pressure reduction, and pressure increase is repeated several times (generally, three to four times per wheel) per second. The three phases do not necessarily follow one another in sequence.

Special conditions

The electronic control unit continuously evaluates wheel sensor signals and responds with a control strategy appropriate to the given situation.

Normally, the front wheels of passenger cars are controlled individually, while the rear wheels are controlled in keeping with the so-called "select low" principle: the rear wheel showing the greatest tendency to lock determines pressure modulation for both rear wheels. The resulting reduced brake force utilization at

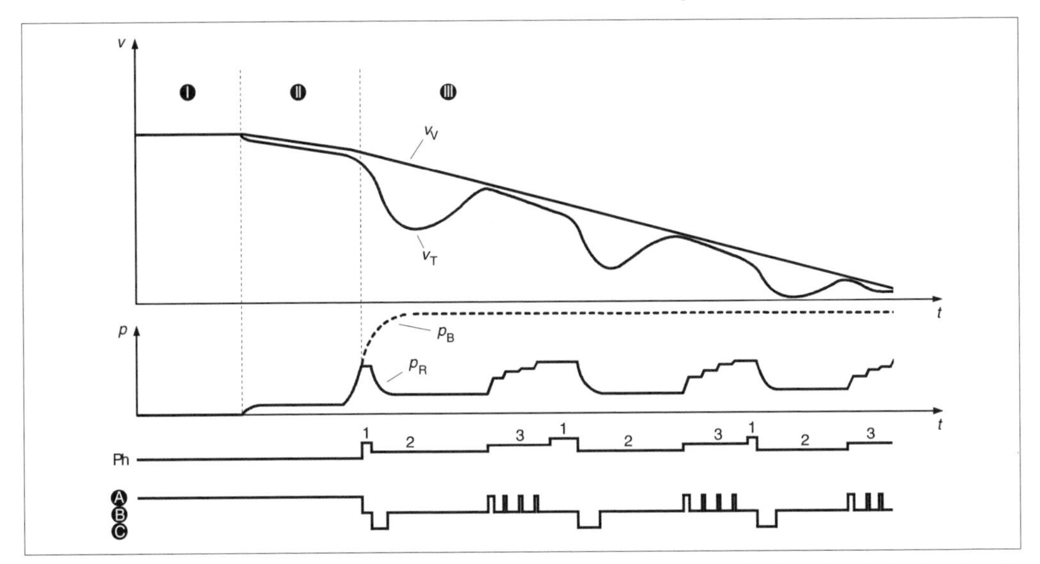

Fig. 7.2-42 Braking event with ABS (single wheel). t, Time; v, speed; p, pressure; Ph, phase; I, unbraked motion; II, partial braking; III, ABS braking; v_V, vehicle speed; v_T, tire circumferential speed; p_B, brake pedal pressure; p_R, wheel brake pressure; A, pressure build; B, pressure hold; C, pressure decay.

the rear wheels leads to greater lateral force and therefore increased vehicle stability.

For special roadway or driving situations, special algorithms have been developed to satisfy the respective needs. These situations include, for example, icy roads, so-called "split μ" surfaces (lateral traction differences), vehicle spins, use of spare tires, and many others.

7.2.5.7 Electronic brake force distribution (EBD)

EBD is an add-on software algorithm for ABS software. It permits optimized brake force distribution between front and rear axles under partial braking to maintain vehicle stability and to improve rear axle traction coefficient utilization during partial braking. To achieve this, the actual vehicle deceleration and lateral acceleration, calculated from the four wheel rotational speeds, are examined. If slight overbraking of the rear axle is detected, further pressure increase is prevented by closing the inlet valve; for more extreme overbraking, pressure may be reduced. In underbraking, pressure to the rear wheel brakes is returned to the master cylinder pressure level in pulses to achieve the best possible utilization of the available traction potential.

7.2.5.8 Electronic Stability Braking System (ESBS)

ESBS, or CBC (Cornering Brake Control, as it is sometimes known), is an expansion of the ABS control algorithm. It recognizes driving situations (particularly in cornering) from wheel speed traces. ESBS optimizes braking slip, thereby improving brake force distribution. Simultaneously, it provides yaw moment compensation. This is achieved by deliberate setting of differing brake forces on either side of the vehicle, which generates a correcting yaw moment leading to vehicle stabilization and improved steerability.

Advantages are apparent above all in cornering at the limit and in lane changes, as well as in maximum-effort braking (i.e., with fully activated ABS control) and especially in partial braking.

7.2.5.9 Antislip regulation (ASR)

7.2.5.9.1 ASR functionality

Antislip Regulation (also known as Automatic Stability Regulation or ASC—Antislip Control) or ASR prevents unnecessary spinning of drive wheels by means of specific brake application (BASR: Brake Antislip Regulation) and/or with engine management intervention (MASR: Motor Antislip Regulation). ASR is both hardware and software based on ABS. ASR advantages in overview include:

— Assurance of vehicle stability for rear-drive applications or steerability for front-drive vehicles
— Limited slip differential function
— Increased propulsion force
— Warning upon reaching physical stability limits by means of an instrument panel light (e.g., on slippery roads)
— Reduced tire wear

7.2.5.9.2 Brake antislip regulation

With differential road surface traction (split μ), the propulsive force (traction) available on the side with more grip cannot be fully utilized because of the differential unit located between the two driven wheels, which limits torque on the side with more grip to the (lower) torque limit of the opposite side.

With the aid of wheel speed sensors, BASR recognizes when the traction limit of the wheel with lower grip has been exceeded. Wheel spin is reduced by appropriate brake application. Simultaneously, the applied brake torque acts as additional support for the differential and is therefore available as drive torque on the opposite side.

Brake ASR is primarily active during drive-off.

BASR cannot act on wheel brakes for an unlimited time. This would result in dangerous brake overheating. For this reason, an electronic controller limits brake application duration according to the load capacity of the brake system.

7.2.5.9.3 Motor antislip regulation

To keep brake loads as low as possible, ASR throttles engine torque in the process of exercising brake control in lower speed ranges. At higher speeds (above ~40 km/h) brake control seldom intervenes; instead, engine control reduces torque in a timely manner to stabilize vehicle handling.

7.2.5.9.4 Engine drag torque control (MSR)

High engine drag torque levels (engine braking, "overrun") may result in increased slip at the drive wheels and therefore vehicle instability— in extreme cases even without brake actuation. This is especially true for rear-wheel-drive vehicles. Under such conditions, engine drag torque is reduced by modulated throttle application, thereby permitting handling stability to be maintained.

7.2.5.10 Electronic stability program (ESP)

ESP is an electronic handling stability control system. It consists of wheel slip control functions (ABS, EBD, ASR) and yaw moment control (GMR). ESP stabilizes vehicle longitudinal and transverse dynamic behavior independently of pedal actuation with the aid of brake application and engine intervention.

Yaw moment control is an electronic control system for improving vehicle lateral dynamic behavior through brake and engine intervention. Given input parameters of wheel rotational speeds, steering wheel angle, and master cylinder pressure, the desired vehicle behavior is calculated on the basis of a computer model. Actual behavior is sensed by means of yaw rate and lateral acceleration.

Especially in the case of very rapid steering movements, the vehicle is no longer capable of translating steering wheel motion into the desired directional change. The result is understeer or oversteer, possibly to the point of spinning. GMR recognizes deviation of actual from intended vehicle behavior and actively intervenes to stabilize the vehicle.

The primary method of counteracting understeer is to brake the inside rear wheel; for oversteer, the outside front wheel. This selective, active brake application on one side of the vehicle generates longitudinal forces on one side and thereby the desired yaw moment. Reduction

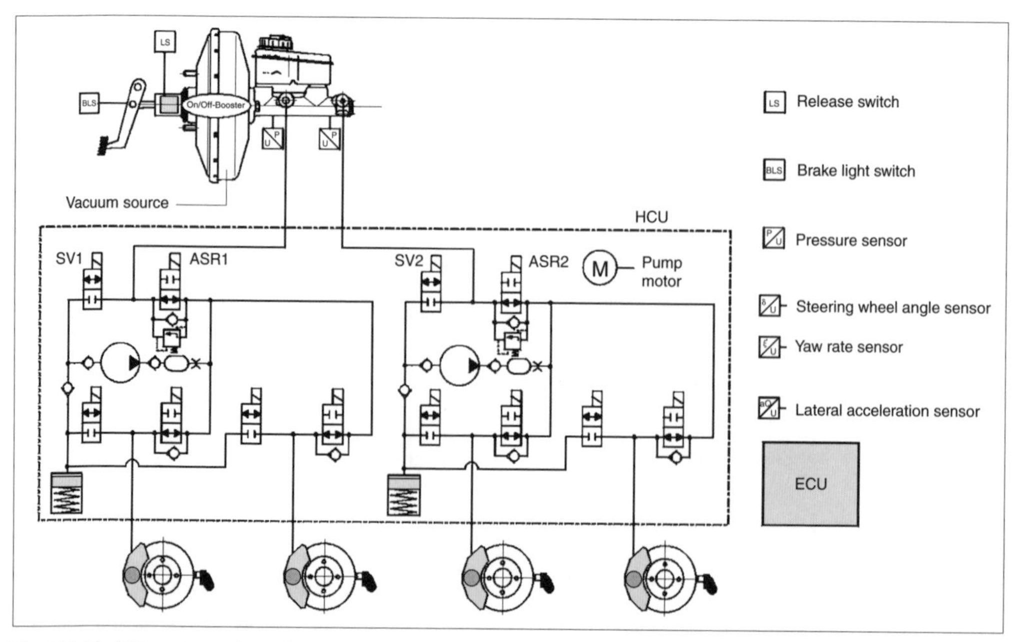

Fig. 7.2-43 ESP system schematic.

of lateral forces in response to added longitudinal forces also serves to reinforce the effects of longitudinal forces.

Excessive drive torque can also be reduced by means of the engine management interface. Expanded ABS/ASR hydraulics with an integrated electronic control unit lie at the core of an ESP system. With the aid of these hydraulics it is possible to generate pressure at any wheel independent of any brake pedal actuation—that is, to brake selectively and actively.

In extremely cold conditions, the pump alone cannot draw sufficient brake fluid volume. Several different preload systems may be used to provide adequate pump volume.

The *on/off booster* or active booster (see also analog booster) may be actuated externally, assuring the necessary hydraulic preload under extremely cold conditions.

Another solution is use of an *electric preload pump.* This draws brake fluid from the reservoir and feeds it to the tandem master cylinder. The master cylinder contains an orifice which causes a stagnation pressure, which appears as preload pressure upstream of the hydraulic pump, thereby assuring the necessary feed volume.

7.2.6 Assistance systems

7.2.6.1 Brake assistant

7.2.6.1.1 Electronic brake assistant

Brake assistant (BA) is an electronic system which provides braking support in dangerous situations, specifically emergency braking. As soon as it recognizes an emergency braking situation in which the driver is acting too slowly, the BA control unit automatically intervenes in the braking process. This situation is recognized by evaluating the brake pedal actuation characteristic. In such cases, brake assistant intervention reduces braking distance by building full braking pressure as rapidly as possible.

The *electronic braking assistant system* consists of a vacuum brake booster which can be activated electrically by means of a solenoid valve (Fig. 7.2-25) and control electronics.

Basic conditions for activation are:

1. Brake light switch activated
2. Release switch not activated

Brake pedal travel is measured indirectly as active brake booster diaphragm travel. If the actuation speed derived from membrane travel exceeds a preset threshold dependent on pedal travel and vehicle speed, the solenoid valve is activated, providing the maximum available brake power boost.

At any time, a decrease in brake pedal pressure causes the release switch in the booster to strike its stop, terminating the emergency braking action and closing the solenoid valve.

7.2.6.1.2 Mechanical brake assistant

To avoid the cost of electronic control, a brake booster was developed which provides properties comparable to the electronic brake assistant but achieves panic recognition, valve activation, and release by purely mechanical means. In this case, brake booster inertia is used to good advantage; in rapid, panicked brake actuation, the control valve overshoots a specific opening lift. By exceeding this opening lift, the control valve is arrested and remains open, even if pedal pressure is reduced.

Advantages of the mechanical brake assistant are:

- Boost variation as a function of actuation speed
- Strong assist in panic braking (infinite ratio)
- Pressure modulation via pedal travel during BA function
- Conventional brake function not compromised ("rounded" threshold pressure for rubber reaction disc)
- Boost function not compromised by loss of one brake circuit
- Production compatibility (integrated in booster, conventional tandem master cylinder)
- Low-cost robust design

7.2.6.1.3 Hydraulic brake assistant (HBA)

A further means of achieving brake assistant functionality is offered by ESP systems, whose existing sensor

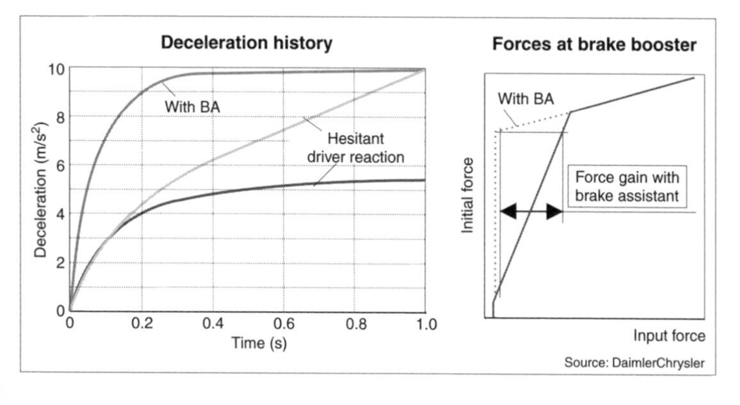

Fig. 7.2-44 Brake assistant.

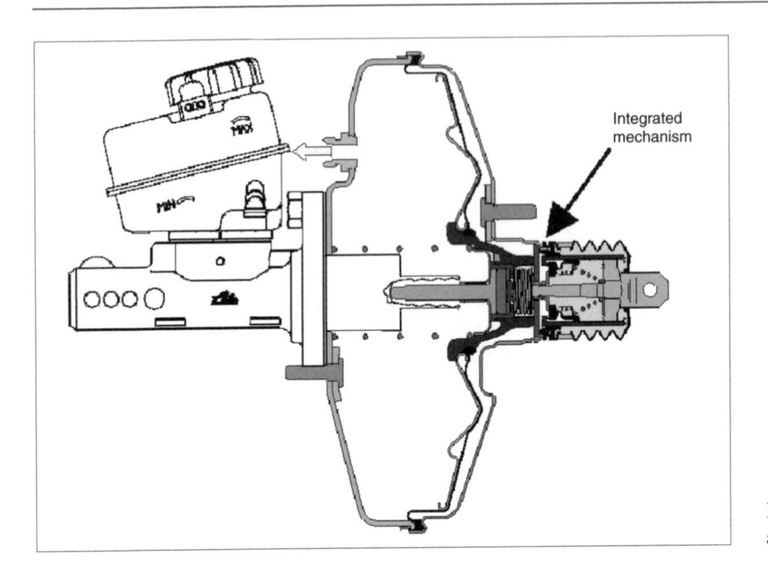

Fig. 7.2-45 Mechanical brake assistant.

suite and hydraulic control unit can be employed (Fig. 7.2-46).

Panicked brake pedal actuation is recognized by means of pressure sensor signals. If a parameterized critical pressure gradient is exceeded, the ASR cutoff valves are closed, electric diverter valves are opened, and the pump is activated. Pressure applied by the brake pedal is raised to wheel locking pressure levels by the pump. In this pressure regulation method, wheel brake pressure follows that of the tandem master cylinder. This provides the possibility of modulating wheel brake pressure within the BA mode. Shutoff of the BA function occurs as pressure drops below a minimum level.

The hydraulic brake assistant is an example of a system in which vacuum brake booster function is supported by a hydraulic control unit (HCU). Further applications are shown in Fig. 7.2-46. The booster vacuum runout point may be recognized by means of vacuum

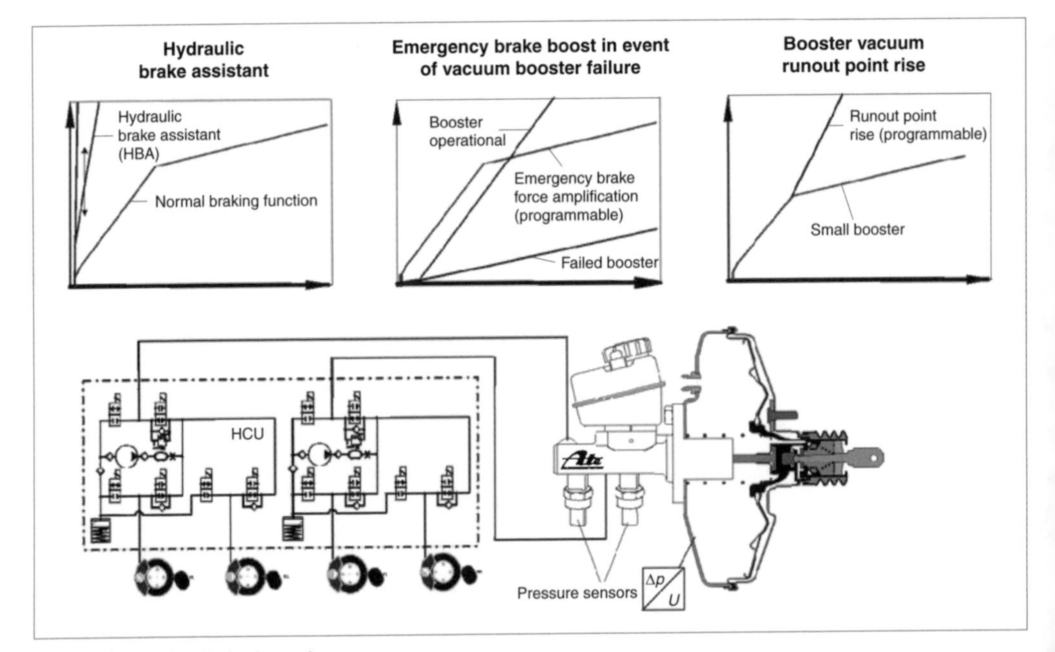

Fig. 7.2-46 Hydraulic brake assistant.

pressure sensing, and a further increase in wheel brake pressure is applied by the HCU as needed. This is useful for vehicles with low vacuum pressure and/or limited vacuum booster installation space. In extreme cases of total loss of vacuum, a backup brake boost function is provided by the HCU alone.

7.2.7 Development of future brake systems

7.2.7.1 Electric parking brake (EPB) and active parking brake (APB)

An electric parking brake replaces a conventional, mechanical parking brake (hand brake) with a switch in the cabin, an electronic control unit, and wheel brake actuators (electric motor and transmission). In general, two actuators are used, each working on one rear brake (duo-servo or combined caliper), either directly or by means of a Bowden cable. To ensure functionality, an additional energy storage unit may be installed that is capable of providing energy for at least one parking brake application (safe condition) in the event of vehicle battery failure. Parking brake components are held in their applied position purely by their own self-locking action. The electronic control unit contains the following algorithms:

- Control of brake application forces
- Pad wear recognition
- Alarm function
- Activation of warning and check light on instrument panel
- Safety logic
- Diagnostic functions

In addition to ignition key position, the electronic control unit detects driver command for parking brake application by means of a rocker switch as well as driving condition by means of an interface to the superposed ABS/ESP control unit.

The basic EPB function is parking brake actuation or release while the vehicle is stationary. In response to a driver-operated button, parking brake application and release (on/off function) take place over a defined time span and maximum force.

Activation of the antilock function during EPB actuation while the vehicle is in motion represents integration of a service brake function within the parking brake function, thereby moving toward the concept of the overriding functionality of an active parking brake (APB). The spectrum of functionality ranges from a simple parking brake or parking pawl (for automatic transmissions) to modulated startup on hills (so-called "hill holder function") all the way to antitheft immobilizer and parking assistance in combination with a distance sensor package.

7.2.7.2 Electrohydraulic brakes (EHB)

Along with the need for smaller form factor because of difficult engine compartment packaging, other desirable functions include improved brake response and therefore modulation ability, setting brake pedal characteristics

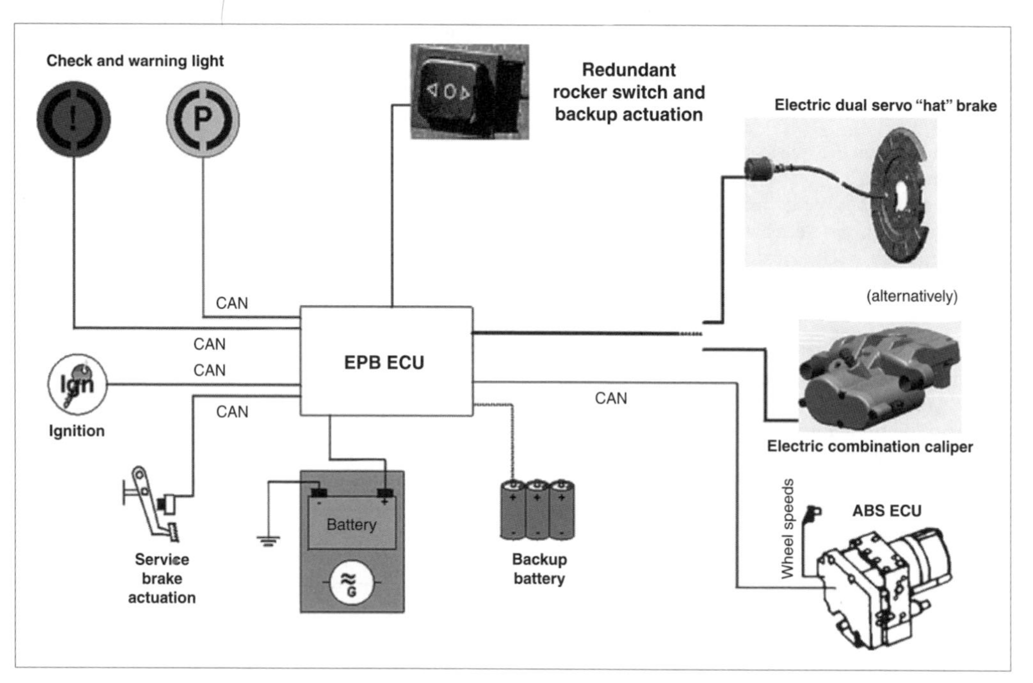

Fig. 7.2-47 Electric parking brake system layout.

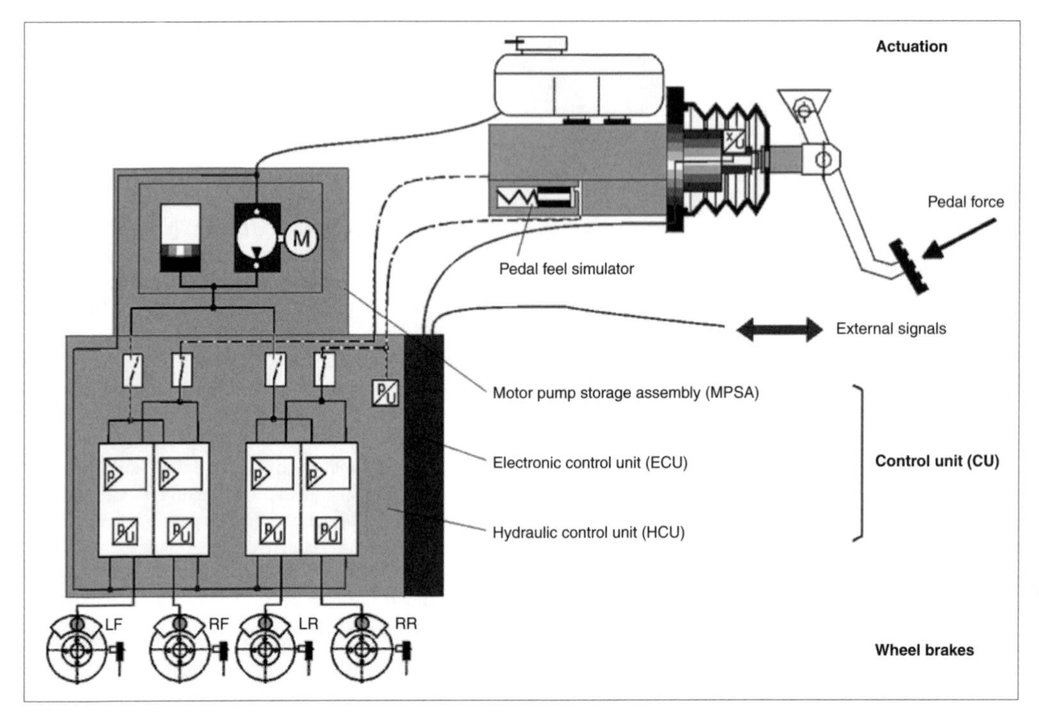

Fig. 7.2-48 EHB schematic with representation of system components.

to meet vehicle manufacturer's expectations, and suppressing pedal kickback during wheel slip control events.

An EHB system is decoupled from the actuating assembly (brake pedal) both in its normal braking mode as well as in wheel slip control mode, and therefore is free of pedal feedback. It consists of the following assemblies (Fig. 7.2-48):

- Actuating unit (in place of the classic actuation mechanism: e.g., vacuum booster) with an individually applicable "pedal feel simulator" and driver command sensor (electronic pedal module)
- Control unit (CU), located anywhere within the vehicle
- Four conventional wheel brakes

Measured pedal travel and pressure generated in the simulator represent the desired degree of vehicle deceleration. These signals are transmitted ("by wire") to an electronic control unit and processed along with other sensor signals which describe the driving situation and external brake applications (e.g., wheel rotational speeds, yaw rate, lateral acceleration). The electronic control unit calculates the individual input parameters to provide optimum wheel brake pressures for corresponding braking behavior and vehicle stability.

The hydraulic control unit and motor pump storage assembly (MPSA) generate braking energy based on these input parameters. The hydraulic connection between the tandem master cylinder and hydraulic control unit is interrupted by isolation solenoid valves. Based on values stored in the preprogrammed memory unit, brake pressure at the wheel is set by means of control valves.

This presents a series of benefits, including:

- Shorter braking and stopping distances through improved pedal arrangements and integration of additional functionality
- Optimum braking and stability behavior
- Optimum pedal feel
- Quiet operation
- No pedal kickback (vibrations) during wheel slip control events
- Improved crash behavior

Further advantages for the vehicle manufacturer include:

- Improved packaging and simplified assembly through deletion of the vacuum booster and a control unit freely positionable within the engine compartment
- Use of standardized component assemblies
- Simple realization of external brake intervention by means of external signals
- No vacuum requirement, therefore ideally suited for new, induction-loss-optimized and thus more economical combustion engines
- Simple networkability with future traffic guidance systems

All brake interventions and wheel slip control functions conceivable today (e.g. ABS, EBD, ASR, ESP, BA, ACC, etc.) may be achieved by means of a single standardized control unit, without any added hardware complexity. (ABS, Antilock Brakes; EBD, Electronic Brake Distribution; ASR, Antislip Regulation; BA, Braking Assistant; ACC, Adaptive Cruise Control).

7.2.7.3 Electromechanical brakes (EMB)

A further step is represented by electromechanical brakes, also known as "dry brake-by-wire," because they do not use brake fluid. As in the case of EHB (electrohydraulic brakes), the basic principle is feedback-free coupling of the brake pedal (driver command input) to the brakes. EMB requires a redundant signal and energy network.

In contrast to EHB, EMB generates brake forces directly at the wheels by means of purely electromechanical wheel brakes in place of conventional calipers.

Actuation consists of an electronic brake pedal, which may be configured as part of an adjustable pedal module also encompassing the throttle pedal and central ECU.

The electronic brake pedal is composed of the pedal feel simulator and sensors to determine driver commands. Pedal travel and force signals are processed by the central ECU, along with other external signals describing the vehicle situation and external interventions (e.g., wheel rotational speeds, yaw rate, lateral acceleration). The electronic control unit calculates the individual input parameters to provide optimum brake clamping forces for corresponding braking behavior and vehicle stability. The appropriate electrical information is transmitted to the wheel brake modules by means of a redundant bus system ("by wire").

Each of the electromechanical wheel brake modules consists of a wheel brake ECU and an electromechanical actuator (motor/transmission unit), which applies the necessary clamping force.

The parking brake may be realized in the form of an arresting mechanism integrated in the actuator (electromechanical combined caliper). This is activated by means of a hand-operated switch and purely electrical signal coupling.

Brake modulation and brake torque variation during wheel slip control events (e.g., by ABS, ASR, ESP) are accomplished as they are in electrohydraulic brakes.

This presents a series of benefits, including:

- Optimum braking and stability behavior
- Optimum pedal feel
- With more advantageous pedal position, less time required to move driver foot, therefore shorter braking or stopping distances
- Absolutely silent operation, without pedal pulsation in ABS mode
- Adjustable pedal module
- Improved crash behavior

Further advantages for the vehicle manufacturer include:

- Smallest possible packaging and assembly complexity ("plug 'n' play" instead of "fill and bleed")
- No vacuum requirement, therefore ideally suited for new, induction-loss-optimized and therefore more economical combustion engines
- Simple networkability with future traffic guidance systems

EMB realizes all currently conceivable brake interventions and wheel slip control functions (e.g., ABS, EBD, ASR, ESP, BA, ACC, etc.). Its operation is abso-

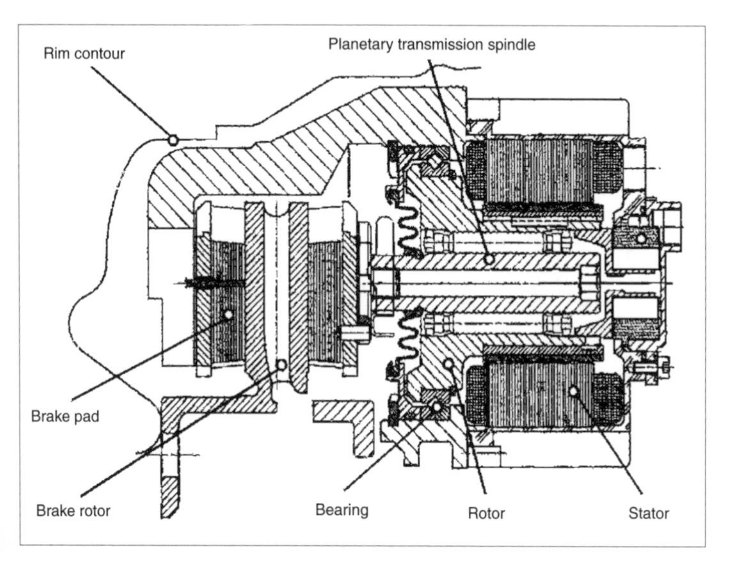

Fig. 7.2-49 EMB cross section.

lutely silent, and, in comparison to hydraulic brake systems, it uses a smaller number of components.

References

Bosch. *Fahrsicherheitssysteme*. Vieweg Verlag, Wiesbaden, 1998.
Bremsbeläge für Straßenfahrzeuge. Verlag Moderne Industrie, 1990.
Bremsen-Handbuch. Bartsch Verlag, Ottobrunn, 1986.
Bremsen-Handbuch, Elektronische Brems-Systeme. Autohaus-Verlag, Ottobrunn, 1995.
Breuer, B. Lecture notes. "Kraftfahrzeuge I," 1997, "Kraftfahrzeuge II," Darmstadt Technical University, 1992.
Burckhardt, M. *Bremsdynamik und PKW-Bremsanlagen*. Vogel Fachbuch, Würzburg, 1991.
Burckhardt, M. *Fahrwerktechnik: Radschlupf-Regelsysteme*. Vogel Fachbuch, Würzburg, 1993.
Buschmann, H., and P. Koessler. *Handbuch der Kraftfahrzeugtechnik*. Wilhelm Heyne Verlag, Munich, 1973.
Strien, H. "Auslegung und Berechnung von PKW-Bremsanlagen." Alfred Teves GmbH, 1980.
Wallentowitz, H. Forschungsgesellschaft Kraftfahrwesen Aachen mbH (fka), "Längsdynamik von Kraftfahrzeugen." Printed lecture notes for "Kraftfahrzeuge I," 1998; "Vertikal-/Querdynamik von Kraftfahrzeugen." Printed lecture notes for "Kraftfahrzeuge II," 1997. "Aufbau von Kraftfahrzeugen." Printed lecture notes for "Kraftfahrzeuge III," 1998.
Zomotor, A. *Fahrwerktechnik: Fahrverhalten*. Vogel Fachbuch, 1993.

7.3 Tires

7.3.1 Introduction

In 1845, Robert William Thomson of Scotland was granted British patent 10990 for an elastic, air-filled tire of rubberized fabric and leather. Thomson was far ahead of his time; his patent, which included grooved tread to prevent slip and which foresaw application to rail vehicles, sank into obscurity.

The pneumatic tire was invented for the second time in 1888 by John Boyd Dunlop. Because it was previously known, Dunlop could only patent it in Britain. By that time, the bicycle was undergoing meteoric development. Compared to iron-tired "boneshakers" and interim use of solid rubber or hollow (unpressurized) tires, introduction of the pneumatic tire represented a decisive step toward increased comfort and reduced rolling resistance. Building on experience gained with "pneumatics" for bicycle applications, tire manufacturers continued development to satisfy the considerably more demanding conditions imposed by automobiles. Tire manufacturers had to conduct their own experiments to convince skeptical automobile pioneers of the pneumatic tire's advantages.

It should also be mentioned that both Dunlop's and Thomson's inventions benefited from a discovery made by American chemist Charles Goodyear in 1839: vulcanization of natural rubber through the use of sulfur.

Development of the pneumatic tire did not follow a linear course to our modern vehicle tires; if anything, it progressed on multiple paths, sometimes parallel, sometimes influencing one another: vulcanizers, reinforcements, tire structures, aspect ratios (Fig. 7.3-1), etc.

The decisive step on the path to modern tires was the introduction of the radial tire, with advantages in terms of life expectancy, rolling resistance, traction coefficient, and handling. Today, radial tires are largely considered standard fare for passenger cars, and "radialization" of commercial vehicles is proceeding at a rapid pace. In the industrialized nations, bias-ply tires have become virtually extinct and therefore will not be considered in this chapter.

7.3.2 Tire construction

A tire is a complex composite construction, made of materials with the most varied physical properties (Fig. 7.3-2).

Radial tires consist of the following main construction elements:

- Bead
- Carcass
- Belts
- Tread

The bead ensures that the tire is securely mounted to the rim. To achieve this, one or more strain-resistant steel

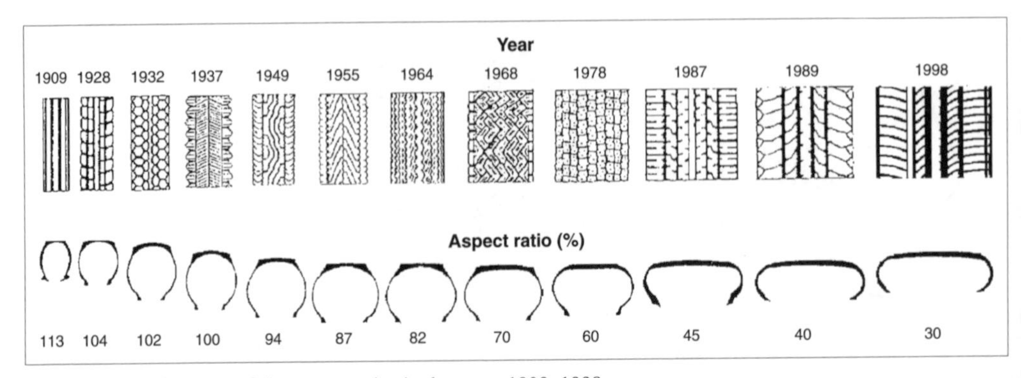

Fig. 7.3-1 Development of tire aspect ratios in the years 1909–1998.

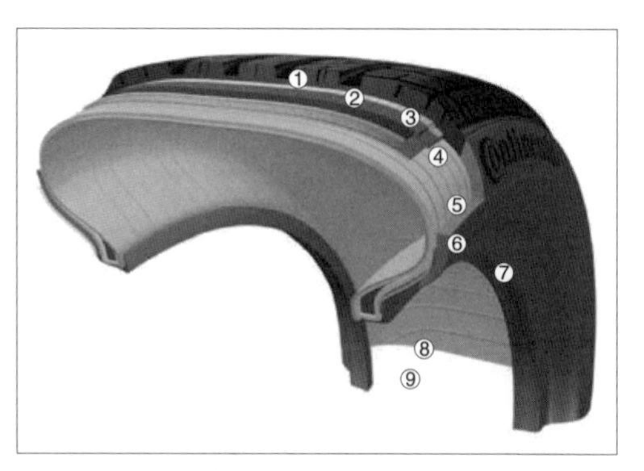

Fig. 7.3-2 Typical passenger car tire construction with components (1) tread, (2) cap ply, (3) nylon belt, (4) steel cord belts, (5) carcass, (6) inner liner, (7) sidewall, (8) bead filler, (9) bead.

cords are embedded in the bead. In tubeless tires, the bead also serves to seal the enclosed air volume against the surrounding environment.

The main reinforcement of a tire is its woven substructure or carcass. The carcass consists of one or more woven belts anchored to a bead.

In radial tires, cord fibers are angled about 90° to the direction of rotation. The stiffness necessary for the tire to function properly is achieved using additional belts laid atop the carcass.

The tread encircles the periphery of the carcass and must transmit forces between the vehicle and roadway. The rubber compound and tread pattern design are chosen to satisfy specific requirements. A wear-resistant sidewall protects the carcass against external influences.

Tread patterns exhibit a wide variety of designs, strongly dependent on their intended use. In winter tires, for example, a definite trend away from coarse lug or block patterns and toward softer, highly siped patterns has taken place over the past few decades (Fig. 7.3-3).

7.3.3 Demands on tires

The tire assumes a prominent role within the vehicle/roadway system. As a connecting link between vehicle and road surface, it transmits all forces and moments. Its ability to transmit these forces strongly affects overall vehicle handling, comfort, and safety.

A vehicle's dynamic properties are significantly affected by tire behavior. Wheel load, springing, damping, suspension kinematics and elasticities, engine performance, and speed as well as intended service have all had great influence on tire design (Fig. 7.3-4).

In a pneumatic tire, a gas or gas mixture, enclosed at higher than atmospheric pressure, is the load-carrying element. The enveloping structure, by its form, construction, and materials, largely determines the service properties of a tire. These service properties are established by the motor vehicle industry, the consumer, and, increasingly, by regulatory agencies. These requirements are not always ideally reconcilable with each other. Opposing demands inevitably lead to conflicting goals, which the tire industry must constantly work to solve.

Tires undergo continuous development in the difficult competition to meet auto industry demands and to achieve tire makers' own ambitious goals. As a result, production tires currently available on the market represent a well-balanced compromise addressing all aspects of vehicle safety, comfort, steering behavior, vehicle stability, and economy, but also more stringent demands for environmental responsibility in a manner acceptable to the consumer.

7.3.3.1 Service properties

Development of passenger car and commercial vehicle tires is decisively influenced by continuously changing (and increasing) vehicular demands on the part of car and truck manufacturers.

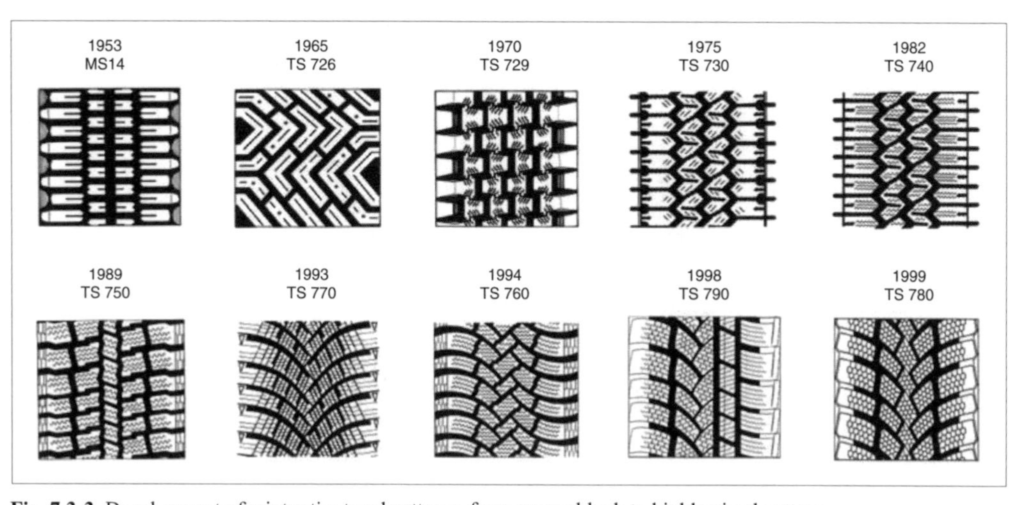

Fig. 7.3-3 Development of winter tire tread patterns, from coarse block to highly siped pattern.

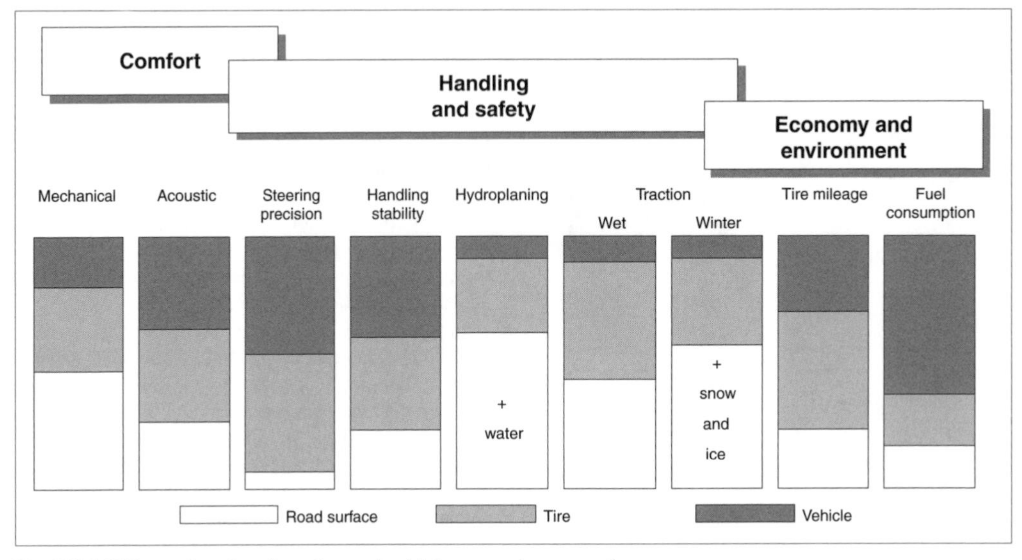

Fig. 7.3-4 Effects of road surface, tire, and vehicle on service properties.

Fig. 7.3-4 illustrates typical effects of tire design factors on service properties.

For the consumer, service properties describe individual tire characteristics and must always be regarded in conjunction with vehicle, road, and driver. To determine service properties, tests are conducted evaluating subjective as well as objective criteria (Table 7.3-1). Service properties are useful for evaluations based on respective customer expectations. As boundary conditions change in the course of time, demands for satisfaction of certain service properties also undergo gradual change.

Over the past few decades, automotive development, in conjunction with expansion of the road network, has led to increased demands on passenger car and commercial vehicle tires. This is also readily apparent in the number of service properties which must be examined within the engineering framework of tire approval.

In 1960, passenger car tires had to satisfy ten major criteria, examined on actual vehicles or in laboratory testing. These included, among others, some very special individual characteristics typical of the time, such as the ability of the tire to guide the vehicle out of streetcar tracks (literally, to avoid "tramlining") without sudden sideways displacement—a vital criterion for the bias-ply tires of the day. Other criteria, such as cornering behavior, were more universal.

In the years following, finer division of the evaluation framework on the part of tire manufacturers proved useful in order to carry out targeted optimization of tires, particularly with respect to special vehicle-specific problems. Parallel to this, the switch from bias-ply to radial tires resulted in elimination of some evaluation criteria which were no longer relevant for radials.

Today, an evaluation catalog containing more than 40 individual criteria has proven useful. Fig. 7.3-5 provides an overview. The following generally familiar vehicle dynamics criteria (Chapter 7.5) are regarded in purely tire-specific terms: that is, only the effect of the tire on overall vehicle behavior is examined.

Table 7.3-1 Effect of tire parameters on service properties

Effect	Property					
	Wet traction	Noise	Mileage	Rolling resistance	Hydroplaning	Quality
Shape/contour	•	•	•	●	●	•
Tread pattern	●	●	•	•	●	•
Material	●	•	●	●	•	•
Construction	•	•	●	•	•	•
Manufacturing process				•		●

● Significant effect
• Discernible effect

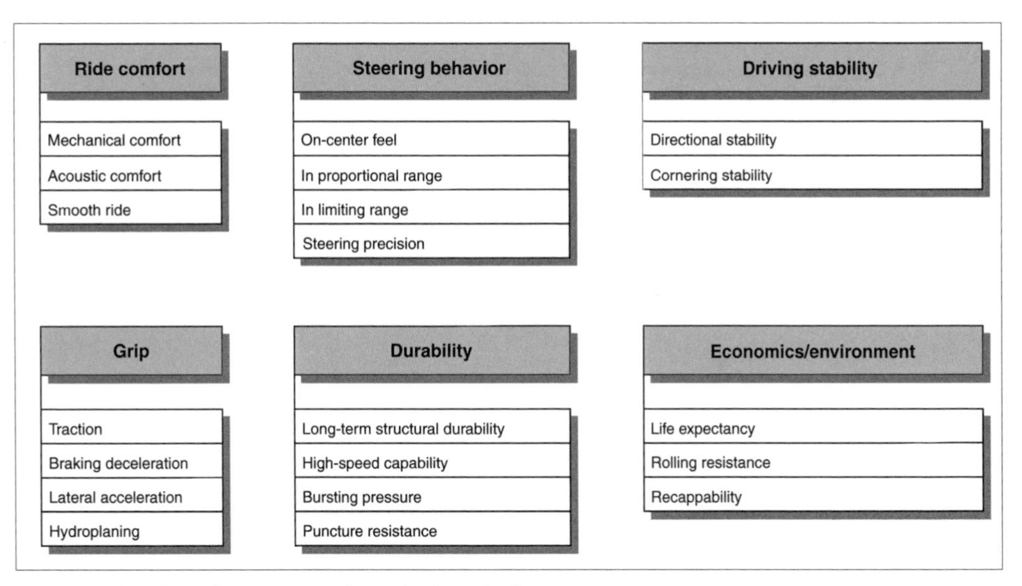

Fig. 7.3-5 Overview of passenger car tire evaluation criteria.

Today, subjective evaluation of a motor vehicle is still the yardstick for whether a tire is suited to a specific vehicle. Nevertheless, great effort is expended in conducting more objective evaluations.

The advantage of objective evaluation is greater reproducibility and better explainability of results by means of physical description of phenomena. If the physics behind a tire property are known, targeted optimization of the desired property may be implemented.

For ride comfort evaluation, measurement of passenger car rear axle vertical acceleration while traversing a cleat discloses a correlation between subjective evaluation and peak acceleration in the frequency range below 100 Hz (Fig. 7.3-6).

With its second vertical mode (at ~75 Hz), the winter tire excites considerably greater vertical rear axle acceleration than the summer tire, whose vertical mode frequency is ~10 Hz higher. This vehicle-specific excitation is clearly perceptible within the vehicle as a drumming noise. With the summer tire, stronger excitation of the fore-and-aft acceleration mode is apparent at ~55 Hz.

Through knowledge of tire resonant modes and measured vehicle acceleration frequency spectra, observed frequency peaks may be explained and the necessary design changes implemented.

In tire/road noise as well, interaction with the road surface is decisive. Fig. 7.3-7 shows that the noise generation bandwidth as a result of various road surfaces is greater than the bandwidth attributable to tire size and tread pattern.

To evaluate vehicle handling within the proportional regime—that is, with lateral acceleration up to 0.4 g—subjective evaluation may be augmented by also examining the vehicle's transfer functions, whereby these functions may be determined by means of various driving maneuvers (Fig. 7.3-8). For a good subjective evaluation, the following are desirable:

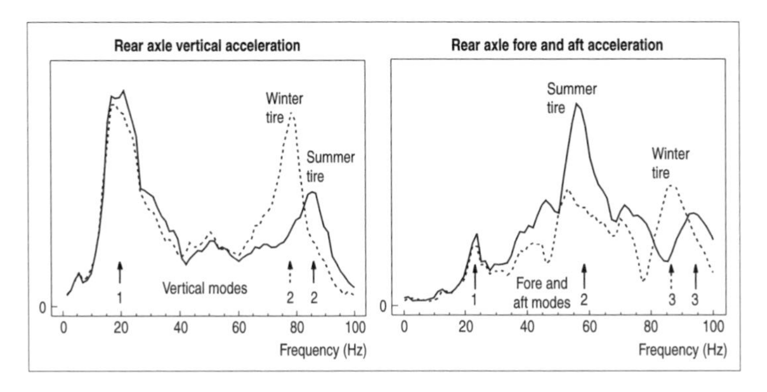

Fig. 7.3-6 Comparison of winter and summer tire comfort in traversing an impact strip.

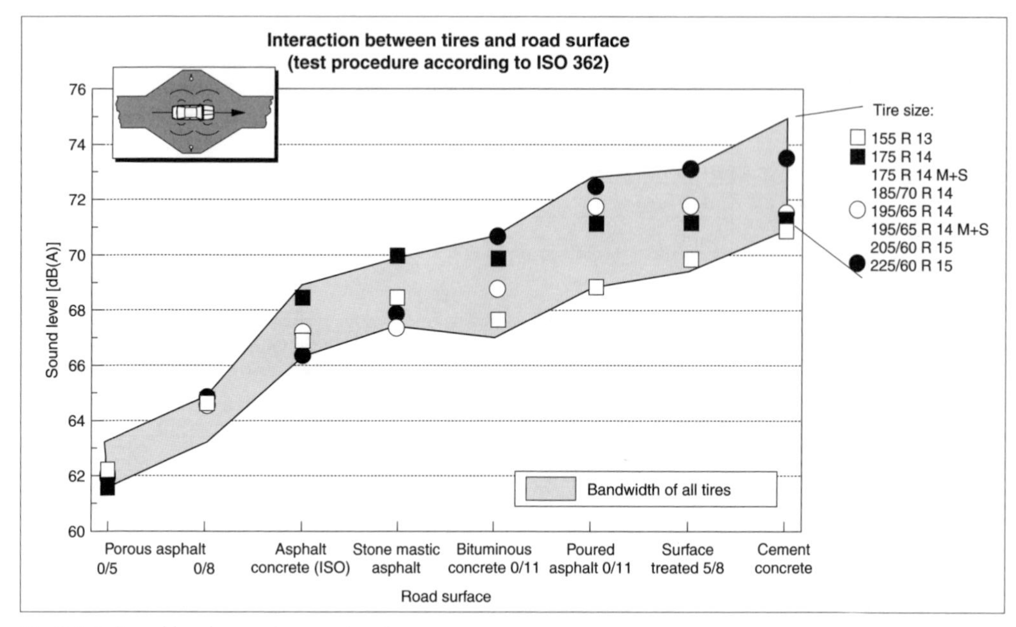

Fig. 7.3-7 Sound level on various road surfaces.

- A large lateral acceleration frequency range
- A high degree of yaw damping
- Small phase shift

At the present level of development, quality tires have high durability reserves and therefore increasing resistance to misuse. Nevertheless, failures resulting from external damage and operator error cannot be eliminated completely.

Definitive analysis of the "failure history" of destroyed tires is problematical. Tires have an integrated memory for the results of misuse. Tire specialists can detect evidence (such as sidewall scuffing and inner liner discoloration) of long-term operation at low inflation pressures.

Tire economics are less influenced by manufacturing or disposal costs than by the cost of operation on the vehicle. Tire rolling resistance must be continuously overcome by application of work supplied by the engine.

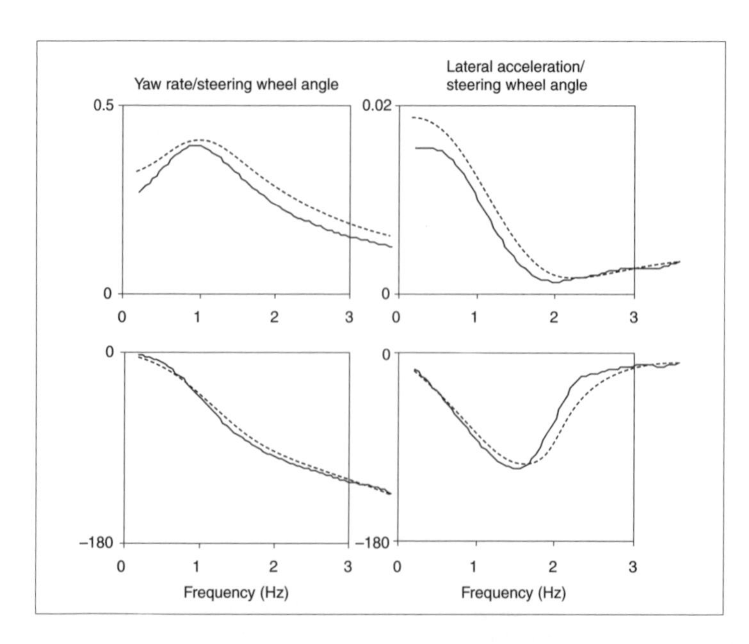

Fig. 7.3-8 Phase and amplification transfer functions for a typical passenger car with subjectively very good (dotted line) and marginally good (solid line) tires.

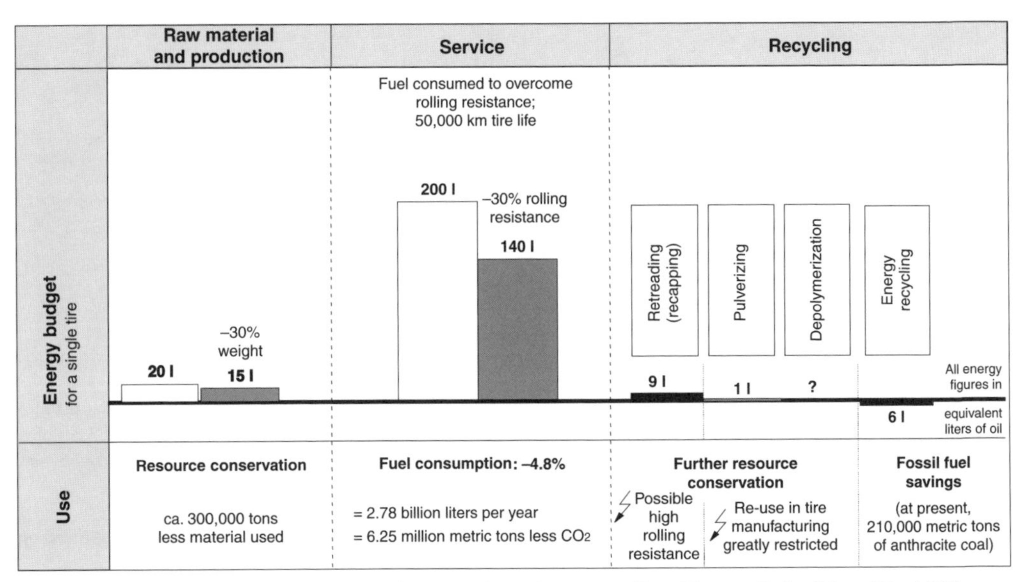

Fig. 7.3-9 Tire energy budget, from production through service to recycling (German Federal Republic, 1997).

The lower a tire's rolling resistance, the lower the vehicle's fuel consumption.

Fig. 7.3-9 illustrates this relationship and its effects on the German economy. By reducing tire rolling resistance by 30%, fuel consumption may be decreased by about 4.8%, which represents a savings of about 60 L per tire over the course of its life. On a national economic level, a rolling resistance reduction of this magnitude would save about 2.5 billion liters of fuel per year.

Compared to new tires, recapped tires usually exhibit about a 10% rolling resistance disadvantage. Consequently, the practice of recapping is questionable from an energy budget standpoint, as usually a greater amount of fuel is used during operation of a recapped tire than is saved by reusing the carcass.

7.3.3.2 Regulatory requirements

Regulatory requirements imposed on tires include sidewall markings with appropriate codes as well as approval marks and approval number in keeping with the ECE 30 standard. Tires so marked have the letter "E" within a circle and the code number of the approving country as well as a multicharacter (e.g., (E12) 028355) approval number.

7.3.3.3 Tires and rims—standardization

Standardization of tires and rims is covered by ETRTO (European Tire and Rim Technical Organization) and DIN (Deutsches Institut für Normung) standards. Tires are marked to show:

- Tire width (e.g., 195)
- Aspect ratio, as a percentage of width (e.g., 65)
- Construction code—R for radial, D for diagonal (bias) ply tires
- Rim diameter in inches (e.g., 15)
- Load rating (e.g., 91, signifying 650 kg maximum load)
- Speed rating symbol (e.g., T = 190 km/h)

This particular example results in a tire designation of 195/65 R 15 91 T.

Other information includes:

- Tread wear indicators (TWI): six transverse ribs molded across tread grooves which appear with 1.6 mm of tread depth remaining
- A DOT (U.S. Department of Transportation) stamp
- The words "TUBELESS" or "TUBE TYPE"
- Production week (e.g., 409 = 40th week of 1999)
- "reinforced" to indicate reinforced tires
- M+S for winter tires.

7.3.4 Tire–roadway force transfer

A tire must provide force transfer between vehicle and roadway, not only on the most diverse types of pavement (asphalt, concrete, cobblestones) but also under all weather conditions and at all vehicle speeds.

Consequently, traction behavior of a tire is a key point for tire development engineers. Factors influencing traction behavior are, above all, tire type and condition, roadway type and condition, operating conditions, and operational misuse.

7.3.4.1 Load-carrying ability

The load-carrying behavior (F_Z) of an ideal membrane is described by its internal pressure (p_i) × contact area

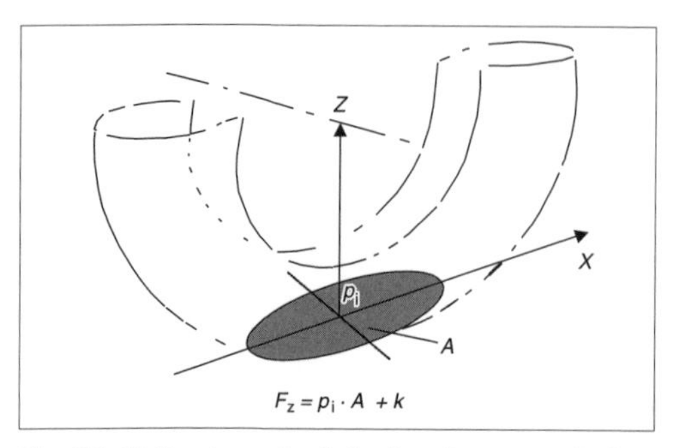

Fig. 7.3-10 Load-carrying behavior of a pneumatic tire.

(*A*); see Fig. 7.3-10. The stiff shell structure of tires adds an additional structural load-carrying component (*k*) of about 10–15%.

7.3.4.2 Traction behavior, generation of horizontal forces

Traction behavior of tires is largely determined by the two frictional partners involved—rubber and road surface.

Adhesion determines force transmission in the regime of smaller sliding speeds (e.g., the front portion of the tire contact patch during ABS braking), while hysteresis determines force transmission at higher sliding speeds (e.g., braking with locked wheels).

The rubber compound developer is readily able to concentrate his efforts on traction in the "grip" or "slide" regimes; that is, the best "ABS tire" is not necessarily the best "locked wheel tire."

The ratio of loss modulus to storage modulus, tan δ, is a measure of energy losses due to deformation of rubber as a viscoelastic material. As an extremely abbreviated explanation, it suffices to say that the developer may associate certain temperature ranges of the tan δ curve with certain properties of a rubber compound based on a temperature/frequency equivalency principle. Under operating conditions, these properties are characterized by a specific temperature profile and mechanical frequency profile (Fig. 7.3-11).

The compromise between good fuel economy through low rolling resistance (small tan δ at 60°C) with short stopping distances (large tan δ in the friction regime; i.e., less than 20°C) should be well balanced on a high level.

Table 7.3-2 Effect of various temperature ranges on tire properties

Temperature range (°C) Frequency: 10 Hz	Tire property
−60 to 0 (adhesion friction)	Winter traction on ice and snow ABS braking
−10 to +20 (hysteresis friction)	ABS braking Locked brakes
+20 to +70	Rolling resistance

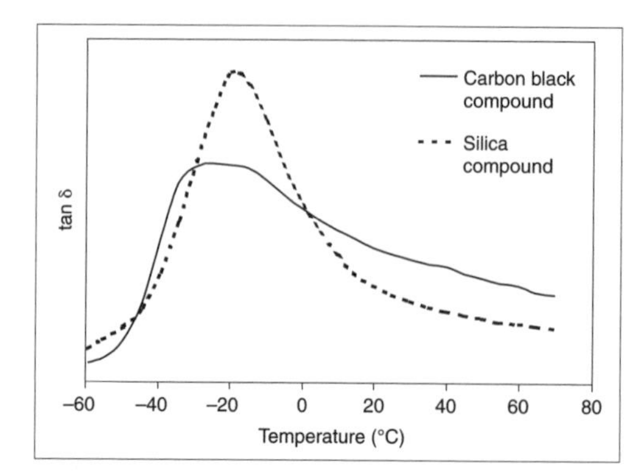

Fig. 7.3-11 Typical trace of loss factor tan δ as a function of temperature for two tire compounds (samples taken from tires, measurements at 10 Hz and constant load amplitude).

The tire developer must find a suitably high compromise level for maximum traction coefficient by means of good grip given the appropriate physical demands of braking or acceleration and fuel-saving low rolling resistance by means of low energy dissipation given the physical demands of rolling. A significant step toward satisfying these demands is introduction of silicate technology (Fig. 7.3-12).

7.3.4.3 Acceleration and braking, tangential forces

Fig. 7.3-13 shows an example of a braking event without ABS. We will examine not the braking system but rather the tire contribution. The representation clearly shows that tires may be loaded throughout the entire regime, up to 100% slip. Notably, in braking with locked wheels, the tires are required to absorb all of the energy which is normally handled by the brake system in ABS braking.

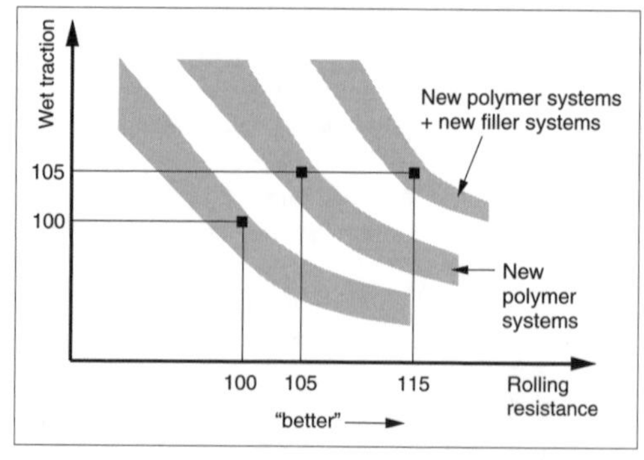

Fig. 7.3-12 Potential of new compound concepts to improve wet traction/rolling resistance compromise (rolling resistance >100% represents lower energy dissipation, therefore "better").

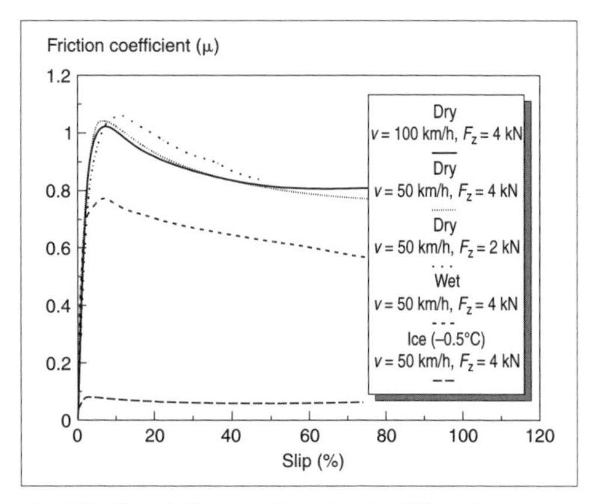

Fig. 7.3-13 Friction coefficient μ for different road surface and service conditions.

With increased slip under braking, sliding zones arise within the tire contact patch (Fig. 7.3-14). Beginning from the trailing end of the contact patch, the sliding zone grows on the right side toward the leading edge of the patch. Just before reaching maximum slip, nearly the entire contact patch is in a sliding condition.

Along with dependence on vehicle design and speed, tire type and road surface roughness have a determining influence.

The representations show the dependence of μ /slip curves, achievable braking accelerations, and braking distances on roadway, tire, and operating conditions as influencing factors.

Various combinations of tire tread patterns and tread compounds may result in very different traction on snow (Fig. 7.3-15). The combination shows that in terms of winter properties, the primary deciding factor is a tread compound which remains more elastic at low temperatures than typical summer compounds do.

7.3.4.4 Slip angle, forces, and moments

For vehicle handling dynamics, the magnitudes and characteristics of forces to be overcome are of decisive importance in providing a pleasant and safe driving experience. With increasing tire slip angle, lateral force increases, as a function of tire normal force, up to a maximum value at slip angles between 5 and 15° (Fig. 7.3-16).

Contact patch deformation and the beginning of contact patch sliding processes between tire and roadway generate a tire aligning moment. This aligning moment attempts to return the steering wheel to its straight-ahead position. It reaches a maximum when the slip angle curve begins to depart significantly from its linear rise and generally becomes negative with continued increase in slip angle.

7.3.4.5 Tires in lateral and longitudinal slip

Along with the question of traction, tire characteristics are also especially interesting in terms of vehicle controllability at the handling limit. Experienced performance-oriented drivers use a higher, narrower limit for higher cornering speeds; more conventional drivers are better served by a broad handling limit, since they seldom encounter borderline situations.

A tire offers its maximum traction potential in only one direction (Fig. 7.3-17). As traction maxima in tangential and transverse directions differ to a certain degree, the generally applied ("Kamm's") traction circle becomes an ellipse.

The relationship between tangential and transverse forces for various slip angles is shown in Fig. 7.3-18.

7.3.4.6 Tire uniformity

Without guidance from a vehicle's suspension, tires would not run in a straight line. Every tire made today generates constant lateral forces which cause a devia-

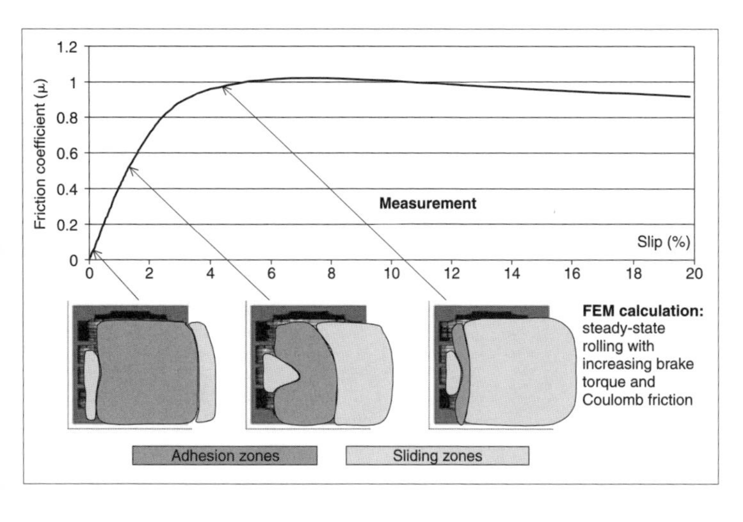

Fig. 7.3-14 Traction and sliding zones in tire contact patch for various levels of wheel slip (from FEM calculations). Left: contact patch leading edge; right: contact patch trailing edge.

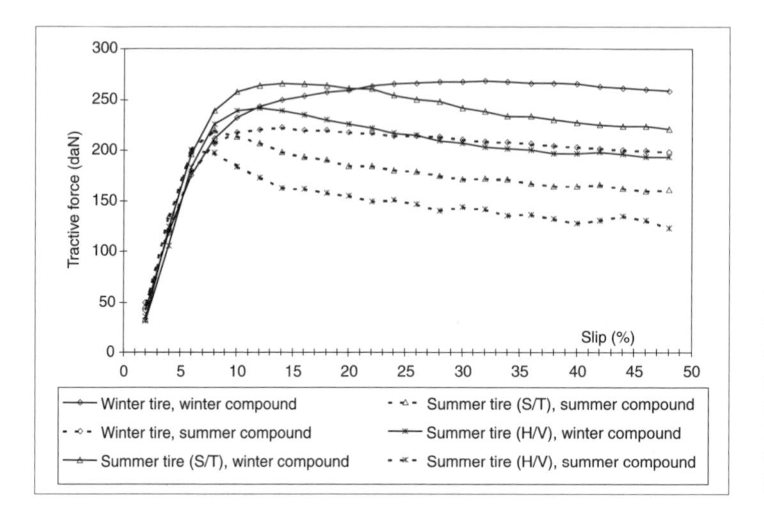

Fig. 7.3-15 Effect of tire and tread compound on friction coefficient for traction on snow. Letters refer to tire speed rating: S ≤ 180 km/h, T ≤ 190 km/h, H ≤ 210 km/h, V ≤ 240 km/h, W ≤ 270 km/h, Y ≤ 300 km/h, ZR ≤ 240 km/h.

tion from the ideal straight-line condition. These forces are composed of rotation-dependent structural side force F_S and rotation-independent conicity F_K.

Structural side force is caused by tire interior structure, while conicity force results from tire geometry.

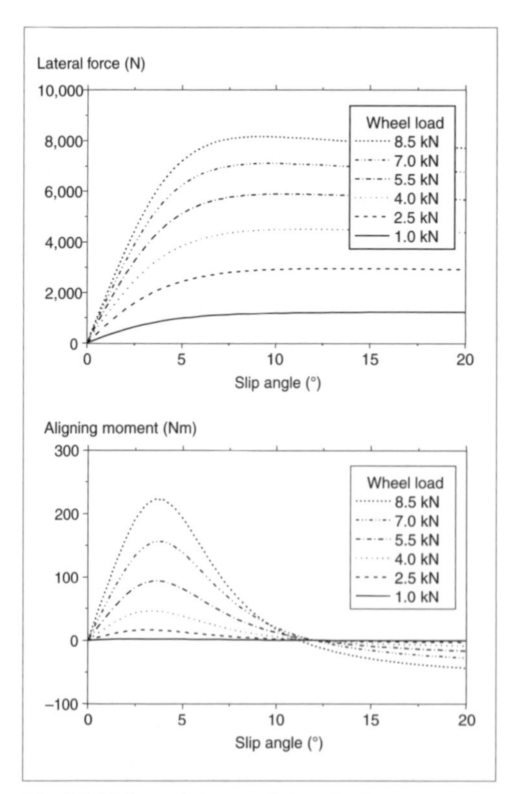

Fig. 7.3-16 Lateral force and tire aligning moment as functions of slip angle for a typical passenger car tire with various wheel loads.

In practice, complaints of pulling to one side may often be addressed by "flipping" one or more tires on their respective rims (Fig. 7.3-19).

7.3.5 Tires as an integral component of the overall vehicle system

Increasingly, tire development is undertaken within the overall vehicle system or subsystems, supported by simulation tools. The objective is to optimize the virtual

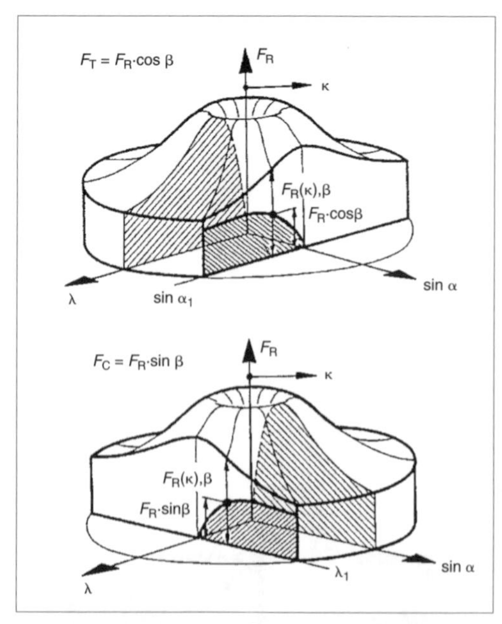

Fig. 7.3-17 "Friction cake" (after Prof. Weber) as a general representation for resultant tire guiding forces. F_T, Tangential force; F_C, lateral force (cornering force); F_R, friction force; κ, slip; α, slip angle; λ, longitudinal slip.

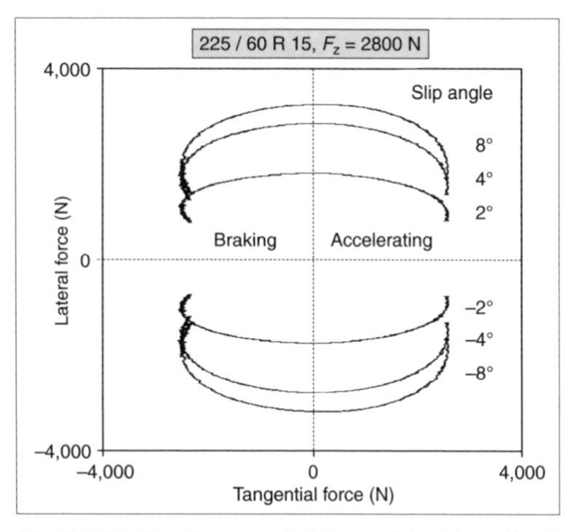

Fig. 7.3-18 Traction potential for combined lateral and longitudinal friction forces.

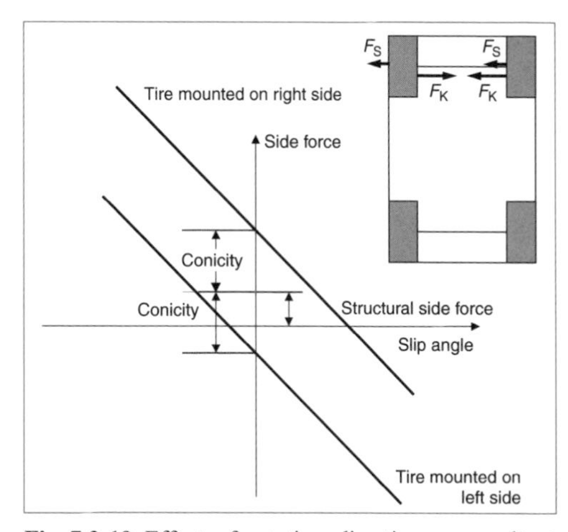

Fig. 7.3-19 Effect of rotation direction on resultant structural side force and conicity F_K.

tire on a virtual vehicle during the development stage to the point where only a few experimental prototypes with a high likelihood of success rapidly lead to improved vehicle behavior. Close development partnerships and joint projects among the tire and vehicle industries, with a primary goal of tuning tires and vehicle concepts, are becoming increasingly common.

7.3.5.1 Tire mechanics and material properties

Mathematical description of rubber behavior, in view of its strongly nonlinear and time-variant characteristics, is a demanding assignment. Fig. 7.3-21 shows a tensile test in which strain is increased by 10% in each of five stages. In each of these stages, strain is run up and down ten times. The first cycle of a load stage always exhibits the highest force, which diminishes with each successive cycle, until an equilibrium condition is reached. Modeling this material behavior requires complex simulations. The model may, for example, consist of an array of springs, dampers and friction elements which describe the elastic, viscous, and plastic behavior of rubber for various strain rates (Fig. 7.3-22)

Individual components may be physically explained as follows.

The general behavior of rubber may be described by its elastic behavior. This is usually modeled by a nonlinear spring, which takes into account a nonlinear stress-strain curve as well as incompressibility.

The viscous component arises from the speed-dependent (i.e., the frequency-dependent) stiffness of rubber, and, by means of the temperature-frequency equivalent, also describes its thermal behavior. The higher the frequency or the lower the temperature, the more stiffly rubber reacts to external loads.

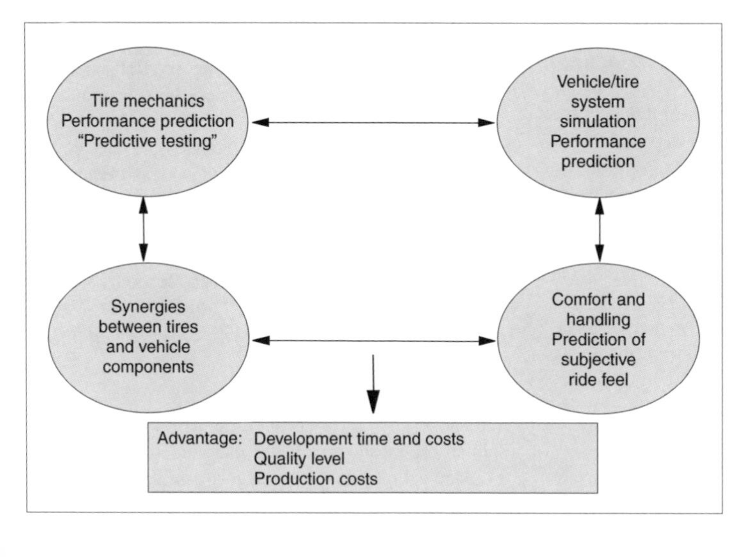

Fig. 7.3-20 Tire development as an integral component of the vehicle/tire/roadway/driver system.

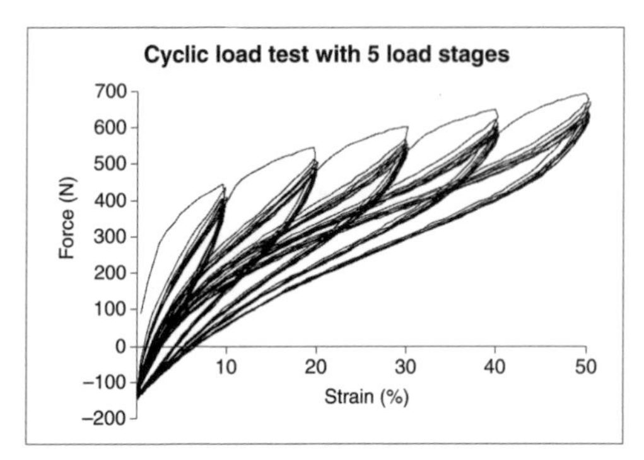

Fig. 7.3-21 Experimental strain behavior examination of carbon black-filled rubber sample subjected to five stages of cyclic loading.

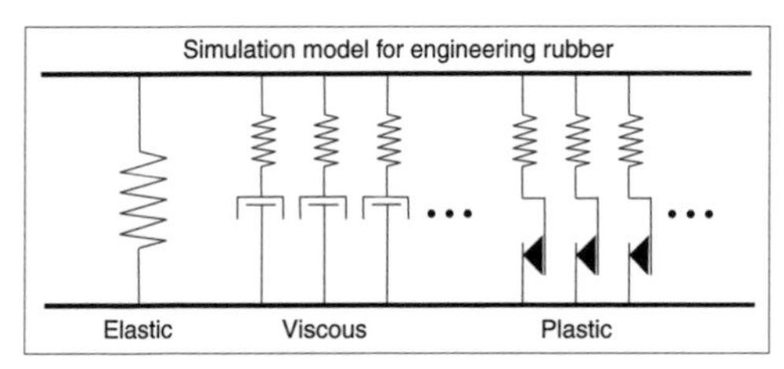

Fig. 7.3-22 Simulation model for rubber behavior.

Matching the storage modulus (G′) and loss modulus (G″) across a frequency range of about five decades (Fig. 7.3-23) is possible using ten Maxwell elements, each consisting of a spring and damper.

The plastic component can account for internal structure of the material. One hypothesis claims that under deformation, polymer chains slide across filler material surfaces, resulting in frictional hysteresis. This behavior is apparent in Fig. 7.3-24, in that even at low strain rates, force during increasing strain (loading) is greater than during decreasing strain (unloading).

Matching is accomplished by Prandtl elements which describe the plastic material component (Fig. 7.3-22). The curve shape and rubber hysteresis typical of filled networks are accurately reproduced.

Introduction of complex rubber descriptions in the form of material laws for tire computation permits a variety of tire property predictions. One example is tire deformation in response to slip angle under a steady-state rolling condition, including information on force and friction conditions in the contact patch in order to optimize traction in the presence of lateral forces during braking and acceleration (Fig. 7.3-25).

Further FEM applications include predicting durability, rolling resistance, temperature distribution, wear, hydroplaning, characteristic curves, etc.

7.3.5.2 Tire models

Tire models serve to qualitatively or quantitatively represent and predict tire properties. Depending on require-

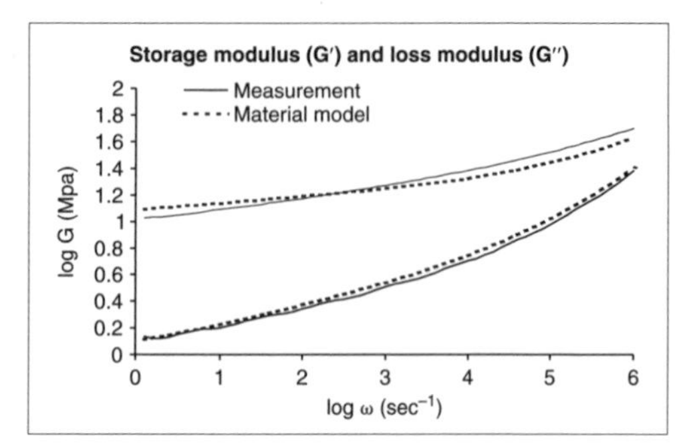

Fig. 7.3-23 Matching material laws to experimental data across a five-decade frequency bandwidth.

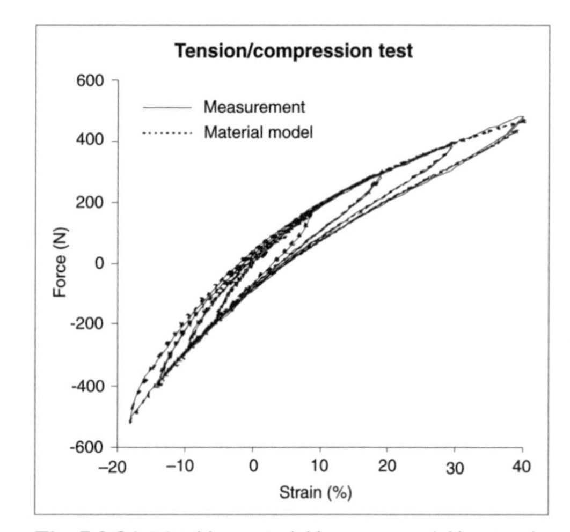

Fig. 7.3-24 Matching material laws to material hysteresis under static conditions with various strain amplitudes.

ments, they may possess different levels of complexity beginning with simple mathematical representations all the way to detailed dynamic FEM models (Fig. 7.3-26).

An example of a simple MBS (multibody system) tire model consisting of springs, masses and dampers may be seen in Fig. 7.3-27. The rim and deformed belt region are apparent as are forces on the axle and in the contact patch.

Tire parameters needed for calculation may be obtained through special experimental measurements or calculated by means of complex tire models (e.g., FEM). The model is therefore capable of "driving" along an uneven "road" and capturing the resulting axle loads to provide a representation of a coupled vehicle model. Ground contact is sensed by so-called "brushes" (not shown) and the resulting contact forces calculated accordingly.

7.3.5.3 Overall models

Exchange of models between the tire and automotive industries, as shown by the example of Continental, broaden vehicle competence by expanding the range of automotive products.

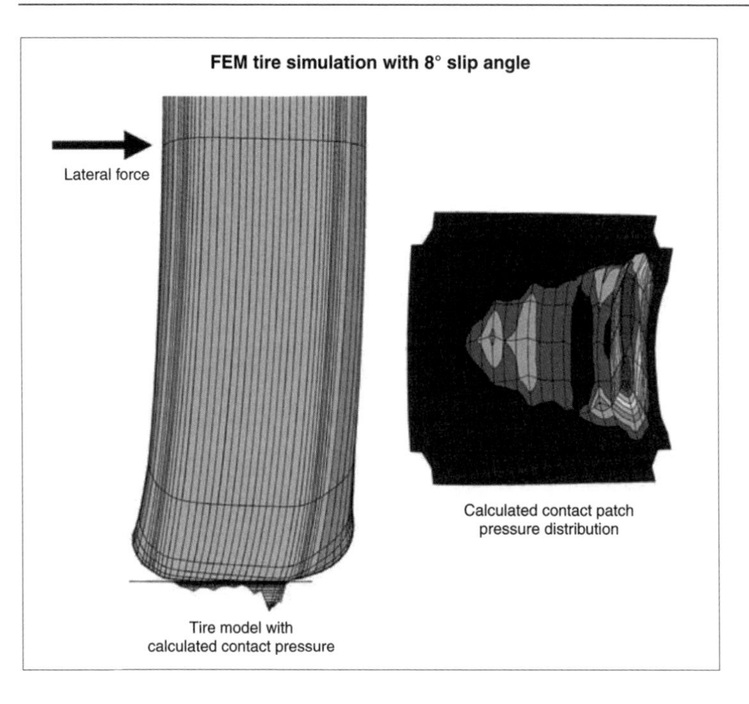

Fig. 7.3-25 Tire simulation using finite element methods (FEM) for a steady-state rolling tire subjected to slip angle, with calculated contact pressure and rim forces.

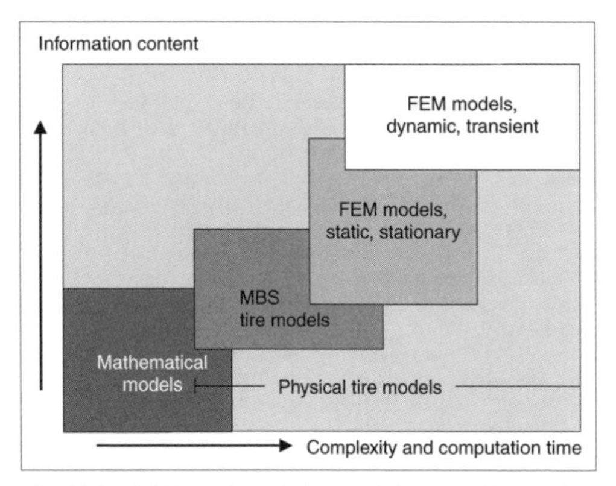

Fig. 7.3-26 Hierarchy of tire models according to information content and complexity.

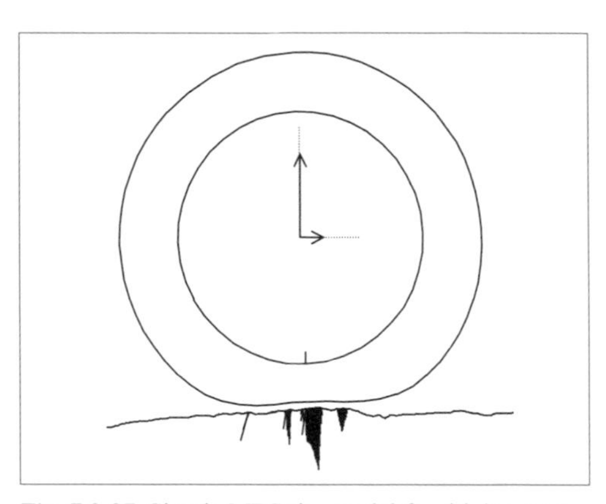

Fig. 7.3-27 Simple MBS tire model for driving over a rough road surface; representation of axle and contact patch forces showing magnitude and direction.

Increasingly, multibody systems are being used to simulate the complex interplay of vehicle components and tires (Fig. 7.3-28). In simulations on flat road surfaces and with constant friction coefficient, horizontal dynamic tire models coupled to vehicle models provide highly accurate representations of tire behavior. In the future, complex tire models capable of providing tire forces and moments on uneven surfaces with varying μ will be applied, allowing conclusions to be drawn regarding longitudinal and vertical dynamics.

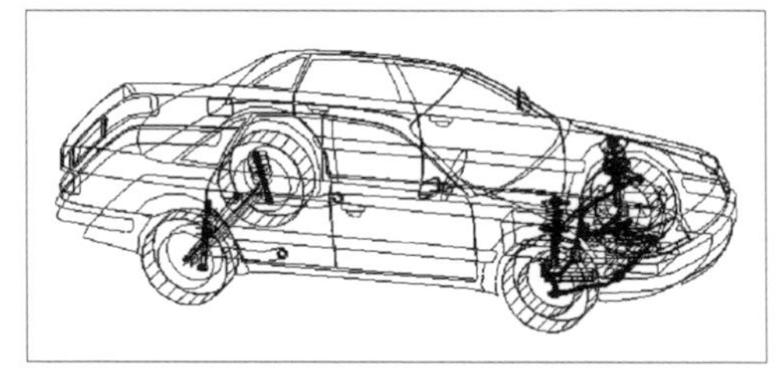

Fig. 7.3-28 Complex vehicle model for dynamic simulation.

7.3.5.4 Description of vehicle handling

The final authority in comfort and handling evaluation is the driver. An understanding of the physics underlying subjective evaluation by actual drivers is a prerequisite for targeted development, especially in the case of computational simulations. Tire developers are placed in the position of objectively evaluating the effects of constructive and materials measures on vehicle system behavior.

7.3.5.5 Synergies between tires and other system components

Recognition of the interplay between tires and other components leads to synergies in properties and costs with regard to the system as well as to components such as tires.

7.3.6 Future tire technologies

The tire, now about 100 years old—younger than the automobile itself—has considerable future potential. The path described in 7.3.5 will lead to new technological approaches.

Among the classic service properties, rolling resistance in particular, handling safety and ride comfort will be at the forefront. Increasingly, demands are being made for adequate run-flat capability in the event of air loss.

7.3.6.1 Run-flat tire systems

Rims with integral surfaces to support flat tires represent a new systems approach. In contrast to self-supporting tire carcasses, it is likely that these will use a membrane approach in which the proportion of load carried by the carcass is as small as possible.

Systems compatible with modern tire/wheel systems are the self-supporting carcass (already on the market), as well as light, conventionally mounted support rings with limited comfort properties under run-flat conditions.

7.3.6.2 Tire-related accessory products

Tire-related accessory products may find application within the tire and fender well—for example, to dampen noise by means of absorbers and resonators. Tires would be designed to radiate noise toward the dampening inner fender shrouding.

7.3.6.3 Tire-tuned suspension system components

Matching tires and suspension components may, for example, be accomplished through the shock strut top mount, which would decouple tire-specific noise and vibration excitation from introduction into the chassis.

7.3.6.4 Additional tire functions

In the future, using suitable sensors, tires will detect forces, speeds, and acceleration, and provide these data as input signals to handling dynamics control systems. Moreover, current research is aimed at determining traction between tires and roadway in real time.

7.3.6.5 Material development

The primary efforts in material development involve improving the compromise between increased traction and reduced rolling resistance. Current research includes work with adaptive materials whose properties change in response to environmental conditions—for example, by means of partially reversible networks.

References

ETRTO. Standard Manual, Brussels, 1999.

ETRTO. Engineering Design Information, Brussels, 1998.

Ammon, D., M. Gipser, J. Rauh, and J. Wimmer. "Effiziente Simulation der Gesamtsystemdynamik Reifen-Achse-Fahrbahn." In: Reifen, Fahrwerk, Fahrbahn, VDI-Berichte No. 1224, 1995.

Bachmann, T. "Literaturrecherche zum Reibwert zwischen Reifen und Fahrbahn." Fortschritt-Berichte VDI Reihe 12: Verkehrstechnik/ Fahrzeugtechnik No. 286, VDI-Verlag.

Becker, A., and B. Seifert. "Simulation von Abrieb und von Reifenkennwerten für Handling mit einem stationär rollenden FE-Reifenmodell." 6. Fachtagung Reifen—Fahrwerk—Fahrbahn, VDI-Berichte 1350, 1997.

Becker, A., V. Dorsch, M. Kaliske, and H. Rothert. "A material model for simulating the hysteretic behavior of filled rubber for rolling tires." *Tire Science and Technology*, Vol. 26, No. 3, 1998.

Böhm, F., and H.-P. Willumeit. *Proceedings* of the 2nd International Colloquium on Tyre Models for Vehicle Dynamic Analysis, University of Berlin, Swets & Zeitlinger Publishers, 1997.

Bussien, *Automobiltechnisches Handbuch*. Technischer Verlag Herbert Cram, Berlin, 1978.

Clark, S.K. "Mechanics of Pneumatic Tires." U.S. Department of Transportation National Highway Traffic Safety Administration, Washington, D.C. DOT HS 805952, August 1981.

Dieckmann, T. "Der Reifenschlupf als Indikator für das Kraftschlusspotential." Dissertation, University of Hannover, 1992.

Eichhorn, U. "Reibwert zwischen Reifen und Fahrbahn—Einfluss und Erkennung." Dissertation, Darmstadt Technical University, Fachgebiet Fahrzeugtechnik, Fortschritt-Berichte VDI-Reihe 12 No. 222, VDI-Verlag, Düsseldorf, 1994.

Eichler, M. "A ride comfort tyre model for vibration analysis in full vehicle simulation." Vehicle System Dynamics Supplement 27, Swets & Zeitlinger B.V., Lisse, the Netherlands, 1997, pp. 109–122

Ernst, G.K. "Nassgriff: Dauerauftrag für die Reifenentwicklung." Conference proceedings, Haus der Technik, Essen, 11/94.

Gabler, A., E. Straube, and G. Heinrich. "Korrelationen des Nassrutschverhaltens gefüllter Vulkanisate mit ihren viskoelastischen Eigenschaften." Kautschuk—Gummi—Kunststoffe 46, No. 12/93.

Gipser, M. "Reifenmodelle für Komfort- und Schlechtwegsimulation." 7th Annual Automobile and Engine Technology Colloquium, Aachen, 1998.

Haken, K.-L., U. Essers, and U. Wohanka. "Neue Erkenntnisse zum Einfluss der Reifenbreite und Querschnittsform auf die Reifeneigenschaften und ihre Berücksichtigung bei der Fahrwerksauslegung." VDI conference "Reifen—Fahrwerk—Fahrbahn," Hannover, 1993.

Hein, H.R., M. Ellmann, and M. Hatzmann. "Die Veränderungen in der Automobilindustrie treffen uns alle." *KGK Kautschuk Gummi Kunststoffe* No. 4/96.

Hein, H. R., and M. Hatzmann. "All-Season-Reifen in der PKW-Erstausrüstung—Eine Möglichkeit zur Anpassung der Fahrwerke an US-spezifische Fahrgewohnheiten, Straßen- und Umweltbedingungen." VDI-Berichte No. 1088 (1993), Reifen—Fahrwerk—Fahrbahn. Conference proceedings of VDI-Gesellschaft Fahrzeugtechnik, VDI-Verlag, pp. 215–242.

Heinrich, G. "The dynamics of tire tread compound and their relationship to wet skid behaviour." *Progress in Colloid & Polymer Science* 90, 1992, pp. 16–26.
Huinink, H., and C. Schröder. "Dynamische Interaktion Bremse—Reifen—Straße." XVIII. Internationales μ-Symposium Bremsen-Fachtagung, Progress Reports, VDI Series 12 No. 373, VDI-Verlag, Düsseldorf, 1999.
Käding, W., E. Kalb, K. Müller, E. Drähne, and C. Schröder. "Untersuchung des Einflusses von Reifen- und Fahrzeugparametern auf die Fahrzeugreaktion bei kleinen Querbeschleunigungen (0,1-0,2 g) am Daimler-Benz-Fahrsimulator." VDI-Tagung Reifen—Fahrwerk—Fahrbahn, Hannover, 1993, VDI-Berichte 1088, VDI-Verlag.
Kaliske, M. "Zur Theorie und Numerik von Polymerstrukturen unter statischen und dynamischen Einwirkungen." Dissertation, University of Hannover, 1995.
Kaliske, M., and H. Rothert. "Internal material friction of rubber modelled by multiplicative elasto-plastic approach." *Proceedings* of the 4th International Conference on Computational Plasticity, Barcelona, 1995.
Mitschke, M. *Dynamik der Kraftfahrzeuge, Antrieb und Bremsung*. Volume A, Third edition, Springer Verlag Berlin, 1995.
Pacejka, H.B., and I. J.M. Besselink. "Magic Formula Tyre Model with Transient Properties." Swets & Zeitlinger B.V., Lisse, the Netherlands, 1997, pp. 234–249.
Reimpell, J. *Fahrwerktechnik: Reifen und Räder*, Vogel Verlag, 1982.
Schröder, C., and S. Chung. "Influence of tire characteristic properties on the vehicle lateral transient response." 8th Annual Meeting and Conference on Tire Science and Technology, Akron, Ohio, March, 1994.
Strzelcyk, M., et al. "Erweiterte Mobilität und Sicherheit über die gegenwärtigen Systemgrenzen hinaus—Innovative Systeme für Pannenlauf und Reifendruckkontrolle." VDI-Berichte 1494, 1999, pp. 221–233.
Wallentowitz, H. "Fahrwerkstechnologie im nächsten Jahrtausend." Tag des Fahrwerks, October 5, 1998, at the Institut für Kraftfahrwesen, Aachen (ika).
Weber, R. "Reifenführungskräfte bei schnellen Änderungen von Schräglauf und Schlupf." Postdoctoral thesis, Department of Mechanical Engineering, University of Karlsruhe, 1981.
Zegelaar, P. W. A. "The dynamic response of tires to brake torque variations and road unevennesses." Dissertation, Delft University, 1998.

7.4 Suspension design

7.4.1 Suspension kinematics

Suspension kinematics determine the spatial motion of wheels in response to spring and steering movement. Analysis of suspension kinematics plays an important role in automobile design because of the importance of wheel (and therefore tire) position with respect to the road surface. It is applied at the beginning of the development process as the suspension concept is established. As development proceeds, suspension kinematics continue to be analyzed using CAD systems, as component clearance issues are examined. In advanced stages of development, the effects of suspension kinematics on vehicle handling are evaluated using complete vehicle models in computational simulations, and later the analysis is optimized through actual driving tests.

Several kinematic parameters have been standardized (ISO 8855/DIN 70000); complete descriptions may be found in [1, 2].

7.4.1.1 Kinematics of vertical wheel motion

Tire camber angle γ, which describes the inclination of the wheel plane to the vertical as measured on a level surface, affects transmissibility of lateral forces between tire and road surface. In the case of a (customarily) negative camber, the top of the wheel is inclined toward the vehicle centerline, and the wheel would tend to roll in a circular path like a cone on its side. If it is kept from doing so by the suspension, a side force toward the vehicle centerline is generated. Therefore, the lateral force transmission potential of a tire under negative camber (within the available traction limit) is greater than for positive camber. For extremely low-profile tires, zero camber is desirable to prevent running on the tire tread shoulders [3]. For independent suspension, camber change in jounce should at least partially compensate for body roll in cornering in order to avoid positive camber at the outside wheel. Beam axles offer the advantage that camber relative to the road remains constant in cornering.

Toe angle λ is positive if the wheels in plan view are pointed inward at the front (toe-in); otherwise, the term toe-out is used. By means of a small positive static toe angle, on the order of a few minutes of arc, small shear forces are generated in the tire contact area during straight-ahead vehicle motion. These shear forces produce opposing left and right lateral forces which take up suspension elasticities and eliminate any possible play in the tie rods and steering box. The vehicle reacts more quickly to steering angle inputs and exhibits better straight-line stability, as preload eliminates having the steering train pass through a "dead zone" or zero point.

Toe-in may change with wheel bump travel (jounce or rebound) in response to springing motion, thereby affecting the vehicle's self-steering behavior; the term "roll steer" (or "ride steer") is used to describe this phenomenon. The relevant effect on vehicle handling is toe-in change in jounce—in other words, at the outside wheel which carries the greater wheel load. As the rear axle generally cannot be directly influenced by the driver, the steering angles generated by suspension vertical travel or by external forces (see Section 7.4.2, Elastokinematics), although small, are decisive for overall vehicle handling behavior. For example, a small rear axle toe-in tendency in jounce may generate an understeer effect. On rough pavement, large toe changes over the range of wheel travel lead to low-frequency steering torque variations which negatively affect vehicle tracking and should, therefore, be avoided.

Spatial movement of the suspension, especially the tire contact patch and wheel center, may be projected onto the wheel longitudinal central plane (side view) and wheel transverse plane (rear view). The point in each plane which represents the instantaneous point about which the suspension moves is termed the center of rotation. Both centers, longitudinal instantaneous center (or "torque arm center") and transverse instantaneous center (Figs. 7.4.1 and 7.4.2), may be regarded as

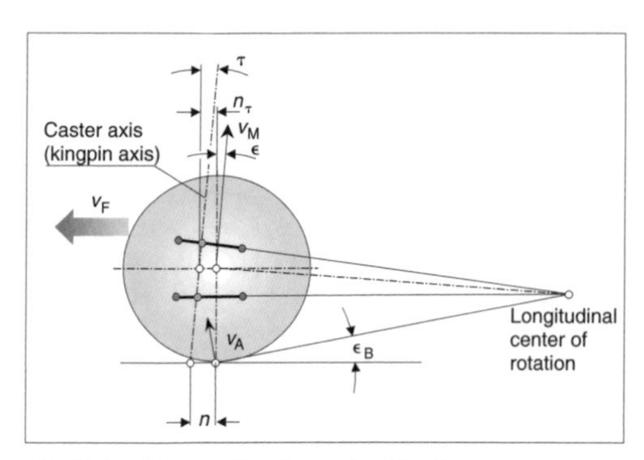

Fig. 7.4-1 Front axle schematic side view.

substitute suspension pickup points. Rays drawn from these instantaneous centers are perpendicular to the velocity vectors of the suspension points under consideration. If one imagines replacing the suspension with a rectilinear motion for the tire contact patch perpendicular to the appropriate ray, then a force perpendicular to this rectilinear path will not result in any motion of the contact patch. Those components of longitudinal and lateral forces passing through these instantaneous centers are taken up directly by the suspension links; the remaining components result in spring movement and therefore body movement. The angles of the longitudinal instantaneous center rays to the horizontal plane determine the proportion of forces taken by suspension arms.

Under braking, a resultant force composed of braking force and dynamic weight transfer is applied to the tire contact patch. If the resultant passes through the torque arm center, the effect is termed 100% antidive, as no spring motion is incurred. Brake antidive is dependent upon wheelbase, center of gravity height, and brake force distribution; the brake force support angle ε_B, between the ray that joins torque arm center and the contact patch (the "side view swing arm") and the horizontal plane, is axle-specific. As braking moment is typically taken by the suspension, one may imagine wheel and suspension members as a rigid assembly. The entire unit and therefore also the tire contact patch rotate

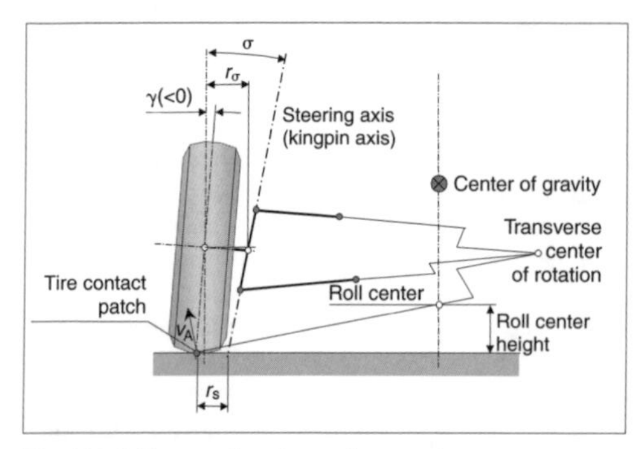

Fig. 7.4-2 Front axle schematic rear view.

about the suspension's torque arm center; for angle ε_B, therefore, motion of the tire contact patch as acting point of the brake force is the important factor.

The ray from the torque arm center to the wheel center establishes the wheel's instantaneous velocity; if this has a rearward component, this is referred to as a positive wheel travel angle. In contrast to longitudinal compliance from suspension rubber bushings, wheel travel angle is no longer of any great significance for ride comfort. It does, however, represent the traction force support angle ε if propulsion is by means of an articulated longitudinal driveshaft. In this case, the suspension is not subjected to any moments, and as far as the suspension is concerned, drive force is applied to the center of the wheel. The offset moment from the wheel center to the tire contact patch is the drive torque and is taken by the final drive unit. Therefore the wheel is free to move with respect to the suspension, and the tire contact patch no longer rotates around the suspension's torque arm center, but rather moves parallel to the wheel center. For the traction force support angle, the important factor is motion of the wheel center as the point of action of drive forces. Antisquat geometry results from negative front wheel travel angle and positive rear wheel travel angle. For rigid (live) axles, however, drive torque is taken by the axle housing; one could imagine wheel and suspension rigidly connected, and traction force support angle is identical to braking force support angle.

The transverse center describes tire contact patch motion relative to the body in the plane transverse to the wheels. For reasons of symmetry, rays from both axle sides intersect at the vehicle centerline to locate the roll center. The line joining the front and rear axle roll centers is termed the roll axis; the instant the vehicle is steered into a turn, it rolls about this axis. The distance between the center of gravity and roll axis is the lever arm for inertia forces (the roll couple) and determines the roll moment, which is counteracted by the suspension springs. A roll center at the pavement surface results in large spring travel and large roll angles. If the roll centers are at the center of gravity height, the springing does not react to any force components, and no roll angle results. With independent suspension, however, high roll centers are associated with appreciable track changes in jounce and, therefore, with lateral slip at the tire contact patch. Therefore, in view of traction and lateral force considerations, roll center heights in excess of 150 mm are rare today. (Beam axles, with their constant track, may exhibit higher roll centers). Roll axis inclination permits influencing distribution of roll support between the front and rear axles. For example, a roll axis inclined upward to the rear (rear axle roll center higher than front) represents greater roll support at the rear, which leads to greater wheel load differences, reduces rear lateral force potential, and thus acts to promote oversteer. The sum of wheel load differences (weight transfer) of course depends solely on center of gravity height and track and is constant for any roll center height; only the front to rear distribution of weight transfer may be affected by roll axis location.

If roll centers change in response to vertical wheel travel, the roll axis location changes with vehicle (cargo) load. In order to prevent vehicle loading from having an undesirable effect on handling, front and rear roll center height changes must be matched, with the objective of avoiding excessive changes in roll axis inclination and, therefore, changes in front-to-rear roll support.

Roll centers and thus the roll axis are defined for symmetrical springing in the plane of vehicle symmetry. In the event of one-sided spring deflection, an intersection of both rays from the transverse instantaneous centers and tire contact patches can be graphically determined, but this may not fall on the vehicle centerline. Instead, it may be displaced laterally; however, this means that the resulting displaced roll axis no longer has any physical significance. Spring jounce and rebound travel are decisive in determining body motion in response to lateral acceleration.

In cornering, with pronounced roll angles, opposing spring travel results in different left- and right-side suspension forces. Which proportion of resultants is taken up by suspension links and which by springs depends on the height of the transverse instantaneous center—that is, its height change in response to spring travel. If this change is small or zero (e.g., for trailing arm suspensions), forces introduced into the springing are smaller for the outside wheel than the inside: that is, the outside corner of the vehicle is compressed less in jounce than the inner corner is extended in rebound. This causes an increase in center of gravity height, which in turn increases the roll moment to be counteracted. This undesirable effect is termed "jacking." It may be reduced by sufficient roll center height change (dependent on roll center height) or avoided entirely, as this will reduce the asymmetric spring force distribution between inner and outer corners.

7.4.1.2 Steering kinematics

The task of the steering system is to enable vehicle guidance as well as provide driver feedback regarding driving and road conditions. These properties may largely be achieved by targeted design of the steering axis about which the wheel swings in response to steering motion—the kingpin axis.

In Fig. 7.4-2, the kingpin inclination σ is shown as the angle of the steering axis to the vertical in front elevation, and the kingpin offset (or steering radius) r_S as the horizontal distance between the center of the tire contact patch and the intersection of the kingpin axis with the ground. In the side view, shown in Fig. 7.4-1, the caster angle τ is similarly defined as the angle of the kingpin axis to the vertical in side elevation and the caster offset n as the horizontal distance between the center of the tire contact patch and the intersection of the steering axis with the ground. Caster angle and caster offset may be chosen independently of each other if one introduces a kingpin offset at the wheel center $n\tau$, a horizontal displacement of the steering axis from the wheel center as seen in side elevation.

As a result of caster and kingpin inclination, the vehicle body is lifted by steering movements. Gravity therefore tends to return the steering to center. Caster angle will produce negative camber at the inward-turned wheel on the outside corner, which has a beneficial effect on tire lateral force potential. Kingpin offset and caster offset combine to generate a wheel load moment arm about the steering axis. The resulting moments cause the steering to return to center if the inner and outer forces change asymmetrically, which is indeed generally the case in the Ackermann condition (see Section 7.4.5). In rapid cornering, the effect of lateral force predominates and, displaced by the caster offset (and tire contact patch trail—"pneumatic trail"), also acts on the kingpin axis to return the steering to center [4].

In jounce, the wheel carrier turns around its longitudinal axis as seen in side elevation, and both caster angle and offset change. In order to limit these changes and therefore avoid unduly affecting the self-centering action, the longitudinal axis must be sufficiently far removed, which limits camber angle. The kingpin offset represents the lever arm for braking forces; for negative values (center of tire contact patch inside the kingpin axis intersection with ground plane), a steering angle is generated which opposes any yaw moment arising from asymmetrical braking forces. However, feedback at the steering wheel is also opposite the actual disturbing force.

The kingpin offset at wheel center, r_σ, as seen in rear elevation, is the lever arm for an impact force acting on the unbraked wheel. This dimension is also known as the disturbing force moment arm. The offset moment between tire contact patch and wheel center therefore acts as an accelerating moment on wheel rotational speed; the suspension is not subjected to this moment.

In propulsion by means of articulated halfshafts, the inclination angle between the inner and outer shaft components must be considered. At an inclination angle of 0°, the traction force radius [1] is identical to the disturbing force moment arm. For driveshafts angled upward at their inner ends, the drive force moment arm is longer, while for inner ends angled downward, the drive force moment arm is shorter than the disturbing force moment arm.

References

[1] Matschinsky, W. *Radführungen der Straßenfahrzeuge*, second edition, Springer Verlag, 1998; English edition *Road Vehicle Suspensions*, Professional Engineering Publishing Ltd., 2000.

[2] Reimpell, J. *Fahrwerktechnik: Grundlagen*, third edition, Vogel Buchverlag, 1995.

[3] Haken, K.-L., et al. "Besondere Eigenschaften von Niederquerschnittsreifen bei trockener und nasser Fahrbahn und ihre

Auswirkung auf die Fahrdynamik." Stuttgarter Symposium Kraftfahrwesen und Verbrennungsmotoren, K8.1, 1995.
[4] Zomotor, A. (ed.), and J. Reimpell. *Fahrwerktechnik: Fahrverhalten*, second edition, Vogel Buchverlag, Würzburg, 1991.

7.4.2 Elastokinematics

Elastokinematics describe the motion of wheels with respect to the body while taking into account elastic deformation of suspension components arising from wheel forces [1].

Elasticity is deliberately built into suspension components with the aid of elastomeric mounts (usually in the form of rubber bushings). In order to achieve the ride comfort demanded of modern passenger cars, the suspension must permit longitudinal compliance, customarily realized by means of rubber mounts. Furthermore, to isolate vehicle occupants from noise, rubber mounts are necessarily interposed between metal components. Additionally, elasticities are unavoidable, because, along with rubber mounts, even steel and aluminum components exhibit significant elasticity. Under load, these cause toe and camber changes. Suspension geometry and elasticities must be tuned to each other in such a way as to optimize handling properties caused by elastic deformation and the resulting steering behavior. Because elasticity-induced steering angles are relatively small, they may be applied above all to improve handling at small slip angles and at higher speeds.

The following will elucidate several key relationships regarding elasticity. Because many individual effects and the spatial arrangement of the entire suspension system must be considered, precise suspension design is only possible by means of simulation. At present, most simulations employ MBS models as well as nonlinear FEM models as needed (e.g., twist-beam axles; see Section 10.2).

7.4.2.1 The effect of component elasticity

"Component elasticity" is understood to mean unavoidable elasticity of necessary metal components. Deformation of suspension components as well as of rubber mounts principally affects vehicle handling; both types of deformation may therefore be deliberately applied [2]. It should be noted that during the course of a vehicle's life, rubber will take a set and harden while metal components retain their elasticity. For vibration isolation, it is advantageous to realize elasticity by means of rubber mounts (see Section 7.4.2.2) because the material damping of rubber is considerably greater than that of steel or aluminum.

Fig. 7.4.3 shows a greatly exaggerated representation of a rear suspension subframe under load. Under such loads, displacement of kinematic points may amount to several millimeters and therefore may not be disregarded in comparison to rubber mount deformation.

In double wishbone and strut-type suspensions, A-arms contribute greatly to overall elasticity. The transverse link of a strut suspension may exhibit additional longitudinal compliance of several millimeters. Contribution of the wheel carrier to elastokinematics is strongly dependent on the suspension principle. In strut suspensions, the shock absorber acts as a wheel locating element and is therefore loaded in bending. Shock absorber bending loads not only result in angularity of the piston and shaft guides, but add a camber contribution under lateral loads and possibly also contribute to steering angle. Although it does not affect suspension elastokinematics per se, the auto body itself affects handling by virtue of its overall torsion and bending behavior. More significant influence is exerted, however, by local elasticities. Usually, these can only be regarded individually, because loading at one body pickup point also results in deformation at other points. The deciding factor is whether overall deformation results in a steering angle. While straight suspension links may generally be regarded as rigid components, the elasticity of links that are angled or cranked for clearance reasons must be considered. Ball joints also exhibit relevant elasticity, which may increase during the vehicle's service life.

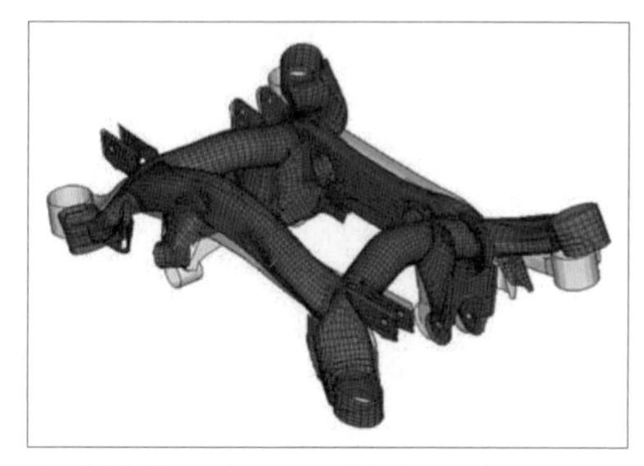

Fig. 7.4-3 Finite element model of rear subframe.

As vehicles must also be optimized with respect to cost and weight, it cannot be demanded that all components be arbitrarily rigid. Moreover, components and the entire suspension system must be cleverly designed so that elasticities do not have any negative effects or, better yet, so that they provide certain benefits [3].

In vehicle development, the first step is application of simulation methods to examine permissible elasticities. These are distributed between rubber mounts, ball joints, and suspension components. Next, individual components are designed. Using finite element methods (FEM), component strength is examined and actual elasticities determined. In suspension and overall vehicle simulations, component and system behavior is evaluated, taking into account all relevant elasticities. Once components have actually been manufactured, test rigs check the stiffness of suspension components and rubber mounts as well as the elastic behavior of an individual axle or the entire vehicle. These test results are compared with prior calculations, possibly to gain more precise insight into applied simulation methods. Finally, actual road testing determines whether the interplay of all effects, in particular kinematics and elastokinematics, actually provides the desired handling behavior under all conceivable conditions.

7.4.2.2 Elastomeric mounts

Elastomeric mounts fulfill a variety of suspension tasks. They are expected to:

- Permit angular motion, in place of ball joints
- Provide compliance in suspension pickup points, as well as damp and isolate vibrations originating in the wheel/tire system and engine/transmission units, to increase occupant comfort
- Act as construction elements for elastokinematic layout to achieve the desired toe behavior in response to external longitudinal, transverse, and vertical forces

These functional requirements are often mutually contradictory.

Steering function: Compared to a ball joint, an appropriately designed elastomeric joint provides more cost-effective joint functionality. Simultaneously, transmission of forces often introduces desired elasticity and damping. Another advantage of elastic joints is their lack of any breakaway torque, particularly in the presence of high radial preloading (e.g., wheel load). This advantage may, however, turn into a disadvantage for large angular displacement with simultaneously high radial joint stiffness. The resulting elastic restitution moment may in this case compromise the desired joint function, or high torsional loading may require a compromise in radial stiffness layout.

Vibration isolation and damping: For disturbances in the form of impacts, elastic mounts permit a more gradual rise in the transmitted force as well as short-term storage of introduced impact energy with associated absorption through elastomeric damping. From this, it is apparent that in addition to their elastic properties, elastomeric mounts also play a significant role in damping.

Dynamic spring rate has a significant effect on vibration isolation behavior of an elastic mount and is therefore an important evaluation and design parameter. Due to the viscoelastic nature of elastomers, spring rate and damping in response to dynamic loads are largely dependent on frequency, excitation amplitude, and temperature (Fig. 7.4-4). Increasing frequency will increase dynamic spring rate, loss angle, and damping, among other factors. This phenomenon is related to a progressive conditional change of the material toward glasslike behavior, as the effect of increased frequency may be compared to that of decreasing temperature. Similarly, increasing excitation amplitude is associated with increased stiffness and loss angle. Finally, the temperature of the elastomeric spring element affects spring rate as well as damping properties of the elastomeric mount, and it influences the service life.

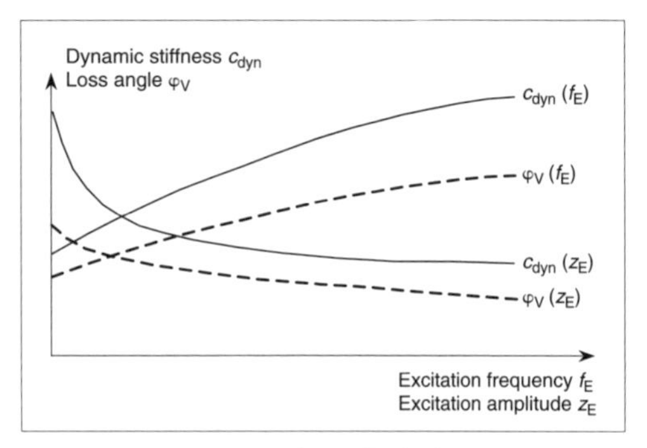

Fig. 7.4-4 Frequency and amplitude dependence on elastomer properties.

A requirement for good isolating action is a sufficiently large ratio of excitation frequency to system resonant frequency. If the excitation frequency is less than $\sqrt{2}$ times the system resonant frequency, isolation is no longer possible (Fig. 7.4-5). In practice, a ratio of mount stiffness to local body stiffness of at least 10:1 has proven effective.

Isolation of vibrating components: Isolating excitation by the wheel/tire system, engine/transmission, final drive, suspension, and so on from the vehicle body or occupants represents one of the most important contributions of elastomeric mounts. By means of soft mounting of the vibrating assembly, an attempt is made to keep forces generated within the mounting element as a function of vibration amplitude at low levels or to permit these to be transmitted to the body only in greatly reduced form. In terms of vibration engineering, this reduction of excitation forces is regarded as insulation or filtering. In this function of elastomeric mounts, damping is not desirable, as it leads to an increase in transmitted force. Because structure-borne sound propagates as waves in solid and liquid media, sound isolation by elastic mounts is accomplished by reflection (Fig. 7.4-6).

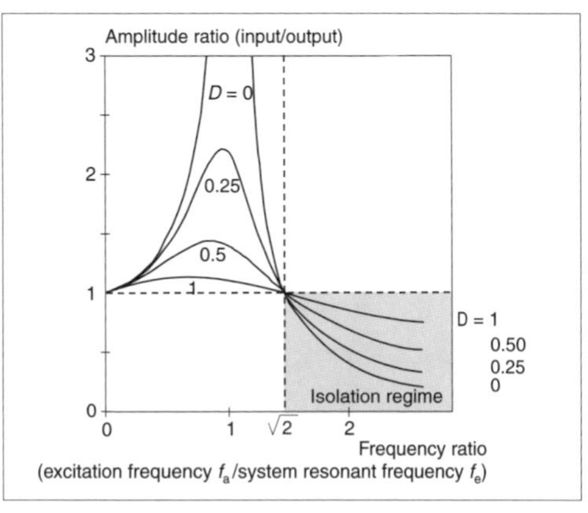

Fig. 7.4-5 Effect of damping and frequency ratio for single-mass oscillator with base excitation.

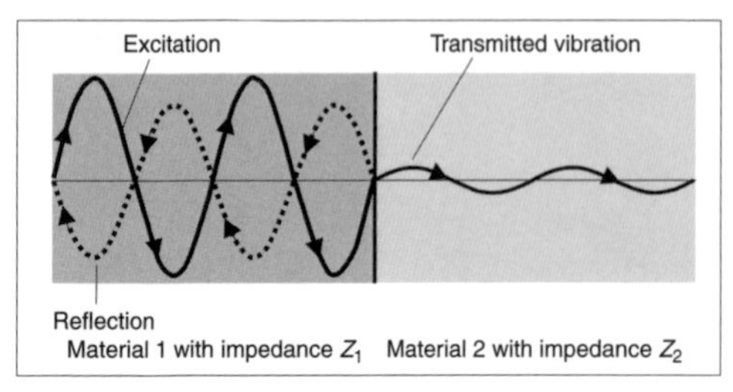

Fig. 7.4-6 Sound insulation by reflection at material interface.

If a wave meets the interface of two different materials with vibration impedances Z_1 and Z_2, it will be partially reflected—that is, its propagation will be hindered. The greater the difference in impedances, p, compared to the ratio Z_1/Z_2, the greater the reflection.

Elastomeric materials possess a low modulus of elasticity and low density; these result in highly effective structure-borne sound damping properties. As it is often impossible in operation to avoid passing through system resonances, it is very important to provide elastic mounts with sufficient damping to remove kinetic energy from the vibrating system by converting it to heat. In this way, vibration peaks resulting from brief passages through resonant zones are kept within limits. Loss angle [4] and damping factor describe the damping behavior of a vibratory system. The loss angle indicates by how many degrees the force resulting from the combination of elastic and damping components precedes elastic deformation; the damping factor is the dimensionless ratio of actual damping to critical damping [5].

Elastomers also have an unfortunate property in that as a result of damping, they stiffen with increasing frequency, so that their isolating effect is reduced at high frequencies. This is overlaid on the previously mentioned stiffness increase with decreased amplitude, so that here, too, the limits of purely elastomeric mounts become apparent. With increasing Shore hardness, material damping generally increases due to the growing proportion of filler material [6].

Configurations: Basically, the rubber of elastomeric mounts may be loaded in shear or tension/compression. Due to the constant volume of the material, greater stiffness may be realized in tension/compression than in shear. These two loading modes permit construction of directionally dependent mounts. In addition to purely elastomeric damping, the range of options may be greatly expanded by use of hydraulically damped mounts (Fig. 7.4-7). Vibration excitation changes the working chamber volume, thereby generating pressures above or below nominal. The fluid mass in the passage and the expansion wall stiffness form a vibratory system, which may be made to resonate at a tuned frequency, and, by means of inertia and flow losses, achieves a high degree of damping. Two grids, on either side of a movable membrane, ensure that hydraulic damping is circumvented for small excitations, thereby improving acoustics. In automobile construction, hydraulically damped mounts have proven themselves in longitudinal wheel motion and engine/transmission mounting applications [7, 8].

7.4.2.3 Effects of external forces

Vertical, longitudinal, and transverse forces acting on a wheel change the wheel orientation as defined by toe and camber. Those changes which are advantageous in certain driving situations will be covered briefly in this section. A more complete treatment can be found in Refs. [9] and [10]. The exact manner in which wheel orientation changes may be influenced by design depends on suspension configuration, location of suspension links, and their elasticities, and these will be illustrated by example. Greater depth is offered by Refs. [2] and [3].

Camber and toe changes over the course of *wheel vertical motion* are primarily a property of suspension kinematics. The significance for handling behavior is described in Section 7.4.1.1. Both parameters are changed by suspension component elasticity. Therefore, elasticity changes must be considered in suspension kinematic layout.

Fig. 7.4-8 shows the transverse link arrangement of a twin-link strut suspension. Longitudinal and transverse

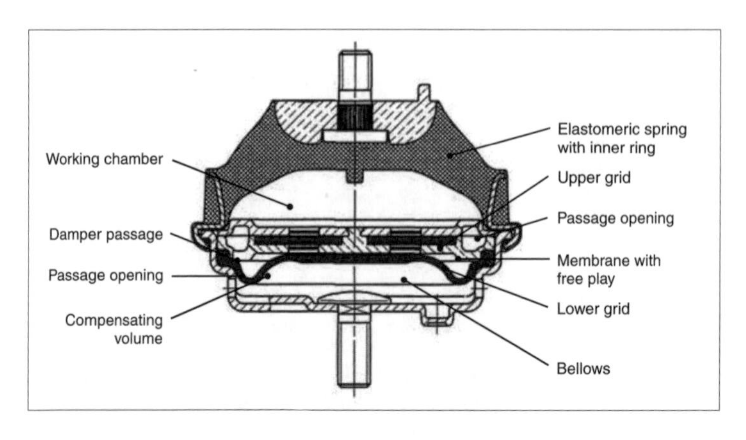

Fig. 7.4-7 Single chamber decoupled hydraulic mount.

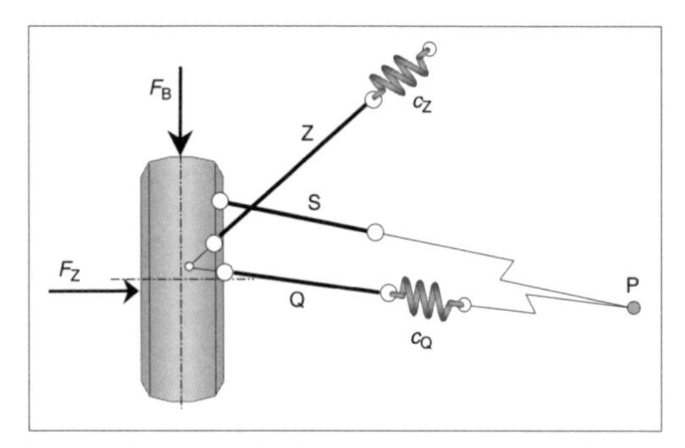

Fig. 7.4-8 Schematic plan view of transverse link mounts of a twin-link strut suspension.

wheel carrier movement is controlled by transverse link Q and semitrailing link Z; steering is by tie rod S. As the suspension strut must be angled inward, tire normal force results in tension on link Q and, because of elasticity c_Q, the wheel-side transverse link point moves outward. The upper strut mount moves inward; this increases negative camber. The direction of toe change depends on whether the steering system is attached ahead of or behind the wheel center. If, as in the illustration, it is ahead of center, the suspension will deflect to give increased toe-in because of the above-described change in the transverse link pickup point in response to tire normal force.

Longitudinal force elastokinematics present several different requirements. Experience has shown that in braking as well as in straight-line acceleration, handling stability is improved if both axles deflect to provide slight toe-in. However, transmitted acceleration forces are greater if tire slip angles are kept as small as possible. In braking on differential-coefficient (split μ) surfaces, differing left and right braking forces produce a yaw moment which attempts to turn the vehicle toward the side with greater friction coefficient. If only the front axle goes into toe-in under braking, the steering angle change generates an opposing yaw moment which stabilizes the vehicle. The same is true if the rear axle shifts into toe-out.

On the other hand, conditions are different if the throttle is closed in a turn (weight transfer) or if braking occurs in a turn. In both maneuvers, wheel loads change first: weight transfers to the front and away from the rear axle. With higher wheel loads and, initially, unaltered slip angles, the front tires generate greater lateral force while the rear axle, with less load, generates less, so that in summation, a yaw moment is generated which tends to turn the vehicle into the turn. This yaw moment is reduced if the front suspension is designed to go into toe-out under braking while the rear axle goes into toe-in. The left and right wheels change their steering angles uniformly, but because the wheel load on the outside wheels is greater, their steering angle change produces greater effect and generates an outward-turning yaw moment.

In general, elastokinematics of both axles are important for vehicle handling. Under braking, front axle elastokinematic steering angle change is greater than that of the rear axle because braking force at the front is considerably greater than at the rear. Furthermore, front steering angles have a greater effect because of forward weight transfer. In acceleration, longitudinal elastokinematics can only be utilized at the driven axle.

These configurations also demonstrate that there is no single optimum elastokinematic layout. Moreover, tuning is always an axle-specific synthesis of the demands imposed by various driving conditions. Tuning must also take into consideration overall vehicle layout objectives and must always be regarded in conjunction with other parameters such as tire behavior, springing, antiroll (stabilizer) bars, and so forth. In reality, vehicles are usually designed so that during braking, the front axle goes into slight toe-out and the rear axle into slight toe-in. If the suspension concept permits it, driven rear axles are also laid out to go into toe-in under acceleration.

In the twin-link strut suspension shown in Fig. 7.4-8, longitudinal compliance is realized by means of the tension rod mount, with stiffness c_Z. For a freely rolling wheel, c_Z is small. With the appearance of a braking force F_B, the wheel carrier first rotates about momentary axis P: that is, in the arrangement shown, the axle goes into slight toe-out. With increasing braking force, the tension rod mount hits its stop (c_Z becomes large). Then the transverse link mount with its considerably greater stiffness, c_Q, comes into play. Because the transverse link is in compression under braking and the tie rod force is comparatively small, the suspension shown goes into greater toe-out; see Fig. 7.4-9.

An important aspect of elastokinematics is the change in kinematic properties as a result of elastic deformation. Fig. 7.4-10 schematically shows a double A-arm suspension in side view. The kingpin axis S represents the line joining the two wheel-side A-arm pivot points for zero longitudinal force (arms shown as dotted lines); the associated caster offset n is also indicated. Longitu-

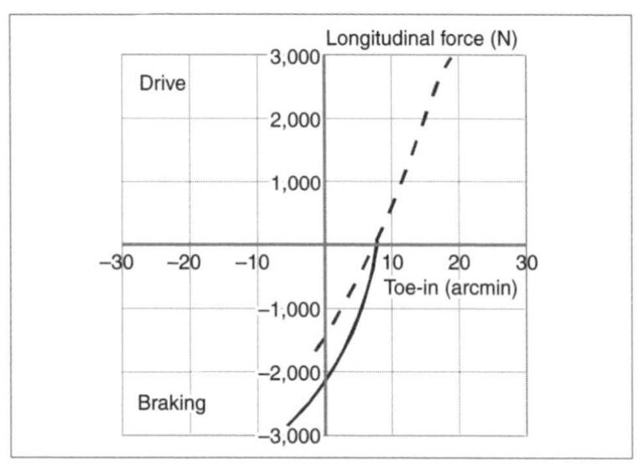

Fig. 7.4-9 Typical toe-in response to drive and braking forces for strut suspension.

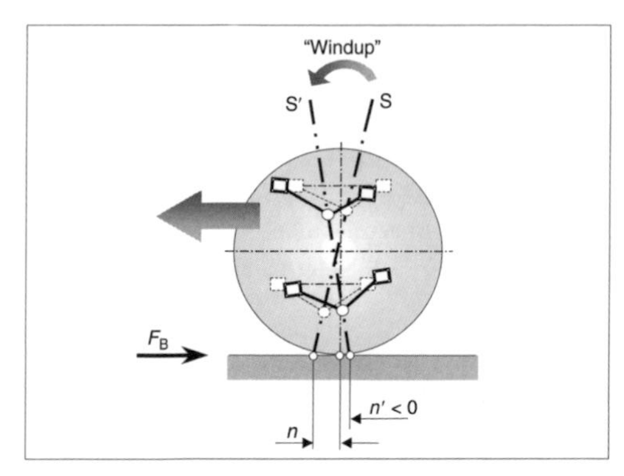

Fig. 7.4-10 Double A-arm suspension principle under braking (side view).

dinal compliance is also achieved by body-side rubber mounts. Because braking torque must be taken up by a pair of reacting forces at the A-arms, the lower arm moves rearward while the upper arm moves forward. Consequently, the wheel carrier "winds up": that is, it rotates about the vehicle transverse axis. The kingpin axis S rotates to S', the caster offset n becomes the negative caster offset n'.

An additional effect in suspension windup is that the tie rod inclination changes, which alters toe-in response to vertical wheel motion. This means that during braking, toe change over wheel movement is skewed compared to that without braking (Fig. 7.4-11).

To limit the wheel carrier windup effect, it is important to provide a large "baseline" for the transverse arms. For a double A-arm suspension, it is therefore advantageous to mount the upper arm above the wheel. If the upper arm must be located within the wheel, longitudinal compliance, an important contribution to comfort, may be realized by an elastically mounted subframe, which permits stiffer transverse arms. It is also desirable to react against braking torque by means of an

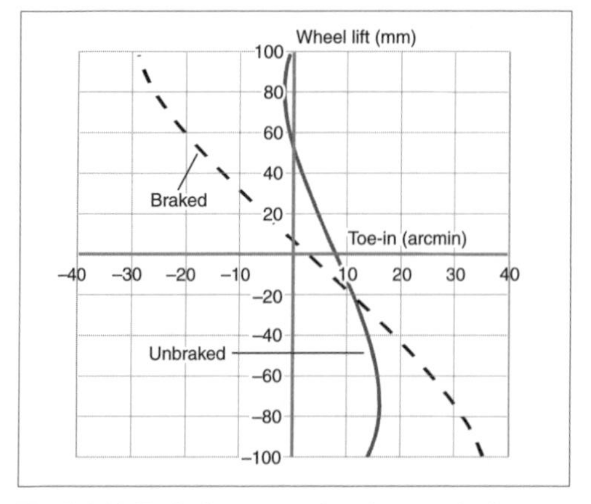

Fig. 7.4-11 Typical response of toe-in over wheel movement for double A-arm suspension ("ride steer" curves).

integral link attached to the lower arm, so that no longitudinal forces are applied to the upper transverse link. This is achieved in the case of integral link rear suspension (see Fig. 7.4-18).

Lateral force also changes toe and camber due to suspension elasticity. As shown in Section 7.4.1.1, independent suspension has a decrease in outside wheel camber in response to body roll angle, which can be only partially compensated for by suspension kinematics. An inward-directed force can only increase camber loss in all currently implemented suspensions. In view of this, the suspension shown in Fig. 7.4-8 requires a stiffly mounted transverse link.

Steering angle change in response to lateral force affects the vehicle's steering behavior. Also, in general, lateral force elastokinematics should produce slightly understeering behavior (see Section 7.1). It is therefore necessary that the front suspension go toward toe-out and the rear toward toe-in in response to an inward-directed lateral force. Wheel forces generated by lateral acceleration will then produce a steering angle which diminishes lateral acceleration. The rear axle steering angle simultaneously reduces the vehicle slip angle (body yaw angle).

Other requirements arise from consideration of conditions such as crosswind and road camber; the vehicle should not exhibit undesirable response to these external forces. Above all, achieving crosswind insensitivity by means of elastokinematics is an attractive option, if an inward-directed force results in increased front axle toe-in.

In general, both axles, insofar as is possible, are designed so that they understeer in response to lateral forces. The interplay of both axles is decisive in determining handling behavior. Definitive tuning is a response to handling targets and must be regarded in conjunction with many other tuning parameters such as springing, stabilization, aerodynamics, etc., but also with kinematic and elastokinematic effects arising from wheel travel and longitudinal forces.

In the suspension shown in Fig. 7.4-8, the transverse link Q is loaded in compression by a lateral force F_S. If the steering is mounted at the front, the suspension will move toward toe-out due to elasticity c_Q (Fig. 7.4-12). If the steering is mounted at the rear, it goes into toe-in. In this case, very high transverse link stiffness is required.

A somewhat different picture appears in the case of a strut suspension if the steering tie rod is connected behind and well above the transverse link. An inward-directed lateral force produces a bending moment in the strut. Bending of the strut increases the (positive) camber of the outside wheel, which results in the top of the wheel carrier being pushed outward above the transverse link. As the wheel carrier is held by the tie rod, the result is a tendency toward toe-out.

A further important aspect for steering behavior in response to lateral force is torsional elasticity of the steering column as part of overall steering elasticity.

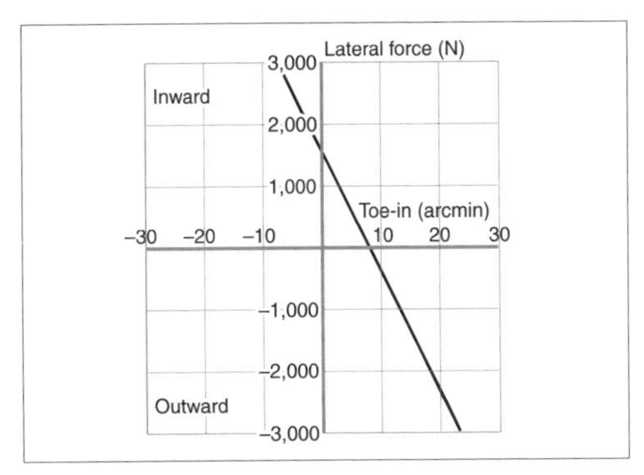

Fig. 7.4-12 Typical toe-in response to lateral force at constant wheel travel of a strut suspension with front-mounted steering arms.

Because of ever-present front suspension caster offset and pneumatic trail, lateral force always produces a steering column torque. This results in twisting of the steering column, which results in an understeering steering angle at the wheels, even with a rigidly held steering wheel.

At the rear axle, it is advantageous to have the wheels go into toe-in in response to an inward-directed side force. In the case of suspension systems which emphasize cost and weight advantages, such as trailing arm, semitrailing arm, and twist beam (or H-arm) suspensions, component elasticities tend to act opposite to this requirement. The steering behavior of these components may be partially compensated for by means of angled rubber mounts between body and rear suspension or twist beam. For more complex double A-arm, multilink, or integral link suspensions, side force steering behavior can be set at will; see Section 7.4-3 as well as Refs. [3] and [11].

The considerations of this chapter show how vehicle handling can be affected by means of elasticity. This is always achieved by wheel forces causing changes in steering angle and camber. The forces themselves are the unavoidable result of vehicle operation. In laying out conventional suspensions, the entire spectrum of possible forces must be considered, which always results in compromises in design synthesis. If, additionally, the attempts to positively influence a vehicle design by means of control systems are successful, the emphasis of elastokinematic tuning may be modified.

References

[1] ISO 8855.

[2] Bantle, M., and H.-H. Braess. "Fahrwerksauslegung und Fahrverhalten des Porsche 928." *ATZ* 79, No. 9, 1977, pp. 369–378.

[3] Matschinsky, W. *Radführungen der Straßenfahrzeuge*, second edition, Springer Verlag, 1998; English edition *Road Vehicle Suspensions*, Professional Engineering Publishing Ltd., 2000.

[4] Hamaekers, A. "Entkoppelte Hydrolager als Lösung des Zielkonfliktes bei der Auslegung von Motorlagern." *Automobilindustrie* 5/85.

[5] Mitschke, M. *Dynamik der Kraftfahrzeuge, Band B: Schwingungen*, third edition, Springer Verlag, 1997.

[6] Hinsch, P. "Einige Untersuchungen zum Messen dynamischer Moduln von Elastomeren." *Kautschuk + Gummi Kunststoffe* 42/89, No. 9, pp. 752–756.

[7] Holzemer, K. "Theorie der Gummilager mit hydraulischer Dämpfung." *ATZ* 87/1985, No. 10, pp. 545-551.

[8] Spurk, J.H., and R. Andrae. "Theorie des Hydrolagers." *Automobilindustrie* 5/85, pp. 553–560.

[9] Zomotor, A. *Fahrwerktechnik: Fahrverhalten*, second edition, Vogel Buchverlag, 1991.

[10] Mitschke, M. *Dynamik der Kraftfahrzeuge, Band C: Fahrverhalten*, second edition, Springer Verlag, 1990.

[11] Matschinsky, W., U. Pfundmeier, and R. Salaiman. "Die Integral-Hinterachse für das Coupe 850i." *ATZ* 92, No. 10, 1990, pp. 554–561.

[12] Matschinsky, W., et al. "Komfortable Sicherheit—Das Fahrwerk der neuen 7er-Baureihe von BMW." *ATZ* 96, No. 11, 1994, pp. 646–651.

[13] Reimpell, J. *Fahrwerktechnik: Grundlagen*, third edition, Vogel Buchverlag, 1995.

7.4.3 Suspensions

Suspensions, composed of wheel carrier (also called hub, spindle, steering knuckle, or suspension upright), wheel bearings, links (including joints and rubber mounts), springs, and dampers (shock absorbers), provide wheel location and motion and reaction against external forces. According to their degrees of freedom, suspensions can be classified as rigid-axle suspension, independent suspension, or compound suspension [1, 2]. Beam axles can compress their springs on one or both sides, therefore possessing two degrees of freedom. In the case of independent suspension, symmetrical and antisymmetrical spring deflections are identical and thus represent only a single degree of freedom. A compound suspension can exhibit various degrees of lateral coupling, depending on design, and therefore lies somewhere between beam axles and independent suspension in terms of properties. For steered axles, an additional degree of freedom is added: rotation of the wheel about a kingpin located near its vertical axis.

Selection of suspension concepts is influenced by many boundary conditions, such as application spectrum (speed range, axle loads, etc.), available installation space, driveline configuration, cost, weight, and so on (See Table 7.4-1). In general, possibilities for optimizing kinematic design parameters increase with the number of links and joints. For example, the front suspension steering axis may be located by the wheel-side ball joints of two A-arms. Location of the joints is not completely free, however, because of space requirements for brakes and wheel clearance required for steering. Therefore compromises must be made with respect to kingpin inclination, caster offset, and kingpin offset, (steering radius). If both actual ball joints are replaced by the virtual centers of rotation of two radius arms, the kingpin axis may be located independently of space restrictions. This virtual kingpin axis may be designed to change deliberately in response to suspension travel and steering lock, which opens additional avenues of suspension layout freedom.

Table 7.4-1 Several suspension concept selection criteria

	Beam axles	Independent suspension			Compound axles
		Planar	Spherical	Spatial	
Kinematic layout potential	–	0	+	++	0
Longitudinal compliance	–	0	+	++	–
Cost	+	0	–	– –	+
Space	– –	0	0	+	–
Weight	–	0	+	+	0
Insensitivity (wheel loads, off-road capability, tolerances, etc.)	++	0	–	–	0

++, appreciably better; +, better; 0, baseline: planar independent suspension; –, negative tendency; – –, unfavorable.

Space considerations also play a part in establishing tire and rim dimensions. With a view to lightweight construction as well as low unsprung and rotating masses, the lightest—that is, smallest and narrowest—wheel/tire combinations would be desirable; to reduce fuel consumption through low rolling resistance, narrow large-diameter tires would be ideal. But for appearance and handling reasons, large rims with wide tires are chosen, which also offer space for generously dimensioned brake discs and calipers. This advantage is offset by greater space requirements in the wheel well, which in turn affects the overall vehicle package (steering lock, turning circle, trunk volume, etc.) and generally results in greater weight. The weight problem is countered by application of innovative materials and processes: for example, stamped aluminum wheels replacing steel wheels, lightweight forged alloy wheels (increasingly being used as standard equipment), and light alloy cast wheels using aluminum foam cores, magnesium, and fiber-reinforced plastics. Further weight savings potential as well as customer-relevant packaging advantages can be found by eliminating the full-size spare tire, which can be justified by the increasing rarity of flat tires. In its place, a narrow high-pressure collapsible spare is often provided, which ideally has the same outside diameter as the standard equipment tire, especially in order to avoid compromising control system functions. In the case of very restricted spaces—for example, two-seat sports cars—one may find no more than an emergency kit containing sealant and a compressor or inflator bottle in place of a spare tire. A different, possibly pioneering approach is taken by special wheel/tire systems (Section 7.3) with run-flat properties, which can still cover considerable distances even after total loss of inflation pressure. Already in production are "self-supporting tires" (EMT, or "Extended Mobility Technology") with reinforced sidewalls, mounted on conventional J-rims with extended humps. New systems contain an added support ring between tire and rim, which assures run-flat capability; Michelin's PAX (tire with unique bead lock) and the Conti CWS (Clamped Wheel System) are examples.

In cases of misuse or accidents, suspensions must satisfy special requirements: in the event a component is overloaded, it should undergo significant deformation before ultimate failure. This is intended first to call attention to overload-induced damage before failure and prevent further use. Second, it addresses product liability concerns: deformation provides evidence that prior to failure, the component was subjected to misuse and overloading through an accident-like situation. In this way, causality for accident and component failure is established: First a collision, then failure, and not the other way around. The materials used for wheel location elements must therefore exhibit yield elongation of at least 6%.

7.4.3.1 Beam axles

Because of their robust nature, beam axles are often found as the front and rear axles of off-road and commercial vehicles (Fig. 7.4-13). The axle beam rigidly joins the left and right wheels and, in the case of driven ("live") axles, contains the final drive. Drive and braking torques react against the axle housing; therefore, braking and traction force support angles are identical. Location is accomplished in the simplest case by leaf springs, otherwise by trailing arms or A-arms. Lateral location is usually by means of a so-called Panhard rod,

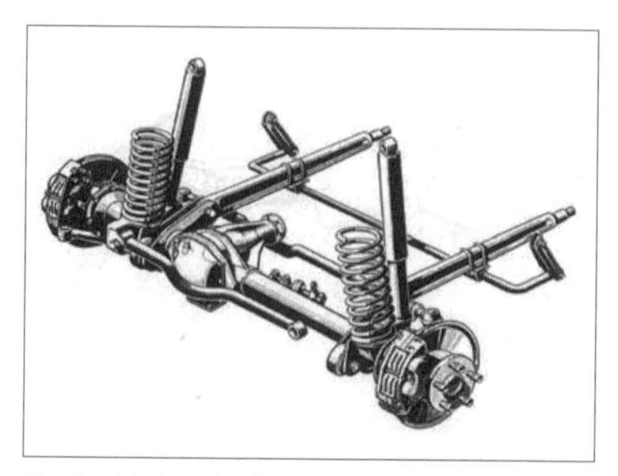

Fig. 7.4-13 Steerable live axle (Mercedes-Benz G-Model).

which is made as long as possible to minimize lateral shake.

The primary advantage of beam axles is their constant track and camber in both roll and parallel suspension movement, which has a beneficial effect on tire load-carrying ability. Furthermore, it is possible to achieve good antidive and antisquat properties, as well as a high roll center, without a track change, reducing body roll and pitch motions. Driven beam axles ("live axles") do not require halfshaft constant velocity joints; undriven axles can be made at relatively low cost. An important advantage for off-road vehicles is constant ground clearance for the axle housing during suspension compression (jounce).

Disadvantages of beam axles include the high unsprung mass of beam axles and the large space requirement for the axle beam in jounce. Because of undesirable elastokinematic steering angles given one-sided longitudinal forces, only limited longitudinal compliance of rubber mounts in locating links is possible. In one-sided jounce, the opposite wheel is subjected to wheel load and camber changes; furthermore, there is a lateral displacement due to finite Panhard rod length. Finally, the transmissible power for a given axle load is limited because longitudinal driveshaft torque reaction leads to differing wheel loads, causing greater slip at the unloaded wheel. This effect may be avoided through use of a limited-slip differential. In driven beam axles with independently body-mounted final drive units (De Dion suspension), the final drive itself remains free of drive moments and acceleration does not induce wheel load differences.

7.4.3.2 Independent suspension

Independent suspensions can be divided according to their motion geometry; wheel and wheel carrier may swing about a fixed axis, thereby describing motion in a single plane perpendicular to the axis of rotation (planar suspension). If suspension motion causes the momentary center of rotation to swing about a fixed point, the wheel carrier pickup points describe spherical arcs about this central point. This is termed spherical suspension. In the general case of spatial motion, rotation about the momentary axis is overlain by axial shift, and the momentary axis becomes an instantaneous screw momentary helical axis. The suspension's kinematic potential increases with the number of independently selectable parameters. A momentary axis is defined by four parameters (point and direction); therefore, not all five suspension layout parameters (toe and camber changes, antidive, antisquat, and roll center height) are freely selectable. In the case of a spatial suspension with an instantaneous screw axis, an independent fifth parameter in the form of axial shift is added. (For undriven axles, traction force support angle is of course of no importance, so that four selectable parameters are sufficient).

Fig. 7.4-14 Semitrailing arm suspension (BMW Z3).

Planar independent suspension

This class includes all suspensions with a fixed axis of rotation, such as trailing arm, semitrailing arm, and double lateral link suspensions in which the links' axes of rotation are parallel. As an example, a semitrailing rear suspension will be described in greater detail (Fig. 7.4-14).

Camber change as a function of spring travel is selectable to at least partially compensate for camber loss (relative to the road) caused by body roll in cornering. Brake antidive is good, due to the high location of the longitudinal instantaneous center. Driven semitrailing arm suspensions are typically mounted on subframes, which also carry the final drive unit. Good longitudinal compliance and acoustic decoupling is achievable with large-volume rubber mounts.

Because of limited design freedom, toe changes in jounce are unavoidable, as are small changes in roll center height, which cause a jacking effect in cornering. Link bushing and component elasticities in reaction to lateral and longitudinal forces result in undesirable steering angles, promoting oversteer. These can only be compensated for through increased complexity in rear axle subframe mounts. Mounts with different longitudinal and transverse spring rates are installed and oriented to provide an elastic center behind the lateral force action point, rotating the entire axle around the elastic center and moving the outside wheel in the direction of toe-in.

Spherical independent suspension

The central point about which the instantaneous center of rotation and all wheel carrier pickup points swing is often configured as the body-side pickup point of a longitudinal link. If wheel carrier and longitudinal link are a single integral component, two additional transverse links are needed for kinematic determinacy (for reasons of constant toe-in, the relatively long, central-link rear suspension of the BMW 3-Series, Fig. 7.4-15) or three shorter links with transverse freedom of motion for the longitudinal link are needed (Ford Focus control blade rear suspension). Spherical independent suspension may also take the form of designs with trapezoidal link

Fig. 7.4-15 Central link rear suspension (BMW 3-Series).

and camber link if the wheel-side and body-side axes of the trapezoidal link intersect at a common point (Audi [3], Jaguar [4], Porsche 928 "Weissach" rear suspension [5]).

In central-link suspensions, the camber curve may be chosen similarly to that in a semitrailing link suspension; the added design freedom provided by a moving instantaneous axis of rotation permits achieving constant toe-in in jounce. Antisquat and antidive are good; the longitudinal link's large-volume mount permits adequate longitudinal compliance. Careful tuning of axial and radial spring rates and mount orientation in plan view permit tuning for stabilizing toe changes in response to longitudinal and lateral forces. On the other hand, roll center height changes, which produce jacking effects similar to those of semitrailing suspensions, are not freely selectable. Furthermore, the two long lateral links need a relatively large amount of space, while the alternative configuration using three short links increases component complexity.

Spatial independent suspension

In view of their need for steering motions, front suspensions are generally spatial mechanisms. Strut-type suspensions (Fig. 7.4-16), in which the damper (shock absorber) is solidly attached to the wheel carrier and assumes a wheel location function, are very common. Spring and damper are principally coaxial and form a single unit, the suspension strut. Shock absorber friction ("stiction") arising from lateral forces may be reduced by angling the spring relative to the damper axis; this is sometimes termed transverse force compensation.

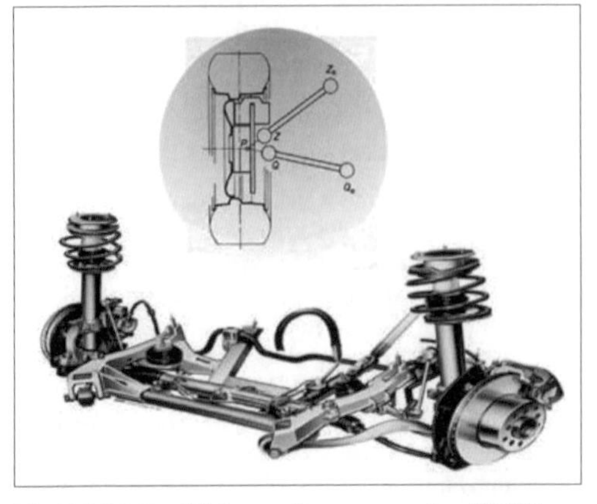

Fig. 7.4-16 Dual link strut front suspension (BMW 5-Series).

Because of its space advantages and attractively low number of components, strut-type (also known as McPherson) suspension has become established as the most popular configuration for front-wheel-drive applications. Kinematically, the strut represents a linear guide which one can imagine as equivalent to an infinitely long A-arm perpendicular to the shock strut and passing through the upper strut mount. Transverse and longitudinal instantaneous centers may then be laid out as in double A-arm suspensions. If the lower A-arm (or "sickle" link) is resolved into two individual lateral links, the result is an instantaneous center which establishes the kingpin axis. This may be used to circumvent spatial restrictions (e.g., brake disc interference).

If the upper A-arm is resolved into two individual lateral links, the result is a five-link suspension as a generalized form of spatial independent suspension. The kingpin axis is now a "virtual" axis, determined by two instantaneous centers. At present, however, such complex solutions are only used on driven axles.

Several different versions of driven spatial rear suspensions are currently on the market: five-link suspensions (DaimlerChrysler multilink suspension, Fig. 7.4-17); three individual links and one A-arm (Porsche 911); and trapezoidal arm with two transverse links and swing link (Audi, BMW 5-Series/7-Series integral suspension, Fig. 7.4-18).

By means of independently selectable parameters, spatial suspensions permit achieving optimum kinematic properties: favorable toe-in and camber curves, good antidive and antisquat, and avoidance of jacking effects through suitable roll center height changes. By virtue of its many rubber mounts, the elastokinematic layout offers the possibility of achieving stabilizing steering angles for all external loads, while simultaneously providing good longitudinal compliance. Because the rear suspension subframe, to which the links are attached, is customarily joined to the body via rubber mounts, final drive noise may be isolated from the vehicle interior. Depending on layout, all of the suspension's longitudinal compliance may be achieved either by the rear subframe mounts or, in part, within the rubber link bushings. The functional advantages are offset by the high degree of complexity associated with the numerous links, joints, rubber mounts, and intricate rear subframe.

7.4.3.3 Compound suspensions

In general, suspension systems exhibit compound effects if one-sided spring compression also affects the opposite wheel or if parallel spring conditions differ from opposing or single-wheel spring conditions. This

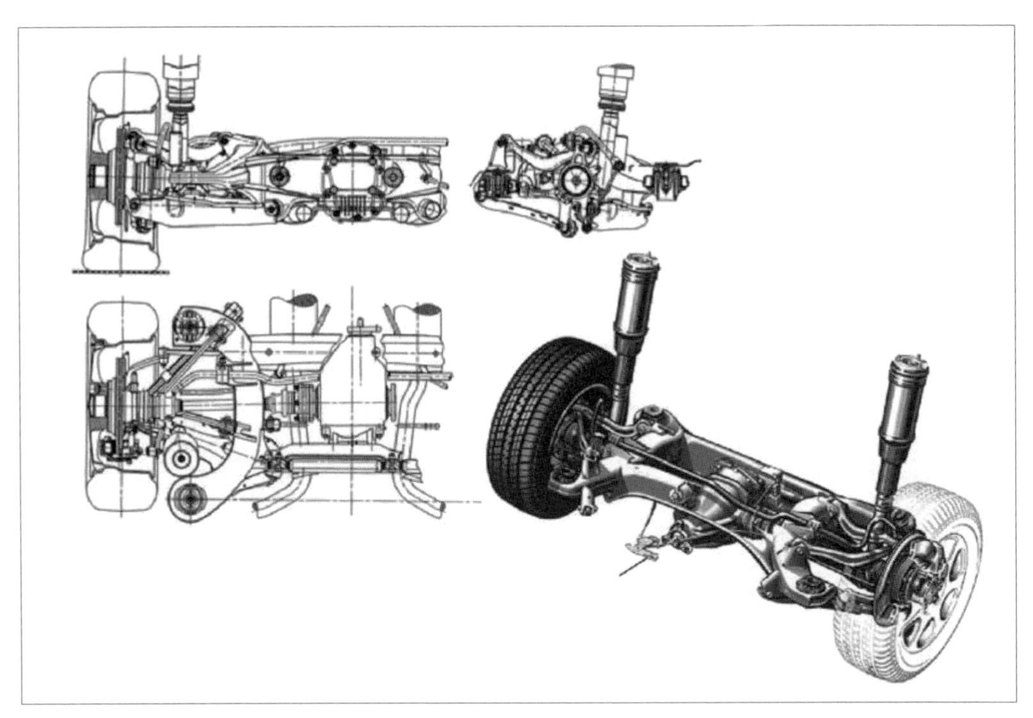

Fig. 7.4-17 Multilink rear suspension (Mercedes E-Class).

was also the case in single-link swing axles, which are no longer used in modern designs. In a more restricted sense, twist beam or H-arm suspension, currently typical practice for rear axles of front-drive passenger cars (Fig. 7.4-19), usually consists of two trailing arms, stiff in torsion and bending, which locate the wheels, and a torsionally "soft" box section connecting both sides.

Depending on location of this box section (closer to the trailing arm bushings or closer to the wheel centers), its kinematic properties may more closely resemble those of a trailing-link suspension or a rigid axle. If the box section is located roughly midway along the trailing arms, jounce on one side produces an instantaneous axis of rotation passing through the trailing arm bushing and the box section shear center, which, for reasons of symmetry, remains stationary. Spatially, this instantaneous axis is located similarly to a semitrailing arm suspension and generates increased toe-in and increased negative camber at the compressed wheel. In the case of simultaneous jounce on both sides, the box section is not loaded in torsion; toe and camber of both wheels remain nearly constant.

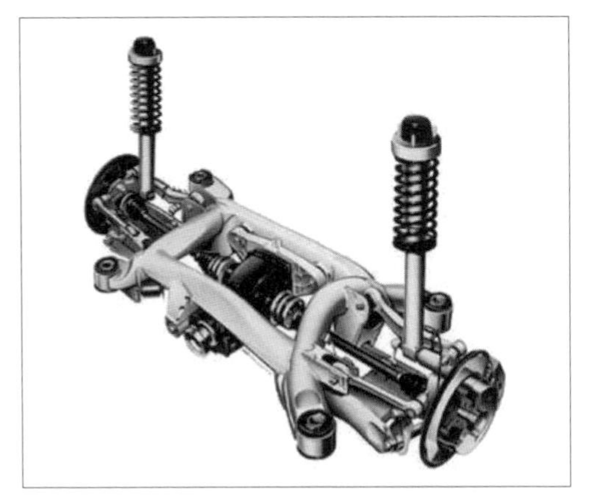

Fig. 7.4-18 Integral rear suspension (BMW 5-Series).

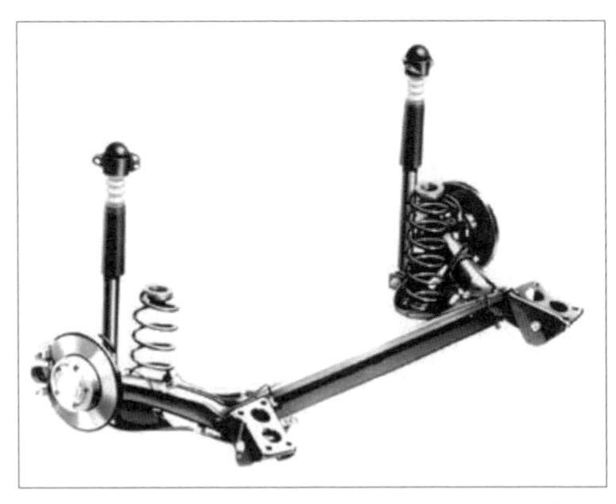

Fig. 7.4-19 Twist beam (H-arm) rear suspension (VW Golf IV).

Along with its economical configuration with only two bearing points, this constancy of toe and camber is the chief advantage of twist beam suspensions. Disadvantages include twist beam space requirements in jounce as well as a tendency toward elastokinematic oversteer in response to longitudinal and lateral forces. However, toe angles may be corrected by means of angled rubber bushings with differing axial and radial spring rates. As in the case of beam axles, longitudinal compliance is limited due to steering effects of one-sided longitudinal forces. Interference between the box section and driveshaft makes twist beam suspensions unsuitable for driven axles.

References

[1] Matschinsky, W. *Radführungen der Straßenfahrzeuge*, second edition, Springer Verlag, 1998; English edition *Road Vehicle Suspensions*, Professional Engineering Publishing Ltd., 2000.
[2] Reimpell, J. *Fahrverktechnik: Grundlagen*. Third edition. Vogel Buchverlag, 1995.
[3] Leitermann, W., et al. "Der neue Audi 200 quattro." *ATZ* 86, 10, 1984, pp. 417–421.
[4] Cartwright, A.J. "The development of a high comfort, high stability rear suspension." IMechE 1986, *Proc. Instn. Mech. Engrs.* Vol. 200, No. D5, pp. 53–60.
[5] Bantle, M., and H.-H. Braess. "Fahrwerksauslegung und Fahrverhalten des Porsche 928." *ATZ* 79, No. 9, 1977, pp. 369–378.

7.4.4 Springing, damping, and stabilizers

"Springing," as an all-encompassing concept for the interplay of load-carrying springs, stabilizers and dampers, performs a series of tasks which are extremely important in determining overall suspension behavior. Springing is expected to:

- Isolate the vehicle body superstructure and thereby, above all, its occupants from unpleasant lift, pitch, and roll vibrations and shocks, thus contributing to mechanical vibration comfort.
- Provide the most consistent possible grip for the tires as a prerequisite for vehicle path stability and for the force transfer between tires and road surface required for propulsion and braking. This is an important aspect of handling safety.
- Make a positive contribution to handling behavior by means of balanced force and moment distribution between the wheels of a single axle, as well as between both axles. This also touches on aspects of ride comfort as well as handling safety.

The layout of individual components is in itself not especially difficult; the actual art lies in balanced tuning of all functions to one another to meet handling and comfort goals. The simplified process shown in Fig. 7.4-20 will be explained in greater detail in the following example. References to sports car layout procedures which deviate from this process will be provided at the end.

The symbols used are derived from the model of Fig. 7.4-21. Tire choice (Section 7.3) establishes c_1. Once a suspension concept is established, m_1 is known, and the available suspension installation space is defined. Experience has shown that for most passenger cars, the coupling mass m_K may be approximately set to zero, representing decoupling of jounce and pitch motion.

7.4.4.1 Load-carrying springs

As their name indicates, springs c_2 primarily fulfill a load-carrying function, namely that of the sprung mass m_2. Because an actual vehicle is operated in laden and unladen conditions, in individual cases m_2 may vary considerably between $m_{2,\text{empty}}$ and $m_{2,\text{laden}}$. It is desirable to keep vibration comfort as constant as possible regardless of load condition, which imposes requirements for adequate spring travel, taking into consideration not only human sensitivity to vibration, but also low (0.7–2 Hz) and as nearly constant spring resonant frequencies as possible [1, 2, 3].

Fig. 7.4-22 shows that body acceleration is only significantly affected by spring c_2 in the vicinity of the superstructure resonant frequency (~1 Hz). With softer springs, these spikes diminish considerably and shift in keeping with decreasing resonant frequencies. In the lower frequency range, this also applies to wheel load oscillations; at higher excitation frequencies, these may be amplified by soft springs.

For a wheel in the half-vehicle model of Fig. 7.4-21, some simplified relationships may be derived:

The superstructure resonant frequency in Hz

$$f_2 = \frac{1}{2\pi}\sqrt{\frac{c_2}{m_2}} \tag{1}$$

or, in terms of the "vibration number" in min^{-1} favored by designers,

$$n_2 = 60 f_2 \tag{2}$$

Static suspension compression due to m_2:

$$z_{21,0} = z_{2,0} - z_{1,0} = \frac{m_2 g}{c_2} \tag{3}$$

and, similarly to Eq. (3),

$$z_{21,0} = \frac{g}{4\pi^2 f_2^2} \tag{4}$$

Eq. (4) quantifies the immediate relationship between resonant frequency and static compression. It is easily determined that the above stated range of favorable perceived vibration requires static compression of up to 500 mm. For package reasons, however, passenger cars normally have only about 200 mm total spring travel; installation of a level control system (Section 7.4.4.4) to compensate for static spring compression must be considered for values of $f_2 < 1$ Hz. Eq. (1) illustrates an additional synthesis boundary condition: To achieve a constant resonant frequency throughout the vehicle

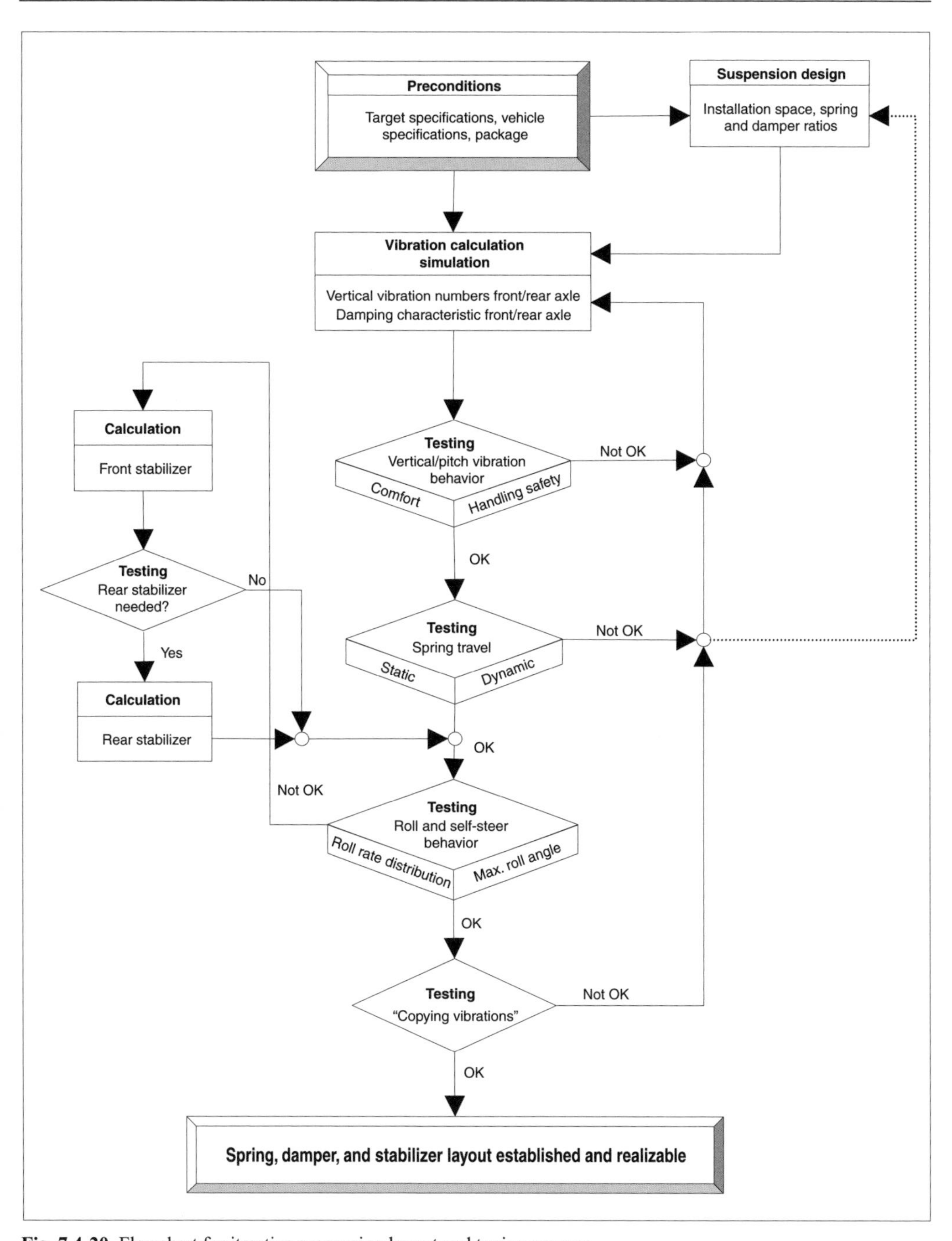

Fig. 7.4-20 Flowchart for iterative suspension layout and tuning process.

load range, progressive springs are needed. While gas springs (see Section 7.4.4.4 and [4]) approach this requirement in principle, steel springs in their most common (straight cylindrical helical) form provide linear response. To smooth bumps, such as those encountered in driving over undulating pavement at an appropriate vehicle speed, auxiliary springs (usually in the form of polyurethane foam components) are installed in such a way that the resulting composite spring characteristic curve provides an overall progressive response (Fig.

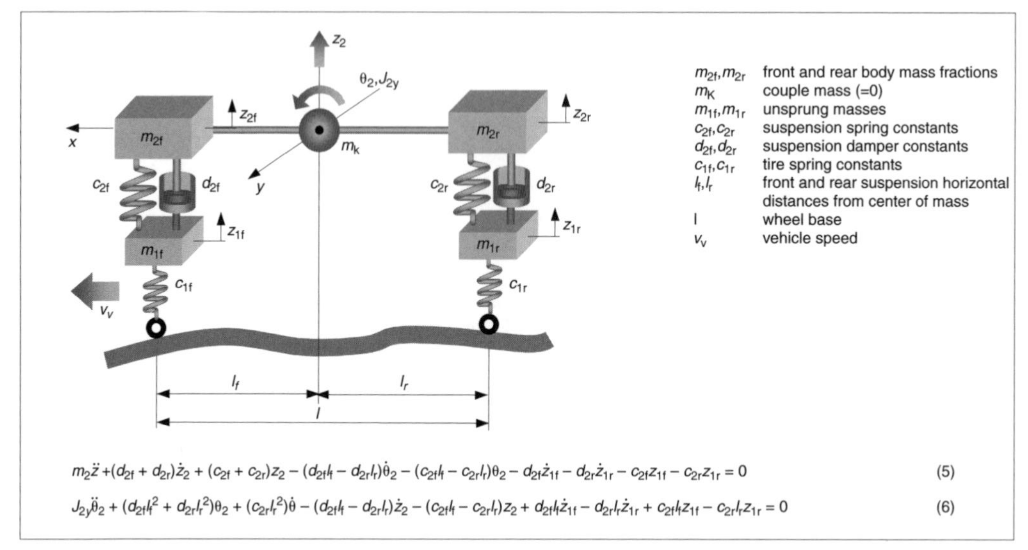

$$m_2\ddot{z} + (d_{2f} + d_{2r})\dot{z}_2 + (c_{2f} + c_{2r})z_2 - (d_{2f}l_f - d_{2r}l_r)\dot{\theta}_2 - (c_{2f}l_f - c_{2r}l_r)\theta_2 - d_{2f}\dot{z}_{1f} - d_{2r}\dot{z}_{1r} - c_{2f}z_{1f} - c_{2r}z_{1r} = 0 \quad (5)$$

$$J_{2y}\ddot{\theta}_2 + (d_{2f}l_f^2 + d_{2r}l_r^2)\theta_2 + (c_{2r}l_r^2)\dot{\theta} - (d_{2f}l_f - d_{2r}l_r)\dot{z}_2 - (c_{2f}l_f - c_{2r}l_r)z_2 + d_{2f}l_f\dot{z}_{1f} - d_{2r}l_r\dot{z}_{1r} + c_{2f}l_f z_{1f} - c_{2r}l_r z_{1r} = 0 \quad (6)$$

Fig. 7.4-21 Half vehicle model for vertical and pitch motion as well as corresponding vibration equations (after [1]).

7.4-23). The illustration reminds us that the near-linear spring component contribution by suspension link rubber bushings must be considered in the layout. Further fine-tuning of the spring characteristic is obtained by suitable travel-dependent spring ratios or for example installation of variable-pitch springs [5]. Progression also helps alternate-side spring compression, as it has a degressive effect on roll angle increase with vehicle load; simultaneously, it generates a jacking effect.

This then establishes an initial spring variant for the strongly load-dependent rear suspension of the case at hand.

Front suspension layout must implement the desired interplay of front and rear suspension. Ref. [1] clearly shows that to give vehicle occupants an impression of comfort (usually described by the effective values of body acceleration measured at the seat, better yet including human perceptions along with seat *K* coefficient [3]), when pitch vibrations are considered, choice of a front/rear suspension tuning relationship depends on seat location. (In this case, pitch vibrations are treated as road irregularities; for antisquat and antidive, see Section 7.4.1).

Because seat location has usually been established well before suspension layout, the designer must be guided by the layout objective: comfort levels are oriented either for the driver or for rear-seat passengers. In the first case, front suspension would be laid out to be softer than at the rear; conversely, the opposite bias would provide appreciably higher rear-seat comfort. The fact that rear-seat–oriented layout generally permits better *K* coefficients to be achieved is attributable to the phase shift between front/rear superstructure accelerations while in forward motion. Because nearly all of the payload on front- or standard (rear)-drive vehicles is applied to the rear axle, rear-seat comfort throughout the payload range is essentially only achievable by means

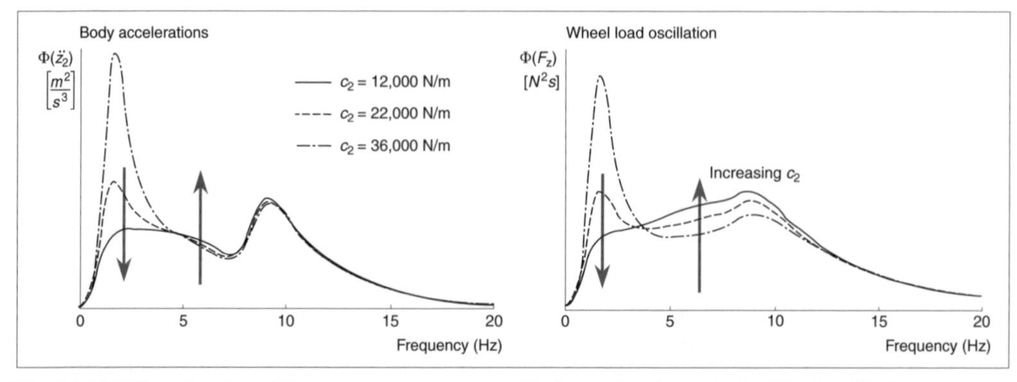

Fig. 7.4-22 Effect of spring stiffness c_2 on power spectra of body acceleration and wheel load oscillation.

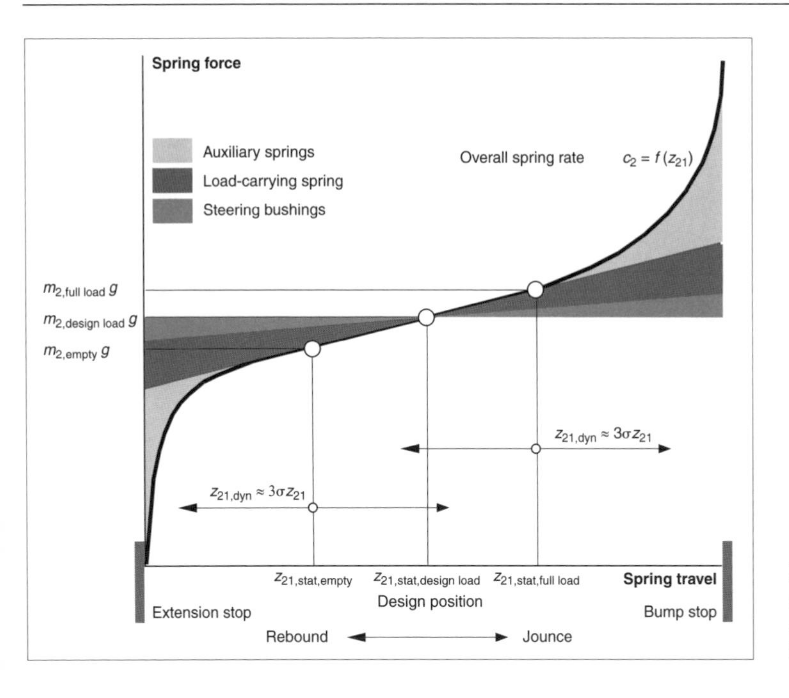

Fig. 7.4-23 Composition of overall spring characteristic for vehicle load-carrying springs.

of load leveling, or the front suspension must be unnecessarily stiffly sprung, which in turn would compromise vertical comfort. In any case, it makes sense to consider spreading front and rear suspension resonant frequencies, not last with regard to behavior in rolling over individual obstacles, as this will rapidly damp any pitch vibrations. Consequently, in practice, f_{2f}/f_{2r} is usually 0.8–0.95.

7.4.4.2 Stabilization

After suspension of both axles has been defined in terms of springs, attention shifts to roll behavior. Because roll oscillations and lift/pitch oscillations may be regarded as decoupled from one another, as shown in [1], the vibration model of Fig. 7.4-24 looks very similar to the single-axle model of Fig. 7.4-21, except that now both axles must be considered, joined by a sufficiently torsionally rigid body of mass m_2 and additionally fitted with front and rear stabilizing springs $c_{S,F}$ and $c_{S,R}$

In cornering with lateral acceleration a_y, a centrifugal force F is applied to center of mass S, which, with roll torque arm h_{RC} leads to the superstructure roll moment

$$M_r = (m_2 \cdot a_y \cdot h_{RC}) \qquad (7)$$

which, after establishment of a roll angle φ, is increased by component $m_2 \cdot g \cdot h_{RC} \cdot \varphi$. The roll torque arm h_{RC} is obtained from the vertical distance between center of gravity and roll axis (Section 7.4.1). The somewhat simplified representation of the roll moment M_r must be in equilibrium with roll spring moment $M_{r,s}$:

$$M_r = m_2 h_{RC}(a_y + g\varphi) = (c_{\varphi f} + c_{\varphi r})\varphi = M_{r,s} \qquad (8)$$

where $c_{\varphi f}$ and $c_{\varphi r}$ are the resulting axle roll spring rates, by whose ratio the torque $M_{r,s}$ is distributed between both axles (assuming torsionally rigid body). A more refined examination may be found in [6].

Unfortunately, roll tendency cannot simply be eliminated by raising the roll center heights to that of the center of gravity ($h_{RC} = 0$). Except in the case of beam axles, this would lead to very unfavorable wheel travel curves with large track changes and therefore traction problems and high tire wear. Also, complete elimination of roll angle, at least at the handling limit, is not even desirable, for driver information reasons.

To maintain small roll angles given normal roll torque arm lengths of about 500 mm and soft springing for comfort reasons, an additional stabilizer (also known as an antiroll bar, antisway bar, sway bar, stabilizer bar) must be installed, which represents a spring acting in parallel to the main vehicle springs, but only in the case of single-sided wheel displacement. The roll spring rate is then obtained by addition of both spring rates, with consideration of front or rear track $s_{f,r}$

$$c_{\varphi f,r} = (c_{2f,r} + c_{sf,r})\frac{s_{f,r}^2}{2} \qquad (9)$$

If the stabilizer rate of the (usually softer sprung, vertically) front suspension is chosen so that the desired maximum roll angle is not exceeded, jacking and self-steering behavior must be examined. As long as only the front suspension is fitted with a stabilizer, its appreciably higher roll stiffness will account for a greater portion of resistance to roll moment. This leads to more

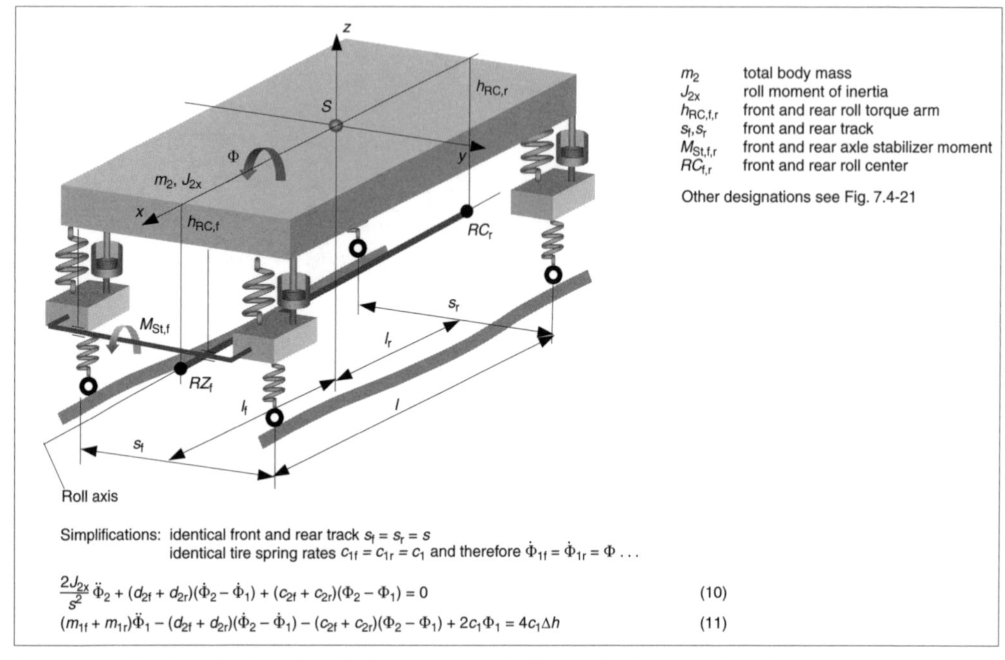

Fig. 7.4-24 Vehicle model for roll oscillations and most significant vibration equations after [1].

pronounced weight transfer between the inner and outer wheels and, above lateral accelerations of 0.3 g, to progressively greater slip angle requirements for the suspension as a result of nonlinearity in tire lateral force characteristics. Because the driver must now add steering input to maintain his course through a turn, we speak of understeer. To a certain extent, this effect is desirable for vehicle stability reasons. However, in the case of excessive understeer, it may be necessary to call on the rear suspension to assist in resisting roll moments. If, for comfort reasons, the rear body spring rate cannot be increased, a stabilizer can be added to the rear axle as well.

7.4.4.3 Damping devices

Even if load-carrying and stabilizing springs provide a good basis for desired lift, pitch, and roll comfort, other measures must still be applied to diminish vibration induced by external forces. This task falls to dampers (also called shock absorbers). Damping not only acts on the body structure, but also makes a very significant contribution to handling safety (Fig. 7.4-25), while load-carrying spring rates have a lesser effect on wheel load variations (Fig. 7.4-22). So for the body degree of damping $D_{2f,r}$ and wheel degree of damping $D_{1f,r}$,

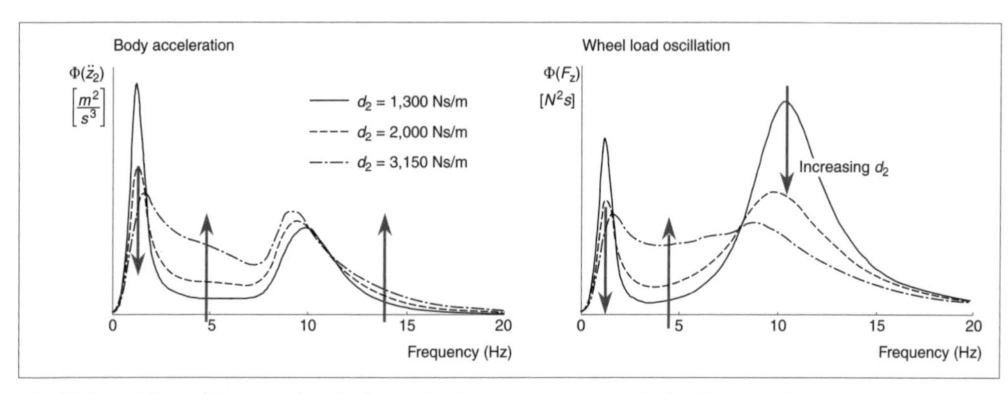

Fig. 7.4-25 Effect of dampers d_2 on body acceleration power spectra and wheel load oscillation.

$$D_{2f,r} = \frac{d_{2f,r}}{2\sqrt{c_{2f,r}m_{2f,r}}}, \quad (12)$$

$$D_{1f,r} = \frac{d_{2f,r}}{2\sqrt{(c_{1f,r}+c_{2f,r})m_{1f,r}}} \approx \frac{d_{2f,r}}{2\sqrt{c_{1f,r}m_{1f,r}}} \quad (13)$$

In practice, values for $D_{2f,r}$ are about 0.2–0.25, and somewhat higher for $D_{1f,r}$, at least for small unsprung masses. Fig. 7.4-26 shows that for stochastic roadway excitation, the main comfort/safety synthesis decision falls somewhere along a line of given body spring constant c_2 by virtue of a choice of a damping constant d_2. Because lift and pitch oscillations always appear in combination and because the entire diagram is dependent on vehicle speed, load, and, above all, the nature of the applied road irregularity spectrum, parallel optimization of damping constants $d_{2f,r}$ at front and rear axles, with variation of the most important simulation input parameters, should be carried out.

Modern shock absorbers—hydraulic telescoping dampers—fall into two types: either single tube (monotube) or dual tube configuration, available in many design variations and with various valve systems.

In the past, one could assume linear damping characteristics (Fig. 7.4-27). If, however, one considers, in terms of comfort [7], response to individual obstacles (e.g., pavement dips, manhole covers) or driver actions (e.g., turning in), then the actual trace of damping force F_d as a function of damper piston speed v_D must undergo additional tuning; the damping characteristic becomes nonlinear.

A useful form of nonlinearity is degression ("falling rate") in extension. This results from an effort to achieve over time a harmonically increasing roll angle in response to sudden avoidance maneuvers. This is a very important contribution to a subjective feeling of safety in such generally uncommon situations. To this end, high damping rates at the inside wheel are required

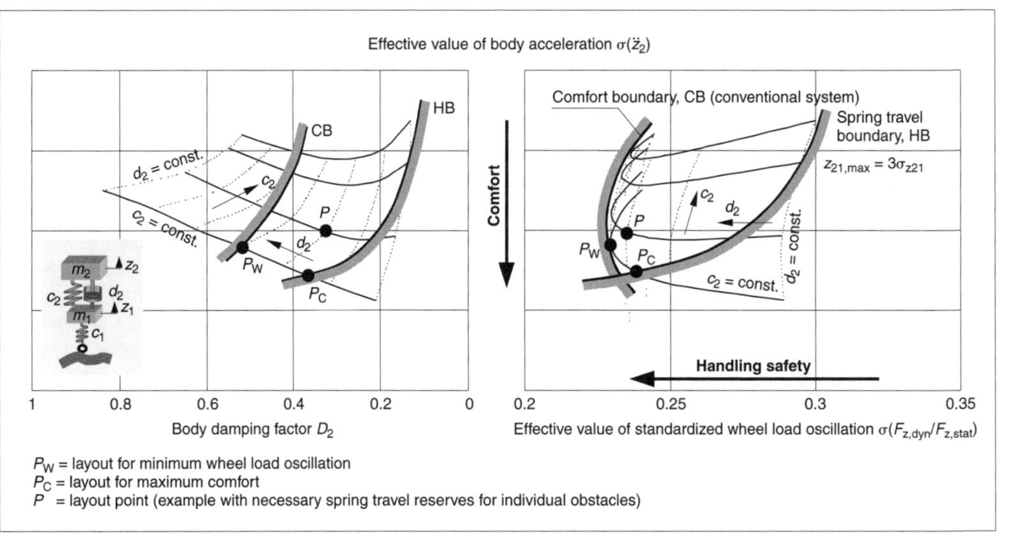

Fig. 7.4-26 Layout limits in comfort/handling safety conflict diagram, as a function of springing and damping, with corresponding damping factors.

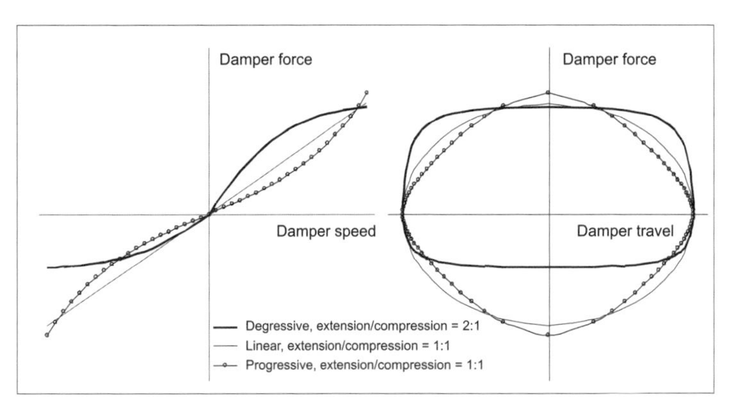

Fig. 7.4-27 Examples of characteristic curves of damper force over damper speed and effects on work diagram.

even at very low values of v_D. This represents a steeper rise of damper characteristic in extension than would be indicated by linear calculation ($d_{2ext} > d_{2lin}$); at higher values of v_D, this would lead not only to increasing discomfort but also to overdamping of the wheels and therefore to greater wheel load oscillations. Consequently, above $v_D \approx 0.15$ m/s, the curve is dropped somewhat below d_{2lin}; the curves in the working diagram become "fuller," and damping work at low v_D is increased.

Another form in current usage is that of the "kinked" damping characteristic, with different damping rates for compression (d_{2C}) and extension (d_{2X}), in which primarily the compression damping rate is dropped to levels as low as $d_{2C}/d_{2X} = 1/3$. The intention is to improve transient "spring feel" in driving over ramp-shaped individual obstacles. This may be achieved within limits, but statistically, with increasing damping rate asymmetry, comfort decreases and wheel load variation increases. Therefore the compression/extension ratio should not be lowered unnecessarily; instead, the positive effect of adequate suspension longitudinal compliance should first be exhausted.

The previous paragraphs have shown that in view of vehicle targets, devising successful overall chassis tuning represents a complex synthesis process, which the designer approaches either empirically, with a great deal of experience, or, better yet, with the aid of modern simulation tools. If simulation methods are applied, better but more complex comfort evaluation criteria may be applied. If initially only the K value for vertical acceleration was mentioned, Ref. [8] shows that the more meaningful overall vibration level K_{ovl} may be calculated:

$$k_{ovl} = \sqrt{K_1^2 + K_2^2 + \ldots + K_n^2} = \sqrt{\sum_{i=1}^{n} K_i^2} \quad (14)$$

Comfort is less a consideration in the case of sports car suspension layout. In such cases, high value is placed both on small spring travel and roll angles, which translate into appreciably stiffer springs, correspondingly stiffer shock absorbers, and higher stabilizer rates, as well as on good tire grip (ultimate cornering speeds). If comfort is ignored, these parameters are limited only by concern for minimal wheel load oscillations.

7.4.4.4 Vertical dynamics systems

Significant potential for improvement is offered by vertical dynamics systems as compared to passive components: that is, control systems which optimize vertical forces over time and according to need. In tuning passive springs, dampers, and stabilizers, the best that can be achieved is a compromise between vehicle behavior (handling, agility) and ride comfort; in other words, suspensions may be tuned to be sporty or comfortable, but not both arbitrarily sporty and comfortable at the same time.

Within small time frames, vertical dynamics systems provide demand-appropriate, vehicle-condition-dependent vertical forces between wheel and body in order to resolve conflicting handling and comfort goals (within the bounds of physical possibility; see Section 7.4.4.3). In terms of Fig. 7.4-26, this represents a breakthrough of the boundary curve CB in the direction of increased comfort and handling safety. In terms of hardware, these systems require sensors, a control unit, and actuators, as well as an energy supply and, on the software side, an appropriate control strategy. In contrast to safety- or slip-oriented control systems, vertical dynamics systems are not only effective at the handling limits of a car but also in the normal driving regime, and they are permanently perceptible and "experienceable."

The following is a brief description of three categories:

- load leveling (at present, usually achieved by means of air springs)
- adjustable damper systems
- active springs/active stabilizers

In *load leveling*, changes in vehicle loading cause an operating medium to be added or withdrawn from the suspension struts so that vehicle ride height remains constant, therefore always keeping an optimum reserve of spring travel available. This characteristic line may be laid out without regard for ride height changes caused by loading. Both points lead to improved vibration comfort, which is experienced not only at maximum payload. Load leveling by means of air suspension is becoming increasingly popular for reasons of weight and cost advantage, gradually replacing hydropneumatic suspension. In addition, air suspension has an advantage in providing a largely constant body resonant frequency throughout the load range (Fig. 7.4-28). All-wheel-drive vehicles have used air suspension at all four corners in production since 1992; now rear suspension air springs [9] and four-wheel air springs [10] are becoming increasingly popular for sedans as well.

Air springs used in load leveling systems are low-

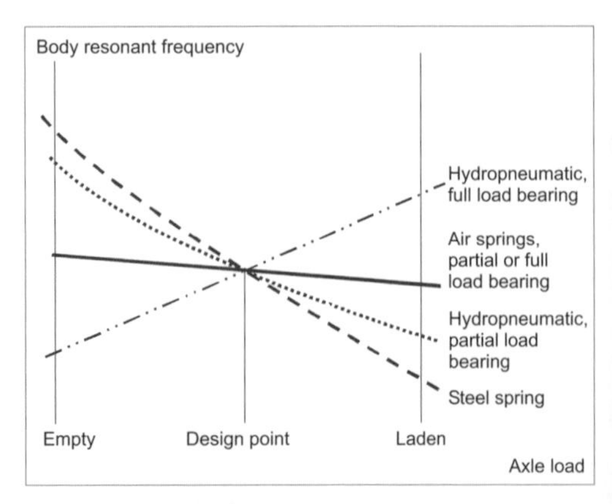

Fig. 7.4-28 Load effect on body resonant frequency for various load-leveling systems.

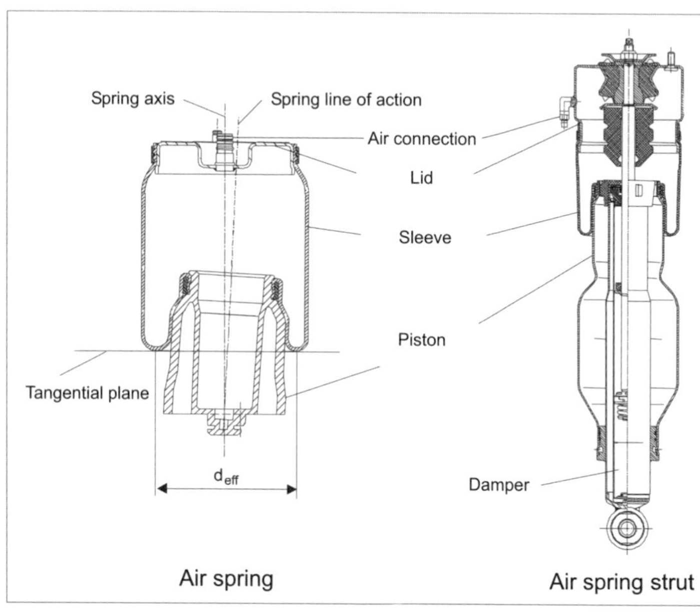

Fig. 7.4-29 Air spring configuration and example of air spring strut.

pressure, constant-volume gas springs. Energy needed for static ride height control for different loading conditions or different selected target ride height is generally supplied by a compressor unit. Of the two air spring varieties—convoluted (bellows) or rolling lobe (sleeve) type—only the rolling lobe has established itself for passenger car applications. Principal air spring elements include lid (bead plate), sleeve, and piston, with mounting elements (Fig. 7.4-29). The sleeve is an elastomeric hose containing a vulcanized reinforcing layer. The latter consists of two or more crossed fabric plies, but for special applications it may use only a single axially oriented fabric layer. A supporting element in the form of external sleeve location is also required. Air springs may be configured as stand-alone springs or as struts (including shock absorber).

In the general case, the (spring travel dependent) gas spring rate of an air spring is composed of two elements:

$$c_G = c_V + c_A \tag{15}$$

with volume stiffness

$$c_V = p_a + p_o \cdot n \cdot \frac{A_{eff} \cdot A_g}{V_o} \tag{16}$$

and surface stiffness

$$c_A = p_o \cdot \frac{\delta A_{eff}}{\delta z_{21}} \tag{17}$$

where

A_{eff} effective cross-section area: the area encompassed by the tangential surface line of contact to the sleeve fold (Fig. 7.4-19, left)

A_g geometric cross-section area (theoretical)

$\delta A_{eff}/\delta z_{21}$ change in effective area over spring travel

n isotropic exponent (spring speed dependent, 1.00–1.38)

p_o air spring overpressure

p_a ambient (atmospheric) pressure

V_o interior volume in nominal layout position

If the effective spring area remains constant throughout the spring travel range—for example, for a cylindrical rolling sleeve on a cylindrical piston—the spring surface stiffness becomes zero and the gas spring rate is solely determined by volume stiffness.

For speed-dependent thermodynamic changes to the working medium, air, during spring travel, distinction is made between dynamic (adiabatic) spring processes ($n = 1.38$) and quasi-static (isothermal) spring processes ($n = 1.00$); for example, see Ref. [11].

For very small excitations, the properties of the rolling sleeve wall, itself a stiffness component, c_{RS}, along with purely thermodynamic spring properties become apparent [12]. This manifests itself as exponential stiffening of the spring with decreasing spring travel, which, depending on sleeve details, can easily exceed the gas spring rate and therefore must be included in the overall air spring rate $c_{AS,ovl}$:

$$c_{AS,ovl} = c_G + c_{RS} \tag{18}$$

Because small spring travel is largely encountered on good roads up to high speeds, comfort loss resulting from sleeve stiffening becomes particularly apparent and is perceived as unpleasant. This negative effect can be greatly reduced by application of single-ply axial sleeve springs with external location and very thin sleeve wall thickness (Fig. 7.4-30).

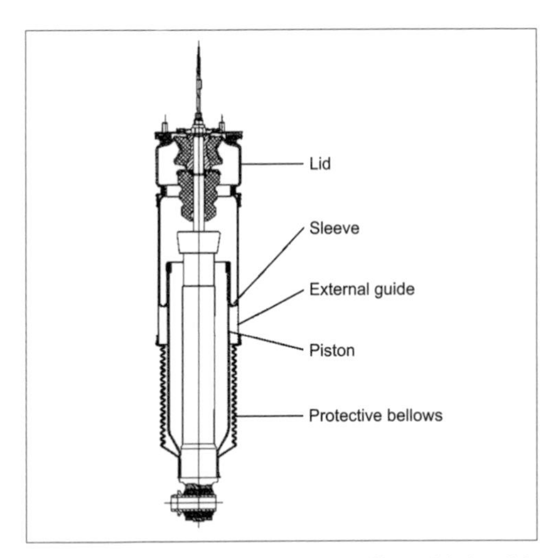

Fig. 7.4-30 Strut variation: externally guided, with single-layer axial sleeve spring.

Electronically *adjustable damping systems* change vertical forces by means of altering damping forces in response to sensed vehicle condition. Good insight into potential improvement in vibration comfort via adjustable damping systems, as well as into their basic function, is again provided by Fig. 7.4-25. At the left, the illustration shows vertical body vibration power spectra for various damper characteristic curves, each of which was locked in while the vehicle was driven over a medium secondary road. The vehicle's body acceleration power spectrum is, as already mentioned, a useful, if simplified, objective evaluation criterion for vibration comfort. It may be seen that in the frequency range from about 2 to 30 Hz, the soft damping characteristic provides better comfort; from 0.3 to about 1.5 Hz, the stiff characteristic reduces acceleration amplitudes, thereby improving vibration comfort.

The fact that handling dynamics and dynamic wheel load variations must be considered in addition to comfort (Fig. 7.4-25, right) gives rise to the following basic system functions for variable damping systems: for primary body vertical vibration frequencies— around 1.2 Hz—as well as dominant longitudinal and transverse vehicle motion, stiff damping characteristics are set by four variable dampers, which quiet or reduce body movements. For primary wheel-frequency vertical vibrations —about 12 Hz—medium damping rates which still provide satisfactory wheel damping are more advantageous. For excitations between these two vehicle resonant frequencies, very soft damping rates should be set for good ride comfort. To ascertain driving condition, variable damping systems use a sensor suite consisting of three vertical accelerometers, a steering angle signal, and front-wheel ABS signals. The latest variable damping systems employ a control strategy based on the so-called "skyhook principle" [8], which offers functional advantages in the case of body-frequency road excitation.

In 1987, the first variable damping system was introduced in Europe under the name EDC (Electronic Damping Control) [13]. This EDC generation was based on three-stage adjustable dampers with selectable and automatically adjustable damping characteristic curves. Along with three-stage versions, in recent years multi-stage dampers have entered the market, such as Mercedes-Benz with individual wheel adjustable four-stage dampers and Toyota with approximately 15-stage dampers. The next generation of adjustable damper systems [8] will consist of map-controlled dampers capable of realizing an infinite number of characteristic curves, a new level of ride comfort (quiet body and isolation), and driving enjoyment.

Active springs have been offered on the Japanese market since 1990 [14, 15], while *active stabilizers* have been available in Europe since 1995 [16] and in a new configuration since 1998. Active springs can provide nearly any desired vertical forces in the body resonant frequency regime, while active stabilizers provide corresponding vertical moments. Customer-relevant functional benefits differ, depending on system details. In general, the complex technology of these hydromechanical systems prevents their widespread market introduction.

Solutions offered by Japanese manufacturers are based on a hydropneumatic active spring (replacing conventional spring and damper), a hydraulic supply, a control unit, and several sensors. The hydraulic supply consists of an axial or radial piston pump, supply and pulsation accumulator, oil reservoir, cooler, high-pressure filter and lines (see Fig. 7.4-31). This supply provides a constant pressure network: that is, the supply pressure is kept as constant as possible at the maximum pressure of, for example, 160 bar. Such constant pressure networks require considerable engineering effort in order to achieve low sound levels. The hydropneumatic active spring is based on hydropneumatic suspension consisting of a differential cylinder, a gas pressure accumulator (acting as a spring element), and an orifice (acting as a damping element), as well as a control valve, which can admit or release high-pressure oil to or from the differential cylinder. Active intervention is accomplished by means of the control valve as an add-on measure. In contrast to a fully active spring [17], the hydropneumatic active spring operates without supply or withdrawal of pressurized oil. This leads to advantages in vibration comfort and energy budget compared to the fully active spring. The required sensor complexity is considerable. To determine driving condition, Nissan employs two lateral, one longitudinal, and three vertical acceleration sensors as well as four height sensors. Toyota adds five pressure sensors. The control unit evaluates the supplied sensor information and controls the four hydropneumatic active springs according to need. In view of the effort involved and in particular the added energy consumption, actual customer benefit must be regarded critically. Consequently, such complex active springs have

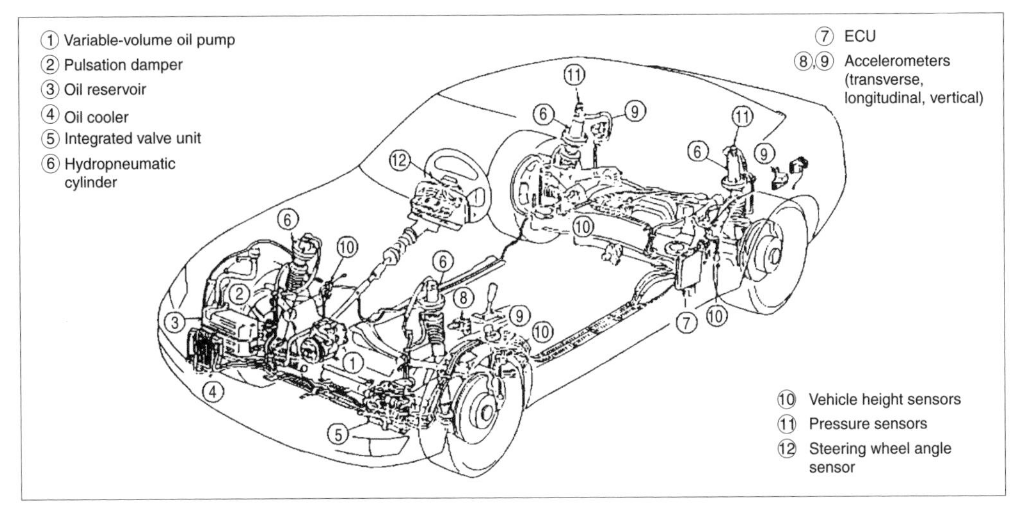

Fig. 7.4-31 Hydropneumatic active spring in 1991 Toyota Soarer.

not achieved any real penetration in the Japanese market.

A new type of active spring was introduced in September 1999 by Mercedes-Benz in its S-class coupe, under the name ABC (for Active Body Control) [18]. Its technical complexity is, however, comparable to that of the systems offered by Japanese manufacturers. The basic difference between ABC and previously described active springs lies in the design of the four suspension struts. Fig. 7.4-32 shows a cross section of the upper portion of an ABC strut. The sleeve of a plunger cylinder is braced against the coil spring that supports the body structure. Admission or removal of pressurized oil by a control valve causes the plunger cylinder sleeve to move, thereby loading the coil spring in response to sleeve travel and providing the desired additional forces. As in the case of previously described active springs, these forces are intended to significantly reduce body motion in the roll, pitch, and lift degrees of freedom and enable ride leveling. In addition to the coil spring, a passive gas pressure shock absorber with a soft characteristic curve is integrated in each suspension strut in order to provide sufficient wheel travel damping. As in the case of hydropneumatic active springs, the ABC system employs a constant pressure network, requiring multiple control valves, and a sensor suite similar to that of Ref. [14] is installed. Added to this are the sensors to control the plunger cylinders. In contrast to hydropneumatic active springs, the ABC solution enjoys two basic advantages: lower friction forces in the flow of forces between wheel and body as well as greater freedom in establishing the spring characteristic. Because of ABC's considerable complexity, the market for these systems will be confined to a very limited vehicle segment, at least for the immediate future.

Fig. 7.4-32 DaimlerChrysler AG Active Body Control strut.

By contrast, thanks to low cost and especially positive effects on the driving experience, ACE (Active Cornering Enhancement) in the new Land Rover Discovery II is likely to enjoy better market opportunities in this vehicle class. During cornering with ACE, active front and rear stabilizers generate vertical moments to counteract vehicle body roll moment. The body remains horizontal in cornering; particularly in off-road driving with a high center of gravity, annoying body roll ("roll rock") does not occur. Other off-road advantages arise because the active stabilizers do not generate vertical moments in response to low-frequency asymmetrical excitation at a given vehicle axle. The hydraulic system schematic of Fig. 7.4-33 indicates the main elements of ACE. These include the oil tank, radial piston pump,

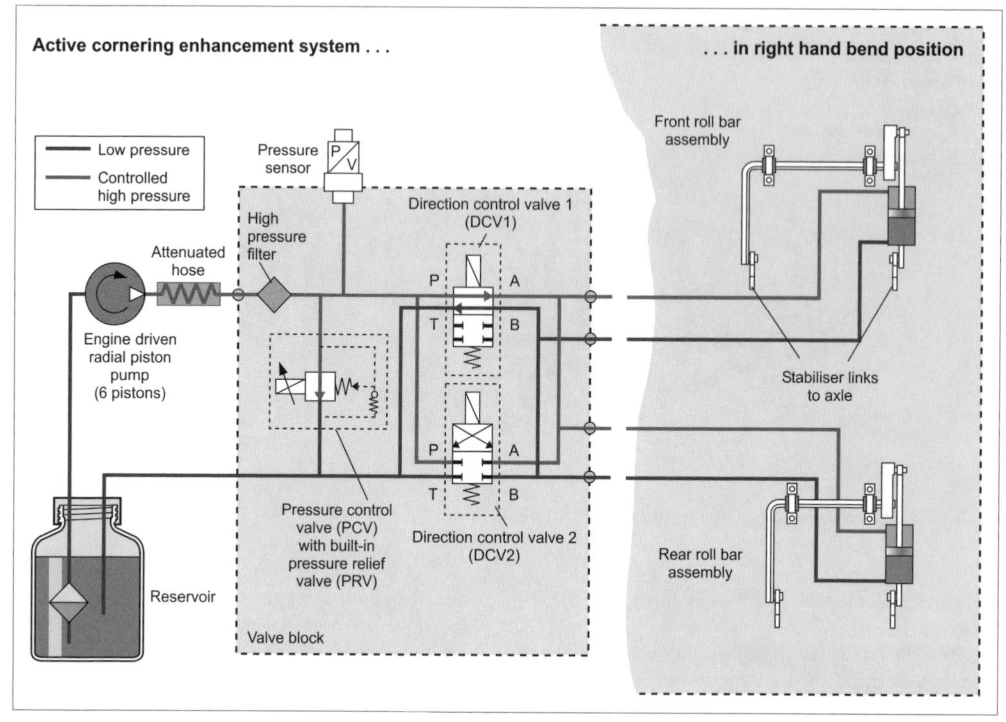

Fig. 7.4-33 Hydraulic schematic for Land Rover Discovery II Active Cornering Enhancement.

valve block, and, schematically, active stabilizers at the front and rear axles. The active stabilizer solution realized by ACE is illustrated in Fig. 7.4-34: a differential cylinder and lever arm take the place of a stabilizer arm. If the upper or lower cylinder chamber is subjected to pressure, this pressure acting on the piston effective area generates a piston force which, multiplied by the effective lever arm, produces an active torsional moment in the stabilizer. This active torsional moment produces an active vertical moment around the roll axis proportional to the applied cylinder pressure. To achieve the desired pressure increase, a pressure control valve is installed in the valve block, which sets pressure in proportion to applied electrical current. Directional valves in the valve block are switched in response to detection of left or right turns or straight-line motion. For demand-appropriate system control, the control unit requires two accelerometers and one pressure sensor. The ACE hydraulic supply system is configured as a circulating system with a constant oil flow volume, and higher pressures are achieved only during cornering.

Fig. 7.4-34 Land Rover Discovery II active stabilizer.

7.4.4.5 Prognosis

Over the past two decades, astounding progress in suspension quality has been achieved. This continued development of suspension elements is attributable to steadily improved understanding, on the part of the motor vehicle and supplier industries, of the complex interrelationships involved, as well as to ever more powerful development tools, materials, and manufacturing methods. Although most present-day suspension systems are still conventional in nature—that is, composed of passive components—we may justifiably speak of a high level of "mechanical intelligence."

Even if future clever engineers introduce additional detail improvements, after more than 100 years of automotive development it must be recognized that the further development potential of conventional-element suspensions does not hold out much promise for customer-relevant, innovative breakthroughs.

In the attempt to answer the question of what we may expect of future suspension development, a bit of self-evaluation may help: Strictly speaking, given the complexity of the task, at present no one is in any position to say whether a newly developed outstanding suspension design actually represents the attainable maximum under a given set of boundary conditions or whether it is "only" the current best effort of the attending development team. Consequently, in the course of advancing virtualization of the development process, the opportunity should be taken to apply the currently available, system-encompassing knowledge base of causal relationships, along with powerful computers and programs, as well as more intensive inclusion of other scientific disciplines, as important aids in the search for individual global optimization of any new suspension synthesis. In this process, continued examination of the necessary limits of cost- and weight-driving stiffness demands on suspension components, local chassis pickup points, and global body properties will remain attractive insofar as increasingly precise knowledge of flexibility in these areas may be more completely integrated in the suspension layout process.

That said, the next technological revolution in the suspension sector will belong to new, powerful suspension control systems with futuristic system configurations and networking, in which the "preview" concept—that is, temporal and spatial anticipation of lane alignment and road irregularities—will presumably experience its translation into productive implementation. The theory behind this "systems universe" has been in preparation for decades (see [19, 20, and 21]) and in part has been explored on a prototype basis. It is only now, however, that technology for the necessary actuators, sensors, bus systems, and electronics have advanced to the point where production-appropriate implementation in keeping with economic aspects can be considered. It is possible that given such boundary conditions, the pressure for compromise in conventional (not actively controlled) suspension components will diminish or be favorably displaced; regardless, even then, the currently attained level of "mechanical intelligence" will remain the indispensable basis of suspension tuning.

References

[1] Mitschke, M. *Dynamik der Kraftfahrzeuge, Band B: Schwingungen*. Third edition. Springer Verlag, 1997.

[2] Griffin, M.J. *Handbook of Human Vibration*. Academic Press, London, 1990.

[3] VDI 2057. "Einwirkung mechanischer Schwingungen auf den Menschen," Sheets 1–3, 1987.

[4] Gold, H. "Eigenschaften einer ausschließlich mit Gas (Luft) arbeitenden Feder-Dämpfer-Einheit." VDI-Berichte 546, VDI-Verlag, 1984.

[5] Brüninghaus GmbH. "Technische Daten Fahrzeugfedern, Teil 1: Drehfedern." Stahlwerke Brüninghaus Werdohl, 1969.

[6] Matschinsky, W. *Radführungen der Straßenfahrzeuge*, second edition, Springer Verlag, 1998; English edition *Road Vehicle Suspensions*, Professional Engineering Publishing Ltd., 2000.

[7] Hennecke, D. "Zur Bewertung des Schwingungskomforts von PKW bei instationären Anregungen." Fortschr.-Ber. VDI Reihe 12 No. 237. VDI-Verlag, Düsseldorf, 1995.

[8] Konik, D., et al. "Electronic Damping Control with Continuously Working Damping Valves (EDCC)." AVEC '96, Aachen, 1996.

[9] Breuer, N., and J. Kock. "Die pneumatische Niveauregelanlage des Audi A6." ATZ/MTZ Sonderausgabe 38, 1995.

[10] Scheerer, H., and M. Römer. "Luftfederung mit adaptivem Dämpfungssystem im Fahrwerk der neuen S-Klasse." 7th Aachen Automobile and Engine Technology Colloquium, 1998.

[11] Schützner, E.-Chr. "Thermodynamische Analysen von Luftfedersystemen." VDI-Berichte 1153, 1994, pp. 113–135.

[12] Dreyer, W., and C. Oehlerking. "Untersuchungen von Luftfeder-Rollbälgen für Personenkraftwagen." *ATZ* (88), No. 10, 1986, p. 535 ff.

[13] Hennecke D., et al. "Anpassung der Dämpferkennung an den Fahrzustand eines PKW." VDI-Bericht 650: Reifen, Fahrwerk, Fahrbahn. Hannover, 1987.

[14] Fukushima, N., and K. Fukayama. "Nissan Hydraulic Active Suspension." In: "Fortschritt der Fahrwerkstechnik 10, Aktive Fahrwerkstechnik." Brunswick, 1991.

[15] Tanaka, H., et al. "Development of a vehicle integrated control system." FISITA '92, London, 1992.

[16] Goroncy, J. "Citroen Xantia Activa mit neuem Fahrwerk." *ATZ* 7/897, 1995.

[17] Williams, D., and P. Wright. "Vehicle suspension arrangements." Group Lotus Car Companies, EP 0 142 947 of 23. 10. 1984.

[18] "ABC-System für Mercedes SL." DaimlerChrysler press information, 1999 Geneva Auto Show.

[19] Gipser, M. "Verbesserungsmöglichkeiten durch aktive Federungselemente aus theoretischer Sicht." VDI-Berichte No. 546, 1984.

[20] Acker, B., W. Darenberg, and H. Gall. "Aktive Federung für Personenwagen." *Ölhydraulik und Pneumatik* 33, Vol. 11, 1989.

[21] Wallentowitz, H., and D. Konik. "Actively influenced suspension systems— Survey of actual patent literature." EAEC Conference, 1991.

[22] Mitschke, M. *Dynamik der Kraftfahrzeuge, Band C: Fahrverhalten*. Second edition. Springer Verlag, 1990.

[23] Zomotor, A. *Fahrwerktechnik: Fahrverhalten*. Second edition. Vogel Buchverlag, 1991.

[24] Parsons, K.G.R., et al. "The Development of ACE for Discovery II." SAE 2000-01-0091. Steering and Suspension Technology Symposium 2000 (SP–1519).

7.4.5 Steering

Guidance of road vehicles is almost exclusively under the control of the driver by means of the steering system. The degree of precision with which, on the one hand, a driver can direct a personally desired course or a course determined by road alignment and traffic situation and, on the other hand, the vehicle maintains such a course is of considerable importance to traffic safety. The driver must always be secure in the knowledge that the vehicle will react reliably to his commands. The more rapidly the driver is able to recognize intentional as well as unintentional course deviations, the better the course-holding precision and the more rapidly and precisely the vehicle will react to driver steering inputs.

For the steering system developer, this gives rise to a number of requirements and recommendations to achieve customer-oriented steering layout:

- Small turning circle, low parking forces, small steering angles at the steering wheel
- Light steering forces, good steering feel, precision, good straight-ahead stability, adequate directness, spontaneous response
- Pronounced road contact, feedback of tire/roadway traction
- Independent return to center, stabilizing behavior in all driving maneuvers
- Suppression of disturbing forces and moments resulting from road surface, crosswind, drive, brakes, tire types, etc.; no pronounced tendency to self-resonance
- Fulfillment of crash requirements for occupant protection
- Freedom from wear and needs for maintenance, as well as no negative effect on tire wear behavior

7.4.5.1 Steering kinematics

Parameters of steering kinematics

The basic properties of a steering system are defined by four geometric properties which describe the relationships between position and orientation of the wheel steering axis (kingpin axis) within the vehicle:

- Kingpin offset
- Caster offset
- Kingpin inclination
- Caster angle

Added to these are parameters referenced to the wheel centers, such as kingpin offset at wheel center and caster offset at wheel center. These are redundant and result from layout of the kingpin axis. All parameters are discussed in Section 7.4.1 (Figs 7.4-1, 7.4-2).

Parameters such as kingpin offset, caster offset, and so on represent effective lever arms for external forces acting on the steering. The very simple definitions of these parameters in standardization (literature) is not sufficiently exact for computation. The reader is therefore referred to Ref. [1]. The significance of these parameters for steering kinematics in suspension design and preferred magnitudes for modern vehicles will be discussed in relation to the most important operating conditions and the resulting external forces.

Kingpin offset—braking forces

Kingpin offset, also known as scrub radius, leads one to believe that the tire contact patch of a steered wheel would roll on a path defined by this radius. This is true, however, only for the special case of a perfectly vertical kingpin axis. In the presence of caster and camber angles, this scrub radius is increased, and the path of the tire contact patch describes a spiral [1]. Kingpin offset therefore does not represent the radius of a circular path, but rather can only be regarded as an effective lever arm for longitudinal forces around the kingpin axis, acting on the tire contact patch at the level of the road surface if braking or drive moments are reacted against the wheel carrier, and therefore wheel and wheel carrier may be temporarily regarded as rigidly joined. This applies, for example, to typical modern outboard (i.e. located within the wheel) brake systems and similarly for drive motors located within wheels (e.g., electric vehicles with hub motors).

Once, before the advent of power steering, optimum kingpin offset values were driven by consideration of steering moment for a stationary vehicle, leading to large offsets. In the event of asymmetrical braking torques (pulling brakes, especially in the case of drum brakes) or road friction coefficient issues, this led to problems in maintaining vehicle course and eventually prompted introduction of *negative kingpin offset*; see Refs. [2] and [3] for examples. Various manufacturers still promulgate different concepts with regard to kingpin offset magnitude and sign, usually as a function of drive concept (Table 7.4-2).

Negative kingpin offset is familiar from its stabilizing effect in braking on differential friction coefficient ("split μ") surfaces. In the quasi-static case of a split μ braking event, higher brake forces on the side with higher friction coefficient generate a yaw moment toward that side, resulting in a corresponding course change. In a suspension layout incorporating negative kingpin offset, the high-friction side produces a steering moment toward the low-friction side, which counteracts the differential brake force yaw moment (Fig. 7.4-35). For the case described here, given a stationary point of application, a layout may be found which leads to balanced

Table 7.4-2 Kinematic parameters of current vehicles (as designed); from [4, 5, 6, 7, 8]

	Audi A4	BMW 3-series	MB A-class	MB S-class	Porsche Boxster
Kingpin offset (mm)	−6.9	4.8	−20.68	−0.63	−7
Caster offset (mm)	21.6	17.0	13.8	31.7	41
Kingpin offset at wheel center (mm)	9.9	83.3	44.12	26.4	83
Caster angle (°)	3.13	5.8	2.83	9.2	8
Kingpin inclination (°)	3.59	15.36	14.1	6	17.9

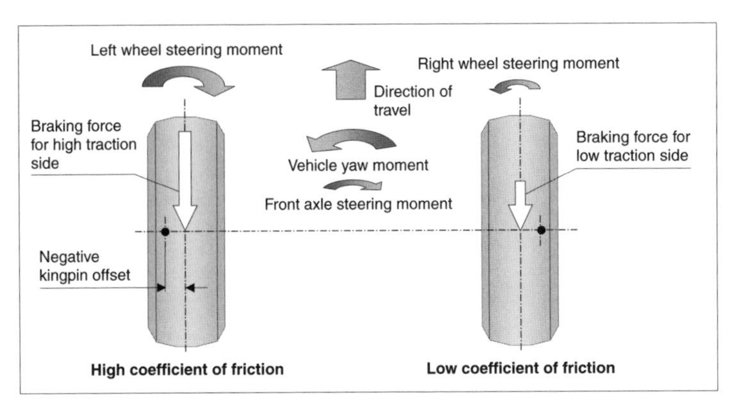

Fig. 7.4-35 Front axle forces and moments for negative kingpin offset and split μ braking.

forces in the steering system without any driver intervention and which results, therefore, in largely undisturbed straight-line vehicle motion.

For front-drive (and all-wheel drive) vehicles with strut suspension, negative kingpin offset is a welcome opportunity to reduce kingpin offset at the wheel center, thereby reducing the disturbing force moment arm which is so important in terms of drive effects.

It should be noted that for negative kingpin offset, the steering wheel will pull toward the side with a lower friction coefficient (i.e., normally toward the outside of the lane), therefore prompting the driver for corrective action. However such action would reinforce the vehicle's initial yaw reaction. One argument against negative kingpin offset is such erroneous reaction and the resulting (unintentional) steering corrections on the part of the driver as an element in the steering control loop.

Kingpin offset should therefore deviate from zero as little as possible to suppress the effect of varying braking forces and to provide the driver with unambiguous information regarding his vehicle's operating condition.

Caster offset, lateral forces, and pneumatic trail

Caster offset results if the tire contact patch is located behind the intersection of the kingpin axis and the road surface (Fig. 7.4-1). The caster offset dimension represents part of the effective lever arm of a lateral force at the road surface acting perpendicular to the projected plane of a turned wheel. It should be noted that tire lateral force is displaced by the amount of pneumatic trail (Section 7.3) and therefore the effective lever arm is composed of caster offset plus pneumatic trail. For effective lever arms >0, lateral forces result in an aligning moment and therefore contribute to stable straight-line stability.

Kingpin offset at wheel center (disturbing force moment arm), rolling resistance, and drive forces

All external forces and force components on a free-rolling wheel can only be transmitted to the steering axis through the wheel bearings and wheel carrier. In the case of nondriven wheels, these are rolling resistance forces, impact forces, and imbalance forces of the tire/rim. These forces act as disturbing forces on the suspension through the kingpin offset at the wheel center (disturbing force moment arm). In general, kingpin offset at wheel center also acts as an effective lever arm for drive, drag, or braking forces transmitted by halfshafts. This statement, however, applies only for the special case of zero halfshaft inclination.

Caster offset at wheel center

Because no lateral forces act at the center of a wheel, the distance between the wheel center and kingpin axis, as seen in side elevation, has no significance as a moment arm. In terms of forces, it is irrelevant for suspension developers whether, in selecting a given kingpin offset and inclination, this distance turns out positive or negative. This is, however, not the case for the spatial movement of the wheel and associated components. For example, in the case of driven axles, this distance leads to greater halfshaft length changes in response to steering movements. Limits are therefore imposed on the distance between wheel center and steering axis for front-drive applications. A positive effect on steered wheel space requirements results from a negative (forward) displacement; with the steering axis (kingpin axis) located behind the wheel center, the inside wheel in a turn does not swing as far inward in the critical area near the firewall. Naturally, the space requirement for the outside wheel becomes correspondingly greater. But because the inside wheel steering angle is greater than the outside wheel angle, this serves to average out the lateral space requirements for steered wheels, providing more space for engine and body structure.

The following quantities have no directly visible design parameters but are very helpful in evaluating various concepts or of decisive importance for optimum steering system function.

Wheel load lever arm, vertical force, restoring torque due to weight

If caster and kingpin inclination angles are zero—that is, the steering axis is vertical—the wheel and body do not experience any height changes as a result of steer-

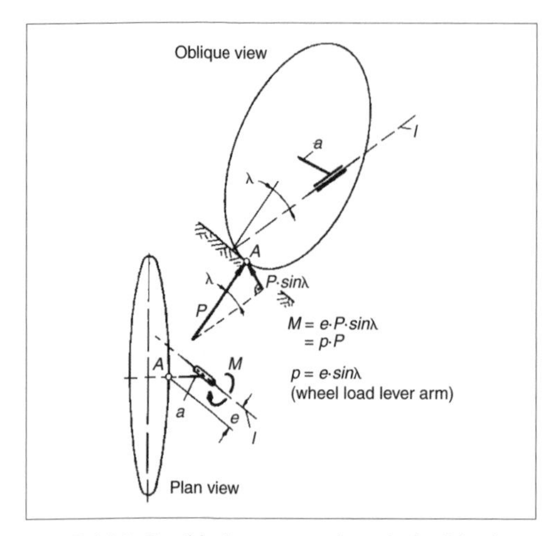

Fig. 7.4-36 Graphical representation of wheel load lever arm (after [9]).

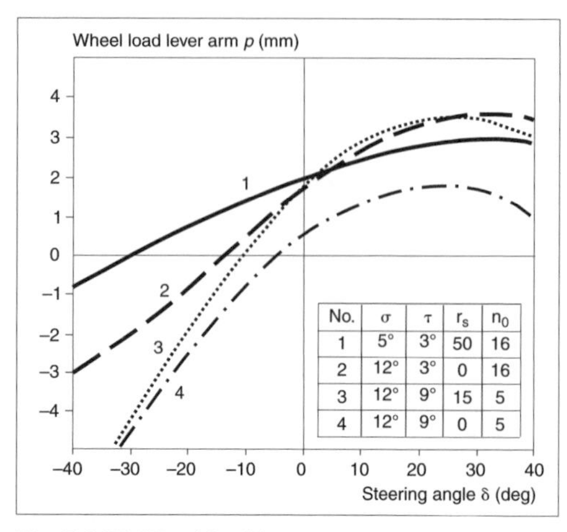

Fig. 7.4-37 Wheel load lever arm p over steering angle δ, for various steering geometries (after [1]). (σ, Kingpin inclination; τ, caster; r_s, kingpin offset; caster offset n_0 at $\delta = 0$)

ing motion and the wheel load lever arm (Fig. 7.4-36), is equal to zero. If the kingpin axis is inclined, the wheel rises or falls with respect to the vehicle body in response to steering inputs; the wheel load is apparently supported by a lever arm of nonzero length. This lifting motion transfers energy to or from the body (and suspension). This energy acts on the steering wheel through the steering linkage.

In mathematical terms, the wheel load lever arm is the derivative of wheel vertical motion over steering angle:

$$p = -\frac{dz}{d\delta} \tag{19}$$

The wheel load lever arm (Fig. 7.4-37) should be nearly zero in the straight-ahead steering position in order to avoid steering moments resulting from vertical force variations (one-sided road surface irregularities). Moreover, the lever arm at the outside wheel should be smaller with steering lock in order to provide gravity return for the steering. This is because on the larger, inside-wheel steering angle, a positive wheel load lever arm lifts the body, while the outside wheel lowers the body, and only a difference in the sum of left and right body lift or drop movements can elevate the center of gravity, promoting a steering return to center. The derivative of wheel load lever arm over steering angle—in other words, the second derivative of wheel lift over steering angle in the straight-ahead position (Eq. 20)—is the "restoring force lever arm due to weight" [1, 9]. Alternatively, this lever may be regarded as a pendulum whose mass is that of the entire front of the car and which produces a corresponding steering aligning moment:

$$\frac{dz}{d\delta} = r_\sigma \tan\sigma - (n \cdot \tan\tau) \quad \text{(for a fixed kingpin axis)} \tag{20}$$

In straight-line motion, restoring torque due to weight is the only quantity (for zero toe-in and camber) which returns steering to center.

Additionally, on suspensions with virtual kingpin axes, the gravity centering effect may be increased or decreased by the swinging motion of the kingpin axis over steering angle [9] or by means of a helical motion along the kingpin axis. Independently of steering kinematics, steering aligning moments may be affected by actuation of spring elements (stabilizer, springs, cam plates, etc.) directly at the wheel carrier. It should be noted that these will affect the steering in the event of asymmetrical wheel load changes.

Traction force radius, drive torque, drag torque

For suspensions with fixed kingpin axes (e.g., conventional upper and lower A-arms), the kingpin offset at the wheel center or the disturbing force moment arm are invariant over steering angle and vertical wheel motion and consequently are always identical at both inside and outside wheels. Given identical drive torque at both wheels, if we ignore steering angle differences even for nonzero disturbing force moment arms, the driving torque on both wheels should be balanced and the steering remain unaffected. Nevertheless, one may observe drive effects in the steering of front-drive vehicles. In Ref. [10], these observations led to a definition of traction force radius. Analogously to kingpin offset (scrub radius), this moment arm must be regarded as that parameter which becomes effective in relation to the steering axis if, with blocked drive, longitudinal forces at the tire contact patch act on the wheel. This traction force radius is represented in Fig. 7.4-38, without derivation, in the very simple case of a fixed kingpin axis and planar kinematics. A line parallel to the bisector of the angle between wheel axis and halfshaft, drawn through

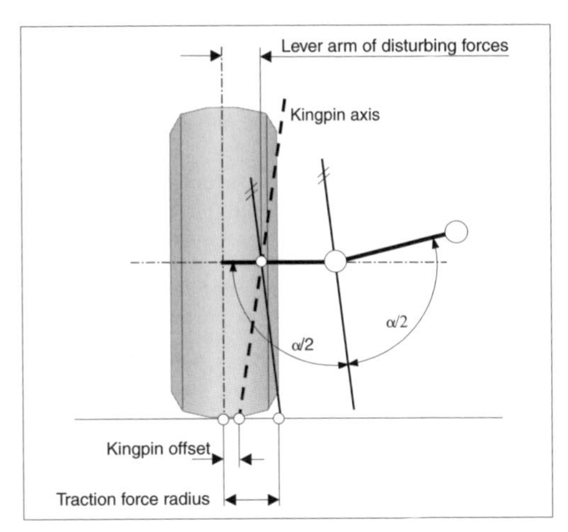

Fig. 7.4-38 Traction force radius of a driven axle with fixed kingpin axis.

the intersection of wheel axis and kingpin axis, intersects the road surface at a distance from the tire contact patch which represents the traction force radius. From the illustration, it may readily be determined that the kingpin offset at wheel center is equal to the traction force radius only for zero halfshaft angularity and, furthermore, that traction force radius decreases for a wheel in jounce and increases for a wheel in rebound. For unequal halfshaft lengths, this leads to differences between right and left wheels not only in cornering but also in parallel suspension jounce or rebound motion and therefore to drive effects on steering.

Effect of disturbing forces and moments

Disturbing forces and moments arising from wheel imbalance, tire radial force oscillations (elastic nonuniformity), or brake torque oscillations generate rotational frequencies at the suspension, and therefore to speed-dependent vibration effects in the steering, in various planes: vertical steering wheel motion, steering wheel rotational oscillations, etc. Proper selection of the above-mentioned kinematically effective moment arms may reduce or largely suppress the effects of ever-present disturbing forces, moments, and resonant frequencies to the point where they are no longer perceptible to the driver. Other important parameters are stiffness of wheel location elements and orientation of the elastokinematic steering axis relative to the unsprung mass center of gravity. Directionally dependent steering box efficiencies may also be used to good advantage in that a higher efficiency from top to bottom— that is, from steering wheel to road wheel—improves steering feel, while lower efficiency from bottom to top helps suppress disturbances. Numerous contributions such as, for example, Ref. [11] show that phenomena causing driver complaints of “unsettled” or “nervous” steering represent a common problem which must be taken seriously.

Determination of ideal steering angle at the wheel

Previously, we have discussed steering kinematics without considering how both wheels ideally should be steered in order to fulfill the primary task of any steering system: to safely guide the vehicle on a driver-determined course. A vehicle has perfect kinematic steering if perpendiculars to the velocity vectors of all wheels meet at a single point. For cornering at low lateral acceleration and small lateral forces, and therefore small slip angles (see Section 7.3), for purposes of simplification we may take the extensions of the wheel axes to determine the optimum steering angle for each turning radius (Fig. 7.4-39). Such a geometric law was first described by Lankensperger and Ackermann. For two-track vehicles with

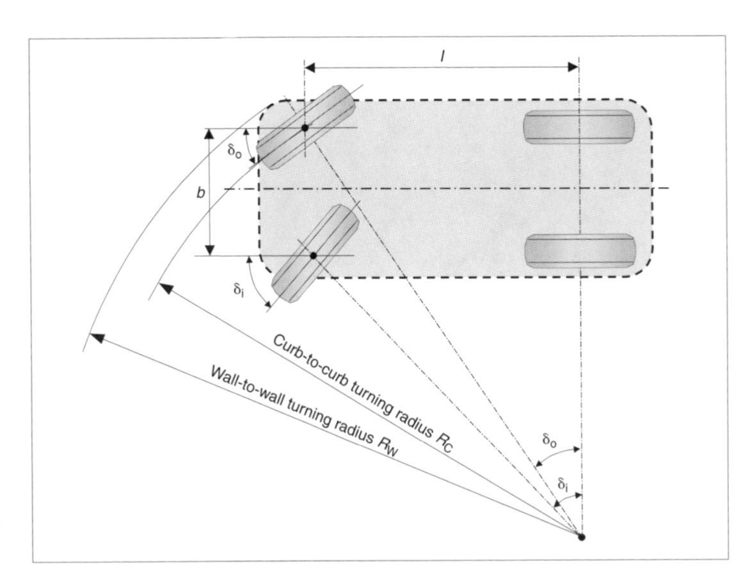

Fig. 7.4-39 Ackermann condition for inner and outer steering angles and definition of wall-to-wall and curb-to-curb turning circles.

two axles (of which the front axle is steered), according to Ackermann, the following rule for the steering angle relationship between the two steered wheels holds:

$$\cot\delta_a = \cot\delta_i + \frac{b}{l} \tag{21}$$

According to the Ackermann principle, the steering linkage should be configured so that any desired curvature, up to the design-limited minimum turning circle, may be achieved with freely turning wheels. Steering errors—that is, deviations from the ideal steering angle relationship (Ackermann function) of Eq. 21—must by force of circumstances lead to tire sideslip. If the steering error is, for example, such that the steering angle of the inside wheel is smaller than the mathematically correct steering angle, we speak of a deviation toward parallel steering angles. This steering angle error is significant for steering return from large steering angles and at slow speeds, for which the necessary lateral force to generate aligning moment would be lacking. The forced slip angle acts like too much toe-in and produces an inward-directed lateral force at both wheels (Fig. 7.4-40). In the case of suspensions with large kingpin inclinations and caster angles (strut suspensions), lateral force on the inside wheel acts on a greater caster offset, while lateral force on the outside wheel acts on a smaller caster offset and therefore cannot compensate for the turning-in moment. Indeed, if the outside wheel already has negative caster offset (kingpin axis *behind* contact patch), the turning-in moment of the inner wheel is reinforced by the outside wheel and the steering "turns in"—it tightens by itself.

From these considerations, the actual realized steering function on strut suspensions may deviate only minimally from the idealized Ackermann function (<3°). This is more easily achieved with front-mounted rack and pinion steering than with steering boxes mounted behind the wheel axis. One factor in favor of a slightly parallel steering angle is the smaller angle of the inside wheel and therefore the smaller space requirement in the wheel well. For equal, average steering ratios (see below), a layout tending toward parallel steer leads to quicker turn-in, as the outside wheel—which accounts for more of the vehicle guidance function—builds up lateral force more quickly due to the superposed toe-in angle.

Effect of steering box configuration on steering linkage

Driver-initiated steering motion is transmitted to the wheels by means of steering gear ("steering box") and steering linkage. As a rule, the steering linkage must also accommodate vertical wheel motion. For independent suspension, this implies articulated tie rods, whose position and length are chosen to provide the desired turn-in behavior for given spring movement and to achieve the desired steering function for a given steering input.

With rack and pinion steering, steering wheel rotational motion is transmitted directly to a steering rack, giving linear motion. Tie rods serve as coupling links between the rigidly mounted steering gear box and steering arms attached to the wheel carriers. Linear motion of the steering rack is translated into rotational motion of the wheel around the kingpin axis (steering angle).

In view of their function, tie rods should be oriented more or less perpendicular to the vehicle longitudinal axis. For rack and pinion steering, this means that two basic orientations are possible for the outer tie rod ends: forward or aft pointing (Fig. 7.4-41). At this point it should also be mentioned that to initiate correct turning motion, in the case of forward-mounted tie rods the pinion of a rack and pinion steering box must be located behind the rack, and for rear-mounted tie rods ahead of or directly on the rack.

Pitman arm steering (or "drag link steering") converts steering wheel rotary motion by means of a steer-

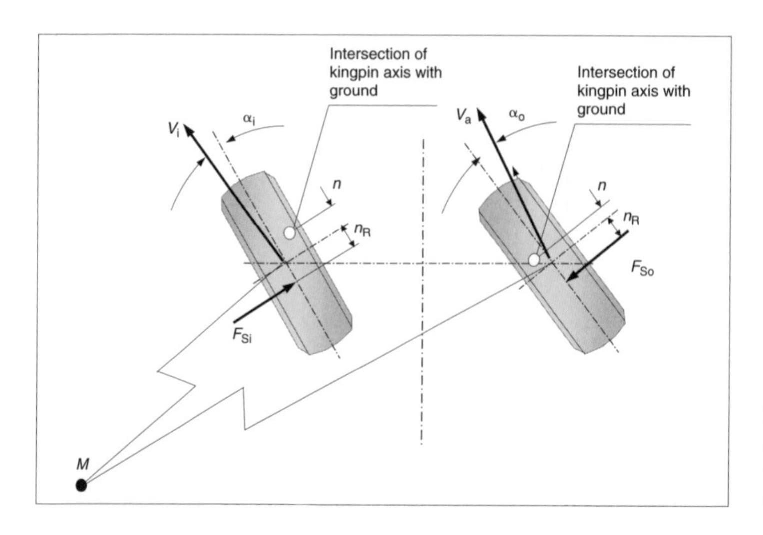

Fig. 7.4-40 Effective forces and lever arms for turned wheels and steering angle error.

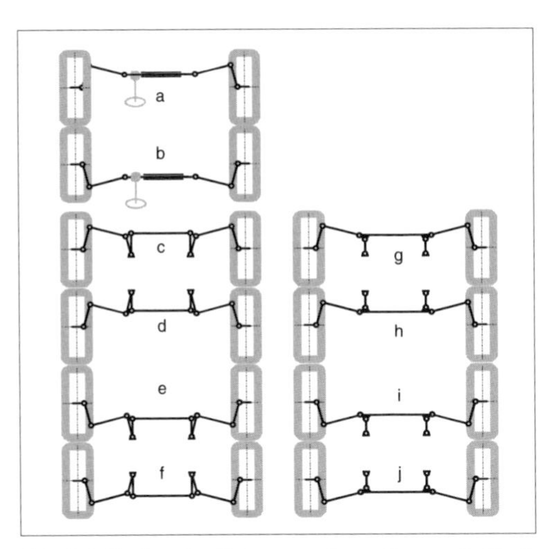

Fig. 7.4-41 Steering linkage configurations. a–b, Rack and pinion steering; c–f, Pitman arm steering, tie rods connected to Pitman (or idler) arm; g–j, Pitman arm steering, tie rods connected to center link.

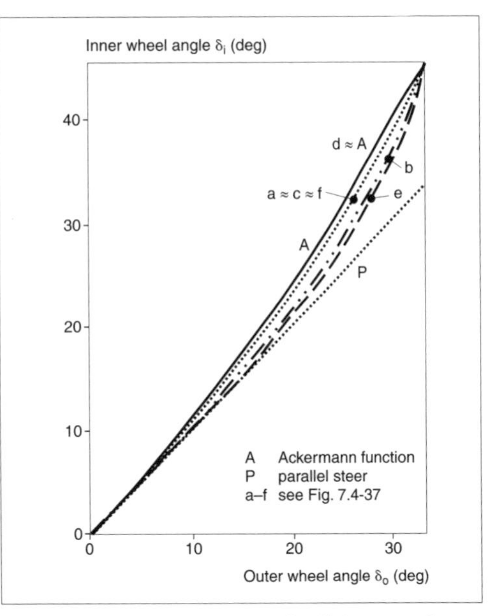

Fig. 7.4-42 Steering functions for linkage configurations of Fig. 7.4-41 (after [1]).

ing box, Pitman arm, center link (also known as "drag link"), idler or slave arm, and tie rods to the steering arms. This fact, combined with the above discussion of tie rod orientation, provides two more possible steering configurations: Pitman arm pointed forward or rearward (Fig. 7.4-41, c–f). A transversely mounted center link connects opposite sides of the vehicle. This transmits steering commands to the idler arm, a lever mounted symmetrically to the Pitman arm. A further doubling of possible design configurations results from the manner in which the inner tie rod ends are actuated: directly by means of the Pitman arm or by the center link (Fig. 7.4-41, g–j). It should be noted that for tie rod actuation via the center link, all four joints must lie on a common line of action; otherwise, the center link must be mounted axially parallel to the Pitman arm to prevent rotation.

Decisive factors in choosing one of these variations are the spatial possibilities within the vehicle and the desired function. Fig. 7.4-42 shows characteristic steering functions for steering linkage arrangements with fixed kingpin axes, as commonly found in passenger cars.

Experience has shown that the best results are obtained with front-mounted outer tie rod ends. Combined with an opposite-rotating Pitman arm, this permits the best approximation of ideal steering function. Unfortunately, this arrangement is nearly impossible in a passenger car, as a forward-mounted steering box and Pitman arm would cause considerable space problems.

Effect of overall vehicle and suspension concepts on steering

Unintended steering angles as a result of vertical wheel movement while the vehicle is in motion must be avoided, or their magnitude and direction must be such that vehicle guidance and stabilization is assured under all conditions. This function is performed by the steering linkage. Using as an example strut suspension, which enjoys wide application in modern road vehicles, we will illustrate various steering linkage configurations in combination with steering box designs. In keeping with the desire for optimum synthesis of function, cost, and weight in all vehicle classes, the trend is toward rack and pinion steering; therefore, this design will be addressed first.

All previously mentioned steering linkage configurations were discussed only in terms of orientation in a single plane. For optimum decoupling of steering from spring motion, however, the proper tie rod length for various vertical positions must be determined. For front-drive vehicles, transversely mounted engine, strut suspension, and rack and pinion steering, there are, for example, only two realizable positions: below the body structure at the level of the wheel axis or above the body structure and tires, in both cases behind the front axle. Since strut suspensions replace the upper transverse link with sliding motion of the damper, the center of rotation of the imaginary replacement link lies on the perpendicular to the strut axis through the upper mount, at infinity. In Fig. 7.4-43, determination of tie rod length for this case is shown using a simplified (planar) method by *Bobillier*. If a tie rod location at the level of the lower transverse link is chosen, the tie rod length L_L approaches that of the lower transverse link. The higher the tie rod location, the longer it must be (L_U) to achieve optimum bump steer. For this reason, strut suspensions require that tie rods actuated by the outer ends of a rack and pinion steering box can only be realized with the lower location option. High-mounted tie rods re-

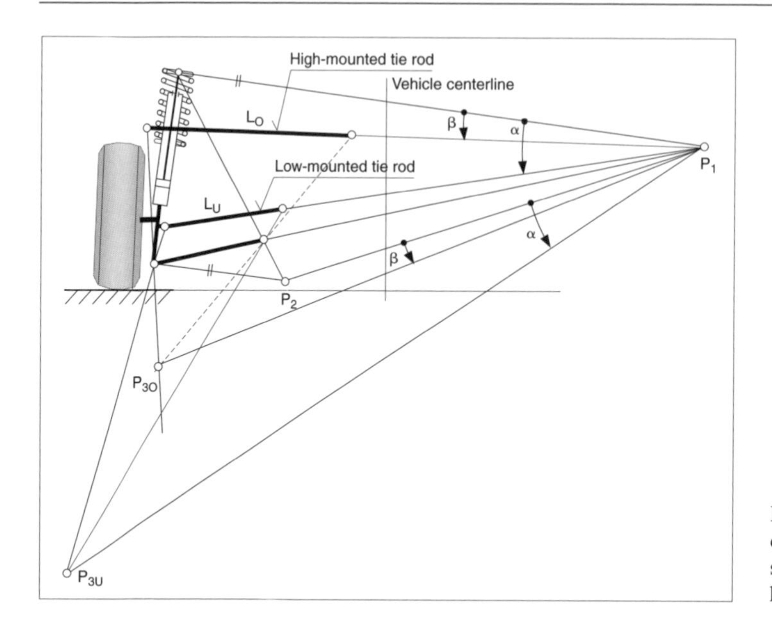

Fig. 7.4-43 Determination of tie rod length for strut suspensions with low- or high-mounted tie rod.

quire centrally actuating steering boxes (see Fig. 7.4-49, bottom).

Steering angle limitations

In the steering system design process, even at as early a stage as determining optimum steering function and the resulting linkage parallelogram or triangle, it is advantageous to determine steering angle limits imposed by linkage members. Inner and outer tie rod transmission angles are important to assure proper function of the steering linkage. If these angles are too small, it means that the linkage may be fully extended under the effect of external forces and therefore no longer capable of assuring exact wheel location, or in extreme cases that the linkage may even go into a dead point ("toggle point") condition. To evaluate the transmission angle, for spatial (nonplanar) suspensions, the effective moment arms must be perpendicular to the axis of rotation for all possible wheel positions. Without mechanical steering lock limits, the transmission angle of passenger car tie rods should never be $<25°$.

For front-drive vehicles, additional steering lock limitations imposed by halfshafts should be examined. Advances in flexible joint technology now permit extreme

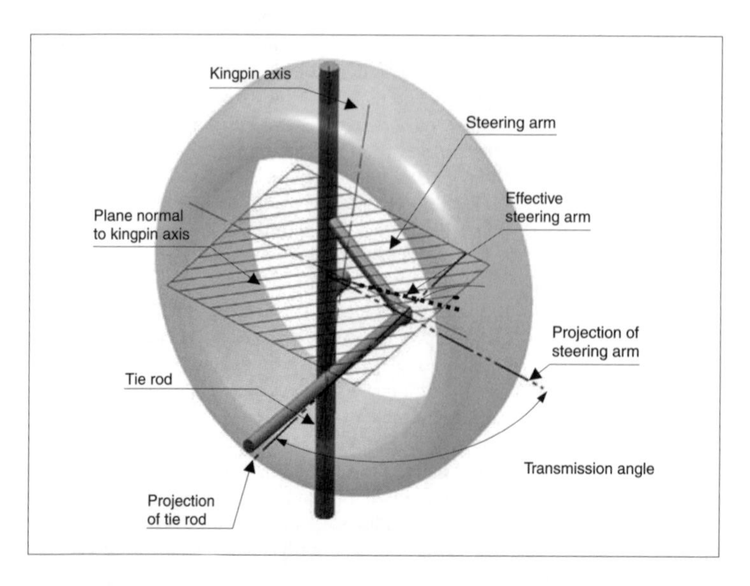

Fig. 7.4-44 Transmission angle between steering arm and tie rod.

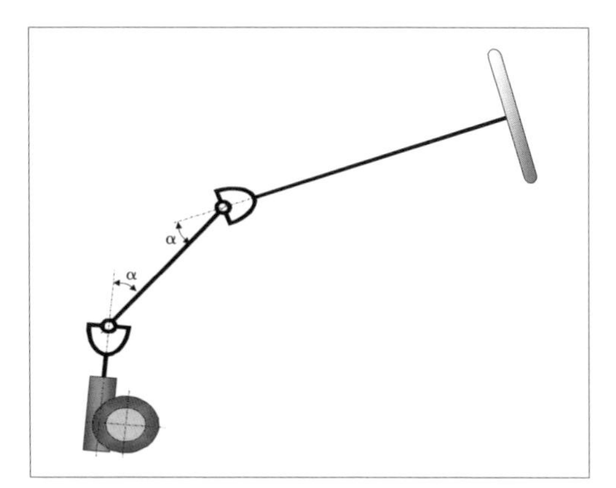

Fig. 7.4-45 Example of a "planar" steering column with two Hooke joints in a "W" arrangement with equal bending angles.

halfshaft angles of nearly 50°. This means that optimized front-drive concepts are at no significant disadvantage compared to conventional drive.

Overall steering ratio

Steering ratio, as the relationship between steering wheel rotation angle and (road) wheel angle, has a lower limit represented by the directness of steering response at high vehicle speeds; here, values below 14:1 are rare. The upper limit follows from the acceptable level steering effort in parking; this is therefore intimately related to availability of power steering assist and seldom exceeds a ratio of 20:1. Choice of steering ratio is also affected by the targeted vehicle market positioning (sporty/agile or emphasis on comfort) [12]. In design terms, the overall steering ratio represents the product of steering linkage ratio and steering box ratio. The steering linkage ratio must be determined by averaging steering angles for inside and outside wheels. For known effective steering arms (Fig. 7.4-44), this may be determined from the steering arm/Pitman arm ratio.

Steering shaft

Very few applications achieve a simple, straight connection (steering shaft) between the steering wheel and steering box. Customary practice is articulated steering columns with one or two joints. For small bending angles (up to ~5°), Hooke joints may be replaced by a flex disc ("rag joint"). At angles >15°, steering columns with a single Hooke joint exhibit rotational nonuniformity, U, readily perceived by the driver:

$$U = \frac{\omega_{2\max} - \omega_{2\min}}{\omega_1} = \tan\beta \cdot \sin\beta \qquad (22)$$

In such cases, an intermediate shaft and second Hooke joint should be inserted (Fig. 7.4-45). The rotational nonuniformity after the first joint may be canceled by the second joint. To achieve this, the following conditions must both be met:

— The joint angles of both joints must be identical.
— The intermediate shaft forks must be simultaneously coplanar with planes A and B formed by the input and output shafts (Fig. 7.4-46, [13, 14]).

For planar steering shaft assemblies, this means that the forks of the center shaft must be coplanar. For steering spindles canted with respect to each other, the forks must be offset (twisted) by the included angle φ between the angled planes of both joints. Both Hooke joints may be angled in the same direction (W arrangement) or opposite directions (Z arrangement).

Redirection of steering torque in the joint creates an additional moment, M_Z, in the fork (Fig. 7.4-47). This added moment gives rise to bearing forces which load the shafts in bending. For practical purposes, this means that joint angles >30° should be generally avoided, as the resulting high transverse forces combined with ex-

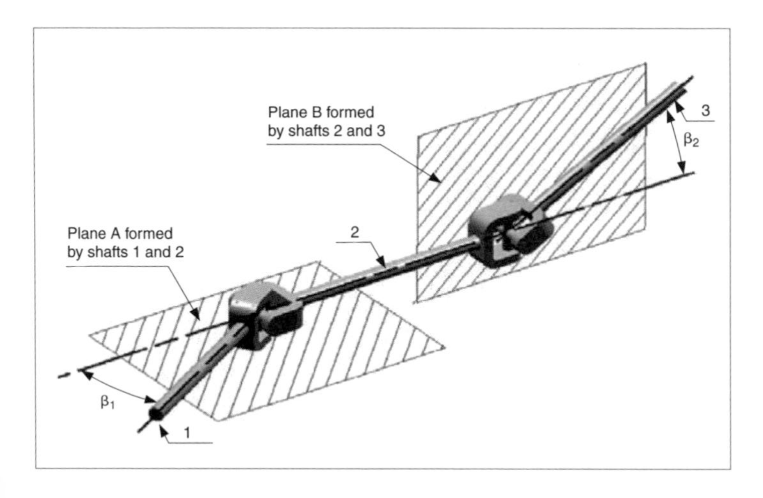

Fig. 7.4-46 Example of a "cranked" steering column with two Hooke joints in a "Z" arrangement with complete compensation.

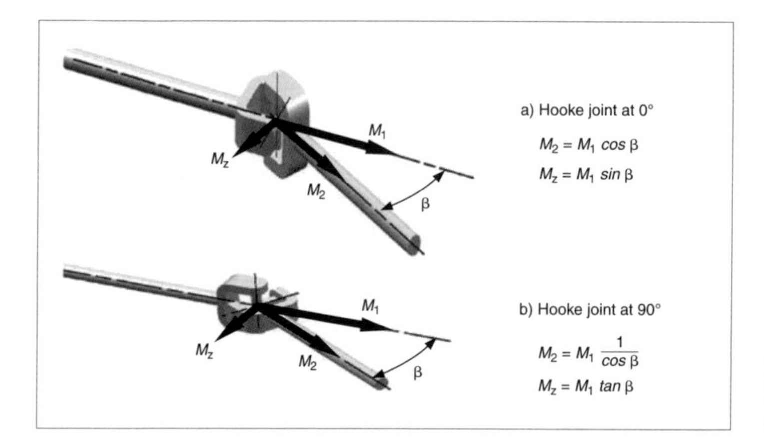

Fig. 7.4-47 Special cases for Hooke joint orientation 0° (a) and 90° (b).

isting bearing friction and finite bearing stiffness will result in perceptible steering moment variations, even for steering shaft designs with equal joint angularity.

References

[1] Matschinsky, W. *Radführungen der Straßenfahrzeuge*, second edition, Springer Verlag, 1998; English edition *Road Vehicle Suspensions*, Professional Engineering Publishing Ltd., 2000.
[2] Braess, H.-H. "Ideeller negativer Lenkrollhalbmesser." *ATZ* 77, 1975, pp. 203–207.
[3] Braess, H.-H. "Beitrag zur Fahrtrichtungserhaltung des Kraftwagens bei Geradeausfahrt unter besonderer Berücksichtigung des Lenkrollhalbmessers." *ATZ* 67, 1965, pp. 218–221.
[4] Paefgen, F.J., U Hackenberg, and E. Müller. "Der neue Audi A4, Teil 2 Fahrwerk und Antriebsstrang." *ATZ* 96, 1994, pp. 734–748.
[5] Hartig, F., and M. Reichel. "Das Fahrwerk des neuen 3er." Special issue of *ATZ* as a supplement to Vol. 5, 1998.
[6] "Die neue A-Klasse von Daimler-Benz." Special issue of *ATZ*, 1997.
[7] Jeglitzka, M., H.-G. Riedel, S. Wolfsried, and U. Zech. "Das Fahrwerk der Mercedes-Benz S-Klasse, leicht und agil." *ATZ* and *MTZ* special issue, 1998, pp. 142–153.
[8] Hentsche, P., and G. Wahl. "Das Fahrwerk des Porsche Boxster." *ATZ* and *MTZ* special issue, 1996, pp. 34–50.
[9] Matschinsky, W., C. Dietrich, and E. Winkler. "Die Doppelgelenk-Federbeinachse der neuen BMW-Sechszylinderwagen der Baureihe 7." *ATZ* 79, 1977, pp. 357–365.
[10] Matschinsky, W. "Bestimmung mechanischer Kenngrößen von Radaufhängungen." Dissertation, TU Brunswick, 1992.
[11] Dödlbacher, G., and H.-G. Gaffke. "Untersuchung zur Reduzierung der Lenkungsunruhe." *ATZ* 80, 1978, pp. 317–322.
[12] Weir, D. H., and R. J. di Marco. "Correlation and Evaluation of Driver/Vehicle Directional Handling Data." SAE 780010, 1978.
[13] Reinecke, W. "Konstruktions-Richtlinien für die Auslegung von Gelenkwellenantrieben." *MTZ* Jahrgang 19, No. 10, 1958, pp. 349–352.
[14] Eugen Klein KG Gelenkwellen; company literature.
[15] Reimpell, J. *Fahrwerktechnik Grundlagen*, Third edition, Vogel Buchverlag, 1995.

7.4.5.2 Steering boxes and linkages

The task of steering boxes and linkages is to translate driver input, in the form of steering wheel rotation, into defined road wheel steering angles at the inside and outside wheels (in the case of passenger cars, front wheels). The most commonly encountered steering box configurations are described below.

Recirculating ball steering

For decades, passenger cars and light trucks have used the proven concept of recirculating ball steering gear (Fig. 7.4-48), preferably with hydraulic assist.

Rotational motion at the steering wheel is applied to the steering box input shaft via the steering column. The lower end of this input shaft is in the form of a worm, whose rotation moves the steering gear piston axially by means of a continuous train of balls. In the hydraulically assisted version, one side of the piston, depending on direction, is subjected to oil pressure, thereby hydraulically supporting mechanically introduced axial motion of the piston. Piston movement rotates a geared sector shaft, mounted perpendicular to the piston.

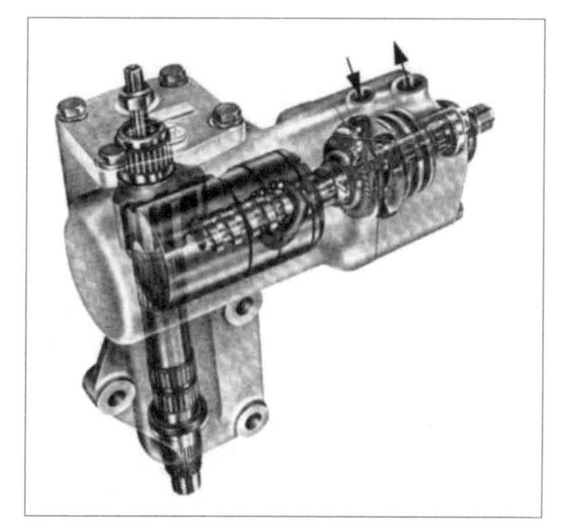

Fig. 7.4-48 Hydraulically assisted recirculating ball steering box with rotary valve (source: ZF).

The Pitman arm is attached to the sector shaft and in turn moves the steering linkage by means of a ball joint. The steering linkage consists of a center link, with ball joints at either end, connecting to tie rods. To guide the steering linkage, an idler arm is mounted symmetrically to the Pitman arm on the right side of the vehicle. The outer tie rod ends are attached to the wheel carriers. To adjust front wheel toe-in, the tie rod length may be varied by means of adjusting threads. Vertical adjustment of the Pitman arm at the steering box can in some cases be used to compensate for vertical location tolerances of the inner tie rod ends. This permits designers to achieve identical toe-in curves for left and right wheels throughout the spring travel range, which ensures straight-line stability during simultaneous deflection of both front springs. The primary steering linkage layout criteria are the desired suspension kinematics, the clearance requirements, and the forces to be transmitted (resistance to buckling).

The advantages of recirculating ball steering are a high degree of steering comfort combined with low impact sensitivity, thanks to higher overall system elasticity. Furthermore, variable layout possibilities for the steering linkage parallelogram offer designers greater freedom in realizing desirable toe-in curves and overall steering ratios, with low tie rod and steering box loads. Higher elasticity, however, may be a disadvantage in terms of on-center steering response and feel. The decisive factors in the recent decreasing popularity of recirculating ball steering for passenger cars are its considerable weight and cost disadvantages compared to rack and pinion steering.

Rack and pinion steering

Rack and pinion steering (Fig. 7.4-49) is the most common system found in modern passenger cars. Increasingly, it is fitted with hydraulic assist, even in smaller vehicles. It is gradually displacing recirculating ball steering, even in high-end vehicles.

In rack and pinion steering, steering wheel movement is transmitted by a (usually helical) pinion to cause lateral motion of a mating rack. A spring-loaded yoke opposite the point of rack engagement ensures minimal backlash between pinion and rack throughout the rack travel range.

Tie rods are attached to the rack by means of ball joints. The ball joints are sealed against dirt and water intrusion by bellows of rubber or synthetic material. Tie rods may be attached to the rack on both ends ("end take-off"), at the center ("center take-off") or both at one end. Tie rod outer ends are in turn attached to the wheel carriers by ball joints. To adjust toe-in, tie rod length is variable by means of threaded adjustment.

For mechanical layout of a rack and pinion steering box, two important parameters in addition to strength and space considerations are rack diameter and rack length. Rack diameter is largely determined by loading in the so-called curb pressure test. In this, no permanent rack deformation is permitted given a blocked front wheel at the permissible vehicle front axle load, full hydraulic steering assist, and 80 Nm of steering wheel torque. Rack length for hydraulically assisted steering with end-mounted tie rods is at least six times rack travel (2× rack gearing + 2× hydraulic piston travel + 1× travel between end of rack and inner rack seal + 1× travel between axial joint to outer rack seal). Added to this are axial space requirements for seals and runout of rack gear cuts. If this required rack length cannot be fit into the car, compromises in turning circle (shorter rack travel) or vehicle handling (shorter tie rods, therefore altered toe-in curves in jounce and rebound) must be expected.

Better conditions for sufficiently long tie rods are offered by the previously mentioned central tie rod ("center take off") attachment. Another possibility is mounting the hydraulic boost cylinder externally, parallel to the rack.

Advantages of rack and pinion steering compared to recirculating ball steering are:

— Lower cost, lower weight thanks to simpler construction, and elimination of steering intermediate linkages and levers

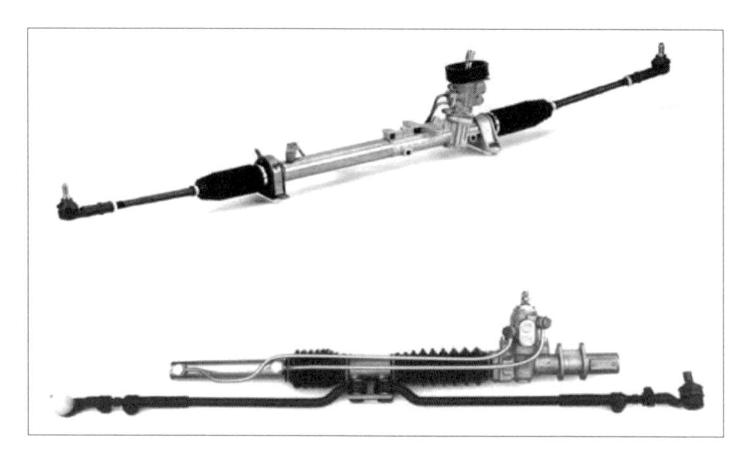

Fig. 7.4-49 Hydraulically assisted rack and pinion steering gear (Source: ZF).

— Smaller space requirements, but with lower-mounted steering and, under some conditions, extra space required below engine oil pan
— Lower steering elasticity and therefore more direct steering response

Disadvantages are greater sensitivity to impacts and greater tendency toward transmission of steering disturbances.

7.4.5.3 Steering assist

The previous section mentioned that even in small cars, manual steering is being replaced by assisted steering, in which steering forces that must be applied by the driver are reduced by means of hydraulic or electric steering aids.

Nevertheless, manual steering offers (mechanical) steering assist in the form of steering gear ratio and steering linkage ratio. For larger steering box ratios, forces at the steering wheel are correspondingly reduced, but at the price of greater steering wheel rotation angles. Vehicles using such designs may easily convey an impression of ungainliness; agility and steering feel are increasingly lost. Up to a certain point, this disadvantage may be compensated in manual rack and pinion steering gear by means of variable steering ratio, whereby rack gearing is produced using a special gear rolling process to provide gearing with variable pitch (modulus) and pressure angle. The pinion has normal gearing. Gearing ratios are designed so that steering is sufficiently direct on center and becomes increasingly indirect toward the steering limit stops in order to reduce parking effort. However, the effect of power assist cannot be attained using this method. Such a variable ratio

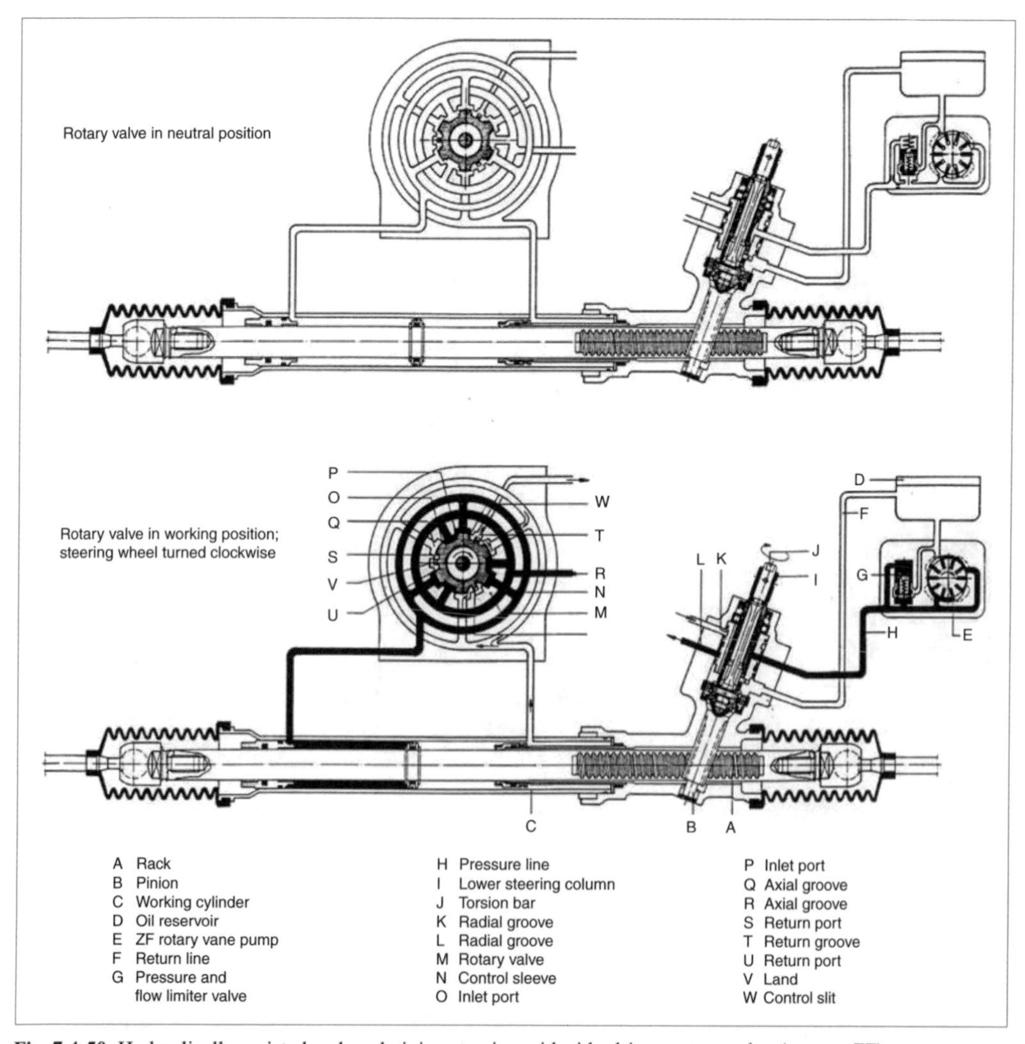

Fig. 7.4-50 Hydraulically assisted rack and pinion steering with side-driven rotary valve (source: ZF).

between piston and sector shaft is also possible in recirculating ball steering systems.

Hydraulic steering assist

In most power-steering–equipped passenger cars currently on the market, assist is provided by hydraulic power steering. The operational principles of hydraulic power assist will be explained using the example of power-assisted rack and pinion steering with a rotary valve (Fig. 7.4-50).

A hydraulic pump, *E*, usually driven from the engine by means of a V-belt (in modern cars, usually a rotary vane pump), provides the necessary pressurized oil by means of a high-pressure flexible hose *H* to a rotary steering valve mounted in the steering box. In the straight-ahead position, a constant flow of oil passes through the rotary valve, which is in its neutral position (open center), and back through a return line to the oil (power steering fluid) reservoir. Pressure in both chambers of the power cylinder is equal and represents the circulation pressure of ~2–5 bar due to flow losses in the steering system. In this position, there is no steering assist.

As the steering wheel is turned in a clockwise direction, the steering rack and therefore the power piston are displaced to the right. As piston movement is to be supported by pressurized oil, this must be directed into the left power cylinder space. Three control grooves in the rotary valve are shifted clockwise, and the inlet ports (*P*) for oil supply are opened further. The inlet ports (*O*), however, close and prevent pressurized oil supply to the axial grooves (*Q*) in the control sleeve. With the valve in its working position, pressurized oil flows through the inlet ports (*P*) to the control sleeve lower radial groove (*L*) and from there to the left power cylinder space, providing power piston movement for hydraulic assistance. The closed control ports (*W*) prevent oil from flowing back to the reservoir. Oil from the right power cylinder space is displaced and flows through the upper radial groove (*K*) in the control sleeve to the return grooves (*S*) in the rotary valve. The lands (*V*) prevent direct admission of oil to the control sleeve return grooves (*T*), which are always open to the oil reservoir. If the steering wheel is turned counterclockwise, hydraulic assist is applied to the right working cylinder space.

The required degree of hydraulic assist depends on the worst-case maximum required tie rod force, obtained at maximum wheel lock with maximum vehicle front end load and maximum friction with worst possible tire choice plus service brake actuation. Other parameters are, as already mentioned, overall steering ratio (steering box and steering linkage ratios) and the maximum permissible driver-applied steering wheel torque (steering effort). On the basis of the necessary assist force and the maximum pressure supplied by the pump, one can calculate the required piston surface area and, given steering rack diameter—which is usually set on the basis of strength-of-materials considerations—calculate the required working piston diameter and steering box diameter at the cylinder.

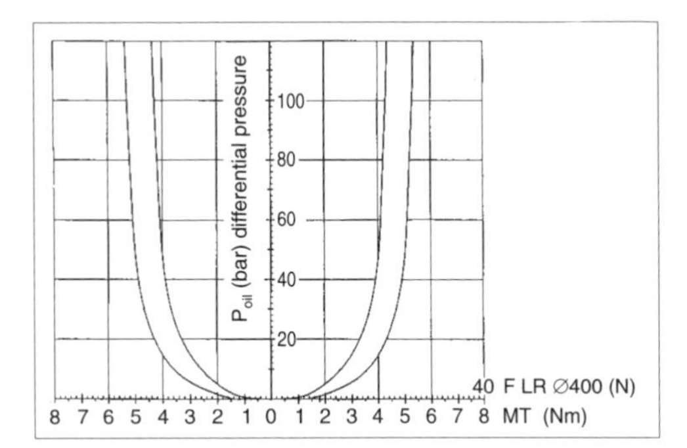

Fig. 7.4-51 Power steering valve characteristic curve (source: ZF).

Another important steering system layout parameter is the volumetric oil flow rate to be supplied by the pump. This establishes the maximum steering angular velocity—that is, how quickly the driver can turn the steering wheel without hydraulic "stiffening" of the steering system.

As already mentioned, the task of the steering valve is to direct pressurized oil to the appropriate working chamber according to steering direction, therefore amplifying force on the appropriate side of the working piston, with the objective of reducing the required torque at the steering wheel. Dependence of applied steering effort on available oil pressure is represented in the so-called valve characteristic curve (Fig. 7.4-51), which is a direct measure of available steering assist.

In most cases, with hydraulically assisted power steering, the pressure difference between the two working chambers is measured. For recirculating ball steering boxes, measuring pressure differential is difficult; therefore, oil pressure is measured at the steering box input. Essentially, the same valve characteristic curve results, but it is displaced by the circulation pressure within the system.

The valve characteristic curve may be adapted to application-specific requirements for individual steering wheel tuning (steering effort, steering feel) by means of modifying valve control surfaces and changing the valve input torsion bar stiffness. One possible means of changing the characteristic curve while the vehicle is in motion is offered by speed-sensitive power steering. The ZF company has developed its "Servotronic" electronically controlled power steering for such applications.

Servotronic works purely in response to vehicle speed; that is, it controls steering assist only in response to speed indicated by an electronic speedometer, independently of engine speed. A microprocessor evaluates the speed signal and establishes the degree of hydraulic assist and therefore steering wheel effort. This parameter is transmitted to the steering system rotary valve in the form of electrical impulses processed by an electrohydraulic converter, which alters hydraulic assist in response to vehicle speed. The unique layout of steering characteristics in such a system (Fig. 7.4-52) results in

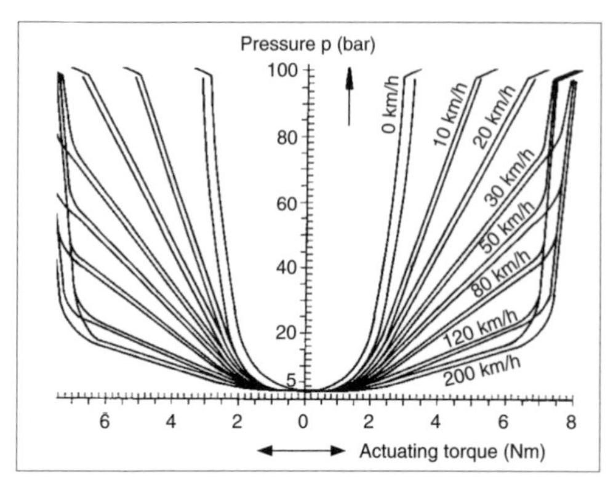

Fig. 7.4-52 Servotronic valve characteristic curves (source: ZF).

very low steering effort in parking and with the vehicle stationary but the steering feel becomes increasingly similar to that of a purely mechanical steering system as speed climbs. This permits precise, accurate steering at high speeds. It should be noted that at no point is oil pressure or volumetric flow rate reduced, and therefore the steering assist function is always assured, even in emergency situations such as rapid steering corrections (steering wheel angular velocities up to 800°/s). These properties provide an extraordinarily high degree of steering precision and safety while simultaneously achieving the greatest steering comfort, thus largely resolving the goal conflict between low parking effort and high steering precision at high speeds.

It should be mentioned that variable steering ratios, discussed above in connection with manual steering boxes, can of course also find application in power-assisted steering, but with reversed objectives. Manual variable-ratio steering boxes have less direct steering toward full steering lock; but in the case of power-assisted variable-ratio steering, in order to reduce parking effort, the ratio at greater steering angles becomes smaller and therefore more direct because power assist eliminates the need to consider low parking effort. This results in very few turns being needed lock-to-lock ($\sim\pm1.5$ turns), which provides optimum vehicle handling in parking and reversing movements.

Electrohydraulic steering assist

A special version of hydraulic steering assist is represented by electrohydraulic assist. In this, the oil pump (rotary vane or gear pump) is not engine driven but is driven rather by an electric motor.

The individual components—electric motor, oil pump, oil reservoir, and electronic control module—may be incorporated in a single compact unit, the so-called power pack (Fig. 7.4-53). The power steering box remains largely unaltered (with the exception of a modified valve characteristic curve).

Advantages of such a power pack include the resulting compactness of the power steering unit, in some cases permitting preassembly of the steering system. Of course, electrohydraulic steering also enables power assist even with the engine shut off (e.g., in coasting, at stoplights or railroad crossings); in the event the engine dies, steering assist is still available.

The electric motor may be controlled in response to demand so that the pump is only activated while steering. This permits lower energy consumption and therefore improved fuel economy. Speed-sensitive power steering, achieved at relatively high added cost in the case of purely hydraulic steering assist, is simply and economically realized by means of electronic motor control in the case of electrohydraulic assist.

Disadvantages of this configuration are, at present, higher system costs than conventional hydraulic power steering and the added weight of the electric motor, which is only partially compensated for by the elimination of oil suction and return lines. Furthermore, due to attainable power limitations, application of electrohydraulic steering is not yet possible in vehicles with high front-end loads.

Electric steering assist

In purely electrically driven steering systems, the necessary steering assist is provided directly by an electric motor which may be located at any of several points in the steering system. At present, four different locations enter consideration, each with unique advantages and disadvantages:

— The servo unit, consisting of an electric motor, worm drive, electronics, and sensors, is integrated within the steering column (Fig. 7.4-54, a).

Advantages:	Limited space requirements; steering column and power assist unit assembled as a ready-to-install unit.
Disadvantages:	Possible problems with crash behavior; noise, if servo unit is placed within the vehicle cabin; assist torque is transmitted by

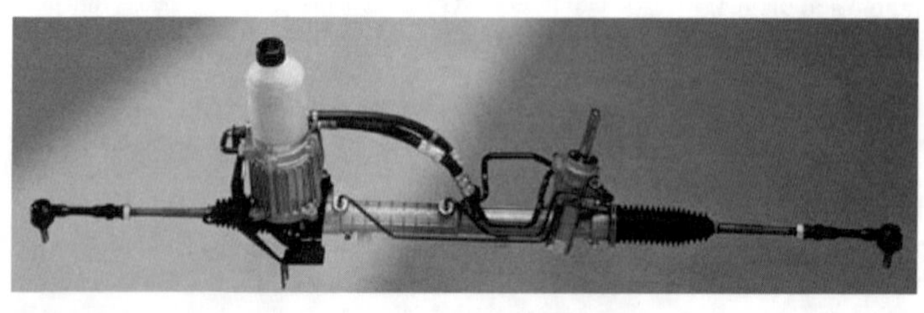

Fig. 7.4-53 Rack and pinion steering gear with electrohydraulic assist unit ("Powerpack"; source: *TRW*).

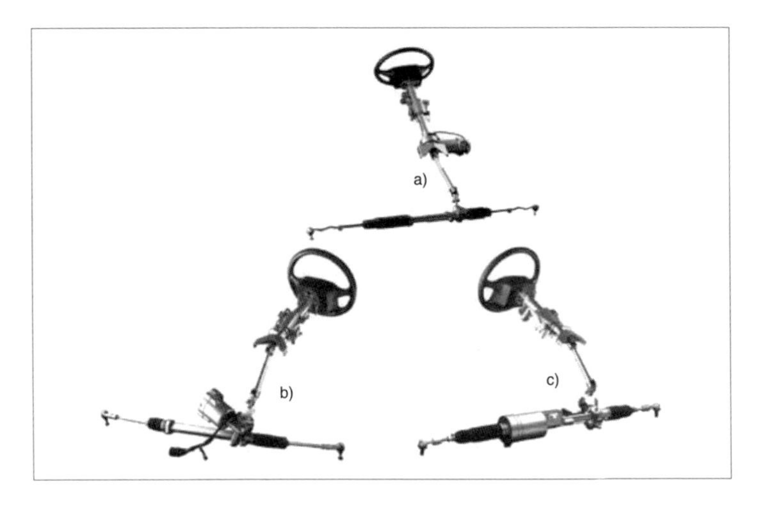

Fig. 7.4-54 ZF Servotronic configuration options (source: ZF).

steering shaft, therefore only suitable for smaller vehicles with low steering forces.

— The servo unit acts on the steering pinion (Fig. 7.4-54, b).
  Advantages: Steering box and servo unit are a single assembly, power assist acts directly on the pinion, therefore higher steering torque is possible.
  Disadvantages: Greater space demand within the engine compartment, possible thermal problems.

— A variation on this configuration is so-called dual pinion architecture, in which (for a left-hand-drive vehicle) the servo unit is located where the steering column would attach to the steering box on a right-hand-drive vehicle; for a right-hand-drive application, these are reversed. One advantage is that space normally reserved for a right-hand-drive steering box configuration is used for the left-hand-drive application's power assist unit. Disadvantages include cost multiplication and possible tolerance problems due to requirement for doubled pinions and rack gearing.

— The power assist unit acts directly on the steering rack (Fig. 7.4-54, c). The electric motor is concentric with the rack and transmits torque by means of a recirculating ball drive.
  Advantages: Assist force is generated precisely at the spot where it is needed. This permits realization of greater rack forces.
  Disadvantages: Higher cost, possible tolerance problems due to recirculating ball drives, possible thermal problems, tight fit to oil pan or front cross-member.

In general, it may be said that with the current state of the art of electric steering, limited as it is by 12-V electrical systems, as yet no luxury-class vehicles can be fitted with electric steering assist. Conversion to higher-voltage electrical systems (e.g., 42 V) may result in a definite push for new development (see Section 5.6).

Also, considerably more complex safety measures and higher overall costs compared to conventional hydraulic assist pose obstacles to rapid adoption. On the other hand, electric steering offers an advantage over hydraulic solutions in terms of greater fuel savings through demand-appropriate control and a high level of service friendliness, as oil level and sealing checks customarily applied to hydraulic systems are eliminated. Furthermore, electric steering offers rapid and manifold variation possibilities with respect to steering tuning merely by changing software; for example, the assist force may be programmed to exert control in response to speed, load, or steering angle.

In the event the abovementioned problems are solved, nothing stands in the way of greater proliferation of electric steering, even in heavier vehicles.

References

[1] Duminy, J. "Contribution of modern steering systems to improve driving stability." Third EAEC Conference, Strasbourg, 1991.

[2] Junker, H. "Moderne Lenkungstechnologie. Von den Anforderungen zur technischen Realisierung." *Automobil-Industrie*, Vol. 4/90, pp. 379–389.

[3] Stoll, H. *Fahrwerktechnik: Lenkanlagen und Hilfskraftlenkungen*. Jörnsen Reimpell, editor. First edition, Vogel Buchverlag, 1992.

[4] "ZF-Kugelmutter-Hydrolenkungen für PKW und leichte Nutzfahrzeuge." Published by the Steering Technology Department, ZF Friedrichshafen AG, 1995.

[5] "ZF-Zahnstangen-Hydrolenkungen mit konstanter und variabler Übersetzung." Published by the Steering Technology Department, ZF Friedrichshafen AG, 1993.

[6] "ZF-OCE-Lenksystem, Elektrohydraulisches Lenksystem für PKW und Transporter." Published by the Steering Technology Department, ZF Friedrichshafen AG, 1997.

[7] "ZF-Servolectric, die elektrische Servolenkung." Published by the Steering Technology Department, ZF Friedrichshafen AG, 1998.

7.4.6 Active steering systems

7.4.6.1 Introduction

Section 7.1 discussed the various mechanisms for generating tire lateral force, and the possibility was presented of affecting these forces by means of small steering angle changes applied by a vehicle dynamic control system. In contrast to design-mandated kinematic and elastic steering angle corrections, these are not reactions to external forces and moments, but rather are active in nature and may be represented as functions of several different input and vehicle dynamic parameters. In this manner, it is possible to create control strategies which result in specific vehicle dynamic behavior that can deviate appreciably from familiar behavior (up to the limits imposed by the laws of physics) and that can stabilize vehicle behavior in every possible situation.

If we ignore individual points of steering assist detail development, which may be attributed to the class of active steering systems (Section 7.4.6.2.1), production versions of handling-dynamic steering systems have heretofore been applied exclusively to rear axles. The reason for this was that improved transient handling at higher speeds could be achieved by influencing the buildup of rear axle lateral forces over time and by reducing the vehicle slip angle (body yaw angle) using a nonfeedback control system without consideration of vehicle dynamic parameters, or indeed with no more than a mechanical connection between rear-wheel steering and front-wheel steering.

Active front-axle steering systems that permit a front-wheel steering angle which deviates significantly from that commanded by the driver were realized at an early stage for research purposes [1, 2]. Not until the mid-1990s was this idea revived by vehicle manufacturers and suppliers for possible installation in production vehicles [3, 4]. Work on such systems is currently still in the research and early development phases.

The same is true of systems which seek to keep the vehicle in its lane by means of active steering moments, so-called "heading control systems" [5].

Along with weight, cost, and assurance against faults, one of the main tasks is to devise active steering system interventions in such a way that the driver either perceives them as positive, does not perceive them at all, or at least does not perceive them as bothersome. Further, it must be assured that in no case does interaction of driver and active steering system lead to deterioration of overall system behavior. The applicable regulatory standards with regard to steering systems require an unambiguous time/distance relationship between road wheel and steering wheel angles, and therefore the standards still represent an obstacle for market introduction of active systems.

7.4.6.2 Active front-wheel steering

Beginning in the mid-1990s, active front-wheel steering moved into the foreground under the "steer-by-wire" concept. This occurred in the wake of "fly-by-wire" making its way from aircraft to automobiles. In many cases, the "steer-by-wire" label does not actually mean interruption of any direct mechanical connection between steering wheel and road wheels but rather a specific solution effort using electromechanical actuators. However, this restrictive interpretation is at odds with aviation usage of the term. For this reason, the more general concept of "active front steering" was chosen here, as it encompasses all active steering systems that can be applied to improve the interplay of driver and vehicle.

The following sections classify familiar active front-wheel steering solutions in terms of function. Because of their relatively recent development and their potential for subsequent rapid evolution of active front-wheel steering, no claims of comprehensive treatment in this discussion are made.

7.4.6.2.1 Active power steering

Characteristic of such steering systems is superposition of a controlled torque over driver-induced steering wheel torque. Such superposed torque is a function of additional handling parameters. Modulated by a controller in response to the situation at hand, it may support driver action to a greater or lesser degree, act opposite to driver actions, or control steering independently of the driver. As in current steering systems, a largely fixed geometric relationship between steering wheel angle and road wheel steering angle is retained.

This category of steering systems includes:

- Steering systems with variable (e.g., speed-sensitive) actuation forces. Historical examples using hydromechanical controls include steering systems used by Citroën on its SM, CX, and XM models. An electronic controller may be found in ZF's Servotronic system.
- Steering systems which recognize the "open loop" mode and provide active return to center along with damping of steering wheel/vehicle oscillations. Early complex hydromechanical systems are known from Citroën's SM, CX, and XM models; realization of these functions is a relatively simple option on electric and electrohydraulic power steering [6].
- Steering systems with situation-relevant application of appropriate moments in the sense of driver assistance [5]. Such "heading control systems" are predominantly still in research or early development, as sensible application is closely tied to progress in developing systems that recognize vehicle surroundings.

7.4.6.2.2 Superposed steering systems

Superposed steering systems are characterized by an additional steering angle which may be applied as needed by an actuator on top of a driver-input steering angle. This additional angle is defined by a controller and serves to increase vehicle agility and stability. There is also the possibility of compensating for disturbing forces and moments and to achieve a road wheel angle/steering wheel angle gradient as a function of road speed and steering wheel angle.

Electric or hydraulic actuators may be employed. In superposed steering systems, a direct mechanical connection between steering wheel and steered road wheels is retained so that in the event of actuator failure, conventional steering function remains intact. This property greatly simplifies meeting safety requirements, compared to the pure "steer by wire" systems described below.

A further property, generally regarded as positive, is direct, proportional transfer of aligning moments to the steering wheel. This eliminates the expense and effort in providing an artificial aligning moment, as is the case in pure steer-by-wire systems. This aspect has a significant effect on system weight and energy consumption.

Along with differentiation resulting from various control strategies, distinction may be made between the following superposed steering system configurations:

- Superposition of translational motion in the tie rod area, steering rack, or steering box. Translation by means of an actuator—for example, movement of the steering box housing—is added to steering rack relative motion (Fig. 7.4-55). Disadvantages are the increased space requirement for translation of the steering box and the associated change in steering geometry as well as a requirement for powerful actuators, since actuator intervention is not amplified by power steering.
- Superposition of rotational motion between steering wheel and steering box (Fig. 7.4-56). This configuration offers the advantage of permitting an unlimited superposed steering angle, taking advantage of power steering support and without altering steering geometry.

In terms of the possibility of using a controller to generate a steering angle independent of driver input, superposed active steering systems are indistinguish-

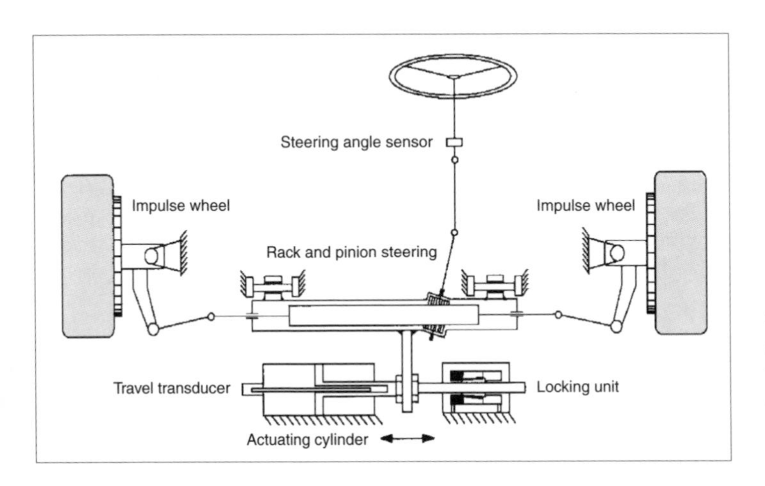

Fig. 7.4-55 Superposition steering based on steering box translation principle, hydraulic actuator, mechanically locked in the event of hydraulic or electronic failure [7].

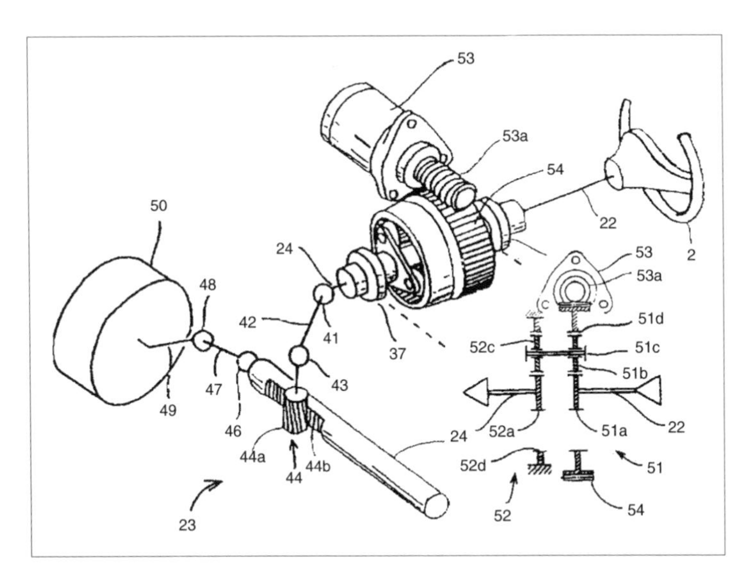

Fig. 7.4-56 Superposition steering based on angular superposition (rotation) by means of a planetary transmission, second planetary transmission to compensate for steering ratio change [Honda, German patent application DE 4326355).

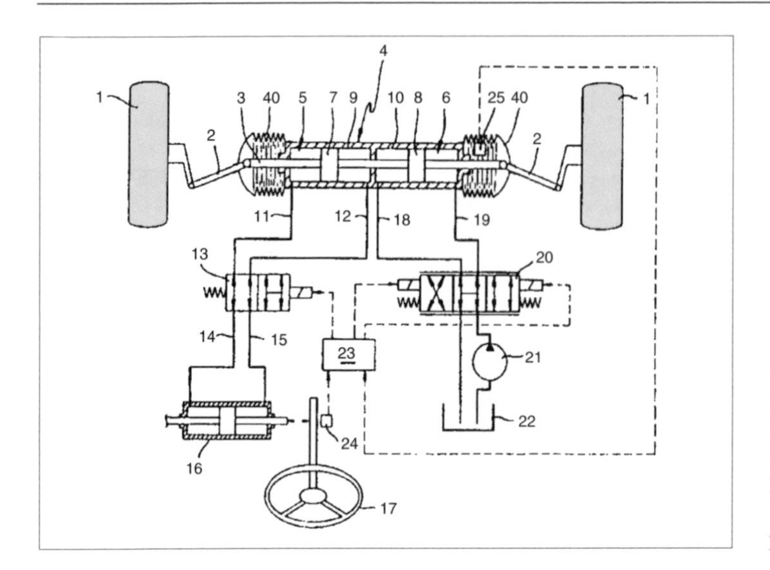

Fig. 7.4-57 Principle schematic of electrohydraulic steering system with hydraulic steering torque simulator and hydrostatic manual backup [DaimlerChrysler, European patent EP 0818382 A2].

able from steer-by-wire systems, which are treated in the next section.

7.4.6.2.3 Steer-by-wire systems

Systems included in this general classification have no mechanical connection between the steering wheel and the steered wheels; road wheel steering angles are entirely generated by a controlled actuator (Fig. 7.4-57).

As in the case of superposed steering, the fixed relationship between steering wheel angle and road wheel angle is eliminated. Furthermore, there is no mechanical transmission of road wheel aligning moment, providing a great degree of freedom in setting steering moments. This freedom, however, is bought at the cost of an active steering torque simulator and its associated control system.

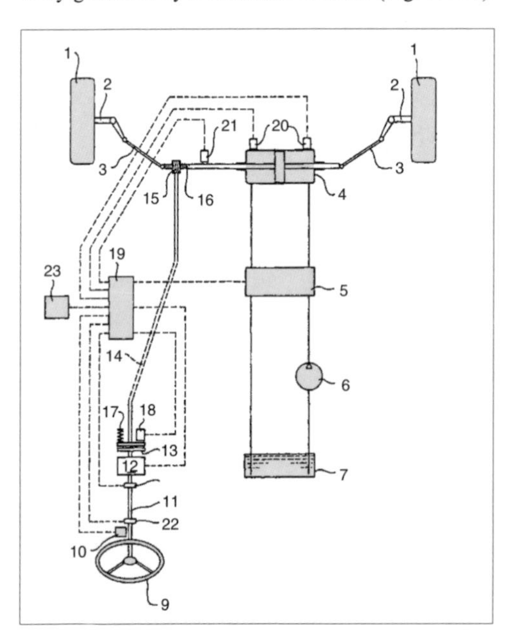

Fig. 7.4-58 Principle schematic of electrohydraulic steering system with electric steering torque simulator and mechanical manual backup; the steering torque simulator may be used to support the manual backup function [DaimlerChrysler, German patent DE 19755044-C1].

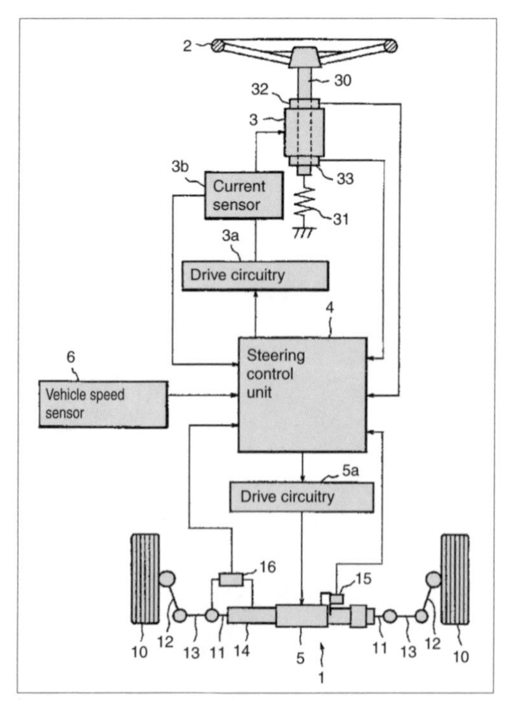

Fig. 7.4-59 Principle schematic of electromechanical steering system with electric steering torque simulator [Koyo Seiko, German patent DE 19806458 A1].

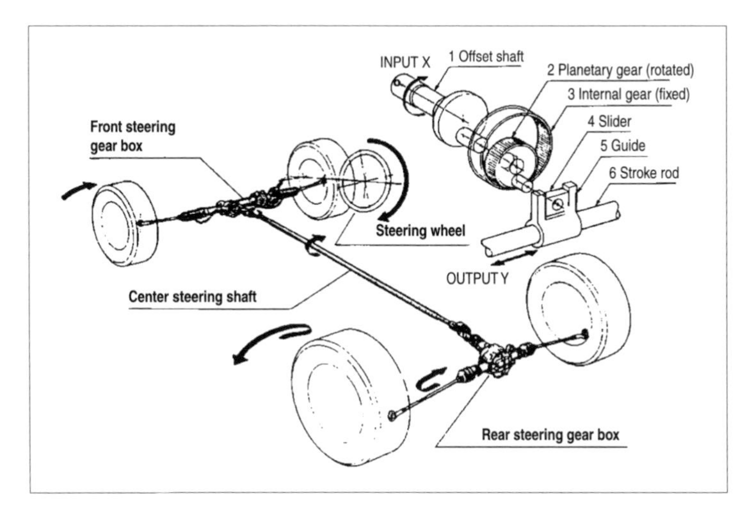

Fig. 7.4-60 Rear-wheel steering with mechanical coupling to front suspension [10].

If the steering wheel no longer serves to deliver at least part of the steering effort supplied by the driver to the steering system, other forms of steering actuation enter consideration. Some researchers are experimenting with automotive implementation of actuation systems familiar from aircraft practice. For motor vehicle applications, increased sensitivity to unintentional effects due to inertia of the driver's arms and the driver's entire body must be taken into consideration. This might introduce disturbing steering motions unintended by the driver. For this reason, as exemplified in [8], a hydraulically damped twist knob, insensitive to external disturbing forces, is preferred.

Historically, the steering wheel replaced the steering tiller, as it represents an ergonomic solution for increasing the travel over which manual force is applied and, for sufficiently low actuating forces, provides the necessary steering work even in the event of power assist failure.

If in laying out a steer-by-wire system one chooses an appreciably more direct steering ratio than in present-day power steering systems, in order to eliminate any risk associated with total system failure such a system must be redundant in terms of control, actuation, and energy supply. This increases cost, weight, space requirements, and complexity. Use of a manual backup with less direct steering ratio represents one alternative. An interesting solution set employing an electric steering moment simulator is shown in Fig. 7.5-58. If the system is switched to its manual backup mode, the simulator assumes the function of an electric power assist system.

Steer-by-wire systems may be classified according to the following characteristics:

- Common actuator for both wheels of a single axle or individual wheel steering. Because of its higher complexity, the latter can only be justified in special situations.
- Electromechanical (Fig. 7.4-59) or hydrodynamic (Fig. 7.4-57) actuators.
- The nature of aligning moment generation. Hydraulic or electric motors may be used for this purpose.
- Existence and type of manual backup. Mechanical backups employing a clutch (Fig. 7.4-58) and hy-

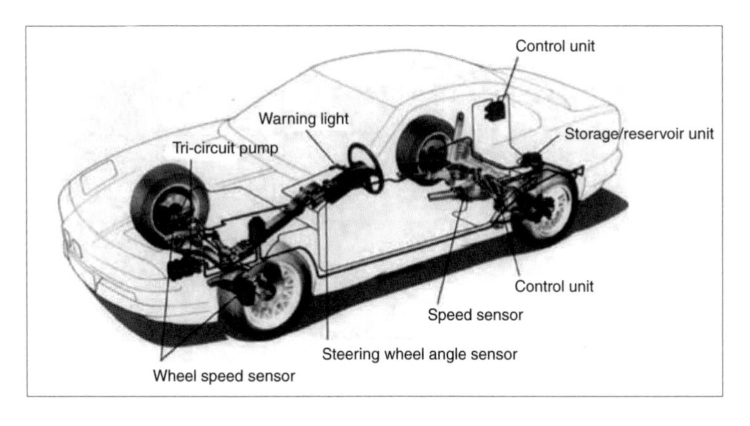

Fig. 7.4-61 Rear-wheel steering of BMW 850 Csi, 1990; overall system configuration [12].

drostatic systems (Fig. 7.4-57) are known from the literature.

- Control regulating algorithms for transfer functions between steering wheel and front-wheel steering angle, as well as vehicle dynamic conditional parameters.
- Stabilizing function strategy.
- Steering aligning moment strategies for optimizing driver/vehicle/environment system performance and for driver's positive subjective evaluation.

7.4.6.3 Active rear-wheel steering

Compared to active front-wheel steering, active rear-wheel steering systems offer possible advantages in decreasing the magnitude of vehicle slip angle as well as exerting an immediate influence on the time history of rear axle side force buildup. The small rear-wheel steer angle (2–3°) needed for improved vehicle dynamics is easier to manage in the event of a system failure. Therefore, it comes as no surprise that vehicle dynamic rear-wheel steering systems have been offered by several automotive manufacturers for quite some time. From a safety aspect, active rear-wheel steering may be regarded similarly to superposed steering systems in that if a fault is detected, returning to a neutral state suffices, or, in extreme cases, the system may be simply shut down. This presupposes that the rear wheels may be locked in the neutral position. With hydraulic systems, this may be achieved by means of preload springs [9]. For electromechanical actuators, self-locking drives are available. In electromechanical systems, a return to the straight-ahead position in the event of failure in position control or energy supply is only possible by means of an unacceptable degree of complexity.

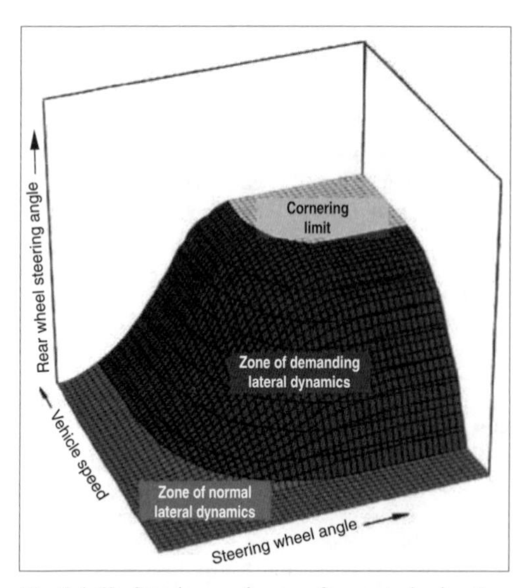

Fig. 7.4-62. Steering angle map for control of active rear-wheel steering as a function of steering wheel angle and vehicle speed [11].

7.4.6.3.1 Rear-wheel steering without vehicle dynamic control

Early systems (Honda Prelude, 1987) used a simple mechanical coupling to join rear-wheel steering to the front-wheel steering system [10]. Strictly speaking, this only represents a special case of all-wheel-steering, a long-established principle on special-purpose vehicles. But while the latter systems serve to reduce turning circle diameter and, for multiaxle vehicles, reduce tire wear, the steering system shown in Fig. 7.4-60 primarily addresses vehicle dynamic goals. Rear wheel/front wheel "same steer" affects buildup of rear axle lateral forces during transient maneuvers. This results in shorter vehicle lateral acceleration response time for steering angle changes while improving stability as excessive yaw rate overshoot is reduced.

Coupling rear axle steering movement to the front axle was achieved by means of a mechanical drive in such a manner that for large steering angles, as are only

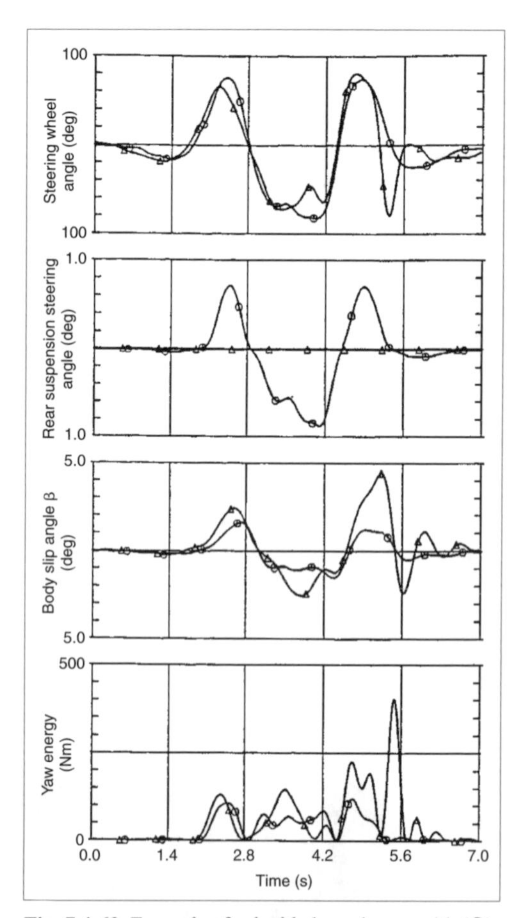

Fig. 7.4-63 Example of a double lane change with (○) and without (△) rear-wheel steering. The improvement in steering effort, body slip angle, and yaw rate is especially remarkable for the second lane change [11].

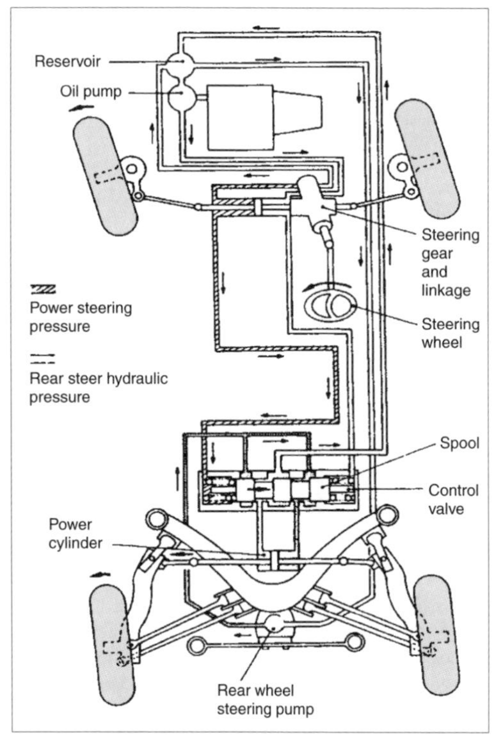

Fig. 7.4-64 Principle schematic, 1988 Mitsubishi rear-wheel steering [13].

encountered at low speeds, the rear axle is in "opposite steer." In this way, rear axle steerability is used to reduce turning circle diameter.

Fig. 7.4-61 illustrates an important advance—electronically controlled hydromechanical rear-wheel steer-

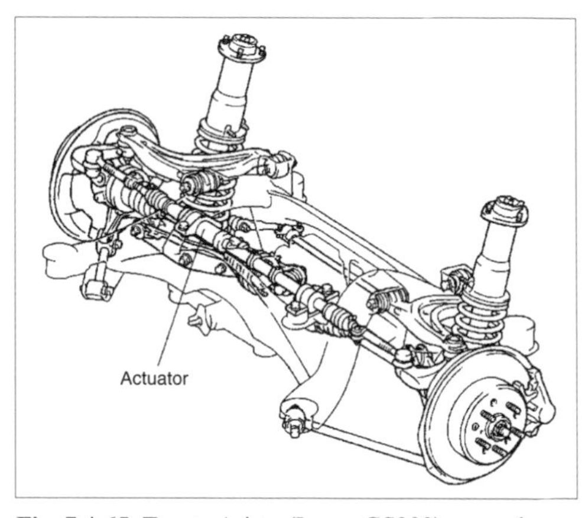

Fig. 7.4-65 Toyota Aristo (Lexus GS300) rear axle configuration with rear-wheel steering [Toyota: technical information, Aristo model, 1998].

ing (BMW, 1991). The rear-wheel steering angle is defined by a speed and steering angle dependent map (Fig. 7.4-62) [11]. To provide more agile response at low speeds, a time delay circuit delays steering angle buildup at the rear axle. As there is no feedback of dynamic handling parameters to the control unit, the BMW steering system, in terms of its vehicle dynamic function, represents a simple, unregulated control system. Fig. 7.4-63 shows the effect of active rear-wheel steering in a double lane change maneuver.

7.4.6.3.2 Rear-wheel steering systems with vehicle dynamic control

A special feature is shown by the rear-wheel steering system of Fig. 7.4-64 (Mitsubishi, 1988) in that it represents a purely hydromechanical control system. Front

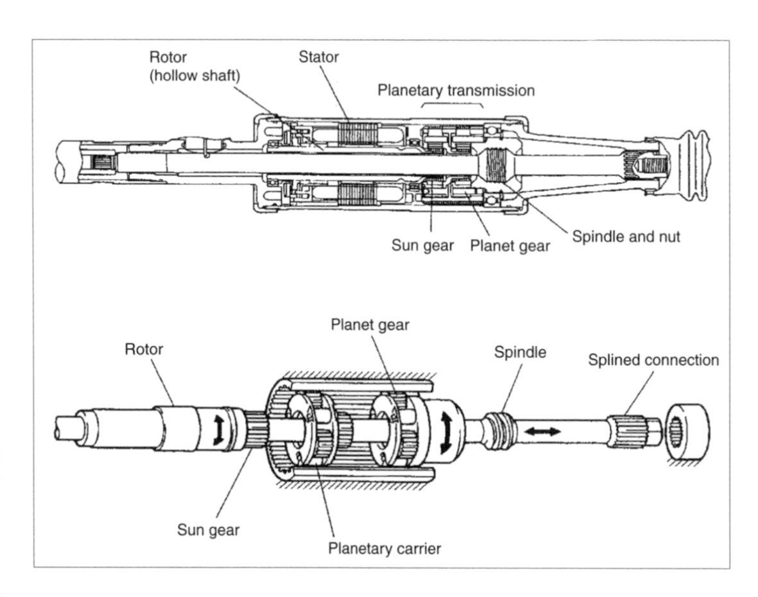

Fig. 7.4-66 Electromechanical actuator for Toyota Aristo active rear-wheel steering [Toyota: technical information, Aristo model, 1998].

axle power steering system pressure activates a hydraulic valve in the rear-wheel steering system, which determines rear axle steering angle direction (tending toward understeer) and influences its magnitude. Rate of steering angle change and front axle lateral force level are sensed from power steering system pressure. A rear axle driven pump supplies the valve with oil flow volume proportional to vehicle speed. This provides the desired additional vehicle speed dependence of rear axle steering angle.

With advancing sensor development, in particular yaw rate sensors, Japanese manufacturers introduced rear-wheel steering systems employing electronic vehicle dynamic control. In this way, external disturbances and deviations from target behavior stored in the control unit may be eliminated.

Introductions of new or further developed electrohydraulic rear-wheel steering systems are being planned by several European manufacturers. Development has been delayed by the economic decline in the automotive sector in the early 1990s, expectations of future mechatronic systems, and market orientation toward more comfort, passive safety, and immediately perceptible customer utility.

In view of the weight and cost advantages expected of mechatronic systems, more widespread introduction of rear-wheel steering systems is quite possible. Fig. 7.4-65 illustrates the Toyota Aristo (Lexus GS300) rear suspension with its active rear-wheel steering. Construction of the actuator with its electric motor, step-down drive, nut and spindle is shown in Fig. 7.4-66. Synergies may result from combination of this type of electrical actuator and electrical power steering systems currently under development.

References

[1] Segel, L. "The Variable Stability Automobile—Vehicle Concept and Design." SAE 650658.
[2] Mitschke, M., and K. Niemann. "Anforderungen an die Dynamik eines Kraftfahrzeugs auf kurvenreichen Fahrbahnen." *ATZ*, 1974, p. 299.
[3] Krämer, W. "Improved Driving Safety by Electronic Steering Assistance." EAEC 6th European Congress, 1997.
[4] Akita, T., M. Yamazaki, T. Kikkawa, and T. Yoshida. "User Benefits of Active Front Steering Control System: Steer-By-Wire." ISATA, 1999.
[5] Donges, E., and K. Naab. "Regelsysteme zur Fahrzeugführung und -stabilisierung in der Automobiltechnik." *AT Automatisationstechnik*, 1996/5, pp. 226–236.
[6] Badawy, A., J. Zuraski, F. Bolourchi, and A. Chandy. "Modeling and Analysis of an Electric Power Steering System." SAE 1999-01-0399 or 982878, in: SP-1438.
[7] Fleck, R. "Konstruktion einer aktiven Lenkung." Master's thesis, TU Munich, IVK 2/1996.
[8] Pauly, A. "Lenkmaschine zur Untersuchung des instationären Fahrverhaltens von Kraftfahrzeugen." *ATZ* 79/1977, Vol. 7, pp. 307–310.
[9] Schneider, R. "Konzipierung einer marktgerechten Stelleinheit als Aktuator einer aktiven Hinterradlenkung für Personenkraftwagen." Dissertation, RWTH Aachen, 1995.
[10] Shoichi, S., M. Tateomi, and F. Yoshimi. "Operational and Design Features of Steer Angle Dependent Four Wheel Steering System." Eleventh ESV Conference, 1987.
[11] Donges, E., R. Aufhammer, P. Fehrer, and T. Seidenfuss. "Funktion und Sicherheitskonzept der Aktiven Hinterachskinematik von BMW." *ATZ* 92/1990, Vol. 10, pp. 580–587.
[12] Donges, E. "Aktive Hinterradkinematik für den BMW 850i, *Automobil Revue* No. 38, 1991.
[13] Yamaguchi, J. "Global Viewpoints," *Automotive Engineering*, April 1988, pp. 96–113.

7.5 Evaluation criteria

7.5.1 Subjective evaluation of ride and handling properties

Even today, despite availability of highly developed measuring technology and CAE development methods, subjective evaluation remains an important tool for fine-tuning of vehicle ride and handling properties.

This may be attributed to the ability of veteran experimental engineers to adequately classify complex relationships as well as handling behavior in a comparatively brief time span. Moreover, due to their complexity, some criteria, such as comfort properties, cannot as yet be measured objectively to a sufficient degree [1].

As a rule, an evaluator simultaneously gains a comprehensive impression of overall vehicle properties, in which the viewpoint of the so-called "normal" driver—that is, the future customer—is also taken into consideration. An evaluation is usually performed on the basis of a series of driving maneuvers on special test tracks with known turns, friction coefficients, and so on. This is supplemented by evaluation drives under "customer conditions" on public roads.

Ultimately, this evaluation provides the experienced test engineer with an overall impression of the vehicle's actual transverse, longitudinal, and vertical dynamics. Evaluation spanning the rather broad field of overall ride and handling properties is done by means of evaluation worksheets, which are logically divided into various categories; for example:

- Ride and handling properties/stability
- Steering properties/suspension kinematics
- Braking properties
- Control system behavior (longitudinal/transverse)
- Comfort properties
- Towing properties
- Drive and traction

Definitive assignment of a multitude of individual criteria to appropriate subject areas is a necessary precondition for goal-oriented evaluation and therefore continued advancement of the vehicle development levels under investigation. For subjective vehicle evaluation, a questionnaire for the appropriate categories is established in which the evaluator ranks vehicles according to a specific scale. For evaluation of individual criteria, an open, unipolar grading scale with ten levels (0 = unsuitable, 10 = outstanding) has proven effective. This also offers the possibility of establishing an average for several judgment criteria or evaluators.

To reduce scatter of the results, evaluators must meet demanding standards with regard to their judgmental ability. They must have adequate specialized knowledge to evaluate and differentiate the properties under examination. Furthermore, they must have outstanding perceptive memory in order to retain all impressions through the final evaluation phase.

If evaluation is conducted by a group, it is possible to recognize and eliminate extreme variations from the average.

7.5.2 Objective ride and handling evaluation

Although the driver of a motor vehicle will in general always judge ride and handling properties subjectively, objective evaluation is becoming increasingly important. In a systematic vehicle development process as well as in comparing several vehicles to each other, capture of objective data has become an indispensable tool. These parameters or characteristic functions may be determined by means of measurements conducted during actual vehicle operation or, increasingly, from simulation results using a virtual computer model. The associated driving maneuvers are largely standardized and unified, with most described in ISO/DIN standards. These mainly concern so-called "open loop" test procedures for transverse dynamics. Vertical dynamic methods are not yet as widespread or standardized. The following descriptions are therefore limited to transverse dynamic vehicle properties.

"Open loop" methods were chosen in order to exclude driver-induced influences from test data or simulation results; that is, during the driving maneuver in question, driver activity is limited to keeping control elements such as steering wheel or throttle pedal as constant as possible during measurement or even releasing them entirely. A comprehensive overview of transverse dynamic test procedures may be found in Ref. [2].

Evaluation is based on vehicle motion within the coordinate system of Fig. 7.5-1. The main measurement and evaluation criteria are:

- Longitudinal and transverse velocities $\dot{x}$ and $\dot{y}$, also used to determine the vehicle slip angle
  $$\beta = \arctan\left(\frac{\dot{y}}{\dot{x}}\right)$$
- Longitudinal and transverse accelerations $\ddot{x}$ and $\ddot{y}$
- Yaw angle and yaw rates ψ and $\dot{\psi}$
- Roll and pitch angles φ and θ

The driver inputs used for evaluation are:

- Steering wheel angle and steering wheel moment δ_{SW} and M_{SW}
- Gas and brake pedal travel and/or forces
- Brake pedal pressure

Gyrostabilized platforms (or similar devices) have proven valuable for capturing vehicle motion data (accelerations and angles), while optical correlation sensors

Table 7.5-1 An example of a list of individual criteria for evaluating handling properties/stability

Subjective vehicle evaluation

Vehicle: **Load:** ○ Empty/driver ○ 2 pers. ○ 3 pers ○ Full

Category: ○ compact ○ lower midrange ○ upper midrange ○ luxury ○ sports car ○ roadster ○ off-road vehicle ○ other **Status**: Model year

Tires

Course

Road conditions ○ dry ○ damp ○ wet ○ variable **Driving style:** ○ gentle ○ forced ○ sporty

	ATZ evaluation	1	2	3	4	5	6	7	8	9	10	Comments
A. Handling characteristics												
Straight-line stability	Smooth road											
	Long-wave road											
	Short-wave road											
Reaction to:	Grooved pavement											
	Road camber											
	Crosswind											
Turn-in behavior	Quasi-static											
	Dynamic											
Windup												
Precision												
Over/understeer behavior	Small radius											
	Large radius											
Traction												
Power-off reaction												
Braking in turns												

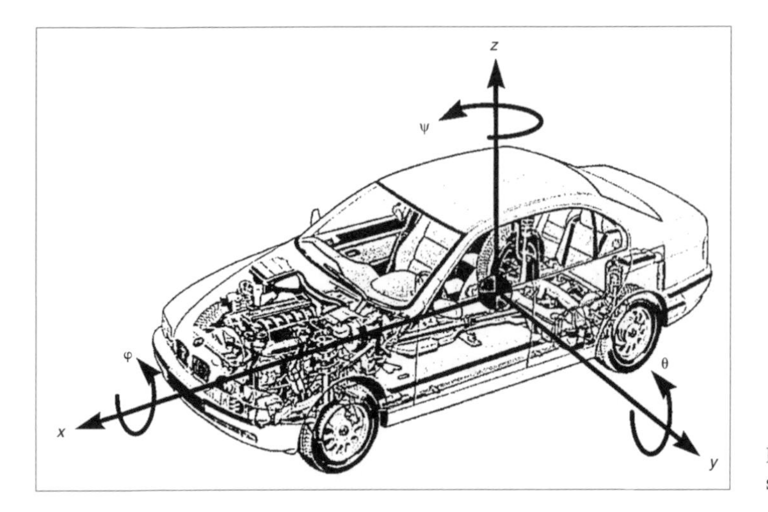

Fig. 7.5-1 Vehicle coordinate system.

are usually employed for speed measurement. Gyrostabilization is necessary to establish coordinate reference planes with respect to an earth-based frame of reference, thereby permitting measurement of vehicle translational acceleration independently of roll and pitch angles. Measurement of steering wheel angles and moments is usually done by means of special instrumented steering wheels. Test data may be processed online by means of apparatus installed in the vehicle. An example of a test setup is shown in Fig. 7.5-2. Data acquisition is driven by means of the display screen, and test results are directly represented onscreen.

In addition, there are of course a large number of test rigs and setups in the automobile industry which are applied for evaluation and optimization of individual properties and above all for specific components. So, for example, in evaluating vehicle dynamics, it is important to know how toe-in and camber of individual wheels respond to various maneuvers—for instance, braking in corners. A test setup for determining these quantities while the vehicle is in motion is shown in Fig. 7.5-3. Wheel-mounted lasers are aimed at body-mounted sensors which register changes in wheel angles. By means of such plots of vehicle motion over time, important suspension elastokinematic functions can be analyzed and their effects evaluated.

7.5.2.1 Straight-line motion

In evaluating vehicle straight-line travel, basic distinction must be made between two different boundary conditions:

1. "Undisturbed" straight-line motion at constant speed, in which disturbances are only introduced into the vehicle by the (presumably flat) road surface
2. Disturbed straight-line motion, in which crosswind, braking, and drive forces, or forces imparted by a trailer, are applied to the vehicle

Fig. 7.5-2 Vehicle dynamic test setup with instrumented steering wheel installed in vehicle.

Fig. 7.5-3 Apparatus for measuring camber and toe-in on moving vehicle.

For condition 1, there are as yet no generally adopted test procedures, first because vehicle motions are very small and therefore difficult to interpret and second because driver influence is always present in extended straight-line motion, so that definitive vehicle evaluation is not possible without added effort. Solution sets may be found in capturing steering angle spectra in drives over a specific course, time to depart from a specific lane with the steering wheel held immobile, or in determination of "yaw rate error." According to Ref. [3], yaw rate measurements during straight-line motion are split into one component due to steering wheel angles (as determined by a "single track" or "bicycle" model) and a remaining component. This remaining component represents the so-called "yaw rate error," which is a measure of deviation from straight-line motion. A close relationship between steering input, vehicle lateral acceleration phase delay, and straight-line qualities is described in Ref. [4]. Vehicles with small phase delays and thus quick response are characterized by good control properties and are therefore evaluated as "good" in terms of straight-line motion. One suitable method of evaluating steering feel in straight-line motion is a "weave test," in which a sinusoidal course is driven with a steering frequency of 0.2 Hz. Conclusions regarding steering properties in terms of ability to perform course corrections are obtained from the resulting plots of steering wheel moment over steering wheel angle near center.

Numerous test maneuvers may be applied to condition 2. For example, vehicle crosswind sensitivity is often evaluated in a crosswind test facility. Fans generate an artificial crosswind of about 60 to 80 km/h, usually perpendicular to the direction of travel. The fan Section typically has a total length of 15–40 m. Main evaluation criteria are yaw angle, yaw rate, and lateral deviation measured while driving past the fan array. An example of test results for various vehicle speeds is shown in Fig. 7.5-4. Compared to natural wind, however, such an artificial test has several inadequacies (angle of attack, gusts), which have led to development of test procedures under natural conditions. In Ref. [5], direction and magnitude of wind disturbances are measured by means of a weathervane mounted on the vehicle roof, and these are related to the resulting vehicle reactions (yaw rate).

Tests to evaluate braking properties in straight-line motion are primarily directed toward determining stop-

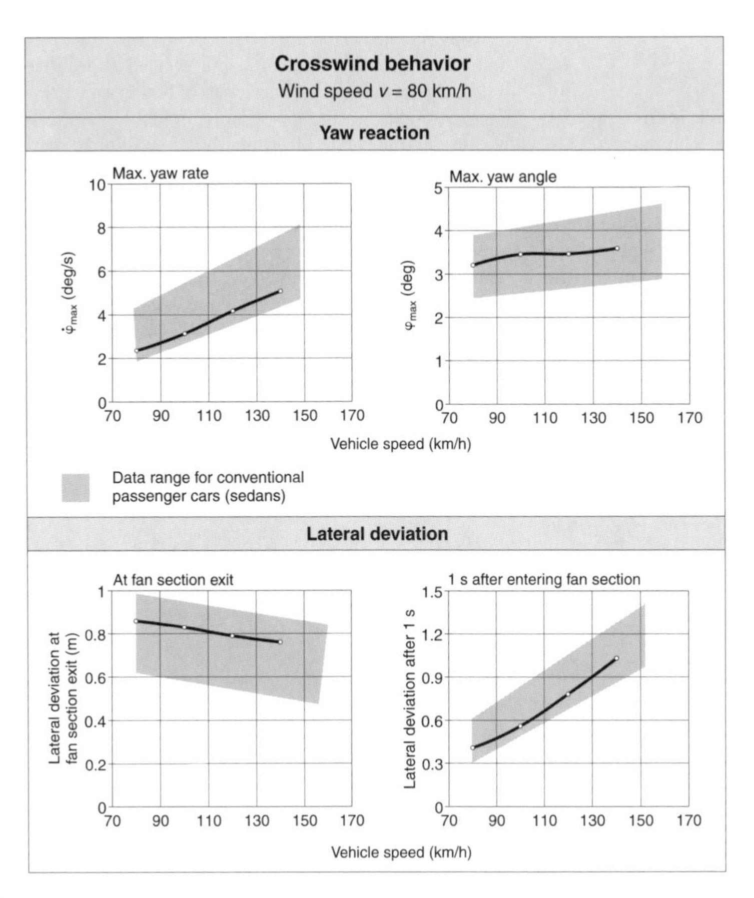

Fig. 7.5-4 Evaluation parameters measured in a crosswind test facility as a function of vehicle speed.

ping distances and directional stability (resulting yaw rates) on various surfaces with different friction coefficients. In addition, the pitch angles resulting from braking and acceleration are determined to provide a measure for the vehicle's antidive and antisquat properties. Normally, good suspensions provide pitch angles $\leq 1°$.

Combining a passenger car with a trailer can result in appreciable loss of vehicle stability. The steering impulse test is used to determine vehicle stability limits. While in undisturbed straight-line motion, vehicle yaw oscillations are excited by means of a short steering impulse. The difference in vehicle and trailer yaw motions—the so-called articulation angle—may be used to determine the degree of yaw damping. Yaw damping for various vehicle speeds may be determined by interpolation or extrapolation using the speed at which the stability limit is reached—that is, where yaw damping is zero. Typical stability limits for a passenger car towing a trailer of identical weight lie between 80 and 140 km/h, depending on trailer configuration.

7.5.2.2 Cornering behavior

The main driving maneuvers used to evaluate cornering behavior are:

— Steady-state cornering
— Power-off reaction in a turn
— Braking or accelerating in a turn
— Hydroplaning

In steady-state cornering, the vehicle is driven at various speeds on a circular track of constant radius (skidpad). Test results are typically given in terms of lateral acceleration (Fig. 7.5-5). Steering wheel angle as a function of lateral acceleration is a measure of the vehicle's under- or oversteer behavior. An increase over lateral acceleration is typical of an understeering car; the cornering limit (typically with lateral acceleration >7 m/s^2) is communicated to the driver by a pronounced increase in steering wheel angle (and usually a decrease in steering wheel moment, not shown). Vehicle slip and body roll angles are measures of comfort and safety and, in this example, represent the requirements of a modern passenger car.

Power-off reactions are measured by suddenly releasing the gas pedal while in steady-state cornering on a circular course with a specific radius. Engine braking torque ("motoring") introduces a deceleration which results in a course deviation due to axle weight transfer and elastokinematic suspension changes. The magnitude of this course deviation is evaluated on the basis of yaw rate and lateral acceleration data. In the interest of a more compact representation and in view of human reaction times, values of these parameters after one second have proven to be useful criteria; in other words, changes in vehicle motion parameters are evaluated one second after the beginning of the event (in this case, release of the throttle pedal). Sample results are shown in Fig. 7.5-6. The change in yaw rate after one second is plotted over initial lateral acceleration. For low levels of initial lateral acceleration, virtually no course change is discernible; yaw rate change is actually slightly negative due to the resulting decrease in vehicle speed. Only at higher levels of lateral acceleration is positive yaw rate change detectable: that is, the vehicle's path tightens.

Braking and acceleration during cornering are evaluated by means of a similar maneuver. In addition, after releasing the throttle, various levels of brake pedal pressure are applied or the gas pedal is further depressed to achieve braking or acceleration. Representation of one-second values is similar to that of the power-off reaction test; the abscissa contains an added variable—longitudinal deceleration or acceleration (Fig. 7.5-7). The entire diagram is valid for a specific initial lateral acceleration, in this example the cornering limit.

One very special driving maneuver is examination of hydroplaning behavior. This requires a skidpad equipped with segments that can be wetted. The vehicle is driven

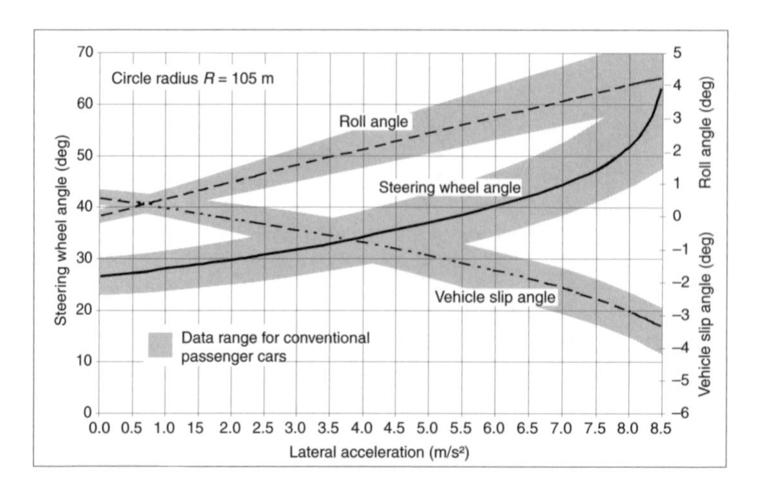

Fig. 7.5-5 Steering, vehicle slip, and roll angles as functions of lateral acceleration in steady-state cornering with a radius of 105 m.

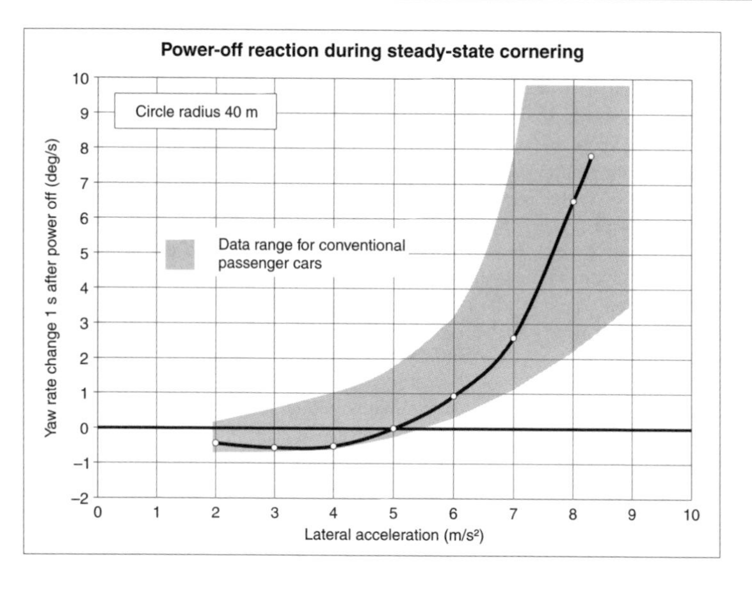

Fig. 7.5-6 One-second values for yaw rate change in response to power-off during steady-state cornering, 40-m radius.

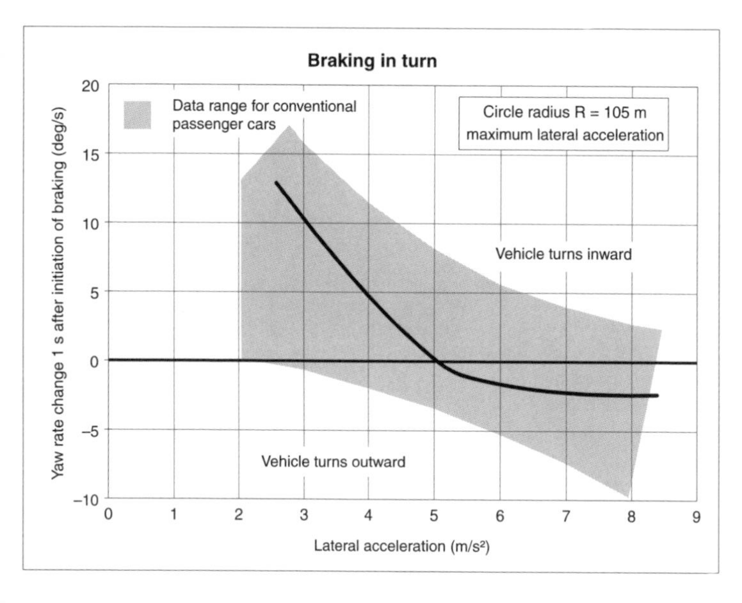

Fig. 7.5-7 One-second values for yaw rate change in response to braking in turn, circle radius 105 m, from maximum lateral acceleration.

through the wet segment at various speeds and therefore various levels of lateral acceleration, with the steering wheel held immobile and with constant throttle opening. Changes in yaw rate and lateral acceleration are again used as measures of course deviation.

7.5.2.3 Transient behavior

Transient behavior describes vehicle handling characteristics in response to transition from straight-line motion into a turn or to a sudden course change. A typical evaluation procedure uses a step steer input. From the straight-ahead position, the steering wheel is quickly turned through a specific angle and then held immobile. Vehicle reaction, primarily measured in terms of yaw rate, lateral acceleration, and vehicle slip angle, is a measure for response speed, handling stability under these conditions, and steering "directness." With a large time lag between steering input and rise in yaw rate, the vehicle feels sluggish and unwilling to corner. If large amplitudes and drawn-out oscillations are observed in transition to stationary values of yaw rate and lateral acceleration, vehicle stability is compromised. A vehicle's "yaw rate gain" is the ratio of yaw rate to steering wheel angle and describes how much steering wheel angle the driver must apply to achieve a given vehicle yaw reaction. Very direct steering is characterized by large yaw gain.

Other open-loop tests are characterized by different steering angle input parameters. Distinction is made between single-cycle sinusoidal input and continuous sinusoidal input, triangular impulse, and random steering angle input (see ISO 7401). In single-sine input, as in stepsteer, the delay times between steering input and vehicle reaction (yaw rate and lateral acceleration) are evaluated; the other tests involve frequency domain evaluation of vehicle motion parameters. Customarily, frequencies up to 4 Hz are evaluated. The gain function for yaw rate and lateral acceleration as well as the associated phase shift provide insight into a vehicle's lateral dynamic properties. Constant amplification, if possible up to a frequency of 2 Hz, and minimal phase shift are criteria for well-harmonized transitional handling behavior.

Closed-loop methods may also be applied. The driver's task is to negotiate a given course as quickly as possible. The most familiar is probably the ISO 3888 double lane change (only the gates are standardized; actual execution is too driver-dependent and therefore not amenable to standardization). Experienced drivers in modern passenger cars negotiate the ISO gates at about 110 km/h. In addition, there are numerous slalom tests. Pylon spacing is usually 18, 30, or 36 m. Objective evaluation criteria are time through the slalom and recorded magnitudes of vehicle motion. In evaluating slalom tests, it must be remembered that results are largely dependent on the interplay of pylon spacing, wheelbase, vehicle length and resonant yaw frequency.

In 1997, the so-called "moose test" (also called the "elk test") became very popular as a test of rollover stability. Because of the very wide lanes used in this test, driver influence is readily apparent, and as a result the test has not gained recognition as an objective, reproducible evaluation method. A committee of VDA experts (Verband der Deutschen Automobilindustrie—German Automobile Industry Association) has reworked the moose test and restricted gate width (dependent on vehicle width) to the point where driver influence is kept as small as possible. An unambiguous definition of gas pedal actuation has been added. The "VDA test" is conducted with the throttle closed as of 10 m before the first gate [6]. This is an effort to more closely approach actual incidents in which the driver will presumably release the throttle pedal in an emergency situation.

7.5.2.4 Prognosis

As described, a series of objective test procedures has been established. However, these only permit conclusions regarding special handling properties in "artificial" driving situations with precisely defined boundary conditions. Evaluation of overall vehicle handling properties is not possible on the basis of the limited results generated by these tests. Additional test procedures must be developed which on the one hand include vertical dynamics and on the other take into consideration the driver and the human capability to act as a control system.

A number of research projects have been conducted to determine which physical motion parameters are actually perceived by humans during vehicle operation and which are judged undesirable or poor on the basis of their intensity. All of these investigations have endeavored to find the best possible correlation between driver evaluations and measured parameters in specific driving maneuvers. One of the most recent studies [7] (financed by the Automotive Technology Research Association and the German Federal Highway Administration), in which numerous ordinary civilian and professional test drivers negotiated the ISO lane change maneuver and a secondary country road course, reported that the time delay between steering angle input and the buildup of yaw rate and lateral acceleration as well as the magnitude of vehicle slip angle largely determined driver assessment of the vehicle.

These relationships will be examined in greater detail in the future. In this way, it will be possible to describe objectively a large range of vehicle handling properties, which in turn will generate continuous improvement in systematic methods for suspension development, and to apply these advancements toward satisfying demands for continued progress. Nevertheless, in the final analysis, given the current state of the art, subjective evaluation of overall vehicle handling properties is still the deciding factor.

References

[1] Hennecke, D. "Zur Bewertung des Schwingungskomforts von PKW bei instationären Anregungen." Fortschr. Ber. VDI Reihe 12, No. 237, 1995.

[2] Zomotor, A., H.-H. Braess, and R. Rönitz. "Verfahren und Kriterien zur Bewertung des Fahrverhaltens von Personenkraftwagen." ATZ 12/97 and 3/98.

[3] Dettki, F. "Methoden zur Bewertung des Geradeauslaufs von PKW." VDI-Bericht 1335, 1997, pp. 385–405.

[4] Loth, S. "Fahrdynamische Einflussgrößen beim Geradeauslauf von PKW." Dissertation, TU Brunswick, 1997.

[5] Schaible, S. "Fahrzeugseitenwindempfindlichkeit unter natürlichen Bedingungen." Dissertation, RWTH Aachen, 1998.

[6] VDA press release by the ad hoc Vehicle Safety Working Group (Arbeitsgruppe Fahrzeugsicherheit), 1. 4. 98.

[7] Riedel, A., and R. Arbinger. "Subjektive und objektive Beurteilung des Fahrverhaltens von PKW." FAT (Forschungsvereinigung Automobiltechnik e.V.) Schriftenreihe No. 139.

7.6 Fuel system

7.6.1 Regulatory and customer-specific requirements

7.6.1.1 Regulatory requirements

Increasing environmental consciousness and demands for the safety of traffic participants force developers of fuel supply systems to deal with existing as well as future regulatory requirements and to address these regulations in development of future products. The increasingly stringent (German) Product Liability Law will only reinforce this trend.

Regulatory requirements are national as well as international in scope; national requirements (see Section 2.2) usually conform to international regulations by virtue of unaltered adoption.

StVZO §45 and §46 (German Federal Republic). The fuel tank, fuel line, and vehicle installation requirements covered in this very comprehensive standard may be found reflected, using the same words, in the regulations of other nations. Significant changes are very rare, and none are expected in the near future.

Because of fuel flammability hazards in the event of an accident, the fuel tank should be spatially separated from the engine, and should not be installed immediately behind the vehicle's front apron nor be a part of the bodywork. The tank should not exhibit any leaks at an overpressure of 0.3 bar. Escaping or evaporating fuel cannot be allowed to collect where it can be ignited by hot components such as the exhaust system or by electrical devices or switches. Fuel lines may consist of elastic metal hoses or other fuel-resistant hoses capable of withstanding relative motions of vehicle and components. Flashback of flames in these lines is not permitted. For gasoline-powered vehicles, fuel supply must be interrupted if the engine is shut off.

ECE-R 34. Fuel system components must be installed in such a way that vibration, torsional and bending motion, and collisions cannot result in damage. Escaping fuel may not enter the vehicle interior nor be ignited by hot components. Other points address design location in the vehicle and possible electrostatic charging of the container. Along with internal pressure tests using 0.3 bar overpressure (6.1), the standard also defines frontal impact against a fixed barrier at 48.3–53.1 km/h and a rear impact at 35–38 km/h against a movable barrier or pendulum (Appendix 4). These impact tests permit a maximum leakage rate of 30 g/min (see also FMVSS 301). Appendix 5 concerns itself solely with plastic fuel tanks and establishes the following requirements:

- Impact test with a pyramidal steel pendulum massing 15 kg, with a 1 m pendulum and at –40°C,
- Internal pressure test for 5 h with 0.3 bar overpressure at 53°C
- Fuel integrity with 12-week storage at 40°C (limit 20 g/24 h)
- Fuel resistance
- Fire test, in which the fuel tank is subjected to defined flames for 2 min
- Resistance at 95°C for 1 h
- Product marking

The oft-cited FKT guideline (Fachausschuss Kraftfahrzeugtechnik—Committee for Motor Vehicle Technology) of §45 StVZO was adopted largely unchanged by the ECE regulation in 1972.

FMVSS 301. This American standard covers fuel system integrity following crash tests and is reflected in nearly all individual national standards. The standard describes a frontal fixed-barrier impact (S6.2) at 30 mph as well as side impact with a movable barrier at 20 mph (S6.3) and a static rollover test (S6.4). The containers are 90–95% full and spillage may not exceed one ounce by weight (28.4 g) of fluid from the time of impact until the vehicle comes to rest (barrier tests) or more than 5 oz (142 g) within 5 min (rollover test).

SHED Test (CARB and EPA). For the SHED test, California and American emissions standards are considered the most stringent and are generally adopted by other countries in their own regulations, but only after extended delays. In the SHED test ("Sealed Housing Evaporative Determination"), total vehicle emissions are measured in a test chamber. This test, including vehicle preparation and conditioning, consumes two days; the actual emissions test takes two hours. The applicable emissions limit is 2 g/test. This limit applies to the entire vehicle, not just the fuel tank and its peripherals.

A fuel supply system developer can only hope to estimate the fuel system contribution to total vehicle (evaporative) emissions and use this estimate as a limit for a so-called mini-SHED test, in which only the fuel container and ancillaries are tested for emissions. For Europe, a SHED test is also ratified as part of ECE-R 83.

OBD II (Onboard Diagnostics, Stage II). Officially designated as Section 1968.1 of Title 13, California Code of Regulations (CCR), this regulation has been implemented in California as of 1994. The purpose of this regulation is to reduce emissions by means of continuous monitoring for possible faults in the catalytic converter, engine, exhaust gas recirculation, and evaporative emissions system, as well as other emissions relevant components, and, in the event of a fault, to indicate these by means of a warning light on the vehicle instrument panel. OBD is intended not only to detect total malfunction of components, but also to detect whether any limits are exceeded. The OBD II regulations are a continuation of OBD I in that, as a result of technical progress, more components can be monitored. A repair shop can examine the vehicle system for faults using a standard diagnostic device ("Scan Tool" or "Scanner"). For the fuel supply system, this regulation mandates introduction of a new monitoring system for the entire air and fuel vapor venting system with respect to HC emissions. This may be accomplished by means of a vacuum or pressure decay test. This portion of OBD II regulations has been implemented as of the 1995 model year.

7.6.1.2 Customer-specific requirements

Naturally, along with regulatory requirements, customer demands have a considerable influence on product execution. Regulatory demands may often be regarded as basic requirements; in many cases, customer requirements exceed these and vary from manufacturer to manufacturer. These demands are generally outlined in technical delivery terms ("technische Lieferbedingungen," TL), quality specifications ("Qualitätsvorschriften," QV), test specifications ("Prüfvorschriften," PV) and experimental guidelines ("Versuchs-

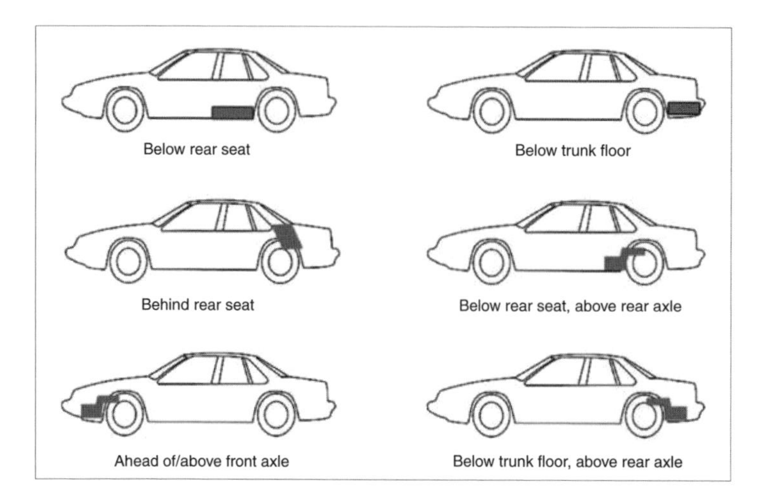

Fig. 7.6-1 Fuel tank location in sedans.

richtlinien," VR), manufacturer-specific standards, and specifications. These may be roughly divided into functional and materials requirements.

7.6.2 Location in vehicle

Location of the fuel container within the vehicle (Fig. 7.6-1) takes into consideration the following:

- Location of engine/transmission unit (front, rear, or mid-engine)
- Body variations (sedan, wagon, roadster, etc.)
- Safety aspects (e.g., installation outside the primary crush zones, etc.)

For vehicles with raised passenger compartments or underfloor engines (Mercedes-Benz A-Class, minivans, etc.), fuel tank locations behind the front axle and below the floorpan have been realized. Customarily, the fuel container is mounted outside the vehicle interior. If located behind the rear seats in the trunk, suitable safety measures must be provided—for example, a sealed bulkhead of steel or suitable materials between the tank and cabin. Combination of such a tank with a fold-down seat is not possible, but a trunk passthrough using a "ski sack" is realizable if measures for interior protection and appropriate fuel tank configuration are implemented.

7.6.3 System variations

Fuel system venting under operating conditions has a significant influence on fuel supply system layout, as does handling of volume increases due to thermal effects. The two following unpressurized systems enjoy the widest application [1]:

7.6.3.1 External overflow volume (e.g., BMW 3-Series)

This system is based on an auxiliary fuel overflow container (Fig. 7.6-2), which ideally is centrally located above the fuel tank. As such a location is not possible for most vehicles (compromise of vehicle interior volume), the fuel overflow container is instead located in the rear wheel well area. The required volume should represent 3–5% of fuel tank volume. The fuel system is vented through this overflow container.

7.6.3.2 Internal overflow volume (e.g., Audi A4)

A compact fuel supply system may be realized using this variation (Fig. 7.6-3), but a larger overflow volume

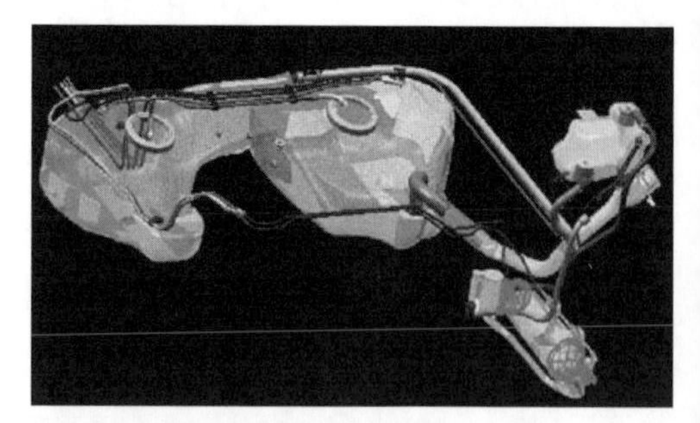

Fig. 7.6-2 Fuel system with external overflow container (BMW 3-Series E46).

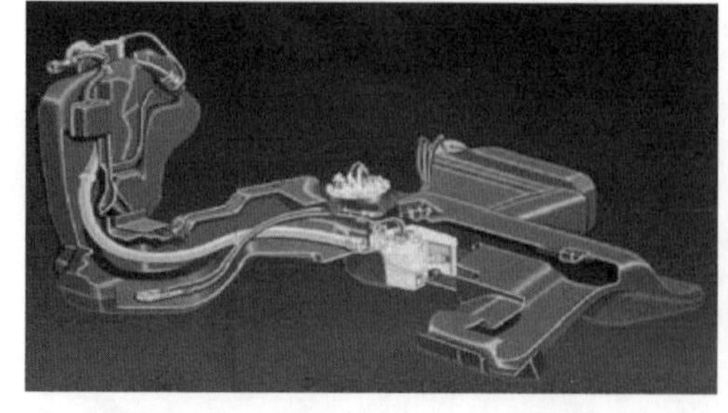

Fig. 7.6-3 Fuel system with internal overflow container (Audi A4).

of up to 10% is needed. The necessary overflow volume is provided by controlling fuel fill quantity by means of valves or by closing the fuel fill opening in response to fuel level. In part, the filler pipe volume may be utilized as overflow volume, in which case filling of the pipe with fuel during refueling must be prevented. One possibility is offered by control of a valve by means of the fuel filler cap.

Layout criteria

Layout of the fuel tank venting system is conducted in accordance with both static and dynamic test procedures. In the static test, maximum vehicle inclinations (exact values are manufacturer-specific) are checked to ensure that no fuel spillage takes place. The dynamic procedure (vehicle or hardware simulation) represents the customer-specific driving cycle, including extreme boundary conditions (e.g., high ambient temperatures with winter fuels). To prevent escape of fuel vapors, connection to atmosphere is through an activated charcoal filter. Depending on regulatory requirements, fuel vapors generated during refueling may need to be recaptured within the vehicle (US regulations covering "Onboard Refueling Vapor Recovery," ORVR).

7.6.4 Fuel tank

Modern fuel tank materials are selected to satisfy many conditions. These include:

- Crash stability and resistance to environmental effects
- Impermeability
- Resistance to aging
- Good workability [2]
- Manufacturability of demanding geometries
- Recycling

The following schematic (Fig. 7.6-4) illustrates the applied processes; a basic distinction is made between two categories:

1. Metal fuel tanks
2. Plastic fuel tanks

7.6.4.1 Metal fuel tanks

Metal fuel tanks have enjoyed a long history. The first automobiles and motorcycles were equipped with metal tanks.

Today, highly complex fuel tank geometries are usually realized using St14Zi/Ni, as well as alloy steel (e.g., X5CrNi1810), and alloy aluminum sheet (primarily on motorcycles).

After the forming process, tank shells are joined to form a gas-tight container by means of an automatic welding process. The most common welding processes are:

- 2-D/3-D seam welding (Fig. 7.6-5)
- Laser welding (Fig. 7.6-6)
- Induction welding
- Soldering

The methods shown in Fig. 7.6-7 may be used to join filler necks or other components to fuel tank shells.

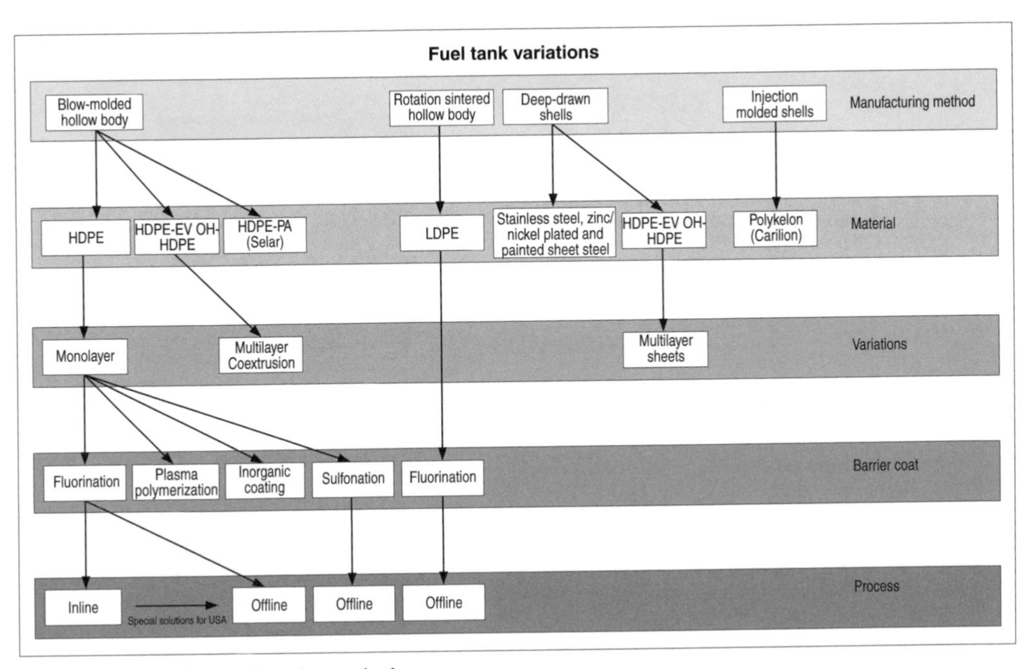

Fig. 7.6-4 Fuel tank manufacturing methods.

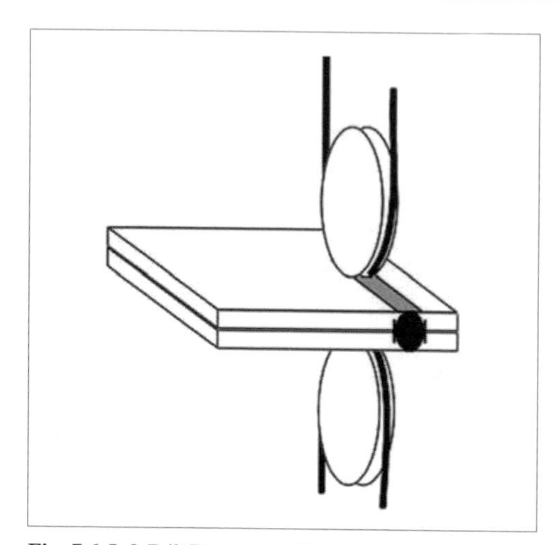

Fig. 7.6-5 2-D/3-D seam welding.

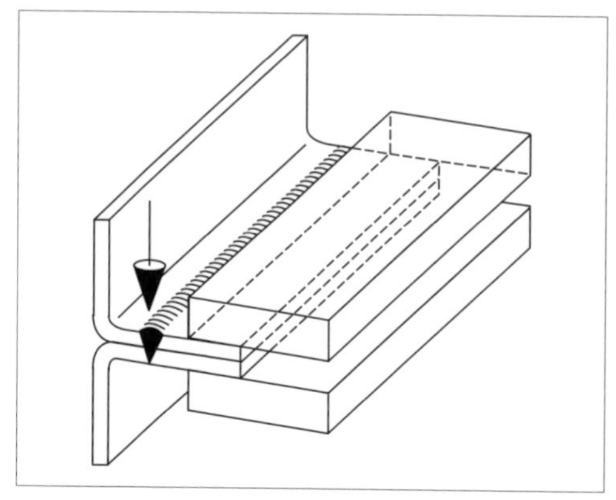

Fig. 7.6-6 Laser welding.

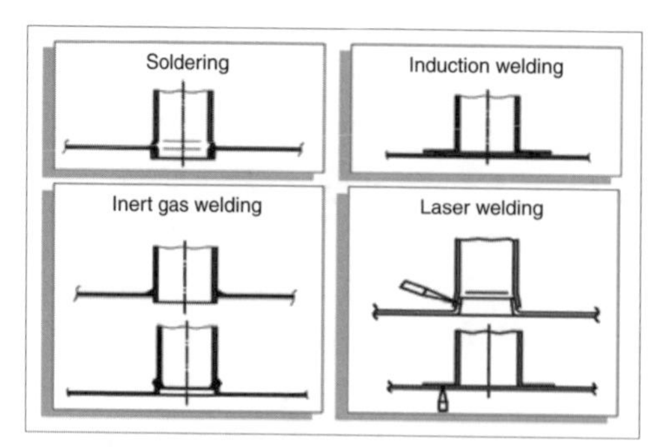

Fig. 7.6-7 Welding methods for filler necks.

7.6.4.2 Plastic fuel tanks

The automobile industry began development of plastic fuel tanks in the 1970s. Due to increasingly complex fuel tank geometries, an extrusion blow molding process [3] underwent further development in cooperation with the plastics industry and producers of plastic materials, and appropriate new materials were developed. The material in question is high density polyethylene (HDPE), modified to resist effects of all commercially available fuels.

The most widely distributed materials and their respective manufacturers are:

Lupolen 4261 A	(ELENAC)
Hostalen GM 7746	(Hoechst)
Finathene MS 456	(Fina)
Eltex RSB 714	(Solvay)

Diffusion barriers were developed to meet increasingly stringent emissions regulations. First, the tank is treated with fluorine gas (F_2). This may be carried out during the extrusion blow molding process (inline fluorination) or after the blow molding and weld completion process (offline fluorination).

In fluorination, the container is first rendered inert, then filled with an F_2/N_2 gas mixture. This produces a polar surface, which appreciably reduces solubility and therefore permeation of nonpolar molecules in the base polymer. Compared to untreated HDPE fuel tanks, gasoline permeability losses are reduced by more than 99%. The 1990s saw development of a process formerly used only by the packaging industry—blow molding using coextruded plastic materials (see Figs. 7.6-8 and 7.6-9)—for the manufacture of plastic fuel tanks. Various welding processes may be used to attach filler necks, brackets, and so on. to such coextruded blown tanks. Some of the most familiar are:

- Vibration welding
- Ultrasonic welding
- Heated element welding
- Laser welding

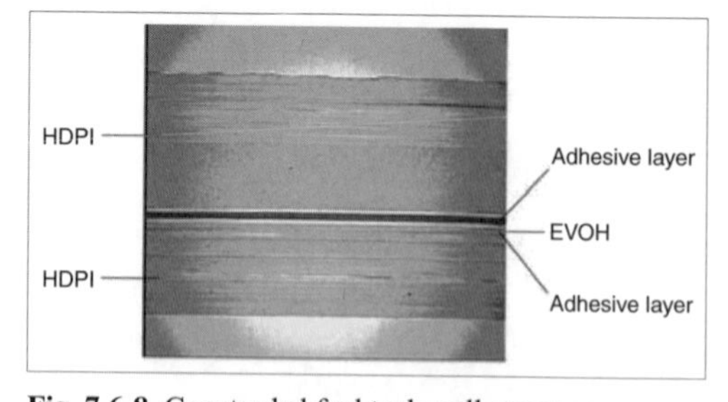

Fig. 7.6-8 Coextruded fuel tank wall structure.

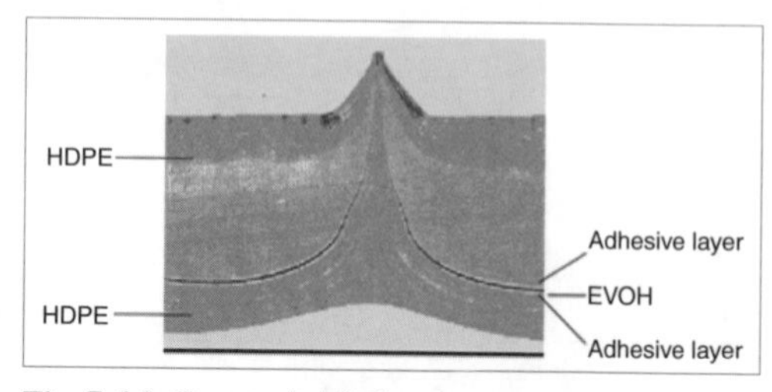

Fig. 7.6-9 Coextruded fuel tank seam structure.

7.6.5 Fuel feed systems

The basic task of the fuel feed system is to supply the engine with sufficient fuel under all operating conditions.

A large number of different influential parameters lead to entirely different fuel supply system configurations. While in the past fuel was circulated in a large loop from the fuel tank to the fuel rail and back to the tank, at present there is a noticeable trend toward shorter circulation loops or nonreturning systems.

7.6.5.1 Fuel transfer

To explain fuel feed requirements, the interaction of fuel pump and swirl pot or fuel tank must be examined. This combination has a considerable influence on fuel intake behavior and therefore fuel transfer from tank to combustion engine.

7.6.5.2 Electric fuel pump (EFP) and its location

The current state of the art in fuel pumps includes single-stage and two-stage pumps based on various pumping principles (ring gear/trochoid, side channel, peripheral channel, rotary vane or roller cell, and screw [helical] pumps).

Side and peripheral port pumps are classified as flow pumps, while the others belong to the positive displacement pump category.

At higher system pressures of $>$1–6.5 bar, trochoidal pumps have proven themselves in service. Side channel and peripheral channel pumps are available in several variations. These are characterized by their continuous pressure increase, very low pulsations, and very smooth operation. However, in terms of maximum pressure and efficiency, they are at a disadvantage compared to positive displacement pumps. Therefore, flow pumps are often used as a prestage, in combination with the aforementioned trochoidal pumps, in applications which may encounter more severe hot feed situations.

Rotary vane and roller cell pumps are no longer as common as they once were. To minimize noise and reduce wear, these must use materials selected for their excellent sliding properties.

In the past few years, screw pumps have been developed to the point of mass production and are preferred for inline installations. Their advantage in comparison to other positive displacement pumps is their outstanding behavior with hot fluids, even without a preliminary stage, making this pump concept an optimum solution for installations outside the fuel tank.

In dimensioning the fuel pump, two aspects demand special attention:

1. During cold starting, despite reduced electrical system voltage (i.e., lower pump speed), the EFP is still expected to provide the engine with sufficient fuel at a specified system pressure (usually below nominal operating pressure).
2. For systems with fuel return lines, the maximum EFP volumetric flow rate must always be appreciably greater than the engine's maximum fuel demand for the following reasons:
   - Often, at least one jet pump is driven by the return flow; for proper function, this requires a flow of 15–25 L/h.
   - The fuel pressure regulator of midrange vehicles requires a minimum flow rate of about 15 L/h to provide a stable regulation range.
   - The hot fuel feed loss (up to 50% of nominal feed rate) must be considered as a function of the selected EFP.

Pressure limiting valve and system pressure check valve

The pump body may include a pressure limiter and a system pressure check valve. If, for example, a fuel line is pinched shut, the pressure limiting valve prevents pressure from rising above the permissible fuel line bursting pressure. Fuel is diverted to the pump suction side.

After engine shutdown of a hot vehicle, the system pressure check valve maintains fuel pressure between pump and pressure regulator for a certain time, allowing fuel to cool to normal temperatures. If pressure were not maintained, engine heat could cause fuel in the lines to vapor-lock, which would make immediate restarting (hot start) more difficult. Therefore, the system pressure check valve must be absolutely tight.

Pump locations. In principle, two installation concepts apply to electric fuel pumps: inline and in-tank installations. Inline pumps are attached to the vehicle body, outside the fuel tank, in the fuel feed line by means of noise-damping brackets. Among the most important EFP characteristics are fuel integrity as well as resistance to salt water and operating fluids such as motor oil, antifreeze, brake fluid, and so on. Due to its distance from the fuel tank, the EFP must provide good suction. A significant advantage of inline pump location is simpler access in the event repairs are needed. In-tank pumps are mounted either on a closing flange or in the swirl pot; in either case, they are vibrationally decoupled from the tank by means of damping rubber mounts. The distance between the intake-side filter and tank floor is an important determinant of suction performance. For ideal exploitation of the tank volume, the gap to the tank floor should be kept as small as possible, independent of tank deformation. This may be achieved by means of a telescoping pickup—that is, a spring mounting. Vertical travel of the in-tank unit may be restricted by a raised tank floor.

7.6.5.3 Requirements for electrical/electronic system integration

As representative for all electrical components, requirements for electrical and electronic integration will be examined using the fuel supply system as an example.

Assuring electromagnetic compatibility (Section 8.4) demands ever greater engineering effort. Injection of voltage spikes into power supply, signal, data, and control

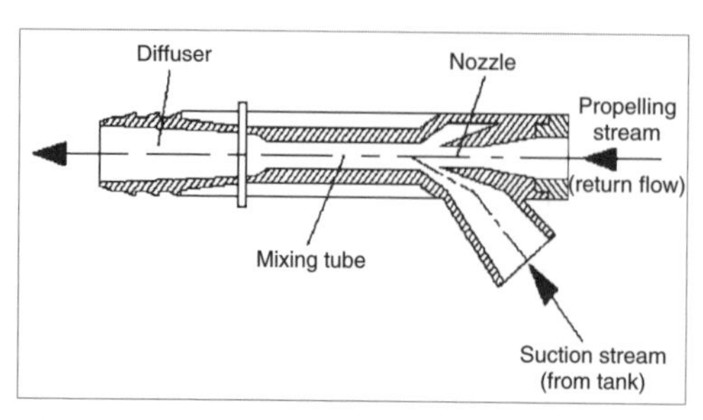

Fig. 7.6-10 Jet pump.

circuits cannot be allowed to have any negative consequences. Furthermore, electrical components should be adequately shielded to prevent interference in all of the usual radio frequency bands. Along with appropriate electrical wiring insulation, assurance must be provided against electrical misconnection (i.e., prevention of reversed polarity) and the resulting functional impairment.

7.6.5.4 Jet pump (venturi pump)

Fig. 7.6-10 shows the principle features of a jet pump (venturi pump) [4] in cross section. This is a passive pump without any moving mechanical parts and is driven by the return fuel flow. Jet pumps are used in multichamber fuel tanks to transfer fuel from outlying chambers to the main tank volume, which contains the swirl pot and EFP. In the case of single-chamber tanks, the jet pump is usually attached directly to or integrated within the swirl pot. Nozzle dimensions are matched to the lowest return fuel flow rate.

7.6.5.5 Swirl pot

Designs of currently implemented swirl pots take on many different shapes. As a whole, however, they serve a single objective: Even in extreme situations, small amounts of fuel remaining in the tank should be made available to the fuel pump pickup to assure problem-free engine operation.

"Extreme situations" are taken to mean, for example:

- Grades (climbing or descending) in the direction of travel
- Tilting perpendicular to the direction of travel
- Maximum vehicle acceleration and braking
- Extended turns taken at high lateral acceleration

The first two points address fuel pickup behavior in normal operation as well as possible longer-term parking. A suitable installation location would be one where fuel readily collects. This must be taken into consideration in designing the fuel tank contours.

Depending on design, additional functions can be integrated in the swirl pot. Demands on the swirl pot must always be regarded in conjunction with the EFP and fuel tank shape.

The swirl pot can be attached to the fuel tank by the following means:

- Welding to the tank
- Subsequent attachment via bayonet fitting, snap fitting, etc.
- Pressing the swirl pot against the fuel tank assembly using spring elements

7.6.6 Fuel filtering

To reduce wear and maintain fuel supply and engine component functions, contaminants must be removed from fuel. Such contaminants may be classified as "original" dirt, found in fuel at service stations or in new cars, and dust, paint and metal particles, rust, and, for diesel fuels, water, tar, and paraffins.

Initial filtration is normally performed by a strainer at the fuel pump pickup. For this, care must be taken not to select an excessively small mesh, as this could be detrimental to fuel pump suction behavior. Supporting elements prevent collapse of this filter "sock." The second and more effective filtration stage is achieved by separate filters installed in the fuel supply line. Designs for these differ considerably, depending on the fuel used—diesel or gasoline. Fuel residence time and degree of filtering must be matched to surrounding conditions.

Because of their higher system pressures, filters for gasoline engines use a metal filter housing. Filters are generally resistant to fuels containing methanol.

Diesel fuel filters are considerably more complex. At low temperatures—that is, below the fuel cloud point—paraffin contained in the fuel may crystallize out, thereby plugging the filter. This is prevented by heating systems which automatically activate in certain preset temperature ranges. Water separators with electric water level sensors and bleed systems to ensure good starting performance after service operations are installed as needed. These accessories may be integrated in the filter or installed ahead of or behind the filter.

Fuel pressure regulators are installed to provide constant system pressure at the fuel injectors independent of the withdrawn fuel quantity. These are made of steel or stainless steel and installed either in the fuel rail or, in shorter fuel circulation systems, directly at the pressure regulator outlet. For stable operation, fuel pressure regulators require a minimum flow rate of about 15 L/h.

7.6.7 Fuel volume measurement system

A variety of different physical principles, with different configurations, may be applied to determine the volume of fuel in the tank:

- Direct fuel level measurement
- Fuel weight determination
- Fuel volume determination
- Measurement of mechanical loads
- Measurement of electrical quantities
- Optical measurement
- Ultrasonic measurement

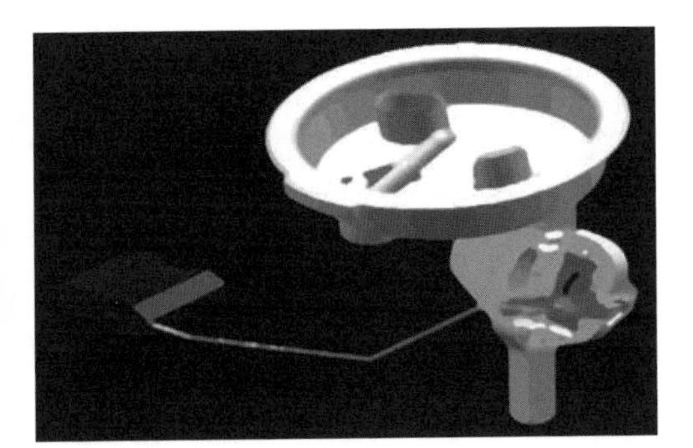

Fig. 7.6-11 Lever arm fuel sender with bottom support (BMW 3-Series E36).

Direct fuel level measurement has become the predominant method and is used on nearly all vehicles because of accuracy and resolution, adaptation to various fuel tank geometries, life expectancy, signal processing, installation and service friendliness, insensitivity to various operating conditions, and, not last, costs.

7.6.7.1 Lever arm sender

Lever arm senders (Fig. 7.6-11) employ a float attached to the end of a wire lever. Appropriate design of the lever and various floats allows considerable influence on measuring range. Floats are available in various geometries (e.g., rectilinear, spherical), made using various manufacturing methods (blown, foamed), and composed of various materials. The float level is converted into an electrical signal by means of a potentiometer, which is sent to the combination instrument on the vehicle's instrument panel. Signal damping to counteract the effects of fuel sloshing is accomplished at the lever using mechanical damping elements or electrically by means of software in the combination instrument.

Attachment methods for fuel level senders are divided into bottom-supported ("bottom sender") and top-supported ("top sender") configurations. Deciding factors for selection of sender mounting type are the tank's ability to retain its shape as well as the desired sensing accuracy throughout the fuel level range.

7.6.7.2 Tube-type senders

In tube-type fuel level senders (Fig. 7.6-12), the float is in the form of a ring and slides along a guide rod in a vertically oriented tube. Contacts attached to the float bear against a rod wound with resistance wire or fitted with printed circuit conductance paths. A hydraulic labyrinth damps out float movement and the resulting signal variations. The sender tube may include an integral deliberate break point designed to fail in the event of extreme fuel tank deformation, thereby preventing the sender mounting flange from being forced out of the tank wall and causing a leak. The sender tube is normally attached to a mounting flange which, like many feed or sensing units, is installed in an appropriate fuel tank opening.

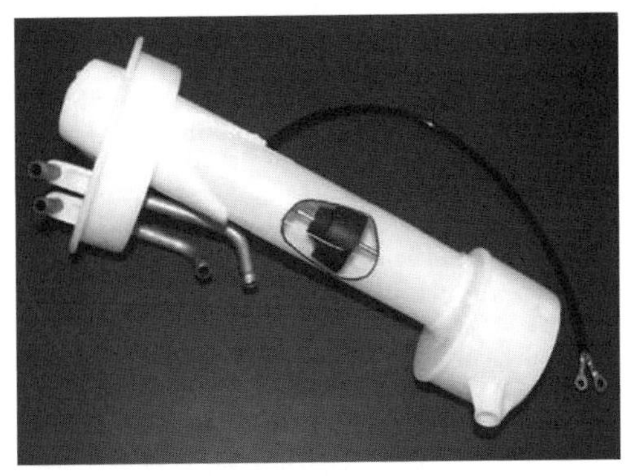

Fig. 7.6-12 Tube-type fuel level sender (sectioned), BMW 5-Series E34.

7.6.8 Special solutions and auxiliary fuel tanks

Due to higher performance engines installed in the sporty production car segment (e.g., BMW M series), the need often arises for larger fuel tanks. Furthermore, regular production tank volumes may be inadequate for certain export markets in view of their fuel supply infrastructure. One solution is installation of an auxiliary fuel tank (Fig. 7.6-13), usually in the trunk area. Filling is accomplished in series with the (regular production) main tank, or, in some cases, in parallel from below (BMW M3, E30). Because of the complexity of a comprehensive fuel volume measurement system, most applications dispense with adding a sensing unit in the auxiliary tank. Most auxiliary tanks are made of metal; plastic tanks are only used in special cases, outside the vehicle interior, due to material-specific permeation of hydrocarbons (objectionable odors).

7.6.9 Fuel supply systems for alternative energy sources

7.6.9.1 Alternative fuels

As described in Section 5.8, methanol and rapeseed oil methyl ester (RME, also called biodiesel) may in prin-

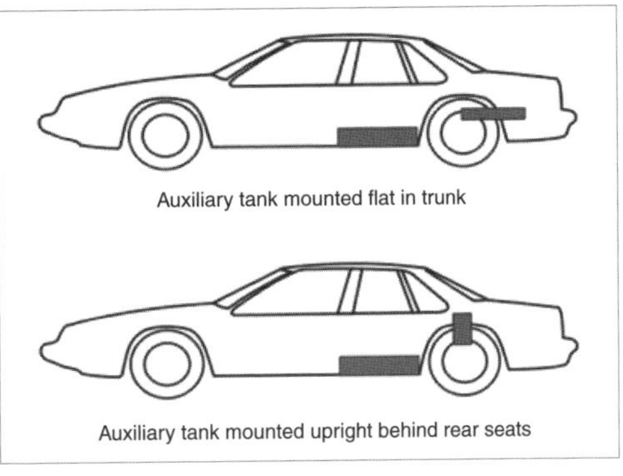

Fig. 7.6-13 Location of auxiliary fuel tanks.

ciple be regarded as alternative fuels. Admixtures of up to 5% RME in diesel fuel have been commonly used in France for years. Methanol and RME may be stored in fuel supply systems as already described above if material selection takes into account the specific chemical properties of methanol and RME. As also shown in Section 5.8, natural gas and, in the longer term, hydrogen are suitable mobile energy sources. Storage of both of these energy carriers is in gaseous form, in pressurized tanks at 200 bar or liquefied at very low temperatures in so-called cryotanks. Natural gas is differentiated as CNG (compressed natural gas) and LNG (liquefied natural gas). The latter is in its liquid state at about –160°C, while hydrogen may be stored in its liquid form at about –250°C.

7.6.9.2 Customer demands and regulatory requirements

It may be assumed that customer demands regarding handling and storage of alternative fuels such as natural gas or hydrogen will be based on previous experience in dealing with gasoline or diesel fuels. Similarly short refueling times and comparable vehicle range will be expected, and shortcomings will hardly be accepted. Furthermore, it can be assumed that such systems cannot cost any more than diesel or gasoline fuel supply systems.

In terms of safety, hydrogen and natural gas storage result in special requirements to meet modern safety demands. The applicable measures are more extensive than for gasoline and diesel fuels, which were introduced more than a century ago and for which extensive experience is available.

With particular regard to modern safety demands, a multitude of regulations addresses pressurized and cryogenic storage systems. For example, in contrast to gasoline and diesel fuel supply systems, vital components of pressure vessels and cryotanks must be licensed by appropriate regulatory agencies. Licensing requirements still vary from country to country, even within the European Union. Unified standards are currently being established [5].

7.6.9.3 Location in the vehicle

Vehicle-installed pressure tanks as well as cryotanks are derived from comparable stationary tanks or transport containers as employed in logistics of technical gases. Tanks are usually cylindrical. Because current natural gas or hydrogen-fueled vehicles are derived from existing gasoline-powered vehicles, integration of cylindrical tanks in existing vehicle concepts is usually difficult. In addition to space requirements, concerns regarding crash safety and installability as well as weight distribution within the vehicle must be addressed. Most such vehicle concepts also retain their conventional fuel tanks, so that such dual fuel vehicles [6] retain their everyday utility when natural gas or hydrogen refueling stations are unavailable [7].

The tank is often installed in the trunk; a reduction in trunk space is part of the bargain with these fuels.

At present, most alternative fuel vehicles operate as part of a fleet (e.g., service vehicles for utility companies). The BMW 316 g (Fig. 7.6-14), the first natural gas vehicle to receive pan-European type approval, was specifically conceived for such applications. The rear seats were deleted to provide more trunk space to house the pressure tank.

Customers experience few restrictions if the storage tanks can be hidden within the vehicle structure. This is the case for the Fiat Multipla Bipower (Fig. 7.6-15), in which several pressure tanks are integrated within the vehicle floor. In the Volvo S80 (Fig. 7.6-16), elimination of the spare tire and a reduction in fuel tank size provide space for a pressure tank.

The appreciably greater density of liquefied natural gas or liquid hydrogen implies that tank volumes may be reduced in comparison to pressurized gas containers, thereby simplifying installation. If tank volume is unaltered, cryotanks permit nearly tripled vehicle range compared to pressurized storage.

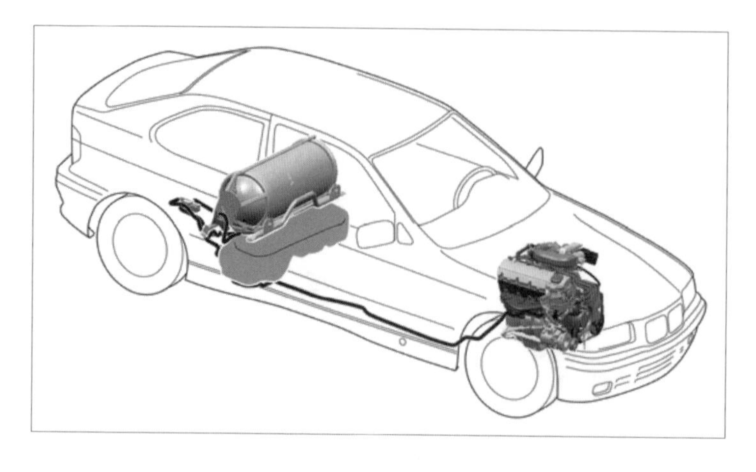

Fig. 7.6-14 BMW 316 g.

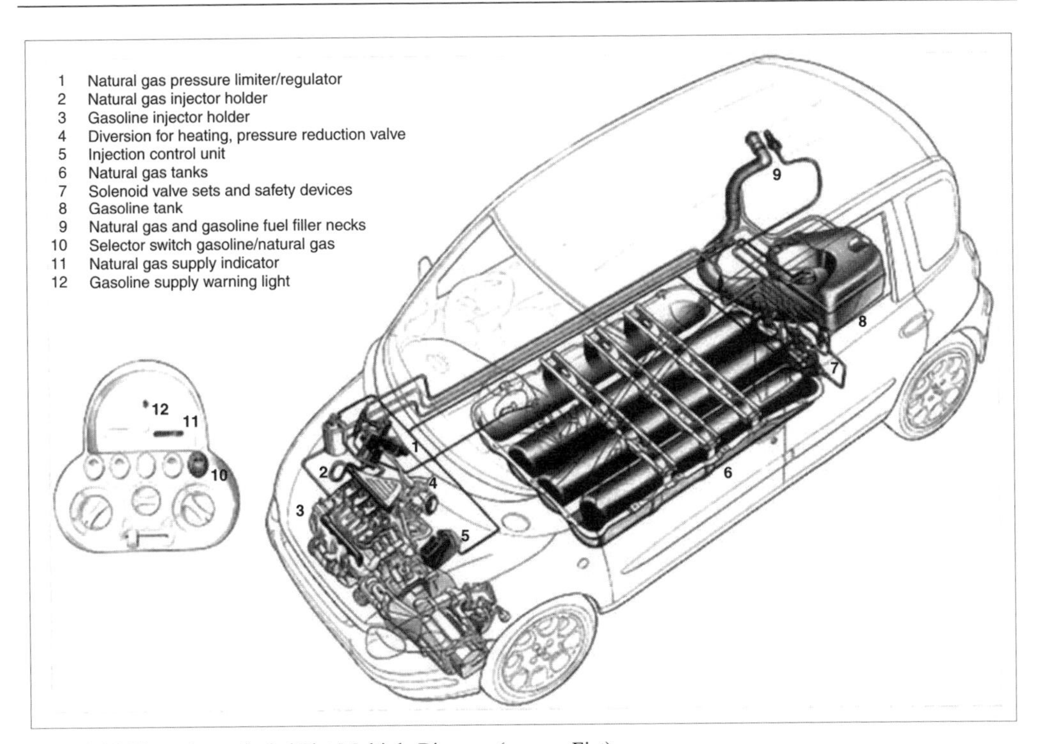

Fig. 7.6-15 Natural-gas–fueled Fiat Multipla Bipower (source: Fiat).

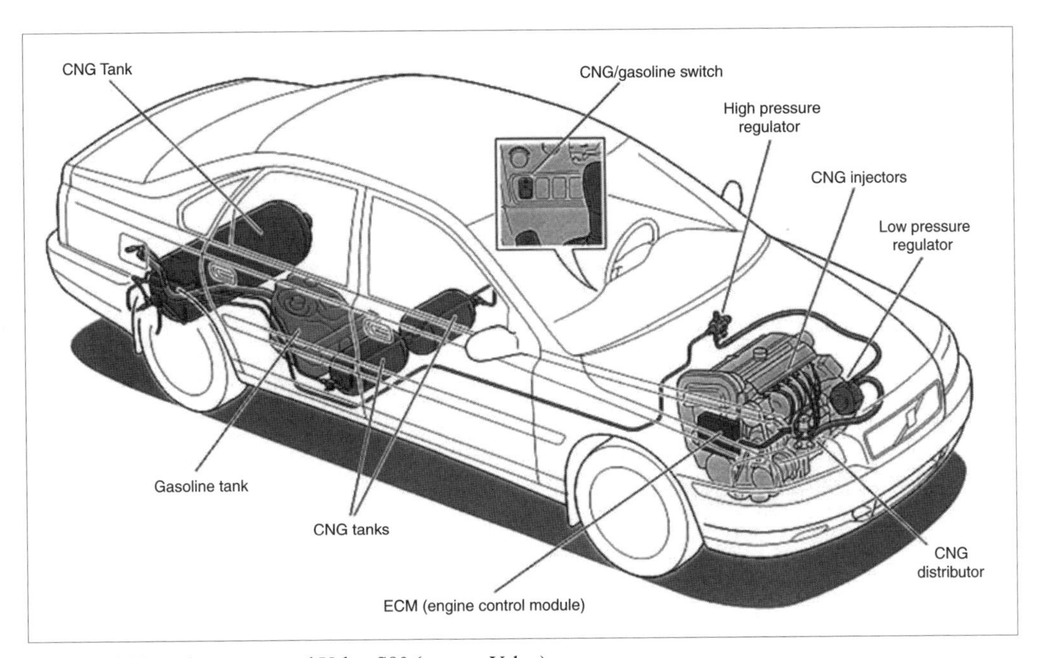

Fig. 7.6-16 Natural-gas–powered Volvo S80 (source: Volvo).

Fig. 7.6-17 Carbon-fiber–reinforced aluminum tank (source: Dynetek).

7.6.9.4 Fuel tanks

Pressurized storage tanks for natural gas or hydrogen are generally operated at a pressure of 200 bar.

Based on steel tanks, as have been used in industry for decades, the past few years have seen development of tanks using new, lighter, and more corrosion resistant materials. Tanks currently available may be made of fiberglass reinforced steel, carbon-fiber–reinforced aluminum, or composite fiber materials entirely. The weight of a carbon-fiber–reinforced aluminum tank is only one-third that of an all-steel tank.

Despite recurring experiments with various geometries [8] to better meet packaging demands, cylindrical tanks were and remain the sole configuration currently in use.

In the event of fire, a solder safety plug installed in the tank prevents tank pressure from rising above the permissible limit by allowing the tank to vent gas. Combined with tough tank materials, this ensures that the pressure tank cannot suddenly burst.

To prevent large volumes of gas from escaping in the event a line ruptures during operation, pressure tanks are fitted with an excess flow valve.

Cryotanks consist of two concentric shells (Fig. 7.6-18) separated by vacuum superinsulation. Vacuum, reflecting aluminum foil, and layers of insulating plastic reduce heat transfer to less than 1 W.

Even at such low levels of heat input, liquefied natural gas or liquid hydrogen will boil, causing pressure in the cryotank to rise. Once the allowed pressure limit is reached, gas escapes by means of an overflow valve [9].

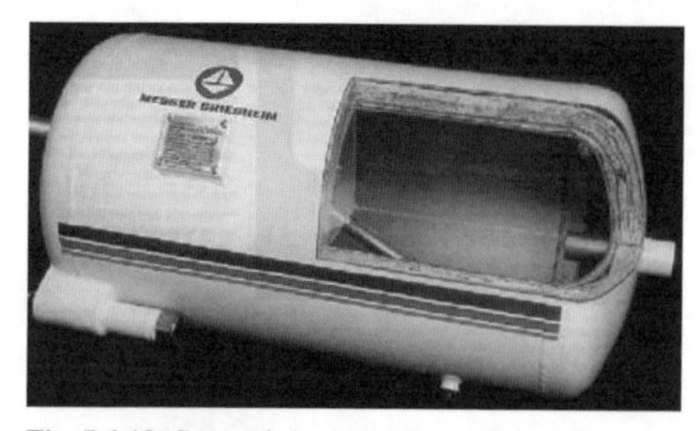

Fig. 7.6-18 Cryotank (source: Messer Griesheim).

Gas may be vented to atmosphere, used by a burner or fuel cell, or temporarily stored using other means.

Cryotanks are also fitted with a safety valve which opens the tank should maximum permissible pressure be reached in the event of fire [10].

7.6.9.5 Fuel supply systems and volume measurement

In pressure tanks as well as cryotanks, gaseous fuel is fed by the pressure differential between the tank and the engine fuel mixing system. Pumps are rarely used and only in certain LNG-powered commercial vehicles (Fig. 7.6-19).

Pressure in cryotanks may be maintained by in-tank heaters which provide controlled evaporation of liquefied natural gas or hydrogen. Alternatively, fuel evaporated outside the tank in a heat exchanger may be fed back into the tank.

Because the density of gaseous fuels is strongly dependent on temperature, simple pressure measurement alone cannot determine the degree of fuel tank filling. Gas temperature must be measured along with pressure and fuel quantity calculated accordingly.

Density of liquefied gases is also temperature depen-

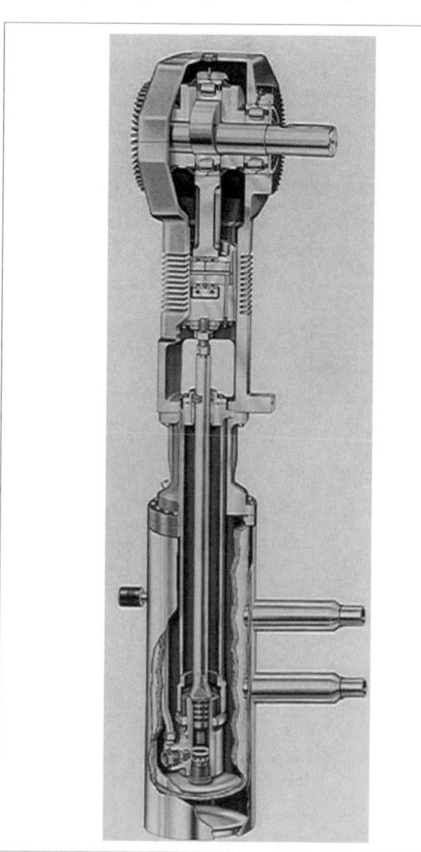

Fig. 7.6-19 High-pressure cryopump (source: CVI).

dent, but these changes are smaller than those of pressure tanks. Movable level indicators as used in gasoline and diesel tanks are not used at the extremely low temperatures found in cryotanks. Fuel level in cryotanks is determined by capacitive sensors or by change in resistance of electrical resistors mounted at various levels.

References

[1] Meinig, U., and J. Heinemann. "Neue Anforderungen für Tankentlüftungssysteme." Automobiltechnische Zeitschrift Vol. 03/99.

[2] Saechtling, H.-J. "Verarbeitungsverfahren." *Kunststoff Taschenbuch*, 26th edition. Carl Hanser Verlag, Munich, 1995, pp. 305–317.

[3] Sievert, H., and M. Thielen. "Trends beim Coextrusionsblasformen." Kunststoffe 88, Carl Hanser Verlag, Munich, 1998, pp. 1218–1221.

[4] "Strahlpumpen." *Techniker-Handbuch*, Vieweg Verlag, Braunschweig, 1995, p. 1184.

[5] Draft of the Economic Commission for Europe (ECE), TRANS/WP.29/1998/33, 1999.

[6] California Environmental Protection Agency, Manufacturers Advisory Correspondence MAC #99-01, 1999.

[7] Hämmerl, A., F. Kramer, P. Langen, G. Schulz, and T. Schulz. "BMW-Automobile für den wahlweisen Benzin- oder Erdgasbetrieb." Automobiltechnische Zeitschrift, Vol. 12/95.

[8] Wozniak, J.J. "The Johns Hopkins University Applied Physics Laboratory, Advanced Natural Gas Vehicle Project." 1998.

[9] *Gase-Handbuch.* Messer Griesheim, 3rd edition, 1989.

[10] Pehr, K. "Experimentelle Untersuchungen zum Worst-Case-Verhalten von LH-Tanksystemen." VDI-Berichte No. 1201, VDI-Verlag, Düsseldorf, 1995, pp. 57–72.

[11] Klee, W., et al. "Barrieretechnologien: Ein Beitrag zur Emissionsreduzierung von Kraftstoffanlagen." In: *Kunststoffe im Automobilbau*, VDI-Gesellschaft Kunststofftechnik, 2000, pp. 309–335.

8 Electrical and electronic components and systems

8.1 Illumination

8.1.1 Certification

External lighting units attached to a vehicle must be certified by a regulatory authority. Applicable regulations consist of installation and operational requirements. Increasingly, national regulations are being replaced by international standards. Worldwide applications may be classified as:

- ECE applications (for left- and right-hand traffic)
- Nations which have adopted large portions of the ECE regulations but have not yet adopted these in full (e.g., Japan) or have only partially adopted them (e.g., China, Australia). EC Directive 76/756/EEC covers installation of lights [1].
- SAE applications are valid in the United States and, with modifications, in Canada [2].

Some countries recognize ECE as well as SAE regulations.

In particular, attempts are under way to harmonize ECE and SAE regulations so that identical illumination equipment may be installed worldwide [3]. However, differences will remain between right- and left-hand traffic.

National certification authorities within the ECE domain issue approval marks consisting of the letter E and a number. For example, E1 represents Germany (ECE, Economic Commission for Europe) while e1 is Germany (EC, European Community). In the United States, DOT markings are required. Certification marks should if possible be applied to the outside of headlamps and other lamps [4].

8.1.2 Lighting concepts

Units—Visible light composes only a small part of the electromagnetic spectrum, with wavelengths ranging from 380 to 700 nm.

Measurement procedure—For headlamps and the 25 m measuring distance often used in Europe, the relationship 1 lx ≈ 625 cd applies. Measuring equipment must be corrected to match color sensitivity of the human eye—the V(λ) function [5]. Our greatest daylight visual sensitivity is to green light with a wavelength of 555 nm.

Range—It is customary to define a distance for which illumination intensity reaches a defined value. For low-beam headlamps, this is often taken as the point at the edge of the road where illuminance is equal to 1 Lux.

Isolux representations—For lighting evaluation, representations of equal illuminance (isolux) or luminous intensity (isocandela) can be used, as well as wall representations (usually for only one headlamp), or they can be converted to road representations.

Visual range—Subjective distance, dependent on many factors, over which an object may still be recognized.

Glare—Physiological glare is measurable illuminance in two-way traffic, while psychological glare represents the subjective degree of interference and may depend on light color and area of the light source.

Table 8.1-1 Photometric units

Photometric units	
Physical quantity	SI units
T luminous flux	lm (lumen)
I luminous intensity	cd (candela)
L luminance	cd/m^2
Q luminous energy	lm · s
P power	W
J light source efficacy	lm/W
E illuminance	lx = lm/m^2 (Lux)

8.1.3 Headlamps

Headlamps serve to illuminate a vehicle's surroundings. Currently approved headlamps for multitrack vehicles are installed in pairs.

8.1.3.1 Historical development

At the dawn of the automobile, night driving was rather uncommon. Around 1905, acetylene gas, produced by a carbide generator mounted on the vehicle, allowed an adequate headlamp design, which, however, was often perceived as too glaring. Electrical illumination eventually established itself after World War I.

Asymmetric low beams, introduced in 1957, illuminated a vehicle's own lane without dazzling oncoming traffic. Headlight differences between right- and left-hand traffic originated at this time.

In the United States after World War II, only sealed beam headlamps were permitted. Halogen lamps were introduced worldwide in the 1960s; xenon headlamps were marketed beginning in 1991.

Beginning about 1960, light output of most headlamps increased dramatically. Along with increased performance on the part of light sources, reflector technology advanced by leaps and bounds (Fig. 8.1-1). Nevertheless, surveys indicate that some drivers regard driving at night and especially at night in rain as particularly unpleasant. In the future, this situation will be improved by vehicles which automatically adjust their light distribution to traffic conditions [5, 6, 7, 8].

8.1.3.2 Headlamp types

Low beam headlamps—with modern traffic density, these are the most commonly used headlamps, accounting for 95–97% of headlamp use.

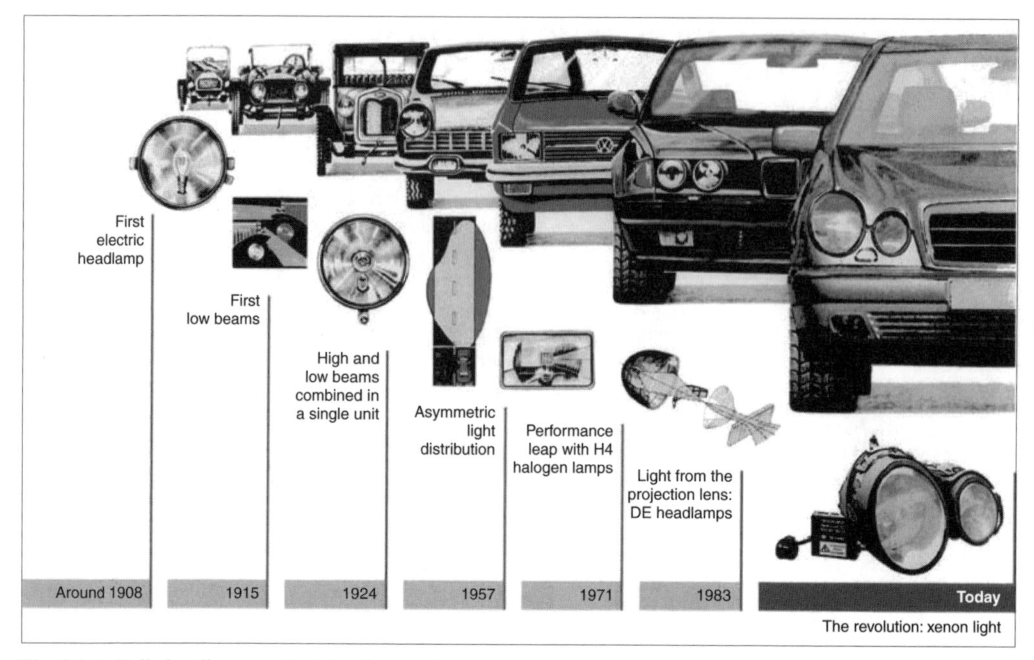

Fig. 8.1-1 Hella headlamp product development.

High beam headlamps—operated alone or in conjunction with low beams.

Auxiliary lamps—to reinforce high beams; may be steered with the front wheels.

Fog lamp—in Germany, these may only be used in fog, rain, or snow, and they provide very limited range. (In Switzerland and Norway, they are used as "cornering lamps".)

Backup lamps—not really headlamps per se, even though commercial vehicles often use fog lamps for this purpose.

Work lamps—may not be used for road illumination while driving.

Searchlamps—like work lamps.

"Flash to pass" (signaling headlamps)—in vehicles with retractable headlamps, auxiliary fixed headlamps are often installed to permit quick signaling; for certification purposes these are high beams.

Two- or four-headlamp systems

If the same reflector is used for low and high beams, the system in question is a two-headlamp system. If low and high beam functions are separate, the result is a four-headlamp system. Whether additional functions such as auxiliary high beams or fog lamps are present does not change this classification, which is important for photometric requirements in the United States and Japan.

Right- or left-hand traffic

In Europe, Great Britain and Ireland require left-hand asymmetric headlamps; otherwise, all other European countries use right-hand asymmetric headlamps. For travel from the Continent to Great Britain and vice versa, there should be a means of masking off the asymmetric low beam "light finger " or of rotating the entire headlamp assembly by 15°. For reflector-type headlamps, this is accomplished by taping the headlamp lens; for projection headlamps, internal shields or shutters must be actuated.

8.1.3.3 Reflector technology

Only that part of the light produced by the light source which is reflected and directed by the reflector provides useful roadway illumination.

Paraboloid reflector

Light originating at the focus of a paraboloidal reflector is rendered parallel, forming a good foundation for high beams. A light source located outside the focus generates a converging bundle of rays. The upper, angled portion of the bundle provides low beams, while the lower part must be blocked by shields for European beam patterns. By using two filaments (as in the H4 bulb), both high and low beams may be generated using the same paraboloid reflector. Actual beam dispersion is produced by prisms and cylindrical lens elements molded into the headlamp lens [4].

Ellipsoid projector

An ellipsoidal reflector has two foci; a large portion of light from a source at one focus is concentrated at the other focus. A shield in the shape of the desired light/dark pattern located near this second focus may be projected onto the roadway by an aspheric projection lens. Advantages of this system are freedom from glare, thanks to as sharp a light cutoff as desired, and small frontal area without performance compromise. Problems include high operating temperature, color fringes at the light/dark boundary, and high light intensity at the lens which oncoming traffic may regard as bothersome. Beginning in the mid-1980s, special reflector materials, modified lens shapes, and additional illumination in areas surrounding the lens have made projection systems practicable. Headlamp manufacturers use their own abbreviations for special reflector shapes—for example, DE for three-axis ellipsoid at Hella or PES for polyellipsoid at Bosch. Projection systems are used for low beams, fog lamps and, more rarely, for high beams [5, 7].

Free-form reflector

Introduction of free-form surfaces represented a definitive advancement in reflector technology, beginning with projection systems in 1988 [7]. Compared to ellipsoidal reflectors, which are already very efficient, usable light output increased by 43% (Super DE, Fig. 8.1-2). Free-form technology is based on point-by-point reflector surface calculations using CAD systems (CAL = computer aided lighting).

Through the use of precision lamps (HB4, H7, or H11), a free-form reflector for low beams may be shaped to utilize the entire reflector surface, representing a gain of about 80% compared to a paraboloid reflector. Free-form technology also permits dispensing with optical profiles on the lens (which is really no longer a lens but a clear window; Fig. 8.1-3). The reflector may be aesthetically shaped—for example, smooth, faceted, or segmented.

Bulb shields

Bulb shields mounted in front of the light source provide suitable means of blocking directly exiting light as well as light reflected by decorative surfaces (that is, surfaces not needed to direct light beams). Their inside surfaces should have a matte finish, preferably black. For decorative purposes, brightwork caps may be fitted over these.

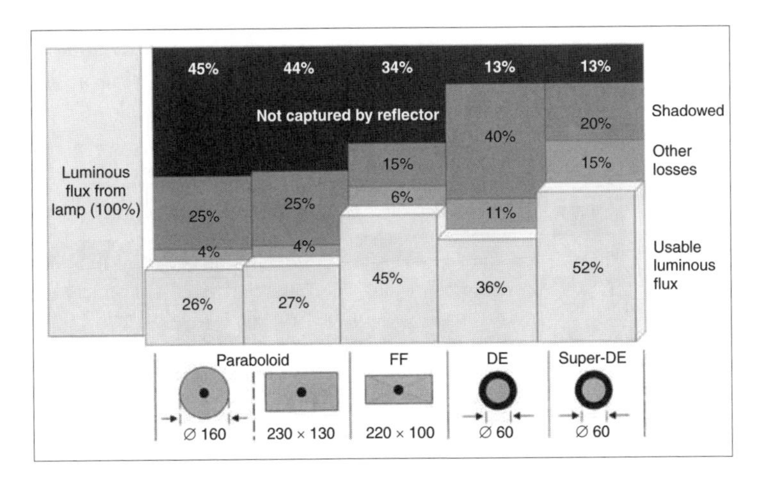

Fig. 8.1-2 Low beam headlamp systems: typical luminous flux distribution.

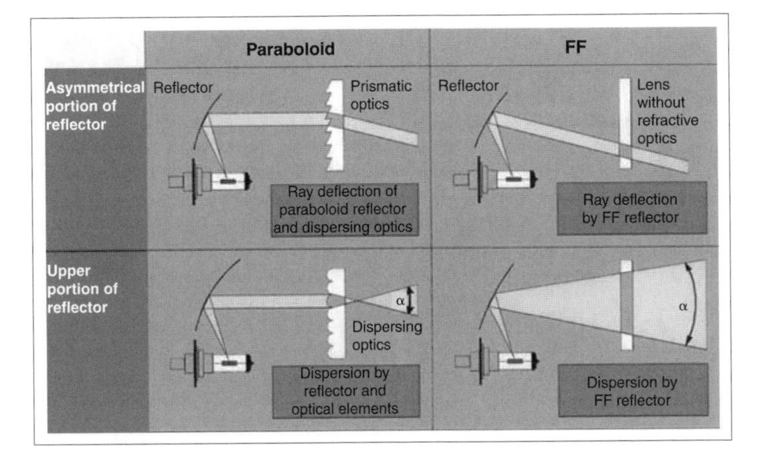

Fig. 8.1-3 Ray tracing for paraboloid and free-form headlamps.

Reflector materials

Traditional reflectors were made of pressed sheet metal, predominantly steel, more rarely brass or aluminum [3]. The surface is smoothed by lacquer and vacuum metallized with a coating of pure aluminum. With an included layer of corrosion protection, reflectivity is about 87%. Cast metal reflectors (magnesium, aluminum, and, earlier, zinc) are used in applications with small dimensions and high temperatures (fog lamps, projectors). Today, thermosetting plastics (BMC, bulk molding compound) are used (e.g., LPP, low profile polyester) with good heat resistance and shape retention, but requiring an undercoat before metallizing.

Mirror surface

Reflector surfaces must satisfy demanding requirements. Surface roughness must not exceed ~1/10,000 mm. In the past, reflectors were often destroyed by condensed liquids. Introduction of ventilated headlamps and improved corrosion protection has solved this problem.

8.1.3.4 Front lens

In current designs, a transparent lens mounted in front of the reflector usually follows body contours. Lenses for paraboloid reflectors require optical profiles to achieve proper light distribution given the ray bundles provided by the reflector. In the case of free-form reflectors and projectors, optical profiles provide more homogeneous light distribution. To achieve a bright or "technical" look, clear lenses or such with few decorative profiles may be used.

Lenses for low beams may not be angled much beyond about 25°; otherwise, the light/dark cutoff would show unacceptable distortion. However, modern aerodynamically styled headlamps, with their rake and tilt angles, achieve total angles of 60° or more (relative to a plane perpendicular to the vehicle axis). Clear lenses permit larger angles. Maximum combined rake and side angles are determined by the ratio of reflectivity to transmissivity. At 0° (lens perpendicular to vehicle axis), transmission is about 92% for glass and 85% for clear plastic (Fig. 8.1-4).

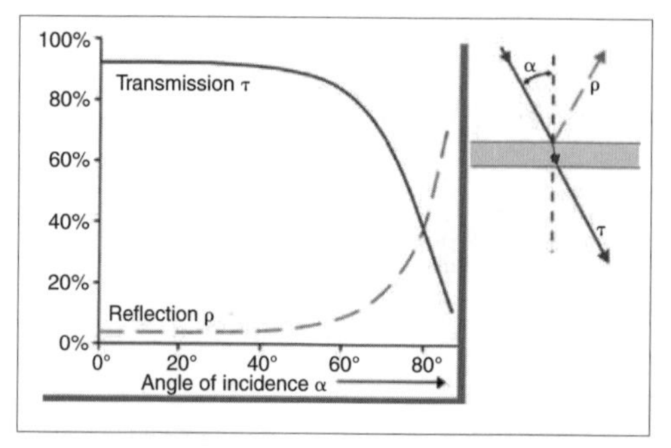

Fig. 8.1-4 Reflectance and transmittance as functions of angle of incidence.

Until 1983, European manufacturers used pressed (or, more rarely, rolled) glass lenses exclusively. The pressing process imposes design limitations, and rapid tool wear results in coarse tolerances. Positive aspects include scratch resistance and relatively good thermal resistance, both of which may be increased by chemical or thermal hardening.

The first European approval for plastic lenses according to the new ECE regulations was granted in 1993 [7, 9]. Plastic lenses were introduced in the United States and Japan somewhat earlier [10]. Advantages compared to glass are lower weight and increased design freedom. The predominant base material is polycarbonate, protected by an outer hard coat which must be applied in cleanroom conditions.

Light soiling of lenses first increases dazzle; more soiling reduces light output. Headlamp cleaning systems (Fig. 8.1-5) represent an important safety element and, until adoption of EC regulations, were mandatory in Scandinavia. For glass lenses, conventional wipers as well as water jets are effective. For plastic lenses, only noncontact water jets are suitable, and these have become the generally accepted solution. In addition to rigidly mounted jets, telescoping jets are preferred; most are simply actuated by water pressure. Jets are activated in conjunction with windshield washers if the lamps are switched on [3, 4, 7].

8.1.3.5 Headlamp aiming

Reflectors must be aligned correctly when the car is assembled, after any repairs, and, possibly, after bulb changes.

For low beams, vertical adjustment is based on the horizontal beam cutoff and horizontal adjustment on the break point of the (usually) 15° angle rise. Alignment may be by means of a mark on a board, an optical headlamp aiming device, or an electronic headlight aiming device. High beams, if not combined with low beam units, are set for maximum illumination. Basic low beam adjustment may be between 1% and 1.6% downward and is marked on the headlamp itself [1].

In the United States, purely mechanical adjustment by means of so-called "aimers" was derived from three aiming pads molded into the lens of sealed beam (SB) headlamps, which for decades were the only type approved for use [2].

Since 1990, installation of vertical (usually bubble level) and horizontal aimers has been approved for all American headlamps. This finally permitted installation of headlamps with fixed front lenses for the American market as well [2, 10].

Beginning in 1997, American regulations also permit "visual aiming," similar to European practice, for a left or right horizontal light cutoff [2]. If horizontal alignment is not precisely defined (as is the case in the United States), horizontal adjustment may be eliminated entirely.

Headlamp adjustment is based on an unladen vehicle.

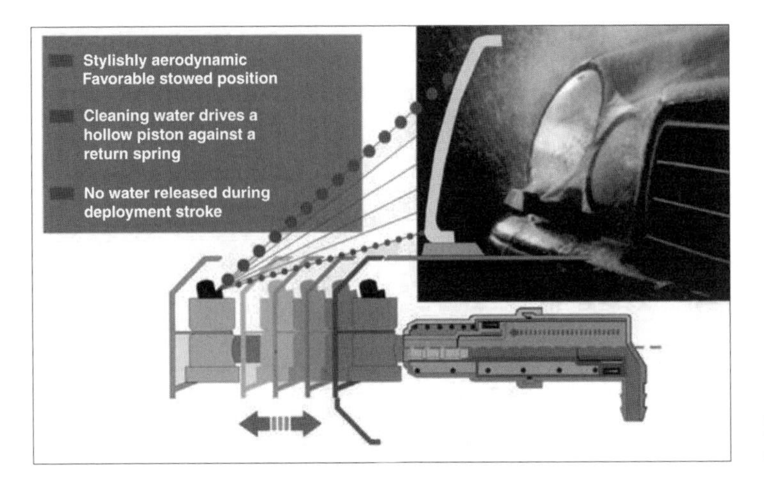

Fig. 8.1-5 Headlamp cleaning system.

Loading causes the rear of the vehicle to settle and can result in considerable glare to oncoming traffic.

Headlamp range adjustment, remotely controlled by the driver, has been required for all new vehicle models in Germany as of 1991 and throughout Europe as of 1998 [1].

Hydraulic headlamp range adjustment is sometimes used in commercial vehicles; pneumatic adjustment, using manifold vacuum, is seldom used in modern applications [7]. Electrical adjustment has become the most common method.

Because correct manual operation of headlamp range adjustment is not always assured, ECE Regulation 48 prescribes automatic regulation for xenon and other high-intensity light sources, which automatically compensates for load changes. Dynamic leveling, although not required but very effective nevertheless, compensates for vehicle motion under hard braking or acceleration. Such leveling systems must react in fractions of a second, and they require a much larger range of reflector adjustment. Stepper motors or powerful DC motors with position feedback are used for these applications [7].

To control automatic headlamp leveling, orientation of the vehicle body relative to the road surface must be determined. Suspension position sensors are used for this purpose. The most common sensors are angular displacement sensors which use linkages to detect jounce or rebound motion of front and rear suspension.

In the future, optical sensors within the headlamps themselves may be used to triangulate the road surface [6, 7].

8.1.3.6 Light sources for headlamps

Only light sources approved by international regulations may be used as replaceable lamps. Worldwide application for many light sources has been made possible by deregulation in the United States and by incorporation of American bulbs in European regulations (Fig. 8.1-6).

Incandescent bulbs radiate invisible infrared light (i.e., heat) along with visible light. Compared to daylight, incandescent bulbs produce yellow-white light. Until 1992, some countries (especially France) mandated yellow lights. Demands on motor vehicle light bulbs are appreciably greater than demands on household bulbs due to shock, vibration, and voltage variations. The pinnacle of conventional incandescent bulb engineering is the dual filament bulb, meeting ECE Regulation 2. This lamp, introduced in 1924, permits generation of low beam and high beam light using a single reflector [4]. Originally, an integral shield in the form of a small concave mirror produced symmetrical low beams; asymmetric low beams were introduced somewhat later (1957). Incandescent bulbs are available for 6-, 12-, and 24-V applications.

The halogen cycle prevents darkening of the lamp envelope and regenerates the incandescent filament, so that halogen bulbs have about twice the light output, higher color temperature, and doubled service life in comparison to ordinary incandescent bulbs [11, 12]. Halogen lamps are covered by ECE Regulation 37 for single-filament bulbs and ECE Regulation 8 for H4 dual filament bulbs.

The following designations are applied to various bulb configurations:

- H1 for low beams, fog lamps, and high beams
- H2 for auxiliary lamps (to be deleted)
- H3 preferred for fog lamps
- H4 for low and high beams using a single reflector, less commonly used for combined fog/high beam units, approved in the United States as HB2.

H4 lamps were first used in 1971 and since then have dominated European headlamp technology. A successor lamp is currently under development.

It was not until 1995 that a more precise alternative to the H1 bulb, the H7, became available for use with free-form reflector technology.

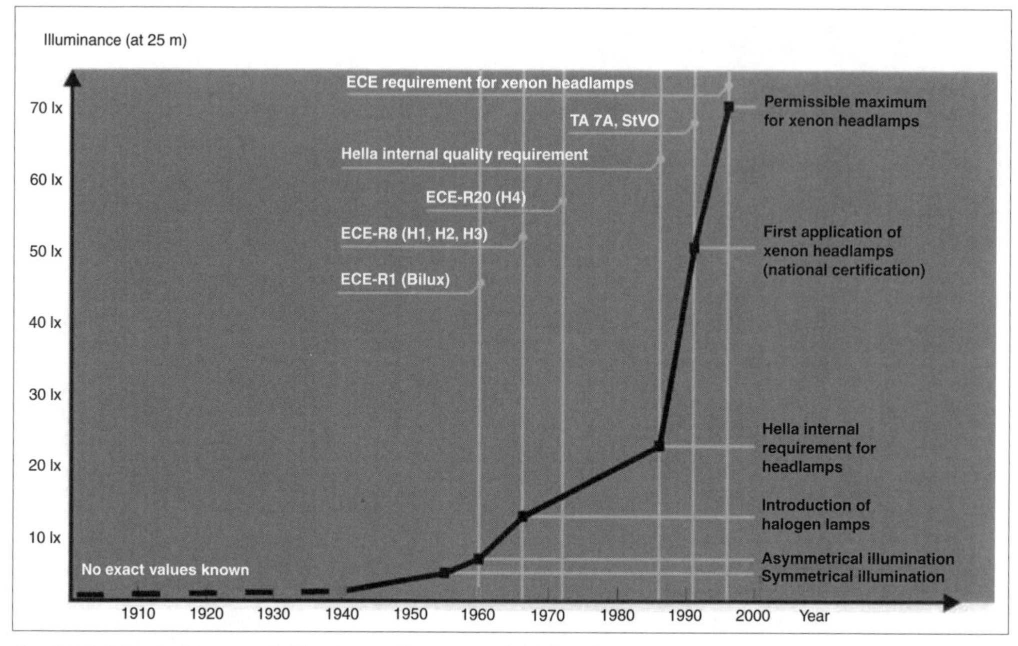

Fig. 8.1-6 Historical increase in illuminance for motor vehicle headlamps.

H8, H9, H11, and H13 are new halogen lamp types with sealing bases; preferred applications are fog lamps (H8), high beams (H9), low beams (H11), and combined high/low beams (H13).

In addition to these types, originally American bulbs HB1, HB3, and HB4, as well as HB5 (replacement for HB1) are finding more widespread application in Europe. Their advantage lies in their sealed base configuration, which permits more compact headlamp units and easier bulb changes.

Bulb life is indicated by various characteristic numbers: Tc represents the number of hours for a 63.6% failure rate and B3 the number of hours for a 3% failure rate for a population of test bulbs. Incandescent bulbs are especially sensitive to overvoltage; on the other hand, actual life of high-quality bulbs is higher than the specifications.

Table 8.1-2 Life expectancy of various halogen bulbs: specified and actual life expectancy

	Specified life expectancy, Tc (DIN 60810) (hours)	Realistic life expectancy, Tc (hours)
H1	400	960
H3	400	990
H4	700	1,050
H7	550	630
H7LL = long life	930	1,000

8.1.3.7 Xenon lights

Halogen bulb technology has attained its zenith. Significant progress can only be achieved through introduction of a completely new light source. This is known as xenon light. The corresponding bulbs are known as gas discharge lamps (GDL) or high intensity discharge (HID) lamps.

Xenon light is about 2.5 times as intense as halogen light, and it is closer to daylight (Fig. 8.1-7). It is produced by an arc, about 4 mm long, in a pea-sized discharge vessel, which contains small quantities of rare earth salts. While halogen lights produce a continuous spectrum with little UV and a great deal of IR radiation, xenon light consists almost entirely of visible light and very little IR. First-generation xenon lamps (D1) had a high UV component. In second-generation xenon lamps (D1S/D1R and D2S/D2R), an additional glass filter jacket reduces UV radiation to the point where plastic headlamp lenses can be used [9, 11, 12].

First-generation D1 lamps had soldered connections to their control electronics; thus, they were not replaceable. The second generation has a connecting socket; the D2S configuration is suitable for projection systems and high beams, while the D2R is suitable for reflection-type low beam headlamps.

In contrast to halogen lamps, xenon lamps require control electronics (ballast) for operation in 12-V, 24-V and, eventually, 42-V electrical systems. Among other functions [7], these complex electronics perform the following tasks:

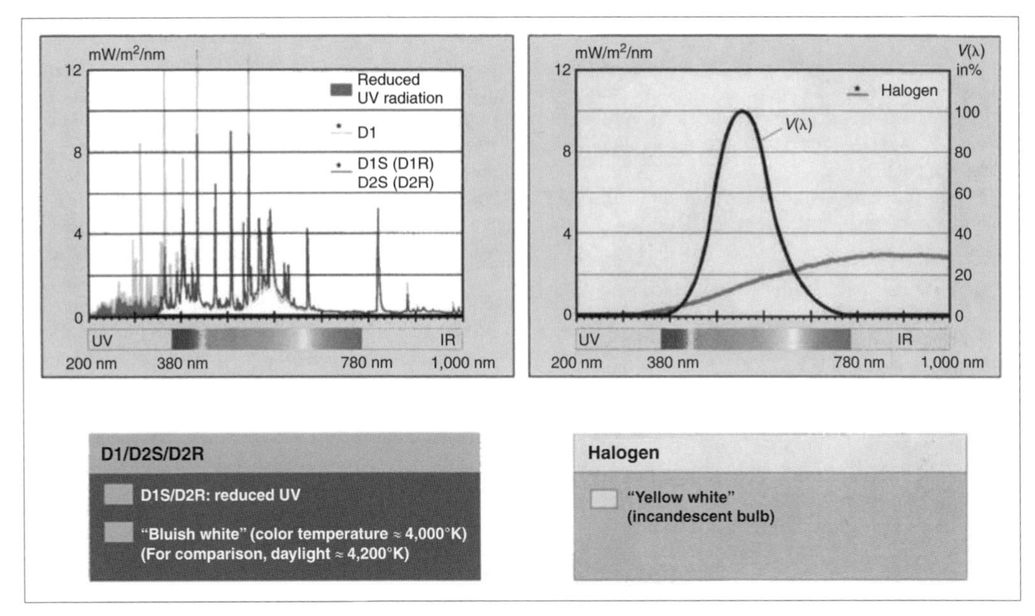

Fig. 8.1-7 Xenon lamps in comparison to halogen lamps.

- Produce high ignition voltage, from 18,000 up to 28,000 V
- Supply high current to evaporate halides (up to 17 A)
- Provide continuous operation with only about 35 W of energy input (in other words, considerably less than the approximately 60 W of halogen lamps)
- Automatic reignition; e.g., after being extinguished by severe shocks
- Monitoring of short circuits (e.g., more than 30 mA leakage to ground) to protect against manual contact
- Voltage stabilization in the event of inconstant supply voltage; e.g., from 9 to 18 V (for nominal 12-V systems).

Xenon lamp control electronics may consist of separate high voltage and low voltage components or a combined electronic package. As a result of their complicated startup procedure, xenon lamps only reach full output after about three seconds. To provide adequate light output immediately upon switching on, the discharge vessel is filled with xenon gas. The life expectancy of a xenon light source depends on the number of times it is switched on rather than the illumination time. In most cases, life expectancy exceeds that of the vehicle itself.

The first xenon headlights were delivered in 1992 in both Germany and the United States. Because of their high UV output, first-generation lamps could only be installed in projection systems with additional metal shielding. High light density of the relatively small (60 mm diameter) projection lens was perceived by drivers of other vehicles as irritating, but their outstanding illumination performance was regarded as a decisive safety benefit. In 1996, with ECE Regulations 98 and 99, worldwide applications became possible for ECE regions in connection with headlamp cleaning systems and automatic leveling (Fig. 8.1-8). The second lamp generation, D2R, was introduced in 1995 for reflection systems, and D2S for projection systems.

Fig. 8.1-8 Xenon headlamp bulbs, as approved by ECE Regulation 99.

In 1989, a research vehicle demonstrated a system with low and high beams using a xenon source and mechanical switching [9]. Such "bi-xenon" systems reached production in 1999 in projection as well as reflection systems [5, 7]. Since then, the bi-xenon projection system has found widespread application.

In the future, use of high-intensity white LEDs is anticipated for applications including front lighting.

8.1.4 Light evaluation

The disproportionate number of nighttime accidents shows that such driving conditions constitute a higher risk. To an increasing degree, visibility using headlamps is being recognized and evaluated as an important safety element [3].

There are as yet no comprehensive quality criteria for headlamp performance. In general, good low beams should:

- Illuminate the road evenly, without dark patches or "hot spots"
- Provide good visual range at the edge of the road
- Emphasize the vehicle's own lane ("guide light")
- Provide adequate beam spread
- Avoid dazzling the oncoming traffic
- Produce minimal backscatter in fog
- Avoid lighting the vehicle's own lane too brightly, out to about 40 meters; reflections from wet roads produce extreme glare for oncoming traffic.

Compared to low beams, high beams enjoy only secondary importance in central Europe, but are given much higher priority in northern Europe (moose!). Lighting performance is determined by headlamp properties and vehicle-related factors. In headlamps, reflector size as well as lens rake and tilt angle play a deciding role; additional factors include lamp type and especially the actual supply voltage. Vehicle factors include installation height, driver seating position, windshield rake, and possible windshield tint or metallization. Magazine comparison tests of various lights are controversial but very popular. Conducted fairly, they may serve to increase awareness of good headlamp illumination and assist advanced lighting technology to achieve a breakthrough. A good lighting test includes a static test, preferably in a "light tunnel," and driving tests under various environmental conditions. Virtual reality methods permit simulation of actual driving tests.

8.1.5 Daytime running lamps

In order to make moving vehicles more obvious even in daylight, many countries require or permit daytime running lamps.

Scandinavian countries and Canada require lights to be on under all conditions. Other countries—for example, Poland—require driving with lights on during certain times of the year. Campaigns are under way in other countries (e.g., Austria) to institute daytime running lamps. At present, Germany requires only motorcycles to use lights during the day.

Daytime running lamps may be realized by low beam headlamps. Because this consumes a relatively large amount of electrical energy and therefore fuel (100 W of electrical power amounts to ~0.12 L per 100 km), other means are also permitted. One method uses dedicated daytime lamps or dimmed (6 V) high beams, as permitted in Canada.

8.1.6 Auxiliary headlamps

According to ECE Regulation 19, a pair of fog lamps is permissible in support of low beams or without low beams but operable in conjunction with marker lamps. Hallmarks of good fog lamps are large beam spread to illuminate the edges of the road and a sharp horizontal beam cutoff [3].

High beams may be reinforced by hardwired or selectable long-range lights. Regulations for these vary regionally. In central Europe, a total reference number of 37.5 (relative to the lights' maximum) may not be exceeded.

Additional lamps may not be installed for purposes of roadway illumination. Searchlamps are movable units which assist in finding street addresses and road signs. In some countries, "alley lights" are permitted on taxis for lateral road illumination. Work lamps use halogen bulbs, and, increasingly, xenon lights.

8.1.7 Signal lamps

Unlike headlamps, signal lamps do not serve to illuminate the roadway (exception: backup lamp), but instead are used to mark vehicle extremities, license plates, and to indicate driver intentions. Here, too, only those functions and light sources specifically described by regulations may be used. Signal lamps appeared late in the history of motor vehicles and were often limited to combination tail lamps/license plate lamps. Directional changes were first signaled on a voluntary basis by rotating arrows, hand signals, and, later, by illuminated semaphores.

Today, signal lamps are design elements with large surface areas which, given homogeneous illumination, take on a wide variety of appearances and, in their "cold" state, can display an almost unlimited range of colors.

Vehicle boundaries are delineated at the front by two white "clearance and sidemarker" lamps and at the rear by two red taillamps ("rear position lamps"). Directional changes and visual warnings are signaled by front, center and rear flashing lamps; braking is indicated by rear and high-mounted center lamps (CHMSL—"center high mounted stop lamp").

Rear license plate illumination and, in some cases, side marker lamps complete the signaling lamp package.

The backup lamp is a hybrid, serving two functions. It illuminates the road behind the vehicle while reversing, and it also indicates to other traffic the driver's intention to reverse.

Rear fog lamps are required for new (European) vehicles and tolerated in the United States. For reasons of practicality, "reflex reflectors" are often combined with tail lamps; triangular reflectors are reserved for trailers.

The intensity of front turn signal lamps depends on their distance from low or high beam lamps. For styling reasons, front turn signals often show a "white" or brilliant appearance even when not activated. Rear directional signals are yellow, or in the United States, often red. Technical illumination requirements are less stringent than for front lamps. This also applies to side-mounted turn signals for new vehicles.

Modern front marker lamps (position lamps) are usually integrated in the headlamp reflectors. In combination with low beams, these create the vehicle's "night design." At the rear, red taillamps must be less than 400 mm from the outermost edges of the car in order to convey a feeling for vehicle width. Additional front and rear marker lamps are required on tall and wide commercial vehicles.

In the United States, taillamps may serve in a multi-role function as brake lamps and direction indicators [2]. In Europe, dual-function tail lamps are permitted if they use dual-filament bulbs, although individual functions are preferable.

Brake lamps should always indicate the onset of any braking event.

Beginning in 1980, Germany permitted a pair of high-mounted brake lamps for passenger cars only. One advantage was visibility through the windshields and backlights of several vehicles and therefore earlier warning of a need for braking [3, 4]. As a result of excessive light intensity, stray light in the vehicle interior, and questionable mechanical attachment, the concept was soon discredited in Europe. In the United States, successful field trials were conducted with single high-mounted stop lamps [10]. In 1985, high-mounted stop lamps became mandatory in the United States [2]; they were tolerated in Europe as of 1991 and became mandatory beginning in 1998 (European Community).

The rear license plate must be illuminated, but depending on body details, two license plate lamps might be required.

One or two red rear fog lamps, with light output greater than that of brake lamps, are intended to make the vehicle more visible in poor visibility conditions. A distance of 100 mm to brake lamps must be maintained.

Backup lamps are intended to illuminate the car's surroundings while reversing and also indicate engagement of reverse gear. For commercial vehicles, fog lamps are tolerated as backup lamps. In Sweden, front side-mounted backup lamps are permissible (Saab).

Parking lamps are optional and are white toward the front, red to the rear. Often, a front position lamp and a taillamp are used as parking lamps.

8.1.7.1 Light sources for signal lamps

Ordinary incandescent bulbs are the dominant choice for signal lamp applications. Customary connections are made by means of a bayonet socket or wedge base. Bayonet coding prevents misapplication. If light sources are replaceable, only such units as are codified in ECE regulations may be used. To simplify bulb changes, especially for rear lamps, the bulbs are incorporated in bulb carriers.

Halogen bulbs are only gradually finding their way into signal lamp applications, although they have advantages in terms of color (white), energy consumption, and intensity. One example of a halogen bulb is the H6W bulb, which is increasingly finding application in front marker lamps.

Compared to incandescent bulbs, light emitting diodes (LEDs) offer many advantages, in particular:

- Rapid switching response, combined with monochromatic red light, results in approximately 170 ms shorter reaction time.
- Lower energy consumption.
- Long service life.
- Small space requirements.

Disadvantages include high price and a change in photometric performance at higher temperatures. The best application for LEDs is brake lamp functions, especially for high-mounted stop lamps [7, 9, 10]. Neon tubes have properties similar to LEDs but add complexity due to their control electronics and difficult electromagnetic compatibility issues [9, 10].

8.1.7.2 Configurations

Signal lamps may be executed as individual functions or combined, with a single lamp serving several functions. At the front, marker lamp, turn signal, and, for the United States, side marker lamp (as well as passive side reflector) are usually combined with the headlamp to form a single light unit. At the side, individual functions predominate for side turn signal lamps and, if present, side marker lamps. At the rear, functions are usually combined in a common rear light unit at each side of the vehicle or, in some cases, divided between two light units per side. Taillamp units are often manufactured using complex injection molding techniques and multicolor molding machines (up to four colors). Colored signal functions may be achieved by means of color filters. Most recently, as in the case of headlamps, brightwork light assemblies with clear covers have become popular.

In the past, lamps have been divided into two categories: those with (parabolic) reflectors, which utilize direct light as well as that directed by the reflector, and lamps without reflectors which distribute directly emitted light by means of prisms in the cover lens [4]. More recently, other principles such as ellipsoid reflectors, faceted reflectors, and free-form reflectors have come into use.

Lens optics may take one of several forms to effect more homogeneous light distribution [7]. Named for their inventor, French physicist Augustin Jean Fresnel, Fresnel lenses are stepped, compound refractive optics with good directional qualities. In flat configurations, these provide illumination through a large included angle [4].

Similar to Fresnel lenses, a lamp reflector may consist of concentric reflecting segments. In contrast to deep paraboloidal reflectors, such a flat mirror reflector requires little installation depth. Holographic reflector structures [5, 7, 11, 12] represent an avenue for future development.

Similar to headlamp applications, free-form reflectors permit continuous light distribution without need of any optical profiles on the lens.

Miniature lamps permit construction of light function arrays. Horizontally extended arrays have been introduced for high-mounted stop lamps [7].

Illumination functions may be achieved using light pipes, in which colored light is fed into transparent rods and directed by prisms arranged transversely to the rod axis [7].

8.1.8 Lighting styling

Whether illuminated or in their "cold" condition, vehicle lighting units have become major styling effects. Consequently, they are accorded great care in styling and are increasingly subject to automotive fashion trends and the desire for individual expression.

In all such efforts toward individual design, the desire of every auto manufacturer to maintain a "family resemblance" is readily apparent. One widespread desire is represented by a "bright" appearance, realized by headlamps without dispersing optics and taillamps with clear lenses. A relatively new aspect is "night design," which with low beams activated or with position lamps only presents a pleasing and preferably easily recognized image.

Lighting equipment is particularly suitable for quickly creating a different vehicle appearance without extensive body modifications, for special models as well as facelifts. Xenon headlamps [9], for example, are particularly well suited for this task.

Clear lenses, highly reflective metallization, and faceted reflectors impart a sparkle to headlamps and taillamps ("jewel look").

Taillamps in particular may be executed in colors other than the customary signal colors, often for a surprising effect. This is achieved by a special filter technology. Dark taillamps (as well as headlamps) are popular [7].

In principle, headlamp night design only applies to low-beam operation (including position lamps), although fog lamps are often placed in accordance with styling considerations rather than functional reasons. Shape and location of headlamps and taillamps should provide ready recognition by day as well as at night.

Interior lighting is becoming increasingly important. Its purpose is to assist in orientation during vehicle entry and egress. While the vehicle is in motion, reading lamps should not distract the driver, and overall, interior lighting should impart a feeling of comfort while making all safety-relevant items sufficiently recognizable. The number of interior light sources is increasing. One solution is offered by light guide systems such as CELIS [7, 9].

8.1.9 Future developments

Because lighting equipment with very few exceptions is subject to international regulations, new concepts must first be expressed in appropriate regulations before they can be applied in production. Concept cars are one means of displaying such ideas in visual form.

Numerous concept cars have sported light guide headlamps, with a rigid or flexible light guide interposed between light source and the point where light emanates from the car.

The AFS (Adaptive Frontlighting System) is currently achieving international standardization. This system independently and optimally adjusts to actual traffic conditions. Instead of the current system of invariant low-beam light for all traffic situations, AFS provides multiple options:

- **Town light** for low speeds.
- **Cross country light** with additional cornering light.
- **Motorway light** as long-range illumination for high-speed driving.
- **Adverse weather light** reduces glare for oncoming traffic while simultaneously increasing illumination of roadway edge.
- **Bend light** to illuminate curves may be combined with any of these lights.

Other functions, including special overhead marker lights, are undergoing testing.

Closely related to AFS are headlamps with built-in "intelligence" which independently react to traffic situations and analyze vehicle surroundings using sensors, monitor their own performance, and restore performance in the event of malfunctions.

Headlamps with low-beam bend lights entered series production in 2002.

Intelligent solutions are being developed for rear signal illumination, similar to headlamps, which, for example, may modify brightness to match visibility conditions or provide a logical means of indicating the intensity of a braking event.

Application of UV light for headlamps is still in the experimental stage. In such a system, an intense bundle of UV rays would illuminate oncoming lanes without generating glare but would cause suitable materials to fluoresce (reradiate visible light).

Another invisible form of light is infrared, which is better able to penetrate fog than visible light. Special cameras and displays can display a "thermal image." Such systems, typically called "night vision systems," also prevent oncoming traffic from being dazzled.

References

[1] European Community Directive E76/756/EEC.
[2] NHTSA: FMVSS 108.
[3] Bockelmann. *Auge—Brille—Auto*. 2nd edition, Springer Verlag, 1987. ISBN 3-540-16429-4.

[4] Bosch. *Kraftfahrtechnisches Taschenbuch*. Bosch, 20th (German) edition, 1987. *Bosch Automotive Handbook*, 5th English edition. Robert Bosch GmbH, 2000. Distributed by SAE.
[5] Bosch, publications of the former "K2" department.
[6] *PAL Symposium Proceedings*, Progress in Automobile Lighting. Darmstadt University of Technology, September 23–24, 1997.
[7] Hella KG, company publications.
[8] Internet, http://www.hella.com.
[9] ATZ, various publications including illumination topics, technical optics.
[10] SAE, annual symposia including illumination topics.
[11] Philips, research reports.
[12] Osram, company literature.

8.2 Sensors, actuators, and systems technology

8.2.1 Introduction

In place of the classic, localized component view, this chapter will consider the automobile from an overall functional viewpoint.

Seen from this perspective, the function of driving a vehicle represents translation of driver commands by means of a system of interrelated and coupled open- and closed-loop control processes. Functions are no longer tuned to optimize components but rather to consider overall vehicle, driver, and traffic-relevant constraints.

This perspective has been made possible by the availability of powerful, cost-effective technology in the form of modern microelectronics and systems engineering and by release into the marketplace of the financial means necessary to satisfy desired gains in vehicle function, customer utility, and compatibility with regulatory requirements.

A typical example is provided by the braking function. For a long period of time, this function was satisfied by a small number of variations (drum or disc brakes, with active redundancy). More recently, this system has quickly developed into an integrated vehicle dynamics control system, with rapid, complex corrective intervention in the brake system as well as powertrain control, with appreciable customer benefits—all while retaining the mechanical/hydraulic basic configuration.

Nearly all other vehicle functions have developed at a similar pace, and this trend continues. The basic pattern is:

- Beginning with luxury vehicles, a mechanical function is initially equipped with electrical supplementary or corrective functions, permitting optimization of the basic function and providing customer benefits or attraction.
- In subsequent stages, the mechanical function is entirely reproduced by an electrical function.
- The system penetrates all vehicle classes; this is economically attractive due to economies of scale and transfer of existing solutions to other models.
- Development then proceeds along a broad front, and application-specific know-how is created worldwide; as a result, new functions are created, with additional, substantial customer benefits.

Definitions

- Sensors
  Capture and measurement of a condition or time-variant behavior by means of a physical or chemical effect and conversion to information (generally, electrical information) which can be processed at some other location in the vehicle. Required elements: sensors.
- Actuators
  Conversion of (generally electrical) information into physical effects in order to deliberately affect vehicle condition or vehicle components. Required elements: actuators.
- Analog to digital conversion
  Conversion of (generally electrical) quantities into a number for further numeric processing by a computer program. Required elements: A/D converter.
- Digital to analog conversion
  Conversion of a number into (generally electrical) quantities for input into the vehicle's electrical/electronic system. Required elements: D/A converter.
- Interface
  The totality of interactions at the component boundaries which are necessary and sufficient for function as well as integration of the component within the vehicle.
- Software
  Ordered sequence of program commands and operations within a microprocessor.
- Hardware
  Electronics (processor, memory, I/O interfaces, etc.) required to run the software.
- Components
  Functional elements with at least one (core) function.
- System
  An assembly of several components whose functionality is derived from component core functions and from additional coordinated or collective (system) functions.

8.2.2 General

The system of sensors, actuators, and electronics (the electrical/electronic, or E/E system) serves to collect, distribute, and generate information and to actively affect vehicle processes. Like the nervous system of a living organism, it permeates the entire automobile and, like a nervous system, is the central element giving rise to functions (Fig. 8.2-1).

Automotive sensors, actuators, and system technology are, on the one hand, based on the possibility of bidirectional transfer of electrically encoded data regarding physical and chemical quantities and, on the other hand, on control technology which, given knowledge of conditions and the ability to influence a control path,

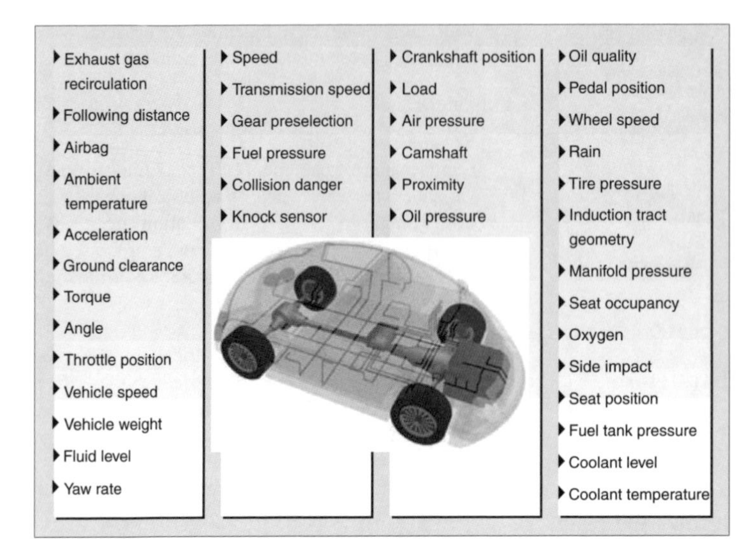

Fig. 8.2-1 The electrical and electronic system permeates the automobile as the nervous system of a living organism does. Shown are sensors to capture the most vital parameters (after an illustration by Delphi).

can modify their behavior to best fulfill given expectations (quality criteria).

Electrical and electronic system configuration is the result of functional (vehicle specific) manufacturing technology and logistical optimization (shared parts concepts, purchasing strategies, etc.) on the part of the vehicle manufacturer.

Electrical and electronic system elements are:

— *Sensors, actuators*
  as interfaces to functional units or the outside world
— *Control units (ECU)*
  as central elements of functional processing, located at network nodes
— *Network (Buses)*
  wiring to conduct energy; electrical and optical buses to transfer information
— *Connectors*
  plugs and sockets for functional integration during the vehicle manufacturing process

Classification of these elements follows according to:

— *Element function*
  What is the element's purpose?
— *Interface to process or unit*
  How is the function generated? What are the associated boundary constraints? How does the function change in response to operating conditions?
— *Interface to installation space*
  How is the element installed within the vehicle? What pressures on manufacturing processes are imposed by the installation situation (logistics, functional check, etc.)?
— *Connection to the information network*
  How is information processed? How is it distributed within the vehicle?
— *Connection to one of the vehicle's energy networks*
  Which network will be used? How severely will it be loaded? How will this affect the vehicle's energy budget?
— *Risk management strategies*
  How will function of the element and the vehicle be assured? How will risks to driver (mobility/quality) and manufacturer (product liability) be minimized? (Safety, maintenance, repair, limp-home strategies, maintenance strategies, certification, documentation, etc.)

In modern automobiles, sensors, actuators, and system technology are the classic domain of the electrical and electronics supplier industry:

— Cost reduction (economies of scale) are achievable by serving the entire global market.
— Marketable products have resulted (small, compact, necessary, etc.).
— Added value is achieved by generic or standardized solutions.
— Product differentiation is possible by optimized production steps in manufacturing and packaging.
— Economical, concentrated problem-solving competence is achieved in particular through simultaneous development of hardware and process-related software and modern electronic manufacturing technologies.
— Nevertheless, there is generally no perceptible differentiation apparent to the customer. It is known that today's customer no longer perceives isolated unit performance, but instead bases his purchase decision mainly on integrated higher level performance (brand, configuration, design, comfort, noise, etc.)

8.2.3 Sensors

Sensors measure or detect conditions and generate data and information.

Even though the general concept of "sensor" encompasses many applications, this section will consider only sensors equipped with an electrical interface. As a consequence of networking (see below), there is a steadily growing demand for vehicle-wide information availability, which in general may only be satisfied economically by complete integration within the electrical/electronic system.

Even with these restrictions, depending on a vehicle's equipment level, a host of information is captured: Position, angular rotation, rotational speed, velocity, electrical current, seat occupancy, temperature, pressure, mass airflow, oxygen partial pressure, driver commands, field strength, airborne sound, structure-borne sound, crash events, stresses, fluid levels, acceleration, torque, yaw rate, vehicle position, access approval, rain intensity, oil quality, fuel alcohol content, vehicle following distance, air quality, and so forth.

Sensor classification

— *Process interfaces*
  Nearly all physiochemical effects may serve as process interfaces; functional limitations are determined by underlying physical or chemical laws.
— *Vehicle interfaces*
  Sensors are generally installed directly on components and are therefore subject to component conditions (installation space, temperature, vibration, accessibility, weight, etc.); installation and testing of installed performance is conducted within the framework of the vehicle assembly process.
— *Information network interfaces*
  Hand-off of raw signals to the wiring network (via connectors); in the case of fault- prone or vital information, provide conditioned and secure signals (e.g., shielding, pulse-width–modulated signals, encoding).
— *Energy network connection (power supply)*
  Sensors are in general supplied by the onboard electrical system (and in some cases with a 5-V reference voltage). The effect of sensors on the vehicle electrical network is generally minimal, as losses are on the order of mW to a few watts.
— *Risk management*
  Proof of assured function and information availability through qualification of packaging and connection technology under given installation conditions (risk minimization through reduction of failure probability). Software recognition of signal path availability by means of limiting value consideration and signal plausibility, substitution of information by default values in the event of implausible values (risk minimization by damage limitation in the event of a fault); for especially critical signals, multiple measurements (increased information availability through active redundancy), incorporation of sensors in vehicle maintenance program within the framework of onboard diagnostics (risk minimization through early warning).

Table 8.2-1 Examples of typical modern sensors

Magnetic sensors	Rotational speed sensor Position sensor Driver control elements
Electrical sensors	Position sensor Angle sensor Switches
Micromechanical sensors	Structure-borne sound, pressure Acceleration Rotational rates, torque
Chemical sensors	Oxygen partial pressure Air quality Exhaust gas composition
Thermal sensors	Mass airflow Exhaust mass flow
Radio/acoustic sensors	Following/parking distance Relative velocity

Application example: Mass airflow meter (Fig. 8.2-2)

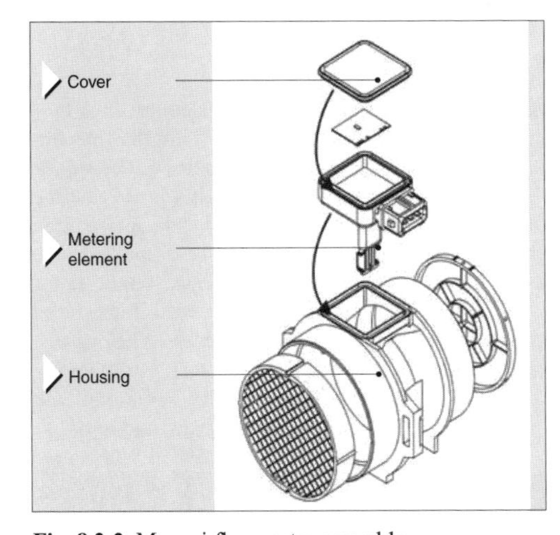

Fig. 8.2-2 Mass airflow meter assembly.

The mass of air inducted by the engine is a fundamental quantity; mass airflow is determined by means of a thermal process (Fig. 8.2-3). A heated resistance element in the airstream is kept at a constant temperature difference with respect to induction air temperature by means of control loop control. The required heating current depends on the mass airflow past the element. To compensate for reverse flow effects, a second heating element is installed downstream; in the event of reverse flow, it ensures that no sensor heating is needed and therefore backflowing air is not metered.

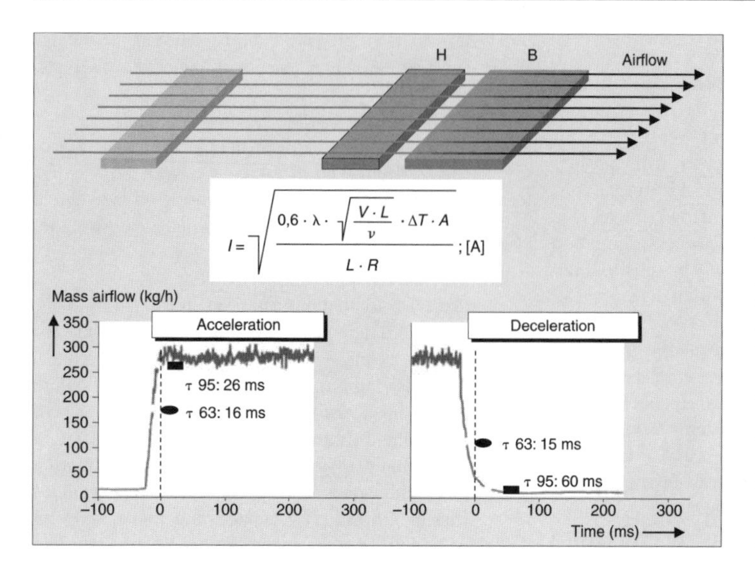

Fig. 8.2-3 Top: operating principle of a thermal mass airflow sensor. Heating element H is kept at a constant temperature difference with respect to air temperature. Heating current I required to accomplish this is directly dependent on mass airflow. Booster B prevents response in the event of reverse flow. Bottom: step responses.

To remain resistant to drossing, the element is installed on a glass substrate. Adaptation is achieved by means of the metering section diameter; protective screens are installed upstream and downstream of the sensor. Once packaged, the element is suitable for automotive service (Fig. 8.2-4).

8.2.4 Actuators

Actuators can serve either of two functions: first, they may actively alter the condition of a unit, thereby converting information into action (control applications). Actuators transfer information to units, thereby altering the vehicle condition. Second, they prepare information and transfer it outside the system, for example, to the driver in HMI (human/machine interface) applications.

Here, we will consider only electrically interfaced actuators: fuel injectors, EGR valves, ABS valves, starters, electric motors, stepper motors, throttle positioners, spark plugs, displays, valve actuators, leak detection pump systems, pumps, airbags, and so on.

Fig. 8.2-4 Mass airflow sensor as a finished product.

Actuator classification

— *Process interfaces*
In open or closed loop control applications: conversion to a mechanical quantity (position), generally by means of electromagnetic effects.
In HMI applications: conversion to a signal (e.g., direct indication in a display); function and functional limits are established by underlying physical or chemical laws.

— *Vehicle interfaces*
Actuators are generally attached to the component and therefore subject to that component's installation conditions (open or closed loop control applications) or installed as optimized-design display units in the vehicle interior (MMI applications). In the former case, installed performance is tested during component assembly; in the latter, during vehicle assembly.

— *Information network interfaces*
Actuation is accomplished through the wiring harness; fault-prone information may be conditioned and transmitted via secure channels (shielding, pulse-width–modulated signals, encoding).

— *Energy network connection (power supply)*
Power is supplied by the onboard electrical system; in measurement/feedback/control applications, this is often also used to control a cascaded control loop in a different energy network (e.g., electrohydraulic system, airbag). In some cases, operation may result in considerable (pulse) electrical system loads (engine starter motor: 1–5 kW).

— *Risk management*
Proof of assured function and functional safety through qualification of packaging and connection technology under given installation conditions (reduced failure probability, "design to last"); recogni-

tion of function through sensor feedback plausibility (diagnostics) and shutoff in the event of implausible values and cascaded paths (damage limitation); diverse redundancy for critical or safety-relevant signals; incorporation in vehicle maintenance (see OBD).

Application example: Electric motor

Fig. 8.2-5 Electric motors for vehicle installation.

Motor vehicles employ a wide variety of electric motors for comfort functions (movement, positioning) as well as control processes (starter, electronically controlled throttle).

Performance spectrum

Starter:	several kilowatts (peak load)
E-gas (drive-by-wire) motor:	several tens of watts (peak load)
Seat/window motors:	several hundred watts (peak load)
Stepper motor for indicator:	several watts (continuous load)

Fig. 8.2-6 shows one configuration in the 100-W power category; the greatest possible standardization has resulted in reduced cost.

By means of precision manufacturing, the functional, torque-relevant air gap parameter is kept below 0.5 mm.

Fig. 8.2-6 Electric motor with connector.

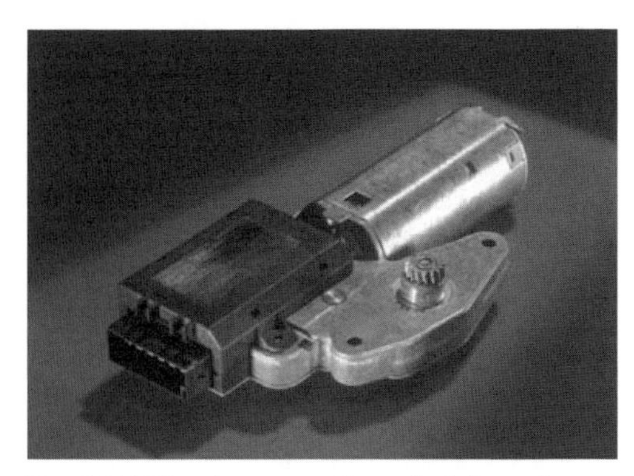

Fig. 8.2-7 Electric motor with attached electronics.

Other available variations include attached electronics, offering “on-site intelligence” (Fig. 8.2-7).

8.2.5 Wiring

The function of vehicle wiring is to transmit signals and information (as a data bus) or electrical energy; modern wiring harnesses are highly complex structures (Fig. 8.2-8).

— *Process interfaces*
  Electrical and electromagnetic properties determine wire functions (material properties, dimensions) and function dependence on loads and environmental conditions.
— *Vehicle interfaces*
  Wiring is generally installed during vehicle assembly and therefore follows the logical assembly sequence of body parts. Individual components may use local wiring harnesses with multipin connectors (dependent on architecture and partitioning) to connect these components to the electrical and electronic system.
— *Information network interfaces*
  Input and output via connectors; wires with suitable material properties. Protection against interference and external influences by means of insula-

Fig. 8.2-8 Wiring harness (section).

tion, shielding, or special configuration (e.g., twisted pair wires for databuses). For high data rates, optical fibers are gaining importance as a new information transfer technology.

— *Energy network connection (power supply)*
  Connection to the electrical network, conduction of electrical power (current). Compromise between weight, function/power losses, and ease of assembly (bending radii, passthroughs, etc.)
— *Risk management*
  Largely standardized materials and insulation, testing of integration performance with respect to operational electromagnetic vulnerability (EMV) within the framework of vehicle certification. Standardization of data protocols (see below).

8.2.6 Connector technology

The function of connectors is to provide separable conductive connections between individual elements of the vehicle electrical/electronic system (Fig. 8.2-9).

— *Process interfaces*
  Wires are connected by at least one line contact in the connector on the conductor cross section. For sensor or actuator connectors, environmental conditions are those of the installed units. Functional behavior is influenced by surface effects (corrosion), geometry, and the total number of connection/disconnection cycles (risk of preinstallation damage).
— *Vehicle interfaces*
  Installed on components as well as wiring harness; significant mutual effects with manufacturer-specific shared components or logistic concepts.
— *Information network interfaces*
  Best possible loss-free and interference-proof transfer of information. High functional demands due to large signal range (mV to kV) and requirement for reproducible integration performance (corrosion, geometry, etc.)
— *Energy network connection*
  Best possible loss-free energy conduction. High functional demands due to large power range (mW to kW) and requirement for reproducible integration performance.
— *Risk management strategies*
  Greatest possible (manufacturer specific) standardization within the vehicle. Combination by functional and topological boundary conditions. High cost pressure due to large number contained in vehicle.

8.2.7 Basics of signal processing and transmission

Analog signals are generally combined with input circuits to define absolute voltage levels (pull-up or pull-down resistors, amplifiers, etc.) and to damp rapid variations (RC lowpass filter; Fig. 8.2-10).

Measurement of analog signals is accomplished by means of an analog to digital converter, which provides a numerical output signal with the information content of the input signal.

Depending on input processing, determination of fault sources in the electrical signal path is possible by means of signal plausibility (broken or short circuits).

Pulse-width–modulated signals are compared to threshold values by means of a comparator circuit and counted by a timer. The actual measured value is the duty cycle τ :

$$\tau = \text{(open duration)/(total duration)}$$

Pulse-width–modulated signals permit especially simple monitoring of the transmission path.

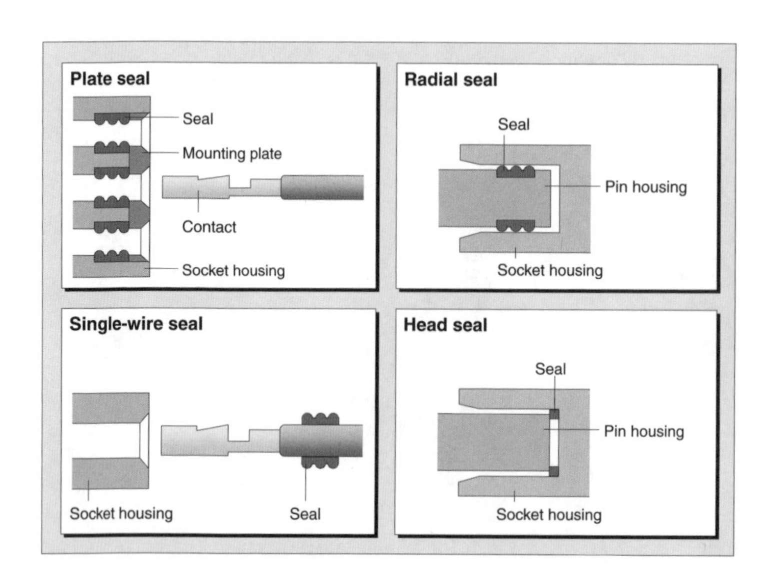

Fig. 8.2-9 Connector and seal configurations (after Tyco Electronics EC).

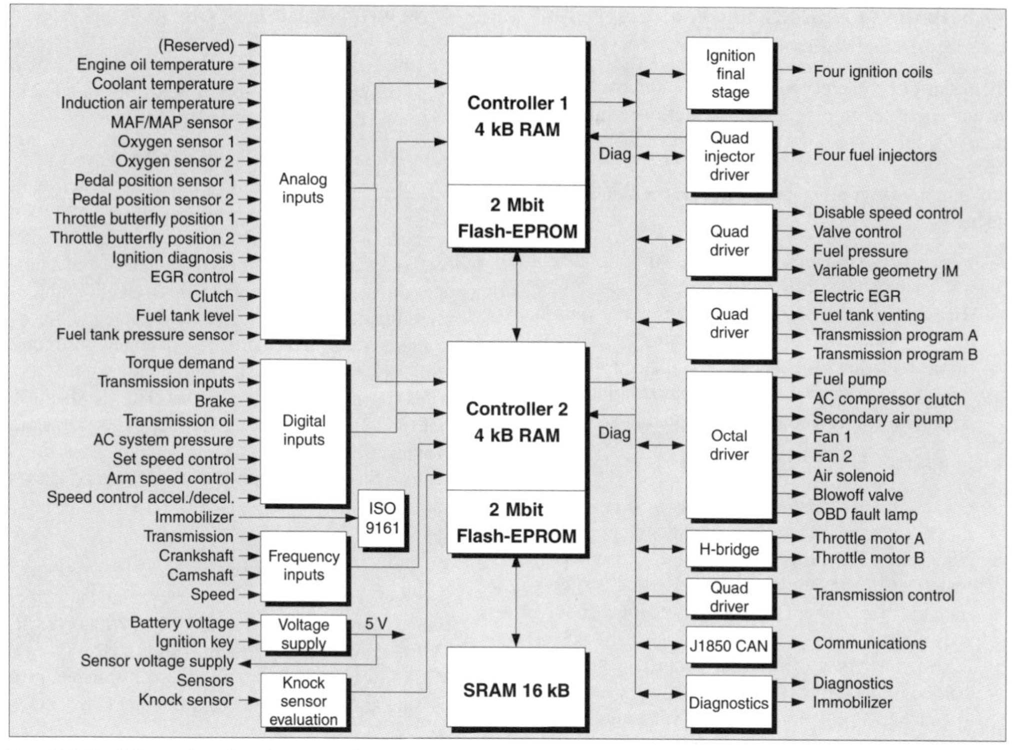

Fig. 8.2-10 Schematic of engine control system. Left: input signals; right: output signals; center: information processing. AC, air conditioning; accel., acceleration; decel., deceleration; IM, intake manifold.

Depending on electrical properties of the signal path, transmission, and evaluation components, a digitized representation is created within the electronic system which differs from the raw input signal:

- By time delay (damping, sampling, sampling time, period, etc.)
- By measurement precision (timer frequency, additional tolerance-limited elements in the signal loop)
- By information resolution (quantization, roundoff error)

Electronic information processing is generally carried out by a control program, either time-dependent (by sampling at constant intervals) or event-dependent (as an interrupt). In addition, the processed information is fed to the bus system (see below) so that it can be used by other processes.

Fault evaluation requires analysis of the entire signal path and precise examination of all relevant influences. The same is true for the reverse process chain (actuator control).

Bus systems play an increasingly important role in distribution of digitized information within the automobile. Serial buses encode information in a pulse-width–modulated signal which is sequentially (serially) fed to a special conductor system consisting of a relatively small number of conductors. Access to the bus system, conversion of information including fault management, serving drivers, etc. is accomplished by special components and protocols. Presently implemented solutions are:

- CAN (Controller Area Network) protocols
- Audio buses
- VAN (Vehicle Area Network) protocols

For parallel buses, information is fed to multiple parallel conductors; digital signals (0/1) are switched to and evaluated from these conductors. The "width" of a bus (number of conductors) generally represents the resolution of transmitted information.

In automotive applications, serial buses are generally used to transmit information over greater distances; they are cost effective and reliable (effective error correction algorithms). For particularly demanding data rates, assured transmission, and freedom from interference, optical fibers may be used.

Parallel buses are primarily used for high-speed signal transmission within control units. Because of their multiple conductors (up to 32) and rapid switching between two digital voltage levels, they require complex measures to assure EMC compatibility as well as crossover to other signal paths.

8.2.8 Basics of feedback and control systems and electronics

By means of computer programs, automotive electronics and control units process sensor information to generate new information, which is fed back to the vehicle through actuators.

Solution of user problems always proceeds in several stages:

— Making available information and possibilities for influence (availability of sensors and actuators)
— Representation of user problem as part of a control loop algorithm
— Implementation in vehicle electrical/electronic system, function verification (problem solution), and compatibility with other vehicle functions

The most important elements are:

— Control loops: control of processes specified by functional requirements (e.g., maintaining constant idle speed). The process to be controlled is represented as a mathematical model using systems analysis tools and controlled by execution of software code.
— Function: every systems analysis provides a control program (software module) which is compiled to generate a program capable of running in a microprocessor. Sequential control combines all modules to form a software package.
— Data set: parameterization of the software module for actual optimization of the concrete control path.

Fig. 8.2-11 shows an application example of modern engine control torque structure:

— All engine interventions relevant to cylinder filling and efficiency are calculated relative to the parameter "torque."
— Pedal information is interpreted as torque demand and fed into the system.
— Engine demands arising from vehicle management (e.g., from suspension control functions) are fed in as torque demands, independent of platform.

Along with the best possible matching of quality criteria in test situations, the most important criterion at the unit and vehicle level is verification of integration performance—that is, proof that quality criteria are met under test conditions (conformity, compliance, functional depth at the unit level) and proof that vehicle usability is ensured under all other imaginable combinations of operational, driver, and environmental conditions (compatibility, application range at the vehicle level).

To this end, modern methods such as HIL (hardware in the loop) and SWIL (software in the loop) as well as standardization efforts play an increasingly important role.

Methods that can be applied to accelerate processes include:

— Rapid prototyping
— Operating systems for standardized interfaces between user programs and hardware, with the objective of reutilizing software modules (Windows CE, OSEK-compatible operating systems, etc.)
— Workflow models and tools adapted from data processing and commercial programming (Java, etc.)

8.2.9 System integration and platforms

The following major trends affect engineering and technology of modern electrical and electronic systems:

— Critical installation conditions (temperatures, vibration, etc.), increasing power losses.
— Continuing tuning of manufacturing to optimize added value changes partitioning of vehicle manufacturing; components and modules are combined

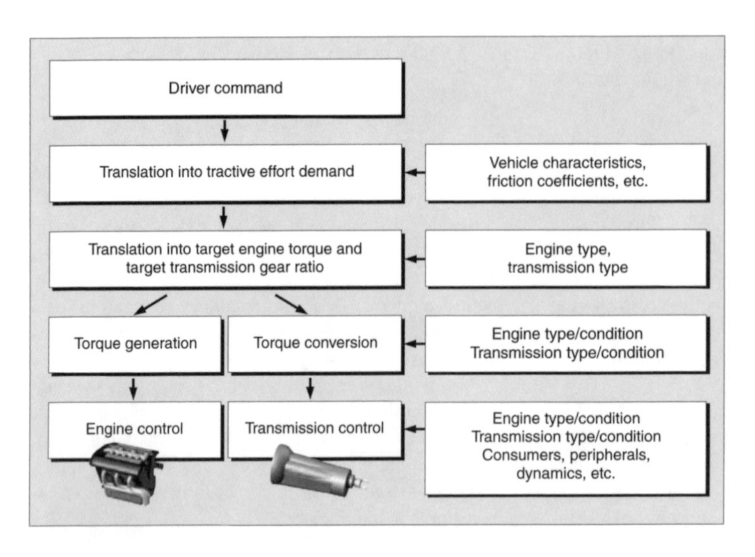

Fig. 8.2-11 Modern drivetrain management torque control architecture.

to form systems, which are preassembled and tested before installation. Testing of functions is becoming more important than testing signals against specifications.

— Continuing cost pressure requires further standardization of core elements (higher production volume basis) and forward integration into applications (work flows, etc.)
— Continued pressure on vehicle component functions demands increased data processing power. Function complexity will continue to rise and verification of integration performance at the vehicle level will become increasingly difficult.
— Disproportionate increase in electronic capabilities combined with decreasing prices permits integration of functions formerly distributed across several control units into a single control unit. As a result, programs become more complex and less coherent.
— Vehicle manufacturer platform strategies: at present the most effective means of increasing vehicle value requires consolidation of platform contents and ongoing functional decoupling of body-relevant functions and services for later integration (noise and vibration management).
— "Leaning" of production processes and the resulting consequences for component testing demand fusion of complex electronics with the units they control (e.g., engine control and engine).
— Change management, variation management, cost of complexity, and product liability require overarching manufacturer cooperation.
— "Leaning" of vehicle manufacturing plants and throughput optimization demands for material flows to match value creation processes.

In practice, these trends manifest themselves in combining components to create functional systems of component groups.

Examples:

— Modular climate control. Nearly all operator control functions, including electronics, are combined into a single unit. The package includes not only the tested (assembled) unit, but also component functions (Fig. 8.2-12).
— Leak detection pump (LDP) as an intelligent actuator capable of conducting fuel tank diagnostic functions nearly autonomously, thereby encompassing a wide range of fuel tank variations (shapes, volumes, installation conditions, etc.) (Fig. 8.2-13).

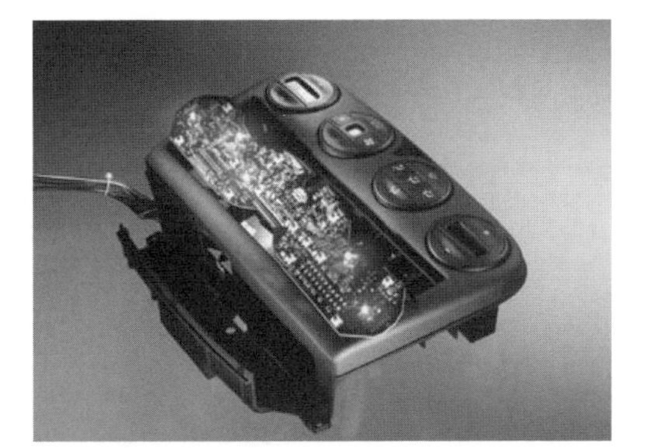

Fig. 8.2-12 Mechanical and electronic integration: climate control system.

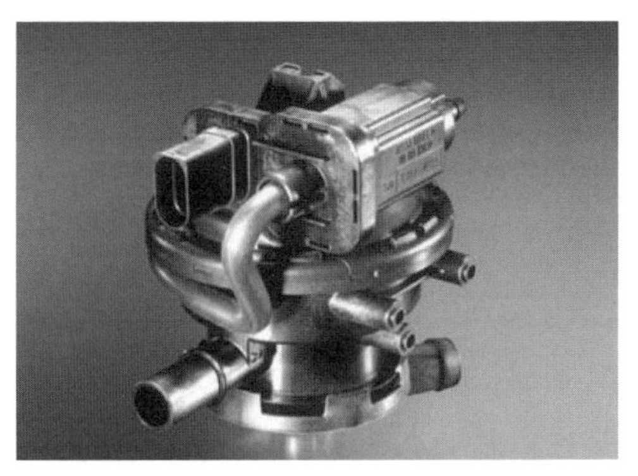

Fig. 8.2-13 Pump with attached electronics for autonomous fuel tank leak detection (OBD II functionality; see Section 7.6).

8.2.10 Open questions and current problems

Onboard Diagnostics (OBD)

OBD systems were introduced to meet regulatory requirements and as part of a traffic risk minimization strategy. Results of electronic tests (plausibility tests, process chain diagnosis) permit one to draw conclusions regarding control loop conditions.

In the course of such diagnostic checks, the ability of the control loop to be influenced as well as its response behavior in noncritical test situations are measured. These situations are insufficient to cause driveability-relevant faults, to exceed preset limits, or to alter product properties.

Fault messages are affected by the entire signal path—that is, even by electrical and mechanical connections and all associated units.

Defective components are not immediately apparent from fault messages, because:

— The software simulates the entire functional process chain, not individual fault mechanisms.
— Models for plausibility tests only apply to the test situation.
— Mutual interactions between functions are in general only known during vehicle testing near nominal conditions (all elements functioning correctly) or only in the case of simple faults in standard situations (vehicle certification).

Rapid and reliable identification of defective components in need of replacement by a repair shop or the necessary control system or parameterization interventions require expert knowledge regarding abnormal component condition, use of models for fault conditions, and inclusion or diagnosis of nonelectrical components (e.g., catalytic converters), which in general can only be performed by selective testing (controlled diagnosis).

One will note the remarkable similarity to trends in medical diagnosis:

- Increase in psychosomatic complaints resulting from recognition of environmental complexity, failure of complexity-reducing recognition processes
- Noninvasive testing
- Minimally invasive repair procedures

Validation and financing of integration services

Modular vehicle manufacturing and modular software exhibit properties inherent in complex or chaotic systems, particularly as the degree of coupling increases as a result of leaning. Precisely this management of the degree of coupling represents the most important consideration on the part of electrical and electronic systems architects:

- Excessively weak coupling between modules wastes potential savings and prevents differentiation by means of integration.
- Stronger coupling tends to show unforeseen overall system behavior and an overproportional rise in testing and validation effort, which cannot be financed by manufacturing revenues.

On the vehicle level as well, strategies to defend value creation are foreseeable.

Fault tolerance

Vehicle operation must be assured even in the face of component or integration process faults. Corresponding risk minimization strategies, particularly in partitioning, establishment of functions, and parameterization, have yet to be devised.

Cost of ownership reduction

There is a worldwide trend to include in onboard check routines such elements outside the electrical and electronic system which may cause breakdowns or drive up costs. These may include:

- Increased oil change intervals
- Wear measurement or estimation
- Road condition recognition, traffic monitoring

To this end, new sensors will quickly establish themselves, in particular additional chemical and radar sensors.

Driver condition and comfort

Within the framework of further customer orientation, the driver is being included in vehicle management to an increasing degree; first, he is, after all, the objective of vehicle function, and second, his physiological condition (reaction time, perception ability, visual acuity, color recognition, health, etc.) represents a significant risk factor.

References

Books

Conzelmann, G., and U. Kiencke. *Mikroelektronik im Kraftfahrzeug.* Springer Verlag, Berlin, 1995. ISBN 3540501282.
Engels, H. *CAN-Bus.* Franzis Verlag, Feldkirchen. ISBN 377235145.
Gretzmeier, F., W. Staudt, and S. Blüml. *KFZ-Elektrik, KFZ-Elektronik, Lehrbuch und Arbeitsbuch zur Kraftfahrzeugtechnik.* Vieweg Verlag, Wiesbaden, 1999. ISBN 3528149159.
Herner, A. *KFZ-Elektronik. Service-Fibel.* Vogel Verlag, Würzburg, 1999. ISBN 3802317416.
Lawrenz, W. *CAN: Controller Area Network. Grundlagen und Praxis.* Hüthig Verlag, Heidelberg, 1999. ISBN 3778527347.
Ribbens, W.B. *Understanding Automotive Electronics.* SAE (Society of Automotive Engineers), 1999. ISBN 07680002117.
Schnell, G. *Bussysteme in der Automatisierungstechnik. Grundlagen und Systeme der industriellen Kommunikation.* Vieweg Verlag, Wiesbaden, 1999. ISBN 3528265698.
Walliser, G. *Elektronik im Kraftfahrzeugwesen.* Expert Verlag, Renningen, 1997. ISBN 3816914152.

Trade Journals

Automobiltechnische Zeitschrift ATZ, Stuttgart.
Motortechnische Zeitschrift MTZ, Stuttgart.

Regularly held conferences and progress reports

Baden-Baden: VDI-Conferences Elektronik im Kraftfahrzeug.
Detroit: Convergence (SAE).

8.3 Electrical system/CAN

8.3.1 Introduction

It is the task of the vehicle electrical system to provide an assured supply of electrical energy to consumers installed in the vehicle. This includes the ability to start the engine at any time, to provide sufficient current during vehicle operation, and to supply electrical consumers with power for a limited time after the vehicle is shut down. The electrical system encompasses all energy supply, storage, and distribution components.

8.3.1.1 Modern electrical system architecture

Modern electrical systems are designated 12-V systems, after their nominal battery voltage. To assure adequate battery charging, the voltage provided by a generator (alternator) is actually about 14 V. Standard architecture consists of an engine-driven generator to provide energy and a battery to store energy.

8.3.1.2 Development trends for future electrical systems

In the future, addition of new electrical consumers will increase electrical power demands to levels which mandate new, more efficient electrical systems. This represents introduction of higher voltage systems, new power distribution components, and new generator and battery concepts, as well as overall control concepts to coordinate power generation and consumption. New safety-relevant systems demand reliable electrical systems which assure a supply of energy despite electrical system malfunctions.

8.3.2 Claw pole alternator

8.3.2.1 Construction and principle of operation

Motor vehicles use alternating current generators (alternators) to supply the electrical system. Because of its robust construction and cost-effective manufacture, the claw pole alternator (Fig. 8.3-1) has come to dominate this application; its behavior is similar to that of salient pole synchronous machines.

Claw pole alternators have a stamped sheetmetal stator package with a normal three-phase winding. The rotating field induces three-phase alternating current in this stator winding. Because the battery requires direct current for charging, the alternator must be connected to the electrical system through a *rectifier*. The generated alternating current is rectified by a full-wave bridge rectifier. This diode bridge also serves as a *reverse current block*: to prevent discharging the battery through the alternator, current can only flow from the alternator to the battery (Fig. 8.3-2).

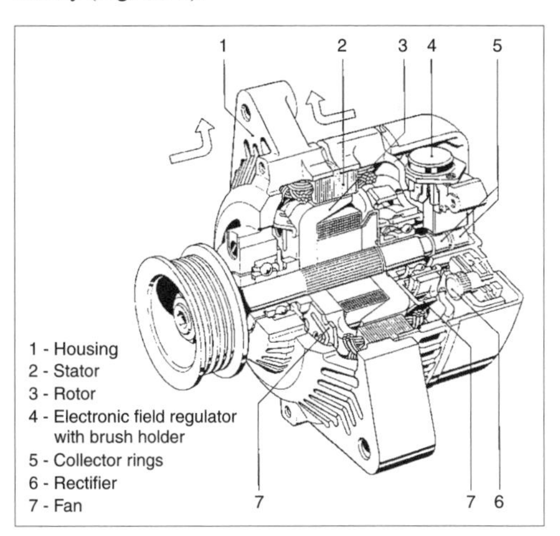

Fig. 8.3-1 Compact alternator construction.

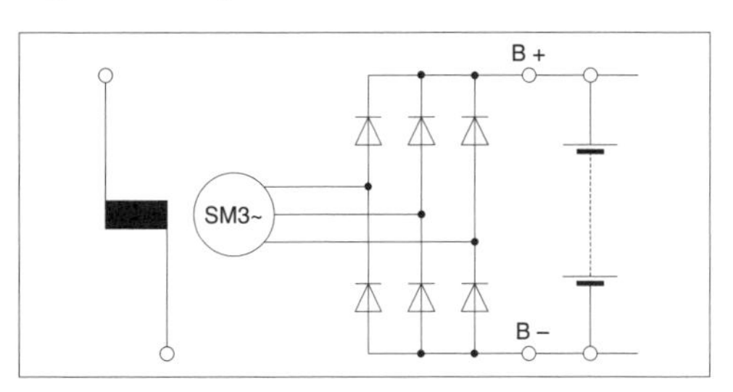

Fig. 8.3-2 Six-diode full-wave bridge rectifier circuit.

The machine differs from conventional synchronous machines only in its rotor design. The alternator rotor consists of two pole plates which interlock like the claws of a dog clutch. The pole system is excited by a DC winding located concentrically between the pole plates. This rotor configuration offers the advantage of mechanically simple and very robust design. The excitation winding usually turns with the rotor; excitation current is maintained by collector rings and carbon brushes. There also exist special alternator configurations without collector rings. In these, only the claw geometry rotates, while the excitation winding remains stationary relative to the stator. Here, though, a larger air gap is required in the excitation circuit, and such machines are somewhat larger and heavier in comparison to machines with rotating exciters. An alternating field builds up in the air gap due to the rotor's alternating pole claws.

8.3.2.2 Performance and efficiency behavior

Alternator power at constant output voltage depends on excitation current and alternator speed. It is customary to plot output current (output performance) over rotation speed. Maximum output is limited by maximum excitation current. Output power decreases with increasing temperature, as resistance of the excitation winding is temperature dependent. At constant electrical system voltage (e.g., excitation voltage 14 V), maximum excitation current drops at higher temperatures due to the exciter's rising ohmic losses.

At low rotational speeds, the alternator's induced voltage is less than electrical system voltage; no current flows across the rectifier bridge to the electrical system. Power output begins once the alternator reaches battery voltage, at the zero-watt speed (n_0). At high speeds, the alternator works near its short-circuit point. Current reaches its maximum value and cannot increase any more. Output power is therefore limited to a maximum value (Fig. 8.3-3).

Electrical system energy generation by means of an

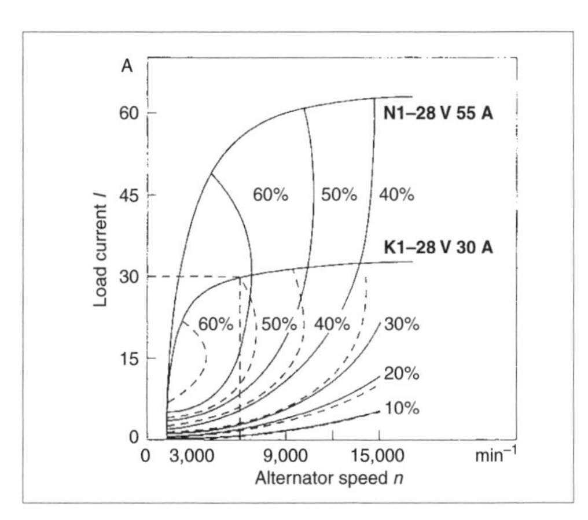

Fig. 8.3-3 Efficiency maps for Bosch K1 and N1 alternators.

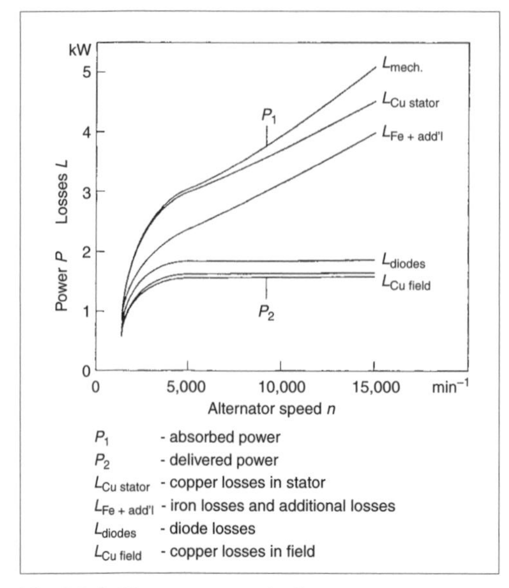

Fig. 8.3-4 Alternator power budget.

alternator necessarily has associated losses. Significant loss factors in air-cooled machines are *copper losses* in the stator winding, *mechanical losses*, and, not to be ignored, *iron losses*. Iron losses arise from alternating currents in the iron laminations of the stator as well as ripple effects within the machine at the pole claw surfaces.

Mechanical losses (primarily, a combination of friction losses and power required to drive the fan) and iron losses rise sharply as alternator speed increases. This explains the drastic drop in alternator efficiency at high speeds (Fig. 8.3-4). Copper and rectification losses, however, depend only on current load and are therefore dependent on output power, not speed.

8.3.2.3 Voltage regulation

The task of voltage regulation is to maintain electrical system voltage at a constant level despite varying alternator speeds and loads. To this end, the voltage regulator of a claw pole alternator uses excitation current as the control variable. Modern voltage regulators are two-point regulators which react to a tolerance range around the target voltage. If alternator voltage exceeds the upper limit of the tolerance band, the regulator shuts off the excitation circuit, excitation current passes through a "free wheeling" or "decay" diode and diminishes according to its excitation time constant. If voltage drops below the lower limit of the tolerance band, the regulator switches on, the excitation winding is connected to the alternator terminals, and excitation current rises. Excitation current therefore oscillates in response to electrical system loads around the required average voltage.

Such voltage regulation can only work in the alternator's part-load range. If electrical system load exceeds the maximum possible alternator output, system voltage may drop below the tolerance band. This is especially true at low alternator speeds. In this operating regime, alternator output drops sharply and power for heavy loads must in part be drawn from the battery.

To provide better battery charging performance at low temperatures, the target voltage is raised for low operating temperatures. Conversely, to prevent overcharging, at high temperatures the target voltage is reduced. Target electrical system voltage is therefore temperature dependent.

The regulator function, along with switch gear and free-wheeling diode, are contained in a single housing (Fig. 8.3-5).

Overvoltage protection

Electrical system overvoltage can damage consumers connected to the network as well as the alternator itself.

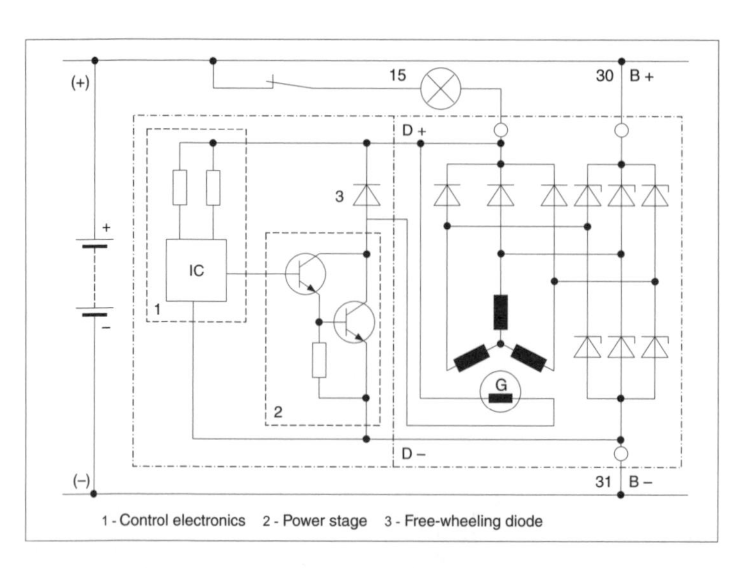

Fig. 8.3-5 Circuit diagram of alternator with electronic regulator.

Therefore, it is necessary to limit maximum voltage. For electrical systems which include a battery, the latter can absorb such overvoltages to a certain extent, but emergency operation without a connected battery is usually a system requirement. With a fully charged battery, electrical system voltage may rise to unacceptable levels, as the battery is no longer capable of accepting additional charge.

Overvoltages may arise from various causes. One important cause is *load dumping*. If the electrical system is subjected to a large power draw and a large consumer is suddenly shut off, system voltage will rise. The alternator has not yet adjusted to the new load. The voltage regulator can only react at a speed represented by the excitation time constant (typically in the range of 100 ms). Excess alternator output must be taken up by the battery, and this leads to a voltage increase. During this time, system voltage must be limited to a maximum value by some suitable means. For 12/24-V systems, this is usually accomplished by a bridge rectifier equipped with *zener diodes*. The zener voltage is chosen so that the diodes break down at about 28 V. This effectively limits maximum electrical system overvoltage and offers effective protection against high-energy overvoltages for the entire electrical system.

Zener diodes cannot be used in electrical systems with higher nominal voltages. In transitioning to new, higher voltage electrical systems, other appropriate electronic measures must be taken to limit overvoltages.

8.3.3 Batteries

8.3.3.1 Battery characteristics

In vehicle electrical systems, the battery's task is to store electrical energy generated by the alternator, thereby supplying power to electrical consumers—for example, at engine idle or with the engine shut off. In particular, it must be capable of providing short-term large currents to start the combustion engine, especially at low ambient temperatures; for this reason it is sometimes called a starter battery. Lead-acid batteries (accumulators) are generally used for this purpose.

The most important characteristics of a battery are its rated voltage and rated capacity. In lead-acid accumulators, rated voltage is a multiple of the 2 V individual cell voltage. In actuality, idle voltage of a cell typically varies between 1.94 and 2.14 V, corresponding to discharged and fully charged states. Typical rated voltage for motor vehicle electrical system batteries is 12 V for passenger cars and 24 V for commercial vehicles, or, in future, 36 V in dual-voltage electrical systems. Rated capacity is defined by DIN standards as that discharge current which can be withdrawn at a constant rate for 20 hours to result in a discharged voltage of 1.75 V/cell. This current is designated I_{20}. Another important characteristic is cold cranking amps, *CCA* (or I_{-18} in DIN terminology). According to DIN standards, battery terminal voltage under I_{-18} must be at least 1.5 V/cell 30 seconds after the beginning of discharge and at least 1 V/cell 150 seconds after beginning of discharge. For example, cold cranking amps of a 36 Ah battery are about 150 A. This parameter is an important consideration in starter motor layout.

8.3.3.2 Operational behavior of lead-acid batteries

Discharge

Fig. 8.3-6 shows a typical trace of battery voltage during constant-current discharge. Shortly after the beginning of discharge, voltage drops to a value which changes little until the battery is completely discharged. Only then does voltage collapse as a result of consumption of acid and/or active electrode material. Also important is the dependence of usable charge on discharge current. With increasing discharge rates, the actual value deviates from the nominal capacity based on 20-hour discharge. This relationship is shown in Fig. 8.3-7 and may be approximated by the so-called Peukert equation:

$$I^n \cdot t_D = const \qquad (1)$$

where I = current, t_D = discharge time, n = Peukert exponent, ≅ 1.2 to 1.5

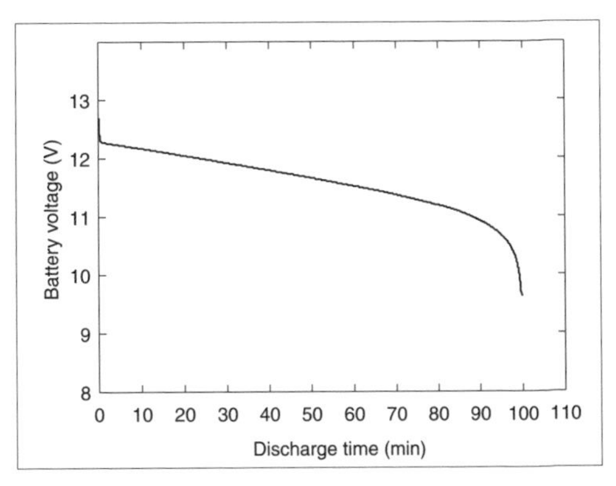

Fig. 8.3-6 Battery voltage drop during constant-current discharge.

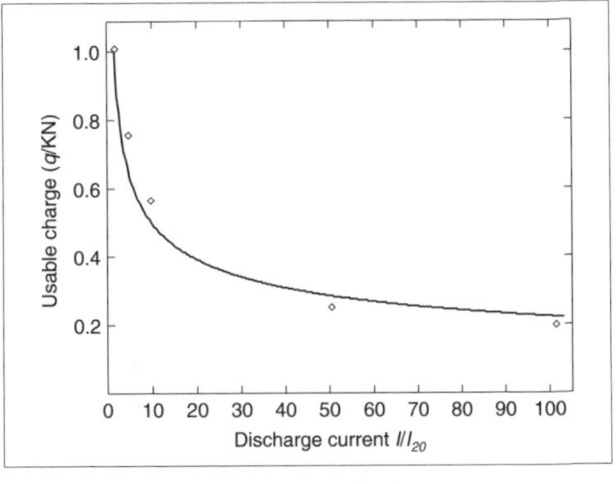

Fig. 8.3-7 Usable charge over discharge current (Peukert exponent = 1.3).

The starting condition, calling for currents on the order of 200 to 300 A, would represent usable capacity of about 45% of rated capacity. However, this applies only to a constant flow of starter current to complete discharge. Because the starting process seldom takes longer than one minute, the associated discharge, relative to rated capacity, is negligibly small. In other words, for the starting process, the battery briefly supplies high current, requiring high energy density, but the actual energy required for starting and the associated energy density are actually relatively small.

Charging

Battery charging may be classified as constant current or constant voltage charging. In vehicle electrical systems, batteries are charged under limited-voltage conditions; that is, when a voltage limit—set by the voltage regulator at some level below the so-called gassing voltage—is reached, charging current drops back to prevent damage by overcharging, which would result in increased water electrolysis and plate corrosion. As gassing voltage is strongly temperature-dependent, charging voltage is about 14.1 ± 0.3 V at 20°C, with a temperature gradient of –(7 . . . 10) mV/°C. At an ambient temperature of 50°C, for example, charging voltage would be set at 13.8 V.

8.3.3.3 Dual battery systems

In conventional motor vehicle electrical systems, starter operation, especially at low temperatures, imposes the most severe loads on batteries in terms of energy density. One requirement for the necessary high discharge currents is low internal resistance. In starter batteries, this is achieved by small plate spacing and a large number of plates. The amount of energy required to start the engine is, however, small. Therefore, a battery with low capacity but high energy density would suffice for starting. However, if electrical consumers are to be supplied with energy over longer periods of time even while generator output is low, a battery with correspondingly higher capacity and energy density would be required.

These contradictory requirements for automotive batteries can be satisfied by implementation of a dual-battery electrical system (Fig. 8.3-8). By using two separate batteries to supply the starter and as a buffer for other consumers, both batteries can be optimized for their respective missions. This permits weight and volume savings of up to 40% compared to conventional single-battery systems. For example, lead-acid batteries with spiral-wound electrodes are suitable as starter batteries; compared to conventional batteries, these have greater electrode surface area per amp-hour and very short current paths. This permits short-term (time frame about 10 s) energy densities of 2,000 W/kg, compared to 300 W/kg for conventional battery technology. By uncoupling the starter current circuit from the remaining electrical system, a charging controller can keep the starter battery charged at a high voltage level to guarantee starting capability, and voltage drops during starting do not affect the voltage for other consumers.

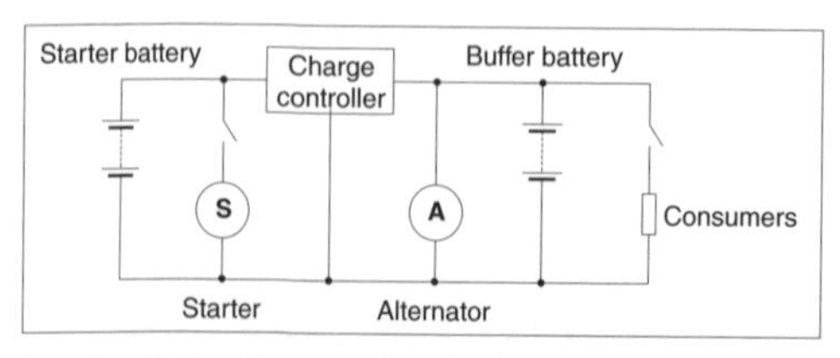

Fig. 8.3-8 Dual battery electrical system.

8.3.4 Future electrical system architectures

8.3.4.1 Dual-voltage electrical system 42 V/14 V

Trends

Objectives in developing future vehicle generations include reduced fuel consumption ("three liter car"), improved active and passive safety, and increased comfort. These objectives are to be achieved using, among other things, new functions such as automatic start/stop, regenerative braking, and new or additional electric consumers. For example, replacement of engine-driven auxiliaries such as power steering pump or water pump by demand-activated, electrically powered components yields improved overall efficiency and therefore fuel savings. Comfort-adding consumers include such features as auxiliary heaters and preheaters. Electromechanical brakes and electric power steering enable realization of functions which improve vehicle dynamics and lead to improved safety.

Demands

For the reasons cited above, electrical power demands will rise sharply in the future. It is estimated that demand will increase from the current level of about 2 kW to somewhere between 6 and 7.5 kW. Depending on equipment level and driving cycle, the modern electrical system contributes up to 1.7 L/100 km to total vehicle fuel consumption (Fig. 8.3-9). In other words, a present-day electrical system would consume well over 3 L/100

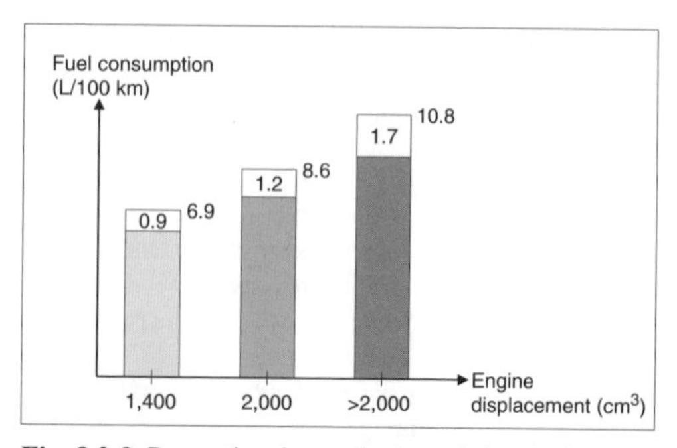

Fig. 8.3-9 Proportional contribution of electrical power generation to total vehicle fuel consumption.

Table 8.3-1

Consumer	Peak output	Continuous output
Heated catalyst	2,000 W	20 . . . 40 W
Heated windshield	1,500 W	100 . . . 200 W
Active yaw stabilization	2,500 W	150 . . . 1,000 W
Electromechanical brake	2,000 W	100 W
Electrohydraulic power steering	1,000 W	100 . . . 250 W
Electric valve actuation	3,400 W	1,000 . . . 3,400 W

km just to satisfy future electrical power demands. Improvement of electrical system efficiency is therefore unavoidable. This may be achieved by improved generating efficiency, optimized power distribution, and optimization of continuous consumers.

In response to introduction of new safety-relevant systems such as "x-by-wire" systems (brake-by-wire, steer-by-wire), future electrical systems must provide a high degree of reliability, as faults in power supply would lead to failure of these systems with possibly fatal consequences.

Table 8.3-1 provides an overview of possible new electrical consumers and their power demands. It is apparent that with an unaltered system voltage of 14 V, steady-state maximum current of 150 A or more would become commonplace. Electrical current of this magnitude cannot be handled in a reasonable manner using present-day alternators, batteries, wires, and connectors. Furthermore, transmission of greater current represents greater ohmic losses in wiring ($P_{loss} = I^2 \cdot R_{wiring}$).

Along with consumers requiring higher system voltage, vehicles also contain electrical consumers which are dependent on the existing 14-V system. These include all light bulbs, whose life expectancy decreases sharply with increased voltage.

Concepts

One solution to this problem is offered by introduction of higher system voltage to reduce current required for a given electrical power level. Simultaneously, 14 V must be retained. This implies that future vehicle generations will have at least two voltages available onboard—the customary 14-V system and a 42-V system. A system voltage of 42 V is advantageous because semiconductor processes currently entering the market have breakdown voltages of 60–70 V. Furthermore, there is an existing standard covering DC touch voltage protection which requires certain protective measures for voltages above 60 V. Other voltage levels are possible according to demand, but voltages above the permissible touch voltage will not be generally available.

Fig. 8.3-10 shows an example of 42-V/14-V electrical system architecture. Actual system voltages are 42 V and 14 V respectively; the batteries have rated voltages of 36 V and 12 V respectively. The 42-V network is supplied by an alternator. A DC/DC converter provides power for the 14-V network. Each voltage level is buffered by its own battery, with the 36-V battery providing starting power. Continuous current consumers are supplied by the 12-V battery. This division of battery duties provides increased starting assurance, as discharge of the starting battery by continuous consumers is no longer possible.

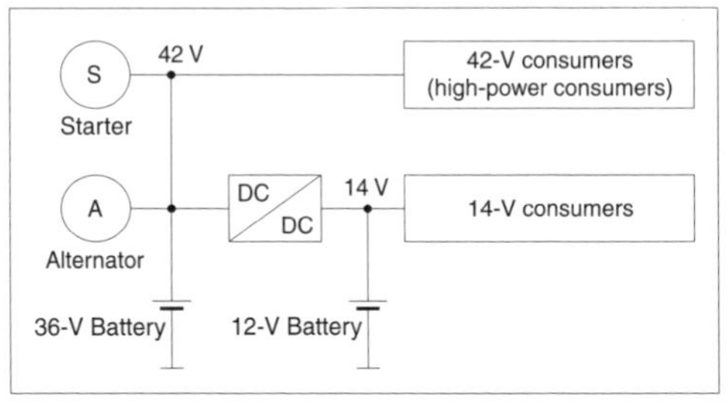

Fig. 8.3-10 42-V/14-V electrical system architecture.

Highly reliable electrical networks include redundant supply to safety-critical systems. Electrical network faults must be recognized and cannot be permitted to cause failure of power supply to such systems.

8.3.4.2 Signal and power distribution

In the future, new electrical consumers and functionalities will be added to vehicle bodies and chassis, joining such standard consumers as lighting, heating systems, etc., currently installed in modern vehicles. To meet demands for increased availability, cost reduction, and expanded functionality, new system architectures must be considered. An increase in the number of functions is most apparent in decentralized subsystems, so-called signal and power distributors, which are interconnected by bus systems to form overarching systems. Extensive networking—for example, by means of CAN—as well as spatial proximity of sensor or actuator units to their respective components of responsibility present a variety of advantages. Among other things, wiring harness complexity is greatly reduced. This results in weight and cost reduction.

Increasingly, with advancing development of semiconductor components, fuses and relays will be replaced by intelligent switches—"smart power." Transition from pure switching functions to electrically controlled loads requires application of semiconductor switches. This permits economical realization of a variety of new functions, such as timed loads, soft start of electric motors, and torque limitation.

Sensing of loads, their diagnosis, central protection against polarity errors and load dumping (i.e., protection against the effects of sudden interruption of current supply to high-draw consumers) and, not last, control unit self-diagnosis are vital functions demanded by automobile manufacturers. Early diagnosis (analysis) of connected components is conducted and, in the event of a fault (load dumping, interrupted connection, short circuits, overtemperature, over/undervoltage, etc.), these are reported to the driver, written to fault code memory, and an appropriate limp-home strategy is implemented. With appropriate wiring of signal and power distributors, safety-relevant systems such as lighting remain operational despite failure of entire modules (Fig. 8.3-11). For example, a defective headlamp may be replaced by a brightness-controlled fog lamp, or a defective taillamp by a pulsed brake lamp. This ensures driving safety and mobility. Software implemented in signal and power distributors is capable of communicating via CAN, switching consumers on command, and providing sensor signals to other bus participants. Physical quantities such as pressure, temperature, voltage, current, etc., are captured by sensors, filtered, converted to a standard format and distributed throughout the CAN bus. An overarching system (software), using its awareness of the load situation, may intervene in energy distribution and apply direct, delayed, or pulse-width–modulated switching to consumers. The number of installed signal and power distributors is vehicle-specific and very strongly dependent on vehicle equipment level.

8.3.4.3 Energy management

Due to high levels of installed electrical power in future vehicles, it is possible that power demands may occasionally exceed generating capacity. This may mean that in the medium term, the battery charge condition required for a restart is not met. Simultaneously, spikes in consumer power demand which cannot be satisfied by the alternator cause electrical system voltage variations. *Electrical energy management* is tasked with preventing these unacceptable conditions by means of suitable coordination of power consumption, power generation, and energy storage. Specifically, this means coordinated control of electrical consumers, alternator, batteries, and vehicle powerplant. Energy management makes it possible to dispense with worst-case dimensioning of electrical system components (battery, alternator, DC/DC converter), thereby saving cost and weight. Coordinated energy management function ensures that critical situations remain under control.

Dynamic interplay between alternator and combustion engine is improved within the framework of energy

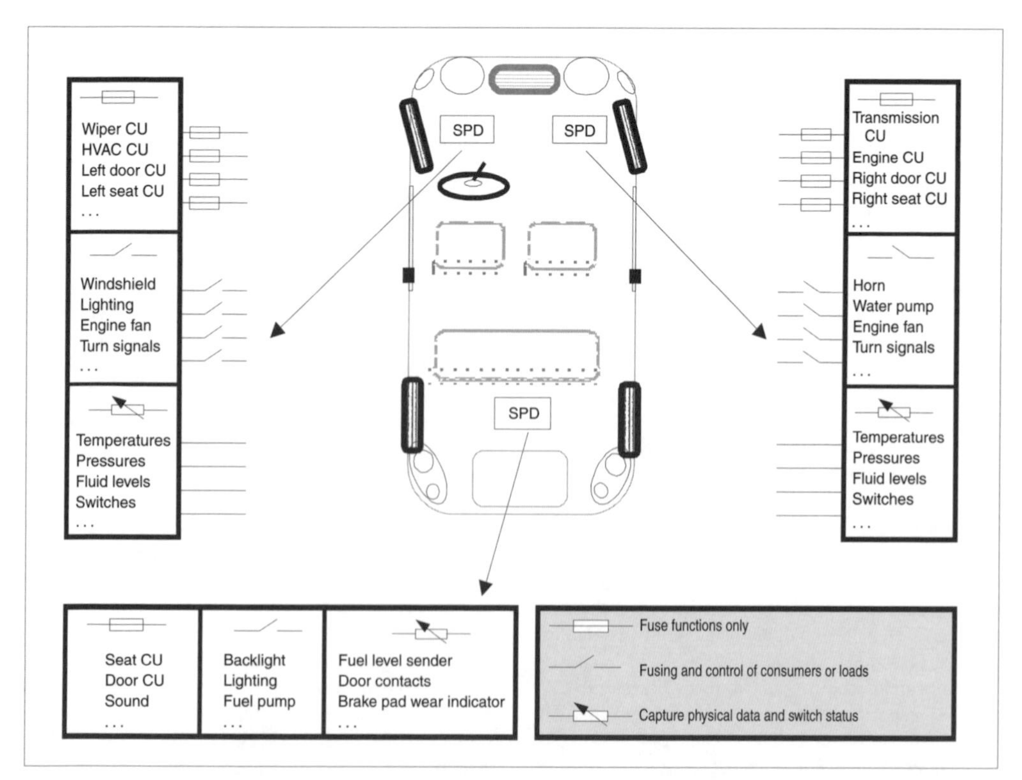

Fig. 8.3-11 Signal and power distribution. SPD, signal and power distribution; CU, control unit.

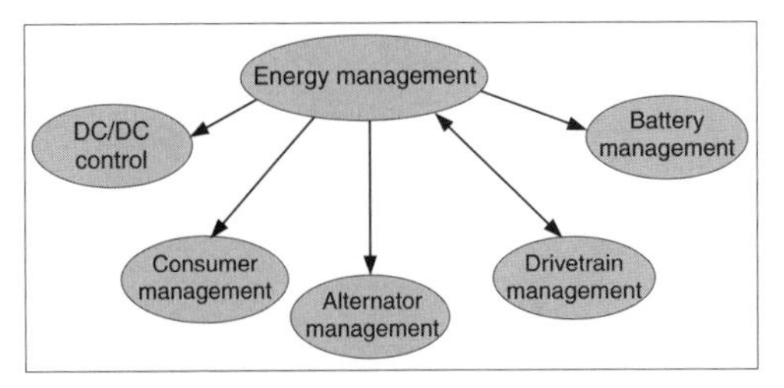

Fig. 8.3-12 Energy management for dual-voltage electrical system.

management. Due to electrical system voltage regulation, step-function activation of large electrical loads results in rapid changes in alternator power generation and therefore associated torque loads on the powerplant. Such dynamic loading results in combustion engine speed fluctuations; at idle, these could even cause engine stalling. Modern vehicles therefore extend the dynamics of electrical power generation over a longer time period, in that alternator excitation is ramped up rather than applied as a step function. The drawback of ramped excitation is that spikes in electrical consumer demand cause electrical system voltage drops. Slowed alternator excitation may be dispensed with if, just before electrical load switching, the required alternator torque demand is passed to the engine control system, enabling the combustion engine to anticipate and compensate for demand.

For inclusive coordination, energy management requires information regarding electrical power demand and potential electrical power generation. Fig. 8.3-12 shows an example of an energy management structure for a dual-voltage electrical system. The energy management system itself acts to coordinate individual components; it coordinates demands from batteries, consumers, drivetrain, DC/DC converter and alternator, and responds accordingly by establishing an overall strategy.

Examples of functions that can be realized through energy management include electrical consumer control as a function of electrical power demand and power generation potential, altering engine speed to improve power generation as a function of battery condition, and matching alternator load to driving situations. This means, for example, reduction of alternator load during acceleration or increasing alternator load during braking (regenerative braking) in response to battery condition.

8.3.5 CAN (Controller Area Network)

8.3.5.1 Motivation

Modern motor vehicles are equipped with a multitude of electronic control units which require extensive communications dialogue to realize ever more complex functions. Conventional data transmission by means of individual dedicated signal lines runs up against the limits of possibility, first because the wiring harness would become nearly unmanageable in scope and second because the limited number of connector pins is increasingly becoming a limiting factor in control unit development. The solution to this problem may be found in installation of vehicle-specific serial bus systems, of which CAN has developed into a standard.

8.3.5.2 Fields of application

Application of CAN for serial communications may be classified according to various functional requirements:

Real time applications to control vehicle motion—for example, networking of Motronic, transmission control, and vehicle dynamic control. Characteristics include typical transmission rates in the range of 125 kbit/s to 1 Mbit/s to achieve the necessary real time behavior.

Multiplex applications for feedback and control tasks in the fields of comfort and body electronics. Examples include climate control, central locking, light module, and seat adjustment. Transmission rates are in the range of 10 kbit/s to 100 kbit/s.

Mobile communications applications which connect a central display and control unit to components such as navigation system, telephone, or car radio. Networking is primarily used to unify actuation processes as much as possible and to encompass status information to minimize driver distraction. Data transmission rates are less than 125 kbit/s; direct transmission of audio and video data up to 20 Mbit/s is, however, not possible with CAN.

Diagnostic applications based on CAN attempt to use the existing network for diagnosis of participating control units. This would eliminate diagnosis by means of the so-called "K-Line" (ISO 9141), which is still in widespread service. A unified data rate of 500 kbit/s is planned.

8.3.5.3 Bus configuration

CAN buses use a linear bus structure (Fig. 8.3-13), which exhibits a lower likelihood of overall system failure compared to other logical structures (ring or star topology). In the event of failure of one participant, bus access to other participants remains unaffected. Stations along the bus, which may consist of control units, sensors, or actuators, operate on the multimaster principle. This means that bus access control is decentralized and is the responsibility of participating, functioning stations; overall central management of the network configuration is not required.

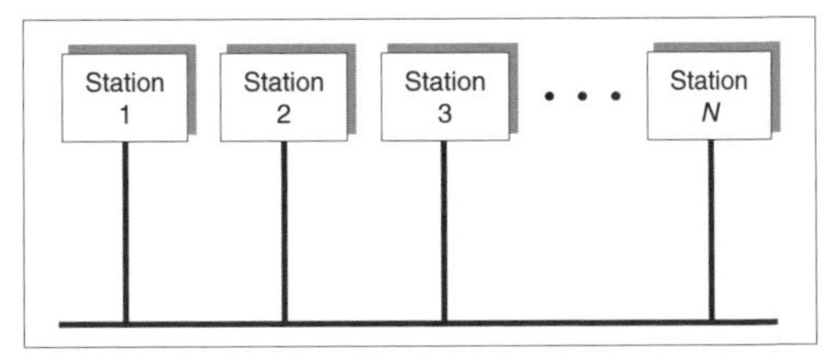

Fig. 8.3-13 Linear bus structure.

8.3.5.4 Bus assignment and addressing

On the basis of the multimaster principle, bus access is not tied to station characteristics. Instead, CAN defines these by means of message content. To keep messages from colliding on the bus, upon transmission each message is given an identifier which defines content as well as priority *(content-relative addressing)*. The smaller the binary identifier number, the higher the message priority.

The CAN protocol is based on the two logical conditions, *dominant* (logical 0) and *recessive* (logical 1). Once messages are ready for transmission, any station may begin sending if the bus is free. A potential conflict in bus access is resolved by *bitwise arbitration* of the respective identifiers. The "wired AND" bus connection ensures that dominant bits sent by one station overwrite the recessive bits of other stations and that ultimately the station with the lowest identifier (equivalent to highest priority) prevails (Fig. 8.3-14). Transmitting stations which "lose" in arbitration automatically become receivers and attempt to resend their message as soon as the bus is free again.

Because CAN addressing is not tied to station characteristics, each receiving station must decide for itself whether any message on the bus is relevant to it. For this reason, the identifier not only includes message priority but also message content. An *acceptance test* (Fig. 8.3-15) on the part of the receiving station ensures that only those messages whose identifiers are in the list of acceptable messages are received. In summary, rejection of a system based on station addresses and use instead of content-based addressing leads to a highly flexible system which permits easier mastery of diverse equipment variations. New stations may be implemented, insofar as they are receivers, without modifying existing stations. If new stations are installed as message transmitters, it may be necessary to expand acceptance tests of existing stations.

8.3.5.5 Message format

The CAN protocol permits two different formats: a *standard format* using an 11-bit identifier and an *extended format* with a 29-bit identifier. Both formats are mutually compatible and may be used on the same existing network. The message frames of both formats differ only in identifier length and are composed of seven successive fields (Fig. 8.3-16).

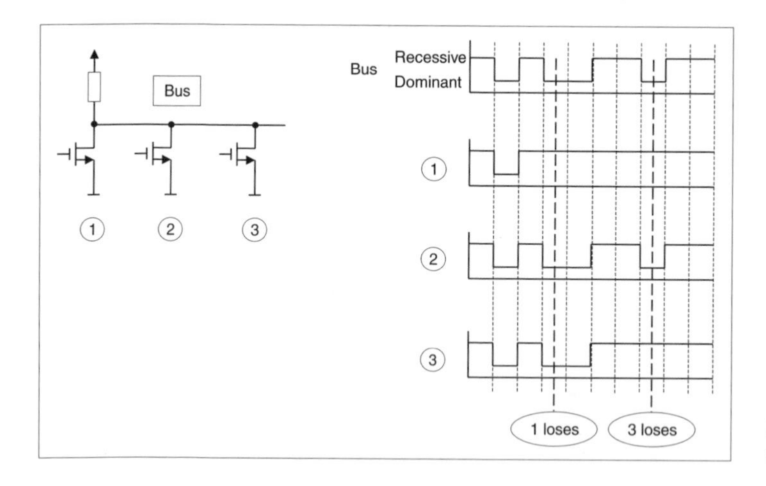

Fig. 8.3-14 Bitwise arbitration.

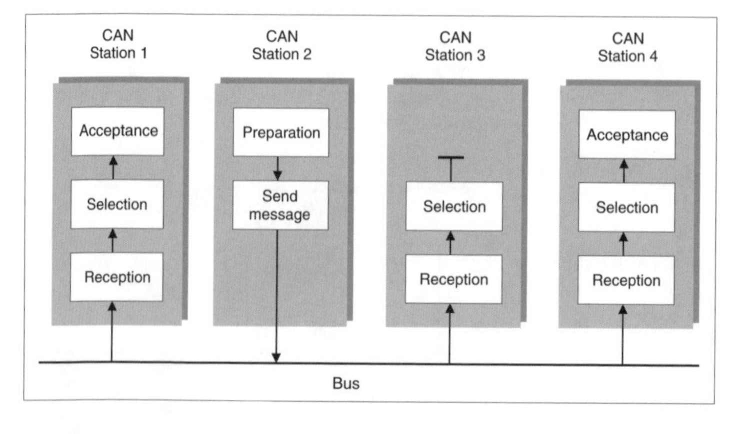

Fig. 8.3-15 Acceptance test.

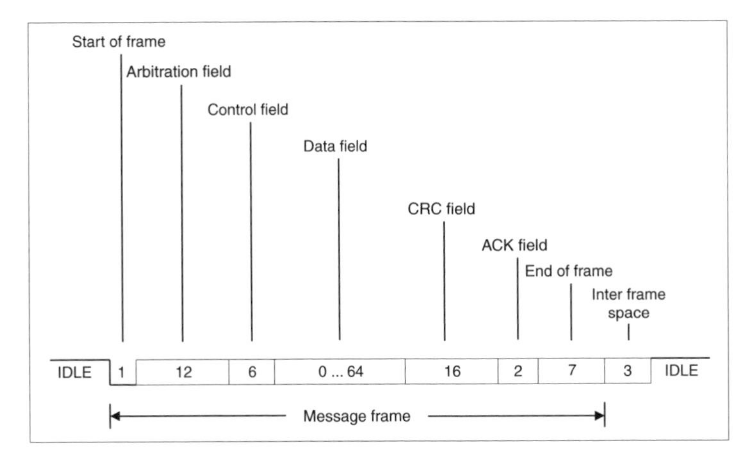

Fig. 8.3-16 Message frame.

Start of Frame uses the dominant bit to indicate the beginning of a message to an idle recessive bus and synchronizes all stations. The *Arbitration Field* consists of the previously described identifier and a control bit. In transmitting this field, the transmitter performs a check at each bit to determine whether it is still authorized to transmit or whether a station with a higher priority has accessed the bus. The following control bit, known as the *RTR Bit* (Remote Transmission Request), identifies whether the message consists of a data frame or a request for data (remote frame) from another transmitter.

The *Control Field* includes the *IDE Bit* (Identifier Extension Bit), which is used to differentiate between standard format (dominant, IDE = 0) and extended format (recessive, IDE = 1), followed by one bit for future expansion. The remaining four bits of this field describe the number of data bytes in the following data field.

The *Data Field* contains between 0 and 8 bytes of data information. A data field with a length of 0 bytes is used to synchronize distributed processes.

The *CRC Field* (Cyclic Redundancy Check) serves to assure frame integrity by recognizing possible transmission errors.

The *ACK Field* (Acknowledgment) allows receivers to confirm receipt of messages. The field encompasses the ACK Slot and ACK Delimiter. The former is transmitted recessive; in the event of correct message receipt, the receivers rewrite this as dominant. It is immaterial whether the message was intended for these receivers, in the sense of the acceptance test; correct receipt is acknowledged regardless.

End of Frame marks the end of a message.

Interframe Spacing consists of three bits which separate successive messages. Thereafter, the bus remains in its recessive idle state until some other station initiates bus access.

8.3.5.6 Error recognition and handling

The CAN protocol implements various control mechanisms to recognize errors; these may be classified as checks conducted at the message frame level or at the bit level. At the frame level, these are:

CRC (Cyclic Redundancy Check), in which the sender inserts redundant check bits which are determined by message information content. Based on the received information bits, the receiver recalculates the check sum and compares this with the CRC sequence inserted by the transmitter.

The *Frame Check* recognizes frame errors by means of verification of frame structure by the receiver. Frame Check examines fields with a defined, invariant format as well as frame length.

ACK Check is the receiver's confirmation of a received frame. Lack of confirmation may indicate recognized transmission errors, a fault in the acknowledgment field itself, or a nonexistent station.

Bit-level mechanisms are:

Monitoring, in which the transmitter simultaneously observes the bus level and thereby recognizes differences between transmitted and received bits. This permits detection of local as well as global errors and distinguishing between the two by means of statistical evaluation.

Code Check examines coding of individual bits to ensure compliance with the following stuffing rule, which states that between Start of Frame and End of Frame of the CRC Field, no more than five successive bits may carry the same polarity. After five identical bits, the transmitter inserts an opposite-polarity bit, which is removed by the receiver after the message is accepted.

If any given CAN station detects a network error, the current transmission is interrupted by transmission of an Error Flag. An Error Flag consists of six successive dominant bits which deliberately violate the stuffing rule or frame check. Sending an Error Flag prevents other stations from accepting the message and ensures netwide data consistency. Interrupted data transmissions are automatically reinitiated by the transmitter; bus access is determined by the previously described arbitration mechanism.

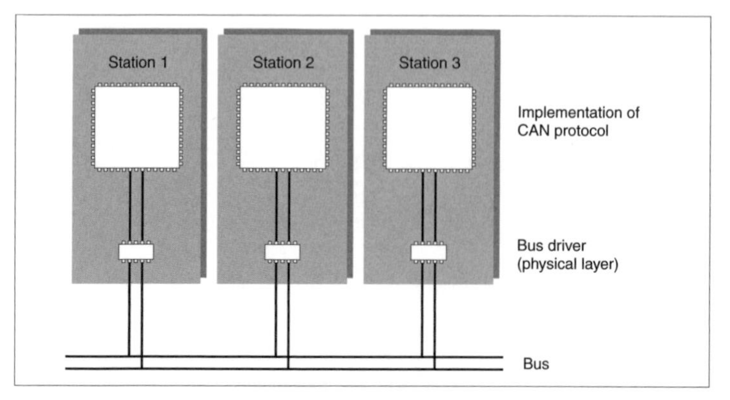

Fig. 8.3.17 System node structure.

Defective stations may impose considerable load on the bus by continuing to retransmit their own interrupted messages or by continuously interrupting valid messages originating from other stations. To prevent blocking the bus, the CAN protocol implements mechanisms to differentiate between stochastically and continuously appearing errors, thereby localizing failure of individual stations. By means of statistical evaluation of its own errors, the affected station is able to recognize its own failure and switch to an operating mode which does not compromise communications by other stations in the network.

8.3.5.7 Implementation

In establishing a CAN node in a control unit, distinction must be made between the actual CAN implementation *CAN Protocol*) and user-specific bus connection by means of driver components (*Physical Layer*; Fig. 8.3-17).

Modern CAN implementations exist for a multitude of microcontrollers of the most varied performance classes as well as special CAN controllers. All of these components share a unified implementation of the CAN protocol with respect to message frame and error handling. This assures that any given CAN implementations are mutually compatible and can communicate with one another without restrictions. What differs is the extent to which the CPU is relieved of chores related to message preparation, acceptance testing, and storage and management of relevant data.

Similarly, there exists an entire series of components which permit physical connection to the bus and which exhibit application-specific differences. In terms of *Transfer Rate*, distinction is made between high speed (>125 kbit/s) and low speed (≤125 kbit/s); accordingly, driver components differ with respect to rise time, output power, and impedance. Depending on EMC (electromagnetic compatibility) requirements, the bus line and therefore *bus coupling* via driver components may be executed as twisted pair, shielded, or unshielded two-wire or single-wire lines. A further difference is in component *fault tolerance* with respect to breaks and shorts of one of the bus lines to fixed potentials. Depending on configuration, the driver may shut down entirely or fall back on single-wire communication.

8.3.5.8 Standardization

The CAN protocol has been standardized by ISO as well as SAE for various fields of application. In terms of transfer rate, distinction is made between low-speed and high-speed applications and between standard format and extended format message frames (Table 8.3-2).

Moreover, application of CAN for diagnostic purposes is proposed under ISO 15765 (Draft International Standard), although approval has not yet been issued.

8.3.5.9 Summary

The following overview summarizes performance characteristics of the CAN protocol:

Topology:	Bus configuration with only one logical bus line
Physical realization:	Single- or dual-wire, shielded or unshielded
Geometric reach:	max. 40 m at maximum transfer rate
Transfer rate:	<10 kbit/s to 1 Mbit/s
Data length:	0 to 8 bytes/message
Identifier length:	11 bits (standard format), 29 bits (extended format)
Message length:	47 bits to 130 bits (150 bits for extended format)

Table 8.3-2 Standardization

Organization	Transfer rate	
	Low speed (≤125 kbits/s)	High speed (>125 Kbytes/s)
ISO	ISO 11519-3 CAN	ISO 11898-2 CAN
SAE passenger cars		J 2284 CAN (standard format)
SAE truck and bus		J 1939 CAN (extended format)

References

Robert Bosch GmbH (ed.). *Kraftfahrtechnisches Taschenbuch*, 24. Auflage. VDI-Verlag, 1995. ISBN 3-528-13876-9.
English edition: *Bosch Automotive Handbook*, 5th edition, Society of Automotive Engineers, 2000. ISBN 0-7680-0669-4.
Ehlers, K. Position paper of the "Bordnetzarchitektur" forum, "Elektronik im Kraftfahrzeug" conference, Baden-Baden, September 1996. VDI Berichte No. 1287, 1996.
Schöttle, R., D. Schramm, and J. Schenk. "Zukünftige Energiebordnetze im Kraftfahrzeug." "Elektronik im Kraftfahrzeug" conference, Baden-Baden, September 1996. VDI Berichte No. 1287, 1996.
Graf, A., D. Vogel, J. Gantioler, and F. Klotz. "Intelligente Leistungshalbleiter für zukünftige Kfz-Bordnetze." "Elektronik im Kraftfahrzeug" conference, Munich, June 3–4, 1997.
Walther, M., R. Schöttle, and K. Dieterich. "Future Electrical Power Supply System." SAE *Proceedings*, Convergence 1998 (October).
Dais, S., and J. Unruh. "Technisches Konzept des seriellen Bussystems CAN—Teil 1 und 2." ATZ Automobiltechnische Zeitschrift 94, Vol. 2, 1992.
Kaiser, K.-H., H.-J. Mathony, and J. Unruh. "Serielles Datenbussystem CAN." "Elektronik im Kraftfahrzeug" conference, Haus der Technik, Essen, June 1992.
Kaiser, K.-H., and W. Schröder. "Maßnahmen zur Sicherung der Daten beim CAN-Bus und deren Beiträge zur Datensicherheit." ATZ Automobiltechnische Zeitschrift 96, Vol. 9, 1994.

8.4 Multimedia systems for motor vehicles

New signal processing and digital data transmission technologies are promoting rapid changes in communication systems as well as entirely new applications. This trend impacts motor vehicles as readily as it does any other field. In the past, the world of vehicle communications consisted solely of analog radio with its audio service extended by audio data formats such as the Compact Cassette (CC) and Compact Disc (CD). These purely audio services were augmented by additional information services such as Radio Data Service (RDS), which permits textual station identification in the radio display. Digital broadcast systems, developed in the past few years, permit transmission of any type of data at high bit rates. This is currently transforming the radio of the past into a true multimedia system of the future, that will enable transmission of audio data (music, news) as well as textual information (news) and video data (from individual images to television transmission). Combining such systems with mobile communications media such as GSM-based cell phones enables interaction with a central server, thereby creating new, interactive services.

The utility of such multimedia systems lies in their ability to provide vehicle occupants with a wide range of entertainment as well as information (e.g., traffic information, news, sports, tourist information). As they combine information and entertainment, these are often referred to as infotainment systems. When coupled with other vehicle systems, these result in complex driver information systems providing high utility. An example is combination of an infotainment system providing traffic information with a navigation system. The result is a dynamic navigation system which informs the driver of an optimum route in response to actual traffic situations, safely and quickly guiding him to his destination.

8.4.1 Analog broadcast systems

Terrestrial-based radio and television systems currently in service employ analog signal transmission. The broadcast signal—in the case of radio an audio signal (music or speech)—is modulated from its analog form by a high-frequency carrier and then wirelessly transmitted to a receiver. The receiver reverses the process, separates the received signal from its carrier, transforms it back to its low-frequency range, and, after suitable amplification and tone manipulation, sends it to loudspeakers.

FM

Seen on a worldwide basis, frequency-modulated broadcast signals (FM) have developed into the most widespread format. The advantage of FM lies in its large bandwidth with good transmission and sound quality. Disadvantages include relatively limited transmission range and the related requirement for high transmitter power.

AM

Amplitude-modulated (AM) broadcasts are now less widespread than in the past. The advantage of AM transmissions is their good range with low transmitter power. However, sound quality is relatively modest by modern standards.

AM long- and shortwave

Long- and shortwave receivers no longer enjoy any great significance. Longwave is used in a few sparsely settled countries; shortwave, because of its very good range, is valued by listeners who enjoy receiving their "home" stations overseas. By modern standards, both services provide only limited sound quality as well as high sensitivity to interference.

8.4.1.1 Mobile radio reception interference

A trait shared by all broadcast services is the fact that they were not originally conceived for use by fast-moving receivers such as typical auto radios. In addition to distorting effects already affecting stationary receivers, such as insufficient transmitter range, overlapping signals, and undesirable signal reflections, mobile receivers are subject to their own unique distortion sources. Variable multipath reception, permanent field strength variations caused by vehicle motion, and interference from vehicle electronics are only a few of these car radio specific influences. Demands on the car radio as a receiver are therefore in many respects very different from those imposed on a stationary receiver. It is often the case that experienced auto radio receiver manufacturers achieve performance far surpassing that of stationary receivers.

8.4.1.2 Supplementary broadcast services

In addition to music and speech transmission, there is significant interest in offering additional radio services. For analog broadcasts, introduction of such services is limited by the remaining available transmission bandwidth as well as its analog basis. Examples of services introduced in the 1970s include:

ARI

In 1974, Blaupunkt, in conjunction with the ADAC (Allgemeiner Deutscher Automobil Club) and ARD (Arbeitsgemeinschaft der Rundfunkanstalten Deutschlands—German Broadcasting Organizations Working Group) broadcast stations developed and offered the first broadcast supplementary service. This auto radio information (ARI) service permits targeted reception of traffic information by means of specific program identification on the part of participating FM stations, appropriate audio volume increase in the event of a muted radio, and mixing of traffic announcements in the event another audio source (e.g., compact cassette) is in use. Functionality of this broadcast service was later integrated with the Radio Data System (RDS) and expanded by the TMC (Traffic Message Channel) jointly developed by Philips and Blaupunkt (see below).

RDS

A further expansion of FM signal capability is the Radio Data System (RDS). RDS is a unified, standardized transmission method for information services accompanying broadcast programs. The classic FM transmitter with its analog audio service is supplemented by a very narrow bandwidth digital data channel. This channel may be used to transmit digital data—for example, station identifier text or artist name/selection title—for the radio display. Because FM radio is an analog system, and the available bandwidth is completely used by the audio signal, the added digital RDS data channel has an extremely small bandwidth (on the order of 100 bits/s).

8.4.1.3 Auto radio receivers for analog broadcast services

In the early 1930s, the first auto radios were independently and simultaneously developed and marketed by Motorola in the United States and Blaupunkt in Germany. Initially, these consisted of multiple heavy and very complex components. Through continuous advancement of electronics and by means of new technologies, installation space was quickly reduced to a format standardized by DIN and ISO (approximately 180 × 52 × 160 mm).

While the first auto radios could only receive longwave signals and reproduce these with limited sound quality, modern auto radios include: receivers for AM, FM, and longwave; drive systems for audio media (CD, cassette); four-channel amplifiers; RDS data decoders; traffic information storage; and many other features. Sound quality of a high-grade auto radio is in every respect comparable to that of a stationary (home) audio system. One of the most important elements is the receiver module, as this is subjected to special demands in a mobile environment (see Section 8.4.1.1). Future-oriented receiver concepts, such as the Digiceiver family developed and implemented by Blaupunkt, operate on a digital basis. Although these are referred to as "digital receivers," it should be noted that these receivers digitize broadcast signals at a very early stage of the signal processing sequence (e.g., as of the intermediate frequency) and then apply digital signal processing methods. The broadcast system is of course still analog; the receiver uses its antenna to receive an analog audio signal.

In addition to broadcast services, auto radios are also expected to enable playback of various audio storage media. At present, the most widespread medium is the Compact Cassette (CC), although market share of auto radios incorporating Compact Disc (CD) drives continues to grow. The advantage of CD lies in outstanding sound quality and a high degree of drive uniformity compared to cassettes.

Another audio format is the Mini Disc (MD). The MD resembles the CD, but a data compression technique is employed to reduce the amount of data required. In the marketplace, MD is developing only slowly in comparison to CD.

Like the CD, the Digital Versatile Disc (DVD) is also based on an electro optical storage principle. Thanks to the high storage capacity of this medium, DVD is particularly suitable for applications requiring large volumes of data, such as storage of video information.

In the future, storage chips (flash memory) will gain increasing significance as audio storage media. Such storage chip cards are already being used in small mobile players and are suitable for downloading music from the Internet. This too involves data compression, in this case using so-called MP3 encoding.

These various storage media and their corresponding playback devices are also subject to the unique demands of the motor vehicle environment. Along with the possibility of large temperature fluctuations, both media and hardware are exposed to moisture and electromagnetic interference. Modern auto radio components must be particularly capable of dealing with these demands and are therefore in many respects quite different from stationary devices.

8.4.2 Digital broadcast systems

Digital broadcasting systems differ significantly from analog systems. The source signal (e.g., music) is digitized and then subjected to two processing steps which have no parallels in analog data transmission. These are so-called source coding and channel coding. In source coding, modern mathematical processes are used to

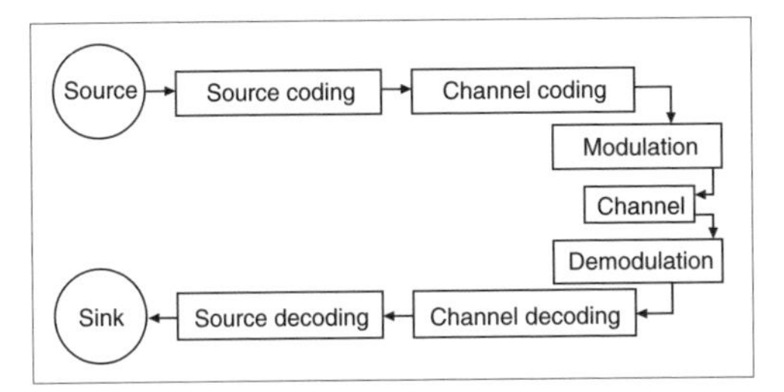

Fig. 8.4-1 Digital data transmission principle.

compress data by removing redundant or irrelevant (i.e., of lesser importance to the receiver) information from the source signal. Channel coding applies special mathematical processes to alter the data stream to minimize distortion by the carrier channel, which is in practice always highly distorted. Only after source and channel coding is the signal modulated to the carrier and transmitted to the receiver, where the process is reversed (see Fig. 8.4-1). Thanks to this process, digital data transmission has higher bandwidth efficiency, is more resistant to interference, and is more energy efficient than analog transmission methods.

Digital broadcast systems are currently being introduced. These are, for example, already in use for satellite broadcasting (satellite TV). Conversion of analog audio broadcasting to a digital system (DAB—Digital Audio Broadcasting) is already under way. In northern Europe (England, Scandinavia), 60–80% of listeners are within range of digital broadcast services. Approximately 30% of central European residents (Germany, France) are within range of digital audio. Implementation is least developed in southern and eastern Europe. Although the label "Digital Audio Broadcasting" implies audio services alone, broadcast services are not limited to these. A wide variety of services is offered, such as news, traffic announcements, and stock market quotes (text services). Graphical services such as maps of traffic jams, CD covers, and parking information in map form are included. These graphics and text services may be utilized by radios equipped with a suitable display.

8.4.2.1 Digital Audio Broadcasting

DAB represents the most important development in broadcast technology since the introduction of FM. This new digital broadcast system was developed as part of the European Eureka 147 project and offers the following features:

- Assured, interference-free reception for mobile as well as stationary applications
- Sound quality comparable to CD
- Frequency economy through audio data compression and shared-frequency networks
- Performance economy through suitable transmission methods and encoding
- Suitability for international, national, regional, and local broadcasting
- Multimedia capability that ensures future compatibility

The system developed by the Eureka 147 DAB project and standardized by ETSI (European Telecommunications Standards Institute) is to date the only system which meets these requirements and which has been endorsed by the ITU (International Telecommunications Union).

Typically, five to seven audio broadcast programs and several digital channels are combined into an ensemble. A program ensemble—for example, all programs provided by a single ARD broadcast facility—is transmitted nationally by a single-frequency network, enabling all of these programs to be received anywhere on the same frequency. Band III (175–239 MHz) is reserved for national broadcasts, while local programming is available on L-Band (1453–1491 MHz). By means of flexible ensemble partitioning, it is possible to transmit other data besides audio programs, such as video/television or Internet web pages. These DAB system applications have been introduced as the DMB (Digital Multimedia Broadcasting) concept. Thanks to DAB's good mobile reception qualities, it is possible to provide drivers with intelligent telematics services (DAB/ITS). An interactive medium is created with the aid of a back channel such as GSM, permitting user access to the Internet or other data services with high data rates and interactive user interfaces. The narrow-band GSM channel requests a service (e.g., by transmitting an Internet web page URL request), while the broadband broadcast medium actually supplies the data (e.g., a graphical web page in HTML format with embedded JPEG images).

The DAB system is based largely on three components: (1) audio data reduction in accordance with MPEG-1 (ISO 11172-3) or MPEG-2 (ISO 13818-3), each Layer II; (2) Coded Orthogonal Frequency Division Multiplexing (COFDM); and (3) flexible division of transmission capacity to a multitude of partial channels capable of transmitting audio and data programs at various data rates, independently of one another and with varying degrees of fault protection (DAB Multiplex). Construction and interrelationships between various DAB components are shown in Fig. 8.4-2, which illustrates schematically how DAB signals are generated.

Data for individual audio programs is reduced using audio encoders. Next, the redundancy required for receiver-side error correction is added by channel coding. Several data services may be combined by a packet multiplexer; this packet multiplexer then also feeds to channel coding. This channel-coded data of partial channels is combined using Fast Information Channel (FIC) data, which contains the structure of multiplexed and program information and is modulated by COFDM. This is generally done digitally by means of fast fourier transforms (FFT). The signal undergoes digital/analog conversion, is mixed to the appropriate transmission frequency, and is amplified, filtered, and transmitted.

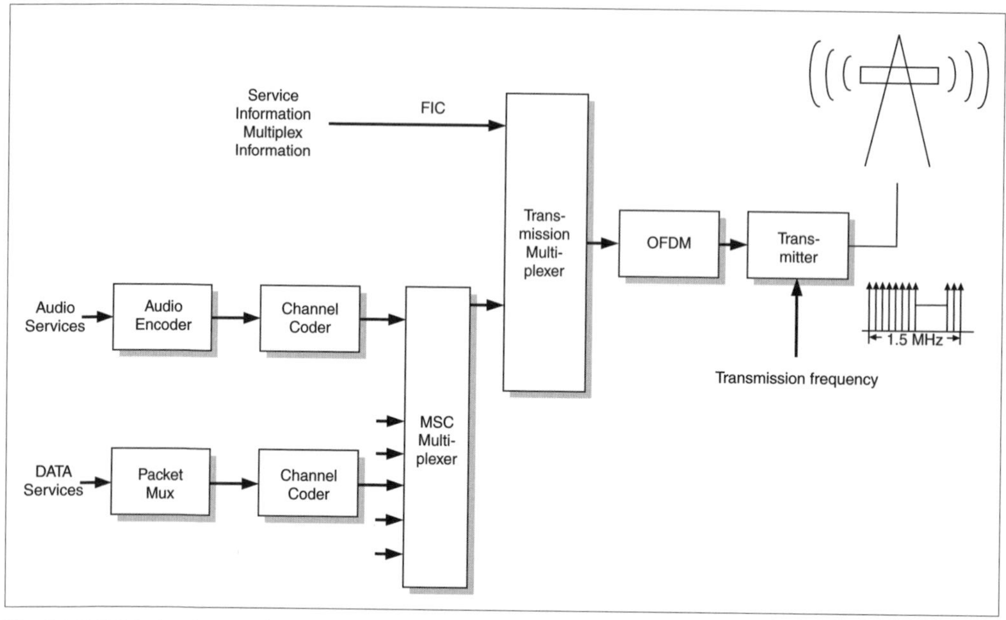

Fig. 8.4-2 DAB signal generation concept.

8.4.2.2 Digital Radio Mondiale

Digital Radio Mondiale (DRM) is a new broadcast standard for long-, medium-, and shortwave transmissions, currently in the process of being standardized. Using methods similar to DAB, signal transmission for these bands is also largely distortion-free; application of the latest audio compression techniques assures good audio quality for music and spoken word programming. This permits taking advantage of greater range while winning audiences through high broadcast quality.

8.4.2.3 Auto radio receivers for digital broadcast services

Because DAB and DMB digital radio services are still in their introductory phase, there are relatively few DAB receivers on the market. Mobile application imposes the same high demands on these as it does on analog broadcast receivers. Because digital radio does not yet enjoy the same coverage area as analog radio, DAB units on the market must, at minimum, also permit reception of analog FM signals. Naturally, the customer also expects audio media drives (CD player, for example), giving the DAB device at least the appearance of a classic auto radio. If the DAB device is also expected to provide graphical services (e.g., traffic jam maps, parking space information, CD covers, weather maps), it must also be equipped with an appropriately sized color-capable display. Due to relatively small market volumes at present, many of the DAB units now being offered consist of a conventional auto radio unit to receive analog broadcast signals coupled with a so-called add-on converter for digital broadcast services and a color display screen. Most recently, highly integrated 1-DIN DAB units have become available: for example, the Blaupunkt DAB Woodstock models. One example of this technology is a chipset developed by Bosch/Blaupunkt consisting of highly integrated components (D-FIRE II and CF 800) that enable construction of the world's smallest DAB module.

8.4.3 Multimedia networking

Penetration of multimedia systems into the automotive environment has several consequences: first, an ever larger number of elements from the classical computer and communications fields gains entry to the vehicle. Examples include mobile two-way radio components (GSM units or GSM modules) and or computerlike architecture in navigation and driver-information systems. A well-equipped vehicle can no longer be served by the installation volume available in a 1-DIN bay, originally standardized to accommodate a conventional car radio. A double-sized 2-DIN bay must be provided, or devices must be installed decentrally and connected by a network. Second, strongly divergent product cycles of computers and communication devices (in extreme cases, amounting to only a few months) compared to motor vehicles (more than ten years) result in a desire to make new devices easily retrofittable to existing vehicles. Standardized networking by means of wiring, in some cases wireless networks, is therefore necessary. Due to the extreme high data rates of multimedia systems even with good data compression, a CD-quality

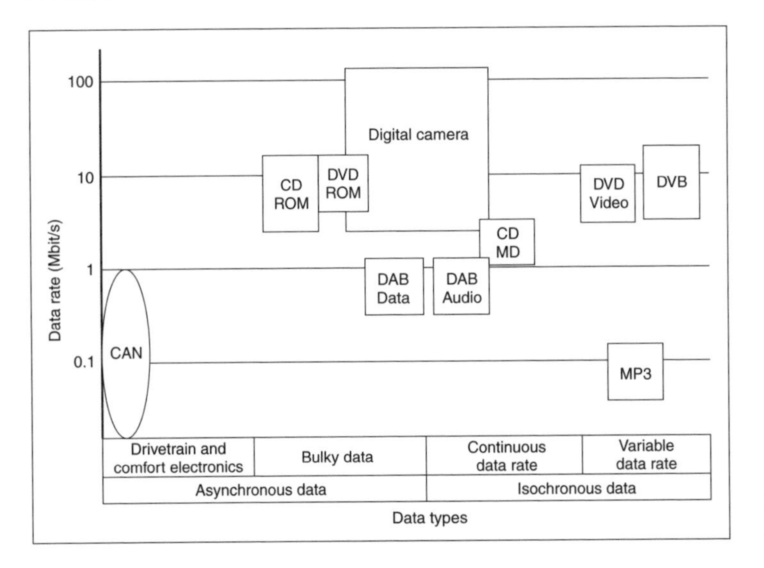

Fig. 8.4-3 Motor vehicle data types and data rates.

audio signal requires a data rate of about 100 kbit/s; an MPEG 2-encoded DVD video signal requires more than 10 Mbit/s. Classical serial databuses familiar in automotive practice, such as the CAN bus (Controller Area Network), cannot be used. Consequently, development of special multimedia buses was initiated several years ago.

These buses are required to transmit control data as well as audio and video signals; they differ greatly in terms of their characteristics (bandwidth, variable or constant bit rate, compressed or uncompressed data, etc.; see Fig. 8.4-3). The variety of different data types imposes special demands on any networking concept.

8.4.3.1 Evolution of motor vehicle multimedia networking

Networked driver information systems have been applied to motor vehicles since about the mid-1990s. In these systems, control data is transmitted separately from audio and video signals, the actual useful data. Control data is transmitted by various bus systems (e.g., CAN, Controller Area Network) or proprietary solutions. Useful data may be classified as digital or analog signals. Along with conventional transmission of analog audio and video signals via copper wires, digital audio data

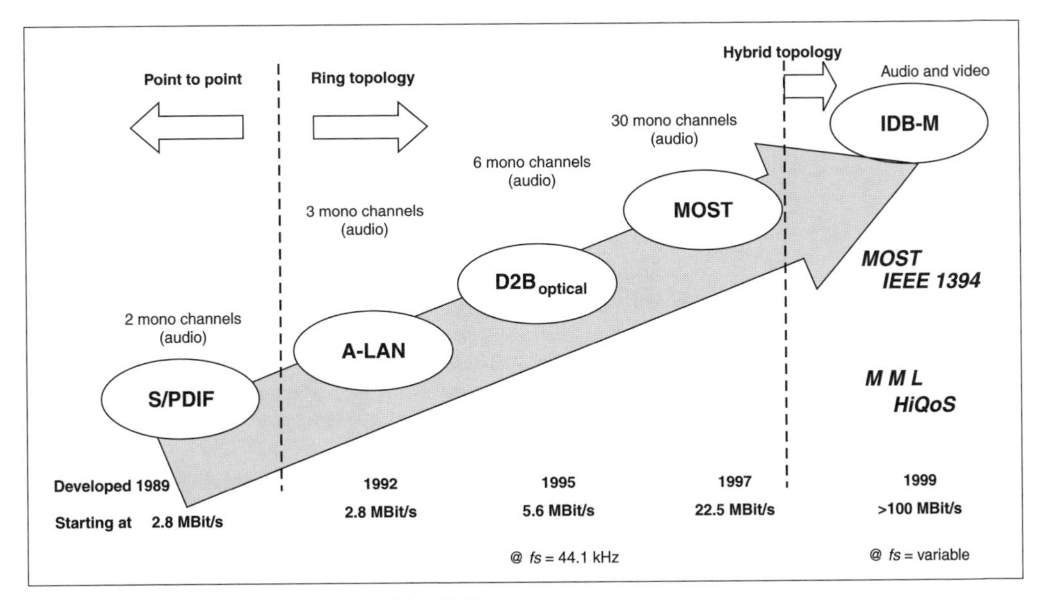

Fig. 8.4-4 Development of automotive multimedia buses.

may be transmitted in the *S/PDIF format* by optical means (fiber optics; see Fig. 8.4-4).

In a further development stage, a bus system in development in Europe since 1995 permits transmission of both control and useful data using a single medium. The result of this development program is the *D2B Optical* bus system. As its name indicates, this uses an optical databus with synchronous transmission and a bandwidth of 5.6 Mbit/s (with a sampling frequency fs = 44.1 kHz). It is suitable for transmission of control data and several channels of audio data. Use of this system still requires high-bandwidth data (e.g., video) to be transmitted using a separate point-to-point connection.

In the future, ever more data will be available in digital form. *MOST* (Media Oriented Systems Transport) was developed to meet the resulting demand for greater bandwidth. As a further development of D2B Optical, this effort quadruples bandwidth (to 22.5 Mbit/s). The MOST databus was developed specifically for automotive applications, along with appropriate chip sets, bus management (e.g., bandwidth, power and configuration management), as well as protocols and message catalogs for device control.

Further development of driver information systems as well as expansion of applications such as video-based driver support systems (e.g., reversing camera, vehicle environment sensors) are reflected in demands imposed on future bus systems. Along with the need for added bandwidth (about 150 Mbit/s by 2005), demands for system availability will also increase (failure safety, data consistency, etc). MOST, as well as automotive development of the FireWire (IEEE 1394) standard with bandwidths of 50 to 150 Mbit/s, are being considered for this future multimedia bus standard (Fig. 8.4-4: IDB-M, *Intelligent Data Bus—Multimedia*).

References

Lauterbach, T. *Digital Audio Broadcasting—Grundlagen, Anwendungen und Einführung von DAB*. Franzis-Verlag, Feldkirchen, 1996.
Werle, H. *Technik des Rundfunks*. R. v. Decker, Heidelberg, 1989.
Kammeyer, K.D. *Nachrichtenübertragung*. Teubner, Stuttgart, 1992.
Robert Bosch GmbH. *Kraftfahrtechnisches Handbuch*, 23rd edition. VDI-Verlag, Düsseldorf, 1999.
English edition: *Bosch Automotive Handbook*, 5th edition, Society of Automotive Engineers, 2000. ISBN 0-7680-0669-4.

8.5 Electromagnetic compatibility

In the past few years, assurance of electromagnetic compatibility (EMC) has become increasingly important. Although historically most vehicle functions were primarily mechanical/hydraulic or electromechanical, use of electronic systems, microprocessor-equipped control units, electronic sensors, and actuators in modern control and feedback systems is increasing. "Electromagnetic compatibility" is generally understood to mean that an electronic device exhibits adequate resistance to interference; that is, it is not unduly affected by outside sources and its own output of interference is limited to locations where interference-free broadcast radio reception is possible within the vehicle and in its immediate vicinity.

8.5.1 Ensuring EMC inside the vehicle

Within the vehicle, the electronic system layout must ensure that vehicle electronics systems operate correctly in proximity to each other without interference and without mutual influence. In dealing with this EMC problem, a multitude of interference mechanisms must be considered. This topic includes interference in the form of voltage spikes resulting from switching processes, whose amplitude may be many times the 12-V or 24-V system voltage. Such interference, generated by electrical system components themselves, reach input and output connections of other electronic components via signal and supply lines. By means of suitable experimental methods (laboratory test facilities), electrical and electronic components may be characterized as to their tendency to cause interference and subjected to standardized test impulses to examine their interference resistance. Final testing in actual cars examines mutual system compatibility.

The topic of self-suppression also ensures that the vehicle electrical system's own (broadcast) radio receivers can operate free of interference and that such electronic systems are not influenced by other transmitters operating inside the vehicle. This becomes increasingly important as the number of vehicle-installed mobile communications systems increases. Currents and voltages appearing in the vehicle electrical system exhibit a signal-shape–dependent frequency spectrum. These disturbances may reach broadcast receiver inputs directly through the wiring harness or in the form of radiation picked up by the vehicle's own broadcast reception antennas; the receiver cannot differentiate these from actual broadcast signals. Disturbance spectra generated by electronic components must therefore be limited to the point where no unacceptably high interference signals appear at the radio receiver's antenna inputs. Digital circuit timing signals are especially critical sources, resulting in narrowband interference. In the course of testing, several laboratory measurement methods are applied to evaluate interference generated by individual components. Finally, overall interference behavior of the production vehicle equipped with rigidly mounted receivers and antennas must be examined.

8.5.2 Resistance to interference from external electromagnetic fields (radiated immunity)

Because vehicles operate in an unknown electromagnetic environment, it must be ensured, for example, that powerful broadcast radio transmitters cannot affect vehicle functions. In operation, a vehicle and therefore its electronic components may be subject to appreciable electromagnetic fields. Broadcast signals are captured by vehicle structures, such as the wiring harness, through which they may reach input and output terminals of electronic devices. As in radio receiver circuits, semiconductor components could demodulate these signals, resulting in undesirable shifts in signal levels. Electronic circuits could interpret altered signal voltages as intended signals, resulting in functional faults such as such shutoff of safety systems as ABS or ASR, engine power losses, or erroneous gauge readings which irritate the driver. Appropriate test procedures in the laboratory and, later, on production cars examine resistance to such disturbances.

Because of the nature of its operation and the fact that any possible effects by external transmitters may endanger life and limb, the automobile assumes a special position in terms of system demands (field strengths). Motor vehicles are therefore required to exhibit very high radiated immunity (resistance to electromagnetic field interference). While a blanket field strength standard of at most 10 V/m is specified for other electrical and electronic products in personal or industrial applications, the "Motor Vehicle EMC Directive" 95/54/EC, relevant for vehicle type approval, establishes a radiated immunity limit of 30 V/m. In practice, automobile manufacturers have their own appreciably more stringent demands for interference resistance of their own products.

8.5.3 Protection of radio reception outside the vehicle

Along with requirements for resistance to interference, Directive 95/54EC also establishes requirements to ensure interference-free broadcast radio reception for nonmobile receivers. Interference radiated by a motor vehicle as a whole must be limited to the point where specified interference field strengths are not exceeded at a certain range. Special attention must be given to correct layout of the high-voltage ignition system, as this is generally responsible for the greatest interference field strengths outside the vehicle.

8.5.4 Standards and directives

Relevant EMC test procedures are described in national and international standards. Table 8.5-1 presents an over-

Table 8.5-1 International EMC standards for motor vehicles

Designation	Title
Radiated interference	
CISPR 12	Limits and methods of measurement of radio interference characteristics of vehicles, motor boats, and spark-ignited engine-driven devices
CISPR 25	Limits and methods of measurement of radio disturbance characteristics for the protection of receivers used onboard vehicles
Interference resistance	
ISO 7637	Road vehicles—Electrical disturbance by conduction and coupling Part 1: Definitions and general considerations Part 2: Passenger cars and light commercial vehicles with nominal 12-V supply voltage—Electrical transient conduction along supply lines only Part 3: Vehicles with nominal 12-V or 24-V supply voltage—Electrical transient transmission by capacitive and inductive coupling via lines other than supply lines
ISO/TR 10605	Road vehicles—Electrical disturbance from electrostatic discharges
ISO 11451	Road vehicles—Electrical disturbances by narrow-band radiated electromagnetic energy—Vehicle test methods Part 1: General and definitions Part 2: Off-vehicle radiation source Part 3: Onboard transmitter simulation Part 4: Bulk current injection (BCI)
ISO 11452	Road vehicles—Electrical disturbances by narrow-band radiated electromagnetic energy—Component test methods Part 1: General and definitions Part 2: Absorber lined chamber Part 3: Transverse electromagnetic mode (TEM) cell Part 4: Bulk current injection (BCI) Part 5: Stripline Part 6: Parallel plate antenna Part 7: Direct radio frequency (RF) power injection

view of the most important international standards. In Germany, vehicle type certification in terms of EMC is covered by §55a of the StVZO (German motor vehicle certification code), which mandates adherence to the requirements of European Community directive 95/54/EC.

8.5.5 EMC assurance

In the past, interference suppression measures were usually applied externally to components requiring suppression in the form of interference filters or additional shielding. In modern production vehicles, such methods are no longer economically viable. Moreover, EMC assurance must be included in the development process. For electronic components, EMC requirements must be considered in circuit conception, component selection, housing design, and printed circuit layout. For electric motors and electromechanical actuators, radiated interference must be minimized by appropriate constructive design and suitable internal switching of interference suppression components. For this reason, automobile manufacturers and suppliers have established test facilities to examine individual components, systems, and entire vehicles [1].

References

[1] Rohde & Schwarz. “EMV-Prüfzentren in jeder Größe—präzise, vollautomatisch, universell.” Vol. 164 (1999/IV).

[2] Gonschorek, K.-H., and H. Neu (ed.). “Die elektromagnetische Umwelt des Kraftfahrzeugs.” FAT-Bericht No. 101, 1993.

[3] Kempen, S., et al. “EMV in zukünftigen 42-V-Kraftfahrzeugbordnetzen.” VDI-Berichte 1547, 2000, pp. 499–509.

[4] Lindl, B., and J. Scheyhing. “EMV—die Entstörung von Kraftfahrzeugen.” ATZ, 1999, pp. 292–301.

9 Materials and manufacturing methods

9.1 A look back

Materials and technologies employed in automobiles reflect the available state of the engineer's art. The material or method best suited for a given set of demands is the one ultimately chosen. An excursion through the materials of early automobiles also serves as a retrospective look at the materials technology of the time.

Automobiles of the nineteenth century had to get by using a variety of materials which, from a modern perspective, would be regarded as a rather limited selection. Chassis and body materials adopted from (horse-drawn) coachbuilding practice as well as iron and nonferrous metals adopted from stationary engine construction for the powerplant formed the basis for new and continued materials development—ultimately driven by the automobile itself.

Initially, automobile body structures offered little protection against the weather; interior equipment in the modern sense was nonexistent.

Brake lining material—applied to brake blocks—consisted of leather pads. Frames were made of steel tubing; in the early years of the twentieth century, these were replaced by frames combining wood and steel and then by frames made of steel pressings, which provided increased strength with lower weight.

Body structures were initially made mostly of wood, with occasional examples of aluminum or wood/sheet metal combinations. Engine hoods, for example, were often made of a single sheet of aluminum.

Until the 1930s, wood remained the most important auto body material. Depending on the type of wood, this might require multiple layers of filler, which took several days to harden. This was in turn covered by several layers of color coat. Wooden boards were used for instrument panels; this heritage is still reflected in the term "dashboard." As was common in horse-drawn carriages, early automobiles rolled on wooden-spoked wheels.

Frequent parts failure as a result of fatigue (e.g., crankshafts or springs) led to extreme oversizing of vehicle components. General Motors founded its first materials laboratory in 1911 to gain a better understanding of metals.

In 1907, for the front axle of his Model T, Henry Ford employed an expensive vanadium steel alloy for its higher strength compared to conventional steels.

Immediately after the end of World War I, designers endeavored to keep vehicle weight to a minimum. Among other things, this led to increased use of aluminum—for example, for crankcases, transmission cases, and body components.

Even earlier, automobile designers primarily in the United States had made use of die-cast aluminum components—for example, to join frame elements. The American automobile industry quickly became the largest user of die-cast parts.

In rare cases, sand casting was employed even for large, thin-walled parts such as roofs or doors.

There were, however, problems, as shown by the example of zinc die casting, in which parts swelled, lost their shape, or turned brittle. The reason was lead and tin contamination of the alloy, which caused electrolytic processes and compromised functionality of the part.

Dodge and Pontiac in 1913 and 1915 respectively first used all-steel bodies in regular automotive production. Bending and torsional stiffness were higher than in the wood and sheet metal composite designs which preceded them. To build these bodies, which were based on a patent by Edward G. Budd and consisted of up to 300 individual sheet metal parts, newly developed oxy-acetylene welding apparatus was imported from France. Spot-welding machines were also available at this early stage. Such an automobile body needed about 1,100 spot welds. In the succeeding years, the number of parts was reduced (temporarily, as it turns out), and with them the number of spot welds. By contrast, modern cars typically employ about 4,000 spot welds.

Metal forming was hand craftsmanship performed by highly trained specialists and therefore correspondingly expensive. This may also be seen as the reason why all-steel bodywork only began to come into its own for series production in the 1920s and 1930s. A prerequisite was development of powerful auto body sheet metal presses to enable mass production of large numbers of sheet metal parts. Ford, for example, built its Model T using wood-framed bodywork until production ceased in 1927. In that same year, Krupp presented a special deep-drawing thin gauge sheet steel especially intended for auto body parts.

A driveable ladder frame, to which a body was later attached, continued as the dominant design configuration.

There were exceptions, such as the Lancia Lambda, presented in October 1922, which had a vertically extended, boxlike frame and can therefore be regarded as the first vehicle with a partially self-supporting body. Lancia held the patent for a vehicle without conventional frame.

In the 1930s, aluminum replaced cast iron as the piston material of choice in most engines. For durability reasons, however, aluminum cylinder heads were less successful at the time.

Also in the 1930s, elastomeric materials began to find increasing application. Chrysler, for example, used rubber mounts to improve acoustic and vibration comfort of its four-cylinder engine. Moreover, elastomeric body seals were first used.

Because vehicle styling became increasingly important as a sales tool, manufacturers began to apply exterior

Table 9.1-1 Examples of successful innovation and failed attempts

Year	Company, model/product	Notes	
1922	Weymann body	Leather-covered body panels to reduce rattles	↓
1924	Du Pont Duco	Nitrocellulose lacquer; reduced painting time from 3 weeks to about 2 days and guaranteed a better surface finish than previous lacquers	↑
1926	Safety Stutz	Splinterproof safety glass with fine steel mesh embedded in glass panels	↑
1936	Volkswagen	Magnesium castings, mainly for transmission case and crankcase, with resulting significant weight savings	↑
1940s/1950s	Crosley Hotshot	Engine of pressed and stamped sheet metal parts	↓
1941	Opel	Dip primer	↑
1945	Gregoire AF	Aluminum body with cast aluminum load-bearing structural components	↓
Late 1940s	Michelin	Steel belted radial tires	↑
1950s	Lloyd 300	Hardwood framework with leatherette-covered plywood panels	↓
1954	Panhard Dyna 54	Spot welded aluminum unit body	↓
1954	GM Corvette	Steel skeleton frame with fiberglass reinforced exterior body parts	→
1957	Lotus Elite Coupe	Fiberglass reinforced plastic body. Three large-area parts bonded to make single assembly.	→
1962	Glas	Camshaft driven by rubber timing belt	↑
1977	Porsche	Hot dip galvanized steel body permitted 6-year, later 10-year, corrosion protection warranty	↑
1980	GM Corvette	Fiberglass reinforced leaf spring	→
1981	DMC (DeLorean Motor Company)	Central steel frame, fiber-reinforced plastic structure with bolted-on stainless steel exterior panels	↓
1983	Polimotor Research Inc.	Four-cylinder gasoline engine with stationary parts (e.g., engine block) made of phenolic composite material	↓
1994	Audi A8	All-aluminum "space frame" body	↑

↑ – concept established itself in this or derivative form
→ – was at least partially followed up
↓ – was not able to establish itself

Table 9.1-2 Motor sports as pacesetter for new materials and technologies

Year	Company/vehicle	Notes
1895	Michelin	Vehicle equipped with pneumatic tires, in Paris-Bordeaux road race
1899	Dürrkopp	Development of small sports car fitted with aluminum bodywork for weight reduction
1900	Maybach/Daimler	Daimler delivered for Jellinek, who named it after his daughter Mercédès, had an engine made largely of aluminum and magnesium as well as a brass honeycomb radiator
1934	Auto-Union	16-cylinder engine with cast aluminum crankcase and cylinder heads
1962	Porsche	Formula 1 engine with titanium connecting rods
1963	Porsche 904 GTS	First German production car with fiberglass exterior skin
1967	Porsche 910/8	Aluminum tube frame with secondary function of frame tubes as oil lines
1971	Porsche 917	Magnesium tube frame
1981	Hercules/McLaren/Lotus	First use of carbon fiber reinforced plastics as load-bearing structures of Formula 1 vehicles

decoration. Ford used stainless steel for radiator shells, door handles, and radiator grilles. With the same intent, the 1930s saw greater use of chrome plated parts as styling elements.

In 1955, the average automobile used about 5 kg of plastics, mostly as decorative parts such as hubcap emblems or horn buttons. Between 1960 and 1970, the average contribution of plastics rose from 11 to 45 kg. Most of this represented nonstructural interior parts.

Introduction of plastics in the motor vehicle was not

without its problems. At times, plastics applications suffered from cheap, unfinished appearance. Only in the 1960s were plastics generally accepted, and they began to lose their former image of "economical" alternative materials.

There are many reasons why fiber composite materials have yet to make a significant penetration into automotive manufacturing. These include higher materials costs compared to aluminum and steel, more complex manufacturing methods with, in some cases, long cycle times and complex testing and quality assurance procedures.

References

[1] Bragg, G. "Materials: Key to 100 Years of Automotive Progress." *Automotive Engineering*, Vol. 104, No. 12, 1996.
[2] Bott, H. "Fortschritte im Automobilbau am Beispiel der Rennwagentechnik." *100 Jahre Automobil*, VDI-Berichte No. 595, VDI-Verlag, Düsseldorf, 1986.
[3] Fersen, O. v. (ed.). *Ein Jahrhundert Automobiltechnik*. VDI-Verlag, Düsseldorf, 1986.
[4] Hediger, F., O. von Fersen, and M. Sedgwick. *Klassische Wagen: 1919-1939*. Hallwag AG, Bern and Stuttgart, 1988.
[5] Gloor, R. *Nachkriegswagen: 1945-1960 Personenautos*. Hallwag AG, Bern and Stuttgart, 1980.
[6] Gloor, R. *Personenwagen der 60er Jahre*. Hallwag AG, Bern and Stuttgart, 1984.

9.2 Modern motor vehicle materials

This section provides an overview of the very extensive subject of materials and manufacturing methods. As many aspects have already been mentioned in previous chapters, we will here consider materials which have been recently developed and applied to production or for which production applications are imminent. The same applies to the closely related topics of manufacturing, joining, and recycling processes.

9.2.1 Materials share in automotive construction

Analysis of individual materials shows a significant change in weight percentages of various materials [1].

Above all, ever more strident demands for lower fuel consumption, despite steadily rising safety and comfort expectations, increasingly lead to lightweight design concepts and therefore to appreciably greater application of plastics and lightweight alloys.

Body

Steel has been, and remains, the leading auto body material. Aluminum-intensive body concepts have been limited to luxury-class vehicles, but in association with the drive for a "3-liter car," they will make inroads into the compact and small car classes (Audi A2).

Plastics play an especially dominant role in vehicle interiors. For exterior applications, plastics are primarily found in nonstructural areas. In the future, magnesium may find more use as a carrier material for lids and doors. By contrast, use of fiber-reinforced composites for structural body elements appears to remain restricted to limited production and experimental vehicles, and, despite many interesting efforts, a breakthrough to large-volume production seems unlikely in the near future.

Suspension

Lightweight alloys account for an ever greater share of suspension components, especially aluminum in the form of cast or wrought alloys. Accordingly, the share of steel and ferrous materials in suspensions is decreasing steadily.

Magnesium is not yet being used for typical suspension applications (swing arms, links, etc.), due to the problem of corrosion and lower values for mechanical properties. Magnesium wheels are technically feasible,

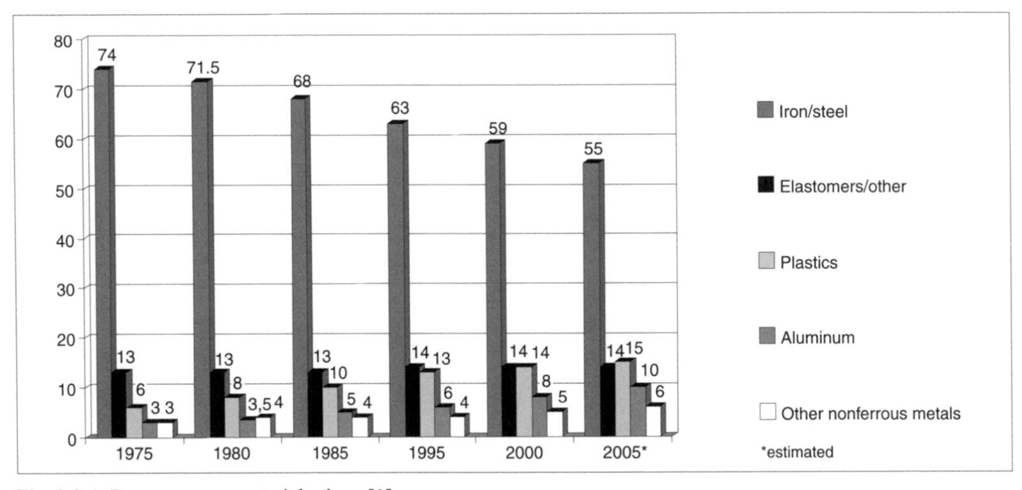

Fig. 9.2-1 Passenger car materials share [1].

but remain limited to small production runs in the sports car sector.

Drivetrain

Drivetrains employ a multitude of materials, making this the application most deserving of the label "multimaterial design" [2].

High-silicon aluminum or cast gray iron crankcases are as common as glass fiber reinforced intake manifolds or forged steel crankshafts. Ceramic catalytic converter materials are as much state of the art today as so-called ceramic preforms for low-wear cylinder surface inserts in aluminum crankcases. Magnesium primarily competes with fiber-reinforced plastics and aluminum (intake system, valve covers), but it is again under discussion as a material for future crankcase designs.

In summary, it should be remembered that while the proportion of steel in drivetrains has steadily decreased, it retains its historically dominant role in automotive construction.

The rapid rise in use of plastics over the past few years will gradually weaken, while use of lightweight metal alloys will grow steadily.

This trend is already apparent in sports cars and luxury vehicles and, with the aid of optimized manufacturing processes and cost-effective material development, appears to be transferable to the mid-range and small car segments.

9.2.2 Requirements and conflicting goals

In response to increasing traffic density and social and ecological changes, the past few years have seen increasingly complex and demanding requirements imposed on the automobile (see Chapter 2). However, extremely diverse performance characteristics give rise to a series of conflicting goals which must be evaluated and weighed against each other. One hundred percent satisfaction of these goals is therefore possible for very few cases. It is the task of engineering development to find the best possible synthesis in meeting these goals.

It is immediately obvious that demand for lightweight construction along with economy, recyclability, and process-assured manufacturability is shared by all component groups and that from the current perspective, therefore, materials choices are based on this central core. The catch phrase "multimaterial design" will thus determine the main avenues of materials engineering development as well as manufacturing and, above all, joining technology.

In the course of this development, materials selection on the basis of classical materials properties will increasingly be displaced by integral consideration of material + construction method + processes [2].

References

[1] "Innovativer Werkstoffeinsatz in Kraftfahrzeugen—Entwicklungstendenzen, Chancen und Ansätze in NRW." Institut für Kraftfahrwesen, Institut für Bildsame Formgebung, Institut für Kunststoffverarbeitung der RWTH, Aachen, January 1997.

[2] Friedrich, H.E. "Leichtbautechnologien im PKW—die Konkurrenz von Werkstoffen und Bauweisen nimmt zu." 5. Euroforum—Werkstofftagung für die Automobilindustrie [5th Euroforum Materials Conference for the Automotive Industry], Bonn, February 17–18, 1997.

9.2.3 Advances in materials performance characteristics

9.2.3.1 Strength and manufacturing methods

9.2.3.1.1 Steels

This section will concentrate on examining recent developments in the field of thin-gauge sheet steel. A drive toward lightweight design on the one hand and increased strength demands arising from more stringent

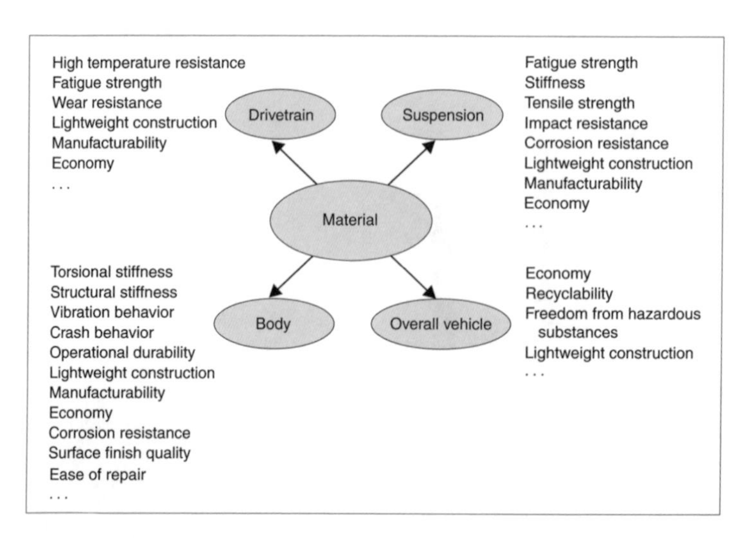

Fig. 9.2-2 Demands on various assemblies in motor vehicle construction.

crash requirements on the other have necessitated development of higher-strength steel sheet (minimum yield strength >180 MPa). Most recently, the industry has seen intensive development of ultrahigh-strength sheet steels, in some cases with tensile strengths in excess of 1,000 MPa. Simultaneously, driven by ever more demanding design, there is increased pressure for improved formability. Even though it must be assumed that these qualities come at higher cost, there may be overall cost benefits due to potential secondary savings. New developments in the field of stainless steels should also be mentioned.

Mild steels

In response to the aforementioned more restrictive demands on deep drawing or yield drawing ability, a new steel, DC 06 (IF 18, microalloyed extra deep-drawing grade) has been developed for use alongside the long-familiar thin-gauge mild steel sheet. By means of further development of the vacuum treating process, it was possible to establish ever lower carbon and nitrogen content and to fix these using appropriate alloying elements.

High-strength steels

Microalloyed steels

Proven, long-serving steels used by many auto manufacturers include microalloyed high-strength steels (alloying elements niobium, vanadium, and titanium). These steels were conceived in the mid-1970s and have undergone continuous development since then. Today, high-strength steels may account for up to 50% by weight of bodies-in-white. Since development of these steel types began, it has been possible to achieve steadily increasing tensile strengths compared to unalloyed steels, while simultaneously improved working properties. Today, these steels are available with minimum yield strengths up to 700 MPa.

Bake hardening steels

Another time-honored technology is bake hardening sheet steel. This designation encompasses steels which are resistant to aging at room temperatures, but in the course of the paint baking process typically employed in modern automotive manufacturing, these undergo controlled carbon aging in addition to prior work hardening during the forming process. This bake hardening process increases yield strength by ~40 MPa. The particular advantage of these steels lies in the fact that due to their low yield strength in the as-delivered condition, they exhibit good cold forming properties, yet are capable of achieving high strength in finished form. Minimum yield strength in the as-delivered condition is in the range of 180–300 MPa.

IF steels

So-called high-strength interstitial free, or IF steels, may at present be regarded as specialty steels. With yield strengths in the range of 180–260 MPa, these materials exhibit especially high deformation capability but are only available in the stated yield strength range.

Isotropic steels (nonscalloping quality strip steel)

These steels have unidirectional flow properties in the plane of the sheet and therefore a low tendency to form "ears" in deep drawing, even while exhibiting high strength. Minimum yield strength of these sheets in the as-delivered condition is in the 210 to 280 MPa range. These steels also exhibit the bake hardening effect after forming.

Multiphase steels

For structural elements and crash-relevant vehicle structures, high and ultrahigh-strength steels—in some cases with tensile strengths >1,000 MPa—are becoming increasingly interesting. The basic problem of increased material strength is an associated loss of formability. This may limit the material's potential applications.

This has led to new concepts—the so-called multiphase steels, in which strength increases are based on structural hardening. Within a matrix of soft ferritic components, harder components consisting of one or more other phases are distributed as uniformly as possible.

This development began with dual phase steels (DP), followed by TRIP (transformation-induced plasticity) steels. More recently, these have been joined by ultra-high-strength complex phase (CP) steels. These cover the strength range from 800 to 1,000 MPa. CP steels have been successfully applied to side impact beams; other conceivable applications include longitudinal members and crossmembers, bumpers, and reinforcing components.

Manufacture and structure of multiphase steels

As already mentioned, the special properties of multiphase steels are derived from their microstructure. One or more hardening phases must be embedded in the soft ferrite matrix as uniformly as possible. Starting from an austenite/ferrite dual-phase structure created in hot-rolled steels immediately after rolling or in cold-rolled steels in the continuous annealing process, subsequent rapid cooling suppresses the usual pearlite formation as much as possible and causes martensite formation of the remaining carbon-enriched austenite. In this way, for example, DP steel may consist of 10–20% martensite. Tensile strengths of 500–600 MPa are achievable; other developments suggest DP steels with even greater strength may be possible.

In TRIP and CP steels, austenite transformation deliberately takes place at higher temperatures in the bainite range. The structure of TRIP steel therefore consists of the main ferrite component plus precipitated bainite and a small amount of metastable retained austenite. This retained austenite changes to martensite in the subsequent deformation process (e.g., deep drawing). Fig.

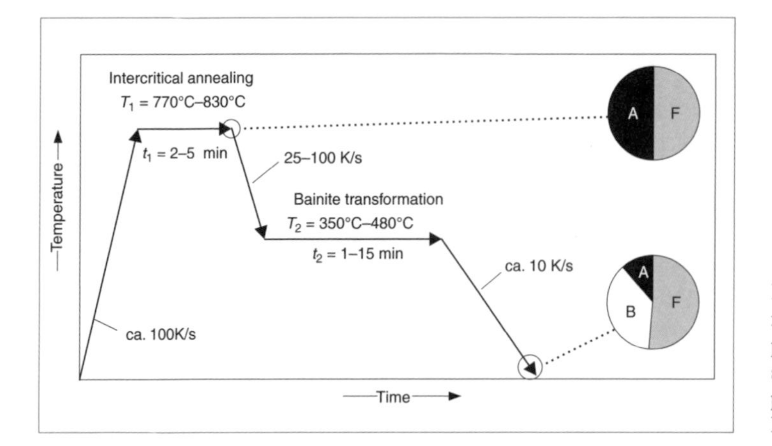

Fig. 9.2-3 Schematic representation of annealing process for cold-rolled TRIP steels indicating resulting phases: A, retained austenite; F, ferrite; B, bainite [1].

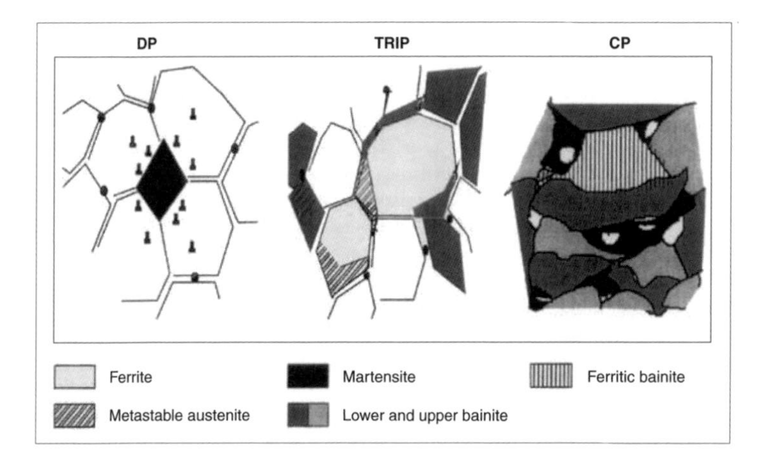

Fig. 9.2-4 Tensile strength increase by structural hardening [2].

9.2-3 shows a schematic of the annealing process for cold-rolled TRIP steel and the phases generated along the way. TRIP steels are currently available with tensile strengths of 700–850 MPa.

Similar to the foregoing examples, CP steels also exhibit very fine microstructure. In addition, these exhibit precipitation hardening by fine carbide and/or nitride precipitation. Tensile strength of these steels lies in the range of 800–1,000 MPa.

Properties of multiphase steels

Good forming behavior of all of these steels is based on overwhelming concentration of plastic deformation on the softer matrix, especially at the beginning of the forming process. In classical tensile tests, good hardening properties are expressed by a high value of n (strain hardening exponent) or a low yield strength to tensile strength ratio R_e/R_m. For conventional high-strength microalloyed wares, the yield point may be more than 90% of tensile strength. In DP steels, however, it is only about 70% of tensile strength. This makes the forming process much easier. The risk of partial thinning or even tearing in difficult forming processes is reduced by the high degree of strain hardening exhibited by these steels. Hardening permits high component strength after forming despite low initial yield strength.

Still lower yield strength to tensile strength ratios with good strain hardening values are achieved with TRIP steels. Formability of these steels relative to their tensile strength in the 700–850 MPa range is unsurpassed.

CP steels are currently produced as hot-rolled sheet with a minimum thickness of about 1.5 mm; cold rolling manufacturing processes are still in development. Previously, materials in this strength class usually had to be hot formed and subsequently heat treated. The advantage of CP steels lies in their cold forming ability without subsequent heat treating, and the associated potential for cost savings.

The bake hardening effect described above may also be observed in multiphase steels. In the case of bake hardening of CP steels, the higher the material's original strength, the greater the degree of hardening after prior forming (BH2). Galvanizing during the manufacturing process—that is, directly on the coil—is possible for these materials and makes sense from a corrosion standpoint, especially in the thinner gauges.

Stainless steels

"Stainless steel" is a general term for noncorroding steels which are characterized by increased resistance to chemical attack. In general, these materials have a chromium content of at least 10.5%. Even higher Cr content and other alloying elements (nickel, molybdenum) result in even better corrosion resistance. Stainless steels are divided into ferritic stainless steels and austenitic stainless steels. In recent years, the latter group has proven most interesting for vehicle construction. Austenitic materials exhibit extremely good work hardening and far surpass ferritic stainless steels in terms of achievable strain. Extremely high degrees of deformation may be realized. Together with the aforementioned favorable work hardening properties, this results in high component strength. Newer materials concepts, in which a portion of expensive alloying elements is replaced by nitrogen, may result in price reductions for stainless steels, which are currently much more expensive than conventional steels. Application to crash-relevant areas of vehicles is also conceivable, as is manufacture of attached ("hang-on") parts or suspension components, which would no longer need corrosion protection in the form of paint.

References

[1] Bleck, W. "TRIP-Stähle—Eine neue Klasse hochfester, kaltumformbarer Stähle." Studiengesellschaft Stahlanwendung, 30 Jahre Forschung für die Stahlanwendung, Düsseldorf, 1998.
[2] Engl, E., and J. Drewes. "From 450-MPa Dualphase to 1100-MPa Complex-Phase Steels: Development of a New Generation of Automotive Sheet Steels." IBEC Conference 1997, Automotive Body Materials.

Cast iron

Cast iron is an alloy of iron and carbon containing at least 2% carbon as well as other alloying elements, above all silicon. Cast iron may be classified as:

- Cast iron with lamellar graphite (DIN EN 1561)
- Cast iron with spheroidal graphite (DIN EN 1563)
- Cast iron with vermicular graphite
- Ductile cast iron (DIN EN 1562)
- Austenitic cast iron (DIN EN 1564)
- Chilled cast iron (white cast iron)
- Wear-resistant alloyed cast iron (DIN 1695)
- Bainitic cast iron (DIN EN 1564)

Cast iron with lamellar graphite, also known as gray cast iron, is by far the most common casting material. Various qualities are available, ranging in tensile strength from 100–200 MPa (EN-GJL 100) to 350–450 MPa (EN-GJL-350).

Reasons to choose gray cast iron include relatively low cost per kilogram, ease of casting, good casting properties, good workability, high compression and cyclic bending strength relative to tensile strength, insensitivity to high and low temperature extremes as well as temperature changes, notch insensitivity, good damping properties, and good limp-home properties. Vehicle applications include brake discs, engine blocks, and, in the form of chilled cast iron, camshafts.

Cast iron with spheroidal graphite, also known as nodular cast iron, has an advantage over gray cast iron in higher tensile strength and, above all, greater ductility. Grades with minimum tensile strengths of 350 MPa (EN-GJS-250-22) to tensile strength >900 MPa (EN-GJS-900-2) are used. Nodular iron automotive applications include differential housings, connecting rods, spindles, flywheels, release levers, and crankshafts.

Bainitic cast iron is subdivided [1] into

- Bainitic-austenitic nodular cast iron
- Bainitic nodular cast iron
- Austempered ductile iron (ADI)
- Ausferritic cast iron (austenitic-ferritic cast iron)

The following will discuss the ADI and austenitic-ferritic forms of cast iron, which represent new developments in the field of bainitic cast iron.

This new type of cast iron exhibits an attractive combination of strength, toughness, damping qualities, and wear resistance. ADI is made by heat treatment of high-quality nodular iron castings. The casting is heated to a temperature of 840–950°C, held at this temperature for austenization, and then rapidly cooled to the bainite transformation temperature of 230–450°C [2]. The bulk of ADI consists of accicular (needle-shaped) ferrite in a matrix of high-carbon, stabilized austenite ("ausferrite"). This structure is only remotely related to bainite as found in steels.

ADI is available in four grades, with minimum tensile strengths of 800 MPa (EN-GJS-800-8) to 1,400 MPa (EN-GJS-1400-1). Typical applications include axle housings, wheel hubs, starter ring gears, and crankshafts.

References

[1] *Konstruieren +Gießen* 22 (1997) No. 3
[2] *Gießerei Praxis* No. 3/4, 1996

9.2.3.1.2 Light alloys

High-strength aluminum alloys

Development of alloys exhibiting both high strength and great toughness has concentrated on suspension components. Furthermore, casting techniques usually employed for suspension components provide the necessary workability properties of these materials. The following will briefly describe and characterize two representative alloys.

"Magsimal-59" (made by Aluminium Rheinfelden GmbH), chemical designation AlMg5Si2Mn, is a new, innovative cast material for die casting and squeeze casting (see Section 9.2-5). In its as-cast condition (F temper), it exhibits high strength and ultrahigh failure strain. Table 9.2-1 lists its material composition [1].

In many cases, heat treatment of Magsimal-59 castings is not required. Still, strength properties are nearly

Table 9.2-1 Chemical composition of alloys AlMg5Si2Mn (mass fractions in %)

	Si	Fe	Cu	Mn	Mg	Zn	Ti	Other
minimum	2.0			0.5	5.0			
maximum	2.5	0.15	0.05	0.8	6.0	0.08	0.20	0.06

identical to those achieved with T6 temper, with outstanding elongation properties.

Mechanical properties of die cast and squeeze cast components depend on cooling rates and achievable dendrite spacing in individual casting sections; properties are therefore dependent on section thickness and mold temperature. This effect is illustrated in Fig. 9.2-5, in which mechanical properties for F temper are shown as a function of casting Section thickness and local solidification time [1].

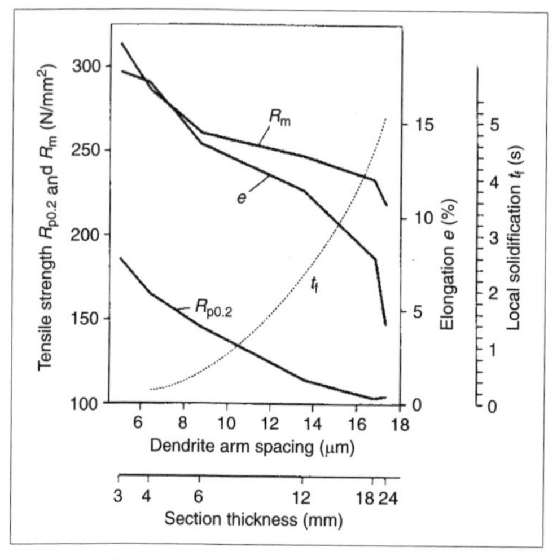

Fig. 9.2-5 Mechanical properties of alloy AlMg5Si2Mn (Magsimal-59) in the as-cast condition, as functions of dendrite spacing, casting section thickness, and local solidification time.

In the as-cast condition, good yield limit ($R_{p0.2}$), tensile strength (R_m) and elongation (e_5) are obtained in sections up to 4 mm thick. Increasing Section thickness up to 12 mm results in decreases in tensile strength, elongation, and, above all, yield limit [1].

Given a 5% failure probability, long-term tensile strength in F temper is 100 MPa (Fig. 9.2-6). By comparison, permanent mold cast AlSi7Mg T6, solution heat treated and artificially aged, has a long-term tensile strength of 93 MPa under the same test conditions [1].

As heat treatment of Magsimal-59 may be eliminated, parts made of this alloy may be produced more economically. Applications include steering wheel frames, suspension subframes, cast nodes for "space frame" bodies, and hubs for spoked wheels with cast-in steel and brake rings [1].

Silafont 36 (also manufactured by Aluminium Rheinfelden GmbH), chemical designation AlSi9MgMnSr, also known as Aluminum Association 365.1, is a ductile die-casting alloy with low iron content, a further development of well-known material AlSi9Mg.

The nominal silicon content is given as 10.5%, making this alloy easily castable and providing good mold filling capability. Iron content is set as low as possible in order to minimize the fraction of usually plate-shaped AlFeSi phases. These phases are the main cause of poor tensile strength and elongation characteristics.

To improve shape retention and reduce tendency to adhere to the mold, manganese content is raised to about 0.65%. Manganese has the same effect in reducing sticking tendency as iron. To ensure that the silicon phase is finely dispersed in the molten condition, the alloy is permanently modified through addition of strontium. Mechan-

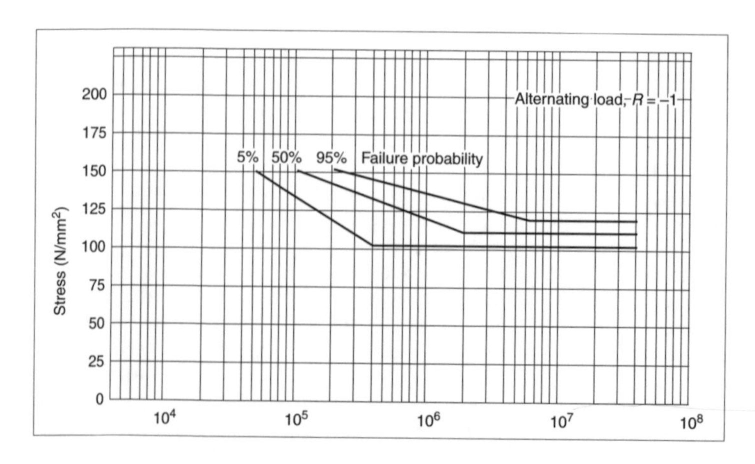

Fig. 9.2-6 Fatigue curves (Wöhler curves) for alloy AlMg5Si2 (Magsimal-59) in as-cast condition. Note: idealized representations; in practice, a slight drop in long-term tensile strength is observed.

Table 9.2-2 Mechanical properties of "Silafont 36" as a function of temper

Material condition (temper)	$R_{p0.2}$ [MPa]	R_m [MPa]	e_5 [%]	HB 5/250-30
F	120–150	250–290	5–10	75–95
T4	95–140	210–260	15–22	60–75
T5	155–245	275–340	4–9	90–110
T6	210–280	290–340	7–12	100–110
T7	120–170	200–240	15–20	60–75

ical properties are established by the magnesium fraction [2]. Lower magnesium content implies high elongation with low tensile strength, while high magnesium content results in high tensile strength and low elongation.

Mechanical properties for various tempers are shown in Table 9.2-2 [3].

Tensile strength and elongation properties may be deliberately modified by varying aging temperature and time (Fig. 9.2-7) [4].

With 5% failure probability, maximum long-term tensile strength is obtained in the as-cast F condition followed by T4 and T6 temper (Fig. 9.2-8). However, these differences are not especially large, which indicates that long-term strength is not a function of heat treat condition. In F temper, long-term tensile strength of 89 MPa is achieved [4].

This new alloy is used for a variety of castings (e.g., steering wheel frames, cast nodes for "space frames," dashboard member and door frames) [4].

Magnesium alloys

Magnesium enjoys a long tradition as a structural material. Taking advantage of air-cooled engine designs, magnesium crankcases and valve covers were state of the art for years on the VW (original Beetle) and Porsche (911). Only with the introduction of more power-

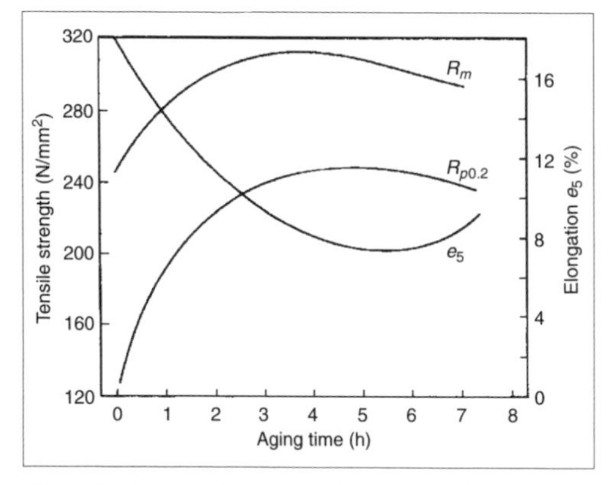

Fig. 9.2-7 Mechanical properties as a function of aging time. Mg content 0.3%, solution heat treated at 490°C/3h, quenched in water, artificially aged at 170°C.

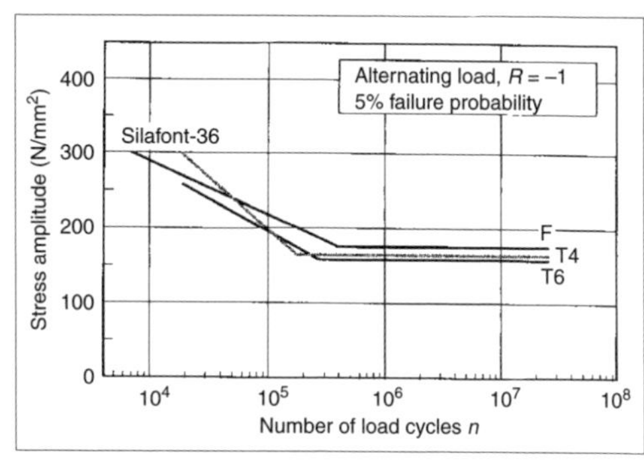

Fig. 9.2-8 Wöhler (fatigue) curves of Silafont 36 as a function of heat treatment. Note: idealized representations; in practice, a slight drop in long-term tensile strength is observed.

ful engines and associated higher thermal and mechanical loads did aluminum finally displace magnesium in these applications. Added to these advances were corrosion problems in water-cooled engines.

Development of heat-resistant as well as extremely pure and therefore corrosion-resistant Mg alloys was an important development consideration in realizing the great lightweight design potential of this material. It should be noted, however, that the problem of anodic corrosion of these alloys has not been solved, and therefore appropriate surface protection and joining technology must be considered.

The following will examine four different magnesium alloys.

AZ91 HP (MgA19Zn1 HP) and *AZ81 HP* (MgA18Znl HP) are magnesium materials with good strength properties, good corrosion resistance, and very good castability. The suffix HP (high purity) represents drastic reduction in heavy metals content (iron, nickel, and copper) and therefore improved corrosion resistance compared to alloys listed in DIN 1729 [5].

The mechanical material properties shown in Table 9.2-3 for alloys AZ91 HP and AZ81 HP are achievable as a function of casting method and heat treat condition [5].

These alloys are primarily used for die cast components such as various covers, including valve covers, etc.

AM60 HP (MgA16 HP) and *AM50 HP* (MgA15 HP) are likewise high purity Mg alloys. They are characterized by high strength and ductility combined with good castability and cold formability.

Table 9.2-4 lists the achievable mechanical material properties as functions of casting method and heat treat condition.

These alloys are eminently suitable for manufacture of seat components (e.g., DaimlerChrysler), instrument panel carriers, wheels, convertible top mechanisms (Porsche Boxster, 911 Carrera) and steering wheel frames, all of which are die cast. Fiber-reinforced and particle-reinforced aluminum and magnesium composites are currently under development but for cost rea-

Table 9.2-3 Mechanical materials properties of cast magnesium alloys

Alloy	Casting process	Temper	Tensile strength [MPa]	0.2% limit [MPa]	Strain at fracture [%]	Brinnell hardness [HB 5/250]	Cyclic bending strength [MPa][b]	Melting point [°C]
AZ 91 HP[a]	Die cast		200–250	150–170	0.5–3.0	65–85	50–70	420–600
	Permanent mold	−F	160–220 (120)	110–130 (90)	2–5 (1)	55–70	70–90	
		−T6	240–300 (170)	150–190 (130)	2–7 (1.5)	60–90	80–100	
	Sand cast	−F	160–220 (130)	90–120 (80)	2–5 (1)	50–65	70–90	
		−T6	240–300 (170)	150–190 (140)	2–7 (1.5)	60-–90	80–100	
AZ 81 HP[a]	Die cast		200–240	140–160	1–3	60–85	50–70	425–615
	Permanent mold	−F	160–220 (130)	90–110 (80)	2–6 (1)	50–65	70–90	
	Sand cast	−F	160–220 (130)	90–110 (80)	2–6 (1)	50–60	70–90	

[a] Unified to DIN 1729 Sheet 2, values in parenthesis are minimum values. Values applicable for section thicknesses to 15 mm.
[b] Rotating bending test to DIN 50113 with 50×10^6 load reversals.

Table 9.2-4 Mechanical materials properties as functions of casting process and heat treat condition

Alloy	Casting process	Temper	Tensile strength [MPa]	0.2% limit [MPa]	Strain at fracture [%]	Brinnell hardness [HB 5/250]	Cyclic bending strength [MPa] [b]	Melting point [°C]
AM 60 HP[a]	Die cast		190–230	120–150	4–8	55–70	50–70	445–630
	Sand cast	−F	180–240 (140)	80–110 (80)	8–12 (4)	50–65 (50)	70–90	
		−T4	190–250 (150)	90–110 (90)	8–15 (6)	50–65 (50)	70–90	
AM 50 HP[a]	Die cast		180–220	110–140	5–9	50–65	50–70	440–625

[a] Unified to DIN 1729 Sheet 2, values in parenthesis are minimum values. Values applicable for section thicknesses to 15 mm.
[b] Rotating bending test to DIN 50113 with 50×10^6 load reversals.

sons are not yet being applied in volume production. Objectives of these development programs are increased tensile strength, yield strength, modulus of elasticity, and hot strength.

References

[1] Hielscher, U., et al. *Gießerei* 85, No. 3, 1998, pp. 62–65.
[2] Hielscher, U., et al. *Gießerbrief,* No. 5, 1997, p. 1.
[3] Silafont-36 processing data sheet, Aluminium Rheinfelden GmbH.
[4] Hielscher, U., et al. *Gießerei* 82, No. 15, 1995, pp. 517–523.
[5] Materials data sheet, Hydro Magnesium.

Metal foams in automotive construction

Metal foams have been known since the 1950s, but until recently it has not been possible to manufacture these in sufficient quantity and with consistent quality. In general, these are classified as powder metal, molten metal, and special processes, of which the powder metal variations are regarded as the most suitable. In these, a blowing agent is mixed with metal powder and consolidated (hot isostatic pressing—HIP, extrusion, etc.). Subsequently, the resulting blank may be reproducibly foamed by heating to a temperature close to the melting point. In addition to the familiar aluminum foams, other metals and alloys such as zinc, tin, bronze, brass, and lead may be foamed. Titanium and steel foams are still under development. Currently manufactured metal foams are almost exclusively of the closed-cell variety. Metal foams permit manufacture of a multitude of shape variations, such as foamed hollow shapes, composite materials, and plates. Plates can be deep-drawn before foaming, which allows even greater possible shape variations (Fig. 9.2-9). Joining of metal foams can be accom-

Fig. 9.2-9 Deep-drawn and subsequently foamed sheet metal [6].

plished with the aid of laser welding processes and also by means of bonding, bolting, or riveting. Brazing is unsuitable for metal foams due to potential anodic corrosion problems in the foam pores [1, 2, 3, 4, 5].

Mechanical properties (tensile strength, modulus of elasticity) rise sharply with increasing foam density. Electrical and thermal conductivity of foams is markedly reduced compared to solid alloys, while the coefficient of thermal expansion remains unchanged. In plastic deformation of metal foams, stress levels remain nearly constant over a large deformation distance, closely approximating an ideal absorber. Common aluminum foams can absorb ~80–90% as much energy as an ideal absorber. For this reason, metal foams can be used for energy-absorbing front, side, and rear impact elements. Another application for metal foams results from the fact that foams greatly improve buckling and crush behavior of hollow sections and can therefore be used as reinforcements. Corrosion behavior of metal foams is strongly dependent on the metals used and their location in the vehicle. Aluminum foams suitable for use in motor vehicles can be viewed as noncritical in this regard as they have a close oxide surface coating which provides adequate insulation for different metal combinations.

References

[1] Banhart, J., J. Baumeister, and M. Weber. "Metallschaum—ein Werkstoff mit Perspektiven." *ALUMINIUM* 70, 1994.

[2] "Aluminiumschäume—Konstruktionswerkstoffe mit großem Potential." *ALUMINIUM* 73, 1997.

[3] Banhart, J., J. Baumeister, M. Weber, and A. Melzer. "Aluminiumschaum—Entwicklungen und Anwendungsmöglichkeiten." *Ingenieur Werkstoffe,* 7, 1998.

[4] Baumeister, J. "Fortschritte der Herstellung und Eigenschaften von metallischen Schäumen." *Proceedings* of the 31st ISATA Conference, "Materials for Energy-Efficient Vehicles," 1998.

[5] "Industrial Production Welding of Aluminium Sandwich Plates." IBEC Advanced Technologies & Processes, 1997.

[6] Alulight. Manufacturer information, Mepura mbH, Ranshofen, Germany.

Titanium alloys

Titanium alloys are employed for aerospace as well as motor sports applications. The main incentive for use of titanium is weight reduction.

The area of greatest potential titanium application is without doubt the drivetrain, as weight savings—that is, reduction in moving/rotating masses, can yield fuel economy benefits. Possible production applications for titanium alloys include connecting rods, cam followers, gears, valves, valve seats, valve heads, bolts, differential housings, various driveshafts, and so forth.

The advantages of titanium alloys include high static

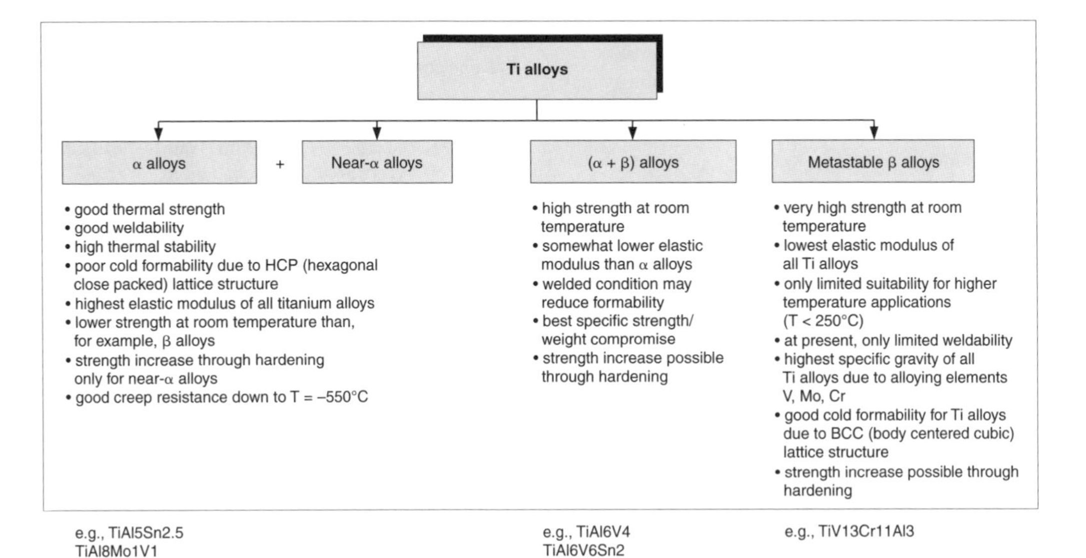

Fig. 9.2-10 Titanium alloys.

Table 9.5-5 Properties of various titanium alloys [1]

Property	Units	α Ti (Ti48Al2Cr)	α+β Ti (TiAl6V4)	γ Ti
Density	[g/cm³]	3.7–3.9	4.5	4.6–4.9
Elastic modulus	[GPa]	155–180	105–115	90–105
Strain limit	[MPa]	400–750	850–950	1,200–1,350
Tensile strength	[MPa]	400–850	900–1,100	1,200–1,600
Elongation at fracture (room temp)	[*e* %]	1–4	10–16	12–25

and dynamic strength combined with low density, good thermal and fatigue strength, and very good corrosion properties.

Disadvantages are difficult formability, difficult machining, greater notch sensitivity than steel, and unfavorable tribological behavior.

Titanium alloy parts are often nitrided to improve wear and fatigue properties.

Another means of improving tribological properties of titanium alloys is surface coating. In addition to the classical electrochemical or chemical processes, plasma or ion beam-based high vacuum processes (CVD, chemical vapor deposition and PVD, physical vapor deposition), combined with galvanic processes, are finding increasing application [1].

Titanium alloys are classified as shown in Fig. 9.2-10.

New intermetallic γ-TiAl(Cr,Mo,Si) alloys are currently being developed for high strength/high temperature applications such as valves, turbochargers, connecting rods, and piston pins [2]

Table 9.2-5 lists mechanical properties of α and α + β alloys in comparison to γ-TiAl alloys (Ti48Al2Cr) [1].

Despite modern manufacturing methods, the high cost of extracting titanium and the high manufacturing costs make production applications of titanium in passenger cars and trucks unlikely in the near future.

However, continued development of titanium extraction and advanced manufacturing techniques should permit introduction of titanium in volume production in the medium to long term.

References

[1] Repenning. “Titan—ein Werkstoff für tribologische Beanspruchungen.” *Ingenieur Werkstoffe* 2/98, p. 14 ff.

[2] Knippscheer and Frommetier. “Intermetallische Leichtbaulegierungen für Motorkomponenten.” *Ingenieur-Werkstoffe* 3/98. p. 32 ff.

Sandwich composites

Metal/plastic/metal composites (Fig. 9.2-11) present an interesting new development for auto body applications. These materials achieve a breakthrough of existing material limitations; the combination of familiar materials and their positive properties have resulted in a composite material with optimum stiffness, weight, and acoustic properties.

Compared to sheet steel offering the same bending strength and to aluminum sheet (1 mm), such sandwich sheets (0.2 mm Al/0.8 mm polypropylene/0.2 mm Al) are about 60% and 35% lighter, respectively [1].

Use of various outer layer materials and different core materials together with choice of appropriate material thicknesses permit properties to be matched to application requirements. As a rule, the core consists of a layer of polypropylene ~0.8 mm thick, bonded on both sides to 0.2 mm aluminum or steel.

Early experiments in the form of a spare tire well in the ULSAB project (steel/plastic/steel) and a front hood study on the VW Lupo (Al/plastic/Al) underscore the great potential for lightweight design offered by this composite material.

As with any new material, its specific properties must be considered in processing and manufacturing.

Consequently, welding techniques commonly used in body manufacturing and their associated thermal loads are not suitable joining methods due to the thermoplastic core layers used in these materials.

Possible joining methods are mechanical joining (self-piercing riveting, crimping, clinching) and adhesive bonding.

In deep drawing, process parameters must be adapted to material properties which differ from those of conventional steel or aluminum sheet.

The same applies to the entire manufacturing layout,

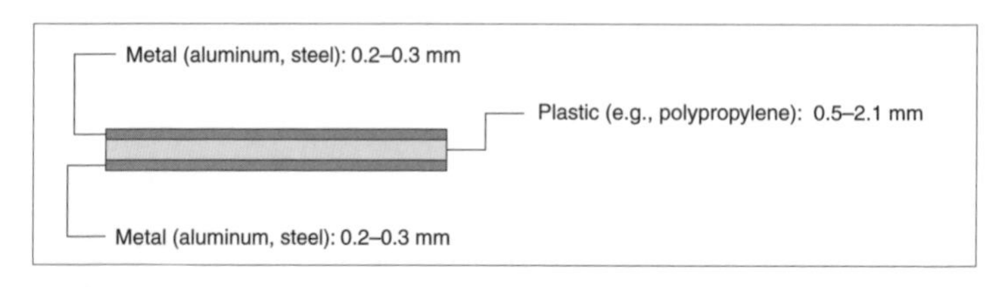

Fig. 9.2-11 Metal/plastic/metal composite.

where present-day cataphoretic dip priming or paint topcoat temperatures might make online painting of such sandwich sheets impossible (melting temperature of polypropylene core layer ~163°C; cataphoretic dip primer temperature ~160–185°C).

Nevertheless, first efforts are promising and lead to expectations of further progress.

Reference

[1] Hoogovens Hylite BV company literature.

9.2.3.1.3 Noble metals

The elements silver (Ag) and gold (Au) as well as ruthenium (Ru), rhodium (Rh), palladium (Pd), osmium (Os), iridium (Ir), and platinum (Pt) are designated noble metals. As they are electrochemically highly unreactive— that is, they exhibit a high degree of corrosion and oxidation resistance—they are used in the chemical and electronic industries.

Electronics components

Noble metals, in particular gold and silver alloys, are employed in various electronic components which are used in motor vehicle construction.

Gold and silver alloys are primarily used in applications where:

- Current flow is low (<5 mA).
- Low electrical resistance is required.
- Contacts must be protected against corrosion.
- Contacts must be protected against wear (caused by relative motion on a microscopic scale).
- A high level of contact assurance is required for electronic safety components

Examples of noble metal applications in the auto industry include:

- Gold-plated contacts in airbag modules
- Gold bond wire for contacts on integrated circuit (IC) circuit boards
- Silver alloys, used for their low electrical resistance in low-current applications in conductor paths as well as in relays
- Sensor technology (wheel speed sensors, outside air temperature sensor, oxygen sensors, etc.)
- Catalysts for combustion engines (platinum, palladium, and rhodium).

Reference

Degussa. *Edelmetall-Taschenbuch*, 2nd edition. Hüthig Verlag, Heidelberg, 1995.

9.2.3.1.4 Plastics

Of all the many developments in the field of plastics (highly crystalline polypropylene, PPO/PA blends, etc.) perhaps the most important are development of fiber-reinforced thermoplastics and long glass fiber technology.

Fiber-reinforced thermoplastic surfaces

This family of materials, in practice often consolidated as "organometals" [1], closes the gap between, on the one hand, short- and long-fiber–reinforced injection molded types and, on the other, glass-mat–reinforced thermoplastic (GMT) and sheet molding compound (SMC) materials.

The main matrix materials for these applications are polypropylene (PP), polyester (PET), and polyamide (PA). Woven or knitted reinforcements are made of glass fibers, synthetic fibers (e.g., aramid), and carbon fibers, alone or in combination. Multiple woven layers can be laid down with various fiber orientations (e.g. +45°/−45°, 0°/90°).

For cost reasons, current attention is focused mainly on glass fibers. Depending on matrix materials, fiber type, fiber content, and type of weave, mechanical properties of these prepregs are easily varied to meet a given set of demands (Fig. 9.2-12) [2].

Properties of these materials may be summarized as follows [3, 4]:

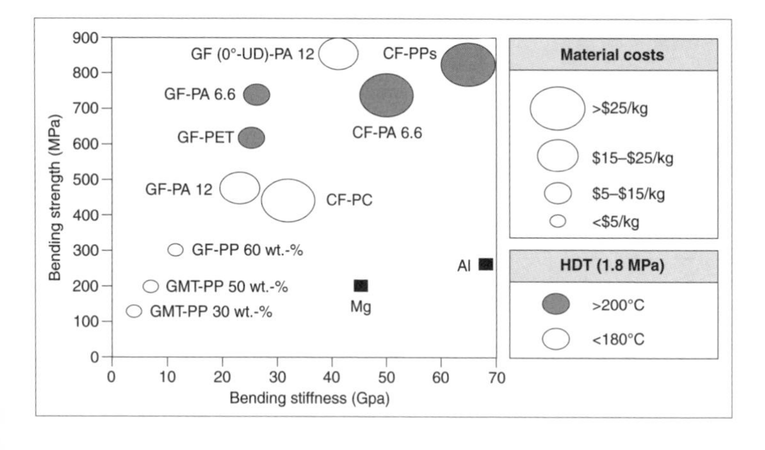

Fig. 9.2-12 Property spectrum of thermoplastic composites [2].

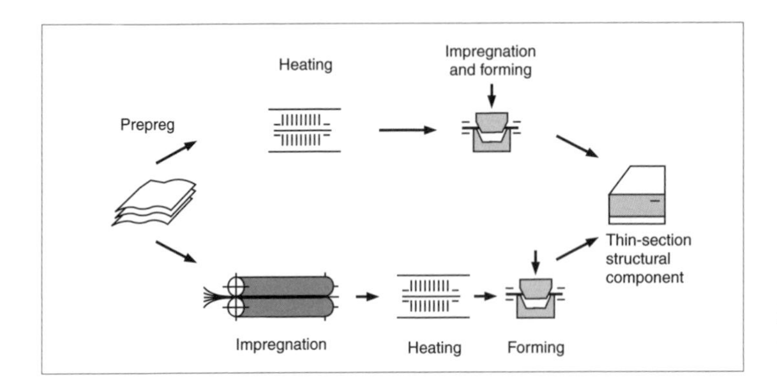

Fig. 9.2-13 Possible production variations for manufacture of thin-section structural components [5].

— Low density
— High strength and elastic modulus
— Corrosion resistance
— High energy absorption/toughness
— Weldable
— Recyclable
— Deep drawable/thermoformable

Prepregs or preconsolidated blanks may be formed by heating beyond the melting point of their respective matrices followed by deep drawing and pressing to form finished parts (Fig. 9.2-13).

Because of their high fiber content, achievable surface quality is poor, ruling out any visually demanding ("class A surfaces") exterior skin applications.

From a recycling standpoint, these materials are easily recycled by grinding and then reusing them as short-fiber–reinforced injection molding material or as a core layer for sandwich composites [6].

Disadvantages include the high cost of blanks and the difficulty in achieving simple, assured damage recognition.

References

[1] Mehn, R. "Leichtbau- und Verarbeitungspotential bei Einsatz von glasgewebeverstärkten Thermoplasten (GF-T) im Automobilbau." 20th Euroforum Conference: New Materials for the Automobile Industry, Frankfurt, 1995.
[2] Breuer, U., and M. Ostgathe. *Halbzeug- und Bauteilherstellung—Umformverfahren-, Faserverbundwerkstoffe mit thermoplastischer Matrix.* expert Verlag, 1997.
[3] Jauss, M. et al. "Neue Hochleistungsverbundwerkstoffe: Sandwichstrukturen mit Rezyklaten, Mit Kunststoffen zu neuen Produkten." Symposium, Karlsruhe, May 13–15, 1997. Fraunhofer Institut für Chemische Technologie (ICT), DWS Werbeagentur und Verlag GmbH, Karlsruhe, 1997.
[4] Mayer, C. "Wirtschaftliche Herstellung von textilverstärkten thermoplastischen Halbzeugen." 28th Internationale AVK Conference, October 1-2, 1997, AVK Frankfurt, 1997.
[5] Neitzel, M., et al. "Gewebeverstärkte Thermoplaste im Karosseriebau." *Autohausspezial* 14/15/1996.
[6] Ostgathe, M., et al. "Organobleche aus Thermoplastpulver." *Kunststoffe* 86 (1996), Carl Hanser Verlag, Munich.

Long-fiber thermoplastics (LFT)

A new material group is developing between the established material systems (GMT, SMC, and short fiber reinforced thermoplastics): so-called long-fiber thermoplastics (LFT).

This new material is marked by high stiffness with very good energy absorption, fully in keeping with the demands of "lightweight design." Long-fiber reinforcement provides an appreciable increase in Izod/Charpy notched impact strength and thermal shape stability. Due to glass fiber content and related poor surface quality, visible "class A" exterior applications are not possible, and interior applications in visible areas are only conditionally possible [1]. Typical applications include front module carriers (front aprons) and laminated instrument panel carriers.

At present, three different LFT material types have been applied successfully. These are classified according to fiber length and the associated processes (Table 9.2-6).

To retain the granulate's good mechanical properties in the finished component, if possible without losses, the most important requirement for LFT processing is

Table 9.2-6 Classification of thermoplastic fiber-reinforced composite systems [2]

	Short-fiber–reinforced thermoplastics		Long-fiber–reinforced thermoplastics		
Fiber length in component (mm)	<1	1–5	5–25	5–25	>10
Raw material/blanks	Short fiber granulate	LFT granulate	LFT granulate	Direct LFT Direct process	GMT mat technology
Processing method	Injection molding		Extrusion		
Tendency toward anisotropy	Very high	Low	High	High	Low

fiber protection—that is, keeping fibers as long and as intact as possible.

One processing variation which is gentle on fibers is extrusion. Plastic extrusion screw geometry appropriate for LFT and melt introduction into open tooling by means of a wide slotted nozzle greatly decrease unacceptable fiber damage in comparison to injection molding.

In conjunction with more heat-resistant materials (e.g., PA, PBT, etc.) and different fiber reinforcement (e.g., carbon fiber), long-fiber–reinforced thermoplastics will capture new applications which are currently still reserved for metallic materials.

References

[1] Bärkle, E. "Status quo in der PP-Spritzgießverarbeitung." "Fortschritte mit Polypropylen im Kfz-Bereich" conference, June 17–18, 1998, Süddeutsches Kunststoffzentrum, Würzburg.

[2] Kuhlmann, G. H. "Hinterpressen und Kompressionsformen von unverstärkten und faserverstärkten Kunststoffen." AVK-TV conference, Baden-Baden, 1998.

Elastomers

The elastomer field has also witnessed important new developments. One example is rubber mixtures for air conditioning hoses. Environmental considerations prompted a switch to refrigerant R134a and, with it, polyethylene glycol lubricant and an increase in operating temperature ($\approx +10°C$). Extensive testing has shown that different elastomers and mixtures would be required for these conditions. Tests indicate that chlorinated or brominated butyl rubber as well as HNBR are outstanding base polymers for rubber mixtures used in motor vehicle air conditioning systems. Elastomers based on these materials meet the most stringent auto manufacturer specifications [1].

Another new development is rubber mixtures for fuel system hoses. Environmental considerations have resulted in extremely strict regulatory requirements for fuel permeation. Extensive tests have shown that permeation reduction can only be achieved by an added barrier layer. FPM has proven to be an outstanding barrier. Test results have shown that application of an inner FPM layer greatly reduced permeation compared to standard hose materials.

New rubber mixtures are also being developed by the tire industry (Section 7.3). Thermoplastic elastomers (TPE) are becoming increasingly important; these will take their place between elastomers and thermoplastics. The automotive industry is often the scene of competition between elastomers and TPE. This concentrates on bellows or boots for axles and driveshafts, while in the engine department, TPE-E seeks to displace conventional elastomeric materials [2]. The advantages of TPE over elastomers lie in weight reduction, recyclability, and manufacturing considerations. Thermoplastic workability of TPE often provides cost advantages. Increasingly, TPE's disadvantages with respect to media resistance or thermal form stability are being addressed by new material developments or TPE-specific design.

References

[1] Harmsworth, N. "Elastomere in Schläuchen für Klimaanlagen." Presentation, Bayer AG, Leverkusen, 1998.

[2] Raue, F. "Elastomer vs. TPE." *Kunststoffe* 88, 12, 1998, pp. 2279–2283.

Textiles and leather

Comfort and a feeling of well-being are the central themes associated with application of automotive textile materials and leather. Textiles are not limited to vehicle interiors; they are also vital aids in meeting various functional demands.

Depending on area of application and intended purpose, a variety of demands with respect to function as well as workmanship must be satisfied (Table 9.2-7) [1].

Natural fibers, popular in the past, have been largely replaced by man-made fibers. The dominant types are polyester and polyamid fibers, which in recent years have been supplemented by much cheaper PP fibers. Natural fibers such as wool are still found in some seat applications.

In addition to the nature of the actual fibers, the type of fabric and its overall structure are deciding factors for its service properties. A rough overview of modern textile materials, using the Mercedes-Benz S-Class as an example, is shown in Table. 9.2-8 [1]. Multilayer convertible top materials of polyester or polyester/polyacrylic blends, with an intermediate layer of chloroprene rubber, complete this overview.

Leather

Leather may be regarded as the classic material for motor vehicle interior design and individualization (Fig. 9.2-14).

Due to high material costs, demanding manufacturing methods, and lamination on components, leather is primarily used in upper mid-range and luxury vehicles. In the past, special attention had been given to improving shrinkage behavior at high temperatures as well as optimized fogging and odor behavior. Shrinkage has been greatly reduced with the introduction of chromium-free

Table 9.2-7 Requirements profile for textile materials [1]

Function	Manufacturing/workability
Stretchability	Deformability
Heatfast/lightfast	Pressure sensitivity
Breathability	Sewability
Scuff resistance	Weldability
Stiffness	Gluing ability
Low flammability	Vulcanisability
Climate resistance	
Fogging/odor	

Table 9.2-8 Examples of textile applications in Mercedes-Benz S-Class [1]

Component	Material application	Textile product	Material composition (by weight)
Seats	Upholstery fabric (seating surface)	Laminated flat weave fabric	55% polyester + 45% wool; polyurethane foam; polyamid or polyester
		Velour fabric	70% polyester + 30% wool in pile
	Trim material (cover)	Laminated rough weave	Polyester; polyurethane foam; polyamid or polyester
	Topper pad	Needled felt	50% wool + 25% cellulose fibers + 25% synthetic fibers
	Intermediate pad	Woven fabric	50% cellulose + 50% polyester
	Foundation pad	Rubberized hair pad	22.5% coir (coconut fiber) + 22.5% hogshair + 55% latex binder
	Safety belts	Webbing	Polyester
Headliner	Prefab headliner covering material	Laminated loop knit velour	Polyester; polyurethane foam; reinforcing Section
Door panels	Center section covering	Laminated flat weave fabric	55% polyester + 45% wool; weldable polyurethane foam
	Bottom section covering, door pocket	Tufted velour carpet	Polyamid on polyester felt or woven polyester

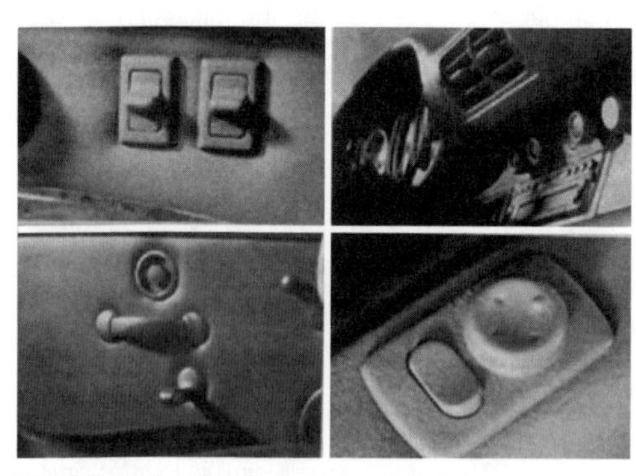

Fig. 9.2-14 Leather-covered interior parts.

tanning and special drying processes. Fogging and odor optimization are achieved by use of low-fogging, low-odor materials in the tanning process.

Today, to obtain the best possible color matching and a flawless surface, leather is coated with a layer of dye about 25 μm thick. Future developments are concentrating on customer desires for natural leather—that is, a stronger emphasis on leather's natural characteristics (odor, haptics, appearance) with the least possible compromises in everyday utility (soiling, colorfastness, wear, etc.).

References

[1] Eissler, E. et al. "Einsatz textiler Materialien bei der Mercedes-Benz S-Klasse." Textilien im Automobilbau, VDI-Kongress, October 30–31, 1991, pp. 18–40, VDI-Gesellschaft Textil und Bekleidung, Düsseldorf 1991.

[2] Francke, G. "Leder—ein klassischer Interieurwerkstoff." "Kunststoffe im Automobilbau" conference, Mannheim, March 25–26, 1998, pp. 337–349. VDI-Verlag GmbH, Düsseldorf, 1998.

9.2.3.1.5 Glazing

Demands on modern automobile glazing are manifold, ranging from customer aesthetic demands to complex protective functions. Accordingly, there is a wide variety of different glass types available, alone and in combination with other materials. The primary trend is toward various tinting and coating systems with flexible, myriad properties.

A general distinction is made between tempered safety glass (TSG) and laminated safety glass (LSG). Tempered safety glass is currently installed in vehicles in thicknesses of ~3.15 mm. Under pressure to reduce weight, efforts are under way to reduce thickness to 2.5 or even 2.1 mm.

In contrast to single-layer tempered glass, laminated safety glass is much more versatile. However, thickness and weight reductions as well as integration of various functions dominate consideration of LSG. Current LSG thicknesses are in the range of 5–6 mm, composed of two sheets of glass each 2.1–2.6 mm thick with an intermediate plastic film. Efforts are under way to reduce thickness to 2.1–1.6 mm, but, as in the case of TSG, this is limited by mechanical property requirements and stone impact protection. Movable side windows currently have a total thickness of 6 mm and in the course of lightweight development will be reduced to 5 mm. Thickness of fixed side glass could even be reduced to 3.5 mm. Due to the layered structure of LSG, it is possible to incorporate other functions, such as special coatings or antennas. In some cases, metal oxide coatings are applied to reflect 60–70% of invisible solar thermal radiation, thereby reducing vehicle interior heating. Metal oxide layers may also be operated as windshield heaters. Along with metal oxide coatings, infrared protective layers (Audi A8, Mercedes S-Class) are also used to prevent uncontrolled heating of the vehicle interior. These layers can also be used as radio and TV antennas. The familiar "printed" circuits are replaced by fine wire structures which are nearly imperceptible to the unaided eye. Outer surfaces of TSG as well as LSG may be given hydrophobic (water repellent) coatings to prevent soiling of waterborne dirt particles. Potential applications are primarily side glazing and backlights [1].

Another possibility is a combination of glass and polycarbonate, in which a thin layer of glass is cemented to a lighter but less scratch-resistant layer of polycarbonate. This combination permits further weight reduction. One disadvantage is reduced torsional stiffness. Such systems are still limited to smaller two-dimensional formats (approximately DIN A4 or U.S. letter size). The next planned development goal is manufacture of large, curved surfaces [2].

A further development in vehicle glazing is the scratch-resistant all-plastic polycarbonate windshield [3]. Along with possible weight savings of about 15–40%, advantages of this technology include very good (cold) impact protection plus great design freedom and a high degree of integration capability thanks to injection molding technology [4]. This technology has already been applied in the form of headlight lenses (covers) [3].

Along with higher component costs, factors weighing against use of polycarbonate for window applications include still unsatisfactory scratch resistance and as yet unknown aging behavior of laminated structures (scratch-resistant coating/adhesive/polycarbonate). Mechanical properties, noise behavior, interior, heating, and recycling problems must also be considered. Despite intensive development, it is likely that application of all-plastic glazing will initially be limited to fixed smaller areas (e.g., rear quarter windows), with a medium-term goal of substituting this material for movable door glass, backlights, and glass sunroofs.

References

[1] Assmann, K. "Automobilverscheibung—sicherer Durchblick mit Polycarbonat." 10 *Kunststoffe* 86, 1996, pp. 366–368.

[2] Assmann, K. "Leichte Scheibensysteme für Automobile." VDI-Z Spezial 1995 Ingenieurwerkstoffe, September III/95.

[3] Anon. "Dauerhaft geschützt – Kratzfestlackierungen von Platten und Fertigteilen." 8 *Kunststoffe* 86, 1996, pp. 1111–1112.

[4] Schottner, G., and G. Abersfelder (ed.). *Fahrzeugverglasung*. Expert Verlag, 1995.

9.2.3.2 Wear protection

According to DIN 50320 (December 1979), wear is defined as "subjecting the surface of a body to stress by contact with and relative motion against a solid, liquid or gaseous body." In engineering, wear is normally undesirable. In certain exceptions, for example during the "break-in" period for moving parts, wear processes may be desirable from a technical standpoint [1].

Finding ever more wear-resistant materials to extend service life and/or prevent failure, reduce fuel consumption, and optimize noise behavior has been a perennial objective, especially in drivetrains.

The primary means of improving wear properties of metallic materials are alloying and/or surface treatment. Another possibility for improving wear resistance as well as component strength is application of sintered metals, composites, and use of ceramic materials.

The last two of these variations will be introduced and described briefly here.

MMC (metal matrix composites)

Metal matrix composites usually employ light-alloy materials. Their objective is to improve local wear behavior and tribological properties in critical areas. One example is presented by crankcases of GD-AlSi9Cu3, partially reinforced by formed hollow cylindrical silicon inserts ("Lokasil").

These highly porous hollow cylindrical silicon shapes, so-called "preforms" (Fig. 9.2-15, a and c), are subjected

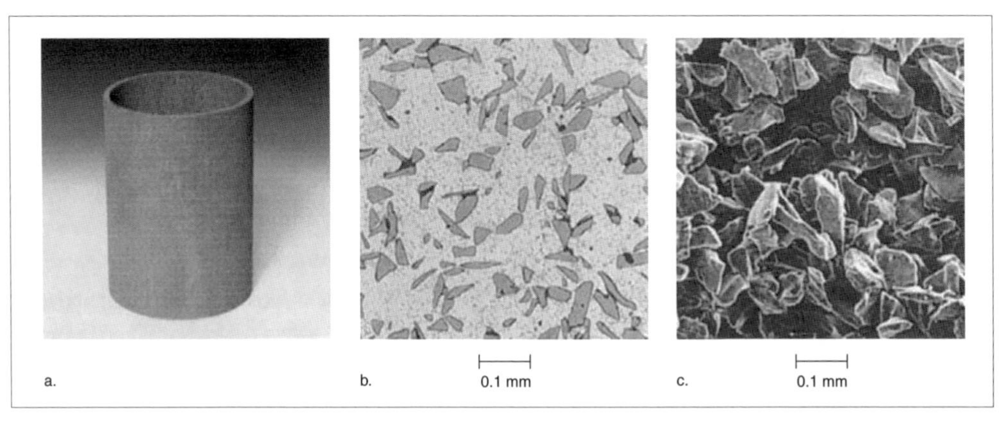

Fig. 9.2-15 a) Lokasil II preform; **b)** microstructure of composite material; average silicon grain size 45–55 μm; **c)** preform structure with ~25% volume silicon.

to pressure during the casting process (squeeze casting), infiltrated by and embedded within a cost-effective secondary aluminum alloy. The resulting local composite material is tribologically comparable to hypereutectoid alloys. The embedded silicon has an average grain size of 45–55 μm (Fig. 9.2-15) [2, 3].

These cast-in silicon preforms provide targeted improvement of cylinder wall surface tribological properties, without requiring that the entire crankcase be cast of a wear-optimized, sophisticated, and therefore expensive primary aluminum alloy. Compared to cast-in gray iron liners, this variation permits a more compact design thanks to closer cylinder spacing. The homogeneous bond between metal matrix and preform provides greatly improved heat transfer properties. A consistent thermal expansion coefficient also prevents out-of-round conditions during the life of the engine, thereby promoting efficient combustion with low emissions.

Ceramics

Three subgroups are recognized within the field of ceramics:

Silicate ceramics

Porcelain, stoneware, glasses, steatite, magnesium aluminum silicate (MAS, cordierite); because of its high resistance to thermal shock and low coefficient of expansion, MAS is of interest for engineering applications (e.g., catalytic converter monoliths) despite its low strength.

Oxide ceramics

Single-phase, single-component metal oxides based on aluminum, zirconium, beryllium, magnesium, titanium, titanates, ferrites.

Nonoxide ceramics

Ceramics mainly based on carbon, silicon, nitrogen, and boron, and sialones (Si, Al, O, and N compounds).

Oxide and non-oxide ceramics and multiphase dispersion ceramics are classified as engineering ceramics (engineering, industrial, or structural ceramics) [4].

The advantages of technically relevant ceramics are their high compressive strength and hardness, high wear, and thermal resistance, as well as good corrosion resistance and low weight.

Disadvantages include low ductility, high scatter of material properties, in some cases low thermal shock resistance, complex manufacturing and processing, and complex methods necessary to join ceramics to other materials.

One possible application for ceramics in motor vehicles is silicon nitride valves. Friction power may be reduced by as much as 40% with these valves, while simultaneously improving wear behavior and minimizing noise emissions [5].

Obstacles to production application include high production costs and inadequate process assurance in manufacturing.

Ceramic surface treatments such as PVD and CVD coatings are finding increased application, as are thermally sprayed coatings. These coatings are applied to metallic components in order to increase hardness and therefore wear and thermal resistance at the surface. They are thin, as a rule, and do not normally require any subsequent processing.

Such surface treatments offer advantages in lightweight design, as the bulk material of a component is no longer defined by highly stressed boundary areas [5].

References

[1] Gräfen. *VDI Lexikon Werkstofftechnik*, 1991.

[2] Gasthuber and Krebser. "Strukturkeramik zur Reduzierung der dynamischen Massen im Ventiltrieb des Kolbenmotors." DaimlerChrysler internal report, 1991.

[3] Kaniut. "Werkstoffe im Automobilbau, Teil 1: Werkstofftypen." Information provided by Zentralwerkstofftechnik (Central Materials Technology), Mercedes-Benz, 1987.

[4] Stenzel, Rehr, Prinz, and Schulze. Abschlussbericht January 1998, BMBF 03 N 30 11: "Monolithische Motorblöcke aus einem Leichtmetall-Verbundwerkstoff."

[5] Institut für Kraftfahrwesen, Institut für Bildsame Formgebung, Institut für Kunststoffverarbeitung der RWTH Aachen, Abschlussbericht January 1997: "Innovativer Werkstoffeinsatz in Kraftfahrzeugen."

[6] Köhler, E., F. Ludescher, J. Niehues, and D. Peppinghaus. "Lokasil-Zylinderlaufflächen." Special publication by KS Aluminium-Technologie AG.

9.2.3.3 Corrosion protection

DIN 50900 defines corrosion as follows: "Corrosion is the reaction of a metallic material with its environment, causing a measurable change of the material which may lead to impaired function of the metallic component or an entire system. In most cases the reaction is electrochemical in nature; in some cases it may be chemical or metallophysical in nature."

The concept of corrosion protection must be distinguished from corrosion resistance, which is used as a rough qualitative measure of a material's ability to withstand attack by a corrosive medium. A completely corrosion-resistant coating—for example, gold—is not absolutely necessary for corrosion protection. The gold coating may be porous, thereby promoting corrosion of the less-noble base material, without itself being attacked. Zinc coated on steel exhibits the opposite behavior; in the event of penetrating damage, zinc goes into solution, protecting the substrate by its own corrosion.

Material coating systems

We will not attempt to describe the entire spectrum of surface protection methods. Instead, we will provide an overview of properties which may undergo desirable or undesirable alteration as a result of coatings. In addition, some characteristics typical of certain processes will be pointed out, along with suggestions for process selection [1]. Surface treatment of a material attempts to optimize that material for an application or indeed to make an application possible.

This applies to classical coating methods as well as conversion processes such as etching, oxidizing, and implanting. Important application goals are corrosion protection surface functions, oxidation protection, wear protection, sliding capability, heat barriers, covering, and achieving a specific surface effect. Often, multiple goals must be met simultaneously.

Demands on an anticorrosion surface coating can be decorative and functional or only functional. This distinction is important, as decorative coatings in general require more involved pretreatment and therefore affect coating costs. In the interest of a technically and economically optimized development process, the requirements of any later surface coating must be considered from the outset.

Electrolytically deposited metallic coatings

Metal deposition by means of electroplating can take place in an electrolyte container or without such a container by means of brush plating (used to selectively apply plating, usually for repair purposes) [2]. The component is electrically connected as a cathode. Of the many electroplatable metals, the following are used for corrosion protection: chromium, nickel, copper, zinc, silver, tin, gold, and lead. Of the galvanically depositable binary metal alloys, ZnCo, ZnFe, and ZnNi have achieved a measure of importance. These layers are chromatable, and, thanks to their appreciably better corrosion resistance, they replace chromated zinc coatings for suspension and engine fastening elements and components. Additional organic thin-layer coatings further increase corrosion resistance and thermal resistance. Such surfaces are also used in contact with magnesium components [3].

For high-strength steels (>1,000 MPa) and electrolytic deposition, hydrogen embrittlement is always a concern [4]. Hydrogen embrittlement must be considered in connection with severely cold-worked sheet metal components. Furthermore, hydrogen is generated by cathodic dip degreasing and inorganic acid etching. These pretreatment processes are often applied prior to surface coating processes which in themselves do not produce hydrogen damage. This also applies to galvanic deposition of aluminum. Hydrogen embrittlement does not arise in electrodeposition using aprotic (i.e., water-free) solutions. Aluminizing produces an exceedingly thin layer, which can be chromated, and, given coating thicknesses in excess of ~20 μm, can be color anodized. It is suitable for hose clamps, for example.

Galvanically deposited plating is not true to contour. As a result of higher electrical current, corners and sharp projections receive thicker coatings. Because of limited dispersion ability, plating of complex geometries is often only possible with the aid of additional or internal electrodes. Good rinsing ability must be considered during component design. Lost electrolyte increases process costs; electrolyte remaining on the part may lead to corrosion. It is advisable to follow design recommendations in publications issued by the various specialist organizations [5].

Electroless metal deposition

Nickel is the metal of choice for corrosion protection using electroless plating. Other depositable metals are copper, silver, and tin. Layers are deposited with the help of chemical reduction partners from aqueous metal salt solutions, hence the alternative designation "chemical nickel" [6]. One advantage is the generation of true contours, in contrast to galvanic coatings. The process permits coating of inside surfaces. Steel and aluminum can be plated; graphite, ceramics, and plastics can be plated if a metal coating is first applied to promote a chemical reaction. Outstanding corrosion resistance is provided by passive protection: that is, on steel and aluminum, the plating cannot have any pores or cracks exposing base metal. More extreme corrosive conditions require plating thicknesses of at least 25 μm on steel and more on aluminum. Stress crack corrosion sensitivity of high-strength steels can be reduced by electroless nickel plating or in some cases eliminated entirely.

Another positive property is the tribological behavior of electroless nickel plating. By embedding PTFE, SiC, or diamond particles, it may be applied as a dispersion coating for many applications. Depending on base material, heat treatment, and plating thickness, electroless nickel plating may reduce fatigue strength by as much as 60%.

Anodizing

Anodizing is the process of converting a metal surface layer into an oxide coating by means of anodic oxidation. For aluminum, this is usually done in sulfuric acid or chromic acid solutions. Components with overlapping sections from which the electrolyte can not be removed cannot be anodized using sulfuric acid; materials with more than 7.5% alloying elements can not be anodized using chromic acid processes. Anodized layers exhibit outstanding corrosion behavior after sealing and can be colored. Anodized layers can be used as a pretreatment prior to painting, but for adhesion purposes they cannot be sealed.

"Hard anodizing," using cooled electrolytes with specific compositions, permits layer thicknesses of 50–100 μm. These can provide electrical and thermal insulation and can briefly withstand temperatures of several thousand degrees with good corrosion and wear behavior. Such layers are hard and brittle and, when subjected to mechanical stresses, form cracks. Fatigue strength is greatly reduced by crack propagation into the base material.

Painting

Paints are protective layers based on organic substances, usually polymers. Various additives such as zinc chromate, lead oxide, and other toxic pigments are no longer permitted or are associated with extensive occupational protection requirements. These active corrosion protection pigments were capable of covering questionably pretreated surfaces. Independent of substrate, paint chemistry, and application method, materials-appropriate surface preparation and pretreatment are prerequisites for corrosion-resistant painted coatings. At this point it should be noted that when subjected to extremely corrosive conditions, paint does not provide any corrosion protection for magnesium components. Another point that should be considered is the settling behavior of organic coatings. Preload forces may be compromised (creep, especially when subjected to thermal loads). Other coating processes, which will not be discussed in detail here, include:

— Thermal spray coating (flame spraying, arc spraying, detonation coating, and plasma spraying)
— PVD (physical vapor deposition) and CVD (chemical vapor deposition) coatings
— Hot dip coatings (Al, Zn, Sn, Pb, and their alloys)
— Cladding (rolled cladding, bi-metal)
— Diffusion coatings

References

[1] AGG, Arbeitsgemeinschaft der Deutschen Galvanotechnik, Düsseldorf.

[2] Boss, M., Singen. "Wasserstoffversprödung: Verstehen der Ursachen ermöglicht Gegenmaßnahmen." *Metalloberfläche* 47, 1993, pp. 24–27.

[3] Hoch, H. "Tampon-Galvanisieren." *Handbuch der Galvanotechnik*, Vol. 11, Section 20.1., Carl Hanser Verlag, Munich, 1966.

[4] Koeppen, H.-J., and G. Laudien. "Bewertung von Oberflächenschutzsystemen für Schrauben." *JOT*, 1998/9, pp. 74–81.

[5] Simon, H., and M. Thoma. *Angewandte Oberflächentechnik für metallische Werkstoffe*. Carl Hanser Verlag, Munich, 1985.

[6] "Verfahren mit Zukunft: Chemisch vernickeln." *Oberfläche+ JOT*, Vol. 10, 1982.

9.2.4 Advances in joining technology

9.2.4.1 Welding and brazing

The following section will examine selected developments in the field of joining technology as applied to sheet metal for body applications (Section 9.2.3.1.1).

Resistance spot welding

In the field of steel materials, resistance spot welding was and remains the dominant joining method. In recent years, so-called medium-frequency welding (1,000 Hz) has made a strong entry into the field of body shell construction. This technology permits realization of transformer welding tongs suitable for robot applications, with greatly reduced weight and smaller form factor compared to conventional 50 or 60 Hz AC welding tongs. Furthermore, a more favorable ratio of effective current to peak current has greatly reduced the tendency to sparking in spot welding. Improved control systems and intelligent controls permit secure welding even of high-strength and coated sheetmetal. Weld bonding, in which an adhesive is additionally applied to the joining plane, can result in further improvement of such joints in terms of load-carrying ability as well as fatigue resistance.

Laser welding

Laser welding, first used in vehicle construction in conjunction with tailored blanks (see Section 6.1.4), is increasingly being used for body assembly operations. Laser welding has a positive effect, especially on high-strength steel sheet metal, in that entire surfaces are effectively welded together. At present, both CO_2 and solid state (Nd-YAG) lasers are in use. From a design standpoint, laser-welded joints permit reduction of flange width. Also, joints with single-sided access (e.g., in the roof area) can be welded with high confidence. Improved laser performance and improved beam quality also permit use of welding filler rod. This permits realization of different joint geometries, including, for example, fillet welds, while simultaneously improving gap filling ability.

Brazing

As variations of conventional brazing, MIG, plasma, and laser brazing have gained entry to the automobile body assembly field. The major difference between welding and brazing is that the latter uses a bronze filler rod which does not require melting of the base metals to create a joint. Nevertheless, high joint strength can be achieved, in some cases exceeding base material strength. With its lower temperature levels, brazing results in less effect on zinc coatings and, in some cases, reduced hardening in the heat-affected zone. Moreover, gap filling ability with the used filler material is better and heat distortion less. Still, in view of the different electrochemical potentials of the materials used, especially in the case of galvanized sheet metal, adequate corrosion protection must be provided in the joint area—for example, by means of flawless application of cataphoretic dip primer.

High-frequency welding

High-frequency welding of tailored blanks is currently being introduced. In this process, precut flat sheet metal components are clamped together, separated by a defined gap. High-frequency current is fed to the stampings directly by means of clamping rails. Because of electrodynamic effects, welding current is concentrated at the edges of the individual stampings. After the necessary temperature is reached, current is cut off and the heated joint edges are pressed together. Extremely short weld times are possible (1 s/m). Requirements for seam preparation are less demanding than for laser welding, and nonlinear seams are easily realized. A variation on this process envisions welding of preformed components—

that is, contoured, three-dimensional seams. In this way, it is conceivable that, for example, identical front stampings could be joined to different rear variations to form automobile side panels. One disadvantage is that the upsetting process results in a raised weld bead which must be removed [1].

Reference

[1] Schmidt, M. "Hochfrequenzschweißen in der Automobilindustrie." Industriekolloquium "Fertigen in Feinblech," Universität Clausthal, 12/98.

9.2.4.2 Mechanical joining methods

Increasingly, pierce riveting and clinching offer alternatives to resistance spot welding. Joining by means of a cold forming process is a general description of manufacturing methods in which parts to be joined and/or supplementary parts are locally, in some cases completely, reformed [1]. These mechanical methods permit joining formed sheet metal parts, both coated and uncoated. Another advantage is that mechanical joining methods are not limited to steel applications. Moreover, such methods permit joining of metallurgically incompatible materials.

Pierce riveting

In pierce riveting, parts are joined by rivet elements without the aid of preformed holes (Fig. 9.2-16). Drilling or punching of rivet holes, a necessary step in conventional riveting, is replaced by an appropriate cutting process performed by the rivet itself.

Pierce riveting with semihollow rivet

Pierce riveting with semihollow rivets produces joints in a single, continuous punching and forming process (Fig. 9.2-16). The sheet metal parts to be joined are laid on a die. The rivet setting unit is placed on the work and commences its stroke, fixing the parts to be joined. As the stroke continues, a rivet is fed to the work.

In the subsequent joining process, the pierce rivet penetrates the upper, punch-side sheet metal component and plastically deforms the lower, die-side sheet metal component as it simultaneously plastically spreads the lower part open to form a locking head. The shape of the closed head is largely determined by the contour sunk into the die.

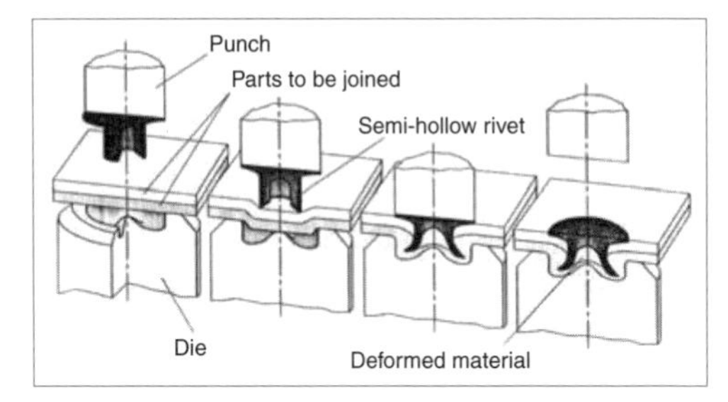

Fig. 9.2-16 Semihollow pierce riveting process.

Plastic deformation of the materials being riveted forms a collar, which in turn gives the rivet element its locking head. Material punched out of the upper sheet-metal layer fills the hollow rivet shaft and is permanently locked in place [2].

Achieving large spread of the rivet shaft is an important geometric parameter. This has a significant influence on transmitted shear and tension forces. Upsetting the pierce rivet produces gap-free mechanically locked joints. Pierce rivets are stressed both radially and axially, applying clamping forces to the joint.

Pierce riveting with solid rivet

In pierce riveting using solid rivets, the rivet element serves as a single-use punch. However, it is not deformed in the process. Solid pierce rivet materials must be harder than the parts to be joined. Materials used in practice are steel, copper, aluminum, and stainless steel, with various coatings.

Clinching

Clinching is a joining process using material deformation without any additional joining elements [3, 4]. Clinching processes may be characterized (Fig. 9.2-17):

- According to joining element configuration: clinching with and without a cutting component
- According to tooling kinematics: single and multistage clinching

Clinching with cutting component

In clinching with a cutting component, the joining element is formed by local effects of a combined shear cutting and clinching process, plus a cold upsetting process. Metal displaced out of the metal plane is upset; expansion forms a combined clamped/interference joint. Clinching and cutting limit the joining process. Depending on the nature and arrangement of cutting or of clinching and upsetting, processes may be classified as single- or multistage clinching systems [2]:

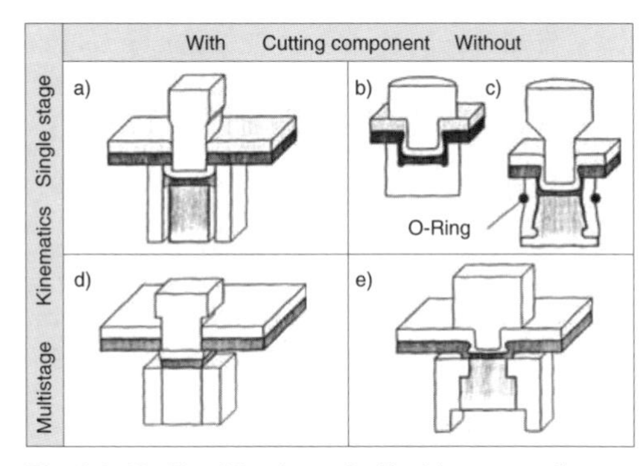

Fig. 9.2-17 Classification of clinching according to kinematic and joining element configuration.

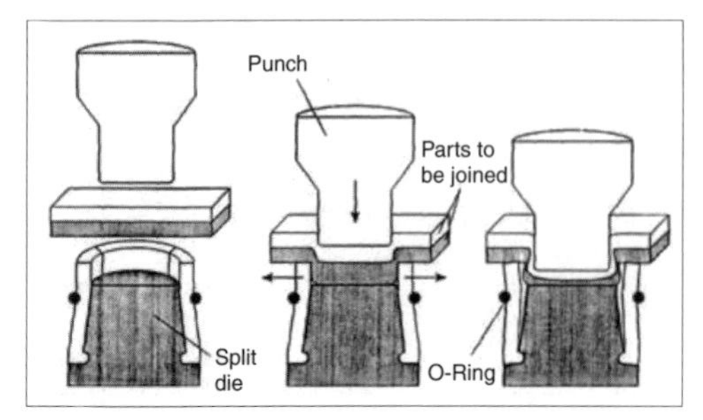

Fig. 9.2-18 Operation sequence, single-stage clinching without cutting component.

— Singlestage clinching with cutting component: single-stage clinching is defined as a process in which the joint is established by an uninterrupted stroke of a single tooling component. This process variation may be performed by single-acting joining presses.
— Multistage clinching with cutting component: multi-stage clinching is defined as a process in which the joint is established by motion of a series of tooling components. The parts to be joined are partially cut and upset at the joint by multipart tools.

Clinching without a cutting component

In recent years, other joining elements have been developed on the basis of clinching with a cutting component. A characteristic feature of these joining elements is that joint strength is increased not only by an enlarged shear area, but also by reducing the cutting component in favor of the upsetting component.

In clinching without a cutting component, a combined sinking and clinching process (sinking limits the joint area) as well as a cold forming process (the clinched material volume is upset) produces a combination clamped/interference joint through cold flow. For single- or multistage clinching without a cutting component, appropriate tooling systems with and without moving dies have been developed [2]:

— Single-stage clinching without cutting component (Fig. 9.2-18):
In single-stage clinching without cutting component, the basic idea is to achieve different cold flow behavior between the die-side and punch-side parts. Only then is a verifiable undercut and therefore a clamped/interference joint possible.
— Multistage clinching without cutting component:
This is characterized by temporal separation between the punching and upsetting phases. As a result, the pressure required to establish a joining element is about 20% lower than in single-stage clinching.

References

[1] Budde, L., and R. Pilgrim. *Stanznieten und Durchsetzfügen*. Verlag Moderne Industrie, 1995.
[2] Budde, L., M. Bold, and O. Hahn. "Grundsatzuntersuchungen zum Festigkeitsverhalten von Durchsetzfügeverbindungen aus Stahl." FAT Schriftenreihe No. 89, Frankfurt, 1991.
[3] Hahn, O. "Fügen durch Umformen." Studiengesellschaft Stahlanwendung, Dokumentation 707, 1996.
[4] Hahn, O. "Neue Lösungen mit Stahl beim Automobil-Leichtbau." Studiengesellschaft Stahlanwendung, 1997.
[5] Hahn, O., and D. Gieske. "Ermittlung fertigungstechnischer und konstruktiver Einflüsse auf die ertragbaren Schnittkräfte an Durchsetzfügeelementen." FAT Schriftenreihe No. 116, Frankfurt, 1995.
[6] Hahn, O. "Untersuchungen zur Prozesssicherheit von selbstlochenden/-stanzenden Nietverfahren beim Fügen von oberflächenveredelten Feinblechen." Forschungsbericht des LWF Paderborn, 1995.

9.2.4.3 Bonding

Bonding is gaining importance as a joining method, both as an independent method and especially in combination with welding and mechanical joining methods [1, 2].

At present, the main point of application for this technology within the auto industry is clearly in the body area. However, bonding technology can be applied in all areas of vehicle construction to solve a variety of problems and demands [1, 3]:

— Joining dissimilar materials
— Sound and vibration damping
— Gas and fluid sealing
— Increasing component stiffness
— Corrosion protection (crevice corrosion, anodic corrosion)
— Low-distortion joining through elimination of thermal loads
— Force distribution across surfaces

A multitude of different requirements has produced a corresponding number of available adhesive systems (Fig. 9.2-19).

For automotive applications in the field of structural bonding, heat-setting single component epoxy adhesives predominate. Their advantages are primarily good mechanical properties and easily automated bonding process, as well as favorable working characteristics.

If sealing, noise, or vibration damping measures must be considered, PUR and PVC adhesives have proven effective.

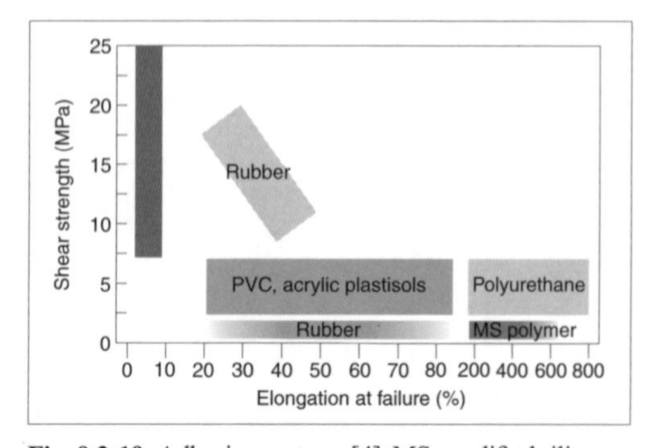

Fig. 9.2-19 Adhesive systems [4]. MS, modified silicone.

The surface condition of the parts to be joined plays a deciding role in all adhesive bonds. For metallic materials, major concerns are contamination caused by forming (drawing oils, etc.) and transportation as well as surface corrosion products.

On the other hand, for polymer materials, molecular structure and the related surface polarity are of decisive importance. In many cases, cleaning (degreasing) or, in the case of plastics, pretreatment (flame treatment, plasma treatment) is required.

Depending on joint requirements, mechanical properties may be further improved by additional chemical (bonding primer) or mechanical (roughing, media blasting) processes [5].

Adhesive-appropriate design has a positive effect on bonded joints. In this regard, mastery of shear and tension loads is most easily assured; peeling loads are the most unfavorable [6]. Several design suggestions are shown in Fig. 9.20, a and b.

Because all adhesive systems involve polymer materials, temperature and aging effects must be considered in the design of bonded joints in addition to design influences and surface conditions [6, 7].

For these reasons (load distribution, temperature, aging), a combination of adhesive bonding technology with familiar thermal or mechanical joining methods is an obvious solution.

Weld bonding, for example, can greatly increase torsional stiffness while reducing the number of spot welds; the same applies for clinch bonding [4].

Spot weld bonding [2] as well as combined adhesive bonding/mechanical joining have proven themselves in many production applications.

Independently of whether bonding technology is used as a stand-alone joining method or in combination with other methods, consistent, reproducible joint quality requires the following:

- Bond-appropriate flange layout and gap dimensions
- Defined adhesive quantity and composition by means of:
  - Robotic application systems
  - Continuous process monitoring (e.g., photographic online monitoring)

Taking these aspects into consideration, many automobile manufacturers have successfully integrated structural bonding in existing manufacturing lines. Other developments mainly seek to achieve better—that is, application-specific—understanding of aging processes and further optimization of automation and process control. In view of future disposal of salvage vehicles, the topic of recycling—that is, separation of bonded joints—must be taken into consideration as part of research and development tasks.

References

[1] Bischoff, J. "Fügetechniken im Vergleich." *Adhäsion, Kleben & Dichten*, Volume 38, 4/94, pp. 10–16.

[2] Jost, R. "Punktschweißkleben in der Serienfertigung." 6. Paderborner Symposium Fügetechnik, Druckerei Reike GmbH, Paderborn, 1998, pp. 106–112.

[3] Hennemann, O. D. "Fügetechniken—Basis für den modernen Leichtbau." VDI-Bericht 1235, "Neue Werkstoffe im Automobilbau," VDI-Verlag GmbH, Düsseldorf, 1995, pp. 71–79.

[4] Basner, G. "Untersuchung des Einflusses von unterschiedlichen Verzinkungsverfahren auf das Tragverhalten von Klebverbindungen aus verzinkten Stahlblechen." *Strukturelles Kleben*, Laboratorium für Werkstoff- und Fügetechnik der Universität-Gesamthochschule-Paderborn (LWF), 1985, pp. 115–139.

[5] Kötting, G. "Klebetechnik fördert Leichtbau." *Automobil-Produktion*, June 1998, pp. 108–110.

[6] Wuich, W. "Metallklebverbindungen mit Kunststoffklebern." *Kunststoffberater* 7/8, 1984, pp. 27–30.

[7] Habenicht, G. *Kleben–Grundlagen, Technologie, Anwendungen.* Springer Verlag, Berlin, Heidelberg, New York, Tokyo, 1986, pp. 308–313.

[8] Kleinert, H. "Klebtechnik." *SCHWEISSEN & SCHNEIDEN* 49, Vol. 6, 1997, pp. 386–391.

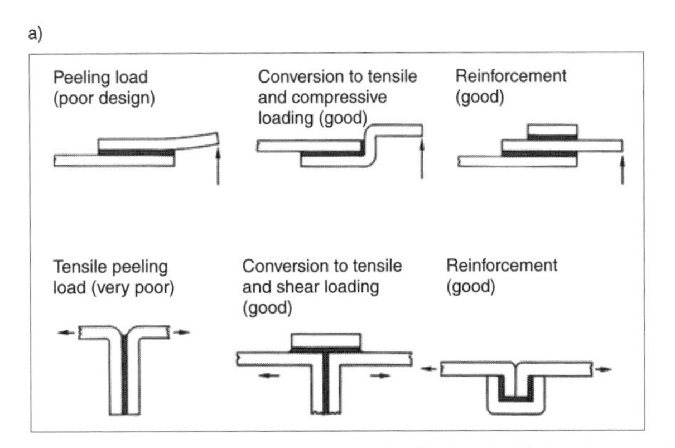

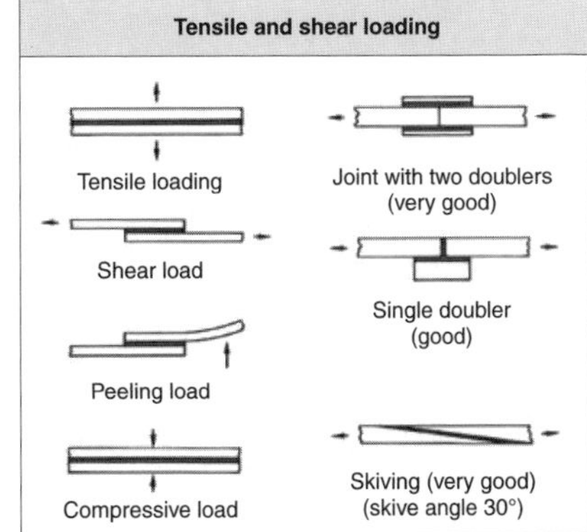

Fig. 9.2-20 a) and b) Design examples for bonded joints (source: Loctite).

9.2.5 Advances in metal forming technology

9.2.5.1 Internal hydroforming of metals

The internal hydroforming process is based on expansion of a body until it conforms to tooling contours in response to the effects of forces in a fluid medium which is in direct contact with the workpiece. Forces exerted by the fluid medium may be supplemented by other forces—for example, axial and transverse forces—which have a supporting function in the forming process.

Internal hydroforming may be classified according to the four different configurations shown in Fig. 9.2-21 [1]:

— Reforming: also termed "calibration" in the literature. A preformed workpiece is made to conform to the forming tooling solely by means of applied internal pressure.
— Expansion: material walls are stretched, thereby expanding the surface. Forming is accomplished solely by means of internal fluid pressure. The degree of expansion that can be achieved depends on ductility of the chosen material in the direction of expansion.
— Expansion swaging: axial movement supplies additional material to the deformation zone. This permits forming of geometrically larger elements compared to simple expansion operations. Location and number of such formed elements are limited in that axial forces cannot be transferred over longer distances, extreme bends, or Section changes.
— Joggling: shifting the axis of a hollow body over part of its length by means of a laterally acting tooling element. By means of an additional transverse force, workpieces may be simultaneously bent and expanded in a single internal hydroforming process.

Engineering advantages of internal hydroforming [2]:

— Complex geometry achievable in a single component.
— Elimination of weld seams for joining.
— High component shape and dimensional accuracy.
— Increased component strength and stiffness by cold forming of material.
— The same tooling can be used for tubes of different wall thickness and different materials.

Engineering disadvantages of internal hydroforming:

— Complex, long components are formed by quenching; therefore only limited circumferential differences are possible.
— Complex prototype tooling.
— Relatively long cycle times.
— Complex facility technology.

In auto body applications, primary IHF advantages are optimum space utilization by elimination of weld flanges and therefore weight savings combined with high component stiffness. Single-piece, continuous longitudinal members or roll bars are impressive examples of advantages offered by this technology.

Hydroforming technology has also proven itself in suspension components, specifically in an IHF-formed rear suspension design.

The classic drivetrain application for IHF is exhaust manifolds in any imaginable shape and size. Complex manifold shapes have been the primary motivation for choice of IHF technology.

Hydromechanical forming

In contrast to conventional mechanical deep drawing, in hydromechanical deep drawing the workpiece is formed by fluid pressure in a drawing chamber. In the process, the deep-drawn component is subjected to triaxial stresses; this has a highly beneficial effect on drawing behavior. In comparison to conventional deep drawing, improvements in drawing depth, surface quality, and less severe wall thickness reduction in critical areas can be realized. A further development of hydromechanical forming is the "Active Hydro-Mec" process developed by SMG Engineering GmbH. Engine hoods, roofs, or door skins made by conventional methods exhibit only limited dent resistance, caused by the low level of deformation and therefore lack of work hardening in these areas [3].

In the "Active Hydro-Mec" process, the sheet metal blank is advanced by controlled application of a pressure medium, which results in work hardening. Follow-

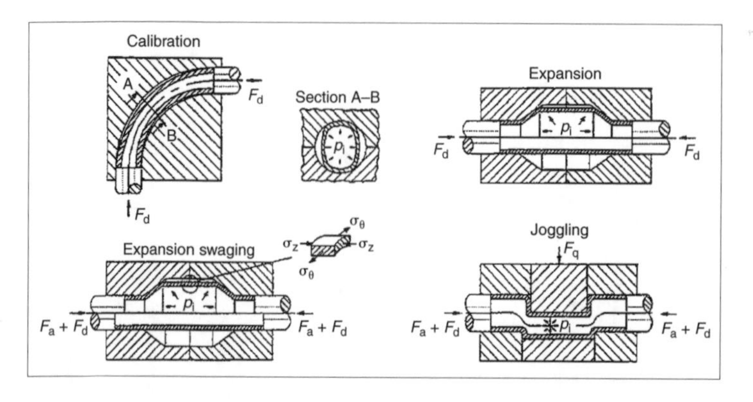

Fig. 9.2-21 Principal internal hydroforming (IHF) process variations.

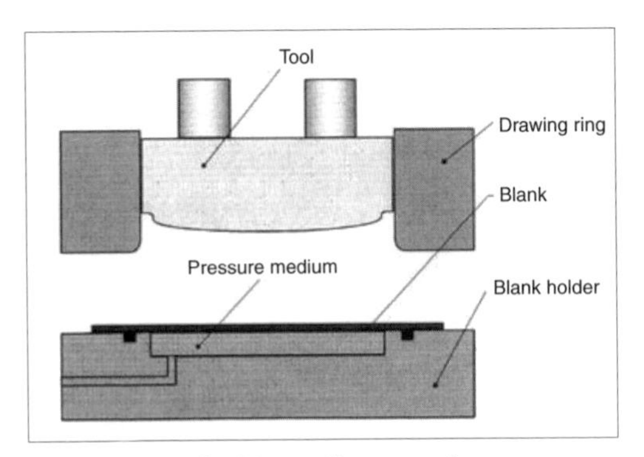

Fig. 9.2-22 Hydro-Mec tooling, opened.

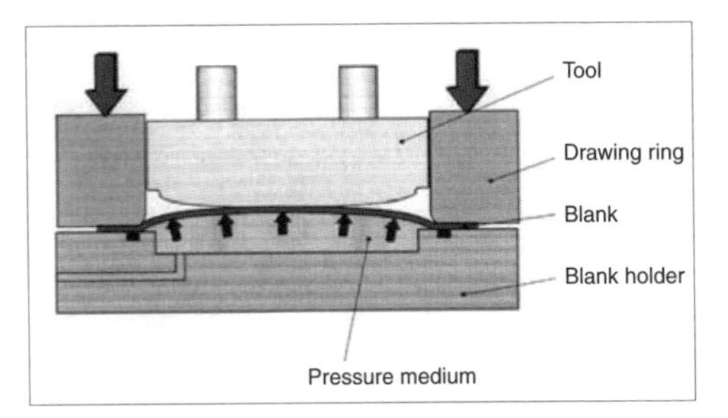

Fig. 9.2-23 Hydro-Mec tooling, closed, and blank advancement.

ing this, the shape is formed against a single-sided contour tool. Figs. 9.2-22 and 9.2-23 show the Active Hydro-Mec process [3].

References

[1] Käsmacher, H. "Innenhochdruckumformen—eine Alternative in der Fertigungstechnik." Europäische Forschungsgesellschaft für Blechverarbeitung e.V., EFB-Tagungsband No. T16, Hannover, 1996.

[2] Lichtenberg, S. "Möglichkeiten und Grenzen des Umformens von Stahlwerkstoffen mit hydraulischen Wirkmedien." Studiengesellschaft Stahlanwendung, 1996.

[3] Schuler. *Handbuch der Umformtechnik.* Springer Verlag, 1996.

Casting technology

Current efforts in casting technology have mainly concentrated on development of new processes and optimization of existing techniques with the primary goal of being able to produce heat-treatable, weldable aluminum die castings. The deciding property of such components is the lowest possible gas porosity (max. 5 cm^3/ 100 g).

As a result of these developments, three new processes have come on the market in the past few years: squeeze casting, thixocasting, and vacuum die casting.

Squeeze casting

Squeeze casting combines the economic advantages of die casting with the special demands of heat-treatable, weldable castings. The squeeze casting process is characterized by a specific speed profile for the casting piston, which displaces air from the casting chamber and feeder head. Monitored, controlled-casting piston speed, combined with optimized geometry of casting chamber and passages, results in laminar and therefore low-gas form filling (0.5–2.0 m/s section flow velocity).

An example of the squeeze casting process is shown in Fig. 9.2-24 [1].

Thixocasting

Prior to the actual thixocasting process (also known as "semisolid metal" (SSM) casting), special bar stock is manufactured and cut to form slugs.

SSM alloy bar stock is produced by a combination of extrusion and electromagnetic stirring. The resulting microstructure (dendrite formation) gives these alloys their thixotropic behavior in the semifluid condition.

The slugs are then heated to a specific temperature—for example, by means of induction coils; at this temperature, aluminum is present in the solid as well as liquid state. These softened slugs are then inserted in the die casting chamber and pressed into the mold (Fig. 9.2-25).

The heated slugs retain their external shape without any external forces and only flow like liquids when subjected to pressure.

At present, the alloy AlSi7Mg is available for thixo-

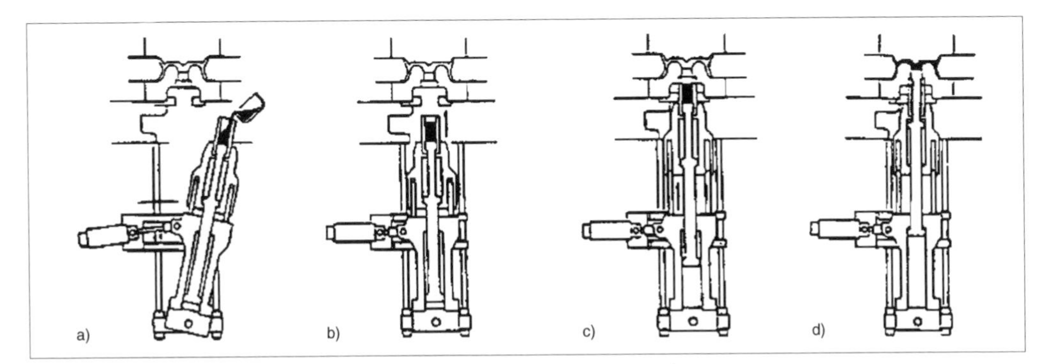

Fig. 9.2-24 Indirect squeeze casting cycle: **a)** metering, **b)** swing to vertical, **c)** dock to mold, **d)** fill mold and solidify.

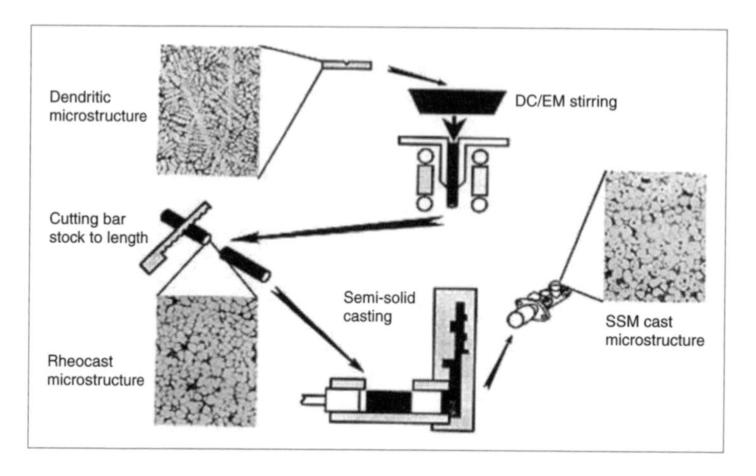

Fig. 9.2-25 Thixocasting sequence [2].

casting. Other aluminum and magnesium alloys are in development.

The resulting castings are free of gas porosity and shrinkage defects and are heat treatable and weldable. The following mechanical properties are achieved with artificially aged AlSi7Mg components:

Yield strength $R_{p0.2}$: >230 MPa
Tensile strength R_m: >290 MPa
Strain e_5: >11%

Vacuum die casting

Pressure tight, heat-treatable, weldable castings with high mechanical strength and ductility may also be produced by means of vacuum die casting.

The key feature of this process is evacuation of the die casting mold, passages, and filling chamber, reaching absolute pressures well under 100 mbar. Cycle times are only slightly longer, thereby retaining the economic advantages of die casting.

Evacuation reduces gas porosity to a level where it becomes possible to heat-treat the resulting castings. Depending on heat treatment parameters, component geometry, and cooling conditions in the mold, artificially aged AlSi10Mg castings can achieve the following mechanical properties:

Yield strength $R_{p0.2}$: >180–260 MPa
Tensile strength R_m: >230–320 MPa
Strain e_5: >6–14%

From a materials viewpoint, this process is not subject to any restrictions; currently available aluminum as well as magnesium alloys can be used.

References

[1] Kaufmann, H. "Endabmessungsnahes Gießen: Ein Vergleich von Squeeze-casting und Thixocasting." *Gießerei,* No. 11, 1994.

[2] Young, K.P. "Semi-Solid Metal Guss—eine neue Technologie von Bühler." No publication date.

Forging

Forging is understood to mean any of several manufacturing processes which DIN 8583 generally classifies as members of the group of pressure-forming processes. This group includes open die forging, drop forging, embossing, and piercing. Forging processes are accordingly classified as open die or closed die processes.

Open die forging requires simple tooling, in general not restricted to the shape of the workpiece. Closed die forging is done using workpiece-related tooling.

Forging is normally done in a temperature range in which the material undergoes recovery and recrystallization processes. This serves to increase material formability and reduce stresses. For some alloys, the forging temperature is tightly controlled to prevent undesirable phase transformations [1].

Forging steels

Both alloy and unalloyed steels are used for forging; the most commonly used types are heat-treatable and case hardenable steels. The ability to change form—and therefore forgeability—decreases as carbon content increases.

In forging, a billet is sheared off, heated to forging temperature, plastically deformed, and subsequently heat-treated. Heat-treating can be done from forging heat (thereby saving energy) or as a separate stage.

Forged surfaces are cleaned by media blasting. Functional surfaces of most closed-die forgings are subsequently machined.

The advantages of forged components compared to castings are, for example, good mechanical properties due to the resulting microstructure and less scatter in production due to lower likelihood of internal material faults.

An important area of application for forgings is highly stressed and safety relevant components—for example, connecting rods, crankshafts, axle pivots, spindles, transmission gears, and so on.

Despite these advantages, use of forged components is decreasing. Attempts are under way to delay this trend by continued development of forging processes. These include precision forging, warm forging, flashless forging, and forging in an inert gas atmosphere to prevent scale formation and decarbonization.

In addition, development continues on heat treatment processes utilizing residual heat. Heat treatment directly from forging temperatures requires alloy compositions appropriate for the chosen cooling conditions [2].

References

[1] Gräfen. *VDI Lexikon Werkstofftechnik*, 1991.
[2] Kaniut. "Werkstoffe im Automobilbau, Teil 1: Werkstofftypen." Information provided by Zentralwerkstofftechnik (Central Materials Technology), Mercedes-Benz, 1987.

9.2.5.2 Polymers

In addition to the introduction of ever more sophisticated, cost-cutting machinery, entire manufacturing systems have been re-evaluated. Continued development of standard technologies—for example, injection molding—has created entirely new manufacturing processes.

Multicomponent molding of bicolored taillight lenses and hard/soft combinations are just two examples of innovative manufacturing technologies which have become the state of the art in the automotive industry.

The application spectrum may be further expanded if one combines various materials by means of clever manufacturing technology, as exemplified by the following three processes.

One-shot technology

One-shot technology is understood to mean joining of any type of decor material (fabric, leather, foam, colored film, etc.) and simultaneous molding of the substrate in a single manufacturing step (Fig. 9.2-26, a and b).

Current applications are mainly found in vehicle interiors, such as manufacturing of interior trim pieces (A/B/C pillars, instrument panels, etc.) with high visual and haptic quality levels. This technology is still an exception rather than the rule for exterior applications.

Injection molding with in-mold decoration (IMD), injection compression molding (ICM), deposit compression molding, or in-mold lamination (IML) of structural foam or blow-molded components can be used, depending on decor material and component size. Decor material cannot be of any arbitrary configuration, but rather must be matched to particulars of the manufacturing process at hand [2].

Main advantages of this process are:

— Reduced production costs (by up to 60%) [3]
— Improved environmental compatibility by eliminating adhesives
— Simultaneously reduction in fogging and vehicle interior emissions.

One disadvantage is tooling costs up to 30% higher, which are offset by elimination of laminating equipment. In summary, it can be said that "one shot technology" is firmly established in the automobile industry and is used wherever value is placed on environmentally friendly manufacturing and emissions-optimized components and where sufficiently large production

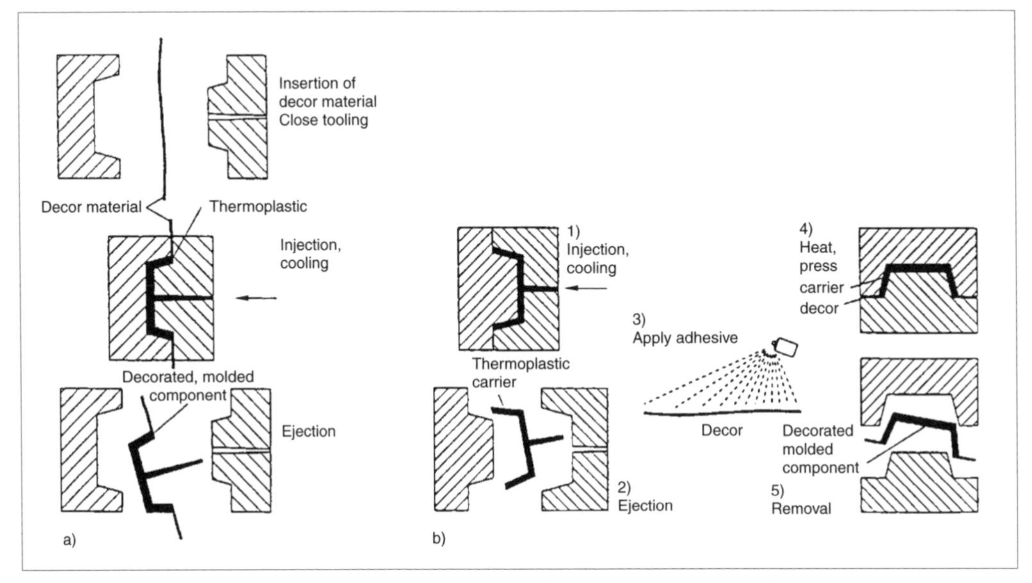

Fig. 9.2-26 a) and b) Manufacturing sequence for "one shot" technology, using in-mold decoration (IMD) as an example, compared with conventional laminating technology [1].

runs justify a somewhat more complex manufacturing process and higher tooling costs.

References

[1] Bürkle, E. "Hinterpressen und Hinterprägen—eine neue Oberflächentechnik." Presented at "Innovative Spritzgießtechnologien—ein Beitrag zur Konjunkturbelebung," Fachtagung Süddeutsches Kunststoffzentrum, Würzburg, December 1–2, 1993.

[2] Michaeli, W., and S. Galuschka. "Hinterspritztechnik, Teil 1: Eine Analyse der Randbedingungen." *Plastverarbeiter,* 44th year, No. 3, 1993, pp. 102–106.

[3] Mischke, J., and G. Bagusche. "Hinterspritzen von Textilien, Teppichen und Folien." *Kunststoffe* 81, No. 3, 1991, pp. 199–203.

Gas injection technique (GIT)

The gas injection technique is employed to generate deliberate, defined voids in injection molded components (Fig. 9.2-27). This constitutes a form of two-component injection molding in which the second component is not a polymer but rather a gas (generally nitrogen).

The void is formed with the aid of a pressurized (up to 300 bar) inert gas [2].

Initially, melt fills 50–100% of the mold (depending on geometry and process) (Fig. 9.2-28, a and b).

Gas is injected in the second stage. Gas pressure forces the melt front farther forward, completely filling the cavity, or excess material is forced back to a side chamber (overflow) or back into the screw antechamber [2].

Gas may be introduced directly into the formed part, in the sprue, or through the injection nozzle.

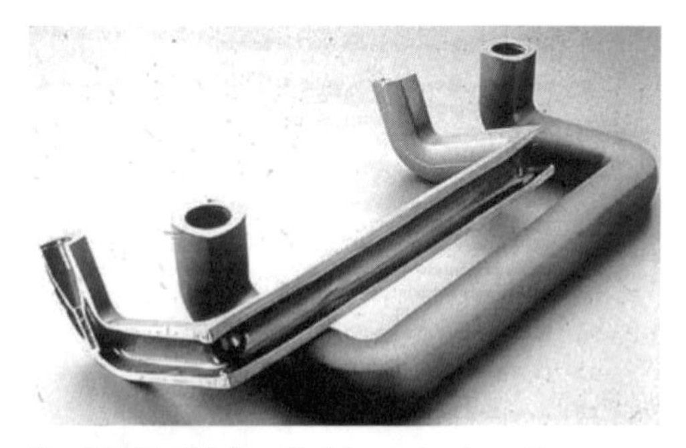

Fig. 9.2-27 Thick-walled formed polyamide component [1].

If one examines parts made by this method, three typical application groups are apparent [4]:

1. Tubular components; examples: exterior trim strips, roof grab handles, coolant distribution components
2. Flat components with reinforcing ribs; example: housing covers
3. Shell-like, thin-walled formed parts incorporating thick-walled sections; examples: outside mirror housings, center consoles, door pockets

To achieve optimum component quality, GIT-specific design rules must be applied for all three groups [2]. It should be pointed out that there are many process variations and patents associated with this technology. Before deciding on the gas injection technique, it is therefore imperative to ascertain whether licensing fees will apply.

References

[1] Theissig, W., and R. Schmid. "Anwendung des GID- (Gas Innendruck)- Verfahrens für Komponenten im Automobil." Presented at "Innovative Spritzgießtechnologien—ein Beitrag zur Konjunkturbelebung," Süddeutsches Kunststoffzentrum, Würzburg, December 1–2, 1993.

[2] Ehritt, J., and K. Schröder. *Gasinnendruck- und Zweikomponenten-Spritzgießverfahren*. Hüthig GmbH, Heidelberg, 1995.

[3] Johannaber, F. "Gas-Injektions-Technik beim Spritzgießen (GIT)." Presented at "Innovative Spritzgießtechnologien—ein Beitrag zur Konjunkturbelebung," Süddeutsches Kunststoffzentrum, Würzburg, December 1–2, 1993.

[4] "Gasinjektionstechnik—Verfahrenstechnik, Anlagentechnik, Gestaltungsregeln." Institut für Kunststoffverarbeitung in Industrie und Handwerk, RWTH Aachen, 1995.

Hybrid technology

Hybrid technology represents an interesting subset within the plastics field. Steel sheetmetal, previously used for many applications, promises stability but is relatively heavy. Plastics are characterized by light weight but often exhibit insufficient strength and stiffness [1, 2]. Hybrid technology, on the other hand, combines the advantages of different, competing materials in a single molded component. The breakthrough of this technology was marked by production application of a front-end carrier module in the Audi A6 (Fig. 9.2-29). Door carrier modules for mounting window lifts, window guides, and audio speakers are also conceivable, as are trunklid or instrument panel structures [3].

To manufacture a hybrid component, a deep drawn and perforated sheet steel component is placed in an in-

Table 9.2-9 Advantages and disadvantages of GIT [2]

Advantages	Disadvantages
+ Greater design freedom in part layout + Shorter cycle times for thick-walled molded parts + Increased stiffness without added weight + Less distortion, lower internal stresses + Fewer sink marks + Reduced clamping force	– Higher facility costs – More demanding process control – More complex quality assurance – Multiple cavities possible, but complex and sensitive

a)

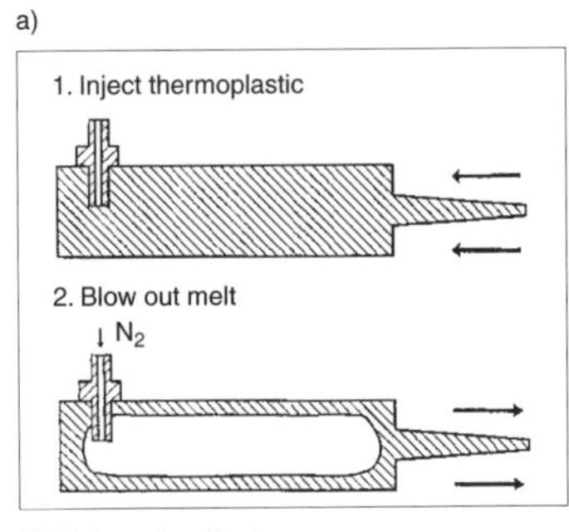

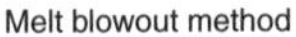
Melt blowout method

b)

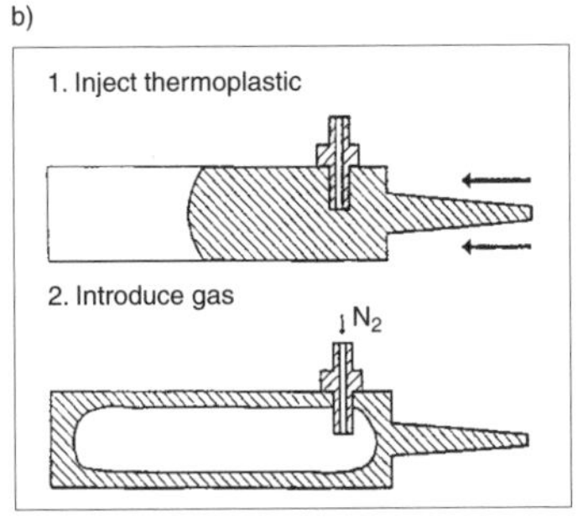

Gas introduced directly into tooling

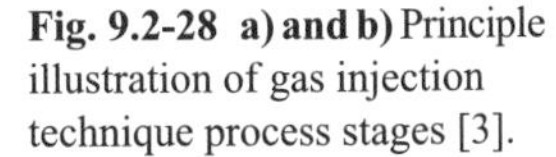
Fig. 9.2-28 a) and b) Principle illustration of gas injection technique process stages [3].

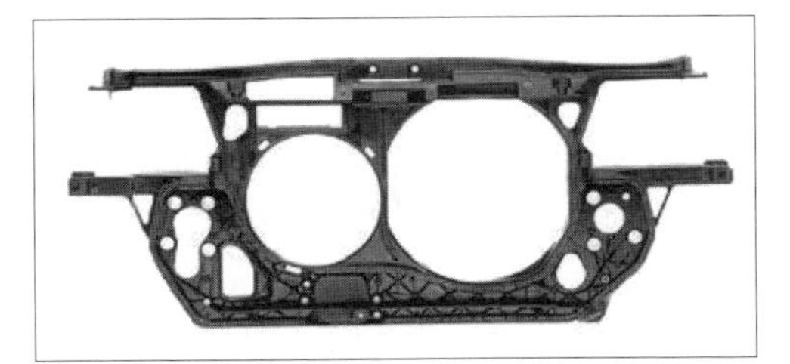

Fig. 9.2-29 Audi A6 hybrid front end carrier [4].

jection molding die and invested by a suitable plastic. The plastic melt penetrates the perforated holes in the sheet metal and upon setting forms a highly stressable, force- and form-fit composite. The metal does not require a bonding agent or any other additional treatment [5]. This new technology permits very light-gauge design for structural metal components, as any danger of deformation is eliminated. Under various load conditions (bending, compression, torsion), metal/plastic composite Sections exhibit greater strength than closed or open steel profiles (Fig. 9.2-30, a and b) [6].

Hybrid technology permits weight savings of up to 40% compared to all-metal structures of equal strength [5].

The usual disadvantages of composite components with respect to recyclability are not expected with hybrid components. Scrapped hybrid components can be quickly pulverized by a hammer mill, separated by means of screens and magnets, and subsequently recycled without undue problems.

In summary, it can be said that metal/plastic composite components represent an optimum solution wherever stability, strength, and assured function are required, and/or where low assembly weight provides added advantages. Compared to same-strength metal designs, hybrid components offer appreciable cost and weight advantages. These advantages are all the greater as more functions are added to such components.

References

[1] Goldbach, H., and J. Hoffner. "Hybridbauteile in der Serienfertigung." *Kunststoffe,* Vol. 87, 1997.

[2] "Ein Werkstoff-Traumpaar für das angestrebte Drei-Liter-Auto. Hybrid- Technologie vereinigt die Vorteile von Metall und technischen Thermoplasten." *VDI Nachrichten*, Vol. 50, Number 35, 1996.

[3] "Kunststoff-Stahlblech-Verbundbauweise." *Materialwissenschaft und Werkstofftechnik,* Vol. 29, No. 1, 1998.

[4] "Innovative Systemlösungen mit Bayer-Werkstoffen für den Audi A6." Bayer brochure, 11/98.

[5] "Stark im Verbund. Hybridtechnologie integriert Zusatzfunktionen beim Herstellprozess." *Produktion*, Landsberg, Vol. 37, No. 42, 1998.

[6] "Metallstrukturen werden durch Kunststoff stärker, leichter und wirtschaftlicher." Bayer brochure, 7/95.

9.2.6 Advances in environmental compatibility

Over the past 20 years, increasing environmental awareness in society and politics has had an impact on the automotive industry. Along with engines optimized for fuel economy, other ecological aspects such as recyclability, use of nonhazardous materials, environmentally friendly manufacturing methods, and so on, all play a major role in an overall view of any vehicle development program.

Replacement of environmentally hazardous constituents and materials

Conversion to environmentally tolerable materials without compromising service properties or unacceptably increasing manufacturing costs has presented significant challenges for engineering development programs. The list of affected components and materials ranges from brake pads to paints to adhesives and seat foam. Vehicle interiors are especially affected, as odor- and emissions-optimized materials represent a major contribution to a feeling of well-being and therefore customer satisfaction.

Innovations and advances currently under development will be examined using several concrete examples (see Table 9.2-10).

In order to meet ecological and societal expectations, application of environmentally friendly materials and substances will continue to command attention.

The automobile industry has addressed this ecopolitical challenge in the form of VDA 232-101, which lists substances subject to declaration and establishes limits for declarable substances as well as forbidden substances.

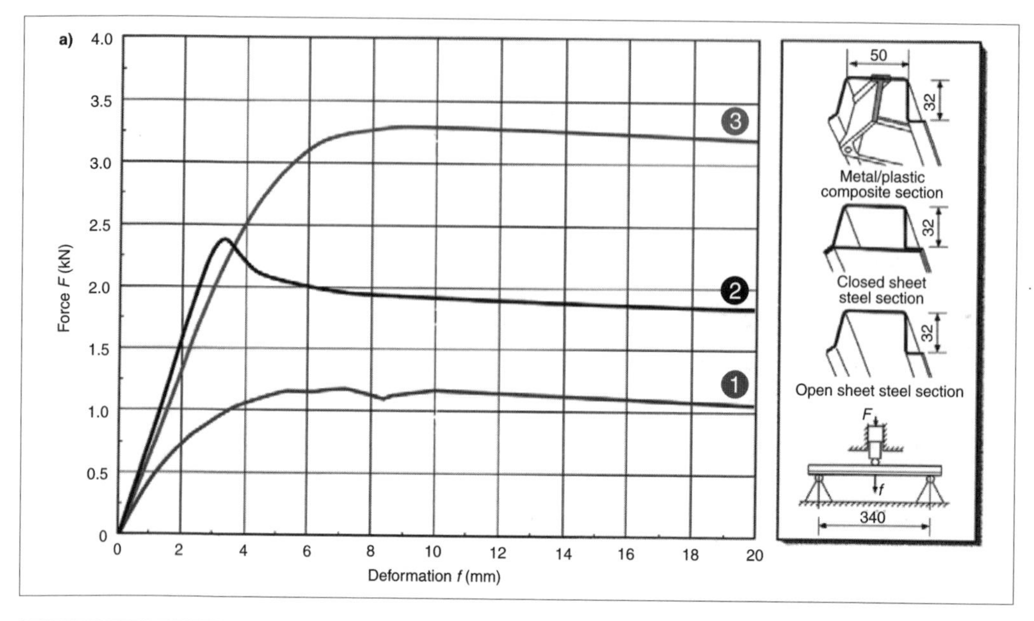

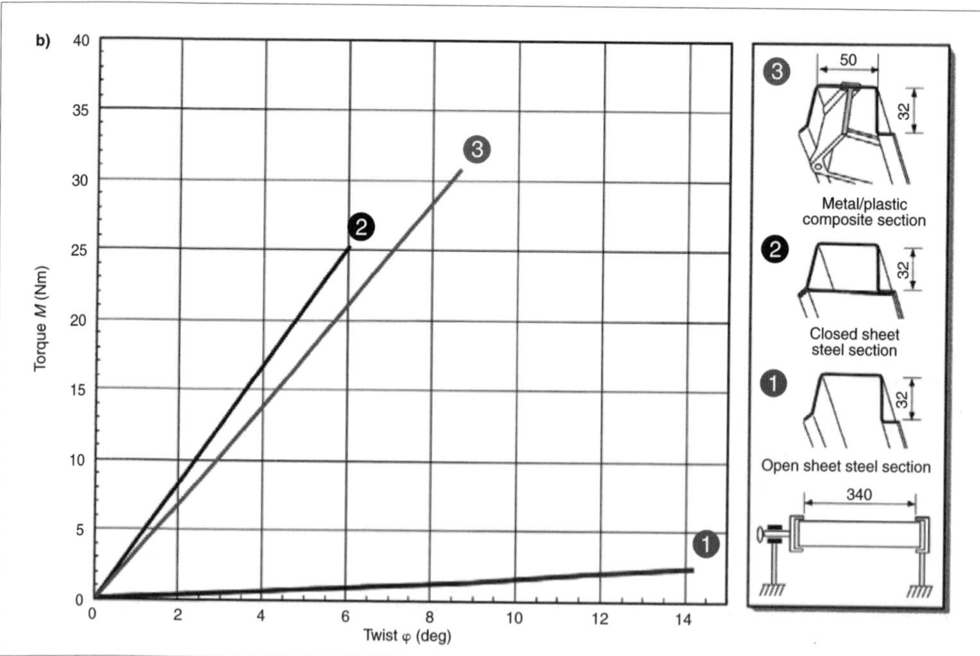

Fig. 9.2-30 a) and b) Behavior of metal/plastic composites under bending and torsional loads, compared to various steel profiles [6].

Application of renewable resources

In addition to familiar automotive materials based on renewable resources—for example, WF/PP (wood-flour–filled polypropylene) or wood-fiber–reinforced thermosetting plastics—for some time the industry has seen increased use of materials based on natural fibers in various matrices.

Renewable resources are primarily used as support components in vehicle interiors, carriers for trunk lining, or for enclosed cargo spaces of commercial vehicles.

Along with familiar applications in vehicle interiors,

Table 9.2-10 Advances in environmental compatibility

Keyword	Environmentally compatible production technology
– Asbestos	Brake and clutch materials, gaskets, etc.
– HCFC	CFC-free refrigerants for air conditioning; foaming agents
– Cadmium	Cadmium-free plastics stabilizers
– Auto body paints	Waterborne primers and fillers; water-based topcoats
– Chromium	Chromium-free leather tanning
– Solvent-based adhesives	Water-based adhesives; one-shot technology (see 9.2.5.2)
– Fogging/interior emissions	Application of TPO (thermoplastic olefin) skin/ foam core cockpit laminations; use of odor- and emissions-optimized polymers

exterior parts based on renewable resources are currently being examined. Various materials can be employed as polymer matrices (Table 9.2-11):

Fiber preparation is by means of conventional or slightly modified textile machinery [1, 2].

Current thermoplastic-based material systems usually employ hybrid fleece consisting of a mixture of natural and polypropylene fibers (mixing ratio 30/70 to 50/50).

To compensate for larger quality variations in natural fibers, mixtures of different fibers are often established. Additionally, mixtures are able to provide custom mechanical properties [3, 4] (Figs. 9.2-31 and 9.2-32).

For thermosetting plastics used as matrix materials,

Table 9.2-11 Polymer matrices

Thermoplastics	Thermosetting plastics	Biopolymers
Polypropylene	Epoxy resin systems	Latex dispersions
Polyester	Unsaturated polyester resin systems	Cellulose-based resins
	Polyurethane systems	Starch-based resins/ polymers
		Shellac

polyurethane systems have recently taken their place alongside more familiar polyester and epoxy resin systems.

Today, the primary fibers used are from the bast family: flax, sisal, and kenaf.

Of the hard fiber group, sisal is the most commonly used [5].

Depending on component geometry and the material system employed, familiar manufacturing methods such as extruding and forming may be applied.

Injection molding of natural-fiber–reinforced materials is not yet being used in production but is currently being investigated. One problem is the appreciably higher stresses imposed on fibers during manufacturing, which may cause fiber damage [6].

A new process, already being applied in production, is the polyurethane spray process. Robots spray both sides of natural fiber mats with highly reactive polyurethane. The coated material is immediately placed in a press and cured by temperature and pressure [7, 8].

Despite these disadvantages, natural fibers represent a sensible extension of the materials spectrum and may displace traditional materials in many automotive applications [9].

Improving recycling ratios

Already, about 75% by weight of scrapped vehicles (model year 1987 and older) is being recycled. This applies primarily to all metallic materials such as steel, cast iron, and ferrous and nonferrous metals, which are shredded, sorted and subsequently recycled (steel mills, remelters).

The remaining 25% by weight, the so-called "shredder light fraction" (SLF), ends up as landfill. Roughly one-third of this consists of plastics as well as glass, elastomers, wood, and dirt. In the future, the recycling ratio will be significantly increased by regulatory agencies on a Europe-wide and, therefore, also on a national scale.

One means of reaching this goal is by establishing closed material cycles, analogous to metal recycling. In the automobile industry, a generally applicable formula has proven useful in roughly determining "recycling suitability" (RS) of a component or assembly [10]:

$$RS = \frac{\text{Costs (equivalent new material + disposal)}}{\text{Costs (disassembly, logistics, processing costs)}}$$

If this metric exceeds 0.8, the component or assembly in question may be economically subjected to a closed material cycle.

Table 9.2-12 Advantages and disadvantages of natural fibers

Advantages	disadvantages
+ Low density compared to glass fibers	– Water absorbent (as much as ~15%)
+ Good acoustic properties	– Odor behavior
+ Benign failure behavior	– Contaminants (fungicides, pesticides)
+ Co_2 neutral combustion	– Quality variations
+ Residue-free thermal recycling	

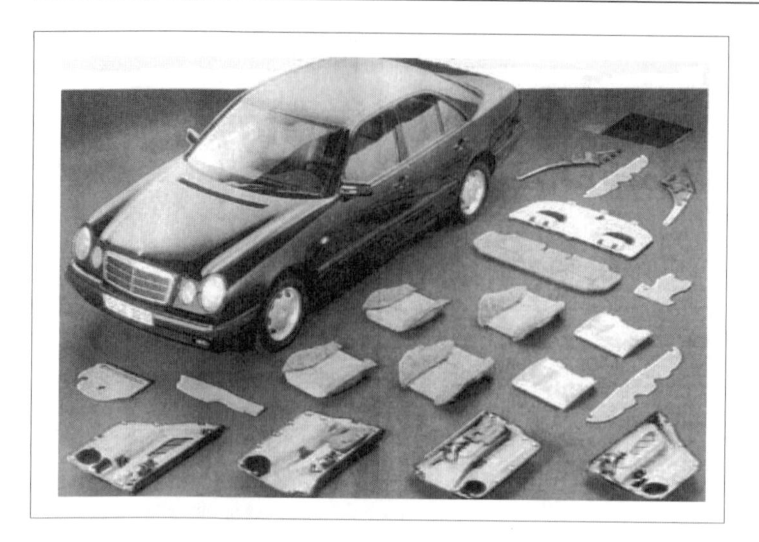

Fig. 9.2-31 Application possibilities for natural-fiber–reinforced materials using the Mercedes-Benz E-Class as an example (source: Daimler-Chrysler).

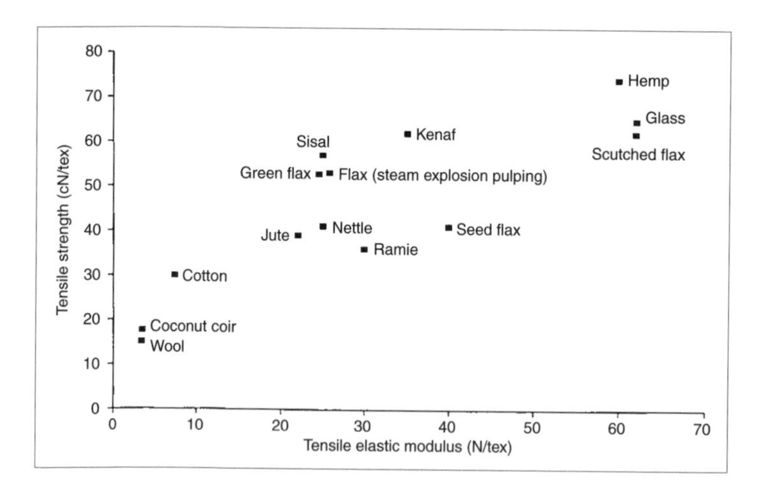

Fig. 9.2-32 Comparison of mechanical properties of various fibers (source: Reutlingen Technical College, Institute of Applied Research).

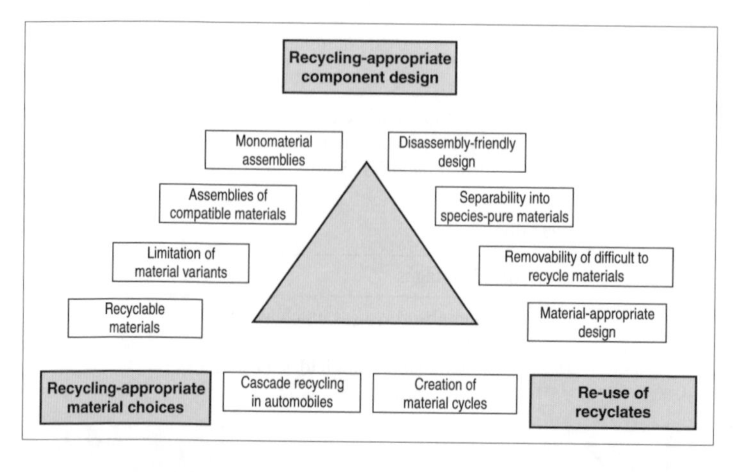

Fig. 9.2-33 Recycling triangle [10].

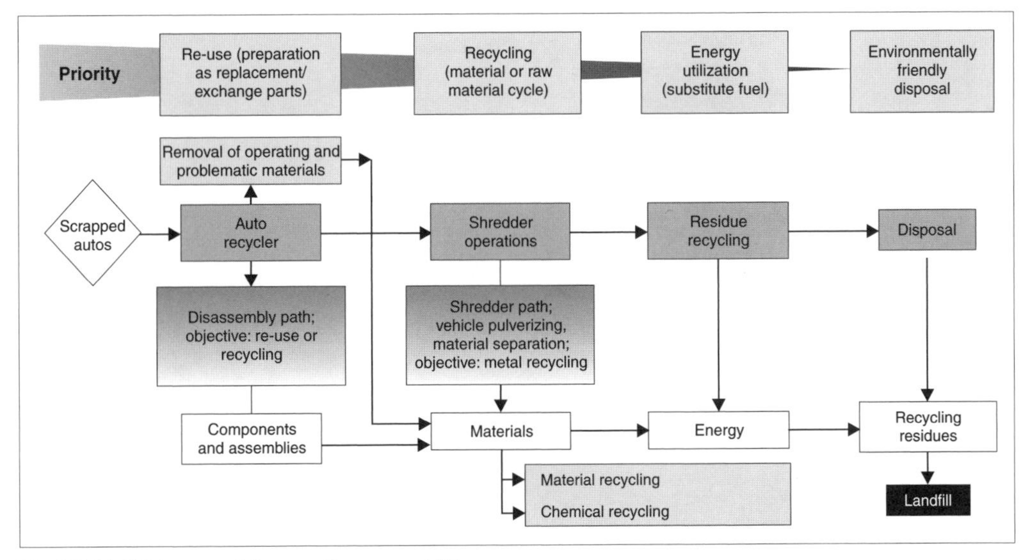

Fig. 9.2-34 VDA automobile recycling concept [8].

Elementary quantities are disassembly costs which are required to separate and sort individual materials.

All automobile manufacturers have at their disposal disassembly analyses, used in conjunction with structured disassembly and evaluation of the entire vehicle. These analyses provide important information regarding current economically achievable recycling rates of existing vehicles and indicate optimization potential for new vehicle designs. Together with internal recycling standards or with VDA Standard 2243 and the European recycling standard "Design for Recycling" (DFR), these form the foundation of environmentally responsible and economically realizable vehicle development. The main points of these standards may be represented by a so-called "recycling triangle" (Fig. 9.2-33) [10].

Measurable success is apparent by the significant rise in the portion of recycled plastics, currently about 12% of the entire plastic content of a vehicle, as well as increased approval of recyclates. Recycling is also evidenced by increasing efforts to find "single-material solutions," which have been successfully implemented, for example, on instrument panels for the Porsche Boxster and 911 Carrera models in the form of an "all-PP" variant.

Energy utilization

The high level of plastic content in SLF (shredder light fraction) automatically points toward thermal recycling. Strictly speaking, in most cases energy recycling involves material recycling; primary energy needed for combustion, in the form of oil, gas, coal, and so on can be replaced or reduced by SLF or used as reducing agents in steel manufacturing ("metallurgical recycling").

The main obstacle to these processes lies not so much in their technical mastery but rather in lack of ecopolitical acceptance, which in many cases has prevented their implementation.

A summary overview of the automobile recycling concept developed by VDA is shown in Fig. 9.2-34 [8].

Ecology budget/energy budget

In principle, when establishing an ecology budget, all material and energy streams impacting the environment must be considered, balanced against each other, weighted, and finally evaluated [7].

The available models are extremely complex and time-consuming to the point where the automotive industry has instead concentrated on the energy budget (Fig. 9.2-35), which significantly affects the ecology budget [8].

On a basic level, an energy budget is related to a CO_2 budget. Compared to all other emissions sources, this is the largest contributor to the greenhouse effect and therefore represents an appropriate instrument for evaluating different materials in connection with special components or assemblies (Fig. 9.2-36). It is conceivable that future system definitions may be expanded to entire component groups (body, drivetrain, etc.) or even entire vehicles [9].

Although many results are already available, knowledge of the most important subject, that of the "right" estimation and evaluation of effects, is still in its initial stages [9]. The necessary transparency and visualization as prerequisites for general acceptance of this method, as well as objective comparability, are not yet assured.

However, for internal comparisons, there is no compelling reason not to use such methods; (energy) budgeting is already capable of making valuable contributions toward materials and technology selection within the framework of vehicle development.

References

[1] Colberg, H., and M. Sauer. "Spritzgiessen naturfaserverstärkter Kunststoffe." *Kunststoffe* 7, 12, Carl Hanser Verlag, Munich, 1997.

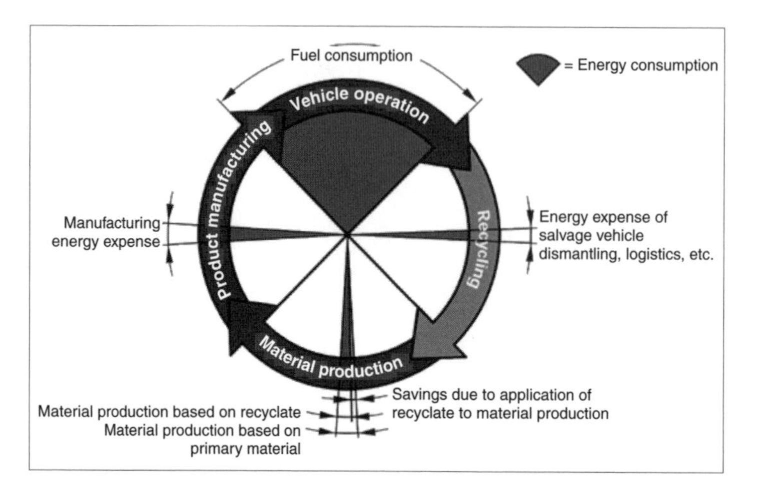

Fig. 9.2-35 Energy budget of a material cycle [8].

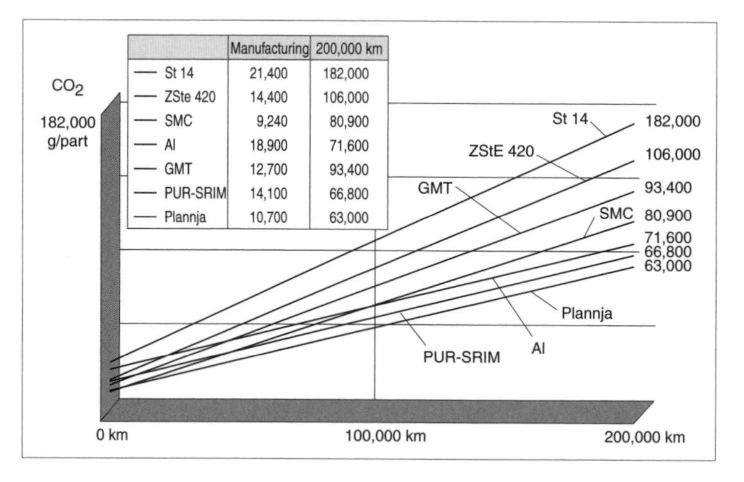

Fig. 9.2-36 CO_2 emissions from manufacture and use of various bumper carriers [9].

[2] "Starke Leichtgewichte." *Plastverarbeiter* 7/1998, Hüthig Verlag, Heidelberg.

[3] Müller, H., and K.-W. Fries. "PURe Natur im Automobil." *Kunststoffe* 88, 4, Carl Hanser Verlag, Munich, 1998.

[4] Hanselka, H. "Automobiler Leichtbau durch den Einsatz von Naturfaserverbundwerkstoffen." Presented at VDI Congress "Kunststoffe im Automobilbau," 1995.

[5] Mast, P., and G. Horsch. "Demontage und Verwertung von Kunststoffbauteilen aus Automobilen." Forschungsvereinigung Automobiltechnik e.V., FAT Schriftenreihe No. 100, Frankfurt, 1993.

[6] Mast, P., et al. "Ressourcenschonung durch Recycling von Kunststoffbauteilen aus Automobilen." *Kunststoffe im Automobilbau—Anwendung und Wiederverwendung*, VDI-K, VDI Verlag, Düsseldorf, 1991.

[7] Braess, H.-H. "Das Automobil von der Produkt- zur Systemoptimierung—Ziele und Aufgaben des Life-Cycle-Managements." *ATZ Automobiltechnische Zeitschrift* 101, 12, 1999.

[8] Schäper, S., et al. "Materialrecycling von aluminiumintensiven Altfahrzeugen am Beispiel des AUDI A8." VDI-Bericht 1235, VDI-Verlag GmbH, Düsseldorf, 1995.

[9] Saur, K., et al. "Fahrzeugdesign und Umweltrelevanz—Erfahrungen mit Life Cycle Engineering." "Werkstoffe im Automobilbau 98/99," special issue of *ATZ* and *MTZ*.

[10] Company brochure, R+S—Stanztechnik, Offenbach, 1996.

[11] Schüssler, A., "Autoinnenteile aus Naturfaservliesen." *Kunststoffe* 88, 7, Carl Hanser Verlag, Munich, 1998.

[12] Fölster, T., and T.P. Schlösser. "Einsatz nachwachsender Rohstoffe im PKW-Innenraum." Presented at VDI-Kongress "Kunststoffe im Automobilbau," Düsseldorf, 1994.

[13] Flemming, M., S. Roth, and G. Ziegmann. *Faserverbundbauweisen, Fasern und Matrices*. Springer Verlag, Munich, 1997.

[14] Classification of fibers according to DIN 60001 T1.

9.3 Where do we go from here?

Today, the automobile industry faces far-reaching changes, as it has so many times in the past. Wide-ranging cost discussions, unleashed only a few years ago in response to the Japanese export offensive, have since been supplemented by two additional trends. The first of these is increasing individualization of customer requirements; the second is even greater consideration of environmental compatibility. Other immediate developments are outgrowths of these two major trends.

More individuality means more model variations and/or smaller production runs of individual models. In

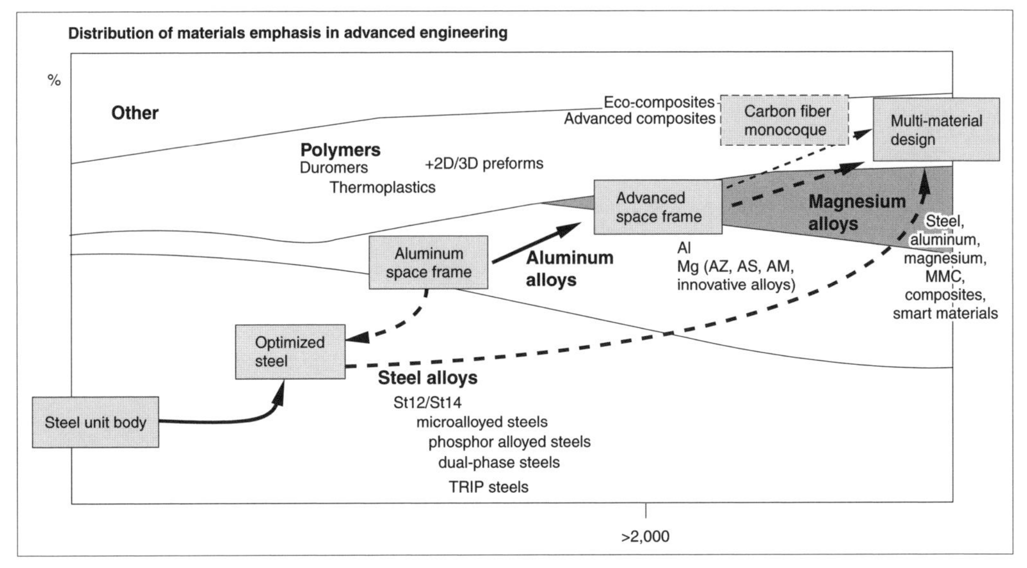

Fig. 9.3-1 Possible evolution of motor vehicle materials and construction methods [2].

the future, to assure cost-effective production of such variations, it will be necessary to implement even more pronounced modularization of individual assemblies. In terms of materials and manufacturing methods, this implies concepts designed around low production runs and specific requirements, synonymous with the growing trend for composite construction and highly flexible, cost-effective manufacturing methods [1].

The international nature of the automobile industry and production that is distributed among multiple manufacturing sites demand flexible, site-independent materials and manufacturing concepts.

Acceptance of new materials will therefore depend in part on their unrestricted availability, on highly developed and process-assured manufacturing technology, as well as on freely selectable manufacturing locations.

In the future, the field of environmental compatibility will be strongly influenced by the demand for significant CO_2 reduction; this will be closely associated with lightweight design concepts for body, drivetrain, and suspension, as well as new propulsion concepts.

A possible scenario for future emphasis on various materials technologies is shown in Fig. 9.3-1 [2].

The trend toward composite construction, as evidenced by modularization, is appreciably reinforced by demands for lightweight construction and CO_2 reduction. However, difficulties associated with such scenarios are soon apparent.

Our currently customary, heat-intensive joining methods fail wherever different materials with different properties and behavior are combined. Development of new, more suitable joining methods therefore takes on elementary importance in meeting the goals outlined earlier.

This also applies to simulation methods. Reliable predictions of crash behavior as well as component life expectancy and manufacturing-friendly component layout are only possible through development of appropriate models. These are basic prerequisites for rapid, cost effective, and efficient vehicle development and production.

Another difficulty associated with composite construction could be economical recyclability of such structures. The drive toward closed material cycles encompassing eventual scrap vehicle disposal is therefore just as important as development of new recycling technologies at the very beginning of new vehicle development.

The vision of responsible designers will tend not to concentrate on individual assemblies and components but rather will focus to an increasing extent on an overall consideration of the entire automobile.

Competition between individual materials groups as well as within a materials family will increase but also lead to currently unimaginable combinations and "alliances." Successful interdisciplinary cooperation between materials scientists, engineers, manufacturing experts, and designers will in future provide the key to success. Simultaneously, this competition will increase opportunities for recognizing potential lightweight design and cost benefits more quickly and, more important, for implementing these, thereby providing lasting assurance of the future of automotive mobility.

References

[1] "Innovativer Werkstoffeinsatz in Kraftfahrzeugen—Entwicklungstendenzen, Chancen und Ansätze in NRW." Institut für Kraftfahrwesen, Institut für Bildsame Formgebung, Institut für Kunststoffverarbeitung der RWTH Aachen, January 1997.

[2] Friedrich, H.E. "Leichtbautechnologie im PKW—die Konkurrenz der Werkstoffe nimmt zu." Euroforum conference "Stahl, Aluminium und Magnesium im Wettbewerb," Bonn, February 17–18, 1998.

10 The product creation process

10.1 Simultaneous engineering and project management in the product creation process

10.1.1 Introduction

An automobile manufacturer that intends to remain competitive must ensure its presence in all markets. To reach this goal, demands of different markets must be precisely analyzed, and these markets must be supplied with products meeting their respective customer expectations. To avoid rushing past actual customer desires, each development department should endeavor to reduce development time. The traditional development process is characterized by sequential development stages, resulting in long development times spanning five years or more. Simultaneous engineering offers great potential for reducing these times. "Simultaneous" in this sense means that all individual steps of preliminary development and production development are initiated as early as possible, and insofar as possible, are carried out in parallel (Fig. 10.1-1).

The following will describe the simultaneous engineering process, using the example of combustion engine development for volume production.

10.1.2 Simultaneous engineering

Along with associated reduction in development time, information exchange between various engineering and manufacturing disciplines is supported. This improvement in communications is the deciding advantage in optimizing the course of individual processes.

In principle, development of a new engine is integrated within a vehicle program, even if it only represents installing a new engine or new engine family in an existing vehicle. It is always the overall vehicle which must satisfy customer desires, regulatory requirements, and corporate philosophy.

From this, we may derive a basic principle: planning the course of a program must be given priority and carried out on the higher vehicle level. If such process planning is not available, it must first be developed as a whole. This means that program control at the vehicle level is indispensable, and so-called vehicle control points must be given priority over the engine program. These points ensure that at any time within the vehicle development process, all necessary work has been carried out and the required results or goals achieved. Corresponding specific control points are planned and worked off for drivetrain subsystems such as engine and transmission.

Fig. 10.1-2 presents an overview of the significance of vehicle control points, counting backward from production startup (A):

K Program strategic direction outlined
J Program strategic direction established
I Vehicle dimensions, weight, etc. established
H Program approved, financing underwritten
G Interior and exterior design approved
F Analytical product approval
E First vehicle available for tests
D Product approval
C Test phase completed
B Production startup approved
A Production startup

The corresponding program control points for engine, transmission, and so on are worked off simultaneously with vehicle control points (lower portion of Fig. 10.1-2).

9 New technologies (off-the-shelf solutions) identified; timing, engineering and manufacturing resources, and cost plan defined

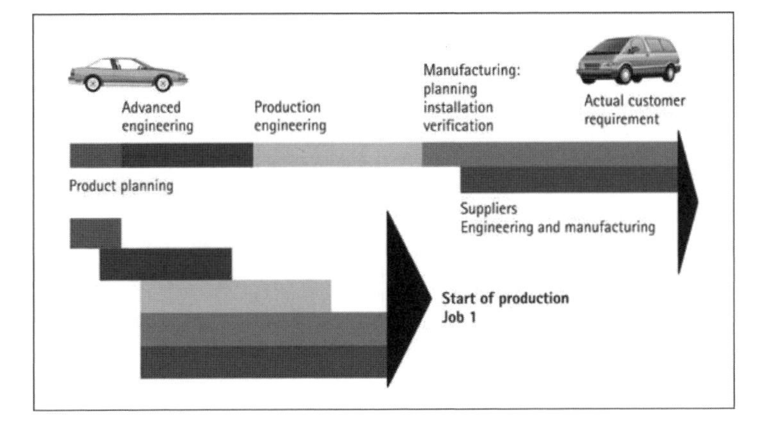

Fig. 10.1-1 Simultaneous engineering.

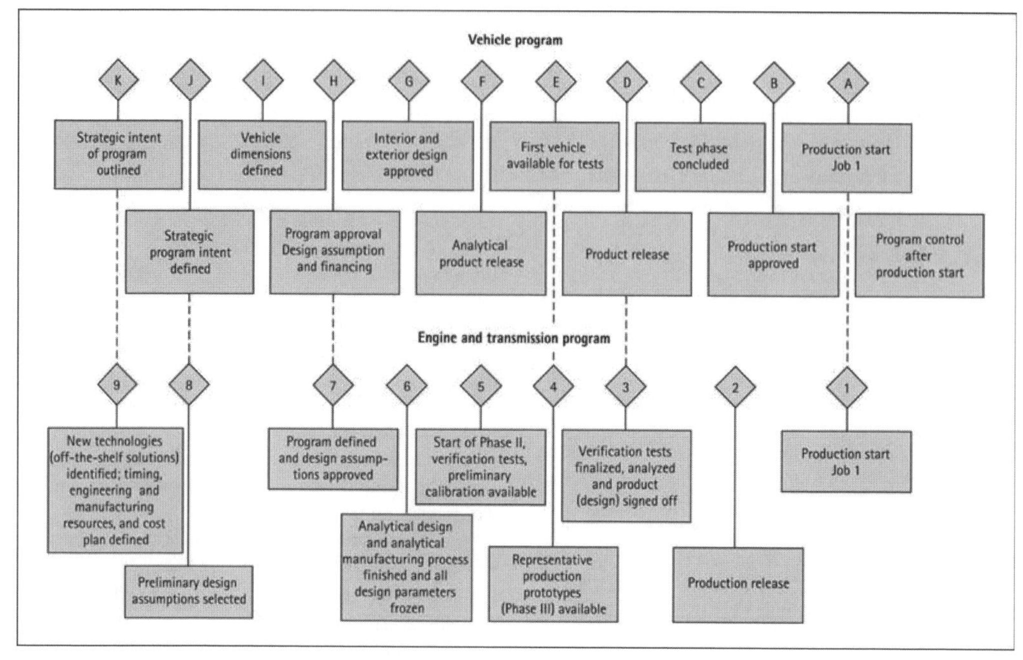

Fig. 10.1-2 Summary of main program checkpoints.

8 Preliminary design assumptions selected
7 Program defined and design assumptions approved
6 Analytic design and analytical manufacturing process finished and all design parameters frozen
5 Start of Phase II, verification tests, preliminary calibration available
4 Representative production prototypes (Phase III) available
3 Verification tests finalized, analyzed, and product (design) signed off
2 Production release
1 Production start Job 1

Vehicle program point K is simultaneous with engine program point 9, J with 8, H with 7, E with 4 and A with 1.

10.1.3 Advanced engineering

10.1.3.1 Establishing specifications and program planning

Planning takes place at the very outset of an engine program. It is characterized by a description of demands imposed on the product to be created and leads to an initial summary of goals (strategic program direction). Program planning is the product of an ongoing strategic business process. The start of a vehicle program is based on analysis of customer demands, market competition, environmental guidelines and regulatory requirements, and company-internal business planning. Interplay of these factors leads to a new vehicle program. Establishment of engine-specific goals is only achieved in connection with vehicle goals. Only when all vehicle goals combined with vehicle weight, aerodynamic drag, rolling resistance, acceleration demands, and transmission design are optimized can engine-specific goals be established. These consist of the required torque and efficiency throughout the entire engine performance map. From these, a specific engine displacement, number of cylinders, valve train concept, and so on are established. Targets are then subdivided for individual components. For example, to meet bearing friction and durability targets, the crankshaft will have an optimum diameter for main and connecting rod bearings.

Beyond this, company strategic considerations such as image, resource utilization, available existing or new production sites, costs (investment and unit costs), and returns must be considered.

Corporate expectations in the specifications include a timeline for product introduction. It should be noted that a product (engine) is introduced at exactly the point in time at which the market demands it. Resource utilization includes, among other things, knowledge of how many employees are available for a specific product development. If, for example, one needs 200 engineers and currently has only 100 available within the company, it will not be possible to realize the product in the allotted time frame.

Costs are especially decisive. The corporation needs a return in order to allocate funds for new investments. This is especially important because the cycle times for new products are becoming ever shorter: that is, each new program must also include preoptimization of the cost structure.

Project scope and the temporal sequence of vehicle planning are determined by the business plan. Consequently, it is generally the case that simultaneous production startup of several new vehicles cannot be scheduled or executed. Furthermore, existing production sites must be considered in terms of their continued utilization and capacity. For example, once a specific vehicle program is selected, the next step is establishing the vehicle segment strategy. This describes the primary reasons and targets for the vehicle program and merges into the preliminary target catalog.

Under the general headings of "function" and "quality," engine target specifications are derived from the preliminary vehicle specifications. In the past, "quality" was understood to mean reconciling development results with requirements as laid out in drawings and specifications. Today, the concept is considerably broader and is primarily defined by the customer. The customer determines exactly what is meant by "quality;" it is the customer who demands products and services to meet his needs and expectations at an appropriate price. The customer is able to experience, feel, and touch quality. An important prerequisite is customer satisfaction; this may be ascertained by surveying various customers regarding their vehicles.

For engine development, engine specifications represent important quality goals:

— Torque and horsepower
— Fuel economy
— Exhaust emissions
— Vibration and acoustics
— Weight and dimensions

These must be optimized in consideration of the following financial aspects:

— Price target
— Development costs
— Investment costs
— Unit costs
— Profitability

Project scope may be established on the basis of objectives outlined in the target specifications. For this, previously developed ideas and basic concepts are pressed into service.

Another reason for starting a new vehicle program may be tightening of exhaust emissions standards, which demand wide-ranging engine and exhaust gas aftertreatment changes.

Along with the ongoing business strategy process, program planning is supported by a continuous technology process (Fig. 10.1-3). Such advanced engineering leads to a level of refinement which assures feasibility and potential of the technology in question. Examples include variable valve timing or gasoline direct injection. From this, it may be seen that a vehicle or engine program takes advantage of certain technologies generated by research and advanced engineering. Advantages of this ongoing process include:

- Shorter development times
- Less development risk (elimination of surprises)
- High product quality

In order to process alternative development concepts, basic decisions must be made in the early planning stages. For example, it must be decided which engine concept will be developed: gasoline or diesel, number and arrangement of cylinders, as well as transmission type. An early decision is of immense importance and has far-reaching consequences; in total, the handicap incurred by a single bad decision cannot be made good by many subsequent good decisions. If, for example, in developing a "three-liter" car (fuel economy of 3 L per 100 km) one were to choose a 1.8-L four-cylinder engine at the beginning of the development process, no amount of "good" technology would compensate for this fundamental error. Every competitor who has chosen a three- or four-cylinder engine displacing less than one liter will have an advantage.

The planning process of the earliest development phase (advanced engineering) includes establishing a project plan which has staged data approval for manufacturing planning. Furthermore, time frames for proto-

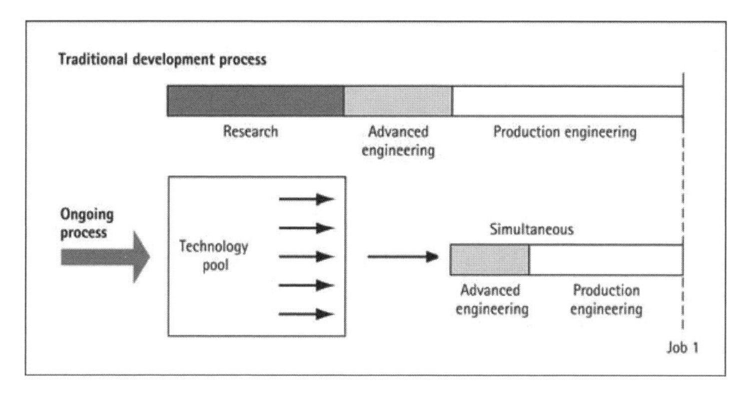

Fig. 10.1-3 Comparison of development processes.

type construction as well as testing program plans for planned variations are established. This means that all design alternatives, such as different camshaft profiles, must be included in the testing program plan along with corresponding verification tests, as approval can only take place on the basis of a verification program. Production startup preparations are made accordingly.

If one considers the vehicle as a whole, in-house manufacturing currently accounts for 20–40% of the vehicle: that is, 60–80% consists of bought-in components which are transported to the assembly plant. Purchasing planning takes place in parallel to initial project planning; this means that suppliers are "signed up" for the development program at this early stage. Supplier manufacturing processes are therefore optimized and integrated in the overall development process.

System and subsystem suppliers are included very early in the development process, immediately after vehicle goals are defined. Subsystems are divided into various levels. Subsystem level 1 might, for example, be the entire drivetrain; Subsystem level 2 would be the engine or transmission. At the latest, selection of all suppliers must be finalized no later than program approval. This commitment includes contractual agreements between manufacturer and supplier to ensure quality, scheduling, and function and cost targets. At this time, along with vehicle targets, important manufacturing constraints must be defined. These include selecting an engine and/or vehicle assembly site. Supplier selection and commitment are significantly affected by production site selection.

After establishing project scope, the program is assigned to a project team which guides the entire development process.

Program planning describes which information, targets, and technologies on the part of the customer and the company are included in the specifications. The first part of the specifications contains overall vehicle objectives. After the preliminary vehicle specifications are established, it is necessary to divide these among engine, transmission, and chassis subsystems. This partition may be described by the concept of "negotiation." Detailed process scheduling, resource determination, and commitment are based on subsystem target catalogs.

10.1.3.2 Process planning

The most important instrument for implementing simultaneous engineering is process planning. This contains the foundation of the entire sequential development process. The sequence provides information on planning of individual development process activities and should not be confused with manufacturing process planning. To plan the development process, it is necessary to take into consideration changes over time in freedom to make modifications. "Modification freedom" is understood to mean the possibility for restructuring during the development phase. The process plan begins after or on the basis of program planning and includes concept selection, the design phase, and verification (Fig. 10.1-4). The latter represents examination of project goals on the basis of prototype testing. After these stages are completed, the development process ends with start of production ("Job 1").

During the definition phase, the development engineer enjoys a relatively high degree of modification freedom. During the concept selection phase, choices can still be made between "turnkey" technologies which had been identified during program startup. The farther along one goes in the development process, the less decision freedom is available. Immediately before start of production, this freedom—as well as any need for it—is nonexistent.

Over time, the need for modification decreases (Fig. 10.1-5). An important, if obvious, requirement is the fact that test and simulation results must bring the engineer ever closer to program goals. As need for modification decreases, greater discipline is needed on the part of developers in order to reach Job 1. Even if a competitor presents a completely different concept during this phase, there can be no deviation from the selected program path; the point of no return has been passed.

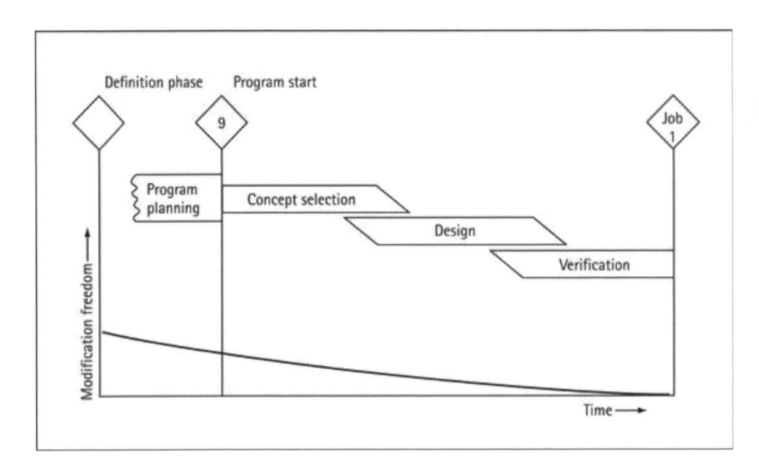

Fig. 10.1-4 Process plan.

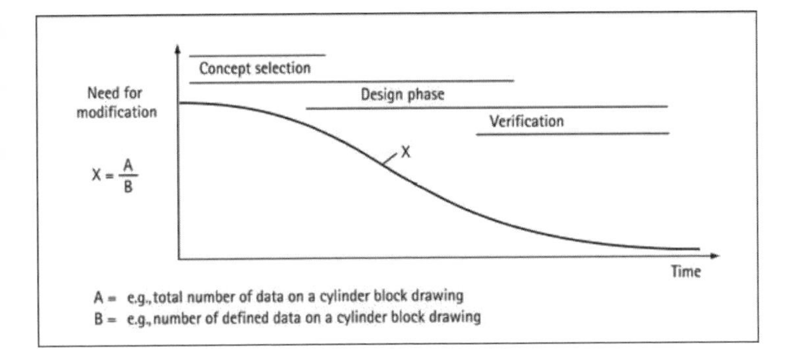

Fig. 10.1-5 Need for modification.

Manufacturing-based modification freedom is restricted by planning and creation of manufacturing facilities (Fig. 10.1-6). At the beginning of the development process, planning, design, and execution of manufacturing facilities demand a small but growing number of finalized, so-called "frozen" product definitions, features, or parameters. In order to acquire the necessary manufacturing machinery and tooling, at some point in time the manufacturing engineer requires concrete data. The resulting "boundary curve" must be observed by the development engineer.

Modification freedom is only limited not only by manufacturing but also by vehicle development (Fig. 10.1-7). The overall vehicle must be taken into consideration during engine development. While engine development proceeds toward predetermined development goals, vehicle development is working on optimization of the overall system. At the beginning of the development process, vehicle development requires a small number of finalized, invariable engine characteristics; this number grows over the course of the program. These characteristics include not only fuel consumption, torque, and power, but also size, acoustics, and cooling requirements. Installation space intended for a four-cylinder engine cannot later be changed to accommodate a six-cylinder inline engine.

Continuous contact with vehicle development is just as important as contact with manufacturing; both of these require frozen transferred parameters from engine development. In a simultaneous development program, modification curves affect one another and result in limitations on modification freedom as a result of production preparations and overall system development.

To harmonize engine and vehicle development plan-

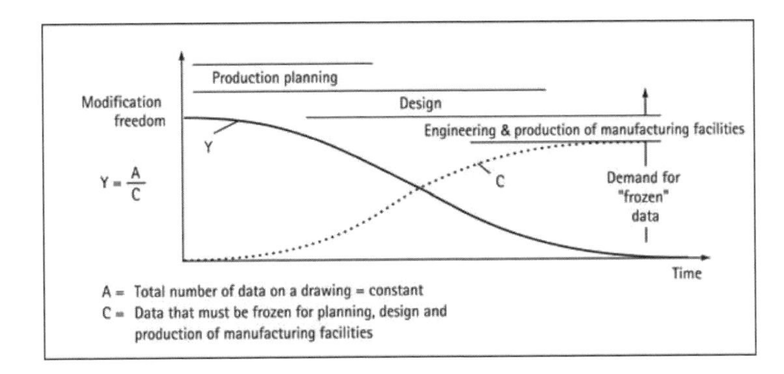

Fig. 10.1-6 Restriction imposed on modification freedom by planning and production of manufacturing facilities.

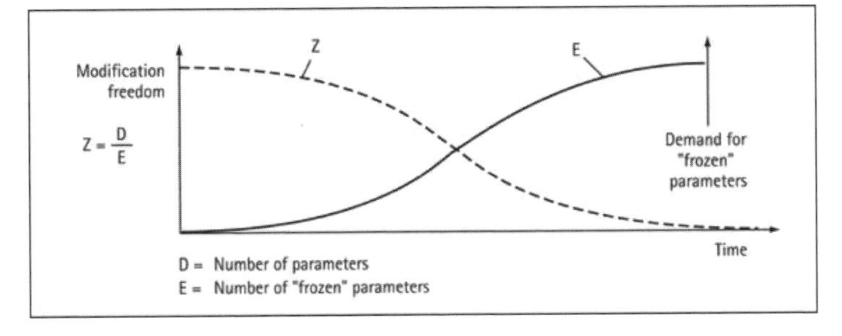

Fig. 10.1-7 Restriction imposed on modification freedom by vehicle development.

ning, the engine is developed during the design phase and built in prototype form. These engines are installed in prototype vehicles and subjected to vehicle testing along with verification tests on engine dynamometers. These engine and vehicle prototypes are built using production tooling, which will later be used at production startup.

In establishing the process plan, it makes sense to begin with production startup (Job 1) and consider all process stages in reverse order (Fig. 10.1-8). The first vehicles are the subjects of a "functional build" prior to production startup; this checks and ensures that production workers are indeed able to assemble components as the designer intended. Engine development is built up analogous to the overall production system. It is apparent that engine production startup must take place before vehicle production startup. One important condition is that production startup for individual components must always be scheduled before the overall system. For the engine development engineer, Job 1 is preceded by validation of the engine assembly process, which in turn is preceded by creation, installation, and activation of the engine assembly line. To permit manufacturing engineers to execute these tasks in a timely manner, design data must be sequentially frozen during development. This also means that there must be a certain point in time (control point 4) after which modifications are no longer permitted, even though the development engineer would like to improve certain features of his product which over the course of development and verification are shown to be less than optimal.

In order to optimize procurement and testing time, prototypes are built using various fabrication methods (different fabrication times) but to the same drawings. These early prototypes (Phase 1) are made using near-production design, employing free-form fabrication methods (rapid prototyping) based on 3-D solid body models and CAM (computer-aided manufacturing). The following prototypes (Phase 2) are based on near-production manufacturing methods. The third phase demands that all engine components, including all components required for engine installation in the vehicle (interface components—for example, engine mounts) are made using processes representative of production methods. Prototype parts are used according to their planned availability for component, system, engine, and vehicle testing. All three prototype phases support a test phase whose sole purpose is verification.

The end of production development is marked by validation of the manufacturing process. Simultaneous planning, procurement, creation, erection, and testing of manufacturing facilities as well as parallel startup of production tooling generation are timed to optimally support completion of production tooling after final approval (control point 2). Working backward, this establishes time frames for process planning and procurement of production tooling as well as engine and vehicle test phases.

How will a process be controlled or monitored? With the aid of program control or monitoring, the current status of the development process it is readily apparent at all times—that is, which requirements have been met

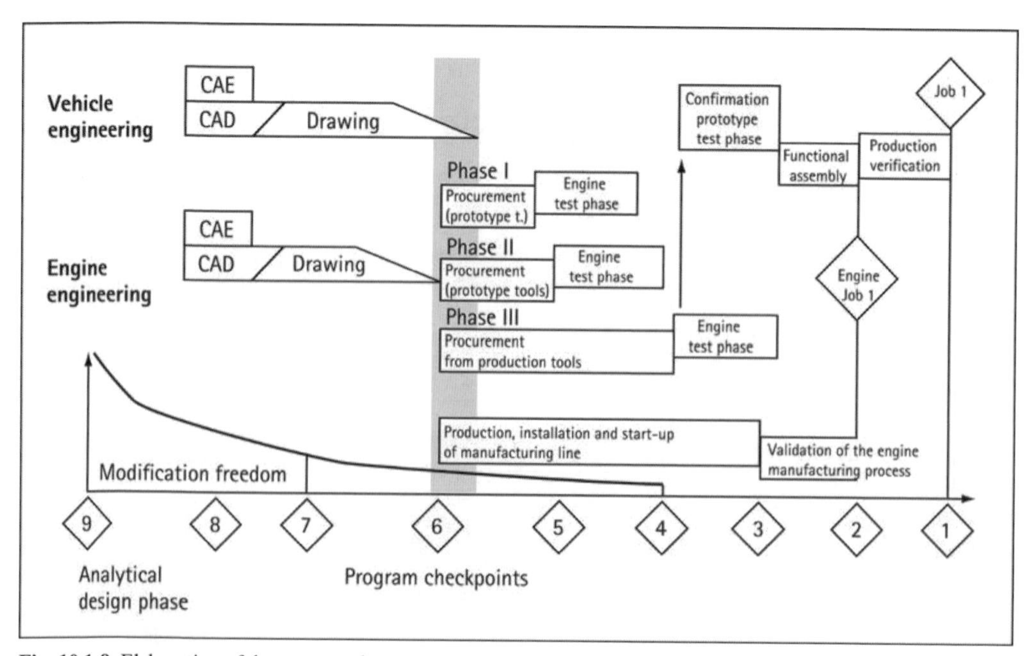

Fig. 10.1-8 Elaboration of the process plan.

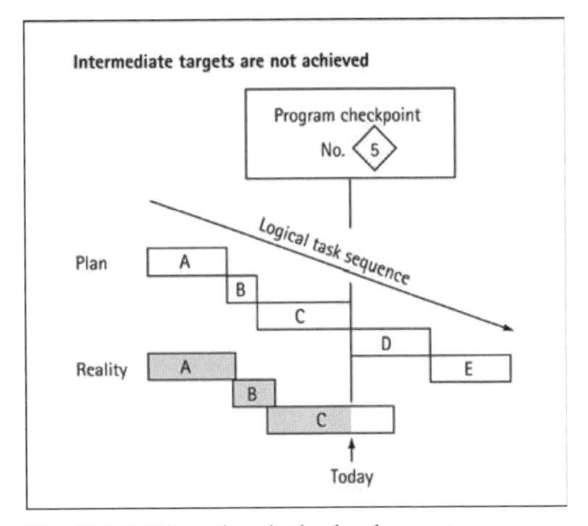

Fig. 10.1-9 Disruptions in the development process due to schedule deviations.

and which still call for attention. Program control points 1 through 9 of Fig. 10.1-8 define specific measures: that is, deliveries or concrete external requirements at a predetermined time.

Work or resource planning must also be included in the overall development plan. If the necessary means are lacking at the critical time, the program cannot be executed to satisfy individual approval criteria. Rough planning of development work may be carried out on the basis of program control points. Detailed near-term planning is served by so-called resource profiles. These in turn are divided into profiles for individual areas of responsibility, such as design or procurement (of prototype parts). In keeping with the control point in question, resources for analytical design (designers and CAD workstations) as well as for CAE applications (CAE specialists and their computer equipment), and later, test rig (dynamometer) planning must be included.

In the course of overall planning, interruptions to the development process cannot be ignored. Theory and practice seldom match; there will hardly be a program completely unaffected by unforeseen external disturbances. Fig. 10.1-9 shows, for example, control point 5, in which it was determined that an intermediate goal had not been met. The plan provides logical sequential execution of individual activities from A through E. In reality, it appears that activity A took longer than expected, which in turn affected the following activities. At the time of control point 5, activity C is not yet complete, even though the plan already envisions the beginning of activity D.

There are several possible ways to eliminate or, if this is not possible, minimize disturbances to the development process. The alternatives shown not only offer relief, but also carry with them certain disadvantages (Table 10.1-1). We can delay all subsequent activities as under option 1 and accept, for example, that new vehicles will not reach market in the spring but rather in the following fall; this scenario is quite possible but in general not acceptable to the company's marketing division. Advertising, press releases, and other activities are concentrated in the spring; if the deadline is not met, major losses are likely. As several projects run concurrently in engine development, delay of activities in other departments would trigger chain reactions which would in turn endanger other projects.

Nonadherence to the logical sequence (option 2) may carry with it a high rate of errors. If already ordered parts must be scrapped, financial losses are significant, especially when one considers that a prototype engine may cost between €50,000 and €150,000. Similarly, additional costs for tooling modifications (in extreme cases, completely new tooling) may be incurred. In the worst case, the result—as in option 1—may be a delay in Job 1. A reduction in quality must always be rejected, as it would be fundamentally impossible to meet predetermined quality criteria.

The safest option appears to be replanning of subsequent activities to take place in shorter time frames or in different sequences. In most cases, however, the initial plan has already been optimized for time, so that there are few opportunities for carrying this out. Despite the

Table 10.1-1 Fault management in the product creation process

Options	Consequences
1. Subsequent activities are correspondingly delayed.	– Job 1 is delayed. – Delay in activities by parallel departments. – New resource profiles no longer match resource availability.
2. Logical sequences are not maintained. Work is carried out in parallel.	– Increased risk of recognizing errors too late, need to repeat tests, scrapping already-ordered tooling or prototype parts, in extreme cases delay in Job 1. – Reduced quality.
3. Replanning of subsequent activities (shorter, different sequence), to eventually regain the original schedule.	– Often impossible (if the original plan was already optimized for time). – Increased resource demand (workers, material, fixtures). – Possibly reduced quality.

need for more workers, facilities, and material, this too may result in compromised quality.

What happens when goals are changed during the program? The development process should be flexible, capable of reacting to customer demands or desires generated by competitors. New objectives interrupt the development process and demand modification outside the agreed-upon goals. The consequences are that test results, calculations, and work already completed by various departments become unusable. For example, if midway through development of a transmission, engine torque is raised appreciably—for instance, through increased displacement or supercharging—the transmission department might not be able to react in time for Job 1 without certain consequences unless reserves are already built into the plan.

From this we have the requirement that even during the program planning phase, goals should be defined to take into account future demands, eliminating the need for possible modifications beyond the agreed-upon freedom of modification.

10.1.3.3 Simultaneous planning of internal and external development and manufacturing

Manufacturing strategy should be established within program planning. In the course of product development—from program startup through program approval to investment approval—quality, cost, and functionality objectives lead to feasibility analysis. The manufacturing strategy must agree with this analysis; only then do results enter the production system.

Various criteria must be considered in establishing manufacturing strategy. The primary goal is unwaveringly good manufacturing quality of the product coupled with high flexibility and low costs. To achieve this goal, employee training levels, worker motivation, and other internal corporate factors must be taken into account. A plant in Brazil will be built up differently than a plant in Germany or in Asia.

The parameters which flow into manufacturing strategy include production volume—that is, large or small volume production. Large-volume production may be defined as more than 300,000 units per year with no upper limit. Volumes of 100,000 to 200,000 units per year characterize medium-sized production; anything below 100,000 units is termed small-volume production. Other parameters include the number of work days per year, and the number of working hours per day. In Germany, present conditions are 225 working days per year with 22.5 working hours per day in three-shift operation. If the production plant is located elsewhere, these values will change to reflect local conditions.

The concept of "uptime" is a measure of efficiency which relates the number of parts actually produced to the theoretically possible number. With a cycle time of one minute per part, in theory 60 parts could be produced per hour. If only 50 units are made, uptime is 83.3%.

Other factors enter into manufacturing technology selection. These include so-called "drivers" such as environment, quality, costs, and deadlines. The objective arising out of the "environment" parameter may, for example, read as "optimum working and environmental conditions." To achieve this, short-term or long-term measures may be implemented. A short-term measure might be ergonomic design of workstations. If ergonomic considerations are ignored, illness or injury are preprogrammed. A long-term measure might be coolant-free machining operations, which would imply significantly reduced environmental impact by eliminating the need for coolant disposal.

Similarly, the objectives "short deadlines," "low production costs," and "durability and customer satisfaction" may all be achieved using respective short- or long-term measures.

On the basis of the simultaneous engineering process—that is, immediate communications between manufacturing, suppliers, and product development—these "drivers" determine design characteristics and therefore manufacturing technology.

Volume productivity is understood to mean manufacture of product variations within the maximum manufacturing capacity. If, for example, two- and four-valve engines are built on the same line, flexibility must be sufficient to ensure that their relative proportions are immaterial. For maximum annual production of 650,000 units, this might represent 100,000 2-V engines and 540,000 4-V engines or vice versa.

Production flexibility means that new product variants may be introduced at minimal cost and with minimum planning time. For example, this may mean rapid conversion from manufacturing a four-cylinder engine to making three-cylinder engines.

The most important component of internal as well as external development and manufacturing planning is early simultaneous cooperation between development and manufacturing engineers. As early as the layout phase, the designer must consider technical requirements posed by manufacturing. For example, let us compare different aspects of cylinder head layout as seen by the designer and the manufacturing process engineer. Corresponding to internal development and manufacturing planning, early cooperation between engine development/manufacturing and supplier or subcontractor manufacturing is of major significance. Increasingly, system suppliers, offering system modules, are included in the development process. One example is combining intake manifold, injection system, air filter, actuators, sensors, and engine control module into a single integrated induction module. From design to manufacturing, these systems are created entirely at the system supplier, with system functions and, in particular, interfaces with the engine system defined and developed in close cooperation with the auto manufacturer. The system supplier is entrusted with overall responsibility and therefore also bears responsibility for productivity of submodules. To ensure that suppliers can keep pace, their development capacities must match—or be made to match—the high expectations of the auto industry.

10.1.3.4 Analytical design and analytical manufacturing process

The first step in realizing individual specifications takes advantage of analytical design. This is understood to include application of CAD, CAE, and FEM for design and optimization—in this case, of the engine. The first designs are carried out almost exclusively on the basis of 3-D CAD models (Fig. 10.1-10), which permit rapid conversion to an FEM model and therefore FEM simulation. The results allow immediate component optimization with respect to mechanical or thermal loads, but also for mechanical behavior (resonant frequencies, noise radiation).

Analytically optimized designs provide a foundation for Failure Mode and Effect Analysis (FMEA). This technique examines all conceivable and possible failure modes to which component or system functions may be subjected, analyzes the existing condition, evaluates, and follows improvement measures. This may mean that the design must be changed and/or appropriate verification tests must be planned and executed. Manufacturing process FMEAs are set up analogously. Process FMEAs examine failure modes which might occur within the manufacturing process to affect component function. This must be carried out as early as possible—based on the initial process planning—to ensure that any necessary design changes are introduced into the analytical design phase. FMEAs are established through teamwork and continuously updated throughout the development process.

Analytical development of the manufacturing process takes place simultaneously with analytical design. In contrast to conventional layout of manufacturing processes based on experience and cost-intensive testing, it is possible with the help of CAE to simulate and optimize nearly all individual manufacturing processes. Using a simulated casting process as an example, optimization of the design in terms of castability can be undertaken even before the first prototype is cast (Fig. 10.1-11). In this way, critical agglomerations and other weak points may be recognized at an early stage and avoided by means of fine-tuning the casting pattern.

The analytical design phase (all subsystems) provides the feedback needed to complete specifications or targets. At this time, the final specifications are in hand and serve to support program approval.

10.1.4 Production development

Production development follows advanced development; the project team that had been responsible for product advanced development is expanded. Some team members from advanced development (engineering and manufacturing) return to their former departments to take on new projects. Those members remaining in the team accompany the program past production startup to ensure that all information, knowledge, and experience from advanced development are taken into account in production development. Project management is especially important in ensuring successful execution of production development; its role will be examined in the following section.

10.1.4.1 Managing simultaneous development and manufacturing processes

Certain preliminary actions need to be performed prior to manufacturing startup. For example, machinery must be ordered about two years earlier; delivery times of 1.5 years are common. Equally decisive is a precisely defined investment effort. Not only must the necessary capital be available at the proper time, but the required expense must already have been planned and managed.

Simultaneous development is significant in the effort to shorten processes and avoid mistakes. This cooperative, goal-oriented effort, extending beyond individual departments with parallel activities in individual sectors, is encompassed under the concept of simultaneous engineering. Simultaneous engineering also represents timely recognition of resulting development costs or their reduction by means of appropriately timed expression of alternative ideas. Product development does not cease with manufacturing; instead, the product is mon-

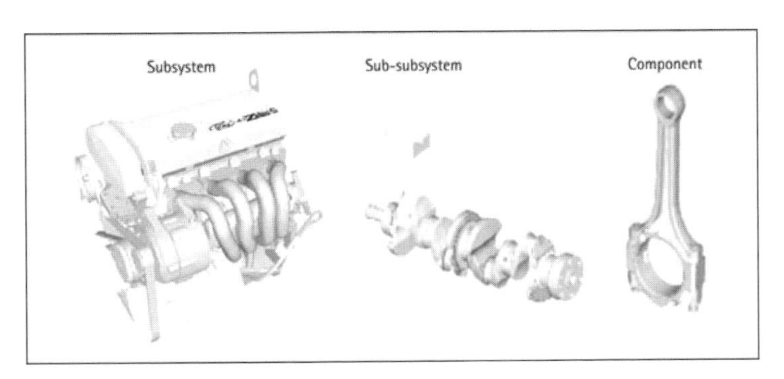

Fig. 10.1-10 Data structure of a solid model.

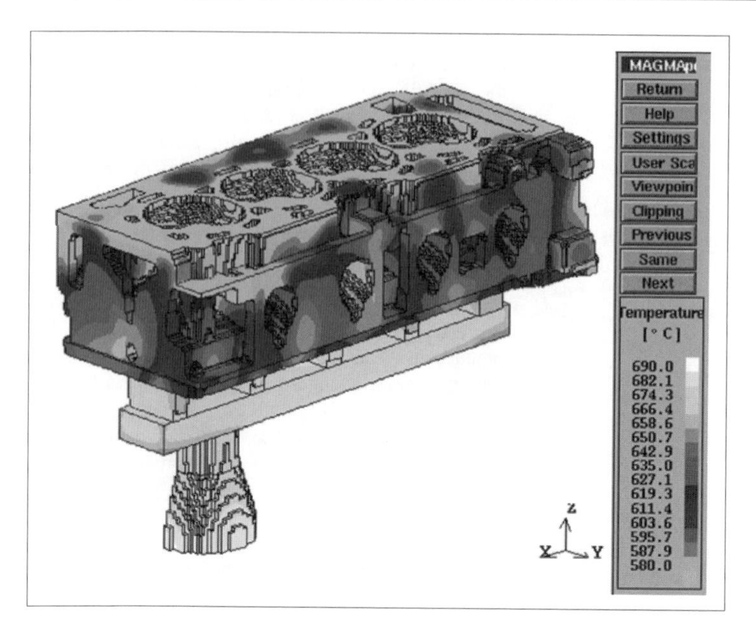

Fig. 10.1-11 Computational simulation of the casting process.

itored until the end of its life. The development team, working within a limited timeframe, operates independently and with intensive customer contact. From the very beginning, system suppliers are included in its operations. Only in this way can development times and costs be reduced and quality improved.

Projects must be guided or coordinated. One form of project coordination consists of a staff with its project leader, who may call on workers from other departments as needed. These, however, are not subordinated to the project leader, but rather, like the project leader himself, they report to respective specialist departments. This form of project coordination is suitable for smaller projects with lower levels of complexity and shorter duration.

The most common organizational structure is matrix project organization (Fig. 10.1-12). The guiding principle is that out of an available pool of engine technology specialists, individual workers may be recruited as needed for development of various vehicle classes. For example, these specialists might all be members of the "combustion engines" department and be completely variable within this framework. All of these specialists would report to the project leader responsible for that department.

For highly complex projects with interdisciplinary content, the so-called pure project organization of Fig. 10.1-13 is implemented. Its primary difference from previously presented organizational structures is that a project leader works with a team which is entirely dedicated, for the duration of a project, to its fulfillment. In this organizational structure, workers are entirely subordinate to the project leader.

Once project organization has been finalized, the objective is to achieve appropriate task distribution and establish communications between project members. To answer the five Ws (who, what, when, where, why), certain responsibilities are precisely defined. For example, the project leader decides what is to be done. Individual departments determine who will carry out the defined tasks and where and when this will happen. All departments carry on a dialogue with each another; all project participants can and must communicate with one another on a continual basis.

Production development is the continuation of analyt-

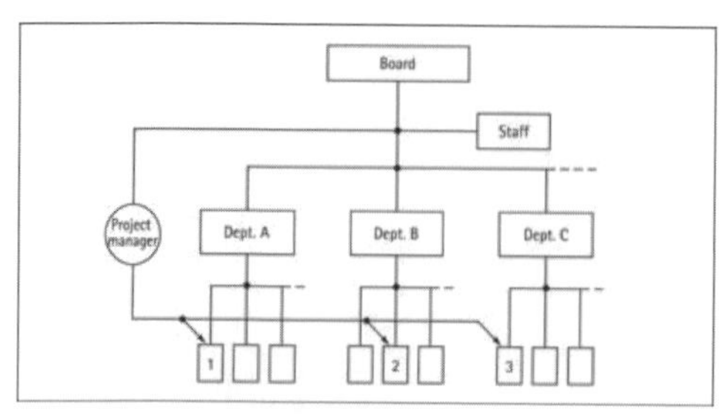

Fig. 10.1-12 Matrix project organization.

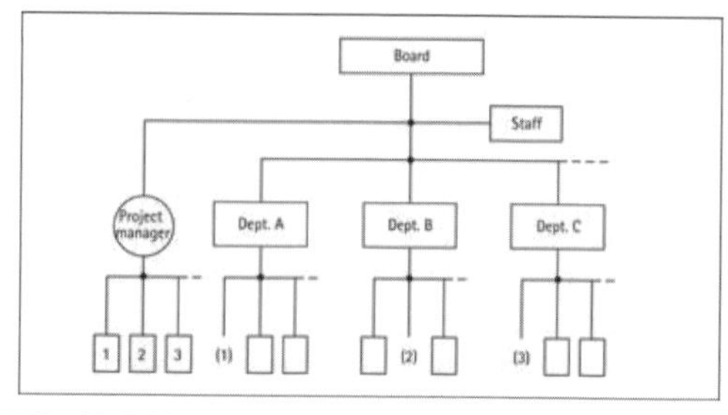

Fig. 10.1-13 100% project organization.

ical design, which envisions staged data release based on verification testing, continuing until final approval. Staged release means that vital dimensions needed for manufacturing are released early in the project—for example, cylinder spacing and cylinder head stud locations; individual details are released later (in stages). This facilitates initiation of manufacturing planning, machine layout, and procurement. Release of financial means is a component of program approval.

Such staged data release must be oriented toward satisfying the needs of engineering as well as manufacturing. The question of which data may be set by engineering and when, and which by manufacturing in order to initiate machinery procurement at the appropriate time can only be answered by application of the simultaneous development and manufacturing process. In practice, this means that the development engineer responsible for a specific component as leader of an interdisciplinary working group must cooperate with his counterpart manufacturing engineer, relevant purchasers, and suppliers. This ensures that all information is available and activities are carried out at the appropriate time. Simultaneous development and manufacturing process management works in a similar manner with all components and overall systems.

10.1.4.2 Reduced testing effort through fewer prototype testing phases

Development time in both advanced and production development can be greatly shortened by reduction in the number of prototype phases. Procuring prototypes is still a very time-intensive process. Prototype fabrication and testing phases as well as procurement represent "dead time" for the designer, as he must await test results before conducting further design optimization. Due to the iterative nature of design, customarily consisting of three successive phases—design, prototype fabrication, and testing and analysis—the development process has historically consumed a relatively large amount of time. By contrast, a time-optimized development process should contain only a design phase and a prototype testing phase: that is, all prototypes are fabricated to the same design status from the same drawings.

Reducing prototype phases to a single iteration is becoming increasingly practicable, as engineering tools such as computer-aided engineering (CAE) have reached a level where prototype testing in the earliest phase (advanced development) and increasingly in the succeeding production development phase can be eliminated. By applying finite element methods to components, dynamic vibration behavior or thermal and mechanical loading can be simulated even before the first prototype is available. At the same time, a first analysis of noise behavior can be carried out. Iterative optimization using model calculations can be conducted on the basis of these results. In the past, such optimization involved comprehensive testing of components, systems, and engines installed in the vehicle, with an enormous associated effort (consuming about one year) in design, procurement, measurement, assembly, instrumentation, test execution, and evaluation; this is now possible in the space of only a few weeks. Nevertheless, experimental testing using actual hardware is indispensable for final evaluation of a combustion engine design.

Even in development segments that demand test development using actual hardware alongside CAE, great time savings can be achieved by coupling CAE and computer-aided manufacturing (CAM). That is, first prototypes can be made extremely quickly (rapid prototyping) on the basis of 3-D CAM models. The first verification prototypes are made using rapid prototyping and freeform fabrication methods in order to permit an early start on proof-of-concept and verification testing. These are followed by prototypes made from the same drawings using production tooling.

Procurement of prototype parts begins during the design phase. Staged data release supports company-internal manufacturing planning and machinery procurement, and it supports to the same degree similar activities on the part of suppliers. For cylinder head tooling, this means freezing major dimensions so that planning and release of the foundry patterns can be initiated and then completed on the basis of subsequent staged data releases (from development and manufacturing).

Verification tests are conducted on the basis of prototypes. Calibration takes place in parallel to design and manufacturing development—that is, matching engine control to the overall vehicle. This calibration in turn can be divided into several phases. Starting with simulation models, in which the interplay of engine, transmission, and vehicle takes place on a purely analytical level, highly dynamic engine/transmission dynamometers are used to carry out additional fine-tuning. Final tuning is conducted in an actual vehicle.

In production development, production processes are precisely matched to design details. To this end, engineering feasibility and optimization of component and system manufacturing are evaluated on a continuous basis. The most important goal is efficient manufacturing and meeting quality goals: in other words, assured function of all components, both individually as well as in assemblies installed in the system.

The conclusion of production development is marked by validation of manufacturing processes. For final approval, the first production engines are installed in production vehicles and subjected to the required tests. Evaluation of results is conducted at the vehicle control point. The "production startup approval" control point is reached about a month later.

10.1.5 Outlook

Market changes, engineering advances, technological revolutions, and product design as well as manufacturing and logistic processes will determine automotive development to an ever increasing degree. All process changes are based on customer requirements, environmental effects, and measures to increase productivity of a corporate entity. A growing development trend is indicated by integration and modularization of components.

Innovation is imperative. Central modules can be regarded as one example of a leap in innovation. Modules which currently still perform for the most part individual functions such as engine control, transmission control, or adaptive suspension will increasingly be integrated, with all functions combined into an overall electronic module networked by CAN bus systems.

Integration will also manifest itself on the hardware side. Examples include system modules such as the oil module, which in addition to oil filtration and cooling, for example, also integrates the oil pump, pressure control valve, and perhaps also the water pump into a single system. To ensure that system suppliers can keep pace with the high demands imposed by the auto industry, their development capacities must be increased.

Rising customer expectations will result in increased development and manufacturing complexity. This in turn will increase pressure to cut development costs and time. Project decisions must be made as early as possible in order to reconcile initially contradictory demands. New computer simulation models guide the first efforts toward finding solutions. Development of virtual prototyping permits manufacturers to bring theory and practice closer together. In this way, decisive steps toward process optimization may be taken at a very early stage.

Further improvements may be found in development methodology. Simultaneous engineering teams not only accompany basic work as it makes its way through suppliers, in-house development, and manufacturing, but they also remain on call throughout the entire service life of a product. Throughout their stewardship, they introduce continuous improvements to the product and the manufacturing and development processes. With the shift of additional development activities to suppliers, mastery of system integration will become a significant competitive factor for automotive companies.

Software utilization is currently undergoing far-reaching changes; eventually, it will displace current hardware-oriented efforts directed toward meaningful differentiation. With the aid of new software programs, existing products may be upgraded and therefore brought up to reflect the latest state of the art. Furthermore, within project management, emphasis on analytical product and process development as well as software applications will permit improved differentiation and therefore improved design in keeping with customer expectations.

The future will see not only integration of new functions. To an even greater degree, striking a balance between the multitude of product characteristics and therefore their optimization will play an increasingly important role. Any company able to present customers with such optimization in terms of technology, longevity, reliability, economy, and environmental protection will reap decisive competitive advantages. Accordingly, the automobile industry must increasingly concentrate on its core competencies—that is, on system integration—while suppliers must see their core competencies in terms of developing such so-called systems or integration modules. The auto industry must ensure that integration of various modules into an overall system with appropriate emphasis on customer expectations is achieved by the best means possible.

Alongside integration, and therefore core competency in the overall system, a further change will be represented by the increased global presence of individual firms, a change which cannot be achieved without adaptation of corporate structures. Productivity and efficiency in development and manufacturing must be further increased; improved flexibility enables more rapid response to sudden shifts in market demands. Here, CAE will form the deciding factor. In addition, more than in the past, it will become necessary for firms to form alliances, either as unified entities or only to have individual shared components developed by suppliers. Efforts in terms of anticipated joint ventures can already be seen in platform strategies or so-called common envelope strategies which define a shared powerplant environment. A decisive factor is application of as many shared vehicle or engine family components as possible. In the longer term, not only will carmakers install identical components in various vehicle lines, but this practice will extend to components shared between various auto manufacturers. If specific parts are required to meet customer expectations and therefore do not permit application of shared-component policies, flexible data banks may prove useful. Such data banks allow rapid and automatic creation of all major components using parameter variations stored on the computer, with simultaneous data transmission to manufacturing.

On the manufacturing level, it is apparent that present-day facilities will in principle be retained. Nevertheless, in the course of general engineering progress, machinery will undergo continuous development, particularly in terms of being tailored to the requirements of individual projects. An important prerequisite for application of flexible manufacturing systems is optimization of their respective machinery concepts, in which a trend toward single- and dual-spindle high-speed dry machining methods is apparent. Choice of linear or rotational work concept depends on the application at hand and the economic realities.

High dynamics—that is, high speed and acceleration in manufacturing as well as in execution of auxiliary functions—reduce productive and nonproductive times and therefore unit times and unit costs. Higher utilization through shorter tool change times, as well as less time spent in fault correction and minimization, are fundamental requirements. Short conversion times for flexible installations will support these efforts. Intelligent tools with automatic dimensional compensation and hydraulic or centrifugal controls are finding increased application. Higher cutting speeds are dependent upon material, process, and the required quality. Cutting speed will not be limited; it will become a relative parameter, to be set according to the chosen process. Reduced use of cutting fluids will become more important; dry machining may be regarded as the ultimate goal.

In general, it can be seen that a further increase in productivity is possible through appropriate measures. Today's limitations cannot be regarded as evidence that operational envelopes have been completely exhausted; instead, continuous development in all sectors leads us to expect additional progress.

References

Bauer, C.O., and R. Arnold. *Qualität in Entwicklung und Konstruktion*. Third completely revised edition. TÜV Rheinland, Cologne, 1992.

Bochtler, W. "Modellbasierte Methodik für eine Integrierte Konstruktion und Arbeitsplanung: Ein Beitrag zum Simultaneous Engineering." RWTH Aachen, 1996.

Eversheim, W., et al. "Die Arbeitsplanung im geänderten produktionstechnischen Umfeld. Teil l: Integration von Arbeitsplanung und Konstruktion." VDI-Z, 1995.

Eversheim, W., et al. "Informationsgerechte Ablaufgestaltung in Konstruktion und Arbeitsplanung." wi-Produktion und Management 86, 3, 1996, pp. 81–85.

Eversheim, W. *Prozessorientierte Unternehmensorganisation. Konzepte und Methoden zur Gestaltung "schlanker" Organisationen*. Second edition. Springer, Berlin, Heidelberg, New York, 1996.

Eversheim, W. *Organisation in der Produktionstechnik. Bd. 1: Grundlagen*. Third completely revised and expanded edition (VDI book). Springer, Berlin, Heidelberg, New York, 1996.

Eversheim, W. *Organisation in der Produktionstechnik. Bd. 2: Konstruktion*. Third completely revised edition (VDI book). Springer, Berlin, Heidelberg, New York, 1998.

Eversheim, W. *Organisation in der Produktionstechnik. Bd. 3: Arbeitsvorbereitung*. Third completely revised edition (VDI book). Springer, Berlin, Heidelberg, New York, 1997.

Hartmann, D. "Modell zur qualitätsgerechten Konstruktion." VDI-Fortschrittsbericht. No. 260, 1996.

Hütten, H. *Motoren—Technik, Praxis, Geschichte*. Motorbuch, Stuttgart, 1985.

Hammer, M., and J. Champy. *Reengineering the Corporation; a Manifesto for Business Revolution*. Harper Collins, New York, 1993.

Küntscher, V. *Kraftfahrzeugmotoren—Auslegung und Konstruktion*. Third, extensively revised, edition. Verlag Technik, Berlin and Munich, 1955.

Lincke, W. *Simultaneous Engineering—Neue Wege zu überlegenen Produkten*. Hanser, Munich, 1995.

Madauss, B. J. *Handbuch Projektmanagement*. Fifth revised and expanded edition, Schaeffer-Poeschel, Stuttgart, 1994.

Masing, W. *Handbuch der Qualitätssicherung*. Hanser, Munich, 1980.

Anon. "FMEA—eine wirksame Methode zur präventiven Qualitätssicherung." VDI-Z 132, 10, 1990.

Anon. "Strategien für die Produktion im 21. Jahrhundert." BMFF, 1994.

Neumann, A. *Quality Function Deployment—Qualitätsplanung für Serienprodukte. Berichte aus der Produktionstechnik*. Shaker, Aachen, 1996.

Pamel, J. *Kooperation mit Zulieferern*. Gabler Verlag, Wiesbaden, 1993.

Pfeifer, T. *Praxishandbuch Qualitätsmanagement*. Hanser, Munich, 1996.

10.2 Computation and simulation

10.2.1 Introduction

Computer-aided processes (CAx) are not new to the automobile industry but for various reasons have garnered a special position in recent years. Auto industry customers profit from this technology indirectly, in that they buy a product whose functionality has been optimized and costs minimized. In the view of auto manufacturers, the motives behind these methods are no less meaningful; they represent a powerful tool, ensuring the competitive viability of their products.

What exactly are the primary advantages of this technology? The most significant benefit is that vehicle functionality (comfort, performance, weight, etc.) can be evaluated or optimized by means of simulation using models created on the basis of geometrical and physical data. This means that the first actual vehicle prototype is built much later and exhibits a much higher level of development. Along with improved quality, this results in significant time savings in the development process. As is generally true whenever new methods are implemented, this technology has a strong influence on work flow and organizational structures in product development. The preconditions which lead to optimal results using simulation methods can be summarized as follows:

- Assured processes, matched to the needs of automotive development
- Data management to ensure availability for model creation and in implementing development process results
- Procedures and organizational structures which support these requirements

For various reasons, methods and procedures have developed very rapidly, and with the availability of powerful computers, they have become available for automotive product development processes. Today, process flows are being restructured to achieve optimum value creation from application of simulation technology.

10.2.2 Organizational aspects

Along with allocation of appropriate resources in the form of very cost-intensive software and hardware, to a certain extent introduction of simulation methods demands careful planning of how this technology will be applied. Optimum value creation using simulation methods is achieved only if its strengths are fully integrated in the product creation process. This also means that individual procedures within this overall process are matched to the requirements of developing a "virtual product." From a simulation technology point of view, the typical product development process can be divided into three phases.

Concept phase

Once a vehicle concept is selected (Section 4.2), the design details that determine major criteria (crash behavior, comfort, fuel economy, etc.) are established. Therefore, simulations play a special role in concept development, as they help to solidify major design features required for package optimization and assurance of functionality. This serves to avoid serious errors which could only be corrected with great difficulty or at considerable expense later in the development process. At this stage, however, a problem arises in that the preconditions for application of simulations, specifically exact descriptions of component geometries and properties, are incomplete at best.

One possible solution is to use the simulation model to generate a suitable progenitor model. Intensive efforts are currently under way on special software systems to support such processes. Another solution is to derive rough as well as more detailed models from the available data and carry out concept evaluation based on these. Because of its uncertain predictive value, this method is only suitable for special situations. In any case, early availability of data required for simulation is a major challenge in establishing these procedures.

Production development

On the whole, vehicle component development and fine-tuning in keeping with the established concept takes place during this phase. The primary task of simulation is to optimize required functionality (comfort, crash behavior, performance, etc.) and major parameters such as weight. Additionally, other examinations are conducted, such as simulation of assembly procedures, deep drawing, and casting processes, for purposes of manufacturing process assurance. At this point, simulation methods draw on CAD data, which by this time exhibit a high degree of design detail. Using these data structures, powerful software systems generate finite element meshes (largely automatically) or conduct examinations of flow processes. The key element is effective data management, which encompasses input quantities (geometry, parameters) as well as models and results.

Vehicle testing

In the ideal situation, completed vehicle designs meet established requirements, and only a single test series is needed for approval. Although reality often approaches this expectation, invariably some problems will become apparent in testing. These problems are to be solved with the minimum expenditure of time and resources. The value of simulations is that these problems—in contrast to experimental methods – are well capable of revealing the causes of faulty behavior and can therefore make decisive contributions toward developing solutions. It is, however, a prerequisite that simulation models, established during previous phases, must be current and well documented. Process organization must ensure that simulation technology is integrated with experimental department processes [1].

10.2.3 Methods and procedures

Table 10.2-1 shows fields of application for numerical simulation in automobile development, along with their impact on the development process. The following expositions will describe major areas of application;

Table 10.2-1 Fields of application and significance of simulation technology

	A	B	C
Body			
Stiffness	x		
Strength, stresses		x	
Acoustics		x	
Operational durability			x
Modal analysis	x		
Crash	x		
Occupant simulation	x		
Deep drawing	x		
Casting simulation		x	
Joining technology		x	
Seals (door, etc.)	x		
Powertrain			
Strength, stresses	x		
Modal analysis	x		
Operational control dynamics		x	
Noise radiation			x
Gas exchange, one-dimensional		x	
Cylinder gas flow	x		
Mixture formation		x	
Combustion			x
Cooling jacket flow	x		
Oil circulation		x	
Overall vehicle			
Handling dynamics		x	
Performance/fuel economy	x		
Vibrations (linear)	x		
Vibrations (nonlinear)		x	
Life expectancy			x
Exterior airflow		x	
Interior airflow	x		
Engine compartment airflow	x		
Climate control			x
Heat management			x
Control technology		x	
Electronics			
Hardware-in-the-loop	x		
EMC		x	
Control unit layout			x
Suspension			
Kinematics, elastokinematics	x		
Strength, stresses	x		
Life expectancy, service durability		x	
Simulated active suspensions	x		

A, In service, predictive capability assured;
B, In service, in support of development;
C, Limited application due to incomplete methodology.

individual applications which are limited to special situations and which, for various reasons, are as yet not applied on a broader scale are not reflected here. An overview of topics involving current applications of simulation technology is offered by regularly held conferences, such as in Refs. [2 and 3], as well as earlier convocations.

10.2.3.1 Structural calculation

The first computer-aided processes employed by the auto industry were based on finite element methods and served to optimize static properties of vehicle components, such as the body. These methods, familiar to every engineer at least in principle, permit calculation of component load response almost regardless of configuration. At present, all questions regarding vehicle structural layout are handled by application of such methods as:

- Stiffness
- Strength
- Joining technology
- Component design for minimum weight (structural optimization)
- Dynamics, vibration comfort, acoustics
- Vehicle safety (crash, occupant simulation)
- Metal forming

Stiffness

Many vehicle components must be dimensioned to provide the maximum possible stiffness. This is especially true for the body. The necessary finite element model results from discrete representations of all load-bearing components (Fig. 10.2-1). Using available CAD data, representation of actual geometry by the computational model can be as detailed as desired, and powerful software systems automatically generate the discretization, with resolution limited only by the available computing power. Accuracy of the results depends on the quality of the finite element model's reproduction of actual geometry as well as on assumptions with regard to loads, boundary conditions, and material properties. Moreover, there is a discretization error, which results from finite element dimensions and stress gradients. To draw conclusions regarding local stiffness, it is sometimes necessary to apply a finer mesh than would be needed for global stiffness issues. As may be seen in the illustration, the degree of discretization even permits taking into account spot welds on body panel flanges. This is capable of providing results whose accuracy is within the limits of measuring accuracy. Today, predictive quality of these methods is so good that body stiffness evaluation may be carried out without resorting to actual testing.

Along with the body, certain chassis components—for example, steering and engine components—are optimized for stiffness using the same methods.

Strength

Strength is one of the essential questions which arise in dimensioning of structural components. Finite element methods have proven to be the only usable means for establishing dimensions. In terms of accuracy, however, certain limitations are imposed that are rooted in the nature of the method. For example, accuracy of stress determination is dependent on the degree of discretization, described above; actual peak stresses could only be determined by unreasonably tight, unachievable meshes. A further uncertainty lies in the type of loading. For most components, this depends on dynamic effects of vehicle operation, which is difficult to reproduce. As this is largely a matter of changing loads, any conclusions regarding component strength can only be made on the basis of life expectancy calculations. Tools for such calculations are available, but their predictive ability is limited because results are dependent on parameters which are inherently difficult to quantify, such as local material properties and manufacturing methods.

Nevertheless, today all highly stressed components are designed using this method. To achieve adequate accuracy, models with sufficiently fine mesh must be employed to permit at least comparative results to be obtained, which in turn enable optimization of force

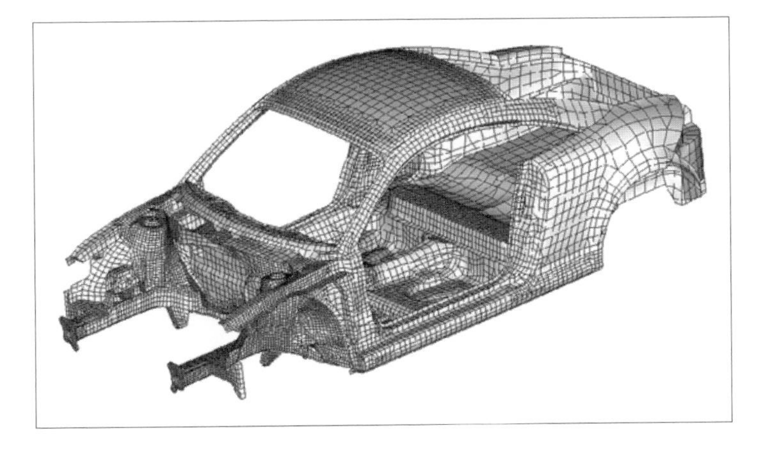

Fig. 10.2-1 Finite element model of an automobile body.

flows. Furthermore, related methods are commercially available; these permit determining stress peaks even for larger elements. These methods are based on variable polynomial order displacement functions for finite elements in a given mesh, known under the concept of "P-convergence" [4]. With this method, accuracy of results depends not on the fineness of the mesh, but rather on the polynomial order of displacement functions applied to elements. This permits achieving great precision despite considerably coarser mesh. Application of commercially available programs is, however, limited to solid body structures. Similar results may be achieved using boundary methods; instead of discretizing an entire structure, only its surface is considered. Using these methods, precise stress analysis of complex solid bodies may be conducted with relatively little effort. However, in contrast to P-convergence, boundary methods have not established themselves, primarily because of their long computation times for larger models.

Pragmatically, the loading problem is solved by subjecting the component in question to loading under laboratory conditions; these conditions are based on experience and are usually set at such a level that failure is not expected even under the most unfavorable conditions. This puts the design on the "safe side," but at the cost of overdimensioning, as the design does not take into account actual forces encountered in vehicle service.

By various investigations, efforts are under way to input road profiles into simulations and apply nonlinear vehicle models to ascertain forces and therefore actual loads. At present, the main obstacle to routine application of this approach is the effort required to establish and validate complex vehicle models. Demanding conditions are imposed on these, as nonlinear dynamic behavior of all components such as dampers, hydraulic mounts [5], tires, etc., must be reproduced. Analogous to currently employed experimental methods, the results of such a simulation include actual loads experienced in a vehicle installation. To evaluate stresses, life expectancy calculations must be included. Before such methods can be routinely applied, several questions regarding methodology must be answered. Still, even with the present body of knowledge, results are already available to a limited extent.

Joining technology

A special case of strength calculation involves dimensioning of joining elements, especially auto body spot welds. On the one hand, the number of spot welds is a cost factor; on the other, these elements are especially critical in terms of strength. Theoretical load analysis is particularly difficult because in addition to metallurgical effects, these joining elements are subjected to stress concentrations at the weld nugget as well as various load conditions such as tension and shear, which must be handled differently. Strength evaluation can therefore only be conducted through consideration of known quantities which have been determined by actual experiment. These are in the form of Wöhler curves (fatigue curves) of samples with comparable weld joints, used to evaluate characteristic stress values, so-called structural stresses. Realizing these methods for practical applications is the object of intense research; more detailed treatment of these relationships as seen from a computational standpoint, with literature references, may be found in Ref. [6].

Structural optimization

The objective of structural calculation is always to establish shapes or particular parameters of components, as well as powerplant and suspension, in such a manner that certain quantities such as weight and stress are minimized. As a rule, this is achieved by individual calculations which examine the effects of various input parameters. Mathematical optimization methods permit determination of minima (or maxima) of a particular quantity as functions of input parameters. The quantity to be optimized is termed the target function; the variable parameters are termed design parameters (or design variables). Only one target function can be specified (usually weight); other quantities (e.g., stress) must be defined as secondary conditions.

These techniques are based on finite element methods; characteristically, their algorithms draw on sensitivity matrices, whose elements are derivatives of the target functions with respect to the design parameters. Efficient solution methods are available for determination of relative minima (absolute minima may only be found for special cases) [7]. The calculations themselves are highly time-intensive, especially given a large number of design parameters. Depending on complexity of the task, one of three types of optimization may be selected.

Parameter optimization

Only parameters with no effect on component form (i.e., wall thickness, elastic modulus) are permitted as design variables. This method may be efficiently applied to optimize stiffness of vehicle components such as the body.

Shape optimization

Additionally, design variables appear which describe the shape of a component. This permits the design to be changed; however, unless one rules out repositioning of nodes, the finite element mesh must be regenerated, which can complicate the process.

Topology optimization

All structural calculations are based on an originating design which is improved or optimized. In contrast to such methods, topology optimization [8] needs only the available space; the algorithm finds the "optimum" necessary structure, without resorting to sensitivity matrices. This is then available in a form resembling a finite element mesh and naturally does not take into consideration questions of manufacturing, costs, etc. This method is particularly suited to the concept phase and offers an outstanding basis for component design.

Experience has shown that this method permits development of designs which lead to considerable weight savings without sacrificing quality.

Dynamics

To borrow a line from the ancient Greek philosopher Heraclitus, who said "Everything flows," it may be said that on a motor vehicle, "Everything vibrates." Vibrations (see Section 3.4) are excited by various processes within the vehicle, including unbalanced inertia forces in the powertrain, road effects, aerodynamic forces, exhaust tip noise, and so forth. Different simulation tools are applied in keeping with various phenomena. The most commonly used tools are based on finite element methods, with which structural layout questions can largely be answered, such as determination of body resonant frequencies. Along with investigations of individual components, linearized models are also used to conduct overall vehicle analyses.

One important field of application is acoustic simulation. Seen from the point of view of simulation processes, a distinction must be drawn between *structure-borne noise*, *sound pressure distribution* within the vehicle interior (caused by standing waves), and *radiated noise*. The first of these covers questions regarding sound distribution paths within the vehicle structure; one example of structure-borne noise investigation is the question of the effect of the driveshaft on vibration transmission behavior of neighboring parts.

To calculate sound pressure distribution, the body of air inside the vehicle must be coupled to the vehicle structure. The difficulty in calculating sound pressure is in part attributable to the large number of resonant frequencies within the frequency range of interest. The degree of complexity of the required computational models is therefore extremely high. Along with a clean representation of vehicle stiffness and mass (in the frequency range between 0 and 100 Hz, a body-in-white has about 200 resonant frequencies), damping behavior must be accurately described. This involves equipment details, all rubber elements, sound insulation mats, and so on [9]. The modeling problem greatly restricts the range of frequencies which must be analyzed. If parameters cannot be correlated by means of test data, 70 Hz represents a cutoff point beyond which a disproportionate effort must be expended to obtain assured results.

At higher frequencies (above ~150 Hz), deterministic calculation of sound pressure level within the vehicle is not possible: on the one hand, finite element models cannot be resolved sufficiently, and their verification using test data is not possible; on the other hand, unavoidable scatter in geometry (e.g., differences in material thickness), damping, absorption, and so on cause indeterminate variations in results at higher frequencies. In contrast to a deterministic process, statistical energy analysis (SEA) [10] describes response behavior by means of statistical quantities such as mean and standard deviation. The structure to be examined is divided into subsystems, for which average energy across a predetermined frequency range is calculated. Subsystems are described by modal quantities—that is, mass, resonant frequency, damping—and are connected via stiffness and damping properties of their respective couplings. To model a vehicle's interior acoustics, the body of air represents one subsystem; average sound pressure is derived from the average energy of this system. One application to an overall vehicle is described in Ref. [11].

The primary source of noise radiation is the engine/transmission unit, with additional contributions from components such as the exhaust system and driveshaft. Computational methods are based on the theory of a spherical radiator, which permits relatively rapid results for approximate values but at significant cost in terms of computing resources.

Vehicle safety

In contrast to implicit finite element methods, as used for strength, stiffness, and dynamic analysis, the explicit finite element method has established itself in recreation of crash processes. From a mathematical standpoint, this approach is based on a finite difference method in the time domain and a finite element method in space. Dynamic equilibrium is established at every point in time for every point in the finite element structure. Given certain assumptions—diagonal mass matrix and only indirectly predetermined damping properties—it is not necessary to solve the system of equations for the stiffness matrix typical of the implicit method. This makes it possible to solve higher-order nonlinear problems within acceptable computing time. This method combines the advantages of classical (implicit) finite element methods with the methods of multibody dynamics [12]. Because very small time increments must be used to achieve convergence, expansion of the explicit finite element method is currently limited by its significant demands on computing time.

Currently available software systems for crash calculation are so well developed (see Section 6.5) that their application, usually during the design stage, permits implementation of structural layout measures to provide optimum passive safety. Such early application greatly reduces the need for later experimental work. Because vehicle crash behavior has been greatly improved by application of simulation technology, this approach has enjoyed widespread application on a worldwide basis. Frontal impact as well as side and rear impact given specific conditions are calculated. Availability of numerical models for the standardized dummies used in actual crash testing permits including occupants and pedestrians in computational simulations. The high level of safety offered by vehicles currently on the market is the result of systematic application of simulations. Analogous to evaluation of actual tests, acceleration history, energy absorption, local deformation, and so on are considered in evaluating simulated results. In addition, simulation results offer an opportunity to uncover the causes of undesirable behavior in the testing phase insofar as these are attributable to structural causes, thereby accelerating the development process. Detail questions, however, such as binding of doors, fuel leakage, and so forth, remain the preserve of actual experimental testing.

Metal forming (deep drawing)

Simulation of the deep drawing process for auto body parts is an established part of the development process. This permits determination of parameters such as blank dimensions, drawing beads, hold-down forces, and so on so that tearing or formation of drawing folds during the forming process may be largely eliminated. Application of such methods yields considerable time and cost savings, as equivalent results determined by experimental means can only be achieved with expensive tooling and great effort. Fig. 10.2-2 shows the result of such a simulation.

Both implicit and explicit finite element methods are applied. Because of the great demand for computing time, approximations must be used. Implicit methods dispense with exact consideration of sheet metal bending properties; this is acceptable, with limitations, for simulation of deep drawing processes with thin-gauge sheet metal. Explicit methods may be disturbed by dynamic effects contained in the solution process. Productive application is predicated on inclusion of all phenomena such as material behavior, friction, drawing fold formation, springback, and so on. For details regarding capture and treatment of important effects as well as current research, the reader is referred to Ref. [13].

10.2.3.2 Flow calculation

Numerical re-creation of flow phenomena is one of the most demanding applications of simulation methods. On the one hand, complex physical effects must be considered; on the other hand, numerics of differential equations place high demands on computing resources. Early efforts were based on potential theory, which permitted engineering solutions to be arrived at with relatively little effort. Today, with the availability of more powerful computers and corresponding software, numerical methods based on the Navier-Stokes equations have established themselves. These encompass, along with convective quantities, viscous effects, although assumptions must be made regarding turbulence and wall effects. As these are often very important for accurate simulation of real processes, great importance is attached to the search for more powerful turbulence models. In most fields of application, the effects of heat transfer must be considered along with determination of the flow field, which increases simulation model complexity and results in added numerical effort. Furthermore, certain material constants, themselves difficult to determine, must be considered.

Along with complex three-dimensional methods in which the fluid volume is divided into discrete cells, one-dimensional methods have been established for certain applications such as pipe flow. These yield rapid results and are therefore well suited for examining parameter variations.

One-dimensional methods are primarily applied to engine investigations; the main fields of three-dimensional simulations are vehicle exterior flow (aerodynamics), interior flow with regard to climate control questions, and unthrottling of engine compartment flow. Engine internal flow processes represent an additional field of application.

One-dimensional flow calculation

One-dimensional flow calculations are applied to flow in pipes or similar passages. Flow coefficients, derived from approximations or through experiment, account for pressure losses as a result of throttling, bends, branching, and so on. These methods are therefore very close to actual experiment and are especially suitable for supporting test bench investigations.

A typical application is simulation of the gas exchange process, a firmly entrenched component of engine development practice. The process serves to simulate dynamic gas flow processes within engine intake and exhaust tracts using simplified assumptions for pro-

Fig. 10.2-2 Result of deep-drawing process simulation for a passenger car side panel.

cesses within the cylinder. Combustion effects are also taken into account; in the simplest case, these are simulated using Vibe's Law [14]. Along with geometric data such as tract dimensions (lengths and diameters of passages, chamber volumes, etc.) as well as boundary conditions such as wall temperatures, the parameters required to construct a model (such as pressure loss coefficients or turbocharger characteristic curves) are usually determined experimentally. The results consist of time-dependent flow parameters such as pressure, flow velocity, gas temperature, and mass flow at any point in the tract system, as well as integral engine characteristics such as charging efficiency, volumetric efficiency, end gas fraction, indicated mean effective pressure, and indicated power. Analysis of these results yields insights into gas dynamic processes within the engine as well as hints toward possible means of optimizing the above-engine characteristics.

One field of application for gas exchange calculations is layout of induction and exhaust systems. This examines the effects of changes in tract length or diameter or configuration of tract intersections on engine parameters such as volumetric efficiency, end gas fraction, or torque. Its one-dimensional calculation approach demands comparatively modest computation time, so that many variations can be examined in short order to find an optimum configuration. Other common questions are, for example, those of finding optimum valve timing, with selection of suitable valve lift curves for cylinder intake and exhaust valves, or evaluation of turbocharger concepts.

Another example of one-dimensional simulation is investigation of engine oil circulation with the goal of assuring oil supply to all components with minimum pumping power. Special attention must be given to accurate assessment of journal bearing properties.

Vehicle exterior flow

For obvious reasons, the primary objective of exterior flow simulation is to minimize aerodynamic drag while taking into consideration aerodynamic lift properties. Due to flow separation, description of wall behavior and turbulence are important, and because of weaknesses in this regard, the predictive ability of such simulations is fraught with uncertainty. Nevertheless, this tool has become a firmly embedded component of the development process. On the one hand, additional questions such as pressure distribution on the surface can be answered with sufficient accuracy. On the other, modern software tools permit trend determinations on the basic body with inclusion of the underbody, which allows computer-based evaluation of vehicle concepts.

Current work on turbulence models leads to the expectation that simulation methods might someday provide a "virtual wind tunnel" to determine lift and drag coefficients to the required degree of accuracy.

Processes based on the Navier-Stokes equations require a computational grid which, in the case of exterior flow, consists of several million cells. Generating these grids is so complex that utility of such methods in the development process is severely limited.

Considerable advantages are offered by a process based on a development of the Lattice Boltzmann method [15]. In this, the fluid is modeled by idealized particles, which move in a discrete cubic grid at discrete velocities and in discrete time intervals. Movement results in collisions between the particles; these are resolved by a rigidly defined set of collision rules which alter the condition (direction, speed, energy) of particles. Suitable formulation of these collision rules ensures that macroscopic behavior of the fluid exactly matches the Navier-Stokes equations. Macroscopic fluid properties such as pressure, velocity, density, and so on are determined through statistical evaluation of particle concentrations and properties. For low Reynolds number flows ($Re <$ 10,000), all length scales of importance to flow behavior may be directly resolved. In the case of higher Reynolds numbers, an additional simple turbulence model is applied. Effects of the turbulent boundary layer are taken into consideration by modeling wall shear stress, which includes local velocity and a friction coefficient. This last is determined by application of an expanded logarithmic law of the wall, which includes pressure gradients to better account for pressure-induced flow separation.

Generation of the cubic calculation is entirely automatic, independent of the geometric complexity of the body in question. Calculation proceeds in discrete time steps and is unconditionally stable—that is, every calculated time step represents a discrete physical time interval. By this means, nonsteady-state flow processes are correctly represented.

Vehicle interior flow, climate comfort

The objective of vehicle interior flow simulation is to optimize climate comfort inside the vehicle. Along with flow effects, heat management must be included in such investigations. Because of the resulting complexity, current practical applications are limited to examination of isothermal flow fields; for thermally affected flows, complex physical effects such as solar insolation as well as human physiology must be considered. The latter are described by "comfort models" which simulate occupant perceptions [16, 17].

With the available flow calculation systems, it is possible to examine arbitrarily bounded bodies—that is, including the passenger cabin and air ducts. Flow from the heating unit, through ducts and vents, to the passenger may be analyzed. This permits the layout of air ducts to achieve maximum flow and establishes preconditions for optimum windshield defrosting, for example. An overview of various methods and their application in automotive development can be found in Ref. [3].

Application of simulation technology opens the possibility of achieving significantly increased climate comfort within the vehicle, as such optimization is much more time-intensive if classical experimental methods are employed.

Cylinder gas exchange processes

Gas flowing into the cylinder and gas motion during the intake stroke largely determine the course of the subsequent combustion process and therefore of power generation, pollutant formation, acoustics, and so on. Inlet flow conditions in the induction system may be analyzed using the same techniques which are applied to vehicle external flow. However, transient processes must be reproduced; the main difficulty is matching the computational grid to changing cylinder volume, which adds considerable complexity to modeling the problem. Today, development of engines with internal mixture formation makes extensive use of numerical methods, as intensive research is currently under way on direct injection for both diesel and gasoline engines. For optimal combustion process layout, experimental methods fail due to their inability to capture complex flow conditions within the cylinder. Simulation methods, however, are able to provide detailed information regarding flow development. Proper combination of both methods opens possibilities for efficient development programs.

Evaluation of mixture formation also requires reconstruction of the injection process, including droplet formation mechanisms, droplet disintegration, vaporization, and so forth. This is provided by so-called spray models, which reflect extremely complex physical effects [18]. Fig. 10.2-3 shows how simulation offers insight into spray formation and its considerable effect on mixture formation. Although the models on which such investigations are based are currently usable, these contain assumptions and approximations which must be parameterized for each individual application. Intensive research is currently under way to refine these models and assure their predictive power.

Precise capture of mixture formation phenomena is a prerequisite for simulation of nonhomogeneous combustion, as found in direct injection gasoline and diesel engines. Energy released as a result of chemical reactions and their intensity and temporal history depend on various factors such as local turbulence. Simulation is carried out on the basis of modeling assumptions which at present are not yet fully validated and are therefore still the subject of research activities.

10.2.3.3 Simulation techniques in support of experimental methods

Simulation and computation are most useful in the earliest phase of vehicle development, when they can be applied during the conceptual process to examine often complex physical relationships. Theoretical methods are also valuable in the testing phase, as soon as phenomena whose investigation requires inclusion of the actual vehicle must be captured. Typical fields of application are processes known under the expressions "Hardware-in-the-Loop" (HiL) or "Software-in-the-Loop" (SiL). One example of HiL is testing of engine control units: the actual control unit is part of a control loop which simulates engine behavior. The applied engine models resemble those discussed above, except that they are considerably "rougher," as they must be capable of real-time response. Another application is found in layout of systems to improve vehicle dynamics (in the simplest case, ABS), in which the control unit is integrated within the simulated vehicle control loop.

Simulation methods which are not yet fully developed are often applied during the testing phase in order to obtain predictive results. These include, for example, assurance of electromagnetic compatibility (EMC). This topic is gaining particular importance as the number of electronic components installed in vehicles increases. Simulation may help to minimize effects of electromagnetic fields emanating from disturbing elements, such as the ignition system, on electronic components,

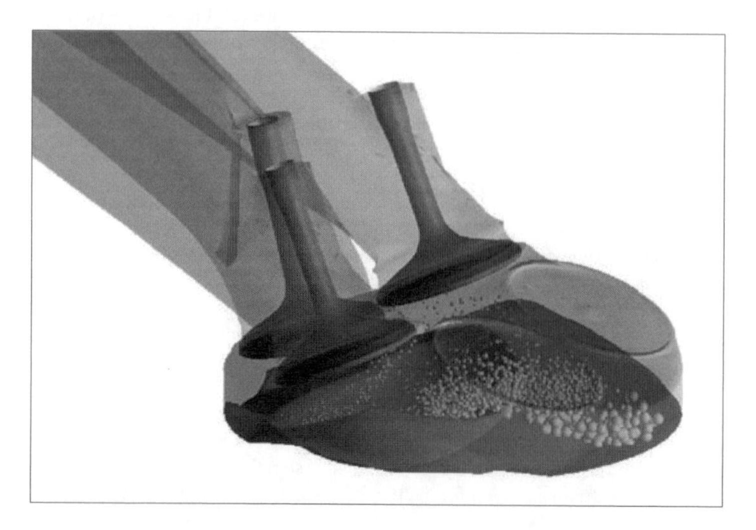

Fig. 10.2-3 Spray propagation inside combustion chamber as determined by cylinder inlet flow simulation.

for example the antenna. Along with this field simulation, transmission line simulation attempts to optimize transmissive behavior of wiring harnesses. Although theoretical solutions for such problems are available [19], their application in vehicles is in its infancy; current problem solutions may be found in [3].

Every investigation using simulation methods, as well as those employing experimental techniques, generally yields a multitude of results given an arbitrarily large number of input parameters. One example is analysis of elastokinematics in development of a vehicle suspension. Wheel kinematics are defined by suspension link pickup point coordinates as well as elasticity of mounts. The results consist of steering angle response, toe curves, and similar quantities. The question to be answered is: how can one reduce the number of input parameter variations needed to discern the desired kinematics, or how can an optimum combination of these parameters be found? This question does not only arise in the case of suspension kinematic layout, but also for many other problem sets, such as operation of engine dynamometers or crash test optimization.

One solution method, known under the "Design of Experiment" (DOE) concept [20], consists of finding a relationship between desired resulting quantities and input parameters by means of trial functions. This permits reducing the number of possible parameter variations to only those which represent a desired system behavior.

Another approach which differs in form but largely follows the same path is application of neural networks [21]. The relationship between input and output parameters is not achieved using trial functions, but rather by special methods for solving systems of equations. Here, too, results of prior simulations (or actual measurements) serve as a foundation.

Future methods permit capturing the effects of input parameter scatter. The results of individual calculations, based on slight modification of input quantities, are evaluated using statistical methods. Expanding deterministic results to stochastic results permits more realistic evaluation—for example, in questions of system robustness. Furthermore, comparison with measurement—which always takes place stochastically—is closer to reality. Important fields of application include higher-order nonlinear simulations, such as vehicle collisions [22, 23].

References

[1] Braess, H.H. "Selbstorganisation und Selbstoptimierung—Chancen einer fraktalen Entwicklung der Mess- und Versuchstechnik im Automobilbau." Presented at "Mess- und Versuchstechnik im Automobilbau 1995." VDI-Gesellschaft Fahrzeug- und Verkehrstechnik; VDI Berichte 1189, VDI-Verlag GmbH, Düsseldorf.

[2] "Berechnung und Simulation im Fahrzeugbau 1996" conference. VDI-Gesellschaft Fahrzeug- und Verkehrstechnik; VDI Berichte 1283, VDI-Verlag GmbH, Düsseldorf.

[3] "Berechnung und Simulation im Fahrzeugbau 1998" conference. VDI-Gesellschaft Fahrzeug- und Verkehrstechnik; VDI Berichte 1411, VDI-Verlag GmbH, Düsseldorf.

[4] Babuska, L., B. Szabo, and I.N. Katz. "The p-Version of the Finite Element Method." *SIAM J. Numerical Analysis*, Vol. 18, No. 3, June 1981.

[5] Weber, B. "Identification of Modelparameter with ADAMS/Design of Experiments (DOE) and ADAMS/Optimization." Proc. ADAMS Users' Conference, Paris, 1998.

[6] Radaj, D., and S. Zhang. "Anschauliche Grundlagen für Kräfte und Spannungen in punktgeschweißten Überlapp-verbindungen." *Konstruktion* 48, Springer Verlag, 1996.

[7] Vanderplaats, G.N. *Numerical Optimization Techniques for Engineering Design—with Applications*. McGraw-Hill, Inc.

[8] Maute, K., E. Ramm, and S. Schwarz. "Adaptive Topologie- und Formoptimierung bei linearem und nichtlinearem Strukturverhalten." Proc. NAFEMS Seminar zur Topologieoptimierung, Aalen, September 23, 1997.

[9] Dirschmid, W. "Parameter Identification for Noise Prediction in Car Structures." SAE 901754.

[10] Lyon, R.H., and R.D. de Jong. *Statistical Energy Analysis*. Second Edition, Butterworth-Heinemann, Boston, 1995.

[11] Geissler, P. "Structure Borne Sound on the Surface of Automotive Mufflers." Proc. Internoise 90, Gothenburg, 1990.

[12] Bathe, K.-J., and E.L. Wilson. *Numerical Methods in Finite Element Analysis*. Prentice-Hall, Inc., Englewood Cliffs, New Jersey.

[13] Döge, E. (ed.): "Umformtechnik an der Schwelle zum nächsten Jahrtausend." 16. Umformtechnisches Kolloquium, Hannover February 1999. ISBN 3-00-003848-5.

[14] Correira da Silva, L.L. "Simulation of the Thermodynamic Processes in Diesel Internal Combustion Engines." SAE 931899.

[15] Chen, H., C. Teixeira, and K. Molvig. "Digital Physics Approach to Computational Fluid Dynamics: Some Basic Theoretical Features." International Journal of Modern Physics C, World Scientific Publishing Company.

[16] Currie, J. "Numerical Simulation of the Flow in a Passenger Compartment and Evaluation of the Thermal Comfort of the Occupants." SAE 970529.

[17] Maue, J., D. Wahl, T. Breitling, and W. Rössner. "Theoretische Untersuchungen zum thermischen Komfort von Fahrzeuginsassen." Haus der Technik, Tagung Thermische Behaglichkeit in Fahrzeugen, 1995.

[18] Islam, M., E. Blümcke, and A.D. Gosman. "Numerical and Experimental Studies of High-Pressure Diesel Fuel Injection." Second Symposium Towards Clean Diesel Engines, Paul Scherrer Institute and ETH Zürich, May 1998.

[19] Jin, J. *The Finite Element Method in Electromagnetics*. John Wiley & Sons, New York, 1993.

[20] Knight, J.W. "Design of Experiments and Simultaneous Engineering." Society of Automotive Engineers, International Congress, Detroit, February 1989.

[21] Stricker, R., and T. Fleischhauer. "Effective Optimization of Complex Systems Using 'Neural Know-How Recycling' from Simulation and Test, by Way of Example Applied to Petrol Engine Management Calibration." *Proceedings* of the 2nd International Conference on Adaptive Computing in Engineering Design and Control, Plymouth, UK, 1996.

[22] Marczyk, J., M. Holzner, et al. "Stochastic Automotive Crash: A New Frontier in Virtual Prototyping." PAM 97 User Conference, Prague, October 1997.

[23] Seiffert, U., and P. Scharnhorst. "Die Bedeutung von Berechnungen und Simulationen für den Automobilbau." *ATZ*, 1989, pp. 241–246 and pp. 303–306.

10.3 Measurement and testing technology

10.3.1 Brief retrospective

The many shortcomings of early, at first entirely hand-built, automobiles demanded comprehensive design improvements as well as development of appropriate

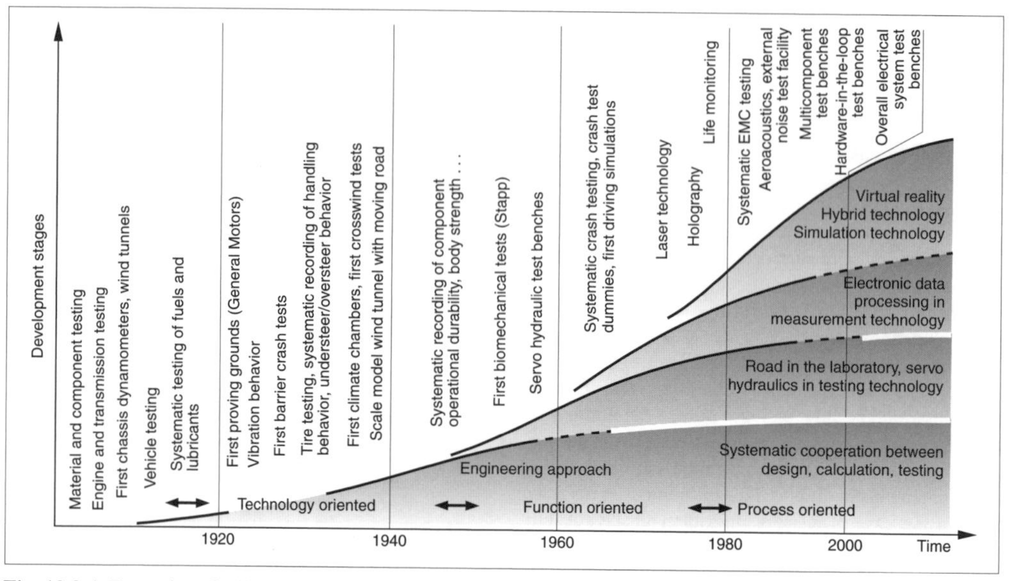

Fig. 10.3-1 Examples of milestones in automotive measurement and testing technology.

test methods (see [1]). Initial emphasis on assurance of some minimum level of usability was soon followed by customer demands for performance (including load-carrying ability, acceleration, top speed), safety, comfort, and economy. After World War II, highly systematic efforts were initiated to address demands arising from difficult operating conditions (heat, cold, high altitude; e.g., [2]), as well as additional requirements such as environmental protection and collision injury mitigation. Over the course of decades, these efforts led to a comprehensive suite of measurement and testing technologies; typical examples are shown in Fig. 10.3-1, from Ref. [3, 1995].

10.3.2 Basic considerations regarding automotive measurement and testing technology

In research and advanced development, experiments are conducted to recognize and verify the potential of various technologies and concepts. In production development and production, tests are conducted to assure product and process maturity (simulations) and comparisons drawn (actual vs. target specifications; see Ref. [3, 1999]). In other words, despite—or indeed because of—comprehensive calculation and simulation methods, measurement and testing technologies are indispens-

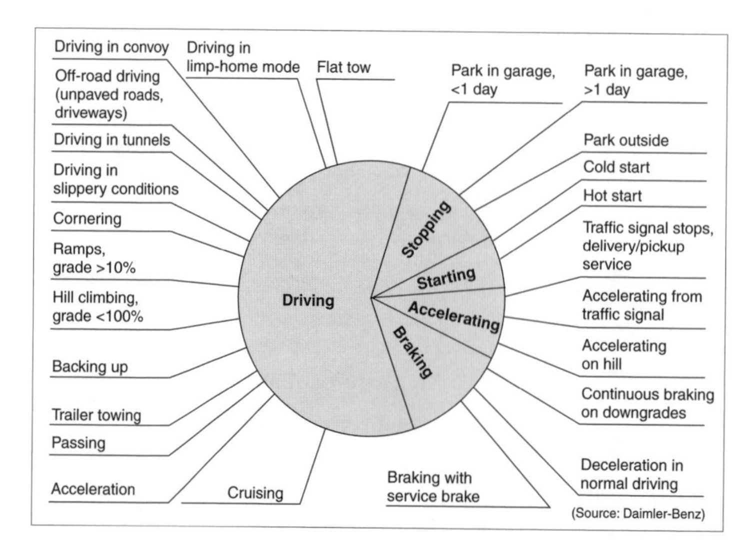

Fig. 10.3-2 Vehicle operating conditions.

able in all phases of the product creation process. Their tasks include:

— Support the goal establishment process, monitor goal attainment
— Determine development, production, regulatory constraints
— Provide a foundation for design decisions
— Fine-tuning and optimization of components and systems
— Validation and verification of computational models and simulation methods, as well as determination of experimentally derived parameters
— Data and time functions for test bench inputs

The significant basis of both the experimental domain and design is the complete range of all vehicle operating conditions and service requirements (Fig. 10.3-2), their frequency and degree of difficulty (customer, road, regional effects, load collectives, etc.) including all interior and exterior vehicle loads (Fig. 10.3-3). A systematic approach to testing technology may be derived from all of these factors. Such a systematic approach encompasses the type of testing (Fig. 10.3-4) as well as assignment of test and measurement methods to specific tasks (Fig. 10.3-5).

In order to establish a testing method, it is imperative that suitable testing objects are available or can be created. These may range from material samples to experimental components, functional assemblies (e.g., engine), driveable but incomplete prototype vehicles, and complete production vehicles. To these must be added establishment of specific test conditions as well as corresponding testing methods and evaluation criteria (for example, see Fig. 10.3-6). The latter are either direct quantitative adoptions of the target specifications (e.g., fuel economy, acceleration ability) as empirically/scientifically derived quantities describing complex behavior or strongly subjective effects (see below), or they are from the biomechanics of collision events (Section 6.5).

As a result of driving and operational errors, vehicles may be subjected to considerably greater loads than those encountered in approved operation. Therefore, from a

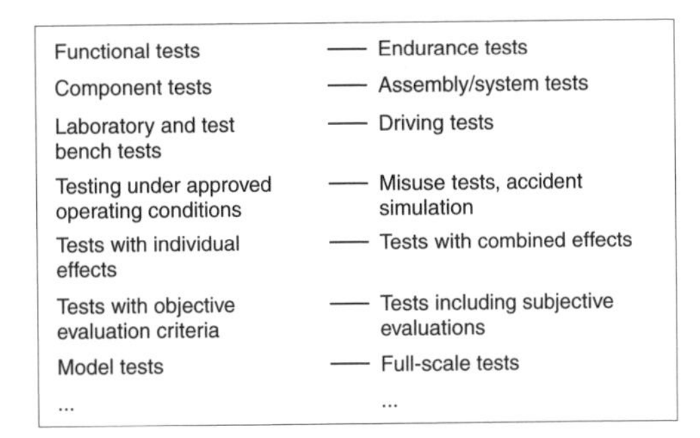

Functional tests	— Endurance tests
Component tests	— Assembly/system tests
Laboratory and test bench tests	— Driving tests
Testing under approved operating conditions	— Misuse tests, accident simulation
Tests with individual effects	— Tests with combined effects
Tests with objective evaluation criteria	— Tests including subjective evaluations
Model tests	— Full-scale tests
...	...

Fig. 10.3-4 Experimental methods in automotive engineering.

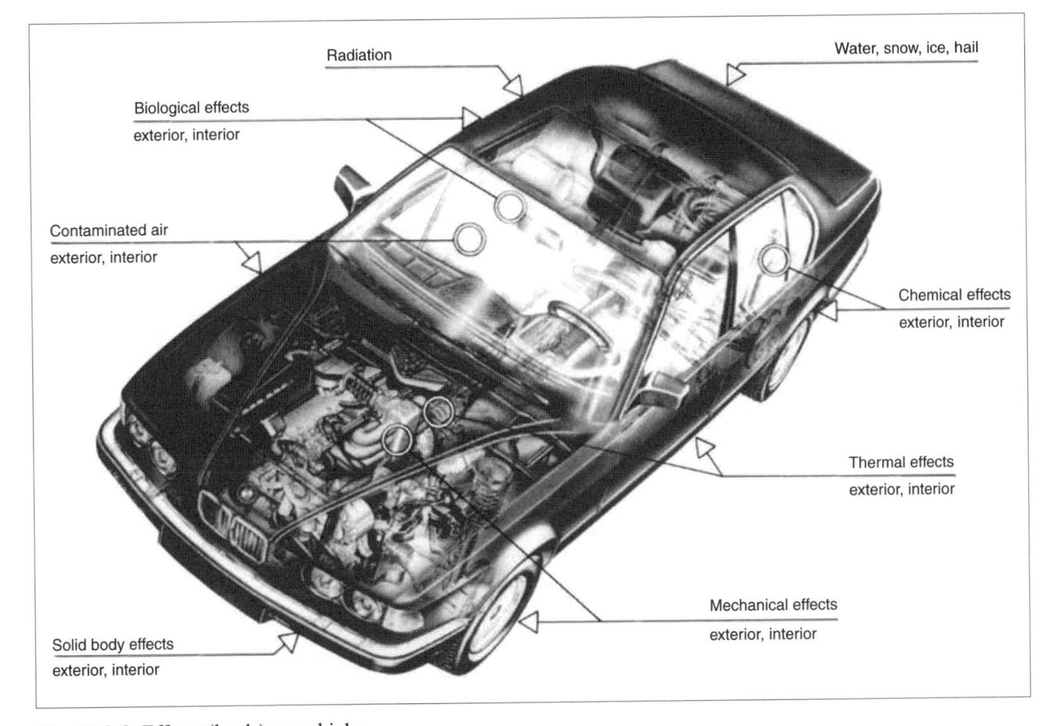

Fig. 10.3-3 Effects (loads) on vehicle.

Testing assignment \ Testing method		Actual road testing	Proving grounds	Physical test benches/ dynamometers	"Dummies in the loop"	"Hardware in the loop"	"Man in the loop"	Virtual prototyping
Functional and endurance testing	Overall vehicle in its environment	• Vehicle overall function, driving experience • Long-distance capability • Endurance testing • Vehicle in extreme conditions (arctic, tropic, high altitude . . .)	Vehicle overall function and driving experience under defined operating conditions	• Aerodynamics • Climate control • Exterior noise • Interior noise • Vibration	• Crash behavior • Climate control behavior • Acoustics	Vehicle dynamics simulation under defined operating conditions, up to the vehicle handling limits	Driving simulation	• Design (exterior, interior) • Ergonomic functions • Accessibility for maintenance and repair
	Aggregate systems	• Complex assemblies and functions under changing and extreme operating conditions • . . .	• Drivetrain tuning • Suspension tuning (longitudinal, transverse, vertical dynamics • . . .	• Engine, transmission, drivetrain • Suspension and braking systems • Body-in-white • . . .	Restraint systems	• Drivetrain and suspension control systems • On-board electrical system	• Ergonomics • Driver assistance systems	Prototype construction, assembly
	Materials, components, operating materials	Behavior of materials, components and operating materials under actual operating conditions	Behavior of materials, components and operating materials under defined operating conditions (including misuse)	Behavior of materials, components and operating materials	In vivo and in vitro experiments	Recognition of component functional faults	Olfactory testing of materials	Component geometry

Fig. 10.3-5 Examples of fields of application for experimental technology in automotive engineering.

	Death Valley summer	Arctic winter
Latitude	35° N	68° N
Solar altitude	78°	2°
Solar radiation intensity, direct	1,000 W/m²	150 W/m²
diffuse	90 W/m²	20 W/m²
Vehicle orientation	Front toward sun	Front toward sun
Ambient temperature	40°C	–20°C
Engine compartment temperature	85°C	+30°C
Trunk temperature	65°C	–10°C
Ground temperature	40°C	–20°C
Relative humidity	10%	90%
Vehicle speed	0, 32, 64, 96 . . . km/h	0, 32, 64, 96 . . . km/h
Blower stages	0 to III	0 to III

Fig. 10.3-6 Test conditions for "Death Valley summer" and "Arctic winter" scenarios.

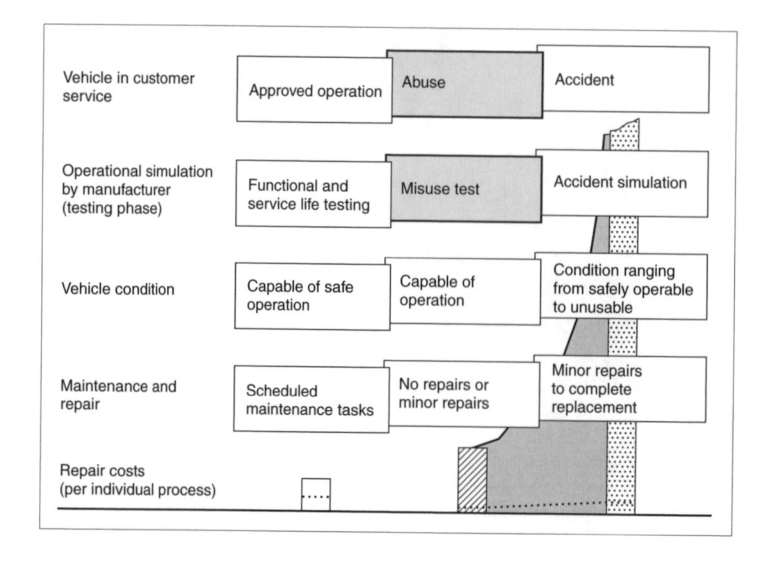

Fig. 10.3-7 Regarding definitions of vehicle abuse and misuse tests.

large number of possible abuses, appropriate misuse tests must be developed and applied (Fig. 10.3-7, from Ref. [3, 1987, 1999]). On the one hand, results of such tests should not be used to generate designs with unacceptable disadvantages in normal vehicle service, but on the other hand those results should ensure that the vehicle remains operable (in some cases, with restricted operability) if abused.

Due to the wide variety of testing tasks, measurement technology has become an independent specialty within the automotive engineering field. Today, a host of stationary and mobile devices and methods permit manifold, often very precise and highly dynamic measurements of physical quantities (e.g., [4]) which can be classified in various fields, notably:

Mechanics

- Component travel, relative speed, acceleration
- Force, moment, deformation, stress, pressure distribution
- Air mass, flow conditions
- Frequency, amplitude, and other vibration- and acoustic-related quantities

Thermodynamics, heat, and energy technology

- Temperature, pressure
- Heat and heat transfer
- Characteristic parameters of combustion processes
- Energy consumption

Electrical technology

- Voltage, current, resistance, frequency
- Electromagnetic fields

Optics

- Illumination intensity
- Light distribution

Materials, components, operating materials

- Physical and chemical properties
- Geometric and technical characteristics

Tribology

- Characteristic quantities of friction and lubrication (lubrication film thickness)
- Wear processes

Accident-mitigating safety

- High speed photography
- Biomechanical characteristics

Exhaust gas technology

- Exhaust gas analysis
- Online exhaust measurements

While comprehensive and rational measurement techniques can be applied to assemblies and entire vehicles, mobile measurement technology applied to moving vehicles must take into consideration space and weight limitations and sometimes other restrictions (Fig. 10.3-8, from Ref. [3]). Nevertheless, many endurance test vehicles are virtual rolling laboratories to record, for example, all possible critical temperatures and loads. Special demands are imposed on noncontact measurement techniques in hard-to-reach areas, such as moving or hot components, as well as on the evaluator's ability to make visible complex processes, such as in compact, highly stressed assemblies.

For years, computer-based measurement systems have been considered state of the art. These permit optimized, monitored testing and complex online evaluation (time and frequency domain, power spectra, modal analysis, etc.) including "quick look" representations.

Before measured values are stored and/or processed, they must be captured free of interference, filtered, conditioned, and transmitted. For driving in extreme conditions (climate, road conditions), this may require that specific measures be taken. In terms of test operation, special organizational efforts are required for planning and execution of test drives with vehicle fleets in distant locations at very specific times of the year (for examples of test locations, see Fig. 10.3-9).

Early on, such factors as the great effort demanded by testing, the conditions which often could not be held constant, problems with measurement technology, and so on led to attempts to supplement testing of assembled units by also moving as many complete vehicle tests as possible "from the road to the lab" (e.g., [1, 3]). Even before World War II, large wind tunnels were built to examine vehicle aerodynamic properties; chassis dynamometers were built to test the entire drivetrain as installed in the vehicle. Later examples include climate-controlled dynamometers as well as operational endurance testing of suspensions and entire vehicles by means of complex servo-hydraulic test rigs that could reproduce in real time any loads measured on actual roads (Fig. 10.3-10; see also Ref. [5]). One principal requirement for such realistic testing methods was development of powerful real-time computers and software systems.

The best efforts of development and manufacturing planning departments notwithstanding, production startup and customer deliveries of new models are only approved after a sufficient number of vehicles have proven their functional ability, reliability, and impression of quality in actual service. Even so, given the high complexity of these products and processes, it is quite possible that not everything works exactly as planned.

10.3.3 Some selected examples

Development of extremely fuel-efficient and low-emissions combustion engines requires precise knowledge of the complex cause-and-effect chain consisting of gas exchange, fuel injection, atomization, vaporization, mixture formation, ignition, combustion, energy conversion,

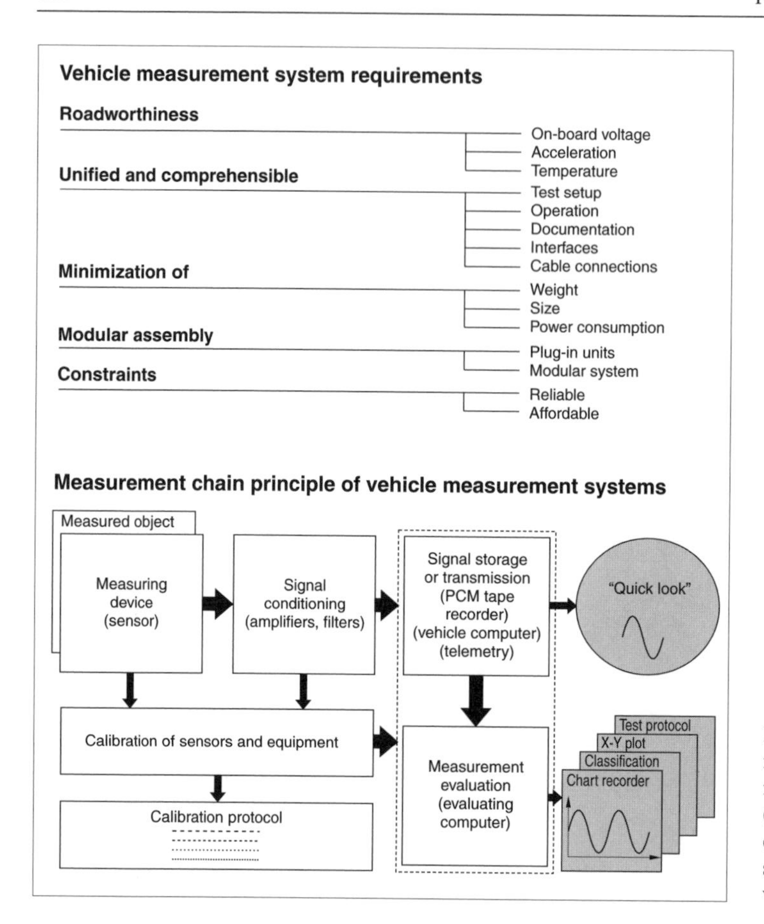

Fig. 10.3-8 Principles and requirements of vehicle measurement systems. (a) Principle of a measurement chain for vehicle measurement systems; (b) requirements of a vehicle measurement system.

Test locations	Emphasis	Special effects
Nürburgring	Suspension, engine, body	Durability testing with extreme loading
Mediterranean area	Suspension, body	Durability testing with extreme loading
Eastern Europe	Suspension, body	Durability testing with high loading
USA/Death Valley	Summer operation	Extreme heat
Arctic	Winter operation	Extreme cold
Upper Bavaria	Overall vehicle	Customer-like conditions
Miramas (proving ground)	Suspension, engine, body	Durability testing with extreme loading
	Engine, drivetrain, cooling system, aerodynamics	High speed testing with durability
Country-specific testing USA, South Africa, Japan, Southeast Asia, Sweden		

Fig. 10.3-9 Overall vehicle testing (example).

and reaction kinetics. In the past decade, to record and understand the associated effects, various laser techniques have been brought to bear. In particular, they have helped make flow fields, spray propagation, spray/wall interactions, flame propagation and particle formation visible, with high temporal and spatial resolution. They have also enabled dynamic exhaust emissions and oil consumption examinations, down to the ppb range (see Refs. [3, 4]). Since the 1970s, laser-induced double-pulse holography has been used to measure vibration

Fig. 10.3-10 Multicomponent chassis test rig for simulation of operational loads.

modes of optically accessible structural surfaces (see Ref. [3]). Another application of laser technology, laser Doppler anemometry, involves visualization of more complex flow conditions such as aerodynamic flow around a vehicle body in the wind tunnel (see Ref. [6]) or in the water jacket of a combustion engine.

The fact that parts or components which appear simple to the uninitiated may often be required to meet demanding test requirements is demonstrated by windshield wipers, in which material properties, geometric details, force distribution, and aerodynamic effects all influence wiping quality, tribology, chatter, and life expectancy.

A further example, the testing sequence for safety belts (Fig. 10.3-11), illustrates the extensive testing effort which is essential in assuring function and longevity of modern vehicles.

Functionality, safety, and reliability of electronic controls and feedback systems cannot be determined using classical methods alone. Therefore, new test system methods as well as automated test environments and fault simulations were developed in order to permit examination of control units and associated software (see Ref. [3]).

Coupling of various methods and technologies is clearly demonstrated by hybrid technologies such as "Hardware-in-the-Loop" or "Software-in-the-Loop," which are increasingly being applied to improve development efficiency (see Section 10.3.4; for examples, see Fig. 10.3-12).

One special attribute of automotive technology is that an entire family of properties cannot be entirely evaluated on an objective basis. An important and increasingly successful subset of automotive testing technology seeks relationships between subjective evaluations and measurable quantities and to derive quasi-objective evaluation criteria from these relationships that can also be applied in evaluating simulation results and in development of "digital cars" (Section 10.2) [7]. Another avenue of investigation involves development and application of anthropomorphic test dummies which are suited for specific tasks, such as crash behavior (Section 6.5), vibration behavior, climate comfort, or audio quality.

10.3.4 On the efficiency of measurement and testing technology

The earlier that cogent test results are available, the sooner a designer can issue his releases and the sooner manufacturing planning can initiate complex investments. However, this had and still has its limits, particularly during the prototype phases. Many "hand-crafted" components are delayed and/or improperly made; test results based on "unfinished" prototypes arrive too late and, to make matters worse, are inconclusive.

For these reasons, auto manufacturers began to develop intensified service life and reliability tests which telescope many years and miles of driving into shorter timespans and distances. In the 1970s, corrosion tests were added to the repertoire (see Refs. [1, 3, 8]). Initially, special "obstacle courses" (incorporating "Belgian blocks," etc.) were built on proving grounds, providing high mileage compression factors. Later, these were supplemented by the previously mentioned computer-controlled simulation test benches, capable of fully reproducing road conditions around the clock regardless of weather and without driver influence. Additionally, time periods of lower-intensity loading may be eliminated (for data reduction processes, for example). It should be mentioned that appropriate time-compression test methods for purely time-dependent aging processes, such as for rubber and plastic components, are not yet available.

In product development, models and sample parts

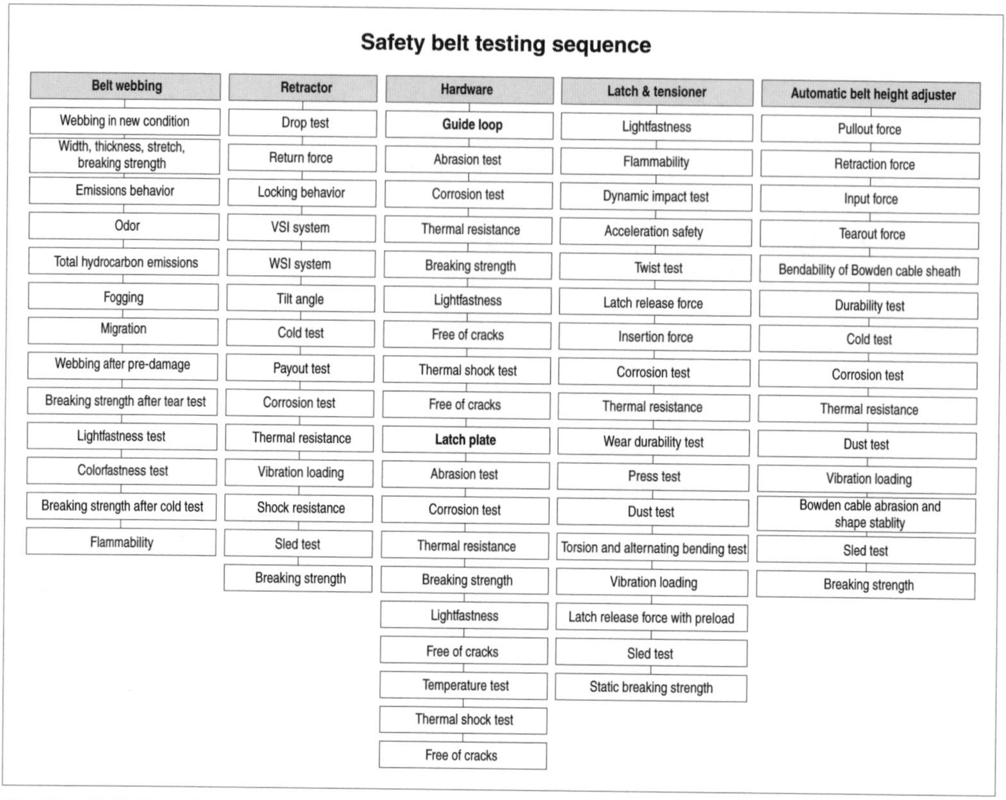

Safety belt testing sequence

Belt webbing	Retractor	Hardware	Latch & tensioner	Automatic belt height adjuster
Webbing in new condition	Drop test	**Guide loop**	Lightfastness	Pullout force
Width, thickness, stretch, breaking strength	Return force	Abrasion test	Flammability	Retraction force
Emissions behavior	Locking behavior	Corrosion test	Dynamic impact test	Input force
Odor	VSI system	Thermal resistance	Acceleration safety	Tearout force
Total hydrocarbon emissions	WSI system	Breaking strength	Twist test	Bendability of Bowden cable sheath
Fogging	Tilt angle	Lightfastness	Latch release force	Durability test
Migration	Cold test	Free of cracks	Insertion force	Cold test
Webbing after pre-damage	Payout test	Thermal shock test	Corrosion test	Corrosion test
Breaking strength after tear test	Corrosion test	Free of cracks	Thermal resistance	Thermal resistance
Lightfastness test	Thermal resistance	**Latch plate**	Wear durability test	Dust test
Colorfastness test	Vibration loading	Abrasion test	Press test	Vibration loading
Breaking strength after cold test	Shock resistance	Corrosion test	Dust test	Bowden cable abrasion and shape stablity
Flammability	Sled test	Thermal resistance	Torsion and alternating bending test	Sled test
	Breaking strength	Breaking strength	Vibration loading	Breaking strength
		Lightfastness	Latch release force with preload	
		Free of cracks	Sled test	
		Temperature test	Static breaking strength	
		Thermal shock test		
		Free of cracks		

Fig. 10.3-11 Safety belt testing sequence.

are an important aid to product design and process planning. In principle, CAD/CAM technology offers the possibility for very rapid creation of sample components based on design data; these are often the connecting link which enables parallelization of product and process design ("simultaneous engineering"; see Section 10.1). Rapid prototyping offers great potential for shortening the product creation process while simultaneously improving product quality. Possibilities include geometric prototypes (in particular during the early phase of process planning) as well as functional prototypes which, depending on material, permit investigation of various functions (see Ref. [9]). Another type of rapid prototyping involves integrated conceptual and evaluation processes for software systems and for design of human–machine interactions (e.g., Ref. [10]).

Although vehicles are becoming increasingly complex and ever more parameters need to be optimized, it must be the objective of every test strategy to conduct as few tests as possible and yet obtain the maximum amount of information. It follows that at a minimum, not all parameters can be investigated separately; multiparameter tests providing detailed information must be conducted. To this end, systems analysis must be applied to determine the most important parameters and their mutual interactions, a process which is all the more precise with greater availability of a priori knowledge, empirical as well as results of calculation and simulation. This analysis is followed by deterministic or statistical test methods best suited for the concrete task at hand (see Refs. [3, 11]). One example is model-supported engine map optimization of modern gasoline engines [3, 1999]. In the early 1980s, only a few parameters sufficed to optimize the two control inputs of ignition and injection timing; at present, several hundred characteristic curves or maps must be optimized for at least eight input parameters, and this for different variations of installed engine, transmission, and exhaust gas treatment. The required development effort ("multiparameter testing environments") would be completely unmanageable using classical testing methods.

The drive for increased efficiency has also led to development of company-internal and, since about 1990, industry-wide activities to form a foundation for hardware- and software-independent standardization of automation, measurement, and evaluation systems (in particular, ASAM, Association for Standardization of Automation and Measuring Systems [3]). Today, standards are indispensable; they enable comparison and communication of test results and binding agreement on methods and tools in development and delivery contracts [3, 1999]. Last but not least, one element of effi-

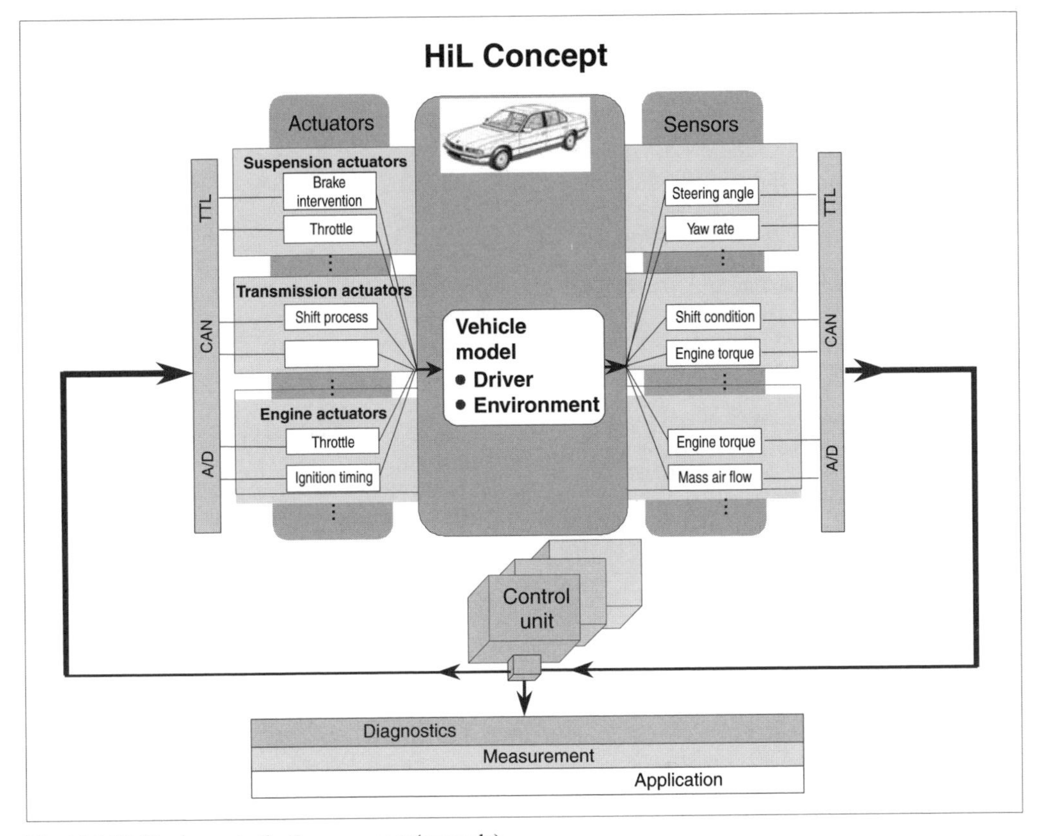

Fig. 10.3-12 Hardware-in-the-Loop concept (example).

cient testing operations is good documentation and, if possible, generalization, in order to minimize the number of new tests. Contributions to this end include new methods of knowledge management, neural networks or, generally, "know-how recycling" [3, 1997].

Still, despite ever more powerful computational and simulation techniques ("digital car"), a large part of vehicle development will be carried out in testing; in powerplant development, for example, two-thirds to three-quarters of development effort is expended in testing.

Even with continually improving and more complete testing methods and techniques, one must always bear in mind that even "proper tests" [3, 1999] are only suitable for detecting faults; they cannot guarantee absolute freedom from faults.

References

[1] "100 Jahre Automobil." VDI-Bericht 595, 1986.
[2] Troesch, M. "Die Alpenstrassen als Prüffeld für Automobile." Annual issue of *Automobil Revue*, Bern, 1955, pp. 101–105.
[3] "Mess- und Versuchstechnik im Automobilbau." VDI-Berichte 632 (1987), 681 (1988), 741 (1989), 791 (1990), 893 (1991), 974 (1992), 1189 (1995), 1335 (1997), 1470 (1999).
[4] Klingenberg, H. *Automobil-Messtechnik.* Springer-Verlag. *Band A: Akustik* (1988), *Band B: Optik* (1994), *Band C: Abgasmesstechnik* (1995).
[5] Weibel, K.-P. "Mehr als 10 Jahre Betriebsfestigkeitsprüfung mit den Mehrkomponenten-Prüfständen der BMW AG." *ATZ*, 1987, pp. 193–198.
[6] Third Stuttgart Symposium on Motor Vehicles and Combustion Engines, February 23–25, 1999. Expert Verlag, 1999.
[7] Multiple authors. "Subjektive Fahreindrücke sichtbar machen." Haus der Technik, Essen, November 30–December 12, 1998.
[8] Langer, J.W. "Entwicklung einer Simulation zur Abprüfung einer 12-Jahre-Gewährleistung gegen Durchrostung." "Steel and Automotive Body" symposium, Cannes, June 17, 1999.
[9] Geuer, A. "Einsatzpotential des Rapid Prototyping in der Produktentwicklung." Dissertation, TU Munich, 1996.
[10] DIN EN ISO 13407 (draft). "Human-centred design processes for interactive systems." Beuth Verlag, Berlin, 1998.
[11] Oelschlegel, F. "Versuche effizient planen." *QZ* 1996, pp. 556–559.

10.4 Quality assurance

Today, quality, as the ability to meet established or prerequisite demands, like reliability (which may be interpreted as "long-term quality"), is more or less a given feature of modern motor vehicles.

In the first decades of automobile construction, quality was largely "checked" after the fact; by the end of World War II, it was systematically controlled—under the motto "prevention instead of inspection." Today, the strategic, all-encompassing effort is known as TQM—"total quality management"—with the objective of "zero defects" or "do it right the first time" instead of checking after the fact and reworking. While this goal may admittedly only be approached asymptotically, by means of manifold and continuous improvements, TQM integrates all links in various value creation chains (vehicle manufacturer, supplier, development partner, customer service, recycling, etc.). A model for interactions between all activities is provided by the "quality circle" (Fig. 10.4-1). Years ago, this comprehensive task, as well as increasingly complex products and processes, prompted the VDA (Verband der Automobilindustrie—the [German] Association of Automobile Manufacturers) to collate all associated quality assurance points. Since then, international standards describe requirements imposed on transcorporate quality assurance systems. Throughout the product phases of any automobile, quality assurance may be divided into three areas of concentration: development, production, and field. All associated activities may be transparently described in the form of process chains, with defined customer/supplier relationships, beginning with establishment of design specifications. These process chains include all pertinent releases—such as experimental, planning, procurement, and production releases—as well as all agreements with value creation partners for capital investments, manufacturing and operating means and materials, as well as services. Further details are assured by continuous process planning and control, including systematic problem solutions in the event that major goals are not achieved or appearance of unexpected obstacles appear.

The task of quality assurance as part of the product creation process has undergone substantial change. Where once a large number of employees carried out the aforementioned checking functions, today quality assurance is an integral part of the product creation process of new vehicle models and powerplants, from establishment of target specifications to styling, prototypes, and production, in part with their own approval tests. Field experience is systematically evaluated for model development measures as well as application to new models; in larger firms, meeting all requirements for the entire range of makes and models is assured.

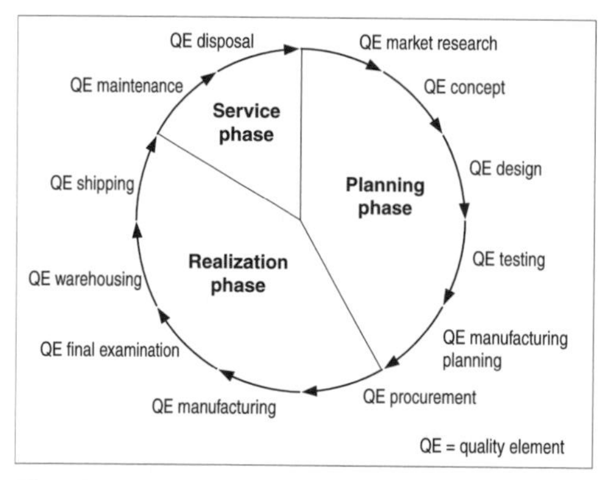

Fig. 10.4-1 Quality assurance throughout the entire product life cycle: the Quality Circle [2].

Often, only small deviations are sufficient to drastically compromise quality and reliability of a complex technical system; some faults appear only sporadically, and long-term quality is not always easily checked. Consequently, for years the automobile industry has employed systematic methods in support of analysis, development, and model development.

Even though there are no cut and dried formulas for solving all problems, in practice, good—indeed outstanding—solutions may be found by application of several principles. These include:

- In terms of product design:
  Avoiding "borderline" designs; striking a balance between innovative and proven concepts and design elements
- In terms of systems technology:
  Comprehensive and systematic consideration of all relevant influencing factors; application of scientifically grounded methods
- In terms of marketing:
  Exceeding customer expectations

Vehicle breakdowns must be regarded very critically, giving this topic especially great importance.

Finally, aside from all technical and scientific criteria, quality must be perceptible to the customer by means of his senses; quality must be seen, felt, smelled, and heard (or perhaps *not* heard). This includes subdivisions such as design quality, workmanship, and acoustic quality, all of which lead to the customer's sense of well-being.

One characteristic of modern automobile design is that specific quality improvements are partially offset by an increase in the number of components and their complexity. Additionally, the scope of software is continually expanding and development of quality and reliability methods for software is still in a state of flux. And because human beings are involved in every aspect of the automobile, it must unfortunately be accepted that things will not always work as planned.

References

[1] VDA (publisher). *Qualitätskontrolle in der Automobilindustrie—Band 4: Sicherung der Qualität vor Serieneinsatz.* Third edition, Frankfurt am Main, 1996.
[2] Masing, W. (ed.). *Handbuch der Qualitätssicherung*. Hanser Verlag, 1988.
[3] Braunsperger, M. "Qualitätssicherung im Entwicklungsablauf." Dissertation, TU Munich, 1992.
[4] VDI/VDE 3542. "Zuverlässigkeit und Sicherheit komplexer Systeme." VDI/VDE-Gesellschaft Mess- und Automatisierungstechnik, 1995.
[5] DIN EN ISO 9000-9004. "Qualitätsmanagement-Systeme." Deutsches Institut für Normung, Berlin, 1999.
[6] Uehlinger, K., and W. Allmen. *TQM live. Das Handbuch der Erfolgskompetenz*. SmartBooks Publishing AG, 1999.
[7] Kuhlang, P., et al. "Software-Entwicklung entlang der Prozesskette—ISO 9000—Zertifizierung auf der Basis von TQM-Grundsätzen." *QZ* 1999, pp. 286–292.

11 Traffic and the automobile—How can this go on?

11.1 Traffic and the automobile: torn between knowledge, desire, and reality

"Traffic serves society; traffic is not an end in itself, but rather is the result of deliberate human action." Those are the first words of the memorandum on traffic issued by the VDI (Verein Deutscher Ingenieure, Association of German Engineers) in 1993 [1], which has since undergone additional development. It may be assumed that this statement (examined at greater length in Chapter 1) will remain valid in the foreseeable future. As all indications are that world population and human activities—regional differences notwithstanding—will continue to increase, the topic of traffic, with all its implications, will in the long run retain its importance, indeed will be given priority treatment.

As is often pointed out (e.g., [2, 3]), within the overall transportation framework, the automobile is accorded great importance for various reasons, in particular because:

- The automobile is uniquely capable of fulfilling human societal demands for individual and special mobility.
- Other means of transportation offer alternatives and additions only in selected (albeit important) segments.
- Even in the past, marked, sometimes appreciable, reduction of drawbacks and reinforcement of advantages have been achieved.
- Potentials for additional, sometimes major, improvement will be available in the future.

Nevertheless, time and again, the future of the automobile is discussed in an atmosphere of skepticism. One reason is the (undoubtedly objective) fact that several specific advances—for example, in fuel economy and CO_2 emissions—have been offset by increased traffic volume. There are also subjective reservations against the automobile: traffic jams are labeled "tin avalanches" or "portable parking lots" and parking congestion in city centers is criticized.

Therefore, certain basic questions are raised repeatedly that might lead to possible approaches and support for traffic problem solutions; for example:

- Can traffic growth in principle be decoupled from economic growth, for example, by means of virtual mobility?
- How much mobility does a city need? How much can it tolerate?

Because accidents are in large part caused by human error, at first glance at least the following questions appear plausible:

- Can accidents be prevented by means of automated traffic?
- Might there be technologies which completely eliminate the most serious accident consequences?

Understandably, resource conservation and environmental protection also raise questions such as:

- Is a practicable "one-liter car" (fuel consumption 1 L/100 km) possible?
- Is a completely recyclable car possible?

Because of the strongly networked nature of traffic, tied as it is to so many aspects of modern life, as well as the consequences of physical law and engineering reality, such questions are bound up with a bewildering array of conflicting goals. There are no simple answers and no patent solutions. To this must be added psychological factors and questions of definition regarding the ambivalence of traffic: is a vehicle entering an area of traffic congestion an instigator or an injured party, perpetrator or victim? The slogan "Mobility yes, traffic no" shows that this ambivalence should actually be regarded as schizophrenia; no person is willing to dispense with individual advantages, but by the same token no one is willing to accept the associated disadvantages.

Still, the following will attempt to indicate future possibilities while observing necessary preconditions and constraints.

11.2 Structure of the overall traffic system

In the past, various modes of transportation—roads, waterways, aviation, rail—developed more or less independently of one another. In addition, for the open system of road transportation, different means of transportation have evolved to meet the various mobility demands of transporting personnel and goods. Networking of transportation means and modes has heretofore only been achieved on an infrastructural level, and in those cases it has been completely achieved only in partial segments. One reason for this is the fact that no common political and administrative institution has been created for traffic as a whole; in Germany, for example, the Bundesverkehrsministerium—the German Federal Traffic Administration—is responsible for federal freeways and highways, but not for urban traffic. Unincorporated towns and their surrounding areas have in many cases formed planning associations, but without any sovereign authority. Housing developments and economic zones are created for local and regional political reasons, but their effects on mobility are not always taken into consideration. Additional problem areas are found in shortcomings of infrastructure and transit organizations. Even today, public transportation systems still tend to think in terms of transportation "from stop to stop" instead of the customer-relevant "from house to house." And many a traffic policy-maker appears not to have grasped the fact that regardless of the quality of any telematics system, permanent traffic jams and bottlenecks cannot be eliminated unless capable bypass routes are available.

Beginning in the 1970s, regional transit authorities have been formed incorporating several different means of transportation (usually commuter rail, light rail, and buses). Since about the end of the 1980s, one goal has been systematic networking of the means of transportation, as no single type of vehicle is capable of satisfying all demands in an optimum manner. This suggests that each means of transportation should be implemented in keeping with its own unique advantages. For years, great value has been given to information networking by means of telematics; by contrast, physical networking of traffic and transit systems, with minimal transition resistance, is usually not given adequate consideration.

The 1992 United Nations Conference on Environment and Development, held in Rio de Janeiro, has charged all of the world's responsible parties with the long-term task of making all human activities "lasting or sustainable" [4]. The associated concept of "sustainable development" has to date been only partially defined in scientific and operational terms; it might rather be regarded as a sort of categorical imperative that the current human population must behave in such a way that succeeding generations enjoy at least the same opportunities for life and survival as the present generation. As a single or potential ideal life situation cannot be defined, one is for the present limited to description of necessary and desirable changes in condition. To this end, in accordance with the core conclusion of the 1987 Brundtland Commission report (which in turn formed a basis for the Rio conference), along with ecological considerations, economic and social dimensions of human activities must be *equitably* included.

In principle, several approaches to this goal are possible; these are best applied in combination. As a starting point, the basic structure of a mobility model [5], containing three spheres of influence—human, organization, and technology—each regarded in the broadest sense and accorded equal weight, will be found useful (Fig. 11.2-1).

Such a basic structure permits detail development and concrete applications using "top-down" as well as "bottom-up" approaches:

1. Conceptual *"top-down" approach*: a certain *basic mobility consensus* is imperative at the beginning of conception; this acknowledges the need for mobility and traffic and their importance for society in general. In view of the ensuing consequences, it would make sense for individual democratic areas of sovereignty and levels of decision making—that is, communities, regions, provinces, and nations—to formally bring their individual consensuses into agreement. It goes without saying that various social value expectations including evaluation principles of different basic rights [6] must be considered. On this basis, synchronized blueprints for economic, environmental, transportation, and land use policy as well as energy policy ("expanded energy consensus") are formulated. These guidelines must at least make an effort to include the Rio Principle. Because all segments involve long-term plans, these blueprints must also be conceived and managed in such a way as to transcend legislative periods. (Those who believe that this objective is unrealistic in the modern environment of day-to-day politics are reminded that in the nineteenth century, over the course of decades, a widespread, all-encompassing railway network was realized at great financial expense and organizational effort, which not least formed the basis for the later welfare and rise of Germany as well as other nations). Concrete examples for communities and regions may include: desired population development and associated housing construction, location of new businesses, venues for large events and convocations, and tourism infrastructure. For the federal government, primary considerations include assurance of international competitiveness or establishment of long-term importance in the "concert of powers," with all its associated consequences. As traffic increasingly transcends national boundaries, these blueprints must be matched to one another to the greatest degree possible within supranational, national, and regional frameworks. Blueprints for transportation modes and infrastructures may be derived from the blueprints of all traffic-generating parameters.
2. The more pragmatically oriented *"bottom-up"* approach: this deals with intelligent, best matched exploitation of necessary and known improvement potential in subsidiary areas—that is, in infrastructures, transportation modes, telematics, organization, and so on. Examples include elimination of permanent bottlenecks and accident concentrations, engineering improvement of passenger cars and public transportation systems, and systematic introduction of parking guidance systems.

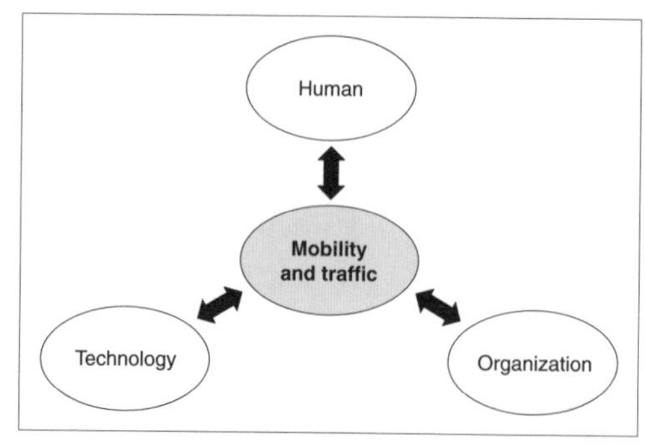

Fig. 11.2-1 Basic structure of a mobility model.

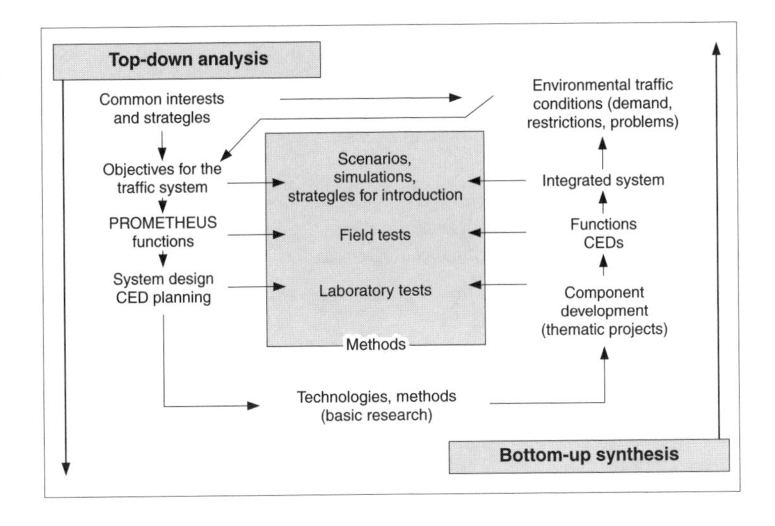

Fig. 11.2-2 Systems approach (after PROMETHEUS).

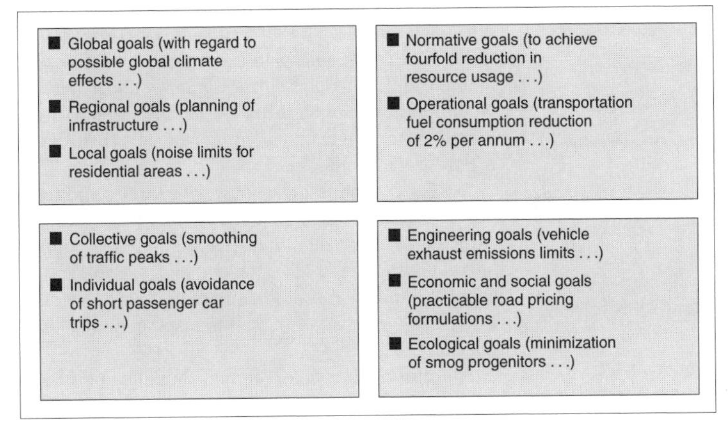

Fig. 11.2-3 Sustainable mobility: foundations of a system of goals.

Expressed in a different way, the top-down approach is necessary to generate fundamentally new concepts, to ensure that various programs fit together, to foster international accord, and so forth. The bottom-up approach is similarly indispensable in applying specialized competencies to practice, to remain closely tied to practical applications, to turn affected parties into participants, and so on.

It should be plausible that both of these rational approaches can be combined into an overall concept from which *concrete goals* for future traffic can be derived, as was demonstrated in the PROMETHEUS program (Fig. 11.2-2, [7]).

As shown in Fig. 11.2-3, a comprehensive system of goals should contain very different categories and dimensions.

For successful application, both approaches and the overall concept derived therefrom require *boundary conditions*, which are best conceived in conjunction with the necessary mobility consensus. These represent technical, ecological, economic, and social boundary conditions, from which all involved parties, such as industry, transit operators, mobility consumers and, not last, political decision-makers and transportation authorities can and must derive their collective and individual decisions in terms of structures, processes, and resource applications (Fig. 11.2-4). An especially important boundary condition involves the willingness of society as a whole to provide the necessary financial means to ensure long-term welfare, even in times of strained public coffers, and, not last, to provide for transportation infrastructure. One possibility toward this end is so-called "public–private partnership" [8].

It is plausible that goals and boundary conditions must take into consideration (very different) transportation-relevant stress limitations (Fig. 11.2-5).

In the process, requirements of so-called "strong sustainability" (e.g., inability to achieve absolute limits in nature) must be given precedence over aspects of "weak sustainability," such as acreage consumption in less sensitive areas.

Additionally, it should be pointed out that sustainability indicators with regard to emissions, resource consumption, or safety (efforts described in Refs. [9, 10])

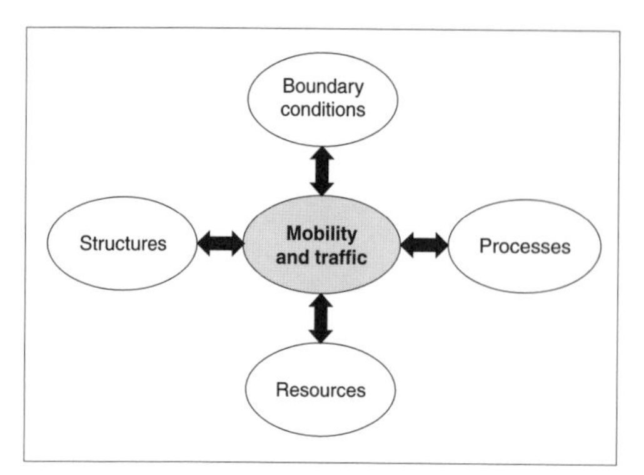

Fig. 11.2-4 Basic factors of traffic concepts.

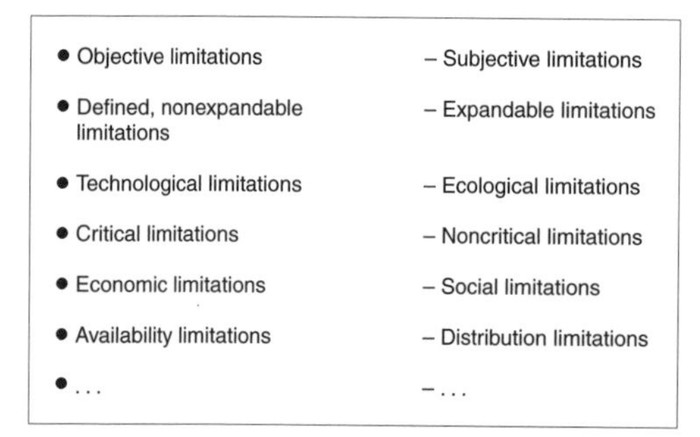

Fig. 11.2-5 Categories of transportation—relevant stress limitations.

may of course be compared to the familiar efficiency principle, in particular with:

- Efficiency as a classical engineering concept
- Profitability as an economic concept
- Resource productivity as an ecological principle

Demanding traffic concepts can only be expected to have high likelihood of success if all participants concentrate on meeting goals rather than engaging in "cherry picking." To achieve this, there should be as many market economic incentives as possible motivating participants to act in their own self-interest. Additional political/regulatory restrictions and requirements should be limited to necessary special situations—for example, in the event of:

- Existing restrictive local or topographic limitations, such as city centers
- Critical ecological impact

or if for other reasons market economic functions are ineffective.

One possible means of achieving this is to conceive the entire traffic system and its processes as a feedback and control system. An initial, purely technical approach to this end was developed as the PROMETHEUS program (Fig. 11.2-6, [7]).

It is apparent that even before embarking on a trip, a traffic participant can decide on a means of reaching his destination. A precondition for this is complete, current information regarding traffic situations and anticipated traffic load developments of various means of transportation. If the traffic participant is already under way, he would profit from controlled traffic flow, which, if required, would direct him along a detour or to a different means of transportation.

Market economic incentives, with locally and temporally variable pricing, extend beyond purely technical controls. Currently, variable parking costs already operate in this sense [11]. Beyond this, variable traffic capacity utilization cost ("road pricing") is being considered in the sense of overall management of scarce resources (e.g., [12]).

All efforts under consideration should be unified in an overall control concept in such a way that given freedom of choice of participants, unstable or undesired growth as well as a detrimental decline in traffic can be avoided. Boundary conditions should be configured so as to avoid competitive distortion by transportation

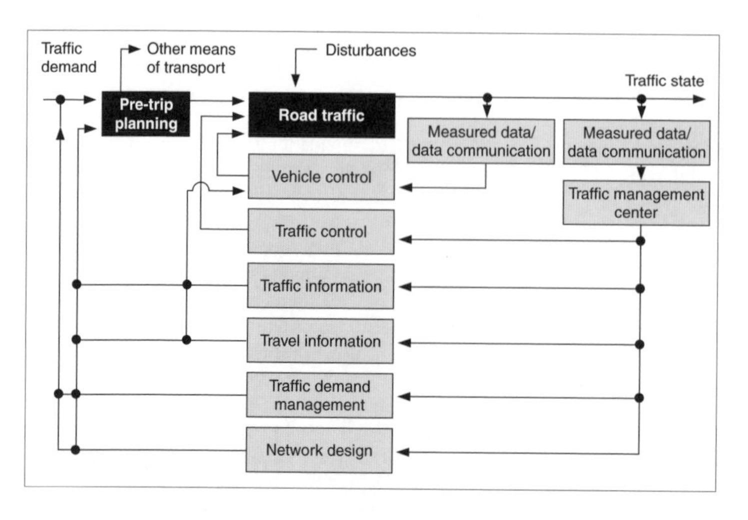

Fig. 11.2-6 Information flows in road traffic (after PROMETHEUS).

modes and service providers, so that all responsible parties do their best to assure sustainable transportation, and so that traffic participants accept the offered transportation means best suited to their needs, whether it be public transportation or individual transportation (including pedestrian paths and bicycles).

Suitable boundary conditions would appear to include, for example, technological limitations, energy prices, and special features of tax policy. These in turn impact the complex control loop in the sense that they affect local and temporal traffic flow, conception of transportation modes and infrastructures, offerings on the part of transit organizations, and so on.

Even with staged application of such an overall effort, it should, for example, be possible to:

- Convince commuters in urban areas to switch to (more attractive) public transportation
- Intensify application of more energy-efficient and environmentally responsible transportation technologies
- Intensify establishment of more efficient transportation organizations and cooperative traffic management systems
- Encourage political decision-makers to more expeditious action
- Eliminate or at least reduce traffic congestion and its detrimental consequences (including the high economic cost of congestion [13])
- In the long term, promote traffic-reducing housing development, economic, and leisure structures
- Motivate traffic participants to more conscientious handling of that valuable commodity known as mobility [14]

Overall, in other words, it should be possible to achieve an altered understanding for mobility that is accepted by all. In contrast to many other efforts, such a process would not treat concrete numbers for reduction of energy consumption and emissions, or for achieving specific modality splits, and so on as more or less political targets, established at the outset, but rather as values derived from combined ecological, economic, and social factors. It is especially important that concrete tasks be formulated in concert and assigned to various responsible parties. This will be illustrated by the following examples:

- Public agencies must above all provide for establishment of "proper" boundary conditions and guiding strategies which transcend individual transportation modes on national, regional, and local levels. Without doubt, more than a few politicians would have to find within themselves greater discipline to accelerate decision-making processes, instead of putting off problems and holding others responsible. Additional tasks of national policy include supranational consensus building and harmonization of international law. The European Union in particular provides opportunities as well as responsibilities for supranational agreements.
- Even in a free economy, business and industry must place a high value on the principles of sustainability, even in the face of prevailing market conditions, and configure their products and services to be as sustainable as possible.
- Transportation organizations must become more customer-oriented.
- It is the task of science to eliminate knowledge deficits with respect to sustainability, ecological stress limitations, complex effects, and so on as quickly as possible and, to a greater degree, present interdisciplinary concepts, evaluation and optimization processes.
- Participants in both passenger and merchandise traffic should, as already discussed, act in keeping with overall goals, supported by incentives and boundary conditions.

All of this will succeed if it is possible for the educational system to convey at an early age the complex relationships of traffic as well as the various advantages and disadvantages of different means of transportation for individuals, the economy, and the environment.

Given this and other consequences, "sustainable mobility" would no longer be a categorical imperative or a set of restrictions, but rather a comprehensive approach for a transgenerational action concept in which areas of action can be sequentially planned and executed.

11.3 Traffic technology management

Since ancient times, traffic and transportation system technologies have been almost exclusively developed and improved using "bottom-up" approaches. Given a framework of sustainable overall concepts (Section 11.2), this is insufficient; moreover, vital impulses must be applied using "top-down" approaches, with the result that emphasis is placed not on technologies for their own sake but rather on their contributions to problem solutions. Traffic technologies may therefore be regarded as one component of the traffic concept management level (Fig. 11.3-1).

Fig. 11.3-2 provides an overview of the associated technological branches.

A prerequisite for traffic of just about every sort is availability of high-grade useful energy. This topic was considered from a technical viewpoint in Section 5.8. In the future, when crude oil derivatives are supplemented and replaced by other energy carriers, this will require a societal energy consensus as part of the mobility consensus treated in Section 11.2. As the energy supply for roadborne traffic must be assured on a largely universal basis, it is necessary to configure complex energy generation and supply structures in conjunction with other energy consumption sectors.

Although infrastructure technologies for moving and stationary traffic will be touched upon only briefly within the scope of this chapter (Fig. 11.3-2), physical networking of transportation modes with one another

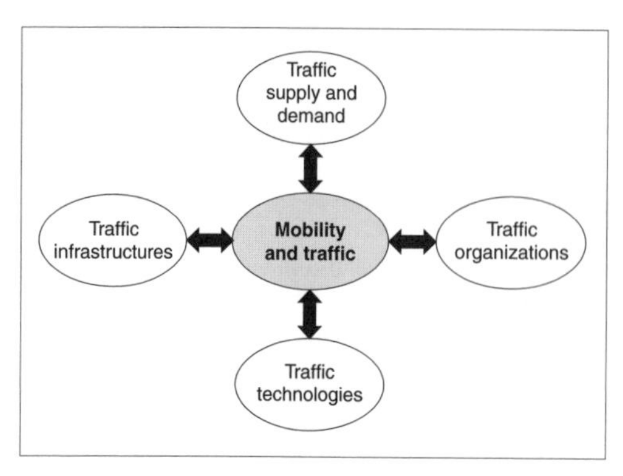

Fig. 11.3-1 Traffic concept management levels.

represents an important subtask for vehicle technology. A familiar if somewhat uncommon example is a two-wheeled vehicle serving as an "auxiliary vessel" for a motor home, as a means of transportation between the motor home and its vicinity. This concept, at least in principle, can be expanded in that while avoiding or minimizing added weight and space requirements, an independently mobile part of the "main vehicle" can be decoupled and reattached. Classical solutions to networking of road and rail are so-called "combined freight traffic" (e.g., [15]) as well as loading of passenger vehicles on railway cars. Various improvements would increase the frequency of use of such methods—for example, in terms of time and space requirements for loading and unloading, improved transport quality, and the net-to-gross relationship of the transportation effort—and therefore increased overall economy.

For years, a successful example of physical networking in rail traffic has been the Karlsruhe (Germany) streetcar system, which operates on German Federal Railway trackage once outside the city (see Ref. [16]).

Information networking of transportation modes and transportation systems, known as telematics, has for years supplied important segments with information, communications, traffic control, and navigation. This has been discussed in detail in Sections 6.4.2 and 8.3 (see also Refs. [17–19]). At this point we will only point out that comprehensive application of information networking (as indicated in Fig. 11.3-2) is also especially useful for physical networking: advanced and above all timely information (including announcement of delays and congestion) is of great importance to traffic participants who might wish to transfer to a different means of transportation. This can be realized by means of mobility centers, with extensive connections to data and associated mobile systems such as the "Personal Travel Assistant" (PTA, [20]).

The "transportation modes" section of Fig. 11.3-2 need only be mentioned in passing, as the topic of this handbook is further development of the automobile. Nevertheless, after considering the topic of sustainability, it makes sense to examine once again the various possibilities of vehicle design philosophies.

Classical goals in developing new products of any kind can be characterized as "design to customer," with subsidiary goals such as "design to cost," "design to quality," or "design to manufacturability." For years, designers have had to consider the "design to environment" principle, with its attendant reductions in emissions and waste, suitability for material cycles, and so on. For the long-term task of "sustainability," the principle of "design to life cycle optimization" seems to apply, as all overall optimization of properties, material and energy streams, product life expectancy, manufacturing and recycling processes, manufacturing location decisions, infrastructures, and so can be subsumed within that concept [21].

With regard to development of other modes of trans-

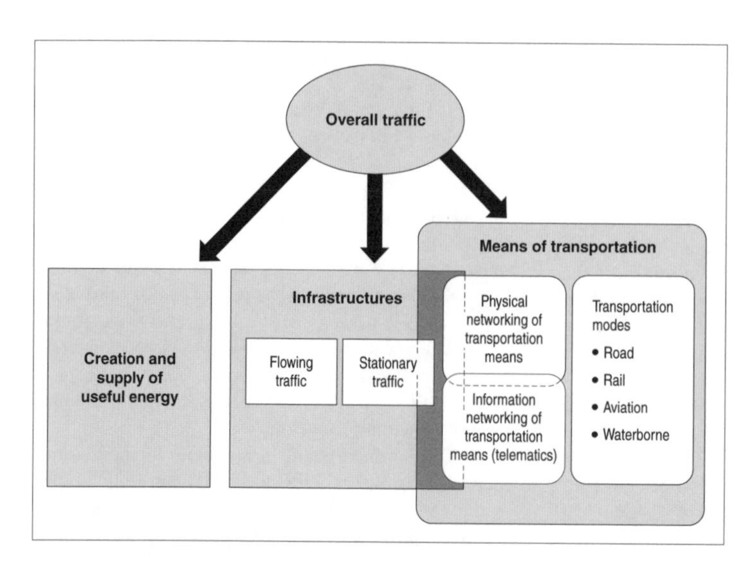

Fig. 11.3-2 Traffic technology disciplines.

portation, also indicated as keywords in Fig. 11.3-2, the reader is referred to the corresponding literature.

11.4 Management of traffic organizations

Within the framework of this book, we will briefly examine the not yet discussed fourth traffic organization management level of Fig. 11.3-1. The task of this sector is to match traffic supply and demand as closely as possible, while fully considering available resources (infrastructure, transportation modes, telematic systems, etc.) as well as ecological and economic boundary conditions. This is above all a political task (with consequences as outlined in Section 11.2), building on the societal mobility consensus and composed of regulatory and institutional prerequisites for traffic management (processes, fare systems, etc.) and transport operators (structures, business models). As has been mentioned several times, integrated traffic systems are indispensable, as no single transportation mode is capable of optimally satisfying all transport tasks while meeting all evaluation criteria [22]. A direct consequence is the need for establishing cooperative systems and organizations in which not only transportation modes are technically networked, but in which various sovereignties as well as public and private transportation organizations work together (Fig. 11.4-1, [23]).

Several concrete examples:

- Mobility centers must provide current information covering several transportation modes (schedules, delays, available capacity for flowing and stationary traffic, fares, etc.).
- "Smart card" systems, with chip cards valid on multiple transportation carriers, simplify use and transfers between different carriers.
- Dynamic, adaptive guidance strategies, also extending over multiple carriers, provide optimum routing, detour, and transfer recommendations.
- To incorporate motorized individual traffic and provide individual availability of vehicles with collective utility attributes, systems such as car-pooling and ride-sharing with intelligent ride centers can be implemented.

A management task vital for all traffic participants is systematic problem management—that is, the task of avoiding, as much as possible, road congestion or collapse of rail or air transport (i.e., due to weather conditions) and to minimize the resulting consequences. This is predicated on strategic measures (e.g., congestion recognition systems) as well as rapid, directed operational management.

In general, traffic management means:

Cooperation between
- Traffic technology
- Traffic organizations
- Traffic policy
- Communication and information

Cooperative traffic management means:

Cooperation of
- Different transportation modes through informational and physical networking
- Various responsible parties
  - Legislative bodies, authorities
  - Public and private transportation organizations
  - Various branches of industry (traffic paths, traffic modes, telematics . . .)
  - Cooperatively acting and thinking traffic participants

As well as
- Economic forces and regulatory policy
- Space utilization and traffic tolerance
- . . .

Fig. 11.4-1 Traffic management: more than just applied telematics.

References

[1] Verein Deutscher Ingenieure. Memorandum Verkehr, Düsseldorf, 1993.
[2] Deutsche Shell AG (publisher). "Mehr Autos—Weniger Emissionen—Szenarien des PKW-Bestands und der Neuzulassungen in Deutschland bis zum Jahr 2020." Hamburg, September 1999.
[3] Waschke, T., et al. "Perspektiven der mobilen Gesellschaft." AVL conference, "Motor und Umwelt 99," September 2–3, 1999, Graz, Austria. Proceedings, pp. 67–101.
[4] Bundesministerium für Umwelt, Naturschutz und Reaktorsicherheit. "Umweltpolitik—Agenda 21," United Nations Conference on Environment and Development, Rio De Janeiro, 1992; Bonn, 1997.
[5] BMW Institut für Mobilitätsforschung. Mobilitätsmodell, Berlin, 1999.
[6] Ronellenfitsch, M. "Mobilität—Vom Grundbedürfnis zum Grundrecht?" DAR 1992, p. 321 ff.
[7] Braess, H.-H., and G. Reichart. "PROMETHEUS—Vision des intelligenten Automobils auf intelligenter Straße—Versuch einer kritischen Würdigung." *ATZ* 1995, pp. 200–205 and pp. 330–343.
[8] Tegner, H. "Risikoallokation zwischen öffentlicher Hand und Privaten im privatfinanziertem Straßenbau." VDI Bericht 1372, 1998 (Gesamtverkehrsforum Braunschweig), pp. 117–128.
[9] Brodmann, U. W., et al. "Messung der Nachhaltigkeit des Verkehrs." *Int. Verkehrswesen* 1 and 2/1999, pp. 23–24.
[10] Meckel, H., and E. Plinke. "Praktikabler Maßstab für Nachhaltigkeit." *Prognos Trendletter* 2/1999, pp. 10–12.
[11] Fiala, E. "Fahrzeug-Mensch-Verkehr." VDI Bericht 948, 1992, pp. 1–13.

[12] Schütte, C. "Road Pricing in der Praxis—Ein konkretes Preiskonzept für Deutschland." Dissertation, TU Berlin, 1998.
[13] Sumpf, J., and D. Frank. "Abschätzung der volkswirtschaftlichen Verluste durch Stau im Straßenverkehr." BMW study, 1994, revised 1998.
[14] Feldhaus, S. *Wege in eine mobile Zukunft—Grundzüge einer Ethik des Verkehrs*. ABERA- Verlag Meyer & Co, Hamburg, 1998.
[15] Müller, W. "Erfahrungen zur Kooperation im kombinierten Verkehr und Ansätze zu ihrer Verbesserung." In: "Technologieansätze zur besseren Kooperation und Vernetzung der Verkehrsträger." BMBF-Statusseminar, May 21–22, 1995; Report, pp. 103–114.
[16] Müller-Hellmann, A. "Voraussetzung und Grenzen der Integration im Nahverkehr, Ref. [15], pp. 133–144.
[17] Bundesministerium für Verkehr. Strategy paper "Telematik im Verkehr." Bonn, 1993.
[18] *Proceedings* of the Annual World Congress on Intelligent Transport Systems, 1999 and previous years.
[19] Institut für Eisenbahnwesen und Verkehrssicherung der TU Braunschweig (publisher). *Beiträge aus dem Zentrum für Verkehr der TU Braunschweig*. Publication series, Vol. 60, 1998.
[20] Reichart, G. "MOTIV—Mobilität und Transport im intermodalen Verkehr." Ref. [15], pp. 227–242.
[21] Braess, H.-H. "Das Automobil von der Produkt- zur Systemoptimierung—Ziele und Aufgaben des Life Cycle Managements." *ATZ* 1999, pp. 984–990.
[22] Seiffert, U. "Mobilität—Gesellschaftliche Anforderungen und technologische Optionen der Zukunft." Presentation, 1. RWE-Zukunftstagung Gesellschaft und Technik im 21. Jahrhundert, Essen, October 22, 1998.
[23] Bundesministerium für Bildung, Wissenschaft, Forschung und Technologie. Leitprojekte Mobilität in Ballungsräumen, Bonn, 1998.
[24] Schad, H., and H. Riedle. "Neue integrierte Mobilitätsdienstleistungen." *Der Nahverkehr* (1999) No. 7–8, pp. 8–12.
[25] Freitag, A. "Verbünde und Verkehrsunternehmen in einem liberalisierten Markt." *Der Nahverkehr* (2000) No. 7–8, pp. 11–15.
[26] Kutter, E. "Verkehrsplanerische Eckwerte einer nachhaltigen regionalen Verkehrstrategie." FAZ-Bericht Nr. 154, 2000.
[27] TÜV Energie und Umwelt GmbH (publisher). *Mobilitätsforschung für das 21. Jahrhundert* (under contract to the BMBF and BMVBW), TÜV-Verlag, Cologne.

12 Outlook—Where do we go from here?

"Reinvent the car!" Time and again, vehicle developers are exhorted to discard old avenues of development, indeed prejudices, and to tread entirely new paths. One reason for these calls to revolution is engineers' attachment to the "old" reciprocating piston engine, which many observers regard as an obstacle to innovation.

Yet the history of the automobile is replete with unconventional proposals in all sectors, some in the form of ideas, patent applications, or scientific disclosures, and others which have been tested in demonstration vehicles or even manufactured in limited production. In addition, for decades, concept vehicles of the widest—and wildest—imaginable variety have been displayed. If anything truly better in terms of components and overall concepts had come of this, it would have readily established itself in the world of tough, international competition—or will do so in the future.

In the future, as today, automobiles will be "movable containers," capable of transporting people and goods. But construction of these containers will exhibit appreciable differences in terms of materials (e.g., variable properties) and construction methods (structures, manufacturing processes).

The automobile of the distant future will experience major impact through staged, gradual abandonment of petroleum fuel as a primary energy carrier. To this end, the present state of the art indicates several possible options. It would be especially attractive if, as in the case of electrified railways, traction energy were no longer carried aboard the vehicle. An extreme attempt is represented by inductive energy transmission from the roadway to the vehicle, requiring that "only" electric motors be carried aboard. Realization of this idea, however, is still far removed even for demonstration purposes, to say nothing of application to the entire road network.

The wheel, one of mankind's oldest inventions, still serves almost all land-based vehicles as a support and guidance element. It seems as good as certain that alternative systems such as air cushion or magnetic levitation technologies will not make any headway in the automotive sector. Instead, wheels, as well as many other vehicle components, will become increasingly complex and demanding. Examples include the mechatronic and adaptronic sectors—that is, networking of mechanical base functions with electrical and electronic systems.

Ever increasing demands on very different transportation needs lead to ever more specialized, in particular more variable, vehicle concepts. Still, there will be limits. For example, it seems highly unlikely that we will ever see a car which is short and compact for urban use and which can be "inflated" prior to the annual vacation trip.

However, a different boundary has been crossed in the past few years: the line between a purely mechanical means of transportation and the "intelligent vehicle on intelligent roads." By intelligence, we of course do not mean human creativity, but rather the human ability to receive signals from the environment, to process these, and to abstract measures and recommendations appropriate to the situation.

The expositions of this book have provided a well-grounded overview of the fascinating development of the automobile. It is readily apparent that engineering solutions will find increased application, and simultaneously will have to pay even closer attention to societal needs. The following trends will be discernible in the future:

- Product creation process
  Application of computer-based systems will reduce product development times of new models to an average of 30 months after "styling freeze." Despite application of numerous simulation methods, experimental testing will gain in importance for reasons of quality. "Zero fault" quality demands will bear first fruit.
- Vehicle safety
  By means of increased application of electronics, sensors, and actuators, considerable progress will be made in the field of accident avoidance. It is to be expected that by the year 2010, telematics and driver assistance systems will cut passenger car and truck accidents in half. Special attention will be paid to improved compatibility between all traffic participants, taking into account the "weaker participants" such as pedestrians and cyclists/motorcyclists.
- Emissions, such as HC, CO, NO_X and particulates, will be reduced by means of engine-side measures, improvement of modern catalytic converters, and oxides of nitrogen catalysts and particulate filters to such an extent that with the exception of carbon dioxide, the motor vehicle will cease to be a topic of discussion with regard to vehicular emissions.
- Exterior noise will be reduced to a minimum. This will be achieved by improvements in regulatory measuring techniques (closer to reality) and the necessary improvement in rolling noise between tire and roadway.
- By the year 2010, consumption of fossil fuels (at least in Europe) will be reduced by 25% compared to 2000 and 30% compared to 1996. The main contributors to this reduction will be drivetrain improvements, reduced vehicle weight, and reduced rolling resistance with continued low aerodynamic drag. Improved networking of all transportation modes will allow transportation of persons and goods with overall lower energy consumption.
- Vehicle weight reduction with simultaneous increases in the usual requirements such as safety, environmen-

tal protection, comfort, and adequate performance is a special challenge for the entire industry. Application of alternative materials and new composite structures is on the increase. The winners are aluminum and magnesium. Additionally, joining technologies and new anticorrosion methods have lent their support to weight reduction. The recycling rate, including thermal recycling, has been increased to more than 90%. This was achieved by a significant increase in economic feasibility of recycling processes.

- Gasoline and diesel engines have been able to defend their high level of market penetration. However, numerous changes have been introduced: smaller displacement with supercharging and turbocharging, a smaller average number of cylinders, cylinder shut-off, gasoline direct injection, fully variable valve timing, and, specifically for diesels, unit injectors and common rail systems with piezo controls. Along with five- and six-speed automatic transmissions, electronically controlled, mechanically shifted "manual" transmissions are in service. CVT transmissions will make additional inroads if their efficiency and comfort exceeds that of previous manual and automatic transmissions. By 2015, about 5% of the fleet will employ alternative drivetrains. Along with CNG (compressed natural gas), hybrid vehicles with various propulsion systems, including fuel cells, will exhibit the considerable technological progress of alternative powerplants.
- The biggest changes are exhibited in the fields of electrical and electronic systems. In 2010, electrical systems and electronics will account for about 40% of vehicle value creation. Electrification of numerous systems—steering, brakes, oil and water pumps—and implementation of 42-volt networks for certain electrical consumers is already under way. With these system changes, the customer will experience new diagnostic and service functions. Internal vehicle communications between various functions has led to completely new onboard electrical system architectures. Simultaneously, the importance of software design has increased. One particular area of emphasis is design of user-friendly system operation and functionality.
- Finally, it has become possible to integrate the automobile in the overall traffic system. Physical and software networking of various transportation modes has increased considerably. Telematics and information systems have played a major role in this development.

In principle, it is the job of engineers to approach as closely as possible the immutable limitations represented by the Carnot cycle and to push the envelope of such factors as lightweight design or artificial intelligence as far as possible. To this end, in the future, qualification of engineers will extend beyond pure technology to include societal and environmental competence.

The authors are firm in their belief that despite all possible engineering innovations, passenger cars of the year 2015 will still be driven individually, and the joy of driving will be maintained through increased support by driver-assistance systems.

Author Index

1	**Mobility**	Prof. Dr. Dr. E.h. Hans-Hermann Braess
2	**Requirements and conflicting goals**	
2.1	Product innovation and prior progress	Prof. Dr. Ulrich Seiffert
2.2	Regulatory requirements	Dipl.-Ing. Wolfgang Barth
2.3	New technologies	Prof. Dr. Dr. E.h. Hans-Hermann Braess
3	**Vehicle physics**	
3.1	Basics	Prof. Dr. Ulrich Seiffert
3.2	Aerodynamics	Dr. Burkhard Leie
		Dr. Heinz Mankau
3.3	Heat transfer and thermal technology	Dipl.-Ing. Hans Kampf
3.4	Vehicle acoustics	Dr.-Ing. habil. Raymond Freymann
4	**Shapes and new concepts**	
4.1	Design	Dipl.-Designer Arnold W. Ostle
4.2	Vehicle concepts and packages in passenger car development	Dipl.-Ing. Dipl. Wirtsch.-Ing. August Achleitner
		Dipl.-Ing. Wolfhelm Gorissen
4.3.1	Electric drive systems	Dr. Erwin Wüchner
4.3.2	Fuel cell propulsion for mobile systems	Prof. Dr. Ferdinand Panik
		Dr. Dietmar Beck
		Dr. Michael Reindl
		Dr. Werner Tillmetz
		Dipl.-Ing. Uwe Benz
4.3.3	Hybrid drive	Dr. Jörg Abthoff
4.3.4	Stirling engines, gas turbines, and flywheels	Dipl.-Ing. Karl E. Noreikat
5	**Powerplants**	
5.1.1	Otto-cycle engines, Part 1: Fundamentals	Dr. Hans-Joachim Oberg
5.1.2	Otto-cycle engines, Part 2	Prof. Dr. Dr. E.h. Franz Pischinger
		Dr. Peter Wolters
5.2	Diesel engines	Dr. Karl-Heinz Neumann
		Dr. Klaus-Peter Schindler
5.3	Supercharging	Dr. Klaus-Dieter Emmenthal
5.4	Drivetrain	Dipl.-Ing. Peter Köpf
		Dr. Gerhard Wagner
		Dr. Hans-Jürgen Drexl
		Dr. Hans Wienholt
5.5.1	All-wheel drive concepts	Dipl.-Ing. Herwig Leinfellner
		Ing. Karl Friedrich
		Dipl.-Ing. Heribert Lanzer
5.5.2	Traction and braking control	Dr. Heinz Leffler
		Dr. Willibald Prestl
5.6	Optimization of auxiliary devices	Prof. Dr. Ulrich Seiffert
5.7	The two-stroke engine: Opportunities and risks	Dipl.-Ing. Bert Pingen
5.8	Alternative fuels in comparison	Dipl.-Ing. Detlef Frank
		Dipl.-Ing. Wolfgang Strobl
		Prof. Dr. Volker Schindler
6	**Body**	
6.1.1	Unit body	Dipl.-Ing. Karl-Peter Schütt
		Dipl.-Ing. Lothar Teske
		Dipl.-Ing. Helmut Goßmann
6.1.2	Space frames and cladding	Dipl.-Ing. Wulf Leitermann

6.1.3	Convertibles	Dipl.-Ing. Winfried Bunsmann
		Dr. Werner Gausmann
		Dr. Peter Kalinke
		Siegfried Licher
6.1.4	Hybrid and mixed material body construction	Prof. Dr. Hans-Günther Haldenwanger
6.2	Automobile body materials	Prof. Dr. Hans-Günther Haldenwanger
6.3	Surface protection	Dipl.-Ing. Paul Dragovic
6.4.1	Ergonomics and comfort	Dipl.-Ing. Claus Volker Gevert
6.4.2	Communications and navigation systems	Dr. Jürgen Kässer
6.4.3	Interior comfort/thermal comfort	Dipl.-Ing. Hans Kampf
6.5	Vehicle safety	Prof. Dr. Ulrich Seiffert
7	**Suspension**	
7.1	General	Dr. Axel Pauly
7.2	Brakes	Dipl.-Ing. Gunther Buschmann
		Dr. Hans-Jörg Feigel
		Dipl.-Ing. Klaus Winter
7.3	Tires	Dipl.-Ing. Heinrich Huinink
7.4	Suspension design	Dr. Dieter Hennecke
		Dipl.-Ing. Andreas Gleser
		Dipl.-Ing. Johann Niklas
		Dipl.-Ing. Reinold Jurr
		Dr. Dieter Konik
		Dr. Axel Pauly
		Dipl.-Ing. Karsten Schille
		Dipl.-Ing. Ludwig Seethaler
		Dipl.-Ing. Günther Seuss
7.5	Evaluation criteria	Dr. Erich Sagan
		Dipl.-Ing. Thomas Unterstraßer
7.6	Fuel system	Dipl.-Ing. Michael Meurer
		Franz J. Henkel-Adam
		Dipl.-Ing. Olaf Meyer
8	**Electrical and electronic components and systems**	
8.1	Illumination	Dipl.-Ing. Wolfgang Hendrischk
8.2	Sensors, actuators, and systems technology	Dr. Wolfgang Strauss
8.3	Electrical system/CAN	Dr.-Ing. Michael Walther
8.4	Multimedia systems for motor vehicles	Dr. Werner Poechmüller
		Dipl.-Ing. Wolfgang Baierl
		Dipl.-Ing. Gerald Spreitz
		Dipl.-Ing. Stephan Rehlich
8.5	Electromagnetic compatibility	Dr. Wolfgang Pfaff
9	**Materials and manufacturing methods**	Dr. Ludwig Hamm
		Dipl.-Ing. Volker Freitag
10	**The product creation process**	Dr. Ludwig Hamm
10.1	Simultaneous engineering and project management in the product creation process	Dipl.-Ing. Johannes Hennecken
		Prof. Dr. Rudolf J. Menne
		Dr. Manfred Rechs
10.2	Computation and simulation	Dr. Werner Dirschmid
10.3	Measurement and testing technology	Prof. Dr. Dr. E.h. Hans-Hermann Braess
10.4	Quality assurance	Prof. Dr. Dr. E.h. Hans-Hermann Braess
		Prof. Dr. Ulrich Seiffert
11	**Traffic and the automobile—How can this go on?**	Prof. Dr. Dr. E.h Hans-Hermann Braess
12	**Outlook—Where do we go from here?**	Prof. Dr. Dr. E.h Hans-Hermann Braess
		Prof. Dr. Ulrich Seiffert

Company Index

Adam Opel AG, Rüsselsheim	Dipl.-Ing. Paul Dragovic Dipl.-Ing. Helmut Goßmann Dipl.-Ing. Karl Peter Schütt Dipl.-Ing. Lothar Teske
Audi AG, Ingolstadt	Dr. Werner Dirschmid Prof. Dr. Hans-Günther Haldenwanger
Audi AG, Neckarsulm	Dipl.-Ing. Wulf Leitermann
Behr GmbH & Co. Stuttgart	Dipl.-Ing. Hans Kampf
Blaupunkt Werke GmbH, Hildesheim	Dr. Jürgen Kässer
BMW AG, München	Dipl.-Ing. Detlef Frank Dr. Raymond Freymann Dipl.-Ing. Claus Volker Gevert Dipl.-Ing. Andreas Gleser Franz J. Henkel-Adam Dr. Dieter Hennecke Dipl.-Ing. Reinold Jurr Dr. Dieter Konik Dr. Heinz Leffler Dipl.-Ing. Michael Meurer Dipl.-Ing. Olaf Meyer Dipl.-Ing. Johann Niklas Dr. Axel Pauly Dr. Willibald Prestl Dr. Erich Sagan Dipl.-Ing. Karsten Schille Prof. Dr. Volker Schindler Dipl.-Ing. Ludwig Seethaler Dipl.-Ing. Günther Seuss Dipl.-Ing. Wolfgang Strobl Dipl.-Ing. Thomas Unterstraßer
Continental AG, Hannover	Dipl.-Ing. Heinrich Huinink
Continental Teves, Frankfurt	Dipl.-Ing. Gunther Buschmann Dr. Hans-Jörg Feigel Dipl.-Ing. Klaus Winter
Daimler-Benz AG, Kirchheim-Nabern	Dr. Dietmar Beck Dipl.-Ing. Uwe Benz Prof. Dr. Ferdinand Panik Dr. Michael Reindl Dr. Werner Tillmetz
Daimler-Benz AG, Stuttgart-Untertürkheim	Dipl.-Ing. Karl E. Noreikat Dr. Erwin Wüchner
FEV-Motorentechnik GmbH, Aachen	Prof. Dr. Dr. E.h. Franz Pischinger Dr. Peter Wolters
Ford Werke AG, Köln	Dipl.-Ing. Johannes Hennecken Prof. Dr. Rudolf J. Menne Dipl.-Ing. Bert Pingen Dr. Manfred Rechs
Hella KG, Lippstadt	Dipl.-Ing. Wolfgang Hendrischk

Keyword Index

About the Editors

Professor Dr.-Ing. Dr.-Ing. E.h. **Hans-Hermann Braess**, VDI, 1936– , studied mechanical engineering and automotive technology at Hannover Technical College [Technische Hochschule Hannover]. After one year as a suspension designer at Ford in Cologne, H.-H. Braess worked as scientific assistant at Munich Technical University [Technische Universität München] from 1964 to 1970, where he obtained his doctorate based on a dissertation on "The Theory of Automotive Steering Behavior." From 1970 to 1980, he worked at Porsche AG, Stuttgart, from 1977 as head of research. In 1980, H.-H. Braess joined BMW AG in Munich, where he was director of "Research, Engineering and Scientific Services" (renamed "Science and Research" in 1985) until his retirement in 1996. H.-H. Braess is honorary professor at the Munich and Dresden Technical Universities as well as the Dresden College of Business and Technology [FHTW Dresden] and is author of more than 150 publications, including numerous books. He was or is a member of several professional societies and panels, including the Automotive Technology Research Association [FAT, Forschungsvereinigung Automobiltechnik] of the German Automotive Industry Association [VDA, Verband der Automobilindustrie] and is on the board of trustees of the Deutsches Museum, Munich. The Association of German Engineers [VDI, Verein Deutscher Ingenieure] honored the professional impact of Dr. Braess's long membership, during which he was actively engaged (from 1972 onward) in the VDI Society of Vehicle and Transportation Technology and for technical and scientific cooperative efforts, by bestowing its Benz-Daimler-Maybach Medal in 1993. In 1994, H.-H. Braess was honored with the (American) National Highway Traffic Safety Administration [NHTSA] Safety Award.

Professor Dr.-Ing. **Ulrich Seiffert**, VDI, 1941– , studied mechanical engineering at Brunswick Technical University [Technische Universität Braunschweig]. He earned his doctorate at Berlin Technical University in 1974 with a study of vehicle safety. He joined Volkswagen in 1966 as a researcher; from 1979 to 1988, he was director of corporate research and powertrain development. Thereafter, until the end of 1995, he was on the Volkswagen AG board of directors as vice president of research and development. In 1996, he became principal partner of WiTech Engineering GmbH, and since 1998 he has been spokesman for the Brunswick Technical University Center of Transportation [Zentrum für Verkehr der TU Braunschweig]. He is a member of ACATECH and of the Royal Swedish Academy of Engineering Sciences. For many years, Ulrich Seiffert has been engaged in honorary engineering work on behalf of the VDI: he is a board member of the VDI Society of Vehicle and Transportation Technology as well as chairman of the VDI Traffic Advisory Board. Since 1981, he has lectured at Brunswick Technical University and is a member of the scientific advisory board of the journal *Motortechnische Zeitschrift* (MTZ). Recognized with national and international awards for his professional efforts in the field of vehicle safety and environmental protection, Ulrich Seiffert holds a number of patents related to vehicle safety. In addition to more than 300 technical publications, including numerous SAE papers, Ulrich Seiffert is author of several technical books, including *Automotive Technology of the Future* (SAE, 1991, with Peter Walzer), *Fahrzeugsicherheit—Personenwagen* (VDI-Verlag, 1992), and, with Lothar Wech, *Automotive Safety Handbook* (SAE, 2003).